T0189987

Communications in Computer and Information Science 1239

More information about this series at http://www.springer.com/series/7899

Marie-Jeanne Lesot · Susana Vieira ·
Marek Z. Reformat · João Paulo Carvalho ·
Anna Wilbik · Bernadette Bouchon-Meunier ·
Ronald R. Yager (Eds.)

Information Processing and Management of Uncertainty in Knowledge-Based Systems

18th International Conference, IPMU 2020
Lisbon, Portugal, June 15–19, 2020
Proceedings, Part III

Editors
Marie-Jeanne Lesot
LIP6-Sorbonne University
Paris, France

Susana Vieira
IDMEC, IST, Universidade de Lisboa
Lisbon, Portugal

Marek Z. Reformat
University of Alberta
Edmonton, AB, Canada

João Paulo Carvalho
INESC, IST, Universidade de Lisboa
Lisbon, Portugal

Anna Wilbik
Eindhoven University of Technology
Eindhoven, The Netherlands

Bernadette Bouchon-Meunier
CNRS-Sorbonne University
Paris, France

Ronald R. Yager
Iona College
New Rochelle, NY, USA

ISSN 1865-0929 ISSN 1865-0937 (electronic)
Communications in Computer and Information Science
ISBN 978-3-030-50152-5 ISBN 978-3-030-50153-2 (eBook)
https://doi.org/10.1007/978-3-030-50153-2

This Springer imprint is published by the registered company Springer Nature Switzerland AG
The registered company address is: Gewerbestrasse 11, 6330 Cham, Switzerland

Preface

We are very pleased to present you with the proceedings of the 18th International Conference on Information Processing and Management of Uncertainty in Knowledge-Based Systems (IPMU 2020), held during June 15–19, 2020. The conference was scheduled to take place in Lisbon, Portugal, at the Instituto Superior Técnico, University of Lisbon, located in a vibrant renovated area 10 minutes from downtown. Unfortunately, due to the COVID-19 pandemic and international travel restrictions around the globe, the Organizing Committee made the decision to make IPMU 2020 a virtual conference taking place as scheduled.

The IPMU conference is organized every two years. Its aim is to bring together scientists working on methods for the management of uncertainty and aggregation of information in intelligent systems. Since 1986, the IPMU conference has been providing a forum for the exchange of ideas between theoreticians and practitioners working in these areas and related fields. In addition to many contributed scientific papers, the conference has attracted prominent plenary speakers, including the Nobel Prize winners Kenneth Arrow, Daniel Kahneman, and Ilya Prigogine.

A very important feature of the conference is the presentation of the *Kampé de Fériet Award* for outstanding contributions to the field of uncertainty and management of uncertainty. Past winners of this prestigious award are Lotfi A. Zadeh (1992), Ilya Prigogine (1994), Toshiro Terano (1996), Kenneth Arrow (1998), Richard Jeffrey (2000), Arthur Dempster (2002), Janos Aczel (2004), Daniel Kahneman (2006), Enric Trillas (2008), James Bezdek (2010), Michio Sugeno (2012), Vladimir N. Vapnik (2014), Joseph Y. Halpern (2016), and Glenn Shafer (2018). This year, the recipient of the *Kampé de Fériet Award* is Barbara Tversky. Congratulations!

The IPMU 2020 conference offers a versatile and comprehensive scientific program. There were four invited talks given by distinguished researchers: Barbara Tversky (Stanford University and Columbia University, USA), Luísa Coheur (Universidade de Lisboa, Instituto Superior Técnico, Portugal), Jim Keller (University of Missouri, USA), and Björn Schuller (Imperial College London, UK). A special tribute was organized to celebrate the life and achievements of Enrique Ruspini who passed away last year. He was one of the fuzzy-logic pioneers and researchers who contributed enormously to the fuzzy sets and systems body of knowledge. Two invited papers are dedicated to his memory. We would like to thank Rudolf Seising, Francesc Esteva, Lluís Godo, Ricardo Oscar Rodriguez, and Thomas Vetterlein for their involvement and contributions.

The IPMU 2020 program consisted of 22 special sessions and 173 papers authored by researchers from 34 different countries. All 213 submitted papers underwent the thorough review process and were judged by at least three reviewers. Many of them were reviewed by more – even up to five – referees. Furthermore, all papers were examined by the program chairs. The review process respected the usual conflict-of-interest standards, so that all papers received multiple independent evaluations.

Organizing a conference is not possible without the assistance, dedication, and support of many people and institutions.

We are particularly thankful to the organizers of special sessions. Such sessions, dedicated to variety of topics and organized by experts, have always been a characteristic feature of IPMU conferences. We would like to pass our special thanks to Uzay Kaymak, who helped evaluate many special session proposals.

We would like to acknowledge all members of the IPMU 2020 Program Committee, as well as multiple reviewers who played an essential role in the reviewing process, ensuring a high-quality conference. Thank you very much for all your work and efforts.

We gratefully acknowledge the technical co-sponsorship of the IEEE Computational Intelligence Society and the European Society for Fuzzy Logic and Technology (EUSFLAT).

A huge thanks and appreciation to the personnel of Lisbon's Tourism Office 'Turismo de Lisboa' (www.visitlisboa.com) for their eagerness to help, as well as their enthusiastic support.

Our very special and greatest gratitude goes to the authors who have submitted results of their work and presented them at the conference. Without you this conference would not take place. Thank you!

We miss in-person meetings and discussions, yet we are privileged that despite these difficult and unusual times all of us had a chance to be involved in organizing the virtual IPMU conference. We hope that these proceedings provide the readers with multiple ideas leading to numerous research activities, significant publications, and intriguing presentations at future IPMU conferences.

April 2020

Marie-Jeanne Lesot
Marek Z. Reformat
Susana Vieira
Bernadette Bouchon-Meunier
João Paulo Carvalho
Anna Wilbik
Ronald R. Yager

Organization

General Chair

João Paulo Carvalho	INESC-ID, Instituto Superior Técnico, Universidade de Lisboa, Portugal

Program Chairs

Marie-Jeanne Lesot	LIP6, Sorbonne Université, France
Marek Z. Reformat	University of Alberta, Canada
Susana Vieira	IDMEC, Instituto Superior Técnico, Universidade de Lisboa, Portugal

Executive Directors

Bernadette Bouchon-Meunier	LIP6, CNRS, France
Ronald R. Yager	Iona College, USA

Special Session Chair

Uzay Kaymak	Technische Universiteit Eindhoven, The Netherlands

Publication Chair

Anna Wilbik	Technische Universiteit Eindhoven, The Netherlands

Sponsor and Publicity Chair

João M. C. Sousa	IDMEC, Instituto Superior Técnico, Universidade de Lisboa, Portugal

Web Chair

Fernando Batista	INESC-ID, Instituto Superior Técnico, Universidade de Lisboa, Portugal

International Advisory Board

Program Committee

Keeley Crockett	Manchester Metropolitan University, UK
Giuseppe D'Aniello	University of Salerno, Italy
Bernard De Baets	Ghent University, Belgium
Martine De Cock	University of Washington, USA
Guy De Tré	Ghent University, Belgium
Sébastien Destercke	CNRS, UMR Heudiasyc, France
Antonio Di Nola	University of Salerno, Italy
Scott Dick	University of Alberta, Canada
Didier Dubois	IRIT, RPDMP, France
Fabrizio Durante	Free University of Bozen-Bolzano, Italy
Krzysztof Dyczkowski	Adam Mickiewicz University, Poland
Zied Elouedi	Institut Supérieur de Gestion de Tunis, Tunisia
Francesc Esteva	IIIA-CSIC, Spain
Dimitar Filev	Ford Motor Company, USA
Matteo Gaeta	University of Salerno, Italy
Sylvie Galichet	LISTIC, Université de Savoie, France
Jonathan M. Garibaldi	University of Nottingham, UK
Lluis Godo	IIIA-CSIC, Spain
Fernando Gomide	University of Campinas, Brazil
Gil González-Rodríguez	University of Oviedo, Spain
Przemysław Grzegorzewski	Systems Research Institute, Polish Academy of Sciences, Poland
Lawrence Hall	University of South Florida, USA
Istvan Harmati	Széchenyi István Egyetem, Hungary
Timothy Havens	Michigan Technological University, USA
Francisco Herrera	University of Granada, Spain
Enrique Herrera-Viedma	University of Granada, Spain
Ludmila Himmelspach	Heirich Heine Universität Düsseldorf, Germany
Eyke Hüllemeier	Paderborn University, Germany
Michal Holčapek	University of Ostrava, Czech Republic
Janusz Kacprzyk	Systems Research Institute, Polish Academy of Sciences, Poland
Uzay Kaymak	Eindhoven University of Technology, The Netherlands
Jim Keller	University of Missouri, USA
Frank Klawonn	Ostfalia University of Applied Sciences, Germany
László T. Kóczy	Budapest University of Technology and Economics, Hungary
John Kornak	University of California, San Francisco, USA
Vladik Kreinovich	University of Texas at El Paso, USA
Ondrej Krídlo	University of P. J. Safarik in Kosice, Slovakia
Rudolf Kruse	University of Magdeburg, Germany
Christophe Labreuche	Thales R&T, France
Jérôme Lang	CNRS, LAMSADE, Université Paris-Dauphine, France
Anne Laurent	LIRMM, UM, France
Chang-Shing Lee	National University of Tainan, Taiwan

Martin Štěpnička	IRAFM, University of Ostrava, Czech Republic
Umberto Straccia	ISTI-CNR, Italy
Olivier Strauss	LIRMM, France
Michio Sugeno	Tokyo Institute of Technology, Japan
Eulalia Szmidt	Systems Research Institute, Polish Academy of Sciences, Poland
Marco Tabacchi	Università degli Studi di Palermo, Italy
Vicenc Torra	Maynooth University, Ireland
Linda C. van der Gaag	Utrecht University, The Netherlands
Barbara Vantaggi	Sapienza University of Rome, Italy
José Luis Verdegay	University of Granada, Spain
Thomas Vetterlein	Johannes Kepler University Linz, Austria
Susana Vieira	Universidade de Lisboa, Portugal
Christian Wagner	University of Nottingham, UK
Anna Wilbik	Eindhoven University of Technology, The Netherlands
Sławomir Zadrożny	Systems Research Institute, Polish Academy of Sciences, Poland

Additional Members of the Reviewing Committee

Raoua Abdelkhalek
Julien Alexandre Dit Sandretto
Zahra Alijani
Alessandro Antonucci
Jean Baratgin
Laécio C. Barros
Leliane N. Barros
Libor Behounek
María José Benítez Caballero
Kylc Bittner
Jan Boronski
Reda Boukezzoula
Ross Boylan
Andrey Bronevich
Petr Bujok
Michal Burda
Rafael Cabañas de Paz
Inma P. Cabrera
Tomasa Calvo
José Renato Campos
Andrea Capotorti
Diego Castaño
Anna Cena
Mihir Chakraborty
Yurilev Chalco-Cano
Manuel Chica
Panagiotis Chountas
Davide Ciucci
Frank Coolen
Maria Eugenia Cornejo Piñero
Cassio P. de Campos
Gert De Cooman
Laura De Miguel
Jean Dezert
J. Angel Diaz-Garcia
Graçaliz Dimuro
Paweł Drygaś
Hassane Essafi
Javier Fernandez
Carlos Fernandez-Basso
Juan Carlos Figueroa-García
Marcelo Finger
Tommaso Flaminio
Robert Fullér
Marek Gagolewski
Angel Garcia Contreras
Michel Grabisch
Karel Gutierrez

Allel Hadjali
Olgierd Hryniewicz
Miroslav Hudec
Ignacio Huitzil
Seong Jae Hwang
Atsushi Inoue
Vladimir Janis
Balasubramaniam Jayaram
Richard Jensen
Luis Jimenez Linares
Katarzyna Kaczmarek
Martin Kalina
Hiroharu Kawanaka
Alireza Khastan
Martins Kokainis
Ryszard Kowalczyk
Maciej Krawczak
Jiri Kupka
Serafina Lapenta
Ulcilea Leal
Antonio Ledda
Eric Lefevre
Nguyen Linh
Nicolas Madrid
Arnaud Martin
Denis Maua
Gilles Mauris
Belen Melian
María Paula Menchón
David Mercier
Arnau Mir
Soheyla Mirshahi
Marina Mizukoshi
Jiří Močkoř
Miguel Molina-Solana
Ignacio Montes
Serafin Moral
Tommaso Moraschini
Andreia Mordido
Juan Antonio Morente-Molinera
Fred Mubang
Vu-Linh Nguyen
Radoslaw Niewiadomski
Carles Noguera
Pavels Orlovs
Daniel Ortiz-Arroyo
Jan W. Owsinski
Antonio Palacio
Manuel J. Parra Royón
Jan Paseka
Viktor Pavliska
Renato Pelessoni
Barbara Pękala
Benjamin Quost
Emmanuel Ramasso
Eloisa Ramírez Poussa
Luca Reggio
Juan Vicente Riera
Maria Rifqi
Luis Rodriguez-Benitez
Guillaume Romain
Maciej Romaniuk
Francisco P. Romero
Clemente Rubio-Manzano
Aleksandra Rutkowska
Juan Jesus Salamanca Jurado
Teddy Seidenfeld
Mikel Sesma-Sara
Babak Shiri
Amit Shukla
Anand Pratap Singh
Damjan Skulj
Sotir Sotirov
Michal Stronkowski
Andrea Stupnánová
Matthias Troffaes
Dana Tudorascu
Leobardo Valera
Arthur Van Camp
Paolo Vicig
Amanda Vidal Wandelmer
Joaquim Viegas
Jin Hee Yoon
Karl Young
Hua-Peng Zhang

Special Session Organizers

Javier Andreu	University of Essex, UK
Michał Baczyński	University of Silesia in Katowice, Poland
Isabelle Bloch	Télécom ParisTech, France
Bernadette Bouchon-Meunier	LIP6, CNRS, France
Reda Boukezzoula	Université de Savoie Mont-Blanc, France
Humberto Bustince	Public University of Navarra, Spain
Tomasa Calvo	University of Alcalá, Spain
Martine Ceberio	University of Texas at El Paso, USA
Yurilev Chalco-Cano	University of Tarapacá at Arica, Chile
Giulianella Coletti	Università di Perugia, Italy
Didier Coquin	Université de Savoie Mont-Blanc, France
M. Eugenia Cornejo	University of Cádiz, Spain
Bernard De Baets	Ghent University, Belgium
Guy De Tré	Ghent University, Belgium
Graçaliz Dimuro	Universidade Federal do Rio Grande, Brazil
Didier Dubois	IRIT, Université Paul Sabatier, France
Hassane Essafi	CEA, France
Carlos J. Fernández-Basso	University of Granada, Spain
Javier Fernández	Public University of Navarra, Spain
Tommaso Flaminio	Spanish National Research Council, Spain
Lluis Godo	Spanish National Research Council, Spain
Przemyslaw Grzegorzewski	Warsaw University of Technology, Poland
Rajarshi Guhaniyogi	University of California, Santa Cruz, USA
Karel Gutiérrez Batista	University of Granada, Spain
István Á. Harmati	Széchenyi István University, Hungary
Michal Holčapek	University of Ostrava, Czech Republic
Atsushi Inoue	Eastern Washington University, USA
Balasubramaniam Jayaram	Indian Institute of Technology Hyderabad, India
Janusz Kacprzyk	Systems Research Institute, Polish Academy of Sciences, Poland
Hiroharu Kawanaka	Mie University, Japan
László T. Kóczy	Budapest University of Technology and Economics, Hungary
John Kornak	University of California, San Francisco, USA
Vladik Kreinovich	University of Texas at El Paso, USA
Henrik Legind Larsen	Legind Technologies, Denmark
Weldon Lodwick	Federal University of São Paulo, Brazil
Maria Jose Martín-Bautista	University of Granada, Spain
Sebastia Massanet	University of the Balearic Islands, Spain
Jesús Medina	University of Cádiz, Spain
Belén Melián-Batista	University of La Laguna, Spain
Radko Mesiar	Slovak University of Technology, Slovakia
Enrique Miranda	University of Oviedo, Spain

Ignacio Montes	University of Oviedo, Spain
Juan Moreno-Garcia	University of Castilla-La Mancha, Spain
Petra Murinová	University of Ostrava, Czech Republic
Vílem Novák	University of Ostrava, Czech Republic
David A. Pelta	University of Granada, Spain
Raúl Pérez-Fernández	University of Oviedo, Spain
Irina Perfilieva	University of Ostrava, Czech Republic
Henri Prade	IRIT, Université Paul Sabatier, France
Anca Ralescu	University of Cincinnati, USA
Eloísa Ramírez-Poussa	University of Cádiz, Spain
Luis Rodriguez-Benitez	University of Castilla-La Mancha, Spain
Antonio Rufian-Lizana	University of Sevilla, Spain
M. Dolores Ruiz	University of Granada, Spain
Andrea Stupnanova	Slovak University of Technology, Slovakia
Amanda Vidal	Czech Academy of Sciences, Czech Republic
Aaron Wolfe Scheffler	University of California, San Francisco, USA
Adnan Yazici	Nazarbayev University, Kazakhstan
Sławomir Zadrożny	Systems Research Institute Polish Academy of Sciences, Poland

List of Special Sessions

Fuzzy Interval Analysis

Antonio Rufian-Lizana	University of Sevilla, Spain
Weldon Lodwick	Federal University of São Paulo, Brazil
Yurilev Chalco-Cano	University of Tarapacá at Arica, Chile

Theoretical and Applied Aspects of Imprecise Probabilities

Enrique Miranda	University of Oviedo, Spain
Ignacio Montes	University of Oviedo, Spain

Similarities in Artificial Intelligence

Bernadette Bouchon-Meunier	LIP6, CNRS, France
Giulianella Coletti	Università di Perugia, Italy

Belief Function Theory and Its Applications

Didier Coquin	Université de Savoie Mont-Blanc, France
Reda Boukezzoula	Université de Savoie Mont-Blanc, France

Aggregation: Theory and Practice

Tomasa Calvo	University of Alcalá, Spain
Radko Mesiar	Slovak University of Technology, Slovakia
Andrea Stupňanová	Slovak University of Technology, Slovakia

Aggregation: Pre-aggregation Functions and Other Generalizations

Humberto Bustince	Public University of Navarra, Spain
Graçaliz Dimuro	Universidade Federal do Rio Grande, Brazil
Javier Fernández	Public University of Navarra, Spain

Aggregation: Aggregation of Different Data Structures

Bernard De Baets	Ghent University, Belgium
Raúl Pérez-Fernández	University of Oviedo, Spain

Fuzzy Methods in Data Mining and Knowledge Discovery

M. Dolores Ruiz	University of Granada, Spain
Karel Gutiérrez Batista	University of Granada, Spain
Carlos J. Fernández-Basso	University of Granada, Spain

Computational Intelligence for Logistics and Transportation Problems

David A. Pelta	University of Granada, Spain
Belén Melián-Batista	University of La Laguna, Spain

Fuzzy Implication Functions

Michał Baczyński	University of Silesia in Katowice, Poland
Balasubramaniam Jayaram	Indian Institute of Technology Hyderabad, India
Sebastià Massanet	University of the Balearic Islands, Spain

Soft Methods in Statistics and Data Analysis

Przemysław Grzegorzewski	Warsaw University of Technology, Poland

Image Understanding and Explainable AI

Isabelle Bloch	Télécom ParisTech, France
Atsushi Inoue	Eastern Washington University, USA
Hiroharu Kawanaka	Mie University, Japan
Anca Ralescu	University of Cincinnati, USA

Fuzzy and Generalized Quantifier Theory

Vilém Novák	University of Ostrava, Czech Republic
Petra Murinová	University of Ostrava, Czech Republic

Mathematical Methods Towards Dealing with Uncertainty in Applied Sciences

Irina Perfilieva	University of Ostrava, Czech Republic
Michal Holčapek	University of Ostrava, Czech Republic

Statistical Image Processing and Analysis, with Applications in Neuroimaging

John Kornak	University of California, San Francisco, USA
Rajarshi Guhaniyogi	University of California, Santa Cruz, USA
Aaron Wolfe Scheffler	University of California, San Francisco, USA

Interval Uncertainty

Martine Ceberio	University of Texas at El Paso, USA
Vladik Kreinovich	University of Texas at El Paso, USA

Discrete Models and Computational Intelligence

László T. Kóczy	Budapest University of Technology and Economics, Hungary
István Á. Harmati	Széchenyi István University, Hungary

Current Techniques to Model, Process and Describe Time Series

Juan Moreno-Garcia	University of Castilla-La Mancha, Spain
Luis Rodriguez-Benitez	University of Castilla-La Mancha, Spain

Mathematical Fuzzy Logic and Graded Reasoning Models

Tommaso Flaminio	Spanish National Research Council, Spain
Lluís Godo	Spanish National Research Council, Spain
Vílem Novák	University of Ostrava, Czech Republic
Amanda Vidal	Czech Academy of Sciences, Czech Republic

Formal Concept Analysis, Rough Sets, General Operators and Related Topics

M. Eugenia Cornejo	University of Cádiz, Spain
Didier Dubois	IRIT, Université Paul Sabatier, France
Jesús Medina	University of Cádiz, Spain
Henri Prade	IRIT, Université Paul Sabatier, France
Eloísa Ramírez-Poussa	University of Cádiz, Spain

Computational Intelligence Methods in Information Modelling, Representation and Processing

Guy De Tré	Ghent University, Belgium
Janusz Kacprzyk	Systems Research Institute, Polish Academy of Sciences, Poland
Adnan Yazici	Nazarbayev University, Kazakhstan
Sławomir Zadrożny	Systems Research Institute Polish Academy of Sciences, Poland

Contents - Part III

Soft Methods in Statistics and Data Analysis

Image Understanding and Explainable AI

Fuzzy and Generalized Quantifier Theory

Mathematical Methods Towards Dealing with Uncertainty in Applied Sciences

Interval Uncertainty

Discrete Models and Computational Intelligence

Current Techniques to Model, Process and Describe Time Series

Mathematical Fuzzy Logic and Graded Reasoning Models

Formal Concept Analysis, Rough Sets, General Operators and Related Topics

Soft Methods in Statistics and Data Analysis

Imprecise Approaches to Analysis of Insurance Portfolio with Catastrophe Bond

Maciej Romaniuk(✉)

Systems Research Institute, Polish Academy of Sciences,
Newelska 6, 01-447 Warsaw, Poland
mroman@ibspan.waw.pl

Abstract. In this paper, imprecise approaches to model the risk reserve process of an insurer's portfolio, which consists of a catastrophe bond and external help, and with a special penalty function in the case of a bankruptcy event, are presented. Apart from the general framework, two special cases, when parameters of the portfolio are described by L-R fuzzy numbers or shadowed sets, are discussed and compared. In a few examples based on the real-life data for these two types of impreciseness, some important characteristics of the portfolio, like the expected value and the probability of the ruin, are estimated, analysed and compared using the Monte Carlo simulations.

Keywords: Risk process · Fuzzy numbers · Shadowed sets · Insurance portfolio · Numerical simulations

1 Introduction

Due to the global warming effect, changes in land use, and other effects related to humans' activities, number and severity of natural catastrophes (like tsunamis, earthquakes, floods etc.) are still increasing. Rising demand for related compensations is a serious problem for the insurers, so they have to introduce new financial and insurance instruments to compensate for their losses. A catastrophe bond (abbreviated as a cat bond, see, e.g., [7,11,12,14]) is an example of such an instrument, which transfer risks from an insurance market to financial markets (which is known as "securitization of losses"). But these new instruments require more complex approaches to the classical problem: how to calculate the probability of the insurer's ruin? Simulations (like the Monte Carlo method) can be applied to analyse even very complex insurer's portfolios (see, e.g., [18,20]), which consist of many different instruments (i.e. layers), like the classical risk process, issued catastrophe bonds, a reinsurance contract, etc. A numerical approach is a very convenient tool also in other areas (see, e.g., [9,15,16]).

Moreover, future characteristics of natural catastrophes and other aspects of the insurer's market (e.g., level of possible external help after a serious natural catastrophe) are not known precisely. Therefore, an imprecise setting (related to, e.g., fuzzy numbers or shadowed sets) is a natural approach to overcome problems with such partially unknown parameters (see, e.g., [18,23]).

M.-J. Lesot et al. (Eds.): IPMU 2020, CCIS 1239, pp. 3–16, 2020.
https://doi.org/10.1007/978-3-030-50153-2_1

The literature devoted to insurance mathematics, especially to the problem of the ruin probability of the insurer, is abundant (see, e.g., [2,7,10]). However, the imprecise approach in this area is still underdeveloped (see, e.g., [5] for a review of the literature). The following paper can be treated as a further development of the ideas presented in [18–20,22]. Then, a generalized form of the classical risk process is considered and additional layers of the insurance portfolio (like a catastrophe bond and an external help) under an assumption about the value of money in time are analysed using the MC simulations. Moreover, the imprecise setting is used to describe some parameters of the considered models.

A contribution of this paper is fourfold. First, a general simulation framework which enables us to use many, not only one, imprecise parameters (with values based, e.g., on the experts' opinions) of the considered models for the insurer's portfolio, is presented. Second, this general framework is discussed in a more detailed way for two practically important types of imprecise values: fuzzy numbers (especially L-R fuzzy numbers) and shadowed sets. Third, a generalization of the classical risk reserve process for the complex insurer's portfolio, which consists of an issued catastrophe bond and a (possible) external (e.g., governmental or from other institution) help, together with an embedded penalty function (applied in the case of a bankruptcy event), is presented. Moreover, this generalization uses the assumption about the value of money in time. Fourth, the two imprecise approaches (i.e. for L-R fuzzy numbers and shadowed sets) are numerically compared in a few examples. In these examples, the Monte Carlo (MC) simulations are used to calculate various characteristics of the portfolio, which are important for the insurer (like the expected value of this portfolio or the probability of the insurer's ruin in finite time).

This paper is organized as follows. In Sect. 2, the classical risk process is generalized to take into account the previously mentioned additional layers and the assumption about the value of money in time. In Sect. 3, necessary definitions concerning fuzzy numbers and shadowed sets are recalled. In Sect. 4, the general approach for imprecise settings of the considered models is presented and then applied for L-R fuzzy numbers and shadowed sets in simulations discussed in Sect. 5. The paper is concluded in Sect. 6 with some final remarks.

2 Model of Insurance Portfolio

The classical risk reserve process R_t is defined as a model of the financial reserves of an insurer depending on time t, i.e.

$$R_t = u + pt - C_t^* \tag{1}$$

where u is an initial reserve of the insurer, p is a rate of premiums paid by the insureds per unit time and C_t^* is a claim process of the form

$$C_t^* = \sum_{i=1}^{N_t} C_i \tag{2}$$

where $C_1, C_2, \ldots$ are iid random values of the claims. Usually, these claims have the same values as the losses caused by the natural catastrophes U_i, i.e. $C_i = U_i$, so we have an additional process of the losses

$$U_t^* = \sum_{i=1}^{N_t} U_i. \tag{3}$$

In [20], the claims were modelled as a deterministic or random part of the losses, i.e. $C_i = \alpha_{\text{claim}} Z_i U_i$, where $\alpha_{\text{claim}} \in [0, 1]$, $Z_i \sim U[c_{\min}, c_{\max}]$, and Z_i, U_i are mutually independent variables. Then, α_{claim} describes a deterministic share of the considered insurer in the whole insurance market (for the given region) and Z_i models a random part of the claim C_i in the loss U_i. In the following, we assume that α_{claim} is given in the imprecise way to model share of the insurer in the market in a more real-life way, e.g., when this share is not exactly stated, its level varies depending on a region of a possible natural catastrophe or its source, etc. It can lead to some hedging problems (see, e.g., [19,20] for additional details).

The process of a number of the claims (and the losses) $N_t \geq 0$ is usually driven by some Poisson process, e.g., a non-homogeneous Poisson process (NHPP). In this paper, we assume an intensity function for this NHPP of the form

$$\lambda_{\text{NHPP}}(t) = a + bt + c \sin\left(2\pi\left(t + d\right)\right) + m \exp\left(\cos\left(\frac{2\pi t}{\omega}\right)\right) \tag{4}$$

with linear, cyclic and exponential parts, which was proposed in [6] and applied in [22]. To generate values from (4), the thinning method (see, e.g., [18]) is then used. Because of using (4), we assume that the premium in (1) is a constant, fixed value for some deterministic moment T, and

$$p(T) = (1 + \nu_p)\, \mathbb{E} C_i \int_0^T \lambda_{\text{NHPP}}(s) ds \tag{5}$$

where ν_p is a safety loading (security loading) of the insurer (usually about 10%–20%, see, e.g, [2]).

In this paper, the classical risk reserve process (1) is generalized to take into account additional financial and insurance instruments (like a catastrophe bond), a penalty function, and the value of money in time (which is modelled using a stochastic interest rate, i.e., the one-factor Vasicek model, given by $dr_t = \kappa\left(\theta - r_t\right) dt + \sigma dW_t$). We enrich the classical insurance portfolio with two additional layers: a catastrophe bond and an external (e.g., governmental or from another institution) help. There are also other possible layers (like a reinsurance contract, see, e.g., [22]).

A catastrophe bond is an example of a complex financial-insurance instrument. When it is issued, the insurer pays an insurance premium p_{cb} in exchange for coverage, when a triggering point (usually some catastrophic event, like an earthquake) occurs. The investors purchase insurance-linked security for cash. The above-mentioned premium and cash flows are usually managed by an SPV (Special Purpose Vehicle), which also issues the catastrophe bonds. The investors hold the issued assets, whose coupons and/or principal depend on the occurrence of the mentioned triggering point. If such a catastrophic event occurs during the specified period, then the SPV compensates the insurer with a payment $f^i_{\text{cb}}(U^*_T)$ and the cash flows for the investors are changed. Usually, these flows are lowered, i.e. there is full or partial forgiveness of the repayment of principal and/or interest. However, if the triggering point does not occur, the investors usually receive the full payment from a cat bond (see, e.g., [7,18,20]).

We assume that the mentioned insurance premium p_{cb} is proportional to both a part α_{cb} of the whole price of the single catastrophic bond I_{cb}, and to a number of the issued bonds n_{cb}, i.e. $p_{\text{cb}} = \alpha_{\text{cb}} n_{\text{cb}} I_{\text{cb}}$. To calculate the price I_{cb} of a catastrophe bond, the martingale method together with the MC simulations are applied (see, e.g., [11,12,18]).

The third layer is an external (e.g., governmental or related to another enterprise from the insurer's consortium) help of the value $f_{\text{hlp}}(U^*_T)$. We assume, that this help is supplied only if the losses surpass some given minimal limit A_{hlp}, and it is also limited to some maximal value B_{hlp}. In some way, this help is similar to a reinsurance contract, but, contrary to such an instrument, there is no additional payment for this help and its limits can be modelled in an imprecise way (see Sect. 4). Moreover, we assume that only part α_{hlp} of this help lowers the expenditures of the insurer, because this help can be directed to both the insureds and the insurer. In practice, the formal requirements and the legal regulations can state that such external help is not allowed. But special financial help from the government was directed to banks (which were "too big to fail") during the global financial crisis in 2007–08 and this can happen again under some extreme circumstances. There is also a list, which was published by the Financial Stability Board, of "specially treated" insurance companies (also "too big to fail"). Moreover, it may be profitable even for smaller insurers to analyse the whole set of possible scenarios during their stress tests to check if with and without the external help their portfolios lead to similar levels of probabilities of insolvency.

And in real-life situations, bankruptcy can be very dangerous for the insurer. Apart from simple lack of necessary funds, this event could lead to additional financial problems, bankruptcy of other enterprises, problems with the law, reputational damages, etc. Therefore, we introduce a special penalty function $f_{\text{pen}}(R_T)$ which is related to the bankruptcy event itself or its value (see, e.g., [22]). Then we have a constant penalty function

$$f_{\text{pen}}(R_T) = b_{\text{pen}} \mathbb{1}(R_T < 0) \tag{6}$$

or, e.g., a linear penalty function (related directly to a level of the bankruptcy)

$$f_{\text{pen}}(R_T) = a_{\text{pen}} \max\{0, -R_T\} + b_{\text{pen}} \mathbb{1}(R_T < 0), \tag{7}$$

where $a_{\text{pen}}, b_{\text{pen}}$ are respective parameters.

Taking into account the previously assumed value of money in time and the additional layers, the classical risk reserve process (1) should be modified into the form

$$R_T = \mathrm{FV}_T(u - p_{\mathrm{cb}}) + \mathrm{FV}_T(p(T)) - \mathrm{FV}_T(C_T^*) + n_{\mathrm{cb}} f_{\mathrm{cb}}^i(U_T^*) + \alpha_{\mathrm{hlp}}\, \mathbb{1}(U_T^* \geq A_{\mathrm{hlp}}) f_{\mathrm{hlp}}(U_T^*)\,, \quad (8)$$

where $\mathrm{FV}_T(.)$ is future value of the cash flow in time T, and together with the penalty function we have

$$R_T^* = R_T - f_{\mathrm{pen}}(R_T). \quad (9)$$

In the following, we will be interested in estimation of some important characteristics of the introduced insurer's portfolio, like $\mathbb{E}\, R_T$ (the mean of the generalized risk reserve process in T), $\Pr(R_T < 0)$ (probability of the ruin in the finite time horizon T), $\mathbb{E}\, R_T^*$ (the mean of the generalized risk reserve process with the penalty function).

3 Fuzzy Numbers and Shadowed Sets

In Sect. 4, the general approach to the insurer's portfolio based on both crisp and imprecise settings is discussed. There are many mathematical setups which can be used to model impreciseness. We focus on trapezoidal fuzzy numbers (TPFNs), L-R fuzzy numbers (left-right fuzzy numbers, LRFNs) and shadowed sets (SHSs), but other types of fuzzy sets can be also applied (see, e.g., [1,3,8,11,13,15,16,18] for additional definitions, notation and applications of fuzzy numbers and shadowed sets). In the case of TPFN, its membership function has the form

$$A(x) = \begin{cases} \frac{x-a_1}{a_2-a_1} & \text{if } a_1 \leqslant x \leqslant a_2, \\ 1 & \text{if } a_2 \leqslant x \leqslant a_3, \\ \frac{a_4-x}{a_4-a_3} & \text{if } a_3 \leqslant x < a_4, \\ 0 & \text{otherwise,} \end{cases} \quad (10)$$

where $a_1, a_2, a_3, a_4 \in \mathbb{R}$, and $a_1 \leqslant a_2 \leqslant a_3 \leqslant a_4$. A trapezoidal fuzzy number $\tilde{a}$ will be further denoted as $[a_1, a_2, a_3, a_4]$, and its α-level cuts are given by intervals $\tilde{a}[\alpha] = [a_L[\alpha], a_U[\alpha]]$. If $a_2 = a_3$, then $\tilde{a}$ is said to be a triangular fuzzy number (TRFN) and we have $\tilde{a} = [a_1, a_2, a_4]$. The operations on fuzzy numbers are defined as in [3].

An SHS S in a universe of discourse X is a set-valued mapping $S : X \to \{0, [0,1], 1\}$ interpreted as follows: all elements of X for which $S(x) = 1$ are called a core of S and they embrace all elements that are fully compatible with the concepts conveyed by S, all elements of X for which $S(x) = 0$ are completely excluded from the concept described by S, and all elements of X for which $S(x) = [0,1]$, called a shadow, are uncertain. The usage of the unit interval for the shadow shows that any element from this shadow could be excluded or exhibit partial membership or could be fully allocated to S. In the following we consider $X = \mathbb{R}$, then a SHS will be denoted by $[s_1, s_2, s_3, s_4]_{\mathrm{SH}}$, where its core is given by the interval $[s_2, s_3]$, its shadow by $(s_1, s_2) \cup (s_3, s_4)$, and its support by $[s_1, s_4]$.

There exist important links between concepts of a fuzzy set and a shadowed set. Based on the initial fuzzy set, a corresponding shadowed set, that captures "the essence" of this fuzzy set, reduces computational efforts related to a membership function (because only two "cuts" are necessary, instead of all possible $\alpha \in [0,1]$), and simplifies the interpretation, can be constructed (see, e.g., [13]). This resultant shadowed set is created from the initial fuzzy set using an elevation of some membership values ("close to 1" or "high enough") and with a reduction of others (which are "close to 0" or "low enough"). In [8], the respective procedure, which is related to the optimization of two weighting functions, was introduced.

4 Imprecise Approaches

In [19,20,22], the models of the insurer's portfolios with only some parameters given by fuzzy numbers have been described. Now we present a generalization of this approach where all respective parameters are described by imprecise values (if such an assumption is appropriate in the considered real-life case) or crisp (i.e. real) values (when the respective parameters are precisely stated, e.g., in trade agreements). In Sect. 5, we focus on TRFNs/TPFNs and SHSs to model the mentioned impreciseness but other types of fuzzy numbers (like LRFNs) can be also applied.

Let us suppose, that our aim is to calculate the value of some function $f(x)$ for x, e.g., $\mathbb{E}\, R^*_T$ (using its respective estimator, i.e. the average $\bar{R}^*_T$ based on the MC approach) for a. To approximate a fuzzy value $\tilde{f}(\tilde{x})$ for a fuzzy counterpart $\tilde{x}$, monotonicity of $f(x)$ should be checked. If $f(x)$ is a non-decreasing function (or a non-increasing, respectively), then for the given $\alpha \in [0,1]$, the respective α-cut of $\tilde{f}(\tilde{x})$ is calculated as the crisp interval $[f(x_L[\alpha]), f(x_U[\alpha])]$ (or $[f(x_U[\alpha]), f(x_L[\alpha])]$, respectively), following the Zadeh's extension rule (see also, e.g., [21]). If monotonicity of $f(x)$ is more complex or, e.g., unknown, then this interval is given as $[\min\{f(x_L[\alpha]), f(x_U[\alpha])\}, \max\{f(x_L[\alpha]), f(x_U[\alpha])\}]$, which complicates the necessary calculations. This step is repeated for all desired values of α (usually we start from $\alpha = 0$ with some increment $\Delta\alpha > 0$ till $\alpha = 1$), with an additional approximation of "missing" α-level sets based on some (rather simple, e.g., linear) function. In the same manner, probability of the ruin in the finite time T can be estimated using $q_{\text{ruin}} = n_{\text{ruin}}/n$ (i.e. ratio of trajectories with final ruin to number of all trajectories of the portfolio) and its fuzzy counterpart $\tilde{q}_{\text{ruin}}$.

This approach can be generalized, if more than only one parameter x is fuzzified, e.g., if we would like to find $\tilde{R}^*_T$ (i.e. the fuzzy counterpart for $\bar{R}^*_T$) for both $\tilde{a}$ and $\tilde{b}$ (i.e. the fuzzy counterparts of a and b). A respective example for $\tilde{R}^*_T$ and some parameters of the considered models is summarized in Table 1, where the plus sign indicates non-decreasing dependency and the minus sign – non-increasing dependency. In the case of $\tilde{q}_{\text{ruin}}$, these dependencies are reverse.

Table 1. Examples of dependencies between $\mathbb{E} R_T^*$ and some parameters of the models.

a	b	α_{claim}	ζ_{LN}	a_{pen}	b_{pen}	A_{hlp}	B_{hlp}
$-$	$-$	$-$	$-$	$-$	$-$	$-$	$+$

The similar approach can be also applied if the parameters are described by SHSs, but then only two "levels" (i.e. $\alpha = 0, 1$) for the respective simulations should be used, and the whole idea is related to interval calculations. The application of SHSs instead of fuzzy numbers significantly reduces the necessary time of the MC simulations, but it can also lead to different results (see Sect. 5). This approach can be profitable if monotonicity of the considered function for some parameters is too complex to asses analytically.

5 Numerical Simulations

To make our simulations more realistic, we focus on the values of the parameters, which have been fitted to real-life data in the literature. But because many future aspects of the considered models are uncertain, some of these parameters are transformed to imprecise values, i.e. they are given as fuzzy numbers (modelled by TPFNs/TRFNs) or SHSs. Then, to describe each U_i, we apply the lognormal distribution with the parameters $\zeta_{\text{LN}}, \sigma_{\text{LN}}$ fitted in [6] to real-life data (collected by PCS, USA). The parameters of the intensity function (4) have been also fitted in [6]. The parameters of the one-factor Vasicek model are based on estimators from [4]. To describe the limits for a catastrophe bond and external help, quantiles of the cumulated value of the losses for the process (3), denoted by $Q_{C_T^*}(x)$, are also applied (see also, e.g., [20]). A piecewise function is used as the payment function $f_{\text{cb}}^i(U_T^*)$ of the issued catastrophe bond (for additional details about these payments and the pricing method, see [18, 22]).

Some of the above-mentioned parameters are given as imprecise values in the following simulations. To make our analysis clearer, we limit the number of these parameters, however, the approach presented in Sect. 4 is more general.

The above-mentioned impreciseness can have many sources. It can be related to future unknown behaviour of the introduced models (like the parameters in (4) or for the distribution of U_i, so the fuzzified value $\tilde{a}$ reflects our doubts concerning the future possible increase of intensity of the number of the catastrophic events caused by the global warming effect), lack of our knowledge because of the rarity of similar events in the past (then the experts' knowledge has to be applied, e.g., to estimate the parameters of the penalty function or the external help from another enterprise, so $\tilde{A}_{\text{hlp}}$ is related to the unknown minimum level of the possible external help, which can be smaller or bigger depending on political restrictions or available financial resources of another company), rapid changes in the market or lack of full information concerning the given region or the specified segment of the market (e.g., if there is a new competitor on the market or a natural catastrophe causes the losses only in some region or to some kind of properties, hence the share of the insurer in this market can be imprecise). Other parameters can be given exactly as crisp values because, e.g., they are precisely stated in contracts or directly known (like p_{cb}, u or the parameters of a cat bond). We consider one-year time horizon in our simulations (i.e. $T = 1$), and for each example one million simulations are used.

First, we assume that the imprecise values are modelled by TPFNs/TRFNs, so we set

$$\begin{aligned}
u = Q_{C_T^*}(0.25), \nu_p = 0.1, I_{\text{cb}} = 0.809896, \alpha_{\text{cb}} = 0.1, n_{\text{cb}} = 1000, w_1 = 1, \\
K_0 = Q_{C_T^*}(0.7), K_1 = Q_{C_T^*}(0.9), \alpha_{\text{hlp}} = 0.5, p_{\text{hlp}} = 1, \\
\tilde{A}_{\text{hlp}} = \left[Q_{C_T^*}(0.95) - 1000, Q_{C_T^*}(0.95), Q_{C_T^*}(0.95) + 1000\right], \\
\tilde{B}_{\text{hlp}} = \left[Q_{C_T^*}(0.99) - 1000, Q_{C_T^*}(0.99), Q_{C_T^*}(0.99) + 1000\right], \\
\kappa = 0.1779, \theta = 0.086565, r_0 = 0.03, \sigma = 0.02, a = [23.93, 24.93, 25.93], \\
b = [0.025, 0.026, 0.027], c = 5.6, d = 7.07, m = 10.3, \omega = 4.76, \\
\zeta_{\text{LN}} = [18.08, 18.58, 19.08], \sigma_{\text{LN}} = 1.49, \alpha_{\text{claim}} = [0.4, 0.5, 0.6], \\
a_{\text{pen}} = [1, 2, 3, 4], b_{\text{pen}} = [5, 10, 15, 20], \qquad (11)
\end{aligned}$$

then, e.g., the insurer issued 1000 catastrophe bonds, the parameter a in the intensity function (4) is "about its value estimated in [6] plus/minus 1", deterministic share of the insurer in the insurance market α_{claim} is "about 50% plus/minus 10%", the constant value b_{pen} in the penalty function is given by TPFN with the core $[10, 15]$ and the support $[5, 20]$, etc.

During our simulations, the insurer's portfolios without the penalty function and without the external help (Example F1), without the penalty function and with the external help (Example F2), and their counterparts with the linear penalty function (Example F3/Example F4), have been analysed. The simulated values of $\tilde{R}_T^*$ and $\tilde{q}_{\text{ruin}}$ can be found in Fig. 1 and Fig. 2, respectively. They are clearly LRFNs with rather wide supports (especially in the cases without the external help), very narrow cores, left-skewed (in the case of $\tilde{R}_T^*$) or right-skewed ones ($\tilde{q}_{\text{ruin}}$). Interestingly, the existence of the external help or/and the linear penalty function has limited influence on the right arms of $\tilde{R}_T^*$ and the left arms of $\tilde{q}_{\text{ruin}}$ (whereas very significant on their left/right counterparts, respectively). The external help narrows the supports of both $\tilde{R}_T^*$ and $\tilde{q}_{\text{ruin}}$ almost twice, but its influence on the cores of $\tilde{R}_T^*$ is more limited.

The insurer can easily evaluate the impacts of different scenarios on his financial results using the simulated outputs. For example, the insurer can be interested if the increasing intensity of the number of natural catastrophes in future can lead to additional problems with his portfolio. Then, instead of the values of a, b given in (11), TPFNs with wider right arms (because we suspect that the intensity will be rather bigger than smaller)

$$a = [23.93, 24.93, 26.93, 28.93], b = [0.025, 0.03, 0.026, 0.028] \qquad (12)$$

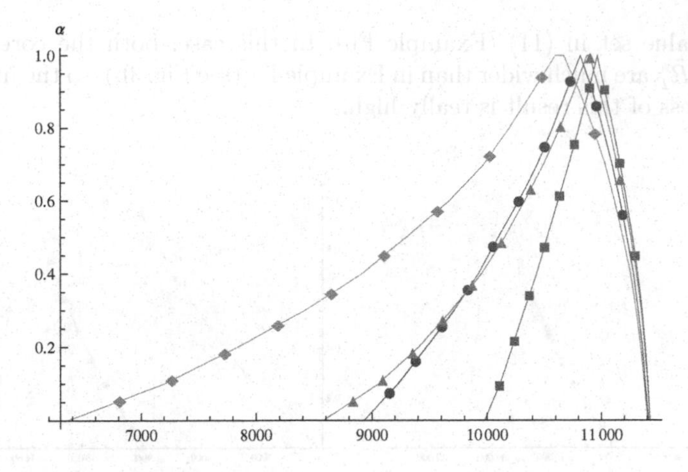

Fig. 1. Simulated values of $\tilde{R}_T^*$ in Example F1 (circles), Example F2 (rectangles), Example F3 (diamonds), Example F4 (triangles).

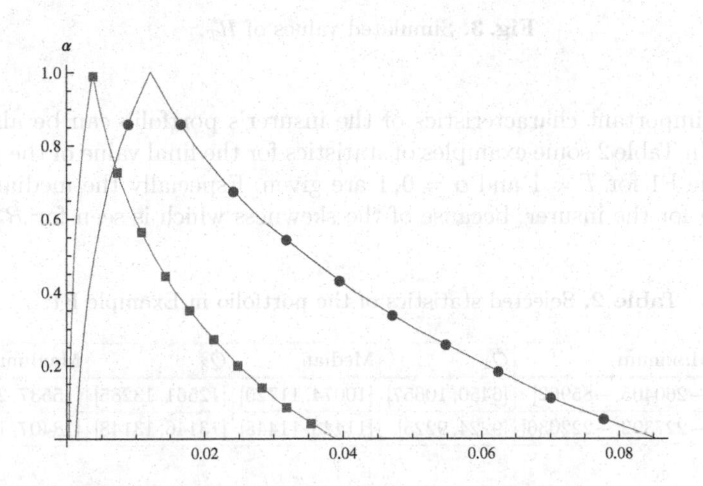

Fig. 2. Simulated values of $\tilde{q}_{\text{ruin}}$ in Example F1/F3 (circles), Example F2/F4 (rectangles).

can be used, together with the setting without the external help and the penalty function (Example F5). The simulated value of $\tilde{R}_T^*$ is similar to the one for Example F1 (see Fig. 3a), but the core in Example F5 is wider. Hence, the obtained results for $\alpha = 1$ are given by the wider interval. On the other hand, the insurer can suspect that his share in the insurance market is given by more imprecise TPFN

$$\alpha_{\text{claim}} = [0.3, 0.4, 0.6, 0.7] \tag{13}$$

than the value set in (11) (Example F6). In this case, both the core and the support of $\tilde{R}^*_T$ are much wider than in Example F1 (see Fig. 3b), so the implicated impreciseness of this result is really high.

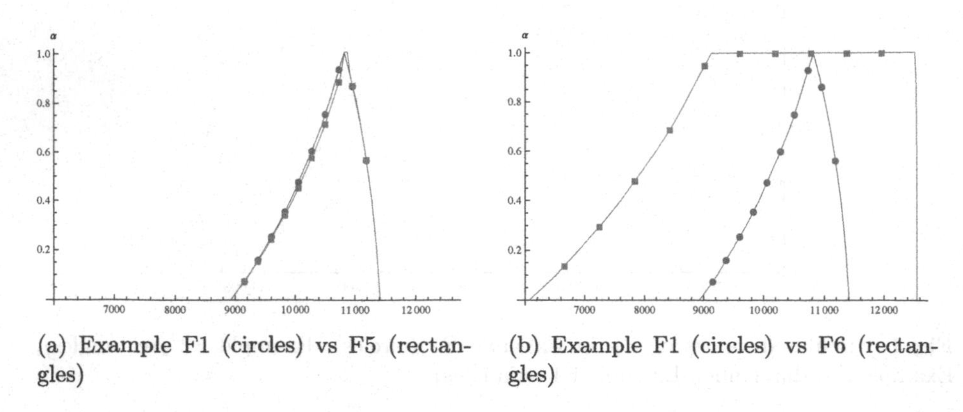

(a) Example F1 (circles) vs F5 (rectangles)

(b) Example F1 (circles) vs F6 (rectangles)

Fig. 3. Simulated values of $\tilde{R}^*_T$.

Other important characteristics of the insurer's portfolio can be also found similarly. In Table 2 some examples of statistics for the final value of the portfolio in Example F1 for $T = 1$ and $\alpha = 0, 1$ are given. Especially the median can be interesting for the insurer, because of the skewness which is seen for $\tilde{R}^*_T$.

Table 2. Selected statistics of the portfolio in Example F1.

	Minimum	Q_1	Median	Q_3	Maximum
$\alpha = 0$	$[-260405, -85962]$	$[6450, 10657]$	$[10074, 11729]$	$[12561, 13285]$	$[15537, 23950]$
$\alpha = 1$	$[-227392, -222036]$	$[9224, 9225]$	$[11443, 11448]$	$[13146, 13148]$	$[18407, 18802]$

Now we compare the simulated outputs when, instead of fuzzy numbers, their SHSs counterparts are applied. To do this, we apply the procedure from [8] to transform the fuzzy numbers in (11) to respective SHSs. We consider similar examples to the previous ones, i.e. without the penalty function and without the external help (Example S1 instead of F1), without the penalty function and with the external help (S2 instead of F1), etc. Then, the simulated values of $\tilde{R}^*_T$ and $\tilde{q}_{\text{ruin}}$ can be found in Fig. 4 and Fig. 5, respectively. The previously obtained plots for Examples F1–F4 are also added there for easier comparison of both approaches.

In general, the obtained SHSs are "similar" to the previously considered LRFNs, especially their skewness can be directly seen. However, the supports of SHSs are much shorter and their cores wider. Therefore, instead of LRFNs with almost the "crisp" cores (like in Example F1 or F2), we obtain rather wide

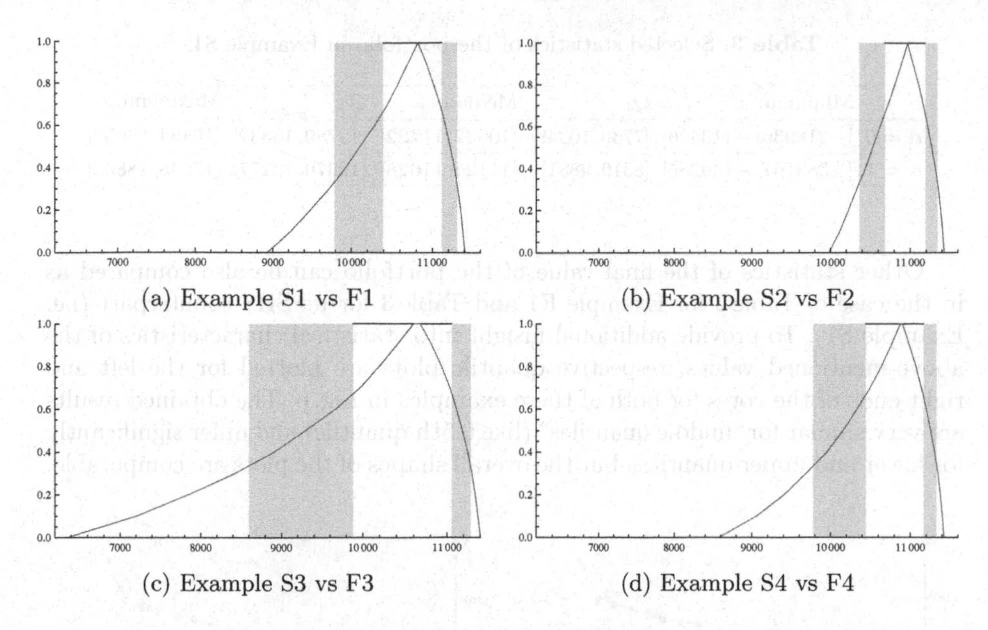

(a) Example S1 vs F1 (b) Example S2 vs F2

(c) Example S3 vs F3 (d) Example S4 vs F4

Fig. 4. Simulated values of $\tilde{R}_T^*$.

intervals for "fully compatible" values. And, on the contrary, the shorter supports can also lead us to different (from the insurer's point of view) evaluation of "fully excluded" values. It can be treated as some drawback of the approach based on SHSs, while its advantage is related to the previously mentioned reduction of time of simulations. However, the general shapes of the calculated SHSs and LRFNs are very similar, so the insurer can obtain a general idea concerning impreciseness of the evaluated values even for the approach based on SHSs. This kind of comparison between both approaches can be also made for other areas of applied sciences (see, e.g., [17]).

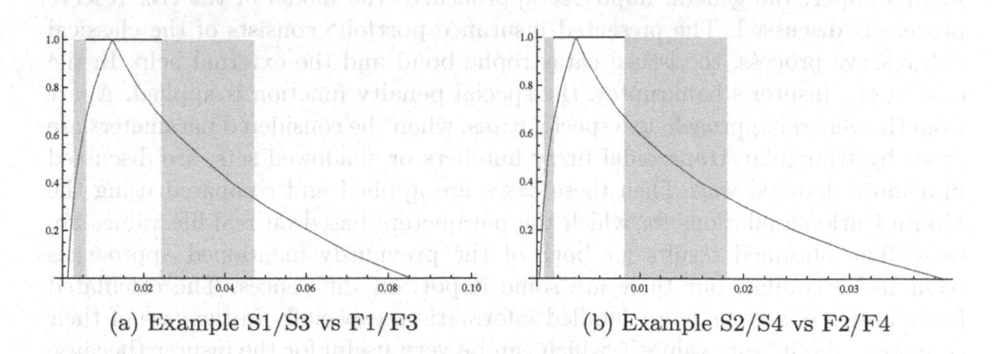

(a) Example S1/S3 vs F1/F3 (b) Example S2/S4 vs F2/F4

Fig. 5. Simulated values of $\tilde{q}_{\text{ruin}}$.

Table 3. Selected statistics of the portfolio in Example S1.

	Minimum	Q_1	Median	Q_3	Maximum
$\alpha = 0$	$[-218936, -113416]$	$[7730, 10349]$	$[10632, 11722]$	$[12780, 13332]$	$[16464, 23950]$
$\alpha = 1$	$[-383767, -149285]$	$[8319, 9881]$	$[11124, 11628]$	$[12970, 13277]$	$[17638, 18802]$

Other statistics of the final value of the portfolio can be also compared as in the case of Table 2 for Example F1 and Table 3 for its SHS counterpart (i.e. Example S1). To provide additional insight into statistical characteristics of the above-mentioned values, respective quantile plots are plotted for the left and right ends of the cores for both of these examples in Fig. 6. The obtained results are very similar for "middle quantiles" (like 0.5th quantile) and differ significantly for lower and upper quantiles, but the overall shapes of the plots are comparable.

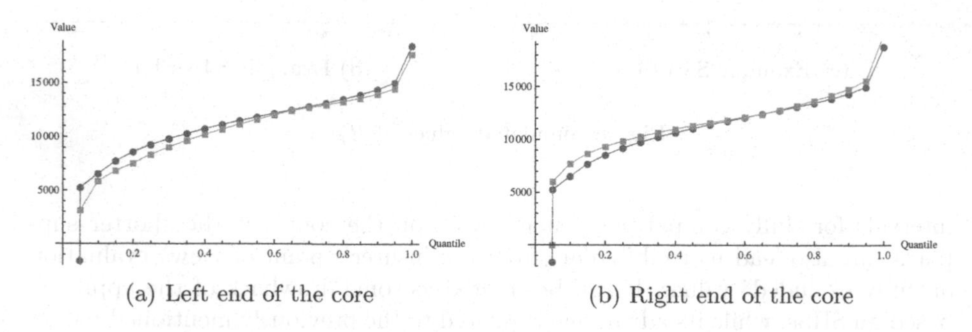

(a) Left end of the core (b) Right end of the core

Fig. 6. Quantile plots for Example F1 (circles) vs S1 (rectangles).

6 Conclusion

In this paper, the general imprecise approach to the model of the risk reserve process is discussed. The presented insurance portfolio consists of the classical risk reserve process, the issued catastrophe bond and the external help. In the case of the insurer's bankruptcy, the special penalty function is applied. Apart from the general approach, two special cases, when the considered parameters are given by triangular/trapezoidal fuzzy numbers or shadowed sets, are discussed in a more detailed way. Then these cases are applied and compared using the Monte Carlo simulations for which the parameters based on real-life values are used. The obtained results for both of the previously mentioned approaches seem to be similar, but there are some important differences. The calculated fuzzy numbers give us more detailed information, especially in the case of their cores (i.e. about "sure values"), which can be very useful for the insurer (because of the above-mentioned "certainty" of these values for $\alpha = 1$). Exact knowledge of the whole membership function of fuzzy number has also a serious drawback related to the expenditure of time of necessary simulations (let's say, 5–10 times more than in the case of the second approach). In the contrary, an application of shadowed sets leads us to a fast, intuitively understandable but rather coarse approximation of the expected output. It can be useful for the insurer as a first attempt to solve the considered problem. Nevertheless, the overall "shapes" of the outputs for both approaches remain very similar.

References

1. Ángeles Gil, M., Hryniewicz, O.: Statistics with imprecise data. In: Meyers, R.A. (ed.) Computational Complexity: Theory, Techniques, and Applications, pp. 3052–3063. Springer, New York (2012). https://doi.org/10.1007/978-1-4614-1800-9
2. Asmussen, S., Albrecher, H.: Ruin Probabilities, 2nd edn. World Scientific, Singapore (2010)
3. Ban, A., Coroianu, L., Grzegorzewski, P.: Fuzzy Numbers: Approximations, Ranking and Applications. Polish Academy of Sciences, Warsaw (2015)
4. Chan, K.C., Karolyi, G.A., Longstaff, F.A., Sanders, A.B.: An empirical comparison of alternative models of the short-term interest rate. J. Finance **47**(3), 1209–1227 (1992)
5. Ghasemalipour, S., Fathi-Vajargah, B.: The mean chance of ultimate ruin time in random fuzzy insurance risk model. Soft Comput. **22**(12), 4123–4131 (2017). https://doi.org/10.1007/s00500-017-2629-0
6. Giuricich, M.N., Burnecki, K.: Modelling of left-truncated heavy-tailed data with application to catastrophe bond pricing. Physica A **525**, 498–513 (2019)
7. Goda, K.: Seismic risk management of insurance portfolio using catastrophe bonds. Comput.-Aided Civ. Inf. Eng. **30**(7), 570–582 (2015)
8. Grzegorzewski, P.: Fuzzy number approximation via shadowed sets. Inf. Sci. **225**, 35–46 (2013)
9. Kulczycki, P., Charytanowicz, M., Dawidowicz, A.L.: A convenient ready-to-use algorithm for a conditional quantile estimator. Appl. Math. Inf. Sci. **9**(2), 841–850 (2015)
10. Ma, J., Sun, X.: Ruin probabilities for insurance models involving investments. Scand. Actuar. J. **2003**(3), 217–237 (2003)
11. Nowak, P., Romaniuk, M.: Catastrophe bond pricing for the two-factor Vasicek interest rate model with automatized fuzzy decision making. Soft Comput. **21**(10), 2575–2597 (2015). https://doi.org/10.1007/s00500-015-1957-1
12. Nowak, P., Romaniuk, M.: Valuing catastrophe bonds involving correlation and CIR interest rate model. Comput. Appl. Math. **37**(1), 365–394 (2016). https://doi.org/10.1007/s40314-016-0348-2
13. Pedrycz, W.: Shadowed sets: representing and processing fuzzy sets. IEEE Trans. Syst. Man Cybern. Part B (Cybern.) **28**(1), 103–109 (1998)
14. Pizzutilo, F., Venezia, E.: Are catastrophe bonds effective financial instruments in the transport and infrastructure industries? Evidence from international financial markets. Bus. Econ. Horiz. **14**(2), 256–267 (2018)
15. Romaniuk, M.: On simulation of maintenance costs for water distribution system with fuzzy parameters. Eksploat. Niezawodn. - Maint. Reliab. **18**(4), 514–527 (2016)
16. Romaniuk, M.: Optimization of maintenance costs of a pipeline for a V-shaped hazard rate of malfunction intensities. Eksploat. Niezawodn. - Maint. Reliab. **20**(1), 46–56 (2018)

17. Romaniuk, M.: Estimation of maintenance costs of a pipeline for a U-shaped hazard rate function in the imprecise setting. Eksploat. Niezawodn. - Maint. Reliab. **22**(2), 352–362 (2020)
18. Romaniuk, M., Nowak, P.: Monte Carlo Methods: Theory, Algorithms and Applications to Selected Financial Problems. Institute of Computer Science Polish Academy of Sciences, Warsaw (2015)
19. Romaniuk, M.: Analysis of the insurance portfolio with an embedded catastrophe bond in a case of uncertain parameter of the insurer's share. In: Wilimowska, Z., Borzemski, L., Grzech, A., Świątek, J. (eds.) Information Systems Architecture and Technology: Proceedings of 37th International Conference on Information Systems Architecture and Technology – ISAT 2016 – Part IV. AISC, vol. 524, pp. 33–43. Springer, Cham (2017). https://doi.org/10.1007/978-3-319-46592-0_3
20. Romaniuk, M.: Insurance portfolio containing a catastrophe bond and an external help with imprecise level—a numerical analysis. In: Kacprzyk, J., Szmidt, E., Zadrożny, S., Atanassov, K.T., Krawczak, M. (eds.) IWIFSGN/EUSFLAT-2017. AISC, vol. 643, pp. 256–267. Springer, Cham (2018). https://doi.org/10.1007/978-3-319-66827-7_23
21. Romaniuk, M.: On some applications of simulations in estimation of maintenance costs and in statistical tests for fuzzy settings. In: Steland, A., Rafajłowicz, E., Okhrin, O. (eds.) SMSA 2019. SPMS, vol. 294, pp. 437–448. Springer, Cham (2019). https://doi.org/10.1007/978-3-030-28665-1_33
22. Romaniuk, M.: Simulation-based analysis of penalty function for insurance portfolio with embedded catastrophe bond in crisp and imprecise setups. In: Wilimowska, Z., Borzemski, L., Świątek, J. (eds.) ISAT 2018. AISC, vol. 854, pp. 111–121. Springer, Cham (2019). https://doi.org/10.1007/978-3-319-99993-7_11
23. Yan, C., Liu, Q., Liu, J., Liu, W., Li, M., Qi, M.: Payments per claim model of outstanding claims reserve based on fuzzy linear regression. Int. J. Fuzzy Syst. **21**, 1950–1960 (2019). https://doi.org/10.1007/s40815-019-00617-x

Random Steinhaus Distances for Robust Syntax-Based Classification of Partially Inconsistent Linguistic Data

Laura Franzoi[1,4](✉), Andrea Sgarro[2,4], Anca Dinu[3,4], and Liviu P. Dinu[1,4]

[1] Faculty of Mathematics and Computer Science, University of Bucharest (Ro), Bucharest, Romania
laura.franzoi@gmail.com

[2] DMG University of Trieste (I), Trieste, Italy

[3] Department of Modern Languages, University of Bucharest (Ro), Bucharest, Romania

[4] Human Language Technologies Research Center, University of Bucharest (Ro), Bucharest, Romania

Abstract. We use the Steinhaus transform of metric distances to deal with inconsistency in linguistic classification. We focus on data due to G. Longobardi's school: languages are represented through yes-no strings of length 53, each string position corresponding to a syntactic feature which can be present or absent. However, due to a complex network of logical implications which constrain features, some positions might be undefined (logically inconsistent). To take into account linguistic inconsistency, the distances we use are Steinhaus metric distances generalizing the normalized Hamming distance. To validate the robustness of classifications based on Longobardi's data we resort to randomized transforms. Experimental results are provided and commented upon.

Keywords: Steinhaus distance · Linguistic classification · Łukasiewicz logic · Fuzzy logic

1 Introduction

The linguist G. Longobardi and his school have an ambitious project on language classification which is based on syntactic rules rather than lexical or phonetic data [1,11–13]: the idea is that syntactic rules have a definitely slower time-drift and so, by being able to reach back deeper into the past, one might obtain precious information on linguistic macrofamilies which have been proposed, but whose adequacy is still a moot point. In a way, one should like to mimic what evolutionary bioinformatics has been able to achieve in the genetic domain of quaternary DNA strings; it is no surprise that tools of bioinformatics, e.g. those used in character-based and distance-based classifications, have been exported into linguistics, cf. e.g. [1,12], a fact we shall comment upon below. Longobardi's approach is defended in [1,11–13], to which the linguistic-minded reader is referred. In linguistics, binary strings are obtained by specifying n linguistic features which can be present = 1

M.-J. Lesot et al. (Eds.): IPMU 2020, CCIS 1239, pp. 17–26, 2020.
https://doi.org/10.1007/978-3-030-50153-2_2

or absent = 0 in a given language L. Since the length n is the same for all strings describing the languages L, $\Lambda, \ldots$ one intends to classify, a distance like Hamming distance, which counts the number of distinct features, appears to be adequate to gauge the dissimilarity between two languages L and Λ (between the corresponding strings), were it not that Longobardi's features are constrained by a complex network of logical implications. This has lead Longobardi's school to the use of a string distance which modifies the Hamming definition so as to get rid of positions corresponding to undefined (and undefinable) positions; the drawback is that the generalized distance they resort to is *not* metric, as instead often required by clustering techniques used to obtain the corresponding classification trees. Due to reasons discussed in next section, also a non-metric generalization the *Jaccard distance* has been used by Longobardi's school: the Jaccard distance "ignores" positions corresponding to features which are absent in both languages L and Λ, and so are linguistically "irrelevant", cf. next section. Now, the original 0–1 Jaccard distance (no inconsistency) can be obtained from the standard Hamming distance by means of a powerful mathematical tool called the *Steinhaus transform*: this transform needs the specification of a *pivot*, which in the Jaccard case is precisely the all-zero string; we stress that the Steinhaus transform of a metric distance is itself metric. In [8] one has already used Steinhaus transforms to deal with old *fuzzy* linguistic data due to Ž. Muljačić, where logical values can be intermediate between 0 and 1: this gave us the idea to represent logical inconsistency by the "ambiguous" value $\frac{1}{2}$ which is equidistant from both *crisp* logical values 0 and 1, cf. Sects. 3 and 4, and to use as pivot the "totally ambiguous" string, i.e. the all-$\frac{1}{2}$ string. The results which we obtain with this Steinhaus transform are surprisingly good, as commented upon in Sects. 3 and 4. The fact that moving from the original Longobardi's non-metric distance to our Steinhaus metric distance leaves the classification tree largely unchanged may be interpreted as a proof of the *robustness* of Longobardi's data: this is in puzzling contrast with results obtained by bootstrapping techniques described and used in [1,11,12] and suggested by bioinformatics, which seemed to show that Longobardi's original classification is not that robust. Below, in Sects. 3 and 4, we argue that this seeming non-robustness might be due to the inadequacy of tools exported from bioinformatics to linguistics. Rather than by bootstrapping, we prefer to validate the robustness of data by *randomly perturbing* our Steinhaus distance, or rather by randomly perturbing the pivot which is used: results are shown and commented upon in Sects. 3 and 4.

2 A Detour: From Muljačić to Steinhaus

In the past, the authors have been working on old and new linguistic data [6–10]; the starting point is the same: languages L, Λ, ... are described by n linguistic features f_i, $1 \leq i \leq n$, which in each language can be either present (1 = true) or absent (0 = false). The usual (crisp) Hamming distance, which counts the number of positions i where the corresponding bits are different, would be to the point, but both in the old Muljačić data and in the new ones due to Longobardi there is a stumbling block, since "ambiguous" situations are possible. Even if in both cases the symbols we will be using[1] are $0, \frac{1}{2}, 1$, the symbol $\frac{1}{2}$, i.e. neither true nor false, neither present nor absent, has a distinct meaning.

In the case of Muljačić [14], $\frac{1}{2}$ can be interpreted as a logical value intermediate between 0 and 1 in a multivalued logic as is *fuzzy logic*, for which cf. e.g. [5]. In ampler generality one may consider strings $\underline{x} = x_1 x_2 \dots x_n$ where each component x_i may belong to the *whole* interval [0,1] allowing for all possible "shadings" of logical values. To define an adequate distance between fuzzy strings, suitably generalizing the usual Hamming distance between crisp strings, the relevant question to be posed is: if x and y are the logical values of feature f in the two languages L and Λ represented by the two strings $\underline{x}$ and $\underline{y}$, is f {present in L *and* absent in Λ} *or* {absent in L *and* present in λ}? Let $\perp$ and $\top$ be the disjunction *or* and the conjunction *and* in the multi-valued logic we choose to use; as for the negation, denoted by an overline, we will always use the 1-complement: $\overline{x} = 1 - x$ (the symbols $\top$ and $\perp$ which we are using for abstract conjunctions and disjunctions remind one of $\wedge$ and $\vee$, and are common when dealing with T-*norms*, cf. e.g. [5]). Assuming additivity w.r.t. to the n features, one gets for the distance $d(\underline{x}, \underline{y})$ between two strings $\underline{x}$ and $\underline{y}$, and so for the corresponding distance $d(\text{L}, \Lambda)$ between languages L and Λ:

$$d(\underline{x}, \underline{y}) = \sum_{1 \le i \le n} (x_i \top \overline{y_i}) \perp (\overline{x_i} \top y_i) \tag{1}$$

In the case of standard fuzzy logic, conjunction *and* and disjunction *or* are computed through *minima* $\wedge$ and *maxima* $\vee$, respectively, $x \top y = x \wedge y = \min[x, y]$, $x \perp y = x \vee y = \max[x, y]$. The distance one obtains from (1) is a fuzzy generalization of the usual crisp Hamming distance; rather than fuzzy Hamming distance as in [15], or even Sgarro distance as in [3], we found it proper to call it *Muljačić distance*. We stress that use of the latter distance has proved to be quite successful in the case of Muljačić data, which, unlike Longobardi's, are genuinely fuzzy. The curious reader is referred to [6,7].

Already in [7] we tried several other logical operators of multi-valued logics, for example Łukasiewicz operators $x \perp y = (x+y) \wedge 1 = \min[x+y, 1]$, $x \top y = (x+y-1) \vee 0 = \max[x+y-1, 0]$. The results were in general uninteresting, since the distances one obtains were metrically unacceptable, cf. [7]; instead, Łukasiewicz case was surprising: as a straightforward computation shows one re-obtains the very well-known *Manhattan distance* or *taxicab distance* or *Minkowski distance*

$$d_T(\underline{x}, \underline{y}) = \sum_{1 \le i \le n} |x_i - y_i|$$

which in this context might even be called *Łukasiewicz distance*. It is precisely this distance that we shall use below, rather than Muljačić; for a more extensive discussion cf. [7,9].

[1] Longobardi instead of $0 \ \frac{1}{2} \ 1$ uses $- \ 0 \ +$.

Before moving to Longobardi's data, we tackle *Steinhaus transforms*. The starting point was Longobardi's observation that positions i where both languages have a zero are linguistically *irrelevant*, and so should be ignored: mathematically, one has to move from Hamming distances to Jaccard distances[2]. What if the strings are not crisp? How should one go from Hamming distances or Muljačić distances to their Jaccard-like counterparts? The answer is precisely the Steinhaus transform, cf. e.g. [3]:

$$\delta_{St}(x,y) = \frac{2\delta(x,y)}{\delta(x,y)+\delta(x,z)+\delta(y,z)} \tag{3}$$

where $\delta(x,y)$ is any metric distance between objects which are not necessarily strings, and where z is a chosen fixed object called the *pivot* of the transformation. As it can be proved, the Steinhaus transform is itself a metric distance; it is normalized to 1, and is equal to 1 when x, z, y form an aligned triple $\delta(x,z)+\delta(z,y) = \delta(x,y)$ for the original distance to be transformed. Now, going back to our strings $\underline{x}, \underline{y}, \ldots$, the Jaccard case corresponds to taking an all-zero pivot string $\underline{z} = 00\ldots0$, in which case the distance from the pivot is nothing else but the *fuzzy weight* $w(\underline{x}) = d(\underline{x},\underline{z}) = \sum_i x_i$ both with Muljačić and the taxicab distance.

The reason why we mentioned here irrelevance is simply that it paves the way to the use of Steinhaus transforms, even if with a different pivot, as we are going to do in the next section.

3 Dealing with Inconsistency

We move to Longobardi's ternary strings[3], where a complex network of logical implications involves features, of the type: if f_2 is false and f_4 is true, then f_6 does not make sense, it is logically inconsistent. In the case of inconsistency we use once more the symbol $\frac{1}{2}$: in the example just given $f_2 = 0$ and $f_4 = 1$ implies $f_6 = \frac{1}{2}$.

The distance used by Longobardi's school is simply a normalized Hamming distance, where the positions where one or both languages have a $\frac{1}{2}$ are *ignored*: in practice, one deals with *shorter* strings, possibly *much* shorter. Since Longobardi's distance is *not* metric, we took a bold step to preserve metricity. In Muljačić case, the numeric value $\frac{1}{2}$ represents suitably total logical ambiguity, but certainly not logical inconsistency, as however we shall now do. In the case of irrelevance an all-0 string did the job and got us rid of positions which are

[2] Actually, in as yet unpublished Longobardi's research this point of view has been relinquished and only inconsistency is taken care of, as we are doing below.

[3] Data we shall work on refer to 38 world languages described by means of 53 syntactic features, cf. [13] and Sect. 4, but Longobardi's group are constantly updating, improving and extending their database.

irrelevant. Forgetting about irrelevance, but of course not about inconsistency, here we shall take a totally ambiguous or rather totally inconsistent pivot string, which is the all-$\frac{1}{2}$ string $\underline{z} = \frac{1}{2}\frac{1}{2}...\frac{1}{2}$; this gets us rid of positions where there is inconsistency in both languages L and Λ, but not, as instead Longobardi's own distance does, of positions where only one of the two is inconsistent: actually, this turns out to be a possible source of weakness in Longobardi's choice, since few positions might survive if far-off languages are compared.

Now, rather than weights, cf. Sect. 2, one has *consistencies*, i.e. distances $d(\underline{x}, \frac{1}{2}\frac{1}{2}...\frac{1}{2})$ from the new pivot[4]: Łukasiewicz consistency turns out to be, as is proper, $\sum_i |x_i - \frac{1}{2}| = \sum_i \left(\frac{1}{2} - f(x_i)\right)$, where $f(x_i) = x \wedge (1-x)$ is often seen as the *fuzziness* of the logical value x, since it is the Euclidean distance from the totally ambiguous fuzzy value $\frac{1}{2}$.

Let us move to Longobardi's data, so as to illustrate our methodology. The tree we obtain, fig. (b), is definitely and surprisingly[5] good, as it is virtually undistinguishable from the original Longobardi's tree (a) [13] based on a non-metric distance, and is linguistically equally sound. The fact that two distinct distances perform so similarly appears to be an indication that data are quite robust. Instead, use of statistical bootstrap techniques as done by Longobardi's school seemed to show that data are not that robust, cf. [1,11–13]. Actually, bootstrapping works quite well with the strings of bioinformatics, whose length is by magnitudes larger than ours, hundreds of thousands vs 53 (also in the case of DNA strings the assumptions of independence between positions, particularly if nearby, is untenable, but this weak point is smoothed out by the huge length of the strings involved). In our case strings are comparatively short and the structure of dependences is pervasive and strong, so the poor performance of bootstrapping might be simply an indication that the network of logical rather than statistical dependences makes the use of bootstrapping inadequate.

Instead, we propose an alternative to check robustness: let us *perturb* the distance, and see what happens. We shall take at random the pivot string $\underline{Z}$ (totally at random with *uniform* distribution on $[0,1]^n$), and check which sort of trees we obtain, taking also into account the taxicab distance between the *observed* random pivot and the "correct" all-$\frac{1}{2}$ pivot (capital letters denote random variables or random n-tuples).

Since the n terms in the random distance $d_T(\underline{x}, \underline{Z}) = \sum_{1 \le i \le n} |x_i - Z_i|$ are independent, not only the expectation, but also the variance is additive, and so

---

[4] Muljačić consistency, based on *maxima* and *minima*, is unusable, being always $\frac{n}{2}$ independent of $\underline{x}$; cf. [7,9].

[5] Farsi (modern Persian) appears to be poorly classified, which is true also with Longobardi's original tree. Also Bulgarian is poorly classified, but Longobardi uses only features relative to the syntax of nouns, and the Bulgarian noun, due to *substratum* influences, is well-known to be an outsider among Slavic languages. Be as it may, the aim of this paper is simply to check mathematical tools and robustness of data, rather than outperforming current classifications; cf. instead [4].

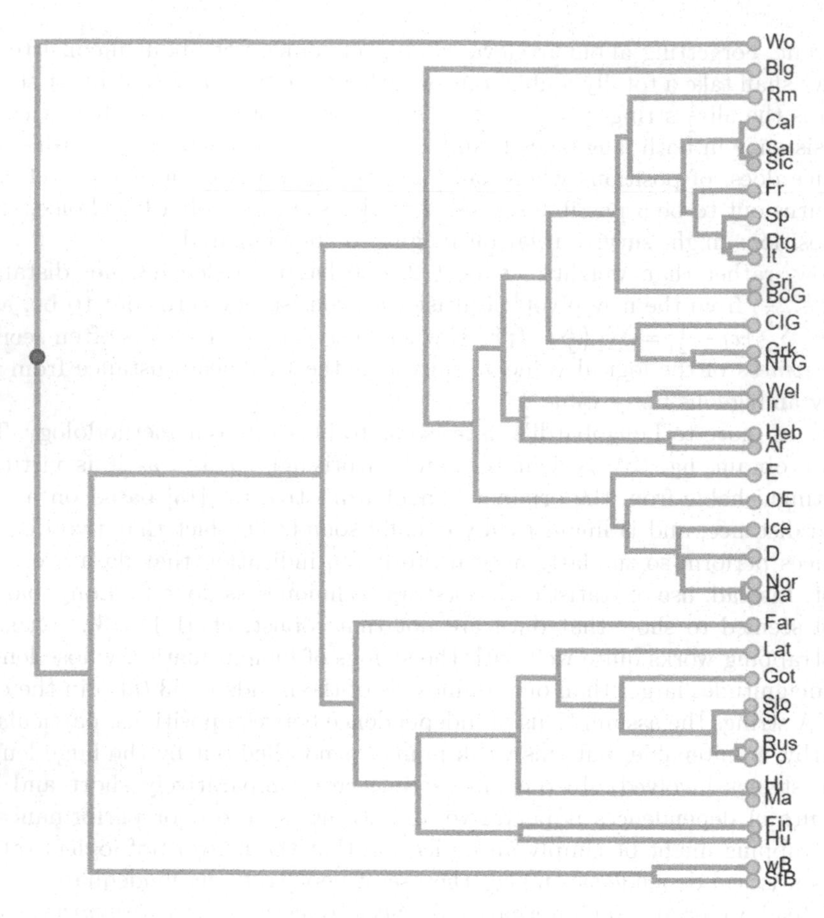

(a) Longobardi's tree

it will be enough to assume $n = 1$. Straightforward computations show that, for given $x \in [0,1]$:

$$\mathrm{E}\big[d_T(x,Z)\big] = x(x-1) + \frac{1}{2}\,, \quad \mathrm{var}\big[d_T(x,Z)\big] = -x^2(x-1)^2 + \frac{1}{12}$$

For XZ uniform on $[0,1]^2$ one has $\mathrm{E}\big[d_T(X,Z)\big] = \frac{1}{3}$, while for x crisp, i.e. $x = 0$ or $x = 1$, the taxicab expectation is $\frac{1}{2}$ and the taxicab variance is $\frac{1}{12}$.

Rather, we are interested in the case when x has the "correct" pivot value $\frac{1}{2}$: then expectation and variance are equal to $\frac{1}{4}$ and $\frac{1}{48}$, respectively, and so, for $n \geq 1$ the standard deviation $\sigma = \sigma\big[d_T(x,Z)\big]$ is approximately $0.144\sqrt{n} \approx 1.05$ with $n = 53$. Since the random distance $d_T(\frac{1}{2} \dots \frac{1}{2}, \underline{Z})$ is the sum of $n = 57$ i.i.d. terms, the central limit theorem allows one to resort to a normal approximation, and so the three intervals of semi-width $i\sigma$, $i = 1, 2, 3$ centered in the expected distance have probability ≈ 0.68, 0.95, 0.997, respectively. Correspondingly, the trees will be called of type α (observed distance inside the first and most probable interval), β (outside the first interval but inside the second), γ (outside the second interval but inside the third), else δ. So, trees of type α, β and γ have approximately probability 0.68, 0.27 and 0.04, respectively.

4 Experimental Results

We reproduce ten trees for Longobardi's data, the first (b) with the correct pivot all-$\frac{1}{2}$, the others, (c) to (k), with a random pivot string; α and β-trees are virtually identical with the unperturbed tree, and in particular preserve the Indoeuropean standard groups; the γ-tree is weaker, e.g. it creates a single large family for Semitic and Celtic languages. For more random trees obtained in successive trials cf. [16].

The languages are 38, namely Sic = Sicilian, Cal = Calabrese as spoken in South Italy, It = Italian, Sal = Salentine as spoken in Salento, South Italy, Sp = Spanish, Fr = French, Ptg = Portuguese, Rm = Romanian, Lat = Latin, ClG = Classical Attic Greek, NTG = New Testament Greek, BoG = Bova Greek as spoken in the village of Bova, Italy, Gri = Grico, a variant of Greek spoken in South Italy, Grk = Greek, Got = Gothic, OE = Old English, E = English, D = German, Da = Danish, Ice = Icelandic, Nor = Norwegian, Blg = Bulgarian, SC = Serbo(Croatian), Slo = Slovenian, Po = Polish, Rus = Russian, Ir = Gaelic Irish, Wel = Welsh, Far = Farsi, Ma = Marathi, Hi = Hindi, Ar = Arabic, Heb = Hebrew or 'ivrit, Hu = Hungarian, Finn = Finnish, StB = standard Basque, wB = Western Basque, Wo = Wolof as spoken mainly in Senegal.

Cf. the supplementary material [16] for more information, inclusive of the 38×53 ternary matrix with the strings of length $n = 53$ associated to the 38 languages.

Final remarks. Unsurprisingly, the trees exhibited perform all very well when compared to ours and the original Longobardi's tree, cf. [1,11–13]; cf. also footnote 5. A finer statistical analysis to gauge tree similarity might require suitable distances between trees like tree edit distances, as we are currently doing in [4]; here we have been more easy-going, since observed similarities are quite obvious to the eye of the linguist. Note that phylogenetic tree distances as we would need, cf. [3], are known to raise nasty computational problems. Unsurprisingly, thinking of Gray's classification tree [2], largely recognized by the linguistic community as a sort of reference benchmark, use of tree distances shows that Longobardi's tree and our own tree have virtually the same distance from Gray's tree, even if the distances used are quite distinct, one of the two not even metric, cf. [4]; once more, this appears to be an indication that Longobardi's data are quite robust. The statistical technique of random perturbation might be readily extended to the generalized Steinhaus distance used in [9], where one copes jointly with irrelevance and inconsistency; cf. however footnote 2.

The idea we are trying to defend in this paper is that, rather than mimicking bioinformatics, evolutionary linguistic should try to create its own new tools. This need has become more and more evident in Longobardi's research, where strings are dramatically shorter than those of bioinformatics and where the distances used, including our own metric distances, are quite different from those used in bioinformatics for distance-based classifications.

Our current work takes into account the new larger data tables provided by Longobardi's school; these data include many non-Indoeuropean languages, so as to get rid of an unwanted prominence of "usual" languages. This much enhances the linguistic significance of the results obtained.

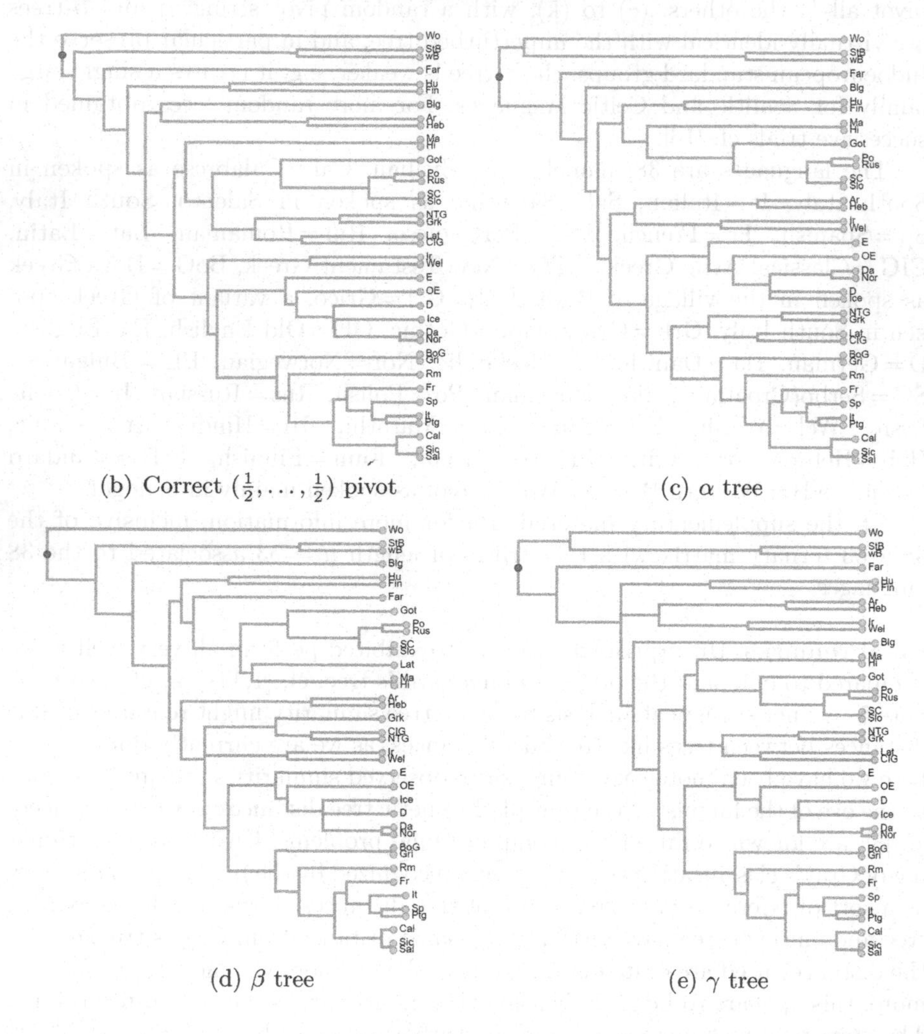

(b) Correct $(\frac{1}{2}, \ldots, \frac{1}{2})$ pivot

(c) α tree

(d) β tree

(e) γ tree

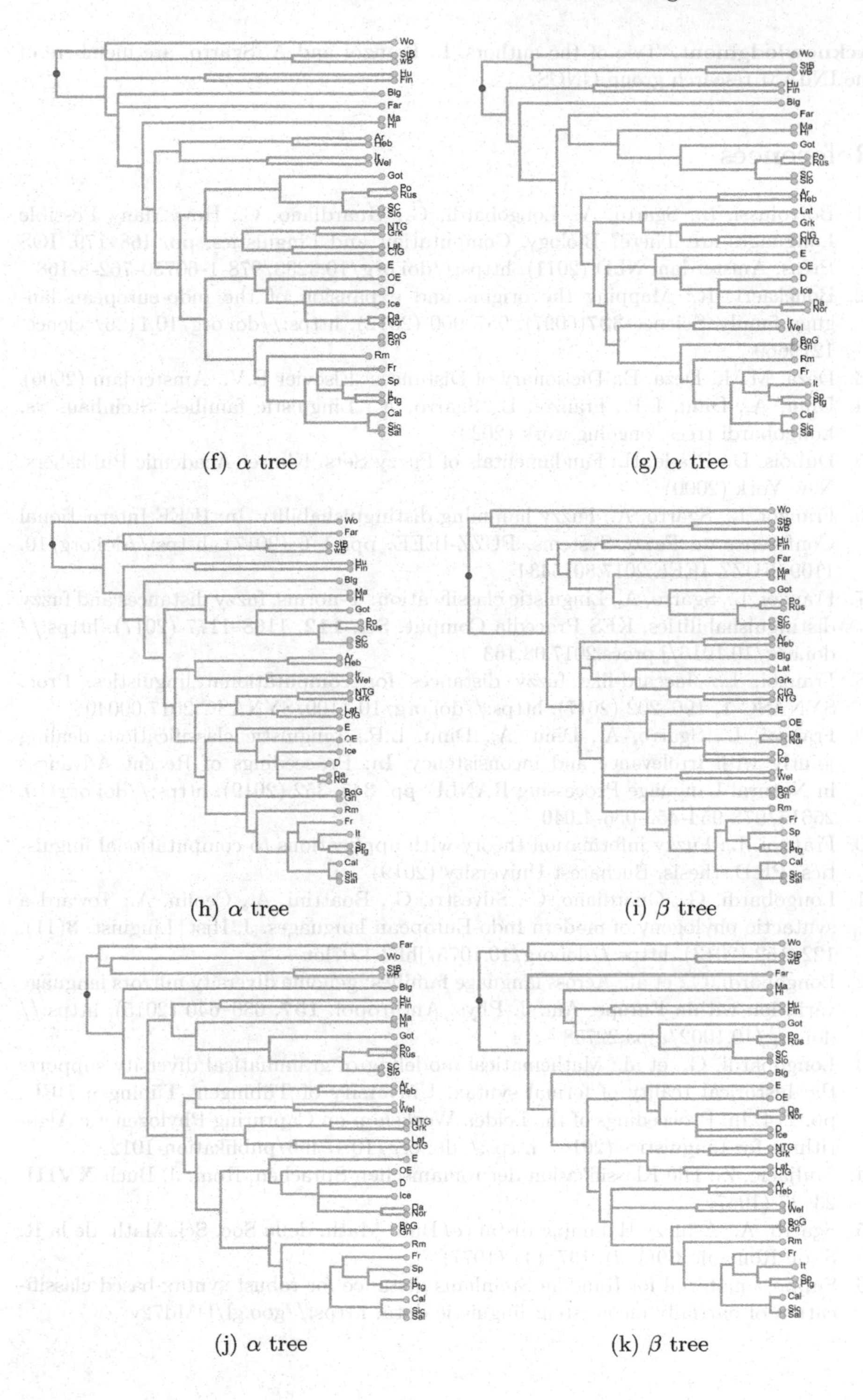

(f) α tree

(g) α tree

(h) α tree

(i) β tree

(j) α tree

(k) β tree

Acknowledgment. Two of the authors, L. Franzoi and A. Sgarro, are members of the INdAM research group GNCS.

References

1. Bortolussi, L., Sgarro, A., Longobardi, G., Guardiano, C.: How Many Possible Languages are There? Biology, Computation and Linguistics, pp. 168–179. IOS Press, Amsterdam NLD (2011). https://doi.org/10.3233/978-1-60750-762-8-168
2. Bouckaert, R.: Mapping the origins and expansion of the indo-european languge family. Science **337**(6097), 957–960 (2012). https://doi.org/10.1126/science.1219669
3. Deza, M.M., Deza, E.: Dictionary of Distances. Elsevier B.V., Amsterdam (2006)
4. Dinu, A., Dinu, L.P., Franzoi, L., Sgarro, A.: Linguistic families: Steinhaus vs. Longobardi trees, ongoing work (2020)
5. Dubois, D., Prade, H.: Fundamentals of Fuzzy Sets. Kluwer Academic Publishers, New York (2000)
6. Franzoi, L., Sgarro, A.: Fuzzy hamming distinguishability. In: IEEE International Conference on Fuzzy Systems, FUZZ-IEEE, pp. 1–6 (2017). https://doi.org/10.1109/FUZZ-IEEE.2017.8015434
7. Franzoi, L., Sgarro, A.: Linguistic classification: T-norms, fuzzy distances and fuzzy distinguishabilities. KES Procedia Comput. Sci. **112**, 1168–1177 (2017). https://doi.org/10.1016/j.procs.2017.08.163
8. Franzoi, L.: Jaccard-like fuzzy distances for computational linguistics. Proc. SYNASC **1**, 196–202 (2017). https://doi.org/10.1109/SYNASC.2017.00040
9. Franzoi, L., Sgarro, A., Dinu, A., Dinu, L.P.: Linguistic classification: dealing jointly with irrelevance and inconsistency. In: Proceedings of Recent Advances in Natural Language Processing RANLP, pp. 345–352 (2019). https://doi.org/10.26615/978-954-452-056-4_040
10. Franzoi, L.: Fuzzy information theory with applications to computational linguistics. Ph.D. thesis, Bucharest University (2019)
11. Longobardi, G., Guardiano, C., Silvestri, G., Boattini, A., Ceolin, A.: Toward a syntactic phylogeny of modern Indo-European languages. J. Hist. Linguist. **3**(11), 122–152 (2013). https://doi.org/10.1075/jhl.3.1.07lon
12. Longobardi, G., et al.: Across language families: genome diversity mirrors language variation within Europe. Am. J. Phys. Anthropol. **157**, 630–640 (2015). https://doi.org/10.1002/ajpa.22758
13. Longobardi, G., et al.: Mathematical modeling of grammatical diversity supports the historical reality of formal syntax, University of Tübingen, Tübingen DEU, pp. 1–4. In: Proceedings of the Leiden Workshop on Capturing Phylogenetic Algorithms for Linguistics (2016). https://doi.org/10.15496/publikation-10122
14. Muljačić, Z.: Die Klassifikation der romanischen Sprachen. Rom. J. Buch **XVIII**, 23–37 (1967)
15. Sgarro, A.: A fuzzy Hamming distance. Bull. Math. de la Soc. Sci. Math. de la R. S. de Romanie **69**(1–2), 137–144 (1977)
16. Support material for Random Steinhaus distance for robust syntax-based classification of partially inconsistent linguistic data. https://goo.gl/DMd72v

Possibilistic Bounds for Granular Counting

Corrado Mencar(✉)

Dipartimento di Informatica, Università degli Studi di Bari "Aldo Moro", 70125 Bari, Italy
corrado.mencar@uniba.it

Abstract. Uncertain data are observations that cannot be uniquely mapped to a referent. In the case of uncertainty due to incompleteness, possibility theory can be used as an appropriate model for processing such data. In particular, granular counting is a way to count data in presence of uncertainty represented by possibility distributions. Two algorithms were proposed in literature to compute granular counting: exact granular counting, with quadratic time complexity, and approximate granular counting, with linear time complexity. This paper extends approximate granular counting by computing bounds for exact granular count. In this way, the efficiency of approximate granular count is combined with certified bounds whose width can be adjusted in accordance to user needs.

Keywords: Granular counting · Possibility theory · Uncertain data

1 Introduction

Data uncertainty may arise as a consequence of several conditions and require proper management [1]. The simplest approach is to ignore uncertainty by estimating a precise value for each observation, but this simplistic approach, though of very simple application, can lead to a distortion in the subsequent processing stages that is difficult to detect. A more comprehensive approach should take into account data uncertainty and propagate its footprint throughout the entire data processing flow. In this way, the results of data processing reveal their uncertainty, which can be evaluated to assess their ultimate usefulness.

Several theories can be applied to represent and process uncertainty, such as Probability Theory [2], which is however a particular case falling in the Granular Computing paradigm. Granular Computing also includes classical Set Theory [3], Rough Sets Theory [4], Evidence Theory [5] and Possibility Theory [6]. The choice of a particular theory depends on the nature of uncertainty; in particular, possibility theory deals with uncertainty due to incomplete information, e.g. when the value of an observation cannot be precisely determined: we will use the term *uncertain data* to denote data characterized by this specific type of uncertainty, therefore we adopt the possibilistic framework in this paper.

A common process on data is counting, which searches for the number of data samples with a specific value. Data counting is often a preliminary step for

M.-J. Lesot et al. (Eds.): IPMU 2020, CCIS 1239, pp. 27–40, 2020.
https://doi.org/10.1007/978-3-030-50153-2_3

different types of analysis, such as descriptive statistics, comparative analysis, etc. It is a fairly simple operation when data are accurate, but it becomes non-trivial when data are uncertain. In fact, the uncertainty in the data should propagate in the count, so that the results are granular rather than precise.

Recently, a definition of granular count through Possibility Theory was proposed [7]. It was shown that the resulting counts are fuzzy intervals in the domain of natural numbers. Based on this result, two algorithms for granular counting were defined: an exact granular counting algorithm with quadratic-time complexity and an approximate counting algorithm with linear-time complexity. Approximate granular counting is appealing in applications dealing with large amounts of data due to its low complexity, but a compromise must be accepted in terms of accuracy of the resulting fuzzy interval. In particular, it is not immediate to know how far is the result of the approximate count from the fuzzy interval resulting from the exact granular count.

In this paper an algorithm is proposed for bounded granular counting, which computes an interval-valued fuzzy set representing the boundaries in which the exact granular count is located. In this way, the efficiency of approximate granular count is combined with certified bounds whose width can be adjusted in accordance to user needs.

The concept of granular count and related algorithms are briefly described in Sect. 2, while the proposal of bounded granular count is introduced in Sect. 3. Section 4 reports some numerical experiments to assess the efficiency of the proposed algorithm, as well as an outline of an application in Bioinformatics.

2 Granular Count

A brief summary of Granular Counting is reported in this Section. Further details can be found in the original papers [7,8].

We assume that data are manifested through *observations*, which refer to some objects or *referents*. The relation between observations and referents—which is called *reference*—may be uncertain in the sense that an unequivocal reference of the observation to one of the referents is not possible. We model such uncertainty with Possibility Theory [6] as we assume that uncertainty is due to the imprecision of the observation, i.e. the observation is not complete enough to make reference unequivocal.

Given a set R of referents and an observation $o \in O$, a possibility distribution is a mapping

$$\pi_o : R \mapsto [0,1]$$

such that $\exists r \in R : \pi_o(r) = 1$. The value $\pi_o(r) = 0$ means that it is impossible that the referent r is referred by the observation, while $\pi_o(r) = 1$ means that the referent r is absolutely possible (though not certain). Intermediate values of $\pi_o(r)$ stand for gradual values of possibility, which quantify the completeness of information resulting from an observation. (More specifically, the lower the possibility degree, the more information we have to *exclude* a referent.) The possibility distributions of all observations can be arranged in a *possibilistic assignment table*, as exemplified in Table 1.

Table 1. Example of possibilistic assignment table. Each row is a possibility distribution π_{o_j}.

	r_1	r_2	r_3
o_1	1	0.3	0.54
o_2	0.8	1	0.6
o_3	1	0	0
o_4	0.864	0.91	1
o_5	1	0	0
o_6	0.5	1	0.64
o_7	1	0.8	1
o_8	0.2	0.5	1
o_9	1	0	0
o_{10}	0.6	1	0.78

2.1 Definition of Granular Count

By using the operators of Possibility Theory, as well as the assumption that observations are non-interactive (i.e. they do not influence each other), the possibility degree, that a subset $O_x \subseteq O$ of $x \in \mathbb{N}$ observations is *exactly*[1] the set of observations referring to a reference $r_i \in R$, is defined as:

$$\pi_{O_x}(r_i) = \min\left\{\min_{o \in O_x} \pi_o(r_i), \min_{o \notin O_x} \max_{r \neq r_i} \pi_o(r)\right\} \tag{1}$$

with the convention that $\min \emptyset = 1$. Informally speaking, Eq. (1) defines the possibility degree that O_x is the subset of all and only the observations of r_i by computing the least possibility degree of two simultaneous events: (i) all observations of O_x refer to r_i, and (ii) all the other observations refer to a different referent.

In order to compute the possibility degree that the number of observations referring to a referent r_i is N_i, we are not interested in a specific set O_x, but in *any set* of x elements. We can therefore define the possibility value that the number of observations for a referent r_i is x as:

$$\pi_{N_i}(x) = \max_{O_x \subseteq O} \pi_{O_x}(r_i) \tag{2}$$

for $x \leq m$ and $\pi_{N_i}(x) = 0$ for $x > m$. Equation (2) provides a granular definition of count. Counting is imprecise because observations are uncertain.

It is possible to prove that a granular count as in Eq. (2) is a fuzzy interval in the domain of natural numbers. A fuzzy interval is a convex and normal fuzzy set on a numerical domain (in our case, it is $\mathbb{N}$). Convexity of a fuzzy set can be established by proving that all α-cuts are intervals, while normality

[1] In the sense that any observation non belonging to O_x does not refer to r_i.

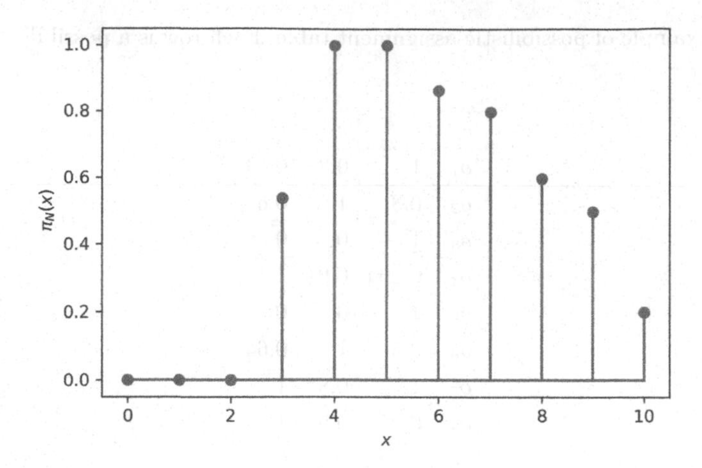

Fig. 1. Exact granular count of referent r_1 as in Table 1

of the granular count is guaranteed because of the normality of the possibility distributions π_o for all $o \in O$. Figure 1 depicts an example of granular count.

2.2 Algorithms for Granular Counting

The direct application of Eq. (2) leads to an intractable counting procedure as all possible subsets of O must be considered. On the other hand, a polynomial-time algorithm can be devised by taking profit of the representation of a granular count as a fuzzy interval. In particular, the granular counting algorithm builds the fuzzy interval by considering the α-cut representation of fuzzy sets. On such basis, two variants of granular counting algorithms can be devised:

- *Exact* granular counting uses all the values of α that correspond to some possibility degree in the possibilistic assignment table;
- *Approximate* granular counting uses the values of α taken from a finite set of evenly spaced numbers over $[0, 1]$. The number of such values depend on a user-defined parameter n_α.

The approximate granular counting is more efficient than the exact version because it does not require to scan the possibilistic assignment table, though at the price of a new required parameter.

Exact granular counting (Algorithm 1) and approximate granular counting (Algorithm 2) share the same core algorithm (Algorithm 3) and only differ by how the set of α-values are computed. In essence, the core algorithm computes the granular count in an incremental way, by reckoning the α-cuts of the fuzzy interval for each α value provided in input.

In brief, the core algorithm works as follows. Given the possibilistic assignment table $\mathbf{R}$, the index i of the referent and the set A of α-cuts, the array $\mathbf{r}$ represents the possibility degrees that an observation refers to r_i, i.e. $\mathbf{r}_j = \pi_{o_j}(r_i)$

Algorithm 1: ExactGranularCount

```
  Data: R, i
  /* R: possibil. assignment table                        */
  /* i: index of referent to count                        */
  Result: N ∈ [0,1]^m
1 A ← {α ∈ R : α ≠ 0};
2 return GranularCount(R, i, A);
```

Algorithm 2: approximateGranularCounting

```
  Data: R, i, n_α
  /* R: possibil. assignment table                        */
  /* i: index of referent to count                        */
  /* n_α: number of α-levels                              */
  Result: N ∈ [0,1]^m
1 ε ← 10^-12;
2 A ← {ε + k · (1-ε)/(n_α-1) : k = 0, 1, ..., n_α - 1};
3 return GranularCount(R, i, A);
```

(line 1), while $\bar{\mathbf{r}}$ represents the possibility degrees that an observation refers to any other referent different from r_i (line 2). N is the array representing the granular count (line 3). The main cycle (lines 4–17) loops over each $\alpha \in A$ and computes the bounds $x_{\min}$ and $x_{\max}$ of the corresponding α-cut (line 5). These bounds are calculated by looping over all observations (lines 6–13), so that $x_{\max}$ is incremented if the possibility degree that the current observation refers to r_i is greater than or equal to α (lines 7–8), while $x_{\min}$ further requires that the possibility degree that the observation refers to any other referent is less than α (lines 9–10). When both $x_{\min}$ and $x_{\max}$ are computed, the degrees of membership of the granular count are updated accordingly (lines 14–16).

For a fixed referent, the time-complexity of exact granular count is $\mathcal{O}\left(nm^2\right)$ (being n the number of referents and m the number of observations), while the time-complexity of approximate granular count drops to $\mathcal{O}\left(m\left(n+n_\alpha\right)\right)$. In consideration that, in typical scenarios, the number of observations is very large (i.e., $m \gg n$), especially in comparison with the number of referents, it is deduced that approximate granular counting is the preferred choice in the case of very large amounts of uncertain data.

3 Bounding the Exact Granular Count

The time-complexity of approximate granular count linearly depends on the number of α values which, in turn, depend on the value of the parameter n_α. On one hand, low values of n_α lead to fast computation of granular counts; on the other hand, low values of n_α may lead to a rough estimate of the possibility degrees of the exact granular count.

Algorithm 3: GRANULARCOUNT

```
   Data: R, i, A
   Result: N ∈ [0,1]^m
   /* m is the number of observations                              */
1  r ← [R_ji] for j = 1,2,...,m;
2  r̄ ← [max_{k≠i} R_jk] for j = 1,2,...,m;
3  N ← [0,0,...,0] (m+1 times);
4  for α ∈ A do
5  |  x_min ← 0; x_max ← 0;
   |  /* Compute α-cut                                              */
6  |  for k = 1,2,...,m do
7  |  |  if R_ki ≥ α then
8  |  |  |  x_max ← x_max + 1;
9  |  |  |  if r̄_k < α then
10 |  |  |  |  x_min ← x_min + 1;
11 |  |  |  end
12 |  |  end
13 |  end
   |  /* Update granular count                                      */
14 |  for x ∈ x_min,...,x_max do
15 |  |  N[x] ← max{N[x], α} ;
16 |  end
17 end
18 return N
```

In Fig. 2 the Jaccard similarity[2] measure between approximate granular count and exact granular count is reported for n_α between 2 and 100: even though similarity values close to 1 are reached for $n_\alpha \gtrsim 20$, for smaller values a significant dissimilarity can be observed. In order to assess whether the discrepancy between approximate and exact granular counts is acceptable for a problem at hand, it is important to identify some bounds for the exact granular count when only an approximate count is available.

3.1 α-Cut Computation

In order to identify such bounds, a closer look at Algorithm 3 is necessary. The algorithm computes the granular count for the i-th referent given a possibilistic assignment table $\mathbf{R}$ and a set A of α-values. The main cycle within the algorithm computes the α-cut of the granular count, which is represented by the array

[2] Given the exact granular count π_N and an approximate count $\widetilde{\pi}_N$, the Jaccard similarity index is computed as

$$S = \frac{\sum_{x=0}^{m} \min\left\{\pi_N(x), \widetilde{\pi}_N(x)\right\}}{\sum_{x=0}^{m} \max\left\{\pi_N(x), \widetilde{\pi}_N(x)\right\}}$$

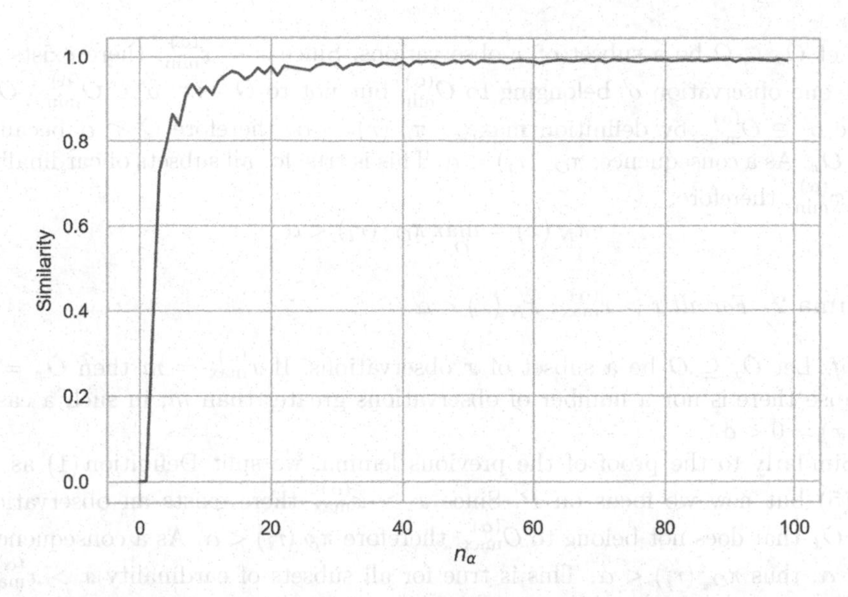

Fig. 2. Similarity of approximate count to exact granular count for referent r_1 in Table 1

N and corresponds to the possibility distribution π_N. For a given value of α, the variable $x_{\max}$ counts the number of observations that refer to r_i with a possibility degree $\geq \alpha$; on the other hand, the variable $x_{\min}$ counts the number of observations that refer to r_i with a possibility degree $\geq \alpha$ *and* refer to any other referent with possibility degree $< \alpha$. As a consequence, $x_{\min} \leq x_{\max}$. Since in our analysis we will consider different values of α, we shall denote the two variables as $x_{\min}^{(\alpha)}$ and $x_{\max}^{(\alpha)}$ respectively.

By construction, the value $x_{\max}^{(\alpha)}$ corresponds to the cardinality of the set

$$O_{\max}^{(\alpha)} = \{o \in O | \pi_o(r_i) \geq \alpha\} \tag{3}$$

while the value $x_{\min}^{(\alpha)}$ is the cardinality of the set

$$O_{\min}^{(\alpha)} = \left\{o \in O | \pi_o(r_i) \geq \alpha \wedge \max_{r \neq r_i} \pi_o(r) < \alpha\right\} \tag{4}$$

with the obvious relation that $O_{\min}^{(\alpha)} \subseteq O_{\max}^{(\alpha)}$. On this basis, it is possible to prove the following lemmas:

Lemma 1. *If $x_{\min}^{(\alpha)} > 0$, then for all $x < x_{\min}^{(\alpha)}$: $\pi_N(x) < \alpha$.*

Proof. By Definition (1), we can write

$$\pi_{O_x}(r_i) = \min\{P, Q\} \tag{5}$$

where $P = \min_{o \in O_x} \pi_o(r_i)$ and $Q = \min_{o \notin O_x} \max_{r \neq r_i} \pi_o(r)$. We focus on Q.

Let $O_x \subset O$ be a subset of x observations. Since $x < x_{\min}^{(\alpha)}$, there exists at least one observation o' belonging to $O_{\min}^{(\alpha)}$ but not to O_x, i.e. $o' \in O_{\min}^{(\alpha)} \setminus O_x$. Since $o' \in O_{\min}^{(\alpha)}$, by definition $\max_{r \neq r_i} \pi_{o'}(r) < \alpha$, therefore $Q < \alpha$ because $o' \notin O_x$. As a consequence, $\pi_{O_x}(r_i) < \alpha$. This is true for all subsets of cardinality $x < x_{\min}^{(\alpha)}$, therefore:

$$\pi_N(x) = \max_{O_x} \pi_{O_x}(r_i) < \alpha$$

Lemma 2. *For all* $x > x_{\max}^{(\alpha)}$*:* $\pi_N(x) < \alpha$.

Proof. Let $O_x \subseteq O$ be a subset of x observations. If $x_{\max}^{(\alpha)} = m$ then $O_x = \emptyset$ because there is not a number of observations greater than m; in such a case, $\pi_N(x) = 0 < \alpha$.

Similarly to the proof of the previous lemma, we split Definition (1) as in Eq. (5) but now we focus on P. Since $x > x_{\max}^{(\alpha)}$ there exists an observation $o' \in O_x$ that does not belong to $O_{\max}^{(\alpha)}$, therefore $\pi_{o'}(r_i) < \alpha$. As a consequence, $P < \alpha$, thus $\pi_{O_x}(r_i) < \alpha$. This is true for all subsets of cardinality $x > x_{\max}^{(\alpha)}$, thus proving the thesis.

Lemma 3. *For all* $x_{\min}^{(\alpha)} \leq x \leq x_{\max}^{(\alpha)}$, $\pi_N(x) \geq \alpha$

Proof. Obvious from Definitions (3) and (4).

The previous lemmas show that, for a given value of α, the exact granular count must satisfy the following relations:

$$\pi_N(x) \in [0, \alpha[\text{ if } x < x_{\min}^{(\alpha)} \vee x > x_{\max}^{(\alpha)}$$
$$\pi_N(x) \in [\alpha, 1] \text{ if } x_{\min}^{(\alpha)} \leq x \leq x_{\max}^{(\alpha)}$$

that is:

Theorem 1. *The interval* $\left[x_{\min}^{(\alpha)}, x_{\max}^{(\alpha)}\right]$ *is the* α*-cut of* π_N.

In Fig. 3 the 0.3- and 0.7- cuts are used to depict the regions that bound the values of π_N. Notice that such regions have been computed without knowing the actual values of the exact granular count.

3.2 Bounds for Exact Granular Count

Thanks to the properties of α-cuts, it is possible to identify tight bounds for an exact granular count by using the results of an approximate granular count. In fact, given two values $\alpha' < \alpha''$, the α''-cut is included in the α'-cut, therefore

$$\left[x_{\min}^{(\alpha'')}, x_{\max}^{(\alpha'')}\right] \subseteq \left[x_{\min}^{(\alpha')}, x_{\max}^{(\alpha')}\right]$$

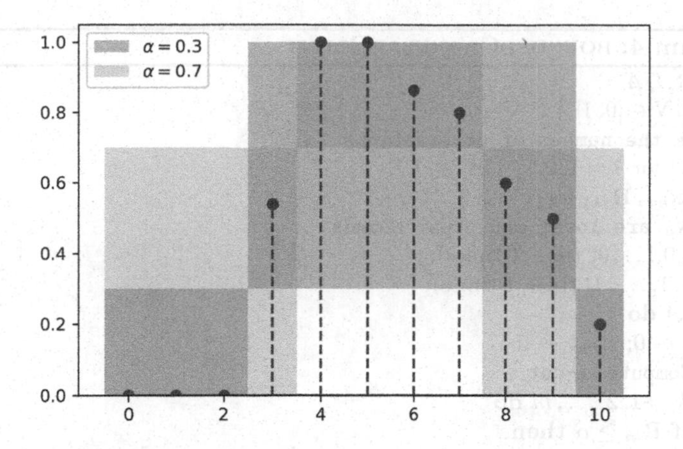

Fig. 3. α values bound the exact granular count π_N

and it is easy to verify the following properties:

$$x < x_{\min}^{(\alpha')} \Rightarrow \pi_N(x) < \alpha'$$
$$x_{\min}^{(\alpha')} \leq x < x_{\min}^{(\alpha'')} \Rightarrow \alpha' \leq \pi_N(x) < \alpha''$$
$$x_{\min}^{(\alpha'')} \leq x \leq x_{\max}^{(\alpha'')} \Rightarrow \pi_N(x) \geq \alpha''$$
$$x_{\max}^{(\alpha'')} < x \leq x_{\max}^{(\alpha')} \Rightarrow \alpha' \leq \pi_N(x) < \alpha''$$
$$x > x_{\max}^{(\alpha')} \Rightarrow \pi_N(x) < \alpha'$$

The previous relations suggest a strategy for computing the bounds of an exact granular count: supposing that, for some x, it is known that $\alpha_1 \leq \pi_N(x) \leq \alpha_2$ and a new value α is considered: if $x_{\min}^{(\alpha)} \leq x \leq x_{\max}^{(\alpha)}$ than it is possible to assert that $\pi_N(x) \geq \alpha$, therefore $\max\{\alpha, \alpha_1\} \leq \pi_N(x) \leq \alpha_2$; if x is outside this interval, then $\pi_N(x) < \alpha$ therefore $\alpha_1 \leq \pi_N(x) \leq \min\{\alpha, \alpha_2\}$.

On the basis of such strategy, it is possible to define a bounded granular counting algorithm to compute bounds of the exact granular count when a set A of α-cuts is given, which is reported in Algorithm 4. In this algorithm the bounds are represented by the arrays N_l and N_u (lines 3–4) and they are updated for each $\alpha \in A$ so as to satisfy the above-mentioned relations (lines 15–22).

The resulting bounded granular count is an Interval-Valued Fuzzy Set (IVFS) [9] which assigns, to each $x \in \mathbb{N}$, an interval $[\pi_{N_L}(x), \pi_{N_U}(x)]$ representing the possibility distribution of $\pi_N(x)$, which may be unknown. In Fig. 4 it is shown an example of bounded granular count by generating the set of α-values as in approximate granular count with $n_\alpha = 5$. It is possible to observe that the core of the granular count (i.e. the set of x values such that $\pi_N(x) = 1$) is precisely identified because the approximate granular counting algorithm includes $\alpha = 1$ in the set of α-values to be considered. Also, since a value of α very close to 0 is also included (namely, 10^{-12}), the impossible counts (i.e. the values of x such

Algorithm 4: BOUNDEDGRANULARCOUNT

```
   Data: R, i, A
   Result: N ∈ [0,1]^m
   /* m is the number of observations                              */
 1 r ← [R_ji] for j = 1,2,...,m;
 2 r̄ ← [max_{k≠i} R_jk] for j = 1,2,...,m;
   /* N_l, N_u are lower and upper bounds                          */
 3 N_l ← [0,0,...,0] (m+1 times);
 4 N_u ← [1,1,...,1] (m+1 times);
 5 for α ∈ A do
 6     x_min ← 0; x_max ← 0;
       /* Compute α-cut                                            */
 7     for k = 1,2,...,m do
 8         if R_ki ≥ α then
 9             x_max ← x_max + 1;
10             if r̄_k < α then
11                 x_min ← x_min + 1;
12             end
13         end
14     end
       /* Update bounds                                            */
15     for x ∈ x_min,...,x_max do
16         N_l[x] ← max{N_l[x], α} ;
17     end
18     for x ∈ 0,...,x_min − 1 do
19         N_u[x] ← min{N_u[x], α} ;
20     end
21     for x ∈ x_max + 1,...,m do
22         N_u[x] ← min{N_u[x], α} ;
23     end
24 end
25 return N_l, N_u
```

that $\pi_N(x) = 0$) are also detected. Finally, since the values of α are equally spaced in $]0,1]$, the lengths of the bounding intervals are constant.

It also possible to set n_α in order to achieve a desired precision. By looking at Algorithm 2, it is possible to observe that the values of α are equally spaced at distance

$$\Delta\alpha = \frac{1-\varepsilon}{n_\alpha - 1}$$

where the value $\Delta\alpha$ coincides with the maximum length of the intervals computed by the bounded granular counting algorithm. By reversing the problem, it is possible to set the value of n_α so that the maximum length is less than a desired threshold β. Since $\varepsilon \approxeq 0$, it suffices to set

$$n_\alpha = \left\lceil \frac{1}{\beta} \right\rceil + 1$$

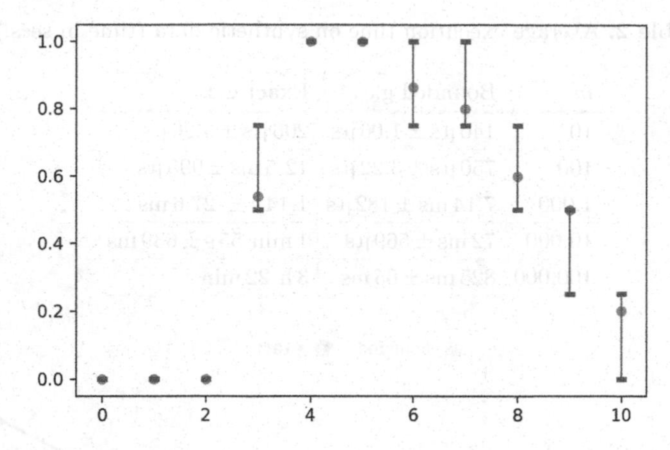

Fig. 4. Bounded granular count of r_1 for $n_\alpha = 5$. Middle dots represent the values of exact granular count.

4 Experimental Results

4.1 Efficiency Evaluation

The evaluation of efficiency has been performed on synthetically generated data. In particular, a number of random possibilistic assignment tables have been generated by varying the number of observations on a geometrical progression with common ratio 10 but keeping the number of referents fixed to three.[3]

For each possibilistic assignment table, both exact and bounded granular counting algorithms (with $n_\alpha = 10$) have been applied on the first referent, and the time required to complete operations has been recorded.[4] Each experiment has been repeated 7 times and average time has been recorded. For each repetition, the experiment has been looped for 10 times and the best timing has been retained. The average execution time is reported in Table 2 and depicted in Fig. 5.

A linear regression in the log-log scale confirms the quadratic trend of the time required for exact granular counting and the linear trend for bounded granular counting algorithm. Noticeably, the change of complexity is most exclusively due to the way the set of α-values have been generated: the selection of all values occurring in the possibilistic assignment table—which is required for exact granular counting—determines a significant reduction of the overall efficiency.

[3] Each possibilistic assignment table has been generated by taking care that each row corresponds to a normal possibility distribution.

[4] Experiments have been executed on a machine equipped by an Intel i7 CPU, 16GiB RAM, Linux SO. Scripts were written in Python 3.7 and executed in Jupyter Notebook. The NumPy library has been used for fast numerical computations, but the scripts were not implemented with the objective of maximizing performance.

Table 2. Average execution time on synthetic data (time in secs.)

m	Bounded g.c	Exact g.c.
10	$140\,\mu s \pm 1.06\,\mu s$	$206\,\mu s \pm 3.26\,\mu s$
100	$750\,\mu s \pm 3.22\,\mu s$	$12.5\,ms \pm 995\,\mu s$
1,000	$7.14\,ms \pm 182\,\mu s$	$1.14\,s \pm\ 27.6\,ms$
10,000	$72\,ms \pm 569\,\mu s$	$1\,min\ 55\,s \pm 639\,ms$
100,000	$825\,ms \pm 65\,ms$	$3\,h\ 22\,min$

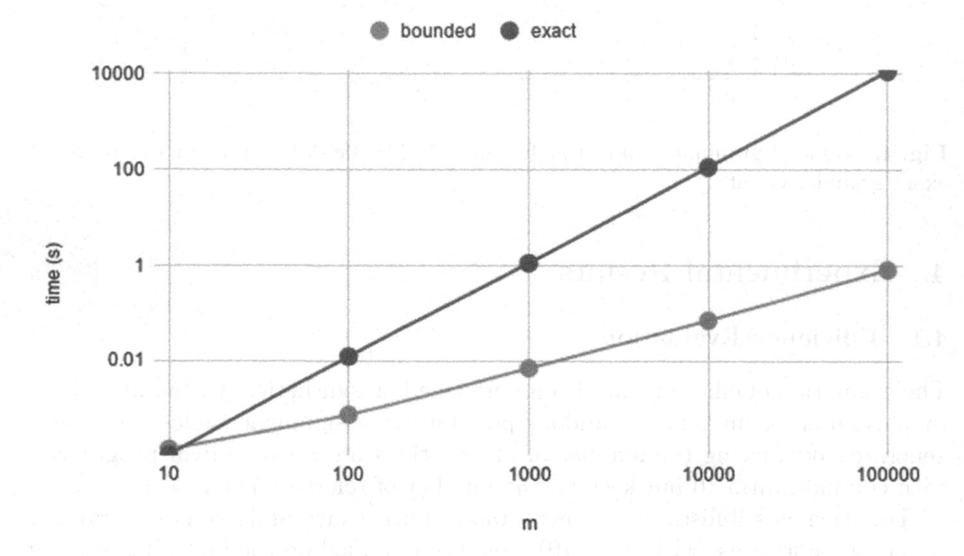

Fig. 5. Average execution time on synthetic data

On the other hand, bounded granular counting takes profit of the advantages of approximate granular counting for light-weight computations but, at the same time, it offers certified bounds on the possibility degrees of the exact granular count.

4.2 Application: Gene Expression Estimation

In Bioinformatics, RNA-Seq is a protocol that allows to examine the gene expression in a cell by sampling fragments of RNA called "reads". When RNA-Seq output is mapped against a reference database of known genes, a high percentage of reads—called *multireads*—map to more than one gene [10]. Multireads are a source of uncertainty in the quantification of gene expression, which should be managed in order to provide significant results. To this end, the mapping procedure provides a quality index that is a biologically plausible estimate of the possibility that a read can be associated to a gene [11]. However, a high

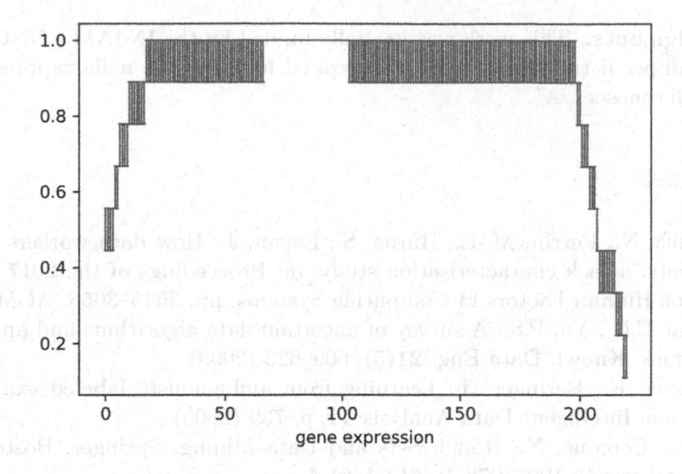

Fig. 6. Bounded granular counting of reads mapping to a sample gene

quality index does not mean certainty in association: two or more genes can be candidate for mapping a read because they can be mapped with similar high quality.

Granular counting finds a natural application in the specific problem of counting the number of reads that are possibly associated to a gene. (Reads are considered as observations, while genes are referents.) However, the amount of data involved in such process may be overwhelming. For example, the public dataset SRP014005 downloaded from NCBI-SRA archive[5], contains a case-control study of the Asthma disease with 55,579 reads mapped on 14,802 genes (16% are multireads). Nonetheless, accurate granular counting can be achieved by the use of the proposed algorithm. As an example, in Fig. 6 the bounded granular count has been computed for gene OTTHUMG00000189570—HELLPAR with $n_\alpha = 10$. It is noteworthy observing how imprecise is the count of this gene, which is due to a large number of multireads (with different quality levels).

5 Conclusions

The proposed bounded granular counting algorithm is an extended version of approximate granular counting where efficient computation is combined with the ability of bounding the exact granular count within intervals whose granularity can be decided by the user. In most cases, it is more than enough that the exact possibility degrees of exact granular count are assured to be within a small range from some approximate values. When such type of imprecision is tolerated, a significant speed-up in calculations can be achieved, thus opening the door of granular counting to big-data problems.

[5] ftp://ftp-trace.ncbi.nlm.nih.gov/sra/sra-instant/reads/ByStudy/sra/SRP/SRP014/SRP014005.

Acknowledgments. This work was partially funded by the INdAM - GNCS Project 2019 "Metodi per il trattamento di incertezza ed imprecisione nella rappresentazione e revisione di conoscenza".

References

1. Boukhelifa, N., Perrin, M.-E., Huron, S., Eagan, J.: How data workers cope with uncertainty: a task characterisation study. In: Proceedings of the 2017 CHI Conference on Human Factors in Computing Systems, pp. 3645–3656. ACM (2017)
2. Aggarwal, C.C., Yu, P.S.: A survey of uncertain data algorithms and applications. IEEE Trans. Knowl. Data Eng. **21**(5), 609–623 (2009)
3. Hüllermeier, E., Beringer, J.: Learning from ambiguously labeled examples. In: Advances in Intelligent Data Analysis VI, p. 739 (2005)
4. Lin, T.Y., Cercone, N.: Rough Sets and Data Mining. Springer, Boston (1996). https://doi.org/10.1007/978-1-4613-1461-5
5. Vannoorenberghe, P.: Reasoning with unlabeled samples and belief functions. In: The 12th IEEE International Conference on Fuzzy Systems. FUZZ 2003, vol. 2, pp. 814–818. IEEE (2003)
6. Dubois, D., Prade, H.: Possibility theory. In: Computational Complexity, pp. 2240–2252. Springer, Boston (2012). https://doi.org/10.1007/978-1-4684-5287-7
7. Mencar, C., Pedrycz, W.: Granular counting of uncertain data. Fuzzy Sets Syst. **387**, 108–126 (2019)
8. Mencar, C., Pedrycz, W.: GrCount: counting method for uncertain data. MethodsX **6**, 2455–2459 (2019)
9. Dubois, D., Prade, H.: Interval-valued fuzzy sets, possibility theory and imprecise probability. In: Proceedings - 4th Conference of EUSFLAT-LFA 2005, pp. 314–319 (2005)
10. Consiglio, A., Mencar, C., Grillo, G., Liuni, S.: Managing NGS differential expression uncertainty with fuzzy sets. In: Angelini, C., Rancoita, P.M.V., Rovetta, S. (eds.) CIBB 2015. LNCS, vol. 9874, pp. 42–53. Springer, Cham (2016). https://doi.org/10.1007/978-3-319-44332-4_4
11. Consiglio, A., Mencar, C., Grillo, G., Marzano, F., Caratozzolo, M.F., Liuni, S.: A fuzzy method for RNA-Seq differential expression analysis in presence of multireads. BMC Bioinform. **17**(S12:345), 167–182 (2016). https://doi.org/10.1186/s12859-016-1195-2

A Fuzzy Model for Interval-Valued Time Series Modeling and Application in Exchange Rate Forecasting

Leandro Maciel[1](✉), Rosangela Ballini[2], and Fernando Gomide[3]

[1] Department of Business, Faculty of Economics, Business and Accounting, University of São Paulo, São Paulo, Brazil
leandromaciel@usp.br

[2] Department of Economics Theory, Institute of Economics, University of Campinas, Campinas, São Paulo, Brazil
ballini@unicamp.br

[3] Department of Computer Engineering and Automation, School of Electrical and Computer Engineering, University of Campinas, Campinas, São Paulo, Brazil
gomide@dca.fee.unicamp.br

Abstract. Financial interval time series (ITS) is a time series whose value at each time step is an interval composed by the low and the high price of an asset. The low-high price range is related to the concept of volatility because it inherits intraday price variability. Accurate forecasting of price ranges is essential for derivative pricing, trading strategies, risk management, and portfolio allocation. This paper suggests a fuzzy rule-based approach to model and to forecast interval-valued time series. The model is a collection of functional fuzzy rules with affine consequents capable to express the nonlinear relationships encountered in interval-valued data. An application concerning one-step-ahead forecast of interval-valued EUR/USD exchange rate using actual data is also addressed. The forecast performance of the fuzzy rule-based model is compared to that of traditional econometric time series methods and alternative interval models employing statistical criteria for both, low and high exchange rate prices. The results show that fuzzy rule-based modeling approach developed in this paper outperforms the random walk, and other competitive approaches in out-of-sample interval-valued exchange rate forecasting.

Keywords: Interval-valued data · Exchange rate forecast · Fuzzy modeling

1 Introduction

Exchange rates play an important role in international trade and in economic competitiveness of a country because they influence the balance of payments.

M.-J. Lesot et al. (Eds.): IPMU 2020, CCIS 1239, pp. 41–53, 2020.
https://doi.org/10.1007/978-3-030-50153-2_4

These rates also have a significant impact on production decision of firms, portfolio allocation, risk management and derivatives pricing [12, 25]. Since the seminal study of Meese and Rogoff [16], the forecasting performance of exchange rate models has turned out to be frequently inferior to the naïve random walk benchmarking. This phenomenon constitutes the "exchange rate disconnect puzzle", which states that exchange rates are largely disconnected from economic fundamentals [1]. Despite the Meese and Rogoff puzzle, the problem of predicting the movement of exchange rates still attracts increasing attention from academy and practitioners [17].

The works of [4,5,19] and [21] give encouraging results for certain statistical forecasting methods regarding the predictability of exchange rates. Their models are shown to outperform random walk in some cases. Additionally, the literature also reports the high performance achieved by computational intelligence techniques for exchange rate forecasting, e.g., using neural networks [12, 22], genetic algorithms [13, 23], genetic programming [25], fuzzy sets [7], and hybrid methods [9, 28].

Despite the recent advances and increasing performance of computational intelligence techniques, the majority of research efforts are devoted to standard forecasting modeling approaches, i.e., the temporal evolution of exchange rates is observed as a single-valued financial time series. For instance, if only the opening (or closing) exchange rate is measured daily, the resulting time series will hide the intraday variability and loose important information [8]. An alternative to alleviate this limitation is when both, the highest and the lowest values of prices are measured at each time step, which results in interval time series (ITS). In particular, considering the high and low values of asset prices, financial ITS modeling and forecasting have received considerable attention in the recent literature with the introduction of several interval-valued time series forecasting methods [11, 14, 26].

This paper introduces an interval fuzzy rule-based model (iFRB) for exchange rate ITS forecasting. The iFRB is a collection of functional fuzzy rules in which the base variables are intervals instead of real numbers. The construction of the iFRB concerns the identification of the rule antecedents, and parameter estimation of the corresponding affine consequents. Rules antecedents are identified using a fuzzy clustering approach for symbolic interval-valued data using the adaptive City-Block distance recently proposed by [6]. The advantage of the adaptive City Block clustering is its ability to accommodate outliers, an essential feature in financial time series forecasting. This is because financial time series values are affected by news and shocks, which reflect in the data as outliers. The parameters of the affine consequents are estimated using a least squares algorithm designed for interval-valued data.

Empirical evaluation of iFRB concerns one-step ahead forecasting the interval-valued Euro/Dollar (EUR/USD) exchange rate for the period from January 2005 to December 2016. The ITS is constructed using actual financial data to extract the daily high and low exchange rate values to assemble the exchange rate intervals. The performance of iFRB is compared with the random walk,

ARIMA and VECM, with the linear and nonlinear interval Holt's exponential smoothing (Holt^I) [15], and with an interval multilayer perceptron neural network (iMLP) [20]. Forecast performance is evaluated using root mean squared error, mean absolute percentage error, and direction accuracy measures, considering statistical tests.

After this introduction, this paper proceeds as follows. Section 2 details the structure and identification of the interval fuzzy rule-based models (iFRB). Forecasting of the EUR/USD exchange rate is addressed in Sect. 3. Finally, Sect. 4 concludes the paper and lists topics for future research.

2 Interval Fuzzy Rule-Based Modeling

2.1 Interval-Valued Time Series

An interval-valued variable X is a closed and bounded set of real numbers indexed by $t \in \Omega$, that is:

$$X_t = [X_t^L, X_t^H] \in \Im, \tag{1}$$

where $\Im = \{[X_t^L, X_t^H] : X_t^L, X_t^H \in \Re, X_t^L \leq X_t^H\}$, $\forall\, t \in \Omega$ is the set of closed intervals of the real line $\Re$. In finance data X_t^L and X_t^H are the daily low and high exchange rate prices for X at time t, respectively.

An interval-valued time series (ITS) is a sequence of interval-valued variables observed in successive time steps t ($t = 1, 2, \ldots, n$) expressed as a two dimensional vector $X_t = [X_t^L, X_t^H]^T \in \Im$, where n denotes the sample size, the number of intervals in the time series.

Processing of interval-valued variables requires interval arithmetic. Interval arithmetic extends traditional arithmetic to operate on intervals. This paper uses the arithmetic operations introduced by [18].

2.2 iFRB Model Structure

The interval-valued fuzzy rule-based model (iFRB) with affine interval consequents consists of a set of fuzzy functional rules of the following form:

$$\mathcal{R}_i: \text{ IF } \mathbf{X} \text{ is } \mu_i \text{ THEN } Y_i = [Y_i^L, Y_i^H], \tag{2}$$

where $\mathcal{R}_i$ is the i-th fuzzy rule, $i = 1, 2, \ldots, c$, c is the number of fuzzy rules. $\mathbf{X} = [X_1, X_2, \ldots, X_p]^T$, $X_j = [X_j^L, X_j^H] \in \Im$, $j = 1, \ldots, p$ is the input, μ_i is the fuzzy set of the antecedent of the i-th fuzzy rule whose membership function is $\mu_i(\mathbf{X}) : \Im \to [0, 1]$, $Y_i = [Y_i^L, Y_i^H] \in \Im$ is the output of the i-th rule, with:

$$\begin{aligned} Y_i^L &= \beta_{i0}^L + \beta_{i1}^L X_1^L + \ldots + \beta_{ip}^L X_p^L, \\ Y_i^H &= \beta_{i0}^H + \beta_{i1}^H X_1^H + \ldots + \beta_{ip}^H X_p^H, \end{aligned} \tag{3}$$

$\{\beta_{i0}^L, \ldots, \beta_{ip}^L\}$ and $\{\beta_{i0}^H, \ldots, \beta_{ip}^H\}$, $j = 1, \ldots, p$, are real-valued parameters of the consequent of the i-th rule associated with the output intervals.

The model output is computed as follows:

$$Y = \sum_{i=1}^{c} \left(\frac{\mu_i(\mathbf{X}) Y_i}{\sum_{j=1}^{c} \mu_j(\mathbf{X})} \right). \quad (4)$$

The expression (4) can be rewritten, using normalized degrees of activation, as:

$$Y = \sum_{i=1}^{c} \lambda_i Y_i, \quad (5)$$

where $\lambda_i = \frac{\mu_i(\mathbf{X})}{\sum_{j=1}^{c} \mu_j(\mathbf{X})}$ is the normalized firing level of the i-th rule.

iFRB modeling requires: i) learning the antecedent part of the model using e.g. an interval fuzzy clustering algorithm, and ii) estimation of the parameters of the affine consequents. Notice that all computations of the iFRB clustering and parameter estimation tasks must consider interval-valued data.

2.3 Antecedent Identification

iFRB antecedent identification uses the adaptive fuzzy clustering algorithm for interval-valued data with City-Block distances [6]. The City-Block distance is more robust to the presence of outliers in the data set than the Euclidean distance. Further, the advantage of using adaptive City Block distance is that the clustering algorithm finds clusters of different shapes and sizes that represents the structures found in data sets better than alternative distances [6].

Let $N = \{1, \ldots, n\}$ be a set of n patterns (each pattern is indexed by t) describing p symbolic interval variables $X_1, \ldots, X_p$ (each variable is indexed by j). Each pattern t is a vector of intervals $\mathbf{X} = [X_1, \ldots, X_p]$, where $X_j = [X_j^L, X_j^H] \in \Im$. Additionally, each prototype $\mathbf{V}_i$ of cluster i, $i = 1, \ldots, c$, is a vector of intervals $\mathbf{V}_i = [V_{i1}, \ldots, V_{ip}]$, where $V_{ij} = [V_{ij}^L, V_{ij}^H] \in \Im$, $j = 1, \ldots, p$.

The interval fuzzy clustering algorithm aims at finding a fuzzy partition of a set of patterns in c clusters and a corresponding set of prototypes $\{\mathbf{V}_i, \ldots, \mathbf{V}_c\}$ that minimize a W criterion that measures how well the clusters and their representatives (prototypes) fits the data set. In this paper W is defined as

$$\begin{aligned} W &= \sum_{i=1}^{c} \sum_{t=1}^{n} (\mu_{it})^m \phi_i(\mathbf{X}_t, \mathbf{V}_i), \\ &= \sum_{i=1}^{c} \sum_{t=1}^{n} (\mu_{it})^m \sum_{j=1}^{p} \theta_{ij} \left(|X_j^L - V_{ij}^L| + |X_j^H - V_{ij}^H| \right), \end{aligned} \quad (6)$$

where $\phi(\cdot)$ is an adaptive City-Block distance that access the dissimilarity between a pair of vectors of intervals. It is defined for each class and is parameterized by vectors of weights $\boldsymbol{\theta}_i = [\theta_{i1}, \ldots, \theta_{ip}]$, $\mathbf{X}_t = [X_{1t}, \ldots X_{pt}]$ is the t-th pattern vector of intervals, $\mathbf{V}_i = [V_{i1}, \ldots, V_{ip}]$ is a prototype vector of intervals of cluster i, μ_{it} is the membership degree of pattern t in cluster i, and m is a fuzzification parameter (usually $m = 2$).

The optimal fuzzy partition is obtained via Picard iterations to find the (local) minimum of W in (6). The algorithm starts with an initial partition and alternates between a representation step and an allocation step until convergence (W reaches a stationary value, often a local minimum) [6]. The representation step sets the best prototypes and the best distances in two stages. The first stage fixes the membership degrees μ_{it} of each pattern t in cluster i and the vector of weights $\boldsymbol{\theta}_i = [\theta_{i1}, \ldots, \theta_{ip}]$. Prototypes $\mathbf{V}_i = [V_{i1}, \ldots, V_{ip}]$, for $i = 1, \ldots, c$ and $j = 1, \ldots, p$ that minimize the clustering criterion W are found solving:

$$\sum_{t=1}^{n} (u_{it})^m \left(|X_j^L - V_{ij}^L| + |X_j^H - V_{ij}^H|\right) \rightarrow \text{Min.} \tag{7}$$

Solution of (7), in turn, results in two minimization problems: find $V_{ij}^L \in \Re$ and $V_{ij}^H \in \Re$ that minimizes, respectively:

$$\sum_{t=1}^{n} (u_{it})^m \left(|X_j^L - V_{ij}^L|\right) \rightarrow \text{ Min. and } \sum_{t=1}^{n} (u_{it})^m \left(|X_j^H - V_{ij}^H|\right) \rightarrow \text{ Min.} \tag{8}$$

Each of these these two problems are equivalent to the minimization of:

$$\sum_{t=1}^{n} |y_t - a z_t|, \tag{9}$$

where $y_t = (u_{it})^m X_j^L$ (respectively, $y_t = (u_{it})^m X_j^H$), $z_k = (u_{it})^m$ and $a = V_{ij}^L$ (respectively, $a = V_{ij}^H$).

Since there is no closed solution for this problem, an heuristic solution can be derived using the following algorithm [6]:

1. Rank (y_t, z_t) such that $\frac{y_{t_1}}{z_{k_1}} \leq \ldots \leq \frac{y_{t_n}}{z_{k_n}}$;
2. For $-\sum_{l=1}^{n} |z_{k_l}|$ add successive values of $2|z_{k_l}|$ and find r such that $-\sum_{l=1}^{n} |z_{k_l}| + 2\sum_{s=1}^{r} |z_{k_s}| < 0$ and $-\sum_{l=1}^{n} |z_{k_l}| + 2\sum_{s=1}^{r+1} |z_{k_s}| > 0$;
3. Set $a = \frac{y_{k_r}}{z_{k_r}}$;
4. If $-\sum_{l=1}^{n} |z_{k_l}| + 2\sum_{s=1}^{r} |z_{k_s}| = 0$ and $-\sum_{l=1}^{n} |z_{k_l}| + 2\sum_{s=1}^{r+1} |z_{k_s}| = 0$, then $a = \frac{\frac{y_{k_r}}{z_{k_r}} + \frac{y_{k_{r+1}}}{z_{k_{r+1}}}}{2}$.

The second stage of the representation step (or weighting step) fixes the membership degrees μ_{it} and the prototypes $\mathbf{V}_i$. The vector of weights $\boldsymbol{\theta}_i = [\theta_{i1}, \ldots, \theta_{ip}]$ minimizing W under $\theta_{ij} > 0$ and $\prod_{j=1}^{p} \theta_{ij} = 1$, for $i = 1, \ldots, c$ and $j = 1, \ldots, p$ is updated using the following expression:

$$\theta_{ij} = \frac{\left\{\prod_{h=1}^{p} \left[\sum_{t=1}^{n} (\mu_{it})^m \left(|X_j^L - V_{ij}^L| + |X_j^H - V_{ij}^H|\right)\right]\right\}^{\frac{1}{p}}}{\sum_{t=1}^{n} (\mu_{it})^m \left[\left(|X_j^L - V_{ij}^L| + |X_j^H - V_{ij}^H|\right)\right]}. \tag{10}$$

Finally, the allocation step defines the best fuzzy partition fixing the prototypes $\mathbf{V}_i$ and the vector of weights $\boldsymbol{\theta}_i$. Next, the membership degrees μ_{it} that minimize W under $\mu_{it} \geq 0$ and $\sum_{t=1}^{c} \mu_{it} = 1$ are found as follows:

$$\mu_{it} = \left[\sum_{h=1}^{c} \left(\frac{\sum_{j=1}^{p} \theta_{ij} \left[(X_{jt}^L - V_{ij}^L)^2 + (X_{jt}^H - V_{ij}^H)^2 \right]}{\sum_{j=1}^{p} \theta_{hj} \left[(X_{jt}^L - V_{hj}^L)^2 + (X_{jt}^H - V_{hj}^H)^2 \right]} \right)^{\frac{1}{m-1}} \right]^{-1} . \quad (11)$$

After fixing the number of clusters c ($2 \leq c < n$), an iteration limit k_{max}, and an error tolerance value ϵ, the algorithm iterates between the representation and allocation steps. The process produces the vector of clusters prototypes $\mathbf{V}_i = [V_{i1}, \ldots, V_{ip}]$ and the respective membership degrees μ_{it} of each pattern t in each cluster i, for $t = 1, \ldots, n$ and $i = 1, \ldots, c$, that locally minimize W. Derivations of expressions (7)–(11) are found in [2] and [6].

2.4 Consequents Identification

In this paper, iFRB consequent parameter identification uses the min-max approach suggested by [3], which is based on the minimization of the errors from two independent linear regression models on the lower and upper bounds of the intervals.

Consider a set of $t = 1, \ldots, n$ samples of $p + 1$ symbolic interval-valued variables $Y_t, X_{1t}, \ldots, X_{pt}$. Each fuzzy rule i, $i = 1, \ldots, c$ corresponds to a linear regression relationship. To keep notation clearer, henceforth we omit the index i related to each cluster or fuzzy rule. The output of iFRB for each fuzzy rule can be rewritten as

$$\begin{aligned} Y_t^L &= \beta_0^L + \beta_1^L X_{1t}^L + \ldots + \beta_p^L X_{pt}^L + \epsilon_t^L, \\ Y_t^H &= \beta_0^H + \beta_1^H X_{1t}^L + \ldots + \beta_p^H X_{pt}^H + \epsilon_t^H, \end{aligned} \quad (12)$$

where ϵ^L and ϵ^H are the corresponding residuals for lower and upper interval bounds equations, respectively.

The sum of the squares of the deviations in the min-max method is [3]:

$$\begin{aligned} S = \sum_{t=1}^{n} (\epsilon_t^L)^2 + (\epsilon_t^H)^2 &= \sum_{t=1}^{n} (Y_t^L - \beta_0^L - \beta_1^L X_{1t}^L - \ldots - \beta_p^L X_{pt}^L)^2 \\ &+ \sum_{t=1}^{n} (Y_t^H - \beta_0^H - \beta_1^H X_{1t}^H - \ldots - \beta_p^H X_{pt}^H)^2, \end{aligned} \quad (13)$$

which is the sum of the lower bound square error plus the sum of the upper bound square error.

The least squares estimates of $\{\beta_0^L, \beta_1^L, \ldots, \beta_p^L\}$ and $\{\beta_0^H, \beta_1^H, \ldots, \beta_p^H\}$ that minimize the expression (13), written in matrix notation, is

$$\hat{\boldsymbol{\beta}} = [\hat{\beta}_0^L, \hat{\beta}_1^L, \ldots, \hat{\beta}_p^L, \hat{\beta}_0^H, \hat{\beta}_1^H, \ldots, \hat{\beta}_p^H]^T = \mathbf{A}^{-1}\mathbf{b}, \quad (14)$$

where $\mathbf{A}$ is a $2(p+1) \times 2(p+1)$ matrix and $\mathbf{b}$ is a $2(p+1) \times 1$ vector:

$$\mathbf{A} = \begin{pmatrix} n & \sum_t x_1^L(t) & \cdots & \sum_t x_p^L(t) & 0 & \cdots & 0 \\ \sum_t x_1^L(t) & \sum_t (x_1^L(t))^2 & \cdots & \sum_t x_p^L(t) x_1^L(t) & 0 & \cdots & 0 \\ \vdots & \vdots & & \vdots & \vdots & & \vdots \\ \sum_t x_p^L(t) & \sum_t x_1^L(t) x_p^L(t) & \cdots & \sum_t (x_p^L(t))^2 & 0 & \cdots & 0 \\ 0 & 0 & \cdots & 0 & n & \cdots & \sum_t x_p^U(t) \\ 0 & 0 & \cdots & 0 & \sum_t x_1^U(t) & \cdots & \sum_t x_p^U(t) x_1^U(t) \\ \vdots & \vdots & & \vdots & \vdots & & \vdots \\ 0 & 0 & \cdots & 0 & \sum_t x_1^U(t) x_p^U(t) & \cdots & \sum_t (x_p^U(t))^2 \end{pmatrix}, \quad (15)$$

and

$$\mathbf{b} = \left[\sum_t y^L(t), \sum_t y^L(t) x_1^L(t), \ldots, \sum_t y^L(t) x_p^L(t), \sum_t y^U(t), \sum_t y^U(t) x_1^U(t), \ldots, \sum_t y^U(t) x_p^U(t)\right]^T. \quad (16)$$

Notice that the least squares estimates of consequent parameters of (14) are computed for each fuzzy rule. Therefore, $\hat{\boldsymbol{\beta}}_i$ are the estimates of the parameters in the consequent of the i-th fuzzy rule.

3 Exchange Rate Forecasting

3.1 Data

The ITS data concerns the exchange rate of the Euro (EUR) against the US Dollar (USD). The sample data are daily interval data for the period from January 3, 2005 to December 31, 2016 with a total of 3,164 and 3,130 observations, respectively[1]. The low and high prices of the exchange rates are the lower and upper bounds in the interval time series.

The data were divided into in-sample and out-of-sample sets. The in-sample set, used for model training, is for the period from January 2005 to December 2012. The remaining four years of data, from January 2013 to December 2016, is the out-of-sample set. The forecasting performance of the methods is assessed based on one-step-ahead forecasts of the out-of-sample data.

3.2 Performance Measures

Evaluation of the forecasting performance of iFRB and selected benchmark approaches are done using the root mean square error (RMSE), and the mean absolute percentage error (MAPE) measures. They are computed as follows:

$$\mathrm{RMSE}^B = \sqrt{\frac{1}{n}\sum_{t=1}^{n}\left(\frac{Y_t^B - \hat{Y}_t^B}{Y_t^B}\right)^2}, \quad (17)$$

[1] Data were collected from the Yahoo Finance website (http://finance.yahoo.com/).

$$\text{MAPE}^B = \frac{100}{n}\sum_{t=1}^{n}\frac{|Y_t^B - \hat{Y}_t^B|}{Y_t^B}, \tag{18}$$

where $B = \{L, H\}$ represents the low and high prices (i.e., the interval bounds), $\mathbf{Y}_t = [Y_t^L, Y_t^H]^T$ and $\hat{\mathbf{Y}}_t = [\hat{Y}_{t^L}, \hat{Y}_{t^H}]^T$ are the actual and predicted intervals exchange rate at t, respectively, n is the sample size, and RMSE^L (MAPE^L) and RMSE^H (MAPE^H) are the RMSE (MAPE) for the ITS lows and highs, respectively.

As stated in [4], the correct prediction of the direction of change can be more important than the magnitude of the error. Therefore, the results are also evaluated using the following measure of direction accuracy:

$$\text{DA}^B = \frac{1}{n}\sum_{t=1}^{n} Z_t^B, \tag{19}$$

where

$$Z_t^B = \begin{cases} 1, & \text{if } \left(\hat{Y}_{t+1}^B - Y_t^B\right)\left(Y_{t+1}^B - Y_t^B\right) > 0, \\ 0, & \text{otherwise.} \end{cases} \tag{20}$$

Statistical significance test of proportions is done to verify if the direction accuracy is significantly different from zero. Rejection of $H_0 : DA = 0$ indicates that the underlying model is superior to the random walk in predicting the direction of changes. One may use $DA = 0.5$ to evaluate the superiority of a model over a random walk, based on the rationale that the random walk "predicts the exchange rate with an equal chance to go up or down", i.e., a 50–50 situation. However, the random walk without drift produces no-change forecasts, since the forecast for each point in time t is the actual value at $t-1$. Hence for a random walk without drift $DA = 0$, the null hypothesis should be $H_0 : DA = 0$, rather than $H_0 : DA = 0.5$ [19].

In addition to the accuracy measurement, significant differences between a pair of forecasting models are evaluated using the Diebold-Mariano test [10] with 5% significance level.

3.3 Results and Analysis

This section details the experiments performed to analyze and to evaluate the interval fuzzy rule-based model (iFRB) for interval-valued EUR/USD exchange rate forecasting. The results are for one-step-ahead forecasts of the out-of-sample data from January 2013 to December 2016.

Concerning exchange rate one-step-ahead forecasting, iFRB inference system is represented as follows:

$$\hat{Y}_{t+1} \approx f_{\text{iFRB}}\left(Y_t, Y_{t-1}, \ldots, Y_{t-l}\right), \tag{21}$$

where $f(\cdot)_{\text{iFRB}}$ represents the nonlinear mapping by iFRB.

iFRB modeling requires the following control parameters: number of fuzzy rules c, and the number l of lagged time series values used as input as in Eq. (21) - exogenous variables can also be included as model input. Simulations were performed on the in-sample data by running the iFRB algorithm for different values of c and l. The best values in terms of RMSE were achieved for $c = 3$ and $l = 2$. All methods were implemented using MATLAB.

Table 1 shows the prediction performance of the models in terms of RMSE, MAPE, and DA. Notice that these metrics are computed individually for both, low (L) and high (H) exchange rate time series. Best results are highlighted in bold. From the point of view of RMSE, iFRB outperforms all competitors in forecasting EUR/USD exchange rate lows and highs. Similar results are found for MAPE as well. Notice that RMSE and MAPE values for highs and lows forecasts of iFRB and random walk models are very similar, which is consistent with the Meese and Rogoff puzzle [16].

Table 1. Performance evaluation of EUR/USD exchange rate forecasting for out-of-sample data (January 2013–December 2016).

Metric	Method					
	RW	ARIMA	VECM	Holt^I	iMLP	iFRB
RMSE^L	0.00483	0.00796	0.00740	0.00834	0.00714	**0.00429**
RMSE^H	0.00530	0.00816	0.00808	0.00869	0.00771	**0.00514**
MAPE^L	0.32927	0.59371	0.54494	0.64138	0.52192	**0.32634**
MAPE^H	0.35945	0.61473	0.58631	0.64781	0.55776	**0.35876**
DA^L	–	0.50928*	0.50557	0.52876*	0.54545*	**0.61114***
DA^H	–	0.52783*	0.53989	0.54824*	**0.59184***	0.57721*

(*) Significantly different from zero at the 5% level for testing a proportion with critical value of 1.96

The forecasting results produced by the interval-valued models iFRB and iMLP achieved better results than the traditional ARIMA, VECM and Holt^I for EUR/USD exchange rate lows and highs (Table 1). It is conceivable to postulate that the reason why ARIMA and VECM models are inferior is that they ignore the possible mutual dependency between the daily highs and lows of the ITS. iFRB and iMLP forecasts are the best among the models, except the random walk. As we move from linear ARIMA, VECM and Holt^I to the nonlinear iFRB and iMLP, the improvement is significant, indicating that modeling non-linearities improve predictive power of interval-valued exchange rates. However, concerning the lower bound of intervals, the differences among iFRB, iMLP and VECM accuracy are lower (RMSE^L and MAPE^L values are slightly distinct).

As shown in [19], dynamic models may outperform random walk in out-of-sample forecasting if forecasting power is measured by direction accuracy and profitability. Table 1 summarizes the results in terms of direction accuracy (DA) and adjusted RMSE (ARMSE) measures for both, low and high EUR/USD exchange rate. Because the random walk without drift predicts no change in the exchange rate, it has zero direction accuracy, and hence a confusion rate of 1, which makes the RMSE and ARMSE equal. In terms of direction accuracy, all the alternative approaches, ARIMA, VECM, Holt^I, iMLP and iFRB are superior to the random walk once the null hypothesis $H_0 : DA = 0$ is rejected for both, exchange rate lows and highs, which means that the models overwhelmingly outperform the random walk in terms of direction accuracy (Table 1).

In addition to goodness of fit, as measured by forecast errors, the models were evaluated using the Diebold-Mariano [10] test statistics for lows and highs of the EUR/USD exchange rate. The results are summarized in Table 2. The test is performed for each pair of models. The null hypothesis of equal predictive accuracy is rejected at the 5% confidence level if $|DM| > 1.96$. From this point of view, for both the lows and highs of the EUR/USD exchange rate, the random walk, iMLP and iFRB approaches can be considered equally accurate ($|DM| <$ 1.96), but they produce statistically superior forecasts against ARIMA, VECM and Holt^I ($|DM| > 1.96$) – see Table 2. The ARIMA, VECM and Holt^I can be considered equally accurate as well, except for EUR/USD highs, in which the VECM model gives statistically more accurate results than ARIMA.

Table 2. Diebold-Mariano statistics of EUR/USD exchange rate low and high prices for out-of-sample forecasts (January 2013–December 2016).

Method	ARIMA	VECM	Holt^I	iMLP	iFRB
Panel A: EUR/USD exchange rate low prices					
RW	−11.595*	−10.920*	−9.920*	−1.323	−1.049
ARIMA	–	5.056*	−3.762*	8.853*	9.352*
VECM	–	–	−4.448*	4.168*	5.271*
Holt^I	–	–	–	4.221*	4.871*
iMLP	–	–	–	–	1.781
Panel B: EUR/USD exchange rate high prices					
RW	−11.999*	−10.078*	−8.781*	−1.532	−1.342
ARIMA	–	1.455	−3.517*	5.852*	6.526*
VECM	–	–	−4.910*	5.253*	5.251*
Holt^I	–	–	–	8.665*	8.917*
iMLP	–	–	–	–	1.098

(∗) Statistically significant at the 5% level

Figure 1 shows the EUR/USD candlesticks based on the observed prices of the exchange rates with the corresponding high-low bands predicted by iFRB for the last three months of data in the out-of-sample sets. Notice that iFRB forecast values follow closely the actual data. Interestingly, the iFRB gives a good fit of the high-low dispersion for both exchange rates, indicating its potential to enhance chart analysis, a tool used by technical traders worldwide.

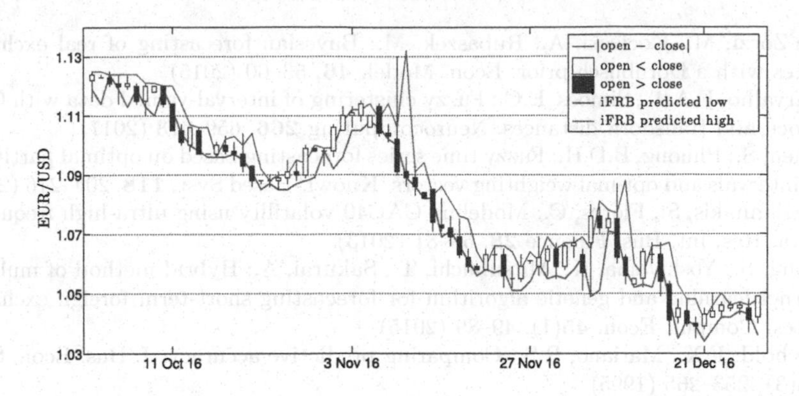

Fig. 1. EUR/USD exchange rates and iFRB high-low forecasts.

4 Conclusion

This paper has suggested an interval fuzzy rule-based model (iFRB) for exchange rate ITS forecasting. Fuzzy rules antecedents are identified with a fuzzy clustering approach for symbolic interval-valued data using adaptive City-Block distances. The parameters of rules consequents are estimated using a least squares algorithm designed for interval-valued data. The iFRB one-step ahead forecasting performance was evaluated in forecasting interval-valued Euro/Dollar (EUR/USD) exchange rate for the period from January 2005 to December 2016. The results show that the iFRB model has higher accuracy than the random walk and alternative approaches for out-of-sample forecasting of interval-valued EUR/USD exchange rate. Future work shall include the automatic selection of the number of clusters in iFRB antecedent identification, performance analysis of medium- and long-term forecasting horizons, and applications in risk management using range-based volatility estimators.

Acknowledgements. The authors acknowledge the Brazilian National Council for Scientific and Technological Development (CNPq) for its support via grants 302467/2019-0 and 304274/2019-4, and the São Paulo Research Foundation (Fapesp).

References

1. Beckmann, J., Schüssler, R.: Forecasting exchange rates under parameter and model uncertainty. J. Int. Money Finance **60**, 267–288 (2016)
2. Bezdek, J.C.: Pattern Recognition with Fuzzy Objective Function Algorithm. Plenum, New York (1981)
3. Billard, L., Diday, E.: Symbolic regression analysis. In: Jajuga, K., Sokołowski, A., Bock, H.H. (eds.) Classification, Clustering, and Data Analysis. Studies in Classification, Data Analysis, and Knowledge Organization, pp. 281–288. Springer, Berlin (2002). https://doi.org/10.1007/978-3-642-56181-8_31
4. Burns, K., Moosa, I.: Enhancing the forecasting power of exchange rate models by introducing nonlinearity: does it work? Econ. Model. **50**, 27–39 (2015)

5. Ca'Zorzi, M., Kociecki, A., Rubaszek, M.: Bayesian forecasting of real exchange rates with a Dornbusch prior. Econ. Model. **46**, 53–60 (2015)
6. Carvalho, F.A.T., Simões, E.C.: Fuzzy clustering of interval-valued data with City-Block and Hausdorff distances. Neurocomputing **266**, 659–673 (2017)
7. Chen, S., Phuong, B.D.H.: Fuzzy time series forecasting based on optimal partitions of intervals and optimal weighting vectors. Knowl.-Based Syst. **118**, 204–216 (2017)
8. Degiannakis, S., Floros, C.: Modeling CAC40 volatility using ultra-high frequency data. Res. Int. Bus. Finance **28**, 68–81 (2013)
9. Deng, S., Yoshiyama, K., Mitsubuchi, T., Sakurai, A.: Hybrid method of multiple kernel learning and genetic algorithm for forecasting short-term foreign exchange rates. Comput. Econ. **45**(1), 49–89 (2015)
10. Diebold, F.X., Mariano, R.S.: Comparing predictive accuracy. J. Bus. Econ. Stat. **13**(3), 253–265 (1995)
11. Froelich, W., Salmeron, J.L.: Evolutionary learning of fuzzy grey cognitive maps for the forecasting of multivariate, interval-valued time series. Int. J. Approximate Reasoning **55**(6), 1319–1335 (2014)
12. Kiani, K.M., Kastens, T.L.: Testing forecast accuracy of foreign exchange rates: predictions from feed forward and various recurrent neural network architectures. Comput. Econ. **32**(4), 383–406 (2008). https://doi.org/10.1007/s10614-008-9144-4
13. Lawrenz, C., Westerhoff, F.: Modeling exchange rate behavior with a genetic algorithm. Comput. Econ. **21**(3), 209–229 (2003). https://doi.org/10.1023/A:1023943726237
14. Lu, W., Chen, X., Pedrycz, W., Liu, X., Yang, J.: Using interval information granules to improve forecasting in fuzzy time series. Int. J. Approximate Reasoning **57**, 1–18 (2015)
15. Maia, A.L.S., Carvalho, F.A.T.: Holt's exponential smoothing and neural network models for forecasting interval-valued time series. Int. J. Forecast. **27**(3), 740–759 (2011)
16. Meese, R.A., Rogoff, K.: Empirical exchange rate models of the seventies: do they fit out of sample? J. Int. Econ. **14**(1–2), 3–24 (1983)
17. Molodtsova, T., Papell, D.H.: Out-of-sample exchange rate predictability with Taylor rule fundamentals. J. Int. Econ. **77**(2), 167–180 (2009)
18. Moore, R.E., Kearfott, R.B., Cloud, M.J.: Introduction to Interval Analysis. SIAM Press, Philadelphia (2009)
19. Moosa, I., Burns, K.: The unbeatable random walk in exchange rate forecasting: reality or myth? J. Macroecon. **40**, 69–81 (2014)
20. Roque, A.M., Maté, C., Arroyo, J., Sarabia, A.: iMLP: applying multi-layer perceptrons to interval-valued data. Neural Process. Lett. **25**(2), 157–169 (2007). https://doi.org/10.1007/s11063-007-9035-z
21. Sarno, L., Valente, G.: Exchange rates and fundamentals: footloose or evolving relationship? J. Eur. Econ. Assoc. **7**(4), 786–830 (2009)
22. Sermpinis, G., Theofilatos, K., Karathanasopoulos, A., Georgopoulos, E.F., Dunis, C.: Forecasting foreign exchange rates with adaptive neural networks using radial-basis functions and particle swarm optimization. Eur. J. Oper. Res. **225**(3), 528–540 (2013)
23. Sermpinis, G., Stasinakis, C., Theofilatos, K., Karathanasopoulos, A.: Modeling, forecasting and trading the EUR exchange rates with hybrid rolling genetic algorithms-support vector regression forecast combinations. Eur. J. Oper. Res. **247**(3), 831–846 (2015)
24. Takagi, T., Sugeno, M.: Fuzzy identification of systems and its applications to modeling and control. IEEE Trans. Syst. Man Cybern. SMC **15**(1), 116–132 (1985)

25. Vasilakis, G.A., Theofilatos, K.A., Georgopoulos, E.F., Karathanasopoulos, A., Likothanassis, S.D.: A genetic programming approach for EUR/USD exchange rate forecasting and trading. Comput. Econ. **42**(4), 415–431 (2013). https://doi.org/10.1007/s10614-012-9345-8
26. Xiong, T., Bao, Y., Hu, Z., Chiong, R.: Forecasting interval time series using a fully complex-valued RBF neural network with DPSO and PSO algorithms. Inf. Sci. **305**, 77–92 (2015)
27. Xiong, T., Li, C., Bao, Y.: Interval-valued time series forecasting using a novel hybrid Holt and MSVR model. Econ. Model. **60**, 11–23 (2017)
28. Yang, H., Lin, H.: Applying the hybrid model of EMD, PSR, and ELM to exchange rates forecasting. Comput. Econ. **49**(1), 99–116 (2017). https://doi.org/10.1007/s10614-015-9549-9

A Method to Generate Soft Reference Data for Topic Identification

Daniel Vélez[1], Guillermo Villarino[2], J. Tinguaro Rodríguez[1,3](✉), and Daniel Gómez[2]

[1] Faculty of Mathematics, Complutense University of Madrid, Madrid, Spain
{danielvelezserrano,jtrodrig}@mat.ucm.es

[2] Faculty of Statistical Studies, Complutense University of Madrid, Madrid, Spain
{gvillari,dagomez}@estad.ucm.es

[3] Interdisciplinary Mathematics Institute, Complutense University of Madrid, Madrid, Spain

Abstract. Text mining and topic identification models are becoming increasingly relevant to extract value from the huge amount of unstructured textual information that companies obtain from their users and clients nowadays. Soft approaches to these problems are also gaining relevance, as in some contexts it may be unrealistic to assume that any document has to be associated to a single topic without any further consideration of the involved uncertainties. However, there is an almost total lack of reference documents allowing a proper assessment of the performance of soft classifiers in such soft topic identification tasks. To address this lack, in this paper a method is proposed that generates topic identification reference documents with a soft but objective nature, and which proceeds by combining, in random but known proportions, phrases of existing documents dealing with different topics. We also provide a computational study illustrating the application of the proposed method on a well-known benchmark for topic identification, as well as showing the possibility of carrying out an informative evaluation of soft classifiers in the context of soft topic identification.

Keywords: Soft classification · Text mining · Topic identification

1 Introduction

In recent years, there has been a significant growth in the volume of available data for companies from different industries regarding their clients. With the aim of being able to exploit such information, the application of mathematical and machine learning methods allowing to identify patterns and relationships useful for decision making purposes has also known an important proliferation.

Frequently, data regarding or coming from clients have an unstructured nature, and particularly it is estimated that approximately 80% of that information is in textual form [1]. Consequently, the analytical field denominated as Text Mining, a multidisciplinary area based on natural language processing (NLP) and machine

Supported by the Government of Spain (grant PGC2018-096509-B-I00) and Complutense University of Madrid (UCM Research Group 910149).

M.-J. Lesot et al. (Eds.): IPMU 2020, CCIS 1239, pp. 54–67, 2020.
https://doi.org/10.1007/978-3-030-50153-2_5

learning techniques focused on extracting value from such texts, has experienced a considerable development. In this field, solutions oriented towards the identification of topics in texts have specially gained relevance, as they enable useful applications for the analysis of contents in social media (Facebook, Twitter, blogs, etc.) or document classification (spam filtering, sentiment analysis, customer complaint classification, etc.) [2].

It is important to note that the manual supervision process of a whole document corpus can be quite costly, which makes the employment of methods allowing to automatically label the texts highly convenient. To some extent, it is possible to manually label just a sample of the available documents, and then apply a supervised method to classify the remaining texts in the corpus from that training sample. However, this forces these new documents to be classified in some of the categories found in the supervised sample, even when their contents do not actually fit into any of them. This fact motivates that unsupervised methods are frequently applied instead of supervised ones.

Nevertheless, even when unsupervised methods are used, it is useful to have some labelled texts available in order to allow the assessment of their performance. In this way, for instance, it is usual to omit at first any knowledge about the topics of the labelled documents, then obtain the topic clusters, and finally validate whether the words that characterize the documents of each cluster are related to the known topics.

Another remark is that it is not realistic to assume that any document has to be always associated to a single topic. To address this situation, many soft classification techniques exist that allow simultaneously assigning a document to a set of topics up to different degrees. However, we have found an important lack regarding the assessment of the performance of such methods, as soft reference corpus for topic identification do not exist or are hardly available. For this reason, in this paper we propose a method to generate topic identification reference documents with a soft nature.

Specifically, the proposed method generates texts that combine phrases of previously available texts associated to different topics. The result is then a corpus of new documents such that, for each of these new documents, soft reference values are provided informing of the relative proportion of phrases it contains from each of the combined topics.

To illustrate the application of the proposed method and the possibilities it enables, a computational study based on a well-known benchmark corpus in topic identification is also included in this work.

This paper is structured as follows. Section 2 is devoted to discuss with more detail the need for soft reference datasets. The proposed method to generate soft reference corpuses for topic identification is then described in Sect. 3, and the computational study illustrating its usage is presented in Sect. 4. Finally, some conclusions and future work are provided in Sect. 5.

2 The Need for Soft Reference Data In topic Identification

The need for soft reference data in some classification contexts has since long been established. In such contexts, the very nature of the classification variable to predict is soft, in such a way that objects to be classified can naturally belong or be assigned to different classes, simultaneously and up to a degree. For instance, in the field of land cover estimation at sub-pixel level, the objects to be classified are pixels in a terrain image, and the classes to be assigned are different terrain types, such as water, wetland, forest, and so on. For many years, the usual image classification approach to this problem consisted in assigning just one class, i.e. a single terrain type, to each pixel [3,4]. However, due to the limited resolution of the images, a pixel may cover an area of several squared meters, in which different terrain types can coexist. For this reason, assigning each image pixel to a single type of cover can be misleading, particularly in applications in which a more precise assessment of land cover is needed, or whenever the spatial resolution of pixels is rather low. In either case, it is important to notice that this kind of crisp assignment omits the complex nature of the problem, and can provide a false appearance of lack of uncertainty regarding the mix of terrain types actually occurring at pixel level.

One could think that a multilabel or label ranking model, respectively associating each pixel to a set or ranking of terrain types, should alleviate this problematic. However, these kinds of solution may be again insufficient, as neither one can fully represent the involved uncertainty regarding the composition of the land cover within each pixel. For instance, if the area associated to a given pixel is composed of a 70% water and a 30% wetland, a label ranking output given by the ordered pair $(water, wetland)$ just informs that *water* is more predominant at this area than *wetland*, but fails to adequately inform about the relative abundance of each cover. Of course, a multilabel (non-ordered) output of the type $\{water, wetland\}$ would be even less informative.

Rather, as this example illustrates, the appropriate, most informative representation estimates the proportion of the pixel area that is covered by each terrain type. Thus, it is a soft output the one that actually allows representing and managing the uncertainty associated to the land cover estimation problem. For this reason, the standard approach to address this problem gradually switched from crisp image classification to linear mixture models [5], and later to more accurate soft classification schemes, as e.g. fuzzy unsupervised and supervised classifiers or neural networks [6,7].

However, it is important to remark that the successful application of soft techniques in the land cover estimation context crucially depended on the availability of contextual soft reference data. Even when a soft supervised approach can be developed without an explicitly soft reference, the proper evaluation of both supervised and unsupervised techniques would have been impossible without adequate soft reference data.

This is also our point concerning some currently developing tasks in text mining, and particularly in topic identification. For instance, this last field is finding an increasing application in the analysis of customer complaint texts. The automatic analysis and understanding of complaint forms is becoming more and more relevant for companies in order to carry out adequate actions to improve their services. Within this context, the usual aim is to classify complaints according to their causes from the basis of the text written by the customers at the complaint forms [8].

It is important to notice the link between this complaint causes classification problem and the previous land cover estimation problem. Particularly, it is possible that a complaint text (object to be classified) refers to more than one cause (classes to be assigned) simultaneously, and each of these causes can have a different weight within the text of a single complaint form. That is, it may be convenient to model the complaint classification problem as a soft classification problem, similarly to the land cover estimation problem described above.

Let us at this point introduce some notation to formalize in a general way the notions being discussed. Let X denote the set of objects to be classified, and let k denote the number of classes being considered. Then, in a soft classification framework each object $x \in X$ has to be assigned to a vector $(c_1(x), \ldots, c_k(x))$, where $c_i(x) \in [0, 1]$ denotes the degree (or proportion) in which object x belongs to class i, $i = 1, \ldots, k$. If the soft class estimations c_i actually represent proportions, it is then usual (see e.g. [9]) to impose the constraint

$$\sum_{i=1}^{k} c_i(x) = 1 \tag{1}$$

A soft classification problem can be addressed in either a supervised or unsupervised way, depending on the available data and the specific context requirements. For instance, in the complaint classification context sometimes an unsupervised approach may be more adequate, since the number k of possible complaint causes may not be perfectly determined *apriori*, and new complaint causes may always arise that are not present in the supervised data. At this respect, let us recall that soft supervised data is not necessarily needed to fit a soft supervised classification model. Even when trained with crisp data, most current supervised classification methodologies produce some kind of soft scores (probabilities, fuzzy degrees, etc.) in an intermediate stage of their process, before applying a decision rule (typically the well-known maximum rule) to map these soft scores into a crisp, single-class output (and see [10] for a discussion on potential drawbacks of such a decision rule).

In either way, whenever a classification problem is modeled in a soft (supervised or unsupervised) form, the crucial question is how the resulting model's performance is going to be evaluated. In this sense, it is important to notice that, if a practical task is modeled in soft terms, but crisp supervised data is used to evaluate the resulting soft model, then many misleading situations may occur.

To see it, suppose two different soft classifiers SC_1 and SC_2 are fitted to the same data in a binary classification task, in such a way that for a given object $x \in X$ the soft scores they respectively assign are $SC_1(x) = (0.51, 0.49)$ and $SC_2(x) = (0.49, 0.51)$. Suppose also that the crisp supervision of x has assessed it actually belongs to the first class, i.e. the correct crisp degrees are $(1, 0)$. Then, if the soft scores $SC_1(x), SC_2(x)$ are mapped into a single class through the maximum-rule, then the first classifier would predict x into the first class, while the second classifier would predict it into the second one. Therefore, one classifier would be evaluated as correctly classifying x while the other would do it wrongly. However, the difference between the real soft outputs $SC_1(x)$ and $SC_2(x)$ is actually not as significant. Even more, suppose the actual soft class-composition of object x before the crisp supervision is $(0.6, 0.4)$. In this situation, both SC_1 and SC_2 would be doing a similar more or less accurate estimation of such a class mix, but imposing a crisp evaluation framework would lead to a totally different assessment of their performance. And if the soft degrees for x and $SC_2(x)$ remain as before, but now it is $SC_1(x) = (0.98, 0.02)$, then SC_1 would be committing a much greater error than SC_2, though however a crisp evaluation would just assess that SC_1 is right and SC_2 is wrong.

These reasons led us to devise a procedure to generate soft reference datasets for topic identification, that can enable the performance of soft classification procedures to be properly evaluated.

3 Soft Reference Documents Generation

This section is devoted to describe the proposed method to automatically generate soft reference documents for the task of topic identification.

Basically, the method departs from a dataset or corpus of documents, each of which is associated in a crisp way to a single topic, and generates an output corpus containing new documents in which the specified topics from the original documents are mixed following different randomly determined proportions.

More specifically, the inputs of the method are the following:

- *InputData*: The name of the input database containing the topic identification corpus. This database has to contain the following fields:
  - *documentID*: An identifier for each of the documents in the database. This identifier is a primary key and, as such, it can not present repeated values.
  - *text*: A free-format character field with the text associated to each document.
  - *topic*: A character variable identifying the topic associated to each document.
- *Topics*: This parameter specifies the topics that will take part in the generation of the documents with mixed topics in the output corpus. These topics names have to be contained in the set of topic names under the *topic* field of the database.
- *OutputCorpusSize*: This parameter specifies the size of the output corpus, that is, the number of documents it has to contain.
- *OutputData*: The name of the output corpus to be generated. This corpus will contain the following fields:
  - *documentID*: An identifier for each of the documents in the output corpus. As before, this identifier is a primary key and, as such, it will not present repeated values.
  - *text*: A free-format character field with the text of each document.

- *topicProportion*[i]: A numerical variable giving the proportion of phrases in each document that are associated to the i-th topic provided in the *Topics* parameter of the method. Therefore, the output corpus will contain as much *topicProportion* variables as different topics have been selected in the *Topics* parameter to be mixed in the output documents.

Regarding the method itself, it proceeds as follows:

1. For each topic i in the *Topics* parameter, a list L_i is generated that stores in each position a different phrase of the documents in the input corpus *InputData* associated to topic i. Therefore, the length of L_i is equal to the total number of phrases of the set of documents in the input corpus associated to topic i. We consider that a 'phrase' is any text between two consecutive periods, or the text between the beginning of a document and the first period.
2. Another list L is generated with as many positions as documents in the input corpus associated to any of the topics in the *Topics* parameter. Let us denote by N the length of this list. The number of phrases of each document in the input corpus associated to any of the topics in the *Topics* parameter is stored at the corresponding position of L. We shall use this list to create documents in the output corpus in such a way that their number of phrases follows a similar distribution to that of the documents in the input corpus.
3. For each $j = 1, \ldots, OutputCorpusSize$, a random number k between 1 and N is generated, and the number of phrases to be placed in the j-th document of the output corpus is assigned as $L[k]$. Then, draw $|Topics| - 1$ random numbers from a uniform $U(0, 1)$ distribution, and sort them in ascending order, so that $u_{(l)}$ denotes the l-th element of the sorted sequence. Assign $u_{(0)} = 0$ and $u_{(|Topics|)} = 1$. For $i = 1, \ldots, |Topics|$, select $(u_{(i)} - u_{(i-1)}) \cdot L[k]$ phrases at random from L_i, and successively write them in the *text* field of the j-th document of the output corpus. Similarly, for each document of the output corpus assign $documentID = j$ and $topicProportion[i] = u_{(i)} - u_{(i-1)}$. This completes the construction of the output corpus *OutputData*.

Figure 1 illustrates the generation of an *OutputData* document from the N documents of the *InputData* corpus that we assume deal with 3 selected topics. List L records the number of phrases of each of these N documents. Lists L_1, L_2, L_3 respectively contain the phrases of the documents associated to each of the 3 topics, and thus it is $N = |L_1| + |L_2| + |L_3|$. The number of phrases to be contained in document 1 of *OutputData* (let us refer to it as *OutDoc*1) is obtained by randomly selecting a value from L, say 8 (second position in L). Then, two random $U(0, 1)$ values are drawn and sorted, and stored as $u_{(1)}, u_{(2)}$. Let say we get $u_{(1)} = 0.125, u_{(2)} = 0.625$. Hence, 12.5% (0.125) of the 8 phrases in *OutDoc*1 are to come from topic 1, 50% (0.625–0.125) from topic 2, and 37.5% (1–0.625) from topic 3. Applying these proportions to the 8 phrases of *OutDoc*1, we get that 1,4 and 3 are the number of phrases to be respectively taken from topics 1 to 3. These number of phrases are then randomly drawn from lists L_1, L_2 and L_3, respectively, and written in *OutDoc*1. Notice that this draw is made with replacement, and thus some of the input phrases (as for instance

$Phrase_{2,|L_2|}$ in Fig. 1) may be repeated in the output documents. The reason behind this selection with replacement is that the number of phrases to select from a topic i may be greater than the number of phrases in the corresponding list L_i.

L	NumPhrases
$Document_1$	5
$Document_2$	8
...	...
$Document_N$	10

L_1	L_2	L_3
$Phrase_{1,1}$	$Phrase_{2,1}$	$Phrase_{3,1}$
$Phrase_{1,2}$	$Phrase_{2,2}$	$Phrase_{3,2}$
...	...	...
$Phrase_{1,\|L_1\|}$	$Phrase_{2,\|L_2\|}$	$Phrase_{3,\|L_3\|}$

Output document	L	$U_{(1)}$	$U_{(2)}$	NumPhrases $Topic_1$	NumPhrases $Topic_2$	NumPhrases $Topic_3$
1	8	0.125	0.625	1	4	3

Output document

$Phrase_{1,2}.Phrase_{2,1}.Phrase_{2,|L_2|}.Phrase_{2,|L_2|}.Phrase_{2,2}.Phrase_{3,|L_3|}.Phrase_{3,1}.Phrase_{3,2}$

1 4 3

Fig. 1. Example of output document generation combining 3 topics.

Therefore, after this method is applied, a new corpus of documents is produced, in such a way that each of the new documents mixes text from the specified topics in different proportions. As these proportions are recorded together with the new text, the new corpus constitutes a soft reference dataset for topic identification. Furthermore, a main feature of the corpus provided by the proposed method is that the soft reference scores (i.e., the proportions of the different topics) assigned to each document are obtained in an objective way, not relying on subjective judgements from human supervisors.

4 Computational Study

In this section, we carry out an small computational experiment on unsupervised soft topic identification with data obtained by applying the soft reference generation method introduced in last section. The aim of this study is to illustrate the application of the proposed method on real data, as well as to provide a comparison of some well-known unsupervised classification techniques on soft reference data.

Therefore, this setting somehow mimics a real-world situation (as that described in Sect. 2 regarding soft complaint classification) in which a corpus of documents is available, in such a way that each document is known to simultaneously deal, up to a (possibly unknown) degree, with a set of topics. From this knowledge, a model is searched that allows quantifying the proportion up to which each document deals with each of the topics being considered. To this aim, we contemplate the following steps:

1. Let k denote the number of topics being considered. Configure a soft reference corpus by applying the method described in Sect. 3, in such a way that each document i in this corpus mixes the k topics in proportions $p_i \in [0,1]^k$.
2. Apply natural language processing (NLP) techniques to translate the unstructured textual information of the documents in the corpus into a matrix with as many rows as documents, and with as many columns as relevant terms, so that it can be processed in the next step.
3. Apply an unsupervised soft classification algorithm to the matrix obtained in the previous step, only assuming that the number of topics k is known. The output of this algorithm then provides an estimation $\hat{p}_i \in [0,1]^k$ of the topics' weights for each document.
4. Obtain an estimation of the performance of the unsupervised algorithm in the task to be performed by comparing the estimations $\hat{p}_i$ with the real proportions p_i.

In the present study we focus on comparing the performance of two unsupervised techniques, the classic k-means (KM) algorithm and the fuzzy k-means (FKM). To this aim, we apply a fuzzyfication step from the final centroids provided by the KM algorithm, so that a soft output is obtained that can be compared in fair terms with that of the FKM. Taking into account that the documents to be clustered present a mix of topics, and thus that the classification variable to predict is soft in nature, the hypothesis we would like to test is whether the FKM outperforms the post-fuzzified KM in providing a more accurate estimation of the actual weights of the topics in the documents to be processed.

4.1 Experimental Setting

Let us now describe the details of the computational study carried out. First of all, we applied the proposed soft reference generation method on the 20-Newsgroup (NG20) dataset [11], which constitutes a well-known benchmark in topic identification. This corpus contains a total of 20,000 documents, with exactly 1000 texts dealing with each of the 20 different topics shown in Table 1.

The proposed method has been setup to provide two different corpus typologies:

- A first kind of corpus contains documents mixing the topics of Atheism (NG1) and Graphics (NG2). Therefore, in this case it is $Topics = \{\text{NG1 NG2}\}$, and both topics are combined in proportions $p_i = (u_i, 1-u_i)$, where u_i is a random $U(0,1)$ value drawn for each document i.
- A second kind combines the topics of Graphics (NG2), Baseball (NG10), and Space (NG15). Then, now it is $Topics = \{\text{NG2 NG10 NG15}\}$, and for each document i a pair of random $U(0,1)$ values u_i, v_i are drawn. Assuming without loss of generality that $u_i < v_i$, the proportions of the mix of the 3 topics are given by $p_i = (u_i, v_i - u_i, 1 - v_i)$.

Table 1. Thematic blocks and topics of the 20-Newsgroup dataset.

Block	Topics [NG]
Alternative	Atheism [1]
Computing	Graphics [2], os.ms-windows [3], sys.ibm [4], sys.mac [5], windows.x [6]
Miscellaneous	Forsale [7]
Recreation	Autos [8], motorcycles [9], baseball [10], hockey [11]
Science	Cryptography [12], electronics [13], medicine [14], space [15]
Social issues	Religion.christian [16]
Talk	Guns [17], Mideast [18], politics.miscellaneous [19], religion.misc [20]

Each of these corpus typologies will be used in a different comparison experiment. To provide a more robust comparison through non-parametric statistical tests, 30 corpus of each typology are produced, each containing $OutputCorpusSize = 1000$ documents.

Once these 60 corpuses were generated through the proposed method, the following NLP steps were applied to each of the 60,000 documents:

- Tokenization [12]: The text of each document is separated in tokens (terms), generating a variable associated to each token.
- Stopwords removal: Non-significant words as conjunctions, determiers, etc. are removed.
- Stemming [13]: Process by which each token is reduced to its root form, so that different inflections are concentrated in a single token independent of number, gender, verbal conjugation, etc. To this aim, the Porter algorithm [14] has been applied.

As a result, for each corpus a matrix with as many rows as documents and as many columns as tokens is produced. Each position of this matrix is filled with the $tf-idf$ metric [15], that represents the relative frequency (term frequency, tf) of each token in a document, penalizing those words that appear with a relatively high frequency in a corpus. This penalization (applied through the so called inverse document frequency, idf) is introduced since a given term should not characterize too strongly a document whenever that term appears frequently in other documents of the corpus. Specifically, the $tf-idf$ metric is defined as follows:

$$tf-idf_{ij} = tf_{ij} \cdot idf_j, \tag{2}$$

where

$$tf_{ij} = \frac{\#\, of\ times\ term\ j\ appears\ in\ document\ i}{total\ \#\, of\ terms\ in\ document\ i} \tag{3}$$

and

$$idf_j = log\left(\frac{total\ \#\, of\ documents}{1 + total\ \#\, of\ documents\ containing\ the\ term\ j}\right). \tag{4}$$

Once these $tf - idf$ matrices are produced, the textual information gets structured in a form allowing the application of unsupervised classification techniques. However, before that, some dimension reduction steps are applied. Firstly, as the corpuses still contain a huge number of tokens (variables), some of which may occur quite infrequently within a corpus, a relevance thresholding step is applied. This consist in removing those tokens for which their accumulated $tf - idf$ in a corpus is lower than a given threshold. This threshold, known as the transition point [16], is set to the 95th percentile of the distribution of accumulated $tf - idf$ of all tokens in the corpus.

Finally, principal component analysis (PCA) has been applied on the remaining tokens in order to further reduce the dimension of the corpus matrices to a few variables. The number of components to use for each corpus typology has been determined through sedimentation graphs. As shown in Fig. 2, the number of components to be used depends on the number of topics considered. Particularly, just the first two components are retained for corpuses of the first typology (NG1/NG2), and three for corpuses of the second typology (NG2/NG10/NG15). Therefore, a 1000×2 matrix is finally obtained for each corpus of the first type, and a 1000×3 matrix is associated to each corpus of the second type.

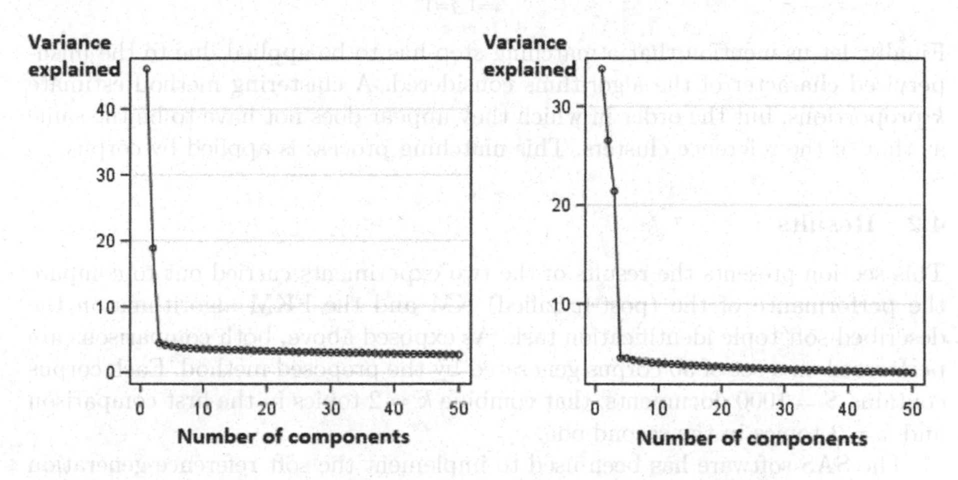

Fig. 2. Sedimentation graphics NG1/NG2 (left) and NG2/NG10/NG15 (right).

The KM and FKM algorithms are then applied on these matrices. The number of clusters k to form are set to the number of topics being combined for each corpus, i.e. $k = 2$ for the first typology and $k = 3$ for the second one. In both methods, 30 random starting centroids are tried for each corpus, allowing a maximum of 100 iterations after each initialization. Only the best result in terms of the intra-cluster variance objective function is keep. In all runs of the FKM the fuzziness parameter was set to $m = 2$.

As KM actually provides a crisp output, but the reference class variable is soft, a fuzzification step is needed in order to allow a fair comparison. This step is applied only after the best final KM centroids (out of the 30 random starts) are obtained. Specifically, the applied post-fuzzification follows the same scheme used in the FKM to obtain degrees from centroids in each iteration. Particularly, as in the FKM, the fuzzified degrees sum up to 1, and thus they can be interpreted as proportions. Therefore, let d_{ij} denote the Euclidian distance between document i and the j-th centroid ($j = 1,..,k$) reached by KM. Then, for the sake of the comparison with the soft reference, we consider the KM estimation of the proportion in which document i deals with topic j to be given by

$$\hat{p}_{ij} = \left(\sum_{l=1}^{k} \frac{d_{ij}^2}{d_{il}^2} \right)^{-1} \tag{5}$$

The performance metric, actually an error measure, will be given by the mean difference between the proportions estimated by the clustering methods $\hat{p}_i = (\hat{p}_{i1}, \ldots, \hat{p}_{ik})$ and the real proportions $p_i = (p_{i1}, \ldots, p_{ik})$ [17]. Thus, for a corpus with S documents and k topics, the error measure is given by

$$err_{Corp} = \frac{1}{Sk} \sum_{i=1}^{S} \sum_{j=1}^{k} |p_{ij} - \hat{p}_{ij}| \tag{6}$$

Finally, let us mention that a matching step has to be applied due to the unsupervised character of the algorithms considered. A clustering method estimate k proportions, but the order in which they appear does not have to be the same as that of the reference clusters. This matching process is applied by corpus.

4.2 Results

This section presents the results of the two experiments carried out to compare the performance of the (post-fuzzified) KM and the FKM algorithms on the described soft topic identification task. As exposed above, both comparisons are performed on a set of 30 corpus generated by the proposed method. Each corpus contains $S = 1000$ documents, that combine $k = 2$ topics in the first comparison and $k = 3$ topics in the second one.

The SAS software has been used to implement the soft reference generation method, as well as for preprocessing the data and perform PCA. The R software (packages *stats* and *fclust*) has been used for fitting the models and compute the errors shown in this section.

Table 2 shows the mean error of the KM and FKM algorithms and its standard deviation for each of the 30 corpuses analyzed. Clearly, in average both algorithms estimate the real proportions of each topic with similar accuracy. However, it is also important to notice that FKM tends to consistently produce slightly lower error rates than KM. Following [18], to rigorously analyze the statistical significance of this behaviour, a Wilcoxon signed rank test is applied on the results in Table 2. The results of this test are shown in Table 3, from which it is possible to conclude that FKM provides corpus error rates with a significantly lower median than KM.

Table 2. Error by corpus (mean ± standard deviation)

	2 Topics		3 Topics	
	FKM	KM	FKM	KM
1	.146 ± .110	.146 ± .109	.120 ± .069	.135 ± .081
2	.147 ± .107	.149 ± .109	.123 ± .070	.127 ± .072
3	.135 ± .100	.138 ± .103	.121 ± .070	.127 ± .075
4	.135 ± .103	.137 ± .104	.130 ± .072	.142 ± .079
5	.129 ± .100	.129 ± .100	.119 ± .068	.120 ± .070
6	.145 ± .106	.147 ± .107	.125 ± .068	.133 ± .075
7	.144 ± .109	.148 ± .112	.123 ± .073	.132 ± .077
8	.153 ± .114	.156 ± .117	.123 ± .071	.121 ± .070
9	.138 ± .103	.138 ± .103	.116 ± .070	.120 ± .070
10	.139 ± .101	.140 ± .103	.117 ± .067	.117 ± .067
11	.140 ± .104	.141 ± .106	.127 ± .074	.133 ± .079
12	.144 ± .111	.144 ± .111	.137 ± .075	.144 ± .079
13	.135 ± .110	.135 ± .109	.122 ± .072	.125 ± .074
14	.128 ± .103	.127 ± .102	.115 ± .070	.113 ± .068
15	.138 ± .104	.140 ± .106	.124 ± .073	.130 ± .077
16	.146 ± .102	.145 ± .102	.119 ± .071	.122 ± .072
17	.146 ± .112	.146 ± .114	.114 ± .069	.122 ± .071
18	.130 ± .098	.132 ± .098	.119 ± .070	.133 ± .080
19	.147 ± .108	.149 ± .111	.123 ± .073	.135 ± .076
20	.140 ± .107	.143 ± .109	.112 ± .066	.124 ± .075
21	.140 ± .106	.140 ± .106	.124 ± .076	.133 ± .078
22	.142 ± .109	.144 ± .111	.121 ± .068	.124 ± .071
23	.136 ± .107	.137 ± .109	.119 ± .076	.123 ± .077
24	.140 ± .101	.141 ± .102	.112 ± .065	.112 ± .064
25	.141 ± .101	.144 ± .104	.115 ± .068	.116 ± .067
26	.141 ± .107	.140 ± .106	.123 ± .069	.210 ± .113
27	.141 ± .103	.144 ± .104	.123 ± .073	.199 ± .106
28	.151 ± .110	.152 ± .111	.125 ± .076	.129 ± .079
29	.146 ± .106	.145 ± .105	.119 ± .070	.125 ± .072
30	.153 ± .106	.159 ± .112	.124 ± .071	.124 ± .072
Mean	**.141 ± .0064**	.143 ± .0071	**.121 ± .0063**	.132 ± .021

In summary, the final centroids produced by the KM and FKM algorithms provide a similar degree of accuracy when estimating the topic composition of the documents in the generated soft reference corpuses. Nevertheless, the inner fuzzification step of the FKM seems to slightly but consistently improve the final

Table 3. Wilcoxon test to compare fuzzy k-means (R^-) against k-means (R^+). For each experiment the differences in error rates for each corpus i are expressed as $FKM_i - KM_i = sign(FKM_i - KM_i)\,|FKM_i - KM_i|$, and sorted in increasing order of the absolute differences. This allows assigning a rank to each difference. The sums of ranks of positive and negative difference are denoted by R^+ and R^-. Under the null hypothesis of equal median error rates these statistics should be similar.

Comparison		R^-	R^+	p-val
FKM vs. KM	2 Topics	431	34	**6.918e-06**
FKM vs. KM	3 Topics	451	14	**2.049e-07**

FKM centroids with respect to those finally achieved by the KM algorithm, which instead performs a crisp cluster assignment in each iteration and has only been fuzzified after its conclusion.

5 Conclusions

A method to generate documents to be used as soft reference data in topic identification tasks has been introduced in this work. The method proceeds by combining phrases of previously available texts associated to different topics, in random but known proportions. As a consequence, a main feature of this method is that it allows generating reference data with objective soft degrees, not relying on subjective judgements from human supervisors. Reference corpuses containing any number of documents that combine a wide variety of topics can be thus obtained through the proposed method. Corpuses of this kind can then be used to allow evaluating the performance of soft topic classification techniques, both supervised and unsupervised. We consider this a relevant contribution as soft reference data for topic identification were nonexistent or hardly available.

A complete computational study was also carried out in this work, illustrating the application of the proposed method on real data and showing the possibility of conducting a proper comparison of the performance of two soft unsupervised classification methods. Particularly, this study allowed to conclude that the inner fuzzification step of the fuzzy k-means algorithm provide slightly but consistently better centroids for soft topic identification than those of the classic k-means algorithm, at least under certain conditions of the soft reference documents.

Future work regarding the proposed method will consider its extension to allow generating documents that combine phrases of different subsets of the specified topics, not necessarily combining all them simultaneously. This will allow more realistic soft reference documents to be generated, particularly for the context of complaint classification. Further work will also be devoted to assess the performance of soft unsupervised methods under various probability distributions of the random numbers used to determine the proportions of phrases of the different topics being combined.

References

1. Chakraborty, G., Pagolu, M., Garla, S.: Text Mining and Analysis. Practical Methods Examples and Case Studies Using SAS, p. 1. SAS Institute Inc., Cary (2013)
2. Berry, M.W., Koban, J.: Text Mining: Applications and Theory. Wiley, Hoboken (2010)
3. Duda, R.O., Hart, P.E.: Pattern Classification and Scene Analysis. Wiley, New York (1973)
4. Swam, P.H., Davis, S.M.: Remote Sensing: The Quantitative Approach. McGraw-Hill, New York (1978)
5. Settle, J.J., Drake, N.A.: Linear mixing and the estimation of ground cover proportions. Int. J. Remote Sens. **14**(6), 1159–1177 (1993)
6. del Amo, A., Montero, J., Fernandez, A., Lopez, M., Tordesillas, J.M., Biging, G.: Spectral fuzzy classification: an application. IEEE Trans. Syst. Man Cybern. Part C **32**(1), 42–48 (2002)
7. Binaghi, E., Brivio, P.A., Ghezzi, P., Rampini, A., Zilioli, E.: Investigating the behaviour of neural and fuzzy-statistical classifiers in sub-pixel land cover estimations. Can. J. Remote Sens. **25**(2), 171–188 (1999)
8. Wang, S., Wu, B., Wang, B., Tong, X.: Complaint classification using hybrid-attention GRU neural network. In: Yang, Q., Zhou, Z.-H., Gong, Z., Zhang, M.-L., Huang, S.-J. (eds.) PAKDD 2019. LNCS (LNAI), vol. 11439, pp. 251–262. Springer, Cham (2019). https://doi.org/10.1007/978-3-030-16148-4_20
9. Ruspini, E.H.: A new approach to clustering. Inf. Control **15**, 22–32 (1969)
10. Villarino, G., Gómez, D., Rodríguez, J.T., Montero, J.: A bipolar knowledge representation model to improve supervised fuzzy classification algorithms. Soft. Comput. **22**(15), 5121–5146 (2018). https://doi.org/10.1007/s00500-018-3320-9
11. Hettich S., Bay S.D.: The UCI KDD archive, pp. 1721–1288. University of California, Department of Information and Computer Science, Irvine, CA (1999). http://kdd.ics.uci.edu
12. Hassler, M., Fliedl, G.: Text preparation through extended tokenization. University Klagenfurt (2006)
13. Jivani, A.G.: A comparative study of stemming algorithms. Department of Computer Science and Engineering, the Maharaja Sayajirao University of Baroda Vadodara, Gujarat, India (2011)
14. Porter, M.F.: An Algorithm For Suffix Stripping, Readings in Information Retrieval, pp. 313–316. Morgan Kaufmann Publishers, Inc., Burlington (1997)
15. Salton, G., Yang, C.S.: On the specification of term values in automatic indexing. J. Doc. **28**(1), 11–21 (1973)
16. Pinto, D., Rosso, P., Jimenez-Salazar, H.: A self-enriching methodology for clustering narrow domain short texts. Comput. J. **54**, 1148–1165 (2011)
17. Gómez, D., Biging, G., Montero, J.: Accuracy statistics for judging soft classification. Int. J. Remote Sens. **29**(3), 693–709 (2008)
18. García, S., Fernández, A., Luengo, J., Herrera, F.: Advanced nonparametric tests for multiple comparisons in the design of experiments in computational intelligence and data mining: experimental analysis of power. Inf. Sci. **180**(10), 2044–2064 (2010)

SK-MOEFS: A Library in Python for Designing Accurate and Explainable Fuzzy Models

Gionatan Gallo, Vincenzo Ferrari, Francesco Marcelloni, and Pietro Ducange(✉)

Department of Information Engineering, University of Pisa, Largo Lucio Lazzarino 1, Pisa, Italy
{gionatan.gallo,vincenzo.ferrari,francesco.marcelloni, pietro.ducange}@unipi.it

Abstract. Recently, the explainability of Artificial Intelligence (AI) models and algorithms is becoming an important requirement in real-world applications. Indeed, although AI allows us to address and solve very difficult and complicated problems, AI-based tools act as a black box and, usually, do not explain how/why/when a specific decision has been taken. Among AI models, Fuzzy Rule-Based Systems (FRBSs) are recognized world-wide as transparent and interpretable tools: they can provide explanations in terms of linguistic rules. Moreover, FRBSs may achieve accuracy comparable to those achieved by less transparent models, such as neural networks and statistical models. In this work, we introduce SK-MOEFS (acronym of SciKit-Multi Objective Evolutionary Fuzzy System), a new Python library that allows the user to easily and quickly design FRBSs, employing Multi-Objective Evolutionary Algorithms. Indeed, a set of FRBSs, characterized by different trade-offs between their accuracy and their explainability, can be generated by SK-MOEFS. The user, then, will be able to select the most suitable model for his/her specific application.

Keywords: Explainable Artificial Intelligence · Multi-objective Evolutionary Algorithms · Fuzzy Rule-Based Systems · Python · Scikit-Learn

1 Introduction

The proliferation of Artificial Intelligence (AI) has a significant impact on society [1]. Indeed, AI has already become ubiquitous in personal life and the modern industry. As regards the latter, we are experiencing the "Industry 4.0 Era", and Machine Learning (ML) and AI play a crucial role among its enabling technologies

This work was supported by the Italian Ministry of Education and Research (MIUR), in the framework of the CrossLab project (Departments of Excellence).

M.-J. Lesot et al. (Eds.): IPMU 2020, CCIS 1239, pp. 68–81, 2020.
https://doi.org/10.1007/978-3-030-50153-2_6

[12]. Models based on ML and AI are learnt from the input data and are generally very accurate. However, in most cases, they are highly non-transparent, *i.e.*, it is not clear which information in the input data causes the generated output. In the context of Industry 4.0, making decisions has a crucial impact, so modern approaches are shifting towards AI models with understandable outcomes.

Recently, a new trend is gaining importance within AI, namely, eXplainable Artificial Intelligence (XAI). XAI methodologies and algorithms aim to make AI-based models and methods more transparent while maintaining high-performance levels of accuracy and precision [5]. Fuzzy Rule-Based Systems (FRBSs) are a category of models strongly oriented towards explainability. FRBSs are highly interpretable and transparent because of the linguistic definitions of fuzzy rules and fuzzy sets, which represent the knowledge base of these models. Moreover, the simplicity of the reasoning method, adopted for providing a decision based on input facts, ensures also a high explainability level of FRBSs [10].

In the last decade, Multi-Objective Evolutionary Algorithms (MOEAs) have been successfully adopted for designing FRBSs from data, leading to the so-called Multi-Objective Evolutionary Fuzzy Systems (MOEFSs) [8,9]. MOEFSs are designed to concurrently optimize the accuracy and explainability of FRBSs, which are two conflicting objectives. Indeed, in general, very accurate models are characterized by low explainability and vice-versa.

Regarding software tools to generate and evaluate XAI models, there are not many options. For example, GUAJE [13] and ExpliClas [2] are examples of tools for designing interpretable models. They also handle FRBSs, but without the boost of MOEAs for optimizing their accuracy and explainability.

In this paper, we propose and discuss SK-MOEFS, a new Python library that helps data scientists to define, build, evaluate, and use MOEFSs, under the Scikit-Learn environment [14]. The latter is an Open Source toolbox that provides state-of-the-art implementations of many well-known ML algorithms. We designed SK-MOEFS according to Scikit-Learn's design principles. Indeed, we exploited the available data structures and methods in the Scikit-Learn library. As a result, the user is allowed, under the same framework, to easily and quickly design, evaluate, and use several ML models, including MOEFSs. The current version of SK-MOEFS includes an implementation of a specific MOEFS, namely PAES-RCS, introduced in [3]. PAES-RCS selects a reduced number of rules and conditions, from an initial set of rules, during the multi-objective evolutionary learning process. Precisely, we implemented PAES-RCS-FDT, which adopts a fuzzy decision tree (FDT) for generating the initial set of rules [7]. We also highlight that SK-MOEFS is an extendable framework that allows easy integration of different types of MOEFSs.

The paper is organized as follows. Section 2 introduces FRBSs and the general multi-objective evolutionary learning scheme for designing them. Afterward, Sect. 3 illustrates the design of SK-MOEFS, focusing on the functionalities provided. Then, we describe in detail the implementation of a specific MOEFS for classification problems in Sect. 4. Section 5 is devoted to show an example of building and evaluating a MOEFS tested with a real-world dataset. Finally, Sect. 6 draws some conclusions.

2 Multi-objective Evolutionary Fuzzy Systems

2.1 Fuzzy Rule-Based Systems

A Fuzzy Rule-Based System (FRBS) is characterized by two main components, namely the Knowledge Base (KB) and the fuzzy inference engine. The KB is composed by a set of linguistic rules and by a set of parameters which describe the fuzzy sets on which the rules are defined. The fuzzy inference engine is in charge of generating a prediction, given a new input pattern, based on the content of the KB.

Let $X = \{X_1, \ldots, X_F\}$ be the set of input attributes and X_{F+1} be the output attribute. Let U_f, with $f = 1, ..., F+1$, be the universe of the fth attribute X_f. Let $P_f = \{A_{f,1}, \ldots, A_{f,T_f}\}$ be a fuzzy partition of T_f fuzzy sets on attribute X_f. Finally, we define the training set $\{(\mathbf{x}_1, x_{F+1,1}), \ldots, (\mathbf{x}_N, x_{F+1,N})\}$ as a collection of N input-output pairs, with $\mathbf{x}_t = [x_{t,1} \ldots, x_{t,F}] \in \Re^F$, $t = 1, \ldots, N$.

In regression problems, X_{F+1} is a continuous attribute and, therefore, $\forall t \in [0..N]$, $x_{F+1,t} \in \Re$. With the aim of estimating the output value corresponding to a given input vector, we can adopt a Fuzzy Rule-Based Regressor (FRBR) with a rule base (RB) composed of M linguistic fuzzy rules expressed as:

$$\begin{aligned} R_m : \textbf{IF } X_1 \textbf{ is } A_{1,j_{m,1}} \textbf{ AND} \ldots \textbf{AND } X_f \textbf{ is } A_{f,j_{m,f}} \textbf{ AND} \\ \ldots \textbf{AND } X_F \textbf{ is } A_{F,j_{m,F}} \textbf{ THEN } X_{F+1} \textbf{ is } A_{F+1,j_{m,F+1}} \end{aligned} \quad (1)$$

where $j_{m,f} \in [1, T_f]$, $f = 1, ..., F+1$, identifies the index of the fuzzy set (among the T_f linguistic terms of partition P_f), which has been selected for X_f in rule R_m.

In classification problems, X_{F+1} is categorical and $x_{F+1,t} \in C$, where $C = \{C_1, \ldots, C_K\}$ is the set of K possible classes. With the aim of determining the class of a given input vector, we can adopt a Fuzzy Rule-Based Classifier (FRBC) with an RB composed of M rules expressed as:

$$\begin{aligned} R_m : \textbf{IF } X_1 \textbf{ is } A_{1,j_{m,1}} \textbf{ AND} \ldots \textbf{AND } X_f \textbf{ is } A_{f,j_{m,f}} \textbf{ AND} \\ \ldots \textbf{AND } X_F \textbf{ is } A_{F,j_{m,F}} \textbf{ THEN } X_{F+1} \textbf{ is } C_{j_m} \textit{ with } RW_m \end{aligned} \quad (2)$$

where C_{j_m} is the class label associated with the mth rule, and RW_m is the rule weight, i.e., a certainty degree of the classification in the class C_{j_m} for a pattern belonging to the fuzzy subspace delimited by the antecedent of the rule R_m. Different definitions of the rule weight RW_m are commonly found in the literature [4]:

Given a new input pattern $\hat{\mathbf{x}} \in \Re^F$, the estimated output value or class label is provided by the FRBR or by the FRBC, respectively, adopting a specific inference engine. In both cases, the output depends on the strength of activation of each rule with the input. Details on the different inference engines can be found in [4].

In the current version of SK-MOEFS, we adopt strong triangular fuzzy partitions. As shown in Fig. 1, each partition is made up of triangular fuzzy sets $A_{f,j}$, whose membership function can be represented by the tuples $(a_{f,j},\ b_{f,j},\ c_{f,j})$, where $a_{f,j}$ and $c_{f,j}$ correspond to the left and right extremes of the support of $A_{f,j}$, and $b_{f,j}$ to its core. Other typologies of FRBSs, such as TSK-FRBSs, FRBSs with DNF rules and FRBSs based on multiple granularities, have also been considered in the MOEFS specialized literature [9]. For the sake of brevity, in this Section we have described only the two types of FRBSs which have been mostly discussed and experimented in the last years, mainly due to their high explainability level. However, the SK-MOEFS toolbox has been designed for allowing the programmer to easily implement multi-objective evolutionary learning schemes for any kind of FRBS, both for regression and classification problems.

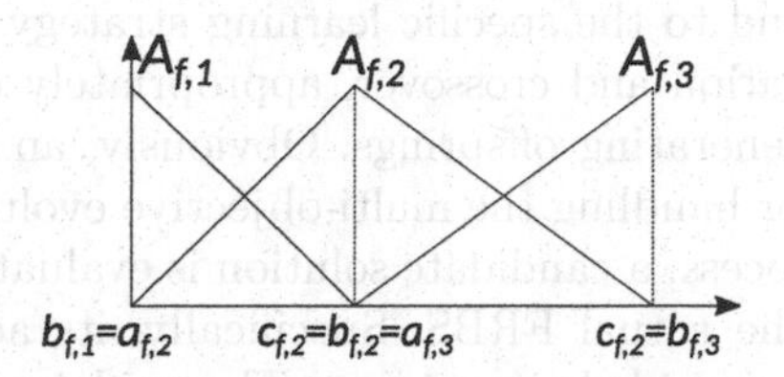

Fig. 1. An example of a strong triangular fuzzy partition with three fuzzy sets.

2.2 Multi-objective Evolutionary Learning Schemes

The FRBS design process aims: i) to determine the optimal set of rules for managing regression or classification problems, and ii) to find the appropriate number of fuzzy sets for each attribute and their parameters. The objective of the design process is to concurrently maximize the system accuracy and, possibly, the model explainability. The accuracy of an FRBR is usually maximized by means of a minimization process of the estimation error of the output values. On the other hand, the accuracy of an FRBC is usually calculated in terms of percentage of correctly classified patterns. As regards the explainability, when dealing with FRBS we usually talk about their *intepretability*, namely the capability of explaining how predictions have been done, using terms understandable to humans. Thus, the simplicity of the fuzzy inference engine, adopted to deduce conclusions from facts and rules, assumes a special importance. Moreover, the intepretability is strictly related to the *transparency* of the model, namely to the capability of understanding the structure of the model itself. FRBSs can be characterized by a high transparency level, whenever the linguistic RB is composed of a reduced number of rules and conditions and the fuzzy partitions have a good *integrity*. The integrity of fuzzy partitions depends on some properties, such as order, coverage, distinguishability and normality [4]. The work in [11] discusses several measures for evaluating the interpretability of an FRBS, taking into consideration semantic and complexity aspects of both the RB and of the fuzzy partitions.

As stated in the Introduction, in the last decade, MOEAs have been successfully adopted for designing FRBSs by concurrently optimizing both their accuracy and explainability, leading to the so-called MOEFSs [9]. Indeed, MOEAs allow us to approach an optimization process in which two or more conflicting objectives should be optimized at the same time, such as accuracy and explainability of FRBSs. MOEAs return a set of non-dominated solutions, characterized by different trade-offs between the objectives, which represents an approximation of Pareto front [6]. Adopting a Multi-Objective Learning Scheme (MOEL) it is possible to learn the structure of FRBSs using different strategies, such as learning only the RB considering pre-defined fuzzy partitions, optimizing only the fuzzy set parameters, selecting rules and conditions, from an initial set of rules, and learning/selecting rules concurrently with the optimization of the fuzzy set parameters. A complete taxonomy of MOEFSs can be found in [8].

In general, an MOEL scheme includes a chromosome coding, which is related to the type of FRBS and to the specific learning strategy, and a set of mating operators, namely mutation and crossover, appropriately defined for acting on the chromosome and generating offsprings. Obviously, an MOEL scheme must use a specific MOEA for handling the multi-objective evolutionary optimization process. During this process, a candidate solution is evaluated decoding its chromosome for building the actual FRBS. Specifically, its accuracy is calculated adopting a training set provided as an input. The explainability is evaluated on the basis of a pre-defined measure, such as the number of rules or the total number of conditions in the RB (also called Total Rule Length (TRL)). At the end of the optimization, a set of FRBSs, characterized by different trade-off between accuracy and intepretability, is returned. In the following sections, we show how to design and implement an MOEL scheme in our SK-MOEFS toolbox.

3 SK-MOEFS Design

Previously, we argued about the importance of MOEFSs in the context of XAI. In this Section, we discuss the design of SK-MOEFS, a Python library for generating explainable FRBSs. SK-MOEFS extends the functionalities of Scikit-Learn[1], a popular Open Source tool for predictive data analysis [14]. Data scientists and researchers deeply adopt Scikit-Learn due to its *ease-of-use*. Moreover, it is highly efficient both in terms of memory occupancy and computational costs. Indeed, Scikit-Learn takes advantage of other Python libraries, such as NumPy, SciPy, and MatplotLib, largely employed in the data analysis field.

Similarly to Scikit-Learn, SK-MOEFS allows us also to adopt the generated models for making predictions and evaluating the models in terms of different metrics. However, since SK-MOEFS creates a collection of different FRBSs, we had to appropriately design data structures and methods for handling more than one model. Indeed, classically, Scikit-Learn algorithms allow the user to define, train, evaluate, and use just one model. Finally, we have also designed the methods for extracting explainability metrics, such as TRL, number of rules, and partition integrity indices [11].

[1] https://scikit-learn.org/.

3.1 Class Hierarchy

To design and implement SK-MOEFS, we followed the official Scikit-Learn guidelines for developers[2].

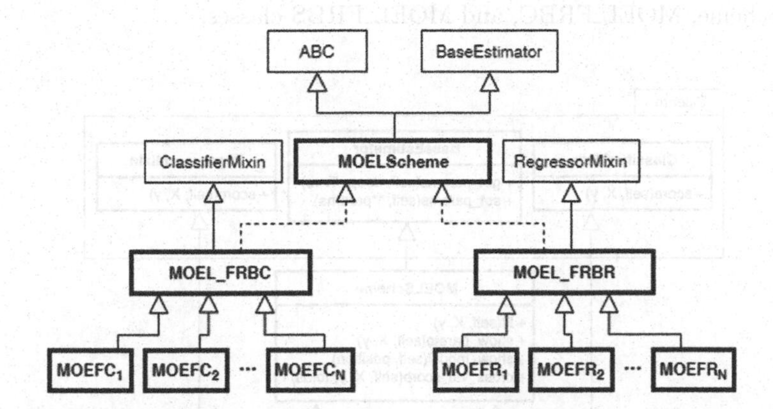

Fig. 2. UML class diagram describing the class hierarchy of SK-MOEFS.

As depicted in Fig. 2, the principal abstract class of SK-MOEFS, that we labeled as MOELScheme, derives from the BaseEstimator class of Scikit-Learn library. Moreover, to define the infrastructure of an abstract class, MOELScheme must extend the ABC class. A MOELScheme represents a general multi-objective evolutionary learning scheme for generating a set of FRBSs characterized by different trade-offs between accuracy and explainability. We recall that the chromosome coding and the mating operators depend on the selected learning scheme. As regards the fitness functions, the accuracy measure depends on the type of problems to be approached (classification or regression), and the explainability measure can be defined in several ways, as discussed in the previous section.

In general, a classifier or a regressor is an instance of a specific class derived by the BaseEstimator and by a ClassifierMixin or RegressorMixin classes: it is an object that fits a model based on some training data and is capable of making predictions on new data.

Since we aim to provide a general scheme for approaching both classification and regression problems by using MOEFSs, we derive two abstract classes from the MOELScheme one, namely MOEL_FRBC and MOEL_FRBR. They define, respectively, the MOEL scheme for Fuzzy Rule-based Classifiers (FRBCs) and the one for Fuzzy Rule-based Regressors (FRBRs). The former includes methods from the ClassifierMixin class and the latter from RegressorMixin class.

Finally, from the MOEL_FRBC and the MOEL_FRBR classes, actual MOEL schemes (we labeled them as Multi-objective Evolutionary Fuzzy Classifier (MOEFC) or Regressor (MOEFR)) can be derived, such as the PAES-RCS that has been implemented and experimented, as discussed in the following sections.

[2] https://scikit-learn.org/stable/developers.

3.2 Description of the Main Methods

Each MOEL scheme must provide the typical Scikit-Learn methods, for both classifiers and regressors adapted explicitly for handling multiple models. In Fig. 3 we show another UML class diagram that describes the main features of MOELScheme, MOEL_FRBC, and MOEL_FRBS classes.

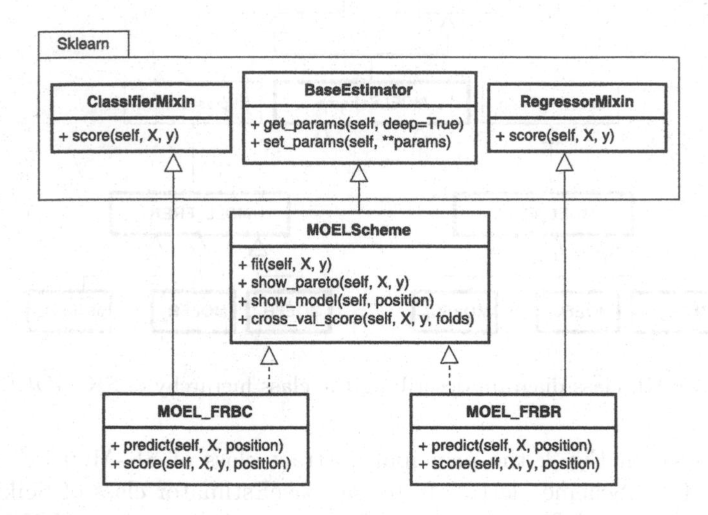

Fig. 3. UML class diagrams describing the main methods of SK-MOEFS

In Scikit-Learn, the methods *fit*, *predict*, and *score* are typically implemented on each classifier or regressor. They allow, respectively, creating a model, using the model for making predictions, and extracting some metrics for evaluating the model. In the following, we describe these and other specific methods that must be implemented for each new MOEL scheme:

- *fit*: this method estimates the model parameters, namely the RB and the fuzzy partitions, exploiting the provided training set. We recall that in Scikit-Learn a training set must be provided in terms of an $N \times F$ NumPy matrix $\boldsymbol{X}$, describing the input patterns in terms of F features, and a vector $\boldsymbol{y}$ with N elements representing the actual label or value associated with a specific input pattern. In the beginning, the method initializes a MOEL scheme according to a specific learning strategy and to the type of problem to be handled, namely classification or regression. Then, an MOEA is in charge of carrying out the learning process, which stops when a specific condition is reached (for example, when the algorithm reaches the maximum number of fitness function evaluations). Finally, it returns an approximated Pareto front of FRBSs, which are sorted by an ascending order per accuracy. The first model, labeled as the FIRST solution, is the one characterized by the highest accuracy and by the lowest explainability. On the contrary, we marked the model with the highest explainability but the lowest accuracy as the LAST solution.

Finally, the MEDIAN model is the middle ground between the two. Indeed, its accuracy is the median among the solutions.

- *predict*: this method is in charge of predicting the class labels or the values associated with a new set of input patterns. It returns a vector of estimated labels or values. Since the MOEL scheme generates multiple models, the method takes as input also an index for selecting the model into the Pareto front. By default, the function adopts the most accurate model (FIRST) for making predictions. Notice that all the learning schemes that extend from MOEL_FRBC or MOEL_FRBR, must implement the predict method to define different and specific behaviors.
- *score*: this method takes as inputs a matrix $\boldsymbol{X}$, which contains data described in the feature space of F values, and the vector $\boldsymbol{y}$ of the r labels or values associated with each input. Moreover, it takes the position of a model belonging to the Pareto front, and it generates the values of the accuracy and explainability measures for that selected model. Also in this case, the FIRST solution is selected by default.
- *show_pareto*: this method extracts and plots the values of the accuracy and the explainability. By default, for each model of the Pareto front generated by an MOEL scheme, it runs the *fit* method on the training set. SK-MOEFS allows the user to provide also a test set; in this case, show_pareto calculates the accuracies considering the additional data. As a result, it returns a plot of the approximated Pareto front, both on the training and the test sets.
- *show_model*: given the position of a model in the Pareto front, this method shows the set of fuzzy linguistic rules and the fuzzy partitions associated with each linguistic attribute. The predefined model of choice is, as always, the FIRST solution.

Finally, since Scikit-Learn provides methods for performing a *k-fold cross-validation* analysis, we re-designed these methods for handling the fact that a MOEL scheme generates a set of solutions. Specifically, we redefined the *cross_val_score* which usually returns an array of k scores, one for each fold. Here, the method returns a $k \times 6$ matrix, where each row contains the accuracy, calculated on the test set, and the explainability of the FIRST, MEDIAN, and LAST solutions. Moreover, when performing cross-validation with MOEFSs [4], we decided to act as follows: first, we compute the mean values of the accuracy and the explainability of the FIRST, MEDIAN and LAST solutions, then we plot them on a graph.

4 An Example of an MOEL Scheme Implementation: PAES-RCS-FDT for Classification Problems

In this Section, we describe the actual implementation of an MOEL scheme for classification problems in SK-MOEFS, namely PAES-RCS-FDT [7]. The implemented algorithm adopts the rule and condition selection (RCS) learning scheme [3] for classification problems. The multi-objective evolutionary learning scheme is based on the (2 + 2)M-PAES, which is an MOEA successfully employed in the context of MOEFSs during the last years. The algorithm concurrently optimizes two objectives: the first objective considers the TRL as explainability measure; the second objective takes into account the accuracy, assessed in terms of classification rate.

In the learning scheme, an initial set of candidate rules must be generated through a heuristic or provided by an expert. In our implementation, the set of candidate rules is generated exploiting the fuzzy multi-way decision trees (FMDT) [15]: each path from the root to a leaf node translates into a rule. Before learning the FMDT, we need to define an initial strong fuzzy partition for each attribute. The adopted FMDT algorithm embeds a discretization algorithm that is in charge of generating such partitions.

During the evolutionary process, the most relevant rules and their conditions are selected. Moreover, each triangular strong fuzzy partition P_f is concurrently tuned, by adapting the positions of the cores $b_{f,j}$.

In PAES-RCS, a chromosome C codifies a solution for the problem. The former is composed of two parts (C_R, C_T), which define, respectively, the RB and the positions of the representatives of the fuzzy sets, namely the cores.

Let J_{DT} be the initial set of M_{DT} rules generated from the decision tree. Compact and interpretable RBs are desirable, so we allow that the RB of a solution contains at most M_{max} rules. The C_R part, which codifies the RB, is a vector of M_{max} pairs $\mathbf{p}_m = (k_m, \mathbf{v}_m)$, where $k_m \in [0, M_{DT}]$ identifies the selected rule of J_{DT} and $\mathbf{v}_m = [v_{m,1}, \ldots, v_{m,F}]$ is a binary vector which indicates, for each attribute X_f, if the condition is present or not in the selected rule. In particular, if $k_m = 0$ the mth rule is not included in the RB. Thus, we can generate RBs with a lower number of rules than M_{max}. Further if $v_{m,f} = 0$ the fth condition of the mth rule can be replaced by a *"don't care"* condition.

C_T is a vector containing F vectors of $T_f - 2$ real numbers: the fth vector $\left[b_{f,2}, \ldots, b_{f,T_f-1}\right]$ determines the positions of the cores of the partition P_f. We recall that using strong fuzzy partitions ensures the partition integrity. Indeed, order, coverage, distinguishability and normality are always ensured. In order to increase the integrity level, we can define constrains on the intervals on which cores can assume valid values. For more details check [3].

In order to generate the offspring populations, we exploit both crossover and mutation. We apply separately the one-point crossover to C_R and the BLX-α-crossover, with $\alpha = 0.5$, to C_T. As regards the mutation, we apply two distinct operators for C_R and an operator for C_T. More details regarding the mating operators and the steps of PAES-RCS can be found in [3]. In the next Section, we will briefly introduce the main parameters that must be set for running PAES-RCS-FDT.

In Fig. 4, we show a detailed UML class diagram describing the main classes and methods that we implemented. First of all, we have derived from the MOEL_FRBC the class MPAES_RCS, which is in charge of handling the rule and condition selection multi-objective learning scheme, by means of (2 + 2)M-PAES algorithm. This class needs the RCSProblem, which is a class derived

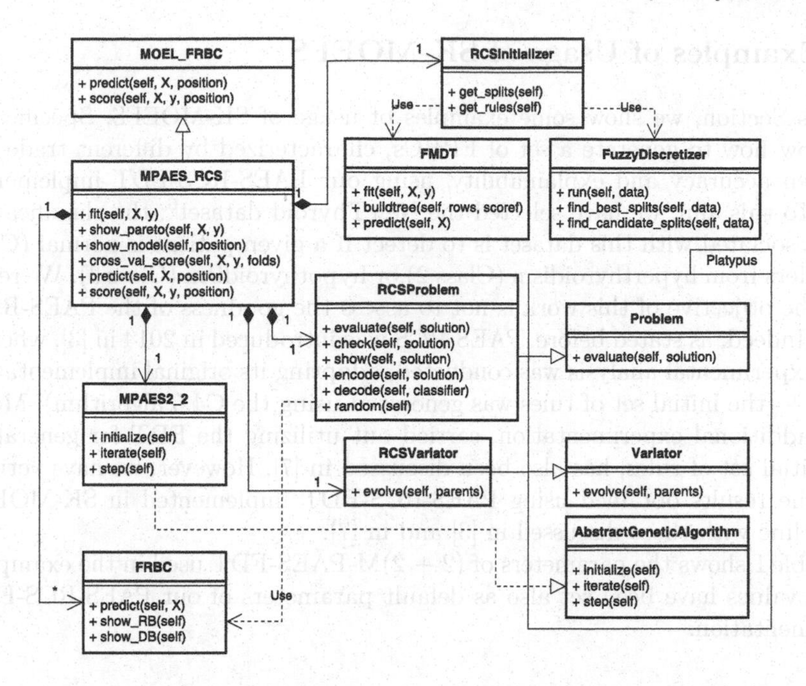

Fig. 4. UML class diagram of PAES-RCS-FDT in SK-MOEFS.

from the Problem class of the Package *Platypus*, a Python framework for multi-objective evolutionary optimization. It defines operations on a possible solution (a chromosome) such as its encoding, feasibility checks, evaluation of objectives, and generation of a random solution. Moreover, the MPAES_RCS class adopts an RCS_Variator, a particular implementation of the Platypus Variator, which includes all the mating operators that we discussed before. Two additional classes are adopted as compositions by the MPAES_RCS class, namely the MPAES2_2 and the RCSInitializer. Specifically, the MPAES2_2 class extends the Abstract-GeneticAlgorithm class of Platypus and implements (2 + 2)M-PAES. Finally, the RCSInitializer implements the methods for the definition of the initial strong fuzzy partitions and for generating the initial set of rules. To this aim, this class uses the fuzzy discretizer (implemented by the FuzzyDiscretizer class) and the Multi-way Fuzzy Decision Tree (implemented by the MFDT class), respectively. Both the discretizer and the algorithm for generating the fuzzy decision tree are described in detail in [15]. More information on the organization of the Platypus framework can be found in the official guide[3] and github repository[4]. The code of SK-MOEFS, including the implementation of PAES-RCS-FDT as a standard Python program, is available on a GitHub repository[5], along with a detailed documentation describing all the classes and its methods.

[3] https://platypus.readthedocs.io/en/latest/.
[4] https://github.com/Project-Platypus/Platypus.
[5] https://github.com/GionatanG/skmoefs.

5 Examples of Usage of SK-MOEFS

In this Section, we show some examples of usage of SK-MOEFS. Specifically, we show how to generate a set of FRBCs, characterized by different trade-offs between accuracy and explainability, using our PAES-RCS-FDT implementation. To this aim, we have selected the NewThyroid dataset[6]: the classification task associated with this dataset is to detect if a given patient is normal (Class 1), suffers from hyperthyroidism (Class 2) or hypothyroidism (Class 3). We recall that the objective of this work is not to assess the goodness of the PAES-RCS-FDT. Indeed, as stated before, PAES-RCS was introduced in 2014 in [3], where a wide experimental analysis was conducted, adopting its original implementation in C++ (the initial set of rules was generated using the C4.5 algorithm). Moreover, additional experimentation, carried out utilizing the FDT for generating the initial set of rules, has also been discussed in [7]. However, we have verified that the results obtained using PAES-RCS-FDT implemented in SK-MOEFS are in line with those discussed in [3] and in [7].

Table 1 shows the parameters of (2 + 2)M-PAES-FDT used in the examples. These values have been set also as default parameters of our PAES-RCS-FDT implementation.

Table 1. Values of the parameters used in the examples

N_{val}	Total number of fitness evaluations	30000
AS	(2 + 2)M-PAES archive size	32
M_{max}	Maximum number of rules for each RB	50
P_{C_R}	Probability of applying crossover operator to C_R	0.1
P_{C_T}	Probability of applying crossover operator to C_T	0.5
P_{MRB_1}	Probability of applying first mutation operator to C_R	0.1
P_{MRB_2}	Probability of applying second mutation operator to C_R	0.7
P_{M_T}	Probability of applying mutation operator to C_T	0.2
T_{max}	Maximum number of fuzzy sets for each attribute	5

```
from skmoefs.rcs import MPAES_RCS, RCSInitializer, RCSVariator
from sklearn.model_selection import train_test_split

X, y = load_dataset('newthyroid')
Xtr, ytr, Xte, yte = train_test_split(X, y, test_size=0.3)
my_moefs = MPAES_RCS(variator=RCSVariator(), initializer=
    RCSInitializer())
my_moefs.fit(Xtr, ytr, max_evals=30000)
my_moefs.show_pareto(Xte, yte)
my_moefs.show_model('median')
```

Listing 1.1: Example for generating and plotting a Pareto front approximation of FRBCs

[6] https://sci2s.ugr.es/keel/dataset.php?cod=66.

In code Listing 1.1, we show an example of usage, in which we first load a dataset from a file and then we divide it into training and test sets. Second, we instantiate an MPAES-RCS object passing to its constructor the RCSVariator and an RCSInitializer. The latter will partition each input attributes into a predefined number of fuzzy sets and will generate the matrix J_{DT}. Afterward, we call the fit method, which returns the fitted model having now a list of the FRBCs characterized by different trade-offs between accuracy and explainability. Then, we call the method for showing the Pareto front approximation (in Fig. 5(a)), both on the training and test sets. Finally, we show the RB and the fuzzy partitions (in Fig. 6) of the MEDIAN solution in the Pareto Front approximation. In this example, we labeled the five fuzzy sets of each partition as Very Low (VL), Low (L), Medium (M), High (H), and Very High (VH). As we can see, the set of linguistic rules allows the user to understand the motivation of a decision: simply speaking, based on the levels of each input attribute describing a new patient, a specific class is associated with him/her. As regards the fuzzy partitions, it is worth noting, especially for the last two, that they moved from the initial uniform shape. However, they are still strong fuzzy partitions, thus ensuring a good integrity level, in terms of order, distinguishability, coverage and normality.

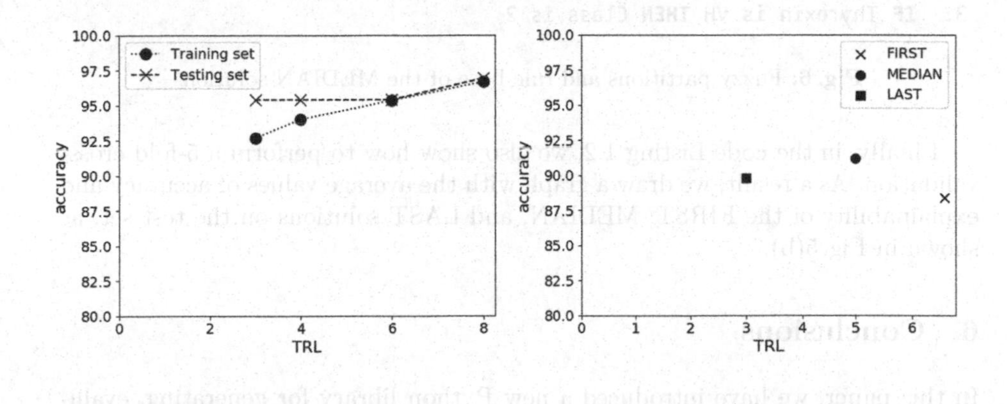

Fig. 5. Two examples of plots

```
from skmoefs.rcs import MPAES_RCS, RCSInitializer, RCSVariator

X, y = load_dataset('newthyroid')
my_moefs = MPAES_RCS(variator=RCSVariator(), initializer=
    RCSInitializer())
my_moefs.cross_val_score(X, y, folds=5)
```

Listing 1.2: Example for performing the cross validation

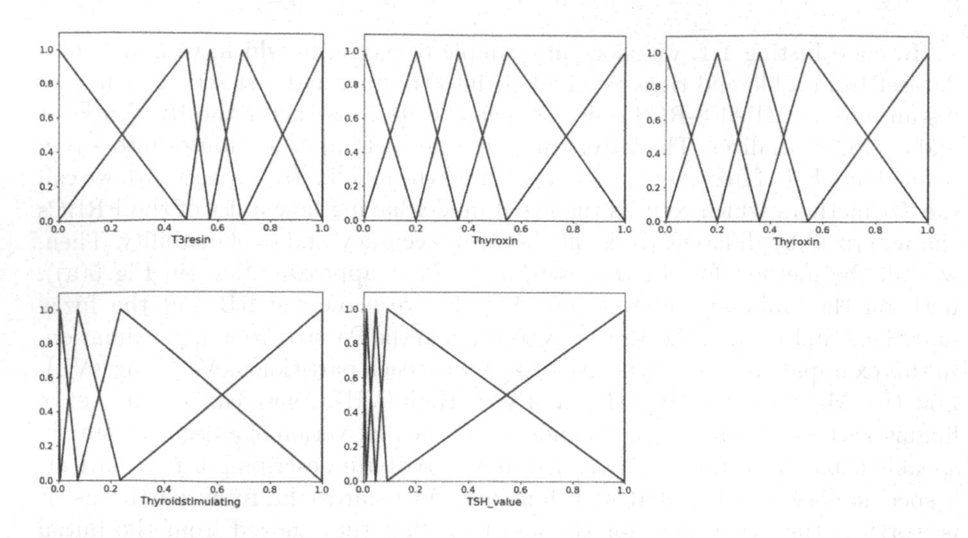


```
RULE BASE
1: IF Thyroxin is VL THEN Class is 3
2:  IF Thyroxin is M  AND Thyroidstimulating is L THEN Class is 1
3:  IF Thyroxin is VH THEN Class is 2
```

Fig. 6. Fuzzy partitions and rule base of the MEDIAN solution

Finally, in the code Listing 1.2, we also show how to perform a 5-fold cross-validation. As a result, we draw a graph with the average values of accuracy and explainability of the FIRST, MEDIAN, and LAST solutions on the test set, as shown in Fig. 5(b).

6 Conclusions

In this paper, we have introduced a new Python library for generating, evaluating and using both accurate, and explainable AI-based models, namely fuzzy rule-based systems (FRBSs). The library, called SK-MOEFS, allows the users to adopt multi-objective evolutionary learning (MOEL) schemes for identifying, from data, the structure of a set of FRBSs, characterized by different trade-offs between accuracy and explainability. Specifically, we designed the overall software infrastructure, i.e. all the class hierarchy, for handling a generic multi-objective evolutionary learning scheme. Moreover, we show an example of an actual implementation of a well known MOEL scheme, namely PAES-RCS-FDT. This scheme, during the evolutionary process, selects rules and conditions from an initial set of candidate classification rules, generated using a fuzzy decision tree. Additionally, the parameters of the fuzzy partitions can be learned concurrently with the set of rules. Finally, we have shown a simple example on how our SK-MOEFS can be used in Python for generating and evaluating a set of fuzzy classifiers on a benchmark dataset.

References

1. Adadi, A., Berrada, M.: Peeking inside the black-box: a survey on explainable artificial intelligence (XAI). IEEE Access **6**, 52138–52160 (2018)
2. Alonso, J.M., Bugarín, A.: ExpliClas: automatic generation of explanations in natural language for weka classifiers. In: 2019 IEEE International Conference on Fuzzy Systems (FUZZ-IEEE), pp. 1–6. IEEE (2019)
3. Antonelli, M., Ducange, P., Marcelloni, F.: A fast and efficient multi-objective evolutionary learning scheme for fuzzy rule-based classifiers. Inf. Sci. **283**, 36–54 (2014)
4. Antonelli, M., Ducange, P., Marcelloni, F.: Multi-objective evolutionary design of fuzzy rule-based systems. In: Handbook on Computational Intelligence: Volume 2: Evolutionary Computation, Hybrid Systems, and Applications, pp. 635–670. World Scientific (2016)
5. Carletti, M., Masiero, C., Beghi, A., Susto, G.A.: Explainable machine learning in industry 4.0: evaluating feature importance in anomaly detection to enable root cause analysis. In: 2019 IEEE International Conference on Systems, Man and Cybernetics (SMC), pp. 21–26. IEEE (2019)
6. Coello, C.A.C., Lamont, G.B.: Applications of Multi-objective Evolutionary Algorithms, vol. 1. World Scientific, Singapore (2004)
7. Ducange, P., Mannarà, G., Marcelloni, F.: Multi-objective evolutionary granular rule-based classifiers: an experimental comparison. In: 2017 IEEE International Conference on Fuzzy Systems (FUZZ-IEEE), pp. 1–6. IEEE (2017)
8. Ducange, P., Marcelloni, F.: Multi-objective evolutionary fuzzy systems. In: Fanelli, A.M., Pedrycz, W., Petrosino, A. (eds.) WILF 2011. LNCS (LNAI), vol. 6857, pp. 83–90. Springer, Heidelberg (2011). https://doi.org/10.1007/978-3-642-23713-3_11
9. Fazzolari, M., Alcala, R., Nojima, Y., Ishibuchi, H., Herrera, F.: A review of the application of multiobjective evolutionary fuzzy systems: current status and further directions. IEEE Trans. Fuzzy Syst. **21**(1), 45–65 (2012)
10. Fernandez, A., Herrera, F., Cordon, O., del Jesus, M.J., Marcelloni, F.: Evolutionary fuzzy systems for explainable artificial intelligence: why, when, what for, and where to? IEEE Comput. Intell. Mag. **14**(1), 69–81 (2019)
11. Gacto, M.J., Alcalá, R., Herrera, F.: Interpretability of linguistic fuzzy rule-based systems: an overview of interpretability measures. Inf. Sci. **181**(20), 4340–4360 (2011)
12. Lu, Y.: Industry 4.0: a survey on technologies, applications and open research issues. J. Ind. Inf. Integr. **6**, 1–10 (2017)
13. Pancho, D.P., Alonso, J.M., Magdalena, L.: Quest for interpretability-accuracy trade-off supported by fingrams into the fuzzy modeling tool GUAJE. Int. J. Comput. Intell. Syst. **6**(sup1), 46–60 (2013)
14. Pedregosa, F., et al.: Scikit-learn: machine learning in Python. J. Mach. Learn. Res. **12**, 2825–2830 (2011)
15. Segatori, A., Marcelloni, F., Pedrycz, W.: On distributed fuzzy decision trees for big data. IEEE Trans. Fuzzy Syst. **26**(1), 174–192 (2018)

Two-Sample Dispersion Problem for Fuzzy Data

Przemyslaw Grzegorzewski[1,2(✉)]

[1] Systems Research Institute, Polish Academy of Sciences, Newelska 6, 01-447 Warsaw, Poland
pgrzeg@ibspan.waw.pl

[2] Faculty of Mathematics and Information Science, Warsaw University of Technology, Koszykowa 75, 00-662 Warsaw, Poland

Abstract. The problem of comparing variability of two populations with fuzzy data is considered. A new permutation two-sample test for dispersion based on fuzzy random variables is proposed. A case-study illustrating the applicability of the suggested testing procedure is also presented.

Keywords: Fuzzy data · Fuzzy number · Fuzzy random variable · Permutation test · Test for dispersion · Test for scale

1 Introduction

Various two-sample statistical tests are designed to determine whether given two populations differ significantly. In such case we assume that the universe of discourse consists of two populations, say X and Y, with cumulative distribution functions F and G, respectively. Then, having a random sample of size n drawn from the X population and another random sample of size m drawn from the Y population, we consider the null hypothesis that these two samples are actually drawn from the same population, i.e. $H_0 : F = G$. One may verify H_0 against the general alternative hypothesis that the populations just differ in some way. The Kolmogorov-Smirnov test or the Wald-Wolfowitz run test are often used in this context (see e.g. [5]). However, they are really useful in preliminary studies only since affected by any type of difference between distributions, they are not very efficient in detecting any specific type of the difference like difference in location or difference in variablity. Other tests, like the Mann-Whitney-Wilcoxon test, the median test, etc. (see e.g. [5]) are particularly sensitive to differences in location when the populations are identical otherwise and hence cannot be expected to perform extremely well against other alternatives.

However, sometimes we need statistical procedures designed to detect differences in variability or dispersion instead of location. Indeed, comparison of

M.-J. Lesot et al. (Eds.): IPMU 2020, CCIS 1239, pp. 82–96, 2020.
https://doi.org/10.1007/978-3-030-50153-2_7

variability might be of interest in many areas including social sciences, biology, clinical trials, engineering, manufacturing and quality control, etc. Moreover, tests for the equality of variances are often required as a preliminary tool for the analysis of variance (ANOVA), dose–response modeling, discriminant analysis, etc.

It is important to emphasize that comparing variability is much harder than comparing measures of location. The famous F test assumes that both underlying populations are normally distributed and is not robust to departures from normality even asymptotically. Thus many nonparametric two-sample tests based on the ranks have been proposed for the scale problem. The best-known tests are the Ansari-Bradley test, the Mood test, the Siegel-Tukey test, the Klotz normal-scores test, the Sukhatme test, etc.

Designing tests for the dispersion problem turns out to be much more difficult in the case of imprecise or vague data which appear quite often in the real-life problems. In particular, human ratings based on opinions or associated with perceptions often lead to data that cannot be expressed in a numerical scale because they consist of intrinsically imprecise or fuzzy elements. Since they are also realizations of some random experiment, we are faced with random fuzzy structures that cannot be analyzed with classical statistical methods. Obviously, one may try to neglect, hide or remove imprecision but the most recommended approach is to consider it as a challenge for modeling and developing new inferential tools.

A general framework for such modeling is given by fuzzy random variables. However, besides mathematical elegence they also bring some fundamental difficulties. For instance, random fuzzy numbers are not linearly ordered so the aforementioned tests based on ranks cannot be directly applied in fuzzy environment. Depending on the context various test constructions have been proposed in the literature (for the overview we refer the reader e.g. to [7,8,11–14,16,18,19,21,26]). However, the dispersion problem with imprecise data has not beed considered very often. Ramos-Guajardo and Lubiano [26] proposed the bootstrap generalization of the Levene test for random fuzzy sets to examine homoscedasticity of k populations. Grzegorzewski [15] introduced two generalizations of the Sukhatme test for interval-valued data.

In this paper we suggest a permutation test for fuzzy data to compare variability of two populations. For motivations we turned back to the classical inference showing that permutation tests, like the bootstrap, require extremly limited assumptions. Indeed, permutation tests are totally distribution-free and require only exchangeability (i.e., under the null hypothesis we can exchange the labels on the observations without affecting the results). Classical permutation test are often more powerful than their bootstrap counterparts (see [9]). Permutation test are exact if all permutation are considered, while bootstrap tests are exact only for very large samples. Moreover, asymptotically permutation tests are usually as powerful as the most powerful parametric tests (see [1]). Keeping this in mind we combine the Pan test [22] and the Marozzi test [20] and then generalize them into the permutation testing procedure that handle fuzzy data.

The paper is organized as follows: in Sect. 2 we recall basic concepts related to fuzzy data modeling and operations on fuzzy numbers. Section 3 is devoted to fuzzy random variables. In Sect. 4 we introduce the two-sample test for the dispersion dedicated to fuzzy data. Next, in Sect. 5 we present some results of the simulation study and the case study with the proposed test. Finally, conclusions and some indications for the futher research are given in Sect. 6.

2 Fuzzy Data

A **fuzzy number** is an imprecise value characterized by a mapping $A : \mathbb{R} \to [0,1]$, called a membership function), such that its α-cut defined by

$$A_\alpha = \begin{cases} \{x \in \mathbb{R} : A(x) \geqslant \alpha\} & \text{if } \alpha \in (0,1], \\ cl\{x \in \mathbb{R} : A(x) > 0\} & \text{if } \alpha = 0, \end{cases} \tag{1}$$

is a nonempty compact interval for each $\alpha \in [0,1]$. Operator cl in (1) denotes for the closure. Thus every fuzzy number is completely characterized both by its memberschip function $A(x)$ or by a family of its α-cuts $\{A_\alpha\}_{\alpha\in[0,1]}$. Two α-cuts are of special interest: $A_1 = \text{core}(A)$ known as the **core**, which contains all values which are fully compatible with the concept described by the fuzzy number A and $A_0 = \text{supp}(A)$ called the **support**, which are compatible to some extent with the concept modeled by A.

The most often used fuzzy numbers are **trapezoidal fuzzy numbers** (sometimes called *fuzzy intervals*) with membership functions of the form

$$A(x) = \begin{cases} \frac{x-a_1}{a_2-a_1} & \text{if } a_1 \leqslant x < a_2, \\ 1 & \text{if } a_2 \leqslant x \leqslant a_3, \\ \frac{a_4-x}{a_4-a_3} & \text{if } a_3 < x \leqslant a_4, \\ 0 & \text{otherwise}, \end{cases} \tag{2}$$

where $a_1, a_2, a_3, a_4 \in \mathbb{R}$ such that $a_1 \leqslant a_2 \leqslant a_3 \leqslant a_4$. A trapezoidal fuzzy number (2) is often denoted as $\text{Tra}(a_1, a_2, a_3, a_4)$. Obviously, $a_1 = \inf \text{supp}(A)$, $a_2 = \inf \text{core}(A)$, $a_3 = \sup \text{core}(A)$ and $a_4 = \sup \text{supp}(A)$, which means that each trapezoidal fuzzy numbers is completely described by its support and core.

A fuzzy number A is said to be a **triangular fuzzy number** if $a_2 = a_3$, while if $a_1 = a_2$ and $a_3 = a_4$ we have the so-called **interval** (or rectangular) fuzzy number. The families of all fuzzy numbers, trapezoidal fuzzy numbers, triangular fuzzy number and interval fuzzy numbers will be denoted by $\mathbb{F}(\mathbb{R})$, $\mathbb{F}^T(\mathbb{R})$, $\mathbb{F}^\Delta(\mathbb{R})$ and $\mathbb{F}^I(\mathbb{R})$, respectively, where $\mathbb{F}^I(\mathbb{R}) \subset \mathbb{F}^\Delta(\mathbb{R}) \subset \mathbb{F}^T(\mathbb{R}) \subset \mathbb{F}(\mathbb{R})$.

Basic arithmetic operations in $\mathbb{F}(\mathbb{R})$ are defined through natural α-cut-wise operations on intervals. In particular, the sum of two fuzzy numbers A and B is given by the Minkowski addition of corresponding α-cuts, i.e.

$$(A+B)_\alpha = \left[\inf A_\alpha + \inf B_\alpha, \sup A_\alpha + \sup B_\alpha\right],$$

for all $\alpha \in [0,1]$. Similarly, the product of a fuzzy number A by a scalar $\theta \in \mathbb{R}$ is defined by the Minkowski scalar product for intervals, i.e. for all $\alpha \in [0,1]$

$$(\theta \cdot A)_\alpha = [\min\{\theta \inf A_\alpha, \theta \sup A_\alpha\}, \max\{\theta \inf A_\alpha, \theta \sup A_\alpha\}] .$$

It is worth noting that a sum of trapezoidal fuzzy numbers is also a trapezoidal fuzzy number: if $A = \text{Tra}(a_1, a_2, a_3, a_4)$ and $B = \text{Tra}(b_1, b_2, b_3, b_4)$ then

$$A + B = \text{Tra}(a_1 + b_1, a_2 + b_2, a_3 + b_3, a_4 + b_4). \tag{3}$$

Moreover, the product of a trapezoidal fuzzy number $A = \text{Tra}(a_1, a_2, a_3, a_4)$ by a scalar θ is a trapezoidal fuzzy number

$$\theta \cdot A = \begin{cases} \text{Tra}(\theta \cdot a_1, \theta \cdot a_2, \theta \cdot a_3, \theta \cdot a_4) & \text{if } \theta \geqslant 0, \\ \text{Tra}(\theta \cdot a_4, \theta \cdot a_3, \theta \cdot a_2, \theta \cdot a_1) & \text{if } \theta < 0. \end{cases} \tag{4}$$

Unfortuntely, $(\mathbb{F}(\mathbb{R}), +, \cdot)$ has not linear but semilinear structure since in general $A + (-1 \cdot A) \neq \mathbb{1}_{\{0\}}$. Consequently, the Minkowski-based difference does not satisfy, in general, the addition/subtraction property that $(A + (-1 \cdot B)) + B = A$. To overcome this problem the so-called Hukuhara difference was defined as follows:

$$C := A -_H B \quad \text{if and only if} \quad B + C = A$$

Although now $A -_H A = \mathbb{1}_{\{0\}}$ or $(A -_H B) + B = A$ hold but the Hukuhara difference does not always exist. Therefore, one should be aware that subtraction in $\mathbb{F}(\mathbb{R})$ generally leads to critical problems and should be avoided, if possible.

At least some of the problems associated with the lack of a satisfying difference in constructing statistical tools for reasoning based on fuzzy observations could be overcome by using adequate metrics defined in $\mathbb{F}(\mathbb{R})$ – for the general overview see [2]. Obviously, one can define various metrics in $\mathbb{F}(\mathbb{R})$ but perhaps the most often used in statistical context is the one proposed by Gil et al. [6] and by Trutschnig et al. [27].

Let λ be a normalized measure associated with a continuous distribution having support in $[0,1]$ and let $\theta > 0$. Then for any $A, B \in \mathbb{F}(\mathbb{R})$ we define a metric D_θ^λ as follows

$$D_\theta^\lambda(A,B) = \left(\int_0^1 \left[(\text{mid}\, A_\alpha - \text{mid}\, B_\alpha)^2 + \theta \cdot (\text{spr}\, A_\alpha - \text{spr}\, B_\alpha)^2 \right] d\lambda(\alpha) \right)^{1/2}, \tag{5}$$

where $\text{mid}\, A_\alpha = \frac{1}{2}(\inf A_\alpha + \sup A_\alpha)$ and $\text{spr}\, A_\alpha = \frac{1}{2}(\sup A_\alpha - \inf A_\alpha)$ denote the mid-point and the radius of the α-cut A_α, respectively.

Both λ and θ correspond to some weighting: λ allows to weight the influence of each α-cut, while by a particular choice of θ one may weight the impact of the distance between the mid-points of the α-cuts (i.e. the deviation in location) in contrast to the distance between their spreads (i.e. the deviation in vagueness). In practice, the most common choice of λ is the Lebesgue measure on $[0,1]$), while the most popular choice is $\theta = 1$ or $\theta = \frac{1}{3}$. It is worth noting that assuming $\theta = 1$ we obtain

$$D_1^{\lambda}(A,B) = \left(\int_0^1 \left[\frac{1}{2}(\inf A_\alpha - \inf B_\alpha)^2 + \frac{1}{2}(\sup A_\alpha - \sup B_\alpha)^2 \right] d\lambda(\alpha) \right)^{1/2}, \quad (6)$$

i.e. the metric which weights uniformly the two squared Euclidean distances and is equivalent to the distance considered in [4,10]. One may also notice that assuming $\theta = \frac{1}{3}$ we obtain

$$D_{1/3}^{\lambda}(A,B) = \sqrt{\int_0^1 \left(\int_0^1 [A_\alpha^{[t]} - B_\alpha^{[t]}]^2 dt \right) d\lambda(\alpha)}, \quad (7)$$

where $A_\alpha^{[t]} = (1-t) \inf A_\alpha + t \sup A_\alpha$, which means that $D_{1/3}^{\lambda}(A,B)$ aggregates uniformly the squared Euclidean distances between the convex combination of points in α-cuts representing A and B.

It should be stressed that whatever (λ, θ) is chosen D_θ^λ is an L^2-type metric in $\mathbb{F}(\mathbb{R})$ having some important and useful properties. It is translational invariant, i.e. $D_\theta^\lambda(A+C, B+C) = D_\theta^\lambda(A,B)$ for all $A, B, C, \in \mathbb{F}(\mathbb{R})$, and it is rotational invariant, i.e. $D_\theta^\lambda((-1) \cdot A, (-1) \cdot B) = D_\theta^\lambda(A,B)$ for all $A, B \in \mathbb{F}(\mathbb{R})$. Moreover, $(\mathbb{F}(\mathbb{R}), D_\theta^\lambda)$ is a separable metric space and for each fixed λ all D_θ^λ are topologically equivalent.

3 Fuzzy Random Variables

Suppose that the result of an experiment consists of random samples of imprecise data described by fuzzy numbers. To cope with such problem we need a model which grasps both aspects of uncertainty that appear in data, i.e. randomness (associated with data generation mechanism) and fuzziness (connected with data nature, i.e. their imprecision). To handle such data Puri and Ralescu [24] introduced the notion of a **fuzzy random variable** (also called a **random fuzzy number**).

Definition 1. *Given a probability space $(\Omega, \mathcal{A}, P)$, a mapping $X : \Omega \to \mathbb{F}(\mathbb{R})$ is called a fuzzy random variable if for all $\alpha \in [0,1]$ the α-cut function X_α is a compact random interval.*

In other words, X is a random fuzzy variable if and only if X is a Borel measurable function w.r.t. the Borel σ-field generated by the topology induced by D_θ^λ.

Puri and Ralescu [24] defined also the Aumann-type mean of a fuzzy random variable X as the fuzzy number $\mathcal{E}(X) \in \mathbb{F}(\mathbb{R})$ such that for each $\alpha \in [0,1]$ the α-cut $(\mathcal{E}(X))_\alpha$ is equal to the Aumann integral of X_α. It is seen that

$$(\mathcal{E}(X))_\alpha = \left[\mathbb{E}(\operatorname{mid} X_\alpha) - \mathbb{E}(\operatorname{spr} X_\alpha), \mathbb{E}(\operatorname{mid} X_\alpha) + \mathbb{E}(\operatorname{spr} X_\alpha) \right].$$

To characterize dispersion of a fuzzy random variable X we can also define (see [17]) the D_θ^λ-Fréchet-type variance $\mathcal{V}(X)$, which is a nonnegative real number such that

$$\mathcal{V}(X) = \mathbb{E}\Big(\big[D_\theta^\lambda(X, \mathcal{E}(X))\big]^2\Big)$$
$$= \int_0^1 \mathrm{Var}(\mathrm{mid}\, X_\alpha) d\lambda(\alpha) + \theta \int_0^1 \mathrm{Var}(\mathrm{spr}\, X_\alpha) d\lambda(\alpha).$$

Given a sample of random fuzzy numbers $\mathbb{X} = (X_1, \ldots, X_n)$ a natural estimator of $\mathcal{E}(X)$ is the average $\overline{X} \in \mathbb{F}(\mathbb{R})$ such that for each $\alpha \in [0,1]$

$$\overline{X}_\alpha = \left[\frac{1}{n}\sum_{i=1}^n \inf(X_i)_\alpha, \frac{1}{n}\sum_{i=1}^n \sup(X_i)_\alpha\right] \tag{8}$$
$$= \left[\frac{1}{n}\sum_{i=1}^n \mathrm{mid}\,(X_i)_\alpha - \frac{1}{n}\sum_{i=1}^n \mathrm{spr}\,(X_i)_\alpha, \frac{1}{n}\sum_{i=1}^n \mathrm{mid}\,(X_i)_\alpha + \frac{1}{n}\sum_{i=1}^n \mathrm{spr}\,(X_i)_\alpha\right],$$

while the estimator of $\mathcal{V}(X)$ is the D_θ^λ-type sample variance $S^2 \in \mathbb{R}$ given by

$$S^2 = \frac{1}{n-1}\sum_{i=1}^n D_\theta^\lambda(X_i, \overline{X})^2. \tag{9}$$

Although aforementioned constructions preserve many properties known from the real-valued inference, one should be aware of the problems typical of statistical reasoning with fuzzy data. As it was noted in Sect. 2, there are problems with subtraction of fuzzy numbers. Similar problems appear in the case of division of fuzzy numbers. Hence, it is advisable to avoid both operations wherever it is possible. Moreover, some difficulties in fuzzy data analysis is caused by the lack of universally accepted total ranking between fuzzy numbers. Another source of possible problems that appear in conjunction of randomness and fuzziness is the absence of suitable models for the distribution of fuzzy random variables. Even worse, there are not yet Central Limit Theorems for fuzzy random variables which can be applied directly in statistical inference.

The disadvantages mentioned above make the straightforward generalization of the classical statistical methodology into the fuzzy context either difficult or, sometimes, even impossible. For instance, in most cases we are not able to find the null distribution of a test statistic based on fuzzy data and, consequently, to find either the critical value or to compute the p-value required for rejection or acceptance of the hypothesis under study. To break through that problem some researchers propose to use the bootstrap [7,8,18,19,21,25,26].

In this paper we suggest another methodology based on permutations. For motivations we turn back to the classical inference which shows that permutation tests, like the bootstrap, require extremly limited assumptions. Bootstrap tests usually rely on assumption that successive observations are independent, while permutation tests require only exchangeability, i.e. under the null hypothesis we can exchange the labels on the observations without affecting the results (obviously, if the observations in a sample are independent and identically distributed then they are exchangeable). In the real-valued framework one can also indicate two advantages of the permutation tests over the bootstrap tests. Firstly, permutation test are often more powerful than their bootstrap counterparts (see [9]). Secondly, permutation test are exact if all permutation are considered, while bootstrap tests are exact only for very large samples. Moreover, asymptotically permutation tests are usually as powerful as the most powerful parametric tests (see [1]). For more information on classical permutation tests we refer the reader to [9,23]. All these reasons indicate that the permutation test applied to fuzzy random variables might be also a competitive tool useful in statistical inference for imprecise data.

4 Permutation Test for Fuzzy Data to Compare Variability

Suppose, we observe independently two fuzzy random samples $\mathbb{X} = (X_1, \ldots, X_n)$ and $\mathbb{Y} = (Y_1, \ldots, Y_m)$ drawn from populations with unknown distributions function F and G, respectively. We want to verify the null hypothesis that both samples come from the same distribution, i.e.

$$H_0 : F(t) = G(t) \quad \text{for all } t \in \mathbb{R}, \tag{10}$$

against the alternative hypothesis that the dispersion of the distributions F and G differ (or against the one-sided alternative that the indicated distribution is more dispersed that the other one).

Most of the tests for scale assume that the distributions under study do not differ in location since possible location differences may mask differences in dispersion. Otherwise, the sample observations should be adjusted by subtrating the respective location parameters, like means or medians. If the true characteristics of location are not known we usually subtract their estimators.

Following remarks of Marozzi [20] on the resampling version of the Pan test [22] and the resampling framework for scale testing described by Boos and Brownie [3], we'll try to eliminate the location effects with sample means. However, keeping in mind problems with subtratiion in fuzzy environment described in Sect. 2, contrary to the crisp case, we do not consider the differences but the distances between sample observations and corresponding sample means calculated as in (8). Therefeore, further on instead of $\mathbb{X} = (X_1, \ldots, X_n)$ and $\mathbb{Y} = (Y_1, \ldots, Y_m)$ we consider the adjusted samples $\mathbb{V} = (V_1, \ldots, V_n)$ and $\mathbb{W} = (W_1, \ldots, W_m)$, respectively, where

$$V_i = D_\theta^\lambda(X_i, \overline{X}), \quad \text{for } i = 1, \ldots, n$$
$$W_j = D_\theta^\lambda(Y_j, \overline{Y}), \quad \text{for } j = 1, \ldots, m.$$

Now let us consider the following test statistics

$$T(\mathbb{X}, \mathbb{Y}) = \frac{\ln \overline{V} - \ln \overline{W}}{\sqrt{\frac{1}{n}\frac{S_V^2}{\overline{V}^2} + \frac{1}{m}\frac{S_W^2}{\overline{W}^2}}} \tag{11}$$

where S_V^2 and S_W^2 denote sample variances of $\mathbb{V}$ and $\mathbb{W}$, respectively, calculated by (9). Obviously, too big or too small values of (11) indicate that the null hypothesis should be rejected since the considered distributions differ in dispersion.

In the original Pan test [22] the decision whether to reject the null hypothesis is based on the test statistic valued with respect to some quantile from the t-Student distribution. However, Marozzi [20] showed that the resampling version of the Pan test should be rather preferred to the original one. In the case of fuzzy data any assumptions on the type of the underlying distribution of the samples are much more dubious than in the crisp case. For this reason we also consider here the permutation version of the Pan test. To carry out such a test we adapt the general idea of permutation tests to our fuzzy context.

The crucial idea of the proposed test construction is that the null hypothesis implies total exchangeability of observed data with respect to groups. Indeed, if H_0 holds then all available observations may be viewed as if they were randomly assigned to two groups but they come from the same population.

Let $\mathbb{Z} = \mathbb{X} \uplus \mathbb{Y}$, where $\uplus$ stands for the vector concatenation, so that the two samples are pooled into one, i.e. $Z_i = X_i$ if $1 \leqslant i \leqslant n$ and $Z_i = Y_{i-n}$ if $n+1 \leqslant i \leqslant N$, where $N = n + m$.

Now, let $\mathbb{Z}^*$ denote a permutation of the initial dataset $\mathbb{Z}$. More formally, if $\nu = \{1, 2, \ldots, N\}$ and π_ν is a permutation of the integers ν, then $Z_i^* = Z_{\pi_\nu(i)}$ for $i = 1, \ldots, N$. Then the first n elements of $\mathbb{Z}^*$ is assigned to the first sample $\mathbb{Z}^*$ and the remaining m elements to $\mathbb{Z}^*$. In other words, it works like a random assignment of elements into two samples of the size n and m, respectively. Each permutation corresponds to some relabeling of the combined dataset $\mathbb{Z}$. Please, note that if H_0 holds then we are completely free to exchange the labels X or Y attributed to particular observations.

As a consequence of elements' exchangeability in $\mathbb{Z}^*$ under H_0 we can estimate the distribution of the test statistic T by considering all permutations of the initial dataset $\mathbb{Z}$ and computing a value of $T(\mathbb{Z}^*)$ corresponding to each permutation. Namely, given $\mathbb{Z} = \mathbb{z}$, where $\mathbb{z} = \mathbb{x} \uplus \mathbb{y}$, we take its permutation $\mathbb{z}^*$ and determine its adjustment with respect to sample means, i.e. we create two samples $\mathbb{v}^* = (v_1^*, \ldots, v_n^*)$ and $\mathbb{w}^* = (w_1^*, \ldots, w_m^*)$ as follows

$$v_i^* = D_\theta^\lambda(z_i^*, \frac{1}{n}\sum_{j=1}^{n} z_j^*), \quad \text{if } i = 1, \ldots, n$$

$$w_j^* = D_\theta^\lambda(z_i^*, \frac{1}{m}\sum_{i=j}^{m} z_j^*), \quad \text{if } i = n+1, \ldots, N.$$

Next, following (11) we compute its actual value corresponding to given permutation $\mathbb{z}^*$, i.e.

$$T(\mathbb{z}^*) = \frac{\ln \overline{v}^* - \ln \overline{w}^*}{\sqrt{\frac{1}{n}\frac{S_{v^*}^2}{(\overline{w}^*)^2} + \frac{1}{m}\frac{S_{w^*}^2}{(\overline{w}^*)^2}}}. \tag{12}$$

Finally, assuming K denotes a fixed number of drawings (usually not smaller than 1000), we calculate the p-value of our test. In the case on the one-sided upperer-tail test, i.e. when verifying $H_0 : F = G$ vs. H_1 stating that F is more disperded than G, we obtain

$$\text{p-value} = \frac{1}{K}\sum_{k=1}^{K} \mathbb{1}\big(T(\mathbb{z}_k^*) \geqslant t_0\big), \tag{13}$$

where each $\mathbb{z}_k^* \in \mathcal{P}(\mathbb{z})$, $\mathbb{z}_k^* = \mathbb{x}_k^* \uplus \mathbb{y}_k^*$, and $t_0 = T(\mathbb{x}, \mathbb{y})$ stands for the test statistic value obtained for the original fuzzy samples $\mathbb{x}$ and $\mathbb{y}$.

For the one-sided lower-tail test, i.e. when verifying $H_0 : F = G$ vs. $H_1 : F$ *is less disperded than* G, we have

$$\text{p-value} = \frac{1}{K}\sum_{k=1}^{K} \mathbb{1}\big(T(\mathbb{z}_k^*) \leqslant t_0\big), \tag{14}$$

while for the two-sided test, i.e. when verifying $H_0 : F = G$ vs. $H_1 : F$ *and* G *differ in dispersion*, we obtain

$$\text{p-value} = \frac{1}{K}\left[\sum_{k=1}^{K} \mathbb{1}\big(T(\mathbb{z}_k^*) \geqslant |t_0|\big) + \mathbb{1}\big(T(\mathbb{z}_k^*) \leqslant -|t_0|\big)\right]. \tag{15}$$

5 Empirical Study

5.1 Simulations

We conducted some simulations to illustrate the behavior of the proposed test. To generate fuzzy samples from a trapezoidal-valued fuzzy random variable $X = \text{Tra}(\xi_1, \xi_2, \xi_3, \xi_4)$, where $\xi_1, \xi_2, \xi_3, \xi_4$ are real-valued random variables such that $\xi_1 \leqslant \xi_2 \leqslant \xi_3 \leqslant \xi_4$, the following characterization appears to be useful (see [19]): $c = \frac{1}{2}(\xi_3 + \xi_2) = \text{mid}_1 X$, $s = \frac{1}{2}(\xi_3 - \xi_2) = \text{spr}_1 X$, $l = \xi_2 - \xi_1$ and $r = \xi_4 - \xi_3$. Conversely, we have $\text{Tra}\langle c, s, l, r\rangle = \text{Tra}(c - s - l, c - s, c + s, c + s + r)$.

In our study we generated fuzzy observations $\mathbb{x} = (x_1, \ldots, x_n)$ and $\mathbb{y} = (y_1, \ldots, y_m)$ by simulating the four real-valued random variables $x_i = \langle c_{Xi}, s_{Xi}, l_{Xi}, r_{Xi}\rangle$ and $y_i = \langle c_{Yj}, s_{Yj}, l_{Yj}, r_{Yj}\rangle$, respectively, with the last three ones random variables in each quartet being nonnegative. In particular, we generated trapezoidal-valued fuzzy random variables using the following real-valued random variables: c_{Xi}, c_{Yj} from the normal distribution and $s_{Xi}, s_{Yj}, l_{Xi}, l_{Yj}, r_{Xi}$ and r_{Yj} from the uniform or chi-square distribution.

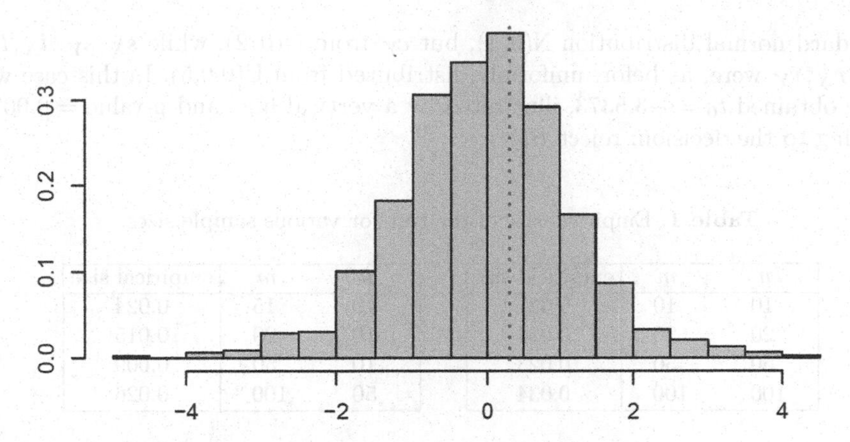

Fig. 1. Empirical null distribution of the permutation test with red vertical line indicatinges the value of the test statistic. (Color figure online)

An illustration how the test works, is shown in Fig. 1 and Fig. 2. Figure 1 shows a histogram made for the test statistic (11) null distribution obtained for two fuzzy samples of sizes $n = 10$ and $m = 12$. Both samples were generated as follows: c_X and c_Y came from the standard normal distribution N(0, 1) and s_X, s_Y, l_X, l_Y and r_X, r_Y from the uniform distribution U(0.0.5). In this case we have obtained $t_0 = 0.3088$, which is illustrated by a vertical line, while p-value = 0.384. A decision suggested by our test is: do not reject H_0.

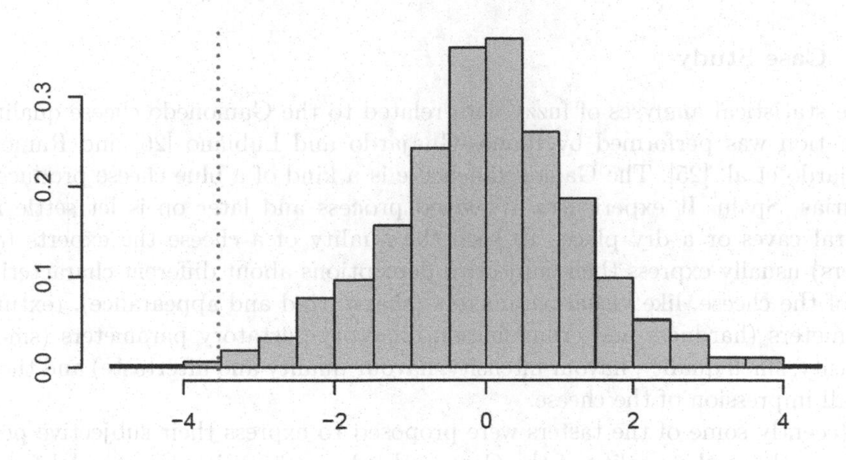

Fig. 2. Empirical null distribution of the permutation test with red vertical line indicatinges the value of the test statistic. (Color figure online)

On the other hand, in Fig. 2 we have a histogram made for the test statistic (11) null distribution obtained for two fuzzy samples of the same samle sizes as before but which differ in dispersion. Namely, c_X was generated from the standard normal distribution N(0, 1), but c_Y from N(0, 2), while s_X, s_Y, l_X, l_Y and r_X, r_Y were, as befor, uniformly distributed from U(0.0.5). In this case we have obtained $t_0 = -3.5373$, illustrated by a vertical line, and p-value = 0.007, leading to the decision: reject H_0.

Table 1. Empirical size of the test for various sample sizes.

n	m	empirical size
10	10	0.021
20	20	0.024
50	50	0.023
100	100	0.034

n	m	empirical size
10	15	0.024
10	20	0.015
10	50	0.009
50	100	0.026

We also examined the proposed permutation test with respect to its size. Therefore, 1000 simulations of the test performed on independent fuzzy samples comming from the same distribution were generated at the significance level 0.05. In each test $K = 1000$ permutations were drawn. Then empirical percentages of rejections under H_0 were determined. The results both for equal and nonequal sample sizes are gathered in Table 1. It is seen that our test is conservative. Moreover, this tendency deepens significantly as the imbalance of the sample sizes increases. These interesting results of the preliminary study of the proposed test properties indicate that further and more extensive study is highly recommended.

5.2 Case Study

Some statistical analyses of fuzzy data related to the Gamonedo cheese quality inspection was performed by Ramos-Guajardo and Lubiano [26] and Ramos-Guajardo et al. [25]. The Gamonedo cheese is a kind of a blue cheese produced Asturias, Spain. It experiences a smoked process and later on is let settle in natural caves or a dry place. To keep the quality of a cheese the experts (or tasters) usually express their subjective perceptions about different characteristics of the cheese, like visual parameters (shape, rind and appearance), texture parameters (hardness and crumbliness), olfactory-gustatory parameters (smell intensity, smell quality, flavour intensity, flavour quality and aftertaste) and their overall impression of the cheese.

Recently some of the tasters were proposed to express their subjective perceptions about the quality of the Gamonedo cheese by using trapezoidal fuzzy numbers. These fuzzy sets were determined in the following way: the set of values considered by the expert to be fully compatible with his/her opinion led to $\alpha = 1$-cut, while the set of values that he/she considered to be compatible with his/her opinion at some extent (i.e., the taster thought that it was not possible that the quality was out of this set) led to $\alpha = 0$-cut of a fuzzy number.

Then these two α-cuts were linearly interpolated to get the trapezoidal fuzzy set representing exppert's personal valuation. For more details on the data aquisition and analysis we refer the reader to Ramos-Guajardo et al. [25].

Table 2. Sample of the opinions of Expert 1 and 2 concerning the overall impression of the Gamonedo cheese (see [25])

Opinion	Expert 1	Expert 2	Opinion	Expert 1	Expert 2
1	(65, 75, 85, 85)	(50, 50, 63, 75)	21	(65, 70, 76, 80)	(60, 64, 75, 85)
2	(35, 37, 44, 50)	(39, 47, 52, 60)	22	(75, 80, 86, 90)	(54, 56, 64, 75)
3	(66, 70, 75, 80)	(60, 70, 85, 90)	23	(65, 70, 73, 80)	(50, 50, 60, 66)
4	(70, 74, 80, 84)	(50, 56, 64, 74)	24	(70, 80, 84, 84)	(44, 46, 55, 57)
5	(65, 70, 75, 80)	(39, 45, 53, 57)	25	(55, 64, 70, 70)	(59, 63, 74, 80)
6	(45, 50, 57, 65)	(55, 60, 70, 76)	26	(64, 73, 80, 84)	(49, 50, 54, 58)
7	(60, 66, 70, 75)	(50, 50, 57, 67)	27	(50, 56, 64, 70)	(55, 60, 70, 75)
8	(65, 65, 70, 76)	(65, 67, 80, 87)	28	(55, 55, 60, 70)	(44, 47, 53, 60)
9	(60, 65, 75, 80)	(50, 50, 65, 75)	29	(60, 70, 75, 80)	(19, 20, 30, 41)
10	(55, 60, 66, 70)	(50, 55, 64, 70)	30	(64, 71, 80, 80)	(40, 44, 50, 60)
11	(60, 65, 70, 74)	(39, 46, 53, 56)	31	(50, 50, 55, 65)	(50, 50, 59, 66)
12	(30, 46, 44, 54)	(19, 29, 41, 50)	32	(50, 54, 60, 65)	(50, 53, 60, 66)
13	(60, 65, 75, 75)	(40, 47, 52, 56)	33	(65, 75, 80, 86)	(50, 52, 58, 61)
14	(70, 75, 85, 85)	(54, 55, 65, 76)	34	(50, 55, 60, 66)	(60, 65, 72, 80)
15	(44, 45, 50, 56)	(59, 65, 75, 85)	35	(40, 44, 50, 50)	(50, 50, 55, 60)
16	(51, 56, 64, 70)	(50, 52, 57, 60)	36	(70, 76, 85, 85)	(30, 34, 43, 47)
17	(40, 46, 54, 60)	(60, 60, 70, 80)	37	(44, 50, 53, 60)	(19, 25, 36, 46)
18	(55, 60, 65, 70)	(50, 54, 61, 67)	38	(34, 40, 46, 46)	(53, 63, 74, 80)
19	(80, 85, 90, 94)	(40, 46, 50, 50)	39	(40, 45, 51, 60)	
20	(80, 84, 90, 90)	(44, 50, 56, 66)	40	(84, 90, 95, 95)	

Here we utilize some data given in [25] to compare the opinions of the two experts about the overall impression of the Gamonedo cheese (the trapezoidal fuzzy sets corresponding to their opinions are gathered in Table 2). Thus we have two independent fuzzy samples of sizes $n = 40$ and $m = 38$ comming from the unknown distributions F and G, respectively. Our problem is to check whether there is a general agreement between these two experts. To reach the goal we verify the following null hypothesis $H_0 : F = G$, stating there is no significant difference between experts' opinions, against $H_1 : \neg H_0$ that their opinions on the cheese quality differ.

Substituting data from Table 2 into formula (11) we obtain $t_0 = 1.355$. Then, after combining samples and generating $K = 10\,000$ random permutations we have obtained the p-value of 0.082. Hence, assuming the typical 5% significant level we may conclude that there is no significant difference between the dispersion of experts' opinion on the overal impression of the Gamonedo cheese. In Fig. 3 one can find the empirical null distribution of the permutation test with red vertical line indicating the value t_0 of the test statistic.

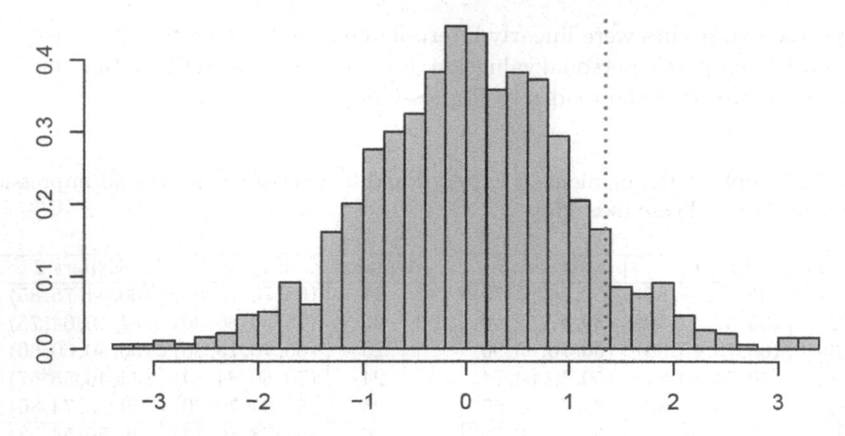

Fig. 3. Empirical null distribution of the permutation test with red vertical line indicating the value of the test statistic.

6 Conclusions

Hypothesis testing with samples which consist of random fuzzy numbers is neither easy nor straightforward. Most of statistical tests developed in this area are based on the bootstrap. In this paper another approach for constructing tests for fuzzy data is proposed. Namely, the two-sample permutation test for dispersion is suggested. Some simulations to illustrate its behavior and to examine its properties are given. Moreover, the case study dedicated to fuzzy rating problem is performed.

The results obtained seem to be promissing, but further research including power studies and a comparison with other tests are still intended in the nearest future. In particular, the behavior of the test under strong imbalance in the sample sizes is worth of further examination. Next, we would like to perform an extensive simulation study to compare the performance of our permuatation test and the bootstrap test for the dispersion.

Moreover, some other topics related to the dispersion problem with fuzzy data seem to be of interest. Firstly, we plan to design other two-sample tests for scale, like the permutation test for fuzzy data based on the classical O'Brien test, as well as a permutation test for the homogeneity of more than two fuzzy samples. Secondly, a permutation test for fuzzy data to compare jointly the central tendency and variability of two populations would be of desirable.

References

1. Bickel, P.M., Van Zwet, W.R.: Asymptotic expansion for the power of distribution-free tests in the two-sample problem. Ann. Stat. **6**(987–1004), 1170–1171 (1978)
2. Blanco-Fernández, A., et al.: A distance-based statistic analysis of fuzzy number-valued data (with Rejoinder). Int. J. Approx. Reason. 55, 1487–1501, 1601–1605 (2014)

3. Boos, D.D., Brownie, C.: Comparing variances and other measures of dispersion. Stat. Sci. **19**(4), 571–578 (2004)
4. Diamond, P., Kloeden, P.: Metric spaces of fuzzy sets. Fuzzy Sets Syst. **100**, 63–71 (1999)
5. Gibbons, J.D., Chakraborti, S.: Nonparametric Statistical Inference. Marcel Dekker, Inc., New York (2003)
6. Gil, M.A., Lubiano, M.A., Montenegro, M., López, M.T.: Least squares fitting of an affine function and strength of association for interval-valued data. Metrika **56**, 97–111 (2002)
7. González-Rodríguez, G., Montenegro, M., Colubi, A., Gil, M.A.: Bootstrap techniques and fuzzy random variables: synergy in hypothesis testing with fuzzy data. Fuzzy Sets Syst. **157**, 2608–2613 (2006)
8. González-Rodríguez, G., Colubi, A., Gil, M.A.: Fuzzy data treated as functional data. A one-way ANOVA test approach. Comput. Stat. Data Anal. **56**, 943–955 (2012)
9. Good, P.: Permutation, Parametric and Bootstrap Tests of Hypotheses. Springer, Heidelberg (2005). https://doi.org/10.1007/b138696
10. Grzegorzewski, P.: Metrics and orders in space of fuzzy numbers. Fuzzy Sets Syst. **97**, 83–94 (1998)
11. Grzegorzewski, P.: Statical inference about the median from vague data. Control Cybern. **27**, 447–464 (1998)
12. Grzegorzewski, P.: Testing statistical hypotheses with vague data. Fuzzy Sets Syst. **112**, 501–510 (2000)
13. Grzegorzewski, P.: Fuzzy tests - defuzzification and randomization. Fuzzy Sets Syst. **118**, 437–446 (2001)
14. Grzegorzewski, P.: K-sample median test for vague data. Int. J. Intell. Syst. **24**, 529–539 (2009)
15. Grzegorzewski, P.: Two-sample dispersion tests for interval-valued data. In: Medina, J., Ojeda-Aciego, M., Verdegay, J.L., Perfilieva, I., Bouchon-Meunier, B., Yager, R.R. (eds.) IPMU 2018. CCIS, vol. 855, pp. 40–51. Springer, Cham (2018). https://doi.org/10.1007/978-3-319-91479-4_4
16. Grzegorzewski, P., Szymanowski, H.: Goodness-of-fit tests for fuzzy data. Inf. Sci. **288**, 374–386 (2014)
17. Lubiano, M.A., Gil, M.A., López-Díaz, M., López-García, M.T.: The λ-mean squared dispersion associated with a fuzzy random variable. Fuzzy Sets Syst. **111**, 307–317 (2000)
18. Lubiano, M.A., Montenegro, M., Sinova, B., de Sáa, S.D.L.R., Gil, M.A.: Hypothesis testing for means in connection with fuzzy rating scale-based data: algorithms and applications. Eur. J. Oper. Res. **251**, 918–929 (2016)
19. Lubiano, M.A., Salas, A., Carleos, C., de Sáa, S.D.L.R., Gil, M.A.: Hypothesis testing-based comparative analysis between rating scales for intrinsically imprecise data. Int. J. Approx. Reason. **88**, 128–147 (2017)
20. Marozzi, M.: Levene type tests for the ratio of two scales. J. Stat. Comput. Simul. **81**, 815–826 (2011)
21. Montenegro, M., Colubi, A., Casals, M.R., Gil, M.A.: Asymptotic and Bootstrap techniques for testing the expected value of a fuzzy random variable. Metrika **59**, 31–49 (2004)
22. Pan, G.: On a Levene type test for equality of two variances. J. Stat. Comput. Simul. **63**, 59–71 (1999)
23. Pesarin, F.: Multivariate Permutation Tests. Wiley, Hoboken (2001)

24. Puri, M.L., Ralescu, D.A.: Fuzzy random variables. J. Math. Anal. Appl. **114**, 409–422 (1986)
25. Ramos-Guajardo, A.B., Blanco-Fernández, A., González-Rodríguez, G.: Applying statistical methods with imprecise data to quality control in cheese manufacturing. In: Grzegorzewski, P., Kochanski, A., Kacprzyk, J. (eds.) Soft Modeling in Industrial Manufacturing. Studies in Systems, Decision and Control, vol. 183, pp. 127–147. Springer, Heidelberg (2019). https://doi.org/10.1007/978-3-030-03201-2_8
26. Ramos-Guajardo, A.B., Lubiano, M.A.: K-sample tests for equality of variances of random fuzzy sets. Comput. Stat. Data Anal. **56**, 956–966 (2012)
27. Trutschnig, W., González-Rodríguez, G., Colubi, A., Gil, M.A.: A new family of metrics for compact, convex (fuzzy) sets based on a generalized concept of mid and spread. Inf. Sci. **179**, 3964–3972 (2009)

Image Understanding and Explainable AI

Transparency of Classification Systems for Clinical Decision Support

Antoine Richard[1,3](✉), Brice Mayag[1], François Talbot[2], Alexis Tsoukias[1], and Yves Meinard[1]

[1] Université Paris-Dauphine, PSL Research University, CNRS, LAMSADE, UMR 7243, 75016 Paris, France
antoine.richard0@dauphine.eu
[2] DSII Bron, Hospices Civils de Lyon, 61 Boulevard Pinel, 69672 Bron, France
[3] GIE Hopsis Lyon, 3 Quai de Célestins, 69229 Lyon, France

Abstract. In collaboration with the Civil Hospitals of Lyon, we aim to develop a "transparent" classification system for medical purposes. To do so, we need clear definitions and operational criteria to determine what is a "transparent" classification system in our context. However, the term "transparency" is often left undefined in the literature, and there is a lack of operational criteria allowing to check whether a given algorithm deserves to be called "transparent" or not. Therefore, in this paper, we propose a definition of "transparency" for classification systems in medical contexts. We also propose several operational criteria to evaluate whether a classification system can be considered "transparent". We apply these operational criteria to evaluate the "transparency" of several well-known classification systems.

Keywords: Explainable AI · Transparency of algorithms · Health information systems · Multi-label classification

1 Introduction

In collaboration with the Civil Hospitals of Lyon (HCL), in France, we aimed to develop and to propose decision support systems corresponding to the clinicians' needs. In 2018, the HCL received more than one million patients for medical consultations. Therefore, the decision has been made to build a decision support system focused on supporting physicians during their medical consultations. After some observations and analyses of medical consultations in the endocrinology department of the HCL [31], we drew two conclusions: physicians mainly need data on patients to reach diagnoses, and getting these data from their information system is quite time-consuming for physicians during consultations. To reduce physicians' workload, we decided to support them by using

M.-J. Lesot et al. (Eds.): IPMU 2020, CCIS 1239, pp. 99–113, 2020.
https://doi.org/10.1007/978-3-030-50153-2_8

a classification system learning which data on patients physicians need in which circumstance. By doing this, we should be able to anticipate and provide the data that physicians will need at the beginning of their future consultations. This can be formalized as a multi-label classification problem, as presented in Table 1 with fictitious data.

In this paper, a "classification system" refers to the combination of a "learning algorithm" and the "type of classifier" produced by this learning algorithm. For example, a classification system based on decision trees can use a learning system such as C4.5 [30], the type of classifier produced by this learning system being a decision-tree. This distinction is necessary because a learning algorithm and a classifier produced by this learning algorithm are not used in the same way and do not perform the same functions.

Table 1. Example of multi-label dataset based on our practical case

X: data known on patient				Y: data on patient needed by physician					
Sex	Age	BMI	Disease	HbA1c	Blood sugar	HDL	LDL	Creatinine	Microalbumin
♀	42	34.23	DT2	1	1	0	0	0	0
♂	52	27.15	HChol	0	0	1	1	0	0
♂	24	21.12	DT1	1	1	0	0	1	1
♀	67	26.22	HChol	0	0	1	1	0	0

However, in the case of clinical decision support systems (CDSSs), a well-known problem is the lack of acceptability of support systems by clinicians [5,19]. More than being performant, a CDSS has first to be accepted by clinicians, and "transparent" support systems are arguably more accepted by clinicians [22, 33]. Mainly because "transparency" allows clinicians to better understand the proposals of CDSSs and minimize the risk of misinterpretation. Following these results, we posit that the "transparency" of support systems is a way to improve the "acceptability" of CDSSs by clinicians.

In the literature, one can find several definition of the concept of "transparency": *"giving explanations of results"* [9,10,15,20,26,28,33,36], *"having a reasoning process comprehensible and interpretable by users"* [1,11,12,24,27,34], *"being able to trace-back all data used in the process"* [2–4,16,40], but also *"being able to take into account feedbacks of users"* [7,40]. Individually, each of the above definitions highlights an aspect of the concept of "transparency" of classification systems, but do not capture all aspects of "transparent" classification systems in our context. In addition, definitions are abstract descriptions of concepts and there is a lack of operational criteria, in the sense of concrete properties one can verify in practice, to determine whether a given algorithm deserves to be called "transparent" or not.

The main objective of this paper is to propose a definition of transparency, and a set of operational criteria, applicable to classification systems in a medical context. These operational criteria should allow us to determine which classification system is "transparent" for users in our use case. Let us specify that, in this paper, the term "users" refers to physicians.

In Sect. 2 we detail the definition and operational criteria we propose to evaluate the transparency of classification systems. In Sect. 3, based to our definition of transparency, we explain why we choose a version of the naive bayes algorithm to handle our practical case. We briefly conclude in Sect. 4, with a discussion on the use of an evaluation of "transparency" for practical use cases.

2 Definition of a "Transparent" Classification System

Even though the concept of algorithm "transparency" is as old as recommendation systems, the emergence and the ubiquity of "black-box" learning algorithms nowadays, such as neural networks, put "transparency" of algorithms back in the limelight [14]. As detailed in Sect. 1, numerous definitions have been given to the concept of "transparency" of classification systems, and there is a lack of operational criteria to determine whether a given algorithm deserves to be called "transparent" or not.

In this paper, we propose the definition below, based on definitions of "transparency" in the literature. Let us recall that our aim here is to propose a definition, and operational criteria, of what we called a "transparent" classification system in a medical context with a user-centered point-of-view.

Definition 1. *A classification system is considered to be "transparent" if, and only if:*

- *the classification system is* ***understandable***
- *the type of classifier and learning system used are* ***interpretable***
- *results produced are* ***traceable***
- *classifiers used are* ***revisable****.*

2.1 Understandability of the Classification System

Although transparency is often defined as *"giving explanations of results"*, several authors have highlighted that these explanations must be "understandable", or "comprehensible", by users [12,26,33]. As proposed by Montavon [28], the fact that something is "understandable" by users can be defined as its belonging to a domain that human beings can make sense of.

However, we need an operational criterion to be sure that users can make sense of what we will provide them. In our case, users being physicians, we can consider that users can make sense of anything they have studied during their medical training. Therefore, we define as "understandable" anything based on notions/concepts included in the school curriculum of all potential users. Based on this operational criterion, we propose the definition below of what we call an "understandable" classification systems.

Definition 2. *A classification system is considered to be understandable by users if, and only if, each of its aspects is based on notions/concepts included in the school curriculum of all potential users.*

Let us consider a classification system based on a set C of notions/concepts, and a set S of notions/concepts included in the school curriculum of all potential users, such than $S \cap C$ can be empty. Defined like this, the "understandability" of a classification system is a continuum extending from $S \cap C = \emptyset$ to $S \cap C = C$.

2.2 Interpretability of Classifiers and Learning System

According to Spagnolli [34], the aim of being "transparent" is to ensure that users are in a position to make informed decisions, without bias, based on the results of the system. A classification system only "understandable" does not prevent misinterpretations of its results or misinformed decisions by users. Therefore, to be considered "transparent" a classification system must also be "interpretable" by users. The criterion of "interpretability" is even more important when applied to sensitive issues like those involved in medical matters. But what could be operational criteria to establish whether a classification system is "interpretable" or not by users?

Let us look at the standard example of a classification system dedicated to picture classification [17]. In practice, the user will use the classifier produced by the learning algorithm and not directly the learning algorithm. Therefore, if the user gives a picture of an animal to the classifier and the classifier says "it's a human", then the user can legitimately ask "Why did you give me this result?" [33]. Here, we have two possibilities: the classifier provides a good classification and the user wants to better understand the reasons underlying this classification, or the classifier provides a wrong classification and the user wants to understand why the classifier didn't provide the right classification.

In the first case, the user can expect "understandable" explanations on the reasoning process that conducted to a specific result. Depending on the classifier used, explanations can take different forms such as "because it has clothes, hair and no claws" or "because the picture is similar to these others pictures of humans". In addition, to prevent misinterpretations, the user can also legitimately wonder "To what extent can I trust this classification?" and expect the classifier to give the risk of error of this result.

In the second case, the user needs to have access to an understandable version of the general process of the classifier and not only the reasoning process that conducts to the classification. This allows the user to understand under which conditions the classifier can produce wrong classifications. In addition, the user can legitimately wonder "To what extent can I trust this classifier in general?". To answer this question, the classifier must be able to provide general performances rates such as its error rate, its precision, its sensitivity and its specificity.

Based on all the above aspects, we are now able to propose the following definition of the "interpretability" of the type of classifier used in the classification system.

Definition 3. *A type of classifier is considered to be "interpretable" by users if, and only if, it is able to provide to users:*

- *understandable explanations of results, including:*
  - *the reasoning process that conducts to results*
  - *the risk of error of results*
- *an understandable version of its general process*
- *its global error, precision, sensitivity and specificity rates.*

Nevertheless, although the classifier can answer the question "Why this result?", it will not be able to answer if the user asks, still to prevent a potential misinterpretation, "How the process of classification have been built? Where does it come from?". Only the learning algorithm used by the classification system can be able to bring elements of a response to users because the function of the learning system is to build classifiers, whereas the function of classifiers is to classify.

Therefore, a "transparent" classification system must be based on a type of classifier "interpretable", as defined in Definition 3, but it must also use an "interpretable" learning algorithm, still to ensure that users are in a position to make informed decisions. A first way to establish whether a learning algorithm is "interpretable" could be to evaluate if users can easily reproduce the process of the algorithm. However, evaluating "interpretability" in this way would be tedious for users. We have then to establish operational criteria of learning algorithms that can contribute to its "interpretability" by users.

First, the more linear it is, the more reproducible it is by users. However, linearity alone is not enough to allow "interpretability". For example, this is the case if the various steps of the algorithm fail to be understandable by users or if branching and ending conditions are not understandable by users. Accordingly, we proposed the following definition of the "interpretability" of a learning algorithm.

Definition 4. *A learning algorithm is considered to be "interpretable" by users if, and only if it has:*

- *a process as linear as possible*
- *understandable steps*
- *understandable branching and ending conditions.*

The use of concept such as "possibility" of the algorithm implies that we cannot tell that a learning algorithm is absolutely "interpretable". By corollary, the assessment algorithm's "interpretability" is quite subjective and dependent on what we consider as "possible" in terms of linearity for an learning algorithm.

2.3 Traceability of Results

Another aspect we have to take into account is the capacity to traceback data used to produce a specific classification. As introduced by Hedbom [18], a user has the right to know which of her/his personal data are used in a classification system, but also how and why. This is all the more true in medical contexts, where the data used are sensitive.

The "understandability" and "interpretability" criteria alone are not enough to ensure the ability to traceback the operations and data used to produce a given result. For example, let us suppose we have a perfectly understandable and interpretable classification system, if this system does some operations randomly, it becomes difficult to traceback operations made from a given result.

By contrast, if a classification system is totally "understandable" and "interpretable", the determinism of classifiers and the learning system is a necessary and sufficient condition to allow "traceability". We can then propose the following definition of the traceability of results.

Definition 5. *The results of a classification system are considered to be "traceable" if, and only if, the learning system and the type of classifier used have a non-stochastic process.*

2.4 Revisability of Classifiers

Lastly, the concept of "transparency" can be associated with the possibility for users to make feedbacks to the classification system to improve future results [40]. When a classification system allows users to make feedbacks that are taken into account, this classification system appears less as a "black-box" system to users.

For example, in the medical context, Caruana et al. [7] have reported that physicians had a better appreciation of a rule-based classifier than of a neural network, in the case of predicting pneumonia risk and hospital readmission. This is despite the fact that neural network had better results than the rule-based classifier. According to the authors, the possibility to modify directly wrong rules of the classifier played a crucial role in the preference of physicians.

However, not all classifiers can be directly modified by users. Another way to take account of users' feedbacks is to use continuous learning algorithms (or online learning). The majority of learning algorithms are offline algorithms, but all can be modified, more or less easily, to become online learning algorithms. In that case, the classifier is considered to be partly "revisable". We then obtain the following definition of "revisability" of the type of classifier used by a classification system.

Definition 6. *A type of classifier used by a classification system is considered to be "revisable" by users if, and only if, users can directly modify the classifier's process or, at least, the learning algorithm can easily become an online learning algorithm.*

3 Evaluation of Different Classification Systems

In this section, we use the operational criteria we have established in Sect. 2 to evaluate the degree of "transparency" of several well-known classification systems. With this evaluation, we aim to determine whether one of these classification systems can be used in our use case, from a "transparency" point of view.

We also evaluate the performances of these algorithms on datasets similar to our use case, to evaluate the cost of using a "transparent" alogrithm in terms of performances.

3.1 "Transparency" Evaluation

Our evaluation of "transparency" has been made on six different classification systems. The BPMLL algorithm (based on artificial neural networks) [42], the MLkNN algorithm (based on k-Nearest Neighbors) [41], the Naive Bayes algorithm (producing probability-based classifiers) [23], the C4.5 algorithm (producing decision-tree classifiers) [30], the RIPPER algorithm (producing rule-based classifiers) [8] and the SMO algorithm (producing SVM classifiers) [25,29].

Figure 1 displays a summary of the following evaluation of our different classification systems. Due to their similarities in terms of "transparency", C4.5 and RIPPER algorithms have been considered as the same entity.

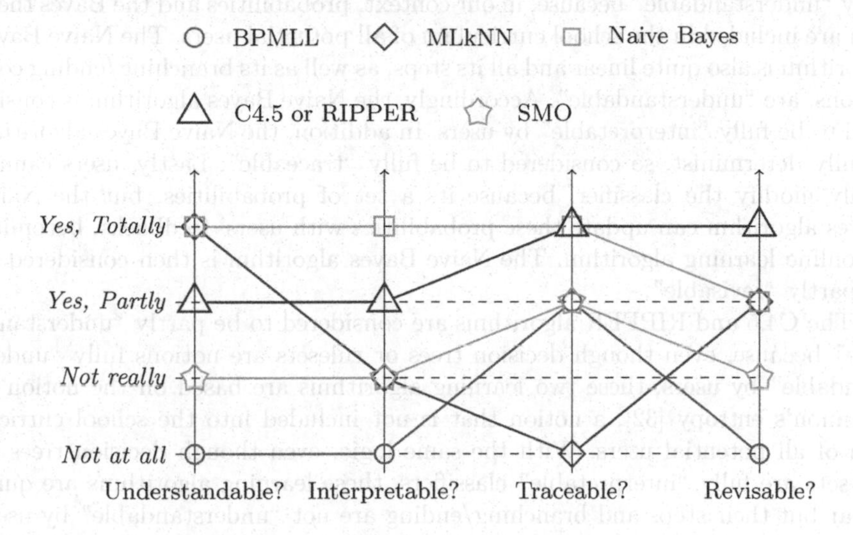

Fig. 1. Graphical representation of the potential "transparency" of different classification systems according to our operational criteria. (Color figure online)

Let us start with the evaluation of a classification system based on the BPMLL algorithm [42] (red circles in Fig. 1). The BPMLL algorithm is based on a neural network and neural networks are based on notions/concepts that are not included in the school curriculum of users such as back-propagation and activation functions. Therefore, the steps of the BPMLL algorithm, as well as its branching/ending conditions, cannot be considered to be "understandable" by users. In addition, the learning process of neural networks is not what might be called a linear process. Accordingly, we cannot consider this classification system to be "understandable" and "interpretable" by users. However, neural networks are generally determinist but, due to their low "understandability", they can only be considered to be partly "traceable". Finally, concerning the "revisability" of such a classification system, users cannot directly modify a wrong part of the classifier process and neural networks are not really adapted to continuous learning due to the vanishing gradient problem [21].

The ML-KNN algorithm [41] (violet diamonds in Fig. 1) is considered to be fully "understandable" because it is based on notions like distances and probabilities. Classifiers produced by the ML-KNN algorithm can produce explanations such as "x is similar to this other example". However, due to nested loops and advanced use of probabilities, the learning algorithm does not fit our criteria of "interpretable". In addition, the k-Nearest Neighbors algorithm [13], used by ML-KNN, is generally not determinist which makes the classification system not "traceable". Nevertheless, although classifiers produced by the ML-KNN algorithm cannot be directly modified by users, ML-KNN can easily be modified to become online learning. Consequently, it is partly "revisable".

The Naive Bayes algorithm [23] (green squares in Fig. 1) is considered to be fully "understandable" because, in our context, probabilities and the Bayes theorem are included in the school curriculum of all potential users. The Naive Bayes algorithm is also quite linear and all its steps, as well as its branching/ending conditions, are "understandable". Accordingly, the Naive Bayes algorithm is considered to be fully "interpretable" by users. In addition, the Naive Bayes algorithm is fully determinist, so considered to be fully "traceable". Lastly, users cannot easily modify the classifier, because its a set of probabilities, but the Naive Bayes algorithm can update these probabilities with users' feedbacks, becoming an online learning algorithm. The Naive Bayes algorithm is then considered to be partly "revisable".

The C4.5 and RIPPER algorithms are considered to be partly "understandable" because, even though decision trees or rulesets are notions fully "understandable" by users, these two learning algorithms are based on the notion of Shannon's entropy [32], a notion that is not included into the school curriculum of all potential users. With the same logic, even though decision trees or rulesets are fully "interpretable" classifiers, these learning algorithms are quite linear but their steps and branching/ending are not "understandable" by users because based on Shannon's entropy. The only difference between C4.5 and RIPPER could be on the linearity of their learning algorithm, because RIPPER may be considered to be less linear than C4.5, so less "interpretable". Accordingly, C4.5 and RIPPER are considered to be partly "interpretable" by users. In addition, the C4.5 and RIPPER algorithms are determinists, so fully traceable, and they are considered to be fully "revisable", because users can modify directly classifiers such as decision trees or rulesets.

Lastly, concerning the SMO algorithm, it is mainly based on mathematical notions, such as a combination of functions, that are not necessarily included in the school curriculum of all potential users. The SMO algorithm is not considered to be really "understandable" and "interpretable" by users. The SMO algorithm is determinist but, due to its low "interpertability" it could be more diffcult to traceback its results. It is then considered to be partly "traceable". In addition, the SMO algorithm can become online [35], but not as easily as ML-kNN or Naive Bayes algorithms (for example), it is not considered to be really "revisable".

Consequently, if we start from the classification system with less operational criteria of "transparency" checked, to the classification system with a majority of operational criteria checked, we obtain: BPMLL, SMO, MLkNN, RIPPER, C4.5 and Naive Bayes. Accordingly, a classification system based on the Naive Bayes algorithm can be considered as the best alternative, from a "transparency" perspective, to treat our medical use case.

3.2 Naive Bayes Algorithm for Multi-label Classification

As developed in Sect. 3.1, the Naive Bayes algorithm can be considered to be "transparent" according to our operational criteria. A common way to apply a one-label classification system to a multi-label classification problem, like in our case, is to use the meta-learning algorithm RAkEL [37]. However, the use of RAkEL, which is stochastic and combine several classifiers, makes classification systems less "interpretable" and "traceable". We proposed then a version of the Naive Bayes algorithm, developed in Algorithm 1, to treat directly multi-label classification problems staying as "transparent" as possible.

Algorithm 1: A Naive Bayes algorithm for multi-label classification

---

Data: a learning dataset $\mathcal{I}$, a set of variables $\mathcal{X}$ and a set of labels $\mathcal{Y}$
Result: sets of approximated probabilities $P_{\mathcal{Y}}$ and $P_{\mathcal{X}|\mathcal{Y}}$

```
   // Computing subsets of numerical variables
1  foreach variable X ∈ 𝒳 do
2  └ Discretize domain of X according to its values in ℐ
   // Counting occurences of 𝒴 and 𝒳 ∩ 𝒴
3  foreach instance I ∈ ℐ do
4  │  foreach label Y ∈ 𝒴 do
5  │  │  y_I ← value of Y for instance I
6  │  │  Increment by one the number of occurences of Y = y_I
7  │  │  foreach variable X ∈ 𝒳 do
8  │  │  │  t^I_X ← the subset of X corresponding to its value in instance I
9  │  │  └  Increment by one the number of occurences of Y = y_I ∩ X = t^I_X
   └  └
10 Compute probabilities P_𝒴 and P_𝒳|𝒴 from computed number of occurences
11 return P_𝒴 and P_𝒳|𝒴
```

To treat numerical variables, the first step of our algorithm is to discretize these numerical variables into several subsets (Algorithm 1, line 2). Discretizing numerical variables allows us to treat them as nominal variables. For each instance of the learning dataset, we get the subset corresponding to the value of each variable for the instance (Algorithm 1, line 8). Then, our algorithm counts occurences of each value of label and variables, and computes their frequency of occurence.

To discretize numerical variables, we first decided to use the fuzzy c-means clustering algorithm [6]. The fuzzy c-means allows to determine an "interpretable" set of subsets T_X of a variable X based on the distribution of observed values in this variable domain. Therefore, the subset t corresponding to a new value $x \in X$ is the subset $t \in T_X$ with the highest membership degree $\mu_t(x)$ (Eq. 1).

$$t_X \leftarrow \underset{t \in T_X}{\arg\, max}\ \mu_t(x) \tag{1}$$

However, we see here that the use of the fuzzy c-means algorithm requires introducing new concepts such as fuzzy sets, membership functions and membership degrees [39]. These concepts are not included into the school curriculum of users, reducing the "transparency" of the classification systems.

Therefore, we propose to use another discretizing method, more "transparent". This method, inspired by histograms, consists in splitting the variable domain into n subsets of equal size. Therefore, the subset t corresponding to a new value $x \in X$ is the subset $t \in T_X$ such as $min(t) \leq x < max(t)$. This method was preferred due to its simplicity and its potential better "transparency".

3.3 The Search for a Right Balance Between Performances and Transparency

Now that we have evaluated the "transparency" of several classifier systems, and we have identified the Naive Bayes algorithm as the most "transparent" alternative in our context, a question still remains: Does "transparency" have a cost in terms of performances?

To answer this question we evaluated classifiers presented at the beginning of this section on performance criteria for different well-known multi-label datasets and a dataset named *consultations* corresponding to our use case. Table 1 is an example based on this dataset. Currently, our dataset contains 50 instances with 4 features (patients' age, sex, BMI and disease) and 18 labels corresponding to data potentially needed by endocrinologists during consultations.

Our aim in this sub-section is to determine if the use of our version of the Naive Bayes algorithm offers suitable performances in our use case. If this is not the case, we won't have the choice but to envisage using a less "transparent" algorithm if it offers better performances.

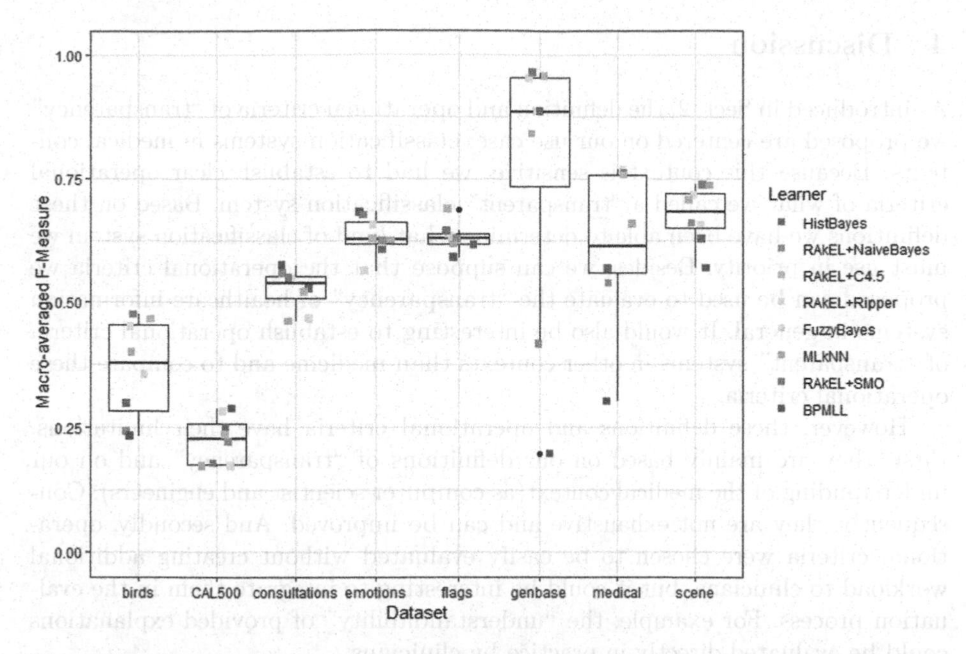

Fig. 2. Distribution of macro-averaged F-measures of several multi-label classification systems for different datasets. Results obtained by cross-validation.

These evaluations were made by using the Java library Mulan [38], which allowed to use several learning systems and cross-validation metrics. The program to reproduce these evaluations can be found on the GitLab of the LAMSADE[1].

Figure 2 shows the distribution of macro-averaged F-measures of classifier systems computed for different multi-label datasets. The F-measure is a harmonic mean of the precision and the recall of evaluated classification systems. These results have been obtained by cross-validation. Classification systems have been ordered by their degree of "transparency" according to the definition developed in Sect. 2. Green for the most "transparent", red for the less "transparent". Although a macro-averaged F-measure alone does not allow a precise evaluation, it allows us to have an overview of classification systems' performances.

We can see that the most "transparent" classification systems (greenest squares in Fig. 2) are not necessarily offering the worst performances. We can also see that, in some cases, "transparent" classification systems can offer performances close to the performances of the less "transparent" ones. In our case, represented by the *consultations* dataset, although the BPMLL algorithm offers the best F-Measure with 0.57, we can see that our version of the Naive Bayes algorithm (HistBayes) offers a quite close F-Measure with 0.53. Note that these results have to be nuanced by the small size of our dataset.

[1] https://git.lamsade.fr/a_richard/transparent-performances.

4 Discussion

As introduced in Sect. 2, the definition and operational criteria of "transparency" we proposed are centered on our use case: classification systems in medical contexts. Because this context is sensitive, we had to establish clear operational criteria of what we called a "transparent" classification system. Based on these definitions we have been able to determine what kind of classification system we must use in priority. Besides, we can suppose that the operational criteria we proposed can be used to evaluate the "transparency" of healthcare information systems in general. It would also be interesting to establish operational criteria of "transparent" systems in other contexts than medicine and to compare these operational criteria.

However, these definitions and operational criteria have their limitations. First, they are mainly based on our definitions of "transparency" and on our understanding of the medical context(as computer scientist and engineers). Consequently, they are not exhaustive and can be improved. And secondly, operational criteria were chosen to be easily evaluated without creating additional workload to clinicians, but it could be interesting to integrate them in the evaluation process. For example, the "understandability" of provided explanations could be evaluated directly in practice by clinicians.

Nevertheless, we claim that establishing clear operational criteria of "transparency" can be useful for decision-makers to determine which systems or algorithm is more relevant in which context. These operational criteria of "transparency" must be balanced with performance criteria. Depending on the use case, performances could be more important than "transparency". In our case, the medical context requires to be as "transparent" as possible. Fortunately, as developed in Subsect. 3.3, in our case being "transparent" had not a lot of impact on performances and did not implies the use of a less "transparent" classification system with better performances.

Acknowledgment. This paper was made in collaboration with employees of the Civil Hospitals of Lyon (France). Thanks to all of them. Special thanks to Pr. Moulin and Dr. Riou for their suggestions and instructive discussions.

References

1. Abdollahi, B., Nasraoui, O.: Transparency in fair machine learning: the case of explainable recommender systems. In: Zhou, J., Chen, F. (eds.) Human and Machine Learning. Human–Computer Interaction Series, pp. 21–35. Springer, Cham (2018). https://doi.org/10.1007/978-3-319-90403-0_2
2. Akkermans, H., Bogerd, P., Van Doremalen, J.: Travail, transparency and trust: a case study of computer-supported collaborative supply chain planning in high-tech electronics. Eur. J. Oper. Res. **153**(2), 445–456 (2004). https://doi.org/10.1016/S0377-2217(03)00164-4

3. Amiribesheli, M., Hosseini, M., Bouchachia, H.: A principle-based transparency framework for intelligent environments. In: Proceedings of the 30th International BCS Human Computer Interaction Conference: Fusion!, p. 49. BCS Learning & Development Ltd. (2016). https://doi.org/10.14236/ewic/HCI2016.68
4. Ananny, M., Crawford, K.: Seeing without knowing: limitations of the transparency ideal and its application to algorithmic accountability. New Media Soc. **20**(3), 973–989 (2018). https://doi.org/10.1177/1461444816676645
5. Berner, E.S.: Clinical Decision Support Systems, 3rd edn. Springer, Heidelberg (2016). https://doi.org/10.1007/978-3-319-31913-1
6. Cannon, R.L., Dave, J.V., Bezdek, J.C.: Efficient implementation of the fuzzy C-means clustering algorithms. IEEE Trans. Pattern Anal. Mach. Intell. **2**, 248–255 (1986). https://doi.org/10.1109/TPAMI.1986.4767778
7. Caruana, R., Lou, Y., Gehrke, J., Koch, P., Sturm, M., Elhadad, N.: Intelligible models for healthcare: predicting pneumonia risk and hospital 30-day readmission. In: Proceedings of the 21th ACM SIGKDD International Conference on Knowledge Discovery and Data Mining, pp. 1721–1730. ACM (2015)
8. Cohen, W.W.: Fast effective rule induction. In: Twelfth International Conference on Machine Learning, pp. 115–123. Morgan Kaufmann (1995)
9. Cramer, H., et al.: The effects of transparency on trust in and acceptance of a content-based art recommender. User Model. User-Adap. Inter. **18**(5), 455 (2008). https://doi.org/10.1007/s11257-008-9051-3
10. Datta, A., Sen, S., Zick, Y.: Algorithmic transparency via quantitative input influence: theory and experiments with learning systems. In: 2016 IEEE Symposium on Security and Privacy (SP), pp. 598–617. IEEE (2016)
11. Dinka, D., Nyce, J.M., Timpka, T.: The need for transparency and rationale in automated systems. Interact. Comput. **18**(5), 1070–1083 (2006)
12. Doran, D., Schulz, S., Besold, T.R.: What does explainable AI really mean? A new conceptualization of perspectives. arXiv preprint arXiv:1710.00794 (2017)
13. Dudani, S.A.: The distance-weighted K-nearest-neighbor rule. IEEE Trans. Syst. Man Cybern. **4**, 325–327 (1976)
14. Goebel, R., et al.: Explainable AI: the new 42? In: Holzinger, A., Kieseberg, P., Tjoa, A., Weippl, E. (eds.) CD-MAKE 2018. LNCS, vol. 11015, pp. 295–303. Springer, Cham (2018). https://doi.org/10.1007/978-3-319-99740-7_21
15. Göritzlehner, R., Borst, C., Ellerbroek, J., Westin, C., van Paassen, M.M., Mulder, M.: Effects of transparency on the acceptance of automated resolution advisories. In: 2014 IEEE International Conference on Systems, Man, and Cybernetics (SMC), pp. 2965–2970. IEEE (2014). https://doi.org/10.1109/SMC.2014.6974381
16. Groth, P.: Transparency and reliability in the data supply chain. IEEE Internet Comput. **17**(2), 69–71 (2013). https://doi.org/10.1109/MIC.2013.41
17. Gunning, D.: Explainable artificial intelligence (XAI). Defense Advanced Research Projects Agency (DARPA), nd Web 2 (2017)
18. Hedbom, H.: A survey on transparency tools for enhancing privacy. In: Matyáš, V., Fischer-Hübner, S., Cvrček, D., Švenda, P. (eds.) The Future of Identity in the Information Society. Privacy and Identity 2008. IFIP Advances in Information and Communication Technology, vol. 298, pp. 67–82. Springer, Heidelberg (2008). https://doi.org/10.1007/978-3-642-03315-5_5
19. Heeks, R., Mundy, D., Salazar, A.: Why health care information systems succeed or fail. Inf. Syst. Public Sector Manage. (1999)
20. Herlocker, J.L., Konstan, J.A., Riedl, J.: Explaining collaborative filtering recommendations. In: Proceedings of the 2000 ACM Conference on Computer Supported Cooperative Work, pp. 241–250. ACM (2000)

21. Hochreiter, S.: The vanishing gradient problem during learning recurrent neural nets and problem solutions. Int. J. Uncert. Fuzziness Knowl.-Based Syst. **6**(02), 107–116 (1998)
22. Holzinger, A., Biemann, C., Pattichis, C.S., Kell, D.B.: What do we need to build explainable AI systems for the medical domain? arXiv preprint arXiv:1712.09923 (2017)
23. John, G.H., Langley, P.: Estimating continuous distributions in Bayesian classifiers. In: Eleventh Conference on Uncertainty in Artificial Intelligence, pp. 338–345. Morgan Kaufmann, San Mateo (1995)
24. Karsenty, L., Botherel, V.: Transparency strategies to help users handle system errors. Speech Commun. **45**(3), 305–324 (2005)
25. Keerthi, S., Shevade, S., Bhattacharyya, C., Murthy, K.: Improvements to Platt's SMO algorithm for SVM classifier design. Neural Comput. **13**(3), 637–649 (2001). https://doi.org/10.1162/089976601300014493
26. Kim, T., Hinds, P.: Who should i blame? Effects of autonomy and transparency on attributions in human-robot interaction. In: ROMAN 2006-The 15th IEEE International Symposium on Robot and Human Interactive Communication, pp. 80–85. IEEE (2006). https://doi.org/10.1109/ROMAN.2006.314398
27. Michener, G., Bersch, K.: Identifying transparency. Inf. Polity **18**(3), 233–242 (2013). https://doi.org/10.3233/IP-130299
28. Montavon, G., Samek, W., Müller, K.R.: Methods for interpreting and understanding deep neural networks. Digit. Signal Proc. **73**, 1–15 (2018)
29. Platt, J.: Fast training of support vector machines using sequential minimal optimization. In: Schoelkopf, B., Burges, C., Smola, A. (eds.) Advances in Kernel Methods - Support Vector Learning. MIT Press, Cambridge (1998)
30. Quinlan, R.: C4.5: Programs for Machine Learning. Morgan Kaufmann Publishers, San Mateo (1993)
31. Richard, A., Mayag, B., Meinard, Y., Talbot, F., Tsoukiàs, A.: How AI could help physicians during their medical consultations: an analysis of physicians' decision process to develop efficient decision support systems for medical consultations. In: PFIA 2018, Nancy, France (2018)
32. Shannon, C.E.: A mathematical theory of communication. Bell Syst. Tech. J. **27**(3), 379–423 (1948)
33. Sinha, R., Swearingen, K.: The role of transparency in recommender systems. In: CHI 2002 Extended Abstracts on Human Factors in Computing Systems, pp. 830–831. ACM (2002). https://doi.org/10.1145/506443.506619
34. Spagnolli, A., Frank, L.E., Haselager, P., Kirsh, D.: Transparency as an ethical safeguard. In: Ham, J., Spagnolli, A., Blankertz, B., Gamberini, L., Jacucci, G. (eds.) Symbiotic 2017. LNCS, vol. 10727, pp. 1–6. Springer, Cham (2017). https://doi.org/10.1007/978-3-319-91593-7_1
35. Tax, D.M., Laskov, P.: Online SVM learning: from classification to data description and back. In: 2003 IEEE XIII Workshop on Neural Networks for Signal Processing (IEEE Cat. No. 03TH8718), pp. 499–508. IEEE (2003)
36. Tintarev, N., Masthoff, J.: Effective explanations of recommendations: user-centered design. In: Proceedings of the 2007 ACM Conference on Recommender Systems, pp. 153–156. ACM (2007). https://doi.org/10.1145/1297231.1297259
37. Tsoumakas, G., Katakis, I., Vlahavas, I.: Random K-labelsets for multi-label classification. IEEE Trans. Knowl. Data Eng. **23**(7), 1079–1089 (2011). https://doi.org/10.1109/TKDE.2010.164
38. Tsoumakas, G., Spyromitros-Xioufis, E., Vilcek, J., Vlahavas, I.: Mulan: a Java library for multi-label learning. J. Mach. Learn. Res. **12**, 2411–2414 (2011)

39. Zadeh, L.A., Klir, G.J., Yuan, B.: Fuzzy Sets, Fuzzy Logic, and Fuzzy Systems: Selected Papers, vol. 6. World Scientific, Singapore (1996)
40. Zarsky, T.: Transparency in data mining: from theory to practice. In: Custers, B., Calders, T., Schermer, B., Zarsky, T. (eds.) Discrimination and Privacy in the Information Society. Studies in Applied Philosophy, Epistemology and Rational Ethics, vol. 3, pp. 301–324. Springer, Heidelberg (2013). https://doi.org/10.1007/978-3-642-30487-3_17
41. Zhang, M.L., Zhou, Z.H.: Ml-KNN: a lazy learning approach to multi-label learning. Pattern Recogn. **40**(7), 2038–2048 (2007)
42. Zhang, M., Zhou, Z.: Multi-label neural networks with applications to functional genomics and text categorization. IEEE Trans. Knowl. Data Eng. **18**, 1338–1351 (2006). https://doi.org/10.1109/TKDE.2006.162

Information Fusion-2-Text: Explainable Aggregation via Linguistic Protoforms

Bryce J. Murray[1(✉)], Derek T. Anderson[1], Timothy C. Havens[2], Tim Wilkin[3], and Anna Wilbik[4]

[1] Electrical Engineering and Computer Science Department, University of Missouri, Columbia, MO, USA
bmndc@mail.missouri.edu

[2] College of Computing and Department of Electrical and Computer Engineering, Michigan Technological University, Houghton, MI, USA

[3] School of Information Technology, Deakin University, Geelong, VIC, Australia

[4] Information Systems Group of the Department of Industrial Engineering and Innovation Sciences, Eindhoven University of Technology, Eindhoven, The Netherlands

Abstract. Recent advancements and applications in artificial intelligence (AI) and machine learning (ML) have highlighted the need for explainable, interpretable, and actionable AI-ML. Most work is focused on explaining deep artificial neural networks, e.g., visual and image captioning. In recent work, we established a set of indices and processes for explainable AI (XAI) relative to information fusion. While informative, the result is information overload and domain expertise is required to understand the results. Herein, we explore the extraction of a reduced set of higher-level linguistic summaries to inform and improve communication with non-fusion experts. Our contribution is a proposed structure of a fusion summary and method to extract this information from a given set of indices. In order to demonstrate the usefulness of the proposed methodology, we provide a case study for using the fuzzy integral to combine a heterogeneous set of deep learners in remote sensing for object detection and land cover classification. This case study shows the potential of our approach to inform users about important trends and anomalies in the models, data and fusion results. This information is critical with respect to transparency, trustworthiness, and identifying limitations of fusion techniques, which may motivate future research and innovation.

Keywords: Deep learning · Machine learning · Information fusion · Information aggregation · Fuzzy integral · Explainable artificial intelligence · XAI · Protoform · Linguistic summary

1 Introduction

We live in a world that is recognizing the potential of *artificial intelligence* (AI) and *machine learning* (ML) in everyday settings. These tools have been integrated into many aspects of our daily lives—whether we realize it or not. These

M.-J. Lesot et al. (Eds.): IPMU 2020, CCIS 1239, pp. 114–127, 2020.
https://doi.org/10.1007/978-3-030-50153-2_9

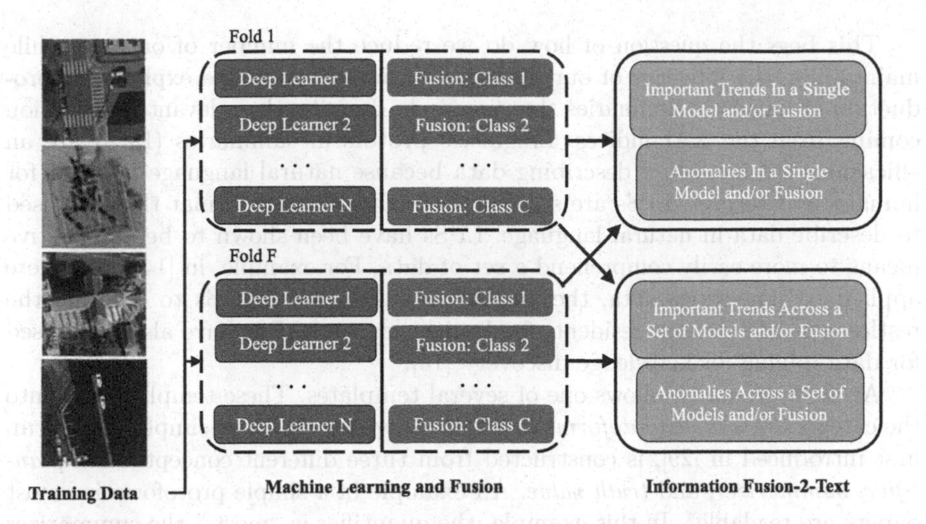

Fig. 1. Graphical illustration of fusion-2-text for the case study explored herein involving object detection and land cover classification in remote sensing. First, multiple machine learning (ML) models are trained in a cross validation context. Next, fusion is used to combat the fact that no single ML architecture is best across all data and classes. However, what have we learned? Fusion-2-text is used to discover a succinct set of important summaries or anomalies for a user in or across models.

tools, which were birthed from academic exercise, are no longer just in academia; they have found home in many different applications. Various AIs are being used to solve real-world problems, or they *simply* make our lives more convenient. Many of these algorithms are built on data-driven methods which scientists, researchers, and engineers have creatively developed and applied mathematics to build. Despite the mathematical foundations, it has become common for these tools to produce solutions that are not understandable. However, many applications require an explanation as to why a machine made a particular decision.

One task in the AI community is data or information fusion. One type of fusion revolves around the *Choquet Integral* (ChI), which can be learned from data. There are many ways to learn the ChI, the reviewer can refer to [18] for a recent review. However, once these parameters are learned, explanations can be derived from the data and learned model. In [9], we exploited properties of the ChI to understand which parameters in a learned model are supported by data. In [17], we exploited this knowledge to produce indices that describe different properties of a learned ChI. Unfortunately, the large quantity of values that our indices produce can be daunting and can lead to information overload. For example, when fusing multiple deep convolutional neural networks (DCNN) for classification, there will be one ChI learned per class—a set of XAI indices for each learned ChI. As a result, there is a need to summarize these results to explain the model at a higher-level as well as reduce the amount of information that they produce.

This begs the question of how do we reduce the number of outputs while maintaining the integrity of our XAI descriptions. Herein, we explore the production of linguistic summaries that concisely describe the relevant information coming from the XAI indices. Linguistic protoform summaries (LPS) are an efficacious mechanism of describing data because natural language is easier for humans to interpret. LPSs are statements with a specific format that are used to describe data in natural language. LPSs have been shown to be an effective means to more easily comprehend a set of data. For example, in [14] LPSs were applied to time-series data, the authors of [26] utilized LPSs to describe the restlessness of eldercare residents for healthcare, and LPSs have also been used for data mining for knowledge discovery [15].

An LPS generally follows one of several templates. These templates fall into the category of *simple protoforms* or *extended protoforms*. The simple protoform, first introduced in [29], is constructed from three different concepts: the *quantifier*, *summarizer*, and *truth value*. An example of a simple protoform is "most papers are readable". In this example, the quantifier is "most," the summarizer is "readable," and the truth value would be computed to determine the degree to which that statement is valid. As time passed, others have extended the simple protoform's template [14,28]. While there are several extensions, one example of an extended protoform (that includes an additional summarizer) is "few papers are readable and noteworthy". Not only has the LPS template been modified, but Yager's original computation of truth values has been scrutinized. In [25], it was shown that Yager's original equations to compute truth may not suited for all membership functions that model the protoforms because they may produce non intuitive summaries. As such, [25] and [12] used the Sugeno integral to overcome shortcomings in Yager's equations. Moreover, the authors of [3] present a holistic view of the development of quantifying sentences and the equations that drive this process.

The main contributions of this paper are as follows. First, we explore the potential for LPSs to reduce the complexity and amount of XAI information for the ChI. To the best of our knowledge, this has not been explored to date. Second, at a high-level, we explore what type of summaries are useful and relevant and should be reported. Third, we propose a way to derive LPSs from two of our data-centric and model-centric indices. While this is only performed on two indices herein, due to space, we discuss how our procedures generalize to other XAI indices. Last, we give a case study for aggregating a set of heterogeneous architecture DCNNs for object detection and land cover classification on a benchmark remote sensing dataset. The benefit of this study is to show actual summaries and assess if they are useful.

The breakdown of this paper is as follows. In Sect. 2, we give a brief overview of the ChI and its optimization, and we identify its data supported parameters. In Sect. 3, we present the XAI indices, Sect. 4 describes how to construct LPSs, and Sect. 5 shows how to construct fuzzy sets with respect to our indices. Last, we present our case study in Sect. 6 and insights are drawn from our data and LPSs. Figure 1 shows the technical breakdown of our fusion-2-text and Fig. 1 shows our remote sensing case study.

2 Choquet Integral

The utility of the ChI has been demonstrated in numerous applications, e.g., [4,6,21,24]. The ChI is a powerful, nonlinear aggregation operator that is parameterized by a fuzzy measure (FM). Let $X = \{x_1, ..., x_N\}$ be a set of N information sources. With respect to a finite domain, the FM, $\mu : 2^X \to \mathbb{R}$, is a function that satisfies: (i) (boundary condition) $\mu(\emptyset) = 0$, and (ii) (monotonicity) if $A, B \subseteq X$, and $A \subseteq B$, $\mu(A) \leq \mu(B)$[1]. It is often convenient to think about the FM as not just free parameters but as a modeling of *interactions* (e.g., possibly correlations) between subsets. The ChI[2] is

$$\int \mathbf{h} \circ \mu = \sum_{j=1}^{N} h_{\pi(j)} \Big(\mu(A_j) - \mu(A_{j-1}) \Big), \tag{2}$$

where π is an ordering of $\mathbf{h} = (h(x_1), ..., h(x_N))$[3]. Furthermore, $h(x_i) = h_i \in \mathbb{R}$ is the input from source i, such that $h_{\pi(1)} \geq h_{\pi(2)} \geq \ldots \geq h_{\pi(N)}$. Last, A_j corresponds to the subset $\{x_{\pi(1)}, ..., x_{\pi(j)}\}$.

The ChI can alternatively be thought about as $N!$ linear convex sums (LCS)[4,5], as each sort of the data yields an LCS. Herein, we follow the nomenclature defined in [17], and we call each sort of the data a *walk* (in the Hasse diagram).

2.1 Optimization

Defining the FM variables in the ChI is not a trivial task and there are many ways to identify them, e.g., [5,10,11,16]. However, in our current data-driven era, it is common place to learn the FM variables. Herein, we use our learning algorithm put forth in [9]. We do not describe the algorithm due to limited page length. The techniques proposed herein extend beyond a specific learner, they are applicable to any ChI derived from data.

2.2 Data Supported Variables

In [9], we established that data supported variables can be identified for the ChI. A variable is called supported if any walk of the data includes it. For example, let $N = 3$ and $h_3 > h_1 > h_2$. The FM variables that are encountered are $\mu(\{h_3\})$, $\mu(\{h_1, h_3\})$ and $\mu(X)$. By considering all the given inputs in the training data, we can easily determine all data supported variables. This fact is important to many of the upcoming indices (Fig. 2).

[1] While not required, it is common in practice to impose $\mu(X) = 1$.

[2] It is important to note that when a FM is set (values are specified), the ChI becomes a specific aggregation operator. For example, consider $\mu(A) = 1, \forall A \in X$, except $\mu(\emptyset)$. As such, the ChI reduces to the max operator.

[3] Hereafter, $h(x_i)$ will be shortened to h_i for simplicity.

[4] When $\mu(X) = 1$.

[5] Who share 2^N weights.

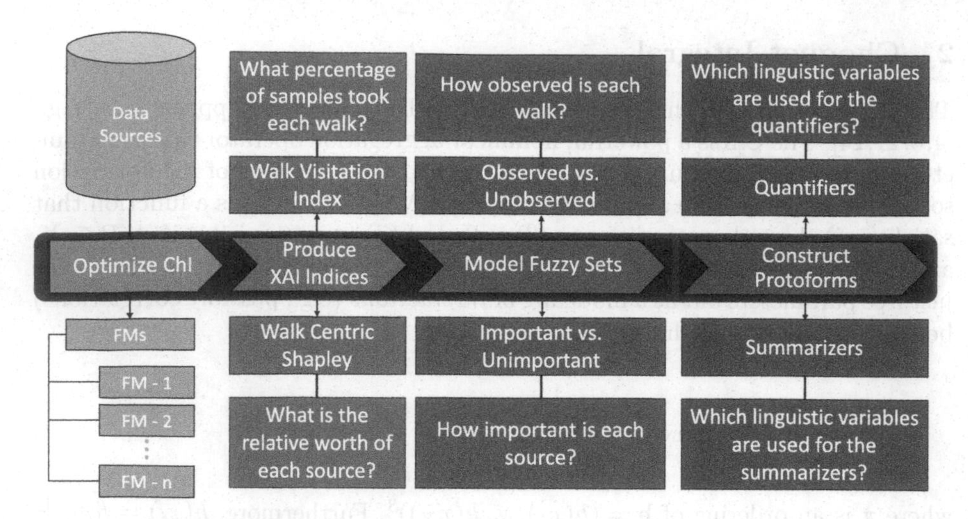

Fig. 2. Illustration of computational stages in our fusion-2-text.

3 Existing Low-Level XAI Indices

In [17], we proposed a set of XAI indices—measures/functions that summarize and highlight important properties about the FM and ChI—in the context of data-driven learning. In [18], we expanded our initial set of indices, including the Shapley and interaction index, to operate more accurately on partially observed domains. In general, our indices can be partitioned into three sets: information about the FM (the sources and their interactions), inquiries about the FM-ChI (e.g., what is the specific aggregation), and inquiries about the data relative to the ChI (e.g., what parts of our model are missing). While each of our indices provide valuable and different insights, we limit the scope herein to one data specific index, walk visitation, and one model specific index, the walk centric Shapley index.

3.1 Walk Centric Shapley

The Walk Centric Shapley (WCS) index is an extension of the Shapley index [20]. The WCS defines the relative worth of each source with respect to its data supported variables. This extension is valuable because the traditional Shapley may be drastically over- or under-estimate the worth of sources as it assumes that the FM is fully observable. The WC Shapley is

$$\bar{\Phi}_\mu(i) = \sum_{K \subseteq X \setminus \{i\}} \zeta_{X,2}(K) \left(\mu(K \cup \{i\}) - \mu(K)\right), \tag{3a}$$

$$\zeta_{X,2}(K) = \frac{(|X| - |K| - 1)!|K|!}{|X|!} \mathbb{1}_{(K \cup \{i\})} \mathbb{1}_{(K)}, \tag{3b}$$

$$\widetilde{\Phi}_\mu(i) = \frac{\bar{\Phi}_\mu(i)}{\sum_{j=1}^{i} \bar{\Phi}_\mu(j)}, \tag{3c}$$

where $K \subseteq X \setminus \{i\}$ denotes all proper subsets from X that do not include source i and $\mathbb{1}$ is an indicator function that is 1 if the FM value is data-supported and 0 otherwise. The Shapley values of μ is the vector $\widetilde{\mathbf{\Phi}}_\mu = [\widetilde{\Phi}_\mu(1), ..., \widetilde{\Phi}_\mu(N)]^t$ where $\sum_{i=1}^{N} \widetilde{\Phi}_\mu(i) = 1$. The WCS values are important because they inform us about the relative worth of each information source.

3.2 Walk Visitation

Understanding the quality of the information sources is merely one aspect of the big XAI picture. It is also important to understand the quality (e.g., completeness) of a learned ChI. Herein, we use the walk visitation metric [17], which describes how many unique walks were taken within the training data. We quickly summarize the index due to limited page count. The index works by sorting all samples (according to their input values), finding which walks were encountered, and dividing the number of times that they were observed by the total number of samples. The goal of this metric is to determine the degree to which each walk was observed. If a probability is zero, then a walk was never seen. Furthermore, if we get a new input for the ChI and its walk was not encountered in training, then one should question the ChI output. In [17], this index was used to derive additional indices, like to what degree should we trust an output of the ChI.

4 Protoforms

Protoform-based linguistic summaries are often an effective liaison between data the data interpreter. As such, deriving linguistic summaries with respect to the XAI indices has the potential to effectively reduce the amount of information by producing concise summaries. Furthermore, as less is often more, there is also the potential to remediate confusion due to complexity, which can improve decision making. While there are multiple LPS templates to follow, the *simple* protoform will suffice for the insights that we are drawing herein. The simple protoform takes the following format,

$$Qy\text{'s are } P. \tag{4}$$

Within the protoform, Q is a linguistic *quantifier*, y is a set of *objects*, and P is the *summarizer*. Both Q and P are modelled by fuzzy sets over the desired domains.

Examples of a quantifier that are suitable for our problem are words like "few", "many", or "most"; whereas examples of a summarizer may be "important" or "observed" when referencing the sources or the walks, respectively. Moreover, an example of a protoform with respect to the XAI indices may be "few walks are observed". With respect to the XAI indices, we produce summaries that describe the importance of each of the sources across all models and how many walks are observed in each model. To do this, we use the vocabulary in Fig. 3.

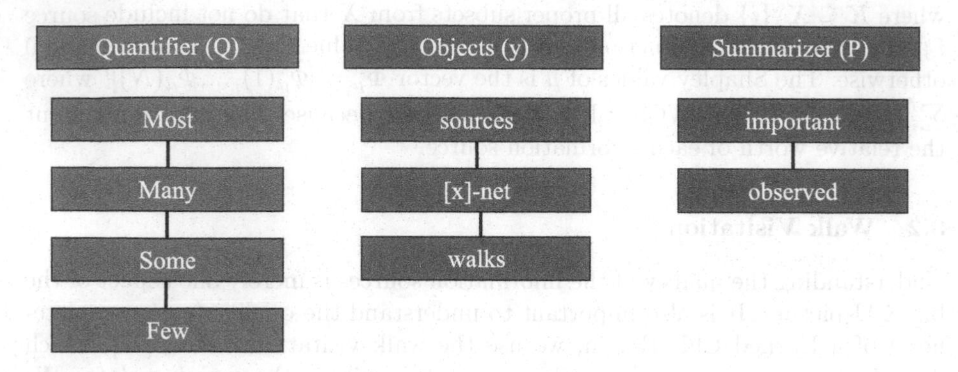

Fig. 3. Vocabulary used herein to produce fusion LPSs.

An LPS has a value of *truth* associated with it. This concept, first introduced by Yager [29], utilizes Zadeh's calculus to use Eq. 5 to determine the truth value, T, associated with the linguistic summary. This equation is as follows,

$$T(Ay's\,are\,P) = A(\frac{1}{N}\Sigma_{i=1}^{N}P(y_i)). \tag{5}$$

However, Eq. 5 may produce non-intuitive results, as noted in [25] and then in [12]. As a result, the fuzzy integral can also be employed to determine the truth value; however, for the scope of our case study, Eq. 5 will suffice.

5 Fuzzy Sets

As mentioned in Sect. 4, $Q(x)$ and $P(x)$ are fuzzy sets. In each case, we use the trapezoidal membership function for the fuzzy sets, and we have empirically determined their parameters. This is acceptable for our initial work, as we have spent a good amount of time working with the ChI and the benchmark data set. However, in future work these sets and their values clearly should be learned from data or a set of experts.

5.1 Walk Centric Shapley

To define the fuzzy sets with respect to the WCS, a fuzzy set must be defined for $P(x)$ (i.e. *important*) and $Q(x)$ (*few, some, many, most*). The fuzzy set is defined by the trapezoidal membership values from the WCS values. In an ideal fusion solution, all Shapley values would be equal to $\frac{1}{N}$, meaning equal contribution. As a result, we consider the source to be *important* if it has a value greater than or equal to $\frac{1}{N}$. Once $P(x)$ is defined, $Q(x)$ can be modelled. The output of averaging $P(x)$ is passed to $Q(x)$, and it is between $[\frac{1}{N}, 1]$ because at least 1 source will be important. As such, the domain of $Q(x)$ is $[\frac{1}{7}, 1]$. Using Eq. 5, we can compute the truth value of each of the statements allowing us to isolate the most relevant summaries that are produced.

5.2 Walk Visitation

Similar to the fuzzy sets that govern the WCS values, fuzzy sets are used to model the walk visitation index. For $P(x)$, the fuzzy sets model the concept of *observed* and *unobserved*. In an ideal case, we desire each walk to have an equal walk visitation index, so with this in mind, we consider the walk to be observed if $z \geq \frac{1}{N!}$. Next, the fuzzy set for $Q(x)$ must be modelled. At least 1 walk will be observed, and it is possible to observe up to however many training samples exist. As such, the domain of $Q(x)$ is $[\frac{1}{M}, 1]$, which is approximately $[0, 1]$. The values of each of the fuzzy sets can be found in Table 1.

Table 1. Trapezoidal membership function parameters.

Walk Visitation Quantifier	a	b	c	d	WC Shapley Quantifier	a	b	c	d
Few	0	0	$\frac{1}{7}$	$\frac{2}{7}$	Few	$\frac{1}{7}$	$\frac{1}{7}$	$\frac{2}{7}$	$\frac{3}{7}$
Some	$\frac{1}{7}$	$\frac{2}{7}$	$\frac{3}{7}$	$\frac{4}{7}$	Some	$\frac{2}{7}$	$\frac{2.5}{7}$	$\frac{3.5}{7}$	$\frac{5}{7}$
Many	$\frac{3}{7}$	$\frac{4}{7}$	$\frac{5}{7}$	$\frac{6}{7}$	Many	$\frac{3}{7}$	$\frac{4.5}{7}$	$\frac{5.5}{7}$	$\frac{6}{7}$
Most	$\frac{5}{7}$	$\frac{6}{7}$	1	1	Most	$\frac{5}{7}$	$\frac{6}{7}$	1	1
Summarizers	a	b	c	d					
Important	0	$\frac{1}{7}$	1	1					
Observed	0	$\frac{1}{7!}$	1	1					

6 Case Study

To show how these indices work in a real-world application, we consider the fusion of a set of 7 different DCNNs for object detection and land classification of remote sensing data. The DCNNs that we fuse are CaffeNet [13], GoogleNet [23], ResNet 50 [7], ResNet 101, DenseNet [8], InceptionResNetV2 [22], and Xception [2]. The dataset is the AID remote sensing data set [27]. This dataset is composed of 10,000 images over 30 different aerial scene types.

The complete training procedure of how these networks were trained can be found in [19]. Furthermore, the complete description of how these DCNNs are fused can be found in [1]. Due to the nature of the multistep classification problem (DCNNs and ChI), it is an important step to determine how to split the data into training and testing. For the DCNNs, five-fold cross validation was used. This means that four folds are used for training, and one fold is used for evaluation. From the evaluation fold, two-fold cross validation is used, due to the limited number of samples in AID. To ensure that each class is approximately balanced in each of the folds, an approximately equal number of samples were chosen from each class. There are multiple ways to perform fusion across the DCNNs. Herein, we train a ChI for each of the 30 classes. As a result, there are $30 \times 7!$ walk visitation values produced, and 30×7 Shapley values produced (a total of 151, 410 values). Using the proposed LPS configuration, we reduce the XAI indices to a few sets of LPSs that are more easily comprehended.

6.1 Source Summaries

XAI Question: How Many Sources are Important?
With respect to each ChI, we can determine how many important sources there are for each class. To produce this set of summaries, we treat each of the 30 ChIs (one per class) as our objects, y. The linguistic summarizer is "important", and our quantifiers are "few", "several", "many", and "most". Figure 4 illustrates the quantity of summaries that are assigned each of the quantifiers.

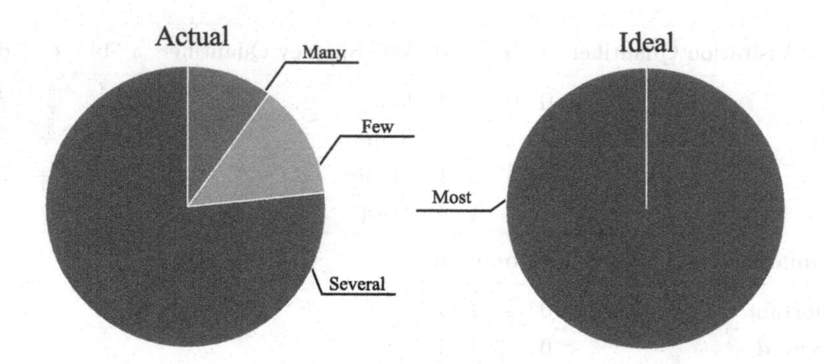

Fig. 4. Actual versus the ideal distribution of quantifier.

In an ideal case, "most" of the sources would be important, meaning all sources are contributing to the fusion solution. However, our experiment produced no summaries of "most sources are important". The majority of the ChIs are summarized by "several sources are important." This begs the question of which sources are important because we now know that not all are important.

XAI Question: How Important is Each Source?
Taking our knowledge from the last set of summaries, it is logical to now isolate the important sources. To do this, we produce a set of summaries specifically for each of the sources. We treat each DCNN across all models as

our objects, y. For example, one set of objects will be *GoogleNets*, such that $GoogleNets = \{GoogleNets_1, GoogleNets_2, ..., GoogleNets_{30}\}$. In Fig. 4, we show the resulting summaries (Fig. 5).

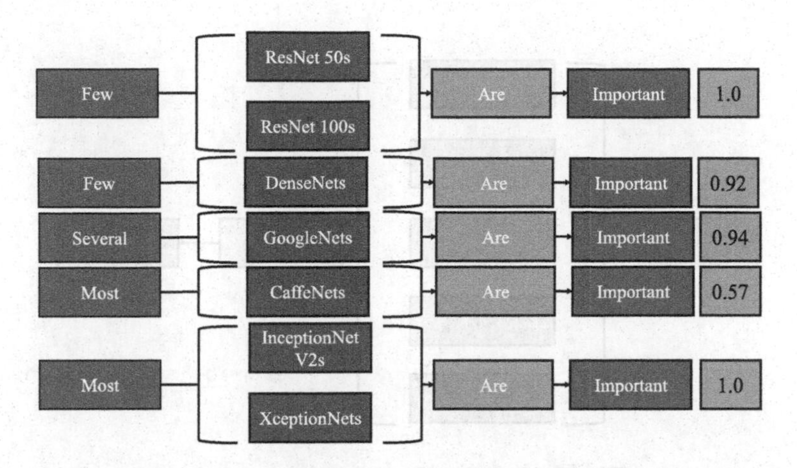

Fig. 5. LPSs describing the importance of each DCNN with truth degree (gray).

From this set of summaries (with strong truth degrees), we can conclude that ResNet 50 and ResNet 100 are not contributing to the fusion solution; however, InceptionNet and XceptionNet are important in *most* of the ChIs, meaning they are strong contributors. This leads us to conclude that ResNet 50 and ResNet 100 can likely be removed, speeding up inference by reducing DCNNs.

XAI Question: How Many Walks are Observed Per Model?
In this case, the summarizer is "observed," the objects are each ChI, and the quantifier is again, "few", "several", "many", and "most". However, for this set, there was only one summary ever found, "few walks are observed." This quickly magnifies the flaw with these models because many of the possible walks have not been observed—meaning the FM values of many of the walks are not actually learned from data. In order to fully learn the ChI, "most" walks must be observed. Observing few walks means that there is not much diversity in the data. This highlights that we may have a dominant walk, or that we only ever observe a relatively low number of walks.

XAI Question: How Observed is Each Walk Across the Data?
Whereas the last summary encapsulated information pertaining to how many walks are observed per model, this set of summaries answers the question of how observed is each walk across the entire data set. We consider each specific walk as an object; for example the walk $[1, 2, 3, 4, 5, 6, 7] = \{[1, 2, 3, 4, 5, 6, 7]_1, [1, 2, 3, 4, 5, 6, 7]_2, ..., [1, 2, 3, 4, 5, 6, 7]_{30}\}$. When producing these summaries,

there would be one summary per walk, so in our example, this would generate 7! summaries, which is far too many for anyone to digest. However, by only evaluating the walks that are observed in "most" ChIs with a high truth degree, we only consider 5 of the walks. The summaries shown in Fig. 6 are produced.

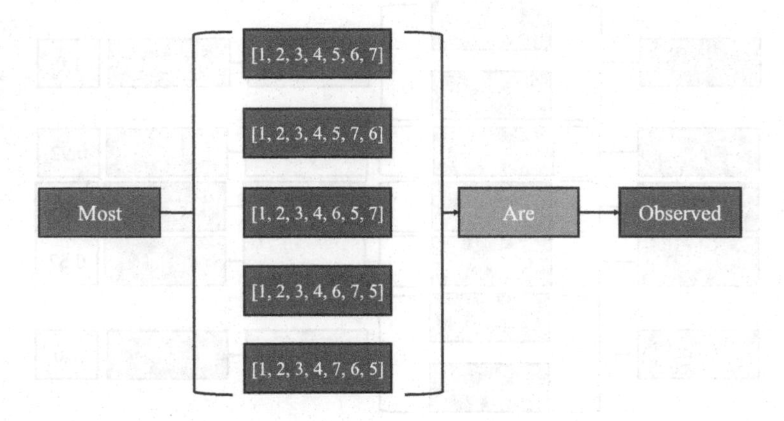

Fig. 6. Most *specific walks* are observed.

While each of these 5 walks have a truth value of 1, they are also the only walks to have the quantifier "most". There are 4,978 walks that are observed a "few" times with a truth degree of 1. This leaves 67 walks that are observed "few", "several", or "most" with a truth values less than 1 for each of them. These summaries clearly highlight that there may be some bias in the data. Specifically, the first 4 sources are typically encountered in the same order, which shows that something is not quite right. This allows us to dig into the data to figure out what might be going on.

6.2 Code

The ability to reproduce an experiment is a cornerstone in the scientific community. As such, we provide the code that produced these summaries at the following repository, https://github.com/B-Mur/ChoquetIntegral.git. Moreover, the data set that was used can be found at the following repository, https://github.com/aminb99/remote-sensing-nn-datasets.

7 Summary and Future Work

Herein, we have proposed and implemented the use of LPSs to reduce a high number of metrics to a short number of concise and more useful summaries. To our knowledge, this is the first work that produces linguistic summaries to explain fusion, without a doubt relative to data-driven learning. Before producing the summaries, the indices produce a large quantity of metrics that are complex to interpret. By producing the summaries, the indices effectively reduce the information that must be digested, while maintaining the integrity of the indices. By first determining only few sources are important, it is a logical step to determine which sources are important. If all sources were important, it would be unnecessary to determine the important sources as they are all important.

The walk visitation summaries tell a similar story. Only few walks are ever observed; this leads us to produce summaries determining the walks are observed across all data (only 5). Before we produce these summaries, these metrics are a raw stream of data that are not intuitive, and interpretable only by those with significant domain experience. However, the summaries allow someone unfamiliar with the indices (and the values they produce) to be practitioners of XAI with their fusion. To our knowledge, has never been done before.

In the future, we hope to generate summaries from the remaining XAI indices to provide more complete and comprehensive insights. By doing this, we will likely produce extended LPSs such that a single, extended LPS contains more information than a simple protoform can provide. We will also explore how to present, in a textual or visual fashion, this information to a human. This foundation also excites us because it is a structured format or language in which information can be extracted and then subsequently computed with. Possibilities including deriving higher-level conclusions about the data, models, and systems, or perhaps using the information to improve the training and/or use of fusion.

References

1. Anderson, D., Scott, G., Islam, M., Murray, B., Marcum, R.: Fuzzy Choquet integration of deep convolutional neural networks for remote sensing. In: Pedrycz, W., Chen, S.M. (eds.) Computational Intelligence for Pattern Recognition. Studies in Computational Intelligence, pp. 1–28. Springer, Heidelberg (2018). https://doi.org/10.1007/978-3-319-89629-8_1
2. Chollet, F.: Xception: deep learning with depthwise separable convolutions. In: 2017 IEEE Conference on Computer Vision and Pattern Recognition (CVPR), pp. 1800–1807, July 2017. https://doi.org/10.1109/CVPR.2017.195
3. Delgado, M., Ruiz, M.D., Sánchez, D., Vila, M.A.: Fuzzy quantification: a state of the art. Fuzzy Sets Syst. **242**, 1–30 (2014). https://doi.org/10.1016/j.fss.2013.10.012. http://www.sciencedirect.com/science/article/pii/S0165011413004247, theme: Quantifiers and Logic
4. Du, X., Zare, A.: Multiple instance Choquet integral classifier fusion and regression for remote sensing applications. IEEE Trans. Geosci. Remote Sens. 1–13 (2018). https://doi.org/10.1109/TGRS.2018.2876687
5. Du, X., Zare, A., Keller, J.M., Anderson, D.T.: Multiple instance Choquet integral for classifier fusion. In: 2016 IEEE Congress on Evolutionary Computation (CEC), pp. 1054–1061, July 2016. https://doi.org/10.1109/CEC.2016.7743905
6. Grabisch, M.: The application of fuzzy integrals in multicriteria decision making. Eur. J. Oper. Res. **89**(3), 445–456 (1996)
7. He, K., Zhang, X., Ren, S., Sun, J.: Deep residual learning for image recognition. arXiv preprint arXiv:1512.03385 (2015)

8. Huang, G., Liu, Z., van der Maaten, L., Weinberger, K.Q.: Densely connected convolutional networks. In: 2017 IEEE Conference on Computer Vision and Pattern Recognition (CVPR), pp. 2261–2269, July 2017. https://doi.org/10.1109/CVPR.2017.243
9. Islam, M.A., Anderson, D.T., Pinar, A.J., Havens, T.C.: Data-driven compression and efficient learning of the Choquet Integral. IEEE Trans. Fuzzy Syst. **PP**(99), 1 (2017). https://doi.org/10.1109/TFUZZ.2017.2755002
10. Islam, M.A., Anderson, D., Petry, F., Elmore, P.: An efficient evolutionary algorithm to optimize the Choquet Integral. Int. J. Intell. Syst. **34**, 366–385 (2018). https://doi.org/10.1002/int.22056
11. Islam, M.A., Anderson, D.T., Pinar, A., Havens, T.C., Scott, G., Keller, J.M.: Enabling explainable fusion in deep learning with fuzzy integral neural networks. IEEE Trans. Fuzzy Syst. 1 (2019). https://doi.org/10.1109/tfuzz.2019.2917124
12. Jain, A., Keller, J.M.: On the computation of semantically ordered truth values of linguistic protoform summaries. In: 2015 IEEE International Conference on Fuzzy Systems (FUZZ-IEEE), pp. 1–8, August 2015. https://doi.org/10.1109/FUZZ-IEEE.2015.7337822
13. Jia, Y., et al.: Caffe: convolutional architecture for fast feature embedding. In: Proceedings of the 22nd ACM International Conference on Multimedia, MM 2014, pp. 675–678. ACM, New York (2014). https://doi.org/10.1145/2647868.2654889
14. Kacprzyk, J., Wilbik, A., Zadrozny, S.: Mining time series data via linguistic summaries of trends by using a modified Sugeno integral based aggregation. In: 2007 IEEE Symposium on Computational Intelligence and Data Mining, pp. 742–749, March 2007. https://doi.org/10.1109/CIDM.2007.368950
15. Kacprzyk, J., Zadrozny, S.: Data mining via protoform based linguistic summaries: some possible relations to natural language generation. In: 2009 IEEE Symposium on Computational Intelligence and Data Mining, pp. 217–224, March 2009. https://doi.org/10.1109/CIDM.2009.4938652
16. Keller, J.M., Osborn, J.: A reward/punishment scheme to learn fuzzy densities for the fuzzy integral. In: Proceedings of International Fuzzy Systems Association World Congress, pp. 97–100 (1995)
17. Murray, B., Anderson, D., Islam, M.A., Pinar, A., Scott, G., Havens, T.: Explainable ai for understanding decisions and data-driven optimization of the Choquet integral. In: World Congress on Computational Intelligence (WCCI), July 2018
18. Murray, B., et al.: Explainable AI for the Choquet integral (accepted). IEEE Trans. Emerg. Top. Comput. Intell.
19. Scott, G.J., England, M.R., Starms, W.A., Marcum, R.A., Davis, C.H.: Training deep convolutional neural networks for land-cover classification of high-resolution imagery. IEEE Geosci. Remote Sens. Lett. **14**(4), 549–553 (2017)
20. Shapley, L.S.: A value for n-person games. Contrib. Theory Games **2**, 307–317 (1953)
21. Smith, R.E., et al.: Genetic programming based Choquet integral for multi-source fusion. In: IEEE International Conference on Fuzzy Systems, July 2017
22. Szegedy, C., Ioffe, S., Vanhoucke, V.: Inception-v4, inception-ResNet and the impact of residual connections on learning. In: AAAI (2016)
23. Szegedy, C., et al.: Going deeper with convolutions. In: Computer Vision and Pattern Recognition (CVPR) (2015)
24. Tahani, H., Keller, J.: Information fusion in computer vision using the fuzzy integral. IEEE Trans. Syst. Man Cybern. **20**, 733–741 (1990)

25. Wilbik, A., et al.: Evaluation of the truth value of linguistic summaries - case with nonmonotonic quantifiers. In: Angelov, P., et al. (eds.) Intelligent Systems. Advances in Intelligent Systems and Computing, vol. 322, pp. 69–79. Springer, Cham (2014). https://doi.org/10.1007/978-3-319-11313-5_7
26. Wilbik, A., Keller, J.M., Bezdek, J.C.: Linguistic prototypes for data from eldercare residents. IEEE Trans. Fuzzy Syst. **22**(1), 110–123 (2014). https://doi.org/10.1109/TFUZZ.2013.2249517
27. Xia, G., et al.: AID: a benchmark data set for performance evaluation of aerial scene classification. IEEE Trans. Geosci. Remote Sens. **55**(7), 3965–3981 (2017). https://doi.org/10.1109/TGRS.2017.2685945
28. Yager, R.R.: Fuzzy summaries in database mining. In: Proceedings the 11th Conference on Artificial Intelligence for Applications, pp. 265–269, February 1995. https://doi.org/10.1109/CAIA.1995.378813
29. Yager, R.R.: A new approach to the summarization of data. Inf. Sci. **28**(1), 69–86 (1982). https://doi.org/10.1016/0020-0255(82)90033-0

Fuzzy and Generalized Quantifier Theory

Graded Decagon of Opposition with Fuzzy Quantifier-Based Concept-Forming Operators

Stefania Boffa(✉), Petra Murinová, and Vilém Novák

Institute for Research and Applications of Fuzzy Modeling, University of Ostrava, NSC IT4Innovations, 30. dubna 22, 701 03 Ostrava 1, Czech Republic
{stefania.boffa,petra.murinova,vilem.novak}@osu.cz

Abstract. We introduce twelve operators called *fuzzy quantifier-based operators*. They are proposed as a new tool to help to deepen the analysis of data in fuzzy formal concept analysis. Moreover, we employ them to construct a graded extension of Aristotle's square, namely the *graded decagon of opposition*.

Keywords: Fuzzy formal concept analysis · Evaluative linguistic expressions · Square of opposition · Łukasiewicz MV-algebra

1 Introduction

Formal Concept Analysis (FCA) is a mathematical theory applied to the analysis of data (see [6]). The input of FCA is a triple called *formal context* that consists of a set of objects, a set of attributes, and a binary relation between objects and attributes. FCA techniques extract a collection of *formal concepts* from every formal context.

Formal concepts are special clusters that correspond to concepts such as "numbers divisible by 5", or "white roses in the garden". *Fuzzy Formal Concept Analysis* (FFCA) generalizes formal concept analysis to include also vague information. The input of FFCA is an L-context (X, Y, I) where L is a support of an algebra of truth values, X is a set of objects, Y a set of attributes, and I is a fuzzy relation $I : X \times Y \longrightarrow L$.

A *fuzzy concept* is a pair (A, B) where A, B are fuzzy sets $A : X \longrightarrow L$, $B : Y \longrightarrow L$. A is called *extent* and it is a fuzzy set of all objects $x \in X$ that have all attributes of B, and B is called *intent* and it is a fuzzy set of all attributes $y \in Y$ being satisfied by all objects of A. Namely, $A(x)$ is the degree to which "x has all attributes of B", and $B(y)$ is the degree to which "the attribute y is satisfied by all objects of A".

The work was supported from ERDF/ESF by the project "Centre for the development of Artificial Intelligence Methods for the Automotive Industry of the region" No. CZ.02.1.01/0.0/0.0/17-049/0008414 and partially also by the MŠMT project NPU II project LQ1602 "IT4Innovations excellence in science".

M.-J. Lesot et al. (Eds.): IPMU 2020, CCIS 1239, pp. 131–144, 2020.
https://doi.org/10.1007/978-3-030-50153-2_10

In this article, we define twelve special operators as a tool to deepen the analysis of data. To explain their function, let us consider the following situation.

Let (X, Y, I) be an L-context, where X is a set of students, Y are their skills, $I(x, y)$ is the degree to which "a student x has the skill y". Thus, given a fuzzy concept (A, B), we know that A is the fuzzy set representing all students with all skills in B.

Let us now ask, how many students share "*almost all* skills in B" ("*most* skills in B", or "*many* skills in B"). Additionally, we may need to classify students with respect to how many skills of B they *do not have* and exactly, to consider the following fuzzy sets of X: students sharing "*few* skills in B", students who do *not* have "*most* skills in B", or students "do *not* have *many* skills in B". Similarly, we can also consider a fuzzy set of Y formed of all skills shared by "almost all" ("most", "many", or "few") students of A, and the fuzzy set of Y made of all skills that are *not shared* by "most" (or "many") students in A. Each of the previous sets is generated by a fuzzy quantifier-based concept-forming operator, that allows us to introduce an extended notion of fuzzy concept.

Fuzzy quantifier-based operators are defined taking into account expressions of natural language *extremely big*, *very big*, and *not small* that are formalized within the *theory of evaluative linguistic expressions* [8]. Finally, starting from the Łukasiewicz MV-algebra, we employ the fuzzy quantifier-based operators to represent a graded decagon of opposition, which is a graded extension of *Aristotle's square* (see Fig. 1).

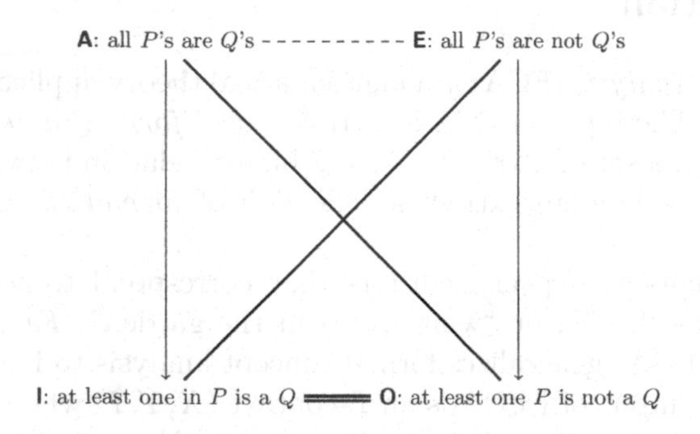

Fig. 1. Aristotle's square. The lines ▬, ═, ⟶, - - - denote that the corresponding propositions are contradictories, sub-contraries, sub-alterns, and contraries, respectively.

The article is organized as follows. Section 2 reviews some basic notions and results regarding MV-algebras, fuzzy formal concept analysis, and the graded square of opposition. Section 3 introduces the fuzzy quantifier-based operators and the corresponding new notions of fuzzy concepts. In Sect. 4, we construct a graded decagon of opposition using the former. Finally, in the last section we discuss further possible development of our results.

2 Preliminaries

This section describes some fundamental notions and results regarding MV-algebras, fuzzy formal concept analysis, and the graded square of opposition.

2.1 MV-Algebras

Definition 1. *A lattice* $\langle L, \vee, \wedge\rangle$ is complete *if and only if all subsets of* L *have both supremum and infimum.*

Definition 2. *A* residuated lattice *is an algebra* $\langle L, \vee, \wedge, \otimes, \rightarrow, \mathit{0}, \mathit{1}\rangle$ *where*

(i) $\langle L, \wedge, \vee, \mathit{0}, \mathit{1}\rangle$ *is a bounded lattice,*
(ii) $\langle L, \otimes, \mathit{1}\rangle$ *is a commutative monoid, and*
(iii) $a \otimes b \leq c$ *iff* $a \leq b \rightarrow c$, *for all* $a, b, c \in L$ *(adjunction property).*

Definition 3 ([3,11]). *An* MV-algebra *is a residuated lattice*

$$\mathcal{L} = \langle L, \vee, \wedge, \otimes, \rightarrow, \mathit{0}, \mathit{1}\rangle$$

where $a \vee b = (a \rightarrow b) \rightarrow b$, *for each* $a, b \in L$. *We will also work with the following additional operations on* L:

(i) $\neg a = a \rightarrow \mathit{0}$ *(negation),*
(ii) $a \oplus b = \neg(\neg a \otimes \neg b)$ *(strong disjunction),*
(iii) $a \leftrightarrow b = (a \rightarrow b) \wedge (b \rightarrow a)$ *(biresiduation).*

Example 1. A special MV-algebra is the *standard Lukasiewicz MV-algebra*

$$\mathcal{L}_{\text{L}} = \langle [0,1], \vee, \wedge, \otimes, \rightarrow, 0, 1\rangle$$

where $a \vee b = \max(a, b)$, $a \wedge b = \min(a, b)$, $a \otimes b = \max(0, a + b - 1)$ and $a \rightarrow b = \min(1, 1 - a + b)$, $\neg a = 1 - a$ and $a \oplus b = \min\{1, a + b\}$, for all $a, b \in L$.

In the following lemma, we list some properties of complete MV-algebras[1] that will be used below.

Lemma 1. *Let* $\mathcal{L} = \langle L, \vee, \wedge, \otimes, \rightarrow, \mathit{0}, \mathit{1}\rangle$ *be a complete MV-algebra. Then the following holds for all* $a, b, c, d, e \in L$:

(a) If $a \leq b$ *and* $c \leq d$, *then* $a \wedge c \leq b \wedge d$.
(b) Let I *be any index set. Then for each* $k \in I$, $\bigwedge_{i \in I} a_i \leq a_k$ *and* $a_k \leq \bigvee_{i \in I} a_i$.
(c) If $a_i \leq b_i$ *for each* $i \in I$, *then* $\bigvee_{i \in I} a_i \leq \bigvee_{i \in I} b_i$.
(d) $a \oplus \neg a = \mathit{1}$ *and* $a \otimes \neg a = \mathit{0}$.
(e) If $a \otimes b \leq e$, *then* $(a \wedge c) \otimes (b \wedge d) \leq e$.
(f) If $a \leq b$ *and* $c \leq d$, *then* $a \otimes c \leq b \otimes d$ *and* $a \oplus c \leq b \oplus d$.

[1] More generally, the properties (a), (b) and (c) hold in any complete lattice.

2.2 Fuzzy Formal Concept Analysis

In this subsection, we recall the definition of two pairs of fuzzy concept-forming operators $(^{\uparrow}, ^{\downarrow})$, and $(^{\cap}, ^{\cup})$ existing in literature. Given a complete residuated lattice L, by a fuzzy set of the universe X we mean a function $A : X \longrightarrow L$. If A is a fuzzy set on X, then we write $A \underset{\sim}{\subset} X$. For each $A, B \underset{\sim}{\subset} X$, we put $\mathcal{S}_X(A, B) = \bigwedge_{x \in X}(A(x) \rightarrow B(x))$, which represents the degree of inclusion of A in B[2].

Definition 4 ([1,12]). *Let (X, Y, I) be an L-context and $A \underset{\sim}{\subset} X$, $B \underset{\sim}{\subset} Y$. We put*

$$A^{\uparrow}(y) = \bigwedge_{x \in X}(A(x) \rightarrow I(x, y)) \quad \text{and} \quad B^{\downarrow}(x) = \bigwedge_{y \in Y}(B(y) \rightarrow I(x, y)),$$

for all $x \in X$ and $y \in Y$.

The $A^{\uparrow}(y)$ and $B^{\downarrow}(x)$ correspond to the truth degrees of the statements "an attribute y is shared by *all* objects of A" and "an object x has *all* attributes of B", respectively.

Definition 5 ([13]). *Let (X, Y, I) be an L-context. If $A \underset{\sim}{\subset} X$ and $B \underset{\sim}{\subset} Y$, then*

$$A^{\cap}(y) = \bigvee_{x \in X}(A(x) \otimes I(x, y)) \quad \text{and} \quad B^{\cup}(x) = \bigwedge_{y \in Y}(I(x, y) \rightarrow B(y)),$$

for all $x \in X$ and $y \in Y$.

The operators $^{\cap}$ and $^{\cup}$ are borrowed from the rough set theory. Namely, $A^{\cap}(y)$ and $B^{\cup}(x)$ correspond to the truth degrees of the statements "an attribute y is shared by at least one object of A" and "an object x has no attributes outside B", respectively.

Each pair $(A, B) \in L^X \times L^Y$ such that $A^{\uparrow} = B$ and $B^{\downarrow} = A$ is called *standard L-concept.* Analogously, each pair $(A, B) \in L^X \times L^Y$ such that $A^{\cap} = B$ and $B^{\cup} = A$ is called *property-oriented L-concept.*

Theorem 1. *The pair of mappings $^{\uparrow} : L^X \rightarrow L^Y$ and $^{\downarrow} : L^Y \rightarrow L^X$ forms an* antitone Galois connection *between X and Y, i.e. $\mathcal{S}_X(A, B^{\downarrow}) = \mathcal{S}_Y(B, A^{\uparrow})$, for each $A \underset{\sim}{\subset} X$ and $B \underset{\sim}{\subset} Y$.*

Theorem 2. *The pair of mappings $^{\cap} : L^X \rightarrow L^Y$ and $^{\cup} : L^Y \rightarrow L^X$ forms an* isotone Galois connection *between X and Y, i.e. $\mathcal{S}_X(A, B^{\cup}) = \mathcal{S}_Y(A^{\cap}, B)$, for each $A \underset{\sim}{\subset} X$ and $B \underset{\sim}{\subset} Y$.*

Definition 6. *Given a set X and a complete residuated lattice L, by a* fuzzy preposet *we mean a pair $(X, \mathcal{R})$ where $\mathcal{R}$ is a fuzzy relation on X that is reflexive, i.e. $\mathcal{R}(x, x) = 1$ for each $x \in X$, and $\otimes$-transitive, i.e. $\mathcal{R}(x, y) \otimes \mathcal{R}(y, z) \leq \mathcal{R}(x, z)$, for each $x, y, z \in X$.*

---

[2] Note that this formula is interpretation of the logical formula $(\forall x)(A(x) \Rightarrow B(x))$ defining classical inclusion between (fuzzy) sets in a model of fuzzy predicate logic.

2.3 Graded Square of Opposition and Fuzzy Concept-Forming Operators

In this subsection, we define graded square of opposition referring to [5], and we enunciate a theorem that shows how this square can be obtained using the fuzzy concept-forming operators introduced in Subsect. 2.2.

Definition 7. *Let P_A and P_B be properties represented by $A, B \underset{\sim}{\subseteq} X$, then we say that*

1. *P_A and P_B are* contraries *if and only if $A(x) \otimes B(x) = 0$ for each $x \in X$,*
2. *P_A and P_B are* sub-contraries *if and only if $A(x) \oplus B(x) = 1$ for each $x \in X$,*
3. *P_A and P_B are* sub-alterns *if and only if $A(x) \to B(x) = 1$ for each $x \in X$,*
4. *P_A and P_B are* contradictories *if and only if $A(x) = \neg B(x)$ for each $x \in X$.*

Definition 8. *In a* graded square of opposition *the vertices* **A**, **E**, **I**, *and* **O** *are fuzzy sets representing the propositions P_{A}, P_{E}, P_{I}, and P_{O} such that the following conditions hold:*

1. *P_{A} and P_{E} are* contraries;
2. *P_{I} and P_{O} are* sub-contraries;
3. *P_{A} and P_{I} are* sub-alterns, *as well as P_{E} and P_{O};*
4. *P_{A} and P_{O} are* contradictories, *as well as P_{E} and P_{I}.*

From now, given the L-contexts (X, Y, I), we suppose that L is the Łukasiewicz MV-algebra, because we will need the *double negation law*, i.e. $\neg\neg a = a$ for each $a \in L$. Moreover, we put $(\neg I)(x, y) = \neg I(x, y)$. In the standard Łukasiewicz algebra, $\neg I(x, y) = 1 - I(x, y)$, for all $x \in X$ and $y \in Y$.

This lemma follows from the results found in [5].

Lemma 2. *Let $A \underset{\sim}{\subseteq} X$ be a normal fuzzy set*[3]*, then*

1. *$A^{\uparrow}_I(y) \otimes A^{\uparrow}_{\neg I}(y) = 0$,*
2. *$A^{\cap}_I(y) \oplus A^{\cap}_{\neg I}(y) = 1$,*
3. *$A^{\uparrow}_I(y) \leq A^{\cap_I}(y)$, and $A^{\uparrow}_{\neg I}(y) \leq A^{\cap}_{\neg I}(y)$,*
4. *$\neg A^{\uparrow}_I(y) = A^{\cap}_{\neg I}(y)$, and $\neg A^{\cap}_I(y) = A^{\uparrow}_{\neg I}(y)$,*

for each $y \in Y$.

Theorem 3. *Let $A \underset{\sim}{\subseteq} X$. If A is normal, then $A^{\uparrow}_I$, $A^{\uparrow}_{\neg I}$, $A^{\cap}_I$ and $A^{\cap}_{\neg I}$ are the vertices of a graded square of opposition, and they represent proprieties that are in relation of contrary, sub-contrary, sub-altern, and contradictory as shown in Fig. 2.*

Observe that we obtain the graded square of opposition defined in [4] when fixing $y \in Y$.

Example 2. Let (X, Y, I) be an L-context, where $X = \{x_1, x_2, x_3, x_4\}$, $Y = \{y_1, y_2, y_3, y_4\}$, and $I(x_1, y_1) = 0.25$, $I(x_2, y_1) = 0.6$, $I(x_3, y_1) = 1$, $I(x_4, y_1) = 0.25$. The graded square of opposition associated to $A = \{x_1, 0.5/x_2, 0.6/x_3, 0.5/x_4\}$ and y_1 is depicted in Fig. 3.

[3] There exists $x \in X$ such that $A(x) = 1$.

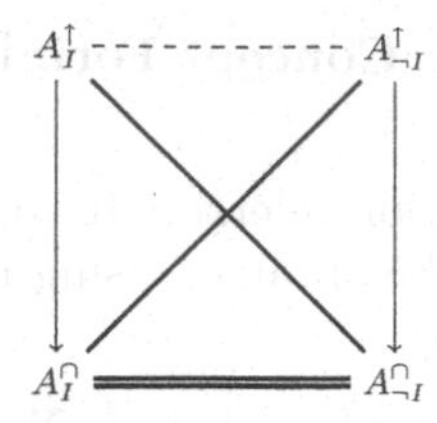

Fig. 2. Graded square of opposition

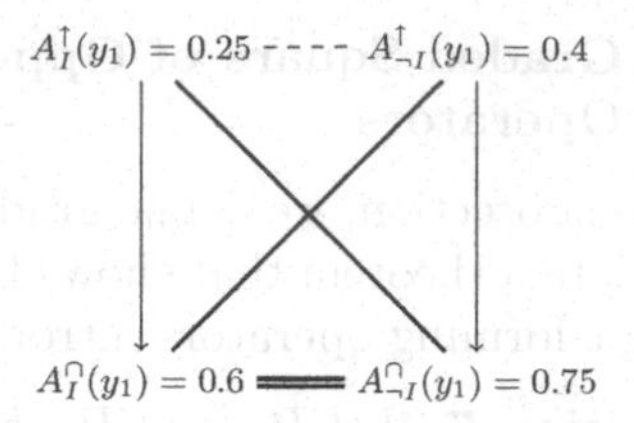

Fig. 3. Example of graded square of opposition

3 Fuzzy Quantifier-Based Operators

In this section, we introduce the *fuzzy quantifier based-operators* extending the notion of fuzzy concept. Our theory is based on the theory of *intermediate quantifiers* presented in [7,9] and elsewhere. The theory is based on the concept of *evaluative linguistic expression.* These are expressions of natural language such as "small, very big, rather medium", etc. In this paper we confine only to "not small", "very big" and "extremely big" and use a simplified model in which we consider only extensions in the (linguistic) context $\langle 0, 0.5, 1\rangle$[4] that are fuzzy sets $\mathsf{BiEx}, \mathsf{BiVe}, \neg\mathsf{Sm}\nu$ depicted in Fig. 4. For justification of this model, see [8,10].

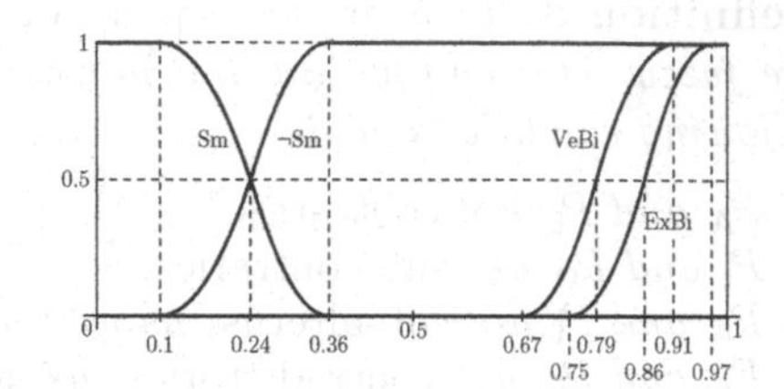

Fig. 4. Shapes of the fuzzy sets BiEx, BiVe, $\neg\mathsf{Sm}\nu$.

Remark 1. It is clear that $\mathsf{BiEx}(x) \leq \mathsf{BiVe}(x) \leq \neg\mathsf{Sm}\nu(x)$ holds for all $x \in [0,1]$.

The *cardinality* of $A \underset{\sim}{\subset} X$ is defined by $|A| = \sum_{x\in X} A(x)$. Furthermore, given $A, B \underset{\sim}{\subset} X$, we consider the following measure that expresses how large the size of A is w.r.t. the size of B (see [9])

$$\mu_B(A) = \begin{cases} 1 & \text{if } B = \emptyset \text{ or } A = B, \\ \frac{|A|}{|B|} & \text{if } B \neq \emptyset \text{ and } A \subseteq B, \\ 0 & \text{otherwise.} \end{cases}$$

For our further reasoning, we need a special operation called *cut of a fuzzy set.* It is motivated by the need to form a new fuzzy set from a given one by extracting several elements together with their membership degrees and putting the other membership degrees equal to 0.

[4] By a linguistic context for evaluative expressions, we understand a triple of numbers $\langle v_L, v_S, v_R\rangle$ that determines an interval $[v_L, v_S] \cup [v_S, v_R]$ in which all values range. For the more detailed explanation, see [10].

Definition 9 ([7]). *Let $A, B \underset{\sim}{\subseteq} X$. The* cut of A with respect to B *is the fuzzy set*

$$(A|B)(x) = \begin{cases} A(x) & \text{if } A(x) = B(x), \\ 0 & \text{otherwise.} \end{cases} \tag{1}$$

Now, we give the definition of positive and negative fuzzy quantifier based-operators that are based on the relation I, and on the functions $\neg$Smv, BiVe and BiEx. Our aim is to capture positive, or negative information in (X, Y, I).

Definition 10 (Fuzzy quantifier-based operators). *Let us consider an L-context (X, Y, I), $A \underset{\sim}{\subseteq} X$, $B \underset{\sim}{\subseteq} Y$, $x \in X$, and $y \in Y$. Let $Ev \in \{\neg\text{Smv}, \text{BiVe}, \text{BiEx}\}$. Then we put:*

(i) Positive fuzzy quantifier based-operators

$$A^{\uparrow}_{I,Ev}(y) = \bigvee_{Z \underset{\sim}{\subseteq} X} (\bigwedge_{x \in X} ((A|Z)(x) \to I(x, y)) \wedge Ev(\mu_A(A|Z))), \tag{2}$$

and

$$B^{\downarrow}_{I,Ev}(x) = \bigvee_{Z \underset{\sim}{\subseteq} Y} (\bigwedge_{y \in Y} ((B|Z)(y) \to I(x, y)) \wedge Ev(\mu_B(B|Z))), \tag{3}$$

(ii) Negative fuzzy quantifier based-operators

$$A^{\uparrow}_{\neg I,Ev}(y) = \bigvee_{Z \underset{\sim}{\subseteq} X} (\bigwedge_{x \in X} ((A|Z)(x) \to \neg I(x, y)) \wedge Ev(\mu_A(A|Z))), \tag{4}$$

and

$$B^{\downarrow}_{\neg I,Ev}(x) = \bigvee_{Z \underset{\sim}{\subseteq} Y} (\bigwedge_{y \in Y} ((B|Z)(y) \to \neg I(x, y)) \wedge Ev(\mu_B(B|Z))), \tag{5}$$

Informal explanation of the formulas in Definition 10 is the following:

(i) $A^{\uparrow}_{I,Ev}(y)$ is the truth degree to which *there exists a cut of A such that "all its objects have the attribute y" and "its size is Ev (not small, very big or extremely big) w.r.t. the size of A"*. Analogous statement holds for $B^{\downarrow}_{I,Ev}(y)$.
(ii) $A^{\uparrow}_{\neg I,Ev}(x)$ is the truth degree to which *there exists a cut of A such that "all its objects do not have the attribute y" and "its size is Ev (not small, very big or extremely big) w.r.t. the size of A"*. Analogous statement holds for and $B^{\downarrow}_{\neg I,Ev}(y)$.

Remark 2. (a) If $Z \underset{\sim}{\subseteq} X$ and $y \in Y$, then $\bigwedge_{x \in X}((A|Z)(x) \to I(x, y)) = (A|Z)^{\uparrow}_{I}(y)$ and $\bigwedge_{x \in X}((A|Z)(x) \to \neg I(x, y)) = (A|Z)^{\uparrow}_{\neg I}(y)$.
(b) If $Z \underset{\sim}{\subseteq} Y$ and $x \in X$, then $\bigwedge_{y \in Y}((B|Z)(y) \to I(x, y)) = (B|Z)^{\downarrow}_{I}(y)$ and $\bigwedge_{y \in Y}((B|Z)(y) \to \neg I(x, y)) = (B|Z)^{\downarrow}_{\neg I}(x)$.

Since BiEx, BiVe and ¬Smv lay behind the definition of the intermediate quantifiers *almost*, *most* and *many* (cf. [9]), formulas $A^{\uparrow}_{I,Ev}(y)$, $A^{\uparrow}_{\neg I,Ev}(y)$, $B^{\downarrow}_{I,Ev}(x)$ and $B^{\downarrow}_{\neg I,Ev}(x)$ can be understood as interpretation of the linguistic expressions summarized in Table 1.

Table 1. Verbal description of the fuzzy quantifier-based operators

Truth degree	Statement
$A^{\uparrow}_{I,\mathsf{BiEx}}(y)$	y is shared by *almost all* objects of A
$B^{\downarrow}_{I,\mathsf{BiEx}}(x)$	x has *almost all* attributes of B
$A^{\uparrow}_{I,\mathsf{BiVe}}(y)$	y is shared by *most* objects of A
$B^{\downarrow}_{I,\mathsf{BiVe}}(x)$	x has *most* attributes of B
$A^{\uparrow}_{I,\neg\mathsf{Smv}}(y)$	y is shared by *many* objects of A
$B^{\downarrow}_{I,\neg\mathsf{Smv}}(x)$	x has *many* attributes of B
$A^{\uparrow}_{\neg I,\mathsf{BiEx}}(y)$	y is shared by *few* objects of A
$B^{\downarrow}_{\neg I,\mathsf{BiEx}}(x)$	x has *few* attributes of B
$A^{\uparrow}_{\neg I,\mathsf{BiVe}}(y)$	y is not shared by *most* objects of A
$B^{\downarrow}_{\neg I,\mathsf{BiVe}}(x)$	*most* attributes of B are not satisfied by x
$A^{\uparrow}_{\neg I,\neg\mathsf{Smv}}(y)$	y is not shared by *many* objects of A
$B^{\downarrow}_{\neg I,\neg\mathsf{Smv}}(x)$	*many* attributes of B are not satisfied by x

In the sequel, new notions of fuzzy concepts are introduced considering additional information generated by the fuzzy-quantifier-based operators.

Definition 11. *Let* $Ev \in \{\neg\mathsf{Smv}, \mathsf{BiVe}, \mathsf{BiEx}\}$ *and* $H \in \{I, \neg I\}$. *For each* $A, \tilde{A} \underset{\sim}{\subset} X$, *and* $B, \tilde{B} \underset{\sim}{\subset} Y$, *we set*

(i) $A^{\Uparrow}_{H,Ev} = (A^{\uparrow}_{H}, A^{\uparrow}_{H,Ev})$ *and* $(B, \tilde{B})^{\Downarrow}_{H,Ev} = B^{\downarrow}_{H}$,
(ii) $(A, \tilde{A})^{\triangle}_{H,Ev} = A^{\uparrow}_{H}$ *and* $B^{\triangledown}_{H,Ev} = (B^{\downarrow}_{H}, B^{\downarrow}_{H,Ev})$.

Definition 12 (Extended fuzzy concepts). *Let* $Ev \in \{\neg\mathsf{Smv}, \mathsf{BiVe}, \mathsf{BiEx}\}$, $A, \tilde{A} \underset{\sim}{\subset} X$, *and* $B, \tilde{B} \underset{\sim}{\subset} Y$. *Then, we say that*

(i) $(A, (B, \tilde{B}))$ *is a* positive concept with *Ev*-attributes *if and only if* $A = (B, \tilde{B})^{\Downarrow}_{I,Ev}$ *and* $(B, \tilde{B}) = A^{\Uparrow}_{I,Ev}$.
(ii) $(A, (B, \tilde{B}))$ *is a* negative concept with *Ev*-attributes *if and only if* $A = (B, \tilde{B})^{\Downarrow}_{\neg I,Ev}$ *and* $(B, \tilde{B}) = A^{\Uparrow}_{\neg I,Ev}$.
(iii) $((A, \tilde{A}), B)$ *is a* positive concept with *Ev*-objects *if and only if* $(A, \tilde{A}) = B^{\triangledown}_{I,Ev}$ *and* $B = (A, \tilde{A})^{\triangle}_{I,Ev}$.
(iv) $((A, \tilde{A}), B)$ *is a* negative concept with *Ev*-objects *if and only if if* $(A, \tilde{A}) = B^{\triangledown}_{\neg I,Ev}$ *and* $B = (A, \tilde{A})^{\triangle}_{\neg I,Ev}$.

The following theorems state that the pairs of operators given by Definition 11 are both Galois connections between fuzzy preposets (see Definition 6). Given a set X, for each $A, B, C, D \subsim X$, we set

$$\mathcal{R}_X((A, B), (C, D)) = \mathcal{S}_X(C, A). \tag{6}$$

Theorem 4. *Let $Ev \in \{\neg\mathsf{Smv}, \mathsf{BiVe}, \mathsf{BiEx}\}$ and $H \in \{I, \neg I\}$. Then,*

(a) the pair of mappings ${}^{\Uparrow}_{H,Ev} : L^X \to L^Y \times L^Y$ and ${}^{\Downarrow}_{H,Ev} : L^Y \times L^Y \to L^X$ is a Galois connection between the fuzzy preposets $(L^X, \mathcal{S}_X)$ and $(L^Y \times L^Y, \mathcal{R}_Y)$, i.e. $\mathcal{S}_X(A, (B, \tilde{B})^{\Downarrow}_{H,Ev}) = \mathcal{R}_Y(A^{\Uparrow}_{H,Ev}, (B, \tilde{B}))$ for each $A \subsim X$ and $B, \tilde{B} \subsim Y$,

(b) the pair of mappings ${}^{\triangle}_{H,Ev} : L^Y \times L^Y \to L^X$ and ${}^{\triangledown}_{H,Ev} : L^X \to L^Y \times L^Y$ is a Galois connection between the fuzzy preposets $(L^Y \times L^Y, \mathcal{R}_Y)$ and $(L^X, \mathcal{S}_X)$, i.e. $\mathcal{R}_Y((A, \tilde{A}), B^{\triangledown}_{H,Ev}) = \mathcal{S}_X((A, \tilde{A})^{\triangle}_{H,Ev}, B)$ for each $A, \tilde{A} \subsim X$ and $B \subsim Y$.

Proof. We prove only item (a), because item (b) can be proved analogously.

Let $A \subsim X$, and $B, \tilde{B} \subsim Y$. By Definition 11(i), $\mathcal{S}_X(A, (B, \tilde{B})^{\Downarrow}_{H,Ev}) = \mathcal{S}_X(A, B^{\downarrow}_H)$. Moreover, by Theorem 1, we know that $\mathcal{S}_X(A, B^{\downarrow}_H) = \bigwedge_{x \in X}(A(x) \to B^{\downarrow}_H(x))$ is equal to $\mathcal{S}_Y(B, A^{\uparrow}_H) = \bigwedge_{y \in Y}(B(y) \to A^{\uparrow}_H(y))$. Eventually, by (6), $\mathcal{S}_Y(B, A^{\uparrow}_H) = \mathcal{R}_Y(A^{\Uparrow}_{H,Ev}, (B, \tilde{B}))$. Then, we conclude that $\mathcal{S}_X(A, (B, \tilde{B})^{\Downarrow}_{H,Ev}) = \mathcal{R}_Y(A^{\Uparrow}_{H,Ev}, (B, \tilde{B}))$. □

4 Graded Decagon of Opposition with Fuzzy Quantifier-Based Operators

In this section, we introduce the definition of graded decagon of opposition, which is a generalization of the graded square of opposition given in Definition 8. Moreover, we construct a graded decagon of opposition using some fuzzy quantifier-based operators.

Definition 13 (Graded decagon of opposition). *A graded decagon of opposition consists of vertices $A_1, \ldots, A_5 \subsim X$, and $N_1, \ldots, N_5 \subsim X$ representing the propositions $P_{A_1}, \ldots, P_{A_5}$, $P_{N_1}, \ldots, P_{N_5}$ such that:*

1. *P_{A_i} and P_{N_j} are* contraries, *for each $i, j \in \{1 \ldots, 4\}$,*
2. *P_{A_5} and P_{N_5} are* sub-contraries,
3. *P_{A_i} and $P_{A_{i+1}}$ are* sub-alterns, *as well as P_{N_i} and $P_{N_{i+1}}$, for each $i \in \{1, \ldots, 4\}$,*
4. *P_{A_1} and P_{N_5} are* contradictories, *as well as P_{A_5} and P_{N_1}.*

The graded decagon of opposition is depicted in Fig. 5.

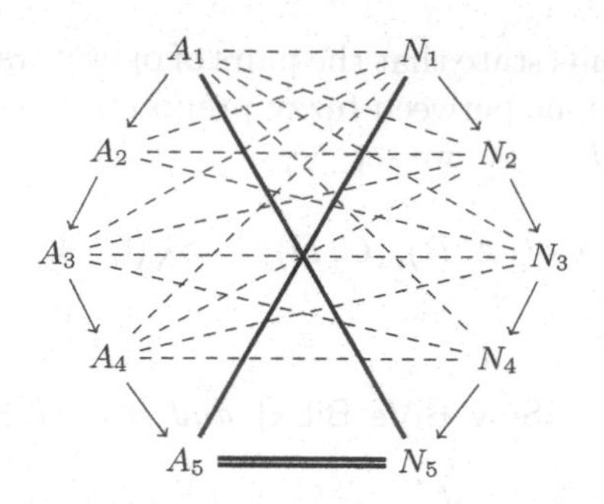

Fig. 5. Graded decagon of opposition

In the sequel, we prove a few lemmas in order to construct a graded decagon of opposition with the fuzzy quantifier-based operators.

Lemma 3. *For each $A \underset{\sim}{\subset} X$ and $y \in Y$, the following properties hold:*

(a) $A_I^{\uparrow}(y) \leq A_{I,\mathsf{BiEx}}^{\uparrow}(y) \leq A_{I,\mathsf{BiVe}}^{\uparrow}(y) \leq A_{I,\neg\mathsf{Smv}}^{\uparrow}(y)$,
(b) $A_{\neg I}^{\uparrow}(y) \leq A_{\neg I,\mathsf{BiEx}}^{\uparrow}(y) \leq A_{\neg I,\mathsf{BiVe}}^{\uparrow}(y) \leq A_{\neg I,\neg\mathsf{Smv}}^{\uparrow}(y)$.

Proof. We give the proof of item (a) only. The proof of item (b) is analogous.

Let $Ev \in \{\neg\mathsf{Smv}, \mathsf{BiVe}, \mathsf{BiEx}\}$. Trivially, $A_I^{\uparrow}(y) = (A|A)_I^{\uparrow}(y) \wedge Ev(\mu_A(A|A))$. By Lemma 1(b),

$$(A|A)_I^{\uparrow}(y) \wedge Ev(\mu_A(A|A)) \leq \bigvee_{Z \underset{\sim}{\subset} X} ((A|Z)_I^{\uparrow}(y) \wedge Ev(\mu_A(A|Z))),$$

namely $A_I^{\uparrow}(y) \leq A_{I,Ev}^{\uparrow}(y)$. By Remark 1, for each $Z \underset{\sim}{\subset} X$,

$$\mathsf{BiEx}(\mu_A(A|Z)) \leq \mathsf{BiVe}(\mu_A(A|Z)) \leq \neg\mathsf{Smv}(\mu_A(A|Z)).$$

Consequently, by Lemma 1(a),

$$(A|Z)_I^{\uparrow}(y) \wedge \mathsf{BiEx}(\mu_A(A|Z)) \leq (A|Z)_I^{\uparrow}(y) \wedge \mathsf{BiVe}(\mu_A(A|Z)) \leq (A|Z)_I^{\uparrow}(y) \wedge \neg\mathsf{Smv}(\mu_A(A|Z)).$$

Finally, by Lemma 1(c), $A_{I,\mathsf{BiEx}}^{\uparrow}(y) \leq A_{I,\mathsf{BiVe}}^{\uparrow}(y) \leq A_{I,\neg\mathsf{Smv}}^{\uparrow}(y)$. □

In some relations, it is necessary to add the assumption that the fuzzy set in concern is non-empty. In classical logic, we add the formula $(\exists x)A(x)$ that assures us that "there exists at least one element x" and speak about *existential import* (or presupposition). In fuzzy logic, the quantifier $\exists$ is interpreted by supremum. This leads us to the following definition.

Definition 14. *Let $A \underset{\sim}{\subset} X$, $y \in Y$, $Ev \in \{\neg\mathsf{Smv}, \mathsf{BiVe}, \mathsf{BiEx}\}$, and $H \in \{I, \neg I\}$. Then the following formulas have existential import:*
(i) $(A^{\uparrow}_{H}(y))^* = \bigwedge_{x\in X}(A(x) \to H(x,y)) \otimes \bigvee_{x\in X} A(x)$,
(ii) $(A^{\uparrow}_{H,Ev}(y))^* = \bigvee_{Z\underset{\sim}{\subset} X}[(\bigwedge_{x\in X}((A|Z)(x) \to H(x,y)) \wedge Ev(\mu_A(A|Z))) \otimes \bigvee_{x\in X}(A|Z)(x)]$.

The existential import is used in the following lemmas.

Lemma 4. *Let $A \underset{\sim}{\subset} X$, $y \in Y$, $Ev \in \{\neg\mathsf{Smv}, \mathsf{BiVe}, \mathsf{BiEx}\}$, and $H \in \{I, \neg I\}$. Then,*

$$(A^{\uparrow}_{H,Ev}(y))^* \leq \left(\bigvee_{x\in X} A(x)\right) \to A^{\cap}_{H}(y).$$

Proof. By Lemma 1(b), the following inequality holds: for each $Z \underset{\sim}{\subset} X$ and $x \in X$

$$(A|Z)^{\uparrow}_{H}(y) \leq (A|Z)(x) \to H(x,y).$$

By the adjunction property, $(A|Z)^{\uparrow}_{H}(y) \otimes (A|Z)(x) \leq H(x,y)$. By Lemma 1(f),

$$(A|Z)^{\uparrow}_{H}(y) \otimes (A|Z)(x) \otimes A(x) \leq A(x) \otimes H(x,y).$$

Hence,

$$(A|Z)^{\uparrow}_{H}(y) \otimes \bigvee_{x\in X}(A|Z)(x) \otimes \bigvee_{x\in X} A(x) \leq \bigvee_{x\in X} A(x) \otimes H(x,y).$$

By Lemma 1(e),

$$((A|Z)^{\uparrow}_{H}(y) \wedge Ev(\mu_A(A|Z))) \otimes \bigvee_{x\in X}(A|Z)(x) \otimes \bigvee_{x\in X} A(x) \leq \bigvee_{x\in X} A(x) \otimes H(x,y).$$

Using the adjunction property, we conclude that $(A^{\uparrow}_{H,Ev}(y))^* \leq (\bigvee_{x\in X} A(x)) \to A^{\cap}_{H}(y)$. □

Lemma 5. *Let $A \underset{\sim}{\subset} X$, $y \in Y$, and $Ev_1, Ev_2 \in \{\neg\mathsf{Smv}, \mathsf{BiVe}, \mathsf{BiEx}\}$. Then,*

$$(A^{\uparrow}_{I,Ev_1}(y))^* \otimes (A^{\uparrow}_{\neg I,Ev_2}(y))^* = 0.$$

Proof. Let $y \in Y$, $x \in X$ and $Z_1, Z_2 \underset{\sim}{\subset} X$. By Definition 4, and by Lemma 1(b),

$$(A|Z_1)^{\uparrow}_{I}(y) \leq (A|Z_1)(x) \to I(x,y), \text{ and } (A|Z_2)^{\uparrow}_{\neg I}(y) \leq (A|Z_2)(x) \to \neg I(x,y).$$

Then, by the adjunction property,

$$(A|Z_1)^{\uparrow}_{I}(y) \otimes (A|Z_1)(x) \leq I(x,y), \text{ and } (A|Z_2)^{\uparrow}_{\neg I}(y) \otimes (A|Z_2)(x) \leq \neg I(x,y).$$

By Lemma 1(e), $((A|Z_1)^{\uparrow}_{I}(y) \wedge Ev_1(\mu_A(A|Z_1))) \otimes (A|Z_1)(x) \leq I(x,y)$, and $((A|Z_2)^{\uparrow}_{\neg I}(y) \wedge Ev_2(\mu_A(A|Z_2))) \otimes (A|Z_2)(x) \leq \neg I(x,y)$.

By Lemma 1(d), (f),

$$((A|Z_1)^{\uparrow}_I(y) \wedge Ev_1(\mu_A(A|Z_1))) \otimes (A|Z_1)(x) \otimes ((A|Z)^{\uparrow}_{\neg I}(y) \wedge Ev_2(\mu_A(A|Z))) \otimes (A|Z_1)(x) = 0,$$

Finally,

$$\bigvee_{Z_1 \underset{\sim}{\subseteq} X} \left((A|Z_1)^{\uparrow}_I(y) \wedge Ev_1(\mu_A(A|Z_1)) \otimes \bigvee_{x \in X} (A|Z_1)(x) \right) \otimes$$

$$\bigvee_{Z_2 \underset{\sim}{\subseteq} X} \left(((A|Z_2)^{\uparrow}_{\neg I}(y) \wedge Ev_2(\mu_A(A|Z_2))) \otimes \bigvee_{x \in X} (A|Z_2)(x) \right) = 0,$$

and hence, $(A^{\uparrow}_{I,Ev_1}(y))^* \otimes (A^{\uparrow}_{\neg I,Ev_2}(y))^* = 0$. □

Lemma 6. *Let $A \underset{\sim}{\subseteq} X$, $y \in Y$, and $Ev \in \{\neg\mathsf{Smv}, \mathsf{BiVe}, \mathsf{BiEx}\}$. Then,*

$$(A^{\uparrow}_{I,Ev}(y))^* \otimes (A^{\uparrow}_{\neg I}(y))^* = 0 \quad and \quad (A^{\uparrow}_I(y))^* \otimes (A^{\uparrow}_{\neg I,Ev}(y))^* = 0.$$

Proof. The proof is similar to that of Lemma 5. □

The following theorem shows that we can obtain a decagon of oppositions starting from our operators.

Theorem 5. *Let (X, Y, I) be an L-context, where L is the standard Łukasiewicz MV-algebra, and let $A \underset{\sim}{\subseteq} X$. If A is normal, then*

$$A^{\uparrow}_I, A^{\uparrow}_{I,\mathsf{BiEx}}, A^{\uparrow}_{I,\mathsf{BiVe}}, A^{\uparrow}_{I,\neg\mathsf{Smv}}, A^{\cap}_I, A^{\uparrow}_{\neg I}, A^{\uparrow}_{\neg I,\mathsf{BiEx}}, A^{\uparrow}_{\neg I,\mathsf{BiVi}}, A^{\uparrow}_{\neg I,\neg\mathsf{Smv}}, A^{\cap}_{\neg I}$$

are the vertices of a graded decagon of opposition, and they represent proprieties that are in relation of contrary, sub-contrary, sub-altern, and contradictory as shown in Fig. 6.

Proof. The proof follows by Theorem 3, Lemma 3, Lemma 4, Lemma 5, and Lemma 6. □

Example 3. Let (X, Y, I) be an L-context, where $X = \{x_1, \ldots, x_{24}\}$, $Y = \{y_1, \ldots, y_{10}\}$, and the L-relation I between the objects of X and the attribute y_1 of Y is defined by Table 2. Let us fix the context $\langle 0, 0.5, 1\rangle$. Then the functions $\neg\mathsf{Smv} : [0,1] \to [0,1]$, $\mathsf{BiVe} : [0,1] \to [0,1]$, and $\mathsf{BiEx} : [0,1] \to [0,1]$ are defined in [10] (cf. also Fig. 4). Furthermore, put

$$A = \{{}^{1}/x_1, \ldots, {}^{1}/x_7, {}^{0.6}/x_8, {}^{0.93}/x_9, {}^{0.5}/x_{10}, {}^{1}/x_{11}, {}^{0.7}/x_{12}, {}^{0.98}/x_{13}, {}^{1}/x_{14}, \ldots, {}^{1}/x_{16}, {}^{0.8}/x_{17}, {}^{1}/x_{18}, \ldots, {}^{1}/x_{20}, {}^{0.5}/x_{21}, {}^{1}/x_{22}, {}^{1}/x_{23}, {}^{0.66}/x_{24}, {}^{1}/x_{25}, {}^{1}/x_{26}\}.$$

Then we obtain the graded decagon of opposition depicted in Fig. 7.

Table 2. The fuzzy relation I between the objects of X and attribute y_1.

I	x_1	x_2	x_3	x_4	x_5	x_6	x_7	x_8	x_9	x_{10}	x_{11}	x_{12}
y_1	0.5	0.15	0.31	0.5	0.66	0.5	0.5	0	0.73	0	0.5	0.8

I	x_{13}	x_{14}	x_{15}	x_{16}	x_{17}	x_{18}	x_{19}	x_{20}	x_{21}	x_{22}	x_{23}	x_{24}	x_{25}	x_{26}
y_1	0.98	0.25	0.5	0.5	0.27	0.5	0.5	0.6	0	0.37	0.5	0.02	0.5	0.6

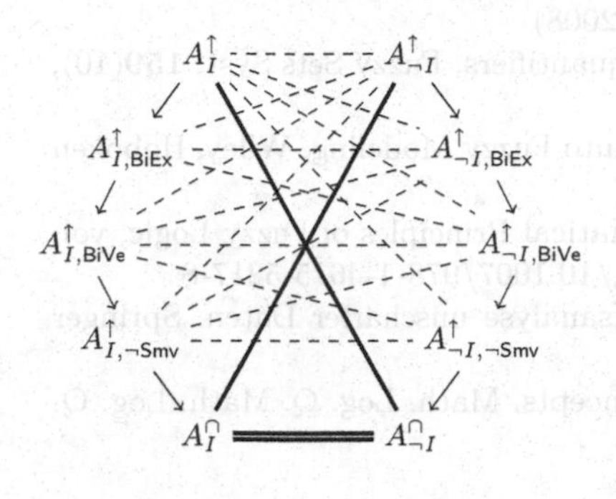

Fig. 6. Graded decagon of opposition

$A^{\uparrow}_{I}(y_1) = 0.15$ · $A^{\uparrow}_{\neg I}(y_1) = 0.04$
$A^{\uparrow}_{I,\mathsf{BiEx}}(y_1) = 0.31$ $A^{\uparrow}_{\neg I,\mathsf{BiEx}}(y_1) = 0.34$
$A^{\uparrow}_{I,\mathsf{BiVe}}(y_1) = 0.37$ $A^{\uparrow}_{\neg I,\mathsf{BiVe}}(y_1) = 0.4$
$A^{\uparrow}_{I,\neg\mathsf{Smv}}(y_1) = 0.5$ $A^{\uparrow}_{\neg I,\neg\mathsf{Smv}}(y_1) = 0.5$
$A^{\cap}_{I}(y_1) = 0.96$ $A^{\cap}_{\neg I}(y_1) = 0.85$

Fig. 7. Example of graded decagon of opposition

5 Future Directions

In this article, a graded decagon of opposition is introduced as a graded generalization of Aristotle's square, and it is constructed using some fuzzy quantifier-based operators. As future work, we intend to analyze more deeply the role that the fuzzy quantifier-based operators could have in fuzzy formal concept analysis. Moreover, fixed an evaluative linguistic expression Ev_1, we will find another evaluative linguistic expression Ev_2 such that the pair of operators $^{\uparrow}_{I,Ev_1}$ and $^{\downarrow}_{I,Ev_2}$ forms a Galois connection. Finally, we would like to propose our operators as fuzzy generalizations of the scaling quantifiers used in Relational concept analysis [2].

References

1. Belohlavek, R.: Fuzzy Relational Systems: Foundations and Principles, vol. 20. Springer, New York (2012). https://doi.org/10.1007/978-1-4615-0633-1
2. Braud, A., Dolques, X., Huchard, M., Le Ber, F.: Generalization effect of quantifiers in a classification based on relational concept analysis. Knowl. Based Syst. **160**, 119–135 (2018)
3. Cignoli, R.L., d'Ottaviano, I.M., Mundici, D.: Algebraic Foundations of Many-Valued Reasoning, vol. 7. Springer, Dordrecht (2013). https://doi.org/10.1007/978-94-015-9480-6

4. Ciucci, D., Dubois, D., Prade, H.: Structures of opposition induced by relations. Ann. Math. Artif. Intell. 76(3–4) 351–373 (2015). https://doi.org/10.1007/s10472-015-9480-8
5. Dubois, D., Prade, H., Rico, A.: Graded cubes of opposition and possibility theory with fuzzy events. Int. J. Approx. Reason. **84**, 168–185 (2017)
6. Ganter, B., Wille, R.: Formal Concept Analysis: Mathematical Foundations. Springer, Heidelberg (2012). https://doi.org/10.1007/978-3-642-59830-2
7. Murinová, P., Novák, V.: The theory of intermediate quantifiers in fuzzy natural logic revisited and the model of "many". Fuzzy Sets and Systems (2020, to appear)
8. Novák, V.: A comprehensive theory of trichotomous evaluative linguistic expressions. Fuzzy Sets Syst. **159**(22), 2939–2969 (2008)
9. Novák, V.: A formal theory of intermediate quantifiers. Fuzzy Sets Syst. **159**(10), 1229–1246 (2008)
10. Novák, V., Perfilieva, I., Dvorak, A.: Insight into Fuzzy Modeling. Wiley, Hoboken (2016)
11. Novák, V., Perfilieva, I., Mockor, J.: Mathematical Principles of Fuzzy Logic, vol. 517. Springer, Boston (2012). https://doi.org/10.1007/978-1-4615-5217-8
12. Pollandt, S.: Fuzzy-Begriffe: Formale Begriffsanalyse unscharfer Daten. Springer (2013)
13. Popescu, A.: A general approach to fuzzy concepts. Math. Log. Q. Math. Log. Q. **50**(3), 265–280 (2004)

Graded Cube of Opposition with Intermediate Quantifiers in Fuzzy Natural Logic

Petra Murinová(✉) and Vilém Novák

Institute for Research and Applications of Fuzzy Modeling,
University of Ostrava, 30. dubna 22, 701 03 Ostrava 1, Czech Republic
{petra.murinova,vilem.novak}@osu.cz
http://irafm.osu.cz/

Abstract. In our previous papers, we formally analyzed the generalized Aristotle's square of opposition using tools of higher-order fuzzy logic. Namely, we introduced general definitions of selected intermediate quantifiers, constructed a generalized square of opposition consisting of them and syntactically analyzed the emerged properties. The main objective of this paper is to extend the graded Peterson's square of opposition into the graded cube of opposition with intermediate quantifiers.

Keywords: Intermediate quantifiers · Fuzzy natural logic · Evaluative linguistic expressions · Generalized peterson square · Graded cube of opposition

1 Introduction

This paper continues the work on intermediate quantifiers. *Fuzzy natural logic* (FNL) is a formal mathematical theory that consists of three theories: (1) a formal theory of evaluative linguistic expressions explained in detail in [1], (2) a formal theory of fuzzy **IF-THEN** rules and approximate reasoning presented in [2,3], and (3) a formal theory of intermediate and generalized fuzzy quantifiers presented in [4–7]. This paper is a contribution to the latter.

Intermediate quantifiers are special linguistic expressions, for example, *almost all, a few, many, a large part of*, etc. which were introduced and deeply studied by Thompson in [8] and later by Peterson in his book in [9]. Peterson introduced a square of opposition as a generalization of the Aristotle's one [10–13]. It consists of five basic intermediate quantifiers.

Formalization of Peterson's square was introduced by Murinová and Novák in [14,15]. The main objective of this paper is to extend this approach to a graded **5**-*cube of opposition* and prove its forming properties.

The work was supported from ERDF/ESF by the project "Centre for the development of Artificial Intelligence Methods for the Automotive Industry of the region" No. CZ.02.1.01/0.0/0.0/17-049/0008414 and partially also by the MŠMT project NPU II project LQ1602 "IT4Innovations excellence in science".

M.-J. Lesot et al. (Eds.): IPMU 2020, CCIS 1239, pp. 145–158, 2020.
https://doi.org/10.1007/978-3-030-50153-2_11

Note that the idea to extend the square of opposition to a cube was already studied by Dubois in [16–18]. The authors introduced a graded Aristotle's square of opposition extended to a cube that associates the traditional square of opposition with a dual one. Ciucci, Dubois, and Prade [19,20] then introduced an application of the graded cube within the possibility theory.

2 Preliminaries

In this section, we will remind the main concepts and properties of the fuzzy type theory (higher-order fuzzy logic) and the theory of evaluative linguistic expressions. The reader can find details in several papers [1,14,21].

2.1 Fuzzy Type Theory

The formal theory of intermediate quantifiers is developed within Łukasiewicz fuzzy type theory (Ł-FTT). The algebra of truth values is a linearly ordered MV_{Δ}-algebra extended by the delta operation (see [22,23]). A special case is the standard Łukasiewicz MV_{Δ}-algebra.

$$\mathcal{L} = \langle [0,1], \vee, \wedge, \otimes, \rightarrow, 0, 1, \Delta \rangle \tag{1}$$

where

$$\wedge = \text{minimum}, \qquad \vee = \text{maximum},$$

$$a \otimes b = 0 \vee (a + b - 1), \qquad a \rightarrow b = 1 \wedge (1 - a + b),$$

$$\neg a = a \rightarrow 0 = 1 - a, \qquad \Delta(a) = \begin{cases} 1 & \text{if } a = 1, \\ 0 & \text{otherwise.} \end{cases}$$

The basic syntactical objects of Ł-FTT are classical (cf. [24]), namely the concepts of type and formula. The atomic types are ϵ (elements) and o (truth values). General types are denoted by Greek letters $\alpha, \beta, \ldots$. We will omit the type whenever it is clear from the context. A set of all types is denoted by *Types*. The (meta-)symbol ":=" used below means "is defined by".

The *language* consists of variables $x_\alpha, \ldots$, special constants $c_\alpha, \ldots$ ($\alpha \in$ *Types*), symbol λ, and parentheses. The connectives (which are special constants) are *fuzzy equality/equivalence* $\equiv$, *conjunction* $\wedge$, *implication* $\Rightarrow$, *negation* $\neg$, *strong conjunction* $\&$, *strong disjunction* ∇, *disjunction* $\vee$, and *delta* Δ.

Formulas are formed of variables, constants (each of specific type), and the symbol λ. Each formula A is assigned a type and we write it as A_α.[1] A set of formulas of type α is denoted by $Form_\alpha$. The set of all formulas is $Form = \bigcup_{\alpha \in Types} Form_\alpha$.

[1] Each formula has a unique type assigned to it. Hence, if α, β are different types then A_α and A_β are different formulas. To increase clarity of explanation, however, we will usually denote different formulas by different letters.

A *model* is $\mathcal{M} = \{(M_\alpha, \doteq_\alpha) \mid \alpha \in Types\}$ where $\doteq_\alpha$ is a fuzzy equality on a set M_α. If $\mathcal{M}$ is a model then $\mathcal{M}(A_o) \in M_o$ is a truth value, $\mathcal{M}(A_\epsilon) \in M_\epsilon$ is some element and $\mathcal{M}(A_{\beta\alpha}) : M_\alpha \to M_\beta$ is a function. For example, $\mathcal{M}(A_{o\alpha}) : M_\alpha \to M_o$ is a fuzzy set and $\mathcal{M}(A_{(o\alpha)\alpha}) : M_\alpha \times M_\alpha \to M_o$ a fuzzy relation. A formula A_o is *true* in T, $T \models A_o$, if it is true in the degree 1 in all models of T.

The fuzzy type theory is *complete*, i.e., a theory T is consistent iff it has a (Henkin) model. We sometimes apply its equivalent version: $T \vdash A_o$ iff $T \models A_o$.

The following special formulas play a role in our theory below:

$$\Upsilon_{oo} \equiv \lambda z_o \cdot \neg\Delta(\neg z_o), \qquad \text{(nonzero truth value)}$$

$$\hat{\Upsilon}_{oo} \equiv \lambda z_o \cdot \neg\Delta(z_o \vee \neg z_o). \qquad \text{(general truth value)}$$

Thus, $\mathcal{M}(\Upsilon(A_o)) = 1$ iff $\mathcal{M}(A_o) > 0$, and $\mathcal{M}(\hat{\Upsilon}(A_o)) = 1$ iff $\mathcal{M}(A_o) \in (0,1)$ holds in any model $\mathcal{M}$.

2.2 Theory of Evaluative Linguistic Expressions

Evaluative linguistic expressions are expressions of a natural language such as *small, medium, big, very short, more or less deep, quite roughly strong, extremely high*, etc. Their theory is the basic constituent of the fuzzy natural logic.

The semantics of evaluative linguistic expressions is formulated in a special formal theory T^{Ev} of Ł-FTT that was introduced in [1] and less formally explained in [25] where also formulas for direct computation are provided.

The evaluative expressions are construed by special formulas $Sm \in Form_{oo(oo)}$ (*small*), $Me \in Form_{oo(oo)}$ (*medium*), $Bi \in Form_{oo(oo)}$ (*big*), and $Ze \in Form_{oo(oo)}$ (*zero*) that can be extended by several selected linguistic hedges. Recall that a *hedge*, i.e., usually (but not necessarily) an adverb such as "very, significantly, about, roughly", etc. is in general construed by a formula $\boldsymbol{\nu} \in Form_{oo}$ with specific properties. To classify that a given formula is a hedge, we introduced a formula $Hedge \in Form_{o(oo)}$. Then $T^{\text{Ev}} \vdash Hedge\ \boldsymbol{\nu}$ means that $\boldsymbol{\nu}$ is a hedge. We refer the reader to [1] for the technical details. We assume that the following is provable: $T^{\text{Ev}} \vdash Hedge\ \boldsymbol{\nu}$ for all $\boldsymbol{\nu} \in \{Ex, Si, Ve, ML, Ro, QR, VR\}$.

2.3 Theory of Intermediate Quantifiers

The theory of intermediate quantifiers is a special formal theory $T^{\text{IQ}}[S]$ of Ł-FTT extending T^{Ev}. A detailed structure of $T^{\text{IQ}}[S]$ and precise definitions can be found in [5,6,14].

As discussed in the Introduction, the semantics of the intermediate quantifiers requires the idea of a "size" of a (fuzzy) set that can be characterized by the concept of a measure.

Definition 1. *Let $R \in Form_{o(o\alpha)(o\alpha)}$ be a formula*[2].

(i) A formula $\mu \in Form_{o(o\alpha)(o\alpha)}$ defined by

$$\mu_{o(o\alpha)(o\alpha)} := \lambda z_{o\alpha}\, \lambda x_{o\alpha}\, (R z_{o\alpha}) x_{o\alpha} \tag{2}$$

represents a measure on fuzzy sets *in the universe of type $\alpha \in Types$ if it has the following properties:*

(M1) $\boldsymbol{\Delta}(x_{o\alpha} \subseteq z_{o\alpha}) \,\&\, \boldsymbol{\Delta}(y_{o\alpha} \subseteq z_{o\alpha}) \,\&\, \boldsymbol{\Delta}(x_{o\alpha} \subseteq y_{o\alpha}) \Rightarrow ((\mu z_{o\alpha}) x_{o\alpha} \Rightarrow (\mu z_{o\alpha}) y_{o\alpha})$,

(M2) $\boldsymbol{\Delta}(x_{o\alpha} \subseteq z_{o\alpha}) \Rightarrow ((\mu z_{o\alpha})(z_{o\alpha} \setminus x_{o\alpha}) \equiv \neg(\mu z_{o\alpha}) x_{o\alpha})$,

(M3) $\boldsymbol{\Delta}(x_{o\alpha} \subseteq y_{o\alpha}) \,\&\, \boldsymbol{\Delta}(x_{o\alpha} \subseteq z_{o\alpha}) \,\&\, \boldsymbol{\Delta}(y_{o\alpha} \subseteq z_{o\alpha}) \Rightarrow ((\mu z_{o\alpha}) x_{o\alpha} \Rightarrow (\mu y_{o\alpha}) x_{o\alpha})$.

(ii) The following formula characterizes measurable fuzzy sets *of a given type α:*

$$\mathbf{M}_{o(o\alpha)} := \lambda z_{o\alpha} \cdot \boldsymbol{\Delta}\neg(z_{o\alpha} \equiv \emptyset_{o\alpha}) \,\&\, \boldsymbol{\Delta}(\mu z_{o\alpha}) z_{o\alpha} \,\&\, (\forall x_{o\alpha})(\forall y_{o\alpha})\boldsymbol{\Delta}((M1) \,\&\, (M3)) \,\&\, (\forall x_{o\alpha})\boldsymbol{\Delta}(M2) \tag{3}$$

where, for the simplicity of expression, we write (M1)–(M3) to stand for the axioms from (i).

Definition 2. *Let $\mathcal{S} \subseteq Types$ be a selected set of types, $P = \{R \in Form_{o(o\alpha)(o\alpha)} \mid \alpha \in \mathcal{S}\}$ be a set of new constants. Let T be a consistent extension of the theory T^{Ev} in the language $J(T) \supseteq J^{Ev} \cup P$. We say that the theory T contains* intermediate quantifiers *w.r.t. the set of types $\mathcal{S}$ if for all $\alpha \in \mathcal{S}$ the following is provable:*

(i)

$$T \vdash (\exists z_{o\alpha}) \mathbf{M}_{o(o\alpha)} z_{o\alpha}. \tag{4}$$

(ii)

$$T \vdash (\forall z_{o\alpha})(\exists x_{o\alpha})(\mathbf{M}_{o(o\alpha)} z_{o\alpha} \Rightarrow (\boldsymbol{\Delta}(x_{o\alpha} \subseteq z_{o\alpha}) \,\&\, \hat{\Upsilon}((\mu z_{o\alpha}) x_{o\alpha})). \tag{5}$$

Formula (5) assures the existence of fuzzy sets in each measurable fuzzy set that have non-trivial measure. Obviously, formulas (4) and (5) can be also introduced as special axioms of T. In the sequel, we will denote a theory that contains intermediate quantifiers due to Definition 2 by T^{IQ}.

For the definition of intermediate quantifiers, we need to define a special operation called *cut of a fuzzy set*, which will be formally defined as follows: Let $y, z \in Form_{o\alpha}$. The cut of y by z is the fuzzy set

$$y|z \equiv \lambda x_\alpha \cdot zx \,\&\, \boldsymbol{\Delta}(\Upsilon(zx) \Rightarrow (yx \equiv zx)).$$

The following lemma can be proved.

[2] This formula can be understood as a procedure providing computation of the output (a value in L) on the basis of a given input—two fuzzy sets. Formula (2) says that the measure is a function.

Lemma 1 ([15]). *Let $\mathcal{M}$ be a model and p an assignment such that $B = \mathcal{M}_p(y) \subsetapprox M_\alpha$, $Z = \mathcal{M}_p(z) \subsetapprox M_\alpha$. Then for any $m \in M_\alpha$*

$$\mathcal{M}_p(y|z)(m) = (B|Z)(m) = \begin{cases} B(m), & \text{if } B(m) = Z(m), \\ 0 & \text{otherwise.} \end{cases}$$

One can see that the operation $B|Z$ takes only those elements $m \in M_\alpha$ from the fuzzy set B whose membership $B(m)$ is equal to $Z(m)$, otherwise $(B|Z)(m) = 0$. If there is no such an element, then $B|Z = \emptyset$. We can thus use various fuzzy sets Z to "picking up proper elements" from B.

The following lemma will play a significant role in proofs of properties of the graded cube of opposition.

Lemma 2. *Let $\mathcal{M}$ be a model and p an assignment such that $B = \mathcal{M}_p(y) \subsetapprox M_\alpha$, $Z = \mathcal{M}_p(z) \subsetapprox M_\alpha$, $Z' = \mathcal{M}_p(z') \subsetapprox M_\alpha$. Then for any $m \in M_\alpha$*

(a) $(B|Z)(m) \otimes (\neg B|Z')(m) = 0$,
(b) $(B|Z)(m) \otimes \neg(B|Z')(m) = 0$.

Proof. (a) Let $p(x) = m$ and $B(m) = 0$. Then the property is trivially fulfilled.

Let $B(m) = Z(m) > 0$ and $\neg B(m) = Z'(m) > 0$. Then from Lemma 1 it follows that $B(m) = (B|Z)(m)$ as well as $\neg B(m) = (\neg B|Z')(m)$. Because $B(m) \otimes \neg B(m) = 1$ holds for any $m \in M_\alpha$, then $(B|Z)(m) \otimes (\neg B|Z')(m) = 0$.

(b) Obviously as (a).

Definition 3. *Let T^{IQ} be a theory containing intermediate quantifiers w.r.t. a set of types $\mathcal{S}$ due to Definition 2. Let $Ev \in Form_{oo}$ be an intension of some evaluative expression. Finally, let $z \in Form_{o\alpha}$, $x \in Form_\alpha$ be variables and $A, B \in Form_{o\alpha}$ be formulas where $T^{IQ} \vdash \mathbf{M}_{o(o\alpha)} B_{o\alpha}$, $\alpha \in \mathcal{S}$. An intermediate generalized quantifier construes the sentence "⟨Quantifier⟩ B's are A" is one of the following formulas:*

$$(Q^{\forall}_{Ev}\, x)(B, A) \equiv (\exists z)[(\forall x)((B|z)\, x \Rightarrow Ax) \wedge Ev((\mu B)(B|z))], \tag{6}$$

$$(Q^{\exists}_{Ev}\, x)(B, A) \equiv (\exists z)[(\exists x)((B|z)x \wedge Ax) \wedge Ev((\mu B)(B|z))]. \tag{7}$$

The following special intermediate quantifiers can be introduced:

A: All B are $A := (Q^{\forall}_{Bi\Delta} x)(B, A) \equiv (\forall x)(Bx \Rightarrow Ax)$,
E: No B are $A := (Q^{\forall}_{Bi\Delta} x)(B, \neg A) \equiv (\forall x)(Bx \Rightarrow \neg Ax)$,
P: Almost all B's are $A := (Q^{\forall}_{Bi\ Ex} x)(B, A)$
B: Almost all B's are not $A := (Q^{\forall}_{Bi\ Ex} x)(B, \neg A)$
T: Most B's are $A := (Q^{\forall}_{Bi\ Ve} x)(B, A)$
D: Most B's are not $A := (Q^{\forall}_{Bi\ Ve} x)(B, \neg A)$
K: Many B's are $A := (Q^{\forall}_{\neg\ Sm} x)(B, A)$

G: Many B's are not $A := (Q^{\forall}_{\neg\, Sm} x)(B, \neg A)$

I: Some B are $A := (Q^{\exists}_{Bi\boldsymbol{\Delta}} x)(B, A) \equiv (\exists x)(Bx \wedge Ax)$,

O: Some B are not $A := (Q^{\exists}_{Bi\boldsymbol{\Delta}} x)(B, \neg A) \equiv (\exists x)(Bx \wedge \neg Ax)$.

3 Graded Cube of Opposition

3.1 From the Graded Square to the Graded Cube of Opposition

The graded Aristotle's square of opposition is formed by two positive and two negative intermediate quantifiers that fulfil the generalized properties of contraries, contradictories, sub-contraries, and sub-alterns. Below, we recall the main definitions from [14].

Definition 4. *Let T be a consistent theory of Ł-FTT, $\mathcal{M} \models T$ be a model, $p \in Asg(\mathcal{M})$ be an assignment, and $P_1, P_2 \in Form_o$ be closed formulas of type o.*

(i) P_1 *and* P_2 *are* contraries in the model $\mathcal{M}$ *if*

$$\mathcal{M}_p(P_1) \otimes \mathcal{M}_p(P_2) = 0. \tag{8}$$

They are contraries in the theory T *if* $T \vdash \neg(P_1 \,\&\, P_2)$. *By completeness, this is equivalent to (8) for every model* $\mathcal{M} \models T$.

(ii) P_1 *and* P_2 *are* subcontraries in the model $\mathcal{M}$ *if*

$$\mathcal{M}_p(P_1) \oplus \mathcal{M}_p(P_2) = 1. \tag{9}$$

They are subcontraries in the theory T *if* $T \vdash (P_1 \,\nabla\, P_2)$. *By completeness, this is equivalent to (9) for every model* $\mathcal{M} \models T$.

(iii) P_1 *and* P_2 *are* contradictories in the model $\mathcal{M}$ *if both*

$$\mathcal{M}_p(\boldsymbol{\Delta} P_1) \otimes \mathcal{M}_p(\boldsymbol{\Delta} P_2) = 0 \text{ as well as } \mathcal{M}_p(\boldsymbol{\Delta} P_1) \oplus \mathcal{M}_p(\boldsymbol{\Delta} P_2) = 1. \tag{10}$$

They are contradictories in the theory T *if both* $T \vdash \neg(\boldsymbol{\Delta} P_1 \,\&\, \boldsymbol{\Delta} P_2)$ *as well as* $T \vdash \boldsymbol{\Delta} P_1 \,\nabla\, \boldsymbol{\Delta} P_2$. *By completeness, this means that (10) hold for every model* $\mathcal{M} \models T$.

(iv) P_2 *is a* subaltern *of* P_1 *(*P_1 *is* superaltern *of* P_2*) in the model* $\mathcal{M}$ *if*

$$\mathcal{M}_p(P_1) \leq \mathcal{M}_p(P_2). \tag{11}$$

P_2 *is* subaltern of P_1 in the theory T *(*P_1 *is* superaltern of P_2 in the theory T*) if* $T \vdash P_1 \Rightarrow P_2$. *By completeness, this means that inequality (11) holds true in every model* $\mathcal{M} \models T$.

All these definitions were introduced as a generalization of the corresponding classical ones. In our previous paper [14], we syntactically proved all the mentioned properties which form the graded Aristotle's square of opposition.

Theorem 1 ([14]). *The following is true in T^{IQ}:*

(a) formulas **A** *and* **O** *are contradictories in T^{IQ},*
(b) formulas **E** *and* **I** *are contradictories in T^{IQ}.*
(c) formulas **A** *and* **E** *are contraries with the presupposition in T^{IQ}.*
(d) the formula **A** *is subaltern of* **I**.

Changing B and A into their negation, $\neg B$ and $\neg A$ respectively, leads to another similar square of opposition **aeio**, provided that we also assume that the fuzzy set $\neg B$ is non-empty. It means that we assume *presupposition*[3] (existential import) which was in detail discussed in our previous papers. The corresponding quantifier with presupposition is denoted by a star. To extend the graded Aristotle's square to a graded cube of opposition, we have to define new formulas as follows:

$$\text{*}\mathbf{a}\text{: All } \neg B \text{ are not } A \qquad (\forall x)(\neg Bx \Rightarrow \neg Ax)\,\&\,(\exists x)\neg Bx, \tag{12}$$

$$\mathbf{e}\text{: All } \neg B \text{ are } A \qquad (\forall x)(\neg Bx \Rightarrow Ax), \tag{13}$$

$$\mathbf{i}\text{: Some } \neg B \text{ are not } A \qquad (\exists x)(\neg Bx \wedge \neg Ax), \tag{14}$$

$$\text{*}\mathbf{o}\text{: Some } \neg B \text{are } A \qquad (\exists x)(\neg Bx \wedge Ax)\,\nabla\,\neg(\exists x)\neg Bx. \tag{15}$$

We can see that the new logical square of opposition (**aeio**) develops from the graded Aristotle's one (**AEIO**) by replacing formulas Bx and Ax by $\neg Bx$ and $\neg Ax$, respectively. It means that the "basic" properties between the intermediate quantifiers inside **aeio** can be proved similarly to **AEIO**.

Theorem 2. *The following is true in T^{IQ}:*

(a) formulas **a** *and* **o** *are contradictories in T^{IQ},*
(b) formulas **e** *and* **i** *are contradictories in T^{IQ}.*
(c) formulas **a** *and* **e** *are contraries with the presupposition in T^{IQ}.*
(d) formula **a** *is subaltern of* **i**.

Proof. The properties (a)–(d) can be proved similarly as the properties of the graded Aristotle's square **AEIO** in [14].

3.2 Relations Between "AEIO" and "aeio"

Lemma 3. *The following is provable:*

(a) $T^{IQ} \vdash \neg(\mathbf{A}\,\&\,\mathbf{e})^*$,
(b) $T^{IQ} \vdash \neg(\mathbf{a}\,\&\,\mathbf{E})^*$.

[3] It is necessary that the universal quantifiers carry a presupposition of existential import for the entailments to their respective particular forms to hold.

Proof. (a) By properties of Ł-FTT we have

$$T^{\mathrm{IQ}} \vdash (Bx \Rightarrow Ax) \Rightarrow (\neg Ax \Rightarrow \neg Bx) \tag{16}$$

as well as

$$T^{\mathrm{IQ}} \vdash (\neg Bx \Rightarrow Ax) \Rightarrow (\neg Ax \Rightarrow Bx). \tag{17}$$

By $T^{\mathrm{IQ}} \vdash (\neg Ax \,\&(\neg Ax \Rightarrow Bx)) \Rightarrow Bx$ and $T^{\mathrm{IQ}} \vdash (\neg Ax \,\&(\neg Ax \Rightarrow \neg Bx)) \Rightarrow \neg Bx$ we obtain the following provable formula

$$T^{\mathrm{IQ}} \vdash (\neg Ax \,\&(\neg Ax \Rightarrow Bx)) \,\&(\neg Ax \,\&(\neg Ax \Rightarrow \neg Bx)) \Rightarrow \bot$$

which, using the properties of Ł-FTT, is equivalent with

$$T^{\mathrm{IQ}} \vdash ((\neg Ax \Rightarrow Bx) \,\&(\neg Ax \Rightarrow \neg Bx)) \Rightarrow (\neg Ax)^2 \Rightarrow \bot \tag{18}$$

From (16), (17) and (18) by transitivity and using MP, we conclude that

$$T^{\mathrm{IQ}} \vdash ((Bx \Rightarrow Ax) \,\&(\neg Bx \Rightarrow Ax)) \Rightarrow ((\neg Ax)^2 \Rightarrow \bot). \tag{19}$$

Finally, by the quantifier properties, we have

$$T^{\mathrm{IQ}} \vdash ((\forall x)(Bx \Rightarrow Ax) \,\&)(\forall x)(\neg Bx \Rightarrow Ax)) \Rightarrow ((\exists x)(\neg Ax)^2 \Rightarrow \bot) \tag{20}$$

which uses the definition of the negation equivalent with

$$T^{\mathrm{IQ}} \vdash \neg((\forall x)(Bx \Rightarrow Ax) \,\&)(\forall x)(\neg Bx \Rightarrow Ax)) \,\&(\exists x)(\neg Ax)^2$$

(b) It can be proved analogously.

Theorem 3. *The following holds true:*

(a) Formulas **a** *and* **E** *are contraries in* T^{IQ}.
(b) Formulas **A** *and* **e** *are contraries in* T^{IQ}.

Proof. It follows from Lemma 3.

Lemma 4. *Let the following be true:*

(a) $T^{IQ} \vdash (\mathbf{i} \nabla \mathbf{O})^*$,
(b) $T^{IQ} \vdash (\mathbf{I} \nabla \mathbf{o})^*$.

Proof. (a) From Lemma 3(a), we have the following provable formula

$$T^{\mathrm{IQ}} \vdash \neg((\forall x)(Bx \Rightarrow Ax) \,\&)(\forall x)(\neg Bx \Rightarrow Ax)) \,\&(\exists x)(\neg Ax)^2.$$

Then by properties of the negation, we have

$$T^{\mathrm{IQ}} \vdash (\exists x)\neg(Bx \Rightarrow Ax) \,\nabla(\exists x)\neg(\neg Bx \Rightarrow Ax)) \,\nabla \neg(\exists x)(\neg Ax)^2$$

which is equivalent with

$$T^{\mathrm{IQ}} \vdash (\exists x)(Bx \,\&\, \neg Ax) \,\nabla(\exists x)(\neg Bx \,\&\, \neg Ax)) \,\nabla \neg(\exists x)(\neg Ax)^2.$$

Finally by properties of & with $\wedge$, we conclude that

$$T^{\mathrm{IQ}} \vdash (\exists x)(Bx \wedge \neg Ax) \,\nabla(\exists x)(\neg Bx \wedge \neg Ax)) \,\nabla \neg(\exists x)(\neg Ax)^2.$$

(b) Analogously as (a).

Theorem 4. *The following is true:*

(a) Formulas **i** *and* **O** *are subcontraries in* T^{IQ}.
(b) Formulas **I** *and* **o** *are subcontraries in* T^{IQ}.

Proof. It follows from Lemma 4.

Lemma 5. *The following is provable:*

(a) $T^{IQ} \vdash \mathbf{A}^* \Rightarrow \mathbf{i} \quad T^{IQ} \vdash \mathbf{a} \Rightarrow \mathbf{I}$,
(b) $T^{IQ} \vdash \mathbf{E} \Rightarrow \mathbf{o}^* \quad T^{IQ} \vdash \mathbf{e} \Rightarrow \mathbf{O}$.

Proof. (a) By the properties of Ł-FTT, we have

$$T^{\mathrm{IQ}} \vdash (\forall x)(Bx \Rightarrow Ax) \Rightarrow (\forall x)(\neg Ax \Rightarrow \neg Bx) \tag{21}$$

as well as

$$T^{\mathrm{IQ}} \vdash (\forall x)(\neg Ax \Rightarrow \neg Bx) \Rightarrow ((\exists x)\neg Ax \Rightarrow (\exists x)(\neg Ax \wedge \neg Bx)). \tag{22}$$

Finally, by (21) and (22), we obtain

$$T^{\mathrm{IQ}} \vdash (\forall x)(Bx \Rightarrow Ax) \,\&\, (\exists x)\neg Ax \Rightarrow (\exists x)(\neg Ax \wedge \neg Bx).$$

(b) is proved analogously.

Theorem 5. *The following is true:*

(a) Formula **A** *is subaltern of* **i**, *and* **I** *is superaltern of* **a**.
(b) Formula **E** *is subaltern of* **o**, *and* **e** *is superaltern of* **O**.

Proof. It follows from Lemma 5.

Example 1 (Interpretation of Cube of opposition in T^{IQ}*).*
(a) Construction of **AEIO** as follows:
Let us consider a model $\mathcal{M} \models T^{\mathrm{IQ}}$ such that $T^{\mathrm{IQ}} \vdash (\exists x)Bx$. Let $\mathcal{M}(\mathbf{A}) = a > 0$ (e.g., $a = 0.2$). Since $[\mathbf{A}, \mathbf{E}]$ are contraries, we have $\mathcal{M}(\mathbf{E}) = e \leq 1 - a$. Because the formulas $[\mathbf{A}, \mathbf{O}]$ are contradictories, it follows from the definition of contradictories that $\mathcal{M}(\boldsymbol{\Delta}\mathbf{A}) = 0$ and so $\mathcal{M}(\boldsymbol{\Delta}\mathbf{O}) = 1$ because $\mathcal{M}(\boldsymbol{\Delta}\mathbf{A}) \otimes \mathcal{M}(\boldsymbol{\Delta}\mathbf{O}) = 0$ and $\mathcal{M}(\boldsymbol{\Delta}\mathbf{A}) \oplus \mathcal{M}(\boldsymbol{\Delta}\mathbf{O}) = 1$. Consequently, **O** is subaltern of **E**. The **I** is subaltern of **A** and thus $\mathcal{M}(\mathbf{I}) = i \geq 0.2$. However, **I** is contradictory with **E** and so $\mathcal{M}(\mathbf{I}) = i = 1$. Finally, **I** is sub-contrary with **O** because $\mathcal{M}(\mathbf{O} \boldsymbol{\nabla} \mathbf{I}) = 1$ and **I** is subaltern of **A**.
(b) Construction of **aeio** as follows:
Let us consider the same model $\mathcal{M} \models T^{\mathrm{IQ}}$ such that $T^{\mathrm{IQ}} \vdash (\exists x)\neg Ax$. Let $\mathcal{M}(\mathbf{a}) = a' > 0$ (e.g., $a' = 0.8$), $\mathcal{M}(\mathbf{e}) = e'$, $\mathcal{M}(\mathbf{i}) = i'$ and $\mathcal{M}(\mathbf{o}) = o'$. Since $[\mathbf{A}, \mathbf{e}]$, as well as $[\mathbf{E}, \mathbf{i}]$ are contraries, we have $\mathcal{M}(\mathbf{e}) = e' \leq 0.2$ and $\mathcal{M}(\mathbf{a}) = a' \leq 0.8$. Similarly, formulas $[\mathbf{i}, \mathbf{O}]$ and $[\mathbf{I}, \mathbf{o}]$ are sub-contraries in T^{IQ} then $\mathcal{M}(i) = i' = 1$ as well as $\mathcal{M}(i) = i' = 1$. Both of these results correspond with contradictories of the pairs $[\mathbf{a}, \mathbf{o}]$ and $[\mathbf{e}, \mathbf{i}]$. Finally, **A** is subaltern of **i** as well as **E** is subaltern of **o**.

These results are summarized in the following scheme. Recall that the straight lines mark contradictories, the dashed lines contraries, and the dotted lines subcontraries. The arrows indicate the relation superaltern–subaltern (Fig. 1).

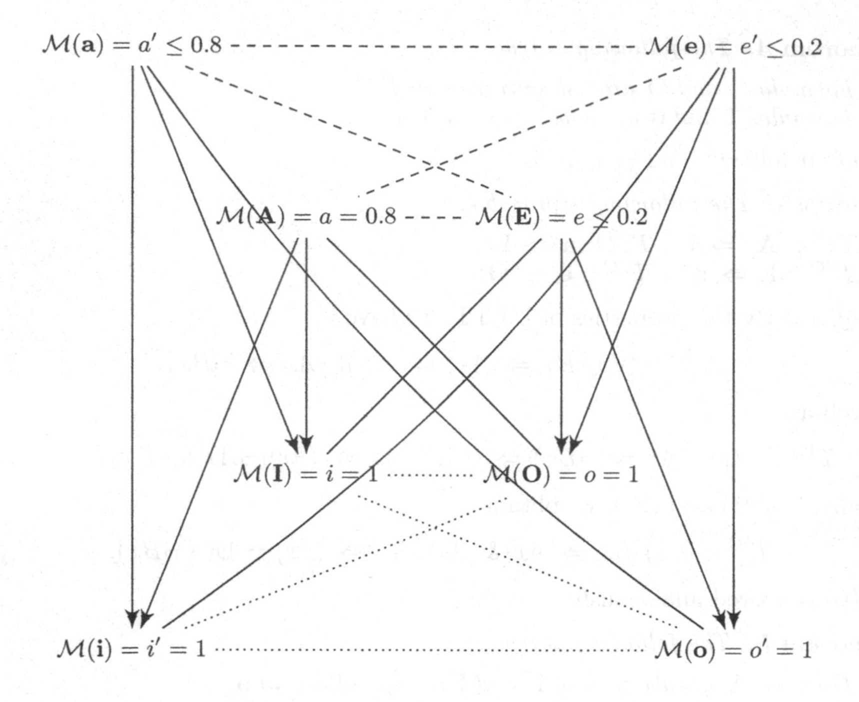

Fig. 1. The example of graded Aristotle's cube of opposition

4 Graded Cube with Intermediate Quantifiers

We continue with an extension of graded the **5**-square of opposition **AEPBT-DKGIO**, which was introduced as a generalization of Peterson's square, to the graded **5**-cube of opposition **aepbtdkgio** with intermediate quantifiers. Below we introduce new forms of intermediate quantifiers as follows:

$$(Q^{\forall}_{Ev}\, x)(\neg B, \neg A) \equiv (\exists z)[(\forall x)((\neg B|z)\, x \Rightarrow \neg Ax) \wedge Ev((\mu(\neg B))(\neg B|z))], \tag{23}$$

$$(Q^{\exists}_{Ev}\, x)(\neg B, \neg A) \equiv (\exists z)[(\exists x)((\neg B|z)x \wedge \neg Ax) \wedge Ev((\mu(\neg B))(\neg B|z))]. \tag{24}$$

Either of the quantifiers (23) or (24) construes the sentence

"⟨Quantifier⟩ not B's are not A".

4.1 Contraries

Lemma 6. *Let $B \in Form_{o\alpha}$ be a formula and $z, z' \in Form_{o\alpha}$ be variables. Then the following is provable:*

$$T^{IQ} \vdash \neg[(\exists z)(\exists z')[(\forall x)((B|z)\, x \Rightarrow Ax) \wedge Ev((\mu B)(B|z))\, \&$$
$$(\forall x)((\neg B|z')\, x \Rightarrow Ax) \wedge Ev((\mu(\neg B))((\neg B|z')))\, \&\, (\exists x)(\neg(Ax))^2]. \tag{25}$$

Proof. The proof of (25) is based on the following provable formulas:

$$T^{\mathrm{IQ}} \vdash ((B|z)x \Rightarrow Ax) \Rightarrow (\neg Ax \Rightarrow \neg(B|z)x)$$

and

$$T^{\mathrm{IQ}} \vdash ((\neg B|z')x \Rightarrow Ax) \Rightarrow (\neg Ax \Rightarrow \neg(\neg B|z')x).$$

Using quantifier properties and by Lemma 2 we obtain from these formulas that

$$T^{\mathrm{IQ}} \vdash (\forall x)((B|z)x \Rightarrow Ax) \,\&\, (\forall x)((\neg B|z')x \Rightarrow Ax) \Rightarrow ((\exists x)(\neg Ax)^2 \Rightarrow \bot). \tag{26}$$

By the adjunction, the properties of $\wedge$, the quantifier properties and the definition of negation, we obtain (25) using the rules of Ł-FTT.

Theorem 6. *The pairs of quantifiers*

(i) $[(Q^{\forall}_{Bi\,Ex}\,x)(B,A), (Q^{\forall}_{Bi\,Ex}\,x)(\neg B,A)]$, *(i.e.,* $[\mathbf{P},\mathbf{b}]$*),*
(ii) $[(Q^{\forall}_{Bi\,Ve}\,x)(B,A), (Q^{\forall}_{Bi\,Ve}\,x)(\neg B,A)]$, *(i.e.,* $[\mathbf{T},\mathbf{d}]$*),*
(iii) $[(Q^{\forall}_{\neg(Sm\,\bar{\nu})}\,x)(B,A), (Q^{\forall}_{\neg(Sm\,\bar{\nu})}\,x)(B,\neg A)]$, *(i.e.,* $[\mathbf{K},\mathbf{g}]$*),*

are contraries in T^{IQ}.

Proof. It follows from Lemma 6 when replacing *Ev* by concrete evaluative linguistic expressions.

Lemma 7. *Let* $B \in Form_{o\alpha}$ *be a formula and* $z, z' \in Form_{o\alpha}$ *be variables. Then the following is provable:*

$$T^{IQ} \vdash \neg[(\exists z)(\exists z')[(\forall x)((\neg B|z)\,x \Rightarrow \neg Ax) \wedge Ev((\mu(\neg B))((\neg B|z))) \,\&\, (\forall x)((B|z')\,x \Rightarrow \neg Ax) \wedge Ev((\mu B)(B|z')) \,\&\, (\exists x)(\neg(Ax))^2]. \tag{27}$$

Proof. Similarly to Lemma 6, the proof of (27) is based on the following provable formula:

$$T^{\mathrm{IQ}} \vdash (\forall x)((\neg B|z)x \Rightarrow \neg Ax) \,\&\, (\forall x)((B|z')x \Rightarrow \neg Ax) \Rightarrow ((\exists x)(Ax)^2 \Rightarrow \bot). \tag{28}$$

By the adjunction, the properties of $\wedge$, the quantifier properties and the definition of negation, we obtain (28) using the rules of Ł-FTT.

Theorem 7. *The pairs of quantifiers*

(i) $[(Q^{\forall}_{Bi\,Ex}\,x)(\neg B,\neg A), (Q^{\forall}_{Bi\,Ex}\,x)(B,\neg A)]$, *(i.e.,* $[\mathbf{p},\mathbf{B}]$*),*
(ii) $[(Q^{\forall}_{Bi\,Ve}\,x)(\neg B,\neg A), (Q^{\forall}_{Bi\,Ve}\,x)(B,\neg A)]$, *(i.e.,* $[\mathbf{t},\mathbf{D}]$*),*
(iii) $[(Q^{\forall}_{\neg(Sm\,\bar{\nu})}\,x)(\neg B,\neg A), (Q^{\forall}_{\neg(Sm\,\bar{\nu})}\,x)(B,\neg A)]$, *(i.e.,* $[\mathbf{k},\mathbf{G}]$*),*

are contraries in T^{IQ}.

Proof. It follows from Lemma 7 when replacing *Ev* by concrete evaluative linguistic expressions.

4.2 Sub-alterns

Lemma 8. *The following is provable in Ł-FTT:*

(a) $T^{IQ} \vdash \mathbf{a} \Rightarrow \mathbf{p}$ $\quad T^{IQ} \vdash \mathbf{e} \Rightarrow \mathbf{b}$,
(b) $T^{IQ} \vdash \mathbf{p} \Rightarrow \mathbf{t}$ $\quad T^{IQ} \vdash \mathbf{b} \Rightarrow \mathbf{d}$,
(c) $T^{IQ} \vdash \mathbf{t} \Rightarrow \mathbf{k}$ $\quad T^{IQ} \vdash \mathbf{d} \Rightarrow \mathbf{g}$,
(d) $T^{IQ} \vdash \mathbf{k} \Rightarrow \mathbf{i}$ $\quad T^{IQ} \vdash \mathbf{g} \Rightarrow \mathbf{o}$.

Proof. This proceeds similarly as in [14] using monotonicity of the corresponding evaluative linguistic expressions.

Below, we introduce a **5**-graded cube of opposition with five basic intermediate quantifiers as a generalization of the graded Peterson's square (Fig. 2).

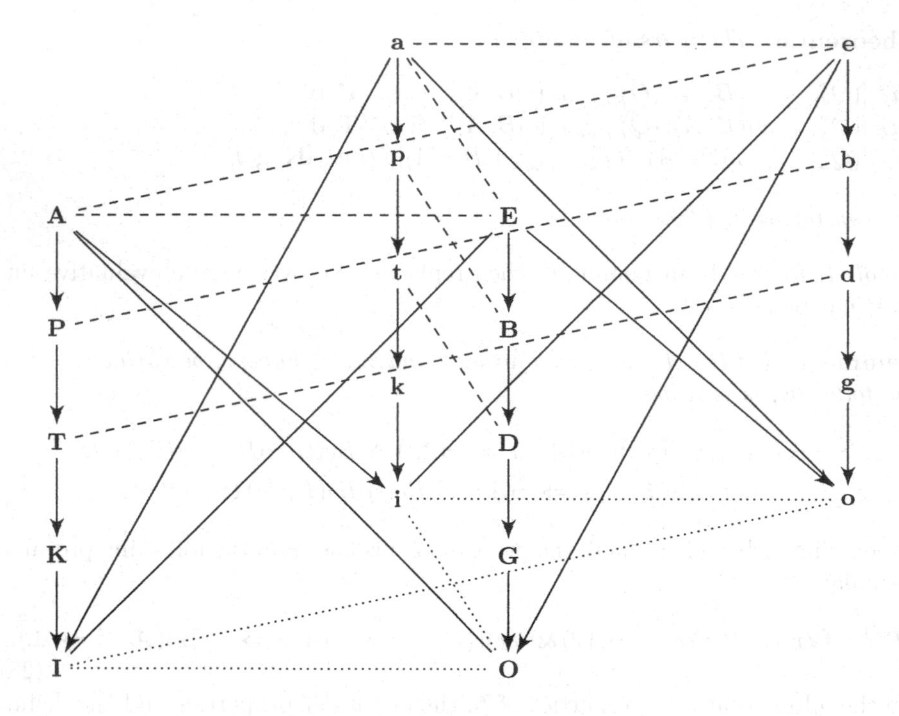

Fig. 2. Graded cube of opposition with generalized intermediate quantifiers

5 Future Applications

As we mentioned above, the idea of this paper was to introduce new forms of generalized intermediate quantifiers forming graded cube of opposition. An idea for future is to apply the theory of syllogistic reasoning introduced in our previous papers [6,26]. Using inferred new forms of valid syllogisms to derive new information which is not included in real data. For example, below we introduce examples of sentences which can be used:

– *Most people who live in an area affected by heavy industry suffer from asthma.*
– *Almost all shares grow with growing economy.*

New idea is to work with examples of natural language linguistic expressions which form graded cube of opposition as follows:

– *Most people who do not smoke have higher lung capacity.*
– *Most children who do not live in an area affected by heavy industry do not suffer from inflammation of the respiratory tract.*

6 Conclusion

In this paper, we extended the theory of the graded classical Aristotle square of opposition to the graded Aristotle cube of opposition. Furthermore, we suggested a generalization of the Peterson's square of opposition to a graded generalized cube, i.e., the cube whose vertices contain intermediate quantifiers.

The future work will focus on a more detailed analysis of the properties of the graded generalized cube of opposition and to extend by new forms of intermediate quantifiers.

References

1. Novák, V.: A comprehensive theory of trichotomous evaluative linguistic expressions. Fuzzy Sets and Systems **159**(22), 2939–2969 (2008)
2. Novák, V.: Perception-based logical deduction. In: Reusch, B. (ed.) Computational Intelligence. Theory and Applications, pp. 237–250. Springer, Berlin (2005)
3. Novák, V., Lehmke, S.: Logical structure of fuzzy IF-THEN rules. Fuzzy Sets and Systems **157**, 2003–2029 (2006)
4. A. Dvořák and M. Holčapek. Type $\langle 1, 1\rangle$ fuzzy quantifiers determined by fuzzy measures on residuated lattices. part i. basic definitions and examples. Fuzzy Sets and Systems, 242:31–55, 20014
5. Novák, V.: A formal theory of intermediate quantifiers. Fuzzy Sets and Systems **159**(10), 1229–1246 (2008)
6. Murinová, P., Novák, V.: A formal theory of generalized intermediate syllogisms. Fuzzy Sets and Systems **186**, 47–80 (2013)
7. Murinová, P., Novák, V.: The structure of generalized intermediate syllogisms. Fuzzy Sets and Systems **247**, 18–37 (2014)
8. Thompson, B.E.: Syllogisms using "few","many" and "most". Notre Dame Journal of Formal Logic **23**, 75–84 (1982)
9. P.L. Peterson. Intermediate Quantifiers. Logic, linguistics, and Aristotelian semantics. Ashgate, Aldershot, 2000
10. L. Miclet and H. Prade. Analogical proportions and square of oppositions. In: A. Laurent et al. (ed.) Proc. 15th Int. Conf. on Information Processing and Management of Uncer- tainty in Knowledge-Based Systems, July 15–19, Montpellier, CCIS, 443:324–334, 2014
11. Pellissier, R.: "Setting" n-opposition. Logica Universalis **2**, 235–263 (2008)
12. Peters, S., Westerståhl, D.: Quantifiers in Language and Logic. Claredon Press, Oxford (2006)

13. Westerståhl, D.: The traditional square of opposition and generalized quantifiers. Studies in Logic **2**, 1–18 (2008)
14. Murinová, P., Novák, V.: Analysis of generalized square of opposition with intermediate quantifiers. Fuzzy Sets and Systems **242**, 89–113 (2014)
15. Murinová, P., Novák, V.: The theory of intermediate quantifiers in fuzzy natural logic revisited and the model of "Many". Fuzzy Sets Syst. **388**, 56–89 (2020)
16. D. Dubois and H. Prade. From blanche's hexagonal organization of concepts to formal concepts analysis and possibility theory. Logica Universalis, pages 149–169, 2012
17. D. Dubois and H. Prade. Gradual structures of oppositions. In: F. Esteva, L. Magdalena, J.L. Verdegay (eds.) Enric Trillas: Passion for Fuzzy Sets, Studies in Fuzziness and Soft Computing, 322:79–91, 2015
18. Dubois, D., Prade, H., Rico, A.: Graded cubes of opposition and possibility theory with fuzzy events. International Journal of Approximate Reasoning **84**, 168–185 (2017)
19. D. Ciucci, D. Dubois, and H. Prade. Oppositions in rough set theory. In: T. Li, H.S. Nguyen, G. Wang, J.W. Grzymala-Busse, R. Janicki, A.E. Hassanien, H. Yu (eds.) Proc. 7th Int. Conf. on Rough Sets and Knowledge Technology (RSKT12), Chengdu, Aug. 17–20, LNCS, 7414:504–513, 2012
20. D. Ciucci, D. Dubois, and H. Prade. The structure of oppositions in rough set theory and formal concept analysis - toward a new bridge between the two settings. In: C. Beierle, C. Meghini (eds.) Proc. 8th Int. Symp. on Foundations of Information and Knowledge Systems (FoIKS14), Bordeaux, Mar. 3–7, LNCS, 8367:154–173, 2014
21. Novák, V.: On fuzzy type theory. Fuzzy Sets and Systems **149**, 235–273 (2005)
22. Cignoli, R.L.O., D'Ottaviano, I.M.L., Mundici, D.: Algebraic Foundations of Many-valued Reasoning. Kluwer, Dordrecht (2000)
23. Novák, V., Perfilieva, I., Močkoř, J.: Mathematical Principles of Fuzzy Logic. Kluwer, Boston (1999)
24. Andrews, P.: An Introduction to Mathematical Logic and Type Theory: To Truth Through Proof. Kluwer, Dordrecht (2002)
25. Novák, V., Perfilieva, I., Dvořák, A.: Insight into Fuzzy Modeling. Wiley & Sons, Hoboken, New Jersey (2016)
26. P. Murinová. Generalized intermediate syllogisms with more premises. In Proc. FUZZ-IEEE 2019, New Orleands, USA, 2019

On the Properties of Intermediate Quantifiers and the Quantifier "MORE-THAN"

Vilém Novák(✉), Petra Murinová, and Stefania Boffa

Institute for Research and Applications of Fuzzy Modeling, University of Ostrava, NSC IT4Innovations, 30. dubna 22, 701 03 Ostrava 1, Czech Republic
{Vilem.Novak,Petra.Murinova,Stefania.Boffa}@osu.cz

Abstract. This paper continues the research in formal theory of intermediate quantifiers. We present some new properties, introduce intermediate quantifiers of type ⟨1⟩, and also new quantifiers MORE-THAN and LESS-THAN.

Keywords: Generalized quantifiers · Fuzzy quantifiers · Fuzzy type theory · Mathematical fuzzy logic

1 Introduction

Quantifiers are special expressions of natural language that characterize quantity of objects having a given property. Typical examples are the classical ones "all, exists", but also "most, almost all, many, few", etc. As they are quite frequently used in common language, they raised interest of logicians who tried to suggest logical models of them. The general theory was initiated by Mostowski [8] and further elaborated by many logicians (cf. [21] and citations therein).

A general and widely accepted definition originated by Lindström in [7] is to take quantifiers as n-ary relations among subsets of powers of a given set M, i.e., subsets of $P(M^{k_1}) \times \cdots \times P(M^{k_n})$. Adoption of this definition to fuzzy logic is the following (cf. [12]): a generalized fuzzy quantifier of type $\langle k_1, \ldots, k_n \rangle$ is a functional $\mathbf{Q}$ that assigns to each non-empty set M a fuzzy relation

$$Q_U : \mathcal{F}(M^{k_1}) \times \cdots \times \mathcal{F}(M^{k_n}) \longrightarrow E \tag{1}$$

where by $\mathcal{F}(\cdot)$ we denote a set of all fuzzy sets on a given universe and E is a support of the algebra of truth values. Note that (1) is an n-ary fuzzy relation over k_i-ary fuzzy relations, $i = 1, \ldots, n$. This definition is semantic which means that (1) interprets a certain formula $Q(A_1^{k_1}, \ldots, A_n^{k_n})$ of a suitable formal logic where $A_1, \ldots, A_n$ are formulas and the exponents $A_i^{k_i}$, $i = 1, \ldots, n$ denote k_i-ary conjunctions of them.

The work was supported from ERDF/ESF by the project "Centre for the development of Artificial Intelligence Methods for the Automotive Industry of the region" No. CZ.02.1.01/0.0/0.0/17-049/0008414 and partially also by the MŠMT project NPU II project LQ1602 "IT4Innovations excellence in science".

M.-J. Lesot et al. (Eds.): IPMU 2020, CCIS 1239, pp. 159–172, 2020.
https://doi.org/10.1007/978-3-030-50153-2_12

An important class of quantifiers are *intermediate* ones, for example *most, few, almost all, a lot of, many, a great deal of, a large part of, a small part of*, etc. Intermediate quantifiers occur in sentences of natural language of the form

$$Q\ B\text{s are } A \tag{2}$$

where Q is a quantifier and B, A are properties of elements. Example of (2) is the sentence "Most (Q) young people (B) are happy (A)". Semantics of intermediate quantifiers lays between the two limit cases: the classical general (universal) $\forall$ and the existential $\exists$ ones (hence the name). From the point of view of (1), intermediate quantifiers are special generalized quantifiers of type $\langle 1, 1\rangle$ (cf. [6,19]).

An in-depth linguistic and logical analysis of intermediate quantifiers was provided by Peterson in [20]. He specified their basic semantic properties, and, using informal tools, demonstrated that 105 generalized syllogisms with five selected intermediate quantifiers should be valid. These results inspired Novák to develop a mathematical model of the meaning of intermediate quantifiers (see [15]). The primary formal tool is higher-order fuzzy logic (namely, Łukasiewicz fuzzy type theory (FTT)). This logic is a generalization of classical higher-order logic (also called λ-calculus), and it was chosen because of its very high explication power. Note that the classical λ-calculus became a standard tool used by linguists when studying the semantic properties of natural language.

The core idea of the mentioned formalization consists in the assumption that intermediate quantifiers can be taken as the classical $\forall$ or $\exists$ quantifiers applied over a universe whose size is characterized by a measure that can be modified and linguistically evaluated. Note that the idea of using the measure in fuzzy quantifiers also occurs in [3,5]).

Using formal language, sentence (2) can be construed by a certain formula $(Qx)(B, A)$ where B and A are subformulas representing properties. This formula is precisely defined in Sect. 4. The theory of intermediate quantifiers is already quite well developed and presented in many papers (see, e.g., [9–11]).

The main objective of this paper is to continue development of the theory of intermediate quantifiers. We introduce simpler quantifiers of type $\langle 1\rangle$ that are time to time needed in some reasoning, and also introduce new intermediate quantifiers MORE-THAN and LESS-THAN, and prove validity of the related generalized syllogisms.

By a fuzzy set in a universe M_α we mean a function $A : M_\alpha \longrightarrow E$ where E is the support of a suitable algebra of truth values. If A is a fuzzy set on M_ϵ then we write $A \subsetapprox M_\alpha$. The kernel of A is the set $\mathrm{Ker}(A) = \{x \mid x \in M_\alpha, A(x) = 1\}$.

2 Preliminaries

2.1 Fuzzy Type Theory

The theory of intermediate quantifiers has been developed in Łukasiewicz fuzzy type theory (Ł-FTT) whose algebra of truth values is a linearly ordered MV-algebra. Note that Ł-FTT is a gneralization of the classical type theory (see [1] and elsewhere).

The basic syntactical objects of Ł-FTT are classical, namely the concepts of *type* and *formula*. Recall that by *type* we understand a certain symbol expressing a kind of objects that are denoted by a formula in concern. The types are recursively formed starting with the atomic types ϵ (elements), and o (truth values). Complex types are defined as follows: if α, β are types then $(\beta\alpha)$ is a type. We denote types by Greek letters and the set of all types by *Types*. Each formula is assigned a type and we write A_α where A is a formula and α a type.

The *language* J of Ł-FTT consists of variables $x_\alpha, \ldots$, special constants $c_\alpha, \ldots$ ($\alpha \in \mathit{Types}$), the symbol λ, and brackets. We will consider the following concrete special constants: $\mathbf{E}_{(o\alpha)\alpha}$ (fuzzy equality) for every $\alpha \in \mathit{Types}$, $\mathbf{C}_{(oo)o}$ (conjunction), $\mathbf{D}_{(oo)}$ (delta operation on truth values) and the description operator $\iota_{\epsilon(o\epsilon)}$.

Formulas are formed of variables, constants (each of specific type), and the symbol λ. A set of all formulas of type α is denoted by Form_α. The set of all formulas is $\mathit{Form} = \bigcup_{\alpha \in \mathit{Types}} \mathit{Form}_\alpha$[1].

If $B \in \mathit{Form}_{\beta\alpha}$ and $A \in \mathit{Form}_\alpha$ then $(BA) \in \mathit{Form}_\beta$. Similarly, if $A \in \mathit{Form}_\beta$ and $x_\alpha \in J$, $\alpha \in \mathit{Types}$, is a variable then $(\lambda x_\alpha A) \in \mathit{Form}_{\beta\alpha}$.

A formal theory $T \subset \mathit{Form}_o$ is a set of formulas of type o (truth values). Provability is defined classically. If A_o is provable in T then we write $T \vdash A_o$.

The algebra of truth values of Ł-FTT is supposed to be a linearly ordered MV-algebra $\langle E, \vee, \wedge, \otimes, \rightarrow, \mathbf{0}, \mathbf{1}, \Delta \rangle$ extended by the operation Δ (see [2,18]). A special case of it is the standard Łukasiewicz MV$_\Delta$-algebra

$$\mathcal{L}_\Delta = \langle [0,1], \vee, \wedge, \otimes, \rightarrow, 0, 1, \Delta \rangle \tag{3}$$

where

$$\wedge = \text{minimum}, \qquad \vee = \text{maximum},$$
$$a \otimes b = 0 \vee (a + b - 1), \qquad a \rightarrow b = 1 \wedge (1 - a + b),$$
$$\neg a = a \rightarrow 0 = 1 - a, \qquad \Delta(a) = \begin{cases} 1 & \text{if } a = 1, \\ 0 & \text{otherwise.} \end{cases}$$

Note that the Δ operation sends all truth values smaller than 1 to 0.

A *model* is $\mathcal{M} = \{(M_\alpha, \doteq_\alpha) \mid \alpha \in \mathit{Types}\}$ where $\doteq_\alpha$ is a fuzzy equality on a set M_α (a binary fuzzy relation on M_α that is reflexive, symmetric, and $\otimes$-transitive). If $\mathcal{M}$ is a model then $\mathcal{M}(A_o) \in M_o$ is a truth value, $\mathcal{M}(A_\epsilon) \in M_\epsilon$ is some element and $\mathcal{M}(A_{\beta\alpha}) : M_\alpha \longrightarrow M_\beta$ is a function. For example, $\mathcal{M}(A_{o\alpha}) : M_\alpha \longrightarrow M_o$ is a fuzzy set and $\mathcal{M}(A_{(o\alpha)\alpha}) : M_\alpha \times M_\alpha \longrightarrow M_o$ a fuzzy relation. A formula A_o is *true* in T, $T \models A_o$, if it is true in the degree 1 in all models of T.

The fuzzy type theory is *complete*, i.e., the completeness theorem stating that a theory T is consistent iff it has a (Henkin) model holds true. We sometimes apply its equivalent version: $T \vdash A_o$ iff $T \models A_o$.

---

[1] To improve readability of formulas, we quite often write the type only once in the beginning of the formula and then omit it. Alternatively, we write $A \in \mathit{Form}_\alpha$ to emphasize that A is a formula of type α and do not repeat its type again.

In the explanation below, we need to characterize that a given formula A_o represents a nonzero truth value, and also a general truth value that is neither equal to 0 nor to 1. The following two formulas will do the job:

$$\Upsilon_{oo} \equiv \lambda z_o \cdot \neg\boldsymbol{\Delta}(\neg z_o), \qquad (z_o \text{ is a nonzero truth value})$$

$$\hat{\Upsilon}_{oo} \equiv \lambda z_o \cdot \neg\boldsymbol{\Delta}(z_o \vee \neg z_o) \qquad (z_o \text{ is a general truth value}).$$

It can be proved that $T \models \Upsilon A_o$ implies that $\mathcal{M}(A_o) > 0$, and $T \models \hat{\Upsilon} A_o$ implies that $\mathcal{M}(A_o) \in (0, 1)$, in any model $\mathcal{M}$ of the theory T.

The following lemma characterizes a few basic properties of $\boldsymbol{\Delta}$ and Υ.

Lemma 1. *Let $z \in Form_{o\alpha}$, $x \in Form_o$. Then*

(a) $\vdash \Upsilon(\boldsymbol{\Delta}x) \equiv \boldsymbol{\Delta}x$
(b) $\vdash ((A_o \Rightarrow B_o) \wedge (A_o \Rightarrow C_o)) \Rightarrow (A_o \Rightarrow B_o \wedge C_o)$
(c) $\vdash ((A_o \Rightarrow B_o) \vee (A_o \Rightarrow C_o)) \Rightarrow (A_o \Rightarrow B_o \vee C_o)$

Proof. (a) is obtained by the following sequence of inferences:

(L.1) $\vdash \Upsilon(\boldsymbol{\Delta}x) \equiv \neg\boldsymbol{\Delta}\neg\boldsymbol{\Delta}x$ (definition of Υ)
(L.2) $\vdash \boldsymbol{\Delta}\bot \equiv \bot$ (properties of $\boldsymbol{\Delta}$)
(L.3) $\vdash \boldsymbol{\Delta}\top \equiv \top$ (properties of $\boldsymbol{\Delta}$)
(L.4) $\vdash \neg\bot \equiv \top$ (properties of FTT)
(L.5) $\vdash \neg\top \equiv \bot$ (properties of FTT)
(L.6) $\vdash \Upsilon(\bot) \equiv \bot$ (L.1–L.5 using Rule (R))
(L.7) $\vdash \Upsilon(\top) \equiv \top$ (L.1–L.5 using Rule (R))
(L.8) $\vdash \Upsilon(\boldsymbol{\Delta}x) \equiv \boldsymbol{\Delta}x$ (L.6, L.7 by [13, Theorem 14])

(b), (c) are proved using the standard means of FTT (cf. also [4, Lemma 2.2.9]).

We will also work with the derived connective

$$\ominus_{(oo)o} \equiv \lambda x_o \, \lambda y_o \cdot x \,\&\, \neg y$$

that is in the standard Łukasiewicz algebra interpreted by the operation $a \ominus b = a \otimes \neg b = \max\{0, a - b\}$. Finally, we define the formula

$$x_o <_{(oo)o} y_o \equiv (x_o \Rightarrow y_o) \,\&\, \neg\boldsymbol{\Delta}(x_o \equiv y_o).$$

Note that a fuzzy set in a universe M_α is in FTT represented by a formula $X_{o\alpha}$. Indeed, let $\mathcal{M}$ be a model. Then $\mathcal{M}(A_\alpha) : M_\alpha \longrightarrow E^2$. In the same way, a λ-formula[3] $\lambda x_\alpha \, B_o$ also represents a fuzzy set.

Our explanation below proceeds mostly on the level of syntax because it is the most general way how to express various kinds of properties, and the results are universally valid in all models. W.r.t. the previous paragraph, we will freely call formulas of type $o\alpha$ "fuzzy sets" instead of more precise "formulas (variables) representing fuzzy sets". For example, we say "a fuzzy set $x_{o\alpha}$" or "a fuzzy set $A_{o\alpha}$". The reader, however, should be aware that fuzzy sets are obtained only in a model after proper interpretation of these formulas.

---

[2] Recall that we identify a fuzzy set with its membership function.
[3] In type theory, λ-formulas are often called λ-terms.

2.2 Evaluative Linguistic Expressions

The theory of intermediate quantifiers is based on the theory of *evaluative linguistic expressions* that are expressions of natural language such as "small, medium, big, very short, more or less deep, quite roughly strong, extremely high", etc. Semantics of them is also formalized using the language of Ł-FTT (see [14]). Less formally, including formulas for the direct computation is their theory presented in [17].

The theory of evaluative linguistic expressions is a special formal theory T^{Ev} of Ł-FTT. Its language J^{Ev} has the following special symbols:

(i) The constants $\top, \bot \in \mathit{Form}_o$ for truth and falsity and $\dagger \in \mathit{Form}_o$ for the middle truth value[4].
(ii) A special constant $\sim \in \mathit{Form}_{(oo)o}$ for an additional fuzzy equality on the set of truth values E.
(iii) A set of special constants $\boldsymbol{\nu}, \ldots \in \mathit{Form}_{oo}$ for linguistic hedges and a set of triples of additional constants $\mathbf{a}_{\boldsymbol{\nu}}, \mathbf{b}_{\boldsymbol{\nu}}, \mathbf{c}_{\boldsymbol{\nu}}, \ldots \in \mathit{Form}_o$ where each triple is associated with one hedge $\boldsymbol{\nu}$. The J^{Ev} is supposed to contain the special constants $\{\mathit{Ex}, \mathit{Si}, \mathit{Ve}, \mathit{ML}, \mathit{Ro}, \mathit{QR}, \mathit{VR}\}$ that represent the linguistic hedges (*extremely, significantly, very, roughly, more or less, rather, quite roughly, very roughly*, respectively).

The logical theory of evaluative expressions contains models of the standard logical and linguistic concepts of *intension* and *extension* (see [14] for the technical details). Evaluative expressions considered in this paper are construed by special formulas of type $oo(oo)$: *Sm* (*small*), *Me* (*medium*), *Bi* (*big*), and *Ze* (*zero*) that can be extended by the linguistic hedges introduced above. For example, *Sm Ve* is a formula whose interpretation is intension of the linguistic expression "very small'. If the concrete expression is not important, we use in the sequel a metavariable *Ev* for intension of an arbitrary evaluative expression.

3 Cuts of Fuzzy Sets

To define intermediate quantifiers, we need a special operation called *cut of a fuzzy set*. It is motivated by the need to form a new fuzzy set from a given one by extracting several elements together with their membership degrees and putting the other membership degrees equal to 0. For example, given a fuzzy set $A = \{^{0.3}/x_1, ^{1}/x_2, ^{0.7}/x_3, ^{0.9}/x_4\}$, we may need to work with its part only, say a fuzzy set $A' = \{^{0.3}/x_1, ^{0.9}/x_4\}$. We thus cut from A the singletons $^{0.3}/x_1$ and $^{0.9}/x_4$ and put them into A'. The elements of A' can be specified by means of some other fuzzy set, say $B = \{^{0.3}/x_1, ^{0.7}/x_2, ^{0.9}/x_4\}$ whose elements of interest (i.e., x_1, x_4) have membership degrees equal to those of A. The resulting fuzzy set A' is thus obtained by a *cut* of A by B, i.e., $A' = A|B$.

[4] The formula $\dagger$ is in the standard Łukasiewicz MV-algebra interpreted by the value 0.5.

This operation is formally defined as follows: Let $y, z \in Form_{o\alpha}$ be variables of type $o\alpha$, $\alpha \in Types$. The cut of $y_{o\alpha}$ by $z_{o\alpha}$ is the fuzzy set

$$y_{o\alpha}|z_{o\alpha} \equiv \lambda x_\alpha \cdot z_{o\alpha}x_\alpha \,\&\, \Delta(\Upsilon(z_{o\alpha}x_\alpha) \Rightarrow (y_{o\alpha}x_\alpha \equiv z_{o\alpha}x_\alpha)).$$

This formula says the following: the fuzzy set $y|z$ is a function that to each x_α assigns a truth value of the conjunction of truth values zx[5] and $\Delta(\Upsilon(zx) \Rightarrow (yx \equiv zx))$ where the latter has the truth value 1, if zx is nonzero and the truth values yx and zx are equal. Otherwise, it has the truth value 0.

The following lemma shows that thus defined operation does precisely what we want.

Lemma 2 ([11]). *Let $\mathcal{M}$ be a model and p an assignment of elements to variables $y_{o\alpha}, z_{o\alpha}$ such that $\mathcal{M}_p(y) = B \underset{\sim}{\subset} M_\alpha$, $\mathcal{M}_p(z) = Z \underset{\sim}{\subset} M_\alpha$. Then for any $m \in M_\alpha$*

$$\mathcal{M}_p(y|z)(m) = (B|Z)(m) = \begin{cases} B(m), & \textit{if } B(m) = Z(m), \\ 0 & \textit{otherwise.} \end{cases}$$

Let us also introduce the following special fuzzy sets: A fuzzy set $X_{o\alpha}$ is *crisp* if it has the property

$$\text{Crisp}_{o(o\alpha)}\, X_{o\alpha} \equiv (\forall u_\alpha)(Xu \equiv \Delta(Xu)).$$

Hence, a crisp fuzzy set has elements with membership degrees equal either to 1 or 0.

Support of a fuzzy set $X_{o\alpha}$ is a set defined by

$$\text{Supp}_{(o\alpha)(o\alpha)}\, X_{o\alpha} \equiv \lambda u_\alpha \cdot \Upsilon(X_{o\alpha}u_\alpha).$$

The universal set is defined by

$$V_\alpha \equiv \lambda x_\alpha \top. \tag{4}$$

Note that interpretation of V_α in a model $\mathcal{M}$ is $\mathcal{M}(V_\alpha) = M_\alpha$.

The following are basic properties of cut.

Lemma 3. *Let $B \in Form_{o\alpha}$, $\alpha \in Types$.*

(a) $\vdash B|B \equiv B$.
(b) $\vdash B|\emptyset \equiv \emptyset$.
(c) $\vdash \text{Crisp}\, \emptyset_{o\alpha}$.
(d) $\vdash \text{Crisp}(\text{Supp}\, B)$.
(e) $\text{Crisp}\, B \vdash (V|B \equiv B)$.
(f) $\text{Crisp}\, B \vdash (\forall z_{o\alpha})\, \text{Crisp}(B|z)$.
(g) $B \subset B' \vdash (\exists z)\neg(B|z \subseteq B'|z)$.

[5] In fact, $z_{o\alpha}x_\alpha$ is a membership degree of x_α in $z_{o\alpha}$.

Proof. (a), (b) are proved in [11], (c), (d) in [16].
(e)

(L.1) $\vdash (V|B)x \equiv Bx \,\&\, \Delta(\Upsilon(Bx) \Rightarrow (Vx \equiv Bx))$ (definition of $|$)
(L.2) $\vdash Bx \equiv \Delta Bx$ (assumption)
(L.3) $\vdash (V|B)x \equiv Bx \,\&\, \Delta(\Upsilon(\Delta(Bx)) \Rightarrow (\top \equiv \Delta(Bx)))$ (L.1, L.2, Rule (R))
(L.4) $\vdash (V|B)x \equiv Bx \,\&\, \Delta(\Delta(Bx) \Rightarrow (\Delta(Bx)))$ (L.3, Lemma 1(a), Rule (R))
(L.5) $\vdash (V|B)x \equiv Bx$ (L.4, properties of FTT)

(g) Let $\mathbf{v}$ be a constant, and $z_{o\alpha}$ a fuzzy set such that $\vdash (B\mathbf{v} < B'\mathbf{v}) \wedge (z\mathbf{v} \equiv B\mathbf{v})$. Then $\vdash (B'|z)\mathbf{v} \equiv \bot$ which implies (g) using the $\exists$-substitution axiom.

4 Intermediate Quantifiers

The theory of intermediate quantifiers is based on the concepts of measure of a fuzzy set, and linguistic evaluation of its size. All technical details not mentioned in this paper can be found in [11].

The measure is defined below. Note that we consider a relative measure, i.e., a measure of a fuzzy set $x_{o\alpha}$ w.r.t. a fuzzy set $z_{o\alpha}$.

Definition 1. *Let $R \in Form_{o(o\alpha)(o\alpha)}$ be a formula*[6].

(i) A formula $\mu \in Form_{o(o\alpha)(o\alpha)}$ defined by

$$\mu_{o(o\alpha)(o\alpha)} := \lambda z_{o\alpha}\, \lambda x_{o\alpha}\, (Rz_{o\alpha})x_{o\alpha} \tag{5}$$

represents a measure on fuzzy sets *in the universe of type $\alpha \in Types$ if it has the following properties:*

(M1) $\Delta(x_{o\alpha} \subseteq z_{o\alpha}) \,\&\, \Delta(y_{o\alpha} \subseteq z_{o\alpha}) \,\&\, \Delta(x_{o\alpha} \subseteq y_{o\alpha}) \Rightarrow ((\mu z_{o\alpha})x_{o\alpha} \Rightarrow (\mu z_{o\alpha})y_{o\alpha})$,

(M2) $\Delta(x_{o\alpha} \subseteq z_{o\alpha}) \Rightarrow ((\mu z_{o\alpha})(z_{o\alpha} \setminus x_{o\alpha}) \equiv \neg(\mu z_{o\alpha})x_{o\alpha})$,

(M3) $\Delta(x_{o\alpha} \subseteq y_{o\alpha}) \,\&\, \Delta(x_{o\alpha} \subseteq z_{o\alpha}) \,\&\, \Delta(y_{o\alpha} \subseteq z_{o\alpha}) \Rightarrow ((\mu z_{o\alpha})x_{o\alpha} \Rightarrow (\mu y_{o\alpha})x_{o\alpha})$.

(ii) The following formula characterizes measurable fuzzy sets *of a given type α:*

$$\mathbf{M}_{o(o\alpha)} := \lambda z_{o\alpha} \cdot \neg\Delta(z_{o\alpha} \equiv \emptyset_{o\alpha}) \,\&\, \Delta(\mu z_{o\alpha})z_{o\alpha} \,\&\, (\forall x_{o\alpha})(\forall y_{o\alpha})\Delta((M1) \,\&\, (M3)) \,\&\, (\forall x_{o\alpha})\Delta(M2) \tag{6}$$

where, for the simplicity of expression, we write (M1)–(M3) to stand for the axioms from (i).

---

[6] This formula can be understood as a procedure for computation of the output on the basis of a given input. In our case, the output is size (element of E) of the measure of a fuzzy set with respect to another one.

Axioms (M1) and (M3) characterize monotonicity of measure; namely that it is isotone w.r.t. $x_{o\alpha}$ and antitone w.r.t. $z_{o\alpha}$. Axiom (M2) characterizes measure of a complement of $x_{o\alpha}$ w.r.t. $z_{o\alpha}$.

We consider a formal theory T^{IQ} in which intermediate quantifiers are definable in the sense of the definition below. The theory must contain the theory of evaluative expressions and measurable fuzzy sets (see [11] for the details).

Definition 2. *Let T^{IQ} be a theory containing intermediate quantifiers and $Ev \in Form_{oo}$ be an intension of some evaluative linguistic expression. Finally, let $z \in Form_{o\alpha}$, $x \in Form_\alpha$ be variables, and $A, B \in Form_{o\alpha}$ be formulas where $T^{IQ} \vdash \mathbf{M}_{o(o\alpha)} B_{o\alpha}$. An intermediate generalized quantifier is one of the following formulas:*

$$(Q^{\forall}_{Ev}\, x_\alpha)(B, A) \equiv (\exists z)[(\forall x)((B|z)\, x \Rightarrow Ax) \wedge Ev((\mu B)(B|z))], \tag{7}$$

$$(Q^{\exists}_{Ev}\, x_\alpha)(B, A) \equiv (\exists z)[(\exists x)((B|z)x \wedge Ax) \wedge Ev((\mu B)(B|z))]. \tag{8}$$

Either of the quantifiers (7) or (8) construes the sentence

"⟨Quantifier⟩ B's are A"

where ⟨Quantifier⟩ is some intermediate quantifier. The formula $B_{o\alpha}$ represents a universe of quantification *and $A_{o\alpha}$ a property of elements of type α.*

Formula (7) has a clear meaning: all elements from a certain part $z_{o\alpha}$ of the universe $B_{o\alpha}$ have the property $A_{o\alpha}$, and size of the cut $B|z$ w.r.t. whole B is linguistically evaluated by Ev. For example the sentence "Most B are A" is construed by the formula $(Q^{\forall}_{Bi\ Ve}\, x)(B, A)$ which, using (7), is equivalent to $(\exists z)[(\forall x)((B|z)x \Rightarrow Ax) \wedge (Bi\ Ve)((\mu B)(B|z))]$. Similarly for (8).

Remark 1. The original definition of intermediate quantifiers introduced, e.g., in [15] considered all fuzzy subsets $z_{o\alpha}$ of $B_{o\alpha}$. When computing intermediate quantifiers on real data, however, it turned out that we obtain counterintuitive results. For example, "Most young women have long hair". If a given woman is young in the degree 0.7, it would be strange to consider her to be young in the degree 0.3. This is the reason why we introduced the operation of cut in (7) and (8).

Theorem 1. *Let $A, B, C, z \in Form_{o\alpha}$ be formulas representing properties of objects and $Ev \in Form_{oo}$ be an intension of some evaluative expression.*

(a) $T^{IQ} \cup \{A \subseteq C\} \vdash (Q^{\forall}_{Ev}\, x_\alpha)(B, A) \Rightarrow (Q^{\forall}_{Ev}\, x_\alpha)(B, C)$,

(b) $T^{IQ} \vdash (Q^{\forall}_{Ev}\, x_\alpha)(B, A) \vee (Q^{\forall}_{Ev}\, x_\alpha)(B, C) \Rightarrow (Q^{\forall}_{Ev}\, x_\alpha)(B, A \cup C)$.

Proof. (a) follows from and $\vdash (\forall x_\alpha)(Bx \Rightarrow Ax) \Rightarrow (\forall x_\alpha)(Bx \Rightarrow Cx)$ and the assumption.

(b) Using Lemma 1(b) and the properties of FTT, we can prove that

$$\vdash (\forall x_\alpha)((B|z)x \Rightarrow Ax) \vee (\forall x_\alpha)((B|z)x \Rightarrow cx) \Rightarrow (\forall x_\alpha)((B|z)x \Rightarrow Ax \wedge Cx)$$

Adding $Ev((\mu B)(B|z))$ to both sides of this implication, we obtain valid implication. Then, using distributivity of $\vee, \wedge$ and the property $\vdash (\exists z)(Pz \vee Qz) \Rightarrow ((\exists z)Pz \vee (\exists z)Qz)$ we obtain (b).

The quantifiers defined above are of type $\langle 1, 1\rangle$. It is possible, however, to introduce also simpler quantifiers of type $\langle 1\rangle$ that make quantification over the whole universe (similarly as the classical quantifiers $\forall$ and $\exists$ do).

Definition 3 (Quantifiers of type $\langle 1\rangle$). *Let $A \in Form_{o\alpha}$, $x \in Form_\alpha$ and $V_{o\alpha}$ be the universal set (4). Let $\vdash \mathbf{M}(V_{o\alpha})$. Then the formula*

$$(Q^{\forall}_{Ev}\, x)A \equiv (Q^{\forall}_{Ev}\, x)(V, A) \equiv (\exists z)[(\forall x)((V|z)\, x \Rightarrow Ax) \wedge Ev((\mu V)(V|z))] \quad (9)$$

is an intermediate quantifier of type $\langle 1\rangle$[7].

Theorem 2. *Let $A, z \in Form_{o\alpha}$ and $Ev \in \{Bi\,\boldsymbol{\nu} \mid \boldsymbol{\nu} \in \{Ex, Si, Ve, ML, Ro, QR, VR\}\}$. Then*

$$(Q^{\forall}_{Ev}\, x)A \equiv (\exists z)[(\forall x)((\mathrm{Supp}\, A|z)x \Rightarrow Ax) \wedge Ev((\mu V)(\mathrm{Supp}\, A|z)).$$

Proof. By Lemma 3(f), $V|z$ is crisp for all fuzzy sets $z \in Form_{o\alpha}$. Let $\mathcal{M}$ be a model, p an assignment, and $z_{o\alpha}$ be such that $\mathcal{M}_p(V|z) \cap (M_\alpha \setminus \mathcal{M}_p(\mathrm{Supp}\, A)) \neq \emptyset$. Then $\mathcal{M}_p((\forall x)((V|z)x \Rightarrow Ax)) = 0$. From it follows that the latter is (in general) non-zero only if $\mathcal{M}_p(V|z) \subseteq \mathcal{M}_p(\mathrm{Supp}\, A)$.

This theorem suggests a simplified way how intermediate quantifiers of type $\langle 1\rangle$ can be computed. Namely, it is sufficient to confine only to the support of parts of A.

5 The Quantifier "MORE-THAN"

This kind of quantifier is studied in the theory of generalized quantifiers [6,19]. An example of such quantifier is the following:

$$More\, girls\, than\, boys\, are\, diligent. \quad (10)$$

Classical model of this quantifier is

$$MT(B, C, A) = 1 \quad \text{iff} \quad |B \cap A| > |C \cap A| \quad (11)$$

where MT is the quantifier and $|\cdot|$ denotes number of elements (in a finite set). This definition does not consider hedging, i.e., modifying by hedges such as

[7] From the point of view of general theory, this quantifier trivially fulfills the property of relativization.

"much" or "a lot of". Introducing them would already require specification of the context w.r.t. which we could specify, how much greater $|B \cap A|$ than $|C \cap A|$ should be. This problem is solved in our definition below, in which we explicitly consider a universe U.

Note that in (10), we do not claim that all girls are more diligent than all boys; only certain part of girls are diligent and the same for boys. We compare sizes of these (fuzzy) sets w.r.t. a certain universe U that can be, e.g., all children at school, or a some more specific part of them.

Definition 4. *Let $A, B, C, z_1, z_2 \in Form_{o\alpha}$ be formulas representing properties of objects and $U \in Form_{o\alpha}$ be a measurable universe, i.e., $T^{IQ} \vdash \mathbf{M}(U)$ holds. Let $T^{IQ} \vdash (B \subseteq U) \wedge (C \subseteq U)$. Then:*

(i) The expression

More B than C are A (in a universe U)

is construed by the formula

$$(MT^{\forall}\, x_\alpha)(B, C, A; U) \equiv (\exists z_1)(\exists z_2)[(\forall x_\alpha)((B|z_1)\, x \Rightarrow Ax) \wedge (\forall x_\alpha)((C|z_2)\, x \Rightarrow Ax) \wedge \Upsilon((\mu U)(B|z_1) \ominus (\mu U)(C|z_2))]. \quad (12)$$

(ii) The modified expression of the form

⟨Hedge⟩ more B than C are A (in a universe U)

where ⟨Hedge⟩ can be, e.g., "much", "a lot", "a little", etc. In formal language, it is construed by the formula

$$MT^{\forall}_{Ev}(B, C, A; U) \equiv (\exists z_1)(\exists z_2)[(\forall x)((B|z_1)\, x \Rightarrow Ax) \wedge (\forall x)((C|z_2)\, x \Rightarrow Ax) \wedge (Ev)((\mu U)(B|z) \ominus (\mu U)(C|z))]. \quad (13)$$

The evaluative expression Ev can be the following:

(i) ⟨Hedge⟩ := *Much*: we put $Ev := Bi\,\bar{\boldsymbol{\nu}}$ (simple "big"),
(ii) ⟨Hedge⟩ := *Very much*: we put $Ev := Bi\,Ve$ ("very big"),
(iii) ⟨Hedge⟩ := *A little*: we put $Ev := Sm\,Ve$ ("very small").

Analogously, we can define other similar kinds of hedges in (ii).

There is one problem with this definition. When analyzing the linguistic expression (10), we see that it internally consists of two propositions:

(a) Girls are diligent.
(b) Boys are diligent.

In our definition we suggest to model both expressions by the implication

$$(\forall x)(Bx \Rightarrow Ax).$$

The classical definition (11), however, would suggest the model

$$(\exists x)(Bx \wedge Ax).$$

In this case, Definition (12) would change into

$$(MT^{\exists}\, x_\alpha)(B,C,A;U) \equiv (\exists z_1)(\exists z_2)[(\exists x)((B|z_1)\, x \wedge Ax) \wedge$$
$$(\exists x)((C|z_2)\, x \wedge Ax) \wedge \Upsilon((\mu U)(B|z) \ominus (\mu U)(C|z))]. \quad (14)$$

We need more investigation to give the definite answer. At this moment, let us only remark that definition (12) corresponds to (7).

Theorem 3. *Let* $A, B, C \in Form_{o\alpha}$ *be formulas. Let* $T^{IQ} \vdash (Q^{\forall}_{Ev} x)(C, A)$ *where* $Ev \in \{\neg\, Sm\,\bar{\nu}, Bi\; Ve, Bi\; Ex\}$ *and* $T^{IQ} \vdash (MT^{\forall}\, x_\alpha)(B,C,A;U)$*. Then* $T^{IQ} \vdash (Q^{\forall}_{Ev} x)(B, A)$.

Proof. **Semantic:** Let $\mathcal{M}$ be a model of T^{IQ}. For better readability, we put $\mathcal{M}_p(z_1) = Z_1$, $\mathcal{M}_p(z_2) = Z_2$, $\mathcal{M}_p(B) = B$, $\mathcal{M}_p(C) = C$, $\mathcal{M}_p(U) = U$, and similarly the other symbols.

By the assumption,

$$\bigvee_{\mathcal{M}_p(z_2)=Z_2 \subsetneq M_\alpha} \mathcal{M}_p((\forall x)((C|z_2)\, x \Rightarrow Ax)) \wedge Ev((\mu C)(C|Z_2)) = 1, \quad (15)$$

$$\bigvee_{\mathcal{M}_p(z_1)=Z_1 \subsetneq M_\alpha}\ \bigvee_{\mathcal{M}_p(z_2)=Z_2 \subsetneq M_\alpha} [\mathcal{M}_p((\forall x)((B|z_1)\, x \Rightarrow Ax)) \wedge$$
$$\mathcal{M}_p((\forall x)((C|z_2)\, x \Rightarrow Ax)) \wedge \Upsilon((\mu U)(B|Z_1) \ominus (\mu U)(C|Z_2))] = 1. \quad (16)$$

Hence, to every a, $0 < a \leq 1$, we have $a \leq (15)$, and also $a \leq (16)$. The latter means that there are $Z_1, Z_2 \subsetneq M_\alpha$ such that

$$(\mu U)(C|Z_2) < (\mu U)(B|Z_1). \quad (17)$$

Furthermore, $B, C \subseteq U$. Let us consider $a = (\mu U)(C|Z_2)$ for some Z_2. Then $a \leq Ev((\mu C)(C|Z_2))$ because $(\mu U)(C|Z_2) \leq (\mu C)(C|Z_2)$ by the properties of measure. But then by (17) and the properties of measure, we obtain $a \leq (\mu B)(B|Z_1)$ for some Z_1 which means that $a \leq Ev((\mu B)(B|Z_1))$ because of the increasing character of the assumed Ev. Since we considered arbitrary $0 < a \leq 1$, we conclude that $\bigvee_{\mathcal{M}_p(z_1)=Z_1 \subsetneq M_\alpha} \mathcal{M}_p((\forall x)((B|z_1)\, x \Rightarrow Ax)) \wedge Ev((\mu B)(B|Z_1)) = 1$.

By this theorem, if we surely know, e.g., that "More B than C are A" and "Most C are A" then it is valid to conclude that surely "Most (Almost all, All) B are A".

From the previous theorem, validity of the following (weak)[8] syllogisms immediately follows.

---

[8] By a strong syllogism with a major premise A_o, a minor premise B_o, and a conclusion C_o we understand provability of the formula $\vdash A_o \,\&\, B_o \Rightarrow C_o$. Note that the syllogism in Corollary 1 is weaker.

Corollary 1.

$$\frac{T^{IQ} \vdash (Q^{\forall}_{Ev}x)(C,A) \qquad T^{IQ} \vdash (MT^{\forall}\, x_\alpha)(B,C,A;U)}{T^{IQ} \vdash (Q^{\forall}_{Ev}x)(B,A)}$$

where $Ev \in \{\neg\, Sm\, \bar{\boldsymbol{\nu}}, Bi\ Ve, Bi\ Ex\}$.

We can also symmetrically consider the quantifier

Less C than B are A (in a universe U)

which can be defined by

$$(LT^{\forall}\, x_\alpha)(C,B,A;U) \equiv (MT^{\forall}\, x_\alpha)(B,C,A;U).$$

Of course, we can also add ⟨Hedge⟩ to this quantifier.

By similar arguments as in Theorem 3, we can prove the following.

Theorem 4. *Let* $A, B, C \in Form_{o\alpha}$ *be formulas. Let* $T^{IQ} \vdash (Q^{\forall}_{Ev}x)(B,A)$ *where* $Ev \in \{^{+}Sm\ Ve, ^{+}Sm\ Si\}$ *and* $T^{IQ} \vdash LT_U(C,B,A)$. *Then* $T^{IQ} \vdash (Q^{\forall}_{Ev}x)(C,A)$.

Corollary 2.

$$\frac{T^{IQ} \vdash (Q^{\forall}_{Ev}x)(B,A) \qquad T^{IQ} \vdash (LT^{\forall}\, x_\alpha)(C,B,A;U)}{T^{IQ} \vdash (Q^{\forall}_{Ev}x)(C,A)}$$

where $Q \in \{A\ few, Several\}$.

We may now ask whether the quantifier MORE-THAN is transitive as in the following example:

More young girls than young boys like yoga.
More young boys than managers like yoga.
———
More young girls than managers like yoga.

Theorem 5. *Let* $A, B, C, D \in Form_{o\alpha}$ *be formulas. Let* $T^{IQ} \vdash (MT^{\forall}\, x_\alpha)(B,C,A;U)$ *and* $T^{IQ} \vdash (MT^{\forall}\, x_\alpha)(C,D,A;U)$. *Then* $T^{IQ} \vdash (MT^{\forall}\, x_\alpha)(B,D,A;U)$.

Proof. **Semantic:** Let $\mathcal{M}$ be a model of T^{IQ}. For better readability, we put $\mathcal{M}_p(z_1) = Z_1$, $\mathcal{M}_p(z_2) = Z_2$, $\mathcal{M}_p(z_3) = Z_3$, $\mathcal{M}_p(A) = A$, $\mathcal{M}_p(D) = D$, $\mathcal{M}_p(B) = B$, $\mathcal{M}_p(C) = C$, $\mathcal{M}_p(U) = U$.

By the assumption,

$$\bigvee_{\mathcal{M}_p(z_1)=Z_1 \underset{\sim}{\subset} M_\alpha} \ \bigvee_{\mathcal{M}_p(z_2)=Z_2 \underset{\sim}{\subset} M_\alpha} [\mathcal{M}_p((\forall x)((B|z_1)\, x \Rightarrow Ax)) \wedge$$

$$\mathcal{M}_p((\forall x)((C|z_2)\, x \Rightarrow Ax)) \wedge \Upsilon((\mu U)(B|Z_1) \ominus (\mu U)(C|Z_2))] = 1. \tag{18}$$

$$\bigvee_{\mathcal{M}_p(z_2)=Z_2 \subsetneq M_\alpha} \bigvee_{\mathcal{M}_p(z_3)=Z_3 \subsetneq M_\alpha} [\mathcal{M}_p((\forall x)((C|z_2)\, x \Rightarrow Ax)) \wedge$$
$$\mathcal{M}_p((\forall x)((D|z_3)\, x \Rightarrow Ax)) \wedge \Upsilon((\mu U)(C|Z_2) \ominus (\mu U)(D|Z_3))] = 1. \quad (19)$$

Similarly as above, to every a, $0 < a \leq 1$, one can see that $a \leq (18)$ and $a \leq (19)$. From these assumptions it follows that there are $Z_1, Z_2, Z_3 \subsetneq M_\alpha$ such that

$$(\mu U)(C|Z_2) < (\mu U)(B|Z_1) \text{ as well as } (\mu U)(D|Z_3) < (\mu U)(C|Z_2) \quad (20)$$

which implies that

$$(\mu U)(D|Z_3) < (\mu U)(D|Z_1). \quad (21)$$

By the properties of L-FTT we have

$$\mathcal{M}_p((\forall x)((B|z_1)\, x \Rightarrow Ax)) \wedge \mathcal{M}_p((\forall x)((C|z_2)\, x \Rightarrow Ax)) \leq$$
$$\mathcal{M}_p((\forall x)((B|z_1)\, x \Rightarrow Ax)) \quad (22)$$

$$\mathcal{M}_p((\forall x)((C|z_2)\, x \Rightarrow Ax)) \wedge \mathcal{M}_p((\forall x)((D|z_3)\, x \Rightarrow Ax))] \leq$$
$$\mathcal{M}_p((\forall x)((D|z_3)\, x \Rightarrow Ax)) \quad (23)$$

Considering arbitrary $0 < a \leq 1$ and using (21), (22), (23) we conclude that

$$\bigvee_{\mathcal{M}_p(z_1)=Z_1 \subsetneq M_\alpha} \bigvee_{\mathcal{M}_p(z_3)=Z_3 \subsetneq M_\alpha} [\mathcal{M}_p((\forall x)((B|z_1)\, x \Rightarrow Ax)) \wedge$$
$$\mathcal{M}_p((\forall x)((D|z_3)\, x \Rightarrow Ax)) \wedge \Upsilon((\mu U)(B|Z_1) \ominus (\mu U)(D|Z_3))] = 1 \quad (24)$$

by the properties of supremum.

6 Conclusion

In this paper, we continued the research in the formal theory of intermediate quantifiers. We proved a few new results, introduced intermediate quantifiers of type $\langle 1 \rangle$, and also new quantifiers MORE-THAN and LESS-THAN. We also proved validity of weak syllogisms with these quantifiers.

The future study will be focused on new forms of generalized syllogisms with the proposed quantifiers. Note that they can be used for a linguistic summarization in human reasoning.

References

1. Andrews, P.: An Introduction to Mathematical Logic and Type Theory: To Truth Through Proof. Kluwer, Dordrecht (2002)

2. Cignoli, R.L.O., D'Ottaviano, I.M.L., Mundici, D.: Algebraic Foundations of Many-Valued Reasoning. Kluwer, Dordrecht (2000)
3. Dvořák, A., Holčapek, M.: L-fuzzy quantifiers of the type $\langle 1 \rangle$ determined by measures. Fuzzy Sets Syst. **160**, 3425–3452 (2009)
4. Hájek, P.: Metamathematics of Fuzzy Logic. Kluwer, Dordrecht (1998)
5. Holčapek, M.: Monadic L-fuzzy quantifiers of the type $\langle 1^n, 1 \rangle$. Fuzzy Sets Syst. **159**, 1811–1835 (2008)
6. Keenan, E., Westerståhl, D.: Quantifiers in formal and natural languages. In: van Benthem, J., ter Meulen, A. (eds.) Handbook of Logic and Language, pp. 837–893. Elsevier, Amsterdam (1997)
7. Lindström, P.: First order predicate logic with generalized quantifiers. Theoria **32**, 186–195 (1966)
8. Mostowski, A.: On a generalization of quantifiers. Fundamenta Mathematicae **44**, 12–36 (1957)
9. Murinová, P., Novák, V.: A formal theory of generalized intermediate syllogisms. Fuzzy Sets Syst. **186**, 47–80 (2012)
10. Murinová, P., Novák, V.: The structure of generalized intermediate syllogisms. Fuzzy Sets Syst. **247**, 18–37 (2014)
11. Murinová, P., Novák, V.: The theory of intermediate quantifiers in fuzzy natural logic revisited and the model of "Many". Fuzzy Sets Syst. **388**, 56–89 (2020)
12. Novák, V.: Antonyms and linguistic quantifiers in fuzzy logic. Fuzzy Sets Syst. **124**, 335–351 (2001)
13. Novák, V.: On fuzzy type theory. Fuzzy Sets Syst. **149**, 235–273 (2005)
14. Novák, V.: A comprehensive theory of trichotomous evaluative linguistic expressions. Fuzzy Sets Syst. **159**(22), 2939–2969 (2008)
15. Novák, V.: A formal theory of intermediate quantifiers. Fuzzy Sets Syst. **159**(10), 1229–1246 (2008)
16. Novák, V.: Topology in the alternative set theory and rough sets via fuzzy type theory. Mathematics **8**(3), 432–453 (2020)
17. Novák, V., Perfilieva, I., Dvořák, A.: Insight into Fuzzy Modeling. Wiley, Hoboken (2016)
18. Novák, V., Perfilieva, I., Močkoř, J.: Mathematical Principles of Fuzzy Logic. Kluwer, Boston (1999)
19. Peters, S., Westerståhl, D.: Quantifiers in Language and Logic. Clarendon Press, Oxford (2006)
20. Peterson, P.: Intermediate Quantifiers. Logic, linguistics, and Aristotelian semantics. Ashgate, Aldershot (2000)
21. Westerståhl, D.: Quantifiers in formal and natural languages. In: Gabbay, D., Guenthner, F. (eds.) Handbook of Philosophical Logic, vol. IV, pp. 1–131. D. Reidel, Dordrecht (1989)

On Semantic Properties of Fuzzy Quantifiers over Fuzzy Universes: Restriction and Living on

Antonín Dvořák and Michal Holčapek(✉)

CE IT4I - IRAFM, University of Ostrava, 30. dubna 22,
701 03 Ostrava 1, Czech Republic
{antonin.dvorak,michal.holcapek}@osu.cz
http://ifm.osu.cz

Abstract. The article investigates important semantic properties of fuzzy quantifiers, namely restriction and living on a (fuzzy) set. These properties are introduced in the novel frame of fuzzy quantifiers over fuzzy universes.

Keywords: Generalized quantifiers · Fuzzy quantifiers · Semantic properties

1 Introduction

In this paper, we continue our investigation of fuzzy quantifiers and their semantic properties [3–5,8]. We are working with the general concept of generalized quantifiers originating from works of Mostowski [12], Lindström [11] and many others. For details, we refer to monograph [13]. For example, a generalized quantifier Q with one argument (so-called type $\langle 1 \rangle^1$) over a set universe M can be understood as a mapping from the powerset of M to the set of truth values $\{0, 1\}$ (i.e., false and true, respectively).

At first [3,8], we investigated a straightforward generalization of these generalized quantifiers, where arguments of fuzzy quantifiers were fuzzy sets and the set of truth values $\{0, 1\}$ has been replaced by some more general structure. Namely, we used a residuated lattice L and defined fuzzy quantifiers on M as mappings from the power set of M to L. However, these generalized quantifiers were still defined over a *crisp* universe M. Gradually, we started to be aware of severe limitations of this approach. For example, it was not possible to define the important operation of *relativization* in a satisfactory way (see [5]). The reason

[1] This notation originated in [11], where quantifiers are understood as classes of relational structures of a certain type (representing a number of arguments and variable binding). It is widely used in the literature on generalized quantifiers [13].

The second author announces a support of Czech Science Foundation through the grant 18-06915S and the ERDF/ESF project AI-Met4AI No. CZ.02.1.01/0.0/0.0/17_049/0008414.

M.-J. Lesot et al. (Eds.): IPMU 2020, CCIS 1239, pp. 173–186, 2020.
https://doi.org/10.1007/978-3-030-50153-2_13

is that in the definition of relativization, the first argument of a quantifier (i.e., a fuzzy set) becomes a new universe for the relativized quantifier. But, only crisp sets have been permitted as universes for fuzzy quantification. To overcome this limitation, in [5] we defined the so-called C-fuzzy quantifiers, where a fuzzy set C served as a universe of quantification. We showed there that relativization can be satisfactorily defined in this frame. However, also the approach of C-fuzzy quantifiers has its limitations. In this contribution, we present initial observations on a more general approach, in which pairs (M, C), where M is a crisp set and C is a fuzzy subset of M, serve as universes for fuzzy quantification. As case studies, important semantic notions of *restriction* and of *living on a fuzzy set* in the setting of fuzzy quantifiers over fuzzy universes are investigated.

This paper is structured as follows: In Sect. 2, we summarize necessary notions on algebras of truth values and on fuzzy sets. Section 3 contains basic definitions of generalized quantifiers, restricted quantifiers and quantifiers living on a set. These notions are then generalized in Sect. 4. Finally, Sect. 5 contains conclusions and directions of further research.

2 Preliminaries

2.1 Algebraic Structures of Truth Values

In this article we assume that the algebraic structure of truth values is a *complete residuated lattice*, i.e., an algebraic structure $L = \langle L, \wedge, \vee, \otimes, \rightarrow, 0, 1\rangle$ with four binary operations and two constants such that $\langle L, \wedge, \vee, 0, 1\rangle$ is a complete lattice, where 0 is the least element and 1 is the greatest element of L, $\langle L, \otimes, 1\rangle$ is a commutative monoid (i.e., $\otimes$ is associative, commutative and the identity $a \otimes 1 = a$ holds for any $a \in L$) and the adjointness property is satisfied, i.e.,

$$a \leq b \rightarrow c \quad \text{iff} \quad a \otimes b \leq c \tag{1}$$

holds for each $a, b, c \in L$, where $\leq$ denotes the corresponding lattice ordering, i.e., $a \leq b$ if $a \wedge b = a$ for $a, b \in L$. A residuated lattice L is said to be *divisible* if $a \otimes (a \rightarrow b) = a \wedge b$ holds for arbitrary $a, b \in L$. The operation of *negation* on L is defined as $\neg a = a \rightarrow 0$ for $a \in L$. A residuated lattice L satisfies the *law of double negation* if $\neg\neg a = a$ holds for any $a \in L$. A divisible residuated lattice satisfying the law of double negation is called an *MV-algebra*. A residuated lattice is said to be *linearly ordered* if the corresponding lattice ordering is linear, i.e., $a \leq b$ or $b \leq a$ holds for any $a, b \in L$.

Obviously, the two elements residuated lattice, i.e., $L = \{0, 1\}$, is a Boolean algebra. We put $\mathbf{2} = L = \{0, 1\}$. Other examples of complete residuated lattices can be determined from left-continuous t-norms on the unit interval:

Example 1. The algebraic structure

$$L_T = \langle [0,1], \min, \max, T, \rightarrow_T, 0, 1\rangle,$$

where T is a left continuous t-norm on $[0, 1]$ and $a \rightarrow_T b = \bigvee\{c \in [0, 1] \mid T(a, c) \leq b\}$, defines the residuum, is a complete residuated lattice (see, e.g., [2,7,10]).

2.2 Fuzzy Sets

Let L be a complete residuated lattice, and let M be a non-empty universe of discourse. A function $A : M \to L$ is called a *fuzzy set* (*L-fuzzy set*) *on* M. A value $A(m)$ is called a *membership degree of* m *in the fuzzy set* A. The set of all fuzzy sets on M is denoted by $\mathcal{F}(M)$. A fuzzy set A is called *crisp* if there is a subset $Z \subseteq M$ such that $A = 1_Z$, where 1_Z denotes the characteristic function of Z. Obviously, a crisp fuzzy set can be uniquely identified with a subset of M. The symbol $\emptyset$ denotes the empty fuzzy set on M, i.e., $\emptyset(m) = 0$ for any $m \in M$. The set of all crisp fuzzy sets on M (i.e., the power set of M) is denoted by $\mathcal{P}(M)$. The set $\mathrm{Supp}(A) = \{m \in M \mid A(m) > 0\}$ is called the *support* of a fuzzy set A.

Let $A, B \in \mathcal{F}(M)$. We say that A *is less than or equal to* B and denoted it as $A \subseteq B$ if $A(m) \le B(m)$ for any $m \in M$. Moreover, A *is equal to* B if $A \subseteq B$ and $B \subseteq A$.

Let $\{A_i \mid i \in I\}$ be a non-empty family of fuzzy sets on M. Then the *union* and *intersection of* A_i are defined as

$$\left(\bigcup_{i\in I} A_i\right)(m) := \bigvee_{i\in I} A_i(m) \quad \text{and} \quad \left(\bigcap_{i\in I} A_i\right)(m) := \bigwedge_{i\in I} A_i(m), \tag{2}$$

for any $m \in M$, respectively. Further, extensions of the operations $\otimes$ and $\to$ on L to the operations on $\mathcal{F}(M)$ are given by

$$(A \otimes B)(m) := A(m) \otimes B(m) \quad \text{and} \quad (A \to B)(m) := A(m) \to B(m), \tag{3}$$

respectively, for any $A, B \in \mathcal{F}(M)$ and $m \in M$. Finally, we introduce the difference of fuzzy sets A and B on M as follows:

$$(A \setminus B)(m) = A(m) \otimes \neg B(m) \tag{4}$$

for any $m \in M$.

3 NL-Quantifiers and Generalized Quantifiers

By *NL-quantifiers*, we in this paper understand natural language expressions such as "for all", "many", "several", etc. For our purposes it is not necessary to delineate the class of NL-quantifiers formally. In fact, we are interested in *general mathematical models* of these NL-quantifiers. For the sake of comprehensibility, we in the following informal explanation consider NL-quantifiers with two arguments, such as "some" in the sentence "Some people are clever."

3.1 Generalized Quantifiers

Generally (see [13]), a model of the NL-quantifier "some" takes the form of a functional (the so-called *global* quantifier) some that to any universe of discourse

M assigns a *local* quantifier some_M. This local quantifier is a mapping that to any two subsets A and B of M assigns a truth value $\mathsf{some}_M(A, B)$. In the following, if we speak about a quantifier Q, we have in mind some *global* quantifier, that is, the functional as above. If we consider only (classical) sets A and B and the truth value of $\mathsf{some}_M(A, B)$ can be only either *true* or *false*, we say that this some is a *generalized quantifier*. If A and B are fuzzy sets and the truth value of $\mathsf{some}_M(A, B)$ is taken from some many-valued structure of truth degrees, we say that this some is a *fuzzy quantifier*.

Definition 1 (Local generalized quantifier). *Let M is a universe of discourse. A* local generalized quantifier *Q_M of type $\langle 1^n, 1\rangle$ over M is a function $\mathcal{P}(M)^n \times \mathcal{P}(M) \to \mathbf{2}$ that to any sets $A_1, \ldots, A_n$ and B from $\mathcal{P}(M)$ assigns a truth value $Q_M(A_1, \ldots, A_n, B)$ from $\mathbf{2}$.*

Definition 2 (Global generalized quantifier). *A* global generalized quantifier *Q of type $\langle 1^n, 1\rangle$ is a functional that to any universe M assigns a local generalized quantifier $Q_M \colon \mathcal{P}(M)^n \times \mathcal{P}(M) \to \mathbf{2}$ of type $\langle 1^n, 1\rangle$.*

Among the most important examples of generalized quantifiers of type $\langle 1\rangle$ belong the classical quantifiers $\forall$ and $\exists$. Their definitions are as follows: $\forall_M(B) = 1$ if and only if $B = M$ and $\exists_M(B) = 1$ if and only if $B \neq \emptyset$ for any $B \in \mathcal{P}(M)$. Formally, these definitions can be also expressed as $\forall_M(B) := B = M$ and $\exists_M(B) := B \neq \emptyset$. The important examples of type $\langle 1, 1\rangle$ generalized quantifiers are all and some, defined as $\mathsf{all}_M(A, B) = 1$ if and only if $A \subseteq B$ and $\mathsf{some}_M(A, B) = 1$ if and only if $A \cap B \neq \emptyset$ for any $A, B \in \mathcal{P}(M)$. Note that the universe M does not appear on the right side of definitions of all and some, which is a difference from the quantifier $\forall_M$, therefore, the truth values of these quantifiers are not directly influenced by their universes. The quantifiers, which possess this essential (semantic) property, are in the generalized quantifier theory said to satisfy the *extension*. Among further essential properties of generalized quantifiers belong the *permutation* and *isomorphism invariance* or the *conservativity*. More about these properties can be found in [9,13].

Let us recall the definition of the relativization of generalized quantifiers [13] mentioned in Sect. 1.

Definition 3. *Let Q be a global generalized quantifier of type $\langle 1^n, 1\rangle$. The* relativization *of Q is a global generalized quantifier Q^{rel} of type $\langle 1^{n+1}, 1\rangle$ defined as*

$$(Q^{\mathrm{rel}})_M(A, A_1, \ldots, A_n, B) := Q_A(A \cap A_1, \ldots, A \cap A_n, A \cap B) \tag{5}$$

for all $A, A_1, \ldots, A_n, B \in \mathcal{P}(M)$. For the most common case of relativization from type $\langle 1\rangle$ to type $\langle 1, 1\rangle$,

$$(Q^{\mathrm{rel}})_M(A, B) = Q_A(A \cap B). \tag{6}$$

It is well known that $\forall^{\mathrm{rel}} = \mathsf{all}$ and $\exists^{\mathrm{rel}} = \mathsf{some}$. In [13], the authors argue that all models of NL-quantifiers of the most common type $\langle 1, 1\rangle$ should be conservative and satisfy the property of extension. It is very interesting that each type $\langle 1, 1\rangle$ generalized quantifier, which possesses the above mentioned properties, is the relativization of a type $\langle 1\rangle$ generalized quantifier.

3.2 Restriction

In the theory of generalized quantifiers, we can distinguish two interesting notions, namely, global generalized quantifiers restricted to a set and local generalized quantifiers living on a set, that play an undoubtedly important rôle in the characterization of generalized quantifiers, but they have not been taken into account for fuzzy quantifiers yet. Note that these notions are considered in [13] for quantifiers of type $\langle 1\rangle$. In this part, we recall (for type $\langle 1\rangle$) and extend (for type $\langle 1^n, 1\rangle$) the concept of the generalized quantifier restricted to a set.

Definition 4. *Let Q be a type $\langle 1\rangle$ global generalized quantifier, and let A be a set. The quantifier Q is* restricted to A *if for any M and $B \subseteq M$ we have*

$$Q_M(B) = Q_A(A \cap B). \tag{7}$$

The set of all type $\langle 1\rangle$ global generalized quantifiers restricted to A is denoted by $RST_{\langle 1\rangle}(A)$.

A natural generalization of the concept of generalized quantifiers restricted to a set for quantifiers of type $\langle 1^n, 1\rangle$ can be provided as follows:

Definition 5. *Let Q be a type $\langle 1^n, 1\rangle$ global generalized quantifier, and let A be a set. The quantifier Q is* restricted to A *if for any M and $A_1, \dots, A_n, B \subseteq M$ we have*

$$Q_M(A_1, \dots, A_n, B) = Q_A(A \cap A_1, \dots, A \cap A_n, A \cap B). \tag{8}$$

The set of all type $\langle 1^n, 1\rangle$ global generalized quantifiers restricted to A is denoted by $RST_{\langle 1^n,1\rangle}(A)$.

One can see that if a global generalized quantifier Q is restricted to a set A, then its evaluations over all considered universes are determined from the local generalized quantifier Q_A.

Lemma 1. *If Q is a type $\langle 1^n, 1\rangle$ global generalized quantifier restricted to a set A, then Q is restricted to any set A' such that $A \subseteq A'$, i.e., $RST_{\langle 1^n,1\rangle}(A) \subseteq RST_{\langle 1^n,1\rangle}(A')$.*

Now we show that global generalized quantifiers restricted to sets can be used to introduce the relativization of a global generalized quantifier Q. More precisely, let Q be a type $\langle 1^n, 1\rangle$ generalized quantifier. Then, for each set A, we can introduce the type $\langle 1^n, 1\rangle$ global generalized quantifier $Q^{[A]}$ as

$$(Q^{[A]})_M(A_1, \dots, A_n, B) = Q_A(A \cap A_1, \dots, A \cap A_n, A \cap B)$$

for any M and $A_1, \dots, A_n, B \subseteq M$. It is easy to show that $Q^{[A]} \in \mathrm{RST}_{\langle 1^n,1\rangle}(A)$. It should be noted that Q and $Q^{[A]}$ are different quantifiers. They become identical if Q is already restricted to A. Now, if we define the global generalized quantifier Q' of type $\langle 1^{n+1}, 1\rangle$ as

$$Q'_M(A, A_1, \dots, A_n, B) = (Q^{[A]})_M(A_1, \dots, A_n, B) \tag{9}$$

for all M and $A_1, \dots, A_n, B \subseteq M$, then it is easy to show that $Q' = Q^{\mathrm{rel}}$, where Q^{rel} has been defined in Definition 3.

3.3 Generalized Quantifiers Living on a Set

A concept related to that of a global generalized quantifier restricted to a set is the concept of a local generalized quantifier living on a set.

Definition 6. *Let M and A be sets. A local generalized quantifier Q_M of type* $\langle 1\rangle$ lives on A *if, for any $B \subseteq M$, we have*

$$Q_M(B) = Q_M(A \cap B). \tag{10}$$

One can see that each local generalized quantifier Q_M, where $Q \in \mathrm{RST}_{\langle 1\rangle}(A)$ (that is, Q is a global generalized quantifier restricted to A and Q_M is the corresponding local quantifier over M), lives on A. Indeed, by (7), we have

$$Q_M(B) = Q_A(A \cap B) = Q_A(A \cap (A \cap B)) = Q_M(A \cap B).$$

Note that A need not be the smallest set on which Q_M lives ($Q \in \mathrm{RST}_{\langle 1\rangle}(A)$). Moreover, the concept of *conservativity* can be introduced in terms of local generalized quantifiers living on sets. Let Q be a global conservative quantifier of type $\langle 1, 1\rangle$.[2] For any M and $A \subseteq M$, define a type $\langle 1\rangle$ local quantifier $(Q[A])_M$ as follows

$$(Q[A])_M(B) = Q_M(A, B) \tag{11}$$

for all $B \subseteq M$. One can see that, as a simple consequence of the conservativity of Q, we obtain that $(Q[A])_M$ lives on A. Vice versa, if each local generalized quantifier $(Q[A])_M$ lives on A for any M, then Q is conservative. Note that Barwise and Cooper expressed the concept of conservativity as we described above using the live-on property [1].

A natural generalization of the concept of a local generalized quantifier living on a set for quantifiers of type $\langle 1^n, 1\rangle$ can be defined as follows:

Definition 7. *Let M and A be sets. The local generalized quantifier Q_M of type* $\langle 1^n, 1\rangle$ lives on A *if, for any $A_1, \dots, A_n, B \subseteq M$, we have*

$$Q_M(A_1, \dots, A_n, B) = Q_M(A \cap A_1, \dots, A \cap A_n, A \cap B). \tag{12}$$

Also in this case, a conservativity of global generalized quantifiers of type $\langle 1^n, 1\rangle$ can be expressed in terms of the live-on property. Indeed, let Q be a global quantifier of type $\langle 1^n, 1\rangle$ for $n \geq 1$. For any M and $A_1, \dots, A_n \subseteq M$, define local quantifier $(Q[A_1, \dots, A_n])_M$ of type $\langle 1\rangle$ as

$$(Q[A_1, \dots, A_n])_M(B) = Q_M(A_1, \dots, A_n, B). \tag{13}$$

[2] Recall that a type $\langle 1, 1\rangle$ global generalized quantifier is *conservative* if $Q_M(A, B) = Q_M(A, B')$ holds for any $A, B, B' \in \mathcal{P}(M)$ such that $A \cap B = A \cap B'$. The conservativity for a type $\langle 1^n, 1\rangle$ generalized quantifier is defined analogously such that $Q_M(A_1, \dots, A_n, B) = Q_M(A_1, \dots, A_n, B')$ holds for any $A_1, \dots, A_n, B, B' \in \mathcal{P}(M)$ such that $A_i \cap B = A_i \cap B'$ for $i = 1, \dots, n$.

It is easy to see that Q is conservative if and only if each local generalized quantifier $(Q[A_1,\dots,A_n])_M$ lives on $A = \bigcup_{i=1}^n A_i$.

The following lemma contains useful facts about generalized quantifiers living on sets (cf. [13, Section 3.2.2]).

Lemma 2. *Let Q be a type $\langle 1^n, 1\rangle$ global generalized quantifier, and let M be a set. Then*

(i) Q_M lives on M.
(ii) Q_M lives on $\emptyset$ if and only if Q_M is trivial (i.e., $Q_M(A_1,\dots,A_n,B) = 0$ for any $A_1,\dots,A_n,B \subseteq M$ or $Q_M(A_1,\dots,A_n,B) = 1$ for any $A_1,\dots,A_n,B \subseteq M$).
(iii) If Q_M lives on C_1 and C_2, then it lives on $C_1 \cap C_2$. Hence, if M is finite, there is always a smallest set on which Q_M lives. This fails, however, when M is infinite.
(iv) $(Q^{[A]})_M$ lives on A and its supersets. If $(Q^{[A]})_M$ is non-trivial, A need not be the smallest set on which $(Q^{[A]})_M$ lives.

4 Fuzzy Quantifiers over Fuzzy Universes

The aim of this section is to introduce the concept of fuzzy quantifiers defined over fuzzy universes. We start with the introduction of (local and global) fuzzy quantifiers over crisp universes, where we demonstrate the limitation of their definitions. This motivates us to introduce the concept of fuzzy universe and define (local and global) fuzzy quantifiers over such universes.

4.1 Fuzzy Quantifiers over Crisp Universes

An immediate generalization of Definitions 1 and 2 consists of replacing classical sets $A_1,\dots,A_n$ and B by fuzzy subsets of M and of using a residuated lattice L instead of the Boolean algebra **2** (see [6,8]).

Definition 8 (Local fuzzy quantifier). *Let M be a universe of discourse. A* local fuzzy quantifier *Q_M of type $\langle 1^n, 1\rangle$ over M is a function $Q_M : \mathcal{F}(M)^n \times \mathcal{F}(M) \to L$ that to any fuzzy sets $A_1,\dots,A_n$ and B from $\mathcal{F}(M)$ assigns a truth value $Q_M(A_1,\dots,A_n,B)$ from L.*

Definition 9 (Global fuzzy quantifier). *A* global fuzzy quantifier *Q of type $\langle 1^n, 1\rangle$ is a functional that to any universe M assigns a local fuzzy quantifier $Q : \mathcal{F}(M)^n \times \mathcal{F}(M) \to L$ of type $\langle 1^n, 1\rangle$. The set of all global fuzzy quantifiers of type $\langle 1^n, 1\rangle$ will be denoted by* $\mathrm{QUANT}_{\langle 1^n,1\rangle}$.

Among the important examples of global fuzzy quantifiers of type $\langle 1\rangle$ are, again, $\forall$ and $\exists$. They are standardly defined as

$$\forall_M(B) := \bigwedge_{m \in M} B(m) \tag{14}$$

and

$$\exists_M(B) := \bigvee_{m \in M} B(m). \tag{15}$$

Important examples of global fuzzy quantifiers of type $\langle 1,1\rangle$ are all and some, defined as

$$\mathsf{all}_M(A,B) := \bigwedge_{m \in M} (A \to B)(m) \tag{16}$$

and

$$\mathsf{some}_M(A,B) := \bigvee_{m \in M} (A \cap B)(m). \tag{17}$$

As we have mentioned in Sect. 1, there is a principal problem to introduce the relativization of a fuzzy quantifier, because one argument of a fuzzy quantifier becomes a universe and a fuzzy set as the universe is not permitted. Therefore, in [8], the relativization of a global fuzzy quantifier Q of type $\langle 1\rangle$ has been proposed:

$$Q^{\mathrm{rel}}_M(A,B) := Q_{\mathrm{Supp}(A)}(A \cap B), \tag{18}$$

where $\mathrm{Supp}(A)$ is used as the universe of the fuzzy quantifier Q instead of A, which is the fuzzy set in the first argument of fuzzy quantifier Q^{rel}_M. Unfortunately, this solution is generally not satisfactory. For example, one can simply derive that

$$(\forall^{\mathrm{rel}})_M(A,B) := \bigwedge_{m \in \mathrm{Supp}(A)} (A \cap B)(m) \neq \mathsf{all}_M(A,B),$$

which is counter-intuitive. Hence, the definition of relativization does not seem to be well established. Another concept called weak relativization was also provided in [8], but again it generally fails. In [5], it was proved that there is no satisfactory definition of relativization of fuzzy quantifiers of type $\langle 1\rangle$. However, an introduction of fuzzy sets as universes for fuzzy quantifiers is not motivated only by relativization of fuzzy quantifiers. The second example can be the concept of restriction of a fuzzy quantifier to a fuzzy set (see Definition 5). Thus, the absence of fuzzy sets as universes for fuzzy quantifiers brings significant limitations in the development of the fuzzy quantifier theory that should possibly cover a wide part of topics studied in the field of generalized quantifiers.

4.2 Fuzzy Universes

Let A be a fuzzy set on N, and let M be a set. Define $A_M : M \to L$ as

$$A_M(m) = \begin{cases} A(m), & \text{if } m \in M \cap N, \\ 0, & \text{otherwise.} \end{cases} \tag{19}$$

The fuzzy set A_M represents A (or its part) on the universe M. Obviously, if $A \in \mathcal{F}(M)$, then $A_M = A$. It is easy to see that $\mathrm{Supp}(A_M) = \mathrm{Supp}(A) \cap M$. Moreover, if A, B are fuzzy sets on N and M is a set, then we have $(A \cap B)_M = A_M \cap B_M$ and $(A \cup B)_M = A_M \cup B_M$.

A pair (M, A), where M is a set and A is a fuzzy set on M, is called a *fuzzy universe*. Obviously, if $M = \emptyset$ in (M, A), then $A = \emptyset$ is the empty function. A fuzzy universe (M, A) is said to be *crisp* if A is crisp and $A = 1_M$. Let (M, A) and (N, B) be fuzzy universes. The basic (fuzzy) set operations for fuzzy universes are defined as follows:

- $(M, A) \cap (N, B) = (K, A_K \cap B_K)$, where $K = M \cap N$;[3]
- $(M, A) \cup (N, B) = (K, A_K \cup B_K)$, where $K = M \cup N$;
- $(M, A) \setminus (N, B) = (M, A_M \setminus B_M)$.

For the sake of simplicity, we write simply $(M, A) \cap (N, B) = (M \cap N, A \cap B)$ and assume that $A \cap B$ is well introduced on the universe $M \cap N$ according to the definition above. A similar notation can be also used for the union and the difference of fuzzy universes.

A non-empty class $\mathcal{U}$ of fuzzy universes is said to be *well defined* if

C1) $(M, A) \in \mathcal{U}$ implies $(M, B) \in \mathcal{U}$ for any $B \in \mathcal{F}(M)$;
C2) $\mathcal{U}$ is closed under the intersection, union and difference.

In what follows, we assume that each class of fuzzy universes is well defined.

Fundamental binary relations for fuzzy universes in a class $\mathcal{U}$ can be introduced as follows. We say that (M, A) *is equal to* (N, B), and denote it by $(M, A) = (N, B)$, if $M = N$ and $A = B$. Moreover, we say that (M, A) *is equal to* (N, B) *up to negligible elements*, and denote it by $(M, A) \sim (N, B)$, if $\text{Supp}(A) = \text{Supp}(B)$ and $A_{\text{Supp}(A)} = B_{\text{Supp}(A)}$. Obviously, $(M, A) \sim (N, B)$ if and only if $(\text{Supp}(A), A_{\text{Supp}(A)}) = (\text{Supp}(B), B_{\text{Supp}(B)})$.[4] Note that if $\text{Supp}(A) = \emptyset = \text{Supp}(B)$, then $A = \emptyset$ on M and $B = \emptyset$ on N. It is easy to see that for any (M, A) and a set $N \supseteq M$, there exists exactly one fuzzy set A' on N such that $(M, A) \sim (N, A')$. This fuzzy set A' is called the *extension of* A *from* M *to* N. We say that (M, A) *is a subset of* (N, B), and denote it by $(M, A) \subseteq (N, B)$, if $(M, A) \cap (N, B) \sim (M, A)$ (or, equivalently, $(M, A) \cup (N, B) \sim (N, B)$).

The following two statements show properties of the equality relation up to negligible elements.

Lemma 3. *If $(M, A) \sim (M', A')$, then $(N, A_N) = (N, A'_N)$ for any set N such that $(N, A_N) \in \mathcal{U}$.*

Theorem 1. *The binary relation $\sim$ on $\mathcal{U}$ is a congruence with respect to the intersection, union and difference of fuzzy universes.*

[3] If $K = M \cap N = \emptyset$, then $A_K \cap B_K$ is the empty mapping.

[4] Note that $(\text{Supp}(A), A_{\text{Supp}(A)}) \notin \mathcal{U}$ in general, but it is not a problem to extend the class $\mathcal{U}$ by such fuzzy universes. Then we can use this equality for the verification that $(M, A) \sim (N, B)$.

4.3 Fuzzy Quantifiers over Fuzzy Universes

In Sect. 4.1, we demonstrated the limitations of the definition of fuzzy quantifiers over set universes. In this subsection we introduce the concept of fuzzy quantifiers defined over fuzzy universes (see Sect. 4.2) in such a way that it overcomes these limitations.

Definition 10 (Local fuzzy quantifier over a fuzzy universe). *Let (M, C) be a fuzzy universe. A* local fuzzy quantifier $Q_{(M,C)}$ *of type $\langle 1^n, 1\rangle$ on (M, C) is a function $Q_{(M,C)} : \mathcal{F}(M)^n \times \mathcal{F}(M) \to L$ that to any fuzzy sets $A_1, \ldots, A_n$ and B from $\mathcal{F}(M)$ assigns a truth value $Q_{(M,C)}(A_1, \ldots, A_n, B)$ from L and*

$$Q_{(M,C)}(A_1, \ldots, A_n, B) = Q_{(M,C)}(A'_1, \ldots, A'_n, B') \tag{20}$$

holds for any $A_1, \ldots, A_n, B, A'_1, \ldots, A'_n, B' \in \mathcal{F}(M)$ such that $A_i \cap C = A'_i \cap C$ for any $i = 1, \ldots, n$ and $B \cap C = B' \cap C$.

One can see that the local fuzzy quantifier $Q_{(M,C)}$, which is defined over a fuzzy universe (M, C), is, in fact, a fuzzy quantifier defined on M that lives on a fuzzy set C (cf. Definition 14). Hence, an analysis of properties of fuzzy quantifiers over fuzzy universes can be practically restricted to fuzzy subsets of the fuzzy set C as it was proposed in [5].

Definition 11 (Global fuzzy quantifier over fuzzy universes). *Let $\mathcal{U}$ be a class of fuzzy universes. A* global fuzzy quantifier Q of type $\langle 1^n, 1\rangle$ *is a functional assigning to any fuzzy universe $(M, C) \in \mathcal{U}$ a local fuzzy quantifier $Q_{(M,C)}$ of type $\langle 1^n, 1\rangle$ such that for any $(M, C), (M', C') \in \mathcal{U}$ with $(M, C) \sim (M', C')$, it holds that*

$$Q_{(M,C)}(A_1, \ldots, A_n, B) = Q_{(M',C')}(A'_1, \ldots, A'_n, B') \tag{21}$$

for any $A_1, \ldots, A_n, B \in \mathcal{F}(M)$ and $A'_1, \ldots, A'_n, B' \in \mathcal{F}(M')$ such that $(M, A_i) \sim (M', A'_i)$ for any $i = 1, \ldots, n$ and $(M, B) \sim (M', B')$.

We should note that condition (21) ensures that if two fuzzy universes are equal up to negligible elements, then the fuzzy quantifiers defined over them are practically identical. More precisely, their evaluations coincide for fuzzy sets that together with their universes are equal up to negligible elements.

Before we provide an example of fuzzy quantifiers defined over fuzzy universes, let us define a binary fuzzy relation of fuzzy equivalence for fuzzy sets on a fuzzy universe. Let $(M, C) \in \mathcal{U}$ be a fuzzy universe. A mapping $\cong_{(M,C)} : \mathcal{F}(M) \times \mathcal{F}(M) \to L$ defined as

$$(A \cong_{(M,C)} B)(m) = \bigwedge_{m \in M} ((A \cap C)(m) \leftrightarrow (B \cap C)(m)) \tag{22}$$

is called a *fuzzy equivalence* on (M, C).

Example 2. Let $\mathcal{U}$ be a family of fuzzy universes. A global fuzzy quantifier over fuzzy universes $\forall$ of type $\langle 1 \rangle$ assigns to any $(M, C) \in \mathcal{U}$ a local fuzzy quantifier over fuzzy universes $\forall_{(M,C)} : \mathcal{F}(M) \to L$ defined for any $B \in \mathcal{F}(M)$ as

$$\forall_{(M,C)}(B) := B \cong_{(M,C)} C,$$

where $\cong_{(M,C)}$ is the fuzzy equivalence (22) on (M, C).[5]

Due to the definition of the fuzzy equivalence on (M, C), we can write

$$\forall_{(M,C)}(B) = B \cong_{(M,C)} C = \bigwedge_{m \in M} ((B \cap C)(m) \leftrightarrow (C \cap C)(m)) =$$
$$\bigwedge_{m \in M} ((B \cap C)(m) \leftrightarrow C(m)) = \bigwedge_{m \in M} ((B(m) \wedge C(m)) \leftrightarrow C(m)) =$$
$$\bigwedge_{m \in M} (C(m) \to B(m)) = \bigwedge_{m \in M} (C \to B)(m), \quad (23)$$

where we used the equality $(a \wedge b) \leftrightarrow b = b \to a$ holding for any $a, b \in L$ in every residuated lattice L. If (M, C) is crisp, then

$$\forall_{(M,C)}(B) = \bigwedge_{m \in M} (C(m) \to B(m)) = \bigwedge_{m \in M} (1 \to B(m)) = \bigwedge_{m \in M} B(m),$$

that is, it coincides with the standard definition of the fuzzy quantifier $\forall_M$ provided in (14).

Finally, we define *relativization* for fuzzy quantifiers defined over fuzzy universes as follows (cf. Definition 3).

Definition 12 (Relativization of fuzzy quantifiers over fuzzy universes). *Let Q be a global fuzzy quantifier of type $\langle 1^n, 1 \rangle$ over fuzzy universes. The* relativization *of Q is a global fuzzy quantifier Q^{rel} of type $\langle 1^{n+1}, 1 \rangle$ over fuzzy universes defined as*

$$(Q^{\text{rel}})_{(M,C)}(A, A_1, \ldots, A_n, B) := Q_{(M, C \cap A)}(A \cap A_1, \ldots, A \cap A_n, A \cap B) \quad (24)$$

for all $A, A_1, \ldots, A_n, B \in \mathcal{F}(M)$. For the most common case of relativization from type $\langle 1 \rangle$ to type $\langle 1, 1 \rangle$,

$$(Q^{\text{rel}})_{(M,C)}(A, B) = Q_{(M, C \cap A)}(A \cap B). \quad (25)$$

4.4 Restriction

As we mentioned in Subsect. 4.1, we are unable to extend the concept of restriction for fuzzy quantifiers defined over crisp universes. In this part, we show that if we employ fuzzy universes for fuzzy quantification, we can introduce this concept

[5] Note the structural similarity of this definition with the definition of the generalized quantifier $\forall_M(B) := B = M$ given below Definition 2.

in an elegant way following the standard definition. In the rest of the paper, by a fuzzy quantifier we mean a fuzzy quantifier over fuzzy universes unless stated otherwise.

The concept of fuzzy quantifiers of type $\langle 1^n, 1\rangle$ restricted to a fuzzy set can be introduced as follows.

Definition 13. *Let Q be a type $\langle 1^n, 1\rangle$ global fuzzy quantifier, and let A be a fuzzy set on N. The fuzzy quantifier Q is* restricted to A *if for any $(M, C) \in \mathcal{U}$ and $A_1, \ldots, A_n, B \in \mathcal{F}(M)$ we have*

$$Q_{(M,C)}(A_1, \ldots, A_n, B) = \\ Q_{(N,A)}(A \cap (C \cap A_1)_N, \ldots, A \cap (C \cap A_n)_N, A \cap (C \cap B)_N). \tag{26}$$

The set of all type $\langle 1^n, 1\rangle$ global fuzzy quantifiers restricted to a fuzzy set A on a universe N is denoted by $FRST_{\langle 1^n,1\rangle}(N, A)$.

Similarly to the classical case of the restriction to a set, a global fuzzy quantifier, which is restricted to a fuzzy set A on a universe N, is determined from the local fuzzy quantifier $Q_{(N,A)}$. One can simply verify that the previous definition of the restriction to a fuzzy set is correct in the sense of Definition 11. Note that such verification is useless for the global generalized quantifiers, because their definition has no requirement on the functionals defining them. Obviously, an equivalent definition could be as follow. A fuzzy quantifier Q is restricted to a fuzzy set A on a universe N if for any $(M, C) \in \mathcal{U}$ and $A_1, \ldots, A_n, B \in \mathcal{F}(M)$ we have

$$Q_{(M,C)}(A_1, \ldots, A_n, B) = Q_{(N,A\cap C_N)}(A \cap A_{1,N}, \ldots, A \cap A_{n,N}, A \cap B_N). \tag{27}$$

The following lemma is a generalization of Lemma 1 for fuzzy quantifiers defined over fuzzy universes.

Lemma 4. *If Q is a type $\langle 1^n, 1\rangle$ global generalized quantifier restricted to a fuzzy set A on N, then Q is restricted to any fuzzy set A' on an arbitrary universe N' such that $(N, A) \subseteq (N', A')$, i.e., $FRST_{\langle 1^n,1\rangle}(N, A) \subseteq FRST_{\langle 1^n,1\rangle}(N', A')$.*

Let us show that the global fuzzy quantifiers restricted to fuzzy sets can be used to introduce the relativization of global fuzzy quantifiers. Let Q be a type $\langle 1^n, 1\rangle$ fuzzy quantifier. Then, for each fuzzy set A on a universe N, we can introduce the type $\langle 1^n, 1\rangle$ global fuzzy quantifier $Q^{[(N,A)]}$ as

$$(Q^{[(N,A)]})_{(M,C)}(A_1, \ldots, A_n, B) = \\ Q_{(N,A)}(A \cap (C \cap A_1)_N, \ldots, A \cap (C \cap A_n)_N, A \cap (C \cap B)_N)$$

for any $(M, C) \in \mathcal{U}$ and $A_1, \ldots, A_n, B \in \mathcal{F}(M)$. It is easy to see that $Q^{[(N,A)]} \in \mathrm{FRST}_{\langle 1^n,1\rangle}(N, A)$. It should be noted that Q and $Q^{[(N,A)]}$ are different quantifiers. They become identical if Q is already restricted to the fuzzy set A on the universe N. Now, if we define the global fuzzy quantifier Q' of type $\langle 1^{n+1}, 1\rangle$ as

$$Q'_{(M,C)}(A, A_1, \ldots, A_n, B) = (Q^{[(M,A)]})_{(M,C)}(A_1, \ldots, A_n, B) \tag{28}$$

for all $(M,C) \in \mathcal{U}$ and $A_1, \ldots, A_n, B \in \mathcal{F}(M)$, then using (27) it is easy to show that $Q' = Q^{\mathrm{rel}}$, where Q^{rel} has been defined in Definition 12.

4.5 Fuzzy Quantifiers Living on a Fuzzy Set

A concept related to that of a global fuzzy quantifier restricted to a fuzzy set is the concept of a local fuzzy quantifier living on a fuzzy set.

Definition 14. *Let $(M,C) \in \mathcal{U}$ and A be a fuzzy set on a universe N. The local fuzzy quantifier $Q_{(M,C)}$ of type $\langle 1^n, 1\rangle$ lives on A if, for any $A_1, \ldots, A_n, B \in \mathcal{F}(M)$, we have*

$$Q_{(M,C)}(A_1, \ldots, A_n, B) = Q_{(M,C)}(A_M \cap A_1, \ldots, A_M \cap A_n, A_M \cap B). \tag{29}$$

If the distributivity of $\wedge$ over $\bigvee$ is satisfied in a residuated lattice L (e.g., L is an MV-algebra), the conservativity of global fuzzy quantifiers of type $\langle 1^n, 1\rangle$ can be expressed in terms of the live-on property.[6] Indeed, let Q be a global fuzzy quantifier of type $\langle 1^n, 1\rangle$ for $n \geq 1$. For any fuzzy universe (M,C) and $A_1, \ldots, A_n \in \mathcal{F}(M)$, define local quantifier $(Q[A_1, \ldots, A_n])_{(M,C)}$ of type $\langle 1\rangle$ as

$$(Q[A_1, \ldots, A_n])_{(M,C)}(B) = Q_{(M,C)}(A_1, \ldots, A_n, B). \tag{30}$$

One can show that Q is conservative if and only if each local generalized quantifier $(Q[A_1, \ldots, A_n])_M$ lives on $A = \bigcup_{i=1}^n A_i$. Note that the distributivity of $\wedge$ over $\bigvee$ ensures the crucial equality $(\bigcup_{i=1}^n A_i) \cap B = \bigcup_{i=1}^n (A_i \cap B)$ important in the proof of a characterization of the conservativity of type $\langle 1^n, 1\rangle$ global fuzzy quantifiers.

The following lemma specifies basic facts about fuzzy quantifiers living on fuzzy sets (cf. Lemma 2).

Lemma 5. *Let Q be a type $\langle 1^n, 1\rangle$ global fuzzy quantifier, let (M,C) be a fuzzy universe from $\mathcal{U}$, and let $C_1 \in \mathcal{F}(N)$ and $C_2 \in \mathcal{F}(N')$ be fuzzy sets. Then*

(i) $Q_{(M,C)}$ *lives on* C.
(ii) $Q_{(M,C)}$ *lives on* $\emptyset$ *if and only if* $Q_{(M,C)}$ *is trivial* ($Q_{(M,C)}(A_1, \ldots, A_n, B) = a$ *for any* $A_1, \ldots, A_n, B \in \mathcal{F}(M)$ *with* $a \in L$).
(iii) *If* $Q_{(M,C)}$ *lives on* C_1 *and* C_2, *then it lives on* $C_{1,N\cap N'} \cap C_{2,N\cap N'}$.
(iv) $(Q^{[(N,A)]})_{(M,C)}$ *lives on* A.

5 Conclusion

In this article, we proposed a novel framework for fuzzy quantifiers which are defined over fuzzy universes. We introduced the concept of a fuzzy universe and define a class of fuzzy universes closed under the operations of intersection, union

[6] A global fuzzy quantifier Q of type $\langle 1^n, 1\rangle$ over fuzzy universes is *conservative*, if for any $(M,C) \in \mathcal{U}$ and $A_1, \ldots, A_n, B, B' \in \mathcal{F}(M)$ it holds that if $A_i \cap B = A_i \cap B'$ for $i = 1, \ldots, n$, then $Q_{(M,C)}(A_1, \ldots, A_n, B) = Q_{(M,C)}(A_1, \ldots, A_n, B')$.

and difference. Moreover, we established a binary relation called the equivalence up to negligible elements, which is essential in the definition of global fuzzy quantifiers over fuzzy universes. Further, we generalized the fuzzy quantifiers defined over crisp universes to fuzzy quantifiers defined over fuzzy universes. The novel conception of fuzzy quantifiers naturally allows us to introduce, for example, the notion of relativization of fuzzy quantifiers, which principally could not be defined in the case of fuzzy quantifiers if only crisp universes are permitted. For an illustration, we investigated the important semantic notions of restriction to a fuzzy set and living on a fuzzy set in our novel framework for fuzzy quantifiers defined over fuzzy universes. It should be noted that the notion of restriction to a fuzzy set also could not be introduced for fuzzy quantifiers defined over crisp universes. Although the presented work is only an initial study, it shows that the fuzzy quantifiers over fuzzy universes enable us to develop the fuzzy quantifier theory in the same fashion as in the theory of generalized quantifiers. In our future research, we will concentrate on investigation of further semantic properties of fuzzy quantifiers over fuzzy universes, e.g., the property of extension and isomorphism invariance.

References

1. Barwise, J., Cooper, R.: Generalized quantifiers and natural language. Linguist. Philos. **4**, 159–219 (1981)
2. Bělohlávek, R.: Fuzzy Relational Systems: Foundations and Principles. Kluwer Academic Publishers, New York (2002)
3. Dvořák, A., Holčapek, M.: **L**-fuzzy quantifiers of type $\langle 1 \rangle$ determined by fuzzy measures. Fuzzy Sets Syst. **160**(23), 3425–3452 (2009)
4. Dvořák, A., Holčapek, M.: Type $\langle 1, 1 \rangle$ fuzzy quantifiers determined by fuzzy measures on residuated lattices. Part I. Basic definitions and examples. Fuzzy Sets Syst. **242**, 31–55 (2014)
5. Dvořák, A., Holčapek, M.: Relativization of fuzzy quantifiers: initial investigations. In: Kacprzyk, J., Szmidt, E., Zadrożny, S., Atanassov, K.T., Krawczak, M. (eds.) IWIFSGN/EUSFLAT 2017. AISC, vol. 641, pp. 670–683. Springer, Cham (2018). https://doi.org/10.1007/978-3-319-66830-7_59
6. Glöckner, I.: Fuzzy Quantifiers in Natural Language: Semantics and Computational Models. Der Andere Verlag, Osnabrück, Germany (2004)
7. Hájek, P.: Metamathematics of Fuzzy Logic. Kluwer Academic Publishers, Dordrecht (1998)
8. Holčapek, M.: Monadic **L**-fuzzy quantifiers of the type $\langle 1^n, 1 \rangle$. Fuzzy Sets Syst. **159**(14), 1811–1835 (2008)
9. Keenan, E.L., Westerståhl, D.: Generalized quantifiers in linguistics and logic. In: van Benthem, J., ter Meulen, A. (eds.) Handbook of Logic and Language, 2nd edn, pp. 859–923. Elsevier, Amsterdam (2011)
10. Klement, E., Mesiar, R., Pap, E.: Triangular Norms, Trends in Logic, vol. 8. Kluwer Academic Publishers, Dordrecht (2000)
11. Lindström, P.: First-order predicate logic with generalized quantifiers. Theoria **32**, 186–195 (1966)
12. Mostowski, A.: On a generalization of quantifiers. Fundamenta Mathematicae **44**, 12–36 (1957)
13. Peters, S., Westerståhl, D.: Quantifiers in Language and Logic. Oxford University Press, New York (2006)

Mathematical Methods Towards Dealing with Uncertainty in Applied Sciences

On the Relationship Among Relational Categories of Fuzzy Topological Structures

Jiří Močkoř(✉)

Institute for Research and Applications of Fuzzy Modeling,
Centre of Excellence IT4Innovations, University of Ostrava, 30. dubna 22,
701 03 Ostrava 1, Czech Republic
Jiri.Mockor@osu.cz
http://irafm.osu.cz/

Abstract. Relational variants of categories of Čech closure or interior L-valued operators, categories of L-fuzzy pretopological and L-fuzzy co-pretopological operators, category of L-valued fuzzy relation, categories of upper and lower F-transforms and the category of spaces with fuzzy partitions are introduced. The existence of relationships defined by functors among these categories are investigated and a key role of a relational category of spaces with fuzzy partitions is described.

1 Introduction

In this paper we want to build on our previous paper [8] and to analyse the relationships between categories that represent various structures generally included under the term fuzzy topological structures. These structures include variants of fuzzy topological spaces, fuzzy rough sets, fuzzy approximation spaces, fuzzy closure operators, fuzzy pretopological operators and their dual terms. In contrast to the original paper [8], however, the relationships between these fuzzy structures are not represented by classical mappings between supports of these fuzzy structures, but in a more general way, i.e., as a fuzzy relations or a fuzzy relations with other properties. This more general approach is based on current trends in the field of fuzzy structures which are based on the application of *fuzzy relations* as morphisms in suitable categories. A typical example of that use of fuzzy relations is the category of sets as objects and L-valued fuzzy relations between sets as morphisms which is frequently used in approximation theory.

The main result of this paper are two theorems describing the existence of functors among categories of these generalized lattice-valued fuzzy topological structures, where a given lattice L is either a complete residuated lattice or a complete MV-algebra. From both these theorems it follows the key position of a relational version of the category of spaces with fuzzy partitions which represents a structure from which all other considered fuzzy topological structures can be derived.

This research was partially supported by the ERDF/ESF project CZ.02.1.01/0.0/0.0/17-049/0008414.

M.-J. Lesot et al. (Eds.): IPMU 2020, CCIS 1239, pp. 189–197, 2020.
https://doi.org/10.1007/978-3-030-50153-2_14

2 Preliminaries

In this section we repeat basic terminology from residuated lattices and we also introduce principal categories and some subcategories we use in the paper. To be self-contained, in this section we repeat most of definitions related to these structures.

We refer to [6,9] for additional details regarding residuated lattices.

Definition 1. *A residuated lattice is an algebra* $\mathcal{L} = (L, \wedge, \vee, \otimes, \rightarrow, 0, 1)$ *such that*

1. $(L, \wedge, \vee, 0, 1)$ *is a bounded lattice with the least element* 0 *and the greatest element* 1*;*
2. $(L, \otimes, 1)$ *is a commutative monoid, and*
3. $\forall a, b, c \in L,\ a \otimes b \leq c \iff a \leq b \rightarrow c.$

A residuated lattice $(L, \wedge, \vee, \otimes, \rightarrow, 0, 1)$ is *complete* if it is complete as a lattice.

The following is the derived unary operations of *negation* $\neg$:

$$\neg a = a \rightarrow 0,$$

A residuated lattice $\mathcal{L}$ is called an *MV-algebra* if it satisfies $(a \rightarrow b) \rightarrow b = a \vee b$. Throughout this paper, a complete residuated lattice $\mathcal{L} = (L, \wedge, \vee, \otimes, \rightarrow, 0, 1)$ will be fixed. For simplicity, instead of $\mathcal{L}$ we use only L if there is no danger of misunderstanding.

Let X be a nonempty set and L^X be a set of all L-fuzzy sets (=L-valued functions) of X. For all $\alpha \in L$, $\underline{\alpha}(x) = \alpha$ is a constant L-fuzzy set on X. For all $u \in L^X$, the *core*(u) is a set of all elements $x \in X$, such that $u(x) = 1$. An L-fuzzy set $u \in L^X$ is called *normal*, if *core*$(u) \neq \emptyset$. An L-fuzzy set $\chi^X_{\{y\}} \in L^X$ is a *singleton*, if it has the following form

$$\chi^X_{\{y\}}(x) = \begin{cases} 1, & \text{if } x = y, \\ 0, & \text{otherwise.} \end{cases}$$

We repeat basic definitions of above mentioned fuzzy topological structures. The original notions of Kuratowski closure and interior operators were introduced in several papers, see [1–5]. In the paper we use a more general form of these operators, called Čech operators or preclosure operators, where the idempotence of operators is not required.

Definition 2. *The map* $i : L^X \rightarrow L^X$ *is called a Čech (L-fuzzy) interior operator, if for every* $\underline{\alpha}, u, v \in L^X$*, it fulfills*

1. $i(\underline{\alpha}) = \underline{\alpha},$
2. $i(u) \leq u,$
3. $i(u \wedge v) = i(u) \wedge i(v).$

We say that a Čech interior operator $i : L^X \to L^X$ is a *strong Čech-Alexandroff interior operator*, if

$$i(\underline{\alpha} \to u) = \alpha \to i(u) \quad \text{and} \quad i(\bigwedge_{j\in J} u_j) = \bigwedge_{j\in J} i(u_j).$$

Definition 3. *The map $c : L^X \to L^X$ is called a Čech (L-fuzzy) closure operator, if for every $\underline{\alpha}, u, v \in L^X$, it fulfils*

1. $i(\underline{\alpha}) = \underline{\alpha}$,
2. $i(u) \geq u$,
3. $i(u \vee v) = i(u) \vee i(v)$.

We say that a Čech closure operator $c : L^X \to L^X$ is a *strong Čech-Alexandroff closure operator*, if

$$c(\underline{\alpha} \otimes u) = \alpha \otimes c(u) \quad \text{and} \quad c(\bigvee_{j\in J} u_j) = \bigvee_{j\in J} c(u_j).$$

We remind the notion of an L-fuzzy pretopological space and L-fuzzy co-pretopological space as it has been introduced in [11].

Definition 4. *An L-fuzzy pretopology on X is a set of functions $\tau = \{p_x \in L^{L^X} : x \in X\}$, such that for all $u, v \in L^X, \alpha \in L$ and $x \in X$,*

1. $p_x(\underline{\alpha}) = \alpha$,
2. $p_x(u) \leq u(x)$,
3. $p_x(u \wedge v) = p_x(u) \wedge p_x(v)$.

We say that an L-fuzzy pretopological space (X, τ) is a *strong Čech-Alexandroff L-fuzzy pretopological space*, if

$$p_x(\underline{\alpha} \to u) = \alpha \to p_x(u) \quad \text{and} \quad p_x(\bigwedge_{j\in J} u_j) = \bigwedge_{j\in J} p_x(u_j).$$

Definition 5. *An L-fuzzy co-pretopology on X is a set of functions $\eta = \{p^x \in L^{L^X} : x \in X\}$, such that for all $u, v \in L^X, \alpha \in L$ and $x \in X$,*

1. $p^x(\underline{\alpha}) = \alpha$,
2. $p^x(u) \geq u(x)$,
3. $p^x(u \vee v) = p^x(u) \vee p^x(v)$.

We say that an L-fuzzy co-pretopological space (X, τ) is a *strong Čech-Alexandroff L-fuzzy co-pretopological space*, if

$$p^x(\underline{\alpha} \otimes u) = \alpha \otimes p^x(u) \quad \text{and} \quad p^x(\bigvee_{j\in J} u_j) = \bigvee_{j\in J} p^x(u_j).$$

We recall the notion of an L-fuzzy partition (see [7] or [10]), which is the basic structure for lattice-valued fuzzy transform, introduced in [10].

Definition 6. *A set $\mathcal{A}$ of normal fuzzy sets $\{A_\alpha : \alpha \in \Lambda\}$ in X is an L-fuzzy partition of X, if*

1. *the corresponding set of ordinary subsets $\{core(A_\alpha) : \alpha \in \Lambda\}$ is a partition of X, and*
2. *$core(A_\alpha) = core(A_\beta)$ implies $A_\alpha = A_\beta$.*

Instead of the index set Λ from $\mathcal{A}$ we use $|\mathcal{A}|$.

We need the following notation. If $R : X \times Y \to L$ is an L-fuzzy relation, then the upper and lower approximation maps $R^\uparrow : L^X \to L^Y$ and $R^\downarrow : L^Y \to L^X$ are defined by

$$t \in L^X, y \in Y, \quad R^\uparrow(t)(y) = \bigvee_{x \in X} t(x) \otimes R(x,y),$$
$$s \in L^Y, x \in X, \quad R^\downarrow(s)(x) = \bigwedge_{y \in Y} R(x,y) \to s(y).$$

Finally, from a set X and a lattice-valued fuzzy partition $\mathcal{A}$ defined on X we can construct examples of upper and lower approximation maps, defined by special fuzzy relations which is derived from a space with a fuzzy partition $(X, \mathcal{A})$. In fact, we can define a fuzzy relation $R : X \times |\mathcal{A}| \to L$ by

$$x \in X, \lambda \in |\mathcal{A}|, \quad R(x, \lambda) = A_\lambda(x),$$

where $\mathcal{A} = \{A_\lambda : \lambda \in |\mathcal{A}|\}$. Fuzzy approximation operators $R^\uparrow$ and $R^\downarrow$ derived from R are then called upper and lower F-transforms based on a fuzzy partition $\mathcal{A}$. Namely, we have

Definition 7. *An upper F-transform on a set X defined by a fuzzy partition $\mathcal{A}$ is a map $F^\uparrow_{X,\mathcal{A}} : L^X \to L^{|\mathcal{A}|}$, such that*

$$\forall u \in L^X, \lambda \in |\mathcal{A}|, \quad F^\uparrow_{X,\mathcal{A}}(u)(\lambda) = \bigvee_{x \in X} A_\lambda(x) \otimes u(x).$$

Definition 8. *A lower F-transform on a set X defined by a fuzzy partition $\mathcal{A}$ is a map $F^\downarrow_{X,\mathcal{A}} : L^X \to L^{|\mathcal{A}|}$, such that*

$$\forall u \in L^X, \lambda \in |\mathcal{A}|, \quad F^\downarrow_{X,\mathcal{A}}(u)(\lambda) = \bigwedge_{x \in X} A_\lambda(x) \to u(x).$$

In the previous paper [8] we discussed the issue of relations, i.e., functors, between categories that have relationships to categories of fuzzy topological structures and whose common features are that morphisms in these categories are special mappings between sets. In this paper, which is a natural continuation of the previous paper [8], we will look at the relationships between analogous categories, where, unlike the original categories, morphisms are defined as special L-valued relations. This shift in understanding relationships between fuzzy objects corresponds to the current trend, where fuzzy relations between different fuzzy structures are of increasing importance.

The categories we will deal with in the paper have the same objects as the categories introduced in [8], the only difference concerns morphisms. Instead of classical maps between sets we use special fuzzy relations as morphisms. To emphasize the link between the original categories in [8], if the original category was labelled **K**, the new category with relational morphisms will be labelled **RK**.

Definition 9. *In what follows by X, Y we denote sets from the standard category* **Set** *and by . a composition of morphisms in this category.*

1. *The category* **RCInt** *is defined by*
   (a) *objects are pairs (X, i), where $i : L^X \to L^X$ is a Čech L-fuzzy interior operator,*
   (b) *$R : (X, i) \to (Y, j)$ is a morphism, if $R : X \times X \to L$ is an L-fuzzy relation and*
$$i.R^{\downarrow} \geq R^{\downarrow}.j.$$
2. *The category* **RCClo** *is defined by*
   (a) *objects are pairs (X, c), where $c : L^X \to L^X$ is a Čech L-fuzzy closure operator,*
   (b) *$R : (X, c) \to (Y, d)$ is a morphism, if $R : X \times Y \to L$ is an L-fuzzy relation, and*
$$R^{\uparrow}.c \leq d.R^{\uparrow}.$$
3. *The category* **RFPreTop** *is defined by*
   (a) *objects are L-fuzzy pretopological spaces (X, τ),*
   (b) *$R : (X, \tau) \to (Y, \sigma)$ is a morphism, where $\tau = \{p_x \in L^{L^X} : x \in X\}$, $\sigma = \{q_y \in L^{L^Y} : y \in Y\}$, if $R : X \times Y \to L$ is an L-fuzzy relation, and for all $x \in X$,*
$$\bigwedge_{z \in Y} (R(x, z) \to q_z) \leq p_x.R^{\downarrow}.$$
4. *The category* **RFcoPreTop** *is defined by*
   (a) *objects are L-fuzzy co-pretopological spaces (X, τ),*
   (b) *$R : (X, \tau) \to (Y, \sigma)$ is a morphism, where $\tau = \{p^x \in L^{L^X} : x \in X\}$, $\sigma = \{q^y \in L^{L^Y} : y \in Y\}$, if $R : X \times Y \to L$ is an L-fuzzy relation, and for all $x \in X, y \in Y$,*
$$q^y.R^{\uparrow} \geq p^x \otimes R(x, y).$$
5. *The category* **RFRel** *is defined by*
   (a) *objects are pairs (X, r), where r is a reflexive L-fuzzy relation on X,*
   (b) *$R : (X, r) \to (Y, s)$ is a morphism, if $R : X \times Y \to L$ is an L-fuzzy relation, and*
$$s \circ R \geq R \circ r,$$
   where $\circ$ is the composition of L-fuzzy relations.
6. *The category* **RSFP** *is defined by*
   (a) *objects are sets with an L-fuzzy partition $(X, \mathcal{A})$,*

(b) $(R,\Sigma):(X,\mathcal{A})\rightarrow(Y,\mathcal{B})$ *is a morphism if* $R:X\times Y\rightarrow L$ *and* $\Sigma:|\mathcal{A}|\times|\mathcal{B}|\rightarrow L$ *are L-fuzzy relations and for each* $\alpha\in|\mathcal{A}|,\beta\in|\mathcal{B}|,x\in X,y\in Y$,

$$A_\alpha(x)\otimes\Sigma(\alpha,\beta)\leq B_\beta(y)\otimes R(x,y),$$
$$A_\alpha(x)\otimes R(x,y)\leq B_\beta(y)\otimes\Sigma(\alpha,\beta).$$

7. *The category* **RFTrans**$^\uparrow$ *is defined by*
   (a) *objects are upper F-transforms* $F^\uparrow_{X,\mathcal{A}}:L^X\rightarrow L^{|\mathcal{A}|}$, *where* $(X,\mathcal{A})$ *are sets with L-fuzzy partitions,*
   (b) $(R,\Sigma):F^\uparrow_{X,\mathcal{A}}\rightarrow F^\uparrow_{Y,\mathcal{B}}$ *is a morphism if* $R:X\times Y\rightarrow L$ *and* $\Sigma:|\mathcal{A}|\times|\mathcal{B}|\rightarrow L$ *are L-fuzzy relations and*

$$\Sigma^\uparrow.F^\uparrow_{X,\mathcal{A}}\leq F^\uparrow_{Y,\mathcal{B}}.R^\uparrow.$$

8. *The category* **RFTrans**$^\downarrow$ *is defined by*
   (a) *objects are lower F-transforms* $F^\downarrow_{X,\mathcal{A}}:L^X\rightarrow L^{|\mathcal{A}|}$, *where* $(X,\mathcal{A})$ *are sets with L-fuzzy partitions,*
   (b) $(R,\Sigma):F^\downarrow_{X,\mathcal{A}}\rightarrow F^\downarrow_{Y,\mathcal{B}}$ *is a morphism if* $R:X\times Y\rightarrow L$ *and* $\Sigma:|\mathcal{A}|\times|\mathcal{B}|\rightarrow L$ *are L-fuzzy relations and*

$$\Sigma^\downarrow.F^\downarrow_{Y,\mathcal{B}}\leq F^\downarrow_{X,\mathcal{A}}.R^\downarrow.$$

Analogously as in the paper [8] we consider the following full subcategories of the above categories:

1. The full subcategory **RsACClo** of **RCClo** with strong Čech-Alexandroff L-fuzzy closure operators as objects.
2. The full subcategory **RsACInt** of **RCInt** with strong Čech-Alexandroff L-fuzzy interior operators as objects.
3. The full subcategory **RsAFPreTop** of **RFPreTop** with strong Čech-Alexandroff L-fuzzy pretopological spaces as objects.
4. The full subcategory **RsAFcoPreTop** of **RFcoPreTop** with strong Čech-Alexandroff L-fuzzy co-pretopological spaces.

3 Relationships Among Relational Categories of L-valued Fuzzy Topological Structures

The main result of the paper are the following two theorems, which use functors to describe the relationships between individual categories from the Definitions 7 and 8. Because each of the category **K** listed in [8]; Theorem 1 and Theorem 2, can be embedded into the corresponding category **RK** and the embedding is done using a graph of the morphisms from **K**, these new theorems generalize results from [8]. In fact, the following simple proposition holds.

Proposition 1. *Let* **K** *be any of categories listed in [8]; Theorem 1 and Theorem 2. Then there exists an embedding functor* $I_{\mathbf{K}}:\mathbf{K}\hookrightarrow\mathbf{RK}$.

From these two theorems it follows that relationships among relational versions of categories associated with F-transforms are analogical to relationships among categories with maps as morphisms.

The first theorem describes the relationships between these categories of L-valued fuzzy topological structures where L is a complete residuated lattice.

Theorem 1. *Let L be a complete residuated lattice. Then the following diagram of functors commutes,*

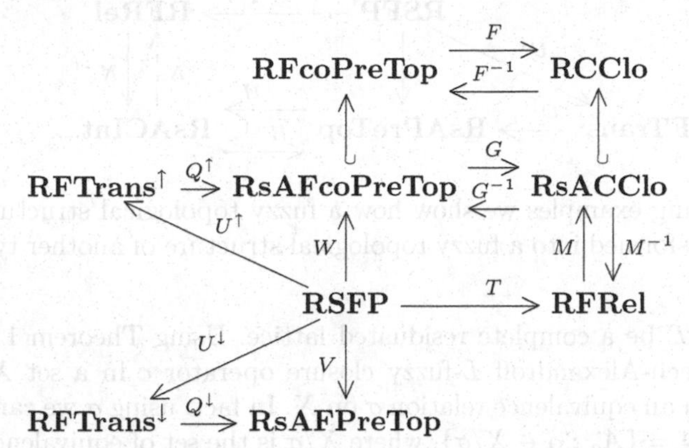

where (F, F^{-1}), (G, G^{-1}) *and* (M, M^{-1}) *are inverse pairs of functors.*

The rather long proof of this theorem will be published elsewhere. We show only how the object functions of the functors are defined. For $(X, \mathcal{A}) \in$ **RSFP** we set

$$T(X,\mathcal{A}) = (X,r), \quad r(x,x') = A_{w_{\mathcal{A}}(x)}(x'),$$

$$W(X,\mathcal{A}) = (X,\{p^x : x \in X\}), \quad p^x(u) = \bigvee_{t\in X} u(t) \otimes A_{w_{\mathcal{A}}(t)}(x),$$

$$V(X,\mathcal{A}) = (X,\{p_x : x \in X\}), \quad p_x(u) = F^{\downarrow}_{X,\mathcal{A}}(u)(w_{\mathcal{A}}(x)),$$

$$U^{\uparrow}(X,\mathcal{A}) = F^{\uparrow}_{X,\mathcal{A}},$$

$$U_{\downarrow}(X,\mathcal{A}) = F^{\downarrow}_{X,\mathcal{A}},$$

where $w_{\mathcal{A}}(t) \in |\mathcal{A}|$ is such that $A_{w_{\mathcal{A}}(t)}(t) = 1$. For other functors we set

$$(X,\{p^x : x \in X\}) \in \mathbf{RFcoPrTop}, F(X,\{p^x : x \in X\}) = (c,u), \quad c(u)(x) = p^x(u),$$

$$(X,r) \in \mathbf{RFRel}, M(X,r) = (X,c), \quad c(u)(x) = r^{\uparrow}(u)(x),$$

$$(X,c) \in \mathbf{RsACClo}, M^{-1}(X,c) = (X,r), \quad r(x,x') = c(\chi^X_{\{x\}})(x').$$

If $\mathcal{L}$ is a complete MV-algebra, we obtain a stronger form of the previous theorem.

Theorem 2. *Let* $\mathcal{L}$ *be a complete* MV*-algebra. Then the following diagram of functors from Theorem 1 and new functors commutes, where* (H, H^{-1}) *and* (N, N^{-1}) *are inverse pairs of functors.*

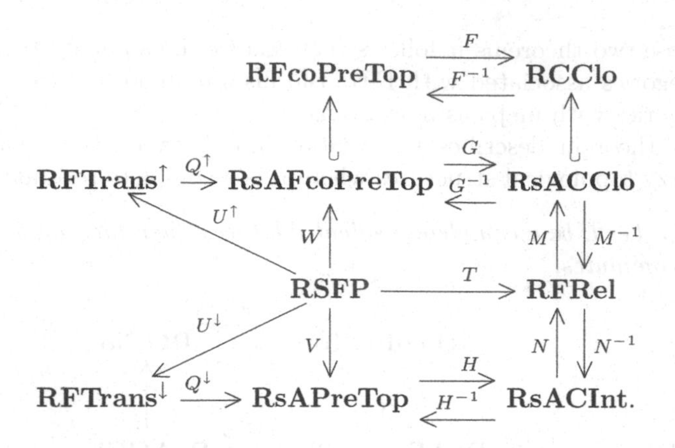

In the following examples we show how a fuzzy topological structure of one type can be transformed into a fuzzy topological structure of another type using these theorems.

Example 1. Let $\mathcal{L}$ be a complete residuated lattice. Using Theorem 1 we show how a strong Čech-Alexandroff L-fuzzy closure operator c in a set X can be constructed from an equivalence relation σ on X. In fact, using σ we can define a fuzzy partition $\mathcal{A} = \{A_\alpha : \alpha \in X/\sigma\}$, where X/σ is the set of equivalence classes defined by σ and $A_\alpha(x) = 1$ iff $x \in \alpha$, otherwise the value is 0. Using functors from the Theorem 1, the strong Čech-Alexandroff L-fuzzy closure operator c in X can be defined by $(X, c) = M.T(X, \mathcal{A})$, i.e., for arbitrary $u \in L^X, x \in X$,

$$c(u)(x) = \bigvee_{t \in X} u(t) \otimes A_{w_{\mathcal{A}}(x)}(t) = \bigvee_{t \in X, (t,x) \in \sigma} u(t).$$

Example 2. Let $\mathcal{L}$ be a complete MV-algebra. Using the Theorem 2 we show how from a strong Čech-Alexandroff L-fuzzy pretopological space (X, τ) a strong Čech-Alexandroff L-fuzzy co-pretopological space (X, ρ) can be defined. In fact, we can put $(X, \rho) = G^{-1}.M.N.H(X, \tau)$. If $\tau = \{p_x : x \in X\}$ then $\rho = \{p^x : x \in X\}$ is defined by

$$p^x(u) = \bigvee_{t \in X} u(t) \otimes \neg p_x(\neg \chi^X_{\{x\}})(t),$$

as it can be verified by a simple calculation. If $R : (X, \tau) \to (Y, \sigma)$ is a morphisms in **RsAFPreTOP**, then R is also a morphism $G^{-1}.M.N.H(X, \tau) \to G^{-1}.M.N.H(Y, \sigma)$ in **RsAFcoPreTop**.

4 Conclusions

The article follows our previous work [8], in which we dealt with the issue of relationships between categories motivated by topological structures. We looked at a more general situation where morphisms in these categories are not mappings, but L-fuzzy relations. In detail, we considered categories and some of their subcategories of Čech closure or interior L-valued operators, categories of L-fuzzy pretopological and L-fuzzy co-pretopological operators, category of L-valued fuzzy relation, categories of upper and lower F-transforms and the category of spaces with fuzzy partitions, where morphisms between objects are based on L-valued relations. As an interesting consequence of these relationships among relational categories, it follows that the category of relational spaces with fuzzy partitions plays a key role, i.e., the objects of this category can be used to create objects of any of the above categories of topological structures.

References

1. Bandler, W., Kohout, L.: Special properties, closures and interiors of crisp and fuzzy relations. Fuzzy Sets Syst. **26**(3), 317–331 (1988). https://doi.org/10.1016/0165-0114(88)90126-1
2. Bělohlávek, R.: Fuzzy closure operators. J. Math. Anal. Appl. **262**, 473–489 (2001). https://doi.org/10.1006/jmaa.2000.7456
3. Bělohlávek, R.: Fuzzy closure operators II. Soft. Comput. **7**(1), 53–64 (2002). https://doi.org/10.1007/s00500-002-0165-y
4. Biacino, V., Gerla, G.: Closure systems and L-subalgebras. Inf. Sci. **33**, 181–195 (1984). https://doi.org/10.1016/0020-0255(84)90027-6
5. Bodenhofer, U., De Cock, M., Kerre, E.E.: Openings and closures of fuzzy preorderings: theoretical basics and applications to fuzzy rule-based systems. Int. J. Gen Syst **32**(4), 343–360 (2003). https://doi.org/10.1080/0308107031000135026
6. Höhle, U.: Fuzzy sets and sheaves. Part I, basic concepts. Fuzzy Sets Syst. **158**, 1143–1174 (2007). https://doi.org/10.1016/j.fss.2006.12.009
7. Močkoř, J.: Spaces with fuzzy partitions and fuzzy transform. Soft. Comput. **21**(13), 3479–3492 (2017). https://doi.org/10.1007/s00500-017-2541-7
8. Močkoř, J., Perfilieva, I.: Functors among categories of L-fuzzy partitions, L-fuzzy pretopological spaces and L-fuzzy closure spaces. In: Kearfott, R.B., Batyrshin, I., Reformat, M., Ceberio, M., Kreinovich, V. (eds.) IFSA/NAFIPS 2019. AISC, vol. 1000, pp. 382–393. Springer, Cham (2019). https://doi.org/10.1007/978-3-030-21920-8_35
9. Novák, V., Perfilijeva, I., Močkoř, J.: Mathematical principles of fuzzy logic, Kluwer Academic Publishers, Boston, Dordrecht. London (1999). https://doi.org/10.1007/978-1-4615-5217-8
10. Perfilieva, I.: Fuzzy transforms: theory and applications. Fuzzy Sets Syst. **157**, 993–1023 (2006). https://doi.org/10.1016/j.fss.2005.11.012
11. Zhang, D.: Fuzzy pretopological spaces, an extensional topological extension of FTS. Chinese Ann. Math. **3**, 309–316 (1999). https://doi.org/10.1142/S0252959999000345

Interactive Fuzzy Fractional Differential Equation: Application on HIV Dynamics

Vinícius Wasques[1], Beatriz Laiate[2], Francielle Santo Pedro[3](✉), Estevão Esmi[2], and Laécio Carvalho de Barros[2]

[1] Department of Mathematics, São Paulo State University, Rio Claro, São Paulo 13506-900, Brazil
vwasques@outlook.com

[2] Department of Applied Mathematics, University of Campinas, Campinas, São Paulo 13083-859, Brazil
beatrizlaiate@gmail.com, {eelaureano,laeciocb}@ime.unicamp.br

[3] Multidisciplinary Department, Federal University of São Paulo, Osasco, São Paulo 06110-295, Brazil
fsimoes@unifesp.br

Abstract. This work presents an application of interactive fuzzy fractional differential equation, with Caputo derivative, to an HIV model for seropositive individuals under antiretroviral treatment. The initial condition of the model is given by a fuzzy number and the differentiability is given by a fuzzy interactive derivative. A discussion about the model considering these notions are presented. Finally, a numerical solution to the problem is provided, in order to illustrate the results.

Keywords: Fuzzy fractional differential equation · Interactive arithmetic · F-correlated fuzzy process · HIV dynamics

1 Introduction

Fractional Differential Equations (FDE) can be seen as a generalization of Ordinary Differential Equations (ODE) to arbitrary non-integer order [14]. The concept of Fuzzy Fractional Differential Equation (FFDE) was introduced by Agarwal *et al.* in [1]. There are several papers that solve FFDEs, for example [26,35].

Here we consider the Fuzzy Fractional Differential Equations (FFDE) under the interactive derivative of Caputo, that is, the differentiability is given by an interactive derivative, as proposed by Santo Pedro *et al.* [26]. We use the fractional interactive derivative to describe a viral dynamics in seropositive individuals under antiretroviral treatment (ART). An HIV population dynamics has already been considered as a process with memory. In this case, it was described by a system of delay-differential equations associated mainly to pharmacological delay, defined as the time interval required to absorption, distribution and penetration of the drug in the target cells of the virus [16].

M.-J. Lesot et al. (Eds.): IPMU 2020, CCIS 1239, pp. 198–211, 2020.
https://doi.org/10.1007/978-3-030-50153-2_15

The dynamics of biological systems usually evolve with some uncertainty, which may be inherent in the phenomenon or result from environmental variation. It seems pertinent modeling a biological system as a process with memory, so it cannot depend on instant time alone. For these reasons, the fractional differential equation is used [2,3].

Our goal in this work is to provide new insight into well-known models of HIV. For this, we will use fractional differential equations, which are used to treat processes with memories [2,3], and the interactive derivative, which considers both correlated processes and variability at the initial condition [5,27]. Current studies consider interactivity in the modeling of biological processes, in particular, in the dynamics of HIV, when assuming the existence of a memory coefficient [15].

This work is structured as follows. Section 2 presents preliminary concepts about fuzzy set theory, as well as the fuzzy derivative for autocorrelated processes. Section 3 presents fuzzy interactive fractional derivatives. Section 4 presents the fuzzy interactive fractional differential equation under the Caputo derivative. Section 5 presents HIV dynamics under Caputo derivative and Sect. 6 presents the final comments.

2 Preliminary

A fuzzy subset A of $\mathbb{R}^n$ is described by its membership function $\mu_A : \mathbb{R}^n \longrightarrow [0,1]$, where $\mu_A(u)$ means the degree in which u belongs to A. The r-levels of the fuzzy subset A are classical subsets defined as:

$$[A]_r = \{u \in \mathbb{R}^n : \mu_A(u) \geq r\} \text{ for } 0 < r \leq 1 \text{ and}$$
$$[A]_0 = \overline{\{u \in \mathbb{R}^n : \mu_A(u) > 0\}}.$$

The fuzzy subset A of $\mathbb{R}$ is a fuzzy number if its r-levels are closed and nonempty intervals of $\mathbb{R}$ and the support of A, $supp(A) = \{u \in \mathbb{R} : \mu_A(u) > 0\}$, is limited [4]. The family of the fuzzy subsets of $\mathbb{R}^n$ with nonempty compact and convex r-levels is denoted by $\mathbb{R}^n_{\mathcal{F}}$, while the family of fuzzy numbers is denoted by $\mathbb{R}_{\mathcal{F}}$.

The Pompeiu-Hausdorff distance $d_\infty : \mathbb{R}^n_{\mathcal{F}} \times \mathbb{R}^n_{\mathcal{F}} \to \mathbb{R}_+ \cup \{0\}$, is defined by

$$d_\infty(A, B) = \sup_{0 \leq r \leq 1} d_H([A]_r, [B]_r), \tag{1}$$

where d_H is the Pompeiu-Hausdorff distance for compact subsets of $\mathbb{R}^n$. If A and B are fuzzy numbers, that is, $A, B \in \mathbb{R}_{\mathcal{F}}$, then (1) becomes

$$d_\infty(A, B) = \sup_{0 \leq r \leq 1} \max\{|a_r^- - b_r^-|, |a_r^+ - b_r^+|\}.$$

From now on, the continuity of a fuzzy function is associated with the metric d_∞. The symbols + and − stands for the traditional (Minkowski) sum and difference between fuzzy numbers, which can be also defined via Zadeh's extension principle [19].

Let $A, B \in \mathbb{R}_{\mathcal{F}}$ and $J \in \mathcal{F}_J(\mathbb{R}^2)$. The fuzzy relation J is a joint possibility distribution of A and B if, [8]

$$\max_v \mu_J(u,v) = \mu_A(u) \text{ and } \max_u \mu_J(u,v) = \mu_B(v), \ \forall\, u, v \in \mathbb{R}.$$

In this case, A and B are called marginal possibility distributions of J.

The fuzzy numbers A and B are said to be *non-interactive* if, and only if, its joint possibility distribution J is given by $\mu_J(u,v) = min\{\mu_A(u), \mu_B(v)\}$ for all $u, v \in \mathbb{R}$. Otherwise, the fuzzy numbers are said to be *interactive* [8,10].

Let A and B be fuzzy numbers with joint possibility distribution J and $f : \mathbb{R}^2 \to \mathbb{R}$. The extension of f with respect to J, applied to the pair (A, B), is the fuzzy subset $f_J(A, B)$ with membership function defined by [7]

$$\mu_{f_J(A,B)}(u) = \begin{cases} \sup\limits_{(w,v)\in f^{-1}(u)} \mu_J(w,v) & \text{if } f^{-1}(u) \neq \emptyset \\ 0 & \text{if } f^{-1}(u) = \emptyset \end{cases}, \tag{2}$$

where $f^{-1}(u) = \{(w, v) : f(w, v) = u\}$.

If J is given by the minimum t-norm, then $f_J(A, B)$ is the Zadeh's extension principle of f at A and B [7].

Theorem 1 [7,12]. *Let $A, B \in \mathbb{R}_{\mathcal{F}}$, J be a joint possibility distribution whose marginal possibility distributions are A and B, and $f : \mathbb{R}^2 \longrightarrow \mathbb{R}$ a continuous function. In this case, $f_J : \mathbb{R}_{\mathcal{F}} \times \mathbb{R}_{\mathcal{F}} \longrightarrow \mathbb{R}_{\mathcal{F}}$ is well-defined and*

$$[f_J(A, B)]_r = f([J]_r) \text{ for all } r \in [0, 1]. \tag{3}$$

Let $A \in \mathbb{R}_{\mathcal{F}}$. The length of the r-level set of A is defined by

$$\text{len}([A]_r) = a_r^+ - a_r^-, \text{ for all } r \in [0, 1].$$

If $r = 0$, then $\text{len}([A]_0) = \text{diam}(A)$.

A strongly measurable and limited integrable fuzzy function is called integrable. The fuzzy integral of Aumann of $x : [a, b] \to \mathbb{R}_{\mathcal{F}}$, with $[x(t)]_r = [x_r^-(t), x_r^+(t)]$ is defined by [13]

$$\begin{aligned} \left[\int_a^b x(t)dt\right]_r &= \int_a^b [x(t)]_r dt = \int_a^b [x_r^-(t), x_r^+(t)]dt \\ &= \left\{\int_a^b y(t)dt | y : [a, b] \to \mathbb{R} \text{ is a measurable selection for } [x(\cdot)]_r\right\}, \end{aligned} \tag{4}$$

for all $r \in [0, 1]$, provided (4) define a fuzzy number.

Let us focus on the special relationship called interactivity. There are several types of joint possibility distributions that generate different interactivities. This manuscript studies the interactivity called linear correlation, which is obtained as follows.

Let $A, B \in \mathbb{R}_{\mathcal{F}} \setminus \mathbb{R}$ and a function $F : \mathbb{R} \to \mathbb{R}$. The fuzzy numbers A and B are called F-correlated if its joint possibility distribution is given by [8]

$$\mu_J(x,y) = \chi_{\{(u,v=F(u))\}}(x,y)\mu_A(x) = \chi_{\{u,v=F(u)\}}(x,y)\mu_B(y) \tag{5}$$

Note that the fuzzy number B coincides with the Zadeh's extension principle of the function F evaluated at the fuzzy number A. If F is invertible, then $A = F^{-1}(B)$ and, in this case,

$$[J]_r = \{(u, F(u)) \in \mathbb{R}^2 | u \in [A]_r\} = \{(F^{-1}(v), v) \in \mathbb{R}^2 | v \in [B]_r\}. \tag{6}$$

Also, if F is a continuous function, then the r-levels of B are given by [4]

$$[B]_r = F([A]_r).$$

The fuzzy numbers are called linearly correlated (or linearly interactive), if the function F is given by $F(u) = qu + r$. Let A and B be F-correlated fuzzy numbers. The operation $B \otimes_F A$ is defined by [7],

$$\mu_{B\otimes_F A}(w) = \begin{cases} \sup\limits_{u\in\Phi_\otimes^{-1}(w)} \mu_A(u) & if\ \Phi_\otimes^{-1}(w) \neq \emptyset \\ 0 & if\ \Phi_\otimes^{-1}(w) = \emptyset \end{cases}, \tag{7}$$

where $\Phi_\otimes^{-1}(w) = \{u | w = u \otimes v,\ v = F(u)\}$, and $\otimes \in \{+, -, \times, \div\}$.

From Theorem 1, the four arithmetic operations of F-correlated fuzzy numbers, for all $r \in [0, 1]$, are given by

$$[B +_F A]_r = \{F(w) + w \in \mathbb{R} | w \in [A]_r\}; \tag{8}$$

$$[B -_F A]_r = \{F(w) - w \in \mathbb{R} | w \in [A]_r\}; \tag{9}$$

$$[B \cdot_F A]_r = \{wF(w) \in \mathbb{R} | w \in [A]_r\}; \tag{10}$$

$$[B \div_F A]_r = \{F(w) \div w \in \mathbb{R} | w \in [A]_r\},\ 0 \notin [A]_0. \tag{11}$$

Moreover, the scalar multiplication of λB, with $B = F(A)$, is given by $[\lambda B]_r = \{\lambda F(w) \in \mathbb{R} | w \in [A]_r\}$.

Proposition 1 [27]. *Let A and B be F-correlated fuzzy numbers, i.e., $[B]_r = F([A]_r)$, with F monotone differentiable, $[A]_r = [a_r^-, a_r^+]$ and $[B]_r = [b_r^-, b_r^+]$, thus, for all $r \in [0, 1]$,*

1) $[B -_F A]_r = \{F(w) - w | w \in [A]_r\} =$

$$\begin{cases} i.\ [b_r^- - a_r^-, b_r^+ - a_r^+]\ if\ \ F'(z) > 1, & \forall z \in [A]_r \\ ii.\ [b_r^+ - a_r^+, b_r^- - a_r^-]\ if\ 0 < F'(z) \leq 1, & \forall z \in [A]_r\,; \\ iii.\ [b_r^- - a_r^+, b_r^+ - a_r^-]\ if\ \ F'(z) \leq 0, & \forall z \in [A]_r \end{cases} \tag{12}$$

2) $[B +_F A]_r = \{F(w) + w | w \in [A]_r\} =$

$$\begin{cases} i.\ [b_r^- + a_r^-, b_r^+ + a_r^+]\ if\ \ F'(z) > 0, & \forall z \in [A]_r \\ ii.\ [b_r^+ + a_r^-, b_r^- + a_r^+]\ if\ -1 < F'(z) \leq 0, & \forall z \in [A]_r \\ iii.\ [b_r^- + a_r^+, b_r^+ + a_r^-]\ if\ \ F'(z) \leq -1, & \forall z \in [A]_r \end{cases} \tag{13}$$

In the first case (12)-*i*., we have $len([A]_r) < len([B]_r)$, and $-_F$ coincides with Hukuhara difference [13], while in (12)-*ii*., we have $len([A]_r) > len([B]_r)$ and, $-_F$ coincides with generalized Hukuhara difference [6,34]. In fact, the generalized Hukuhara and Hukuhara differences are particular cases of an interactive difference [34]. Additionally, $-_F$ coincides with standard difference when $F'(z) \leq -1$, and $+_F$ coincides with standard sum when $F'(z) > 0$.

Assuming A and B linearly correlated fuzzy numbers, that is $F(u) = qu + r$, and $[B]_r = q[A]_r + r$, with $[A]_r = [a_r^-, a_r^+]$ and $[B]_r = [b_r^-, b_r^+]$, (12) and (13) becomes

$$[B -_L A]_r = \begin{cases} i. \ [b_r^- - a_r^-, b_r^+ - a_r^+] & \text{if } q \geq 1 \\ ii. \ [b_r^+ - a_r^+, b_r^- - a_r^-] & \text{if } 0 < q < 1 \\ iii. \ [b_r^- - a_r^+, b_r^+ - a_r^-] & \text{if } q < 0 \end{cases} \tag{14}$$

and

$$[B +_L A]_r = \begin{cases} i. \ [b_r^- + a_r^-, b_r^+ + a_r^+] & \text{if } q > 0 \\ ii. \ [b_r^+ + a_r^-, b_r^- + a_r^+] & \text{if } -1 \leq q < 0. \\ iii. \ [b_r^- + a_r^+, b_r^+ + a_r^-] & \text{if } q < -1 \end{cases} \tag{15}$$

It is worth to notice that $+_L$ coincides with standard sum when q is positive, and $-_L$ coincides with standard difference when q is negative [11]. Moreover, $-_L$ coincides with generalized Hukuhara difference [6] when q is positive and when $q > 1$ it coincides with Hukuhara difference [13]. It is interesting to mention that the authors of [25] used linearly interactive fuzzy numbers to fit an HIV dataset.

2.1 Autocorrelated Fuzzy Processes

Autocorrelated fuzzy processes are similar to autocorrelated statistical processes [5,11,27,28]. These types of fuzzy processes have been carried out in areas such as, epidemiology [30,33] and population dynamics [27,29].

Let $L([a,b], \mathbb{R}_\mathcal{F})$ be the set of all Lebesgue integrate functions from the bounded interval $[a,b]$ into $\mathbb{R}_\mathcal{F}$, and $AC([a,b], \mathbb{R}_\mathcal{F})$ be the set of all absolutely continuous functions from $[a,b]$ into $\mathbb{R}_\mathcal{F}$. A fuzzy process x is defined by a fuzzy-number-valued function $x : [a,b] \longrightarrow \mathbb{R}_\mathcal{F}$. Considering $[x(t)]_r = [x_r^-(t), x_r^+(t)]$, for all $r \in [0,1]$, the process x is δ-locally F-autoregressive at $t \in (a,b)$ (F-autoregressive for short) if there exists a family of real functions $F_{t,h}$ such that, for all $0 < |h| < \delta$ [27],

$$[x(t+h)]_r = F_{t,h}([x(t)]_r), \forall r \in [0,1]. \tag{16}$$

If $x : [a,b] \rightarrow \mathbb{R}_F$ is a F-autoregressive fuzzy process, then the function x is F-correlated differentiable (F-differentiable for short) at $t_0 \in [a,b]$ if there exists a fuzzy number $x'_F(t_0)$ such that [27]

$$x'_F(t_0) = \lim_{h \rightarrow 0} \frac{x(t_0+h) -_F x(t_0)}{h}, \tag{17}$$

where the above limit exists and it is equal to $x'_F(t_0)$ (using the metric d_∞). If x'_F exists, for all $t \in [a,b]$, then we say that x is F-differentiable.

Next theorem provides a characterization of the F derivative by means of r-levels.

Theorem 2 [27]. *Let $x : [a,b] \to \mathbb{R}_{\mathcal{F}}$ be F-differentiable at $t_0 \in [a,b]$, with $[x(t)]_r = [x_r^-(t), x_r^+(t)]$, where the corresponding family of functions $F_{t_0,h} : I \to \mathbb{R}$ is monotone continuously differentiable for each h, for $r \in [0,1]$ and $F_{t,h}$, $\forall t \in [a,b]$. Then,*

$$[x_F^{'}(t_0)]_r = \begin{cases} \left[(x_r^-)^{'}(t_0), (x_r^+)^{'}(t_0)\right] & \text{if} \quad F_{t,h}^{\prime}(w) > 1 \\ \left[(x_r^+)^{'}(t_0), (x_r^-)^{'}(t_0)\right] & \text{if} \, 0 < F_{t,h}^{\prime}(w) \leq 1 \\ \{(x_r^-)^{'}(t_0)\} = \{(x_r^+)^{'}(t_0)\} & \text{if} \quad F_{t,h}^{\prime}(w) \leq 0 \end{cases}.$$

for each $0 < |h| < \delta$, $\delta > 0$, and $\forall w \in [x(t)]_r$.

The process x is called expansive if, the diameter of $x(t)$ is a non-decreasing function at t, and equivalently, x is called contractive if, the diameter of $x(t)$ is a non-increasing function at t.

Theorem 3 [26]. *Let $x \in AC([a,b], \mathbb{R}_{\mathcal{F}})$ be F-differentiable, where $[x(t)]_r = [x_r^-(t), x_r^+(t)]$.*

I. Suppose x is expansive, that is, $len([x(t)]_r)$ is an increase function on $[a,b]$. If function $x_F^{'}$ is Aumann integrable then $(x_r^-)^{'}(t)$ and $(x_r^+)^{'}(t)$ are integrable on $t \in [a,b]$, and

$$\left[\int_a^t x_F^{'}(s)ds\right]_r = \left[\int_a^t (x_r^-)^{'}(s)ds, \int_a^t (x_r^+)^{'}(s)ds\right].$$

II. Suppose x is contractive, that is, $len([x(t)]_r)$ is a decrease function on $[a,b]$. If function $x_F^{'}$ is Aumann integrable then $(x_r^-)^{'}(t)$ and $(x_r^+)^{'}(t)$ are integrable on $t \in [a,b]$, and

$$\left[\int_a^t x_F^{'}(s)ds\right]_r = \left[\int_a^t (x_r^+)^{'}(s)ds, \int_a^t (x_r^-)^{'}(s)ds\right].$$

2.2 Fuzzy Fractional Integral and Fuzzy Fractional Derivative

The Riemann-Liouville fractional integral $I_{a^+}^{\alpha} f$ of a function $f \in (L[a,b], \mathbb{R})$, of order $\alpha \in (0,1]$ is defined by [31],

$$(I_{a^+}^{\alpha} f)(t) = \frac{1}{\Gamma(\alpha)} \int_a^t (t-s)^{\alpha-1} f(s)ds, \quad \text{for } t > a \tag{18}$$

where $\Gamma(\alpha)$ is the gamma function. If $\alpha = 1$, we have $(I_{a^+}^1 f)(t) = \int_a^t f(s)ds$.

The Riemann-Liouville derivative of order $\alpha \in (0,1]$, is defined by [31]

$$(D^{\alpha} f)(t) = \frac{d}{dt} I^{1-\alpha} f(t) = \frac{1}{\Gamma(1-\alpha)} \frac{d}{dt} \int_a^t (t-s)^{-\alpha} f(s)ds, \tag{19}$$

for $t > a$.

Definition 1 [31]. *The Riemann-Liouville derivative of order* $\alpha \in (0, 1]$, *is defined by*

$$({}^{RL}D^{\alpha}_{a+}f)(t) = \frac{d}{dt}I^{1-\alpha}_{a+}f(t) = \frac{1}{\Gamma(1-\alpha)}\frac{d}{dt}\int_a^t (t-s)^{-\alpha}f(s)ds, \ \textit{for}\ t > a. \tag{20}$$

Definition 2 [31]. *Let* $f \in (L[a, b], \mathbb{R})$ *and suppose there exists* ${}^{RL}D^{\alpha}_{a+}f$ *on* $[a, b]$. *The Caputo fractional derivative* $CD^{\alpha}_{a+}f$ *is defined by*

$$({}^{C}D^{\alpha}_{a+}f)(t) = \left({}^{RL}D^{\alpha}_{a+}[f(\cdot) - f(a)]\right)(t), \ \textit{for}\ t \in (a, b]. \tag{21}$$

Besides that, if $f \in AC([a, b], \mathbb{R})$, then

$$({}^{C}D^{\alpha}_{a+}f)(t) = \frac{1}{\Gamma(1-\alpha)}\int_a^t (t-s)^{-\alpha}f'(s)ds, \ \ \forall t \in (a, b] \tag{22}$$

and

$$({}^{\mathrm{RL}}D^{\alpha}_{a+}f)(t) = ({}^{\mathrm{C}}D^{\alpha}_{a+}f)(t) + \frac{(t-a)^{-\alpha}}{\Gamma(1-\alpha)}f(a), \ \ \forall t \in (a, b]. \tag{23}$$

The next section considers the fuzzy process x in the above definitions, instead of the deterministic function f. The idea is to use the concepts of fuzzy integral and fuzzy F-correlated derivative.

3 Fuzzy Interactive Fractional Derivative

The fuzzy integral fractional Riemann-Liouville, of order $\alpha > 0$, of x is defined by

$$[(I^{\alpha}_{a+}x)(t)]_r = \frac{1}{\Gamma(\alpha)}\left[\int_a^t (t-s)^{\alpha-1}x^-_r(s)ds, \int_a^t (t-s)^{\alpha-1}x^+_r(s)ds\right], \ \ t > a. \tag{24}$$

For fuzzy fractional derivative consider $x \in L([a, b], \mathbb{R}_{\mathcal{F}})$ and the fuzzy process

$$x_{1-\alpha}(t) = \int_a^t \frac{(t-s)^{-\alpha}}{\Gamma(1-\alpha)}x(s)ds, \ \ \text{for all } t \in (a, b], \tag{25}$$

where $x_{1-\alpha}(a) = \lim_{t \to a^+} x_{1-\alpha}(t)$ in the sense of Pompeiu-Hausdorff metric. Recall that for all $0 < \alpha \leq 1$, the fuzzy function $x_{1-\alpha} : (a, b] \to \mathbb{R}_{\mathcal{F}}$ defines a fuzzy number.

Definition 3 [26]. *The fuzzy Riemann-Liouville fractional derivative of order* $0 < \alpha \leq 1$ *of* x *with respect to* F*-derivative is defined by*

$$({}^{RL_F}D^{\alpha}_{a+}x)(t) = \frac{1}{\Gamma(1-\alpha)}\left(\int_a^t (t-s)^{-\alpha}x(s)ds\right)'_F = (x_{1-\alpha}(t))'_F, \tag{26}$$

where $\int_a^t (t-s)^{-\alpha}x(s)ds$ *is a* F*-correlated fuzzy process,* F*-differentiable for all* $t \in (a, b]$.

It is important to highlight that, $\int_a^t (t-s)^{-\alpha} x(s)ds$ can be an expansive or contractive fuzzy process. However, it is expansive if $x(\cdot)$ is expansive [31]. So, if $x_{1-\alpha}(\cdot)$ or $x(\cdot)$ is expansive, then

$$[^{RL_F}D^{\alpha}_{a^+}x(t)]_r = \frac{1}{\Gamma(1-\alpha)}\left[\frac{d}{dt}\int_a^t (t-s)^{-\alpha}x_r^-(s)ds, \frac{d}{dt}\int_a^t (t-s)^{-\alpha}x_r^+(s)ds\right]. \tag{27}$$

Thus,

$$[^{RL_F}D^{\alpha}_{a^+}x(t)]_r = \begin{cases} i.\ [D^{\alpha}_{a^+}x_r^-(t), D^{\alpha}_{a^+}x_r^+(t)] & \text{if } x_{1-\alpha}(\cdot) \text{ or } x(\cdot) \text{ is expansive}) \\ ii.\ [D^{\alpha}_{a^+}x_r^+(t), D^{\alpha}_{a^+}x_r^-(t)] & \text{if } x_{1-\alpha}(\cdot) \text{ is contractive} \end{cases}. \tag{28}$$

Definition 4. *Let x be a F-correlated fuzzy process. The fuzzy Caputo fractional derivative $C_F D^{\alpha}_{a^+}x$ with respect to F-derivative is defined by*

$$(^{C_F}D^{\alpha}_{a^+}x)(t) = \left(^{RL_F}D^{\alpha}_{a^+}[x(\cdot) -_F x(a)]\right)(t),\ for\ t \in (a,b]. \tag{29}$$

Thus,

$$(^{C_F}D^{\alpha}_{a^+}x)(t) = \frac{1}{\Gamma(1-\alpha)}\left(\int_a^t (t-s)^{-\alpha}(x(s) -_F x(a))ds\right)'_F. \tag{30}$$

From (26) if $x_{1-\alpha}(\cdot)$ is contractive, then

$$[(^{C_F}D^{\alpha}_{a^+}x)(t)]_r = \begin{cases} \left[(^{C}D^{\alpha}_{a^+}x_r^-)(t), (^{C}D^{\alpha}_{a^+}x_r^+)(t)\right] & \text{if } x(\cdot) \text{ is expansive} \\ \left[(^{C}D^{\alpha}_{a^+}x_r^+)(t), (^{C}D^{\alpha}_{a^+}x_r^-)(t)\right] & \text{if } x(\cdot) \text{ is contractive} \end{cases}. \tag{31}$$

Theorem 4 [26]. *Let $x \in AC([a,b], \mathbb{R}_{\mathcal{F}})$ be a F-correlated fuzzy process, F-differentiable with $[x(t)]_r = [x_r^-(t), x_r^+(t)]$, for $r \in [0,1]$, and $0 < \alpha \leq 1$. In this case,* $[(^{C_F}D^{\alpha}_{a^+}x)(t)]_r =$

$$\begin{cases} \left[\int_a^t \frac{(t-s)^{-\alpha}}{\Gamma(1-\alpha)}(x_r^-)'(s)ds, \int_a^t \frac{(t-s)^{-\alpha}}{\Gamma(1-\alpha)}(x_r^+)'(s)ds\right] & if\ x\ is\ expansive \\ \left[\int_a^t \frac{(t-s)^{-\alpha}}{\Gamma(1-\alpha)}(x_r^+)'(s)ds, \int_a^t \frac{(t-s)^{-\alpha}}{\Gamma(1-\alpha)}(x_r^-)'(s)ds\right] & if\ x\ is\ contractive \end{cases}, \tag{32}$$

for $t \in [a,b]$.

In the fuzzy fractional calculus the derivative that the researchers usually used is the generalized Hukuhara derivative (gH). Our results via F-correlated derivative are similar to those obtained via gH. However, the domains of arithmetic operations via F-correlated process and via gH are different as can be seen in (8)–(11). Although the difference (9) coincides with the difference gH, the multiplication and division operations F-correlated do not coincide with standard arithmetic operations, which are used with gH. These facts imply that the solutions of fuzzy differential equations via gH and via F can be different. For example via numerical simulations [32].

4 Fuzzy Interactive Fractional Differential Equations

Consider the following fuzzy fractional initial value problem given by the F-correlated fractional Caputo derivative of order $\alpha \in (0,1]$

$$\begin{aligned} ({}^{C_F}D^{\alpha}_{a+}x)(t) &= f(t,x(t)), \\ x(a) &= x_0 \in \mathbb{R}_{\mathcal{F}}, \end{aligned} \tag{33}$$

where $f : (a,b] \times \mathbb{R}_{\mathcal{F}} \to \mathbb{R}_{\mathcal{F}}$ is fuzzy continuous function on Pompeiu-Hausdorff metric. The F-correlated fuzzy process $x : [a,b] \to \mathbb{R}_{\mathcal{F}}$ is said to be a solution of (33) if $x \in C([a,b],\mathbb{R}_{\mathcal{F}})$, $x(a) = x_0$ and $({}^{C_F}D^{\alpha}_{a+}x)(t) = f(t,x(t))$, for all $t \in (a,b]$.

For all $r \in [0,1]$, consider $[x_0]_r = [x^-_{0r}, x^+_{0r}]$ and

$$[f(t,x)]_r = [f^-_r(t,x^-_r(t),x^+_r(t)), f^+_r(t,x^-_r(t),x^+_r(t))].$$

Thus, for all $r \in [0,1]$, the solutions $x(\cdot)$ of (33) satisfy [26]

- if x is expansive on $[a,b]$

$$\begin{aligned} ({}^{C_F}D^{\alpha}_{a+}x^-_r)(t) &= f^-_r(t,x^-_r(t),x^+_r(t)); \;\; x^-_r(a) = x^-_{0r} \\ ({}^{C_F}D^{\alpha}_{a+}x^+_r)(t) &= f^+_r(t,x^-_r(t),x^+_r(t)); \;\; x^+_r(a) = x^+_{0r} \end{aligned} \tag{34}$$

- if x is contractive on $[a,b]$

$$\begin{aligned} ({}^{C_F}D^{\alpha}_{a+}x^+_r)(t) &= f^-_r(t,x^-_r(t),x^+_r(t))]; \;\; x^-_r(a) = x^-_{0r} \\ ({}^{C_F}D^{\alpha}_{a+}x^-_r)(t) &= f^+_r(t,x^-_r(t),x^+_r(t)); \;\; x^+_r(a) = x^+_{0r} \end{aligned}. \tag{35}$$

The Fuzzy Fractional Initial Value Problems (FFIVPs) given by (34) and (35) boil down to classical Fractional Initial Value Problems. Hence, numerical solution for the FFIVP can be provided by the method proposed by [22], which is based on the modified trapezoidal rule and the fractional Euler's method, for Caputo fractional derivative. The generalization of this method for FFIVPs can be founded in [17].

Consider a fractional initial (classical) value problem given by

$${}^{C}D^{\alpha}_{a+}x(t) = f(t,x(t)), \quad x(0) = x_0.$$

Let $[0,a]$ be an interval divided in k subintervals $[t_i, t_{i+1}]$ with equal size h. Then the solution $x(t_j)$, for each $t_j \in [0,a]$, is given by

$$\begin{aligned} x(t_j) &= x_0 + M((j-1)^{\alpha+1} - (j-\alpha-1)j^{\alpha})f(t_0,x(t_0)) \\ &+ M\sum_{i=1}^{j-1}((j-i+1)^{\alpha+1} - 2(j-i)^{\alpha+1} + (j-i-1)^{\alpha+1})f(t_i,x(t_i)) \\ &+ M(f(t_j,x(t_{j-1})) + Nf(t_{j-1},x(t_{j-1}))), \end{aligned} \tag{36}$$

where

$$M = \frac{h^{\alpha}}{\Gamma(\alpha+2)} \quad \text{and} \quad M = \frac{h^{\alpha}}{\Gamma(\alpha+1)}.$$

Next an application of this method is applied in a HIV model that describes the viral dynamics of individuals, under antiretroviral treatment.

5 Viral Dynamics for Seropositive Individuals Under Antiretroviral Treatment (ART)

Data obtained in various studies [20,24] suggests that the virus concentration decay in bloodstream is approximately exponential after the patient was placed on a potent antiretroviral drug. One of the simplest models of viral dynamics consider the effect of antiretroviral as Eq. (37)

$$\frac{dv}{dt} = P - cv, \tag{37}$$

where P is the rate of virus production, c is the clearance rate and $v = v(t)$ is the virus concentration. This model assumes that the treatment is initiated at $t = 0$ and that the efficiency of the treatment is partial when $P > 0$, once the drug could not instantly block all viral production [23].

Although Eq. (37) describes the viral dynamics considering the effect of the drugs, the classical differential equation does not take some behaviours of this dynamic into account. For instance, there is a time interval between the infection of the cell and the release of new infectious viral particles, called *virions*. This means that there exists an intracellular delay, which can be modeled by a system of delay differential equation [9]. For this reason consider the gamma distribution. According to Mittler et al. [18] the gamma distribution can be used to describe the delay presented in the HIV dynamic, because the curves of the gamma distribution are more realistic than the curves of normal distribution, since some cells may take a long time to release virus.

The gamma distribution is widely used to deal with fractional differential equations. Due to the well-established fractional calculus theory, here we adopt the Caputo derivative. To this end, consider the intracellular delay given by the difference $t - s$, where $0 < s < t$. The Caputo derivative of v of order $\alpha \in [0, 1]$ is given by

$$D_C^\alpha v(t) = \frac{1}{\Gamma(1-\alpha)} \int_0^t v'(s)(t-s)^{-\alpha} ds, \tag{38}$$

which can be rewritten as

$$D_C^\alpha v(t) = \int_0^t [f(t-s)e^{t-s}]v'(s)ds, \tag{39}$$

where $f(t - s)$ is the gamma distribution of $t - s$, that is, for $0 < s < t$ and $\alpha \in [0, 1]$,

$$f(t-s) = \frac{(t-s)^{-\alpha} e^{-(t-s)}}{\Gamma(1-\alpha)}. \tag{40}$$

Therefore, as a non local operator, the Caputo derivative provides the effect of intracellular delay at the virus concentration. In this case, it is weighted by the exponential function e^{t-s}, which assign more weight to a shorter delay, as depicted in Fig. 1.

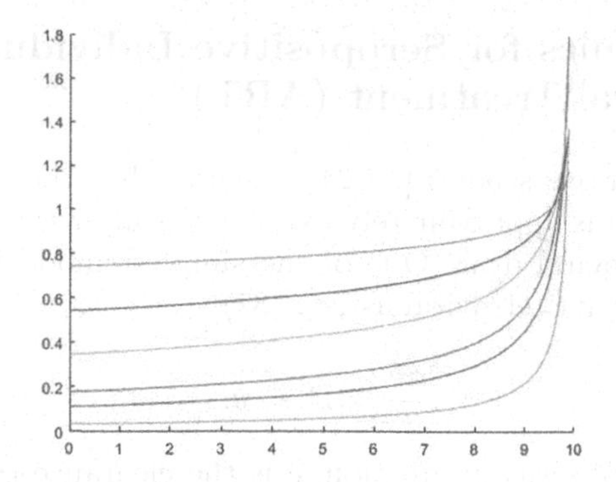

Fig. 1. Representation of wheighted gamma distribution of intracellular delay.

Now, since the initial value of the virus concentration is usually uncertain, the initial condition to this model is described by a fuzzy number, which gives raise to the following Fuzzy Fractional Differential Equation with Caputo derivative

$$\begin{cases} ({}^{C_F}D^{\alpha}v)(t) + cv(t) = P, \\ v(0) = V_0 \in \mathbb{R}_{\mathcal{F}} \end{cases} \tag{41}$$

where $c, P \in \mathbb{R}^+$.

Here we consider two cases for this dynamic. The first one is when the fuzzy process is expansive i.e, the diameter of the process is a non-decreasing function at t, and the second one is when the fuzzy process is contractive, i.e, the diameter of the process is a non-increasing function at t. So, the function $f(t_i, x(t_i)))$ that appears in the formula (36) must be adapted for each case, using the formulas (34) and (35).

Figure 2 illustrates the numerical solution for the FFIVP considering different fuzzy processes. In the case where one expects that uncertainty increases over time, then we must take an expansive process into account, as Subfigure (a) of Fig. 2 depicts. On the other hand, in the case where one expects that uncertainty decreases over time, then we must take a contractive process into account, as Subfigure (b) of Fig. 2 depicts.

Note that the numerical solution for the expansive process assumes negative values. Since we are dealing with the number of infected individuals, the numerical solution obtained from the expansive process is not consistent. This implies that only the contractive process is appropriated for this model. Now, we can still interpret the expansive process for this case. Although it assumes negative values, we verify that the evolution of the disease increases over time. In addition, its width increases, illustrating a chaotic scenario with increasing uncertainty.

Also observe that, in both cases, there is an oscillation in the beginning of the solutions. This is a typical behavior of problems involving FDEs.

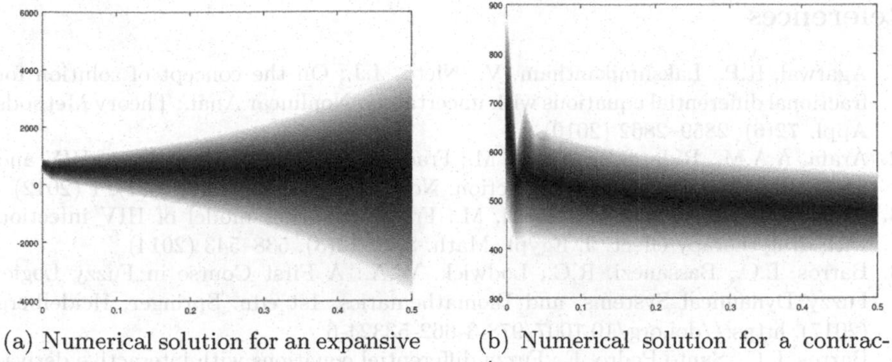

(a) Numerical solution for an expansive process

(b) Numerical solution for a contractive process

Fig. 2. Numerical solution to the HIV model given by (41). The gray lines represent the r-levels of the fuzzy solutions, where their endpoints for r varying from 0 to 1 are represented respectively from the gray-scale lines varying from white to black. The initial condition is given by $v_0 = (470; 670; 870)$, $h = 0.125$ and $\alpha = 0.3$.

6 Final Comments

In this manuscript, we present an HIV viral dynamics model for individuals under antiretroviral treatment. The modeling was done by considering Interactive Fuzzy Fractional Differential Equations (IFFDE), that considers an underlying interactivity in the process and its use is justified by the fact that biological processes have memories in their dynamics [2,3].

Viral load, as an autocorrelated process, considers that there is a memory coefficient in its modeling, this means that the instant of time t is associated to the previous instant time $t-1$. Specifically, the Caputo fractional derivative allows us to take the intracellular delay as a non fixed value into account, by means of the gamma distribution. This distribution assigns more weight to a lower intracellular delay and it carries biological informations, in contrast to the classical derivatives. The FFIVP via Caputo derivative provides solutions related to the value of $\alpha \in [0,1]$, once the bigger the value of α, the faster the viral load decays.

The uncertainty in the number of viral particles produced by each infected cell suggests that the viral load can be represented as a fuzzy number. Through IFFDE, it was possible to describe the phenomenon from two points of view: expansive process (the diameter of the solution is a non-decreasing function in t) and contractive process (the diameter of the solution is a non-increasing function in t), in contrast to other methods given in the literature.

Finally, we present a numerical solution to illustrate the obtained results. In both cases, a decrease in plasma viremia in the bloodstream is obtained, which corroborates the data presented in the literature [21].

Acknowledgments. This research was partially supported by FAPESP under grants no. 2018/10946-2, and 2016/26040-7, and CNPq under grants no. 306546/2017-5 and 42309/2019-2.

References

1. Agarwal, R.P., Lakshmikantham, V., Nieto, J.J.: On the concept of solution for fractional differential equations with uncertainty. Nonlinear Anal.: Theory Methods Appl. **72**(6), 2859–2862 (2010)
2. Arafa, A.A.M., Rida, S.Z., Khalil, M.: Fractional modeling dynamics of HIV and CD4+ T-cells during primary infection. Nonlinear Biomed. Phys. **6**(1), 1 (2012)
3. Arafa, A.A.M., Rida, S.Z., Khalil, M.: Fractional-order model of HIV infection with drug therapy effect. J. Egypt. Math. Soc. **22**(3), 538–543 (2014)
4. Barros, L.C., Bassanezi, R.C., Lodwick, W.A.: A First Course in Fuzzy Logic, Fuzzy Dynamical Systems, and Biomathematics, 1st edn. Springer, Heidelberg (2017). https://doi.org/10.1007/978-3-662-53324-6
5. Barros, L.C., Santo Pedro, F.: Fuzzy differential equations with interactive derivative. Fuzzy Sets Syst. **309**, 64–80 (2017)
6. Bede, B.: Mathematics of Fuzzy Sets and Fuzzy Logic. Springer, Heidelberg (2013). https://doi.org/10.1007/978-3-642-35221-8
7. Cabral, V.M., Barros, L.C.: Fuzzy differential equation with completely correlated parameters. Fuzzy Sets Syst. **265**, 86–98 (2015)
8. Carlsson, C., Fullér, R., Majlender, P.: Additions of completely correlated fuzzy numbers. In: Proceedings of 2004 IEEE International Conference on Fuzzy Systems, vol. 1, pp. 535–539 (2004)
9. Culshaw, R.V., Ruan, S.: A delay-differential equation model of HIV infection of $CD4^+$ T-cells. Math. Biosci. **165**(1), 27–39 (2000)
10. Dubois, D., Prade, H.: Additions of interactive fuzzy numbers. IEEE Trans. Autom. Control **26**(4), 926–936 (1981)
11. Esmi, E., Santo Pedro, F., Barros, L.C., Lodwick, W.A.: Fréchet derivative for linearly correlated fuzzy function. Inf. Sci. **435**, 150–160 (2018)
12. Fullér, R.: Fuzzy reasoning and fuzzy optimization. Turku Centre for Computer Science (1998)
13. Kaleva, O.: Fuzzy differential equations. Fuzzy Sets Syst. **24**, 301–317 (1987)
14. Kilbas, A.A.A., Srivastava, H.M., Trujillo, J.J.: Theory and Applications of Fractional Differential Equations, vol. 204. Elsevier Science Limited, Amsterdam (2006)
15. Laiate B., Santo Pedro, F., Esmi E., Barros L.C.: HIV dynamics under antiretroviral treatment with interactivity. In: Lesot, M.-J., et al. (eds.) IPMU 2020. CCIS, vol. 1239, pp. 212–225. Springer, Heidelberg (2020)
16. Laiate, B., Jafelice, R.M., Esmi, E., Barros, L.C.: An interpretation of the fuzzy measure associated with choquet calculus for a HIV transference model. In: Kearfott, R.B., Batyrshin, I., Reformat, M., Ceberio, M., Kreinovich, V. (eds.) IFSA/NAFIPS 2019 2019. AISC, vol. 1000, pp. 306–317. Springer, Cham (2019). https://doi.org/10.1007/978-3-030-21920-8_28
17. Mazandarani, M., Kamyad, A.V.: Modified fractional Euler method for solving fuzzy fractional initial value problem. Commun. Nonlinear Sci. Numer. Simul. **18**, 12–21 (2013)
18. Mittler, J.E., Sulzer, B., Neumann, A.U., Perelson, A.S.: Influence of delayed viral production on viral dynamics in HIV-1 infected patients. Math. Biosci. **152**, 143–163 (1998)
19. Mizumoto, M., Tanaka, K.: Bounded-sum or bounded-difference for fuzzy sets. Trans. IECE (D) **59**, 905–912 (1976)
20. Nelson, P.W., Mittler, J.E., Perelson, A.S.: Effect of drug efficacy and the eclipse phase of the viral life cycle on estimates of HIV viral dynamic parameters. J. Acquir. Immune Defic. Syndr. **26**, 405–412 (2001)

21. Notermans, D.W., Goudsmit, J., Danner, S.A., DeWolf, F., Perelson, A.S., Mittler, J.: Rate of HIV-1 decline following antiretroviral therapy is related to viral load at baseline and drug regimen. Aids **12**(12), 1483–1490 (1998)
22. Odibat, Z.M., Momani, S.: An algorithm for the numerical solution of differential equations of fractional order. J. Appl. Math. Inf. **26**, 15–27 (2008)
23. Perelson, A.S., Nelson, P.W.: Mathematical analysis of HIV-1 dynamics in vivo. Soc. Ind. Appl. Math. **41**(1), 3–44 (1999)
24. Perelson, A.S., et al.: Decay characteristics of HIV-1-infected compartments during combination therapy. Nature **387**(6629), 188–191 (1997)
25. Pinto, N.J.B., Esmi, E., Wasques, V.F., Barros, L.C.: Least square method with quasi linearly interactive fuzzy data: fitting an HIV dataset. In: Kearfott, R., Batyrshin, I., Reformat, M., Ceberio, M., Kreinovich, V. (eds.) IFSA/NAFIPS 2019 2019. Advances in Intelligent Systems and Computing, vol. 1000, pp. 177–189. Springer, Cham (2019). https://doi.org/10.1007/978-3-030-21920-8_17
26. Santo Pedro, F., Martins, M. M., Wasques, V. F., Esmi, E., Barros, L. C.: Fuzzy fractional under interactive derivative. Fuzzy Sets Syst. (submitted)
27. Santo Pedro, F., Barros, L.C., Esmi, E.: Population growth model via interactive fuzzy differential equation. Inf. Sci. **481**, 160–173 (2019)
28. Santo Pedro, F., Esmi, E., Barros, L.C.: Calculus for linearly correlated fuzzy function using Fréchet derivative and Riemann integral. Inf. Sci. **512**, 219–237 (2020)
29. Santo Pedro, F., Barros, L.C., Esmi, E.: Measure of interactivity on fuzzy process autocorrelated: malthusian model. In: Kearfott, R., Batyrshin, I., Reformat, M., Ceberio, M., Kreinovich, V. (eds.) FSA/NAFIPS 2019 2019. Advances in Intelligent Systems and Computing, vol. 1000, pp. 567–577. Springer, Cham (2019). https://doi.org/10.1007/978-3-030-21920-8_50
30. Santo Pedro, F., Barros, L.C., Esmi, E.: Interactive fuzzy process: an epidemiological model. In: Barreto, G., Coelho, R. (eds.) NAFIPS 2018. Communications in Computer and Information Science, vol. 831, pp. 108–118. Springer, Cham (2019). https://doi.org/10.1007/978-3-319-95312-0_10
31. Van Ngo, H., Lupulescu, V., O'Regan, D.: A note on initial value problems for fractional fuzzy differential equations. Fuzzy Sets Syst. **347**, 54–69 (2018)
32. Wasques, V.F., Esmi, E., Barros, L.C., Bede, B.: Comparison between numerical solutions of fuzzy initial-value problems via interactive and standard arithmetics. In: Kearfott, R.B., Batyrshin, I., Reformat, M., Ceberio, M., Kreinovich, V. (eds.) IFSA/NAFIPS 2019 2019. AISC, vol. 1000, pp. 704–715. Springer, Cham (2019). https://doi.org/10.1007/978-3-030-21920-8_62
33. Wasques, V.F., Esmi, E., Barros, L.C., Sussner, P.: Numerical solutions for bidimensional initial value problem with interactive fuzzy numbers. In: Barreto, G., Coelho, R. (eds.) NAFIPS 2018. Communications in Computer and Information Science, vol. 831, pp. 84–95. Springer, Cham (2018). https://doi.org/10.1007/978-3-319-95312-0_8
34. Wasques, V.F., Esmi, E., Barros, L.C., Sussner, P.: The generalized fuzzy derivative is interactive. Inf. Sci. **519**, 93–109 (2020)
35. Zahra, A., Dumitru, B., Babak, S., Guo-Cheng, W.: Spline collocation methods for systems of fuzzy fractional differential equations. Chaos Solit. Fractals **131**, 109510 (2020)

HIV Dynamics Under Antiretroviral Treatment with Interactivity

Beatriz Laiate[1(✉)], Francielle Santo Pedro[2], Estevão Esmi[1(✉)], and Laécio Carvalho de Barros[1(✉)]

[1] Department of Applied Mathematics, University of Campinas, Campinas, SP 13083-970, Brazil
beatrizlaiate@gmail.com, eelaureano@gmail.com, laeciocb@ime.unicamp.br
[2] Multidisciplinary Department, Federal University of São Paulo, Osasco, SP 06110-295, Brazil
fsimoes@unifesp.br

Abstract. This manuscript presents a model for HIV dynamics of seropositive individuals under antiretroviral treatment described from fuzzy set theory by two different approaches considering interactivity: differential equation with interactive derivative and differential equation with Fréchet derivative. It also establishes an identity between interactive derivative and fuzzy Fréchet derivative. With this identity, we establish when the solutions of the two differential equations coincide. Lastly, we present biological interpretations for both cases.

Keywords: HIV · Antiretroviral treatment · Fuzzy interactive differential equation · Fréchet derivative · Fuzzy interactive derivative

1 Introduction

HIV dynamics considering antiretroviral treatment (ART) has already been studied in several articles [11,16,18,20]. The major target of HIV are CD4+ T cells, a class of immune cells. Antiviral drugs act blocking biological processes involved in life cycle virus into cell cytoplasm. Most common therapies combine protease inhibitors and reverse transcriptase inhibitors. The first ones block HIV protease, so that noninfectious viral particles start being produced by infected T-cells, and the last ones prevent the successfully infection of T-cells. Data obtained in previous studies [17,18] show that under combination of protease inhibitors and RT inhibitors, a viral decline in the bloodstream is followed by the increase of CD4+ population are expected.

Fuzzy set theory applied to HIV dynamics under treatment was already studied using fuzzy rule-based systems [12–14] and Choquet Calculus [15], both considering an intracellular delay assigned maily to the pharmacological delay, defined as the interval of time required for the absorption of the antiviral drugs in the bloodstream. Viral dynamics represented as an interactive process is a new

M.-J. Lesot et al. (Eds.): IPMU 2020, CCIS 1239, pp. 212–225, 2020.
https://doi.org/10.1007/978-3-030-50153-2_16

approach in the literature and provides results subject to new interpretations for already known HIV models.

The existence of memory in biological processes was already considered in previous studies [1,2]. Current studies consider the existence of memory in biological processes, in particular, in the dynamics of HIV, when representing them by fuzzy fractional derivatives with interactivity [25]. Autocorrelated processes take into account the dependence between their states in consecutive instants. This type of approach allows us to describe processes that are hidden or inherent to the phenomenon when considering a memory coefficient $f_{t,h}$. This coefficient changes with the process and is determined by the current moment. Thus, there may exist memories with different properties for different periods of time. Therefore, it is necessary a derivative operator that incorporates the memory of the system, being responsible for its variation.

There are various different theories of fuzzy differential equations for fuzzy-set-valued functions [4,5,21,24], that is, functions $f : [a, b] \to \mathbb{R}_{\mathcal{F}}$, where $\mathbb{R}_{\mathcal{F}}$ is the space of fuzzy numbers, that is, fuzzy subsets of $\mathbb{R}$ whose α-levels are closed intervals in real line [3]. In this work we will represent an HIV intracellular model throught two interactive derivatives defined in an autocorrelated process. In particular, we assume that it is a linearly correlated process [8,23].

Firstly we describe the dynamics via the linearly correlated derivative, based on the difference between fuzzy sets, that is, the difference obtained from possibility distributions of fuzzy sets envolved [26]. This derivative provides two possible behaviors for the solution, expansive or contractive. In the first case, the fuzziness of the solution increase with time while in the second one, it decreases. In this work, the fuzziness is measured accordingly to the diameter of the fuzzy number. The larger the diameter of the fuzzy number, the greater its fuzziness.

On the other hand, fuzzy differential equation via Fréchet derivative is based on the isomorphim $\Psi_A : \mathbb{R}^2 \to \mathbb{R}_{\mathcal{F}(A)}$, where $\mathbb{R}_{\mathcal{F}(A)}$ is the set of all fuzzy numbers linearly correlated to $A \in \mathbb{R}_{\mathcal{F}}$ [10]. This allows us to define the induced sum and scalar multiplication in $\mathbb{R}_{\mathcal{F}(A)}$ given by $B +_A C = \Psi_A({\Psi_A}^{-1}(B) + {\Psi_A}^{-1}(C))$ and $\eta \cdot_A B = \Psi_A(\eta {\Psi_A}^{-1}(B)), \forall B, C \in \mathbb{R}_{\mathcal{F}(A)}$ and $\eta \in \mathbb{R}$. With this operations it is possible to confer a Banach space structure to the space $\mathbb{R}_{\mathcal{F}(A)}$ and, therefore, develop a calculus theory for the family of fuzzy functions linearly correlated to $A \in \mathbb{R}_{\mathcal{F}}$, as it was done in [22].

This work is structured as follows. Section 2 provides the mathematical concepts necessary to understand the development of this work. Section 3 presents HIV dynamics in two different approaches: via L-derivative and Fréchet derivative. Section 4 presents final comments.

2 Mathematical Background

A fuzzy subset A of $\mathbb{R}$ is described by its membership function $\mu_A : \mathbb{R} \to [0, 1]$, where $\mu_A(x)$ is the degree of membership of x in A. The α-cuts of A are subsets of $\mathbb{R}$ given by $[A]_\alpha = \{x \in \mathbb{R} : A(x) \geq \alpha\}$, for $\alpha \in (0, 1]$, and $[A]_0$ is the closure of the support of A, that is, $[A]_0 = \overline{\{x \in \mathbb{R} : A(x) > 0\}}$.

The fuzzy subset A of $\mathbb{R}$ is a fuzzy number if all the α-cuts are closed and nonempty intervals of $\mathbb{R}$ and the support of A is bounded [3]. The set of all fuzzy numbers is denoted by $\mathbb{R}_{\mathcal{F}}$. We define the $\text{diam}(A) = |a_0^+ - a_0^-|$, where a_0^- and a_0^+ are the endpoints of $[A]_0$, for all $A \in \mathbb{R}_{\mathcal{F}}$.

Let $a, b, c \in \mathbb{R}$ such that $a \leq b \leq c$, a triangular fuzzy number A is a well-known example of fuzzy number given by the following membership function:

$$\mu_A(x) = \begin{cases} 0, & x \leq a \,\text{or}\, x \geq b \\ \frac{x-a}{b-a} \wedge \frac{c-x}{c-b} & \text{otherwise,} \end{cases} \tag{1}$$

where $\wedge$ is the minimum operator. In this case, we denote A by the symbol $(a; b; c)$.

Next, we recall some concepts necessary to understand the theory of interactive derivative in the space of fuzzy numbers.

A possibility distribution on $\mathbb{R}^n$ is a fuzzy subset J of $\mathbb{R}^n$ with membership function $\mu_J : \mathbb{R}^n \to [0, 1]$ satisfying $\mu_J(x_0) = 1$ for some $x_0 \in \mathbb{R}^n$. The family of possibility distributions of $\mathbb{R}^n$ will be denoted by $\mathcal{F}_J(\mathbb{R}^n)$.

Definition 1 [6]. *Let $A, B \in \mathbb{R}_{\mathcal{F}}$ and $J \in \mathcal{F}_J(\mathbb{R}^2)$. Then μ_J is a joint possibility distribution of A and B if $\max_y \mu_J(x, y) = \mu_A(x)$ and $\max_x \mu_J(x, y) = \mu_B(y)$, for any $x, y \in \mathbb{R}$.*

In this case, μ_A and μ_B are called marginal possibility distributions of J.

Definition 2 [9]. *The fuzzy numbers A and B are said to be non-interactive if and only if their joint possibility distribution J satisfies the relationship $\mu_J(x, y) = \min(\mu_A(x), \mu_B(y))$ for all $x, y \in \mathbb{R}$. Otherwise, are said to be interactive.*

Definition 3 [6,9]. *The fuzzy numbers A and B are said to be completely correlated if there exist $q, r \in \mathbb{R}$, $q \neq 0$ such that joint possibility distribution is defined by*

$$\mu_C(x, y) = \mu_A(x)\chi_{qx+r=y}(x, y) = \mu_B(y)\chi_{qx+r=y}(x, y) \tag{2}$$

where $\chi_{qx+r}(x, y)$ represents the characteristic function of the line $\{(x, y) \in \mathbb{R}^2 : qx + r = y\}$.

Definition 4 [8]. *Two fuzzy numbers A and B are said linearly correlated if there exist $q, r \in \mathbb{R}$ such that their α-levels satisfy $[B]_\alpha = q[A]_\alpha + r$ for all $\alpha \in [0, 1]$. In this case, we write $B = qA + r$.*

Definition 5. *The four arithmetic operations between linearly correlated fuzzy numbers are defined, in levels, by:*

- $[B +_L A]_\alpha = (q + 1)[A]_\alpha + r, \forall \alpha \in [0, 1]$;
- $[B -_L A]_\alpha = (q - 1)[A]_\alpha + r, \forall \alpha \in [0, 1]$;
- $[B \cdot_L A]_\alpha = \{qx_1^2 + rx_1 \in \mathbb{R} | x_1 \in [A]_\alpha\}, \forall \alpha \in [0, 1]$;
- $[B \div_L A]_\alpha = \{q + \frac{r}{x_1} \in \mathbb{R} | x_1 \in [A]_\alpha\}, \forall \alpha \in [0, 1]$.

The Pompeiu-Hausdorff distance $d_\infty : \mathbb{R}^n_\mathcal{F} \times \mathbb{R}^n_\mathcal{F} \to \mathbb{R}_+ \cup \{0\}$ is defined by

$$d_\infty(A, B) = \sup_{0 \leq \alpha \leq 1} d_H([A]_\alpha, [B]_\alpha), \tag{3}$$

where d_H is the Pompeiu-Hausdorff distance for sets in $\mathbb{R}^n$. If A and B are fuzzy numbers, then (3) becomes

$$d_\infty(A, B) = \sup_{0 \leq \alpha \leq 1} max\{|a^-_\alpha - b^-\alpha|, |a^+_\alpha - b^+_\alpha|\}. \tag{4}$$

The derivative enunciated in this subsection is related to an autocorrelated process $F : [a, b] \to \mathbb{R}_\mathcal{F}$, that is, for h with absolute value sufficiently small, $F(t+h) = q(h)F(h) + r(h)$, for all $t \in [a, b]$, $q(h), r(h) \in \mathbb{R}$. This formula means that $[F(t+h)]_\alpha = q(h)[F(t)]_\alpha + r(h)$, $\forall \alpha \in [0, 1]$.

Definition 6 [8]. *Let $F : [a, b] \to \mathbb{R}_\mathcal{F}$ be a fuzzy-number-valued function and for each h with absolute value sufficiently small, let $F(t_0 + h)$ and $F(t_0)$ with $t_0 \in [a, b]$ be linearly correlated fuzzy numbers. F is called L-differentiable at t_0 if there exists a fuzzy number $D_L F(t_0) \in \mathbb{R}_\mathcal{F}$ such that the limit*

$$\lim_{h \to 0} \frac{F(t_0 + h) -_L F(t_0)}{h} \tag{5}$$

exists and is equal to $D_L F(t_0)$, using the metric d_∞. $D_L F(t_0)$ is called linearly correlated fuzzy derivative of F at t_0. At the endpoints of $[a, b]$ we consider only one-sided derivative.

The next theorem provides a practical formula to calculate the L-derivative of an autocorrelated process.

Theorem 1 [8]. *Let $F : [a, b] \to \mathbb{R}_\mathcal{F}$ be L-differentiable in t_0 and $F_\alpha(t_0) = [F(t_0)]_\alpha = [f^-_\alpha(t_0), f^+_\alpha(t_0)]$, for all $\alpha \in [0, 1]$. Then f^-_α and f^+_α are differentiable in t_0 and for all $|h| < \delta$, for some $\delta > 0$ and*

$$[D_L F(t_0)]_\alpha = \begin{cases} i. \ [(f^-_\alpha)'(t_0), (f^+_\alpha)'(t_0)] & if\, q(h) \geq 1 \\ ii. \ [(f^+_\alpha)'(t_0), (f^-_\alpha)'t_0] & if\, 0 < q(h) \leq 1 \\ iii. \ [(f^-_\alpha)'(t_0), (f^-_\alpha)'(t_0)] & if\, q(h) < 0 \end{cases} \tag{6}$$

where $D_L F(t_0)$ is the L-derivative.

Next, we present important results related to the theory of calculus developed in the space of fuzzy numbers linearly correlated to a given fuzzy number A.

A fuzzy number $A \in \mathbb{R}_\mathcal{F}$ is said to be symmetric with respect to $x \in \mathbb{R}$ if $A(x - y) = A(x + y)$, $\forall y \in \mathbb{R}$, and it is said to be non-symmetric if there exists no x such that A is symmetric. For example, the fuzzy number $A = (-1; 0; 1)$ is symmetric with respect to 0 and the fuzzy number $B = (2; 3; 5)$ is not symmetric.

Given $A \in \mathbb{R}_\mathcal{F}$, we can define the operator $\Psi_A : \mathbb{R}^2 \to \mathbb{R}_\mathcal{F}$ so that $\Psi_A(q, r) = qA + r$, that is, the image of the pair (q, r) is the fuzzy number $\Psi_A(q, r)$ whose

α-cuts are given by $[\Psi_A(q,r)]_\alpha = \{qx + r \in \mathbb{R} | x \in [A]_\alpha\} = q[A]_\alpha + r$. The range of the operator Ψ_A is denoted by $\mathbb{R}_{\mathcal{F}(A)} = \{\Psi_A(q,r) | (q,r) \in \mathbb{R}^2\}$. This operator defines an isomorphism between $\mathbb{R}^2$ and $\mathbb{R}_{\mathcal{F}(A)}$ whenever A is a non-symmetric fuzzy number [10]. Since $\mathbb{R}^2$ is a Banach space, we can conclude that $\mathbb{R}_{\mathcal{F}(A)}$ is also a Banach space.

Let A be a non-symmetric fuzzy number. We say that a fuzzy-number-valued function $f : [a,b] \to \mathbb{R}_{\mathcal{F}(A)} \subseteq \mathbb{R}_{\mathcal{F}}$ is continuous in $\mathbb{R}_{\mathcal{F}(A)}$ when it is continuous with respect to the norm $\| \cdot \|_{\Psi_A}$. These functions are called A-linearly correlated fuzzy processes. The following lemma characterizes this type of function.

Lemma 1 [10]. *Let $A \in \mathbb{R}_{\mathcal{F}}$ be non-symmetric. There exists unique $(q,r) = p : \mathbb{R} \to \mathbb{R}^2$ such that $f = \Psi_A \circ p$.*

Theorem 2 [10]. *Let $B = \Psi_A(q,r) \in \mathbb{R}_{\mathcal{F}}$ for some $A \in \mathbb{R}_{\mathcal{F}}$ and some pair $(q,r) \in \mathbb{R}^2$ with $q \neq 0$. Then the ranges of operators Ψ_A and Ψ_B are identical, that is, $\mathbb{R}_{\mathcal{F}(A)} = \mathbb{R}_{\mathcal{F}(B)}$.*

The next theorem establishes sufficient and necessary conditions to an A-linearly correlated fuzzy process to be continuous.

Theorem 3 [22]. *Let A be non-symmetric and $f = \Psi_A \circ p : [a,b] \longrightarrow \mathbb{R}_{\mathcal{F}(A)}$. The function $f : [a,b] \longrightarrow \mathbb{R}_{\mathcal{F}(A)}$ is continuous if, and only if, $p : [a,b] \to \mathbb{R}^2$ is continuous.*

Since, for $A \in \mathbb{R}_{\mathcal{F}}$ non-symmetric, $\mathbb{R}_{\mathcal{F}(A)}$ is a Banach space, it is possible to define the Fréchet derivative of f as it was done in [10]. The next proposition presents a necessary and sufficient condition to $f : \mathbb{R} \to \mathbb{R}_{\mathcal{F}(A)}$ to be Fréchet differentiable.

Proposition 1 [10]. *Let $A \in \mathbb{R}_{\mathcal{F}}$ be non-symmetric and $f : [a,b] \longrightarrow \mathbb{R}_{\mathcal{F}(A)} \subset \mathbb{R}_{\mathcal{F}}$. The function f is Fréchet differentiable at t if, and only if, $\Psi_A^{-1} \circ f : [a,b] \longrightarrow \mathbb{R}^2$ is Fréchet differentiable at t.*

Theorem 4 [10]. *Let $A \in \mathbb{R}_{\mathcal{F}}$ be non-symmetric, the functions $q, r : \mathbb{R} \to \mathbb{R}$ and $f : \mathbb{R} \to \mathbb{R}_{\mathcal{F}(A)}$ such that $f(t) = \Psi_A(q(t), r(t))$, $\forall t \in \mathbb{R}$. The function f is Fréchet differentiable (F-differentiable) at $t \in \mathbb{R}$ if and only if $q'(t)$ and $r'(t)$ exist. Additionally, the F-derivative of f at t is given by $f'(t,h) = \Psi_A(q'(t)h, r'(t)h)$, $\forall h \in \mathbb{R}$.*

Fuzzy interactive derivatives studied in this paper can be related algebraically throught Theorem 5.

Theorem 5. *Let A be a non-symmetric fuzzy number, $f : \mathbb{R} \to \mathbb{R}_{\mathcal{F}(A)}$ given by $f(t) = \Psi_A(p(t), q(t))$, where p, q are real functions for all $t \in \mathbb{R}$. Then f is Fréchet differentiable if, and only if, f is L-differentiable, where the L-derivative is the interactive derivative [8]. Moreover, the Fréchet derivative of f coincide with the L-derivative of f, that is*

$$\Psi_A(p'(t), q'(t)) = D_L f(t), \forall t \in \mathbb{R}. \tag{7}$$

Proof. Let $f : \mathbb{R} \to \mathbb{R}_{\mathcal{F}(A)}$ be given by $f(t) = p(t)A + q(t)$, for all $t \in \mathbb{R}$ such that $f(t) \neq \mathbb{R}$, $\forall t \in \mathbb{R}$.

For $h \in \mathbb{R}$, we have that $f(t+h) = p(t+h)A + q(t+h)$ and, therefore,

$$f(t+h) = \left(\frac{p(t+h)}{p(t)}\right) f(t) + \left(q(t+h) - p(t+h)\frac{q(t)}{p(t)}\right). \tag{8}$$

Denoting $\tilde{p}(h) = \frac{p(t+h)}{p(t)}$ and $\tilde{q}(h) = q(t+h) - \frac{p(t+h)q(t)}{p(t)}$, we may write

$$f(t+h) = \tilde{p}(h)f(t) + \tilde{q}(h),\ \forall h \in \mathbb{R}. \tag{9}$$

Thus, if $f(t) = \Psi_A(p(t), q(t))$, $\forall t \in \mathbb{R}$, f represents a linearly correlated fuzzy process. So, we have that

$$f(t+h) -_L f(t) = (\tilde{p}(h) - 1)f(t) + \tilde{q}(h) \tag{10}$$

for all $t, h \in \mathbb{R}$. If f is Fréchet differentiable, then the derivatives $p', q' : \mathbb{R} \to \mathbb{R}$ exist for all $t \in \mathbb{R}$, according to Theorem 4. Moreover,

$$\lim_{h\to 0} \tilde{p}(h) = \lim_{h\to 0} \frac{p(t+h)}{p(t)} = 1 \tag{11}$$

and

$$\lim_{h\to 0} \tilde{q}(h) = \lim_{h\to 0} q(t+h) - q(t)\frac{p(t+h)}{p(t)} = 0, \tag{12}$$

once p and q are continuous in t. Therefore, $\lim_{h\to 0}[f(t+h) -_L f(t)] = \lim_{h\to 0}[(\tilde{p}(h) - 1)f(t) + \tilde{q}(h)] = 0$.

For all $\alpha \in [0, 1]$ we have that

$$\lim_{h\to 0} \frac{[f(t+h) -_L f(t)]_\alpha}{h} = \lim_{h\to 0}\left[\left(\frac{\tilde{p}(h) - 1}{h}\right)[f(t)]_\alpha + \left(\frac{\tilde{q}(h)}{h}\right)\right] \tag{13}$$

Note that

$$\lim_{h\to 0}\left(\frac{\tilde{p}(h) - 1}{h}\right) = \lim_{h\to 0} \frac{\left(\frac{p(t+h)}{p(t)} - 1\right)}{h} = \frac{p'(t)}{p(t)} \tag{14}$$

and

$$\lim_{h\to 0} \frac{\tilde{q}(h)}{h} = \lim_{h\to 0} \frac{q(t+h) - \frac{p(t+h)q(t)}{p(t)}}{h} = q'(t) - \frac{q(t)p'(t)}{p(t)} \tag{15}$$

for all $t \in \mathbb{R}$. Therefore,

$$[D_L f(t)]_\alpha = \lim_{h\to 0}\left(\frac{\tilde{p}(h) - 1}{h}\right)[f(t)]_\alpha + \lim_{h\to 0}\left(\frac{\tilde{q}(h)}{h}\right) \tag{16}$$

$$= \frac{p'(t)}{p(t)}[f(t)]_\alpha + q'(t) - \frac{q(t)p'(t)}{p(t)}, \tag{17}$$

for all $t \in \mathbb{R}$ and $\alpha \in [0, 1]$.

Since $[f(t)]_\alpha = p(t)[A]_\alpha + q(t)$ for all $\alpha \in [0, 1]$, we have

$$D_L(f)(t) = p'(t)A + q'(t) = \Psi_A(p'(t), q'(t)),\ \forall t \in \mathbb{R}. \tag{18}$$

The converse implication is immediate, from Theorem 1.

3 HIV Dynamics Under Antiretroviral Treatment (ART)

Data obtained in various studies [17] appear to show that the decay of plasma viraemia in bloodstream is approximately exponential after the patient was placed on a potent antiretroviral drug. One of the simplest models of viral dynamics consider the effect of antiretroviral in viral population as in Eq. (19)

$$\frac{dV}{dt} = P - cV,\ V(0) = V_0 \tag{19}$$

where P is the rate of virus production, c is the clearance rate and $V = V(t)$ is the virus concentration in bloodstream. This model assumes that the treatment is initiated at $t = 0$ and that the efficiency of the treatment is partial, so that $P > 0$. With this assumption, virus decay is not perfectly "exponential", so that the solution of (19) is given by

$$V(t) = V_0 e^{-ct} + \frac{P}{c} - \frac{P}{c} e^{-ct}. \tag{20}$$

Equation (20) means viral load declines whenever $V_0 > \frac{1}{c}$, where $\frac{1}{c}$ is the average life time of the viruses when the efficiency of the treatment is total, that is, $P = 0$.

Next, we will establish two different fuzzy approaches to HIV dynamics described in (19), both considering interactivity into the process V.

3.1 Fuzzy Interactive Differential Equation via L-Derivative

Analysis of models considering CD4+ cell population suggests that, when starting the treatment, viral load is related to parameters such as virus and infected cells elimination rates, as well as and the number of viral particles produced by each infected CD4+ cell [19]. The uncertainty of these rates suggests that the viral load may be well represented when V is a fuzzy number.

We will consider that viral dynamics described in (19) is an autocorrelated fuzzy process. According to Definition 6, this means that for each h with absolute value sufficiently small, $V(t+h) = q(h)V(t) + r(h)$ for all $t \geq 0$, where q, r are real functions. Then, the corresponding Fuzzy Initial Value Problem (FIVP) via L-derivative is given by

$$\begin{cases} V_L'(t) = P -_L cV(t) \\ V(0) = V_0 \in \mathbb{R}_{\mathcal{F}}, \end{cases} \tag{21}$$

where $P, c > 0$ are real constants.

According to Theorem 1, there are two cases to consider:

i) $q \geq 1$:
In this case, the solution of (21) for $V(t) = [v_\alpha^-(t), v_\alpha^+(t)]$, in levels, is given by

$$[V(t)]_\alpha = [k_1^\alpha e^{ct} + k_2^\alpha e^{-ct} + \tfrac{P}{c}, -k_1^\alpha e^{ct} + k_2^\alpha e^{-ct} + \tfrac{P}{c}], \tag{22}$$

where $k_1^\alpha = -\frac{v_\alpha^+(0)-v_\alpha^-(0)}{2}$ and $k_2^\alpha = \frac{v_\alpha^+(0)+v_\alpha^-(0)}{2} - \frac{P}{c}$. As expected in the expansive solution, we have that $\lim_{t\to\infty} \text{diam}(V_0) = \lim_{t\to\infty} |v_0^+(0) - v_0^-(0)|e^{ct} = +\infty$, that is, the fuzziness raises with time, as depicted in Fig. 1.

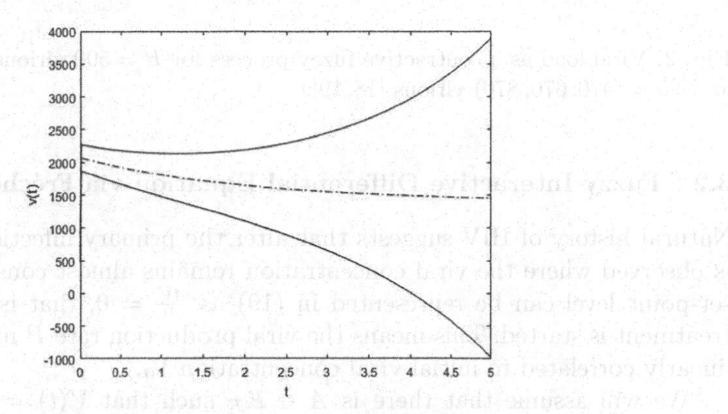

Fig. 1. Viral load as an expansive fuzzy process for $P = 700$ virions/day, $c = 0.5$/day and $V_0 = (470; 670; 870)$ virions [18,19]. Only positive values for V should be considered.

ii) $0 < q < 1$:
In this case, the solution of (21) for $V(t)_\alpha = [v_\alpha^-(t), v_\alpha^+(t)]$, in levels, is given by

$$[V(t)]_\alpha = [(v_\alpha^-(0) - \tfrac{P}{c})e^{-ct} + \tfrac{P}{c}, (v_\alpha^+(0) - \tfrac{P}{c})e^{-ct} + \tfrac{P}{c}]. \tag{23}$$

As expected in the contractive solution, $\lim_{t\to\infty} v_\alpha^-(t) = \lim_{t\to\infty} v_\alpha^+(t) = \frac{P}{c}$ and, therefore, $\text{diam}(V(t)) \to 0$, that is, the fuzziness vanishes with time. We have that $v_\alpha^-(t), v_\alpha^+(t) \geq 0, \forall t \geq 0$ if, and only if, $v_\alpha^-(0), v_\alpha^+(0) \geq \frac{P}{c}(1-e^{ct})$ for all $t \geq 0$, that is, $v_\alpha^-(0), v_\alpha^+(0) \geq \frac{P}{c}$. Viral load declines whenever $v_\alpha^-(0), v_\alpha^+(0) > \frac{P}{c}$ as depicted in Fig. 2. This represents a constraint on the initial condition to the solution to be consistent with the expected immune recovery expected in individuals under ART.

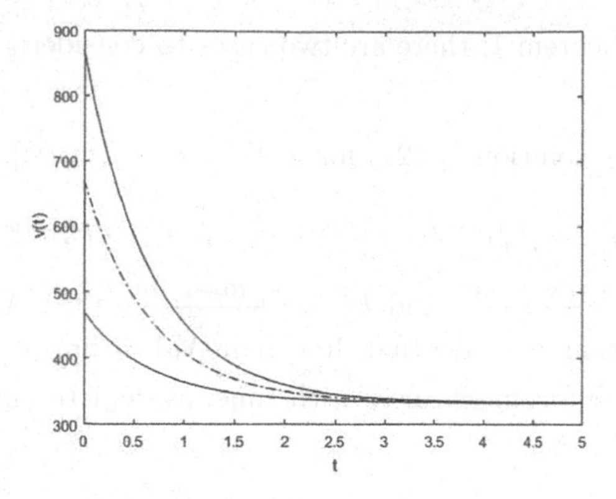

Fig. 2. Viral load as a contractive fuzzy process for $P = 500$ virions/day, $c = 1.5$/day and $V_0 = (470; 670; 870)$ virions [18,19].

3.2 Fuzzy Interactive Differential Equation via Fréchet Derivative

Natural history of HIV suggests that after the primary infection, a large phase is observed where the viral concentration remains almost constant [7,20]. This set-point level can be represented in (19) as $\frac{dV}{dt} = 0$, that is, $P = cV_0$ when treatment is started. This means the viral production rate P may be considered linearly correlated to initial viral concentration V_0.

We will assume that there is $A \in \mathbb{R}_{\mathcal{F}}$ such that $V(t) = \Psi_A(p(t), q(t)) = p(t)A + q(t)$ for all $t \geq 0$. We will also assume that $p, q : \mathbb{R} \to \mathbb{R}$ are differentiable. Then, the FIVP corresponding to (19) is given by

$$\begin{cases} V'(t) = P -_{\Psi_A} cV \\ V(0) = V_0 \in \mathbb{R}_{\mathcal{F}(A)}, \end{cases} \tag{24}$$

where $P \in \mathbb{R}_{\mathcal{F}(A)}$, that is, there are $p_1, p_2 \in \mathbb{R}$ such that $P = p_1 A + p_2 = \Psi_A(p_1, p_2)$, and $c \in \mathbb{R}$ is constant. Note that, in this case, we have that

$$P = \frac{p_1}{p_0}(p_0 A + q_0) + \left(p_2 - \frac{p_1 q_0}{p_0}\right) = \frac{p_1}{p_0} V_0 + \left(p_2 - \frac{p_1 q_0}{p_0}\right), \tag{25}$$

that is, P is linearly correlated to V_0. Theorem 4 ensures that

$$V'(t) = \Psi_A(p'(t), q'(t)) = p'(t)A + q'(t). \tag{26}$$

Then, (24) can be rewritten as

$$\Psi_A(p'(t), q'(t)) = \Psi_A(p_1, p_2) + c \cdot_A (p(t), q(t)) \tag{27}$$

and by the linearity of Ψ_A, we have that

$$\Psi_A(p'(t), q'(t)) = \Psi_A(p_1 - cp(t), p_2 - cq(t)). \tag{28}$$

As $A \in \mathbb{R}_{\mathcal{F}}$ is non-symmetric, the operator Ψ_A is injective, so that (28) is equivalent to two real systems, given by

$$\begin{cases} p'(t) = p_1 - cp(t) \\ p(0) = p_0 \end{cases} \tag{29}$$

and

$$\begin{cases} q'(t) = p_2 - cq(t) \\ q(0) = q_0, \end{cases} \tag{30}$$

whose solutions are given by $p(t) = \left(p_0 - \frac{p_1}{c}\right) e^{-ct} + \frac{p_1}{c}$ and $q(t) = \left(q_0 - \frac{p_2}{c}\right) e^{-ct} + \frac{p_2}{c}$, respectively. Therefore, the solution of (24) is given by

$$V(t) = \left[\left(p_0 - \frac{p_1}{c}\right) e^{-ct} + \frac{p_1}{c}\right] A + \left(q_0 - \frac{p_2}{c}\right) e^{-ct} + \frac{p_2}{c}, \ \forall t \in \mathbb{R}. \tag{31}$$

Since $V(t) = p(t)A + q(t)$, we have that $\text{diam}(V(t)) = \text{diam}(p(t)A) \rightarrow \text{diam}(\frac{p_1}{c}A)$ when $t \rightarrow \infty$. Moreover, if $A \in \mathbb{R}_{\mathcal{F}}$ is such that $0 \in [A]_1$, then we can consider that A is a fuzzy number around 0. In this case, we may expect that viral load to be around $\frac{p_2}{c}$ when $t \rightarrow \infty$. Once $P = p_1 A + p_2$, this result coincides with the classic case when $P \in \mathbb{R}$.

Lastly, we have three cases to consider:

i) $c < \frac{p_1}{p_0}$:
In this case, $p'(t) = -c\left(p_0 - \frac{p_1}{c}\right) e^{-ct} > 0$, that is, $\text{diam}(V(t)) = \text{diam}(p(t)A)$ is an increasing function, as depicted in Fig. 3. Therefore, the fuzziness of V increases with time, as seen in the expansive case described in Subsection B.

ii) $c = \frac{p_1}{p_0}$:
In this case, $p'(t) = 0$, that is, $\text{diam}(V(t)) = \text{diam}(p(t)A) = \text{diam}(\frac{p_1}{c}A)$ is constant with time. Therefore, the fuzziness of V remains constant, as depicted in Fig. 4.

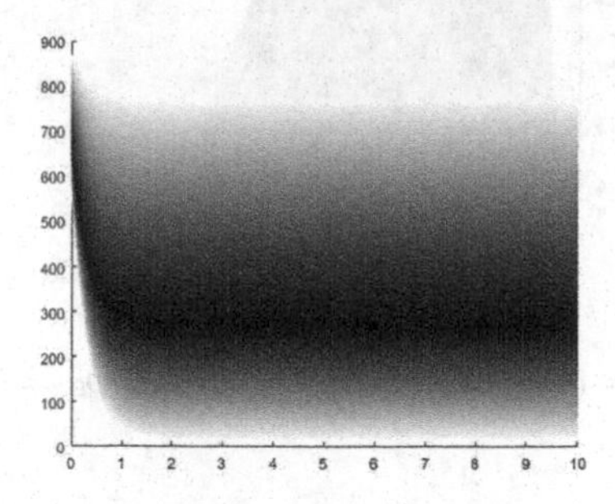

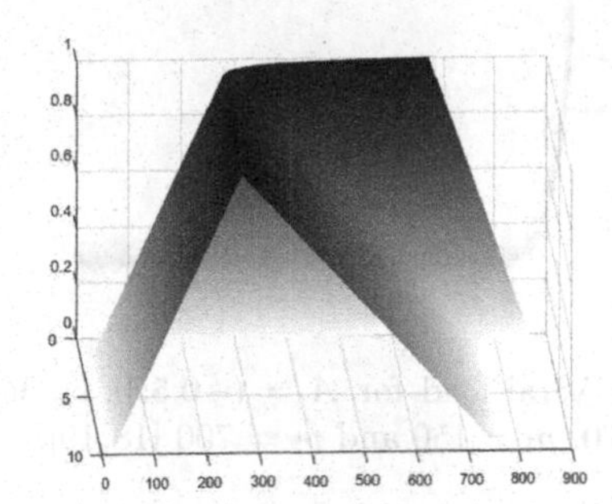

Fig. 3. Viral load for $A = (-0.5; 0; 1)$, $V_0 = (470; 670; 870)$, $c = 3/\text{day}$, $p_0 = 1500$, $q_0 = 670$, $p_1 = 1500$ and $p_2 = 800$ [18,19].

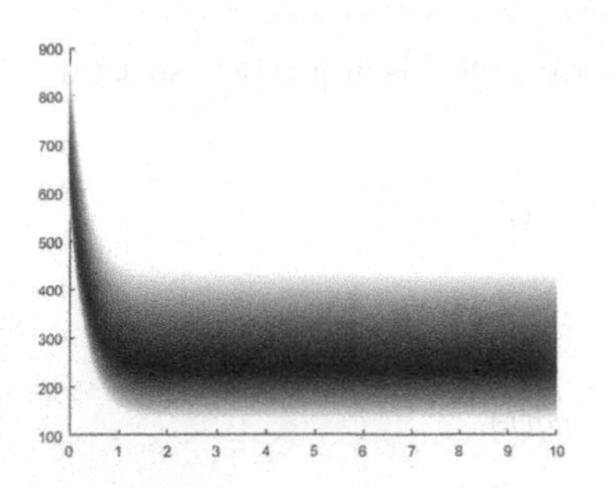
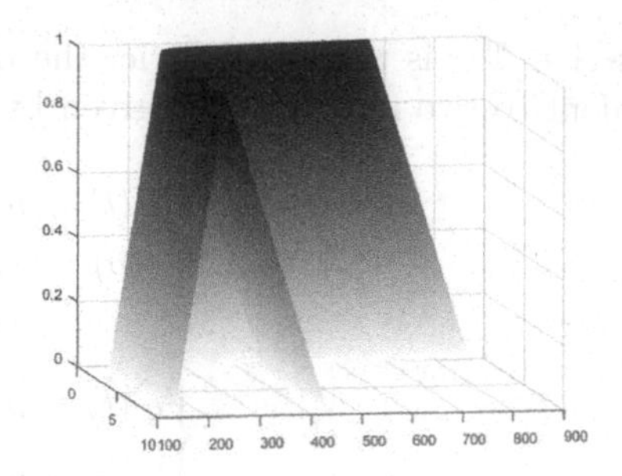

Fig. 4. Viral load for $A = (-0.5; 0; 1)$, $V_0 = (470; 670; 870)$, $c = 3$/day, $p_0 = 1500$, $q_0 = 670$, $p_1 = 600$ and $p_2 = 700$ [18,19].

iii) $c > \frac{p_1}{p_0}$:
In this case, $p'(t) = -c\left(p_0 - \frac{p_1}{c}\right)e^{-ct} < 0$, that is, $\text{diam}(V(t)) = \text{diam}(p(t)A)$ is a decreasing function. Therefore, the fuzziness of V decreases with time, as depicted in Fig. 5.

When viral production rate P is a real constant, that is, when $p_1 = 0$, the representation of P in the space $\mathbb{R}_{\mathcal{F}(A)}$ is $P = \Psi_A(0, p_2) = 0A + p_2 \in \mathbb{R}$. Theorem 5 ensures that, in this case, the Initial Value Problems (21) and (24) are equivalent if the initial conditions are the same.

However, only solutions (23) and (31) may coincide. As we observed previously, if $p_1 = 0$, then $\text{diam}(V(t)) \to 0$ when $t \to \infty$, that is, the fuzziness of the solution vanishes with time, as represented in Fig. 6. For HIV dynamics predicting viral drop, the autocorrelated process described by Fréchet derivative is always contractive if $P \in \mathbb{R}$, being expansive exclusively when P is a fuzzy number.

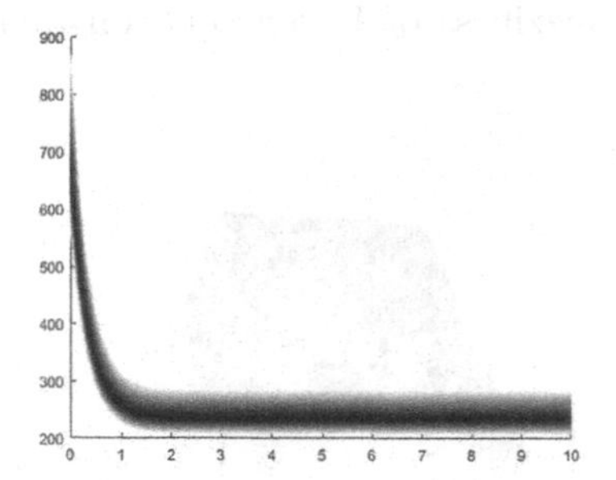
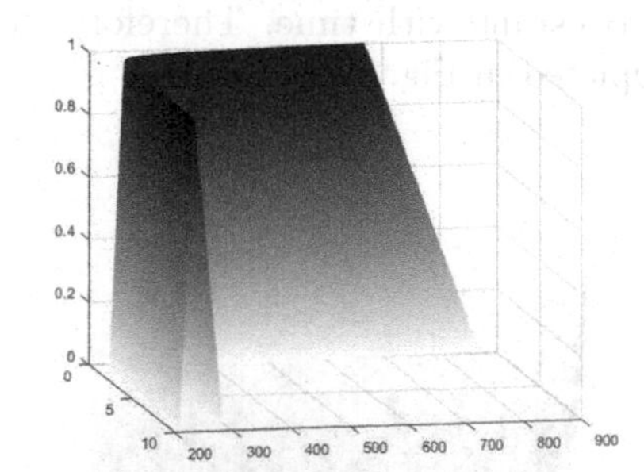

Fig. 5. Viral load for $A = (-0.5; 0; 1)$, $V_0 = (470; 670; 870)$, $c = 3$/day, $p_0 = 1500$, $q_0 = 670$, $p_1 = 150$ and $p_2 = 700$ [18,19].

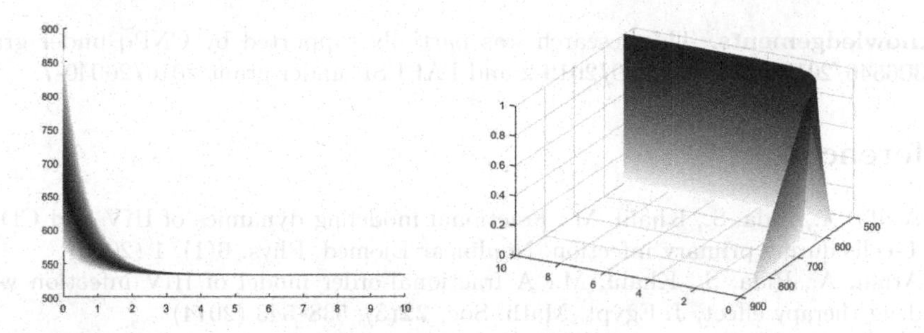

Fig. 6. Viral load for $A = (-0.5; 0; 1)$, $V_0 = (570; 670; 870)$, $c = 1.5/\text{day}$, $p_0 = 200$, $q_0 = 670$, $p_1 = 0$ and $p_2 = 800$ [18,19].

4 Final Comments

In this work we presented an HIV dynamics for individuals under ART as an application of two different approaches from fuzzy set theory: differential equation via interactive derivative and differential equation via Fréchet derivative. Biological processes may be considered as processes with memory, or from the point of view of fuzzy interactivity, autocorrelated processes. Viral dynamics was considered as an autocorrelated process in this manuscript.

Differential equation via interactive derivative provided two different types of solutions: the contractive and the expansive one. The underlying memory coefficient determines if the fuzziness of the solution decreases with time, as in the first case, or increases, as in the second case. Viral dynamics predicted a drop on viral load in bloodstream as was also observed in the two cases. For interactive derivative, viral production rate P was a real constant and it was not related to the stability of the solution.

Diferential equation via Fréchet derivative provided three different types of solution: the contractive, the expansive and also a third kind, that one with constant fuzziness with time. Modelling HIV via Fréchet derivative allowed us to evaluate the viral production rate P as a fuzzy number linearly correlated to $A \in \mathbb{R}_{\mathcal{F}}$. This was adopted due to the set-point that seropositive individuals remain after the primary infection, according to the natural history of HIV. It suggests that the viral production rate is linearly correlated to initial viral load V_0, with the coefficient $\frac{p_1}{p_0}$. The lower this coefficient is in relation to clearance rate c, the lower is the fuzziness of viral load in bloodstream. It also means that the representation of P on the space $\mathbb{R}_{\mathcal{F}(A)}$ determined the fuzziness of the solution V.

Furthermore, for HIV dynamics presented in this work, when P is a real constant, the FIVP determined by differential equation with Fréchet derivative is equivalent to the FIVP determined by differential equation with interactive derivative. In this case, only the solutions (23) and (31) coincide and $\text{diam}(V(t)) \to 0$ when $t \to \infty$. We can conclude that, for HIV dynamics, the interactive process determined by Fréchet derivative has an underlying memory coefficient in $(0, 1)$, that is, only provides solution whose fuzziness vanishes with time.

Acknowledgements. This research was partially supported by CNPq under grant no. 306546/2017-5, and 142309/2019-2 and FAPESP under grant 2016/26040-7.

References

1. Arafa, A., Rida, S., Khalil, M.: Fractional modeling dynamics of HIV and CD4+ T-cells during primary infection. Nonlinear Biomed. Phys. **6**(1), 1 (2012)
2. Arafa, A., Rida, S., Khalil, M.: A fractional-order model of HIV infection with drug therapy effect. J. Egypt. Math. Soc. **22**(3), 538–543 (2014)
3. Barros, L.C.d., Bassanezi, R.C., Lodwick, W.A.: A first course in fuzzy logic, fuzzy dynamical systems, and biomathematics: theory and applications (2017)
4. Bede, B., Gal, S.G.: Generalizations of the differentiability of fuzzy-number-valued functions with applications to fuzzy differential equations. Fuzzy Sets Syst. **151**(3), 581–599 (2005)
5. Bede, B., Stefanini, L., et al.: Generalized differentiability of fuzzy-valued functions. Fuzzy Sets Syst. **230**(1), 119–141 (2013)
6. Carlsson, C., Fullér, R., et al.: Additions of completely correlated fuzzy numbers. In: 2004 IEEE International Conference on Fuzzy Systems (IEEE Cat. No. 04CH37542), vol. 1, pp. 535–539. IEEE (2004)
7. Coutinho, F.A.B., Lopez, L., Burattini, M.N., Massad, E.: Modelling the natural history of HIV infection in individuals and its epidemiological implications. Bull. Math. Biol. **63**(6), 1041–1062 (2001)
8. De Barros, L.C., Santo Pedro, F.: Fuzzy differential equations with interactive derivative. Fuzzy Sets Syst. **309**, 64–80 (2017)
9. Dubois, D., Prade, H.: Additions of interactive fuzzy numbers. IEEE Trans. Autom. Control. **26**(4), 926–936 (1981)
10. Esmi, E., Santo Pedro, F., de Barros, L.C., Lodwick, W.: Fréchet derivative for linearly correlated fuzzy function. Inf. Sci. **435**, 150–160 (2018)
11. Herz, A., Bonhoeffer, S., Anderson, R.M., May, R.M., Nowak, M.A.: Viral dynamics in vivo: limitations on estimates of intracellular delay and virus decay. Proc. Natl. Acad. Sci. **93**, 7247–7251 (1996)
12. Jafelice, R.M., Barros, L., Bassanezi, R.: Study of the dynamics of HIV under treatment considering fuzzy delay. Comput. Appl. Math. **33**(1), 45–61 (2014)
13. Jafelice, R.M., Silva, C.A., Barros, L.C., Bassanezi, R.C.: A fuzzy delay approach for HIV dynamics using a cellular automaton. J. Appl. Math. **2015** (2015)
14. Jafelice, R.M., Barros, L.C., Bassanezi, R.C.: A fuzzy delay differential equation model for HIV dynamics. In: IFSA/EUSFLAT Conference, pp. 265–270 (2009)
15. Laiate, B., Jafelice, R.M., Esmi, E., Barros, L.C.: An interpretation of the fuzzy measure associated with choquet calculus for a HIV transference model. In: Kearfott, R.B., Batyrshin, I., Reformat, M., Ceberio, M., Kreinovich, V. (eds.) IFSA/NAFIPS 2019. AISC, vol. 1000, pp. 306–317. Springer, Cham (2019). https://doi.org/10.1007/978-3-030-21920-8_28
16. Nelson, P.W., Mittler, J.E., Perelson, A.S.: Effect of drug efficacy and the eclipse phase of the viral life cycle on estimates of HIV viral dynamic parameters. J. Acquir. Immune Defic. Syndr. (1999) **26**(5), 405–412 (2001)
17. Notermans, D.W., Goudsmit, J., Danner, S.A., De Wolf, F., Perelson, A.S., Mittler, J.: Rate of HIV-1 decline following antiretroviral therapy is related to viral load at baseline and drug regimen. Aids **12**(12), 1483–1490 (1998)
18. Perelson, A.S., et al.: Decay characteristics of HIV-1-infected compartments during combination therapy. Nature **387**(6629), 188 (1997)

19. Perelson, A.S., Nelson, P.W.: Mathematical analysis of HIV-1 dynamics in vivo. SIAM Rev. **41**(1), 3–44 (1999)
20. Perelson, A.S., Neumann, A.U., Markowitz, M., Leonard, J.M., Ho, D.D.: HIV-1 dynamics in vivo: virion clearance rate, infected cell life-span, and viral generation time. Science **271**(5255), 1582–1586 (1996)
21. Puri, M.L., Ralescu, D.A.: Differentials of fuzzy functions. J. Math. Anal. Appl. **91**(2), 552–558 (1983)
22. Santo Pedro, F., Esmi, E., de Barros, L.C.: Calculus for linearly correlated fuzzy function using Fréchet derivative and Riemann integral. Inf. Sci. **512**, 219–237 (2020)
23. Simões, F.S.P., et al.: Sobre equações diferenciais para processos fuzzy linearmente correlacionados: aplicações em dinâmica de população (2017)
24. Stefanini, L., Bede, B.: Generalized hukuhara differentiability of interval-valued functions and interval differential equations. Nonlinear Anal. Theory Methods Appl. **71**(3–4), 1311–1328 (2009)
25. Wasques, V., Laiate, B., Santo Pedro, F., Esmi, E., Barros, L.C.: Interactive fuzzy fractional differential equation: application on HIV dynamics. In: Proceedings on Uncertainty in Knowledge Bases: 18th International Conference on Information Processing and Management of Uncertainty in Knowledge-Based Systems, Lisboa, Portugal, 15–19 June (2020)
26. Zadeh, L.A.: Fuzzy sets as a basis for a theory of possibility. Fuzzy Sets Syst. **1**(1), 3–28 (1978)

On Categories of L-Fuzzifying Approximation Spaces, L-Fuzzifying Pretopological Spaces and L-Fuzzifying Closure Spaces

Anand Pratap Singh(✉) and Irina Perfilieva

Institute for Research and Applications of Fuzzy Modeling, CE IT4Innovations, University of Ostrava, 30. dubna 22, 701 03 Ostrava 1, Czech Republic
{anand.singh,irina.perfilieva}@osu.cz

Abstract. This paper investigates the essential connections among several categories with a weaker structure than that of L-fuzzifying topology, namely category of L-fuzzifying approximation spaces based on reflexive L-fuzzy relations, category of L-fuzzifying pretopological spaces and category of L-fuzzifying interior (closure) spaces. The interrelations among these structures are established in categorical setup.

Keywords: L-fuzzifying approximation space · L-fuzzifying pretopological space · Čech L-fuzzifying interior (closure) spaces · Galois connection

1 Introduction

Since the introduction of the rough set by Pawlak [11], this powerful theory drawn the attention of many researchers due to its importance in the study of intelligent systems with insufficient and incomplete information. Several generalizations of rough sets have been made by replacing the equivalence relation by an arbitrary relation. Dubois and Prade [3] generalized this theory and introduced the concept of fuzzy rough set. Various types of fuzzy rough approximation operators have been introduced and studied (c.f. [9,17–21]) in the context of fuzzy rough set theory. The most well known introduced fuzzy rough set is obtained by replacing the crisp relations with fuzzy relations and the crisp subset of the universe by fuzzy sets. Further, a rough fuzzy set was introduced in [23] by considering the fuzzy approximated subsets and crisp relations. In [25] Yao, introduced another kind of fuzzy rough set which is based on fuzzy relations and crisp approximated subsets, and is further studied by Pang [10] through the constructive and axiomatic approach. Several interesting studies have been carried on relating the theory of fuzzy rough sets with fuzzy topologies (cf., [2,6,13,16,19,22]). Further, Ying [26] introduced a logical approach to study the fuzzy topology and proposed the notion of fuzzifying topology. In brief, a

M.-J. Lesot et al. (Eds.): IPMU 2020, CCIS 1239, pp. 226–239, 2020.
https://doi.org/10.1007/978-3-030-50153-2_17

fuzzifying topology on a set X assigns to every crisp subset of X a certain degree of being open. A number of articles were published based on this new approach (cf., [4,5,8,24,29,30]). Fang [4,5] showed the one to one correspondence between fuzzifying topologies and fuzzy preorders and Shi [24] discussed the relationship of fuzzifying topology and specialization preorder in the sense of Lai and Zhang [7]. In 1999, Zhang [28] studied the fuzzy pretopology through the categorical point of view and Perfilieva et al. in [12,14] discussed its relationship with F-transform. Further following the approach of Ying [26], Lowen and Xu [8], Zhang [30] discussed the categorical study of fuzzifying pretopology.

Recently, Pang [10] followed the approach of Ying [26] and studied L-fuzzifying approximation operators through the constructive and axiomatic approaches. So far, the relationship among L-fuzzifying pretopological spaces, Čech L-fuzzifying interior (closure) spaces and L-fuzzifying approximation spaces has not been studied yet. In this paper, we will discuss such relationship in more details. It is worth to mention that our motivation is different from Qiao and Hu [15], in which such connection is established in the sense of Zhang [28] rather than L-fuzzifying pretopological setting. Specifically, we established the Galois connection between L-fuzzifying reflexive approximation space and L-fuzzifying pretopological spaces. Finally, we investigate the categorical relationship between Čech L-fuzzifying interior spaces and L-fuzzy relational structure.

2 Preliminaries

Throughout this paper, L denotes a De Morgan algebra $(L, \vee, \wedge, ', 0, 1)$, where $(L, \vee, \wedge, 0, 1)$ is a complete lattice with the least element 0 and greatest element 1 and an order reversing involution " $'$ ". For any $a \subseteq L$, $\bigvee a$ and $\bigwedge a$ are respectively the least upper bound and the greatest lower bound of a. In particular, we have $\bigvee \phi = 0$ and $\bigwedge \phi = 1$.

Let X be a nonempty set. The set of all subsets of X will be denoted by $\mathscr{P}(X)$ and called powerset of X. For $\lambda \in \mathscr{P}(X)$, λ^c is the complement of λ and characteristic function of λ is 1_λ. Let X, Y be two nonempty sets and $f : X \rightarrow Y$ be a mapping, then it can be extended to the powerset operator $f^\rightarrow : \mathscr{P}(X) \rightarrow \mathscr{P}(Y)$ and $f^\leftarrow : \mathscr{P}(Y) \rightarrow \mathscr{P}(X)$ such that for each $C \in \mathscr{P}(X)$, $f^\rightarrow(C) = \{f(x) : x \in C\}$ and for each $D \in \mathscr{P}(Y)$, $f^\leftarrow(D) = f^{-1}(D) = \{x : f(x) \in D\}$.

A map $f : X \rightarrow Y$ can be extended to the powerset operators $f^\rightarrow : L^X \rightarrow L^Y$ and $f^\leftarrow : L^Y \rightarrow L^X$ such that $\lambda \in L^X, \mu \in L^Y, y \in Y$,

$$f^\rightarrow(\lambda)(y) = \bigvee_{x, f(x)=y} \lambda(x), \quad f^\leftarrow(\mu) = \mu \circ f.$$

For a nonempty set X, L^X denotes the collection of all L-fuzzy subsets of X, i.e. a mapping $\lambda : X \rightarrow L$. Also, for all $a \in L$, $\mathbf{a}(x) = a$ is a constant L-fuzzy set on X. The greatest and least element of L^X is denoted by 1_X and 0_X respectively. For the sake of terminological economy, we will use the notation λ for both crisp set and L-fuzzy set. Further, an L-fuzzy set $1_y \in L^X$ is called a *singleton*, if it has the following form

$$1_y(x) = \begin{cases} 1, & \text{if } x = y, \\ 0, & \text{otherwise.} \end{cases}$$

Let X be a nonempty set. Then for $\lambda, \mu \in L^X$, we can define new *L-fuzzy sets* as follows:

$$\lambda = \mu \iff \lambda(x) = \mu(x),\ \lambda \le \mu \iff \lambda(x) \le \mu(x),$$
$$(\lambda \wedge \mu)(x) = \lambda(x) \wedge \mu(x),\ (\lambda \vee \mu)(x) = \lambda(x) \vee \mu(x),$$
$$(\lambda)'(x) = (\lambda(x))', \forall x \in X.$$

Let I be a set of indices, $\lambda_i \in L^X$, $i \in I$. The *meet* and *join* of elements from $\{\lambda_i \mid i \in I\}$ are defined as follows:

$$(\bigwedge_{i\in I} \lambda_i)(x) = \bigwedge_{i\in I} \lambda_i(x),\ (\bigvee_{i\in I} \lambda_i)(x) = \bigvee_{i\in I} \lambda_i(x).$$

Throughout this paper, all the considered categories are concrete. A *concrete category* (or construct) [1] is defined over **Set**. Specifically, it is a pair $(\mathbf{C}, \mathbb{U})$, with $\mathbf{C}$ as a category and $\mathbb{U} : \mathbf{C} \to \mathbf{Set}$ is a faithful (forgetful) functor. We say $\mathbb{U}(X)$ the underlying set for each $\mathbf{C}$-object X. We write simply $\mathbf{C}$ for the pair $(\mathbf{C}, \mathbb{U})$, since $\mathbb{U}$ is clear from the context.

A *concrete functor* between concrete categories $(\mathbf{C}, \mathbb{U})$ and $(\mathbf{D}, \mathbb{V})$ is a functor $\mathbb{F} : \mathbf{C} \to \mathbf{D}$ with $\mathbb{U} = \mathbb{V} \circ \mathbb{F}$. It means, $\mathbb{F}$ only changes structures on the underlying sets. For more on category we refer to [1].

Now we recall the following definition of L-fuzzy relation from [27].

Definition 1 [27]. *Let X be a nonempty set. An* L-fuzzy relation θ *on X is an L-fuzzy subset of $X \times X$. An L-fuzzy relation θ is called* reflexive *if $\theta(x, x) = 1$, $\forall\, x \in X$.*

A set X equipped with an L-fuzzy relation θ is denoted by (X, θ) and is called a *L-fuzzy relational structure*.

Below we define the category **FRS** of L-fuzzy relational structures.

1. The pairs (X, θ) with reflexive L-fuzzy relation θ on X are the *objects*, and
2. for the pairs (X, θ) and (Y, ρ) a *morphism* $f : (X, \theta) \to (Y, \rho)$ is a map $f : X \to Y$ such that $\forall\, x, y \in X$, $\theta(x, y) \le \rho(f(x), f(y))$.

3 L-Fuzzifying Approximation Operators

In this section, we recall the notion of L-fuzzifying approximation operators and its properties presented in [10]. We also define the category of L-fuzzifying approximation space and show that this category is isomorphic to the category of fuzzy relational structures.

Definition 2 [10]. *Let θ be an L-fuzzy relation on X. Then* upper (lower) L-fuzzifying approximation *of λ is a map $\overline{\theta}, \underline{\theta} : \mathscr{P}(X) \to L^X$ defined by;*

$$(\forall \lambda \in \mathscr{P}(X),\ x \in X),\ \ \overline{\theta}(\lambda)(x) = \bigvee_{y \in \lambda} \theta(x, y),$$

$$(\forall \lambda \in \mathscr{P}(X),\ x \in X),\ \ \underline{\theta}(\lambda)(x) = \bigwedge_{y \notin \lambda} \theta(x, y)'.$$

We call $\underline{\theta}$, $\overline{\theta}$ the lower L-fuzzifying approximation operator and the upper L-fuzzifying approximation operator respectively. Further, the pair $(\overline{\theta}, \underline{\theta})$ is called L-fuzzifying rough set and (X, θ) is called an L-fuzzifying approximation space *based on L-fuzzy relation θ.*

(i) It is important to note that, if $\lambda = \{y\} \in \mathscr{P}(X)$ for some $y \in X$, then we have the upper L-fuzzifying approximation $\overline{\theta}(\{y\})(x) = \theta(x, y)$ for each $x \in X$. If $\lambda = X - \{y\} \in \mathscr{P}(X)$ for some $y \in X$, then we have the lower L-fuzzifying approximation $\underline{\theta}(X - \{y\})(x) = \theta(x, y)'$ for each $x \in X$.
(ii) Let X be a nonempty set and θ be reflexive L-fuzzy relation on X. We call the pair (X, θ), an *L-fuzzifying reflexive approximation space.*

Now, we give some useful properties of L-fuzzifying upper (lower) approximation operators from [10]. These properties will be used in the further text.

Proposition 1 [10]. *Let (X, θ) be an L-fuzzifying reflexive approximation space. Then for $\lambda \in \mathscr{P}(X)$ and $\{\lambda_i \mid i \in I\} \subseteq \mathscr{P}(X)$, the following holds.*

(i) $\overline{\theta}(\phi) = 0_X,\ \underline{\theta}(X) = 1_X,$
(ii) $\overline{\theta}(\lambda) = \underline{\theta}(\lambda^c)',\ \underline{\theta}(\lambda) = \overline{\theta}(\lambda^c)',$
(iii) $\overline{\theta}(\lambda) \geq 1_\lambda,\ \underline{\theta}(\lambda) \leq 1_\lambda,$
(iv) $\overline{\theta}(\bigcup_{i \in I} \lambda_i) = \bigvee_{i \in I} \overline{\theta}(\lambda_i),\ \underline{\theta}(\bigcap_{i \in I} \lambda_i) = \bigwedge_{i \in I} \underline{\theta}(\lambda_i).$

Below we give the notion of morphism between two L-fuzzifying reflexive approximation spaces.

Definition 3. *The* morphism *$f : (X, \theta) \to (Y, \rho)$ between two L-fuzzifying reflexive approximation spaces (X, θ) and (Y, ρ) is given by*

$$f^{\leftarrow}(\underline{\rho}(\lambda)) \leq \underline{\theta}(f^{\leftarrow}(\lambda))\ \forall \lambda \in \mathscr{P}(Y).$$

It is easy to verify that all L-fuzzifying reflexive approximation spaces as objects and morphism defined above form a category. We denote this category by **L-FYAPP**.

Theorem 1. *The category* **L-FYAPP** *is isomorphic to the category* **FRS**.

Proof. The proof is divided into two parts. On one hand we can see that both the categories have the identical objects. It only remains to show that both the categories have the identical morphisms. Let $f : (X,\theta) \to (Y,\rho)$ be a morphism in the category **FRS**, then for any $\lambda \in \mathscr{P}(Y)$ and $x \in X$ we have

$$\begin{aligned}\underline{\theta}(f^{\leftarrow}(\lambda))(x) &= \bigwedge_{y \notin f^{\leftarrow}(\lambda)} \theta(x,y)' \geq \bigwedge_{f(y)\notin\lambda} \rho(f(x),f(y))' \\ &\geq \bigwedge_{t\notin\lambda} \rho(f(x),t)' = f^{\leftarrow}(\underline{\rho}(\lambda))(x).\end{aligned}$$

On the other hand, let f be a morphism in the category **L-FYAPP**. Then for all $x, y \in X$ we have

$$\begin{aligned}\rho(f(x),f(y))' &= \bigwedge_{t\notin(Y-\{f(y)\})} \rho(f(x),t)' = \underline{\rho}(Y-\{f(y)\})(f(x)) \\ &= f^{\leftarrow}(\underline{\rho}(Y-\{f(y)\}))(x) \leq \underline{\theta}(f^{\leftarrow}(Y-\{f(y)\}))(x) \\ &= \bigwedge_{y\notin f^{\leftarrow}(Y-\{f(y)\})} \theta(x,y)' \\ &= \underline{\theta}(X-\{y\})(x) = \theta(x,y)'.\end{aligned}$$

Hence we get $\rho(f(x),f(y))' \leq \theta(x,y)'$. Since "$'$" is order reversing, hence $\theta(x,y) \leq \rho(f(x),f(y))$ holds and f is a morphism in the category **FRS**. We denote this isomorphism by $\mathbb{N}$.

4 *L*-Fuzzifying Approximation Space and *L*-Fuzzifying Pretopological Space

This section is towards the categorical relationship among L-fuzzifying pretopological space, Čech (L-fuzzifying) interior space and L-fuzzifying approximation space. We discuss how to generate an L-fuzzifying pretopology by an reflexive L-fuzzy relation and our approach is based on the L-fuzzifying approximation operator studied in L-fuzzifying rough set theory.

Below, we present the definition of L-fuzzifying pretopological space which is similar (but not identical) to that in [8].

Definition 4. *A set of functions $\tau_X = \{p_x : \mathscr{P}(X) \to L \mid x \in X\}$ is called an* L-fuzzifying pretopology *on X if for each $\lambda, \mu \in \mathscr{P}(X)$, and $x \in X$, it satisfies,*

(i) $p_x(X) = 1$,
(ii) $p_x(\lambda) \leq 1_\lambda(x)$,
(iii) $p_x(\lambda \cap \mu) = p_x(\lambda) \wedge p_x(\mu)$.

For an L-fuzzifying pretopology τ_X, the pair (X,τ_X) is called an L-fuzzifying pretopological space.

An L-fuzzifying pretopological space (X,τ_X) is called Alexandroff, *if*

(iv) $p_x(\bigcap_{i\in I}\lambda_i) = \bigwedge_{i\in I} p_x(\lambda_i)$.

With every L-fuzzifying pretopological space $\tau_X = \{p_x : \mathscr{P}(X) \to L \mid x \in X\}$ and each $\lambda \in \mathscr{P}(X)$, we can associate another L-fuzzy set $\phi_\lambda \in L^X$ such that for all $x \in X$, $\phi_\lambda(x) = p_x(\lambda)$. Obviously, $\phi : \lambda \mapsto \phi_\lambda$ is an operator on X.

A mapping $f : (X, \tau_X) \to (Y, \tau_Y)$ between two L-fuzzifying pretopological spaces is called *continuous* if for all $x \in X$ and for each $\lambda \in \mathscr{P}(Y)$, $q_{f(x)}(\lambda) \leq p_x(f^{\leftarrow}(\lambda))$, where $\tau_X = \{p_x : \mathscr{P}(X) \to L \mid x \in X\}$, $\tau_Y = \{q_{f(x)} : \mathscr{P}(Y) \to L \mid f(x) \in Y\}$ and $f^{\leftarrow}(\lambda) = \{x : f(x) \in \lambda\}$. It can be verified that all L-fuzzifying pretopological spaces as objects and their continuous maps as morphisms form a category, denoted by **L-FYPT**.

Now, we define the concepts of Čech L-fuzzifying interior (closure) operators by considering the domain as crisp power set $\mathscr{P}(X)$ rather than L-fuzzy set L^X.

Definition 5. *A mapping* $\hat{i} : \mathscr{P}(X) \to L^X$ *is called a* Čech (L-fuzzifying) interior operator *on* X *if for each* $\lambda, \mu \in \mathscr{P}(X)$, *and* $x \in X$, *it satisfies*

(i) $\hat{i}(X) = 1_X$,
(ii) $\hat{i}(\lambda) \leq 1_\lambda$,
(iii) $\hat{i}(\lambda \cap \mu) = \hat{i}(\lambda) \wedge \hat{i}(\mu)$.

The pair $(X, \hat{i})$ *is called a* Čech (L-fuzzifying) interior space.

A Čech (L-fuzzifying) interior operator $(X, \hat{i})$ *is called* Alexandroff, *if*

(iv) $\hat{i}(\bigcap_{i \in I} \lambda_i) = \bigwedge_{i \in I} \hat{i}(\lambda_i)$.

The map $f : (X, \hat{i}) \to (Y, \hat{j})$ between two Čech L-fuzzifying interior spaces is called *continuous* if for each $x \in X$ and $\lambda \in \mathscr{P}(Y)$, $\hat{j}(\lambda)(f(x)) \leq \hat{i}(f^{\leftarrow}(\lambda))(x)$. It is trivial to verify that all Čech (L-fuzzifying) interior spaces as objects and continuous maps as morphisms form a category. We denote this category by **L-FYIC**. Moreover, we denote the subcategory (full) **L-AFYIC** of **L-FYIC** with Čech Alexandroff L-fuzzifying interior operators as objects.

The notion of Čech (L-fuzzifying) closure operator can be defined using the duality of L.

Definition 6. *A mapping* $c_X : \mathscr{P}(X) \to L^X$ *is called a* Čech (L-fuzzifying) closure operator *on* X *if for each* $\lambda, \mu \in \mathscr{P}(X)$, *and* $x \in X$, *it satisfies*

(i) $c_X(\phi) = 0_X$,
(ii) $c_X(\lambda) \geq 1_\lambda$,
(iii) $c_X(\lambda \cup \mu) = c_X(\lambda) \vee c_X(\mu)$.

The pair (X, c_X) *is called a Čech* L-fuzzifying closure space.

A Čech (L-fuzzifying) closure space (X, c_X) *is called* Alexandroff, *if*

(iv) $c_X(\bigcup_{i \in I} \lambda_i) = \bigvee_{i \in I} c_X(\lambda_i)$.

A mapping $f : (X, c_X) \to (Y, c_Y)$ between two Čech (L-fuzzifying) closure spaces is called *continuous* if for all $x \in X$ and $\lambda \in \mathscr{P}(X)$, $f^{\rightarrow}(c_X(\lambda)) \leq c_Y(f^{\rightarrow}(\lambda))$.

Remark 1. For a De Morgan algebra L, the L-fuzzifying pretopologies, Čech L-fuzzifying interior operators and Čech L-fuzzifying closure operators are generally considered as equivalent and can be defined using the immanent duality of L in the following manner.

$$c_X(\lambda) = (i_X(\lambda^c))^c, \ \forall \lambda \in \mathscr{P}(X)$$

From now on, we will only study the relationship between L-fuzzifying approximation spaces, L-fuzzifying pretopological spaces and Čech L-fuzzifying interior spaces. Since the similar results can be obtained for Čech L-fuzzifying closure spaces.

The following Proposition is an easy consequence of Definitions 4 and 5.

Proposition 2. *The set of functions $\tau_X = \{p_x : \mathscr{P}(X) \to L \mid x \in X\}$ is an L-fuzzifying pretopology on X iff the map $\hat{i}_{\tau_X} : \mathscr{P}(X) \to L^X$ such that for all $x \in X$,*

$$\hat{i}_{\tau_X}(\lambda)(x) = p_x(\lambda), \tag{1}$$

is a Čech L-fuzzifying interior operator. Moreover, if L-fuzzifying pretopology τ_X is Alexandroff, then the map $\hat{i}_{\tau_X}$ is a Čech-Alexandroff L-fuzzifying interior operator.

Theorem 2. *The category* **L-FYPT** *and* **L-FYIS** *are isomorphic.*

Proof. Let $f : (X, \tau_X) \to (Y, \tau_Y)$ is a morphism (continuous map) in **L-FYPT**. We define the functor $\mathbb{G}$ as follows

$$\mathbb{G} : \begin{cases} \textbf{L-FYPT} \longrightarrow \textbf{L-FYIS} \\ (X, \tau_X) \quad \longmapsto (X, \hat{i}_{\tau_X}) \\ \quad f \qquad \longmapsto \quad f, \end{cases}$$

and for all $\lambda \in \mathscr{P}(X)$, $x \in X$, $\hat{i}_{\tau_X}(\lambda)(x) = p_x(\lambda)$. Since $(X, \hat{i}_{\tau_X})$ is the object of category **L-FYIS**, then $f : (X, \hat{i}_{\tau_X}) \to (Y, \hat{j}_{\tau_Y})$ is a continuous map, i.e. $\forall \lambda \in \mathscr{P}(Y)$, $\hat{j}_{\tau_Y}(\lambda)(x) = q_{f(x)}(\lambda) \le p_x(f^{\leftarrow}(\lambda)) = \hat{i}_{\tau_X}(f^{\leftarrow}(\lambda))(x)$.

Conversely, let $f : (X, \hat{i}_{\tau_X}) \to (Y, \hat{j}_{\tau_Y})$ is a continuous map in the category **L-FYIS**. We define the inverse functor $\mathbb{G}^{-1}$ as follows

$$\mathbb{G} : \begin{cases} \textbf{L-FYIS} \longrightarrow \textbf{L-FYPT} \\ (X, \hat{i}_{\tau_X}) \longmapsto (X, \tau_X) \\ \quad f \qquad \longmapsto \quad g, \end{cases}$$

and for all $\lambda \in \mathcal{P}(X)$, $p_x(\lambda) = \hat{i}_{\tau_X}(\lambda)(x)$. Then clearly $\mathbb{G}^{-1}$ is an inverse functor with the inverse $\mathbb{G}$.

In the next proposition, we show that an L-fuzzifying pretopology on X can be represented by an L-fuzzifying lower approximations of sets on X with respect to a reflexive L-fuzzy relation.

Proposition 3. *Suppose that* (X, θ) *be an L-fuzzifying reflexive approximation space. Let for all* $\lambda \in \mathscr{P}(X)$, $x \in X$, *we denote*

$$p_x^\theta(\lambda) = \bigwedge_{y \notin \lambda} \theta(x, y)'. \tag{2}$$

Then $\tau_\theta = \{p_x^\theta : \mathscr{P}(X) \to L | x \in X\}$, *is an L-fuzzifying pretopology on* X.

Proof. For all $x \in X$ and $\lambda \in \mathscr{P}(X)$, from Proposition 1, it can be easily verified that τ_θ as defined in Eq. 2 satisfies the properties (i)–(iii) of lower L-fuzzifying approximation operator.

Proposition 4. *Let* (X, τ_X) *be an L-fuzzifying pretopological space. Then for any* $x \in X$, *we define*

$$\Theta_{\tau X}(x, y) = p_x(X - \{y\})'.$$

Then, Θ_{τ_X} *is a reflexive L-fuzzy relation and* (X, Θ_{τ_X}) *is an L-fuzzifying reflexive approximation space.*

Proof. For all $x \in X$ and from the Definition 4 we have, $\Theta_{\tau_X}(x, x) = p_x(X - \{x\})' \leq 1_{X-\{x\}}(x)' = 0' = 1$. Which shows that Θ_{τ_X} is a reflexive L-fuzzy relation and hence (X, Θ_{τ_X}) is an L-fuzzifying reflexive approximation space.

Proposition 5. *If* $f : (X, \theta) \to (Y, \rho)$ *is a morphism between two L-fuzzifying reflexive approximation spaces. Then* f *is continuous function between two L-fuzzifying pretopological spaces* (X, τ_θ) *and* (Y, τ_ρ).

Proof. The proof directly follows from Proposition 3.

Thus from the Propositions 3 and 5 we obtain a concrete functor τ as follows:

$$\tau : \begin{cases} \textbf{L-FYAPP} \longrightarrow \textbf{L-FYPT} \\ (X, \theta) \quad\quad \longmapsto (X, \tau_\theta) \\ \quad f \quad\quad\quad \longmapsto \ f. \end{cases}$$

Next, we prove a result, which gives a concrete functor Θ : **FYPT** $\to$ **FYAPP**.

Proposition 6. *If* f *is a continuous function between two L-fuzzifying pretopological spaces* (X, τ_X) *and* (Y, τ_Y). *Then* $f : (X, \Theta_{\tau_X}) \to (Y, \Theta_{\tau_Y})$ *is a morphism between two L-fuzzifying reflexive approximation spaces.*

Proof. Let $\lambda \in \mathscr{P}(Y)$ and $x \in X$, we have

$$\begin{aligned} f^{\leftarrow}(\underline{\Theta_{\tau_Y}}(\lambda))(x) &= \underline{\Theta_{\tau_Y}}(\lambda)(f(x)) = \bigwedge_{t\notin\lambda} \Theta_{\tau_Y}(f(x),t)' \\ &= \bigwedge_{t\notin\lambda} q_{f(x)}(Y-\{t\}) = \bigwedge_{f(y)\notin\lambda} q_{f(x)}(Y-\{f(y)\}) \\ &\leq \bigwedge_{y\notin f^{\leftarrow}(\lambda)} p_x(f^{\leftarrow}(Y-\{f(y)\})) \\ &\leq \bigwedge_{y\notin f^{\leftarrow}(\lambda)} p_x(X-\{y\}) = \bigwedge_{y\notin f^{\leftarrow}(\lambda)} \Theta_{\tau_X}(x,y)' \\ &= \underline{\Theta_{\tau_X}}(f^{\leftarrow}(\lambda))(x). \end{aligned}$$

Hence, we have $f : (X, \Theta_{\tau_X}) \to (Y, \Theta_{\tau_Y})$ is a morphism between two L-fuzzifying reflexive approximation spaces (X, Θ_{τ_X}) and (Y, Θ_{τ_Y}). In particular, we obtain a concrete functor Θ as follows:

$$\Theta : \begin{cases} \textbf{L-FYPT} & \longrightarrow \textbf{L-FYAPP} \\ (X, \tau_X) & \longmapsto (X, \Theta_{\tau_X}) \\ \quad f & \longmapsto f. \end{cases}$$

In the next theorem we prove the adjointness between the categories **L-FYAPP** and **L-FYPT**. Now we have the following.

Theorem 3. *Let (X,θ) be an L-fuzzifying reflexive approximation space. Then τ :* **L-FYAPP** $\to$ **L-FYPT** *is a left adjoint of Θ :* **L-FYPT** $\to$ **L-FYAPP**. *Moreover $\Theta \circ \tau(X,\theta) = (X,\theta)$ i.e., Θ is a left inverse of τ.*

Proof: The proof is divided into two parts. At first, we show that for any L-fuzzifying reflexive approximation space (X,θ), $\mathbb{I}_X : (X,\theta) \to (X, \Theta_{\tau_\theta})$ is a morphism between L-fuzzifying reflexive approximation spaces.

For any $\lambda \in \mathscr{P}(X)$ and $x \in X$, we have

$$\begin{aligned} \underline{\Theta_{\tau_\theta}}(\lambda)(x) &= \bigwedge_{y\notin\lambda} \Theta_{\tau_\theta}(x,y)' \\ &= \bigwedge_{y\notin\lambda} (p_x^\theta(X-\{y\})')' \quad \text{(from Proposition 4)} \\ &= \bigwedge_{y\notin\lambda} p_x^\theta(X-\{y\}) \qquad \text{(by involution of “ $'$ ”)} \\ &= \bigwedge_{y\notin\lambda} \underline{\theta}(X-\{y\}) = \bigwedge_{y\notin\lambda} \theta(x,y)' = \underline{\theta}(\lambda)(x). \end{aligned}$$

Hence, $\mathbb{I}_X : (X,\theta) \to (X, \Theta_{\tau_\theta})$ is a morphism between L-fuzzifying reflexive approximation spaces.

On the other hand, for any $\lambda \in \mathscr{P}(X),\ x \in X$, we have

$$\begin{aligned}
p_x^{\Theta_{\tau_X}}(\lambda) &= \underline{\Theta_{\tau_X}}(\lambda)(x) = \bigwedge_{y \notin \lambda} \Theta_{\tau_X}(x,y)' \\
&= \bigwedge_{y \notin \lambda} (p_x(X - \{y\})')' \\
&= \bigwedge_{y \notin \lambda} p_x(X - \{y\}) \qquad \text{(by involution of " } ' \text{ ")} \\
&\geq p_x \bigcap_{y \notin \lambda} (X - \{y\}) = p_x(\lambda).
\end{aligned}$$

Hence, we show that $\mathbb{I}_X : (X, \tau_{\Theta_{\tau_X}}) \to (X, \tau_X)$ is continuous.

Therefore, τ : **L-FYAPP** $\to$ **L-FYPT** is a left adjoint of Θ : **L-FYPT** $\to$ **L-FYAPP** (Fig. 1).

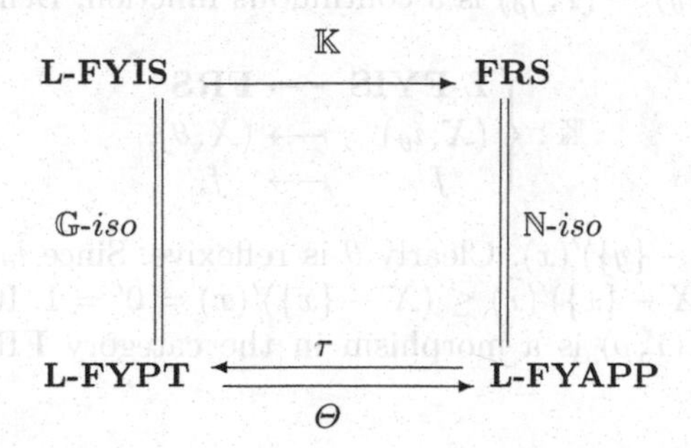

Fig. 1. Commutative diagram of Theorems 1, 2 and 3.

5 L-Fuzzy Relational Structures and Čech L-Fuzzifying Interior Space

In this section, we establish the categorical relationship between the category **FRS** of L-fuzzy relational structures and the category **L-FYIS** of Čech L-fuzzifying interior spaces. Now we have the following.

Proposition 7. *Let $f : (X, \theta) \to (Y, \rho)$ be a morphism in the category* **FRS**, *then $f : (X, \hat{i}_\theta) \to (Y, \hat{j}_\rho)$ is a continuous function (morphism) in the category* **L-AFYIS**.

Proof: Given that $f : (X, \theta) \to (Y, \rho)$ be a morphism in the category **FRS**. We define a functor $\mathbb{F}$ as follows;

$$\mathbb{F} : \begin{cases} \mathbf{FRS} & \longrightarrow \mathbf{L\text{-}AFYIS} \\ (X, \theta) & \longmapsto (X, \hat{i}_\theta) \\ \quad f & \longmapsto \quad f, \end{cases}$$

and $\forall \lambda \in \mathscr{P}(X), x \in X,\ \hat{i}_\theta(\lambda)(x) = \wedge_{y \notin \lambda} \theta(x,y)'$. As $(X, \hat{i}_\theta)$ is the object of category **L-AFYIS**, we need to show that $f : (X, \hat{i}_\theta) \to (Y, \hat{j}_\rho)$ is a continuous function (morphism) in the category **L-AFYIS**. For all $\lambda \in \mathscr{P}(Y), x \in X$ we have

$$\begin{aligned}\hat{j}_\rho(\lambda)(f(x)) = \bigwedge_{z \notin \lambda} \rho(f(x), z)' &\leq \bigwedge_{f(y) \notin \lambda} \rho(f(x), f(y))' \\ &\leq \bigwedge_{y \notin f^{\leftarrow}(\lambda)} \theta(x,y)' = \hat{i}_\theta(f^{\leftarrow}(\lambda))(x).\end{aligned}$$

Hence $\mathbb{F}$ is a functor.

Proposition 8. *Let $f : (X, \hat{i}_\theta) \to (Y, \hat{j}_\rho)$ be a continuous function (morphism) in the category* **L-FYIS**, *then $f : (X, \theta) \to (Y, \rho)$ is a morphism in the category* **FRS**.

Proof: Let $f : (X, \hat{i}_\theta) \to (Y, \hat{j}_\rho)$ is a continuous function. Define a functor $\mathbb{K}$ as follows;

$$\mathbb{K} : \begin{cases} \mathbf{L\text{-}FYIS} \longrightarrow \mathbf{FRS} \\ (X, \hat{i}_\theta) \longmapsto (X, \theta), \\ \quad f \quad \longmapsto \quad f, \end{cases}$$

and $\theta(x,y) = \hat{i}_\theta(X - \{y\})'(x)$. Clearly θ is reflexive. Since $\hat{i}_\theta$ is anti-extensive, hence $\theta(x,x) = \hat{i}_\theta(X - \{x\})'(x) \leq (X - \{x\})'(x) = 0' = 1$. It remains to show that $f : (X, \theta) \to (Y, \rho)$ is a morphism in the category **FRS**, i.e., $\theta(x,y) \leq \rho(f(x), f(y))$, or

$$\begin{aligned}\hat{i}_\theta(X - \{y\})'(x) &\leq \hat{j}_\rho(Y - \{f(y)\})'(f(x)), \\ \text{or, } \hat{i}_\theta(X - \{y\})(x) &\geq \hat{j}_\rho(Y - \{f(y)\})(f(x)).\end{aligned} \tag{3}$$

Since $f : (X, \hat{i}_\theta) \to (Y, \hat{j}_\rho)$ is a continuous function, we have $\hat{j}_\rho(\lambda)(f(x)) \leq \hat{i}_\theta(f^{\leftarrow}(\lambda))(x)$. Therefore for $\lambda = (Y - \{f(y)\})$, we get

$$\begin{aligned}\hat{j}_\rho(Y - \{f(y)\})(f(x)) &\leq \hat{i}_\theta(f^{\leftarrow}(Y - \{f(y)\}))(x) \\ &\leq \hat{i}_\theta(X - \{y\})(x).\end{aligned}$$

Hence (3) holds and $f : (X, \theta) \to (Y, \rho)$ is a morphism in the category **FRS**.

Proposition 9. *Let $(X, \hat{i}_\theta)$ be a Čech Alexandroff L-fuzzifying interior space and $\mathbb{F}$:* **FRS** $\to$ **L-AFYIS**, *$\mathbb{K}$:* **L-AFYIS** $\to$ **FRS** *be the concrete functors. Then $\mathbb{FK}(X, \hat{i}_\theta) = (X, \hat{i}_\theta)$ (Fig. 2).*

Proof: Let $\mathbb{FK}(X, \hat{i}_\theta) = (X, \hat{j}_\rho)$, where,

$$\forall \lambda \in \mathscr{P}(X),\ \ \hat{j}_\rho(\lambda)(x) = \bigwedge_{y \notin \lambda} \hat{i}_\theta(X - \{y\})(x) = \bigwedge_{y \notin \lambda} \theta(x,y)'.$$

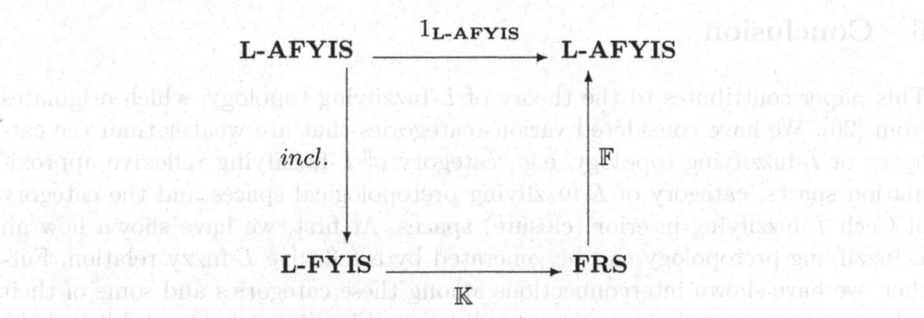

Fig. 2. Commutative diagram of Proposition 9.

As we know that, any arbitrary set $\lambda \in \mathscr{P}(X)$ can be decomposed as

$$\lambda = \bigcap_{y \notin \lambda} (X - \{y\}).$$

Therefore for Čech Alexandroff L-fuzzifying interior operator $\hat{i}_\theta$, we have

$$\hat{i}_\theta(\lambda)(x) = \hat{i}_\theta\left(\bigcap_{y \notin \lambda}(X - \{y\})(x)\right) = \bigwedge_{y \notin \lambda} \hat{i}_\theta(X - \{y\})(x)$$
$$= \bigwedge_{y \notin \lambda} \theta(x, y)' = \jmath_\rho(\lambda)(x).$$

Hence we have $\mathbb{FK}(X, \hat{i}_\theta) = (X, \hat{i}_\theta)$.

In the end, we give a graph to collect the relationships among the discussed categories.

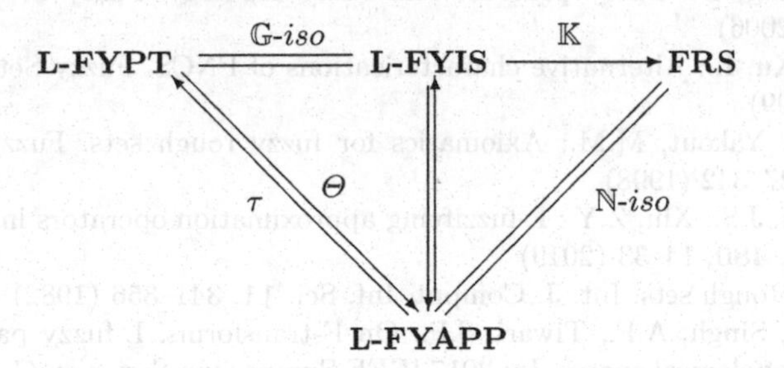

6 Conclusion

This paper contributes to the theory of L-fuzzifying topology, which originates from [26]. We have considered various categories that are weaker than the category of L-fuzzifying topology, e.g., category of L-fuzzifying reflexive approximation spaces, category of L-fuzzifying pretopological spaces and the category of Čech L-fuzzifying interior (closure) spaces. At first, we have shown how an L-fuzzifying pretopology can be generated by a reflexive L-fuzzy relation. Further, we have shown interconnections among these categories and some of their subcategories using the commutative diagram. Finally, we have established the relationship between L-fuzzy relational structure and Čech L-fuzzifying interior space by means of Galois connection.

Acknowledgments. This work is supported by University of Ostrava grant IRP201824 "Complex topological structures" and the work of I. Perfilieva is partially supported by the Grant Agency of the Czech Republic (project No. 18-06915S).

References

1. Adámek, J., Herrlich, H., Strecker, G.: Abstract and Concrete Categories: The Joy of Cats. John Wiley and Sons, New York (1990)
2. Boixader, D., Jacas, J., Recasens, J.: Upper and lower approximations of fuzzy sets. Int. J. Gen. Syst. **29**, 555–568 (2000)
3. Dubois, D., Prade, H.: Rough fuzzy sets and fuzzy rough sets. Int. J. Gen. Syst. **17**, 191–209 (1990)
4. Fang, J., Chen, P.: One-to-one correspondence between fuzzifying topologies and fuzzy preorders. Fuzzy Sets Syst. **158**, 1814–1822 (2007)
5. Fang, J., Qiu, Y.: Fuzzy orders and fuzzifying topologies. Int. J. Approx. Reason. **48**, 98–109 (2008)
6. Hao, J., Li, Q.: The relationship between L-fuzzy rough set and L-topology. Fuzzy Sets Syst. **178**, 74–83 (2011)
7. Lai, H., Zhang, D.: Fuzzy preorder and fuzzy topology. Fuzzy Sets Syst. **157**, 1865–1885 (2006)
8. Lowen, R., Xu, L.: Alternative characterizations of FNCS. Fuzzy Sets Syst. **104**, 381–391 (1999)
9. Morsi, N.N., Yakout, M.M.: Axiomatics for fuzzy rough sets. Fuzzy Sets Syst. **100**(1–3), 327–342 (1998)
10. Pang, B., Mi, J.S., Xiu, Z.Y.: L-fuzzifying approximation operators in fuzzy rough sets. Inf. Sci. **480**, 14–33 (2019)
11. Pawlak, Z.: Rough sets. Int. J. Comput. Inf. Sci. **11**, 341–356 (1982)
12. Perfilieva, I., Singh, A.P., Tiwari, S.P.: On F-transforms, L-fuzzy partitions and L-fuzzy pretopological spaces. In: 2017 IEEE Symposium Series on Computational Intelligence (SSCI), pp. 1–8. IEEE (2017)
13. Perfilieva, I., Singh, A.P., Tiwari, S.P.: On the relationship among F-transform, fuzzy rough set and fuzzy topology. Soft. Comput. **21**(13), 3513–3523 (2017). https://doi.org/10.1007/s00500-017-2559-x

14. Perfilieva, I., Tiwari, S.P., Singh, A.P.: Lattice-valued F-transforms as interior operators of L-fuzzy pretopological spaces. In: Medina, J., et al. (eds.) IPMU 2018. CCIS, vol. 854, pp. 163–174. Springer, Cham (2018). https://doi.org/10.1007/978-3-319-91476-3_14
15. Qiao, J., Hu, B.Q.: A short note on L-fuzzy approximation spaces and L-fuzzy pretopological spaces. Fuzzy Sets Syst. **312**, 126–134 (2017)
16. Qin, K., Pei, Z.: On the topological properties of fuzzy rough sets. Fuzzy Sets Syst. **151**, 601–613 (2005)
17. Radzikowska, A.M., Kerre, E.E.: A comparative study of fuzzy rough sets. Fuzzy Sets Syst. **126**(2), 137–155 (2002)
18. Radzikowska, A.M., Kerre, E.E.: Fuzzy rough sets based on residuated lattices. In: Peters, J.F., Skowron, A., Dubois, D., Grzymała-Busse, J.W., Inuiguchi, M., Polkowski, L. (eds.) Transactions on Rough Sets II. LNCS, vol. 3135, pp. 278–296. Springer, Heidelberg (2004). https://doi.org/10.1007/978-3-540-27778-1_14
19. She, Y.H., Wang, G.J.: An axiomatic approach of fuzzy rough sets based on residuated lattices. Comput. Math. Appl. **58**, 189–201 (2009)
20. Thiele, H.: On axiomatic characterisations of crisp approximation operators. Inf. Sci. **129**(1–4), 221–226 (2000)
21. Thiele, H.: On axiomatic characterizations of fuzzy approximation operators. In: Ziarko, W., Yao, Y. (eds.) RSCTC 2000. LNCS (LNAI), vol. 2005, pp. 277–285. Springer, Heidelberg (2001). https://doi.org/10.1007/3-540-45554-X_33
22. Tiwari, S.P., Srivastava, A.K.: Fuzzy rough sets, fuzzy preorders and fuzzy topologies. Fuzzy Sets Syst. **210**, 63–68 (2013)
23. Wu, W.Z., Zhang, W.X.: Constructive and axiomatic approaches of fuzzy approximation operators. Inf. Sci. **159**(3–4), 233–254 (2004)
24. Yao, W., Shi, F.G.: A note on specialization L-preorder of L-topological spaces, L-fuzzifying topological spaces, and L-fuzzy topological spaces. Fuzzy Sets Syst. **159**, 2586–2595 (2008)
25. Yao, Y.Y.: Combination of rough and fuzzy sets based on α-level sets. In: Rough Sets and Data Mining, pp. 301–321. Springer, Boston (1997). https://doi.org/10.1007/978-1-4613-1461-5_15
26. Ying, M.: A new approach for fuzzy topology (i). Fuzzy Sets Syst. **39**, 303–321 (1991)
27. Zadeh, L.A.: Similarity relations and fuzzy orderings. Inf. Sci. **3**, 177–200 (1971)
28. Zhang, D.: Fuzzy pretopological spaces, an extensional topological extension of FTS. Chin. Ann. Math. **20**, 309–316 (1999)
29. Zhang, D.: L-fuzzifying topologies as L-topologies. Fuzzy Sets Syst. **125**, 135–144 (2002)
30. Zhang, D., Xu, L.: Categories isomorphic to FNS. Fuzzy Sets Syst. **104**, 373–380 (1999)

Measure of Lattice-Valued Direct F-transforms and Its Topological Interpretations

Anand Pratap Singh(✉) and Irina Perfilieva

Institute for Research and Applications of Fuzzy Modeling, CE IT4Innovations, University of Ostrava, 30. dubna 22, 701 03 Ostrava 1, Czech Republic
{anand.singh,irina.perfilieva}@osu.cz

Abstract. The goal is to introduce and study the measure of quality of approximation of a given fuzzy set by its lattice-valued F-transform. Further, we show that this measure is connected with an Alexandroff LM-fuzzy topological (co-topological) spaces. Finally, we discuss the categorical relationship between the defined structures.

Keywords: M-valued partition · Direct F-transforms · Fuzzy inclusion measure · LM-fuzzy (co)topology · Ditopology

1 Introduction

The importance of various kinds of transforms such as Fourier, Laplace, integral, wavelet are well-known in classical mathematics. The main idea behind these techniques consists of transforming an original space of functions into a new computationally simpler space. Inverse transformations back to the original spaces and produce either the original functions or their approximations. The notion of F-transforms were proposed in [17] has now been significantly developed. Through the viewpoint of application purpose, this theory represent new methods which have turned out to be useful in denoising, time series, coding/decoding of images, numerical solutions of ordinary and partial differential equations (cf., [3,13,24]) and many other applications.

In the seminal paper [17], F-transforms were defined on real-valued functions and another type of F-transform was also introduced based on a residuated lattice in the interval [0,1]. A number of researchers have initiated the study of F-transforms, where they are applied to L-valued functions in a space defined by L-valued fuzzy partitions (cf., [15,16,18,19,25]), where L is a complete residuated lattice. Among these studies, a categorical study of L-partitions of an arbitrary universe is presented in [15] and an interesting relationship among F-transforms, L-topologies/co-topologies and L-fuzzy approximation spaces are established in [19]. The relationships between F-transforms and similarity relations are investigated in [16], axiomatic study of F-transforms have been done in [14], while F-transforms based on a generalized residuated lattice are studied in [25].

M.-J. Lesot et al. (Eds.): IPMU 2020, CCIS 1239, pp. 240–253, 2020.
https://doi.org/10.1007/978-3-030-50153-2_18

In the past few years, some studies have been conducted on the theoretical development of lattice-valued F-transforms. Among them, the papers [18–20, 22] are focused on establishing the relationship between the lattice-valued F-transform and various structured spaces, namely, fuzzy (co) topological/pre-(co)topological space, fuzzy approximation space and fuzzy interior/closure spaces. The ground structure for the above-mentioned studies is the lattice-valued F-transform defined on a space with a fuzzy partition [19]. There are many papers [2, 12, 28], where implications are used to evaluate the measure of inclusion in the same lattice L. To our knowledge, the first attempt to measure the degree of roughness of a given fuzzy set in fuzzy rough set theory was undertaken in [6, 7]. In [7], an approach to measure the quality of rough approximation of fuzzy set is discussed. Motivated by this work, we aim at studying the measure of approximation by the lattice-valued direct F-transforms. In other words, we measure the degree of inclusion of lattice-valued F-transforms into the L-fuzzy set.

For this purpose, in Sect. 3, we consider a more general version of lattice-valued F-transform defined on space with an M-valued partition. Specifically, we define the lattice-valued F-transform operators of an L-fuzzy subset of a set endowed with an M-valued partition. These operators, as special cases, contain various rough approximation-type operators used by different authors [10, 11, 21, 26]. Interestingly, as a main result of this contribution, in Sect. 4, we show that the M-valued measure of F-transform operators based on space with M-valued partition determine the Alexandroff LM-fuzzy topological (co-topological) spaces. We also discuss the mentioned relationship through a categorical viewpoint. It is worth mentioning that our work differs from the existing study in the sense that here we consider the lattice-valued F-transform operator based on the M-valued partition represented by M-fuzzy preorder relation and showed that with this M-fuzzy preorder relation, Alexandroff LM-fuzzy topology (co-topology) can be induced by defining the M-valued measure of direct F-transforms. Finally, we give some direction for further research.

2 Preliminaries

Here, we recall the basic notions and terminologies related to residuated lattices, integral commutative cl-monoid, measure of inclusion between two fuzzy sets and fuzzy topological spaces. These terminologies will be used in the remaining text.

Definition 1 [1,8,9]. *An integral commutative cl-monoid (in short, iccl monoid) is an algebra $(L, \leqslant, \vee, \wedge, \otimes)$ where $(L, \leqslant, \vee, \wedge)$ is a complete lattice with the bottom element 0_L and top element 1_L, and $(L, \otimes, 1_L)$ is a commutative monoid such that binary operation $\otimes$ distributes over arbitrary joins, i.e.,*

$$a \otimes (\vee_{i\in I} b_i) = \vee_{i\in I}(a \otimes b_i), \quad \textit{for all } a \in L,\ \{b_i \mid i \in I\} \subseteq L.$$

Given an iccl-monoid $(L, \leqslant, \vee, \wedge, \otimes)$, we can define a binary operation "$\rightarrow$", for all $a, b, c \in L$,

$$a \rightarrow b = \vee\{c \in L \mid c \otimes a \leq b\},$$

which is adjoint to the monoidal operation $\otimes$, i.e.,

$$a \otimes b \leq c \iff a \leq b \rightarrow c.$$

A particular example of residuated lattice is Lukasiewicz algebra

$$\mathscr{L}_L = ([0,1], \vee, \wedge, \otimes, \rightarrow, 0, 1),$$

where the binary operations "$\otimes$" and " $\rightarrow$" are defined as, $a \otimes b = \min(a+b-1, 1)$ and $a \rightarrow b = \max(1 - a + b, 0)$.

The following properties of residuated lattice will be used, for proof refer to [1,5,9].

Proposition 1. *Let* $(L, \leqslant, \wedge, \vee, \otimes, \rightarrow)$ *be a residuated lattice. Then for all* $a, b, c \in L$,

(i) $a \leq b \rightarrow (a \otimes b)$, $a \otimes (a \rightarrow b) \leq b$,
(ii) $a \otimes (b \rightarrow c) \leq (a \rightarrow b) \rightarrow c$,
(iii) $a \otimes (\vee_{j \in J} b_j) = \vee_{j \in J}(a \otimes b_j)$,
(iv) $a \otimes (\wedge_{j \in J} b_j) \leq \wedge_{j \in J}(a \otimes b_j)$,
(v) $\vee_{j \in J}(a_j \rightarrow b) \leq \wedge_{j \in J} a_j \rightarrow b$,
(vi) $a \rightarrow (b \rightarrow c) = (a \otimes b) \rightarrow c = b \rightarrow (a \rightarrow c)$,
(vii) $a \rightarrow \wedge_{j \in J} b_j = \wedge_{j \in J}(a \rightarrow b_j)$,
(viii) $\vee_{j \in J} a_j \rightarrow b = \wedge_{j \in J}(a_j \rightarrow b)$,
(ix) $a \rightarrow \vee_{j \in J} b_j \geq \vee_{j \in J}(a \rightarrow b_j)$.

Given a nonempty set X, L^X denotes the collection of all L-fuzzy subsets of X, i.e. a mapping $\lambda : X \rightarrow L$. Also, for all $a \in L$, $\mathbf{a}(x) = a$ is a constant L-fuzzy set on X. The greatest and least element of L^X is denoted by 1_X and 0_X respectively. For all $\lambda \in L^X$, the *core*(λ) is a set of all elements $x \in X$, such that $\lambda(x) = 1$. An L-fuzzy set $\lambda \in L^X$ is called *normal*, if *core*$(\lambda) \neq \emptyset$.

The notion of powerset structures are well known and have been widely used in several constructions and applications. According to Zadeh's principle, any map $f : X \rightarrow Y$ can be extended to the powerset operators $f^{\rightarrow} : L^X \rightarrow L^Y$ and $f^{\leftarrow} : L^Y \rightarrow L^X$ such that for $\lambda \in L^X, \mu \in L^Y, y \in Y$, $f^{\rightarrow}(\lambda)(y) = \vee_{x, f(x)=y} \lambda(x)$, $f^{\leftarrow}(\mu) = \mu \circ f$.

Definition 2. [4] *Let X be a nonempty set. Then for given two L-fuzzy sets λ, μ and for each $x \in X$, following are the* new L-fuzzy sets *defined:*

$$\lambda = \mu \iff \lambda(x) = \mu(x), \;\; \lambda \leq \mu \iff \lambda(x) \leq \mu(x),$$
$$(\lambda \wedge \mu)(x) = \lambda(x) \wedge \mu(x), \;\; (\lambda \vee \mu)(x) = \lambda(x) \vee \mu(x),$$
$$(\lambda \otimes \mu)(x) = \lambda(x) \otimes \mu(x), \; (\lambda \rightarrow \mu)(x) = \lambda(x) \rightarrow \mu(x).$$

Let I be a set of indices, $\lambda_i \in L^X$, $i \in I$. The *meet* and *join* of elements from $\{\lambda_i \mid i \in I\}$ are defined as follows:

$$(\wedge_{i\in I}\lambda_i)(x) = \wedge_{i\in I}\lambda_i(x), (\vee_{i\in I}\lambda_i)(x) = \vee_{i\in I}\lambda_i(x).$$

In this paper, we work with two independent iccl-monoids $L = (L, \leqslant, \vee, \wedge, \otimes_L)$ and $M = (M, \leqslant, \vee, \wedge, \otimes_M)$. The iccl-monoid L is used as range for L-fuzzy sets (i.e. for the approximative objects), while M is used as the range of values taken by the measure estimating the precision of the approximation. Both the iccl-monoids are unrelated, however for proving some results based on measure of approximation, a connection between them is required, for this purpose we define fixed mappings $\phi : L \to M$ and $\psi : M \to L$ such that the top and bottom elements are preserved i.e., for $\{a_i \mid i \in I\} \subseteq L$, $\{b_i \mid i \in I\} \subseteq M$, $\phi(\wedge_{i\in I}a_i) = \wedge_{i\in I}\phi(a_i)$, $\psi(\vee_{i\in I}b_i) = \vee_{i\in I}\psi(b_i)$. Additionally, we need that $\phi(\alpha \otimes_L a) = \phi(\alpha) \otimes_L \phi(a)$, $\forall \alpha, a \in L$ and $\psi(\beta \otimes_M b) = \psi(\beta) \otimes_M \psi(b)$, $\forall \beta, b \in M$.

Now we recall the following definition of M-fuzzy relation from [27].

Definition 3. [27] *Let X be a nonempty set. An M-*fuzzy relation *δ on X is an M-fuzzy subset of $X \times X$. An M-fuzzy relation δ is called*

(i) reflexive *if $\delta(x, x) = 1$ $\forall x \in X$,*
(ii) transitive *if $\delta(x, y) \otimes \delta(y, z) \leq \delta(x, z)$, $\forall x, y, z \in X$.*

*A reflexive and transitive M-fuzzy relation δ is called an M-*fuzzy preorder.

Here, we define the measure of inclusion between two given L-fuzzy sets as it was introduced in [7].

Definition 4. *Let the iccl-monoids L, M be given and $\phi : L \to M$ be the fixed mapping. The M-*valued measure of inclusion *of the L-fuzzy set λ into the L-fuzzy set μ is a map $\hookrightarrow: L^X \times L^X \to M$ defined as*

$$\lambda \hookrightarrow \mu = \wedge_{x\in X}\phi(\lambda(x) \to \mu(x)), \text{ for all } \lambda, \mu \in L^X.$$

In other words, the measure of inclusion function $\hookrightarrow$ can be defined by $\lambda \hookrightarrow \mu = \phi(\wedge(\lambda \to \mu))$, where the infimum of the L-fuzzy set $\lambda \hookrightarrow \mu$ is taken in the lattice L^X.

Below, we give some useful properties of the map $\hookrightarrow$, which is similar (in some sense) to the properties of residuated lattices.

Proposition 2. [7] *The inclusion map $\hookrightarrow: L^X \times L^X \to M$ satisfies the following properties.*

(i) $(\vee_{i\in I}\lambda_i) \hookrightarrow \mu = \wedge_{i\in I}(\lambda_i \hookrightarrow \mu)$, $\forall\{\lambda_i \mid i \in I\} \subseteq L^X$, $\mu \in L^X$,
(ii) $\lambda \hookrightarrow (\wedge_{i\in I}\mu_i) = \wedge_{i\in I}(\lambda_i \hookrightarrow \mu)$, $\forall \lambda \in L^X, \{\mu_i \mid i \in I\} \subseteq L^X$,
(iii) $\lambda \hookrightarrow \mu = 1 \Leftrightarrow \lambda \leq \mu$,
(iv) $1_X \hookrightarrow \lambda = \phi(\wedge_x \lambda(x))$, $\forall \lambda \in L^X$,
(v) $\lambda \hookrightarrow \mu \leq (\lambda \otimes c \hookrightarrow \mu \otimes c)$, $\forall \lambda, \mu, c \in L^X$,

(vi) $(\wedge_i \lambda_i) \hookrightarrow (\wedge_i \mu_i) \geq \wedge_i(\lambda_i \hookrightarrow \mu_i)$, $\forall\{\lambda_i \mid i \in I\}$, $\{\mu_i \mid i \in I\} \subseteq L^X$,
(vii) $(\vee_i \lambda_i) \hookrightarrow (\vee_i \mu_i) \geq \wedge_i(\lambda_i \hookrightarrow \mu_i)$, $\forall\{\lambda_i \mid i \in I\}$, $\{\mu_i \mid i \in I\} \subseteq L^X$.

We close this section by recalling the following definition of LM-fuzzy topology presented by [23].

Definition 5. [23] *An* LM-fuzzy topology $\mathfrak{T}$ *on universe* X *is a mapping* $\mathfrak{T} : L^X \to M$, *such that for each* $\lambda, \mu \in L^X$, $\{\lambda_i \mid i \in I\} \subseteq L^X$ *it satisfies,*

(i) $\mathfrak{T}(\phi) = \mathfrak{T}(X) = 1$,
(ii) $\mathfrak{T}(\lambda \wedge \mu) \geq \mathfrak{T}(\lambda) \wedge \mathfrak{T}(\mu)$,
(iii) $\mathfrak{T}(\vee_{i\in I}\lambda_i) \geq \wedge_{i\in I}\mathfrak{T}(\lambda_i)$.

For an LM*-fuzzy topology* $\mathfrak{T}$ *on nonempty set* X*, the pair* $(X, \mathfrak{T})$ *is called an* LM-fuzzy topological space. *Further,* $(X, \mathfrak{T})$ *is*

(vi) strong, *if* $\mathfrak{T}(\mathbf{a} \otimes \lambda) \geq \mathfrak{T}(\lambda)$,
(v) Alexandroff, *if* $\mathfrak{T}(\wedge_{i\in I}\lambda_i) \geq \wedge_{i\in I}\mathfrak{T}(\lambda_i)$.

Given two LM-fuzzy topological space $(X, \mathfrak{T}_X)$ and $(Y, \mathfrak{T}_Y)$ a map $f : (X, \mathfrak{T}_X) \to (Y, \mathfrak{T}_Y)$ is called *continuous* if for all $\lambda \in L^Y$, $\mathfrak{T}_X(f^{\leftarrow}(\lambda)) \geq \mathfrak{T}_Y(\lambda)$. We denote by **LM-ToP**, the category of Alexandroff LM-fuzzy topological spaces.

The notion of LM-fuzzy co-topology [23] $\mathscr{T}$ can be defined similarly. Given two LM-fuzzy co-topological space $(X, \mathscr{T}_X)$ and $(Y, \mathscr{T}_Y)$ a map $f : (X, \mathscr{T}_X) \to (Y, \mathscr{T}_Y)$ is called *continuous* if $\lambda \in L^Y$, $\mathcal{T}_X(f^{\leftarrow}(\lambda)) \geq \mathcal{T}_Y(\lambda)$. We denote by **LM-CToP**, the category of Alexandroff LM-fuzzy co-topological spaces.

3 *M*-valued Partition and Direct *F*-transforms

In this section, we recall the notion of M-valued partitions and lattice-valued F-transforms [19]. We also discuss its behaviour based on ordering in spaces with M-valued partition. We begin with the following definition of M-valued partition as introduced in [19].

Definition 6. *Let* X *be a nonempty set. A collection* $\mathcal{P}$ *of normal* M*-valued sets* $\{A_\xi : \xi \in \Lambda\}$ *in* X *is an* M-valued partition *of* X*, if* $\{core(A_\xi) : \xi \in \Lambda\}$ *is a partition (crisp) of* X*. A pair* $(X, \mathcal{P})$*, where* $\mathcal{P}$ *is an* M*-valued partition of* X*, is called a* space with an M-valued partition.

Let $\mathcal{P} = \{A_\xi : \xi \in \Lambda\}$ be an M-valued partition of X. With this partition, we associate the following surjective index-function $i_{\mathcal{P}} : X \to \Lambda$:

$$i_{\mathcal{P}}(x) = \xi \iff x \in core(A_\xi). \tag{1}$$

Then M-valued partition $\mathcal{P}$ can be uniquely represented by the reflexive M-fuzzy relation $\delta_{\mathcal{P}}$ on X, such that

$$\delta_{\mathcal{P}}(x, y) = A_{i_{\mathcal{P}}(x)}(y). \tag{2}$$

In [19], it has been proved that the relation (2) can be decomposed into a constituent M-fuzzy preorder relation $\delta_{\mathcal{P}_\xi}(x, y)$, where for each $\xi \in \Lambda$

$$\delta_{\mathcal{P}_\xi}(x, y) = \begin{cases} A_{i_{\mathcal{P}}(x)}(y), & \text{if } x \in core(A_{i_{\mathcal{P}}(x)}), \\ 1, & \text{if } x = y, \\ 0, & \text{otherwise.} \end{cases}$$

For given two spaces with M-valued partitions, following is the notion of morphism between them.

Definition 7. [15] *Let $(X, \mathcal{P} = \{A_\xi : \xi \in \Lambda)\}$ and $(Y, \mathcal{Q} = \{B_\omega : \omega \in \Omega)\}$ be two spaces with M-valued partitions. A* morphism *in the space with M-valued partitions is a pair of maps (f, g), where $f : X \to Y$ and $g : \Lambda \to \Omega$ are maps such that for each $x \in X$, $\forall\ \xi \in \Lambda,\ \ A_\xi(x) \leq B_{g(\xi)}(f(x))$.*

It can be easily verified that all spaces with M-valued partition as objects and the pair of maps (f, g) defined above as morphisms form a category [15], denoted by **SMFP**. If there is no danger of misunderstanding, the object-class of **SMFP** will be denoted by **SMFP** as well.

Let L, M be iccl-monoids and $\psi : M \to L$ be the fixed mapping. Here, we define the following concept of lattice-valued direct $F^{\uparrow}(F^{\downarrow})$-transforms.

Definition 8. *Let X be a nonempty set and $\mathcal{P} = \{A_\xi : \xi \in \Lambda\}$ be M-valued partition of X. Then for all $\lambda \in L^X$, and for all $x \in X$,*

(i) direct $F^{\uparrow}$-transform *of L-valued function λ is defined by*

$$F^{\uparrow}_{\xi}(\lambda) = \vee_{y \in X}(\psi(A_\xi(y)) \otimes \lambda(y)),$$

(ii) direct $F^{\downarrow}$-transform *of L-valued function λ is defined by*

$$F^{\downarrow}_{\xi}(\lambda) - \wedge_{y \in X}(\psi(A_\xi(y)) \to \lambda(y)).$$

It has been shown in [19] that for M-valued partition $\mathcal{P} = \{A_\xi : \xi \in \Lambda\}$ represented by M-fuzzy preorder relation $\delta_{\mathcal{P}_\xi}$ and with the associated index-function $i_{\mathcal{P}}$, we can associate two operators $F^{\uparrow}_{\mathcal{P}_\xi}$ and $F^{\downarrow}_{\mathcal{P}_\xi}$ corresponding to direct $F^{\uparrow}$ and $F^{\downarrow}$-transforms. Where, for each $\lambda \in L^X$,

$$F^{\uparrow}_{\mathcal{P}_\xi}(\lambda)(x) = F^{\uparrow}_{i_{\mathcal{P}}(x)}(\lambda) = \vee_{y \in X}(\psi(\delta_{\mathcal{P}_\xi}(x, y)) \otimes \lambda(y)), \quad (3)$$

$$F^{\downarrow}_{\mathcal{P}_\xi}(\lambda)(x) = F^{\downarrow}_{i_{\mathcal{P}}(x)}(\lambda) = \wedge_{y \in X}(\psi(\delta^{T}_{\mathcal{P}_\xi}(x, y)) \to \lambda(y)). \quad (4)$$

In [19], it has been proved that the operators defined in Eqs. 3 and 4 connect the $i_{\mathcal{P}}(x)$-th upper (lower) $F^{\uparrow}(F^{\downarrow})$-transform approximation with a closure (interior) operators in corresponding L-fuzzy co-topology (topology).

Remark 1. Initially, the notion of lattice-valued direct F-transforms were introduced in [17], where the fuzzy partition A_ξ was considered as a fuzzy subset of $[0,1]$ (i.e., $A_\xi : X \to [0,1]$) fulfilling the covering property, $(\forall x)$ $(\exists \xi)$, $A_\xi(x) > 0$. Later, this notion were extended to the case where they are applied to L-valued functions in a space defined by L-valued fuzzy partitions, where L is a complete residuated lattice. Here, if we consider the case where $L = M$ and $\psi = id$ is an identity map, then the above definition of lattice-valued direct F-transforms will be similar to the lattice-valued direct F-transforms appeared in [17–20].

The following properties of the operators $F^{\uparrow}_{\mathcal{P}_\xi}$ and $F^{\downarrow}_{\mathcal{P}_\xi}$ were presented in [17,19]. All of them are formulated below for arbitrary $\mathbf{a}, \lambda, \lambda_i \in L^X$.

Proposition 3. *Let X be a nonempty set and $\mathcal{P}$ be the M-valued partition of X. Then for all $\mathbf{a}, \lambda \in L^X$, and $\{\lambda_i \mid i \in I\} \subseteq L^X$, the operators $F^{\uparrow}_{\mathcal{P}_\xi} : L^X \to L^X$ and $F^{\downarrow}_{\mathcal{P}_\xi} : L^X \to L^X$ satisfies the following properties.*

(i) $F^{\uparrow}_{\mathcal{P}_\xi}(\mathbf{a}) = \mathbf{a}$, $F^{\downarrow}_{\mathcal{P}_\xi}(\mathbf{a}) = \mathbf{a}$,
(ii) $\lambda \leq F^{\uparrow}_{\mathcal{P}_\xi}(\lambda)$, $F^{\downarrow}_{\mathcal{P}_\xi}(\lambda) \leq \lambda$,
(iii) $F^{\uparrow}_{\mathcal{P}_\xi}(\vee_{i\in I}\lambda_i) = \vee_{i\in I} F^{\uparrow}_{\mathcal{P}_\xi}(\lambda_i)$, $F^{\downarrow}_{\mathcal{P}_\xi}(\wedge_{i\in I}\lambda_i) = \wedge_{i\in I} F^{\downarrow}_{\mathcal{P}_\xi}(\lambda_i)$,
(iv) $F^{\uparrow}_{\mathcal{P}_\xi}(\mathbf{a} \otimes \lambda) = a \otimes F^{\uparrow}_{\mathcal{P}_\xi}(\lambda)$, $F^{\downarrow}_{\mathcal{P}_\xi}(\mathbf{a} \to \lambda) = a \to F^{\downarrow}_{\mathcal{P}_\xi}(\lambda)$.

Below, we investigate the properties of lattice-valued direct F-transforms with respect to ordering between spaces with M-valued partition. We have the following.

Proposition 4. *Let $(f,g) : (X, \mathcal{P}) \to (Y, \mathcal{Q})$ be a morphism between spaces with M-valued partition and fixed mapping ψ is monotonic decreasing. Then,*

(i) $F^{\uparrow}_{\xi}(f^{\leftarrow}(\lambda)) \leq F^{\uparrow}_{g(\xi)}(\lambda)$,
(ii) $F^{\downarrow}_{\xi}(f^{\leftarrow}(\lambda)) \geq F^{\downarrow}_{g(\xi)}(\lambda)$,
(iii) $F^{\uparrow}_{g(\xi)}(f^{\rightarrow}(\lambda)) \geq F^{\uparrow}_{\xi}(\lambda)$,
(iv) $F^{\downarrow}_{g(\xi)}(f^{\rightarrow}(\lambda)) \leq F^{\downarrow}_{\xi}(\lambda)$.

Proof. (i) Let $\lambda \in L^Y$. From Definition 8, we have

$$\begin{aligned} F^{\uparrow}_{\xi}(f^{\leftarrow}(\lambda)) &= \vee_{x\in X}\psi(A_\xi(x)) \otimes f^{\leftarrow}(\lambda)(x) = \vee_{x\in X}\psi(A_\xi(x)) \otimes \lambda(f(x)) \\ &\leq \vee_{x\in X}\psi(B_{g(\xi)}(f(x))) \otimes \lambda(f(x)) \leq \vee_{y\in Y}\psi(B_{g(\xi)}(y)) \otimes \lambda(y) \\ &= F^{\uparrow}_{g(\xi)}(\lambda). \end{aligned}$$

(ii) Let $\lambda \in L^Y$. From Definition 8, we get

$$\begin{aligned} F^{\downarrow}_{\xi}(f^{\leftarrow}(\lambda)) &= \wedge_{x\in X}\psi(A_\xi(x)) \to f^{\leftarrow}(\lambda)(x) = \wedge_{x\in X}\psi(A_\xi(x)) \to \lambda(f(x)) \\ &\geq \wedge_{x\in X}\psi(B_{g(\xi)}(f(x))) \to \lambda(f(x)) \geq \wedge_{y\in Y}\psi(B_{g(\xi)}(y)) \to \lambda(y) \\ &= F^{\downarrow}_{g(\xi)}(\lambda). \end{aligned}$$

(iii) Let $\lambda \in L^X$. From Definition 8, we obtain
$F^{\uparrow}_{g(\xi)}(f^{\rightarrow}(\lambda))$

$$= \vee_{y\in Y}\left(f^{\rightarrow}(\lambda)(y) \otimes \psi(B_{g(\xi)}(y))\right) = \vee_{y\in Y}\left(\vee_{x:f(x)=y}\,\lambda(x) \otimes \psi(B_{g(\xi)}(y))\right)$$
$$\geq \vee_{x\in X}\lambda(x) \otimes \psi(B_{g(\xi)}(f(x))) \geq \vee_{x\in X}\lambda(x) \otimes \psi(A_{\xi}(x)) = F^{\uparrow}_{\xi}(\lambda).$$

(iv) Let $\lambda \in L^X$. From Definition 8, we obtain
$F^{\downarrow}_{g(\xi)}(f^{\rightarrow}(\lambda))$

$$= \wedge_{y\in Y}\left(\psi(B_{g(\xi)}(y)) \rightarrow f^{\rightarrow}(\lambda)(y)\right) = \wedge_{y\in Y}\left(\psi(B_{g(\xi)}(y)) \rightarrow \vee_{x:f(x)=y}\lambda(x)\right)$$
$$= \wedge_{x\in X}\psi(B_{g(\xi)}(y)(f(x)) \rightarrow \lambda(x) \leq \wedge_{x\in X}\psi(A_{\xi}(x)) \rightarrow \lambda(x) = F^{\downarrow}_{\xi}(\lambda).$$

4 Main Results

In this section, we introduce and study the measure of lattice-valued direct F-transforms of an L-fuzzy set. The defined measure determines the amount of preciseness of L-fuzzy subset into L-fuzzy set. Further, we investigate the topologies induced by the measure of lattice-valued direct F-transform operators. In particular, we show that every measure of lattice-valued direct F-transform operators determine strong Alexandroff LM-fuzzy topological and co-topological spaces.

Definition 9. *Let X be a nonempty set and $\mathcal{P}$ be the M-valued partition of X. Then for an L-fuzzy set $\lambda \in L^X$. The* M-valued measure of direct $F^{\uparrow}$-transform $\mathcal{F}_u(\lambda) = F^{\uparrow}_{\mathcal{P}_{\xi}}(\lambda) \hookrightarrow \lambda$ *and* M-valued measure of direct $F^{\downarrow}$-transform $\mathcal{F}_l(\lambda) = \lambda \hookrightarrow F^{\downarrow}_{\mathcal{P}_{\xi}}(\lambda)$ *are maps $\mathcal{F}_u, \mathcal{F}_l : L^X \rightarrow M$ and defined as,*

$$\mathcal{F}_u(\lambda) = \phi\left(\wedge_{x\in X}(F^{\uparrow}_{\mathcal{P}_{\xi}}(\lambda)(x) \rightarrow \lambda(x))\right), \quad \mathcal{F}_l(\lambda) = \phi\left(\wedge_{x\in X}(\lambda(x) \rightarrow F^{\downarrow}_{\mathcal{P}_{\xi}}(\lambda)(x))\right).$$

To investigate in more detail how the M-valued measure of F-transform operators preserve the inclusion of two sets defined in the real interval [0,1], instead of a general complete residuated lattice L, we will use the specific example of residuated lattices.

Example 1. Consider the case $L = M$, and $\phi = \psi = id$ and the Lukasiewicz algebra

$$\mathscr{L}_L = ([0,1], \vee, \wedge, \otimes, \rightarrow, 0, 1).$$

Then the measure of lattice-valued F-transform operators of an L-fuzzy set is obtained as below,

$$\mathcal{F}_u = \wedge_{x,y\in X}(2 - \lambda(x) + \lambda(y) - A_{i_{\mathcal{P}}(y)}(x)),$$
$$\mathcal{F}_l = \wedge_{x,y\in X}(2 - \lambda(x) + \lambda(y) - A_{i_{\mathcal{P}}(x)}(y)).$$

Proposition 5. *Let X be a nonempty set and $\mathcal{P}$ be the M-valued partition of X. Then for each $\mathbf{a}, \lambda \in L^X$ and for all $\{\lambda_i \mid i \in I\} \subseteq L^X$, the M-valued measure of F-transform operators $\mathcal{F}_u, \mathcal{F}_l : L^X \to M$ satisfy the following properties.*

(i) $\mathcal{F}_u(\mathbf{a}) = 1_M$, $\mathcal{F}_l(\mathbf{a}) = 1_M$,
(ii) $\mathcal{F}_u(\vee_{i\in I}\lambda_i) \geq \wedge_{i\in I}\mathcal{F}_u(\lambda_i)$, $\mathcal{F}_l(\vee_{i\in I}\lambda_i) \geq \wedge_{i\in I}\mathcal{F}_l(\lambda_i)$,
(iii) $\mathcal{F}_u(\wedge_{i\in I}\lambda_i) \geq \wedge_{i\in I}\mathcal{F}_u(\lambda_i)$, $\mathcal{F}_l(\wedge_{i\in I}\lambda_i) \geq \wedge_{i\in I}\mathcal{F}_l(\lambda_i)$,
(iv) $\mathcal{F}_u(\mathbf{a} \otimes \lambda) \geq \mathcal{F}_u(\lambda)$, $\mathcal{F}_l(\mathbf{a} \to \lambda) \geq \mathcal{F}_l(\lambda)$.

Proof. We give only proof for the M-valued measure of direct F transform operator $\mathcal{F}_u$. The proof for operator $\mathcal{F}_l$ follows similarly.

(i) For all $\mathbf{a} \in L^X$, and from Proposition 3, we have,

$$\begin{aligned}\mathcal{F}_u(\mathbf{a}) = F^{\uparrow}_{\mathcal{P}_\xi}(\mathbf{a}) \hookrightarrow \mathbf{a} &= \phi(\wedge_{x\in X}(F^{\uparrow}_{\mathcal{P}_\xi}(\mathbf{a})(x) \to \mathbf{a}(x)))\\ &= \phi(\wedge_{x\in X}(\mathbf{a}(x) \to \mathbf{a}(x))) = 1_M.\end{aligned}$$

(ii) For all $\lambda \in L^X$, and $\{\lambda_i \mid i \in I\} \subseteq L^X$, from Proposition 2, we have

$$\begin{aligned}\mathcal{F}_u(\vee_{i\in I}\lambda_i) &= F^{\uparrow}_{\mathcal{P}_\xi}(\vee_{i\in I}\lambda_i) \hookrightarrow (\vee_{i\in I}\lambda_i)\\ &= \vee_{i\in I}(F^{\uparrow}_{\mathcal{P}_\xi}(\lambda_i)) \hookrightarrow (\vee_{i\in I}\lambda_i) \geq \wedge_{i\in I}(F^{\uparrow}_{\mathcal{P}_\xi}(\lambda_i)) \hookrightarrow (\lambda_i))\\ &= \wedge_{i\in I}\mathcal{F}_u(\lambda_i).\end{aligned}$$

(iii) For all $\lambda \in L^X$, and $\{\lambda_i \mid i \in I\} \subseteq L^X$, from Proposition 2, we have

$$\begin{aligned}\mathcal{F}_u(\wedge_{i\in I}\lambda_i) &= F^{\uparrow}_{\mathcal{P}_\xi}(\wedge_{i\in I}\lambda_i) \hookrightarrow (\wedge_{i\in I}\lambda_i)\\ &\geq \wedge_{i\in I}(F^{\uparrow}_{\mathcal{P}_\xi}(\lambda_i)) \hookrightarrow (\wedge_{i\in I}\lambda_i) \geq \wedge_{i\in I}(F^{\uparrow}_{\mathcal{P}_\xi}(\lambda_i)) \hookrightarrow (\lambda_i))\\ &= \wedge_{i\in I}\mathcal{F}_u(\lambda_i).\end{aligned}$$

(iv) For all $\mathbf{a}, \lambda \in L^X$, from Proposition 3, we have,

$$\begin{aligned}\mathcal{F}_u(\mathbf{a} \otimes \lambda) &= F^{\uparrow}_{\mathcal{P}_\xi}(\mathbf{a} \otimes \lambda) \hookrightarrow (\mathbf{a} \otimes \lambda)\\ &= \phi(\wedge_{x\in X}(\vee_{y\in X}(\psi(A_{i_{\mathcal{P}}(x)}(y)) \otimes (\mathbf{a} \otimes \lambda)(y)) \to (\mathbf{a} \otimes \lambda)(x)))\\ &\geq \phi(\wedge_{x\in X}(\vee_{y\in X}(\psi(A_{i_{\mathcal{P}}(x)}(y)) \otimes \lambda(y)) \to \lambda(x)))\\ &= \phi(\wedge_{x\in X}(F^{\uparrow}_{\mathcal{P}_\xi}(\lambda)(x) \to \lambda(x))) = \mathcal{F}_u(\lambda).\end{aligned}$$

4.1 Measure of Direct $F^{\downarrow}$-transform Operator and LM-fuzzy Topology

In this subsection, we show that the M-valued measure of direct $F^{\downarrow}$-transform induces LM-fuzzy topology.

Theorem 1. *Let $(X, \mathcal{P})$ be a space with an M-valued partition, $\delta_{\mathcal{P}_\xi}$ representing M-fuzzy preorder relation and $F^{\downarrow}_{\mathcal{P}_\xi} : L^X \to L^X$ be the corresponding $F^{\downarrow}$-operator. Then the pair $(X, \mathfrak{T}_{\mathcal{P}})$, where $\mathfrak{T}_{\mathcal{P}} = \{\mathcal{F}_l(\cdot)(x) : L^X \to M \mid x \in X\}$ is such that for every $\lambda \in L^X$ and $x \in X$,*

$$\mathcal{F}_l(\lambda) = \lambda \hookrightarrow F^{\downarrow}_{\mathcal{P}_\xi}(\lambda)$$

is a strong Alexandroff LM-fuzzy topological space.

Proof. Let $\mathcal{P} = \{A_\xi : \xi \in \Lambda\}$ be an M-valued partition and $i_{\mathcal{P}}$ be the corresponding index-function, such that for all $x \in X$, the value $i_{\mathcal{P}}(x)$ determines the unique partition element $A_{i_{\mathcal{P}}(x)}$, where $x \in core(A_{i_{\mathcal{P}}(x)})$. For every $x \in X$, $\lambda \in L^X$, we claim that $F^{\downarrow}_{\mathcal{P}_\xi}(\lambda)(x) = F^{\downarrow}_{i_{\mathcal{P}}(x)}(\lambda)$, where the right-hand side is the $i_{\mathcal{P}}(x)$-th $F^{\downarrow}$-transform component of λ. Indeed, for a particular $x \in X$, $F^{\downarrow}_{i_{\mathcal{P}}(x)}(\lambda)$ is computed in accordance with (4) and by this, coincides with $F^{\downarrow}_{\mathcal{P}_\xi}(\lambda)(x)$. Now it only remains to verify that the collection $\mathfrak{T}_{\mathcal{P}} = \{\mathcal{F}_l(\cdot)(x) : L^X \to M \mid x \in X\}$, where for every $\lambda \in L^X$, $\mathcal{F}_l(\lambda) = \lambda \hookrightarrow F^{\downarrow}_{i_{\mathcal{P}}(x)}(\lambda)$ is a strong Alexandroff LM-fuzzy topological space. It can be seen that the properties (i)–(iv) of Proposition 5 characterize the M-valued measure of direct $F^{\downarrow}$-transform operator $\mathcal{F}_l : L^X \to M$ as strong Alexandroff LM-fuzzy topological space.

Theorem 2. *Let $(f, g) : (X, \mathcal{P}) \to (Y, \mathcal{Q})$ be a morphism between the spaces with M-valued partition, characterized by index-functions $i_{\mathcal{P}}$ and $k_{\mathcal{P}}$, respectively. Let, moreover, Alexandroff LM-fuzzy topologies $\mathfrak{T}_X = \{\mathcal{F}^{i_{\mathcal{P}}}_l : L^X \to M \mid x \in X\}$ on X and $\mathfrak{T}_Y = \{\mathcal{F}^{k_{\mathcal{P}}}_l : L^Y \to M \mid y \in Y\}$ on Y be induced by the $F^{\downarrow}$-transform. Then $f : (X, \mathfrak{T}_X) \to (Y, \mathfrak{T}_Y)$ is a continuous map.*

Proof. Since $(f, g) : (X, \mathcal{P}) \to (Y, \mathcal{Q})$ be an FP-map. Then from Definition 7, for all $\xi \in \Lambda$, $A_\xi(x) \leq B_{g(\xi)}(f(x))$. Now we have to show that for all $\lambda \in L^Y$, $\mathfrak{T}_X(f^{\leftarrow}(\lambda)) \geq \mathfrak{T}_Y(\lambda)$, where

$$\begin{aligned}
\mathfrak{T}_X(f^{\leftarrow}(\lambda)) &= \mathcal{F}^{i_{\mathcal{P}}}_l(f^{\leftarrow}(\lambda)) = f^{\leftarrow}(\lambda) \hookrightarrow F^{\downarrow}_{i_{\mathcal{P}}(x)}(f^{\leftarrow}(\lambda)) \\
&= \phi\left(\wedge_{x\in X}(f^{\leftarrow}(\lambda)(x) \to F^{\downarrow}_{i_{\mathcal{P}}(x)}(f^{\leftarrow}(\lambda)(x)))\right) \\
&= \phi\left(\wedge_{x\in X}(\lambda(f(x)) \to \wedge_{t\in X}(\psi(A_{i_{\mathcal{P}}(x)}(t)) \to \lambda(f(t))))\right) \\
&= \phi\left(\wedge_{x\in X} \wedge_{t\in X} (\lambda(f(x)) \to (\psi(A_{i_{\mathcal{P}}(x)}(t)) \to \lambda(f(t))))\right) \\
&\geq \phi\left(\wedge_{x\in X} \wedge_{t\in X} (\lambda(f(x)) \to (\psi(B_{k_{\mathcal{P}}(x)}(f(t))) \to \lambda(f(t))))\right) \\
&\geq \phi\left(\wedge_{z\in X} \wedge_{y\in X} (\lambda(z) \to (\psi(B_{k_{\mathcal{P}}(x)}(y)) \to \lambda(y)))\right) \\
&\geq \phi\left(\wedge_{z\in X}(\lambda(z) \to \wedge_{y\in X}(\psi(B_{k_{\mathcal{P}}(x)}(y)) \to \lambda(y)))\right) \\
&= \lambda \hookrightarrow F^{\downarrow}_{k_{\mathcal{P}}(z)}(\lambda) = \mathcal{F}^{k_{\mathcal{P}}}_l(\lambda) = \mathfrak{T}_Y(\lambda).
\end{aligned}$$

Thus $f : (X, \mathfrak{T}_X) \to (Y, \mathfrak{T}_Y)$ is a continuous map.

From Theorems 1, 2, we obtain a functor $\mathbb{G}$ as follows:

$$\mathbb{G}:\begin{cases}\mathbf{SMFP} & \longmapsto \mathbf{LM\text{-}ToP}\\ (X,\mathcal{P}) & \longmapsto (X,\mathfrak{T}_X)\\ (f,g) & \longmapsto \ f,\end{cases}$$

where $\mathcal{P}=\{A_\xi:\xi\in\Lambda\}$ is an M-valued partition of X, $\mathfrak{T}_X=\{\mathcal{F}_l^{i_\mathcal{P}}:L^X\to M\mid x\in X\}$ is the induced strong Alexandroff LM-fuzzy topology on X by the M-valued measure of direct $F^\downarrow$-transform operator $\mathcal{F}_l$, and $(f,g):(X,\mathcal{P})\to(Y,\mathcal{Q})$ is a morphism between the spaces with M-valued partition.

4.2 Measure of Direct $F^\uparrow$ -transform Operator and LM-fuzzy Co-topology

Here, we show that the M-valued measure of direct $F^\uparrow$-transform induces LM-fuzzy co-topology.

Theorem 3. *Let $(X,\mathcal{P})$ be a space with an M-valued partition, $\delta_{\mathcal{P}_\xi}$ representing M-fuzzy preorder relation and $F^\uparrow_{\mathcal{P}_\xi}:L^X\to L^X$ be the corresponding $F^\uparrow$-operator. Then the pair $(X,\mathcal{T}_\mathcal{P})$, where $\mathcal{T}_\mathcal{P}=\{\mathcal{F}_u(\cdot)(x):L^X\to M\mid x\in X\}$ is such that for every $\lambda\in L^X$ and $x\in X$,*

$$\mathcal{F}_u(\lambda)=F^\uparrow_{\mathcal{P}_\xi}(\lambda)\hookrightarrow\lambda$$

is a strong Alexandroff LM-fuzzy co-topological space.

Proof. The proof is analogous to Theorem 1.

Theorem 4. *Let $(f,g):(X,\mathcal{P})\to(Y,\mathcal{Q})$ be a morphism between the spaces with M-valued partition, characterized by index-functions $i_\mathcal{P}$ and $k_\mathcal{P}$, respectively. Let, moreover, Alexandroff LM-fuzzy co-topologies $\mathcal{T}_X=\{\mathcal{F}_u^{i_\mathcal{P}}:L^X\to M\mid x\in X\}$ on X and $\mathcal{T}_Y=\{\mathcal{F}_u^{k_\mathcal{P}}:L^Y\to M\mid y\in Y\}$ on Y be induced by the $F^\uparrow$-transform. Then $f:(X,\mathcal{T}_X)\to(Y,\mathcal{T}_Y)$ is a continuous map.*

Proof. Since $(f,g):(X,\mathcal{P})\to(Y,\mathcal{Q})$ be a morphism between space with M-valued partitions. Then from Definition 7, for all $\xi\in\Lambda$, $A_\xi(x)\le B_{g(\xi)}(f(x))$. Now we have to show that for all $\lambda\in L^Y$, $\mathcal{T}_X(f^\leftarrow(\lambda))\ge\mathcal{T}_Y(\lambda)$, where

$$\begin{aligned}\mathcal{T}_X(f^\leftarrow(\lambda))&=\mathcal{F}_u^{i_\mathcal{P}}(f^\leftarrow(\lambda))=F^\uparrow_{i_\mathcal{P}(x)}(f^\leftarrow(\lambda))\hookrightarrow f^\leftarrow(\lambda)\\&=\phi\left(\wedge_{x\in X}(F^\uparrow_{i_\mathcal{P}(x)}(f^\leftarrow(\lambda)(x))\to f^\leftarrow(\lambda)(x))\right)\\&=\phi\left(\wedge_{x\in X}(\wedge_{t\in X}(\psi(A_{i_\mathcal{P}(x)}(t))\otimes\lambda(f(t)))\to\lambda(f(x))\right)\\&=\phi\left(\wedge_{x\in X}\wedge_{t\in X}(\psi(A_{i_\mathcal{P}(x)}(t))\otimes\lambda(f(t)))\to\lambda(f(x))\right)\\&\ge\phi\left(\wedge_{x\in X}\wedge_{t\in X}(\psi(B_{k_\mathcal{P}(x)}(f(t)))\otimes\lambda(f(t)))\to\lambda(f(x))\right)\\&\ge\phi\left(\wedge_{z\in X}\wedge_{y\in X}(\psi(B_{k_\mathcal{P}(x)}(y))\otimes\lambda(y))\to\lambda(z)\right)\\&\ge\phi\left(\wedge_{z\in X}\wedge_{y\in X}(\psi(B_{k_\mathcal{P}(x)}(y))\otimes\lambda(y))\to(\lambda(z)\right)\\&=F^\uparrow_{k_\mathcal{P}(z)}(\lambda)\hookrightarrow\lambda=\mathcal{F}_u^{k_\mathcal{P}}(\lambda)=\mathcal{T}_Y(\lambda).\end{aligned}$$

Thus $f : (X, \mathcal{T}_X) \to (Y, \mathcal{T}_Y)$ is a continuous map.

From Theorems 3, 4, we obtain a functor $\mathbb{H}$ as follows:

$$\mathbb{H} : \begin{cases} \mathbf{SMFP} \longmapsto \mathbf{LM{-}CToP} \\ (X, \mathcal{P}) \longmapsto (X, \mathcal{T}_X) \\ (f, g) \longmapsto f, \end{cases}$$

where $\mathcal{P} = \{A_\xi : \xi \in \Lambda\}$ is an M-valued partition of X, $\mathcal{T}_X = \{\mathcal{F}_u^{i_\mathcal{P}} : L^X \to M \mid x \in X\}$ is the induced strong Alexandroff LM-fuzzy co-topology on X by the measure of direct $F^\uparrow$-transform operator $\mathcal{F}_u$, and $(f, g) : (X, \mathcal{P}) \to (Y, \mathcal{Q})$ is a morphism between the spaces with M-valued partition.

Below, we show that the M-valued measure of F-transform operators $\mathcal{F}_u$, $\mathcal{F}_l$ together with its induced LM-fuzzy topologies $\mathfrak{T}, \mathcal{T}$ can be interpreted as LM-fuzzy ditopology.

Corollary 1. *Let $(X, \mathcal{P})$ be space with M-valued partition, $\mathfrak{T}_\mathcal{P} = \{\mathcal{F}_l(\cdot)(x) : L^X \to M \mid x \in X\}$ and $\mathcal{T}_\mathcal{P} = \{\mathcal{F}_u(\cdot)(x) : L^X \to M \mid x \in X\}$ be the induced strong Alexandroff LM-fuzzy topology and co-topology. Then the triple $(X, \mathfrak{T}_{\mathcal{P}_X}, \mathcal{T}_{\mathcal{P}_X})$ is a strong Alexandroff LM-fuzzy ditopology.*

The following proposition is an easy consequence of Theorems 2 and 4.

Proposition 6. *Let $(f, g) : (X, \mathcal{P}) \to (Y, \mathcal{Q})$ be a morphism between the spaces with M-valued partition. Let, moreover, assume that the assumptions of Theorems 2 and 4 holds. Then $f : (X, \mathfrak{T}_{\mathcal{P}_X}, \mathcal{T}_{\mathcal{P}_X}) \to (Y, \mathfrak{T}_{\mathcal{P}_Y}, \mathcal{T}_{\mathcal{P}_Y})$ is a continuous mapping between two strong Alexandroff LM-fuzzy ditopological spaces.*

From the above two results we obtain a functor $\mathbb{L}$ as follows:

$$\mathbb{L} : \begin{cases} \mathbf{SMFP} \longmapsto \mathbf{LM{-}DIToP} \\ (X, \mathcal{P}) \longmapsto (X, \mathfrak{T}_{\mathcal{P}_X}, \mathcal{T}_{\mathcal{P}_X}) \\ (f, g) \longmapsto f. \end{cases}$$

5 Conclusion

The paper is an effort to show that by defining the M-valued measure of F-transform operators, we can associate Alexandroff LM-fuzzy topological and co-topological spaces. Specifically, we have introduced and studied the M-valued measure of inclusion defined between the lattice-valued $F^\uparrow$ and $F^\downarrow$-transform and L-fuzzy set. The basic properties of defined operators are studied. The M-valued measure of F-transforms defined here are essentially in some sense determine the amount of preciseness of given L-fuzzy set. We have discussed such connections through the categorical point of view.

We believe that the M-valued measure of F-transform operators propose an abstract approach to the notion of "precision of approximation" and naturally arise in connection with applications to image and data analysis etc. Which will be one of our directions for the future work. In another direction, we propose to investigate the categorical behavior of the operators $\mathcal{F}_u, \mathcal{F}_l$ in a more deeper way.

Acknowledgment. The work of first author is supported by University of Ostrava grant IRP201824 "Complex topological structures" and the work of Irina Perfilieva is partially supported by the Grant Agency of the Czech Republic (project No. 18-06915S).

References

1. Blount, K., Tsinakis, C.: The structure of residuated lattices. Int. J. Algebr. Comput. **13**(04), 437–461 (2003)
2. Bustince, H.: Indicator of inclusion grade for interval-valued fuzzy sets. Application to approximate reasoning based on interval-valued fuzzy sets. Int. J. Approximate Reasoning **23**(3), 137–209 (2000)
3. Di Martino, F., Loia, V., Perfilieva, I., Sessa, S.: An image coding/decoding method based on direct and inverse fuzzy transforms. Int. J. Approximate Reasoning **48**(1), 110–131 (2008)
4. Goguen, J.A.: L-fuzzy sets. J. Math. Anal. Appl. **18**, 145–174 (1967)
5. Hájek, P.: Metamathematics of Fuzzy Logic, vol. 4. Springer, Dordrecht (2013). https://doi.org/10.1007/978-94-011-5300-3
6. Han, S.-E., Šostak, A.: M-valued measure of roughness for approximation of L-fuzzy sets and its topological interpretation. In: Merelo, J.J., Rosa, A., Cadenas, J.M., Dourado, A., Madani, K., Filipe, J. (eds.) Computational Intelligence. SCI, vol. 620, pp. 251–266. Springer, Cham (2016). https://doi.org/10.1007/978-3-319-26393-9_15
7. Han, S.E., Šostak, A.: On the measure of M-rough approximation of L-fuzzy sets. Soft Comput. **22**(12), 3843–3855 (2018). https://doi.org/10.1007/s00500-017-2841-y
8. Höhle, U.: M-valued sets and sheaves over integral commutative CL-monoids. In: Rodabaugh, S.E., Klement, E.P., Höhle, U. (eds.) Applications of Category Theory to Fuzzy Subsets. Theory and Decision Library (Series B: Mathematical and Statistical Methods), vol. 14, pp. 33–72. Springer, Dordrecht (1992). https://doi.org/10.1007/978-94-011-2616-8_3
9. Höhle, U.: Commutative, residuated L–monoids. In: Höhle, U., Klement, E.P. (eds.) Non-Classical Logics and their Applications to Fuzzy Subsets. Theory and Decision Library (Series B: Mathematical and Statistical Methods), vol. 32 pp. 53–106. Springer, Dordrecht (1995). https://doi.org/10.1007/978-94-011-0215-5_5
10. Järvinen, J.: On the structure of rough approximations. Fundam. Inform. **53**(2), 135–153 (2002)
11. Järvinen, J., Kortelainen, J.: A unifying study between modal-like operators, topologies and fuzzy sets. Fuzzy Sets Syst. **158**(11), 1217–1225 (2007)
12. Kehagias, A., Konstantinidou, M.: L-fuzzy valued inclusion measure, l-fuzzy similarity and l-fuzzy distance. Fuzzy Sets Syst. **136**(3), 313–332 (2003)

13. Khastan, A., Perfilieva, I., Alijani, Z.: A new fuzzy approximation method to cauchy problems by fuzzy transform. Fuzzy Sets Syst. **288**, 75–95 (2016)
14. Močkoř, J.: Axiomatic of lattice-valued F-transform. Fuzzy Sets Syst. **342**, 53–66 (2018)
15. Močkoř, J., Holčapek, M.: Fuzzy objects in spaces with fuzzy partitions. Soft Comput. **21**, 7269–7284 (2017). https://doi.org/10.1007/s00500-016-2431-4
16. Močkoř, J., Hurtik, P.: Lattice-valued F-transforms and similarity relations. Fuzzy Sets Syst. **342**, 67–89 (2018)
17. Perfilieva, I.: Fuzzy transforms: theory and applications. Fuzzy Sets Syst. **157**(8), 993–1023 (2006)
18. Perfilieva, I., Singh, A.P., Tiwari, S.P.: On F-transforms, L-fuzzy partitions and L-fuzzy pretopological spaces. In: 2017 IEEE Symposium Series on Computational Intelligence (SSCI), pp. 1–8. IEEE (2017)
19. Perfilieva, I., Singh, A.P., Tiwari, S.P.: On the relationship among F-transform, fuzzy rough set and fuzzy topology. Soft Comput. **21**, 3513–3523 (2017). https://doi.org/10.1007/s00500-017-2559-x
20. Perfilieva, I., Tiwari, S.P., Singh, A.P.: Lattice-valued F-transforms as interior operators of L-fuzzy pretopological spaces. In: Medina, J., et al. (eds.) IPMU 2018. CCIS, vol. 854, pp. 163–174. Springer, Cham (2018). https://doi.org/10.1007/978-3-319-91476-3_14
21. Qin, K., Pei, Z.: On the topological properties of fuzzy rough sets. Fuzzy Sets Syst. **151**, 601–613 (2005)
22. Singh, A.P., Tiwari, S.P., Perfilieva, I.: F-transforms, L-fuzzy partitions and L-fuzzy pretopological spaces: an operator oriented view. Fuzzy Sets Syst. Submitted
23. Sostak, A., Brown, L.M.: Categories of fuzzy topology in the context of graded ditopologies on textures. Iran. J. Fuzzy Syst. **11**(6), 1–20 (2014)
24. Stepnicka, M., Valasek, R.: Numerical solution of partial differential equations with help of fuzzy transform. In: The 14th IEEE International Conference on Fuzzy Systems. FUZZ 2005, pp. 1104–1109. IEEE (2005)
25. Tiwari, S.P., Perfilieva, I., Singh, A.P.: Generalized residuated lattices based F-transform. Iran. J. Fuzzy Syst. **15**(2), 165–182 (2018)
26. Tiwari, S.P., Srivastava, A.K.: Fuzzy rough sets, fuzzy preorders and fuzzy topologies. Fuzzy Sets Syst. **210**, 63–68 (2013)
27. Zadeh, L.A.: Similarity relations and fuzzy orderings. Inf. Sci. **3**, 177–200 (1971)
28. Zeng, W., Li, H.: Inclusion measures, similarity measures, and the fuzziness of fuzzy sets and their relations. Int. J. Intell. Syst. **21**(6), 639–653 (2006)

Gold Price: Trend-Cycle Analysis Using Fuzzy Techniques

Linh Nguyen(✉), Vilém Novák, and Michal Holčapek

Institute for Research and Applications of Fuzzy Modelling, NSC IT4Innovations, University of Ostrava, 30. dubna 22, 701 03 Ostrava 1, Czech Republic
{Linh.Nguyen,Vilem.Novak,Michal.Holcapek}@osu.cz

Abstract. In this paper, we apply special fuzzy techniques to analyze the gold price historical data. The main tools are the higher degree fuzzy transform and specific methods of fuzzy natural logic. First, we show how to apply the former for the estimation of the trend-cycle. Then, we provide methodologies for identifying monotonous periods in the trend-cycle and describe them by sentences in natural language.

Keywords: Fuzzy transform · Fuzzy modeling · Financial time series · Data mining

1 Introduction

The crises of the global financial environment in recent years have put many financial markets into complicated situations, e.g., the drop in oil prices, the unusual fluctuation of gold prices, the unprecedented boom of digital currencies. These make it difficult for people who would like to optimize their profit when trading in such financial markets. Facing this situation, any knowledge of how a market behaves is significantly helpful. In this paper, we devote several techniques to mine information behind a market based on its historical data. Our focus is on the gold market.

First, we apply the fuzzy transform (F-transform) technique for estimation of the trend-cycle of the gold price historical data. The F-transform is a fuzzy approximation technique proposed by I. Perfilieva in [21] and later elaborated by several authors in [5,10,13,22] and elsewhere. It has been successfully applied in many branches of applied sciences [6,8,12,14,25,26], and especially, in time series analysis [3,4,17,18,27] that inspires our investigation in this paper. From those contributions, it is known that the trend-cycle of a time series can be estimated by the F-transform technique, successfully. However, the quality of this estimation strongly depends on practical experience in setting parameters of the latter. Therefore, in addition to the application of the fuzzy transform to the trend-cycle estimation, we introduce a technique for choosing parameters to make it possible to achieve an efficient estimation without requiring much practical experience.

M.-J. Lesot et al. (Eds.): IPMU 2020, CCIS 1239, pp. 254–266, 2020.
https://doi.org/10.1007/978-3-030-50153-2_19

To help investors have a better understanding of the behavior of a market (particularly the gold market), we introduce the concept of bull and bear periods on the trend-cycle to characterize the monotonous stages on it. These concepts are inspired through the notions of bull and bear markets in finance [2,19]. For practical purposes, an algorithm is provided for identifying bull and bear periods on the estimated trend-cycle.

Finally, we employ one of the important tasks in mining information from time series that is to extract linguistic characterization of the trend-cycle. Let us note that mining linguistic information or time series summarization has been studied for quite a while, recently. There are several approaches that have been proposed to this issue such as [1,9,15,20,24,28] and elsewhere. Being motivated by fuzzy natural logic techniques used for describing the behavior of the trend-cycle [15], we develop a methodology to extract linguistic characteristics of bull and bear periods on the trend-cycle of the gold price market and represent them by sentences.

The paper is structured as follows. The next section provides a brief introduction to the (higher degree) F-transform and specific tools in fuzzy natural logic. The main contribution of this paper is described in Sect. 3, where we analyze the trend-cycle of the gold price data. This consists of the estimating of the trend-cycle, identifying of monotonous periods on it, and describing its course in natural language (by sentences). The last section is the conclusion.

2 Preliminaries

Let $\mathbb{N}$, $\mathbb{Z}$ and $\mathbb{R}$ denote the set of natural numbers, integers and real numbers, respectively.

2.1 The Higher Degree Fuzzy Transform

The central notion in the theory of fuzzy transform is the fuzzy partition. Standardly, a fuzzy partition is a set of fuzzy sets on $\mathbb{R}$ that satisfy the Ruspini's condition.[1] Together with the development of the fuzzy transform, this concept has several modified versions such as: the generalized (uniform) fuzzy partition [7], the adjoint fuzzy partition [23]. For the practical purposes of this paper, we restrict our consideration on a particularly simple one, called the triangular generalized uniform fuzzy partition.

Definition 1. *Let $t_0 \in \mathbb{R}$, and two positive constants h and r be such that $h/r \in \mathbb{Z}$. Let $\mathcal{A} = \{A_k \mid k \in \mathbb{Z}\}$ be a set of fuzzy sets on $\mathbb{R}$, defined by*

$$A_k(t) = \frac{r}{h} \cdot \max\left\{1 - \left|\frac{t - t_0 - kr}{h}\right|, 0\right\}, \quad k \in \mathbb{Z}.$$

[1] The sum of all membership functions corresponding to fuzzy sets in a fuzzy partition over each point in $\mathbb{R}$ is one.

Then, $\mathcal{A}$ is called a triangular generalized uniform fuzzy partition of the real line $\mathbb{R}$ determined by the triplet (t_0, h, r). Each fuzzy set in this fuzzy partition is called a basic function.

Note that basic functions determined in Definition 1 need not to be normal, because there can be more than one basic function covering their peak, in general (see [7]).

In the sequel, since t_0 does not affect the theoretical results concerning the fuzzy transform, for the sake of simplicity, we restrict our investigation to the fuzzy partitions with $t_0 = 0$. Moreover, we fix $r = h/2$. Therefore, we omit the reference both to t_0 as well as r in the triplet (h, r, t_0) and deal with h only. We call h the *bandwidth* of the fuzzy partition.

The higher degree fuzzy transform (F^m-transform, $m \in \mathbb{N}$) is a fuzzy approximation technique that can provide both local and global approximation of functions. The first phase, called *direct F^m-transform*, transforms a given function to a family of polynomials of degree up to m. These polynomials are orthogonal projections of the given function onto polynomial approximation spaces concerning specific inner products defined with respect to basic functions of a fuzzy partition. Below, we briefly introduce how the direct F^m-transform of a function is computed.

Let f be a locally square Lebesgue integrable function on $\mathbb{R}$, and $\mathcal{A}$ be a fuzzy partition of the latter in the sense of Definition 1. Let $K : \mathbb{R} \to [0, 1]$ be defined by[2]

$$K(t) = \frac{1}{2} \cdot \max\{1 - |t|, 0\}.$$

The *direct F^m-transform of f* with respect to $\mathcal{A}$ is the family

$$\mathrm{F}^m_{\mathcal{A}}[f] = \{F^m_k[f] \mid k \in \mathbb{Z}\},$$

where, for any $k \in \mathbb{Z}$,

$$F^m_k[f](t) = C_{k,0} + C_{k,1}\left(\frac{t - t_k}{h}\right) + \ldots + C_{k,m}\left(\frac{t - t_k}{h}\right)^m, \quad t \in [t_k - h, t_k + h],$$

with $t_k = \frac{kh}{2}$,

$$(C_{k,0}, C_{k,1}, \ldots, C_{k,m})^T = (\mathcal{Z}_m)^{-1} \cdot \mathcal{Y}_{m,k}, \tag{1}$$

where $\mathcal{Z}_m = (Z_{ij})$ is an $(m+1) \times (m+1)$ invertible matrix defined by

$$Z_{ij} = \int_{-1}^{1} t^{i+j-2} K(t)\, dt, \quad i, j = 1, \ldots, m+1,$$

and $\mathcal{Y}_{m,k} = (Y_{k,1}, \ldots, Y_{k,m+1})^T$ is defined by

$$Y_{k,\ell} = \int_{-1}^{1} f(ht + t_k) \cdot t^{\ell-1} K(t)\, dt, \quad \ell = 1, \ldots, m+1. \tag{2}$$

[2] This function is known as a generating function of the fuzzy partition $\mathcal{A}$ (see in [7]).

The polynomial $F_k^m[f]$ is called the *k-th component of the direct F^m-transform* of f. It provides an approximation of the latter on the region covered by the basic function A_k.

The second phase called the *inverse F^m-transform* of f transforms the vector of components to a function defined as the linear-like combination of the direct F-transform components and the corresponding basic functions, i.e.,

$$\hat{f}_{\mathcal{A}}^m(t) = \sum_{k\in\mathbb{Z}} F_k^m[f](t) \cdot A_k(t), \quad t \in \mathbb{R}.$$

Being motivated by fuzzy natural logic techniques used for characterizing behavior of the trend-cycle in time series, we develop a methodology to extract linguistic characteristics of bull and bear periods on the trend-cycle of the gold price market and represent them by sentences. This function provides an approximation to f on its domain $\mathbb{R}$ (global approximation). The quality of this approximation depends on the settings of fuzzy partition $\mathcal{A}$, particularly, the bandwidth h. By contrast, under reasonable setting of the bandwidth, the inverse F^m-transform can suppress high frequencies in the function f. Namely, $\hat{f}_{\mathcal{A}}^m$ is a smoothed function of the latter. In the sequel, when needeing to emphasize the influence of the bandwidth to the inverse function, we use the notation $\hat{f}_h^m$.

2.2 Evaluative Linguistic Expressions: A Formal Theory in Fuzzy Natural Logic

The fuzzy natural logic (FNL) was established as the formal logic aiming at modeling of natural human reasoning which proceeds in natural language. It is an extension of *mathematical fuzzy logic*, and its paradigm extends the classical concepts of *natural logic* suggested by Lakoff in [11]. FNL is a class of several formal theories such as: theory of evaluative linguistic expressions, theory of fuzzy IF-THEN rules, etc. For the purposes of this paper, we only focus on the theory of evaluative linguistic expressions.

Evaluative linguistic expressions are special expressions of natural language that are used to specify the course of development of some processes, to evaluate a phenomenon or to make a decision. These expressions may have a complicated structure, but in this paper, we simply consider evaluative linguistic expressions of the following form:

$$\langle\text{linguistic hedge}\rangle\langle\text{atomic evaluative expression}\rangle, \tag{3}$$

where the atomic evaluative expression comprises any of the *canonical adjectives*: small, medium, big, and the *linguistic hedge* is a specific adverb, for example, extremely, significantly, very, rather, roughly, very roughly. The linguistic hedge makes the meaning of the atomic expression more or less precise. An evaluative expression of the form (3) will be denoted by a script letter, for example, $\mathscr{A}$ or $\mathscr{B}$.

When using an evaluative expression to evaluate values of a variable X, the resulting expression is of the form

$$X \text{ is } \mathscr{A}, \tag{4}$$

is called the *evaluative (linguistic) predication*. To make such a predication meaningful, one must specify the *context* (the state of the world) in which variable X is considered. The context is characterized by a triplet $w = \langle v_L, v_M, v_R \rangle$, where $v_L, v_M, v_R \in \mathbb{R}$ and $v_L < v_M < v_R$. These numbers characterize the minimal, typically middle, and maximal values, respectively, of the evaluated characteristics (e.g., "height", "distance") in the specified context of use. Only when a context is specified to X, the meaning of the predication in (4) is well-defined and interpreted by a fuzzy set on $\mathbb{R}$. Moreover, if a context w is known then one can determine an evaluative linguistic expression $\mathscr{A}$ characterizing a given value x_0 of X by using a function of local perception

$$LPerc(x_0, w) = \mathscr{A}.$$

For the details of this function as well as tools in FNL, we refer to the book [16].

3 Gold Price: Trend-Cycle Model and Analysis

Our investigation employs the Gold Future Historical Data published in the site investing.com. This is the daily data of the price, the open, high, and low prices in the US Dollar of one Troy Ounce of gold. Our methodology is focused on the price only. A similar analysis can be applied to the rest of the data. This section aims at developing techniques based on the fuzzy transform and tools developed in the theory of evaluative linguistic expressions for estimation of the trend-cycle of the data, classifying periods in the trend-cycle into specific stages concerning its monotonousness, and finally, for extraction of the linguistic characterization of the course of the trend-cycle. Established methods are applied to the data from October 1, 2018, to December 3, 2019, of the length 312, as presented in Fig. 1.

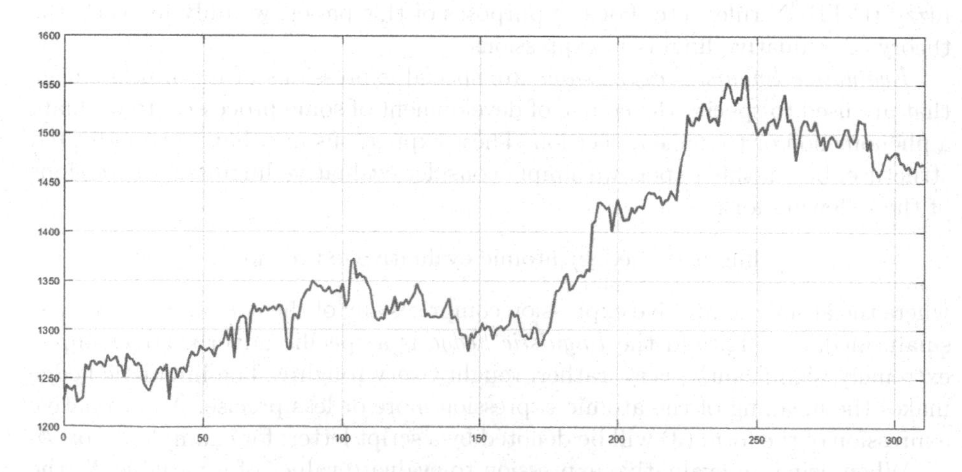

Fig. 1. The daily gold price from October 1, 2018, to December 3, 2019.

3.1 Time Series Model

Let us consider the gold price values as a time series $\{X(t), t \in \mathbb{T}\}$, where $\mathbb{T} = \{0, 1, \ldots, T\}$. Standardly, a time series can be additively (multiplicatively) decomposed into a trend-cycle, a seasonal component and an irregular fluctuation. However, we know that the influence of the seasonality to the gold market is weak, and even, unclear. Therefore, let us assume that the time series X can be decomposed as follows:

$$X(t) = TC(t) + R(t), \quad t \in \mathbb{T}, \tag{5}$$

where TC is the trend-cycle component and $R(t)$ is a realization of a random process characterized by quick decay of its autocorrelation function ρ, i.e.,

$$\int_0^{\ell} |\rho(\tau)| d\tau = o(\ell), \quad \text{for } \ell \to \infty. \tag{6}$$

3.2 Trend-Cycle Estimation

This task has been elaborated in several papers as mentioned in the introduction. Most of these investigations are to show that the (higher degree) F-transform is a good technique for estimating the trend-cycle of a time series. More precisely, the inverse F^m-transform of X provides an estimation of the trend-cycle TC,

$$TC(t) \approx \hat{X}_h^m(t), \quad t \in \mathbb{T}. \tag{7}$$

However, one can see that the quality of this estimation depends on the degree m of the fuzzy transform and the construction of the fuzzy partition (the bandwidth h). In what follows, we describe a methodology for setting these parameters such that the trend-cycle is well estimated by (7) without requiring much practical experience.

- *Choosing of the degree m:* The more the time series changes its course (or has higher volatility), the higher the degree m should be chosen. A rule of thumb says that if the observed trend-cycle is a nearly linear function then we should choose $m = 0$ or $m = 1$; otherwise, choose $m = 2$ or $m = 3$.
- *Choosing of the bandwidth h:* Let $\tilde{R}_h(t) = X(t) - \hat{X}_h^m(t)$ and $\hat{\rho}_h$ be its sample autocorrelation function. The chosen bandwidth h_0 is the value satisfying that
$$\sum_{k=0}^{N_{max}} |\hat{\rho}_{h_0}(k)| = \min \left\{ \sum_{k=0}^{N_{max}} |\hat{\rho}_h(k)| \mid h = 1, 2, \ldots, H_{max} \right\},$$
where N_{max} and H_{max} are two positive integers chosen by the users.

Applying these rules to the gold price data displayed in Fig. 1, we find $m = 2$ with the bandwidth $h_0 = 6$. Namely, the trend-cycle of the considered data is estimated by the inverse F^2-transform with respect to the fuzzy partition determined as in Definition 1 with $h = 6$. The estimated trend-cycle is depicted in Fig. 2.

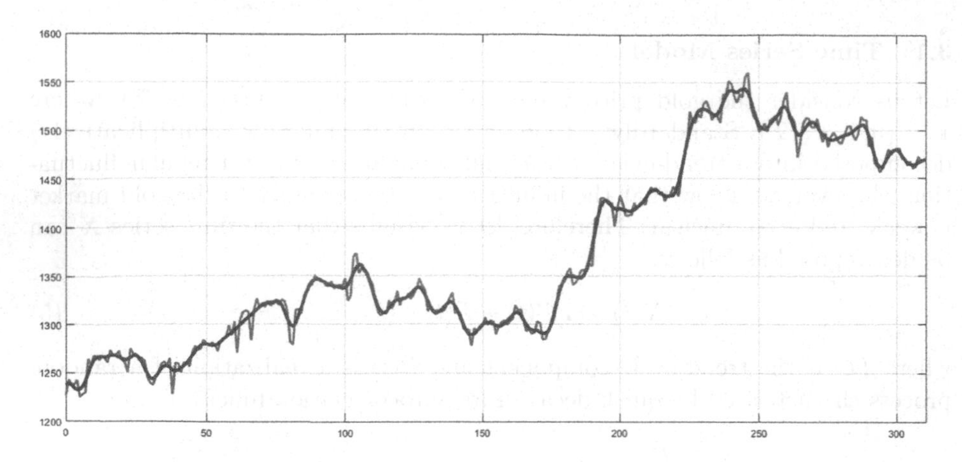

Fig. 2. The trend-cycle estimated by the fuzzy transform (dark/blue line). (Color figure online)

3.3 Identification of Monotonous Periods on the Trend-Cycle

This section provides a method for identification of monotonous periods on the trend-cycle TC of time series X. These periods are determined based on duration and market move constraints. Inspired by the notion of bull and bear markets in finance [2,19], we classify monotonous periods into three categories: bull, bear and neutral ones, corresponding to the increasing, decreasing and stagnating of the trend-cycle. To characterize the monotonousness of the trend-cycle, we use the forward difference defined as follows.

Definition 2. *Let $f(t)$, $t = 1, 2, \ldots, N$ be a discrete function. The* forward difference $\Delta[f]$ of f *is a discrete function determined by*

$$\Delta[f](t) = f(t+1) - f(t), \quad t = 1, 2, \ldots, N-1.$$

Below, we define the bull, bear and neutral periods in the trend-cycle.

Definition 3. *Let $\mathbb{S} = \{s, s+1, \ldots, s+p\}$ be a time period in $\mathbb{T}$. Let β_{bull}, β_{bear} and L_{min} be three positive real numbers. The period $S = \{TC(t) \mid t \in \mathbb{S}\}$ in TC is said to be a* bull (or bear) period *with respect to β_{bull} (or β_{bear}) and L_{min} if the following statements hold true:*

(i) $\Delta[TC](t) > 0$ *(or* $\Delta[TC](t) < 0$*), for any* $t \in \mathbb{S}$,
(ii) *If* $\mathbb{S}' \supseteq \mathbb{S}$ *is a time period satisfying that* $\Delta[TC](t) > 0$ *(or* $\Delta[TC](t) < 0$*), for any* $t \in \mathbb{S}'$*, then* $\mathbb{S}' = \mathbb{S}$,
(iii) $\frac{TC(s+p)-TC(s)}{p} \geq \beta_{bull}$ *(or* $\frac{TC(s+p)-TC(s)}{p} \leq -\beta_{bear}$*) and* $p+1 \geq L_{min}$.

S is said to be a neutral period *if it is neither bull nor bear period.*

In Definition 3, L_{min} is the smallest length of the bull and bear periods, while β_{bull} and β_{bear} are thresholds characterize the steepness of the trend-cycle. From (iii), one can see that β_{bull} and β_{bear} are chosen to categorize long stages with little change of the trend-cycle to the neutral periods. In practice, L_{min} is the shortest time period, chosen by traders based on the time frame of their trading strategy. Knowing bull and bear periods with lengths greater than L_{min} is significantly insightful. For example, if one would like to trade on the gold market weekly based on the given data, he would choose $L_{min} = 5$. Moreover, β_{bull} and β_{bear} are chosen based on the market move and the duration constraints. Let us assume that one is interested to discover bull (or bear) periods in the gold market with at least 20$ jump in at most a half of the month (10 days). In this specific case, he should choose $\beta_{bull} = 2$ (or $\beta_{bear} = 2$).[3]

In what follows, we describe an informal algorithm for identifying the bull and bear periods on a trend-cycle.

Algorithm 1. Identification of the bull, bear and neutral period in the trend-cycle TC of X

Input:
trend-cycle $TC(t)$, $t = 0, 1, \ldots, T$;
three thresholds β_{bull}, β_{bear}, and L_{min};
Processing:
compute $\Delta[TC](t)$, $t = 0, 1, \ldots, T-1$;
classify points $TC(t)$, $t = 0, 1, \ldots, T-1$ into distinct segments based on the sign (positive and negative) of $\Delta[TC](t)$;
evaluate each obtained segment with respect to the thresholds β_{bull}, β_{bear}, and L_{min} to determine bull, bear and neutral periods;
Output:
bull, bear and neutral periods in TC and their position

Apply Algorithm 1 to the considered gold price data where the inputs are the estimated trend-cycle obtained in the previous subsection, $L_{min} = 5$ and $\beta_{bull} = \beta_{bear} = 2$. We obtain the result as in Fig. 3.

3.4 Evaluation of the Trend-Cycle in Natural Language

In this subsection, we describe how the course of the trend-cycle can be evaluated in natural language. Namely, we focus on the bull and bear periods of the trend-cycle TC and introduce a method for generating the linguistic characterization of these stages. Each bull and bear period is characterized by the following sentence,

$$\textit{Period} \text{ is } \langle \textbf{price change evaluation} \rangle \text{ in } \langle \textbf{time period evaluation} \rangle \text{ period}, \quad (8)$$

[3] β_{bull} and β_{bear} can be different. The setting is up to users.

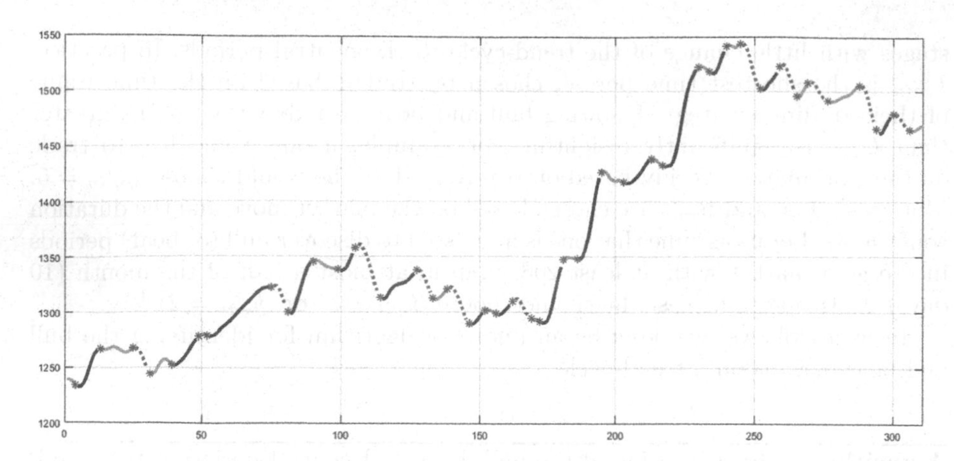

Fig. 3. The bull (dark/blue lines), bear (dotted/red lines) and neutral (grey/green lines) periods on the estimated trend-cycle. (Color figure online)

where

$$\begin{aligned}\langle \textbf{price change evaluation} \rangle &:= \langle \text{evaluative expression 1} \rangle \langle \text{sign} \rangle,\\ \langle \textbf{time period evaluation} \rangle &:= \langle \text{evaluative expression 2} \rangle,\end{aligned}$$

with

$$\begin{aligned}\langle \text{evaluative expression 1} \rangle &:= \langle \text{linguistic hedge} \rangle \langle \text{weak} \mid \text{moderate} \mid \text{strong} \rangle,\\ \langle \text{sign} \rangle &:= \langle \text{increasing} \mid \text{decreasing} \rangle,\\ \langle \text{evaluative expression 2} \rangle &:= \langle \text{linguistic hedge} \rangle \langle \text{short} \mid \text{medium} \mid \text{long} \rangle.\end{aligned}$$

Neutral stages are evaluated as "stagnating".

Let $S = \{TC(t) \mid t = s, s+1, \ldots, s+p\}$ be a bull (bear) period on the trend-cycle TC. From Definition 3, this stage is characterized by two essential components: the change of price and the length of the stage. Let δ_S and l_S be the change of price and the length of S, respectively. In this case, $\delta_S = TC(s+p) - TC(s)$ and $l_S = p+1$. Then,

$$\begin{aligned}\langle \text{evaluative expression 1} \rangle &= LPerc(|\delta_S|, w_1),\\ \langle \text{sign} \rangle &= \begin{cases} \text{increasing}, & \delta_S > 0 \\ \text{decreasing}, & \delta_S < 0 \end{cases},\\ \langle \text{evaluative expression 2} \rangle &= LPerc(l_S, w_2),\end{aligned}$$

where w_1 and w_2 are the contexts corresponding to the change of price and the length of periods, respectively. For details how to set the context, we refer the readers to [16].

In what follows, we apply the proposed technique for extracting the linguistic characteristics of the trend-cycle of the considered gold price data depicted

Table 1. Linguistic characterization of the trend-cycle.

Time period	Identified stage	Linguistic characterization
[0, 3]	neutral	stagnating
[3, 12]	bull	ML moderate increasing in VR short period
[12, 24]	neutral	stagnating
[24, 30]	bear	Ro weak decreasing in Ra short period
[30, 38]	neutral	stagnating
[38, 48]	bull	QR weak increasing in ML medium period
[48, 51]	neutral	stagnating
[51, 74]	bull	Ty moderate increasing in Ra long period
[74, 80]	bear	Ro weak decreasing in Ra short period
[80, 89]	bull	Ra moderate increasing in VR short period
[89, 98]	neutral	stagnating
[98, 104]	bull	ML weak increasing in Ra short period
[104, 113]	bear	Ra moderate decreasing in VR short period
[113, 126]	bull	QR weak increasing in Ty medium period
[126, 132]	bear	QR weak decreasing in Ra short period
[132, 137]	neutral	stagnating
[137, 145]	bear	VR weak decreasing in QR short period
[145, 151]	bull	Si weak increasing in Ra short period
[151, 155]	neutral	stagnating
[155, 161]	bull	Ex weak increasing in Ra short period
[161, 168]	bear	Ra weak decreasing in ML short period
[168, 171]	neutral	stagnating
[171, 180]	bull	Ra moderate increasing in VR short period
[180, 185]	neutral	stagnating
[185, 194]	bull	ML strong increasing in VR short period
[194, 202]	neutral	stagnating
[202, 212]	bull	Ro weak increasing in ML medium period
[212, 216]	neutral	stagnating
[216, 229]	bull	Si strong increasing in Ty medium period
[229, 233]	neutral	stagnating
[233, 239]	bull	QR weak increasing in Ra short period
[239, 244]	neutral	stagnating
[244, 251]	bear	Ra moderate decreasing in ML short period
[251, 258]	bull	ML weak increasing in ML short period
[258, 264]	bear	VR weak decreasing in Ra short period
[264, 268]	neutral	stagnating
[268, 275]	bear	Ra weak decreasing in ML short period
[275, 287]	neutral	stagnating
[287, 293]	bear	Ra moderate decreasing in Ra short period
[293, 299]	bull	Ve weak increasing in Ra short period
[299, 305]	bear	Ra weak decreasing in Ra short period
[305, 311]	neutral	stagnating

in Fig. 3. Since the exact trend-cycle is unknown, we mine the information on its estimation. Namely, we form the linguistic characterization of the bull, bear, and neutral periods (cf. Fig. 3). To do this, we set $w_1 = \langle 9, 43, 92\rangle$ and $w_2 = \langle 5, 14, 26\rangle$. The obtained results are presented in Table 1, where linguistic hedges are abbreviated as follows: Ex (Extremely), Si (Significantly), Ra (Rather), Ve (Very), ML (More or Less), Ro (Roughly), QR (Quite Roughly), Ty (Typically) and VR (Very Roughly). From this table, one can extract sentences characterizing the behavior of the bull and bear stages on the trend-cycle of the given data, for example, "the `bull` period on [51, 74] is `Ty moderate increasing` in `Ra long` period", or "the `bear` period on [137, 145] is `VR weak decreasing` in `QR short` period".

4 Conclusions

We applied special fuzzy techniques, consisting of the F-transform and the FNL techniques, for analyzing the gold price historical data. We focused on three critical tasks in mining information of financial time series, including in estimating the trend-cycle, classifying periods in the trend-cycle into bull, bear and neutral stages, and extracting linguistic characterization of the course of the trend-cycle. In this paper, we restricted our consideration on the gold market. However, since the suggested methodologies are described in general schemes, this makes it possible to apply them to other financial data. This is one of the topics for our future researches.

Acknowledgments. This work was supported by the project GA ČR No. 18-13951S. Additional support was provided also by the project (No. CZ.02.1.01/0.0/0.0/17_049/0008414.) "Centre for the development of Artificial Inteligence Methods for the Automotive Industry of the region".

References

1. Chiang, D.A., Chow, L.R., Wang, Y.F.: Mining time series data by a fuzzy linguistic summary system. Fuzzy Sets Syst. **112**(3), 419–432 (2000)
2. Hoepfner, M.: Retrospective identification of bull and bear markets: a new, but simple algorithm (2018)
3. Holčapek, M., Nguyen, L.: Suppression of high frequencies in time series using fuzzy transform of higher degree. In: Carvalho, J.P., Lesot, M.-J., Kaymak, U., Vieira, S., Bouchon-Meunier, B., Yager, R.R. (eds.) IPMU 2016. CCIS, vol. 611, pp. 705–716. Springer, Cham (2016). https://doi.org/10.1007/978-3-319-40581-0_57
4. Holčapek, M., Nguyen, L.: Trend-cycle estimation using fuzzy transform of higher degree. Iran. J. Fuzzy Syst. **15**, 23–54 (2018)
5. Holčapek, M., Nguyen, L., Tichý, T.: Polynomial alias higher degree fuzzy transform of complex-valued functions. Fuzzy Sets Syst. **342**, 1–31 (2018)
6. Holčapek, M., Tichý, T.: A smoothing filter based on fuzzy transform. Fuzzy Sets Syst. **180**, 69–97 (2011)

7. Holčapek, M., Perfilieva, I., Novák, V., Kreinovich, V.: Necessary and sufficient conditions for generalized uniform fuzzy partitions. Fuzzy Sets Syst. **277**, 97–121 (2015)
8. Hurtík, P., Perfilieva, I., Hodáková, P.: Fuzzy transform theory in the view of image registration application. In: Laurent, A., Strauss, O., Bouchon-Meunier, B., Yager, R.R. (eds.) IPMU 2014. Communications in Computer and Information Science, vol. 443, pp. 143–152. Springer, Cham (2014)
9. Kacprzyk, J., Wilbik, A., Zadrożny, S.: Linguistic summarization of time series using a fuzzy quantifier driven aggregation. Fuzzy Sets Syst. **159**(12), 1485–1499 (2008)
10. Kokainis, M., Asmuss, S.: Higher degree F-transforms based on B-splines of two variables. In: Carvalho, J.P., Lesot, M.-J., Kaymak, U., Vieira, S., Bouchon-Meunier, B., Yager, R.R. (eds.) IPMU 2016. CCIS, vol. 610, pp. 648–659. Springer, Cham (2016). https://doi.org/10.1007/978-3-319-40596-4_54
11. Lakoff, G.: Linguistics and natural logic. Synthese **22**, 151–271 (1970)
12. Loia, V., Tomasiello, S., Vaccaro, A.: Fuzzy transform based compression of electric signal waveforms for smart grids. IEEE Trans. Syst. Man Cybern.: Syst. **47**(1), 121–132 (2016)
13. Nguyen, L., Holčapek, M., Novák, V.: Multivariate fuzzy transform of complex-valued functions determined by monomial basis. Soft. Comput. **21**(13), 3641–3658 (2017). https://doi.org/10.1007/s00500-017-2658-8
14. Nguyen, L., Perfilieva, I., Holčapek, M.: Boundary value problem: weak solutions induced by fuzzy partitions. Discrete Continuous Dyn. Syst.-B **25**(2), 715 (2020)
15. Novák, V.: Linguistic characterization of time series. Fuzzy Sets Syst. **285**, 52–72 (2016)
16. Novák, V., Perfilieva, I., Dvořák, A.: Insight into Fuzzy Modeling. Wiley, Hoboken (2016)
17. Novák, V., Perfilieva, I., Holčapek, M., Kreinovich, V.: Filtering out high frequencies in time series using F-transform. Inf. Sci. **274**, 192–209 (2014)
18. Novák, V., Štěpnička, M., Dvořák, A., Perfilieva, I., Pavliska, V., Vavříčková, L.: Analysis of seasonal time series using fuzzy approach. Int. J. Gener. Syst. **39**, 305–328 (2010)
19. Pagan, A.R., Sossounov, K.A.: A simple framework for analysing bull and bear markets. J. Appl. Econ. **18**(1), 23–46 (2003)
20. Palpanas, T., Vlachos, M., Keogh, E., Gunopulos, D.: Streaming time series summarization using user-defined amnesic functions. IEEE Trans. Knowl. Data Eng. **20**(7), 992–1006 (2008)
21. Perfilieva, I.: Fuzzy transforms: theory and applications. Fuzzy Sets Syst. **157**, 993–1023 (2006)
22. Perfilieva, I., Daňková, M., Bede, B.: Towards a higher degree F-transform. Fuzzy Sets Syst. **180**, 3–19 (2011)
23. Perfilieva, I., Holčapek, M., Kreinovich, V.: A new reconstruction from the F-transform components. Fuzzy Sets Syst. **288**, 3–25 (2016)
24. Sripada, S., Reiter, E., Hunter, J., Yu, J.: Summarizing neonatal time series data. In: 10th Conference of the European Chapter of the Association for Computational Linguistics (2003)
25. Tomasiello, S.: An alternative use of fuzzy transform with application to a class of delay differential equations. Int. J. Comput. Math. **94**(9), 1719–1726 (2017)
26. Vlašánek, P., Perfilieva, I.: Patch based inpainting inspired by the F1-transform. Int. J. Hybrid Intell. Syst. **13**(1), 39–48 (2016)

27. Štěpnička, M., Dvořák, A., Pavliska, V.: A linguistic approach to time series modeling with the help of F-transform. Fuzzy Sets Syst. **180**, 164–184 (2011)
28. Yu, J., Reiter, E., Hunter, J., Mellish, C.: Choosing the content of textual summaries of large time-series data sets. Nat. Lang. Eng. **13**(1), 25–49 (2007)

On PSO-Based Approximation of Zadeh's Extension Principle

Jiří Kupka and Nicole Škorupová(✉)

CE IT4I - IRAFM, University of Ostrava, 30. dubna 22, Ostrava, Czech Republic
{Jiri.Kupka,Nicole.Skorupova}@osu.cz

Abstract. Zadeh's extension is a powerful principle in fuzzy set theory which allows to extend a real-valued continuous map to a map having fuzzy sets as its arguments. In our previous work we introduced an algorithm which can compute Zadeh's extension of given continuous piecewise linear functions and then to simulate fuzzy dynamical systems given by them. The purpose of this work is to present results which generalize our previous approach to a more complex class of maps. For that purpose we present an adaptation on optimization algorithm called particle swarm optimization and demonstrate its use for simulation of fuzzy dynamical systems.

Keywords: Zadeh's extension · Particle swarm optimization · Fuzzy dynamical systems

1 Introduction

Zadeh's extension principle plays an important role in the fuzzy set theory. Mathematically it describes a principle due to which each map $f\colon X \to Y$ induces a map $z_f\colon \mathbb{F}(X) \to \mathbb{F}(Y)$ between some spaces of fuzzy sets $\mathbb{F}(X)$ (resp. $\mathbb{F}(Y)$) defined on X (resp. Y).

In general the calculation of Zadeh's extension principle is a difficult task. This is caused mainly by difficult computation of inverses of the map f. Only some cases, e.g. when f satisfies some assumptions like monotonicity, one can find an easier solution. Consequently, the problem of approximation of the image of a fuzzy set $A \in \mathbb{F}(X)$ under Zadeh's extension z_f has been attracted by many mathematicians. For example, in [2] and [3] the authors introduced a method approximating Zadeh's extension $z_f(A)$ which is based on decomposition of a fuzzy set A and multilinearization of a map f. Later in [1] another method using an optimization algorithm applied to α-cuts of a chosen fuzzy set was proposed and tested. Further some specific representations of fuzzy numbers have also been used. For example, a parametric LU-representation of fuzzy numbers

The support of the grant "Support of talented PhD students at the University of Ostrava" from the programme RRC/10/2018 "Support for Science and Research in the Moravian-Silesian Region 2018" is kindly announced.

M.-J. Lesot et al. (Eds.): IPMU 2020, CCIS 1239, pp. 267–280, 2020.
https://doi.org/10.1007/978-3-030-50153-2_20

was proposed and elaborated in [5] and [15]. The authors claim that the LU-fuzzy representation allows a fast and easy simulation of fuzzy dynamical systems and, thus, it avoids the usual massive computational work. However this method is restricted for fuzzy numbers only. Further, the authors of [14] presented another simple parametric representations of fuzzy numbers or intervals, based on the use of piecewise monotone functions of different forms. And finally, another more general procedure allowing to approximate Zadeh's extension of any continuous map using the F-transform technique was introduced in [9].

We contributed to the problem above in [10] where we introduced an algorithm which can compute Zadeh's extension of a given continuous piecewise linear map and, consequently, to simulate a fuzzy dynamical system given by this map. We first focused on continuous one-dimensional piecewise linear interval maps and piecewise linear fuzzy sets. These assumptions allowed us to precisely calculate Zadeh's extension for the class of piecewise linear fuzzy sets, for which we do not necessarily assume even the continuity. This feature should be considered as an advantage of our approach because discontinuities naturally appear in simulations of fuzzy dynamical systems. Still the algorithm proposed in [10] covered a topologically large, i.e. dense, class of interval maps.

The aim of this contribution is to extend the use of our previous algorithm (from [10]) for a more complex class of maps, namely, for the class of continuous interval maps. In order to do this, we intend to linearize a given map f, which is an approximation task leading to an optimization problem (of the determination of appropriate points of the linearization) minimizing the distance between the original function f and its piecewise linear linearization $\tilde{f}$. As there is no feasible analytical solution of the minimization problem, we can approach it from the perspective of the stochastic optimization. For that purpose, we chose the particle swarm optimization algorithm (PSO) that helps us to find appropriate distributions of points in a given space defining piecewise linear approximations as close as possible to the original function f. PSO is one of the most known stochastic algorithms from the group of swarm algorithms. We considered this algorithm due to several reasons. For example, it was the best among algorithms used for a similar task e.g. in [13]. Of course, there are naturally other options to be considered and deeper analysis in this direction is in preparation. Due to page limit we present several preliminary observations only, although we have prepared much more tests.

The structure of this paper is the following. In Sect. 2, some basic terms and definitions used in the rest of this manuscript are introduced. In Sect. 3, we introduce a modification of the original PSO algorithm applied to the problem mentioned in the previous paragraph and then we demonstrate the linearization procedure on a few examples. Further in Sect. 4 we provide a testing of the proposed algorithm, taking into account mainly its accuracy and the choice of parameters. And in the final section (Sect. 5) the proposed algorithm for approximation of Zadeh's extension is shortly introduced and, finally, the whole process is demonstrated on several examples.

2 Preliminaries

In this subsection we shortly introduce some elementary notions. For more detailed explanation we refer mainly to [8,9] and references therein.

A *fuzzy set* A on a given (compact) metric space (X, d_X), where X is a nonempty space (often called a *universe*), is a map $A\colon X \to [0,1]$. The number $A(x)$ is called a *membership degree* of a point $x \in X$ in the fuzzy set A. For a given $\alpha \in (0,1]$ an α*-cut* of A is the set $[A]_\alpha = \{x \in X \mid A(x) \geq \alpha\}$. Let us remark that if a fuzzy set A is upper semi-continuous then every α-cut of A is a closed subset of X. This helps us later to define a metric on the family of upper semi-continuous fuzzy sets on X which will be denoted by $\mathbb{F}(X)$. Note that if X is not compact then an assumption that every $A \in \mathbb{F}(X)$ has a compact support is required.

Before to define fuzzy dynamical systems it is necessary to define a metric on the family of fuzzy sets $\mathbb{F}(X)$ and such metrics are usually based on the well known Hausdorff metric D_X which measures distance between two nonempty closed subsets of X. For instance, one of the most used metrics on $\mathbb{F}(X)$ is a *supremum metric* d_∞ defined as

$$d_\infty(A,B) = \sup_{\alpha\in(0,1]} D_X([A]_\alpha, [B]_\alpha),$$

for $A, B \in \mathbb{F}(X)$. Considering a metric topology on $\mathbb{F}(X)$ induced by some metric, e.g. by d_∞, we can obtain a topological structure on $\mathbb{F}(X)$.

Thus, let X be a (compact) metric space and $f\colon X \to X$ be a continuous map. Then a pair (X, f) is called a *(discrete) dynamical system.* Dynamics of an initial state $x \in X$ is given by a sequence $\{f^n(x)\}_{n\in\mathbb{N}}$ of forward iterates of x, i.e. $x, f(x), f^2(x) = f(f(x)), f^3(x) = f(f(f(x))), \ldots$ The sequence $\{f^n(x)\}_{n\in\mathbb{N}}$ is called a *forward trajectory* of x under the map f. Now properties of given points are given by properties carried by their trajectories. For instance, a *fixed point* of the function f is a point $x_0 \in X$ such that $f(x_0) = x_0$. A *periodic point* is a point $x_0 \in X$ for which there exists $p \in \mathbb{N}$ such that $f^p(x_0) = x_0$. However, usual trajectories are much more complicated as it is demonstrated on Fig. 1.

In this manuscript we deal with a fuzzy dynamical system which admits the standard definition of a discrete dynamical system and, at the same time, forms a natural extension of a given *crisp*, i.e. not necessarily fuzzy, discrete dynamical system on X. Discrete dynamical systems of the form (X, f) are used in many applications as mathematical models of given processes, see e.g. [11]. Fuzzy dynamical systems studied in this paper are defined with the help of Zadeh's extension which was firstly mentioned by L. Zadeh in 1975 [17] in a more general context. Later in [7] P. Kloeden elaborated a mathematical model of discrete fuzzy dynamical system $(\mathbb{F}(X), z_f)$, which is induced from a given discrete (crisp) dynamical system (X, f). This direction was further elaborated by many mathematicians.

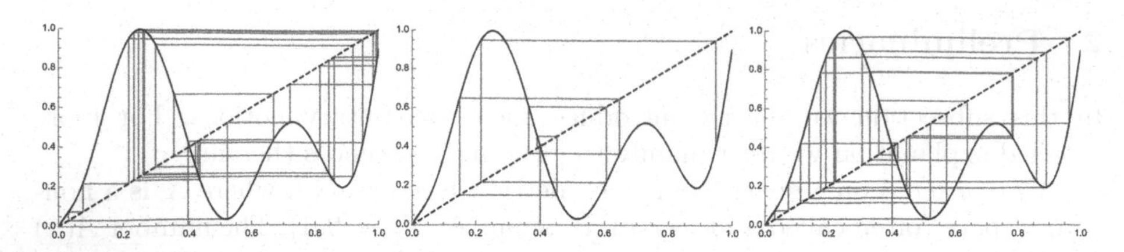

Fig. 1. Trajectories of the dynamical system of a given map, where the initial points are around the value 0.4.

So, formally, let a discrete dynamical system (X, f) be given. Then a continuous map $f\colon X \to X$ induces a map $z_f\colon \mathbb{F}(X) \to \mathbb{F}(X)$ by the formula

$$(z_f(A))(x) = \sup_{y\in f^{-1}(x)} \{A(y)\}.$$

The map z_f is called a *fuzzification* (or *Zadeh's extension*) of the map $f\colon X \to X$. The map z_f fulfils many natural properties, for instance the following equality $[z_f(A)]_\alpha = f([A]_\alpha)$, for any $A \in \mathbb{F}(X)$ and $\alpha \in (0, 1]$, which shows a natural relation to the family of compact subsets (α-cuts) of X. It was proved earlier (see e.g. [7] and [8]) that in the most common topological structures on $\mathbb{F}(X)$ (e.g. for the metric topology induced from the metric d_∞), the continuity of $f\colon X \to X$ is equivalent to the continuity of the fuzzification $z_f\colon \mathbb{F}(X) \to \mathbb{F}(X)$. Consequently a pair $(\mathbb{F}(X), z_f)$ fulfils a formal definition of a discrete dynamical system and this dynamical system is called a *fuzzy dynamical system*. For more information we again refer to [7], [8] and references therein.

3 Particle Swarm Optimization

Particle swarm optimization (abbr. PSO) is an evolutionary optimization algorithm based on stochastic searching in the domain which was originally attributed to R. Eberhart and J. Kennedy in 1995 [4]. The reason of the algorithm, whose behavior is inspired by a social behavior of species, is usually to optimize a given problem. For example, it can be used to find the global optimum of a function of one or more variables. Roughly speaking, population is composed from particles moving around in a given search space according to simple mathematical formulas. In every step of the algorithm, some characteristics (e.g. velocity, the best found solution of every particle, or the best solution in the population) are computed and used to show how the particles move towards the desired solutions. Naturally PSO can stop after a certain number of iterations or after fulfilling some predefined conditions like, for example, accuracy or required size of errors etc. [6,12,16].

3.1 Pseudocode of PSO

The aim of this algorithm is an optimization, i.e. searching for a global optima of a function of one or more variables, which is an essential thing in many applications.

Initially, a finite number of particles x_i is placed into the domain and each particle is evaluated by a given function. Each particle then determines its movement in the domain, with the help of its historical (personal) best position and of the neighbouring particles combined together with some random parameters.

Below we can see the pseudocode of the original version of PSO which can search for a global optima of a given function. Let $f\colon \mathbb{R} \to \mathbb{R}$ be a function for which we search for a global optimum. Now, a number of particles is equal to n and a vector $x = (x_1, \ldots, x_n)$ gives a position of each particle $x_i \in \mathbb{R}$, $p = (p_1, \ldots, p_n)$ is a vector of the best found positions of each particle in its history, p_g is the best position of a particle from the population called *the best neighbour* and v_i is a *velocity* of each particle. Because it is an iterative algorithm, in each iteration, the function f is evaluated in all particles. If the value $f(x_i)$ is better than the previous best value p_i, then this point is rewritten as the best point ($p_i := x_i$) and the value of the function $f(p_i)$ is saved to the value of p_{best_i}. The new position of the particles is then reached by updating the velocity v_i. Now we can explain the parameters which are used in the equation defining the velocity v_i of the next iteration of a particle position. The elements U_{Φ_1}, U_{Φ_2} indicate random points given by a uniform distribution of each component from intervals $[0, \Phi_1]$, $[0, \Phi_2]$, where $\Phi_1, \Phi_2 \in \mathbb{R}$. Parameters Φ_1, Φ_2 are called *acceleration coefficients*, where Φ_1 gives the importance of the personal best value and Φ_2 gives the importance of the neighbors best value. If both of these parameters are too high then the algorithm can be unstable because the velocity could grow up faster. Parameter χ called a *constriction factor*, multiplies the newly calculated velocity and it can affect the movement propagation given by the last velocity value. The original version of PSO works with $\chi = 2/(\Phi - 2 + \sqrt{\Phi^2 - 4\Phi})$, where $\Phi = \Phi_1 + \Phi_2$. The value of this parameter is not changed in time and it has restrictive effect to the result.

1. Initialization (functions, variables, constants, ...)
2. Cycle - for all i calculate $f(x_i)$
3. Comparison
   - compare p_{best_i} and $f(x_i)$
   - if $p_{best_i} \leq f(x_i)$, then $p_i := x_i$ and $p_{best_i} := f(p_i)$
4. Best neighbour
   - find the best neighbour of i and assign it j
   - if $f(p_g) \leq f(x_j)$, then $p_g := x_j$ and $f(p_g) := f(x_j)$
5. Calculation
   - $v_i := \chi(v_i + U_{\Phi_1}(p_i - x_i) + U_{\Phi_2}(p_g - x_i))$
   - $x_i := x_i + v_i$

3.2 The Use of PSO for Linearization

This algorithm is an adaptation of the one-dimensional algorithm introduced above to a higher-dimensional case. We have a map given by a formula $f(x)$ and we want to search for a particle, i.e. a vector of ℓ points, which gives us the best possible linearization of f. These points indicate a *dimension* ℓ of the problem under consideration and it defines a size of each particle $\mathbf{x}$ in the population. Because the idea is to get points (particles) which give us a piecewise linear function, it becomes a problem in which we want to minimalize distances between the original function f and approximating functions given by particles.

The distance between the initial function f and the approximating piecewise linear function is calculated with the help of the following *Manhattan metric* d_M on a finite number of points D in the domain of f. This metric is given by a function $d_M\colon \mathbb{R}^D \times \mathbb{R}^D \to \mathbb{R}$ defined by

$$d_M(\mathbf{x}, \mathbf{y}) = \sum_{i=1}^{D} |x_i - y_i|,$$

where $\mathbf{x} = \{x_1, x_2, \ldots, x_D\}$ and $\mathbf{y} = \{y_1, y_2, \ldots, y_D\}$.

4 Testing

4.1 Parameter Selection

In this subsection we briefly describe the choice of selected parameters and then we study their influence to the accuracy of the proposed algorithm. For each parameter settings the results are calculated 50 times and then evaluated by means of mean and standard deviation.

In general the choice of PSO parameters can have a large impact on optimization performance. In our optimization problem we look for the best setting of the constriction factor χ and acceleration coefficients Φ_1, Φ_2. If both of the parameters Φ_1, Φ_2 are too high then the algorithm can be unstable because the related velocity could grow up faster. There are some recommendations that can be found in the literature. For example, the authors of the original algorithm [4] recommended the values of Φ_1, Φ_2 to be set to 2.05 and they also recommended the following equation $\Phi_1 + \Phi_2 > 4$ to be satisfied. However, parameter setting can be different model by model and therefore we did it for our purpose as well.

In our testing we set the parameter $\chi \in \{0.57, 0.61, 0.65, 0.69, 0.73\}$ and parameters $\Phi_1, \Phi_2 \in \{1.65, 1.85, 2.05, 2.25, 2.45\}$. Thus we have 125 possible combinations of parameters (i.e. 25 combinations for Φ_1, Φ_2 and 5 possibilities for χ). For each of these combinations the proposed algorithm runs 50 times to get the mean and standard deviation from computed outcomes. Some of the initial parameters are fixed - namely, all results are calculated for fixed numbers defining linear parts of approximating function ($\ell = 12$), the number of particles in population ($\mathbf{x} = 25$), the number of iterations of PSO ($I = 100$) and the number of points at which the metric d_M is computed ($D = 80$). These parameters are chosen only for our testing, therefore for the use of this algorithm it should always be considered what the best initial parameters are for our linearized function. For instance the recommended number of linear parts ℓ should definitely be much bigger that the number of monotone parts of the function under consideration to be able somehow cover at least all monotone parts of the approximated function.

4.2 Testing Functions

For the purpose of testing we have chosen functions g_1, g_2 given by the following expressions:

$$\begin{aligned} g_1(x) = 0.9 + (-1 + x)(0.9 + (-0.16 + (5.4 + (-27 + (36 + (510 + \\ (-120 - 2560(-0.9 + x))(-0.1 + x))(-0.6 + x))(-0.2 + x)) \\ (-0.8 + x))(-0.4 + x))x), \end{aligned}$$

$$g_2(x) = 1/25(\sin 20x + 20x \cdot \sin 20x \cdot \cos 20x) + 1/2.$$

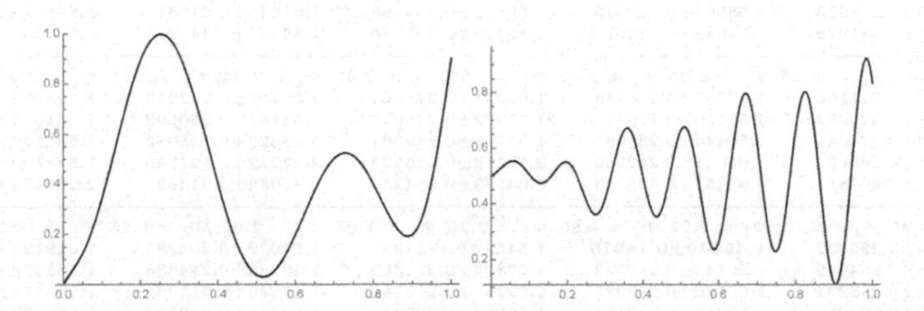

Fig. 2. The graphs of the functions $g_1(x)$ (the left one) and $g_2(x)$ (the right one).

4.3 Parameters Selection

In Tables 1 and 2 below one can see that the choice of parameters χ, Φ_1, Φ_2 can affect the results of the linearization procedure. After some preliminary testing on several functions we can provide some introductory observations. Namely, the best setting was found when the constriction factor χ was taken among values 0.57, 0.61, 0.65 and in a few cases also for the value 0.69. However for χ lying outside of the interval [0.61, 0.73], the results were much worse and did not meet our expectations. Concerning parameters Φ_1, Φ_2, again, our general observation is that better results are obtained when $\Phi_1 \geq \Phi_2$.

It is natural for stochastic methods, that in particular tasks you can get more specific combination of parameters - in the following two tables we emphasize in bold the best 3 combinations of attributes for particular functions g_1 and g_2. More extensive testing is planned as the continuation of this manuscript.

4.4 Examples

In this subsection we demonstrate the use of the proposed algorithm on three nontrivial functions g_1, g_2 and g_3, where the first two functions were defined in Subsect. 4.2, and we consider the best settings of random parameters selected in Subsect. 4.3.

Table 1. The values of mean and standard deviation for the function $g_1(x)$.

x	$\Phi_1 = 1.65, \Phi_2 = 1.65$	$\Phi_1 = 1.65, \Phi_2 = 2.05$	$\Phi_1 = 1.65, \Phi_2 = 2.45$	$\Phi_1 = 2.05, \Phi_2 = 2.05$	$\Phi_1 = 2.05, \Phi_2 = 2.45$
0.57	1.63547±0.238939	1.67338±0.266557	1.65651±0.266593	1.57143±0.213432	1.54162±0.198801
0.61	1.62098±0.281265	1.60798±0.247689	1.59534±0.222001	1.60351±0.226622	1.6108±0.252729
0.65	1.68093±0.257731	1.58062±0.214022	1.59761±0.21376	1.57592±0.224728	1.52531±0.189646
0.69	1.62679±0.227925	1.59228±0.235681	1.9139±0.34237	1.65493±0.276727	1.98223±0.355166
0.73	1.62422±0.214448	1.83744±0.276758	3.10602±0.67275	2.26001±0.571314	4.03971±0.787626

x	$\Phi_1 = 2.45, \Phi_2 = 2.45$	$\Phi_1 = 2.45, \Phi_2 = 2.05$	$\Phi_1 = 2.45, \Phi_2 = 1.65$	$\Phi_1 = 2.05, \Phi_2 = 1.65$	$\Phi_1 = 1.65, \Phi_2 = 1.85$
0.57	1.63376±0.249797	1.53581±0.194398	1.54222±0.197587	1.58635±0.233554	1.59812±0.229893
0.61	1.55107±0.191518	1.55327±0.205339	1.58552±0.22103	1.57803±0.227844	1.60375±0.242386
0.65	1.57619±0.219947	1.57579±0.204595	**1.51653±0.178176**	1.56608±0.222457	1.63805±0.218378
0.69	3.421±0.689219	1.57886±0.239874	1.54552±0.217573	1.53172±0.20908	1.617±0.223034
0.73	5.72016±0.834381	2.50061±0.643045	1.60521±0.212556	1.53968±0.200264	1.54681±0.205785

x	$\Phi_1 = 1.65, \Phi_2 = 2.25$	$\Phi_1 = 1.85, \Phi_2 = 1.85$	$\Phi_1 = 1.85, \Phi_2 = 2.05$	$\Phi_1 = 1.85, \Phi_2 = 2.25$	$\Phi_1 = 1.85, \Phi_2 = 2.45$
0.57	1.59498±0.216631	1.60299±0.243964	1.62142±0.226532	1.56213±0.22595	1.66831±0.217123
0.61	1.62145±0.209883	1.57229±0.184818	1.58244±0.228172	1.57339±0.212051	1.57717±0.229499
0.65	1.63451±0.232793	1.595±0.241515	1.53383±0.189491	1.60258±0.228966	1.58079±0.222188
0.69	1.65213±0.278563	1.58664±0.221945	1.59723±0.234288	1.61548±0.218672	1.86881±0.415271
0.73	1.99725±0.391979	1.6464±0.257361	1.89429±0.301735	2.86462±0.545212	3.5004±0.789913

x	$\Phi_1 = 2.05, \Phi_2 = 2.25$	$\Phi_1 = 2.25, \Phi_2 = 2.25$	$\Phi_1 = 2.25, \Phi_2 = 2.45$	$\Phi_1 = 2.45, \Phi_2 = 2.25$	$\Phi_1 = 2.45, \Phi_2 = 1.85$
0.57	1.53667±0.217165	1.57757±0.208589	1.63313±0.241569	1.56437±0.223043	1.59828±0.211077
0.61	1.57841±0.220252	1.51553±0.19691	1.57972±0.230095	1.54784±0.250593	1.57163±0.223167
0.65	1.56308±0.216044	1.56444±0.235687	1.53595±0.189361	1.62629±0.235062	1.58514±0.220281
0.69	1.70348±0.29063	1.64649±0.242506	2.72718±0.626525	2.27222±0.527745	1.5791±0.202365
0.73	3.2433±0.546781	3.93251±0.809978	5.04399±0.684256	3.92034±0.91482	2.20619±0.495479

x	$\Phi_1 = 2.25, \Phi_2 = 2.05$	$\Phi_1 = 2.25, \Phi_2 = 1.85$	$\Phi_1 = 2.25, \Phi_2 = 1.65$	$\Phi_1 = 2.05, \Phi_2 = 1.85$	$\Phi_1 = 1.85, \Phi_2 = 1.65$
0.57	1.57289±0.220332	**1.49649±0.18915**	1.5457±0.195436	1.56029±0.208285	1.61812±0.239638
0.61	1.56538±0.204659	1.56143±0.209023	1.57824±0.221359	1.59234±0.234838	1.53362±0.194484
0.65	1.60941±0.238231	1.60536±0.219927	1.52257±0.213785	1.57849±0.21733	1.58347±0.247829
0.69	1.54909±0.20935	1.53217±0.192154	1.5495±0.180544	1.58344±0.210707	1.57927±0.2096
0.73	2.61903±0.579384	1.71432±0.33481	**1.50373±0.166186**	1.73587±0.267968	1.53825±0.173687

[a] Compiled in *Mathematica* 12.0 on a laptop with processor 1,8 GHz Intel Core i5.

Table 2. The values of mean and standard deviation for the function $g_2(x)$.

x	$\Phi_1 = 1.65, \Phi_2 = 1.65$	$\Phi_1 = 1.65, \Phi_2 = 2.05$	$\Phi_1 = 1.65, \Phi_2 = 2.45$	$\Phi_1 = 2.05, \Phi_2 = 2.05$	$\Phi_1 = 2.05, \Phi_2 = 2.45$
0.57	4.48464±0.808152	4.33324±0.670205	4.43686±0.774412	4.24214±0.707924	4.43853±0.737978
0.61	4.31768±0.682493	4.49225±0.749727	4.21937±0.63087	4.24245±0.755797	4.30317±0.677813
0.65	4.53499±0.531912	4.14321±0.657339	4.29982±0.670138	4.18762±0.674639	4.64488±0.94961
0.69	4.43666±0.719598	4.18925±0.727941	5.7266±1.06469	4.70045±0.902873	6.64708±1.16885
0.73	4.23356±0.74098	5.4096±1.10863	7.88197±0.951119	6.76307±1.05201	8.72105±0.956886

x	$\Phi_1 = 2.45, \Phi_2 = 2.45$	$\Phi_1 = 2.45, \Phi_2 = 2.05$	$\Phi_1 = 2.45, \Phi_2 = 1.65$	$\Phi_1 = 2.05, \Phi_2 = 1.65$	$\Phi_1 = 1.65, \Phi_2 = 1.85$
0.57	4.20244±0.667743	4.22678±0.549085	4.18617±0.510273	4.4319±0.685112	4.4517±0.706558
0.61	4.20342±0.764773	4.22028±0.520342	4.10346±0.603738	4.23213±0.711974	4.32719±0.636389
0.65	5.42939±1.16048	4.4184±0.802053	**3.99467±0.590382**	4.18327±0.635378	4.26412±0.781714
0.69	8.30996±0.983165	5.46925±0.995767	**3.90281±0.646766**	**4.03962±0.527906**	4.17317±0.555974
0.73	8.31745±1.04308	8.02816±1.0483	6.02647±1.24507	4.58364±0.833236	4.56542±0.978583

x	$\Phi_1 = 1.65, \Phi_2 = 2.25$	$\Phi_1 = 1.85, \Phi_2 = 1.85$	$\Phi_1 = 1.85, \Phi_2 = 2.05$	$\Phi_1 = 1.85, \Phi_2 = 2.25$	$\Phi_1 = 1.85, \Phi_2 = 2.45$
0.57	4.76712±0.819473	4.46511±0.841069	4.19868±0.648816	4.33206±0.617926	4.32131±0.780307
0.61	4.46591±0.645347	4.47089±0.791886	4.33041±0.483959	4.21253±0.815479	4.26108±0.471753
0.65	4.08529±0.640999	4.39325±0.749726	4.08447±0.641931	4.30325±0.79269	4.47726±0.719004
0.69	4.86003±0.80498	4.09351±0.600797	4.61469±0.862227	5.39318±0.93059	6.07631±1.1166
0.73	6.91749±1.08444	5.06067±1.0387	5.79045±0.991023	7.78185±1.09352	8.15509±1.11355

x	$\Phi_1 = 2.05, \Phi_2 = 2.25$	$\Phi_1 = 2.25, \Phi_2 = 2.25$	$\Phi_1 = 2.25, \Phi_2 = 2.45$	$\Phi_1 = 2.45, \Phi_2 = 2.25$	$\Phi_1 = 2.45, \Phi_2 = 1.85$
0.57	4.33195±0.763994	4.23792±0.618683	4.27062±0.71422	4.26326±0.707052	4.31927±0.729022
0.61	4.27319±0.596503	4.29391±0.724273	4.14659±0.812527	4.20238±0.573447	4.18677±0.682581
0.65	4.30942±0.741097	4.25068±0.734095	4.73966±0.798268	4.6961±0.934194	4.10398±0.530543
0.69	5.87982±0.967094	6.35378±1.00249	7.23311±1.22546	7.43348±1.08212	4.84346±0.956979
0.73	8.18012±0.796192	8.64843±0.990123	8.77355±0.937052	8.30777±1.15618	6.78641±1.22497

x	$\Phi_1 = 2.25, \Phi_2 = 2.05$	$\Phi_1 = 2.25, \Phi_2 = 1.85$	$\Phi_1 = 2.25, \Phi_2 = 1.65$	$\Phi_1 = 2.05, \Phi_2 = 1.85$	$\Phi_1 = 1.85, \Phi_2 = 1.65$
0.57	4.26443±0.661071	4.30843±0.643455	4.24803±0.727079	4.16216±0.593539	4.4644±0.712637
0.61	4.22633±0.68801	4.34254±0.67098	4.20053±0.648986	4.25709±0.687653	4.31687±0.55673
0.65	4.23431±0.701167	4.22943±0.691592	4.14209±0.633141	4.12123±0.646354	4.15398±0.577304
0.69	5.41382±1.04234	4.28741±0.7065	4.13273±0.713539	4.23241±0.641686	4.33662±0.707127
0.73	6.98054±1.18231	6.47896±1.11268	5.11513±0.968546	5.39849±1.12682	4.22943±0.885804

[a] Compiled in *Mathematica* 12.0 on a laptop with processor 1,8 GHz Intel Core i5.

Example 1. We have the function g_1 whose graph is depicted in Fig. 2. The initial parameters are set to $\chi = 0.69, \Phi_1 = 2.45, \Phi_2 = 1.65$. In our testing we choose $\ell = 6, 12, 18$, $I = 100$ and $D = 80$. This function g_1 has 5 monotone parts, thus if our intention is to linearize the function g_1 the smallest number ℓ to be considered is 6. Naturally, the higher ℓ we take, the smoother result we obtain. This is demonstrated on the following figure (Fig. 3).

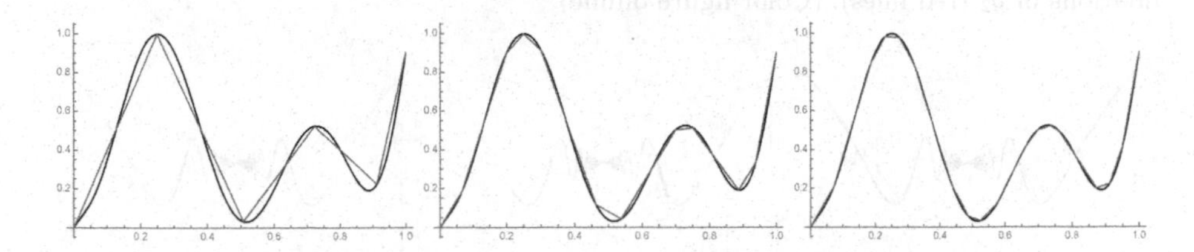

Fig. 3. Graphs of the original function g_1 (black lines) and the piecewise linear functions (red lines) approximating g_1, where $\ell = 6, 12, 18$. (Color figure online)

Example 2. Consider a function g_2 (see Fig. 2) and take the initial parameters $\chi = 0.69, \Phi_1 = 2.45, \Phi_2 = 1.65$, $D = 80$, $I = 100$, $\ell = 15, 18, 25$ (see Fig. 4).

In this example, 14 monotone parts are divided almost equidistantly. Naturally, for better accuracy the number of linear parts should be higher as we can see in the next figure.

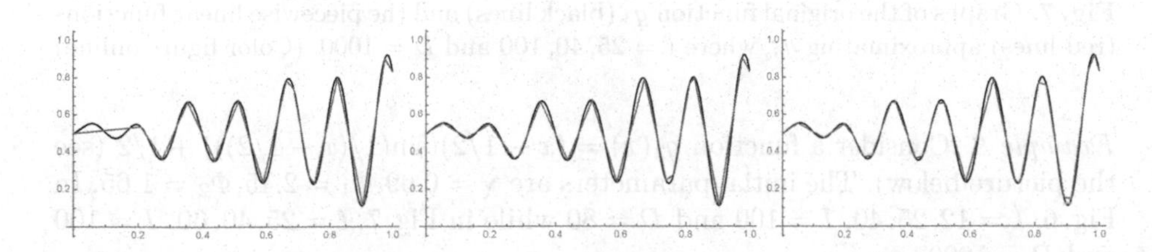

Fig. 4. Graphs of the original function g_2 (black lines) and its piecewise linear approximations (red lines), where $\ell = 15, 18, 25$. (Color figure online)

Another simple observation is that for better accuracy, it need not help to increase the number of linear parts only, but we need to increase also the number D of discretization points accordingly. To demonstrate this, we choose D to be equal to 200, where $\ell = 25$, $\ell = 50$ and $I = 100$ (see Fig. 5).

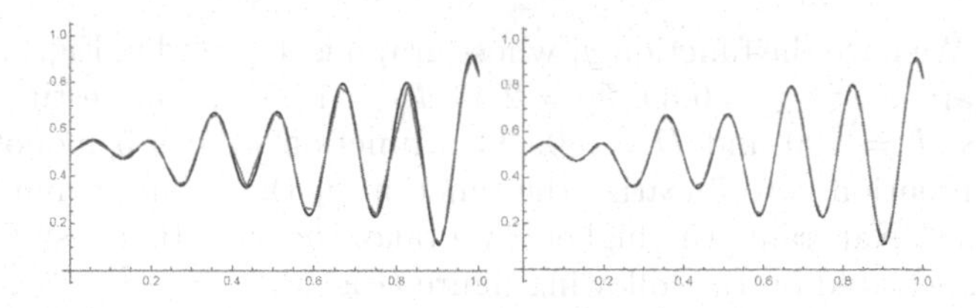

Fig. 5. Graphs of the original function g_2 (black lines) and its piecewise linear approximations of g_2 (red lines). (Color figure online)

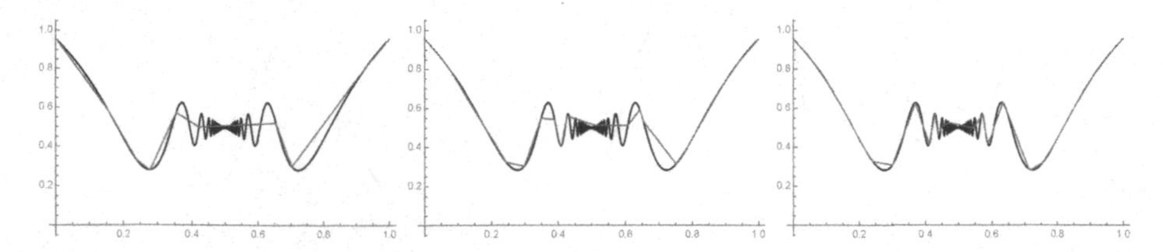

Fig. 6. Graphs of the original function g_3 (black lines) and the piecewise linear functions (red lines) approximating g_3, where $\ell = 12, 25, 40$ and $D = 80$. (Color figure online)

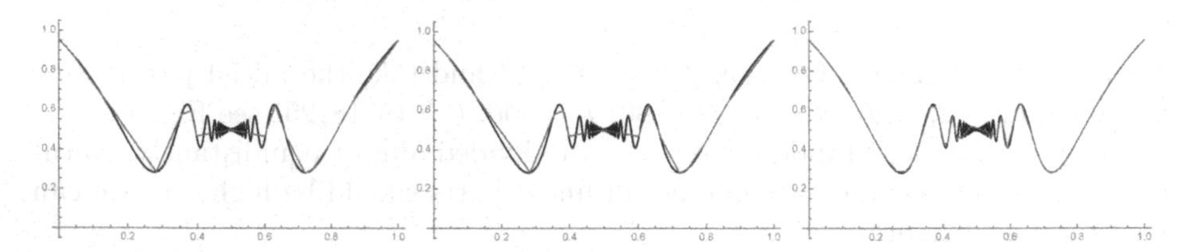

Fig. 7. Graphs of the original function g_3 (black lines) and the piecewise linear functions (red lines) approximating g_3, where $\ell = 25, 40, 100$ and $D = 1000$. (Color figure online)

Example 3. Consider a function $g_3(x) = (x - 1/2)(\sin(1/(x - 1/2))) + 1/2$ (see the picture below). The initial parameters are $\chi = 0.69, \Phi_1 = 2.45, \Phi_2 = 1.65$. In Fig. 6, $\ell = 12, 25, 40$, $I = 100$ and $D = 80$ while in Fig. 7, $\ell = 25, 40, 60$, $I = 100$ and $D = 1000$.

We intentionally consider the function g_3 to demonstrate limits of the proposed algorithm because g_3 has infinitely many monotone parts at arbitrary small neighborhood of the point 1/2. Consequently, in this case it is not possible to approximate all monotone parts correctly. Despite of this drawback we can see in Figs. 6 and 7 that when we increase the number D of discretization points and the number ℓ of linear pairs appropriately, the proposed algorithm works smoothly outside of some neighborhood of the "oscillating" point 1/2.

4.5 Computational Complexity

In this section, we briefly discuss computation complexity of the proposed algorithm. Naturally, the computation time depends on more factors, mainly on the

Table 3. Computing time in seconds.

Function g_1	$D = 80$	$D = 200$	$D = 500$	$D = 1000$
$\ell = 12$	35.92	78.38	189.95	372.82
$\ell = 18$	50.33	111.92	273.8	537.57
$\ell = 25$	65.75	163.33	364.63	734.85
$\ell = 50$	126.28	322.48	775.28	1462.72

[a] Compiled in *Mathematica* 12.0 on a laptop with processor 1,8 GHz Intel Core i5.

number ℓ of linear parts, the number I of iterations, the number D of discretization points and also on computer which is used for compiling. In the table below, we show the time in dependence on number of pairs ℓ and number of discretization points D, which are the most important parameters for the accuracy of this algorithm. The test was executed on function g_1 defined above and with parameters $\chi = 0.69, \Phi_1 = 2.45$, $\Phi_2 = 1.65$ and $I = 100$ (Table 3).

5 Approximation of Zadeh's Extension

5.1 Algorithm

In this subsection we briefly recall an algorithm for calculation of Zadeh's extension of a given function f. For a more detailed description of this algorithm we refer to [10]. The algorithm in [10] was proposed for one-dimensional continuous functions $f\colon X \to X$, i.e. we assume $X = [0, 1]$, but one can consider any closed subinterval of $\mathbb{R}$. The algorithm was proposed for piecewise linear maps f and piecewise linear fuzzy sets A. In this section we demonstrate a generalization of the algorithm from [10] to maps which are not necessarily piecewise linear.

The purpose of the algorithm was to compute a trajectory of a given discrete fuzzy dynamical subsystem $(\mathbb{F}(X), z_f)$, which is obtained as a unique and natural extension of a given discrete dynamical system (X, f).

Thus we consider a continuous map $f\colon [0,1] \to [0,1]$ and a piecewise linear fuzzy set A representing an initial state of induced fuzzy dynamical system $(\mathbb{F}([0,1]), z_f)$. First we use the PSO-based linearization (described in Sect. 5) to get an approximated piecewise linear function $\tilde{f}$, and then we use the algorithm from [10] to calculate a trajectory of the initial state A in the fuzzy dynamical system $(\mathbb{F}([0,1]), z_{\tilde{f}})$. This simple and natural generalization is demonstrated in the following subsection.

5.2 Examples

Let us see two examples of the procedure described in the previous subsection.

Example 4. Let a function f_1 be given by a formula

$$f_1(x) = (-2.9 + (-4.1 + (-15.6 - 14(-0.8 + x))(-0.2 + x))(-0.6 + x))(-1 + x)x$$

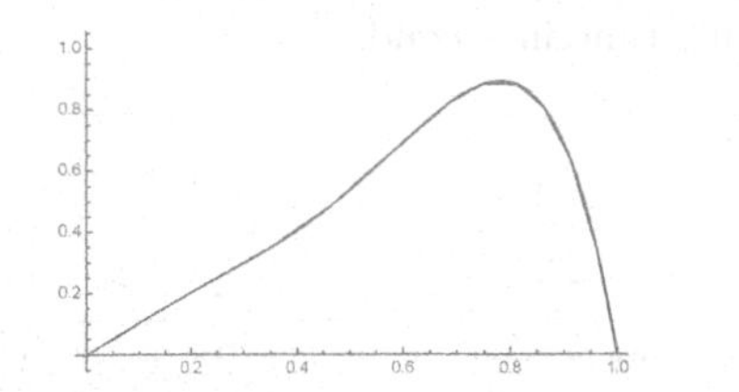

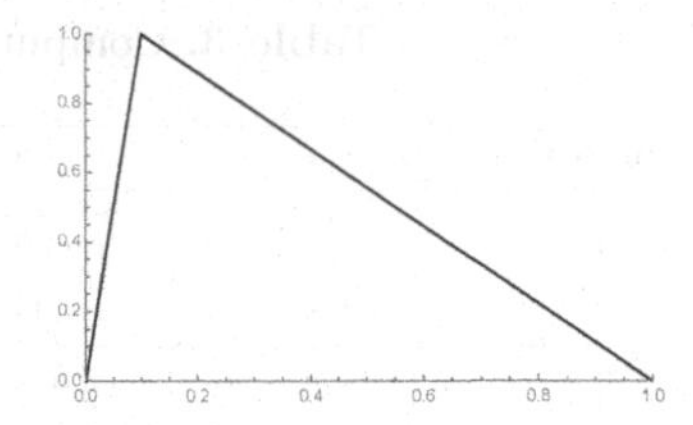

Fig. 8. The graph of a function f_1 (the left figure, black line) and the linearization of f_1 given by PSO (the left figure, red line), the graph of a fuzzy set A (the right figure). (Color figure online)

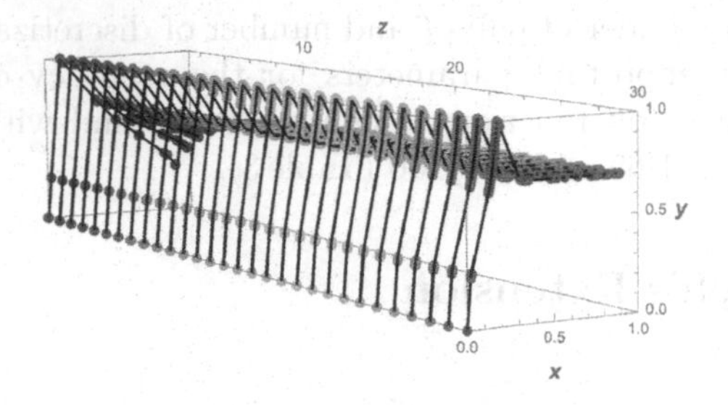

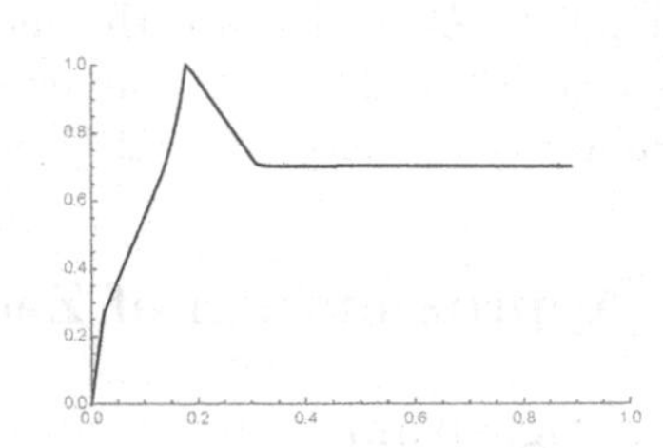

Fig. 9. The graphs of $A, z_{\tilde{f}_1}(A), \ldots, z_{\tilde{f}_1}^{30}(A)$ (the left one) and $z_{\tilde{f}_1}^{30}(A)$ (the right one).

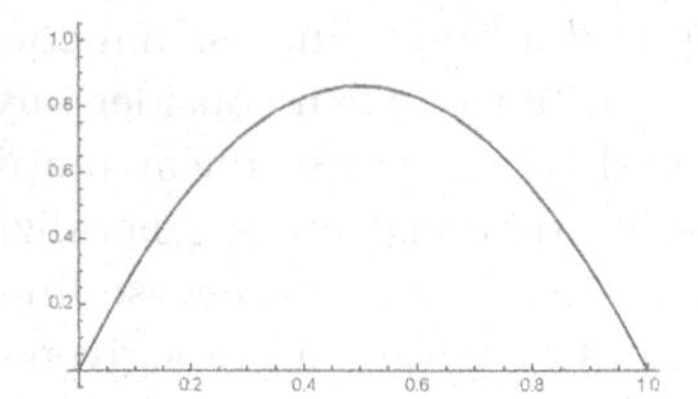

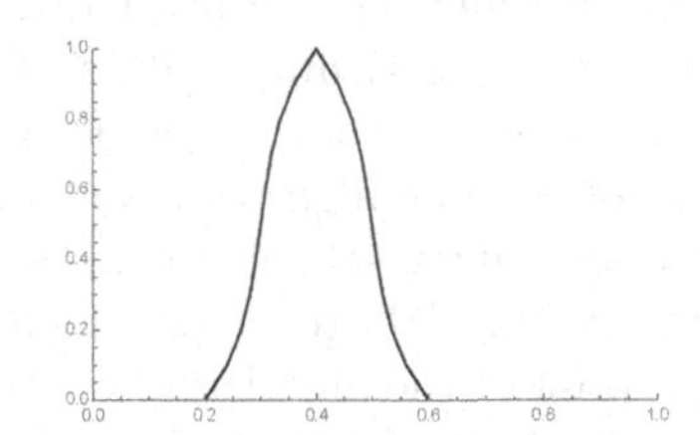

Fig. 10. The graph of a function f_2 (the left figure, black line) and the linearization of f_2 given by PSO (the left figure, red line), the graph of a fuzzy set A (the right figure). (Color figure online)

and let $A(y)$ be a fuzzy set depicted on Fig. 8. As the first step of the algorithm we linearize the function f_1. For that reason, we use the PSO-based algorithm with parameters $\ell = 12, D = 80, I = 100$. After the linearization process we obtain a piecewise linear function $\tilde{f}_1$ and we can compute a plot containing the first 30 iterations of the fuzzy set A (see Fig. 9).

Example 5. Let a function f_2 be given by the following formula $f_2(x) = 3.45(x - x^2)$ and $A(y)$ be a fuzzy set depicted in Fig. 10. Again, we need to linearize the function f_2. To do this we use the PSO-based algorithm with the following parameters $\ell = 18, D = 80, I = 100$.

Finally we can see a plot of the images of the fuzzy set A for the first 30 iterations (see Fig. 11).

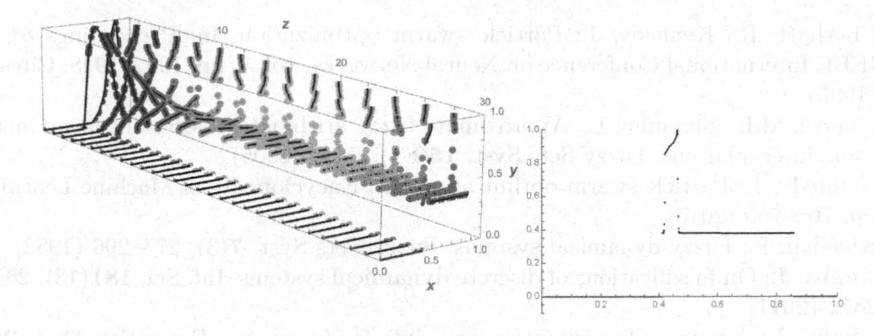

Fig. 11. The graphs of $A, z_{\tilde{f}_2}(A), \ldots, z^{30}_{\tilde{f}_2}(A)$ (the left one) and $z^{30}_{\tilde{f}_2}(A)$ (the right one).

6 Conclusion

In this contribution we generalized our previous algorithm from [10]. The main idea of this algorithm was to calculate Zadeh's extension for a piecewise linear function and a fuzzy set. Because we restricted our attention only to piecewise linear functions the next natural step becomes a generalization of the previous approach to arbitrary continuous functions. Consequently, we took an evolutionary algorithm called particle swarm optimization and we adapt this algorithm to searching for the best possible linearization of a given function. This naturally extends the use of our previous algorithm of an approximation of Zadeh's extension, which now gives us an approximated trajectory of the initial state A in a more general fuzzy dynamical system.

The newly proposed algorithm has been briefly tested from several points of view, mainly parameters selection has been taken into account. In our future work we plan more extensive testing involving also computational complexity of the algorithm given by Big O notation which will deal, for example, with the size of the population, number of iterations, number of linear parts, etc. Another natural step is to provide a deep comparison of the original trajectory derived by the Zadeh's extension with the one given by our algorithm, to provide various comparisons to previously known approaches, and, eventually, to involve, dynamic adaptation of parameters in our PSO-based algorithm. After that, the algorithm should be naturally extended to higher dimensions.

References

1. Ahmad, M.Z., Hasan, M.K.: A new approach for computing Zadeh's extension principle. Matematika **26**, 71–81 (2010)
2. Chalco-Cano, Y., Misukoshi, M.T., Román-Flores, H., Flores-Franulic, A.: Spline approximation for Zadeh's extensions. Int. J. Uncertain. Fuzziness Knowl. Based Syst. **17**(02), 269–280 (2009)
3. Chalco-Cano, Y., Román-Flores, H., Rojas-Medar, M., Saavedra, O., Jiménez-Gamero, M.D.: The extension principle and a decomposition of fuzzy sets. Inf. Sci. **177**(23), 5394–5403 (2007)

4. Eberhart, R., Kennedy, J.: Particle swarm optimization. In: Proceedings of the IEEE International Conference on Neural Networks, vol. 4, pp. 1942–1948. Citeseer (1995)
5. Guerra, M.L., Stefanini, L.: Approximate fuzzy arithmetic operations using monotonic interpolations. Fuzzy Sets Syst. **150**(1), 5–33 (2005)
6. Kennedy, J.: Particle swarm optimization. In: Encyclopedia of Machine Learning, pp. 760–766 (2010)
7. Kloeden, P.: Fuzzy dynamical systems. Fuzzy Sets Syst. **7**(3), 275–296 (1982)
8. Kupka, J.: On fuzzifications of discrete dynamical systems. Inf. Sci. **181**(13), 2858–2872 (2011)
9. Kupka, J.: A note on the extension principle for fuzzy sets. Fuzzy Sets Syst. **283**, 26–39 (2016)
10. Kupka, J., Škorupová, N.: Calculations of Zadeh's extension of piecewise linear functions. In: Kearfott, R.B., Batyrshin, I., Reformat, M., Ceberio, M., Kreinovich, V. (eds.) IFSA/NAFIPS 2019 2019. AISC, vol. 1000, pp. 613–624. Springer, Cham (2019). https://doi.org/10.1007/978-3-030-21920-8_54
11. Lynch, S.: Dynamical Systems with Applications using MATLAB. Springer, Boston (2004). https://doi.org/10.1007/978-0-8176-8156-2
12. Olivas, F., Valdez, F., Castillo, O., Melin, P.: Dynamic parameter adaptation in particle swarm optimization using interval type-2 fuzzy logic. Soft Comput. **20**(3), 1057–1070 (2016)
13. Scheerlinck, K., Vernieuwe, H., De Baets, B.: Zadeh's extension principle for continuous functions of non-interactive variables: a parallel optimization approach. IEEE Trans. Fuzzy Syst. **20**(1), 96–108 (2011)
14. Stefanini, L., Sorini, L., Guerra, M.L.: Parametric representation of fuzzy numbers and application to fuzzy calculus. Fuzzy Sets Syst. **157**(18), 2423–2455 (2006)
15. Stefanini, L., Sorini, L., Guerra, M.L.: Simulation of fuzzy dynamical systems using the LU-representation of fuzzy numbers. Chaos Solitons Fractals **29**(3), 638–652 (2006)
16. Valdez, F.: A review of optimization swarm intelligence-inspired algorithms with type-2 fuzzy logic parameter adaptation. Soft Comput. **24**(1), 215–226 (2020)
17. Zadeh, L.A.: Fuzzy logic and approximate reasoning. Synthese **30**(3–4), 407–428 (1975)

On Uncertain Discontinuous Functions and Quasi-equilibrium in Some Economic Models

Inese Bula[1,2](✉)

[1] Department of Mathematics, University of Latvia,
Jelgavas Street 3, Riga 1004, Latvia
inese.bula@lu.lv

[2] Institute of Mathematics and Computer Science, University of Latvia,
Raina bulv. 29, Riga 1048, Latvia

Abstract. In the paper is studied some properties of uncertain discontinuous mappings, the so-called w-discontinuous mappings. Based on them, the existence of a quasi-equilibrium for a new economic model is proved.

Keywords: Discontinuity · Fixed point theorem · Market equilibrium · Quasi-equilibrium

1 Introduction

One of the basic assumptions in a mathematical modelling of the standard economic model is the continuity of the excess demand function involved. There are reasons to maintain that the necessity of this assumption is caused by the methods provided by mathematics. First of all the fixed points theorems of Brouwer and Kakutani have to be mentioned, since both require the continuity of the maps. They are the main tools for establishing the existence of an equilibrium. However, the necessity of the assumption of continuity has also some economic motivation: in a neoclassical exchange economy due to the strict convexity and strict monotony of the preferences of all consumers the excess demand function is continuous (s. [2], Th.1.4.4).

The paper offers a possibility to substitute the continuity of the excess demand function by the w-discontinuity of this function and therefore to deal, in some extent, with unstable economies. We will examine the properties of w-discontinuous mappings and finally, under some additional conditions, we prove the existence of a generalized equilibrium. The concept w-discontinuity includes uncertainty about the deviation of a function from continuity.

The classical microeconomic models have their origins mainly in the work of L. Walras [18], (1954), a wider discussion of them is presented by K. J. Arrow and G. Debreu [3], (1954) and also by K.J. Arrow and F.H. Hahn [4], (1991). An extended description of the classical model can also be found in textbooks on

M.-J. Lesot et al. (Eds.): IPMU 2020, CCIS 1239, pp. 281–294, 2020.
https://doi.org/10.1007/978-3-030-50153-2_21

microeconomics, for example, H. Varian [17], (1992), D.M. Kreps [14], (1990), W. Nicholson [15], (1992) or R.M. Starr [16], (2011). For a strictly functional-analytic approach we refer to the book of C.D. Aliprantis, D.J. Brown and O. Burkinshaw [2], (1990).

2 w-Discontinuous Mappings and Their Properties

A class of discontinuous mappings is defined as follows. Let (X, d) and (Y, ϱ) be two metric spaces and w a positive number.

Definition 1. *A mapping $f: X \to Y$ is said to be w-discontinuous at the point $x_0 \in X$ if for every $\varepsilon > 0$ there exists δ such that whenever $x \in X$ and $d(x_0, x) < \delta$ follows that $\varrho(f(x_0), f(x)) < \varepsilon + w$.*

The constant w may not be the best possible (smallest) one. Very often, especially in economic applications, there is known only a rough upper estimation for the "jump". Exactly the constant w includes uncertainty about the division of a function from continuity.

A mapping f is called *w-discontinuous in X* if it is w-discontinuous at all points of X.

The notion of w-discontinuous maps is not new. It is already found in [12] as the concept of *oscillation* or as *continuity defect* in [8]. The notion of w-discontinuity (former w-continuity) was introduced by the author in [5].

Example 1. The usual Dirichlet function on $\mathbf{R}$ and also the generalized Dirichlet function $f : \mathbf{R}^n \to \{a, b\}$, a, $b \in \mathbf{R}$, $a \neq b$, defined for all $x = (x_1, x_2, ..., x_n) \in \mathbf{R}^n$ by

$$f(x) = \begin{cases} a, \text{ if all components } x_i \in \mathbf{Q} \\ b, \text{ if there exists } i_0 \text{ such that } x_{i_0} \in \mathbf{R} \setminus \mathbf{Q} \end{cases}$$

are $|a - b|$-discontinuous (and for any $w \geq |a - b|$ also w-discontinuous) functions. ■

If X, Y, V are real normed vector spaces the following properties of w-discontinuous mappings are established (similar as for continuous mappings). For proofs see [7].

Proposition 1. *Let be $f_i : X \to Y$, $\alpha_i \in \mathbf{R}$, $i = 1, \ldots, k$ and $g = \alpha_1 f_1 + \cdots + \alpha_k f_k$. Suppose $w_i > 0$ and that f_i is w_i-discontinuous on the set X for each $i = 1, \ldots, k$. Then $g = \alpha_1 f_1 + \cdots + \alpha_k f_k$ is a $|\alpha_1| w_1 + \cdots + |\alpha_k| w_k$-discontinuous mapping.*

From the Definition 1, which makes sense also for $w = 0$, immediately follows that the 0-discontinuous mappings are exactly the continuous ones.

Corollary 1. *Suppose that $f, g : X \to Y$, f is w'-discontinuous and g is w''-discontinuous. Then $f + g$ and $f - g$ are $w' + w''$-discontinuous mappings. In particular, if one of the mappings (f or g) is continuous, then $f \pm g$ are w'-discontinuous (or w''-discontinuous).*

Corollary 2. *If $f\colon X \to Y$ is w-discontinuous and c is a constant then $c \cdot f$ is a $|c|w$-discontinuous mapping.*

Proposition 2. *Let $f\colon \mathrm{dom} f \to \mathbf{R}$ and $g\colon \mathrm{dom}\, g \to \mathbf{R}$ be w'-, w''-discontinuous functions, respectively. Then the functions $f \wedge g$ and $f \vee g$ are $w' + w''$-discontinuous on $\mathrm{dom} f \cap \mathrm{dom}\, g$.*

Corollary 3. *If f is w-discontinuous and g is continuous then $f \vee g$ is w-discontinuous.*

In order to consider the product of mappings we need the notation of the product in a normed space.

Definition 2 ([13]). *Let X, Y, Z be real normed vector spaces. A mapping $\pi\colon X \times Y \to Z$ is called a product if it satisfies the following conditions: for all $a, b \in X$, $u, v \in Y$ and $\lambda \in \mathbf{R}$ one has*

1. $\pi((a+b, v)) = \pi((a, v)) + \pi((b, v))$
2. $\pi((a, u+v)) = \pi((a, u)) + \pi((a, v))$
3. $\pi((\lambda a, u)) = \lambda \pi((a, u)) = \pi((a, \lambda u))$
4. $\|\pi((a, u))\|_Z \leq \|a\|_X \|u\|_Y$.

A simple example is given by $X = Y = \mathbf{R}^n$, $Z = \mathbf{R}$ and $\pi((x, y)) = \langle x, y \rangle$ – the scalar product in $\mathbf{R}^n$, i.e., $\langle x, y \rangle = \sum_{i=1}^{n} x_i\, y_i$.

Let V, X, Y, Z be real normed vector spaces and let $\pi\colon X \times Y \to Z$ be a product. The product of the mappings $f\colon \mathrm{dom} f \subseteq V \to X$ and $g\colon \mathrm{dom}\, g \subseteq V \to Y$ is understood pointwise, i.e.,

$$(f \cdot g)(v) = \pi(f(v), g(v)), \quad \forall v \in \mathrm{dom} f \cap \mathrm{dom}\, g,$$

where $\mathrm{dom} f, \mathrm{dom}\, g \subseteq V$.

Proposition 3. *Suppose that $f\colon \mathrm{dom} f \to X$ is w'-discontinuous and $g\colon \mathrm{dom}\, g \to Y$ is w''-discontinuous on $\mathrm{dom} f \cap \mathrm{dom}\, g$. Then $f \cdot g$ is a $(w'w'' + w'\|g(x_0)\|_Y + w''\|f(x_0)\|_X)$-discontinuous mapping at every point $x_0 \in \mathrm{dom} f \cap \mathrm{dom}\, g$.*

Corollary 4. *If $f : V \to X$ is w-discontinuous and $g : V \to Y$ is continuous then $f \cdot g$ is a $\|g(x_0)\|_Y\, w$-discontinuous mapping at every point $x_0 \in V$.*

For the division we reconcile with simplified situation, where (X, d) is again a metric space.

Proposition 4. *Let the function $f\colon X \to \mathbf{R}$ be w-discontinuous at the point x_0 and $f(x_0) \neq 0$. If there exists a neighborhood U of x_0 and a number $\alpha_0 > 0$ such that $|f(x)| \geq \alpha_0$ for all $x \in U$ then the function $\frac{1}{f}$ is $\frac{w}{\alpha_0 |f(x_0)|}$-discontinuous at x_0.*

As a special case we get

Corollary 5. *If $f : X \to [1, +\infty[$ is w-discontinuous then $\frac{1}{f}$ is a $\frac{w}{f(x_0)}$- discontinuous mapping for every point $x_0 \in X$*

If the domain of definition for a continuous mapping is compact, then its range is also compact and, in particular, bounded. The boundedness of the range is guaranteed for w-discontinuous mappings as well, however, compactness may not hold.

Example 2. Define $f\colon [0;2] \to [0;2]$ as

$$f(x) = \begin{cases} 1, \text{ if } x \in \{0, 2\} \\ x, \text{ if } \; x \in]0, 2[. \end{cases}$$

The function f is 1-discontinuous and its range $]0, 2[$ is bounded, but not compact. ■

Theorem 1. *Suppose that $A \subset X$ is compact and let $f\colon A \to X$ be w-discontinuous. Then $f(A)$ is bounded.*

The following essential result is proved by O. Zaytsev in [19] and can be considered as a generalization of the Bohl-Brouwer-Schauder fixed point theorem for w-discontinuous mappings.

Theorem 2. *Let K be a nonempty, compact and convex subset in a normed vector space X. For every w-discontinuous mapping $f : K \to K$ $(w > 0)$ there exists a point $x^* \in K$ such that $\| x^* - f(x^*) \| \leq w$.*

3 Market Equilibrium of the Standard Economic Model

We give the description of a simple economic model $\mathcal{E}$ considered by Arrow and Hahn in [4].

Let there be n ($n \in \mathbf{N}$) different goods (commodities) on the market: services and wares, and a finite number of economic agents: households and firms, where each household can be considered as a firm, and, vice versa, each firm can be considered as a household.

Let x_{hi} be the quantity of good i which is needed to the household h. If $x_{hi} < 0$ then $|x_{hi}|$ denotes the quantity of good i which is supplied by the household h. If $x_{hi} \geq 0$ then x_{hi} is the (real) demand of good i by h, including the zero demand. The summation over all households will be indicated by $x_i = \sum_h x_{hi}$ – the total demand of good i, $i = 1, \ldots, n$.

The quantity of good i that is supplied by the firm f will be denoted by y_{fi}. Again, if $y_{fi} < 0$ then $|y_{fi}|$ is the demand (input) of good i by f. If $y_{fi} \geq 0$ then y_{fi} is the supplied quantity (output) of i by f, where the zero supply again is included. The summation over all firms will be indicated by $y_i = \sum_f y_{fi}$ – the supply of good i, $i = 1, \ldots, n$.

The initially available amount (or resources) of good i in all households will be denoted by $\overline{x_i}$. Note that $\overline{x_i}$ must be non-negative.

A market equilibrium, which is one of the most important characteristics of any economy (see f. e. [1,2,4,9,11,16]), describes the economic situation that the total demand of each good in the economy is satisfied by its total supply. This fact is obviously expressed by saying that the difference between the total demand of each good and its total supply is less than or equal to zero. The total supply of good i is understood as the sum of the supply of the good i and the quantity of i which is already available, i.e. the total supply of the good i equals to $y_i + \overline{x_i}$. The excess demand of good i is then defined as $x_i - y_i - \overline{x_i}$, $i = 1, ..., n$.

If economic agents at the market are faced with a system of prices, i.e. with a price vector $p = (p_1, \ldots, p_n)$, where p_i is the price of one unit of good i, then the quantities x_{hi}, y_{fi} and also $x_i, y_i, \overline{x_i}$ depend on p. Now we denote the excess demand of the good i by $z_i(p)$, i.e.

$$z_i(p) = x_i(p) - \Big(y_i(p) + \overline{x_i}(p)\Big).$$

If prices are involved then an equilibrium price (a price system at which an equilibrium is reached) clears the markets.

Further on we frequently make use of the natural order in $\mathbf{R}^n$ introduced by the positive cone

$$\mathbf{R}^n_+ = \{x = (x_1, \ldots, n) \in \mathbf{R}^n \mid x_i \geq 0,\ i = 1, \ldots, n\},$$

i.e. for two vectors $x = (x_1, \ldots, x_n), y = (y_1, \ldots, y_n)$ we write $x \leq y$ iff $x_i \leq y_i$ for all $i = 1, \ldots, n$, we write $x < y$ iff $x \leq y$ and $x_{i_0} < y_{i_0}$ for at least one index i_0. The norm we will use in the space $\mathbf{R}^n$ is defined as

$$\|x\| = \sum_{i=1}^{n} |x_i|,\ x = (x_1, ..., x_n) \in \mathbf{R}^n.$$

This norm is equivalent to the euclidean norm which is introduced by means of the scalar product $\langle x, y\rangle = \sum_{i=1}^{n} x_i\, y_i$. Note that in economic publications the scalar product of two vectors $x, y \in \mathbf{R}^n$ is usually written as $x\, y$.

For the standard economic model the following four assumptions have to be met (see [4]).

Assumption 1. Let $p = (p_1, ..., p_n)$ be an n-dimensional price vector with the prices p_i for one unit of good i as components, $i = 1, 2, ..., n$. For any p let the excess demand for i be characterized by a unique number $z_i(p)$ and so the unique vector $z(p) = (z_1(p), \ldots, z_n(p))$ - the excess demand function with excess demand functions for i as components $(i = 1, 2, ..., n)$ - is well defined.

Assumption 2. $z(p) = z(\lambda p), \quad \forall p > \mathbf{0}$ and $\lambda > 0$.

The Assumption 2 asserts that z is a homogeneous vector-function of degree zero. Economically this means that the value of the excess demand function does not depend on the price system if the latter is changed for all the goods simultaneously by the same portion.

From the Assumption 2 follows that prices can be normalized (see [4], p.20 or [9], p.10). If for some price p one has $z(p) = 0$ then $z(\lambda p) = 0$ for all prices of the ray $\{\lambda p \colon \lambda > 0\}$. Therefore, further on we consider only prices from the $n-1$-dimensional simplex of R^n

$$\Delta_n = \{p = (p_1, p_2, ..., p_n) \mid p_i \geq 0 \text{ and } \sum_{i=1}^{n} p_i = 1\}.$$

We rule out the situations when all the prices are zero or some of them are negative. Note that Δ_n is a compact and convex set in the space $\mathbf{R}^n$ equipped with one of its (equivalent) norms.

Assumption 3 or Walras' Law. $p\, z(p) = 0, \quad \forall p \in \Delta_n$.

Walras' Law can be regarded as an attempt to have a model sufficiently truly reflecting rationally motivated activities of economic agents. According to Walras' Law all the firms and all the households both spend their financial resources completely [9].

Assumption 4. The excess demand function z is continuous on its domain of definition Δ_n.

It means that a small change of a price system will imply only a small change in the excess demand. As a consequence from continuity of z, the standard model can be used only for the description of economies with continuous excess demand functions. Sometimes they are called stable economies.

In economies such prices are important at which the excess demand for each good is nonpositive, i.e. the total supply of each good satisfies at least its total demand.

Definition 3. *A price $p^* \in \Delta_n$ is called an* **equilibrium** *(price) if $z(p^*) \leq 0$.*

If p^* is an equilibrium price then $\sum_{i=1}^{n} z_i(p^*) \leq 0$.

For the standard model of an economy with a finite number of goods and agents such prices always exist as is proved in the following theorem.

Theorem 3 ([4]). *If an economy $\mathcal{E}$ with a finite number of goods and agents satisfies the Assumptions 1–4, then there exists an equilibrium in $\mathcal{E}$.*

4 Economic Models with Discontinuous Excess Demand Functions

If z is the excess demand function for a neoclassical exchange economy, then z is continuous on the set

$$S = \{p \in \Delta_n \mid p_i > 0, i = 1, 2, ..., n\}$$

(see [2], Th.1.4.4 and Th.1.4.6). A neoclassical exchange economy (see [2]) is characterized by a finite set of agents, where each agent i has a non-zero initial endowment ω_i and his preference relation $\succeq_i$ is continuous (a preference relation $\succeq$ is continuous if, given a two sequences $(x^n)_{n=1}^{\infty}$, $(y^n)_{n=1}^{\infty}$ with $\lim_{n\to\infty} x^n = x$, $\lim_{n\to\infty} y^n = y$ and $x^n \succeq y^n$, $n = 1, 2...$, then $x \succeq y$), strictly monotone and strictly convex (on $\mathbf{R}_+^n$) or else his preference relation $\succeq_i$ is continuous, strictly monotone and strictly convex on interior of $\mathbf{R}_+^n$, and everything in the interior is preferred to anything on the boundary and the total endowment $\omega = \sum_i \omega_i$ is strictly positive. If the preference relation $\succeq_i$ is continuous, strictly monotone and strictly convex then the corresponding utility function and the excess demand function are continuous on the set S. We will consider the situation with a discontinuous excess demand function. It is clear that in this case the properties of the preference relations differ from them in the neoclassical exchange economy.

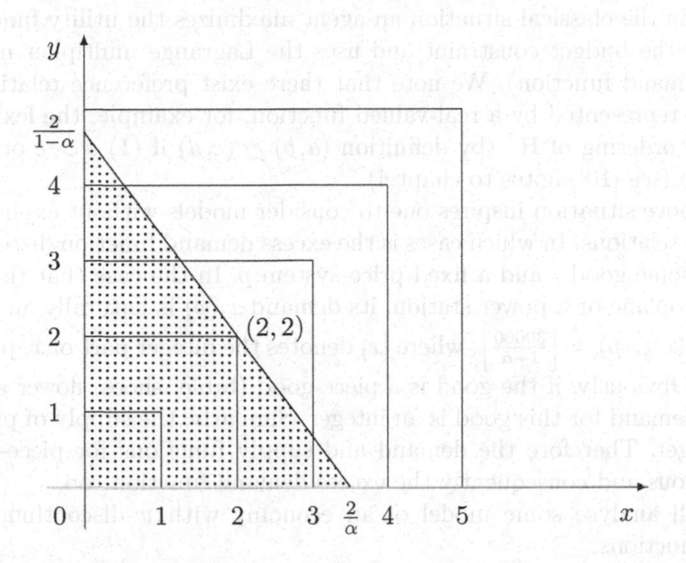

Fig. 1. The indifference curves of utility function $u(x, y) = \max\{x, y\}$ for the values 1, 2, 3, 4 and 5.

For example, consider the preference relation on $\mathbf{R}_+^2$ that is represented by the utility function $u(x, y) = \max\{x, y\}$ and an initial endowment $\omega = (2, 2)$. The utility function is continuous, but it is not strictly monotone (for example, $(2, 2) > (2, 1)$ but $u(2, 2) = 2 = u(2, 1)$) and it is not strictly concave, it is convex. The indifference curves for the values 1, 2, 3, 4 and 5 are illustrated in Fig. 1. Let $p = (\alpha, 1-\alpha)$ be a fixed price vector for some $0 < \alpha < 1$. We maximize the utility function u subject to the budget constraint $\alpha x + (1-\alpha)y = 2\alpha + 2(1-\alpha) = 2$. This line goes through the point $(2, 2)$ and intersects the axis in the points $(0, \frac{2}{1-\alpha})$ and $(\frac{2}{\alpha}, 0)$. From Fig. 1 we see that the maximal vector of u over budget set (the dotted region in Fig. 1) is the point $(0, \frac{2}{1-\alpha})$ if $\alpha > \frac{1}{2}$ and $(\frac{2}{\alpha}, 0)$ if $\alpha < \frac{1}{2}$, respectively. If $\alpha = \frac{1}{2}$ then $\frac{2}{1-\alpha} = \frac{2}{\alpha}$ and therefore we have two maximizing vectors. The demand function in this case is

$$d(p) = d(\alpha, 1-\alpha) = \begin{cases} (0, \frac{2}{1-\alpha}), & \alpha > \frac{1}{2}, \\ \{(0,4),(4,0)\}, & \alpha = \frac{1}{2}, \\ (\frac{2}{\alpha}, 0), & \alpha < \frac{1}{2}. \end{cases}$$

In the point $(\frac{1}{2}, \frac{1}{2})$ the demand multifunction is discontinuous.

In [1] it is proved that in a neoclassical exchange economy the condition $p_n \to p \in \partial S$ with $(p_n)_{n \in \mathbf{N}} \subset S$ implies $\lim_{n\to\infty} ||z(p_n)|| = \infty$. It is also not our case (see Theorem 1). In [1] it is shown that a utility function $u\colon X \to \mathbf{R}$ (X - topological space) representing a continuous preference relation is not necessarily continuous. If we start with an arbitrary chosen discontinuous utility function then we have no mathematical tools for finding the corresponding demand function (in the classical situation an agent maximizes the utility function with respect to the budget constraint and uses the Lagrange multiplier method for finding demand function). We note that there exist preference relations which cannot be represented by a real-valued function, for example, the lexicographic preference ordering of $\mathbf{R}^2$ (by definition $(a, b) \succeq (c, d)$ if (1) $a > c$ or (2) $a = c$ and $b > d$) (see [10], notes to chapt.4).

The above situation inspires one to consider models without explicitly given preference relations. In which cases is the excess demand function discontinuous? Consider some good i and a fixed price system p. In the case that this good is, e.g. an aeroplane or a power station, its demand $x_i(p)$ is naturally an integer. A function like $x_i(p) = \left[\frac{30000}{1+\alpha}\right]$, where $[x]$ denotes the integer part of x, provides an example. Obviously, if the good is a piece-good (table, shoes, flower and other) then the demand for this good is an integer. Similarly, the supply of piece-goods is an integer. Therefore the demand and supply functions for piece-goods are discontinuous and consequently the excess demand function too.

We will analyse some model of an economy with w-discontinuous excess demand functions.

For the economies under consideration we keep the two first assumptions from the standard model and change the two last as follows.

Assumption 4'. The excess demand function z is w-discontinuous on its domain of definition Δ_n.

The w-discontinuity of the excess demand function makes our model available to describe some properties of an unstable economy as well.

It is quite natural that for every price vector $p \in \Delta_n$ there exist at least one good i with the price $p_i > 0$ and such that the demand for them is satisfied, i.e. $z_i(p) \leq 0$.

If for some economy $\mathcal{E}$ with the excess demand vectors $z(p)$, $p \in \Delta_n$ there holds the Walras' Law, i.e. $p\,z(p) = 0$ for any $p \in \Delta_n$, then for each $p \in \Delta_n$ the inequality

$$\gamma_p = \sum_{i:z_i(p)\leq 0} p_i > 0$$

is satisfied. (We write further "$z_i(p) \leq 0$" instead of "$i : z_i(p) \leq 0$" and in similar cases.) Indeed, if for some $p = (p_1, \ldots, p_n) \in \Delta_n$ there would be $\sum_{z_i(p)\leq 0} p_i = 0$, then

$$\sum_{z_i(p)\leq 0} p_i + \sum_{z_i(p)>0} p_i = \sum_{i=1}^{n} p_i = 1$$

would imply the existence of an index i_0 such that $p_{i_0} > 0$ and $z_{i_0}(p) > 0$, which hold then, because of $\sum_{i=1}^{n} p_i = 1$, for some i_0 there must be $p_{i_0} > 0$ and $z_{i_0}(p) > 0$, which yields $p\,z(p) = \sum_{i=1}^{n} p_i z_i(p) \geq p_{i_0} z_{i_0}(p) > 0$, a contradiction to Walras' Law.

Our next assumption requires the existence of a uniform lower bound for the sums $\sum_{z_i(p)\leq 0} p_i$, for all $p \in \Delta_n$.

Assumption 3'. $\gamma = \inf_{p\in\Delta_n} \gamma_p > 0$.

It seems to be clear that it would be hard to find out why an equilibrium exists in our model. But it will be possible if we can estimate the unsatisfied aggregate demand. This leads to the concept of quasi- or k-equilibrium.

Definition 4. *Let k be a positive real. A price vector $p^* \in \Delta_n$ is called a* **k-equilibrium** *if it satisfies the condition*

$$\sum_{z_i(p^*)>0} z_i(p^*) \leq k.$$

The constant $k \in \mathbf{R}_+$ as a numerical value of the maximally possible unsatisfied demand for a given price $p^* \in \Delta_n$ characterizes to what state the economy differs from the market equilibrium (Definition 3).

We can prove now the following

Theorem 4. *Let $\mathcal{E}$ be an economy with n goods that satisfies the Assumptions 1, 2 and the Assumption 3' with some number $\gamma > 0$. Put*

$$w_+ = w_+(n, \gamma) = \frac{1}{2n}\left(-(n+1) + \sqrt{(n+1)^2 + 8n\gamma}\right).$$

If now the Assumption 4' is satisfied with $w \in [0, w_+)$, then the economy $\mathcal{E}$ possesses a k-equilibrium for each $k \geq \frac{nw^2+(n+1)w}{2\gamma-nw^2-(n+1)w}$.

Proof. For $p \in \Delta_n$ define $z_i^+(p) = \max\{0, z_i(p)\}$, $i = 1,...,n$, $z^+(p) = (z_1^+(p),\ldots,z_n^+(p))$,

$$\nu(p) = \langle p + z^+(p), e\rangle = 1 + \sum_{z_i(p)>0} z_i(p) \quad \text{and} \quad t_i(p) = \frac{p_i + z_i^+(p)}{\nu(p)},\ i = 1,...,n,$$

where $e = (1,\ldots,1)$ denotes the vector of $\mathbf{R}^n$ with all components equal to 1. Note that $\|e\| = n$.

Define now a map $T\colon \Delta_n \to \Delta_n$ by $T(p) = \frac{p+z^+(p)}{\langle p+z^+(p),e\rangle}$, then $T(p) = (t_1(p),...,t_n(p))$. Since $0 \le t_i(p) \le 1$ for each i and

$$\sum_{i=1}^n t_i(p) = \frac{\sum\limits_{i=1}^n (p_i + z_i^+(p))}{\nu(p)} = \frac{1 + \sum\limits_{z_i(p)>0} z_i(p)}{\nu(p)} = 1$$

one has $T(p) : \Delta_n \to \Delta_n$.

Now the particular maps which the map T consists of, possess the following properties. The identity map id on Δ_n is continuous, by Assumption 4' the map $z\colon \Delta_n \to \mathbf{R}^n$ is w-discontinuous and by Corollary 3 so is z^+. By Corollary 1 the map $id + z^+$ is w-discontinuous, what by Corollary 4 implies the $w\|e\|$-discontinuity, i.e. the nw-discontinuity of $\nu(p) = \langle p + z^+(p), e\rangle$. Since $\nu\colon \Delta_n \to [1,\infty)$ the function $\frac{1}{\nu}$ is $\frac{nw}{\nu(p)}$-discontinuous as a consequence of Corollary 5. Finally, based on Proposition 3, the map $T(p) = (p + z^+(p))\,\frac{1}{\nu(p)}$ is w_0-discontinuous at a every point $p \in \Delta_n$, where

$$w_0 = w_0(p) = \frac{nw^2}{\nu(p)} + \frac{w}{\nu(p)} + \frac{nw\|p + z^+(p)\|}{\nu(p)} = \frac{nw^2 + w}{\nu(p)} + nw \le nw^2 + (n+1)w \tag{1}$$

and so, the map T is also $nw^2 + (n+1)w$-discontinuous on the set Δ_n.

Since Δ_n is a convex and compact subset in the normed vector space $\mathbf{R}^n$ and $T(p) : \Delta_n \to \Delta_n$ we conclude by means of Theorem 2 that there exists a vector $p^* \in \Delta_n$ satisfying the inequality

$$\|T(p^*) - p^*\| \le nw^2 + (n+1)w.$$

Using the norm in $\mathbf{R}^n$ this yields

$$\begin{aligned}\|T(p^*) - p^*\| &= \left\|\tfrac{p^*+z^+(p^*)}{\nu(p^*)} - p^*\right\| = \sum_{i=1}^n \left|\tfrac{p_i^*+z_i^+(p^*)}{\nu(p^*)} - p_i^*\right| \\ &= \sum_{i=1}^n \left|\frac{p_i^*+z_i^+(p^*)-p_i^*-p_i^*\sum\limits_{z_i(p^*)>0} z_i(p^*)}{\nu(p^*)}\right| \le nw^2 + (n+1)w.\end{aligned}$$

Since $1 + \sum\limits_{z_i(p^*)>0} z_i(p^*) > 0$ one has

$$\sum_{i=1}^n \left| z_i^+(p^*) - p_i^* \sum_{z_i(p^*)>0} z_i(p^*)\right| \le \left(nw^2 + (n+1)w\right)\, \nu(p^*). \tag{2}$$

The left side of inequality (2) can be splitted into two sums

$$\sum_{z_i(p^*)\leq 0}\left|z_i^+(p^*) - p_i^*\sum_{z_i(p^*)>0} z_i(p^*)\right| + \sum_{z_i(p^*)>0}\left|z_i(p^*) - p_i^*\sum_{z_i(p^*)>0} z_i(p^*)\right|$$

$$= \sum_{z_i(p^*)\leq 0} p_i^* \sum_{z_i(p^*)>0} z_i(p^*) + \sum_{z_i(p^*)>0}\left|z_i(p^*) - p_i^*\sum_{z_i(p^*)>0} z_i(p^*)\right|. \quad (3)$$

Using the triangle inequality we get the estimation

$$\left|\sum_{z_i(p^*)>0}\left(z_i(p^*) - p_i^*\sum_{z_i(p^*)>0} z_i(p^*)\right)\right| \leq \sum_{z_i(p^*)>0}\left|z_i(p^*) - p_i^*\sum_{z_i(p^*)>0} z_i(p^*)\right| \quad (4)$$

and further the left hand side of (4) calculates as

$$\left|\sum_{z_i(p^*)>0}\left(z_i(p^*) - p_i^*\sum_{z_i(p^*)>0} z_i(p^*)\right)\right| = \left|\sum_{z_i(p^*)>0} z_i(p^*)\left(1 - \sum_{z_i(p^*)>0} p_i^*\right)\right|$$

$$= \sum_{z_i(p^*)>0} z_i(p^*)\left(1 - \sum_{z_i(p^*)>0} p_i^*\right) = \sum_{z_i(p^*)>0} z_i(p^*) \sum_{z_i(p^*)\leq 0} p_i^*. \quad (5)$$

By means of the equalities (3), (5) and the inequalities (2), (4) we obtain now

$$2\sum_{z_i(p^*)>0} z_i(p^*) \sum_{z_i(p^*)\leq 0} p_i^* \leq \sum_{z_i(p^*)>0} z_i(p^*) \sum_{z_i(p^*)\leq 0} p_i^*$$

$$+ \sum_{z_i(p^*)>0}\left|z_i(p^*) - p_i^*\sum_{z_i(p^*)>0} z_i(p^*)\right| \leq (nw^2 + (n+1)w)\ \nu(p^*).$$

It follows by means of the Assumption 3'

$$2\gamma \sum_{z_i(p^*)>0} z_i(p^*) \leq 2\sum_{z_i(p^*)>0} z_i(p^*) \sum_{z_i(p^*)\leq 0} p_i^* \leq (nw^2 + (n+1)w)\ \nu(p^*).$$

Since $\nu(p^*) = 1 + \sum_{z_i(p^*)>0} z_i(p^*)$ the last inequality yields

$$\sum_{z_i(p^*)>0} z_i(p^*) \leq \frac{nw^2 + (n+1)w}{2\gamma - nw^2 - (n+1)w}, \quad \text{i.e.} \quad \sum_{z_i(p^*)>0} z_i(p^*) \leq k,$$

where k satisfies $k \geq \frac{nw^2+(n+1)w}{2\gamma - nw^2 - (n+1)w}$.

In order to have the number $2\gamma - nw^2 - (n+1)w$ positive the value of w must belong to the interval $[0, w_+)$, where w_+ is the positive root of the equation $w^2 + \frac{n+1}{n}w - \frac{2\gamma}{n} = 0$. □

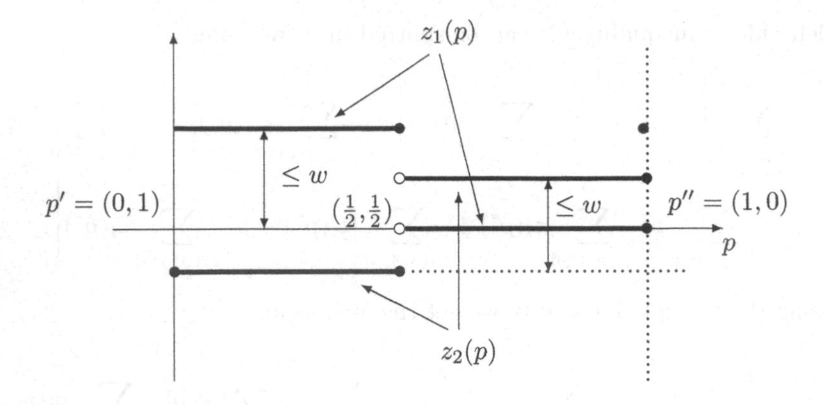

Fig. 2. No classical equilibrium, but k-equilibrium exists.

5 Conclusions

We make some remarks.

1. Let n and $\gamma > 0$ be fixed. Then $w_+ = w_+(n, \gamma)$ is defined as indicated in the theorem. For $w \in [0, w_+)$ put

$$k_0(n, w) = \frac{nw^2 + (n+1)w}{2\gamma - nw^2 - (n+1)w}.$$

The number $k_0(n, w)$ is non-negative as was shown above. Note that a sharper estimation (our estimation is based on the rough inequality $\nu(p) \geq 1$) in (1) would yield a smaller value of $k_0(n, w)$ and, therefore, would give a better result. In view of Theorem 2, however, an estimation has be obtained independently on p.

2. In Fig. 2 for $n = 2$ there is shown a situation without a classical equilibrium. It is clear that there is no $p \in \Delta_2$ which satisfies the inequality $z(p) = (z_1(p), z_2(p)) \leq \mathbf{0}$. The Assumptions 1, 2, 4' are obviously fulfilled. The Assumption 3' also holds. Indeed, represent $p = (p_1, p_2) \in \Delta_2$ as

$$p = (1-t)p' + tp'', \qquad t \in [0, 1],$$

then $t \in [0, \frac{1}{2}]$ implies $z_1(p) > 0$, $z_2(p) < 0$ and so $\gamma_p = p_2$ and $t \in (\frac{1}{2}, 1]$ implies $z_1(p) = 0$, $z_2(p) > 0$ and so $\gamma_p = p_1$. In both cases we get $\gamma_p \geq \frac{1}{2}$ which shows that the Assumption 3' holds with $\gamma = \frac{1}{2}$. Theorem 4 guarantees the existence of a k-equilibrium for $k \geq \frac{2w^2+3w}{1-2w^2-3w}$ if $w < -\frac{3}{4} + \frac{\sqrt{17}}{4}$. Note that Walras' Law is not satisfied.

3. The number $w_+(n, \gamma)$ is positive for each n and fixed $\gamma > 0$. If one takes $w = 0$ then $k_0(n, \gamma) = 0$ and with $k = 0$ there is obtained the classical case. Observe that in this case it is not necessary to use the Walras' Law for establishing a classical equilibrium.

4. Note that in the classical situation it is impossible to carry out any quantitative analysis. On the contrary, the inequalities from Theorem 4

$$w < w_+(n,\gamma) \quad \text{and} \quad k \geq k_0(n,w)$$

give a chance to analyse the behaviour of an economy for different numerical values of the parameters n, w, γ included in our model. From

$$0 \leq w_+(n,\gamma) = \frac{-(n+1) + \sqrt{(n+1)^2 + 8n\gamma}}{2n}$$

$$< \frac{-(n+1) + (n+1) + \sqrt{8n\gamma}}{2n} = \sqrt{\frac{2\gamma}{n}}$$

it follows that $\lim_{n\to\infty} w_+(n,\gamma) = +0$. Since $k_0(n,0) = 0$, the positive number k can be chosen arbitrary small. This shows that the larger the number of goods the better the chance for a classical equilibrium.
5. It is reasonable to put $k_0(n, w_+(n,\gamma)) = +\infty$. If for fixed n and γ the value w is sufficiently close to $w_+(n,\gamma)$, then k is very large. In such a case the existence of an k-equilibrium seems to be of low economic meaning.
6. The results of this paper have been developed in a collaboration with prof. M. R. Weber from the Dresden University of Technology [7].
7. Other application of w-discontinuous mappings is to find a of quasi-equilibrium in economic models that the author has developed in [6] in a collaboration with a student D. Rika.

References

1. Aliprantis, C.D.: Problems in Equilibrium Theory. Springer, Heidelberg (1996). https://doi.org/10.1007/978-3-662-03233-6
2. Aliprantis, C.D., Brown, D.J., Burkinshaw, O.: Existence and Optimality of Competitive Equilibria. Springer, Heidelberg (1990). https://doi.org/10.1007/978-3-642-61521-4
3. Arrow, K.J., Debreu, G.: Existence of an equilibrium for a competitive economy. Econometrica **42**, 265–290 (1954)
4. Arrow, K.-J., Hahn, F.-H.: General Competitive Analysis. Advanced Textbook in Economics, vol. 12. North-Holland Publishing Company, Amsterdam (1991)
5. Bula, I.: On the stability of Bohl-Brouwer-Schauder theorem. Nonlinear Anal. Theory Methods Appl. **26**, 1859–1868 (1996)
6. Bula, I., Rika, D.: Arrow-Hahn economic models with weakened conditions of continuity. In: Game Theory and Mathematical Economics, vol. 71, pp. 47–61. Banach Center Publications (2006)
7. Bula, I., Weber, M.R.: On discontinuous functions and their application to equilibria in some economic model. Preprint of Technische Universiteat Dresden MATH-AN-02-02 (2002)
8. Burgin, M., Schostak, A.: Towards the theory of continuity defect and continuity measure for mappings of metric spaces. Latvijas Universitātes Zinātniskie Raksti, Matemātika **576**, 45–62 (1992)

9. Cornwall, R.R.: Introduction to the Use of General Equilibrium Analysis. Advanced Textbooks in Economics, vol. 20. North-Holland Publishing Company, Amsterdam (1984)
10. Debreu, G.: Theory of Value. Yale University Press, New Haven, London (1959)
11. Hildenbrand, W., Kirman, A.P.: Equilibrium Analysis. Advanced Textbooks in Economics, vol. 28. North-Holland Publishing Company, Amsterdam (1991)
12. Kuratowski, K.: Topology I. Academic Press, New York (1966)
13. Lang, S.: Analysis I. Addison-Wesley Publishing Company, Boston (1976)
14. Kreps, D.M.: A Course in Microecnomic Theory. Harvester Wheatsheaf, London (1990)
15. Nicholson, W.: Microeconomic Theory. Basic Principles and Extensions. The Dryden Press, Fort Worth (1992)
16. Starr, R.M.: General Equilibrium Theory: An Introduction, 2nd edn. Cambridge University Press, New York (2011)
17. Varian, H.R.: Microeconomic Analysis, 3rd edn. W. W. Norton & Company, New York (1992)
18. Walras, L.: Elements of Pure Economics. Allen and Unwin, London (1954)
19. Zaytsev, O.: On discontinuous mappings in metric spaces. Proc. Latv. Acad. Sci. Sect. B **52**, 259–262 (1998)

Nonlocal Laplace Operator in a Space with the Fuzzy Partition

Hana Zámečníková(✉) and Irina Perfilieva

IRAFM, University of Ostrava, 30. dubna 22, 701 03 Ostrava, Czech Republic
p18113@student.osu.cz, irina.perfilieva@osu.cz
http://www.osu.eu/

Abstract. Differential operators play an important role in the mathematical modeling of dynamic processes and the analysis of various structures. However, there are certain limitations in their use. To remove them, nonlocal differential operators have been proposed. In this work, we focus on nonlocal Laplace operator, which has become increasingly useful in image processing. We introduce the representation of F-transform based Laplace operator in a space with a fuzzy partition. Many useful properties of this operator are proposed and their proofs are also included.

Keywords: Nonlocal Laplace operator · Proximity · Basic functions · Fuzzy transform

1 Introduction

Image processing requires quick and efficient processing of large amounts of data. The most important factor here is the speed of processing, which can be generally in conflict with quality. For this reason, the classical metric spaces are gradually being replaced by more general spaces based on the notion of proximity. The direction of research is moving towards nonlocal differential operators defined on these spaces.

Methods based on nonlocal Laplace operator have become widely used in many application fields. Dimensionality reduction (extracting low dimensional structure from high dimensional data) [1] or clustering (automatic identification of groups of similar objects) [2] can be mentioned. This approach also turned out to be successful for image colorization [3], image denoising or segmentation [4].

In our approach we try to extend the theory of fuzzy transforms, that was proved to be useful in image processing. The purpose is to focus on similarities between approach based on the theory of fuzzy transform and framework based on nonlocal Laplace operator, mentioned above. We assume, that the use of nonlocal operators-based methods in spaces determined by fuzzy partition can somehow enhance the research in this direction.

M.-J. Lesot et al. (Eds.): IPMU 2020, CCIS 1239, pp. 295–303, 2020.
https://doi.org/10.1007/978-3-030-50153-2_22

Let f be a real function, $f : \Omega \to \mathbb{R}$, $\Omega \subset \mathbb{R}^n$. *Laplace operator* is a second order differential operator in the n-dimensional Euclidean space, given by the divergence of the gradient of a function f. Equivalently it can be defined by the sum of all the unmixed second partial derivatives of function f in the Cartesian coordinates x_i, $\Delta f = \sum_{i=1}^{n} \frac{\partial^2 f}{\partial x_i^2}$. This formula incorporates the partial derivatives, that should fulfill certain well-known conditions. However, this puts limitations on sets of functions, that can be differentiated as required.

For this reason, the notion of derivative was extended to a nonlocal version by defining [5]:

$$\partial_y f(x) = \frac{f(y) - f(x)}{\tilde{d}(x, y)}, \quad \forall x, y \in \Omega \tag{1}$$

where $\tilde{d}$, $0 < \tilde{d}(x, y) \leq \infty$, defines a positive distance measure between x and y.

Definition 1. *Let function* $w : \Omega \times \Omega \to \mathbb{R}$ *be non-negative* $(0 \leq w(x, y) < \infty)$ *and symmetrical* $(w(x, y) = w(y, x))$*, then* w *is called a measure of proximity. A pair* (Ω, w) *defines a proximity space.*

Remark 1. A proximity space (Ω, w) can be also introduced with the reference to a distance measure, e.g. defining

$$w(x, y) = \tilde{d}^{-2}(x, y). \tag{2}$$

In (Ω, w), a nonlocal derivative can be defined as follows [5]:

$$\partial_y f(x) = (f(y) - f(x)) \sqrt{w(x, y)}. \tag{3}$$

This notion can be extended for functions of several real variables. In a similar way, other operators of vector field can be defined in a nonlocal form.

The nonlocal gradient $\nabla_w f(x) : \Omega \to \Omega \times \Omega$, which is defined as a vector of all partial derivatives, has a form [5]:

$$(\nabla_w f)(x, y) = (f(y) - f(x))\sqrt{w(x, y)}, \quad \forall x, y \in \Omega. \tag{4}$$

Assume vector $\boldsymbol{v} = v(x, y) \in \Omega \times \Omega$, then nonlocal divergence $div_w \boldsymbol{v}(x) : \Omega \times \Omega \to \Omega$ can be written as follows [5]:

$$(div_w \boldsymbol{v})(x) = \int_{\Omega} (v(x, y) - v(y, x))\sqrt{w(x, y)} dy. \tag{5}$$

It is well known that the Laplace operator can be defined by the divergence of the gradient up to a constant multiplication, therefore with the notions mentioned above we can define nonlocal Laplace operator in this case by:

$$\Delta_w f(x) = \int_a^b (f(y) - f(x)) w(x, y) dy, \tag{6}$$

for each function $f : \Omega \to \mathbb{R}$ and $[a, b] \subset \Omega$.

2 Fuzzy Partition

The choice of proper proximity function is extremely important. In this article we proposed one, where proximity is determined by *fuzzy partition* $A_1, \ldots, A_n$.

Definition 2. *Fuzzy sets* $A_1, \ldots, A_n$, *establish a fuzzy partition of a real interval* $[a, b]$ *with nodes* $x_1 < \cdots < x_n$ *if for* $k = 1, \ldots, n$ *holds [6]:*

1. $A_k : [a, b] \to [0, 1]$, $\quad A_k(x_k) = 1$, $\quad A_k(x) > 0$ *if* $x \in (x_{k-1}, x_{k+1})$
2. $A_k(x) = 0$ *if* $x \notin (x_{k-1}, x_{k+1})$, *where* $x_0 = a$ *and* $x_{n+1} = b$
3. $A_k(x)$ *is continuous*
4. $A_k(x)$, *for* $k = 2, \ldots, n$, *strictly increases on* $[x_{k-1}, x_k]$ *and* $A_k(x)$ *strictly decreases on* $[x_k, x_{k+1}]$ *for* $k = 1, \ldots, n-1$,
5. $\forall x \in [a, b]$

$$\sum_{k=1}^{n} A_k(x) = 1. \tag{7}$$

The membership functions $A_1, \ldots, A_n$ *are called basic functions.*

Definition 3. *The fuzzy partition* $A_1, \ldots, A_n$, *for* $n \geq 2$ *is h-uniform [6] if nodes* $x_0 < \cdots < x_{n+1}$ *are h-equidistant, i.e. for all* $k = 1, \ldots, n+1$, $x_k = x_{k-1} + h$, *where* $h = (b - a)/(n + 1)$ *and the following additional properties are fulfilled:*

1. *for all* $k = 1, \ldots, n$ *and for all* $x \in [0, h]$, $A_k(x_k - x) = A_k(x_k + x)$,
2. *for all* $k = 2, \ldots, n$ *and for all* $x \in [x_{k-1}, x_{k+1}]$, $A_k(x) = A_{k-1}(x - h)$.

Definition 4. *If the fuzzy partition* $A_1, \ldots, A_n$ *of* $[a, b]$ *is h-uniform, then there exists [7] an even function* $A_0 : [-1, 1] \to [0, 1]$, *such that for all* $k = 1, \ldots, n$:

$$A_k(x) = A_0\left(\frac{x - x_k}{h}\right), \quad x \in [x_{k-1}, x_{k+1}].$$

A_0 *is called a generating function of uniform fuzzy partition.*

Corollary 1. *Generating function* A_0 *produces infinitely many rescaled functions [8]* $A_H : \mathbb{R} \to [0, 1]$ *with the scale factor* $H > 0$, *so that:*

$$A_H(x) = A_0\left(\frac{x}{H}\right).$$

A (h,H)-uniform partition of $\mathbb{R}$ *is then a collection of translations* $\{A_H(x - k \cdot h), k \in \mathbb{Z}\}$.

3 Fuzzy Transform

Direct Fuzzy transform or *F-transform* is a result of weighted linear integral transformation of a continuous function with weights determined by basic functions.

Definition 5. *Let $A_1, \ldots, A_n$ be basic functions which form a fuzzy partition of $[a, b]$ and f be any function from $C([a, b])$. We say that n-tuple of real numbers $F[f] = (F_1, \ldots, F_n)$ given by [6]*

$$F_k = \frac{\int_a^b f(x)A_k(x)dx}{\int_a^b A_k(x)dx}, \quad k = 1, \ldots, n, \tag{8}$$

is the direct integral F-transform of f with respect to $A_1, \ldots, A_n$.

F-transform establishes a correspondence between a set of continuous functions on $[a, b]$ and the set of n-dimensional vectors. *Inverse F-transform* then converts an n-dimensional vector of components $(F_1, \ldots, F_n)$ into another continuous function:

$$\hat{f}(x) = \sum_{k=1}^{n} F_k A_k(x), \tag{9}$$

which approximates the original one.

4 Proximity Function Determined by Fuzzy Partition

Let $[a, b] \subset \mathbb{R}$, $f \in L^2([a, b])$, $h > 0$ and A_h be a h-rescaled generating function. Assume the measure of proximity as follows:

$$w(x, y) = \frac{1}{h} A_h(x - y). \tag{10}$$

Proposition 1. *Let a generalized h-uniform fuzzy partition of $[a, b]$ be given by the infinite set of basic functions $\{A_y(x) | y \in [a, b]\}$, where $A_y(x) = A_h(x - y)$. Then for all $x \in [a, b]$:*

$$\Delta_w^{FT} f(x) = F_x - f(x), \tag{11}$$

defines nonlocal Laplace operator. F_x denotes the particular x^{th} F-transform component of F-transform $F[f]$ of function f.

Proof.

$$\begin{aligned}
\Delta_w^{FT} f(x) &= \int_\Omega (f(y) - f(x))\omega(x, y)\,\mathrm{d}y \\
&= \int_a^b (f(y) - f(x))\frac{1}{h} A_h(x - y)\,\mathrm{d}y \\
&= \frac{\int_a^b f(y)A_h(x - y)\,\mathrm{d}y}{h} - \frac{f(x)}{h} \int_a^b A_h(x - y)\,\mathrm{d}y \\
&= F_x - f(x).
\end{aligned}$$

□

5 Properties of FT-Laplace Operator Δ_w^{FT}

In this section we propose several properties of the operator Δ_w^{FT}.

Proposition 2. *For all $x, y \in [a, b]$, the operator Δ_w^{FT} admits following properties:*

1. *If $f(x) = \text{const}$, then $\Delta_w^{FT} f(x) = 0$.*
2. *For $w(x, y) > 0$, if $\Delta_w^{FT} f(x) = 0$, then $f(x) = \text{const}$.*
3. *If $f(x_0) \geq f(x)$, then $\Delta_w^{FT} f(x) \leq 0$.*
4. *Similarly for a minimum, if $f(x_1) \leq f(x)$, then $\Delta_w^{FT} f(x_1) \geq 0$.*
5. *Δ_w^{FT} is a positive semi-definite operator, i.e.*

$$\langle -\Delta_w^{FT} f(x), f(x) \rangle \geq 0, \tag{12}$$

where $\langle \cdot, \cdot \rangle$ denotes inner product on L^2, defined as $\langle f, g \rangle = \int_a^b f(x) g(x)\,\mathrm{d}x$.

6. *The following equation holds:*

$$\int_a^b \Delta_w^{FT} f(x)\,\mathrm{d}x = 0. \tag{13}$$

Proof. Property 1:

$$\begin{aligned}\Delta_w^{FT} f(x) &= \int_a^b (f(y) - f(x)) \frac{1}{h} A_h(x - y)\,\mathrm{d}y \\ &= \frac{f(x)}{h} \int_a^b A_h(x - y)\,\mathrm{d}y - f(x) = 0.\end{aligned}$$

Property 2:

$$\begin{aligned}\Delta_w^{FT} f(x) &= \frac{1}{h} \int_a^b f(y) A_h(x - y)\,\mathrm{d}y - f(x) \\ &= \frac{f(c)}{h} \int_a^b A_h(x - y)\,\mathrm{d}y - f(x) \\ &= f(c) - f(x) = 0 \\ \Rightarrow \quad & f(x) = \text{const.} \quad \forall x \in [a, b].\end{aligned}$$

Property 3:

$$\begin{aligned}\Delta_w^{FT} f(x_0) &= \frac{1}{h} \int_a^b (f(y) - f(x_0)) A_h(x - y)\,\mathrm{d}y \\ &= \frac{1}{h} \int_a^b f(y) A_h(x - y)\,\mathrm{d}y - \frac{f(x_0)}{h} \int_a^b A_h(x - y)\,\mathrm{d}y \\ &\leq \frac{f(x_0)}{h} \int_a^b A_h(x - y)\,\mathrm{d}y - \frac{f(x_0)}{h} \int_a^b A_h(x - y)\,\mathrm{d}y = 0\end{aligned}$$

$$\Rightarrow \Delta_w^{FT} f(x_0) \leq 0.$$

Property 4:

$$\begin{aligned}\Delta_w^{FT} f(x_1) &= \frac{1}{h}\int_a^b (f(y) - f(x_1))A_h(x-y)\,\mathrm{d}y \\ &= \frac{1}{h}\int_a^b f(y)A_h(x-y)\,\mathrm{d}y - \frac{f(x_1)}{h}\int_a^b A_h(x-y)\,\mathrm{d}y \\ &\geq \frac{f(x_1)}{h}\int_a^b A_h(x-y)\,\mathrm{d}y - \frac{f(x_1)}{h}\int_a^b A_h(x-y)\,\mathrm{d}y = 0 \\ \Rightarrow \Delta_w^{FT} f(x_1) &\geq 0.\end{aligned}$$

Property 5:

$$\begin{aligned}\langle -\Delta_w^{FT} f(x), f(x)\rangle &= \langle f(x) - \frac{1}{h}\int_a^b f(y)A_h(x-y)\,\mathrm{d}y, f(x)\rangle \\ &= \frac{1}{h}\int_a^b\int_a^b (f^2(x) - f(x)f(y))A_h(x-y)\,\mathrm{d}y\mathrm{d}x \\ &= \frac{1}{2h}\int_a^b\int_a^b [(f^2(x) - f(x)f(y))A_h(x-y) \\ &+ (f^2(x) - f(x)f(y))A_h(x-y)]\,\mathrm{d}y\mathrm{d}x \\ &= \frac{1}{2h}\int_a^b\int_a^b [(f^2(x) - f(x)f(y))A_h(x-y) \\ &+ (f^2(y) - f(y)f(x))A_h(x-y)]\,\mathrm{d}y\mathrm{d}x \\ &= \frac{1}{2h}\int_a^b\int_a^b (f(x) - f(y))^2 A_h(x-y)\,\mathrm{d}y\mathrm{d}x \geq 0 \\ \Rightarrow \langle -\Delta_w^{FT} f(x), f(x)\rangle &\geq 0.\end{aligned}$$

Property 6:

$$\begin{aligned}&\int_a^b \Delta_w^{FT} f(x)\,\mathrm{d}x \\ &= \frac{1}{2}\int_a^b\int_a^b [(f(y) - f(x))A_h(x-y) + (f(y) - f(x))A_h(x-y)]\,\mathrm{d}x\mathrm{d}y \\ &= \frac{1}{2}\Big[\int_a^b\int_a^b ((f(y) - f(x))A_h(x-y)\mathrm{d}x\mathrm{d}y - \int_a^b\int_a^b ((f(x) - f(y))A_h(x-y)\mathrm{d}x\mathrm{d}y\Big] \\ &= 0.\end{aligned}$$

□

6 Application to Image Processing

Image regularization that uses the nonlocal Laplace operator is proved to be very efficient [9]. The regularization of an image function f^0 corresponds to an optimization problem, which can be formalized by the minimization of a weighted sum of two energy terms [9]:

$$\min_{f}\left\{E_w^p(f,f^0,\lambda)=R_w^p(f)+\frac{\lambda}{2}\|f-f^0\|_2^2\right\}, \tag{14}$$

where $R_w^p(f)=\frac{1}{p}\sum_{u\in V}|\nabla_w f(u)|^p$ (details can be found in [9]). When assuming $p=2$, this problem has a unique solution.

In [9], linearized Gauss-Jacobi iterative method was used to solve this problem. Let t be an iteration step, and let $f^{(t)}$ be the solution at the step t. The method is given by the following algorithm:

$$f^{(0)}=f^0 \tag{15}$$

$$f^{(t+1)}(u)=\frac{\lambda f^0(u)+\sum_{v\sim u}\gamma_w^{f^{(t)}}(u,v)f^{(t)}(v)}{\lambda+\sum_{v\sim u}\gamma_w^{f^{(t)}}(u,v)},\quad \forall u\in V. \tag{16}$$

It describes a family of discrete diffusion processes, which is parametrized by the structure of the graph, the edge weights, the parameter λ and the parameter p.

In our case, $p=2$, the equation of the $(t+1)$th step is simplified to:

$$f^{(t+1)}(u)=\frac{\lambda f^0(u)+2\sum_{v\sim u}w(u,v)f^{(t)}(v)}{\lambda+2\sum_{v\sim u}w(u,v)}. \tag{17}$$

The minimization problem and the discrete diffusion processes can be used to regularize any function defined on a finite set of discrete data. This is realized by constructing a weighted graph $G=(V,E,w)$, and by selecting the function to be regularized as a function f^0, defined on the vertices of the graph.

Graph is produced as follows. Each pixel is identified with one vertex and semantically related pixels are connected by edges. The edges weights are computed according to a symmetric similarity function $m: V\times V\to\mathbb{R}$. If between vertices u and v does not exist an edge, then $w(u,v)=0$, otherwise $w(u,v)=m(u,v)$.

Every data $u\in V$ is assigned with a feature vector $F(f^0,u)\in\mathbb{R}^q$. In the simplest case, one can consider $F(f^0,u)=f^0(u)$.

Also the choice of graph topology plays an important role, because different types of graphs are suitable to use for different types of problems.

6.1 Image Denoising

Consider an image damaged by additional noise and the goal of this method is to restore the initial uncorrupted image. In our case RGB noise was added. We analyzed the case of weight function depending on the fuzzy partition for a fixed value of parameter p. The scalar feature vector was used, $F(f^0,u)=f(u)$. And for this configuration, we considered a standard 4-adjacency grid graph.

Various values of parameter λ were tested. For illustration we selected the one best output after 15 iterations of regularization process. Figure 1 shows, that the noise was partly removed and this process caused minimal damage to geometric features.

Fig. 1. Image denoising illustration, original image (first), image corrupted by a noise (second) and recovered image (third). Parameters: RGB noise (noise level 0.4 in each channel, random seed set to 222), 4-adjacency grid graph, $F(f^0, u) = f(u)$, weight function based on fuzzy partition, $\lambda = 2$.

7 Conclusion

A new representation of nonlocal Laplace operator in a space with a fuzzy partition is proposed and analysed. It stems from the theory of fuzzy transform, where the weight assignment is based on a generating function of a fuzzy partition and represents proximity between points. We proved validity of all important properties of this operator and illustrated its usefulness in image denoising.

In the future work, we would like to continue in this direction, we are now focusing on how a new expression of nonlocal Laplace operator can be applied to image processing tasks, specifically we would like to turn the attention to image segmentation and filtering, that are connected to the regularization. This nonlocal approach is significantly computationally simpler, so we expect that compared to classical methods it will be a significantly lower time consuming, which is one of the main priorities in this field.

Acknowledgements. The authors thank the reviewers for their valuable comments and suggestions to improve the quality of the paper. The support of the grant SGS01/UVAFM/2020 is kindly announced.

References

1. Belkin, M., Niyogi, P.: Laplacian eigenmaps for dimensionality reduction and data representation. Neural Comput. **15**(6), 1373–1396 (2003). https://doi.org/10.1162/089976603321780317
2. von Luxburg, U.: A tutorial on spectral clustering. Stat. Comput. **4**(17), 395–416 (2007). https://doi.org/10.1007/s11222-007-9033-z
3. Lezoray O., Ta V. T., Elmoataz A.: Nonlocal graph regularization for image colorization. In: 19th International Conference on Pattern Recognition, Tampa, FL, pp. 1–4 (2008). https://doi.org/10.1109/ICPR.2008.4761617
4. Gilboa, G.: Nonlocal linear image regularization and supervised segmentation. Multiscale Model. Simul. **6**(2), 595–630 (2007). https://doi.org/10.1137/060669358
5. Gilboa, G., Osher, S.: Nonlocal operators with applications to image processing. Multiscale Model. Simul. **7**(3), 1005–1028 (2009). https://doi.org/10.1137/070698592
6. Perfiljeva, I.: Fuzzy transforms: theory and applications. Fuzzy Sets Syst. **157**(8), 993–1023 (2006). https://doi.org/10.1016/j.fss.2005.11.012
7. Perfiljeva, I., Daňková, M., Bede, B.: Towards a higher degree F-transform. Fuzzy Sets Syst. **180**(1), 3–19 (2011). https://doi.org/10.1016/j.fss.2010.11.002
8. Perfilieva, I., Vlašánek, P.: Total variation with nonlocal FT-Laplacian for patch-based inpainting. Soft. Comput. **23**(6), 1833–1841 (2018). https://doi.org/10.1007/s00500-018-3589-8
9. Elmoataz, A., Lézoray, O., Bougleux, S.: Nonlocal discrete regularization on weighted graphs: a framework for image and manifold processing. IEEE Trans. Image Process. **17**(7), 1047–1060 (2008). https://doi.org/10.1109/TIP.2008.924284

A Comparison of Explanatory Measures in Abductive Inference

Jian-Dong Huang(✉), David H. Glass(✉), and Mark McCartney(✉)

School of Computing, Ulster University,
Newtownabbey, Co. Antrim, Northern Ireland BT37 0QB, UK
{jd.huang,dh.glass,m.mccartney}@ulster.ac.uk

Abstract. Computer simulations have been carried out to investigate the performance of two measures for abductive inference, Maximum Likelihood (ML), and Product Coherence Measure (PCM), by comparing them with a third approach, Most Probable Explanation (MPE). These have been realized through experiments that compare outcomes from a specified model (the correct model) with those from incorrect models which assume that the hypotheses are mutually exclusive or independent. The results show that PCM tracks the results of MPE more closely than ML when the degree of competition is greater than 0 and hence is able to infer explanations that are more likely to be true under such a condition. Experiments on the robustness of the measures with respect to incorrect model assumptions show that ML is more robust in general, but that MPE and PCM are more robust when the degree of competition is positive. The results also show that in general it is more reasonable to assume the hypotheses in question are independent than to assume they are mutually exclusive.

Keywords: Inference to the Best Explanation (IBE) · Explanatory reasoning · Hypotheses competition · Abduction

1 Introduction

In modern literature, abduction refers to the study of explanatory reasoning in justifying hypotheses, or Inference to the Best Explanation (IBE) that considers a number of plausible candidate hypotheses in a given evidential context and then compares these hypotheses in order to make an inference to the one that best explains the relevant evidence [1–11].

In conventional studies involving IBE, significant attention has been paid to dealing with the hypotheses being mutually exclusive [12–15], and the measure for identifying the most plausible hypothesis was typically chosen as the maximized posterior probability [16, 17] termed Most Probable Explanation (MPE). However, modern studies have highlighted situations where hypotheses can be in competition even though they are not mutually exclusive and can compete to varying degrees [18, 19]. Meanwhile, alternative measures to MPE have been considered and applied [12, 19–29], such as Maximum Likelihood (ML) and Product Coherence Measure (PCM) [11]. The reality that hypotheses often have various degrees of competition gives rise to the necessity of examining the characteristics of abductive inference in such a context, along with the

M.-J. Lesot et al. (Eds.): IPMU 2020, CCIS 1239, pp. 304–317, 2020.
https://doi.org/10.1007/978-3-030-50153-2_23

characteristics of the functioning and performance of the explanatory measures used to make inferences in these contexts. This then motivates a study to compare the performance of several measures when conducting abductive inference under the assumption that they may be competing to some extent, with an objective of identifying the most suitable measure(s) as the criteria/criterion for the inference to the best explanation, thus benefiting abductive inference research. Therefore, in this study, comparison of the performance and functionality among the measures MPE, ML, and PCM in identifying the best explanation has been carried out. We consider a number of different probability model settings, as an extension of the study of abductive inference under various degrees of competition between candidate hypotheses [19].

2 Competing Hypotheses and Degree of Competition

We start by giving an example to illustrate the competition concept. Suppose a detective has two main suspects in a murder inquiry, Smith and Jones. The detective tries to determine which hypothesis best explains all the relevant evidence by treating the suspects as two competing hypotheses and reasoning abductively. The hypotheses can be represented as:

H_S: Smith committed the murder
H_J: Jones committed the murder

In general, the hypotheses need not be assumed to be mutually exclusive, since both Smith and Jones could have colluded in committing the murder and hence it would be improper to assume that $P(H_S \& H_J) = 0$. Clearly, if the two hypotheses are known to be mutually exclusive (if Smith and Jones could not have colluded), then they are competing hypotheses. In reality, it might be difficult to establish mutual exclusion, but in many cases, it would still be reasonable to treat them as competing hypotheses. Perhaps, for example, in light of the evidence it is very unlikely but not impossible that Smith and Jones colluded.

Glass [19] proposed a definition for competing hypotheses: Let each of H_1 and H_2 be hypotheses and E evidence under consideration and suppose that $P(H_1 \& E)$ and $P(H_2 \& E)$ are greater than zero. Hypotheses H_1 and H_2 are said to be competing hypotheses with respect to evidence E if and only if $P(H_1 | H_2 \& E) < P(H_1 | E)$. Because the competition is a symmetric concept, the formula can also be expressed as $P(H_2 | H_1 \& E) < P(H_2 | E)$.

Schupbach and Glass [18] recently defined a measure of the degree of competition between two hypotheses, H_1 and H_2, with respect to evidence E, as the average degree to which H_1 and H_2 disconfirm each other given E:

$$\mathrm{Comp}(H_1, H_2 | E) = \frac{1}{2} \times [C_1(H_1, \neg H_2 | E) + C_1(H_2, \neg H_1 | E)] \tag{1}$$

where C_l is the likelihood ratio measure of confirmation conditioned on E, that is,

$$C_l(H_1,\ H_2 \mid E) = \log \frac{P\,(H_1 \mid H_2 \& E)}{P\,(H_1 \mid \neg H_2 \& E)} \tag{2}$$

By the definition, Comp has increasingly positive values to the extent that H_1 and H_2 disconfirm one another given E, and increasingly negative values to the extent that H_1 and H_2 confirm one another given E, and zero when H_1 and H_2 are probabilistically independent given E. Note that if H_1 and H_2 are assumed to be mutually exclusive, then $P(H_1| H_2\&E) = 0$ and hence this would be a case with the highest possible degree of competition. H_1 and H_2 can be said to compete with respect to E if Comp > 0. It can then be shown that H_1 and H_2 compete with respect to E if the condition in the definition is met. Since the measure of competition lies in the range $[-\infty, \infty]$, their alternative measure, which lies in the range $[-1, 1]$, has been used for convenience in this study. It is given by [18]

$$\text{dcomp} = \frac{1}{2} \times [C_k(H_1, \neg H_2 \mid E) + C_k(H_2, \neg H_1 \mid E)] \tag{3}$$

where C_k is the confirmation measure proposed by Kemeny and Oppenheim [30] when conditioned on E,

$$C_k(H_1, H_2 \mid E) = \frac{P\,(H_1 \mid H_2 \& E) - P(H_1 \mid \neg H_2 \& E)}{P\,(H_1 \mid H_2 \& E) + P(H_1 \mid \neg H_2 \& E)} \tag{4}$$

3 Probability Model and Experiment Design

Computer simulations were carried out to investigate the performance and functionality of the different measures when incorrectly assuming hypotheses to be mutually exclusive or independent, on making inferences. Each of the experiments concerned generating a probability model involving evidence E and hypotheses H_1, H_2 and a catchall hypothesis, $H_c = \neg H_1 \& \neg H_2$, with a specified degree of competition between H_1 and H_2 given E [19]. This model was stipulated to be the correct model, $Prob_O$, and three different incorrect probability models, modified from that of $Prob_O$, in which H_1 and H_2 were treated as mutually exclusive for two experiments (named MEx1 and MEx2 respectively), and treated as independent for a third experiment (named IND), were used for the inference.

These are intended to represent simplifying assumptions that might be made in practice when the true probability model is unknown. The goal is then to evaluate several versions of abductive inference under these assumptions. Each of the experiments were repeated a large number of times ($N = 10^6$) to sample the distribution over the variables and obtain a meaningful average, and a Degree of Agreement (*DA*) is defined as follows:

Let N_identified be the number of times the hypothesis identified by the incorrect model agrees with the correct model, N_total be the total number of observations, under a given degree of competition dcomp, then

$$DA(\%) = 100 \times \frac{\text{N\_identified}}{\text{N\_total}} \tag{5}$$

Note that the *DA* is a parameter closely linked to the degree of competition, or the *DA* is a function of dcomp, because N_identified is a function of dcomp.

We also use *DA* Index (*DAI*) to represent the average Degree of Agreement over the interval [−1, 1] containing successively (with a certain step) n points of dcomp:

$$DAI = \frac{\sum_{k=1}^{n} (DA)_k}{n} \tag{6}$$

The simulations are an extension of a study in which the measure MPE was used as the standard for hypothesis identification [19] since here ML and PCM are also used. Details of how the correct probability model and the incorrect models were constructed can be found in [19]. Design of the extended experiments can be sketched as follows [11, 19]:

Firstly, for a specified value of the degree of competition, dcomp (Eq. (3)), a probability model was defined and stipulated as the correct model $Prob_0$, involving hypotheses H_1, H_2, a catchall H_c, and evidence E, where H_1 and H_2 are not assumed to be mutually exclusive; the initial parameters in the model are randomly generated from a uniform distribution.

For MEx1, a mutually exclusive probability model $Prob_1$ is obtained from the original model, $Prob_0$, by replacing H_1 with $H_1\&\neg H_2$ and H_2 with $H_2\&\neg H_1$, setting:

$$Prob_1(H_1) = Prob_0(H_1\&\neg H_2) \tag{7}$$

$$Prob_1(H_2) = Prob_0(H_2\&\neg H_1) \tag{8}$$

$$Prob_1(H_1\&H_2) = 0 \tag{9}$$

$$Prob_1(H_c) = 1 - Prob_1(H_1) - Prob_1(H_2) \tag{10}$$

$$Prob_1(E \mid H_1) = Prob_0(E \mid H_1\&\neg H_2) \tag{11}$$

$$Prob_1(E \mid H_2) = Prob_0(E \mid H_2\&\neg H_1) \tag{12}$$

$$Prob_1(E \mid H_c) = Prob_0(E \mid H_c) \tag{13}$$

In the second experiment MEx2, another mutually exclusive probability model, $Prob_2$, is obtained from the original model $Prob_0$,

$$Prob_2(H_1) = Prob_0(H_1) \times \frac{Prob_0(H_1 \vee H_2)}{Prob_0(H_1) + Prob_0(H_2)} \quad (14)$$

$$Prob_2(H_2) = Prob_0(H_2) \times \frac{Prob_0(H_1 \vee H_2)}{Prob_0(H_1) + Prob_0(H_2)} \quad (15)$$

The probabilities for the likelihood terms are set in the same way as for $Prob_1$, and the probability for the catchall hypothesis, H_c, is similarly set to

$$Prob_2(H_c) = 1 - Prob_2(H_1) - Prob_2(H_2) \quad (16)$$

since H_1 and H_2 are assumed to be mutually exclusive [19]; and we have:

$$Prob_2(H_1 \& H_2) = 0 \quad (17)$$

$$Prob_2(E \mid H_1) = Prob_0(E \mid H_1 \& \neg H_2) \quad (18)$$

$$Prob_2(E \mid H_2) = Prob_0(E \mid H_2 \& \neg H_1) \quad (19)$$

In contrast to the models $Prob_1$ and $Prob_2$, the third experiment IND treats H_1 and H_2 as independent:

$$Prob_3(H_1) = Prob_0(H_1) \times \frac{Prob_0(H_1 \vee H_2)}{Prob_0(H_1) + Prob_0(H_2)} \quad (20)$$

$$Prob_3(H_2) = Prob_0(H_2) \times \frac{Prob_0(H_1 \vee H_2)}{Prob_0(H_1) + Prob_0(H_2)} \quad (21)$$

$$Prob_3(H_1 \& H_2) = Prob_3(H_1) \times Prob_3(H_2) \quad (22)$$

$$Prob_3(H_c) = 1 - Prob_3(H_1 \vee H_2) \quad (23)$$

$$\begin{aligned} &Prob_3(E|H_1) = \\ &Prob_0(E|H_1 \& H_2) \times Prob_3(H_2) + Prob_0(E|H_1 \& \neg H_2) \times Prob_3(\neg H_2) \end{aligned} \quad (24)$$

$$\begin{aligned} &Prob_3(E|H_2) = \\ &Prob_0(E|H_2 \& H_1) \times Prob_3(H_1) + Prob_0(E|H_2 \& \neg H_1) \times Prob_3(\neg H_1) \end{aligned} \quad (25)$$

This is because $Prob_3$ provides a compromise between incorrectly treating hypotheses as mutually exclusive and fully taking into account the dependence between them; we need to consider that in some cases hypotheses are modelled as being independent as well.

Secondly, abductive inference was carried out under a given degree of competition dcomp, for the correct models to find the hypothesis which maximizes the selected measure (MPE, ML, PCM) for evidence E. If an inference made by the incorrect model (mutually exclusive or independent) in identifying the hypothesis agreed with the inference made using the correct model, this inference is then counted as a success.

Thirdly, the above process was repeated $N = 10^6$ times and the number of total successful inferences S was obtained. The accuracy, or the degree of agreement between the inference of the incorrect model (mutually exclusive or independent) and the correct model, was defined as S/N (percentage success). The accuracy values reflect how often an incorrect model identifies the same hypothesis as the correct model.

Finally, the process was repeated for a range of values of the degree of competition between −0.9 and 0.9, with a step of 0.1.

There are a number of different measures proposed in the literature, to quantify how well a hypothesis H explains evidence E. For example, the Measure of Explanatory Power proposed by Schupbach and Sprenger [22]; the measure proposed by Crupi and Tentori [24]; the measure by Good [25] and McGrew [26]; the Likelihood Ratio measure [19]; the Overlap Coherence Measure used to rank explanations by Glass [27–29]; and the Product Coherence Measure by Glass [11, 12]. In this study, the following measures have been used for the inference in the computer simulation:

(1) MPE: Most Probable Explanation; selects the hypothesis with the maximum posterior probability, of the hypotheses in light of the evidence,

$$MPE = \underset{H_i, i \in \{1,2,C\}}{\operatorname{argmax}} P(H_i|E) \tag{26}$$

(2) ML: selects the hypothesis with the Maximum Likelihood,

$$ML = \underset{H_i, i \in \{1,2,C\}}{\operatorname{argmax}} P(E|H_i) \tag{27}$$

(3) PCM: selects the hypothesis with the maximum value of the Product Coherence Measure [11]:

$$PCM = \underset{H_i, i \in \{1,2,C\}}{\operatorname{argmax}} [P(H_i|E) \times P(E|H_i)] \tag{28}$$

Arguably, MPE is not a good measure of explanation [20–27]. In the example of the murder suspects, the probability that both Smith and Jones are guilty, $P(H_S\&H_J|E)$, will obviously always be less than or equal to that of the individual hypotheses, $P(H_S|E)$ or $P(H_J|E)$. However, if MPE is used as a measure of explanation, this means that the joint explanation that Smith and Jones committed the murder can never provide a better explanation than the individual explanations that Smith (or Jones) committed the murder. More generally, this means that it only makes sense to use MPE for a fixed number of explanatory variables, but arguably in various contexts, such as explanation in Bayesian networks, it is desirable to compare different numbers of explanatory variables in order to obtain explanations that are neither too simple nor too complex [21].

But MPE is still a useful measure to include for comparison since we are interested in whether inferences made using explanatory measures such as ML and PCM have a high probability of being correct.

As a further extension of [19], the performance of these three measures was examined and compared in the computer simulation of abductive inference for identifying the most probably correct explanation. The computer simulations were carried out with the procedures described earlier. In reality we typically do not know the true model, so we are evaluating how well abductive inference works with different incorrect assumptions, mutually exclusive or independent. Bearing this in mind, within each of the experiments, two groups of comparisons were made:

Group 1: Here the assumption is that explanatory approaches (ML and PCM) should be compared against MPE as the standard to see how good ML and PCM are at inferring hypotheses that are probably true. Thus, with MPE as a standard in the hypothesis identification in the correct model, this group is to find out the degree of agreement of hypothesis identification made from the correct model against that from an incorrect model, with MPE, ML, and PCM respectively as the criterion in the hypothesis identification in the incorrect model. These experiments are repeated for each of the incorrect models (MEx1, MEx2, IND). The inferences with the three measures have been abbreviated as:

- MPE_F versus MPE_T: using MPE criterion in the incorrect (False or _F) model against using MPE criterion in the correct (True or _T) model to infer a hypothesis;
- ML_F versus MPE_T: using Maximum Likelihood (ML) criterion in the incorrect (_F) model against using MPE criterion in the correct (_T) model to infer a hypothesis;
- PCM_F versus MPE_T: using Product Coherence Measure (PCM) criterion in the incorrect (_F) model against using MPE criterion in the correct (_T) model to infer a hypothesis;

Group 2: In abductive inference we do not necessarily need to consider MPE as the only standard. Therefore, in this group, each of the three measures was applied in the inference as the criterion for both incorrect model and the correct model, i.e., taking each of the three measures as the standard, to find out the degree of agreement of hypothesis identification made from the correct model against that from an incorrect model. These experiments can be seen as evaluating the robustness of each of the measures with respect to the incorrect model assumptions (mutually exclusive or independent). Again, these experiments are repeated for each of the incorrect models (MEx1, MEx2, IND). The inferences have been abbreviated as:

- MPE_F versus MPE_T: this is the same as in the Group1;

- ML_F versus ML_T: using Maximum Likelihood (ML) criterion in the incorrect (_F) model against using ML criterion in the correct (_T) to infer a hypothesis;
- PCM_F versus PCM_T: using Product Coherence Measure (PCM) criterion in the incorrect (_F) model against using PCM criterion in the correct (_T) model to infer a hypothesis;

The output of the two groups are expressed by Degree of Agreement, *DA*, and *DAI*, formulated in (5) and (6). It should be noted that this metric should be interpreted in one of two ways, depending on what comparison is being made. For Group 1, the *DA* reflects the Accuracy of the corresponding measure, i.e., given that we presume that MPE is a standard for identifying the true hypothesis, the degree of agreement with the identification by MPE is then viewed as the accuracy of the relevant measure, and the higher the *DA* is, the better the Accuracy the measure possesses. In Group 2, the same measure is used in the incorrect model and the correct model, and in this case the output of the three experiments will show how close the three series of output will be, i.e., the degree of consistency of the same measure under different experiments. In this case we say that the higher the degree of agreement the more *robust* the measure is. Therefore, under the circumstance of Group 2, we say that the Degree of Agreement reflects the Robustness of the measure. Accuracy and Robustness are then used in this work to represent the relevant properties of the explanatory measures.

4 Results

Graphs were plotted to illustrate the results of MEx1, MEx2 and IND, for comparison of the performance (Accuracy and Robustness) of the three explanatory measures, MPE (Most Probable Explanation), ML (Maximum Likelihood), and PCM (Product Coherence Measure).

Figure 1-1 shows that when the MPE is used as the standard, ML and the PCM have lower degree of agreement with the identification using MPE, but PCM is much closer to MPE than ML as found in [12]. When dcomp < 0, the curves drop to below 50%, suggesting that for negative degree of competition all three measures result in poor agreement with the output of using the standard MPE and PCM performs slightly better than MPE.

Figure 1-2 and 1-3 exhibit a similar trend as in Fig. 1-1 when dcomp > 0; but in the range of dcomp < 0, the curves are better ordered from high to low without crossing. As expected, the MPE curve is higher than those of ML and PCM, and noticeably the MPE and PCM curves are all above 50% in the whole range [−0.9, 0.9]. The degree of agreement for the measure ML appears lower, with the value less than 50% in the majority of the interval for all three experiments.

In Fig. 1-3, all the curves for MPE and PCM are above 60% when dcomp > 0, showing that the PCM performs very well compared to ML in identifying the most probably correct explanation, with MPE as the standard. For dcomp < 0, the output of MPE and PCM still have their accuracy greater than 50%. However, the ML curve

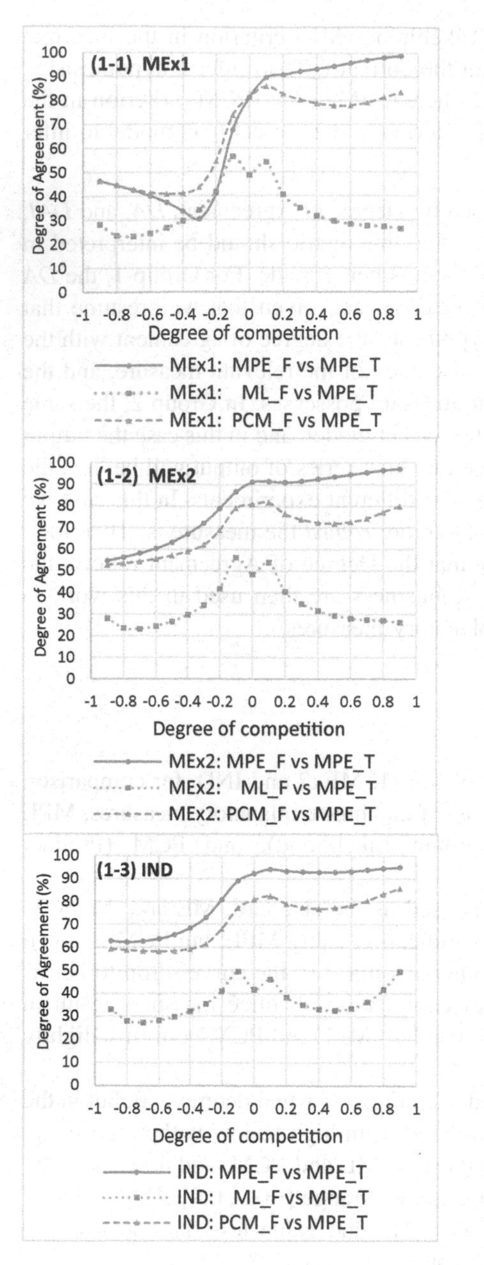

Fig. 1. Degree of agreement (Accuracy) between the output of the incorrect model (in MEx1, MEx2, and IND) using each of the measures and the correct model ($Prob_0$) using MPE to infer a hypothesis.

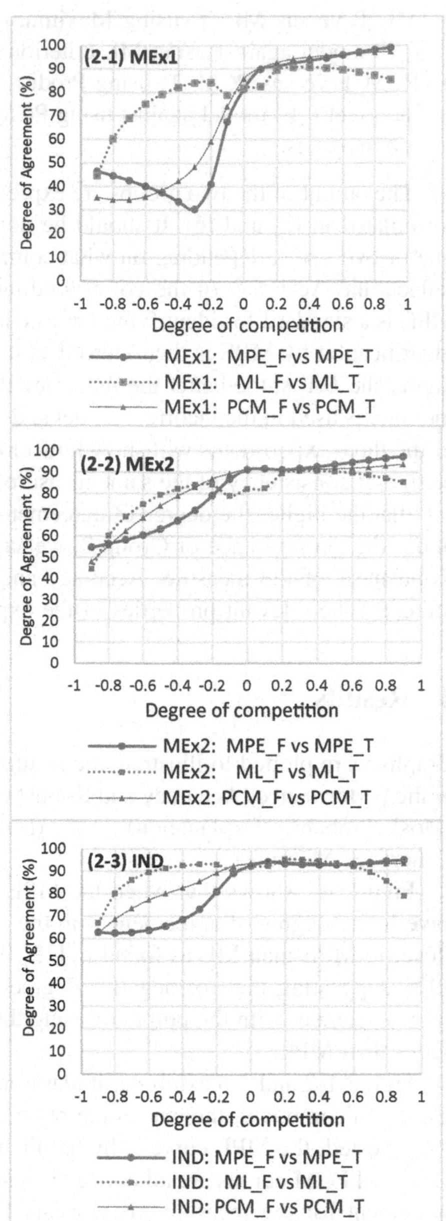

Fig. 2. Degree of agreement (Robustness) between the output of the incorrect model (in MEx1, MEx2, and IND) using each of the measures and the correct model ($Prob_0$) using the corresponding measure to infer a hypothesis.

output drops to below 30% when dcomp is in [−0.8,−0.6]. In all the three curves, PCM performs much better than ML.

In Fig. 1-1 through 1-3, it appears that the curves of the MPE measure are higher than PCM's for dcomp > 0 but the PCM curve is higher than the MPE curve for MEx1 from −0.6 to 0. In the majority of the range [−0.9, 0.9], MPE curves are higher. For MPE itself, it shows good degrees of agreement with the correct model in identifying the hypotheses when dcomp > 0; whilst in the case of dcomp < 0, it results in poor agreement with the output of the correct model.

In general, the experiment results indicate that, with the MPE measure and correct model used as the standard in hypothesis identification (a) for dcomp > 0 (or dcomp > 0.1 for MEx1), the MPE curves are all above 90%; for dcomp < 0, the curves are above 50% for MEx2 and IND. This is the same as one of the results in [19], implying that to presume the hypotheses to be mutually exclusive or independent appears reasonable especially for the situation in which the hypotheses compete to some degree (dcomp > 0), in the context of the experiments; (b) the measure ML has low degree of agreement with MPE in hypotheses identification; and (c) PCM results in closer degrees of agreement with the output of the MPE measure.

However, the above features do not necessarily mean that the ML and PCM are 'worse' measures than the MPE measure. Although MPE is often referred to in the artificial intelligence literature as the most probable explanation, arguably, this is an inadequate definition of 'best explanation' [11, 12, 19] but it nevertheless provides a standard against which to compare the various explanatory measures to determine how good they are at identifying hypotheses that are probably true.

The curves merely reflect the degree of agreement of the identifications made by ML or PCM with MPE. Therefore, a further comparison as illustrated in Fig. 2, examines the performance and Robustness when using the measures of ML and PCM with the correct model as the standard for the same measures with the incorrect models (experiment Group 2). This shall reveal more significant information on the performance of the measures.

It can be seen in Figs. 2-1 to 2-3, that in the range of dcomp > 0, MPE and PCM show a better degree of agreement with the identification of the correct model in the hypothesis identification (greater than 90%), whilst when dcomp < 0, ML performs better than the other two, with the degree of agreement largely above 50%, and for IND it goes up to 90% in (−0.6,−0.2). These features reflect that MPE does not always perform better than the other measures. PCM has a high degree of agreement similar to the MPE when dcomp > 0 (the difference is less than 5%) but ML performs better than MPE and PCM in the majority of the range of dcomp < 0.

Among the three figures of Fig. 2-1 through 2-3, ML has its highest curve in Fig. 2-3 (for the experiment IND), and PCM has its highest one in Fig. 2-3 as well. The curves of PCM are only slightly lower than MPE in all three figures when dcomp > 0, i.e. the PCM curve and the MPE curve are very close when dcomp > 0.

Moreover, curves for PCM are above their MPE counterparts in the majority of the range of dcomp < 0. For dcomp < 0, PCM perform better than MPE whilst when dcomp > 0 the two measures perform similarly.

The PCM curve for IND in Fig. 2-3 is above 60% in the whole range −0.9 to 0.9. It is obvious that, with PCM as the standard, presuming the hypotheses to be mutually exclusive and independent are both reasonable when dcomp > 0; and presuming them as being independent appears more reasonable when dcomp < 0, under the condition of the experiments.

Further, for a quantitative understanding of the properties of the measures, Fig. 3 and 4 give the average values of the Degree of Agreement over the interval [−0.9, 0], [0, 0.9] and the whole range of [−0.9, 0.9], with Fig. 3 reflecting the Accuracy of the measures and Fig. 4 the Robustness of the measures. It can be seen that PCM has much higher Accuracy than ML but ML has slightly higher Robustness.

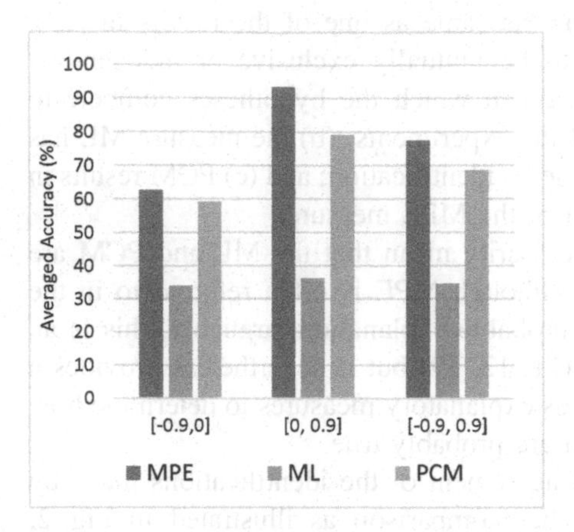

Fig. 3. Averaged *DAI* for the three experiments, showing a comparison of the performance of the measures ML and PCM against the presumed standard MPE for inferring a hypothesis in the experiment group 1, for the situations of dcomp ≤ 0, dcomp ≥ 0 and the whole interval [−0.9, 0.9].

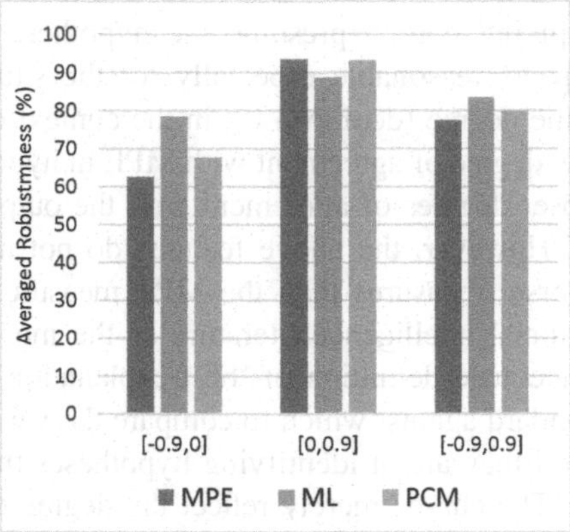

Fig. 4. Averaged Robustness for the output of the three experiments, illustrating a comparison of consistency of the measures for inferring a hypothesis under the incorrect model and correct model in the experiment group 2, for the situation of dcomp ≥ 0, dcomp ≤ 0, and for the whole range of [−0.9, 0.9].

5 Conclusions

Computer simulations have been carried out to investigate the performance, in terms of Accuracy and Robustness, of two measures for abductive inference, Maximum Likelihood (ML), and Product Coherence Measure (PCM). This has been done by comparing them with a third approach, Most Probable Explanation (MPE), for the identification of the best explanation. The results show that:

1. It appears appropriate to represent the functioning characteristics of the measures separately according to the sign of the degree of competition (dcomp) of the hypotheses, which can be calculated in practice using the Eqs. (1) through (4).
2. In terms of Accuracy, the results show that PCM tracks the results of MPE much more closely than ML especially when the degree of competition is positive, hence it is able to infer explanations that are much more likely to be true under such a condition.
3. Experiments on the Robustness of the measures with respect to incorrect model assumptions show that ML is in general more robust, although it is only slightly more robust than PCM. It performs better than MPE and PCM when the hypotheses are not competing (dcomp < 0) and in general. MPE and PCM are more robust and similar to each other when the degree of competition is positive; in general, PCM is more robust than MPE.
4. Presuming the hypotheses in question to be mutually exclusive appears reasonable when the hypotheses are competing (dcomp > 0) but could result in a low degree of agreement (accuracy) when they are not (dcomp < 0).
5. The experimental results also show that it is more reasonable to assume that the hypotheses are independent than to assume that they are mutually exclusive, both in the case of competing hypotheses and non-competing hypotheses.

Overall, the results show that PCM performs much better in terms of accuracy and only slightly worse in terms of robustness than ML. Hence, PCM seems preferable to ML as a measure for abductive inference. One limitation of the current work is that MPE has been used as a standard for determining accuracy. Future work will include simulations that designate hypotheses as true or false and then evaluate all three measures (MPE, PCM and ML) on an equal footing. Also, in the current work the different measures are used to infer the single best hypothesis, but since the hypotheses are not assumed to be mutually exclusive more than one hypothesis could be true. There is scope for comparing single hypotheses such as H_1 or H_2 with conjunctive hypotheses involving two or more hypotheses such as $H_1\&H_2$. Clearly, such a conjunction cannot be more probable than one of its conjuncts, so PCM and ML might be expected to have benefits over MPE in such contexts.

Acknowledgments. This publication was supported by a grant from the John Templeton Foundation (Grant ID 61115). The opinions expressed in this publication are those of the authors and do not necessarily reflect the views of the John Templeton Foundation. The authors would like to thank anonymous reviewers for helpful comments and suggestions.

References

1. Douven, I.: Abduction. In: Zalta, E.N. (ed.) The Stanford Encyclopedia of Philosophy. Metaphysics Research Lab, Stanford University, winter 2016 edition (2016)
2. Lipton, P.: Inference to the best explanation, 2nd edn. Taylor and Francis, London (2004)

3. Josephson, J.R., Susan, G.J.: Abductive Inference: Computation. Philosophy, Technology. Cambridge University Press, Cambridge (1996)
4. Harman, G.H.: The inference to the best explanation. Philos. Rev. **74**(1), 88–95 (1965)
5. Thagard, P.R.: The best explanation: Criteria for theory choice. J. Philos. **75**(2), 76–92 (1978)
6. Ben-Menahem, Y.: The inference to the best explanation. Erkenntnis **33**(3), 319–344 (1990)
7. Douven, I.: Inference to the best explanation made coherent. Philos. Sci. **66**, S424–S435 (1999)
8. Niiniluoto, I.: Defending abduction. Philos. Sci. **66**, S436–S451 (1999)
9. Gabbay, D., Woods, J.: The reach of abduction: Insight and Trial. 289–294 (2005)
10. Douven, I.: Inference to the best explanation, Dutch books, and inaccuracy minimization. Philos. Q. **63**(252), 428–444 (2013)
11. Glass, D.H.: Coherence, explanation, and hypothesis selection. Br. J. Philos. Sci. axy063 (2018). https://doi.org/10.1093/bjps/axy063
12. Glass, D.H.: An evaluation of probabilistic approaches to inference to the best explanation. Int. J. Approximate Reasoning **103**, 184–194 (2018)
13. Fenton, N., Neil, M., Lagnado, D., Marsh, W., Yet, B., Constantinou, A.: How to model mutually exclusive events based on independent causal pathways in Bayesian network models. Knowl.-Based Syst. **113**, 39–50 (2016)
14. Lam, F.C., Yeap, W.K.: Bayesian updating: on the interpretation of exhaustive and mutually exclusive assumptions. Artif. Intell. **53**(2–3), 245–254 (1992)
15. Norman Fenton, M.N., Lagnado, D.: Modelling mutually exclusive causes in Bayesian networks. Submitted to IEEE Transactions on Knowledge and Data Engineering (2011)
16. Pearl, J.: Probabilistic reasoning in intelligent systems: networks of plausible inference. Elsevier (2014)
17. Shimony, S.E.: Explanation, irrelevance and statistical independence. In: Proceedings of the Ninth National Conference on Artificial Intelligence-Volume **1**, 482–487 (1991)
18. Schupbach, J.N., Glass, D.H.: Hypothesis competition beyond mutual exclusivity. Philos. Sci. **84**(5), 810–824 (2017)
19. Glass, D.H.: Competing hypotheses and abductive inference. Ann. Math. Artif. Intel. 1–18 (2019). https://doi.org/10.1007/s10472-019-09630-0
20. Chajewska, U., Halpern, J.Y.: Defining explanation in probabilistic systems. In: Proceedings of the Thirteenth conference on Uncertainty in Artificial Intelligence, pp. 62–71. Morgan Kaufmann Publishers Inc. (1997)
21. Yuan, C., Lim, H., Lu, T.C.: Most relevant explanation in Bayesian networks. J. Artif. Intell. Res. **42**, 309–352 (2011)
22. Schupbach, J.N., Sprenger, J.: The logic of explanatory power. Philos. Sci. **78**(1), 105–127 (2011)
23. Schupbach, J.N.: Comparing probabilistic measures of explanatory power. Philos. Sci. **78**(5), 813–829 (2011)
24. Crupi, V., Tentori, K.: A second look at the logic of explanatory power (with two novel representation theorems). Philos. Sci. **79**(3), 365–385 (2012)
25. Good, I.J.: Weight of evidence, corroboration, explanatory power, information and the utility of experiments. J. Roy. Stat. Soc.: Ser. B (Methodol.) **22**(2), 319–331 (1960)
26. Mc Grew, T.: Confirmation, heuristics, and explanatory reasoning. Br. J. Philos. Sci. **54**(4), 553–567 (2003)
27. Glass, D.H.: Inference to the best explanation: does it track truth? Synthese **185**(3), 411–427 (2012). https://doi.org/10.1007/s11229-010-9829-9

28. Glass, D.H., McCartney, M.: Explanatory Inference under Uncertainty. In: International Conference on Intelligent Data Engineering and Automated Learning, pp. 215–222. Springer, Cham (2014)
29. Glass, D.H.: Coherence measures and inference to the best explanation. Synthese **157**, 275–296 (2007). https://doi.org/10.1007/s11229-006-9055-7
30. Kemeny, J.G., Oppenheim, P.: Degree of factual support. Philos. Sci. **19**(4), 307–324 (1952)

On Integral Transforms for Residuated Lattice-Valued Functions

Michal Holčapek(✉) and Viec Bui

CE IT4I - IRAFM, University of Ostrava, 30. dubna 22,
701 03 Ostrava 1, Czech Republic
michal.holcapek@osu.cz, bqviec@gmail.com
http://irafm.osu.cz

Abstract. The article aims to introduce four types of integral transforms for functions whose function values belong to a complete residuated lattice. The integral transforms are defined using so-called qualitative residuum based fuzzy integrals and integral kernels in the form of binary fuzzy relations. We present some of the basic properties of proposed integral transforms including a linearity property that is satisfied under specific conditions for comonotonic functions.

Keywords: Integral transform · Fuzzy transform · Residuated lattice · Integral kernel · Fuzzy integral

1 Introduction

Mathematical operators known as integral transforms produce a new function $g(y)$ by integrating the product of an existing function $f(x)$ and an integral kernel function $K(x, y)$ between suitable limits. The Fourier and Laplace transforms belong among the most popular integral transforms and are applied for real or complex-valued functions. The importance of the integral transforms is mainly in solving (partial) differential equations, algebraic equations, signal and image processing, spectral analysis of stochastic processes (see, e.g., [2,21,23]).

In fuzzy set theory we often deal with functions whose function values belong to an appropriate algebra of truth values as a residuated lattice and its special variants as the BL-algebra, MV-algebra, IMTL-algebra (see, e.g. [1,5,18]). In [20], Perfilieva introduced lattice-valued upper and lower fuzzy transforms that are, among others, used for an approximation of functions. A deeper investigation of fuzzy transforms properties can be found in [13–17,19,22]. In a recent article [9], we demonstrated that the lower and upper fuzzy transforms can be introduced as two type of integral transforms, where the multiplication based fuzzy integral is applied [3,4]. Namely, for a fuzzy measure space $(X, \mathcal{F}, \mu)$, an integral kernel $K : X \times Y \to L$ and a function $f : X \to L$, where L is a complete

The first author announces a support of Czech Science Foundation through the grant 18-06915S and the ERDF/ESF project AI-Met4AI No. CZ.02.1.01/0.0/0.0/17_049/0008414.

M.-J. Lesot et al. (Eds.): IPMU 2020, CCIS 1239, pp. 318–331, 2020.
https://doi.org/10.1007/978-3-030-50153-2_24

residuated lattice, we proposed the integral transforms given by the following formulas:

$$F^{\otimes}_{(K,\otimes)}(f)(y) = \int^{\otimes} K(x,y) \otimes f(x)\, d\mu,$$

$$F^{\rightarrow}_{(K,\otimes)}(f)(y) = \int^{\otimes} K(x,y) \rightarrow f(x)\, d\mu,$$

where $F^{\otimes}_{(K,\otimes)}$ becomes the upper fuzzy transform if $\mu(A) = 1$ for any $A \in \mathcal{F} = \mathcal{P}(X)$ such that $A \neq \emptyset$ and $F^{\rightarrow}_{(K,\otimes)}(f)$ becomes the lower fuzzy transform if $\mu(A) = 0$ for any $A \in \mathcal{F} = \mathcal{P}(X)$ such that $A \neq X$.[1] Moreover, to get the exact definitions of lower and upper fuzzy transforms, the family of fuzzy sets $\{K(\cdot, y) \mid y \in Y\}$ has to form a fuzzy partition of X (see [20]).

The aim of this article is to introduce further integral transforms for residuated lattice-valued functions and analyze their basic properties for which we consider the residuum based fuzzy integrals that were proposed by Dvořák and Holčapek in [4] and Dubois, Prade and Rico in [3]. Together with the integral transform with the multiplication based fuzzy integral introduced in [9] we get a class of nonstandard integral transforms for the residuated lattice-valued functions based on fuzzy (or also qualitative) integrals that are often used in data processing. Note that the fuzzy integrals aggregate data and, in this way, provide summary information that is not directly visible from data. Obviously, the proposed integral transforms also provide an aggregation of function values, mainly, if the set Y has a significantly smaller size than the set X. This can be used, for example, in hierarchical decision making, classification problem or signal and image processing, where kernels can express relationships between different levels of criteria, object attributes and classes or introduce windows for some kind of filtering, respectively.

The article is structured as follows. In the next section, we recall the definition of complete residuated lattices and the basic concepts of fuzzy set theory and the theory of fuzzy measure spaces. The third section introduces two types of the residuum based fuzzy integrals and shows their basic properties. The integral transforms for residuated lattice-valued functions are established in the fourth section. We present their elementary properties and demonstrate the linearity property under the restriction to comonotonic functions. The last section is a conclusion.

Because of the space limitation almost all proofs are omitted in this article.

2 Preliminary

2.1 Truth Value Structures

We assume that the structure of truth values is a *complete residuated lattice*, i.e., an algebra $L = \langle L, \wedge, \vee, \otimes, \rightarrow, 0, 1\rangle$ with four binary operations and two

[1] Note that we use here the denotation of the integral transforms employed in this article which is slightly different from [9].

constants such that $\langle L, \wedge, \vee, 0, 1\rangle$ is a complete lattice, where 0 is the least element and 1 is the greatest element of L, $\langle L, \otimes, 1\rangle$ is a commutative monoid (i.e., $\otimes$ is associative, commutative and the identity $a \otimes 1 = a$ holds for any $a \in L$) and the adjointness property is satisfied, i.e.,

$$a \leq b \to c \quad \text{iff} \quad a \otimes b \leq c \tag{1}$$

holds for each $a, b, c \in L$, where $\leq$ denotes the corresponding lattice ordering, i.e., $a \leq b$ if $a \wedge b = a$ for $a, b \in L$. A residuated lattice L is said to be *divisible* if $a \otimes (a \to b) = a \wedge b$ holds for arbitrary $a, b \in L$. The operation of *negation* on L is defined as $\neg a = a \to 0$ for $a \in L$. A residuated lattice L satisfies the *law of double negation* if $\neg\neg a = a$ holds for any $a \in L$. A divisible residuated lattice satisfying the law of double negation is called an *MV-algebra.* A residuated lattice is said to be *linearly ordered* if the corresponding lattice ordering is linear, i.e., $a \leq b$ or $b \leq a$ holds for any $a, b \in L$.

Theorem 1. *Let $\{b_i \mid i \in I\}$ be a non-empty set of elements from L, and let $a \in L$. Then*
(a) $a \otimes (\bigvee_{i\in I} b_i) = \bigvee_{i\in I}(a \otimes b_i)$,
(b) $a \to \bigwedge_{i\in I} b_i = \bigwedge_{i\in I}(a \to b_i)$,
(c) $(\bigvee_{i\in I} b_i) \to a = \bigwedge_{i\in I}(b_i \to a)$,
(d) $a \otimes \bigwedge_{i\in I} b_i \leq \bigwedge_{i\in I}(a \otimes b_i)$,
(e) $\bigvee_{i\in I}(a \to b_i) \leq a \to \bigvee_{i\in I} b_i$,
(f) $\bigvee_{i\in I}(b_i \to a) \leq \bigwedge_{i\in I} b_i \to a$.

If L is a complete MV-algebra the above inequalities may be replaced by equalities.

For more information about residuated lattices, we refer to [1,18]. In what follows, we present two examples of linearly ordered lattice.

Example 1. It is easy to prove that the algebra

$$L_T = \langle [0,1], \min, \max, T, \to_T, 0, 1\rangle,$$

where T is a left continuous t-norm (see, e.g., [11]) and $a \to_T b = \bigvee\{c \in [0,1] \mid T(a,c) \leq b\}$, defines the residuum, is a complete residuated lattice. In this article, we will refer to complete residuated lattices determined by the Łukasiewicz t-norm and nilpotent minimum, i.e.,

$$T_{\text{Ł}}(a,b) = \max(a + b - 1, 0),$$
$$T_{nM}(a,b) = \begin{cases} 0, & \text{if } a + b \leq 1, \\ \min(a,b), & \text{otherwise,} \end{cases}$$

respectively. Their residua are as follows:

$$a \to_{\text{Ł}} b = \min(1, 1 - a + b),$$
$$a \to_{nM} b = \begin{cases} 1, & \text{if } a \leq b, \\ \max(1-a, b), & \text{otherwise.} \end{cases}$$

In the first case, the complete residuated lattice will be denoted by $L_{Ł}$. Note that $L_{Ł}$ is a complete MV-algebra called the *Łukasiewicz algebra* (*on* $[0,1]$), where, for example, the distributivity of $\otimes$ over $\bigwedge$ is satisfied.[2] The residuated lattice determined by the nilpotent minimum is an example of a residuated lattice in which the above-mentioned distributivity fails.

Example 2. Let $a, b \in [0, \infty]$ be such that $a < b$. One checks easily that $L_{[a,b]} = \langle [a,b], \min, \max, \min, \rightarrow, a, b \rangle$, where

$$c \rightarrow d = \begin{cases} b, \text{ if } c \leq d, \\ d, \text{otherwise}, \end{cases} \tag{2}$$

is a complete residuated lattice. Note that $L_{[a,b]}$ is a special example of a more general residuated lattice called a *Heyting algebra*.[3]

In the end of this section, we introduce two families of subsets of L from which important algebras of sets are later generated. Let $u : \mathcal{P}(L) \rightarrow \mathcal{P}(L)$ be defined as

$$u(X) = \{x \in L \mid \exists a \in X, a \leq x\} \tag{3}$$

for any $X \in \mathcal{P}(L)$. Obviously, $X \subseteq u(X)$. A set $X \in \mathcal{P}(L)$, for which $u(X) = X$ holds, is called the *upper set* or *upset*. We use $\mathcal{U}(L)$ to denote the family of all upsets in L, i.e., $\mathcal{U}(L) = \{u(X) \mid X \in \mathcal{P}(L)\}$.[4] Similarly, let $\ell : \mathcal{P}(L) \rightarrow \mathcal{P}(L)$ be defined as

$$\ell(X) = \{x \in L \mid \exists a \in X, a \geq x\} \tag{4}$$

for any $X \in \mathcal{P}(L)$. A set $X \in \mathcal{P}(L)$ for which $\ell(X) = X$ holds is called the *lower set* or *loset*. The family of all losets in L is denoted $\mathcal{L}(L)$.

2.2 Fuzzy Sets

Let L be a complete residuated lattice, and let X be a non-empty universe of discourse. A function $A : X \rightarrow L$ is called a *fuzzy set* (*L-fuzzy set*) *on* X. A value $A(x)$ is called a *membership degree of* x *in the fuzzy set* A. The set of all fuzzy sets on X is denoted by $\mathcal{F}(X)$. A fuzzy set A on X is called *crisp* if $A(x) \in \{0, 1\}$ for any $x \in X$. Obviously, a crisp fuzzy set can be uniquely identified with a subset of X. The symbol $\emptyset$ denotes the empty fuzzy set on X, i.e., $\emptyset(x) = 0$ for any $x \in X$. The set of all crisp fuzzy sets on X (i.e., the power set of X) is denoted by $\mathcal{P}(X)$. A constant fuzzy set A on X (denoted as a_X) satisfies $A(x) = a$ for any $x \in X$, where $a \in L$. The sets $\text{Supp}(A) = \{x \mid x \in X \ \& \ A(x) > 0\}$ and $\text{Core}(A) = \{x \mid x \in X \ \& \ A(x) = 1\}$ are called the *support* and the *core* of a fuzzy set A, respectively. A fuzzy set A is called *normal* if $\text{Core}(A) \neq \emptyset$.

[2] Here we mean that $\bigwedge_{i \in I}(a \otimes b_i) = a \otimes \bigwedge_{i \in I} b_i$ holds.

[3] A Heyting algebra is a residuated lattice with $\otimes = \wedge$.

[4] In [8], a type of topological spaces derived from upsets in L was proposed.

Let A, B be fuzzy sets on X. The extension of the operations $\wedge$, $\vee$, $\otimes$ and $\rightarrow$ on L to the operations on $\mathcal{F}(X)$ is given by

$$\begin{aligned} (A \wedge B)(x) = A(x) \wedge B(x) \quad &\text{and} \quad (A \vee B)(x) = A(x) \vee B(x) \\ (A \otimes B)(x) = A(x) \otimes B(x) \quad &\text{and} \quad (A \rightarrow B)(x) = A(x) \rightarrow B(x) \end{aligned} \tag{5}$$

for any $x \in X$. Obviously, $A \wedge B$ and $A \vee B$ are the standard definitions of the intersection and union of fuzzy sets A and B, respectively, but we prefer here the symbols of infimum ($\wedge$) and supremum ($\vee$) over the classical $\cap$ and $\cup$.

Let X, Y be non-empty universes. A fuzzy set $K : X \times Y \rightarrow L$ is called a (binary) fuzzy relation. A fuzzy relation K is said to be *normal*, whenever $\text{Core}(K) \neq \emptyset$, and *normal in the first coordinate*, whenever $\text{Core}(K(x, \cdot)) \neq \emptyset$ for any $x \in X$. Similarly, a fuzzy relation is normal in the second component. A fuzzy relation K is said to be *complete normal* whenever K is normal in the first and the second coordinates. A relaxation of the normality of fuzzy relation is a *semi–normal* fuzzy relation defined as $K \neq \emptyset$, i.e., $K(x, y) > 0$ for certain $(x, y) \in X \times Y$. Similarly one can define *semi-normal in the the first (second) coordinate* and *complete semi-normal* fuzzy relation.

2.3 Fuzzy Measure Spaces

Measurable spaces and functions. Let us consider algebras of sets as follows.

Definition 1. *Let X be a non-empty set. A subset $\mathcal{F}$ of $\mathcal{P}(X)$ is an* algebra of sets on X *provided that.*

(A1) $X \in \mathcal{F}$,
(A2) if $A \in \mathcal{F}$, *then* $X \setminus A \in \mathcal{F}$,
(A3) if $A, B \in \mathcal{F}$, *then* $A \cup B \in \mathcal{F}$.

Definition 2. *An algebra $\mathcal{F}$ of sets on X is a* σ-algebra of sets *if*

(A4) if $A_i \in \mathcal{F}$, $i = 1, 2, \dots$, *then* $\bigcup_{i=1}^{\infty} A_i \in \mathcal{F}$.

It is easy to see that if $\mathcal{F}$ is an algebra (σ-algebra) of sets, then the intersection of finite (countable) number of sets belongs to $\mathcal{F}$. A pair $(X, \mathcal{F})$ is called a *measurable space* (on X) if $\mathcal{F}$ is an algebra (σ-algebra) of sets on X. Let $(X, \mathcal{F})$ be a measurable space and $A \in \mathcal{F}(X)$. We say that A is *$\mathcal{F}$-measurable* if $A \in \mathcal{F}$. Obviously, the sets $\{\emptyset, X\}$ and $\mathcal{P}(X)$ are σ-algebras of fuzzy sets on X.

A beneficial tool how to introduce an algebra or a σ-algebra of sets on X is an algebra (σ-algebra) generated by a non-empty family of sets.

Definition 3. *Let $\mathcal{G} \subseteq \mathcal{P}(X)$ be a non-empty family of sets. The smallest algebra (σ-algebra) on X containing $\mathcal{G}$ is denoted by* $\text{alg}(\mathcal{G})$ *($\sigma(\mathcal{G})$) and is called the* generated algebra *(σ-algebra) by the family $\mathcal{G}$.*

Note that the intersection of algebras (σ-algebras) is again an algebra (σ-algebra), hence, the smallest algebra (σ-algebra) on X containing $\mathcal{G}$ always exists and its unique. Moreover, the generated algebra $\text{alg}(\mathcal{G})$, in contrast to

$\sigma(\mathcal{G})$, can be simply constructed from the elements $\mathcal{G}$ as the set which consists of all finite unions applied on the set of all finite intersections over the elements of $\mathcal{G}$ and their complements. Note that the construction of a generated σ-algebra needs a transfinite approach. In this article, we will consider the algebras generated from the families of upsets $\mathcal{U}(L)$ and losets $\mathcal{L}(L)$.

Let $(X, \mathcal{F})$ and $(Y, \mathcal{G})$ be measurable spaces, and let $f : X \to Y$ be a function. We say that f is $\mathcal{F}$-$\mathcal{G}$-measurable if $f^{-1}(Z) \in \mathcal{F}$ for any $Z \in \mathcal{G}$. The following theorem shows that the verification of $\mathcal{F}$-$\mathcal{G}$-measurability of functions can be simplified if $\mathcal{G}$ is a generated algebra (σ-algebra).

Theorem 2. *Let $\mathcal{G} \subseteq \mathcal{P}(Y)$ be a subset such that $Y \in \mathcal{G}$, and let $(X, \mathcal{F})$ be a measurable space. A function $f : X \to Y$ is $\mathcal{F}$-alg$(\mathcal{G})$-measurable if and only if $f^{-1}(Z) \in \mathcal{F}$ for any $Z \in \mathcal{G}$.*

Proof. ($\Rightarrow$) The implication is a simple consequence of $\mathcal{G} \subseteq \text{alg}(\mathcal{G})$.

($\Leftarrow$) Let $\mathcal{Q} = \{Z \mid f^{-1}(Z) \in \mathcal{F}\}$. Note that $\mathcal{Q}$ is called the preimage algebra on Y. From the definition of the generated algebra $\text{alg}(\mathcal{G})$ by the family $\mathcal{G}$, we find that $\text{alg}(\mathcal{G}) \subseteq \mathcal{Q}$. Hence, we obtain that $f^{-1}(Z) \in \mathcal{F}$ for any $Z \in \text{alg}(\mathcal{G})$, which means that f is $\mathcal{F}$-alg$(\mathcal{G})$-measurable.

Note that the previous theorem remains true if $\text{alg}(\mathcal{G})$ is replaced by $\sigma(\mathcal{G})$. In the following three statements we provide sufficient conditions under which the functions obtained applying the operations to measurable functions remain measurable. For the purpose of this article, we restrict to fuzzy sets and algebras determined by upsets and losets.

Theorem 3. *Let L be linearly ordered, let $(X, \mathcal{F})$ be an algebra, and let $\mathcal{B} \subseteq \mathcal{F}(X)$ be a set of all $\mathcal{F}$-alg$(\mathcal{U}(L))$-measurable fuzzy sets. Then $f \wedge g, f \vee g \in \mathcal{B}$ for any $f, g \in \mathcal{B}$.*

Theorem 4. *Let $(X, \mathcal{F})$ be an algebra, and let $\mathcal{B} \subseteq \mathcal{F}(X)$ be a set of all $\mathcal{F}$-alg$(\mathcal{U}(L))$-measurable fuzzy sets. If $\mathcal{F}$ is closed over arbitrary unions, then*

$$f \otimes g, f \wedge g, f \vee g, \quad f, g \in \mathcal{B}.$$

Theorem 5. *Let L be linearly ordered and dense. Let $(X, \mathcal{F})$ be an algebra, and let $\mathcal{B} \subseteq \mathcal{F}(X)$ be a set of all $\mathcal{F}$-alg$(\mathcal{U}(L))$-measurable fuzzy sets. If $\mathcal{F}$ is closed over arbitrary unions, then*

$$f \to g \in \mathcal{B}, \quad f, g \in \mathcal{B}.$$

The previous theorems become true if the algebra $\text{alg}(\mathcal{U}(L))$ is replaced by $\text{alg}(\mathcal{L}(L))$ and the $\mathcal{F}$-alg$(\mathcal{L}(L))$-measurability is considered.

Fuzzy Measures. The concept of a fuzzy measure on a measurable space $(X, \mathcal{F})$ is a slight extension of the standard definition of the normed measure where the unit interval (or the real line) is replaced by a complete residuated lattice L (e.g., [6,12]).

Definition 4. *A map $\mu : \mathcal{F} \to L$ is called a* fuzzy measure *on a measurable space $(X, \mathcal{F})$ if*

(i) $\mu(\emptyset) = 0$ and $\mu(X) = 1$,
(ii) if $A, B \in \mathcal{F}$ such that $A \subseteq B$, then $\mu(A) \leq \mu(B)$.

A triplet $(X, \mathcal{F}, \mu)$ is called a *fuzzy measure space* whenever $(X, \mathcal{F})$ is a measurable space and μ is a fuzzy measure on $(X, \mathcal{F})$.

Example 3. Let L_T be an algebra from Example 1, where T is a continuous t-norm. Let $X = \{x_1, \dots, x_n\}$ be a finite non-empty set, and let $\mathcal{F}$ be an arbitrary algebra. A *relative* fuzzy measure μ^r on $(X, \mathcal{F})$ can be given as

$$\mu^r(A) = \frac{|A|}{|X|}$$

for all $A \in \mathcal{F}$, where $|A|$ and $|X|$ denote the cardinality of A and X, respectively. Let $\varphi : L \to L$ be a monotonically non-decreasing map with $\varphi(0) = 0$ and $\varphi(1) = 1$. The relative measure μ^r can be generalized as a fuzzy measure μ^r_φ on $(X, \mathcal{F})$ given by $\mu^r_\varphi(A) = \varphi(\mu^r(A))$ for any $A \in \mathcal{F}$.

3 Residuum Based Fuzzy Integrals

In the following part, we introduce two types of fuzzy (qualitative) integrals based on the operation of residuum. The first type of this fuzzy integral was proposed by Dvořák and Holčapek in [4] for fuzzy quantifiers modelling, the second type was proposed by Dubois, Prade and Rico in [3], known also under the name *desintegral*, for the reasoning with a decreasing evaluation scale. A comparison of both fuzzy integrals can be found in [10].

3.1 $\to_{\mathrm{DH}}$–Fuzzy Integral

The integrated functions are fuzzy sets on X. We consider a modified version of the original definition of the residuum based fuzzy integral presented in [4].

Definition 5. *Let $(X, \mathcal{F}, \nu)$ be a complementary fuzzy measure space, and let $f : X \to L$. The* $\to_{\mathrm{DH}}$-fuzzy integral of f *on X is given by*

$$\int_{\mathrm{DH}}^{\to} f \, d\nu = \bigwedge_{A \in \mathcal{F}} \left(\bigwedge_{x \in A} f(x) \right) \to \nu(A). \tag{6}$$

Note that the original definition in [4] and the previous definition of residuum based integrals coincide on MV-algebras. The following statement presents basic properties of $\to_{\mathrm{DH}}$-fuzzy integral.

Theorem 6. *For any $f, g \in \mathcal{F}(X)$ and $a \in L$, we have*

(i) $\int_{\mathrm{DH}}^{\to} f \, d\nu \geq \int_{\mathrm{DH}}^{\to} g \, d\nu$ if $f \leq g$;
(ii) $\int_{\mathrm{DH}}^{\to} a_X \, d\nu = \neg a$;

(iii) $\int_{\mathrm{DH}}^{\rightarrow} a_X \otimes f\, d\nu \leq a \rightarrow \int_{\mathrm{DH}}^{\rightarrow} f\, d\nu$;
(iv) $\int_{\mathrm{DH}}^{\rightarrow} a_X \rightarrow f\, d\nu \geq a \otimes \int_{\mathrm{DH}}^{\rightarrow} f\, d\nu$.

Note that the inequality (iii) of the previous theorem becomes the equality in a complete MV-algebra. An equivalent, and useful from the practical point of view, definition of $\rightarrow_{\mathrm{DH}}$–fuzzy integrals can be obtained for $\mathcal{F}$-alg$(\mathcal{U}(L))$-measurable functions.

Theorem 7. *If $f : X \rightarrow L$ be $\mathcal{F}$-alg$(\mathcal{U}(L))$-measurable, then*

$$\int_{\mathrm{DH}}^{\rightarrow} f\, d\nu = \bigwedge_{a\in L} (a \rightarrow \nu(\{x \in X \mid f(x) \geq a\})) . \tag{7}$$

We say that $f, g \in \mathcal{F}(X)$ are *comonotonic* if and only if there is no pair $x_1, x_2 \in X$ such that $f(x_1) < f(x_2)$ and simultaneously $g(x_1) > g(x_2)$. Note that the Sugeno integral preserves the infimum and supremum for the commonotonic functions, i.e., it is comonotonically minitive and comonotonically maxitive, (see, [7, Theorem 4.44]). A similar result for the residuum based fuzzy integral can be simply derived using the following lemma whose proof can be found in [9].

Lemma 1. *Let L be linearly ordered, and let $f, g \in \mathcal{F}(X)$. Denote $C_f = \{C_f(a) \mid a \in L\}$, where $C_f(a) = \{x \in X \mid f(x) \geq a\}$. Then C_f is a chain with respect to $\subseteq$, and if f and g are comonotonic, then $C_{f\star g}(a) = C_f(a)$ or $C_{f\star g}(a) = C_g(a)$ for any $a \in L$, where $\star \in \{\wedge, \vee\}$.*

Theorem 8. *Let L be linearly ordered, and let $f, g \in \mathcal{F}(X)$ be comonotonic $\mathcal{F}$-alg$(\mathcal{U}(L))$-measurable functions. Then*

$$\int_{\mathrm{DH}}^{\rightarrow} (f \vee g)\, d\nu = \int_{\mathrm{DH}}^{\rightarrow} f\, d\nu \wedge \int_{\mathrm{DH}}^{\rightarrow} g\, d\nu. \tag{8}$$

Note that a dual formula to (8), where the infimum is replaced by the supremum and vice versa, is not true in general even if we restrict ourselves to linearly ordered Heyting algebra (cf. Theorem 3.4 in [9] for the multiplication based fuzzy integral).

3.2 $\rightarrow_{\mathrm{DPR}}$–Fuzzy Integrals

The integrated functions are again fuzzy sets on X.

Definition 6. *Let $(X, \mathcal{F}, \mu)$ be a fuzzy measure space, and let $f : X \rightarrow L$. The* $\rightarrow_{\mathrm{DPR}}$-fuzzy integral of f on X *is given by*

$$\int_{\mathrm{DPR}}^{\rightarrow} f\, d\mu = \bigwedge_{A\in\mathcal{F}} \left(\mu^c(A) \rightarrow \bigvee_{x\in A} f(x) \right). \tag{9}$$

Note that if $A = \emptyset$, then $\mu^c(X \setminus \emptyset) \rightarrow \bigvee \emptyset = 0 \rightarrow 0 = 1$, hence, the empty set has no influence on the value of the $\rightarrow_{\mathrm{DPR}}$–fuzzy integral. The following statement presents basic properties of $\rightarrow_{\mathrm{DPR}}$-fuzzy integral.

Theorem 9. *For any $f, g \in \mathcal{F}(X)$ and $a \in L$, we have*

(i) $\int_{\mathrm{DPR}}^{\rightarrow} f \, d\mu \leq \int_{\mathrm{DPR}}^{\rightarrow} g \, d\mu$ *if* $f \leq g$;
(ii) $\int_{\mathrm{DPR}}^{\rightarrow} a_X \, d\mu = a$;
(iii) $\int_{\mathrm{DPR}}^{\rightarrow} a_X \otimes f \, d\mu \geq a \otimes \int_{\mathrm{DPR}}^{\rightarrow} f \, d\mu$;
(iv) $\int_{\mathrm{DPR}}^{\rightarrow} a_X \rightarrow f \, d\mu \leq a \rightarrow \int_{\mathrm{DPR}}^{\rightarrow} f \, d\mu$.

Note that the inequality (iv) of the previous theorem becomes the equality in a complete MV-algebra. An equivalent formula to (9) under the assumption of $\mathcal{F}$-alg($\mathcal{L}(L)$)-measurability of functions is as follows.

Theorem 10. *Let $(X, \mathcal{F}, \mu)$ be a fuzzy measure space, and let $f : X \rightarrow L$ be $\mathcal{F}$-alg($\mathcal{L}(L)$)-measurable. Then*

$$\int_{\mathrm{DPR}}^{\rightarrow} f \, d\mu = \bigwedge_{a \in L} \left(\mu^c(\{x \in X \mid f(x) \leq a\}) \rightarrow a\right). \tag{10}$$

Lemma 2. *Let L be linearly ordered, and let $f, g \in \mathcal{F}(X)$. Denote $B_f = \{B_f(a) \mid a \in L\}$, where $B_f(a) = \{x \in X \mid f(x) \leq a\}$. Then B_f is a chain with respect to $\subseteq$, and if f and g are comonotonic, then $B_{f \star g}(a) = B_f(a)$ or $B_{f \star g}(a) = B_g(a)$ for any $a \in L$, where $\star \in \{\wedge, \vee\}$.*

Theorem 11. *Let L be linearly ordered, and let $f, g \in \mathcal{F}(X)$ be comonotonic $\mathcal{F}$-alg($\mathcal{L}(L)$)-measurable functions. Then*

$$\int_{\mathrm{DPR}}^{\rightarrow} (f \wedge g) \, d\mu = \int_{\mathrm{DPR}}^{\rightarrow} f \, d\mu \wedge \int_{\mathrm{DPR}}^{\rightarrow} g \, d\mu.$$

4 Integral Transforms for Lattice-Valued Functions

In this section, we propose four types of integral transforms for functions whose function values are evaluated in a complete residuated lattice. For their definitions, we use the residuum based fuzzy integral introduced in Sect. 3. The integral transforms transform fuzzy sets from $\mathcal{F}(X)$ to fuzzy sets from $\mathcal{F}(Y)$.

4.1 $\rightarrow_{\mathrm{DH}}$–Integral Transforms

In this part, we propose two types of integral transform based on $\rightarrow_{\mathrm{DH}}$–fuzzy integral. For their definitions we are inspired by a straightforward generalization of the lower and upper fuzzy transforms in terms of the multiplication based fuzzy integral presented in [9]. We start with the definition of $\rightarrow_{\mathrm{DH}}$–integral transforms merging an integral kernel and a transformed function by the multiplication operation.

Definition 7. *Let $(X, \mathcal{F}, \nu)$ be a complementary fuzzy measure space, and let $K : X \times Y \rightarrow L$ be a semi-normal in the second component fuzzy relation. A map $F^{\otimes}_{(K, \rightarrow_{\mathrm{DH}})} : \mathcal{F}(X) \rightarrow \mathcal{F}(Y)$ defined by*

$$F^{\otimes}_{(K, \rightarrow_{\mathrm{DH}})}(f)(y) = \int_{\mathrm{DH}}^{\rightarrow} K(x, y) \otimes f(x) \, d\nu \tag{11}$$

is called a $(K, \otimes, \rightarrow_{\mathrm{DH}})$–integral transform.

It is easy to see that a complementary measure ν and a semi-normal in the second component fuzzy relation K are parameters of $(K,\otimes,\rightarrow_{\mathrm{DH}})$–integral transform. The fuzzy relation K will be called the *integral kernel*, which corresponds to the standard notation in the theory of integral transforms. Note that the semi-normality in the second component of integral kernels is considered as a natural assumption avoiding the trivial case, namely, if $K(x,y)=0$ for any $x\in X$ and some $y\in Y$, we trivially obtain $F^{\otimes}_{(K,\rightarrow_{\mathrm{DH}})}(f)(y)=1$ as a consequence of $\int^{\rightarrow}_{\mathrm{DH}} 0_X\, d\nu=\neg 0=1$ (see Theorem 6(b)).

Remark 1. If an integral kernel K is normal in the second component for any $y\in Y$ and, moreover, the family of sets $\{\mathrm{Core}(K(\cdot,y))\mid y\in Y\}$ forms a partition of X, the family of fuzzy sets $\{K(\cdot,y)\mid y\in Y\}$ is called a *fuzzy partition* of X, which is a crucial concept in the definition of lower and upper fuzzy transforms [20]. In this article, we significantly relax the concept of fuzzy partition because we require only that $K(\cdot,y)>0$ for any $y\in Y$. Nevertheless, the fulfillment of certain integral transforms properties usually forces to introduce specific conditions for integral kernels (see Theorems 4.4 and 4.7 in [9]).

The following theorem shows basic properties of $(K,\otimes,\rightarrow_{\mathrm{DH}})$–integral transforms.

Theorem 12. *For any $f,g\in\mathcal{F}(X)$ and $a\in L$, we have*

(i) $F^{\otimes}_{(K,\rightarrow_{\mathrm{DH}})}(f)\geq F^{\otimes}_{(K,\rightarrow_{\mathrm{DH}})}(g)$ *if* $f\leq g$;
(ii) $F^{\otimes}_{(K,\rightarrow_{\mathrm{DH}})}(f\wedge g)\geq F^{\otimes}_{(K,\rightarrow_{\mathrm{DH}})}(f)\vee F^{\otimes}_{(K,\rightarrow_{\mathrm{DH}})}(g)$;
(iii) $F^{\otimes}_{(K,\rightarrow_{\mathrm{DH}})}(f)\wedge F^{\otimes}_{(K,\rightarrow_{\mathrm{DH}})}(g)\geq F^{\otimes}_{(K,\rightarrow_{\mathrm{DH}})}(f\vee g)$;
(iv) $F^{\otimes}_{(K,\rightarrow_{\mathrm{DH}})}(a_X\otimes f)\leq a\rightarrow F^{\otimes}_{(K,\rightarrow_{\mathrm{DH}})}(f)$;
(v) $F^{\otimes}_{(K,\rightarrow_{\mathrm{DH}})}(a_X\rightarrow f)\geq a\otimes F^{\otimes}_{(K,\rightarrow_{\mathrm{DH}})}(f)$.

Moreover, if L is a complete MV-algebra, the equality in (iv) holds.

Proof. The first three statements are trivial consequences of the monotonicity of the operation $\otimes$ (i.e., monotonically non-decreasing) and the $\rightarrow_{\mathrm{DH}}$–fuzzy integral (Theorem 6(i)). Using Theorem 6(iii) and the commutativity of $\otimes$, for any $y\in Y$, we have

$$\begin{aligned}F^{\otimes}_{(K,\rightarrow_{\mathrm{DH}})}(a_X\otimes f)(y)&=\int^{\rightarrow}_{\mathrm{DH}} K(x,y)\otimes(a_X(x)\otimes f(x))\,d\nu\\&\leq a\rightarrow\int^{\rightarrow}_{\mathrm{DH}} K(x,y)\otimes f(x)\,d\nu=a\rightarrow F^{\otimes}_{(K,\rightarrow_{\mathrm{DH}})}(f)(y).\end{aligned}$$

Moreover, if L is a complete MV-algebra, the previous inequality becomes the equality and hence, the equality in (iv) holds. Since $K(x,y)\otimes(a\rightarrow f(x))\leq a\rightarrow(K(x,y)\otimes f(x))$, using (i) and (iv) of Theorem 6, one can simply prove (v). □

Let us continue with another type of $\rightarrow_{\mathrm{DH}}$–integral transforms, where the integral kernels are combined with the transformed functions using the residuum operation.

Definition 8. *Let $(X, \mathcal{F}, \nu)$ be a complementary fuzzy measure space, and let $K : X \times Y \to L$ be a semi-normal in the second component fuzzy relation. A map $F^{\to}_{(K,\to_{\rm DH})} : \mathcal{F}(X) \to \mathcal{F}(Y)$ defined by*

$$F^{\to}_{(K,\to_{\rm DH})}(f)(y) = \int_{\rm DH}^{\to} K(x, y) \to f(x)\, d\nu \tag{12}$$

is called a $(K, \to, \to_{\rm DH})$-integral transform.

Note that if K is not semi-normal in the second component, i.e., $K(x, y) = 0$ for some $y \in Y$ and any $x \in X$, we trivially obtain $F^{\to}_{(K,\to_{\rm DH})}(f)(y) = 0$ as a consequence of $\int_{\rm DH}^{\to} 1_X\, d\nu = \neg 1 = 0$ (see Theorem 6(b)). In what follows, we present some basic properties of the $(K, \to, \to_{\rm DH})$-integral transform.

Theorem 13. *For any $f, g \in \mathcal{F}(X)$ and $a \in L$, we have*

(i) $F^{\to}_{(K,\to_{\rm DH})}(f) \geq F^{\to}_{(K,\to_{\rm DH})}(g)$ *if* $f \leq g$;
(ii) $F^{\to}_{(K,\to_{\rm DH})}(f \wedge g) \geq F^{\to}_{(K,\to_{\rm DH})}(f) \vee F^{\to}_{(K,\to_{\rm DH})}(g)$;
(iii) $F^{\to}_{(K,\to_{\rm DH})}(f \vee g) \leq F^{\to}_{(K,\to_{\rm DH})}(f) \wedge F^{\to}_{(K,\to_{\rm DH})}(g)$;
(iv) $F^{\to}_{(K,\to_{\rm DH})}(a_X \otimes f) \leq a \to F^{\to}_{(K,\to_{\rm DH})}(f)$;
(v) $F^{\to}_{(K,\to_{\rm DH})}(a_X \to f) \geq a \otimes F^{\to}_{(K,\to_{\rm DH})}(f)$.

Proof. The first three statements are trivial consequences of the monotonicity of the operation $\to$ (i.e., monotonically non-decreasing in the second component) and the $\to_{\rm DH}$-fuzzy integral (Theorem 6(i)). Since $K(x, y) \to (a \otimes f(x)) \geq a \otimes (K(x, y \to f(x))$, then using (i) and (iii) of Theorem 6, for any $y \in Y$, we obtain

$$F^{\otimes}_{(K,\to_{\rm DH})}(a_X \otimes f)(y) = \int_{\rm DH}^{\to} K(x, y) \to (a \otimes f(x))\, d\nu$$
$$\leq \int_{\rm DH}^{\to} a \otimes (K(x, y) \to f(x))\, d\nu \leq a \to F^{\otimes}_{(K,\to_{\rm DH})}(f)(y).$$

Since $K(x, y) \to (a \to f(x)) = a \to (K(x, y) \to f(x))$, using Theorem 6(iv), one can simply prove (v). □

One could see that, although, the $\to_{\rm DH}$-integral transforms are defined by different operations, i.e., $\otimes$ and $\to$, their basic properties coincide.

We showed in Theorem 8 that under the assumption of the linearity of complete residuated lattices, the $\to_{\rm DH}$-fuzzy integral is a linear operator in the sense that the $\to_{\rm DH}$-fuzzy integral of the supremum of comonotonic functions is the infimum of $\to_{\rm DH}$-fuzzy integrals of these functions. The linearity property of $\to_{\rm DH}$-fuzzy integral can be used to prove the analogous property for $\to_{\rm DH}$-integral transforms.

Theorem 14. *Let L be a linearly ordered and assume that the algebra $\mathcal{F}$ is closed over arbitrary unions. Let $f, g, K(\cdot, y)$ be $\mathcal{F}$-alg$(\mathcal{U}(L))$-measurable for any $y \in Y$. If $K(\cdot, y) \star f$ and $K(\cdot, y) \star g$ are comonotonic for $\star \in \{\otimes, \to\}$, then*

$$F^{\star}_{(K,\to_{\rm DH})}(f \vee g) = F^{\star}_{(K,\to_{\rm DH})}(f) \wedge F^{\star}_{(K,\to_{\rm DH})}(g) \tag{13}$$

4.2 $\rightarrow_{\text{DPR}}$–Integral Transform

Similarly to the previous subsection we propose two types of integral transforms based now on the $\rightarrow_{\text{DPR}}$–fuzzy integral. Again we start with the definition of integral transform, where integral kernels and transformed functions are merged by the multiplication operation.

Definition 9. *Let $(X, \mathcal{F}, \mu)$ be a fuzzy measure space, and let $K : X \times Y \to L$ be a semi-normal in the second component fuzzy relation. A map $F^{\otimes}_{(K,\rightarrow_{\text{DPR}})} : \mathcal{F}(X) \to \mathcal{F}(Y)$ defined by*

$$F^{\otimes}_{(K,\rightarrow_{\text{DPR}})}(f)(y) = \int^{\rightarrow}_{\text{DPR}} K(x,y) \otimes f(x)\, d\mu \tag{14}$$

*is called a $(K, \otimes, \rightarrow_{\text{DPR}})$–*integral transform.

The following theorem shows several basic properties of $(K, \otimes, \rightarrow_{\text{DPR}})$–integral transforms.

Theorem 15. *For any $f, g \in \mathcal{F}(X)$ and $a \in L$, we have*

(i) $F^{\otimes}_{(K,\rightarrow_{\text{DPR}})}(f) \leq F^{\otimes}_{(K,\rightarrow_{\text{DPR}})}(g)$ *if* $f \leq g$;
(ii) $F^{\otimes}_{(K,\rightarrow_{\text{DPR}})}(f \wedge g) \leq F^{\otimes}_{(K,\rightarrow_{\text{DPR}})}(f) \wedge F^{\otimes}_{(K,\rightarrow_{\text{DPR}})}(g)$;
(iii) $F^{\otimes}_{(K,\rightarrow_{\text{DPR}})}(f \vee g) \geq F^{\otimes}_{(K,\rightarrow_{\text{DPR}})}(f) \vee F^{\otimes}_{(K,\rightarrow_{\text{DPR}})}(g)$;
(iv) $F^{\otimes}_{(K,\rightarrow_{\text{DPR}})}(a_X \otimes f) \geq a \otimes F^{\otimes}_{(K,\rightarrow_{\text{DPR}})}(f)$;
(v) $F^{\otimes}_{(K,\rightarrow_{\text{DPR}})}(a_X \rightarrow f) \leq a \rightarrow F^{\otimes}_{(K,\rightarrow_{\text{DPR}})}(f)$;

Proof. Similarly to the proof of Theorem 12 one can simply prove all the statements using the properties of $\rightarrow_{\text{DPR}}$–fuzzy integral presented in Theorem 9. □

Definition 10. *Let $(X, \mathcal{F}, \mu)$ be a fuzzy measure space, and let $K : X \times Y \to L$ be a semi-normal in the second component fuzzy relation. A map $F^{\rightarrow}_{(K,\rightarrow_{\text{DPR}})} : \mathcal{F}(X) \to \mathcal{F}(Y)$ defined by*

$$F^{\rightarrow}_{(K,\rightarrow_{\text{DPR}})}(f)(y) = \int^{\rightarrow}_{\text{DPR}} K(x,y) \rightarrow f(x)\, d\mu \tag{15}$$

*is called a $(K, \rightarrow, \rightarrow_{\text{DPR}})$–*integral transform.

Some of basic properties of $(K, \rightarrow, \rightarrow_{\text{DPR}})$-integral transform are presented in the following theorem.

Theorem 16. *For any $f, g \in \mathcal{F}(X)$ and $a \in L$, we have*

(i) $F^{\rightarrow}_{(K,\rightarrow_{\text{DPR}})}(f) \leq F^{\rightarrow}_{(K,\rightarrow_{\text{DPR}})}(g)$ *if* $f \leq g$;
(ii) $F^{\rightarrow}_{(K,\rightarrow_{\text{DPR}})}(f \wedge g) \leq F^{\rightarrow}_{(K,\rightarrow_{\text{DPR}})}(f) \wedge F^{\rightarrow}_{(K,\rightarrow_{\text{DPR}})}(g)$;
(iii) $F^{\rightarrow}_{(K,\rightarrow_{\text{DPR}})}(f \vee g) \geq F^{\rightarrow}_{(K,\rightarrow_{\text{DPR}})}(f) \vee F^{\rightarrow}_{(K,\rightarrow_{\text{DPR}})}(g)$;
(iv) $F^{\rightarrow}_{(K,\rightarrow_{\text{DPR}})}(a \otimes f) \geq a \otimes F^{\rightarrow}_{(K,\rightarrow_{\text{DPR}})}(f)$;

(v) $F^{\rightarrow}_{(K,\rightarrow_{\text{DPR}})}(a \rightarrow f) \leq a \rightarrow F^{\rightarrow}_{(K,\rightarrow_{\text{DPR}})}(f)$.

Moreover, if L is a complete MV-algebra, the equality in (v) holds.

Again one could notice that although, the $\rightarrow_{\text{DPR}}$–integral transforms are defined by different operations, their basic properties are identical. The following linear property ensured for comonotonic functions is a straightforward consequence of Theorem 11.

Theorem 17. *Let L be a linearly ordered and assume that the algebra $\mathcal{F}$ is closed over arbitrary unions. Let $f, g, K(\cdot, y)$ be $\mathcal{F}$-alg$(\mathcal{L}(L))$-measurable for any $y \in Y$. If $K(\cdot, y) \star f$ and $K(\cdot, y) \star g$ are comonotonic for $\star \in \{\otimes, \rightarrow\}$, then*

$$F^{\star}_{(K,\rightarrow_{\text{DPR}})}(f \wedge g) = F^{\star}_{(K,\rightarrow_{\text{DPR}})}(f) \wedge F^{\star}_{(K,\rightarrow_{\text{DPR}})}(g) \tag{16}$$

5 Conclusion

In this article, we introduced four types of integral transforms, where we used the residuum based fuzzy (qualitative) integrals, namely, the $\rightarrow_{\text{DH}}$–fuzzy integral proposed by Dvořák and Holčapek in [4] and the $\rightarrow_{\text{DPR}}$–fuzzy integral proposed by Dubois, Prade and Rico in [3]. We presented some of the basic properties of the residuum based fuzzy integrals including a linearity property for the comonotonic functions which holds in the linearly ordered complete residuated lattices. Using these properties we provided an initial analysis of elementary properties of proposed integral transforms. The further development of the theory of integral transforms for residuated lattice-valued functions is a subject of our future research, where, among others, we want to focus on the seeking of inverse integral kernels to be able to approximate the original functions from the transformed functions. Our motivation comes from the relationship between the lower and upper fuzzy transforms and their related inverse fuzzy transforms.

References

1. Bělohlávek, R.: Fuzzy Relational Systems: Foundations and Principles. Kluwer Academic Publishers, New York (2002)
2. Debnath, L., Bhatta, D.: Integral Transforms and Their Applications. CRC Press, New York (1914)
3. Dubois, D., Prade, H., Rico, A.: Residuated variants of sugeno integrals: towards new weighting schemes for qualitative aggregation methods. Inf. Sci. **329**, 765–781 (2016)
4. Dvořák, A., Holčapek, M.: **L**-fuzzy quantifiers of type ⟨1⟩ determined by fuzzy measures. Fuzzy Sets Syst. **160**(23), 3425–3452 (2009)
5. Galatos, N., Jipsen, P., Kowalski, T., Ono, H.: Residuated Lattices: An Algebraic Glimpse at Substructural Logics, Studies in Logic and Foundations of Mathematics, vol. 151. Elsevier, Amsterdam (2007)
6. Grabisch, M., Murofushi, T., Sugeno, M. (eds.): Fuzzy Measures and Integrals. Theory and Applications. Studies in Fuzziness and Soft Computing. Physica Verlag, Heidelberg (2000)

7. Grabish, M.: Set Functions, Games and Capacities in Decision Making. Springer, Cham (2016). https://doi.org/10.1007/978-3-319-30690-2
8. Holdon, L.: New topology in residuated lattices. Open Math. **2018**(16), 1104–1127 (2018)
9. Holčapek, M., Bui, V.: Integral transforms on spaces of complete residuated lattice valued functions. In: Proceedings of IEEE World Congress on Computational Intelligence (WCCI) 2020, pp. 1–8. IEEE (2020)
10. Holčapek, M., Rico, A.: A note on the links between different qualitative integrals. In: Proceedings of IEEE World Congress on Computational Intelligence (WCCI) 2020, pp. 1–8. IEEE (2020)
11. Klement, E., Mesiar, R., Pap, E.: Triangular Norms, Trends in Logic, vol. 8. Kluwer Academic Publishers, Dordrecht (2000)
12. Klement, E., Mesiar, R., Pap, E.: A universal integral as common frame for Choquet and Sugeno integral. IEEE Trans. Fuzzy Syst. **18**, 178–187 (2010)
13. Močkoř, J.: Spaces with fuzzy partitions and fuzzy transform. Soft Comput. **21**(13), 3479–3492 (2017). https://doi.org/10.1007/s00500-017-2541-7
14. Močkoř, J.: Axiomatic of lattice-valued f-transform. Fuzzy Sets Syst. **342**, 53–66 (2018)
15. Močkoř, J.: F-transforms and semimodule homomorphisms. Soft Comput. **23**(17), 7603–7619 (2019). https://doi.org/10.1007/s00500-019-03766-1
16. Močkoř, J., Holčapek, M.: Fuzzy objects in spaces with fuzzy partitions. Soft Comput. **21**(24), 7269–7284 (2016). https://doi.org/10.1007/s00500-016-2431-4
17. Močkoř, J., Hurtík, P.: Lattice-valued f-transforms and similarity relations. Fuzzy Sets Syst. **342**, 67–89 (2018)
18. Novák, V., Perfilieva, I., Močkoř, J.: Mathematical Principles of Fuzzy Logic. Kluwer Academic Publishers, Boston (1999)
19. Perfilieva, I., Tiwari, S.P., Singh, A.P.: Lattice-valued F-transforms as interior operators of L-fuzzy pretopological spaces. In: Medina, J., et al. (eds.) IPMU 2018. CCIS, vol. 854, pp. 163–174. Springer, Cham (2018). https://doi.org/10.1007/978-3-319-91476-3_14
20. Perfilieva, I.: Fuzzy transforms: theory and applications. Fuzzy Sets Syst. **157**(8), 993–1023 (2006)
21. Tenoudji, F.: Analog and Digital Signal Analysis: From Basics to Applications. Springer, Switzerland (2016). https://doi.org/10.1007/978-3-319-42382-1
22. Tiwari, S., Perfilieva, I., Singh, A.: Generalized residuate lattice based F-transform. Iran. J. Fuzzy Syst. **18**(2), 165–182 (2015)
23. Yaglom, A.M.: An Introduction to the Theory of Stationary Random Functions. Revised English edn. Translated and edited by Richard A. Silverman, vol. XIII. Prentice-Hall, Inc., Englewood Cliffs (1962)

Optimal Control Under Fuzzy Conditions for Dynamical Systems Associated with the Second Order Linear Differential Equations

Svetlana Asmuss[1,2] and Natalja Budkina[1,3(✉)]

[1] University of Latvia, Riga, Latvia
svetlana.asmuss@lu.lv
[2] Institute of Mathematics and Computer Science, Riga, Latvia
[3] Riga Technical University, Riga, Latvia
natalja.budkina@rtu.lv

Abstract. This paper is devoted to an optimal trajectory planning problem with uncertainty in location conditions considered as a problem of constrained optimal control for dynamical systems. Fuzzy numbers are used to incorporate uncertainty of constraints into the classical setting of the problem under consideration. The proposed approach applied to dynamical systems associated with the second order linear differential equations allows to find an optimal control law at each α-level using spline-based methods developed in the framework of the theory of splines in convex sets. The solution technique is illustrated by numerical examples.

Keywords: Dynamical system · Fuzzy constraints · Optimal control

1 Introduction

Optimal control is the process of determining control and state trajectories for a dynamical system over a period of time to minimize an objective function. In this paper we analyse the special case of the following control theory problem:

$$x'(t) = Mx(t) + \beta u(t), y(t) = \gamma^\top x(t), \quad t \in [a, b], \tag{1}$$

considered with the initial condition

$$x(a) = c. \tag{2}$$

Here x is a vector-valued absolutely continuous function defined on $[a, b]$, M is a given quadratic constant matrix and β, γ are given constant vectors of compatible dimensions. We consider system (1) as the curve $z = y(t)$ generator. The goal is

M.-J. Lesot et al. (Eds.): IPMU 2020, CCIS 1239, pp. 332–343, 2020.
https://doi.org/10.1007/978-3-030-50153-2_25

to find a control law $u \in L_2[a,b]$ which drives the scalar output trajectory close to a sequence of set points at fixed times

$$\{(t_i, z_i) : i = 1, 2, \ldots, n\}, \quad \text{where} \quad a < t_1 < t_2 < \ldots < t_n \leq b, \tag{3}$$

by minimization of the objective functional

$$\int_a^b (u(t))^2 dt. \tag{4}$$

In some applications of such type of control problems, for example, doing trajectory planning in traffic control, we need to be able to generate curves that pass through predefined states at given times since we need to be able to specify the position in which the system will be in at a sequence of times (see, e.g., [1]). In this case we refer to the classical setting of the problem under consideration:

$$\int_a^b (u(t))^2 dt \to \min_{u \in L_2[a,b]:\ x(a)=c,\ y(t_i)=z_i,\ i=1,\ldots,n}, \tag{5}$$

where x and y depend on u by means of (1). It is shown in [1] and the references therein that a number of interpolation and path planning problems can be incorporated into control problem and studied using control theory and optimization techniques on Hilbert spaces with efficient numerical spline-based schemes. Control splines give a richer class of smoothing curves relative to polynomial curves. They have been proved to be useful for trajectory planning in [2], mobile robots in [3], contour modelling of images in [4], probability distribution estimation in [5] and so on.

However, in many situations, it is not really crucial that we pass a trajectory through these points exactly, but rather that we go reasonably close to them, while minimizing the objective functional. Such approach is closely related to the idea of smoothing under fuzzy interpolation conditions. We propose to use fuzzy numbers Z_i, $i = 1, \ldots, n$, in (5) instead of crisp z_i, $i = 1, \ldots, n$, to incorporate uncertainty of location conditions (3) into the model. According to this idea, we rewrite optimisation problem (5) in the following way:

$$\int_a^b (u(t))^2 dt \to \min_{u \in L_2[a,b]:\ x(a)=c,\ y(t_i) \text{is } Z_i,\ i=1,\ldots,n}, \tag{6}$$

where x and y depend on u by means of (1).

In this paper, the main attention is paid to the special case of problem (1):

$$M = \begin{pmatrix} 0 & 1 \\ -q & -p \end{pmatrix}, \quad x = \begin{pmatrix} x_1 \\ x_2 \end{pmatrix}, \quad \beta = \begin{pmatrix} 0 \\ 1 \end{pmatrix}, \quad \gamma = \begin{pmatrix} \gamma_1 \\ \gamma_2 \end{pmatrix}, \quad c = \begin{pmatrix} c_1 \\ c_2 \end{pmatrix}.$$

For this case problem (6) can be rewritten as

$$\int_a^b (g''(t) + pg'(t) + qg(t))^2 dt \longrightarrow \min_{\substack{g \in L_2^2[a,b]:\ g(a)=c_1,\ g'(a)=c_2, \\ y(t_i) \text{is } Z_i,\ i=1,\ldots,n}}, \tag{7}$$

where

$$y(t) = \gamma_1 g(t) + \gamma_2 g'(t),$$
$$u(t) = g''(t) + pg'(t) + qg(t),$$

and g is used to denote x_1.

2 Control Problem at α-Levels

In this paper we suggest a method for construction of solutions of (7) and finding corresponding control laws at each α-level with respect to fuzzy numbers used in the model by applying results from the theory of splines in convex sets.

To rewrite (7) for α-levels we introduce notations to deal with fuzzy numbers Z_i, $i = 1, \dots, n$. Fuzzy real number Z_i is a normal fuzzy subset of $\mathbb{R}$ that satisfies the condition: all α-cuts of Z_i are closed bounded intervals.

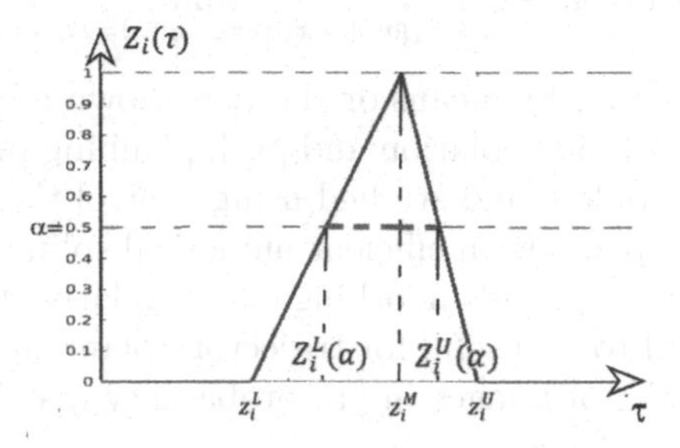

Fig. 1. Triangular fuzzy number

The α-cut ($\alpha \in (0, 1]$) of fuzzy number Z_i is the crisp set $(Z_i)_\alpha$ defined as

$$(Z_i)_\alpha = \{\tau \in \mathbb{R} \mid Z_i(\tau) \geq \alpha\}.$$

If $\alpha = 0$, then α-cut $(Z_i)_0$ can be defined as the support of function Z_i. The constraints "$y(t_i)$ is Z_i, $i = 1, \dots, n$," can be written at α-levels using α-cuts:

$$y(t_i) \in (Z_i)_\alpha, \; i = 1, \dots, n.$$

For each α-level the α-cut of Z_i is the closed interval

$$(Z_i)_\alpha = [Z_i^L(\alpha), Z_i^U(\alpha)].$$

Therefore problem (7) at α-level can be written in the following form:

$$\int_a^b (g''(t) + pg'(t) + qg(t))^2 dt \longrightarrow \min_{\substack{g \in L_2^2[a,b]:\; g(a)=c_1,\; g'(a)=c_2, \\ Z_i^L(\alpha) \leq y(t_i) \leq Z_i^U(\alpha), i=1,\dots,n}}, \tag{8}$$

where

$$y(t) = \gamma_1 g(t) + \gamma_2 g'(t), \; u(t) = g''(t) + pg'(t) + qg(t).$$

We apply triangular fuzzy numbers Z_i (see Fig. 1) given by triples (z_i^L, z_i^M, z_i^U):

$$Z_i(\tau) = \begin{cases} (\tau - z_i^L)(z_i^M - z_i^L)^{-1} & \text{if } \tau \in [z_i^L, z_i^M], \\ (z_i^U - \tau)(z_i^U - z_i^M)^{-1} & \text{if } \tau \in (z_i^M, z_i^U], \\ 0, & \text{otherwise.} \end{cases}$$

Then

$$Z_i^L(\alpha) = z_i^L + \alpha(z_i^M - z_i^L), \; Z_i^U(\alpha) = z_i^U - \alpha(z_i^U - z_i^M) \text{ for all } \alpha \in [0, \; 1].$$

3 Spline-Based Approach

We consider problem (8) as the special case of the following more general conditional minimization problem:

$$||Tg||_{L_2[a,b]} \longrightarrow \min_{\substack{g \in L_2^r[a,b]: \; (Ag)_0 = c_1, \; (Ag)_{n+1} = c_2, \\ Z_i^L(\alpha) \leq (Ag)_i \leq Z_i^U(\alpha), \; i=1,\ldots,n}} , \tag{9}$$

where linear operators $T: \; L_2^r[a,b] \; \rightarrow \; L_2[a,b]$, and $A: L_2^r[a,b] \rightarrow I\!R^{n+2}$ are continuous (here $L_2^r[a,b]$ is the Sobolev space), vector $c \in I\!R^2$ is given and $Z_i^L(\alpha), Z_i^U(\alpha), i = 1, \ldots, n$, are known. We assume that $A(L_2^r[a,b]) = I\!R^{n+2}$. In the case under consideration $r = 2$ and

$$Tg = g'' + pg' + qg = u, (Ag)_i = \gamma_1 g(t_i) + \gamma_2 g'(t_i), i = 1, ..., n,$$
$$(Ag)_0 = g(a), (Ag)_{n+1} = g'(a). \tag{10}$$

The solution of problem (9) will be considered for different α-levels. Value $\alpha = 1$ corresponds to the case when we pass the output trajectory through points (3) exactly (the case $z_i = z_i^M$). In this case problem (9) turns into the interpolating problem. For $\alpha < 1$ problem (9) will be considered applying smoothing splines.

3.1 Interpolating Splines

Problem (9) in the case $\alpha = 1$ corresponds to the following interpolating problem:

$$||Tg||_{L_2[a,b]} \longrightarrow \min_{\substack{g \in L_2^r[a,b]: \; (Ag)_0 = c_1, \; (Ag)_{n+1} = c_2, \\ (Ag)_i = z_i^M, \; i=1,\ldots,n}} . \tag{11}$$

The conditions of existence and uniqueness of solution of (11) and its characterization follow from the well known theorems (see, e.g., Theorems 4.4.2. and 4.5.9. in [6]).

Proposition 1. *Under the assumption, that $kerT \cap ker(A) = \{0\}$ and $kerT$ is finite-dimensional, the unique solution of problem (11) exists. An element $s \in L_2^r[a,b]$, such as $(As)_0 = c_1$, $(As)_i = z_i^M$, $i = 1, \ldots, n$, and $(As)_{n+1} = c_2$, is a solution of (11) if and only if there exists vector $\lambda \in I\!R^{n+2}$ such that*

$$T^*Ts = A^*\lambda. \tag{12}$$

This result implies that a solution of problem (11) is a spline from the space

$$S(T, A) = \{s \in L_2^r[a,b] \mid \forall x \in kerA \quad \langle Ts, Tx \rangle = 0\}.$$

Here and in the sequel the corresponding inner product is denoted by $\langle \cdot, \cdot \rangle$, and $kerA$ is the kernel of operator A.

The view of splines from the space $S(T, A)$ in depending on parameters p and q for the considered case of operators (10) is obtained in [7] using the general theorem (see Theorem 1 in [8]) and applying functional analysis tools. For example, if $p = q = 0$ then elements of $S(T, A)$ are polynomial cubic splines from $C^1[a,b]$, i.e., they are cubic polynomials on each interval $[t_{i-1}, t_i]$, $i = 1, \ldots, n+1$, where $t_0 = a$ and $t_{n+1} = b$.

3.2 Splines in Convex Sets

Problem (9) in the case $\alpha < 1$ corresponds to the following smoothing problem (problem on splines in a convex set):

$$||Tg||_{L_2[a,b]} \longrightarrow \min_{\substack{g \in L_2^r[a,b]:\ (Ag)_0 = c_1,\ (Ag)_{n+1} = c_2, \\ Z_i^L(\alpha) \le (Ag)_i \le Z_i^U(\alpha),\ i=1,\ldots,n}} \tag{13}$$

considered under assumption $Z_i^L(\alpha) < Z_i^U(\alpha)$.

The conditions of existence and uniqueness of solution of (13) follow from the known theorem (see Theorem 7 in [8]).

Proposition 2. *Under the assumption that $kerT$ is finite-dimensional a solution of problem (13) exists. An element $s \in L_2^r[a,b]$, such as $(As)_0 = c_1$, $(As)_{n+1} = c_2$, $Z_i^L(\alpha) \le (As)_i \le Z_i^U(\alpha)$, $i = 1, \ldots, n$, is a solution of (13) if and only if there exists vector $\lambda \in I\!R^{n+2}$ such that*

$$T^*Ts = A^*\lambda \tag{14}$$

and components λ_i, $i = 1, \ldots, n$, satisfy the conditions

$$\begin{array}{ll} \lambda_i = 0, & \text{if} \quad Z_i^L(\alpha) < (As)_i < Z_i^U(\alpha), \\ \lambda_i \ge 0, & \text{if} \quad (As)_i = Z_i^L(\alpha), \\ \lambda_i \le 0, & \text{if} \quad (As)_i = Z_i^U(\alpha). \end{array} \tag{15}$$

Under the additional assumption $kerT \cap ker(A) = \{0\}$ this solution is unique.

This result implies that a solution of problem (13) belongs to the space $S(T, A)$. To find it we can use the method of adding-removing interpolation knots which is considered in details, for example, in [9] or [10]. It is an iterative method. On the k-th step of it we need to solve the following interpolation problem: to construct a spline $s^k \in S(T, A)$ such that the initial conditions $(As^k)_0 = c_1$, $(As^k)_{n+1} = c_2$, and the interpolation conditions written in the form $(As^k)_i = d_i^k$, $i \in I^k$, are satisfied. The set of indices $I^k \subset \{1, \ldots, n\}$ and numbers d_i^k are specified during the iterations. The knots t_i for $i \in I^k$ are considered as interpolation knots on the k-th step. We start with a solution s_1 obtained using only the initial conditions, i.e., $I^1 = \emptyset$. The iterative step from I^k to I^{k+1} is done by adding to I^k all indices $i \in \{1, \ldots, n\}$ such that the restriction $Z_i^L \leq (As^k)_i \leq Z_i^U$ is not satisfied. For the added index i we take $d_i^{k+1} = Z_i^L(\alpha)$ if $Z_i^L(\alpha) > (As^k)_i$, and we take $d_i^{k+1} = Z_i^U(\alpha)$ if $(As^k)_i > Z_i^U(\alpha)$. On the other hand, we remove from I^k all indices $i \in I^k$ such that the rule (15) is not satisfied for the corresponding coefficient of s^k. To finish the k-th step we also denote $d_i^{k+1} = d_i^k$ for $i \in I^{k+1} \cap I^k$. If $I^{k+1} = I^k$ then the algorithm ends and the obtained s^k is a solution of (13).

4 Numerical Solutions

In this paper we consider problem (8) as (9) with operator T and A defined by (10). According to Proposition 1 and Proposition 2 in this case solutions of (8) at each α-level belong to the space $S(T, A)$. The view of splines from the corresponding $S(T, A)$ (i.e., the view of solutions of problem (8)) is obtained in [7]. This view in [7] is given depending on the roots r_1, r_2 of the equation $r^2 + pr + q = 0$:

- Class 1 (exponential splines with polynomial coefficients): $r_1 = r_2 \in \mathbb{R} \setminus \{0\}$.
- Class 2 (exponential splines): $r_1, r_2 \in \mathbb{R}$, $r_1 \neq r_2$.
- Class 3 (polynomial-exponential splines): $r_1, r_2 \in \mathbb{R}$, $r_1 \neq r_2, r_1 \neq 0, r_2 = 0$.
- Class 4 (polynomial splines): $r_1 = r_2 = 0$.
- Class 5 (trigonometric splines with polynomial coefficients): $r_{1,2} = \pm i\eta \neq 0$.
- Class 6 (trigonometric splines with exponential-polynomial coefficients): $r_{1,2} = \zeta \pm i\eta$ with $\eta \neq 0$ and $\zeta \neq 0$.

The simplest case with $p = q = 0$, i.e., $Tg = g''$, corresponds to the classical smoothing problem in the theory of splines according to which a solution of (8) without the initial conditions is a cubic spline. Taking into account the initial conditions we get the following form for solution s of problem (8) for this case:

$$s(t) = c_1 + c_2(t-a) + \frac{\lambda_0}{6}(t-a)_+^3 - \frac{\lambda_{n+1}}{2}(t-a)_+^2 + \sum_{i=1}^{n} \lambda_i\left(\frac{\gamma_1}{6}(t-t_i)_+^3 - \frac{\gamma_2}{2}(t-t_i)_+^2\right)$$

with the following conditions on coefficients

$$\lambda_0 + \sum_{i=1}^{n} \lambda_i(\gamma_1 + \gamma_2) = 0, \quad \lambda_0 a + \lambda_{n+1} + \sum_{i=1}^{n} \lambda_i(\gamma_1 t_i + \gamma_2) = 0.$$

Here and in the sequel the truncated power function is defined as

$$(t - t_j)_+^k = \begin{cases} (t - t_j)^k, \ t \geq t_j, \\ \quad 0, \quad t < t_j. \end{cases}$$

The corresponding control function u could be obtained as $u = s''$.

Two numerical examples corresponding to more complicated cases are considered below for illustration of the proposed technique. Numerical results are obtained by using Maple.

4.1 Example 1: Exponential Splines

We consider the numerical example for the case $p = -3$ and $q = 2$, $\gamma_1 = 1$ and $\gamma_2 = 0$, interval $[a, b] = [0, 0.5]$, the initial conditions are with $c_1 = 1$, $c_2 = 1$. At equally spaced points of interval $[0.1, 0.5]$ with step size 0.1 we take the following fuzzy numbers $Z_i, i = 1, \ldots, 5$: $(4, 5, 6), (1, 2, 3), (5, 6, 7), (2, 3, 4), (6, 7, 8)$. This case corresponds to the case of two nonzero roots of characteristic equation $r_1 = 1, r_2 = 2$, i.e., to the case when solutions belong to the class of exponential splines.

As it is obtained in [7], the class of exponential splines for problem (8) consists of splines

$$s(t) = \mu_1 e^{r_1(t-a)} + \mu_2 e^{r_2(t-a)} + \frac{1}{2(r_1^2 - r_2^2)}\Big(\frac{(\lambda_0 - \lambda_{n+1} r_1)e^{r_1(t-a)}}{r_1}$$
$$- \frac{(\lambda_0 - \lambda_{n+1} r_2)e^{r_2(t-a)}}{r_2} + \sum_{i=1}^{n} \lambda_i\Big(\frac{\gamma_1}{r_1} e^{r_1|t-t_i|} + \gamma_2(e^{r_1(t_i-t)_+} - e^{r_1(t-t_i)_+})$$
$$- \frac{\gamma_1}{r_2} e^{r_2|t-t_i|} - \gamma_2(e^{r_2(t_i-t)_+} - e^{r_2(t-t_i)_+})\Big)\Big). \tag{16}$$

For the solution of (8) the coefficients are expressed by using the following system:

$$(\gamma_1 + \gamma_2 r_1) \sum_{i=1}^{n} \lambda_i e^{r_1 t_i} + (\lambda_0 + \lambda_{n+1} r_1) e^{r_1 a} = 0,$$

$$(\gamma_1 + \gamma_2 r_2) \sum_{i=1}^{n} \lambda_i e^{r_2 t_i} + (\lambda_0 + \lambda_{n+1} r_2) e^{r_2 a} = 0, \tag{17}$$

and the system of interpolating conditions for $g(t_i) = z_i^M, i = 1, \ldots, n$, in case $\alpha = 1$. For $\alpha < 1$ the interpolating conditions are precised by iterations of the method of adding-removing knots.

The corresponding control function u is given by

$$u(t) = \sum_{i=1}^{n} \frac{\lambda_i (t_i - t)_+^0}{r_1 - r_2} (\gamma_1 (e^{r_1(t_i-t)} - e^{r_2(t_i-t)}) + \gamma_2 (r_1 e^{r_1(t_i-t)} - r_2 e^{r_2(t_i-t)})). \tag{18}$$

For the considered case the conditions of the uniqueness of solution are satisfied. This solution is constructed for four α-levels: $\alpha_1 = 0$, $\alpha_2 = 0.25$, $\alpha_3 = 0.5$ and $\alpha_4 = 1$. The solution of problem (8) in this case is the exponential spline (16) with coefficients from Table 1. The control law u is obtained by (18). The corresponding graphs are considered in Fig. 2 and Fig. 3. The values of the objective functional for these α-levels are compared (see Table 2).

Table 1. Coefficients of the solution at α-level for *Example 1*

	$\alpha_1 = 0$	$\alpha_2 = 0.25$	$\alpha_3 = 0.5$	Interpolation ($\alpha_4 = 1$)
μ_1	-5.3984×10^1	-4.0037×10^1	-4.4686×10^1	-5.3984×10^1
μ_2	5.4984×10^1	4.1037×10^1	4.5686×10^1	5.4984×10^1
λ_0	-1.9715×10^4	-1.3245×10^4	-1.5402×10^4	-1.9715×10^4
λ_1	5.5236×10^4	3.5234×10^4	4.1902×10^4	5.5236×10^4
λ_2	-7.4174×10^4	-4.4561×10^4	-5.4432×10^4	-7.4174×10^4
λ_3	7.2526×10^4	4.2014×10^4	5.2185×10^4	7.2526×10^4
λ_4	-4.6963×10^4	-2.6943×10^4	-3.3617×10^4	-4.6963×10^4
λ_5	1.3092×10^4	7.4793×10^3	9.3504×10^3	1.3092×10^4
λ_6	-1.5895×10^2	-1.1711×10^2	-1.3105×10^2	-1.5895×10^2

By comparison of the objective functional at these α-levels from Table 2 we see that the minimum of this functional is obtained for $\alpha_1 = 0$ and the interpolating spline gives the biggest value.

Table 2. Comparison of the values of the objective functional for *Example 1*

α	$\alpha_1 = 0$	$\alpha_2 = 0.25$	$\alpha_3 = 0.5$	Interpolation ($\alpha_4 = 1$)
$\|u\|$	2.8622×10^2	3.7282×10^2	4.5988×10^2	6.34657×10^2

4.2 Example 2: Trigonometric Splines with Polynomial Coefficients

We consider the second numerical example for the case $p = 0$ and $q = 1$, $\gamma_1 = 1$ and $\gamma_2 = 0$, interval $[a, b] = [0, 0.5]$, the initial conditions are given with $c_1 = 1$ and $c_2 = 1$. At equally spaced points in interval $[0.1, 0.5]$ with

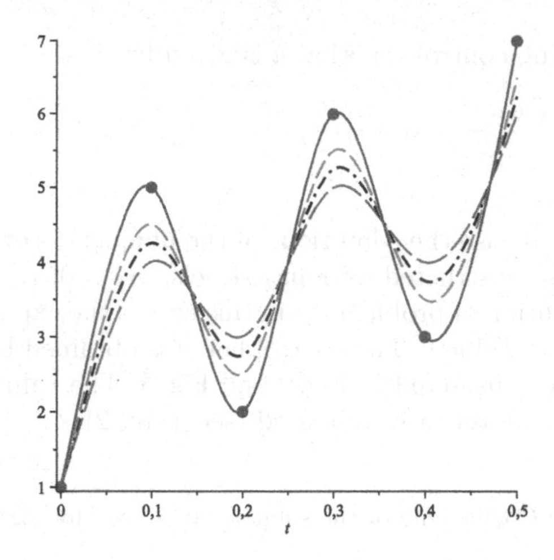

Fig. 2. State trajectories for *Example 1* (solid line for $\alpha = 1$, dash line for $\alpha = 0.25$, dash dot line for $\alpha = 0.5$, long dash line for $\alpha = 0$)

step size 0.1 we take the following values of fuzzy numbers $Z_i, i = 1, \ldots, 5$: $(2,3,4), (0,1,2), (4,5,6), (1,2,3), (3,4,5)$. This case corresponds to the case of complex roots of characteristic equation $r_{1,2} = \pm i$ (for this case $\eta = 1$), i.e., to the case when solutions belongs to the class of trigonometric splines with polynomial coefficients.

As it is obtained in [7], the class of trigonometric splines with polynomial coefficients for problem (8) in the considered case consists of splines

$$s(t) = c_1 cos\,(\eta(t-a)) + \frac{c_2}{\eta} sin\,(\eta(t-a)) + \frac{\lambda_0}{2\eta^3}(\eta(t-a)cos\,(\eta(t-a)) - sin\,(\eta(t-a)))$$

$$+\frac{\lambda_{n+1}}{2\eta}(t-a)sin\,(\eta(t-a)) + \frac{1}{2}\sum_{i=1}^{n}\lambda_i(\frac{\gamma_1}{\eta^3}sin\,(\eta(t-t_i)_+)$$

$$-\frac{(t-t_i)_+}{\eta^2}(\gamma_1 cos\,(\eta(t-t_i)) + \gamma_2\eta sin\,(\eta(t-t_i)))). \qquad (19)$$

The coefficients fulfil the following conditions

$$\begin{aligned} &\sum_{i=1}^{n}\lambda_i(\gamma_1 sin\,(\eta t_i) + \gamma_2\eta cos\,(\eta t_i)) + \lambda_0 sin\,(\eta a) + \lambda_{n+1}\eta cos\,(\eta a) = 0, \\ &\sum_{i=1}^{n}\lambda_i(\gamma_1 cos\,(\eta t_i) - \gamma_2\eta sin\,(\eta t_i)) + \lambda_0 cos\,(\eta a) - \lambda_{n+1}\eta sin\,(\eta a) = 0, \end{aligned} \qquad (20)$$

and the system of the interpolating conditions

$$\gamma_1 s(t_i) + \gamma_2 s'(t_i) = z_i^M, i = 1, \ldots, n,$$

in case $\alpha = 1$. For $\alpha < 1$ the interpolating conditions are specified by iterations of the method of adding-removing knots.

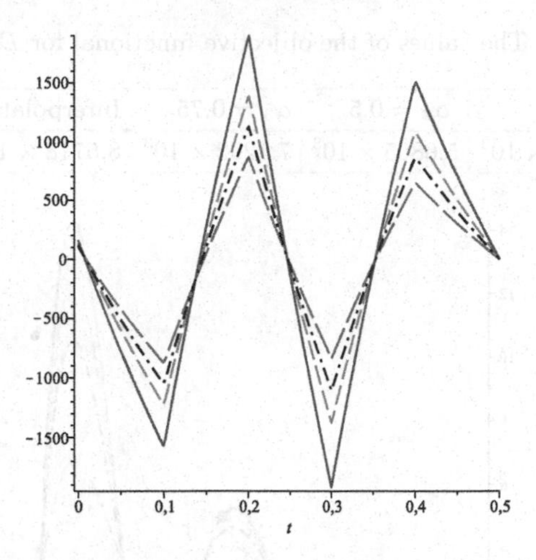

Fig. 3. Control law for *Example 1* (solid line for $\alpha = 1$, dash line for $\alpha = 0.25$, dash dot line for $\alpha = 0.5$, long dash line for $\alpha = 0$)

The corresponding control function u is given by

$$u(t) = \sum_{i=1}^{n} \lambda_i(\frac{\gamma_1}{\eta} sin\, \eta(t_i - t)_+ + \gamma_2(t_i - t)_+^0 cos\, \eta(t_i - t)). \tag{21}$$

For the considered case the conditions on the uniqueness of solution are satisfied. This solution is constructed for four α-levels: $\alpha_1 = 0$, $\alpha_2 = 0.5, \alpha_3 = 0.75$ and $\alpha_4 = 1$. The solution of problem (8) in this case is the trigonometric splines with polynomial coefficients (19) with coefficients from Table 3. The control law u could be obtained by (21). The corresponding graphs are considered in Fig. 4 and Fig. 5. The values of the objective functional for considered α-levels are compared (see Table 4).

Table 3. Coefficients of the solutions at α-levels for *Example 2*

	$\alpha_1 = 0$	$\alpha_2 = 0.5$	$\alpha_3 = 0.75$	Interpolation ($\alpha_4 = 1$)
λ_0	9.0736×10^3	1.8084×10^4	2.2590×10^4	2.7095×10^4
λ_1	2.0746×10^4	4.1800×10^4	5.2327×10^4	6.2854×10^4
λ_2	-3.1238×10^4	-6.1490×10^4	-7.6616×10^4	-9.1742×10^4
λ_3	5.6274×10^4	1.1130×10^5	1.3881×10^5	1.6633×10^5
λ_4	-1.3993×10^5	-2.8177×10^5	-3.5268×10^5	-4.2360×10^5
λ_5	8.6636×10^5	1.7523×10^5	2.1953×10^5	2.63837×10^5
λ_6	4.6398×10^2	8.6472×10^2	1.0650×10^3	1.2654×10^3

Table 4. The values of the objective functional for *Example 2*

α	$\alpha_1 = 0$	$\alpha_2 = 0.5$	$\alpha_3 = 0.75$	Interpolation ($\alpha_4 = 1$)
$\|u\|$	2.8012×10^3	5.6865×10^3	7.1292×10^3	8.5718×10^3

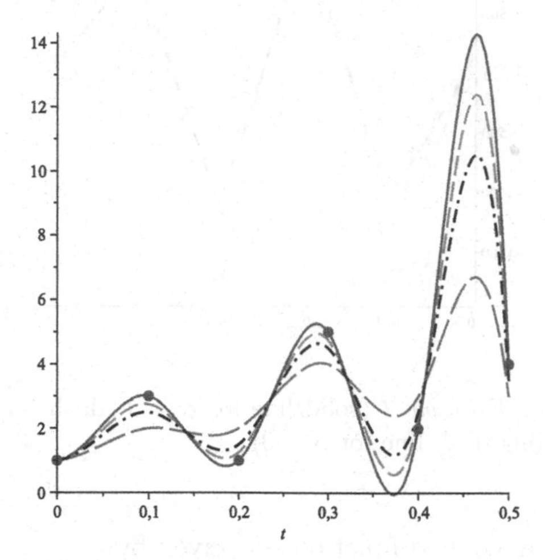

Fig. 4. State trajectories for *Example 2* (solid line for $\alpha = 1$, dash line for $\alpha = 0.5$, dash dot line for $\alpha = 0.75$, long dash line for $\alpha = 0$)

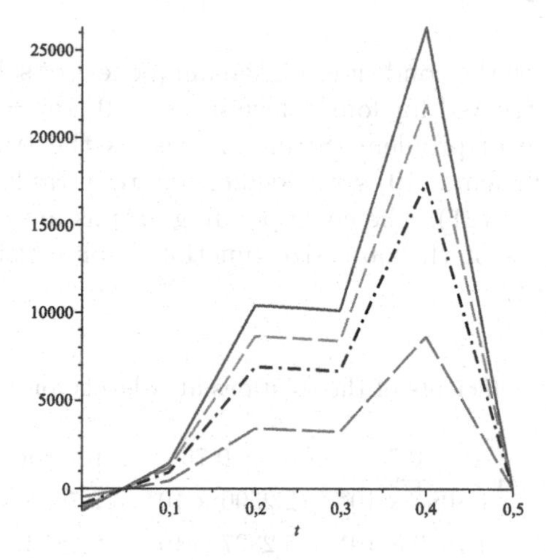

Fig. 5. Control law for *Example 2* (solid line for $\alpha = 1$, dash line for $\alpha = 0.5$, dash dot line for $\alpha = 0.75$, long dash line for $\alpha = 0$)

By comparison of the values of the objective functional from Table 4 we see that the minimum of this functional is obtained for $\alpha_1 = 0$ and the interpolating spline gives the biggest value.

5 Conclusion

The proposed method can be effectively used for dynamical systems associated with linear differential equations (the restriction on the order of equations is not essential) when uncertainty in location conditions is described by fuzzy numbers $Z_i,\ i = 1, \ldots, n$ (the restriction on the triangular type of fuzzy numbers is not essential). It seems natural to incorporate into the model also uncertainty of the sequence of times to be considered. For such purpose fuzzy numbers $T_i,\ i = 1, \ldots, n$, could be used instead of crisp $t_i,\ i = 1, \ldots, n$. In this case the constraints

$$y(t_i) \text{ is } Z_i,\ i = 1, \ldots, n,$$

could be rewritten using IF-THEN rules as

$$\text{IF } t \text{ is } T_i \text{ THEN } y(t) \text{ is } Z_i,\ i = 1, \ldots, n.$$

The future research could be devoted to development of the proposed approach for the following problem

$$\int_a^b (u(\tau))^2 d\tau \to \min_{\substack{u \in L_2[a,b]:\ x(a)=c, \\ \text{IF } t \text{ is } T_i \text{ THEN } y(t) \text{ is } Z_i,\ i=1,\ldots,n}}$$

considered in the context of (1)–(4).

References

1. Egerstedt, M., Martin, C.: Control Theoretic Splines. Princeton University Press, New Jersey (2010)
2. Egerstedt, M., Martin, C.: Optimal trajectory planning and smoothing splines. Automatica **37**, 1057–1064 (2001)
3. Martin, C. F., Takahashi, S.: Optimal control theoretic splines and its application to mobile robot. In: Proceedings of the IEEE International Conference on Control Applications, Taipei, Taiwan, pp. 1729–1732. IEEE (2004)
4. Egerstedt, M., Fujioka, H., Kano, H., Martin, C.F., Takahashi, S.: Periodic smoothing splines. Automatica **44**, 185–192 (2008)
5. Charles, J.K., Martin, C.F., Sun, S.: Cumulative distribution estimation via control theoretic smoothing splines. In: Hu, X., Jonsson, U., Wahlberg, B., Ghosh, B. (eds.) Three Decades of Progress in Control Sciences, pp. 95–104. Springer, Heidelberg (2010). https://doi.org/10.1007/978-3-642-11278-2_7
6. Laurent, P.-J.: Approximation et Optimisation. Hermann, Paris (1972)
7. Asmuss, S., Budkina, N.: Control smoothing splines with initial conditions. In: Proceedings of the 17th Conference on Applied Mathematics, pp. 14–21. STU, Bratislava, Slovakia (2018)
8. Asmuss, S., Budkina, N.: On some generalization of smoothing problems. Math. Model. Anal. **15**(3), 11–28 (2015). https://doi.org/10.3846/13926292.2015.1048756
9. Budkina, N.: Construction of smoothing splines by a method of gradual addition of interpolating knots. Proc. Latv. Acad. Sci. Sect. B **55**(4), 145–151 (2001)
10. Leetma, E., Oja, P.: A method of adding-removing knots for solving smoothing problems with obstacles. Eur. J. Oper. Res. **194**(1), 28–38 (2009). https://doi.org/10.1016/j.ejor.2007.12.020

Statistical Image Processing and Analysis, with Applications in Neuroimaging

High Dimensional Bayesian Regularization in Regressions Involving Symmetric Tensors

Rajarshi Guhaniyogi(✉)

Department of Statistics, SOE 2, University of California Santa Cruz, 1156 High Street, Santa Cruz, CA 95064, USA
rguhaniy@ucsc.edu

Abstract. This article develops a regression framework with a symmetric tensor response and vector predictors. The existing literature involving symmetric tensor response and vector predictors proceeds by vectorizing the tensor response to a multivariate vector, thus ignoring the structural information in the tensor. A few recent approaches have proposed novel regression frameworks exploiting the structure of the symmetric tensor and assume symmetric tensor coefficients corresponding to scalar predictors to be low-rank. Although low-rank constraint on coefficient tensors are computationally efficient, they might appear to be restrictive in some real data applications. Motivated by this, we propose a novel class of regularization or shrinkage priors for the symmetric tensor coefficients. Our modeling framework a-priori expresses a symmetric tensor coefficient as sum of low rank and sparse structures, with both these structures being suitably regularized using Bayesian regularization techniques. The proposed framework allows identification of tensor nodes significantly influenced by each scalar predictor. Our framework is implemented using an efficient Markov Chain Monte Carlo algorithm. Empirical results in simulation studies show competitive performance of the proposed approach over its competitors.

Keywords: Low-rank structure · Symmetric tensor predictor · Shrinkage prior · Spike and slab prior

1 Introduction

This article is motivated by a variety of applications, in which a sample of symmetric tensors is available along with a few scalar variables of interest. Analogous to rows and columns of a matrix, various axes of a tensor are known as tensor modes and the indices of a tensor mode are often referred to as "tensor nodes". A tensor is known to be symmetric if interchanging modes results in the same tensor. Entries in a tensor are known as "tensor cells". In our motivating applications, each sample point is represented by its own symmetric tensor, and the

Supported by Office of Naval Research and National Science Foundation.

M.-J. Lesot et al. (Eds.): IPMU 2020, CCIS 1239, pp. 347–357, 2020.
https://doi.org/10.1007/978-3-030-50153-2_26

tensor nodes are labeled and shared across all sample points through a registration process. The goals of such scientific applications are two fold. First, it is important to build a predictive model to assess the change of the symmetric tensor response as the predictors of interest vary. A more important scientific goal becomes identifying nodes of the symmetric tensor significantly impacted by each predictor.

Although there is a gamut of applications involving symmetric tensors, our work is specifically motivated by scientific applications pertaining to brain connectomics. In such applications the dataset contains brain network information along with a few observable phenotypes (e.g., IQ, presence or absence of any neuronal disorder, age) for multiple subjects. Brain network information for each subject is encoded within a symmetric matrix of dimension $V \times V$, with V as the number of regions of interest (ROI) a human brain is parceled into following a popular brain atlas. The (k,l)th cell of the matrix consists of the number of neuron connections between the k-th and l-th regions of interest (ROI). Thus, each mode of this symmetric matrix (when viewed as a 2-D symmetric tensor) consists of V nodes ($V = 68$ when Desikan-Killany brain atlas is followed [1]), each corresponding to a specific ROI in the human brain. The most important scientific goal here boils down to making inference on brain regions of interest (ROIs) and their inter-connections significantly associated with each phenotypic predictor [9].

One approach is to vectorize the symmetric tensor and cast the modeling problem as a high dimensional multivariate reduced rank sparse regression framework with the vectorized tensor response and scalar predictors. There are adequate literature on frequentist penalized optimization [15], as well as on Bayesian shrinkage [2] which deal with model fitting and computational issues with high dimensional multivariate reduced rank regression models. Although computationally efficient, these approaches treat the cells of the symmetric tensor coefficients as if they were fully exchangeable, ignoring the fact that coefficients that involve common tensor nodes can be expected to be correlated a priori. Ignoring this correlation may appear to be detrimental in terms of model performance. Additionally, such modeling framework does not directly lead to the identification of nodes significantly associated with each predictor.

We develop a symmetric tensor response regression model with a symmetric tensor response and scalar predictors. The symmetric tensor coefficients corresponding to each predictor in this regression is assigned a novel Bayesian shrinkage prior that combines ideas from low-rank parallel factor (PARAFAC) decomposition methods, spike-and-slab priors and Bayesian high dimensional regularization techniques to generate a model that respects the tensor structure of the response. These structures are introduced to achieve several inferential goals simultaneously. The low-rank structure is primarily assumed to capture the interactions between different pairs of tensor nodes, the node-wise sparsity offers inference on various tensor nodes significantly associated with a predictor. The Bayesian regularization structure allows appropriate shrinkage of unimportant cell coefficients towards zero while minimally shrinking the important cell coefficients. All structures jointly achieve parsimony and deliver accurate characterization of uncertainty for estimating parameters and identifying significant tensor nodes. The proposed approach finds excellent synergy with the recent literature on bilinear relational data models [4,7], multiway regression models [6] and other object oriented regression models [3], where low-rank and/or regularization structures are imposed on parameters.

The proposed framework is similar to but distinct from the recent developments in high dimensional regressions with multidimensional arrays (tensors) and other object oriented data. For example, recent literature that builds regression models with a scalar response and tensor predictors [6] is less appealing in this context, since it does not incorporate the symmetry constraint in the tensor. In the same vein, [5] formulate a Bayesian tensor response regression approach that is built upon a novel multiway stick breaking shrinkage prior on the tensor coefficients. While [5] is able to identify important tensor cells, it does not allow detection of tensor nodes influenced by a predictor. Moreover, these approaches have not been extended to accommodate scenarios other than a tensor response with continuous cell entries and do not directly incorporate the symmetry constraint in the tensor coefficient corresponding to a predictor. Also, unlike these approaches, our approach does not assume a low-rank representation of tensor coefficients; hence allowing more flexible structure to analyze impact of the predictors on tensor cells and interaction between tensor nodes. A work closely related to our framework develops shrinkage priors in a regression framework with a scalar response and an undirected network predictor, expressed in the form of a symmetric matrix [3]. However, they treat the tensor as a predictor, whereas we treat it as a response. This difference in the modeling approach leads to a different focus and interpretation. The symmetric tensor predictor regression focuses on understanding the change of a scalar response as the symmetric tensor predictor varies, while regression with symmetric tensor response aims to study the change of the symmetric tensor as the predictors vary.

Rest of the article flows as follows. In Sect. 2 the model and prior distributions on parameters are introduced. Section 2 also briefly discusses posterior computation, where as Sect. 3 presents simulation studies to validate our approach. Finally, Sect. 4 concludes the article with an eye to the future work.

2 Model Development and Posterior Computation

2.1 Model and Prior Distributions

For $i = 1, ..., n$, let $\boldsymbol{y}_i = ((y_{i,(k_1,...,k_D)}))_{k_1,...,k_D=1}^V \in \mathcal{Y} \subseteq \mathbb{R}^{V\times\cdots\times V}$ denote the D-way symmetric tensor response with dummy diagonal entries and $\boldsymbol{z}_i = (z_{i1}, ..., z_{ip})'$ be p predictors of interest corresponding to the ith individual. The symmetric constraint in the tensor response implies $y_{i,(k_1,...,k_D)} = y_{i,(P(k_1),...,P(k_D))}$, with $P(\cdot)$ being any permutation of $\{k_1, ..., k_D\}$. Due to the diagonal entries being dummies in the symmetric tensor response, it is enough to build a probabilistic model for $\boldsymbol{y}_{upper} = \{y_{i,\boldsymbol{k}} : \boldsymbol{k} = (k_1, ..., k_D), 1 \le k_1 < \cdots < k_D \le V\}$.

For the sake of this article, we assume $y_{i,k} \in \mathbb{R}$ and propose

$$y_{i,k} = \gamma_0 + \Gamma_{1,k} z_{i1} + \cdots + \Gamma_{p,k} z_{ip} + \epsilon_{i,k},\ \epsilon_{i,k} \overset{iid}{\sim} N(0, \sigma^2), \tag{1}$$

where $\Gamma_{1,k}, ..., \Gamma_{p,k}$ are the $\boldsymbol{k} = (k_1, ..., k_D)$th cells of the $V \times \cdots \times V$ symmetric coefficient tensors $\boldsymbol{\Gamma}_1, ..., \boldsymbol{\Gamma}_p$ with dummy diagonal entries, respectively. Here $\Gamma_{s,k}$, $s = 1, ..., p$, is the coefficient corresponding to the sth predictor of interest on the $\boldsymbol{k} = (k_1, ..., k_D)$th cell of the symmetric tensor response. The coefficient $\gamma_0 \in \mathbb{R}$ is the intercept in the regression model (4).

To account for association between the tensor response $\boldsymbol{y}$ and predictors $z_1, ..., z_p$, we propose a *shrinkage prior* distribution on $\Gamma_{s,k}$, $s = 1, ..., p$, represented as a location and scale mixture of normal distributions. In particular, the distribution of $\Gamma_{s,k}$ is assumed to be conditionally independent normal distribution with

$$\Gamma_{s,k} \sim N\left(\sum_{r=1}^{R} \lambda_{s,r} \gamma_{s,k_1}^{(r)} \cdots \gamma_{s,k_D}^{(r)}, \kappa_{s,k}\sigma^2\right),\ \boldsymbol{k} = (k_1, ..., k_D),\ 1 \le k_1 < \cdots < k_D \le V. \tag{2}$$

(2) implies $E[\Gamma_{s,k}|\sigma^2, \kappa_{s,k}] = \sum_{r=1}^{R} \lambda_{s,r} \gamma_{s,k_1}^{(r)} \cdots \gamma_{s,k_D}^{(r)}$, i.e., the prior distribution of $\boldsymbol{\Gamma}_s$ is centered on a rank-R PARAFAC [10] decomposed tensor. The PARAFAC or parallel factor decomposition is a multiway analogue to the two-dimensional factor modeling of matrices. In particular it provides a low-rank structure to the mean function of $\boldsymbol{\Gamma}_s$. Note that, in the tensor regression literature, it is a fairly common practice to assume a low-rank structure for $\boldsymbol{\Gamma}_s$ directly [6]. In contrast, the prior distribution in (2) centers on a low-rank PARAFAC/CP representation [10], precluding any additional imposition of a low-rank structure on $\boldsymbol{\Gamma}_s$ a priori. This allows more flexibility in the structure of the coefficients. $\kappa_{s,k}$ is the scale parameter corresponding to each $\Gamma_{s,k}$ controlling the local variability of each coefficient a priori and $\lambda_{s,r} \in \{0, 1\}$, $r = 1, ..., R$, are introduced to assess the effect of the r-th summand on the mean of $\Gamma_{s,k}$. In particular, $\lambda_{s,r} = 0$ implies that the r-th summand of the low-rank factorization is not informative to predict the response.

To develop a data dependent learning of nonzero $\lambda_{s,r}$'s, we propose $\lambda_{s,r} \sim Ber(\theta_{s,r})$, $\theta_{s,r} \sim Beta(1, r^c)$, $c > 1$, a priori. The hyper-parameters in the beta distribution are set so as to penalize the usage of large number of summands in the PARAFAC decomposition, which protects the model from over-fitting. Define $\gamma_{s,v} = (\gamma_{s,v}^{(1)}, ..., \gamma_{s,v}^{(R)})' \in \mathbb{R}^R$ as the tensor node specific vectors (for $v = 1, ..., V$) describing $\boldsymbol{\Gamma}_s$. In the course of identifying important tensor nodes significantly associated with the sth predictor, we note that the v-th node has minimal effect on the sth predictor if $\boldsymbol{\gamma}_{s,v} = \boldsymbol{0}$. Thus, in order to directly infer on $\boldsymbol{\gamma}_{s,v}$, a *spike-and-slab* mixture distribution prior [8] is assigned on $\boldsymbol{\gamma}_{s,v}$ as below

$$\boldsymbol{\gamma}_{s,v} \sim \begin{cases} N(\boldsymbol{0}, \boldsymbol{H}_s), & \text{if } \xi_{s,v} = 1 \\ \delta_{\boldsymbol{0}}, & \text{if } \xi_{s,v} = 0 \end{cases} \tag{3}$$

where $\delta_{\boldsymbol{0}}$ is the Dirac-delta function at $\boldsymbol{0}$ and $\boldsymbol{H}_s$ is a covariance matrix of order $R \times R$. The rest of the hierarchy is completed by setting $\kappa_{s,k} \sim Exp(\zeta^2/2)$, $\xi_{s,v} \sim$

$Ber(\Delta_s)$, $\boldsymbol{H}_s \sim IW(\boldsymbol{I}, \nu)$, $\Delta_s \sim U(0,1)$, where IW stands for the inverse wishart distribution. Finally, γ_0 and σ^2 are assigned $N(\mu_\gamma, \sigma_\gamma^2)$ and IG(a,b) priors respectively, IG corresponding to the inverse gamma density. The posterior distribution of $\xi_{s,v}$ is monitored and the vth tensor node is related to the sth predictor if the posterior probability of $\{\xi_{s,v} = 1\}$ turns out to be greater than 0.5.

A few remarks are in order. Note that, if $\kappa_{s,\boldsymbol{k}} = 0$ for all $\boldsymbol{k}$, then $\boldsymbol{\Gamma}_s$ assumes an exact low-rank decomposition given by, $\Gamma_{s,\boldsymbol{k}} = \sum_{r=1}^{R} \lambda_{s,r}\gamma_{s,k_1}^{(r)} \cdots \gamma_{s,k_D}^{(r)}$. Also, if $\boldsymbol{\gamma}_{s,v} = \boldsymbol{0}$, then a priori $\Gamma_{s,\boldsymbol{k}}$ follows an ordinary Bayesian lasso shrinkage prior distribution [12] for all $\boldsymbol{k}$ with some $k_l = v$. In general, $\Gamma_{s,\boldsymbol{k}}$ a priori can be expressed as $\Gamma_{s,\boldsymbol{k}} = \sum_{r=1}^{R} \lambda_{s,r}\gamma_{s,k_1}^{(r)} \cdots \gamma_{s,k_D}^{(r)} + \Gamma_{s,2,\boldsymbol{k}}$, where $\Gamma_{s,2,\boldsymbol{k}}$ following an ordinary Bayesian lasso shrinkage prior [12].

2.2 Posterior Computation

Although summaries of the posterior distribution cannot be computed in closed form, full conditional distributions for all the parameters are available and correspond, in most cases, to standard families. Thus, posterior computation can proceed through a Markov chain Monte Carlo algorithm. Details of the Markov chain Monte Carlo algorithm with the conditional posterior distributions are provided in the Appendix. We run the MCMC chain for 15000 iterations. With the first 5000 as burn-ins, the posterior inference is drawn on the $L = 10000$ post burn-in draws suitably thinned. In order to identify whether the v-th tensor node is significantly related to the sth predictor, we rely on the post burn-in L samples $\xi_{s,v}^{(1)}, \ldots, \xi_{s,v}^{(L)}$ of $\xi_{s,v}$. Node v is said to be influential if $\frac{1}{L}\sum_{l=1}^{L} \xi_{s,v}^{(l)} > 0.5$. Here $\frac{1}{L}\sum_{l=1}^{L} \xi_{s,v}^{(l)}$ corresponds to the empirical estimate of the posterior probability of $\{\xi_{s,v} = 1\}$. We also assess the point estimates on tensor cell coefficients $\Gamma_{s,\boldsymbol{k}}$ and present uncertainty in the estimation procedure.

3 Simulation Studies

3.1 Simulation Settings

This article illustrates the performance of our proposed approach referred to as the symmetric tensor regression (STR) along with some of its competitors under various simulation scenarios. In fitting our model, we fix the hyper-parameters at $a = 1, b = 1$, $\mu_\gamma = 0$, $\sigma_\gamma = 1$, $\nu = 10$ and $\zeta = 1$. We compare our approach to ordinary least squares (LS), which proposes a cell by cell regression of the response on the predictors. Although a naive approach, LS is included due to its widespread use in neuro-imaging applications. Additionally, we employ the *Higher-Order Low-Rank Regression* (HOLRR) [14] as a competitor. HOLRR provides a framework for higher order regression with a tensor response and scalar predictors. A comparative assessment of these three methods will help evaluate possible gains in inference in our method for taking into account the symmetry in the tensor response.

For the sake of simplicity, we work with $p = 1$ (hence get rid of the subscript s hereon), with the scalar variable of interest z_i's are drawn iid from N(0, 1). We also set $D = 2$. The response is simulated from the following model

$$y_{i,k} = \gamma_0^* + \Gamma_k^* z_i + \epsilon_{i,k},\ \epsilon_{i,k} \overset{iid}{\sim} N(0, \sigma^{*2}), \tag{4}$$

where γ_0^* is the true intercept and $\boldsymbol{\Gamma}^* = ((\Gamma_k^*))_{k_1,k_2=1}^V$ is the true symmetric tensor coefficient. The true intercept γ_0^* is set to be 0.2. To simulate the true symmetric tensor coefficient $\boldsymbol{\Gamma}^*$, we draw V tensor node specific latent variables $\boldsymbol{\gamma}_v^* = (\gamma_v^{*(1)}, ..., \gamma_v^{*(R^*)})'$, $v = 1, ..., V$, each of dimension R^*, from a mixture distribution given by $\boldsymbol{\gamma}_v^* \sim \pi_1^* N_{R^*}(0.61, 0.5\boldsymbol{I}) + (1 - \pi_1^*)\delta_{\mathbf{0}}$. We then construct $\boldsymbol{\Gamma}^*$ under two different simulation scenarios, referred to as *Simulation 1* and *Simulation 2*, as described below.

Simulation 1

Simulation 1 constructs cell coefficients $\Gamma_k^* = \sum_{r=1}^{R^*} \gamma_{k_1}^{*(r)} \cdots \gamma_{k_D}^{*(r)}$. Thus, the coefficient tensor assumes a symmetric rank-R^* PARAFAC decomposition. Note that, if $\boldsymbol{\gamma}_v^* = \mathbf{0}$, then the vth tensor node specific variable has no impact in describing the relationship between $\boldsymbol{y}$ and z. Hence $(1 - \pi_1^*)$ is the probability of a tensor node being not related to z_i. We refer to it as the *node sparsity* parameter. In particular, the node sparsity parameter indicates the proportion of nodes in the tensor response (among the total of V tensor nodes) which are not related to the predictor. Notably, this data generation mechanism in Simulation 1 is quite similar (although not identical) to our fitted model. Hence, the goal of this first simulation is to evaluate the ability of the model to recover the true data-generation mechanism. In particular, we consider four cases under Simulation 1 (see Table 1) by varying the fitted rank of PARAFAC (R), true rank of $\boldsymbol{\Gamma}^*$ (R^*), sample size (n), no. of tensor nodes (V) and the tensor node sparsity (defined before).

Table 1. First six columns present the cases under Simulation 1. The next seven columns present the cases under Simulation 2.

Simulation 1						Simulation 2						
Cases	R	R^*	n	V	π_1^*	Cases	R	R^*	n	V	π_1^*	π_2^*
1	4	2	70	30	0.4	1	4	2	70	30	0.4	0.5
2	3	2	70	60	0.6	2	3	2	70	60	0.6	0.5
3	5	2	100	30	0.5	3	4	2	70	30	0.4	0.7
4	5	3	100	60	0.7	4	4	2	70	30	0.6	0.7

Simulation 2
Under Simulation 2, we first simulate node specific latent variables, similar to Simulation 1. If either of $\gamma_{k_1}^*, ..., \gamma_{k_D}^*$ is $\mathbf{0}$, we set $\Gamma_{\boldsymbol{k}}^* = 0$. Otherwise, $\Gamma_{\boldsymbol{k}}^*$ is simulated from a mixture distribution $\pi_2^* N(0,1) + (1 - \pi_2^*)\delta_0$, where $(1 - \pi_2^*)$ is referred to as the *cell sparsity* parameter and δ_0 refers to the Dirac delta function at $\mathbf{0}$. Unlike Simulation 1, Simulation 2 does not necessarily leads to a low-rank structure of $\boldsymbol{\Gamma}^*$. Hence this simulation is ideal for investigating the performance under model mis-specification. Table 1 shows different cases under Simulation 2 where the model is investigated.

We compare the three competitors in terms of their accuracy of estimating $\boldsymbol{\Gamma}^*$. The accuracy of estimating $\boldsymbol{\Gamma}^*$ is measured by the scaled mean squared error (MSE) defined as $||\boldsymbol{\Gamma}^* - \hat{\boldsymbol{\Gamma}}||^2/||\boldsymbol{\Gamma}^*||^2$, where $\hat{\boldsymbol{\Gamma}}$ corresponds to a suitable point estimate of $\boldsymbol{\Gamma}$, e.g., the posterior mean of $\boldsymbol{\Gamma}$ for STR. $||\cdot||$ refers to the Frobenius norm of a matrix. Additionally, we quantify the uncertainty offered by each these competitors through the coverage and length of 95% credible intervals of $\Gamma_{\boldsymbol{k}}$, averaged over all $\boldsymbol{k}$. Length and coverage of posterior 95% credible intervals for each $\Gamma_{\boldsymbol{k}}$ are available empirically from the post burn-in MCMC samples of $\boldsymbol{\Gamma}$ for our proposed approach. On the other hand, the 95% confidence intervals of frequentist competitors are constructed using a bootstrap approximation. To infer on the performance of STR in terms of identifying tensor nodes significantly associated with z_i, we present True Positive Rate (TPR) and False Positive Rate (FPR) with different choices of the cut-off t for all simulation cases. As mentioned earlier, such measures are not available for our competitors since they are not designed to detect tensor nodes related to the predictor of interest.

3.2 Simulation Results

Scaled MSE, coverage and length of 95% CI for all competitors are presented under Simulations 1 and 2 in Tables 2 and 3, respectively. With no model mis-specification under Simulation 1, STR is showing significantly better performance than LS and HOLRR under all four cases. The low-rank structure of HOLRR facilitates its superior performance over LS for larger V/n ratio. However, as V/n ratio increases, HOLRR loses its edge over LS. All three competitors show over-coverage under Simulation 1, with STR producing substantially narrower credible intervals. Moreover, for a fixed n, the credible intervals tend to be more narrow with increasing V for all competitors.

Even under model mis-specification in Simulation 2, STR outperforms all competitors in cases with smaller V and higher node sparsity (Cases 1 and 3), as seen in Table 3. However, with a larger V in case 2, LS and STR are competitive to each other. Since HOLRR is constructed on variants of sparsity within low-rank principle, it loses edge over LS in terms of MSE in these cases. Comparing MSE of STR between cases 3 and 4, we find that MSE increases as the node sparsity decreases. Similarly, comparing Cases 1 and 3 reveals adverse effect of decreasing cell sparsity on MSE of STR. It is generally found that the effect of node sparsity is more profound than the effect of cell sparsity on the performance.

Table 2. Mean Squared Error (MSE), average coverage and average length of 95% credible interval for STR, LS and HOLRR are presented for cases under Simulation 1, with the lowest MSE in each case is boldfaced.

Case	Competitors			
		SGTM	LS	HOLRR
1	MSE $\times 10^3$	**3.4**	46	42
	Avg. cov	0.99	0.99	0.99
	Avg. length	0.10	1.74	1.68
2	MSE $\times 10^3$	**1.3**	18	12
	Avg. cov	0.99	0.99	0.99
	Avg. length	0.08	2.79	2.63
3	MSE $\times 10^3$	**1.9**	26	35
	Avg. cov	0.99	0.99	0.99
	Avg. length	0.10	2.53	2.37
4	MSE $\times 10^3$	**0.8**	5.1	8.5
	Avg. cov	0.99	1.00	0.99
	Avg. length	0.21	4.82	4.74

Table 3. Mean Squared Error (MSE), average coverage and average length of 95% credible interval for STM, LS and HOLRR are presented for cases under Simulation 2, with the lowest MSE in each case in boldfaced.

Case	Competitors			
		SGTM	LS	HOLRR
1	MSE	**0.09**	0.21	0.30
	Avg. cov	0.96	0.97	0.89
	Avg. length	0.12	0.95	0.78
2	MSE	0.23	0.29	**0.21**
	Avg. cov	0.89	0.98	0.81
	Avg. length	0.14	1.53	0.78
3	MSE	**0.13**	0.22	0.30
	Avg. cov	0.96	0.98	0.92
	Avg. length	0.15	1.09	0.88
4	MSE	0.16	**0.15**	0.32
	Avg. cov	0.92	0.99	0.92
	Avg. length	0.26	1.69	1.29

Moving onto uncertainty characterization, STR shows close to nominal coverage along with its competitors in cases 1, 2 and 4 when V is small. With increasing V, coverage of STR and HOLRR drops below 0.90. Similar to Simulation 1, STR demonstrates sufficiently narrower credible intervals than HOLRR in all cases. LS offers over-coverage with much wider 95% credible intervals in all cases.

Table 4. True Positive Rates (TPR) and False Positive Rates (FPR) in identifying nodes which are significantly related to the predictor under all cases with cut-offs $t = 0.1, 0.5, 0.9$.

Simulation	Accuracy of tensor node identification						
		TPR ($t = 0.1$)	TPR ($t = 0.5$)	TPR ($t = 0.9$)	FPR ($t = 0.1$)	FPR ($t = 0.5$)	FPR ($t = 0.9$)
1	Case 1	1.00	1.00	1.00	0.00	0.00	0.00
	Case 2	1.00	1.00	1.00	0.00	0.00	0.00
	Case 3	1.00	1.00	1.00	0.00	0.00	0.00
	Case 4	1.00	1.00	1.00	0.00	0.00	0.00
2	Case 1	1.00	0.90	0.60	0.00	0.00	0.00
	Case 2	1.00	1.00	0.84	0.00	0.00	0.00
	Case 3	0.98	0.80	0.55	0.24	0.00	0.00
	Case 4	1.00	1.00	0.76	0.24	0.15	0.00

Since LS and HOLRR are not designed to detect nodes significantly related to the predictor, we focus our inference on STR for node detection. To this end, we choose three cur-off values $t = 0.1, 0.5, 0.9$ and present TPR and FPR values for our approach. STR yields the posterior probability of a node being related to the predictor of interest to be very close to 1 or 0 for all reasonable values of cut-off t, depending on whether a tensor node is related or not to the predictor of interest in the truth, respectively. As a consequence, TPR and FPR values (see Table 4) turn out to be close to 1 and 0, respectively, for all the simulation cases, indicating a close to perfect active node detection. The posterior distributions of γ_0 also appear to be centered around the truth (not shown here).

4 Conclusion and Future Work

The overarching goal of this article is to propose a symmetric tensor regression framework with a symmetric tensor response and scalar predictors. The model is aimed at identifying tensor nodes significantly related to each scalar predictor. Unlike the existing approaches, the proposed framework does not assume any low-rank constraint on the symmetric tensor coefficients. Rather, we propose a tensor shrinkage prior which decomposes the symmetric tensor coefficients into low-rank and sparse components a priori. The low-rank component is further assigned a novel hierarchical mixture prior to enable identification of tensor nodes related to each predictor. The sparse component is equipped with Bayesian regularization or shrinkage priors to enable accurate estimation of tensor cell coefficients. Detailed simulation study with data generated under the true model and mis-specified model demonstrates superior performance of our approach compared to its competitors.

Although there is a considerable literature on theoretical understanding of Bayesian shrinkage priors in high dimensional regression, there is a limited literature on theoretical aspects of shrinkage prior on tensor coefficients. In future, we will focus on developing conditions for posterior consistency of the proposed approach under suitable conditions imposed on tensor shrinkage priors. It is also instructive to employ other shrinkage priors on $\Gamma_{s,2,\boldsymbol{k}}$ from the class of global-local shrinkage prior distributions [13] and provide a comparative understanding of their performances. Finally, we would like to extend our approach when each entry in $y_{i,\boldsymbol{k}}$ are categorical or count in nature. Some of these constitute our future work.

Appendix

The full conditional distributions of parameters for implementing the MCMC is given by following.

1. $\gamma_0|- \sim N\left[\frac{\sum_{i=1}^{n}\boldsymbol{1}^T(\boldsymbol{y}_i-\sum_{s=1}^{p}\boldsymbol{\Gamma}_s z_{is})/\sigma^2}{(nq)/\sigma^2+1}, \frac{1}{(nq)/\sigma^2+1}\right]$
2. $\kappa_{s,\boldsymbol{k}}\,|- \sim RGIG\left[\frac{1}{2}, \frac{(\Gamma_{s,\boldsymbol{k}}-\sum_{r=1}^{R}\lambda_{s,r}\gamma_{s,k_1}^{(r)}\cdots\gamma_{s,k_D}^{(r)})^2}{\sigma^2}, \zeta^2\right]$, $s=1,...,p$; $1\le k_1 < \cdots < k_D \le V$
3. Let $\boldsymbol{Z}$ be an $n\times p$ matrix with the ith row as $(z_{i1},...,z_{ip})$. Let $\boldsymbol{y_k} = (y_{1,\boldsymbol{k}},...,y_{n,\boldsymbol{k}})'$, $\boldsymbol{D_k} = diag(\kappa_{1,\boldsymbol{k}},...,\kappa_{p,\boldsymbol{k}})$, $\boldsymbol{m_k} = (\sum_{r=1}^{R}\lambda_{s,r}\gamma_{s,k_1}^{(r)}\cdots\gamma_{s,k_D}^{(r)} : 1\le s\le p)'$ and $\boldsymbol{\gamma_k} = (\Gamma_{1,\boldsymbol{k}},...,\Gamma_{p,\boldsymbol{k}})'$. Let $\boldsymbol{\Sigma_\Gamma} = \sigma^2(\boldsymbol{Z}'\boldsymbol{Z}+\boldsymbol{D_k}^{-1})^{-1}$, $\boldsymbol{\mu_\Gamma} = \boldsymbol{\Sigma_\Gamma}\left[\boldsymbol{Z}'(\boldsymbol{y_k}-\gamma_0\boldsymbol{1})+\boldsymbol{D_k}^{-1}\boldsymbol{m_k}\right]/\sigma^2$. Then $\boldsymbol{\gamma_k}|- \sim N(\boldsymbol{\mu_\Gamma},\boldsymbol{\Sigma_\Gamma})$.
4. $\sigma^2|- \sim IG(a+(nq)/2+(pq)/2, b+\sum_{i=1}^{n}||\boldsymbol{y}_i-\sum_{s=1}^{p}\boldsymbol{\Gamma}_s z_{is}||^2/2 + \sum_{s=1}^{p}\sum_{\boldsymbol{k}}(\Gamma_{s,\boldsymbol{k}}-\sum_{r=1}^{R}\lambda_{s,r}\gamma_{s,k_1}^{(r)}\cdots\gamma_{s,k_D}^{(r)})^2)$
5. Let $\mathcal{K}_v = \{\boldsymbol{k} : \exists\, l,\ s.t.,\ k_l = v\}$ $\boldsymbol{v}_s = (\Gamma_{s,\boldsymbol{k}} : \boldsymbol{k}\in\mathcal{K}_v)$. Define $\boldsymbol{J}$ as a matrix of the order $\#\mathcal{K}_v\times R$ with a representative row $(\prod_{k_l\ne v}\lambda_{s,1}\gamma_{s,k_l}^{(1)},...,\prod_{k_l\ne v}\lambda_{s,R}\gamma_{s,k_l}^{(R)})$. Define, $w_{\gamma_{s,v}} = \frac{(1-\Delta_s)N(\boldsymbol{v}_s|\boldsymbol{0},\sigma^2\boldsymbol{I})}{(1-\Delta_s)N(\boldsymbol{v}_s|\boldsymbol{0},\sigma^2\boldsymbol{I})+\Delta_s N(\boldsymbol{v}_s|\boldsymbol{J}\boldsymbol{H}_s\boldsymbol{J}'+\sigma^2\boldsymbol{I})}$, $\boldsymbol{M}_s = diag(\kappa_{s,\boldsymbol{k}} : \boldsymbol{k}\in\mathcal{K}_v)$ $\boldsymbol{\Sigma}_{\gamma_{s,v}} = \left(\boldsymbol{J}'\boldsymbol{M}_s^{-1}\boldsymbol{J}/\sigma^2+\boldsymbol{H}_s^{-1}\right)^{-1}$, $\boldsymbol{\mu}_{\gamma_{s,v}} = \boldsymbol{\Sigma}_{\gamma_{s,v}}\boldsymbol{J}'\boldsymbol{M}_s^{-1}\gamma_{s,v}/\sigma^2$. Then $\gamma_{s,v}|- \sim w_{\gamma_{s,v}}\delta_{\boldsymbol{0}}+(1-w_{\gamma_{s,v}})N(\boldsymbol{\mu}_{\gamma_{s,v}},\boldsymbol{\Sigma}_{\gamma_{s,v}})$
6. $\boldsymbol{H}_s|- \sim IW(\boldsymbol{I}+\sum_{v:\gamma_{s,v}\ne\boldsymbol{0}}\gamma_{s,v}\gamma_{s,v}', \nu+\{\#v : \gamma_{s,v}\ne\boldsymbol{0}\})$, $s=1,...,p$
7. $\theta_{s,r}|- \sim Beta(1+\lambda_{s,r}, r^c+1-\lambda_{s,r})$, $s=1,...,p$; $r=1,...,R$
8. $\Delta_s|- \sim Beta(1+\sum_{v=1}^{V}\xi_{s,v}, 1+\sum_{v=1}^{V}(1-\xi_{s,v}))$, $s=1,...,p$
9. $\lambda_{s,r}\,|- \sim Ber(p_{\lambda_{s,r}})$, where $p_{\lambda_{s,r}} = \frac{\theta_{s,r}N_{s,r}}{\theta_{s,r}N_{s,r}+(1-\theta_{s,r})D_{s,r}}$, where $N_{s,r} = \prod_{\boldsymbol{k}}N(y_{i,\boldsymbol{k}}|\sum_{r'=1,r'\ne r}^{R}\lambda_{s,r'}\gamma_{s,k_1}^{(r')}\cdots\gamma_{s,k_D}^{(r')}+\gamma_{s,k_1}^{(r)}\cdots\gamma_{s,k_D}^{(r)},\sigma^2)$, $D_{s,r} = \prod_{\boldsymbol{k}}N(y_{i,\boldsymbol{k}}|\sum_{r'=1,r'\ne r}^{R}\lambda_{s,r'}\gamma_{s,k_1}^{(r')}\cdots\gamma_{s,k_D}^{(r')},\sigma^2)$ for $r=1,..,R$.

References

1. Desikan, R.S., et al.: An automated labeling system for subdividing the human cerebral cortex on MRI scans into gyral based regions of interest. Neuroimage **31**(3), 968–980 (2006)
2. Goh, G., Dey, D.K., Chen, K.: Bayesian sparse reduced rank multivariate regression. J. Multivar. Anal. **157**, 14–28 (2017)
3. Guha, S., Rodriguez, A.: Bayesian regression with undirected network predictors with an application to brain connectome data. arXiv preprint arXiv:1803.10655 (2018)
4. Guhaniyogi, R., Rodriguez, A.: Joint modeling of longitudinal relational data and exogenous variables. Bayesian Anal. (2019, in press). https://doi.org/10.1214/19-BA1160
5. Guhaniyogi, R., Spencer, D.: Bayesian tensor response regression with an application to brain activation studies. UCSC Technical report: UCSC-SOE-18-15 (2019)
6. Guhaniyogi, R., Qamar, S., Dunson, D.B.: Bayesian tensor regression. J. Mach. Learn. Res. **18**(5), 1–31 (2017)
7. Hoff, P.D.: Bilinear mixed-effects models for dyadic data. J. Am. Stat. Assoc. **100**(469), 286–295 (2005)
8. Ishwaran, H., Rao, J.S.: Spike and slab variable selection: frequentist and Bayesian strategies. Ann. Stat. **33**(2), 730–773 (2005)
9. Jung, R.E., et al.: Neuroanatomy of creativity. Hum. Brain Mapp. **31**(3), 398–409 (2010)
10. Kolda, T.G., Bader, B.W.: Tensor decompositions and applications. SIAM Rev. **51**(3), 455–500 (2009)
11. Li, L., Zhang, X.: Parsimonious tensor response regression. J. Am. Stat. Assoc. **112**(519), 1131–1146 (2017)
12. Park, T., Casella, G.: The Bayesian lasso. J. Am. Stat. Assoc. **103**(482), 681–686 (2008)
13. Polson, N.G., Scott, J.G.: Shrink globally, act locally. Sparse Bayesian regularization and prediction. Bayesian Stat. **9**, 501–538 (2010)
14. Rabusseau, G., Kadri, H.: Higher-order low-rank regression. arXiv preprint arXiv:1602.06863 (2016)
15. Rothman, A.J., Levina, E., Zhu, J.: Sparse multivariate regression with covariance estimation. J. Comput. Graph. Stat. **19**(4), 947–962 (2010)

A Publicly Available, High Resolution, Unbiased CT Brain Template

John Muschelli[(✉)]

Department of Biostatistics, Johns Hopkins Bloomberg School of Public Health, 615 N Wolfe St, Baltimore, MD 21205, USA
jmuschel@jhu.edu

Abstract. Clinical imaging relies heavily on X-ray computed tomography (CT) scans for diagnosis and prognosis. Many research applications aim to perform population-level analyses, which require images to be put in the same space, usually defined by a population average, also known as a template. We present an open-source, publicly available, high-resolution CT template. With this template, we provide voxel-wise standard deviation and median images, a basic segmentation of the cerebrospinal fluid spaces, including the ventricles, and a coarse whole brain labeling. This template can be used for spatial normalization of CT scans and research applications, including deep learning. The template was created using an anatomically-unbiased template creation procedure, but is still limited by the population it was derived from, an open CT data set without demographic information. The template and derived images are available at https://github.com/muschellij2/high_res_ct_template.

Keywords: CT imaging · CT template · Brain template · Computed tomography

1 Introduction

Many research applications of neuroimaging use magnetic resonance imaging (MRI). MRI allows researchers to study a multitude of applications and diseases, including studying healthy volunteers as it poses minimal risk. Clinical imaging, however, relies heavily on X-ray computed tomography (CT) scans for diagnosis and prognosis. Studies using CT scans cannot generally recruit healthy volunteers or large non-clinical populations due to the radiation exposure and lack of substantial benefit. As such, much of head CT data is gathered from prospective clinical trials or retrospective studies based on health medical record data and hospital PACS (picture archiving and communication system). Most of this research is on patients with neuropathology, which can cause deformations of the brain, such as mass effects, lesions, stroke, or tumors.

Electronic supplementary material The online version of this chapter (https://doi.org/10.1007/978-3-030-50153-2_27) contains supplementary material, which is available to authorized users.

M.-J. Lesot et al. (Eds.): IPMU 2020, CCIS 1239, pp. 358–366, 2020.
https://doi.org/10.1007/978-3-030-50153-2_27

Many clinical protocols perform axial scanning with a high within-plane resolution (e.g. $0.5\,\text{mm} \times 0.5\,\text{mm}$) but lower out-of-plane resolution (e.g. 5 mm). High resolution scans (out of plane resolution $\approx 0.5\,\text{mm}$) may not be collected or reconstructed as the lower resolution scans are typically those read by the clinician or radiologist for diagnosis and prognosis. Recently, a resource of a large number of CT scans were made available, denoted as CQ500 (Chilamkurthy et al. 2018). These scans include people with a number of pathologies, including hemorrhagic stroke and midline shifts. Fortunately, this data also includes people **without indicated pathology** with high resolution scanning, which is what we will use in this study.

The goal of this work is to create an anatomically unbiased, high-resolution CT template of the brain. That is, we wish to create a template that represents the population, regardless of any initial templates we start with. The first, and we believe the only, publicly-available CT template was released by Rorden et al. (2012) (https://www.nitrc.org/projects/clinicaltbx/). That template was created with the specific purpose of creating a template with a similar age range as those with stroke, using 30 individuals with a mean age of 65 years old (17 men). The associated toolbox released contained a high resolution ($1 \times 1 \times 1\,\text{mm}$) template, with the skull on, in Montreal Neurological Institute (MNI) space. Subsequent releases have included skull-stripped brain templates, but only in a lower ($2 \times 2 \times 2\,\text{mm}$) space (https://github.com/neurolabusc/Clinical). This lower resolution template matches what is used in many MRI and functional MRI analyses.

Thus, the current CT templates available are a high-resolution template ($1\,\text{mm}^3$), but not of the brain only (and skull stripping the template performs marginally well), and a low-resolution template of the brain only, both in MNI space. We have used these templates in previous analyses, but would like a brain template that was 1) constructed using an unbiased anatomical procedure, 2) uses more patients, 3) uses high-resolution scans to achieve a higher resolution, and 4) provide an image which dimensions are easily used in deep learning frameworks.

As the CQ500 data was released under a Creative Commons Attribution-NonCommercial-ShareAlike 4.0 (CC-NC-SA) International License, we can release the template under the same license.

2 Methods

All code, analysis, and reporting was done the R statistical programming language (R Core Team 2015) and a number of packages from the R medical imaging package platform Neuroconductor (Muschelli et al. 2019).

2.1 Data

We defined a high-resolution patient scan as having a within-axial resolution of $0.7 \times 0.7\,\text{mm}$ or less, with full coverage of the brain. For example, if the cerebellum was not imaged, that image was discarded. All scans were non-contrast CT scans with a soft-tissue convolution kernel. As CT scans are generally well calibrated across sites and are measured in standard units of Hounsfield Units (HU), no intensity normalization was done. Intensities less than (−1024) HU (the value for air) and greater than 3071 HU were Winsorized (Dixon and Yuen 1974) to those values, as values outside of these are likely artifact or areas outside the field of view.

All data was converted from DICOM files to NIfTI (Neuroimaging Informatics Technology Initiative) using `dcm2niix` (Li et al. 2016) using the `dcm2niir` package (Muschelli 2018). This conversion corrects for any gantry tilt and enforces one fixed voxel size for the image, which is necessary if different areas of the image are provided at different resolutions, which is sparsely seen in clinical CT images.

From the CQ500 data set, 222 subjects had no indication of pathology, of which 141 had a high-resolution scan (if multiple were present, the one with the highest resolution was used). From these 141 people, 130 had "thick-slice" scans where the out-of-plane resolution was greater than 4 mm. We used these 130 scans for construction of the template. The 11 scans were discarded as we wish to perform the same operation using low-resolutions scans to see the effect of initial resolution on template creation, but that is not the focus of this work.

For all images, the head was skull-stripped so that only brain tissue and cerebrospinal fluid (CSF) spaces were kept, using a previously validated method (Muschelli et al. 2015) using the brain extraction tool (BET) from FSL (FMRIB Software Library) (Smith 2002; Jenkinson et al. 2012). We chose an image (patient 100 from CQ500), for template creation. This choice was based on a within-plane resolution close to 0.5×0.5 mm (0.488×0.488 mm), an axial slice size of 512×512, and an out-of-plane resolution of 0.5 mm. The image was resampled to $0.5 \times 0.5 \times 0.5$ mm resolution so that the voxels are isotropic. We would like the image to be square; we padded the image back to 512×512 after resampling, and the image had 336 coronal-plane slices.

2.2 Template Creation

The process of template creation can be thought of as a gradient descent algorithm to estimate the true template image as inspired by the advanced normalization tools (ANTs) software and the R package ANTsR that implements the registration and transformation was used (https://github.com/ANTsX/ANTsR) (Avants et al. 2011). The process is as follows:

1. Let I_i represent the image where i represents subjects. We registered all images to the template, denoted $\bar{T}_k$ where k represents iteration, using an affine registration followed by symmetric normalization (SyN), a non-linear deformation/diffeomorphism, where the composed transformation is denoted as $G_{i,k}$ (Avants et al. 2008). Let the transformed image be denoted as $T_{i,k}$. In other words, $I_i \xrightarrow{G_{i,k}} T_{i,k}$. The transformation $G_{i,k}$ is represented by a 4D warping image. Let T_1 be the original template chosen above and $G_{i,1}$ be the transformation for an image to the original template.

2. Calculate the mean, median, and standard deviation images, where the mean image is $\bar{T}_k = \frac{1}{n}\sum_{i=1}^{n} T_{i,k}$, using a voxel-wise average.
3. Calculate the average warping transformation: $\bar{G}_k = \frac{1}{n}\sum_{i=1}^{n} G_{i,k}$. A gradient descent step size of 0.2 was specified for SyN gradient descent, such that: $\bar{T}_{k+1} = \bar{T}_k \times (-0.2 * \bar{G}_k)$. The median and standard deviation are transformed accordingly.

For each iteration k, we can calculate a number of measures to determine if the template has converged compared to the previous iteration $k-1$. We calculated the Dice Similarity Coefficient (DSC) (Dice 1945) between the mask of iteration k and $k-1$, where the mask for iteration k is defined as $\bar{T}_k > 0$. The DSC measures if the overall shape is consistent across iterations. We also the root mean squared error (RMSE) of voxel intensities, e.g. $\frac{1}{V}\sum\left(\bar{T}_k - \bar{T}_{k-1}\right)^2$, where V is the number of voxels in the volume. The RMSE can be calculated over a series of volumes, either 1) the entire image, 2) over the non-zero voxels in iteration k, 3) in iteration $k-1$, or 4) the union (or intersection) of the 2 masks. Calculation over the entire image gives an optimistic estimate as most of the image are zeroes, and the choice of either iteration k or $k-1$ masks is arbitrary, so we calculated the RMSE over the union of the 2 masks. The RMSE represents if the values of the image are close across iterations.

To define convergence, we would like a high DSC between the masks and a low RMSE. Ideally, the convergence criteria would set a DSC of 1 and a RMSE less than 1 Hounsfield Unit (HU), which would indicate the voxel intensity is changing less than 1 HU on average. As CT scans are measured in integers, this RMSE would likely be as good as possible. We set a DSC cutoff of 0.95 and chose the template with the lowest RMSE. As this procedure is computationally expensive, we ran 40 iterations, which was adequate for achieving stable results (Fig. 1).

Values of the final template that were lower than 5 HU were boundary regions, outside the region of the brain and likely duc to averagc of onc or a small few of images, incongruent with the remainder of the template (Supplemental Figure 1). We did not constrain the DSC and RMSE calculation excluding these regions, but excluded values less than 5 HU from the final template.

After the template was created, we padded the coronal plane so that the template was 512 × 512 × 512. The intention is that these dimensions allow it easier to create sub-sampled arrays that are cubes and multiples of 8, such as 256 × 256 × 256, 128 × 128 × 128, or 64 × 64 × 64 with isotropic resolution.

2.3 Segmentation

Though the template itself is the main goal of the work, many times researchers use or are interested in annotations/segmentations of the template space. The contrast between gray matter and white matter in CT imaging is not as high as T1-weighted MRI. Some areas, such as the cerebellum, corpus callosum, and basal ganglia can be delineated well. Thus, segmentation methods based on intensity may not differentiate gray and white matter adequately. We instead used a multi-atlas registration approach using previously-published set of 35 MRI atlases from Landman et al. (2012), which had whole brain segmentations, including tissue-class segmentations.

We registered each brain MRI to the CT template using SyN and applied the transformation to the associated tissue segmentation and whole brain segmentation from that MRI template. Thus, we had 35 tissue segmentations of the CT template in template space, and the segmentations were combined using STAPLE (Warfield et al. 2004) via the `stapler` package (Muschelli 2019). The whole brain structures were combined using majority vote.

Separating the brain from the cerebrospinal fluid areas (mainly ventricles) are of interest in many applications, such as Alzheimer's disease (Leon et al. 1989; Braak et al. 1999). In addition, we segmented the template using Atropos (Avants et al. 2011), which used a k-means clustering approach with 2 clusters (CSF/tissue) to obtain a CSF mask. Additionally, we registered the MNI T1-weighted template to the CT Template using SyN, and applied the transformation used the ALVIN (Automatic Lateral Ventricle delIneatioN) mask of the ventricles (Kempton et al. 2011). We masked the CSF mask with this transformed ALVIN mask to get a mask of lateral ventricles as well.

3 Results

As we see in Fig. 1A, the DSC quickly increases and reaches a high score, where the horizontal line indicates a DSC of 0.99. The red dot and vertical line indicate the iteration that had the maximum DSC (0.9896). As the DSC is high for all iterations past iteration 15, we chose the template based on the minimum RSE. In Fig. 1B, we see a similar pattern of improving performance, but by lowering the RMSE. The lowest RMSE is noted by the red point with a value of 1.47. Thus, this iteration (iteration 37) is the template we will choose.

The template for this image can be seen in Fig. 2, along with the standard deviation image, and a histogram of the intensities of the template. Areas outside the brain mask were removed for visualization. We see the template is relatively smooth, with values from 5 HU to around 65 HU. The standard deviation image shows high variability around the lateral horns, which may be due to calcifications in a set of patients, which have abnormally high HU values. The high standard deviation areas near the midline are likely due to dense areas of the falx cerebri, including potential falx calcifications.

In Figure 3, we see the template again, with the tissue-class segmentation (Panel B), whole brain structural segmentation (Panel C), and Atropos lateral ventricle segmentation. Overall, we see some differences between the segmentation of the CSF based on Atropos and the multi-atlas labeling approach. We have provided a lookup table for each structure label with its corresponding value in the image.

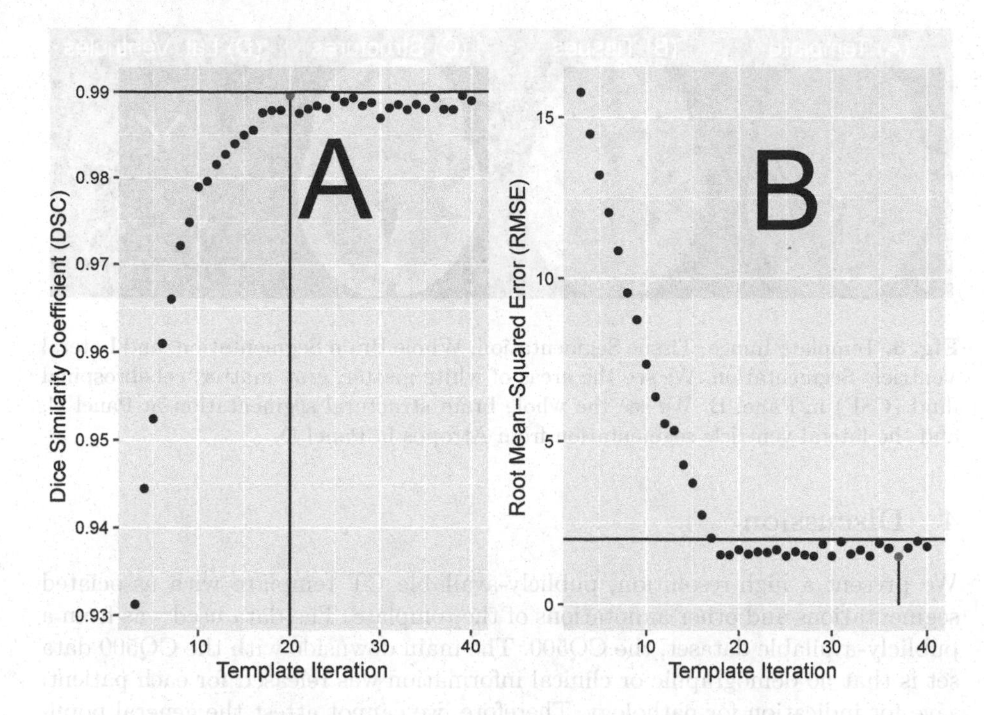

Fig. 1. Convergence of Shape and Intensity of the Template over Iterations. Here we see the Dice Similarity Coefficient (DSC) increase between an iteration and the previous iteration, achieving high degrees of overlap, indicating the shape of the surface of the image is similar and converging (panel A). We also see the root mean-squared error (panel B) drops as the iterations increase and then levels off around 4 Hounsfield units (HU), the horizontal line. The red dot indicates the iteration chosen for the template.

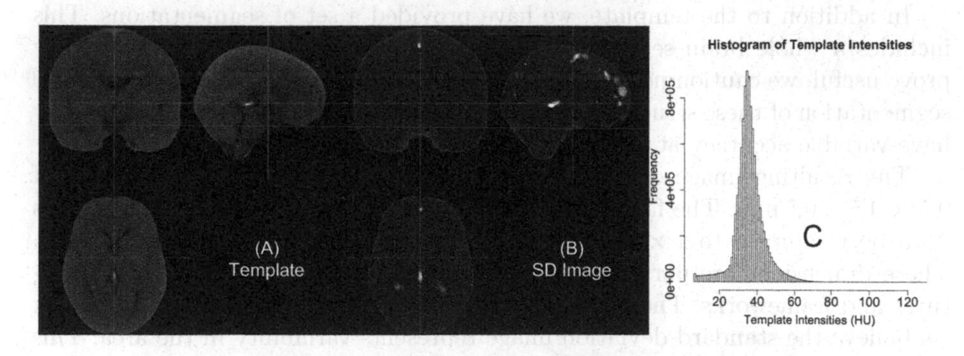

Fig. 2. Template Image, Standard Deviation Image, and Histogram of Intensities. Here we show the template in the left panel, the voxel-wise standard deviation, denoting areas of variability (which include biological and technical variability), and the histogram of the template intensities/Hounsfield Units (HU). Overall the template is smooth and values fall in the range of 5 to 65 HU.

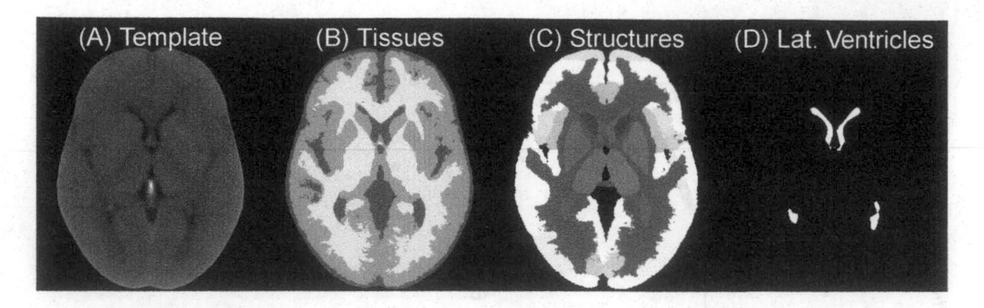

Fig. 3. Template Image, Tissue Segmentation, Whole Brain Segmentation, and Lateral Ventricle Segmentation. We see the areas of white matter, gray matter, cebebrospinal fluid (CSF) in Panel B. We see the whole brain structural segmentation in Panel C, and the lateral ventricle segmentation from Atropos in Panel D.

4 Discussion

We present a high-resolution, publicly-available CT template with associated segmentations and other annotations of the template. The data used was from a publicly-available dataset, the CQ500. The main downside with the CQ500 data set is that no demographic or clinical information was released for each patient, save for indication for pathology. Therefore, we cannot attest the general population of interest for this template. Furthermore, we cannot fully assume these patients were disease-free as a lack of pathology only applies to the categories of interest in the CQ500 dataset (intracranial/subdural/subarachnoid/epidural hematoma, calvarial or other fractures, mass effect and midline shifts). In future work, we hope to prepare age- and sex-specific templates for each population based on hospital scans and records, where we have demographic information and confirmation of lack of neuropathology.

In addition to the template, we have provided a set of segmentations. This includes a whole brain segmentation of over 150 structures. Though this may prove useful, we caution users to how well this template can provide an accurate segmentation of these structures. At least, the accuracy of the segmentation may have variable accuracy at different areas of the brain.

The resulting image dimensions was $512 \times 512 \times 512$, with a resolution of $0.5 \times 0.5 \times 0.5$ mm. The fact that the image dimension is a multiple of 8 allows it to be resampled to $1 \times 1 \times 1$ mm and $2 \times 2 \times 2$ mm and remain as a cube. These dimensions are particularly important in certain deep learning architectures and frameworks. Though most templates are given using the mean image, we believe the standard deviation image represents variability in the area. This variability represents true systematic and biologic variability. One important area of systemic variability is registration errors. Therefore this template allows for the creation of z-score images, where a new image is registered to the mean image, the mean image is subtracted, and then divided by the standard-deviation image, so that voxels represent standard deviations away from the mean voxel. This image may be a useful tool in feature extraction. Thus, we believe this template provides a standard, isotropic space that is conducive to machine learning and can reduce the burden of standardization for medical imaging applications.

CQ500 is Creative Commons Attribution-NonCommercial-ShareAlike 4.0 International License. Therefore, the template is released under the same license. The images are located on https://github.com/muschellij2/high_res_ct_template and can be accessed at https://johnmuschelli.com/high_res_ct_template/template/.

Acknowledgments. This work has been supported by the R01NS060910 and 5U01NS080824 grants from the National Institute of Neurological Disorders and Stroke at the National Institutes of Health (NINDS/NIH).

References

Avants, B.B., Epstein, C.L., Grossman, M., Gee, J.C.: Symmetric diffeomorphic image registration with cross-correlation: evaluating automated labeling of elderly and neurodegenerative brain. Med. Image Anal. **12**(1), 26–41 (2008). https://doi.org/10.1016/j.media.2007.06.004. Special Issue on the Third International Workshop on Biomedical Image Registration (WBIR 2006)

Avants, B.B., Tustison, N.J., Song, G., Cook, P.A., Klein, A., Gee, J.C.: A reproducible evaluation of ANTs similarity metric performance in brain image registration. Neuroimage **54**(3), 2033–2044 (2011). https://doi.org/10.1016/j.neuroimage.2010.09.025

Avants, B.B., Tustison, N.J., Wu, J., Cook, P.A., Gee, J.C.: An open source multivariate framework for n-tissue segmentation with evaluation on public data. Neuroinformatics **9**(4), 381–400 (2011). https://doi.org/10.1007/s12021-011-9109-y

Braak, E., Griffing, K., Arai, K., Bohl, J., Bratzke, H., Braak, H.: Neuropathology of Alzheimer's disease: what is new since A. Alzheimer? Eur. Arch. Psychiatry Clin. Neurosci. **249**(3), S14–S22 (1999). https://doi.org/10.1007/PL00014168

Chilamkurthy, S., et al.: Deep learning algorithms for detection of critical findings in head CT scans: a retrospective study. The Lancet **392**(10162), 2388–2396 (2018)

Dice, L.R.: Measures of the amount of ecologic association between species. Ecology **26**(3), 297–302 (1945). http://www.jstor.org/stable/1932409

Dixon, W.J., Yuen, K.K.: Trimming and winsorization: a review. Statistische Hefte **15**(2–3), 157–170 (1974). https://doi.org/10.1007/BF02922904

Jenkinson, M., Beckmann, C.F., Behrens, T.E.J., Woolrich, M.W., Smith, S.M.: FSL. NeuroImage **62**(2), 782–790 (2012). https://doi.org/10.1016/j.neuroimage.2011.09.015

Kempton, M.J., et al.: A comprehensive testing protocol for MRI neuroanatomical segmentation techniques: evaluation of a novel lateral ventricle segmentation method. Neuroimage **58**(4), 1051–1059 (2011)

Landman, B.A., et al.: MICCAI 2012 Workshop on Multi-Atlas Labeling. CreateSpace Independent Publishing Platform (2012)

de Leon, M.J., et al.: Alzheimer's disease: longitudinal CT studies of ventricular change. Am. J. Roentgenol. **152**(6), 1257–1262 (1989)

Li, X., Morgan, P.S., Ashburner, J., Smith, J., Rorden, C.: The first step for neuroimaging data analysis: DICOM to NIfTI conversion. J. Neurosci. Methods **264**, 47–56 (2016)

Muschelli, J.: dcm2niir: conversion of DICOM to NIfTI imaging files through R (2018). https://www.nitrc.org/plugins/mwiki/index.php/dcm2nii:MainPage

Muschelli, J.: Simultaneous Truth and Performance Level Estimation (2019). https://CRAN.R-project.org/package=stapler

Muschelli, J., et al.: Neuroconductor: an R platform for medical imaging analysis. Biostatistics **20**(2), 218–239 (2019)

Muschelli, J., Ullman, N.L., Mould, W.A., Vespa, P., Hanley, D.F., Crainiceanu, C.M.: Validated automatic brain extraction of head CT images. Neuroimage **114**, 379–385 (2015). https://doi.org/10.1016/j.neuroimage.2015.03.074

R Core Team: R: A Language and Environment for Statistical Computing. R Foundation for Statistical Computing, Vienna, Austria (2015). https://www.R-project.org/

Rorden, C., Bonilha, L., Fridriksson, J., Bender, B., Karnath, H.O.: Age-specific CT and MRI templates for spatial normalization. Neuroimage **61**(4), 957–965 (2012). https://doi.org/10.1016/j.neuroimage.2012.03.020

Smith, S.M.: Fast robust automated brain extraction. Hum. Brain Mapp. **17**(3), 143–155 (2002). https://doi.org/10.1002/hbm.10062

Warfield, S.K., Zou, K.H., Wells, W.M.: Simultaneous truth and performance level estimation (STAPLE): an algorithm for the validation of image segmentation. IEEE Trans. Med. Imaging **23**(7), 903–921 (2004)

Statistical Methods for Processing Neuroimaging Data from Two Different Sites with a Down Syndrome Population Application

Davneet S. Minhas[1], Zixi Yang[1], John Muschelli[2], Charles M. Laymon[1], Joseph M. Mettenburg[1], Matthew D. Zammit[3], Sterling Johnson[3], Chester A. Mathis[1], Ann D. Cohen[1], Benjamin L. Handen[1], William E. Klunk[1], Ciprian M. Crainiceanu[3], Bradley T. Christian[3], and Dana L. Tudorascu[1](✉)

[1] University of Pittsburgh, Pittsburgh, PA 15213, USA
dlt30@pitt.edu
[2] Johns Hopkins University, Baltimore, MD 21205, USA
[3] University of Wisconsin-Madison, Madison, WI 53705, USA

Abstract. Harmonization of magnetic resonance imaging (MRI) and positron emission tomography (PET) scans from multi-scanner and multi-site studies presents a challenging problem. We applied the Removal of Artificial Voxel Effect by Linear regression (RAVEL) method to normalize T1-MRI intensities collected on two different scanners across two different sites as part of the Neurodegeneration in Aging Down syndrome (NiAD) study. The effects on FreeSurfer regional cortical thickness and volume outcome measures, in addition to FreeSurfer-based regional quantification of amyloid PET standardized uptake value ratio (SUVR) outcomes, were evaluated. A neuroradiologist visually assessed the accuracy of FreeSurfer hippocampus segmentations with and without the application of RAVEL. Quantitative results demonstrated that the application of RAVEL intensity normalization prior to running FreeSurfer significantly impacted both FreeSurfer volume and cortical thickness outcome measures. Visual assessment demonstrated that the application of RAVEL significantly improved FreeSurfer hippocampal segmentation accuracy. The RAVEL intensity normalization had little impact on PET SUVR measures.

Keywords: Harmonization · MRI · PET

1 Introduction

Positron emission tomography (PET) and structural magnetic resonance imaging (MRI) are two neuroimaging modalities frequently used in studies of Alzheimer's Disease (AD). [11C] Pittsburgh Compound B (PiB) PET standardized uptake value ratio (SUVR) and T1-weighted MRI-based volumetric measures, such as hippocampal volume and entorhinal cortical thickness, play a crucial role in studying the progression of AD in the elderly (Jack et al. 2017; Schwarz et al. 2016; Villemagne et al. 2011), autosomal dominant AD mutation carriers (Yau et al. 2015), and Down Syndrome (DS) populations (Lao et al. 2017).

M.-J. Lesot et al. (Eds.): IPMU 2020, CCIS 1239, pp. 367–379, 2020.
https://doi.org/10.1007/978-3-030-50153-2_28

Adults with DS are at high risk to reveal AD pathology in part due to the triplication of chromosome 21 encoding the amyloid precursor protein. Adults with DS are typically affected by AD pathology by their 30s, demonstrate AD-associated structural and functional brain changes in midlife, and have a 70–80% chance of clinical dementia by their 60s (Lao et al. 2017; Tudorascu et al. 2019).

Multi-site multimodal neuroimaging studies are becoming increasingly popular in studying neurodegenerative diseases, but there is a lack of statistical methods for combining data acquired across multiple sites and scanners. The use of different scanner types for data acquisition can introduce a significant amount of variability in the image processing, which propagates into the statistical analysis pipeline. Therefore, data harmonization should be considered a key element in the analysis of data from multi-site studies.

Previous research in addressing the multi-site confound and data harmonization has focused on either characterizing differences in derived imaging measures obtained across different scanners (Tummala et al. 2016) or on harmonizing derived measures using statistical models that account for scanner variability (Fortin et al. 2018; Johnson, Li and Rabinovic 2007). In both cases the focus has been on post-image processing derived measures.

Heinen and colleagues (Heinen et al. 2016) have characterized the robustness of measures from 1.5T and 3T MRI on the same subjects, using multiple automated techniques and evaluating accuracy using manual segmentations, and show that all automated methods demonstrate variability in accuracy depending on the segmented compartment/tissue type (Heinen et al. 2016). Pfefferbaum et al. characterized agreement between volume measures acquired at 1.5T and 3T using two techniques (FreeSurfer and SRI24) (Pfefferbaum et al. 2012). Their results demonstrated varying degrees of agreement depending on region and analysis method. Notably, the globus pallidus showed particularly poor agreement with either method. It has also been shown that including acquisition site as a confounding variable does not improve prediction accuracy of a cognitive outcome from imaging data (Rao, Monteiro and Mourao-Miranda 2017).

The focus of our study was to investigate the impact of a T1-weighted MRI intensity normalization method, Removal of Artificial Voxel Effect by Linear regression (RAVEL) (Fortin et al. 2016), on MRI- and PET-based outcome measures. The purpose of RAVEL is to remove unwanted variation from and standardize MRI voxel intensity values. RAVEL consists of two main steps: (1) factor analysis of control voxels within cerebrospinal fluid (CSF) space, where voxel intensities are known to be unassociated with disease status and clinical covariates, to estimate the principal directions of unwanted technical variability; and (2) linear regression at the whole brain level, including both covariates of biological interest and direction of unwanted variation.

We specifically focused on investigating MRI intensity normalization using RAVEL for improving neuroimaging outcomes derived from intensity normalized images. Four major steps were combined to create a novel characterization of the effects of RAVEL normalization on neuroimaging derived outcomes. The processes in place were: 1) RAVEL intensity normalization, 2) MRI segmentation using FreeSurfer v5.3 (FS) and derivation of cortical thickness and Region-of-Interest (ROI) volumes, 3)

PET SUVR quantification using ROIs derived from MRI images processed with and without RAVEL intensity normalization, and 4) neuroradiological evaluation to assess the accuracy of hippocampus segmentation derived with and without RAVEL intensity normalization. The effect of RAVEL normalization on MRI- and PET-based outcome measures was explicitly assessed. However, the harmonization of outcome measures was not, as the assessed cohort did not include same-subject data scanned across multiple scanners and/or sites, which would be necessary to explicitly and accurately assess harmonization.

2 Methods

2.1 Study Participants

Forty subjects with Down syndrome were selected from two sites participating in the Neurodegeneration in Aging Down Syndrome (NiAD) study, 20 subjects from the University of Pittsburgh (UPitt) and 20 from the University of Wisconsin (UWisc). The 20 subjects from each site were selected to span the age range of total study participants (32–54 y/o from UPitt, 30–54 y/o from UWisc), and the frequency of amyloid burden positivity as assessed by PiB PET (D. L. Tudorascu et al. 2018) was matched across sites. In all, 10 subjects with PiB scans indicative of amyloid deposition (PiB (+)) and 30 subjects without indication of deposition (PiB (−)) were used in this work. Amyloid classification was established as part of our existing standard analysis of the data and based on previously published methods from our group (Cohen et al. 2013).

2.2 Image Acquisition

MRI. Sagittally acquired structural T1-weighted MRIs were acquired at both sites. A Siemens MAGNETOM Prisma 3T scanner was employed at the UPitt (TE = 2.95 ms; TI = 900.0 ms; TR = 2300 ms; Weighting = PD; Flip Angle = 11.0° ; Pulse Sequence = GR/IR). A General Electric Discovery MR750 scanner was used at the UWisc (TE = 3.04 ms; TI = 400.0 ms; TR = 7.35 ms; Weighting = T1; Flip Angle = 11.0° ; Pulse Sequence = GR). MRIs from both sites had an in-plane pixel spacing of 1.05 mm and a sagittal thickness of 1.2 mm.

PET. PiB PET imaging was performed on a Siemens mCT Biograph PET/CT scanner (UPitt) and a Siemens EXACT HR + PET scanner (UWisc). For both sites, a nominal dose of 15 mCi PiB was delivered as a bolus injection over approximately 30 s, followed by a saline flush, and subjects were imaged over 50–70 min post-injection. PET images were reconstructed into four 5-min time frames using the manufacturers' software and included corrections for scatter, deadtime, random coincidences, and radioactive decay.

Reconstruction at the UPitt site was via ordered subset expectation maximization (OSEM) using 4 iterations of 12 subsets with no post filtering. Data required for attenuation and scatter correction were derived from a low-dose CT-scan, without contrast, acquired at the start of the scan session. Voxel size of the reconstructed PET image was 1.02 mm × 1.02 mm × 2.03 mm (axial).

PET images at the UWisc site were also produced via OSEM (4 iterations, 16 subsets) with no post filtering. Attenuation data were acquired in transmission scans performed using Ge-68/Ga-68 rotating rod sources. PET image voxel size was 2.57 mm × 2.57 mm × 2.43 mm (axial).

2.3 RAVEL Intensity Normalization

The application of RAVEL to a set of T1-weighted MRIs requires 3 conditions: (Condition 1) MRIs to be in a common anatomical space, (Condition 2) MRIs to be skull-stripped, and (Condition 3) a single control ROI indicating common CSF voxels across all images. Condition 2 is, in fact, not an explicit requirement of the RAVEL process but is for the WhiteStripe WM normalization step which precedes RAVEL [https://cran.rproject.org/web/packages/WhiteStripe/index.html]. The MRI preprocessing steps undertaken to meet these conditions are outlined below.

First, MRIs were processed through the Statistical Parametric Mapping v12 (SPM12) Segment tool (Ashburner and Friston 2005) in order to generate forward and inverse nonlinear registration transforms to SPM12-defined Montreal Neurological Institute (MNI) space and subject-specific GM, WM, and CSF tissue probability maps in both subject native space and MNI space. Incidentally, the application of SPM Segment tool is commonly used as part of the Centiloid process in amyloid PET imaging studies of Alzheimer's disease and aging in Down syndrome (Klunk et al. 2015; Tudorascu et al. 2018).

To satisfy Condition 1, native-space MRIs were warped to MNI space using subject-specific SPM12-defined forward transforms. In addition, MNI-space MRIs were bias corrected using the "N4BiasFieldCorrection" function implemented within the "ANTsR" library (Avants 2019), applied in RStudio version 1.2.1335.

To satisfy Condition 2, a single MNI-space intracranial mask was generated by summing the GM, WM, and CSF tissue prior probability maps provided within SPM12 (TPM.nii), thresholding the sum image at 0.1 and binarizing the result. To remove noncontiguous extracranial voxels from the MNI-space intracranial mask, the "cluster" function within FSL FMRIB Software Library v5.0 (Jenkinson et al. 2012) was used to identify and isolate the largest cluster. All MNI-space bias-corrected MRIs were then masked using the intracranial mask to generate skull-stripped MRIs.

To satisfy Condition 3, MNI-space CSF tissue probability maps were first thresholded at a value of 0.3 and binarized to generate subject-specific MNI-space CSF ROIs. The intersection of all MNI-space CSF ROIs was then calculated to generate a single CSF control ROI indicating all common CSF voxels across subjects.

RAVEL intensity normalization was subsequently performed using the "normalizeRAVEL" function within the "RAVEL" R library (https://github.com/Jfortin1/RAVEL) with MNI-space bias-corrected skull-stripped MRIs as "input.files" input, MNI-space intracranial mask as "brain.mask" input, and MNI-space CSF ROI as "control.mask" input. Default parameters were used. Resulting RAVEL-intensity normalized MRIs were also in MNI space with WM voxel intensities centered on a value of 0, such that GM and CSF voxel intensities were negative. This is a consequence of WhiteStripe normalization, which is run by default within the "normalizeRAVEL" function.

To remove negative voxel intensities from RAVEL-normalized images, the minimum voxel value across all images was added to each RAVEL-normalized MRI. The voxel intensity-shifted MNI-space RAVEL-normalized MRIs were subsequently spatially warped back to subject-specific native space using the SPM12 Segment-derived inverse nonlinear transforms.

2.4 MRI Processing

Native-space unnormalized and RAVEL-normalized T1-weighted MRIs were processed using FreeSurfer v5.3 (FS) to generate regional cortical thicknesses, volumes, and binary ROI images for sampling PiB PET images. Unnormalized MRIs were processed through FS using default parameters. RAVEL-normalized images were processed using the "-noskullstrip" flag given that skull-stripping had already been performed as part of the RAVEL normalization process.

2.5 PiB PET Processing

The multi-frame (5 min/frame) PiB PET images were visually inspected for frame-to-frame motion, and motion corrected when appropriate, using PMOD software version 3.711 (PMOD Technologies LLC, https://www.pmod.com/). Single-frame PET images were generated by averaging the time frames from 50–70 min post-injection. The single-frame PET images were subsequently registered to subject-specific unnormalized T1-weighted MRIs using the "Coregister: Estimate and Reslice" tool within SPM12. Previously described PiB-specific composite ROIs were used to sample single-frame PET images, generating regional PiB radioactivity concentration values (Tudorascu et al. 2018). Regional values were normalized by dividing by the FS-defined cerebellar GM radioactivity, resulting in regional SUVRs. PiB regional SUVR values were derived using both unnormalized MRI-derived FS ROIs and RAVEL-normalized MRI-derived FS ROIs. However, the PiB-PET images were not re-registered to RAVEL-normalized MRIs. The results of this process, for each subject, were two sets of PiB-PET SUVR measures based on FS ROIs: one set from the unnormalized MRI and one set from the RAVEL-normalized MRI. For the purposes of this work, we evaluated PET results using a striatal ROI, a region of early accumulation of amyloid deposition in DS (Tudorascu et al., 2019), and a global ROI that is the union of all our standard amyloid quantitation regions. Amyloid status (PiB (−) or PiB (+)) was not reevaluated as part of this work.

2.6 Statistical Evaluation

Descriptive statistics including means and standard deviations (SD) were computed for each cortical thickness and ROI volume as well as for the global PiB SUVR derived measures with and without the use of RAVEL. Group differences were determined using Cohen's d effect sizes for each measure between PiB (−) and PiB (+) subject groups.

Intraclass correlation coefficients (ICCs) and their corresponding 95% confidence intervals (CI) were calculated to measure agreement between the derived measures from unnormalized MRI and RAVEL-normalized MRI-derived FS cortical thicknesses, volumes, and PiB SUVR values. FS volumes were normalized to FS-reported total intracranial volume (tICV) before statistical assessment to account for sex-specific differences in anatomical volumes (Schwarz et al. 2016).

A neuroradiologist (JMM) visually rated unnormalized and RAVEL-normalized MRI-derived FS hippocampal masks on unnormalized MRIs using a four-point scale to rate the accuracy of the segmentations (1 = poor, 2 = some errors, 3 = good, 4 = excellent). The rater was blinded to subject information, including age, gender, amyloid status and to the origin and methodology from which hippocampal masks were derived. The Wilcoxon signed-rank test was used to assess pairwise differences in rater ranking between unnormalized MRI-derived FS hippocampal masks and RAVEL-normalized MRI-derived FS hippocampal masks.

3 Results

Density plots, smoothed versions of histograms, of MRI voxel intensities are shown by site and amyloid status at multiple preprocessing steps (Raw, after applying n4 correction, after applying White Stripe and, last, after RAVEL) in Fig. 1. It can be observed that after RAVEL intensity normalization the histograms of intensities have a much greater overlap, reinforcing that RAVEL normalization significantly reduces MRI voxel intensity variability.

Descriptive statistics and PiB(−) to PiB(+) effect sizes for volumes, cortical thicknesses, and SUVR are presented in Table 1. The derived PET global SUVR and striatum SUVR measures were almost identical with or without the use of RAVEL in the preprocessing stream. The estimated regional brain volumes obtained using RAVEL were larger for left and right hippocampus in both PiB(−) and PiB (+) groups. For assessing variability, standard deviations were larger for the PiB (+) group, SD = 0.67 for left hippocampus and SD = 0.52 for right hippocampus when RAVEL was used compared to SD = 0.57 for left hippocampus and SD = 0.50 for right hippocampus when RAVEL was not used. In the PiB (-) group, the SD was smaller and almost identical regardless of the use of RAVEL.

However, the cortical thickness values were much more variable, with no clear pattern observed. For example, the means and SDs for the right entorhinal values are almost identical across methods, while left entorhinal values diverged (3.77 mm mean with SD = 0.40 for PiB (−) when RAVEL was not used and 3.69 mm mean with SD = 0.39 when RAVEL was used). Similarly, in the PiB (+) group, the values were 3.37 mm, SD = 0.44 without RAVEL and 3.49 mm, SD = 0.42 with RAVEL. The computed Cohen's d effect sizes between PiB (+) and PiB (−) were 0.95 without RAVEL and 0.45 with RAVEL.

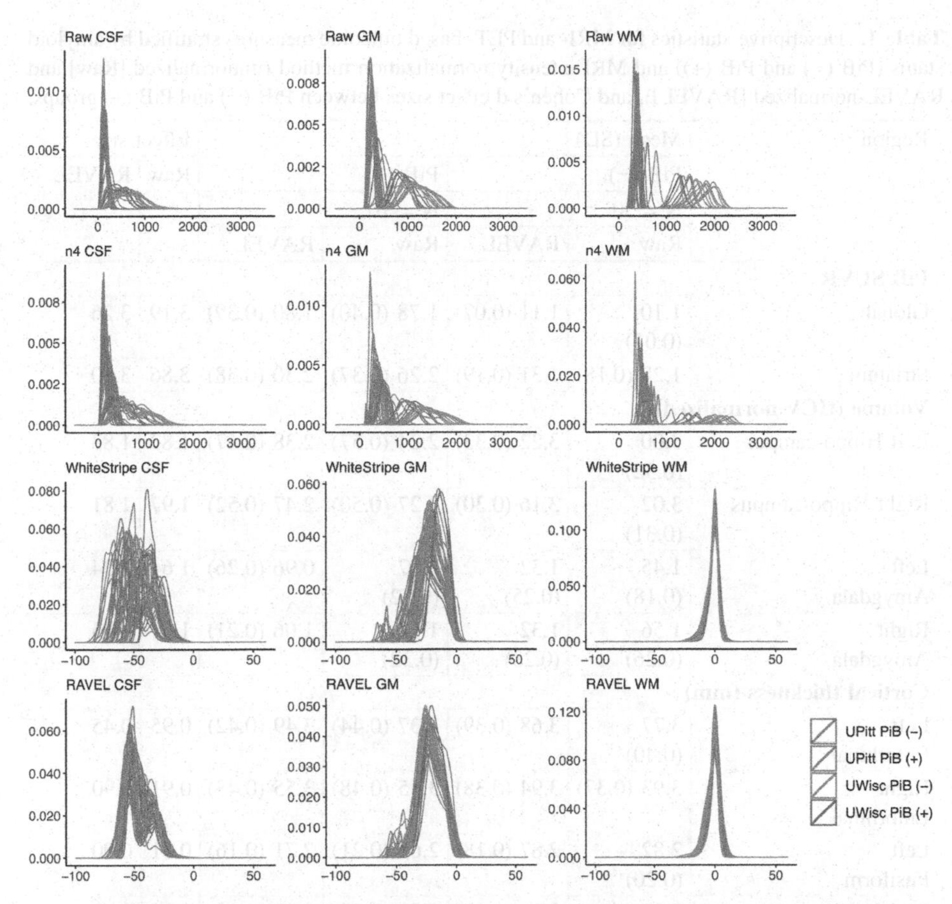

Fig. 1. Density plots of MRI voxel intensities by site (University of Pittsburgh [UPitt] and University of Wisconsin [UWisc]), amyloid classification status (PiB (−) and PiB (+)), and tissue type (cerebrospinal fluid [CSF], grey matter [GM], and white matter [WM]) for multiple steps of the RAVEL intensity normalization process (unnormalized [Raw], bias corrected [n4], WhiteStripe-normalized [WhiteStripe], and RAVEL-normalized [RAVEL]). Results shown for RAVEL are prior to the constant intensity shift described in the text.

ICCs are shown for SUVR, volumes, and cortical thickness between outcomes obtained with and without RAVEL in Table 2. The ICCs show very good agreement between PET SUVR measures, with values greater than 0.9. The ICC values were lower for volumes, with a range of ICC = 0.365 (95% CI: 0.066, 0.604) in the right amygdala up to ICC = 0.830 (95% CI: 0.703, 0.906) for the right hippocampus. The ICCs for the cortical thicknesses were much lower, specifically for middle temporal left (0.228 (95% CI: −.083, 0.5)) and middle temporal right (0.33 (95% CI: 0.028, 0.579)). Areas with highest ICC values for the cortical thickness were left fusiform (0.851 (95% CI: 0.589, 0.864)) and right fusiform (0.871 (95% CI: 0.77, 0.929)).

Table 1. Descriptive statistics for MRI- and PET- based outcome measures stratified by amyloid status (PiB (−) and PiB (+)) and MRI intensity normalization method (unnormalized [Raw] and RAVEL-normalized [RAVEL]), and Cohen's d effect sizes between PiB (−) and PiB (+) groups.

Region	Mean (SD)				Effect size	
	PiB (−)		PiB (+)		Raw	RAVEL
	N = 30		N = 10			
	Raw	RAVEL	Raw	RAVEL		
PiB SUVR						
Global	1.10 (0.07)	1.11 (0.07)	1.78 (0.40)	1.80 (0.39)	3.19	3.26
Striatum	1.28 (0.18)	1.31 (0.19)	2.26 (0.37)	2.30 (0.38)	3.86	3.80
Volume (tICV normalized)						
Left Hippo-campus	3.00 (0.32)	3.22 (0.33)	2.24 (0.57)	2.38 (0.67)	1.83	1.83
Right Hippo-campus	3.02 (0.31)	3.16 (0.30)	2.27 (0.50)	2.47 (0.52)	1.97	1.81
Left Amygdala	1.45 (0.18)	1.32 (0.25)	1.07 (0.32)	0.96 (0.26)	1.64	1.34
Right Amygdala	1.56 (0.26)	1.32 (0.28)	1.13 (0.32)	1.06 (0.21)	1.51	0.93
Cortical thickness (mm)						
Left Entorhinal	3.77 (0.40)	3.68 (0.39)	3.37 (0.44)	3.49 (0.42)	0.95	0.45
Right Entorhinal	3.93 (0.37)	3.94 (0.38)	3.55 (0.48)	3.55 (0.48)	0.91	0.90
Left Fusiform	2.82 (0.20)	2.87 (0.18)	2.65 (0.21)	2.71 (0.16)	0.81	0.90
Right Fusiform	2.95 (0.19)	2.94 (0.18)	2.78 (0.27)	2.82 (0.26)	0.77	0.53
Left Inferior Temporal	2.79 (0.27)	2.93 (0.20)	2.65 (0.22)	2.82 (0.14)	0.50	0.59
Right Inferior Temporal	2.86 (0.26)	2.92 (0.21)	2.75 (0.20)	2.82 (0.19)	0.41	0.47
Left Middle Temporal	2.86 (0.26)	3.09 (0.20)	2.75 (0.21)	2.94 (0.21)	0.43	0.71
Right Middle Temporal	2.80 (0.27)	3.01 (0.25)	2.70 (0.27)	2.86 (0.28)	0.37	0.58

Neuroradiological ratings revealed that RAVEL intensity normalization significantly improved right hippocampal segmentation accuracy (Wilcoxon W-value = 69, p-value = 0.02) as shown in Fig. 2. When RAVEL was used, the right hippocampal segmentation was excellent for 18 subjects as compared to only 10 when RAVEL was not used. A single case is illustrated in Fig. 3, showing the hippocampus segmentation by method with arrows pointing to the areas that were incorrectly segmented without RAVEL normalization. The arrows point to voxels that were incorrectly classified as hippocampus when RAVEL was not implemented (in red).

Table 2. Intraclass correlation coefficients (ICC) between unnormalized and RAVEL-intensity normalized FreeSurfer-based regions of interest outcome measures for both sites (University of Pittsburgh and University of Wisconsin) combined.

Region	ICC (95% CI)
PiB SUVR	
Global	0.998 (0.996, 0.999)
Striatum	0.995 (0.992, 0.998)
Volume	
Left Hippocampus	0.754 (0.583, 0.862)
Right Hippocampus	0.830 (0.703, 0.906)
Left Amygdala	0.616 (0.382, 0.776)
Right Amygdala	0.365 (0.066, 0.604)
Cortical thickness	
Left Entorhinal	0.683 (0.477, 0.819)
Right Entorhinal	0.758 (0.589, 0.864)
Left Fusiform	0.851 (0.738, 0.918)
Right Fusiform	0.871 (0.770, 0.929)
Left Inferior Temporal	0.474 (0.197, 0.682)
Right Inferior Temporal	0.647 (0.425, 0.796)
Left Middle Temporal	0.228 (−0.083, 0.500)
Right Middle Temporal	0.330 (0.028, 0.579)

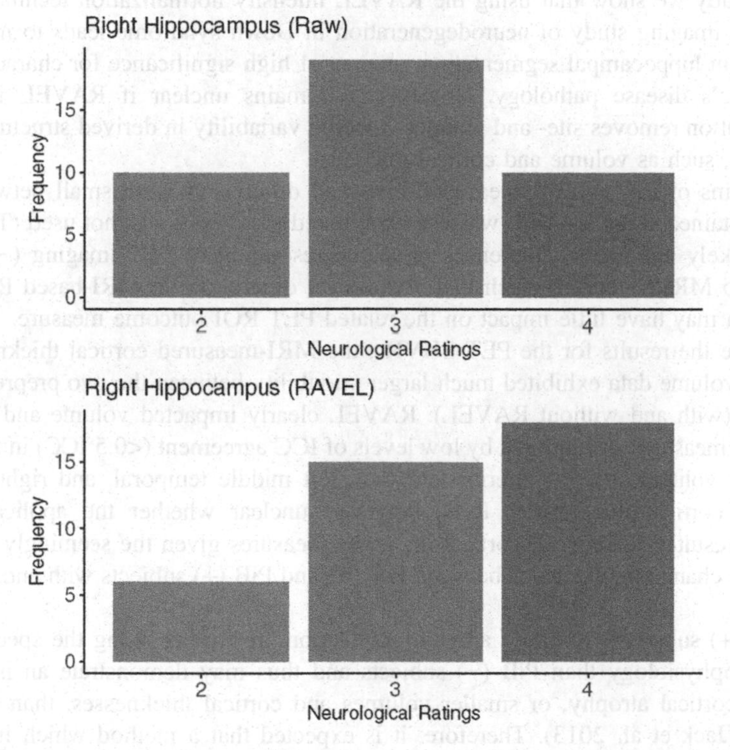

Fig. 2. Visual ratings of FreeSurfer-based right hippocampus segmentations based on unnormalized (Raw) and RAVEL-intensity normalized (RAVEL) MRIs. A four-point rating scale was used, such that 1 = poor, 2 = some errors, 3 = good, and 4 = excellent. (Color figure online)

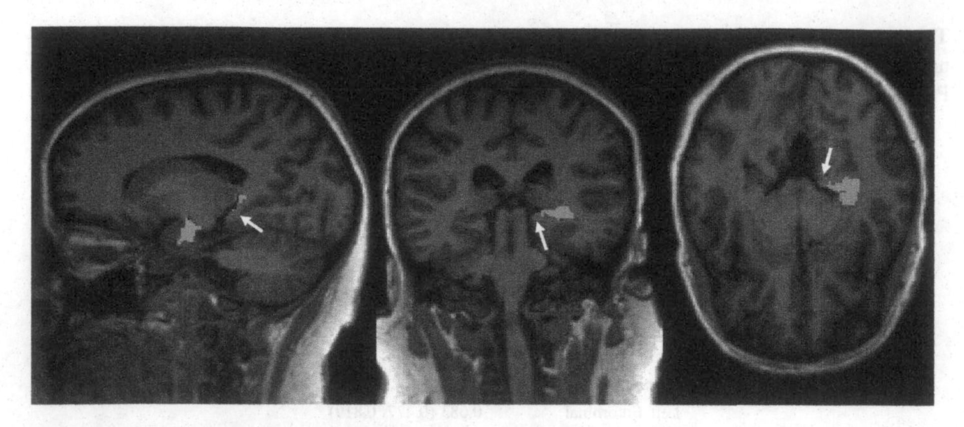

Fig. 3. Sagittal, coronal, and axial slices of a single subject MRI with right hippocampal segmentations overlaid generated without the use of RAVEL (red) and with RAVEL (green). Arrows indicate extraneous non-hippocampus voxels classified as hippocampus prior to RAVEL intensity normalization. (Color figure online)

4 Discussion

In this study we show that using the RAVEL intensity normalization technique in a multi-site imaging study of neurodegeneration in Down syndrome leads to increased accuracy in hippocampal segmentation, an area of high significance for characterizing Alzheimer's disease pathology. However, it remains unclear if RAVEL intensity normalization removes site- and scanner- specific variability in derived structural MRI outcomes, such as volume and cortical thickness.

In terms of PET SUVR measures, observed differences were small between the values obtained when RAVEL was used compared to when it was not used (Table 2). This is likely due to the differences in spatial resolution of PET imaging ($\sim$5 mm) relative to MRI ($\sim$1 mm), such that voxel-level differences in MRI-based ROI segmentation may have little impact on the related PET ROI outcome measure.

Unlike the results for the PET SUVRs, the MRI-measured cortical thickness and regional volume data exhibited much larger variability between the two preprocessing methods (with and without RAVEL). RAVEL clearly impacted volume and cortical thickness measures as exhibited by low levels of ICC agreement (<0.5 ICC) in the right amygdala volume and left inferior temporal, left middle temporal, and right middle temporal cortical thicknesses. It is, however, unclear whether the application of RAVEL resulted in superior cortical thickness measures given the seemingly random pattern of change in effect size between PiB (−) and PiB (+) subjects with and without RAVEL.

PiB (+) subjects exhibiting amyloid deposition are further along the spectrum of AD pathophysiology than PiB (−) subjects and thus may demonstrate an increased level of cortical atrophy, or smaller volumes and cortical thicknesses, than PiB (−) subjects (Jack et al. 2013). Therefore, it is expected that a method which improves MRI-based regional segmentation accuracy would increase the effect sizes of volumes and cortical thicknesses between PiB (−) and PiB (+) groups. Five of the eight cortical thickness regions presented in Table 1 had improved effect sizes with the application of RAVEL, but the greatest difference in cortical thickness effect size was observed in the left entorhinal region with no normalization outperforming RAVEL-intensity normalization. For cortical thickness measurements, the features of the cortical ribbon present challenges in processing that are dependent on size, shape and location within the particular ribbon.

In conclusion, applying RAVEL normalization during preprocessing of image data acquired at different sites using different scanners reduced variability in the distribution of voxel intensities and improved the accuracy of hippocampal segmentation but not the discriminatory power of quantitative FS-based outcome measures. The primary purpose of this work was to assess the effects of RAVEL intensity normalization on MRI- and PET-derived outcome measures. However, the efficacy of RAVEL in removing site- and scanner- specific variability from volume and cortical thickness measures was not assessed. Future work will address this by applying RAVEL on same-subject data acquired on multiple scanners to more accurately assess RAVEL's effect on harmonization of imaging outcome measures than was possible with this cohort. The effect of each preprocessing step (n4 bias correction, skull-stripping, and WhiteStripe intensity normalization) on the accuracy and harmonization of MRI- and PET- based outcome measures of same-subject data across different scanners will also be more thoroughly characterized.

RAVEL can be easily incorporated into the preprocessing stream, and the code is available online (https://github.com/Jfortin1/RAVEL) (Fortin et al. 2016).

Acknowledgement. This research investigation was funded by the National Institute of Aging (R01 AG031110 to B.L.H and B.T.C; U01AG051406 to B.L.H., W.E.K., and B.T.C., U01AG051406-05S1 to C.M.L and D.L.T. and R01 AG063752 to D.L.T.).

References

Ashburner, J., Friston, K.J.: Unified segmentation. Neuroimage **26**(3), 839–851 (2005)

Avants, B.B.: ANTsR: ANTs in R: Quantification Tools for Biomedical Images (2019)

Cohen, A.D., Mowrey, W., Weissfeld, L.A., Aizenstein, H.J., McDade, E., Mountz, J.M., Nebes, R.D., Saxton, J.A., Snitz, B., De Kosky, S., Williamson, J., Lopez, O.L., Price, J.C., Mathis, C.A., Klunk, W.E.: Classification of amyloid-positivity in controls: comparison of visual read and quantitative approaches. Neuroimage **71**, 207–215 (2013). https://doi.org/10.1016/j.neuroimage.2013.01.015

Fortin, J.-P., Cullen, N., Sheline, Y.I., Taylor, W.D., Aselcioglu, I., Cook, P.A., Adams, P., Cooperh, C., Favai, M., Mc Grath, P.J., Mc Innis, M., Phillips, M.L., Trivedi, M.H., Weissmangl, M.M., Shinohara, R.T.: Harmonization of cortical thickness measurements across scanners and sites. NeuroImage **167**, 104–120 (2018). https://doi.org/10.1016/j.neuroimage.2017.11.024

Fortin, J.-P., Sweeney, E.M., Muschelli, J., Crainiceanu, C.M., Shinohara, R.T.: Removing inter-subject technical variability in magnetic resonance imaging studies. NeuroImage **132**, 198–212 (2016). https://doi.org/10.1016/j.neuroimage.2016.02.036

Heinen, R., Bouvy, W.H., Mendrik, A.M., Viergever, M.A., Biessels, G.J., et al.: Robustness of automated methods for brain volume measurements across different MRI field strengths. Plos one **11**(10), e0165719 (2016). https://doi.org/10.1371/journal.pone.0165719

Jack Jr., C.R., Knopman, D.S., Jagust, W.J., Petersen, R.C., Weiner, M.W., Aisen, P.S., Shaw, L. M., Vemuri, P., Wiste, H.J., Weigand, S.D., GLesnick, T., Pankratz, V.S., CDonohue, M., Trojanowski, J.Q.: Tracking pathophysiological processes in Alzheimer's disease: an updated hypothetical model of dynamic biomarkers. Lancet Neurol. **12**(2), 207–216 (2013). https://doi.org/10.1016/s1474-4422(12)70291-0

Jack Jr., C.R., Wiste, H.J., Weigand, S.D., Therneau, T.M., Lowe, V.J., Knopman, D.S., Gunter, J.L., Senjem, M.L., Jones, D.T., Kantarci, K., Machulda, M.M., Mielke, R.O., Roberts, M.M., Vemuri, P., Reyes, D.A., Petersen, R.C.: Defining imaging biomarker cut points for brain aging and Alzheimer's disease. Alzheimers Dement **13**(3), 205–216 (2017). https://doi.org/10.1016/j.jalz.2016.08.005

Jenkinson, M., Beckmann, C.F., Behrens, T.E., Woolrich, M.W., Smith, S.M.: FSL. Neuroimage **62**(2), 782–790 (2012)

Johnson, W.E., Li, C., Rabinovic, A.: Adjusting batch effects in microarray expression data using empirical Bayes methods. Biostatistics **8**(1), 118–127 (2007)

Klunk, W.E., Koeppe, R.A., Price, J.C., Benzinger, T.L., Devous, M.D. Sr., Jagust, W.J., Johnson, K.A., Mathis, C.A., Minhas, D., Pontecorvo, M.J., Rowe, C.C., Skovronsky, D.M., Mintun, M.A.: The Centiloid Project: standardizing quantitative amyloid plaque estimation by PET. Alzheimers Dement, **11**(1), 1–15, e11-14 (2015). https://doi.org/10.1016/j.jalz.2014.07.003

Lao, P.J., Handen, B.L., Betthauser, T.J., Mihaila, I., Hartley, S.L., Cohen, A.D., Bulova, P.D., Tumuluru, R.V., Murali, D., Mathis, C.A., Barnhart, T.E., Stone, C.K., Price, J.C., Devenny, D.A., Mailick, M.R., Klunk, W.E., Johnson, S.C., Christian, B.T.: Longitudinal changes in amyloid positron emission tomography and volumetric magnetic resonance imaging in the nondemented Down syndrome population. Alzheimers Dement (Amst) **9**, 1–9 (2017). https://doi.org/10.1016/j.dadm.2017.05.001

Pfefferbaum, A., Rohlfing, T., Rosenbloom, M.J., Sullivan, E.V.: Combining atlas-based parcellation of regional brain data acquired across scanners at 1.5 T and 3.0 T field strengths. Neuroimage **60**(2), 940–951 (2012)

Rao, A., Monteiro, J.M., Mourao-Miranda, J.: Predictive modelling using neuroimaging data in the presence of confounds. NeuroImage **150**, 23–49 (2017). https://doi.org/10.1016/j.neuroimage.2017.01.066

Schwarz, C.G., Gunter, J.L., Wiste, H.J., Przybelski, S.A., Weigand, S.D., Ward, C.P., Senjem, M.L., Vemuri, P., Murray, M.E., Dickson, D.W., Parisi, J.E., Kantarci, K., Weiner, M.W., Petersen, R.C., Jack, C.R.: A large-scale comparison of cortical thickness and volume methods for measuring Alzheimer's disease severity. NeuroImage Clin. **11**, 802–812 (2016). https://doi.org/10.1016/j.nicl.2016.05.017

Tudorascu, D.L., Anderson, S.J., Minhas, D.S., Yu, Z., Comer, D., Lao, P., Hartley, S., Laymon, C.M., Snitz, B.E., Lopresti, B.J., Johnson, S., Price, J.C., Mathis, C.A., Aizenstein, H.J., Klunk, W.E., Handen, B.L., Christiand, B.T., Cohen, A.D.: Comparison of longitudinal Aβ in nondemented elderly and Down syndrome. Neurobiol. Aging **73**, 171–176 (2019). https://doi.org/10.1016/j.neurobiolaging.2018.09.030

Tudorascu, D.L., Minhas, D.S., Lao, P.J., Betthauser, T.J., Yu, Z., Laymon, C.M., Lopresti, B.J., Mathis, C.A., Klunk, W.E., Handen, B.L., Christian, B.T., Cohen, A.D.: The use of Centiloids for applying [(11)C] PiB classification cutoffs across region-of-interest delineation methods. Alzheimers Dement (Amst) **10**, 332–339 (2018). https://doi.org/10.1016/j.dadm.2018.03.006

Tummala, S., Chu, R., Khalid, F., Dupuy, S., Tauhid, S., Healy, B., Bakshi, R.: Cortical Thickness Measurements from 1.5 t vs. 3t MRI in Healthy Subjects and Patients with Multiple Sclerosis (P4. 179). In: AAN Enterprises (2016)

Villemagne, V.L., Pike, K.E., Chetelat, G., Ellis, K.A., Mulligan, R.S., Bourgeat, P., Ackermann, U., Jones, G., Szoeke, C., Salvado, O., Martins, R., O'Keefe, G., Mathis, C.A., Klunk, W.E., Ames, D., Masters, C.L., Rowe, C.C.: Longitudinal assessment of Abeta and cognition in aging and Alzheimer disease. Ann. Neurol. **69**(1), 181–192 (2011). https://doi.org/10.1002/ana.22248

Yau, W.W., Tudorascu, D.L., McDade, E.M., Ikonomovic, S., James, J.A., Minhas, D., Mowrey, W., Sheu, L.K., Snitz, B.E., Weissfeld, L., Gianaros, P.J., Aizenstein, H.J., Price, J.C., Mathis, C.A., Lopez, O.L., Klunk, W.E.: Longitudinal assessment of neuroimaging and clinical markers in autosomal dominant Alzheimer's disease: a prospective cohort study. Lancet Neurol. **14**(8), 804–813 (2015). https://doi.org/10.1016/S1474-4422(15)00135-0

Bayesian Image Analysis in Fourier Space Using Data-Driven Priors (DD-BIFS)

John Kornak(✉), Ross Boylan, Karl Young, Amy Wolf, Yann Cobigo, and Howard Rosen

University of California, San Francisco, CA 94143, USA
john.kornak@ucsf.edu
https://github.com/bifs

Abstract. Statistical image analysis is an extensive field that includes problems such as noise-reduction, de-blurring, feature enhancement, and object detection/identification, to name a few. Bayesian image analysis can improve image quality, by balancing a priori expectations of image characteristics, with a model for the noise process via Bayes Theorem. We have previously given a reformulation of the conventional Bayesian image analysis paradigm in Fourier space, i.e. the prior distribution (the prior) and likelihood are given in terms of spatial frequency signals. By specifying the Bayesian model in Fourier space, spatially correlated priors, that are relatively difficult to model and compute in conventional image space, can be efficiently modeled as a set of independent processes across Fourier space. The originally inter-correlated and high-dimensional problem in image space is thereby broken down into a series of (trivially parallelizable) independent one-dimensional problems. In this paper we adapt this Fourier space process into a data-driven framework in which the Fourier space priors are built empirically from a database of images and then used to enhance future images. We will describe the data-driven Bayesian image analysis in Fourier space (DD-BIFS) modeling approach, illustrate it's computational efficiency and speed. Finally, we give specific applications of DD-BIFS to improve the quality of arterial-spin-labeling (ASL) perfusion images via a database of human brain positron emission tomography (PET) images.

Keywords: Bayesian image analysis · Data-driven priors · Fourier space

1 Introduction

Bayesian image analysis models provide a solution for improving image quality in image reconstruction/enhancement problems by incorporating *a priori* expectations of image characteristics along with a model for image noise. We have previously presented an approach to reformulating the Bayesian Image

Supported by NIH R01 EB022055 and U19 AG063911.

M.-J. Lesot et al. (Eds.): IPMU 2020, CCIS 1239, pp. 380–390, 2020.
https://doi.org/10.1007/978-3-030-50153-2_29

analysis in Fourier Space (BIFS) [15]. Spatially correlated prior distributions (priors) that are difficult to model and compute in conventional image space can be modeled as independent across locations in Fourier space. The original high-dimensional problem in image space is thereby broken down into a series of one-dimensional problems, leading to easier specification and implementation, and faster computation.

2 BIFS Modeling Framework

Consider x to be a true (or idealized) image that we wish to recover from a sub-optimal image dataset y. (We use the common shorthand notation of not explicitly distinguishing the random variables and the corresponding image realizations [5], i.e., we use lower case x and y throughout.) The Bayesian image analysis paradigm incorporates a priori desired spatial characteristics (e.g., smoothness) via a prior distribution for the true image x and defines the noise degradation process via the likelihood.

Instead of the conventional Bayesian image analysis approach of generating prior and likelihood models for the true image x based on image data y directly, we formulate them via their discrete Fourier transforms representations: $\mathcal{F}x$ and $\mathcal{F}y$. After applying Bayes' Theorem, the posterior becomes

$$\pi(\mathcal{F}x|\mathcal{F}y) \propto \pi(\mathcal{F}x)\pi(\mathcal{F}y|\mathcal{F}x). \tag{1}$$

The key component of the BIFS formulation that leads to its useful properties of easy definition and computational speed, is that we define both the prior and likelihood (and therefore the posterior) to consist of a set of independent distributions across Fourier space locations. Desired spatial correlation in image space is induced by allowing the parameters of the distributions to change over Fourier space [18,20]. This independence definition can be contrasted with conventional Bayesian image analysis using Markov random field (MRF) priors, where Markovian neighborhood structures are used to induce correlation structures across pixels via joint or conditional distributional specifications [4,5,9].

When defining a spatially correlated prior in image space via a set of independent processes across Fourier space, the full conditional posterior at a Fourier space location $k = (k_x, k_y)$, or for volumetric data (k_x, k_y, k_z), now only depends on the prior at that location k, i.e.,

$$\pi(\mathcal{F}x_k|\mathcal{F}y) \propto \pi(\mathcal{F}x_k)\pi(\mathcal{F}y_k|\mathcal{F}x_k), \tag{2}$$

where we use $\mathcal{F}x_k$ as shorthand for $(\mathcal{F}x)_k$. The joint posterior density for the image is then

$$\pi(\mathcal{F}x|\mathcal{F}y) \propto \prod_{k \in K} \pi(\mathcal{F}x_k)\pi(\mathcal{F}y_k|\mathcal{F}x_k), \tag{3}$$

where K is the set of all Fourier space point locations.

2.1 The Data-Driven BIFS Prior (DD-BIFS)

The standard process of generating the BIFS prior distribution given in Kornak [15] is based on choosing a pair of distributions to be applied as priors at each location in Fourier space (one for the modulus and the other for the argument of the complex value signal) and a set of parameter functions to define how the parameters of the distributions vary over Fourier space. In contrast, for the data-driven approach, although we again choose the probability distribution forms across Fourier space, the parameters are estimated empirically from a database of transformed images. That is, all of the images in the database are Fourier transformed, the data at each location in Fourier space are extracted, and the distribution parameters for that Fourier space location are estimated from that data. In this way empirical maps of parameter estimates is generated over space. These parameter estimates are then used as the parameters for the prior at each Fourier space location.

Separate priors and associated parameter functions are defined for each of the modulus and argument of the complex value at each Fourier space location. Working with the modulus and argument provides a more natural framework for defining prior information at specific Fourier space locations (i.e., specific spatial frequencies) than working with the real and imaginary components, because prior information (e.g., smoothness, edges, or features of interest) most strongly relates to the magnitude of the process involved.

2.2 BIFS Likelihood

As for the prior, the BIF likelihood is again modeled separately for the Modulus and Argument of the signal at each Fourier space location. Because we model based on independence across Fourier space points, a range of different noise structures (defined in Fourier space) can readily be incorporated into the likelihood $\pi(\mathcal{F}y_k|\mathcal{F}x_k)$. For example, the combination of independent and identically distributed (*i.i.d.*) zero mean Rayleigh noise/Rician likelihood [19] for the modulus with uniform argument on the circle in Fourier space corresponds to the likelihood model of *i.i.d.* Gaussian noise in image space.

2.3 Posterior Estimation

It is at the posterior estimation stage that the computational gains of the independent BIFS formulation are ultimately realized. Posterior estimation in conventional Bayesian image analysis tends to focus on *maximum a posteriori* (MAP) estimation (minimizing a 0–1 loss function), because it is the most computationally tractable. In the BIFS formulation the MAP estimate can be estimated with added efficiency by maximizing the posterior distribution separately at each Fourier space location and then taking the inverse Fourier transform of the Fourier space MAP estimates, i.e, $x_{\text{MAP}} = \mathcal{F}^{-1}(\mathcal{F}x_{\text{MAP}})$ where $\mathcal{F}x_{\text{MAP}} = \{\mathcal{F}x_{k,\text{MAP}}, k = 1, \ldots, K\}$ and $\mathcal{F}x_{k,\text{MAP}} = \max_{\mathcal{F}x_k} \{\pi(\mathcal{F}x_k|\mathcal{F}y_k)\}$.

This computational efficiency contrasts with conventional Bayesian image analysis, where even the most computationally convenient MAP estimate requires iterative computation methods such as conjugate gradients or expectation-maximization to obtain it.

2.4 Implementation of DD-BIFS

Implementation of DD-BIFS modeling requires the following steps:

1. Fast Fourier transform (FFT) all images in the database that are to be used to build the DD-BIFS prior, i.e., from image space into Fourier space.
2. Choose the distributional form of the prior at each location in Fourier space
3. Estimate the parameters of the prior at each location in Fourier space using the data from the corresponding Fourier space locations across the subjects in the database.
4. Define the likelihood in Fourier space.
5. FFT the dataset to be reconstructed from image space into Fourier space.
6. Combine the DD-BIFS prior and likelihood for the image at each Fourier space location via Bayes' Theorem to generate the DD-BIFS posterior
7. Generate Fourier space MAP estimate by maximizing the posterior at each Fourier space location.
8. Inverse FFT the Fourier space MAP estimate back to image space and display

3 Simulated Dataset Example – Lesion Enhancement

In this example we use simulations to drive the formulation of a DD-BIFS prior which is subsequently applied to independent data. We simulated 1000 256×256 images representing lesions/tumor patterns. The number of lesions was modeled as a Poisson process and the lesions were simulated as randomly positioned truncated Gaussian probability density functions (resembling bumps) with random intensity, and standard deviation on each axis, and with correlation distributed uniformly between -1 and 0, so that the process was anisotropic (i.e. so the spatial autocorrelation was not uniform in all directions).

We use the following model applied at each Fourier space location:

- Gaussian (normal) prior for the modulus: $\text{Mod}(\mathcal{F}x_k) \sim N(\mu_k, \tau_k^2)$
- Uniform prior on the circle for the argument: $\text{Arg}(\mathcal{F}x_k) \sim U(0, 2\pi)$ (uninformative prior)
- Gaussian noise model for the modulus $\text{Mod}(\epsilon_k) \sim N(0, \sigma^2)$
- Uniform noise model for the argument $\text{Arg}(\epsilon_k) \sim U(0, 2\pi)$

where ϵ_k is the complex noise at Fourier space location k. (Note this model is not Gaussian noise in image space, for which we use a Rayleigh noise model/Rician likelihood, for the modulus.)

The prior for the modulus at each Fourier space location was then generated from the empirical estimates of the mean (μ_k) and standard deviation (τ_k) at the corresponding location across the simulated datasets. The global posterior mode was then be obtained by generating the posterior mode at each Fourier space location based on conjugate Bayes for the Gaussian distribution [8] with $\mathrm{Mod}(x_{k,\mathrm{MAP}}) = (\frac{\mu_k}{\tau_k^2} + \frac{y_k}{\sigma^2})/(\frac{1}{\tau_k^2} + \frac{1}{\sigma^2})$ and $\mathrm{Arg}(x_{k,\mathrm{MAP}}) = \mathrm{Arg}(x_{k,\mathrm{MAP}})$ (when the prior for the argument is uninformative).

Figure 1 shows a single new realization (i.e. that was not included in the simulation set) of the process at the top-left, the same image with added noise at the top-right, a parameter function-based BIFS MAP reconstruction on the bottom-left (approximating a pairwise absolute difference MRF prior Bayesian reconstruction); and the DD-BIFS prior on the bottom-right. The parameter function-based BIFS prior does a reasonable job of enhancing lesions, in particular the blurred lesion furthest to the right. However, the DD-BIFS reconstruction clearly improves the enhancement of the simulated lesions beyond that of the parameter function-based BIFS. The DD-BIFS reconstruction is able to better retain the detail of the lesions, in particular, their non-isotropic elongated form.

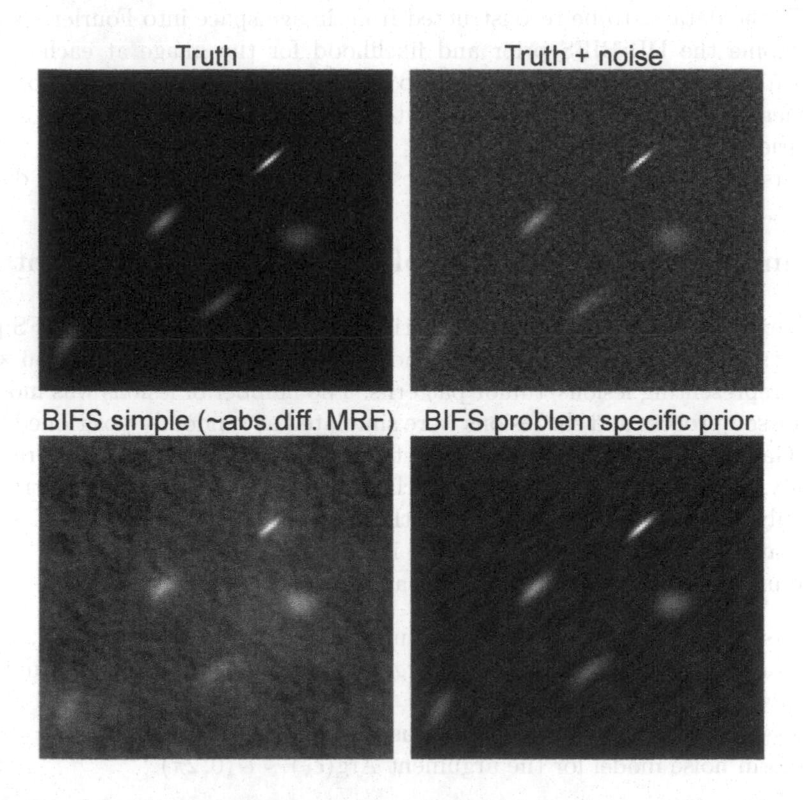

Fig. 1. Simulation study and reconstruction of lesion/tumor patterns. Top-left: new process realization; top-right: with added noise; bottom left: pairwise difference prior approximation BIFS reconstruction; bottom right: simulation-driven BIFS reconstruction.

It should be noted here that primary objective of the study lies with visualization and not direct quantification of image intensities. If the clinical goal is to detect tumors then the objective of reconstruction is to have visually clear tumors that the radiologist can readily identify. Alternatively, if the goal is differentiation of benign vs cancer tumors, the decision may be based on shape characteristics so that accentuating them may help the clinician to visually differentiate between the lesion types.

4 Perfusion/PET Imaging of the Brain Example

The overall objective of this study is to process arterial spin labeling (ASL) perfusion MRI to enhance blood flow patterns in the brain associated with frontotemporal lobar dementia (FTLD). ASL perfusion has been shown to be sensitive to FTLD pathology, showing hypoperfusion in frontal regions, potentially providing a cheaper and radiation free alternative to the conventional (FDG)-PET (fluorodeoxyglucose positron emission tomography) [7,12]. Preliminary data indicates that ASL based perfusion patterns associated with FTLD are coherent with PET, albeit with reduced image quality due to increased noise (Fig. 2). The objective of our BIFS modeling procedure is to improve ASL perfusion maps toward the quality of PET by using a DD-BIFS prior generated from a database of PET images.

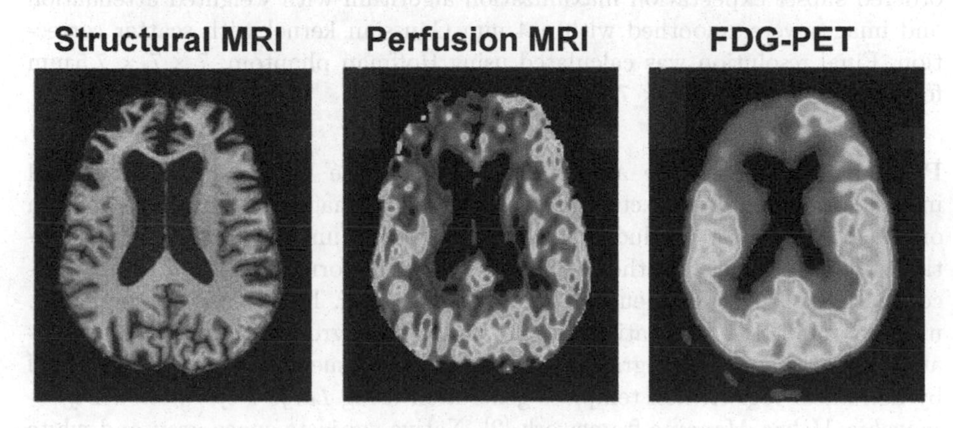

Fig. 2. Comparison ASL perfusion MRI and FDG-PET cerebro-blood-flow (CBF) maps for the same subject.

4.1 Data

PET and ASL perfusion pairs for individuals acquired around the same time are taken from the Frontotemporal Lobar Degeneration Neuroimaging Initiative (FTLDNI), a multi-center biomarker trial aimed at identifying promising markers as surrogate endpoints for FTLD disease progression in clinical trials.

Structural MRI T1 and ASL Perfusion MRI Data: Volumetric MPRAGE sequences at UCSF were used to acquire T1-weighted structural images of the entire brain (Sagital slice orientation; slice thickness = 1.0 mm; slices per slab = 160; in-plane resolution = 1.0 × 1.0 mm; matrix = 240 × 256; TR = 2,300 ms; TE = 2.98 ms; TI = 900 ms; flip angle = 9°). Pulsed ASL (PASL) imaging was acquired using QUIPSSII with Thin-slice TI1 Periodic Saturation (Q2TIPS) sequence incorporated in a PICORE (Proximal Inversion with Control of Off-Resonance Effects) labeling scheme [16]. The periodic saturation pulses started at the postlabeling delay inversion time TI1 = 700 ms after the in-plan presaturation radio infrequency pulse; the readout started at the postlabeling delay inversion time TI2 = 1800 ms. The repetition and echo time were TR/TE = 2500/11 ms. Sixteen slices were acquired, each 6 mm thick with a 7.2 mm center to center distance and a matrix 64 × 56 of 4 × 4 mm^2 in-plane voxel resolution.

PET Data: PET data were acquired at the Lawrence Berkeley National Laboratory on a Siemens ECAT EXACT HR scanner or a Siemens Biograph PET/CT scanner. FDG was supplied by a radiopharmacy (IBA Molecular). Six emission frames lasting 5 min each were acquired starting 30 min post-injection. Attenuation correction was performed using a 10 min transmission scan on the ECAT scanner or a low-dose CT scan on the Biograph, both being acquired prior to PET acquisition. For both scanners, PET data were reconstructed using an ordered subset expectation maximization algorithm with weighted attenuation and images were smoothed with a 4 mm Gaussian kernel with scatter correction. Final resolution was calculated using Hoffman phantom: 7 × 7 × 7.5 mm for ECAT and 6.5 × 6.5 × 7.25 mm for Biograph.

Pre-processing: Before any prepossessing of the images, all T1-weighted images were visually inspected for quality control. Images with excessive motion or image artifact were excluded. T1-weighted images underwent bias field correction using N3 algorithm, the segmentation was performed using SPM12 (Wellcome Trust Center for Neuroimaging, London, UK, http://www.fil.ion.ucl.ac.uk/spm) unified segmentation [10]. A customized group template was generated from the segmented gray and white matter tissues and cerebrospinal fluid by non-linear registration template generation using *Large Deformation Diffeomorphic Metric Mapping* framework [2]. Native subjects space gray and white matter were geometrically normalized to the group template, modulated and then smoothed in the group template. The applied smoothing used a Gaussian kernel with 8 mm full width half maximum. All steps of the transformation were carefully inspected from the native space to the group template. Linear and non-linear transformations between the group template space and International Consortium of Brain Mapping (ICBM) [17] was applied.

Frames of the ASL acquisition were corrected for motion, co-registered with the first frame (M0) using FSL [13]. An automatic quality control process removed tagged/untagged pairs of frames when the relative root mean square (RMS) distance value between two consecutive frames was higher than 0.5 mm.

The subject was dropped if this RMS value was higher than 1 mm. Differential perfusion images were created by subtracting unlabeled from adjacent labeled frames and averaging these subtraction images [1]. Susceptibility artifacts along the phase-encoding direction were corrected for the M0 frame and perfusion map using ANTs [3] restricted to the coronal axis of the patient. Cerebral Blood Flow (CBF) was calculated by applying the Buxton kinetic model to the perfusion map [6]. CBF data was processed to obtain partial volume corrected maps of gray matter perfusion, based on the tissue segmentation using transformation matrix from T1 to M0 as previously described [7,11,14]. Normalized CBF values were obtained by dividing the voxel CBF value by the mean Calacarin CBF value region of interest. Calcarine was selected based on the observation that FTD variants do not impact this area neither the acquisition field of view.

Analyses on partial volume corrected, non-partial volume corrected and normalized perfusion images was performed in MNI space using the structural set of geometric transformations and smoothed with an isotropic 12 mm Gaussian kernel full width half maximum. All CBF images were visually inspected in the native and MNI spaces. Poor quality images out of the field of view, with large susceptibility or motion artifacts were removed from the study.

Finally, voxels in the ASL image were rescaled to match the dynamic range of the PET images. To do this the voxels in the ASL were first ordered by intensity, ignoring their spatial coordinates. Then a 10% random sample was drawn from all the voxels in all the PET images of the database, and they were ordered by intensity. The n'th brightest voxel in the ASL to be processed was reassigned the intensity of the m'th brightest voxel in the PET sample, where $m = \text{round}\left(n * \frac{n_p}{n_a}\right)$, and where n_p and n_a are the sizes of the PET subsample and the ASL image respectively, and n ranges from 0 to $n_a - 1$.

4.2 Reconstruction Results

The same DD-BIFS model set as that used for the lesion simulation study. 101 subjects were used to build the PET prior and 1 individual with corresponding ASL-perfusion scan was reserved for evaluation.

Figure 3 shows the results of the reconstruction of an individual's ASL image. In the top-left panel is the PET image that we would like to emulate and in the top-right the corresponding original ASL perfusion MRI for the same individual. In the bottom left the DD-BIFS prior reconstruction is displayed (with mask applied) and at the bottom right is DD-BIFS prior reconstruction with additional shrinkage on the prior variance (specifically the standard deviation is multiplied by a factor of 0.01). The reconstructed image does adjust the ASL-perfusion MRI so that it emulates the characteristics of the PET, and shrinking the prior variance (essentially putting more weight on the prior) serves to further move the original ASL to an image that emulates what the clinician might see with PET.

Ongoing work on this project will focus on performing full clinical validation of DD-BIFS to assess clinical applicability as a surrogate for PET.

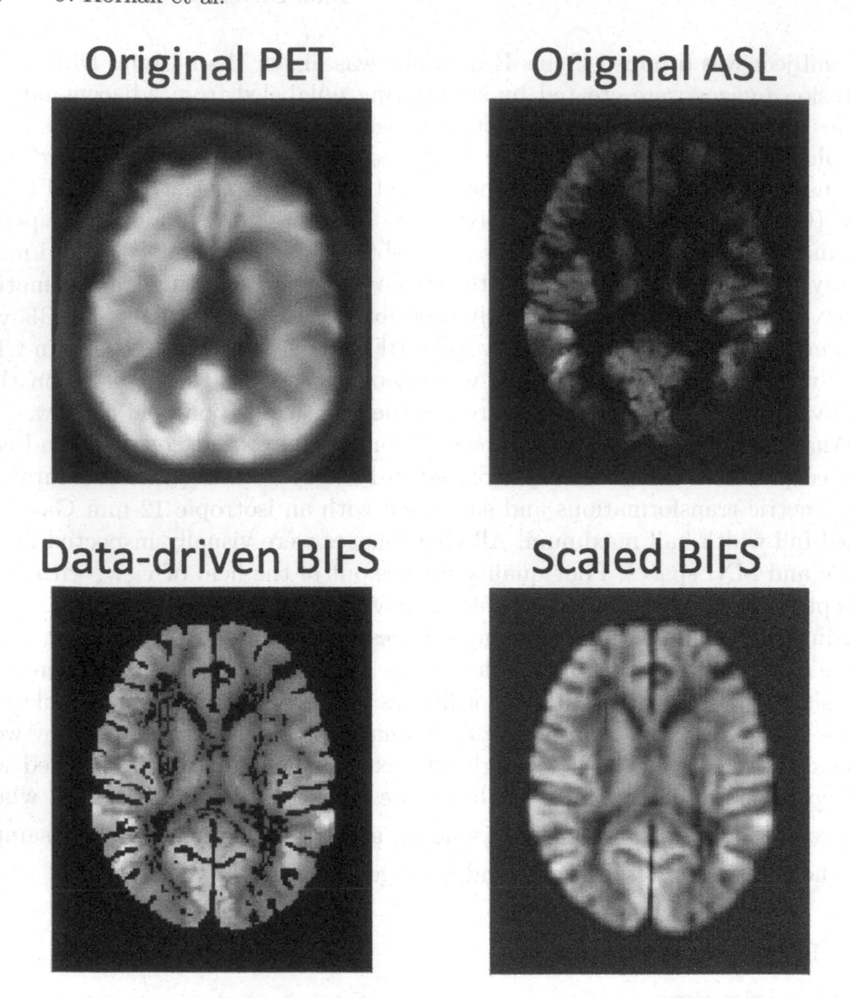

Fig. 3. Individual reconstruction of ASL data to resemble PET. Top-left: Original PET image; top-right: ASL for the same individual; bottom left: DD-BIFS prior reconstruction; bottom right: DD-BIFS reconstruction with additional shrinkage on the prior variance.

5 Conclusion

The DD-BIFS modeling framework provides a powerful new approach to using information available in large databases to improve reconstruction in individual images. In particular, the independence across Fourier space specification allows for fast and efficient computation which can be further improved with parallelization. Additionally, the ability to efficiently use empirical prior information from a database of images without the need for explicit modeling provides a powerful approach to improving image quality. Our preliminary application to the reconstruction of ASL perfusion MRI shows great promise and large-scale validation work is currently under way to determine its applicability in clinical practice.

Acknowledgements. Thanks To: Renaud La Joie and Amelia Strom for providing information on the PET dataset acquisitions.

References

1. Aguirre, G., Detre, J., Zarahn, E., Alsop, D.: Experimental design and the relative sensitivity of BOLD and perfusion fMRI. Neuroimage **15**(3), 488–500 (2002)
2. Ashburner, J., Friston, K.J.: Diffeomorphic registration using geodesic shooting and Gauss-Newton optimisation. NeuroImage **55**(3), 954–967 (2011)
3. Avants, B.B., Tustison, N., Song, G.: Advanced normalization tools (ANTS). Insight J. **2**(365), 1–35 (2009)
4. Besag, J.: Spatial interaction and the statistical analysis of lattice systems (with discussion). J. Roy. Stat. Soc. Ser. B (Methodol.) **36**(2), 192–236 (1974)
5. Besag, J.: Digital image processing: towards Bayesian image analysis. J. Appl. Stat. **16**(3), 395–407 (1989)
6. Buxton, R.B., Wong, E.C., Frank, L.R.: Dynamics of blood flow and oxygenation changes during brain activation: the balloon model. Magn. Reson. Med. **39**(6), 855–864 (1998)
7. Du, A.T., et al.: Hypoperfusion in frontotemporal dementia and Alzheimer disease by arterial spin labeling MRI. Neurology **67**(7), 1215–1220 (2006)
8. Gelman, A., Carlin, J.B., Stern, H.S., Dunson, D.B., Vehtari, A., Rubin, D.B.: Bayesian Data Analysis, vol. 2. CRC Press, Boca Raton (2014)
9. Geman, S., Geman, D.: Stochastic relaxation, Gibbs distributions, and the Bayesian restoration of images. IEEE Trans. Pattern Anal. Mach. Intell. **6**(6), 721–741 (1984)
10. Hallam, G.P., et al.: The neural correlates of emotion regulation by implementation intentions. PLoS ONE **10**(3), e0119500 (2015)
11. Hayasaka, S., et al.: A non-parametric approach for co-analysis of multi-modal brain imaging data: application to Alzheimer's disease. Neuroimage **30**(3), 768–779 (2006)
12. Hu, W.T., Wang, Z., Lee, V.M.Y., Trojanowski, J.Q., Detre, J.A., Grossman, M.: Distinct cerebral perfusion patterns in FTLD and AD. Neurology **75**(10), 881–888 (2010)
13. Jenkinson, M., Beckmann, C.F., Behrens, T.E., Woolrich, M.W., Smith, S.M.: FSL. Neuroimage **62**(2), 782–790 (2012)
14. Johnson, N., et al.: Pattern of cerebral hypoperfusion in Alzheimer disease and mild cognitive impairment measured with arterial spin-labeling mr imaging: initial experience. Radiology **234**(3), 851–859 (2005)
15. Kornak, J.: Bayesian image analysis in fourier space (BIFS). In: JSM Proceedings, Statistical in Imaging Section, Alexandria, VA, pp. 1487–1492. American Statistical Association (2014)
16. Luh, W., Wong, E., Bandettini, P., Hyde, J., et al.: QUIPSS II with thin-slice TI 1 periodic saturation: a method for improving accuracy of quantitative perfusion imaging using pulsed arterial spin labeling. Magn. Reson. Med. **41**(6), 1246–1254 (1999)

17. Mazziotta, J.C., Toga, A., Evans, A., Fox, P., Lancaster, J., Woods, R.: A probabilistic approach for mapping the human brain: the international consortium for brain mapping (ICBM). In: Toga, A.W., Mazziotta, J.C. (eds.) Brain Mapping: The Systems, pp. 141–156. Elsevier, Amsterdam (2000)
18. Peligrad, M., Utev, S.: Central limit theorem for stationary linear processes. Ann. Probab. **34**(4), 1608–1622 (2006)
19. Rice, S.O.: Mathematical analysis of random noise. Bell Syst. Tech. J. **24**(1), 46–156 (1945)
20. Zeger, S.L.: Exploring an ozone spatial time series in the frequency domain. J. Am. Stat. Assoc. **80**(390), 323–331 (1985)

Covariate-Adjusted Hybrid Principal Components Analysis

Aaron Wolfe Scheffler[1(✉)], Abigail Dickinson[2], Charlotte DiStefano[2], Shafali Jeste[2], and Damla Şentürk[3]

[1] Department of Epidemiology and Biostatistics, University of California, San Francisco, USA
aaron.scheffler@ucsf.edu

[2] Department of Psychiatry and Biobehavioral Sciences, University of California, Los Angeles, USA

[3] Department of Biostatistics, University of California, Los Angeles, USA

Abstract. Electroencephalography (EEG) studies produce region-referenced functional data in the form of EEG signals recorded across electrodes on the scalp. The high-dimensional data capture underlying neural dynamics and it is of clinical interest to model differences in neurodevelopmental trajectories between diagnostic groups, for example typically developing (TD) children and children with autism spectrum disorder (ASD). In such cases, valid group-level inference requires characterization of the complex EEG dependency structure as well as covariate-dependent heteroscedasticity, such as changes in variation over developmental age. In our motivating study, resting state EEG is collected on both TD and ASD children aged two to twelve years old. The peak alpha frequency (PAF), defined as the location of a prominent peak in the alpha frequency band of the spectral density, is an important biomarker linked to neurodevelopment and is known to shift from lower to higher frequencies as children age. To retain the most amount of information from the data, we model patterns of alpha spectral variation, rather than just the peak location, regionally across the scalp and chronologically across development for both the TD and ASD diagnostic groups. We propose a covariate-adjusted hybrid principal components analysis (CA-HPCA) for region-referenced functional EEG data, which utilizes both vector and functional principal components analysis while simultaneously adjusting for covariate-dependent heteroscedasticity. CA-HPCA assumes the covariance process is weakly separable conditional on observed covariates, allowing for covariate-adjustments to be made on the marginal covariances rather than the full covariance leading to stable and computationally efficient estimation. A mixed effects framework is proposed to estimate the model components coupled with a bootstrap test for group-level inference. The proposed methodology provides novel insights into neurodevelopmental differences between TD and ASD children.

Keywords: Electroencephalography · Autism spectrum disorder · Functional data analysis · Marginal covariances · Functional principal components analysis · Covariate-adjustments

M.-J. Lesot et al. (Eds.): IPMU 2020, CCIS 1239, pp. 391–404, 2020.
https://doi.org/10.1007/978-3-030-50153-2_30

1 Introduction

Despite the numerous developmental delays observed in children with autism spectrum disorder (ASD) compared to their typically developing peers (TD), the neural mechanisms underpinning these delays are not well characterized. To address this gap, our motivating study collected resting-state electroencephalograms (EEG) on TD and ASD children aged two to twelve years old, making it possible to contrast neural processes between the two diagnostic groups over a wide developmental range. EEG and magnetoencephalography (MEG) characterize cortical and intracortical brain activity, respectively, via the measurement of electrical potentials and their corresponding oscillatory dynamics (i.e. spectral characteristics). Recent studies in cognitive development using both EEG and MEG highlight the peak alpha frequency (PAF), defined as the location of a single prominent peak in the spectral density within the alpha frequency band (6–14 Hz), as a potential biomarker associated with autism diagnosis [7–9]. Specifically, the location of the PAF tends to shift from lower to higher frequencies as TD children age but this chronological shift is notably delayed or absent in ASD children [7,8,12,16]. This trend is observed in our motivating data from a temporal electrode (T8) where the PAF, identifiable as 'humps' in age-specific slices of the group-specific bivariate mean spectral density (across age and frequency), increases in frequency with age for TD children but not for ASD children (Fig. 1(a)).

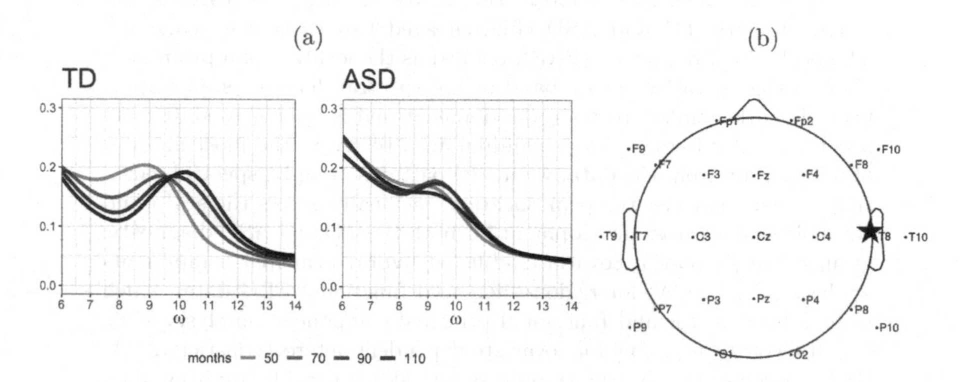

Fig. 1. (a) The group-specific bivariate mean alpha band spectral density (across age and frequency (6–14 Hz)) at ages 50, 70, 90 and 110 months from the T8 electrode. (b) A diagram of the 25 electrode montage used in our motivating data with the T8 electrode marked by a star.

Although the PAF holds promise as a biomarker for neural development in TD and ASD children, emphasis on the identification of a single peak produces considerable drawbacks. Estimation of a subject-electrode specific PAF can be error prone due to the presence of noise and multiple local maxima [6] and measurement of PAF inherently reduces the information from the alpha spectral band to a single scalar summary resulting in a loss of information. To avoid these limitations, we follow Scheffler et al. [15] and consider the entire spectral density across the alpha band as a functional measurement of neural activity. We focus on modeling and contrasting patterns of alpha spectral variation regionally across the scalp and chronologically across development for both the ASD and TD diagnostic groups. While Scheffler et al. [14] proposed a hybrid principal components analysis (HPCA) decomposition that models variation in region-referenced functional EEG data, it does not allow for the covariance structure to change across development as needed in our application. Previous research clearly shows that alpha spectral dynamics differ as a function of age between TD and ASD children and to assume a constant covariance structure across development risks missing important findings. To avoid this misspecification, we propose a covariate-adjusted hybrid principal components analysis (CA-HPCA) that models variation in high-dimensional functional data while simultaneously allowing the patterns of variation to change as a function of subject-specific covariates. CA-HPCA assumes the covariance process is weakly separable conditional on observed covariates, allowing for covariate-adjustments to be made on the marginal covariances rather than the full covariance leading to stable and computationally efficient estimation.

In the simplified context of one-dimensional functional data, existing methods allow for covariate-adjustments to the functional covariance in two ways: (1) both the eigenvalues and eigenfunctions of the functional covariance are allowed to change as a function of observed covariates or (2) the eigenfunctions are assumed to be constant across the covariate dimension but their corresponding eigenvalues (hence principal scores) are covariate-dependent. In the former class, Cardot [2] proposed a non-parametric covariate-adjusted functional principal components analysis (FPCA) in the context of dense functional data and Jiang and Wang [10] extended covariate-adjusted FPCA to noisy or sparse settings by estimating subject-specific scores using conditional expectation. In both cases, covariance estimation is performed non-parametrically by simultaneous smoothing across the covariate and functional domains via kernel methods. By fixing eigenfunctions across the covariate domain, Chio et al. [5] introduced a semi-parametric functional regression model that estimates covariate-dependent principal scores using a single-index model and Backenroth et al. [1] developed a heteroscedastic FPCA for repeatedly measured curves that models eigenvalues as an exponential function of covariate and subject-dependent effects.

Our proposed covariate-adjusted hybrid principal components analysis combines existing one-dimensional methods for covariate-dependent functional heteroscedasticity with recent advances in multi-dimensional FPCA to allow covariate-adjustments in the context of high-dimensional functional data. We briefly explore the methodological contributions of our proposed model and the resulting computational gains. A central theme in FPCA decompositions for multi-dimensional functional data is the use of simplifying assumptions regarding the covariance structure to ease estimation. A flexible approach in modeling two-dimensional functional data is to assume weak separability of the covariance process [4,11] in which the marginal covariances along each dimension are targeted and the full covariance is projected onto a tensor basis formed from the corresponding marginal eigenfunctions. Thus, estimation is reduced from that of the total covariance in four-dimensions to the marginal covariances in two-dimensions for which efficient two-dimensional smoothers exist. Scheffler et al. [14] extended weak separability to region-referenced functional EEG data (similar to our motivating study) by allowing a discrete regional dimension via HPCA. We leverage the simplifying assumptions and computational efficiency of HPCA through the proposed CA-HPCA which introduces covariate-dependence to the marginal covariances rather than the total covariance and allows the marginal eigenvalues and eigenfunctions to change across the covariate domain.

CA-HPCA provides a flexible modeling framework but introduces potential compute burden through the addition of a covariate dimension to estimation of the marginal covariances which for a scalar covariate requires smoothing across three dimensions. Previous methods such as Cardot [2] and Jiang and Wang [10] utilized kernel methods to estimate covariate-dependent marginal covariances but these approaches are computationally intensive and scale poorly with the introduction of additional covariates. To address this challenge, we extend the fast functional covariance smoothing proposed by Cederbaum et al. [3] to allow for covariate-adjustments by including an additional basis along the covariate dimension. Thus, CA-HPCA generalizes covariate-adjustments to high-dimensional functional covariances and substantially reduces the resulting computational burden by applying adjustments to the marginal covariances and introducing covariate-dependence to cutting-edge fast covariance smoothers. A mixed effects framework is proposed to estimate the model components and is paired with parametric bootstrap resampling to perform inference across the covariate domain. The remaining sections are organized as follows. Section 2 introduces the proposed CA-HPCA and Sect. 3 describes the corresponding estimation procedure. Application of the proposed method to our resting state EEG data follows in Sect. 4. Section 5 concludes with a brief summary and discussion.

2 Covariate-Adjusted Hybrid Principal Components Analysis (CA-HPCA)

Let $Y_{di}(a_i, r, \omega)$ be a random function observed in the presence of some continuous non-functional covariate $a_i \in \mathcal{A}$, for subject i, $i = 1, \ldots, n_d$, from group d, $d = 1, \ldots, D$, in region r, $r = 1, \ldots, R$, and at frequency ω, $\omega \in \Omega$. We decompose $Y_{di}(a_i, r, \omega)$ additively such that the expectation and covariance of the process depend on the covariate a_i,

$$Y_{di}(a_i, r, \omega) = \eta_d(a_i, r, \omega) + Z_{di}(a_i, r, \omega) + \epsilon_{di}(a_i, r, \omega),$$

where $\eta_d(a_i, r, \omega) = E\{Y_{di}(a_i, r, \omega)|a_i\}$ denotes the group-region mean function, $Z_{di}(a_i, r, \omega)$ denotes a mean zero region-referenced stochastic process with total variance $\boldsymbol{\Sigma}_{d,T}(a_i; r, \omega; r', \omega') = \text{cov}\{Z_{di}(a_i, r, \omega), Z_{di}(a_i, r', \omega')|a_i\}$, and

$\epsilon_{di}(a_i, r, \omega)$ denotes measurement error with mean zero and variance σ_d^2 that is independent across the regional and functional domains. We assume the group-region means $\eta_d(a, r, \omega)$ are smooth in both the functional domain Ω and the non-functional domain $\mathcal{A}$ though we place no restrictions across the regional domain $\mathbb{R}^R$.

In the proposed CA-HPCA model, we assume that the total covariance $\boldsymbol{\Sigma}_{d,T}(a; r, \omega; r', \omega')$ is weakly separable for each $a \in \mathcal{A}$. Weak separability, a concept recently proposed by Lynch and Chen [11] and adapted by Scheffler et al. [14] to region-referenced functional EEG data, implies that a covariance can be approximated by a weighted sum of separable covariance components and that the direction of variation (i.e. eigenvectors/eigenfunctions) along one dimensions of the EEG data is the same across fixed slices of the other dimension. Note that weak separability is more flexible than strong separability (i.e. separability) commonly utilized in spatiotemporal modeling which requires the total covariance, not just the directions of variation, is the same up to a constant for fixed slices of the other dimensions. Unlike previously applications of weak separability, we assume that the total covariance is weakly separable conditional on observed covariates and the marginal covariance functions vary smoothly along the covariate domain. Let the covariate-adjusted regional and functional marginal covariances be defined as

$$\{\boldsymbol{\Sigma}_{d,\mathcal{R}}(a_i)\}_{r,r'} = \int_{\Omega} \text{cov}\{Z_{di}(a_i, r, \omega), Z_{di}(a_i, r', \omega)\} d\omega = \sum_{k=1}^{R} \tau_{dk,\mathcal{R}}(a_i) \text{v}_{dk}(a_i, r) \text{v}_{dk}(a_i, r'),$$

$$\boldsymbol{\Sigma}_{d,\Omega}(a_i, \omega, \omega') = \sum_{r=1}^{R} \text{cov}\{Z_{di}(a_i, r, \omega), Z_{di}(a_i, r, \omega')\} = \sum_{\ell=1}^{\infty} \tau_{d\ell,\Omega}(a_i) \phi_{d\ell}(a_i, \omega) \phi_{d\ell}(a_i, \omega'),$$

where $\text{v}_{dk}(a, r)$ are covariate-adjusted marginal eigenvectors, $\phi_{d\ell}(a, \omega)$ are covariate-adjusted marginal eigenfunctions, and $\tau_{dk,\mathcal{R}}(a)$ and $\tau_{d\ell,\Omega}(a)$ are their respective covariate-adjusted marginal eigenvalues. Utilizing the covariate-dependent eigenvectors and eigenfunctions, the covariate-adjusted hybrid principal components decomposition (**CA-HPCA**) of $Y_{di}(a_i, r, \omega)$ is given as,

$$\begin{aligned} Y_{di}(a_i, r, \omega) &= \eta_d(a_i, r, \omega) + Z_{di}(a_i, r, \omega) + \epsilon_{di}(a_i, r, \omega) \\ &= \eta_d(a_i, r, \omega) + \sum_{k=1}^{R} \sum_{\ell=1}^{\infty} \xi_{di,k\ell}(a_i) \text{v}_{dk}(a_i, r) \phi_{d\ell}(a_i, \omega) + \epsilon_{di}(a_i, r, \omega), \end{aligned}$$

where $\xi_{di,k\ell}(a_i)$ are subject-specific scores defined through the projection $\langle Z_{di}(a_i, r, \omega),\, \text{v}_{dk}(a_i, r)\phi_{d\ell}(a_i, \omega)\rangle = \sum_{r=1}^{R} \int Z_{di}(a_i, r, \omega)\ \text{v}_{dk}(a_i, r)\phi_{d\ell}(a_i, \omega) d\omega$.

Weak separability of the total covariance at each covariate value implies that the scores $\xi_{di,k\ell}(a_i)$ are uncorrelated leading to the decomposition of the total covariance $\boldsymbol{\Sigma}_{d,T}(a;r,\omega;r',\omega')$ as follows,

$$\begin{aligned}\boldsymbol{\Sigma}_{d,T}(a;r,\omega;r',\omega') &= \text{cov}\{Z_{di}(a,r,\omega), Z_{di}(a,r',\omega')|a\} + \sigma_d^2\delta(a;r,\omega;r',\omega')\\ &= \sum_{k=1}^{R}\sum_{\ell=1}^{\infty}\tau_{d,k\ell}(a)\mathrm{v}_{dk}(a,r)\mathrm{v}_{dk}(a,r')\phi_{d\ell}(a,\omega)\phi_{d\ell}(a,\omega') + \sigma_d^2\delta(a;r,\omega;r',\omega'),\end{aligned}$$

where $\tau_{d,k\ell}(a) = \text{var}\{\xi_{di,k\ell}(a)\}$. Note, the above model assumes that both the marginal directions of variation and their associated tensor weights are allowed to vary across the covariate domain. In practice, the CA-HPCA decomposition is truncated to include only K_d and L_d covariate-adjusted marginal eigencomponents for the regional and functional domains, respectively, with the number of components initially selected on the marginal fraction of variance explained (FVE). One guideline is to include the minimum number of covariate-adjusted marginal eigencomponents in the CA-HPCA expansion that explain at least 90% of variation in their respective covariate-adjusted marginal covariances. The final number of components can be fixed after the subject-specific scores and their associated variance components are estimated in Sect. 3 which allow enumeration of the overall FVE in the observed data not just the marginal covariances.

3 Estimation of Model Components and Inference

The following section outlines the CA-HPCA estimation procedure, provides detailed descriptions of each step, and outlines how to perform inference via parametric bootstrap.

(1) Estimation of group-region mean functions: We estimate the group-region mean function $\eta_d(a,r,\omega)$ for each region via smoothing performed by projection onto a tensor basis formed by penalized marginal B-splines in the regional and functional domains. Smoothing parameter selection and variance components are estimated using restricted maximum likelihood (REML) methods.

(2) Estimation of covariate-adjusted marginal covariances and measurement error variance: We estimate the covariate-adjusted marginal covariances by assuming each two-dimensional marginal covariance varies smoothly over the covariate dimension. For the functional marginal covariance, $\boldsymbol{\Sigma}_{d,\Omega}(a,\omega,\omega')$, we extend the fast bivariate covariance smoother of Cederbaum et al. [3] to include a third covariate dimension $a \in \mathcal{A}$. The resulting trivariate smoother maintains the computational efficiency of Cederbaum et al. [3] while simultaneously allowing the marginal functional covariance to vary smoothly along the covariate dimension. As an added bonus, we also obtain an initial estimate of the measurement error variance $\hat{\sigma}^2_{d,\Omega}$.

Algorithm: *CA-HPCA Estimation Procedure*

---

1. Estimation of group-region mean functions
   (a) Calculate $\hat{\eta}_d(a_i, r, \omega)$ by applying a bivariate penalized spline smoother to all observed data $\{a_i, \omega, Y_{di}(a_i, r, \omega) : i = 1, \ldots, n_d; a_i \in \mathcal{A}; \omega \in \Omega\}$.
   (b) Mean center each observation, $\widehat{Y}^c_{di}(a_i, r, \omega) = Y_{di}(a_i, r, \omega) - \hat{\eta}_d(a_i, r, \omega)$.
2. Estimation of covariate-adjusted marginal covariances and measurement error variance
   (a) Calculate $\widehat{\boldsymbol{\Sigma}}_{d,\Omega}(a, \omega, \omega')$ and $\hat{\sigma}^2_{d,\Omega}$ by applying trivariate penalized spline smoothers to the products, $\{a_i, \omega, \omega', \widehat{Y}^c_{di}(a_i, r, \omega)\widehat{Y}^c_{di}(a_i, r, \omega')$:b $i = 1, \ldots, n_d; a_i \in \mathcal{A}; \omega, \omega' \in \Omega\}$.
   (b) Calculate $\widehat{\boldsymbol{\Sigma}}_{d,\mathcal{R}}(a)$ by smoothing each (r, r') entry across $\mathcal{A}$. For $r \neq r'$, estimate $\{\widehat{\boldsymbol{\Sigma}}_{d,\mathcal{R}}(a)\}_{(r,r')}$ by applying a univariate kernel smoother to $\{a_i, r, r', \widehat{Y}^c_{di}(a_i, r, \omega)\widehat{Y}^c_{di}(a_i, r', \omega) : i = 1, \ldots, n_d; a_i \in \mathcal{A}\}$. For $r = r'$, estimate $\{\widehat{\boldsymbol{\Sigma}}_{d,\mathcal{R}}(a)\}_{(r,r)}$ by applying a univariate penalized smoother to $\{a_i, r, r, \widehat{Y}^c_{di}(a_i, r, \omega)\widehat{Y}^c_{di}(a_i, r, \omega) - \hat{\sigma}^2_{d,\Omega} : i = 1, \ldots, n_d; a_i \in \mathcal{A}\}$.
3. Estimation of covariate-adjusted marginal eigencomponents
   (a) For each unique value of a observed, employ FPCA on $\widehat{\boldsymbol{\Sigma}}_{d,\Omega}(a, \omega, \omega')$ to estimate the eigenvalue, eigenfunction pairs, $\{\tau_{d\ell,\Omega}(a), \phi_{d\ell}(a, \omega) : \ell = 1, \ldots, L_d\}$.
   (b) For each unique value of a observed, employ PCA on $\widehat{\boldsymbol{\Sigma}}_{d,\mathcal{R}}(a)$ and to estimate the eigenvalue, eigenvector pairs $\{\tau_{dk,\mathcal{R}}(a), \mathrm{v}_{dk}(a, r) : k = 1, \ldots K_d\}$.
4. Estimation of covariate-adjusted variance components and subject-specific scores via linear mixed effects models
   (a) Calculate $\hat{\tau}_{dg}(a_i) = \mathrm{cov}\{\hat{\zeta}_{dig}(a_i)\}$ and $\hat{\sigma}^2_d$ by fitting the proposed linear mixed effects model.
   (b) Select G'_d such that $FVE_{dG'} > .8$ for $d = 1, \ldots, D$ and form predictions $\widehat{Y}_{di}(a_i, r, \omega)$.
   (c) Calculate $\hat{\zeta}_{dig}\{a_i$ as the BLUP $\hat{\zeta}_{dig}(a_i) = E\{\zeta_{dig}(a_i)|\boldsymbol{Y}_{di}\}$.
5. Inference via parametric bootstrap.

For fixed slices of the covariate domain, the regional marginal covariance $\{\boldsymbol{\Sigma}_{d,\mathcal{R}}(a)\}_{r,r'}$ is discrete and thus not amenable to trivariate smoothers as the functional marginal covariance above. Therefore, we estimate the raw regional marginal covariance at each covariate-value, remove the measurement variance from the diagonals as in Scheffler et al. [14] and then apply a Nadarya-Watson kernel smoother to the resulting matrices entry-by-entry along the covariate domain. Our kernel smoother is the kernel weighted-average, $\{\widehat{\boldsymbol{\Sigma}}_{d,\mathcal{R}}(a_o)\}_{(r,r')} = \sum_{i=1}^{n_d} \sum_{\omega\in\Omega} K_\lambda(a_o, a_i)\widehat{Y}^c_{di}(a_i, r, \omega)\widehat{Y}^c_{di}(a_i, r', \omega)/|\Omega| \sum_{i=1}^{n_d} K_\lambda(a_o, a_i)$, where $K_\lambda(\cdot\,, \cdot\,)$ is some kernel with smoothness parameter λ and $|\Omega|$ is the number of observed functional grid points. The smoothing parameter λ is selected to minimize the LOSOCV(λ) statistic across all channel pairs (r, r'), $\text{LOSOCV}(\lambda) = \sum_{r=1}^{R} \sum_{r'<r} \text{LOSOCV}(\lambda, r, r')$, $\text{LOSOCV}(\lambda, r, r') = (1/|\Omega| n_d) \sum_{i=1}^{n_d} \sum_{\omega \in \Omega} [\widehat{Y}^c_{di}(a_i, r, \omega) \widehat{Y}^c_{di}(a_i, r', \omega) - \{\widehat{\boldsymbol{\Sigma}}^{(-i)}_{d,\mathcal{R}}(a_i)\}_{(r,r')}]^2$, where $\{\widehat{\boldsymbol{\Sigma}}^{(-i)}_{d,\mathcal{R}}(a_i)\}$ is the smoothed marginal covariance matrix with the ith subject left out. Thus, we introduce two novel covariate-adjusted smoothers that allow for calculation of the covariate-adjusted marginal covariances which are then used for subsequent covariate-adjusted eigendecompositions.

(3) Estimation of covariate-adjusted marginal eigencomponents: To estimate the covariate-adjusted marginal eigencomponents we perform eigendecompositions at each fixed covariate-value as described in Scheffler et al. [14] retaining a common number of K_d and L_d covariate-adjusted eigencomponents.

(4) Estimation of covariate-adjusted variance components and subject-specific scores via linear mixed effects models: We make use of the estimated functional fixed effects and marginal eigencomponents to propose a linear mixed effects framework for modeling covariate-adjusted region-referenced functional data. Under the assumption of joint normality of the covariate-adjusted subject-specific scores and measurement error, the proposed mixed effects framework induces regularization and stability in modeling the data by enforcing a low-rank structure on the covariate-adjusted variance components $\tau_{dg}(a)$. The resulting variance components can be used to select the number of eigencomponents to include in the CA-HPCA decomposition by quantifying the proportion of variance explained and for hypothesis testing and point-wise confidence bands via parametric bootstrap. We present the linear mixed effects modeling framework below.

To make the notation more compact, we replace the double index $k\ell$ in CA-HPCA truncated at K_d and L_d with a single index $g = (k-1) + K_d(\ell - 1) + 1$,

$$Y_{di}(a_i, r, \omega) = \eta_d(a_i, r, \omega) + \sum_{g=1}^{G_d} \xi_{di,g}(a_i) \varphi_{dg}(a_i, r, \omega) + \epsilon_{di}(a_i, r, \omega),$$

where $\varphi_{dg}(a_i, r, \omega) = \text{v}_{dk}(a_i, r)\phi_{d\ell}(a_i, \omega)$, $\zeta_{dig}(a_i) = \langle Z_{di}(a_i, r, \omega), \varphi_{dg}(a_i, r, \omega)\rangle$, $\tau_{dg} = \text{cov}\{\zeta_{dig}(a_i)\}$, and $G_d = K_d L_d$. Let $\mathbf{Y}_{di}(a_i)$ represent the vectorized form of $Y_{di}(a_i, r, \omega)$ for subject i, $i = 1, \ldots, n_d$, observed along with covariate value a_i. Note, an argument for the covariate domain a is included to stress that a subject's covariance is covariate-dependent. Analogous vectorized forms for the covariate-adjusted functional fixed effects, $\eta_d(a_i, r, \omega)$, the region-referenced stochastic process $Z_{di}(a_i, r, \omega)$, multidimensional orthonormal basis $\varphi_{dg}(a_i, r, \omega)$, and the measurement error $\epsilon_{di}(a_i, r, \omega)$ are denoted by $\boldsymbol{\eta}_{di}(a_i)$, $\boldsymbol{Z}_{di}(a_i)$, $\boldsymbol{\varphi}_{dg}(a_i)$, and $\boldsymbol{\epsilon}_{di}$, respectively. Under the assumption that $\boldsymbol{\zeta}_{di}(a_i) = \{\zeta_{di1}(a_i), \ldots, \zeta_{diG_d}(a_i)\}$ and $\boldsymbol{\epsilon}_{di}$ are jointly Gaussian and $\text{cov}\{\boldsymbol{\zeta}_{di}(a_i), \boldsymbol{\epsilon}_{di}\} = \mathbf{0}$ at a fixed value of a_i, the proposed linear mixed effects model is given as

$$\begin{aligned} \boldsymbol{Y}_{di}(a_i) &= \boldsymbol{\eta}_{di}(a_i) + \boldsymbol{Z}_{di}(a_i) + \boldsymbol{\epsilon}_{di} \\ &= \boldsymbol{\eta}_{di}(a_i) + \sum_{g=1}^{G_d} \zeta_{dig}(a_i)\boldsymbol{\varphi}_{dg}(a_i) + \boldsymbol{\epsilon}_{di}, \quad \text{for} \quad i = 1, \ldots, n_d. \end{aligned} \tag{1}$$

The model can be fit separately in each group, $d = 1, \ldots, D$ and the regional and functional dependencies in $\boldsymbol{Y}_{di}(a_i)$ are induced through the subject-specific random effects $\zeta_{dig}(a_i)$ in (1). Given the assumption that the total covariance is weakly separable for fixed values of a, $\text{cov}\{\zeta_{dig}(a_i), \zeta_{dig\prime}(a_i)\} = 0$ for $g \neq g\prime$ and thus the covariance of the subject-specific scores possess a diagonal diagonal structure, $\text{cov}\{\boldsymbol{\zeta}_{di}(a_i)\} = \boldsymbol{T}_d(a_i) = \text{diag}\{\boldsymbol{\tau}_d(a_i)\}$, where $\boldsymbol{\tau}_d(a_i) = \{\tau_{d1}(a_i), \ldots, \tau_{dG}(a_i)\}$. We further assume that $\boldsymbol{T}_d(a)$ evolves smoothly along the covariate domain and target the smooth variance components through their corresponding precision components. Given previous estimates for $\boldsymbol{\eta}_{di}(a_i)$ and $\boldsymbol{\varphi}_{dg}(a_i)$, estimates of the variance components and subject-specific scores are obtained using REML methods [17].

The assumption that the variance components evolve smoothly over the covariate domain resolves several challenges that emerge when modeling the covariate-adjusted subject-specific scores. First, the estimation procedure is able to borrow strength across the covariate-domain when modeling variation, a necessity when specific covariate values may only be observed once as in our motivating data. Second, we are able to project the precision components onto a smooth low-rank basis which induces regularization and control over the speed at which $\tau_g(a)$ is allowed to vary. Alternatively, a projection based approach is less computationally burdensome with estimates of the subject-specific scores obtained directly by numerical integration, $\hat{\zeta}_{dig}(a_i) = \langle Z_{di}(a_i, r, \omega), \hat{\varphi}_{dg}(a_i, r, \omega)\rangle$ and their corresponding variance components calculated empirically, $\hat{\tau}_{dg} = \text{cov}\{\hat{\zeta_{dig}}(a)\}$, but the resulting estimates are unstable due to the limited number of observations at each point along the covariate domain. Therefore, despite the added compute time, our proposed linear mixed effects framework is better suited for providing covariate-adjustments to the region-referenced functional process in a controlled and principled manner.

The estimated variance components are used to choose the number of eigencomponents included in the CA-HPCA decomposition where G'_d denotes a set of eigencomponents such that the total fraction of variance explained ($FVE_{dG'_d}$) is greater than 0.8 in each group $d = 1, \ldots, D$. We recommend starting with a larger number $G_d = K_d L_d$ of eigencomponents in the mixed effects modeling used for the estimation of $\{\tau_{dg}(a_i) : g = 1, \ldots, G_d\}$ and then reducing or adding components as appropriate to fix the final value of G'_d. In order to estimate the group-specific fraction of total variance explained via the G'_d eigencomponents, we consider the quantity, $FVE_{dG'_d} = \int\{\sum_{i=1}^{n_d} \sum_{g=1}^{G'_d} \hat{\tau}_{dg}(a_i)\}da / \int[\sum_{i=1}^{n_d}\{||Y_{di}(a_i, r, \omega) - \hat{\eta}_d(a_i, r, \omega)|| - R\int \hat{\sigma}_d^2 da\}]da$, where $||f(a_i, r, \omega)||^2 = \sum_{r=1}^{R} \int f(a_i, r, \omega)^2 d\omega$. Once G'_d is selected, the subject-specific scores can be obtained using their best linear unbiased predictor (BLUP) as in Scheffler et al. [14].

(5) Inference via parametric bootstrap: Inference in the form of hypothesis testing and point-wise confidence intervals can be performed via a parametric bootstrap based on the estimated CA-HPCA model components. To test the null hypothesis that all groups have equal means in the region r for a fixed covariate value $a \in \mathcal{A}$, i.e. $H_0 : \eta_d(a,r,\omega) = \eta(a,r,\omega)$ for $d = 1, \dots D$, we propose a parametric bootstrap procedure based on the CA-HPCA decomposition. For $b = 1, \dots, B$, the proposed parametric bootstrap generates outcomes based on the estimated model components under the null hypothesis as $Y^b_{di}(a_i, r, \omega) = \hat{\eta}(a_i, r, \omega) + Z^b_{di}(a_i, r, \omega) + \epsilon^b_{di}(a_i, r, \omega) = \hat{\eta}(a_i, r, \omega) + \sum_{g=1}^{G'_d} \zeta^b_{dig}(a_i) \hat{\varphi}_{dg}(a_i, r, \omega) + \epsilon^b_{di}(a_i, r, \omega)$ in region r and as $Y^b_{di}(a_i, r, \omega) = \hat{\eta}_d(a_i, r, \omega) + Z^b_{di}(a_i, r, \omega) + \epsilon^b_{di}(a_i, r, \omega) = \hat{\eta}_d(a_i, r, \omega) + \sum_{g=1}^{G'_d} \zeta^b_{dig} \hat{\varphi}_{dg}(a_i, r, \omega) + \epsilon^b_{di}(a_i, r, \omega)$ in the other regions, where subject-specific scores and measurement error are sampled from $\zeta^b_{dig}(a_i) \sim \mathcal{N}\{0, \hat{\tau}_{dg}(a_i)\}$ and $\epsilon^b_{di}(a_i, r, \omega) \sim \mathcal{N}(0, \hat{\sigma}^2_d)$. The proposed test statistic $T_r(a) = [\sum_{d=1}^{D} \int \{\hat{\eta}_d(a, \omega, s) - \hat{\eta}(a, r, \omega)\}^2 d\omega]^{1/2}$ is based on the norm of the sum of the departures of the estimated group-region shifts $\hat{\eta}_d(a, r, \omega)$ from the estimate of the common shift across groups, $\hat{\eta}(a, r, \omega)$. The common region shift estimate $\hat{\eta}(a, r, \omega)$, under the null, is set to the point-wise average of the group-region shift estimates, $\hat{\eta}_d(a, r, \omega)$, $d = 1, \dots, D$. We utilize the proposed parametric bootstrap to estimate the distribution of the test statistic $T_r(a)$ which can be used to evaluate the null hypothesis along the covariate domain.

To generate point-wise confidence intervals for estimates of $\hat{\eta}_d(a, r, \omega)$, we repeat the above parametric bootstrap procedure but instead generate outcomes from the model $Y^b_{di}(a_i, r, \omega) = \hat{\eta}_d(a_i, r, \omega) + \sum_{g=1}^{G'_d} \zeta^b_{dig} \hat{\varphi}_{dg}(a_i, r, \omega) + \epsilon^b_{di}(a_i, r, \omega)$. At each iteration of the bootstrap, estimate $\hat{\eta}^b_d(a, r, \omega)$ from the simulated data and then form point-wise confidence intervals based on percentiles of the estimated group-region mean functions as a function of a, r and ω across iterations, $\{\hat{\eta}^b_{dg}(a, r, \omega) : b = 1, \dots, B\}$.

4 Application to the Task-Free Paradigm Data

Data structure: In our motivating data application, EEG signals were sampled at 500 Hz for two minutes from a 128-channel HydroCel Geodesic Sensor Net on 58 ASD and 39 TD children aged 25 to 146 months old (diagnostic groups were age matched). EEG recordings were collected during an 'eyes-open' paradigm in which bubbles were displayed on a screen in a sound-attenuated room to subjects at rest [7]. We describe the dataset in our previous work and present an abbreviated description here, though the reader may reference Scheffler et al. [15] for technical details related to pre-processing and data acquisition. EEG data for each subject is interpolated down to a standard 10–20 system 25 electrode montage (R = 25) using spherical interpolation as detailed in Perrin et al. [13], producing 25 electrodes with continuous EEG signal (Fig. 1 (b)). Alpha spectral density (Ω = (6 Hz, 14 Hz)) estimates for each electrode were obtained and the resulting electrode-specific alpha spectral estimates form an instance of region-referenced functional data.

Data analysis results: We present the results from our application of CA-HPCA to the EEG data collected under the 'eyes-open' paradigm. While the main focus of our analysis is to characterize differences in alpha spectral dynamics between TD and ASD children over the course of development via inference on the group-region mean functions, we begin by briefly discussing the eigencomponents produced by the decomposition. The leading four; four and four; six covariate-adjusted regional and functional marginal eigencomponents are collectively found to explain 1.006 and 0.895 of the total FVE ($FVE_{dG'_d}$) in the TD and ASD groups, respectively. In the functional dimension along the covariate domain, the first least leading covariate-adjusted marginal eigenfunctions $\phi_{d1}(a, \omega)$ (Fig. 2(a), top row) display maximal variation at approximately 6 and 10 Hz (in opposing directions), where the location of maximal variation shifts in TD children from higher to lower frequencies as age increases but remains relatively constant in ASD children. The second leading covariate-adjusted marginal eigenfunctions $\phi_{d2}(a, \omega)$ (Fig. 2(a), bottom row) show maximal variation at 6 Hz, 8.5 Hz, 10.5 Hz and 6 Hz, 7.5 Hz in the TD and ASD groups, respectively, where again peak variation moves from higher to lower frequencies as age increases in the TD group but not the ASD group which instead displays shifts in the magnitude of maximal variation across development. The first two leading covariate-adjusted marginal eigenfunctions together explain at least 65% of the variation in the covariate-adjusted functional marginal covariances. In the regional dimension along the covariate domain, the first leading covariate-adjusted marginal eigenvectors $\mathrm{v}_{d1}(a, r)$ (Fig. 2(b), top row) display maximal variation in the central; right temporal; left posterior and central; middle posterior electrodes at younger ages with a shift to right posterior and frontal; right temporal electrodes at older ages in the TD and ASD groups, respectively. The second leading covariate-adjusted marginal eigenvector $\mathrm{v}_{d2}(a, r)$ (Fig. 2(b), bottom row) shows maximal variation in the frontal and right frontal; right temporal electrodes at younger ages which moves to frontal; right posterior (opposing directions) and central; left posterior (opposing directions) at older ages in the TD and ASD groups, respectively. The first two covariate-adjusted marginal eigenvectors together explain at least 70% of the variation in the covariate-adjusted regional marginal covariances.

To test for differences between TD and ASD groups in the alpha spectrum over development, we utilize the parametric bootstrap procedure described in Sect. 3 under the null hypothesis that the TD and ASD group-region mean functions are equal for every electrode r at each age $a = 25, \ldots, 145$ months which takes the form $H_0 : \eta_d(r, \omega, a) = \eta(r, \omega, a)$, $d = 1, 2$. Figure 3(a) displays the results of the hypothesis tests for all electrodes and ages with p-values transformed to the $-\log_{10}$ scale to better stratify results where values greater than $-\log_{10}(0.05) = 1.30$ denote significance at level $\alpha = .05$. Nearly all electrodes show significant differences between diagnostic groups in the alpha spectrum at

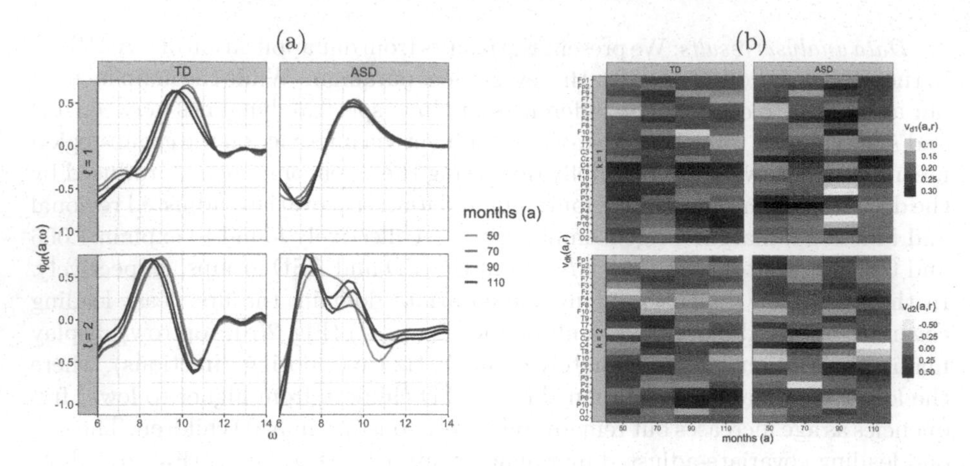

Fig. 2. (a) Estimated first and second leading covariate-adjusted eigenfunctions $\phi_{d1}(a,\omega)$ and $\phi_{d2}(a,\omega)$ at $a = 50, 70, 90, 110$ months. (b) Estimated first and second leading covariate-adjusted eigenvectors $\mathrm{v}_{d1}(a,r)$ and $\mathrm{v}_{d2}(a,r)$ at $a = 50, 70, 90, 110$ months.

some point over development (with the exception of the P3 and P7 electrodes) with the strongest group differences occurring at younger ages (30–50 months) and older ages (100–130 months) in the frontal, central, temporal, and posterior regions.

The greatest differences in the group-region mean functions across development are observed in the T8 and T10 electrodes displayed in Fig. 3(b) along with their 95% point-wise confidence intervals generated as described in Sect. 3. At both electrodes, the TD group displays a well-defined peak in the alpha spectrum that shifts from 8 Hz–10 Hz moving from 50–110 months, whereas the ASD group generally has less clearly-defined peaks that tend to center around 9 Hz throughout development. Differences in the estimated group-region mean functions mirror the results found from the parametric bootstrap procedure with separation of the point-wise confidence intervals occurring at 50; 90; 110 months and 110 months for the T8 and T10 electrodes, respectively. When aggregated, the observations and inferences obtained from the CA-HPCA model components provide evidence for differences in both the mean structure and patterns of covariation between the two diagnostic groups that shift and change over development highlighting the need to provide covariate-adjustments in modeling the high-dimensional EEG data.

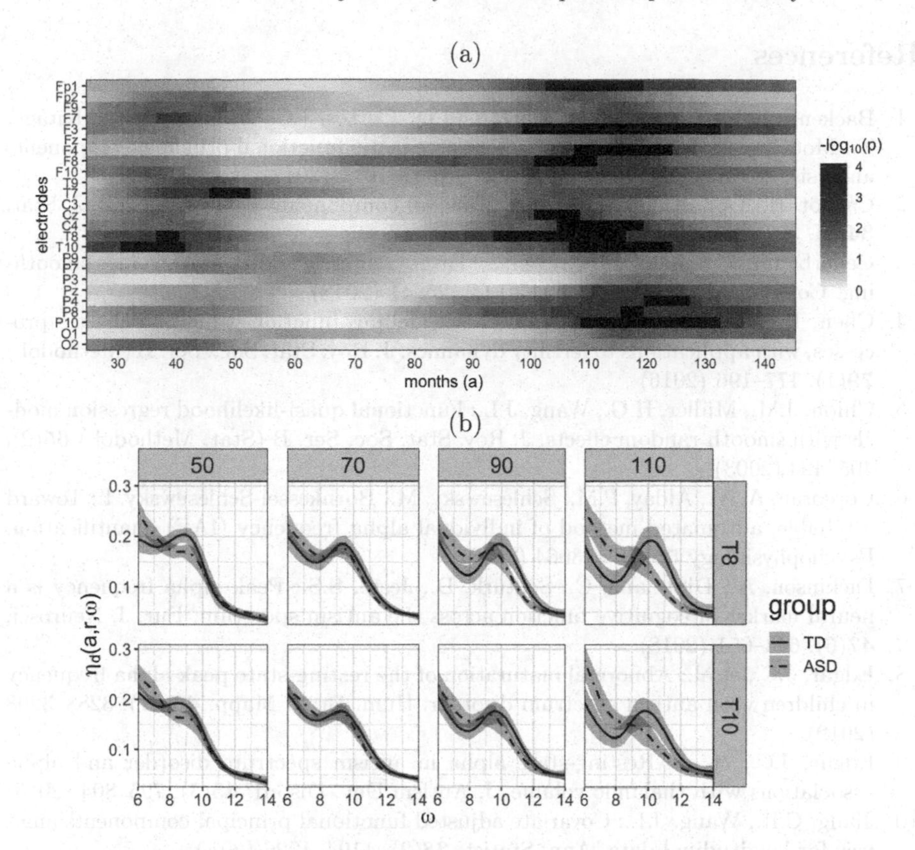

Fig. 3. (a) The $-\log_{10}$ transformed p-values from the hypothesis test for each electrode from the parametric bootstrap test for $a = 25, \ldots, 145$ months. (b) The estimated group-region mean functions $\eta_d(a, r, \omega)$ at ages $a = 50, 70, 90, 110$ months from the T8 and T10 electrodes. Grey shading denotes 95% point-wise confidence intervals for each estimate.

5 Discussion

We proposed a covariate-adjusted hybrid principal components analysis (CA-HPCA) which decomposes region-referenced functional data and accounts for covariate-dependent heteroscedasticity by assuming the high-dimensional covariance structure is weakly separable conditional on observed covariates. The proposed estimation procedure develops computationally efficient fast-covariance smoothers that incorporate covariate-dependence when estimating marginal covariances as well as a mixed effects framework which admits inference along the covariate-domain via bootstrap sampling. The CA-HPCA decomposition was developed to model EEG data over a broad developmental range but may be applied to other settings where high-dimensional data is expected to exhibit differential covariation as a function heterogenous covariates.

References

1. Backenroth, D., Goldsmith, J., Harran, M.D., Cortes, J.C., Krakauer, J.W., Kitago, T.: Modeling motor learning using heteroscedastic functional principal components analysis. J. Am. Stat. Assoc. **113**(523), 1003–1015 (2018)
2. Cardot, H.: Conditional functional principal components analysis. Scand. J. Stat. **34**(2), 317–335 (2007)
3. Cederbaum, J., Scheipl, F., Greven, S.: Fast symmetric additive covariance smoothing. Comput. Stat. Data Anal. **120**(C), 25–41 (2018)
4. Chen, K., Delicado, P., Müller, H.G.: Modelling function-valued stochastic processes, with applications to fertility dynamics. J. Roy. Stat. Soc.: Ser. B (Methodol.) **79**(1), 177–196 (2016)
5. Chiou, J.M., Müller, H.G., Wang, J.L.: Functional quasi-likelihood regression models with smooth random effects. J. Roy. Stat. Soc. Ser. B (Stat. Methodol.) **65**(2), 405–423 (2003)
6. Corcoran, A.W., Alday, P.M., Schlesewsky, M., Bornkessel-Schlesewsky, I.: Toward a reliable, automated method of individual alpha frequency (IAF) quantification. Psychophysiology **55**(7), e13064 (2018)
7. Dickinson, A., DiStefano, C., Senturk, D., Jeste, S.S.: Peak alpha frequency is a neural marker of cognitive function across the autism spectrum. Eur. J. Neurosci. **47**(6), 643–651 (2018)
8. Edgar, J.C., et al.: Abnormal maturation of the resting-state peak alpha frequency in children with autism spectrum disorder. Hum. Brain Mapp. **40**(11), 3288–3298 (2019)
9. Edgar, J.C., et al.: Resting-state alpha in autism spectrum disorder and alpha associations with thalamic volume. J. Autism Dev. Disord. **45**(3), 795–804 (2015)
10. Jiang, C.R., Wang, J.L.: Covariate adjusted functional principal components analysis for longitudinal data. Ann. Statist. **38**(2), 1194–1226 (2010)
11. Lynch, B., Chen, K.: A test of weak separability for multi-way functional data, with application to brain connectivity studies. Biometrika **105**(4), 815–831 (2018)
12. Miskovic, V., et al.: Developmental changes in spontaneous electrocortical activity and network organization from early to late childhood. NeuroImage **118**(Supplement C), 237–247 (2015)
13. Perrin, F., Pernier, J., Bertrand, O., Echallier, J.: Spherical splines for scalp potential and current density mapping. Electroencephalogr. Clin. Neurophysiol. **72**(2), 184–187 (1989)
14. Scheffler, A.W., et al.: Hybrid principal components analysis for region-referenced longitudinal functional EEG data. Biostatistics **21**(1), 139–157 (2018)
15. Scheffler, A.W., et al.: Covariate-adjusted region-referenced generalized functional linear model for EEG data. Stat. Med. **38**(30), 5587–5602 (2019)
16. Valdas-Hernandez, P., et al.: White matter architecture rather than cortical surface area correlates with the EEG alpha rhythm. NeuroImage **49**(3), 2328–2339 (2010)
17. Wood, S.: Generalized Additive Models: An Introduction with R. Chapman and Hall/CRC, London (2017)

Interval Uncertainty

On Statistics, Probability, and Entropy of Interval-Valued Datasets

Chenyi Hu[1(✉)] and Zhihui H. Hu[2]

[1] University of Central Arkansas, Conway, AR, USA
chu@uca.edu
[2] Edge Pursuit LLC, San Francisco, CA, USA

Abstract. Applying interval-valued data and methods, researchers have made solid accomplishments in information processing and uncertainty management. Although interval-valued statistics and probability are available for interval-valued data, current inferential decision making schemes rely on point-valued statistic and probabilistic measures mostly. To enable direct applications of these point-valued schemes on interval-valued datasets, we present point-valued variational statistics, probability, and entropy for interval-valued datasets. Related algorithms are reported with illustrative examples.

Keywords: Interval-valued dataset · Point-valued variational statistics · Probability · Information entropy

1 Introduction

1.1 Why Do We Study Interval-Valued Datasets?

Statistic and probabilistic measures play a very important role in processing data and managing uncertainty. In the literature, these measures are mostly point-valued and applied to point-valued dataset. While a point-valued datum intends to record a snapshot of an event instantaneously in theory, it is often imprecise in real world due to system and random errors. Applying interval-valued data to encapsulate variations and uncertainty, researchers have developed interval methods for knowledge processing. With data aggregation strategies [1,5,21], and others, we are able to reduce large size point-valued data into smaller interval-valued ones for efficient data management and processing. By doing so, researchers are able to focus more on qualitative properties and ignore insignificant quantitative differences.

Studying interval-valued data, Gioia and Lauro developed interval-valued statistics [4] in 2005. Lodwick and Jamison discussed interval-valued probability [17] in the analysis of problems containing a mixture of possibilistic, probabilistic, and interval uncertainty in 2008. Billard and Diday reported regression analysis of interval-valued data in [2]. Huynh *et al.* established a justification on decision making under interval uncertainty [13]. Works on applications of interval-valued data in knowledge processing include [3,8,16,19,20,22], and

M.-J. Lesot et al. (Eds.): IPMU 2020, CCIS 1239, pp. 407–421, 2020.
https://doi.org/10.1007/978-3-030-50153-2_31

many more. Applying interval-valued data in the stock market forecasting, Hu and He initially reported an astonishing quality improvements in [9]. Specifically, comparing against the commonly used point-valued confidence interval predictions, the interval approaches have increased the average accuracy ratio of annual stock market forecasts from 12.6% to 64.19%, and reduced the absolute mean error from 72.35% to 5.17% [9]. Additional results on the stock market forecasts reported in [6,7,10], and others have verified the advantages of using interval-valued data. The paper [12], published in the same volume as this one, further validates the advantages from the perspective of information theory.

Using interval-valued data can significantly improve efficiency and effectiveness in information processing and uncertainty management. Therefore, we need to study interval-valued datasets.

1.2 The Objective of this Study

As a matter of fact, powerful inferential decision making schemes in the current literature use point-valued statistic and probabilistic measures, not interval-valued ones [4] and [17], mostly. To enable direct applications of these schemes and theory on analyzing interval-valued datasets, we need to supply *point-valued* statistics and probability for *interval-valued* datasets. Therefore, the primary objective of this work is to establish and to calculate such point-valued measures for interval-valued datasets.

To make this paper easy to read, it includes brief introductions on necessary background information. It also provides easy to follow illustrative examples for novel concepts and algorithms in addition to pseudo-code. Numerical results of these examples are obtained with a recent version of Python 3. However, readers may use any preferred general purpose programming language to verify the results.

1.3 Basic Concepts and Notations

Prior to our discussion, let us first clarify some basic concepts and notations related to intervals in this paper. An *interval* is a connected subset of $\mathbb{R}$. We denote an interval-valued object with a boldfaced letter to distinguish it from a point-valued one. We further specify the greatest lower bound and least upper bound of an interval object with an underline and an overline of the same letter but not boldfaced, respectively. For example, while a is a real, the boldfaced letter $\mathbf{a}$ denotes an interval with its greatest lower bound $\underline{a}$, and least upper bound $\overline{a}$. That is $\mathbf{a} = \{a : \underline{a} \leq a \leq \overline{a}, a \in \mathbb{R}\} = [\underline{a}, \overline{a}]$. The absolute value of $\mathbf{a}$, defined as $|\mathbf{a}| = \overline{a} - \underline{a}$, is also called the length (or norm) of $\mathbf{a}$. This is the greatest distance between any two numbers in $\mathbf{a}$.

The *midpoint* and *radius* of an interval $\mathbf{a}$ are defined as $mid(\mathbf{a}) = \dfrac{\underline{a} + \overline{a}}{2}$ and $rad(\mathbf{a}) = \dfrac{\overline{a} - \underline{a}}{2}$, respectively. Because the midpoint and radius of an interval $\mathbf{a}$ are point-valued, we simply denote them as $mid(a)$ and $rad(a)$ without boldfacing the letter a. We call $[\underline{a}, \overline{a}]$ the endpoint (or min-max) representation of *a*. We can specify an interval **a** with $mid(a)$ and $rad(a)$ too. This is because of $\underline{a} = mid(a) - rad(a)$ and $\overline{a} = mid(a) + rad(a)$. In the rest of this paper, we use both min-max and mid-rad representations for an interval-valued object.

While we use a boldfaced lowercase letter to indicate an interval, we denote an *interval-valued dataset*, i.e., a collection of real intervals, with a boldfaced uppercase letter. For instance, $\mathbf{X} = \{x_1, x_2, \ldots, x_n\}$ is an interval-valued dataset. The sets $\underline{X} = \{\underline{x}_1, \underline{x}_2, \ldots, \underline{x}_n\}$ and $\overline{X} = \{\overline{x}_1, \overline{x}_2, \ldots, \overline{x}_n\}$ are the left- and right-end sets of $\mathbf{X}$, respectively. Although items in a set are not ordered, the $\underline{x}_i \in \underline{X}$ and $\overline{x}_i \in \overline{X}$ are related to the same interval $x_i \in \mathbf{X}$. For convenience, we denote both $\underline{X}$ and $\overline{X}$ as ordered tuples. They are the *left- and right-endpoints* of $\mathbf{X}$. That is $\underline{X} = (\underline{x}_1, \underline{x}_2, \ldots, \underline{x}_n)$ and $\overline{X} = (\overline{x}_1, \overline{x}_2, \ldots, \overline{x}_n)$. Similarly, the *midpoint and radius* of $\mathbf{X}$ are point-valued tuples. They are $mid(X) = (mid(x_1), mid(x_2), \ldots, mid(x_n))$ and $rad(X) = (rad(x_1), rad(x_2), \ldots, rad(x_n))$, respectively.

Example 1. Provided an interval-valued sample dataset $\mathbf{X_0} = \{[1,5], [1.5, 3.5], [2,3], [2.5, 7], [4,6]\}$. Then, its left-endpoint is $\underline{X}_0 = (1, 1.5, 2, 3, 2.5, 4)$, and right-endpoint is $\overline{X}_0 = (5, 3.5, 3, 7, 6)$. The midpoint of $\mathbf{X_0}$ is $mid(X_0) = \frac{\underline{X}_0 + \overline{X}_0}{2} = (3, 2.5, 2.5, 4.75, 5)$, and the radius is $rad(X_0) = \frac{\overline{X}_0 - \underline{X}_0}{2} = (2, 1, 0.5, 2.25, 1)$.

We use this sample dataset $\mathbf{X_0}$ in the rest of this paper to illustrate concepts and algorithms for its simplicity.

In the rest of this paper, we discuss statistics of an interval-valued dataset in Sect. 2; define point-valued probability distributions for an interval-valued dataset in Sect. 3; introduce point-valued information entropy in Sect. 4; and summarize the main results and future work in Sect. 5.

2 Descriptive Statistics of an Interval-Valued Dataset

We introduce positional statistics for an interval-valued dataset first, and then discuss its point-valued variance and standard deviation.

2.1 Positional Statistics of an Interval-Valued Dataset X

The left-, right-endpoints, midpoint, and radius $\underline{X}, \overline{X}, mid(X)$, and $rad(X)$ are among positional statistics of an interval-valued dataset $\mathbf{X}$ as presented in Example 1. The mean of $\mathbf{X}$, denoted as $\mu_{\mathbf{X}}$, is the arithmetic average of $\mathbf{X}$. Because $\sum_{i=1}^{n} x_i = [\sum_{i=1}^{n} \underline{x}_i, \sum_{i=1}^{n} \overline{x}_i]$ in interval arithmetic[1], we have

$$\mu_{\mathbf{X}} = \frac{1}{n} \sum_{i=1}^{n} x_i = \left[\frac{\sum_{i=1}^{n} \underline{x}_i}{n}, \frac{\sum_{i=1}^{n} \overline{x}_i}{n} \right] = [\mu_{\underline{X}}, \mu_{\overline{X}}] \tag{1}$$

We now define few more observational statistics for $\mathbf{X}$.

[1] For readers who want to know more about standardized interval arithmetic, please refer the IEEE Standards for Interval Arithmetic [14] and [15].

Definition 1. *Let* $\mathbf{X}$ *be an interval-valued dataset, then*

1. *The envelope of* $\mathbf{X}$ *is the interval* $env(\mathbf{X}) = [\min(\underline{X}), \max(\overline{X})]$;
2. *The core of* $\mathbf{X}$ *is the interval* $core(\mathbf{X}) = \bigcap_{i=1}^{n} x_i = [\max(\underline{X}), \min(\overline{X})]$; *and*
3. *The mode of* $\mathbf{X}$ *is a tuple,* $mode(\mathbf{X}) = (\bigcap_{s\in S_j} x_s, k)$, *where* $\bigcap_{s\in S_j} x_s \neq \emptyset$, S_j *is a cardinality* k *subset of* $\{1, 2, \ldots, n\}$, *and for any* $S_i \subseteq \{1, 2, \ldots, n\}$ *if* $\bigcap_{s\in S_i} x_s \neq \emptyset$ *then* $|S_i| \leq k$.

In other words, $\forall x_i \in \mathbf{X}$, x_i is a subset of *env*($\mathbf{X}$), and *core*($\mathbf{X}$) is a subset of x_i. Furthermore, *mode*($\mathbf{X}$) is an ordered tuple. In which, $\bigcap_{s\in S_j} x_s$ is the non-empty intersection of x_s for all $s \in S_j$, such that, the cardinality of S_j is the greatest. For a given $\mathbf{X}$, its mode may not be unique. This is because of that, there may be multiple cardinality k subsets of {1, 2, ..., n} satisfying the nonempty intersection requirement $\bigcap_{s\in S_j} x_s \neq \emptyset$.

Corollary 1. *Let* $\mathbf{X}$ *be an interval-valued dataset, then*

1. *For all* $x_i \in \mathbf{X}$, $x_i \subseteq env(\mathbf{X})$;
2. *The core of* $\mathbf{X}$ *is not empty if and only if* $\max(\underline{X}) \leq \min(\overline{X})$; *and*
3. *The mode of* $\mathbf{X}$ *is* $(core(\mathbf{X}), n)$ *if and only if* $core(\mathbf{X}) \neq \emptyset$.

Corollary 1 is straightforward.

Instead of providing a proof, we provide the mean, envelop, core and mode for the sample dataset $\mathbf{X}_0 = \{[1, 5], [1.5, 3.5], [2, 3], [2.5, 7], [4, 6]\}$. In addition to its endpoints, midpoint, and radius presented in Example 1, we have its mean $\mu_{\mathbf{X}_0} = [2.2, 4.9]$; $env(\mathbf{X}_0) = [1, 7]$; $core(\mathbf{X}_0) = \emptyset$ because of $\max(\underline{X}_0) = 4$ is greater than $\min(\overline{X}_0) = 3$; and $mode(\mathbf{X}_0) = ([2.5, 3], 4)$. Figure 1 illustrates the sample dataset $\mathbf{X}_0$. From which, one may visualize the $env(\mathbf{X}_0)$ and $mode(\mathbf{X}_0)$ by imaging a vertical line, like the y-axis, continuously moving from left to right. The first and last points the line touches any $x_i \in \mathbf{X}_0$ determine the envelop $env(\mathbf{X}_0) = [1, 7]$. The line touches at most four intervals for all $x_i \in \mathbf{X}_0$ between [2.5, 3]. Hence, the mode is $mode(\mathbf{X}_0) = ([2.5, 3], 4)$.

While finding the envelop, core, and mean of $\mathbf{X}$ is straightforward, determining the mode of $\mathbf{X}$ involves the $2n$ numbers in $\underline{X}$ and $\overline{X}$, which divide *env*($\mathbf{X}$) into $2n-1$ sub-intervals in general (though some of them maybe degenerated as points.) Each of these $2n-1$ sub-intervals can be a candidate of the nonempty intersection part in the mode. For any $x_i \in \mathbf{X}$, it may cover some of these $2n-1$ sub-intervals (candidates) consecutively. For each of these candidates, we accumulate its occurrences in each $x_i \in \mathbf{X}$. The mode(s) for $\mathbf{X}$ is (are) the candidate(s) with the (same) highest occurrence. As a special case, if *core*($\mathbf{X}$) is not empty, then $mode(\mathbf{X}) = (core(\mathbf{X}), n)$. We summarize the above as an algorithm.

Algorithm 1: (*Finding the mode for an interval dataset* $\mathbf{X}$)

`Input:` $\mathbf{X}$`: an` n`-element interval dataset.`
`Output:` $mode(\mathbf{X})$.
`If` $\max(\underline{X}) < \min(\underline{X})$

```
    mode(X) = ([max(X_), min(X_)], n)
Else
  # Initialization:
  Concatenating X_ and X‾ as a single list c
  Sort the list c.
  For i from 0 to 2n − 1:
      cand_i = ([c_i, c_{i+1}], count_i = 0)
  End for
  # Counting frequency:
  For each x_i ∈ X
      Find j and k, such that c_j = x_i (lower) and c_k = x_i (upper)
          For l from j to k:
              cand_l.count_l += 1
          End for
  End for
  # Find the mode:
  m = max{cand.count}
  For j from 0 to 2n − 1:
      If cand_j.count_j = m,
          mode(X) = ([c_j, c_{j+1}], m)
  End for
Return mode(X).
```

Algorithm 1 is $O(n^2)$. This is because of that for each interval x_i, it may update the count in each of the $2n - 1$ candidates takes $O(n^2)$.

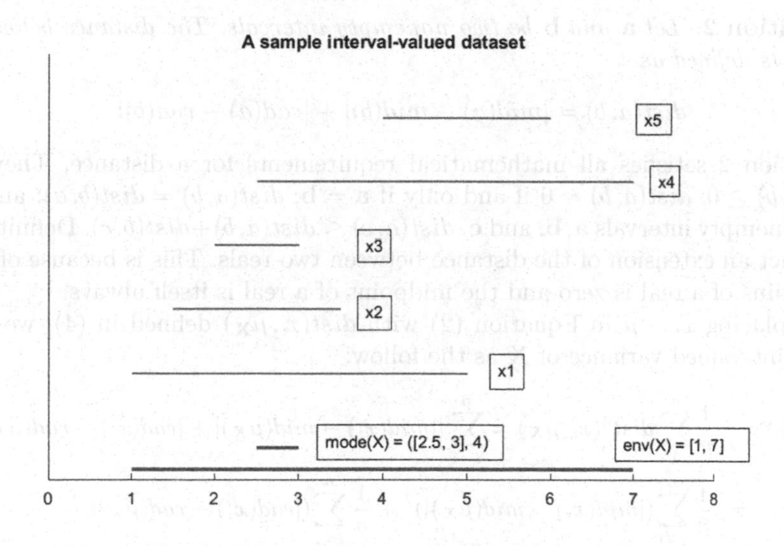

Fig. 1. The sample interval-valued dataset $\mathbf{X}_0$.

2.2 Point-Valued Variational Statistics of an Interval-Valued Dataset

In the literature, the variance of a point-valued dataset X is defined as

$$Var(X) = \frac{1}{n}\sum_{i=1}^{n} |x_i - \mu|^2 \tag{2}$$

in which, the term $|x_i - \mu|$ is the distance between $x_i \in X$ and μ, which is the mean of X.

Using (2) to define a variance for an interval-valued $\mathbf{X}$, we need a notion of point-valued distance between two intervals, $x_i \in \mathbf{X}$ and the interval $\mu_{\mathbf{X}}$. May we simply use $|\mathbf{a} - \mathbf{b}|$, the absolute value of the difference between two intervals $\mathbf{a}$ and $\mathbf{b}$, as their distance? Unfortunately, it does not work.

In interval arithmetic [18], the difference between two intervals $\mathbf{a}$ and $\mathbf{b}$ is defined as the follow:

$$\mathbf{a} - \mathbf{b} = [\min\{\underline{a} - \underline{b}, \underline{a} - \overline{b}, \overline{a} - \underline{b}, \overline{a} - \overline{b}\}, \max\{\underline{a} - \underline{b}, \underline{a} - \overline{b}, \overline{a} - \underline{b}, \overline{a} - \overline{b}\}] \tag{3}$$

Equation (3) ensures $\forall a \in \mathbf{a}, \forall b \in \mathbf{b}, a - b \in \mathbf{a} - \mathbf{b}$. However, it also implies $|\mathbf{a} - \mathbf{b}| = \max\{|a - b|, \forall a \in \mathbf{a}, \forall b \in \mathbf{b}\}$, which is the maximum distance between $a \in \mathbf{a}$ and $b \in \mathbf{b}$.

Mathematically, a distance between two nonempty sets A and B is usually defined as the minimum distance between $a \in A$ and $b \in B$ but not the maximum. Hence, we need to define a notion of distance between two intervals.

Definition 2. *Let* **a** *and* **b** *be two nonempty intervals. The distance between* **a** *and* **b** *is defined as*

$$dist(a, b) = |mid(a) - mid(b)| + |rad(a) - rad(b)| \tag{4}$$

Definition 2 satisfies all mathematical requirements for a distance. They are $dist(a, b) \geq 0$; $dist(a, b) = 0$ if and only if $\mathbf{a} = \mathbf{b}$; $dist(a, b) = dist(b, a)$; and for any nonempty intervals $\mathbf{a}$, $\mathbf{b}$, and $\mathbf{c}$, $dist(a, c) \leq dist(a, b) + dist(b, c)$. Definition 2 is in fact an extension of the distance between two reals. This is because of that the radius of a real is zero and the midpoint of a real is itself always.

Replacing $x_i - \mu$ in Equation (2) with $dist(x_i, \mu_{\mathbf{X}})$ defined in (4), we have the point-valued variance of $\mathbf{X}$ as the follow:

$$\begin{aligned} Var(X) &= \frac{1}{n}\sum_{1}^{n} dist^2(x_i, \mu_X) = \sum_{i=1}^{n} [|mid(x_i) - mid(\mu_X)| + |rad(x_i) - rad(\mu_X)|]^2 \\ &= \frac{1}{n}\sum_{i=1}^{n}(|mid(x_i) - mid(\mu_X)|)^2 + \frac{1}{n}\sum_{i=1}^{n}(|rad(x_i) - rad(\mu_X)|)^2 \\ &+ \frac{2}{n}\sum_{1}^{n}(|mid(x_i) - mid(\mu_X)|)(|rad(x_i) - rad(\mu_X)|). \end{aligned}$$

The expression above has three terms. All of them involve $mid(\mu_X)$ and $rad(\mu_X)$.

Since $\mu_{\mathbf{X}} = \left[\frac{\sum_{i=1}^{n} \underline{x}_i}{n}, \frac{\sum_{i=1}^{n} \overline{x}_i}{n}\right]$, $mid(\mu_{\mathbf{X}}) = \frac{1}{2}\left(\sum_{i=1}^{n} \underline{x}_i/n + \sum_{i=1}^{n} \overline{x}_i/n\right)$ $= \frac{1}{n}\sum_{i=1}^{n}\left(\frac{\underline{x}_i + \overline{x}_i}{2}\right) = \frac{1}{n}\sum_{i=1}^{n} mid(x_i) = \mu_{mid(X)}$. Therefore, the first term in the expression above $\frac{1}{n}\sum_{1}^{n}(|mid(x_i) - mid(\mu_X)|)^2 = \frac{1}{n}\sum_{1}^{n}(mid(x_i) - \mu_{mid(X)})^2 =$ $Var(mid(X))$ according to (2). Similarly, the second term $\frac{1}{n}\sum_{i=1}^{n}(|rad(x_i) - rad(\mu_X)|)^2 = Var(rad(X))$.

The third term is related to the absolute covariance between $mid(X)$ and $rad(X)$. Let $\Delta m_i = mid(x_i) - mid(\mu_X)$ and $\Delta r_i = rad(x_i) - rad(\mu_X)$, then we can rewrite the term $\frac{2}{n}\sum_{1}^{n}(|mid(x_i) - \mu_{mid(X)})|)(|rad(x_i) - \mu_{rad(X)})|)$ as $\frac{2}{n}\sum_{1}^{n}|\Delta m_i \Delta r_i|$.

Summarizing the discussion above, we have the point-valued variance for an interval-valued dataset $\mathbf{X}$ as the follow.

Definition 3. *Let* $\mathbf{X} = (x_1, x_2, \ldots, x_n)$ *be an interval-valued dataset, then the point-valued variance of* $\mathbf{X}$ *is*

$$Var(X) = Var(mid(X)) + Var(rad(X)) + \frac{2}{n}\sum_{i=1}^{n}|\Delta m_i \Delta r_i| \quad (5)$$

Because midpoints and radii of interval-valued objects are point-valued, the variance defined in (5) is also point-valued. Hence, we have the point-valued standard deviation of $\mathbf{X}$ as usual:

$$Std(X) = \sqrt{Var(X)} \quad (6)$$

In evaluating (5) and (6), one does not need interval computing at all. For the sample dataset $\mathbf{X}_0$, we have its point-valued variance $Var(X_0) =$ $var(mid(X_0)) + var(rad(X_0)) + \frac{2}{5}\sum_{i=1}^{5}|\Delta m_i \Delta r_i| = 1.5125 + 0.55 + 1.282 = 3.3445$; and the standard deviation $Std(X_0) = 1.8288$.

It is worthwhile to note that, Eq. (5) is an extension of (2) and applicable to point-valued datasets too. This is because of that, for all x_i in a point-valued X, $rad(x_i) = 0$ and $mid(x_i) = x_i$ always. Hence, $Var(X) = Var(mid(X))$ for a point-valued X.

3 Probability Distributions of an Interval-Valued Population

An interval-valued dataset $\mathbf{X}$ can be viewed as a sample of an interval-valued population. In this section, we study practical ways to find probability distributions for an interval-valued dataset $\mathbf{X}$. Our discussion addresses two different cases. One assumes distribution information for all $x_i \in \mathbf{X}$. The other does not.

3.1 On Probability Distribution of X with Distribution Information for Each $x_i \in \mathbf{X}$

Our discussion involves the concept of a probability distribution over an interval. Let us very briefly review the literature first.

A function $f(x)$ is a probability density function (*pdf*) of a random variable x on the interval $x = [\underline{x}, \overline{x}]$ if and only if $f(x) \geq 0, \forall x \in \mathbb{R}$, and $\int_{-\infty}^{\infty} f(t)dt = \int_{\underline{x}}^{\overline{x}} f(t)dt = 1$. Well-known *pdf*s in the literature include the uniform distribution: $f(x) = \begin{cases} 1 & \text{if } x \in [0,1], \\ 0 & \text{otherwise.} \end{cases}$; normal distribution: $f(x) = \frac{1}{\sigma\sqrt{2\pi}} e^{\frac{(x-\mu)^2}{2\sigma^2}}$; and beta distribution: $f(x) = \frac{x^{\alpha-1}(1-x)^{\beta-1}}{B(\alpha,\beta)}$, where $B(\alpha,\beta) = \frac{\Gamma(\alpha)\Gamma(\beta)}{\Gamma(\alpha+\beta)}$ and both parameters α and β are positive, and $\Gamma(t)$ is the gamma function. There are software tools available to fit point-valued sample data, which means computationally determining the parameter values in a chosen type of distribution. For instance, the Python `scipy.stats` module is available to find the optimal μ and σ to fit a point-valued dataset in a normal distribution, and/or α and β in a beta distribution.

It is safe to assume an availability of a *pdf* for each $x_i \in \mathbf{X}$ both theoretically and computationally. In practice, an interval $x_i \in \mathbf{X}$ is often obtained through aggregating observed points. For instances, in [9] and [11], min-max and confidence intervals are applied to aggregate points into intervals, respectively. If an interval is provided directly, one can always pick points from the interval and fit these points with a selected probability distribution computationally. Hereafter, we denote the *pdf* of $x_i \in \mathbf{X}$ as $pdf_i(x)$.

We now define a notion of *pdf* for an interval-valued dataset $\mathbf{X}$.

Definition 4. *A function $f(x)$ is called a probability density function of an interval-valued dataset $\mathbf{X} = \{x_1, x_2, \ldots, x_n\}$ if and only if $f(x)$ satisfies all of the conditions:*

$$\begin{cases} f(x) \geq 0 \ \forall x \in (-\infty, \infty); \\ \sum_{i=1}^{n} \int_{x_i \in \mathbf{X}} f(t)dt = 1. \end{cases} \tag{7}$$

The theorem below provides a practical way to calculate a *pdf* for $\mathbf{X}$.

Theorem 1. *Let $\mathbf{X} = (x_1, x_2, \ldots, x_n)$ be an interval-valued dataset; and $pdf_i(x)$ be the pdf of x_i provided $i \in \{1, 2, \ldots, n\}$. Then,*

$$f(x) = \frac{\sum_{i=1}^{n} pdf_i(x)}{n} \tag{8}$$

is a pdf of $\mathbf{X}$.

Proof. Because $pdf_i(x) \geq 0 \; \forall i \in \{1, 2, \ldots, n\}$, we have $\sum_{i=1}^{n} pdf_i(x) \geq 0$. Hence, $f(x) \geq 0$. In addition, $\int_{-\infty}^{\infty} pdf_i(t)dt = 1$ for all $i \in \{1, 2, \ldots, n\}$, we have $\sum_{i=1}^{n} \int_{x_i} f(t)dt = \int_{-\infty}^{\infty} \frac{\sum_{i=1}^{n} pdf_i(x)}{n} dx = \frac{\sum_{i=1}^{n} \int_{-\infty}^{\infty} pdf_i(t)dt}{n} = \frac{n}{n} = 1$. Equation (7) satisfied. Hence, the $f(x)$ is a *pdf* of X. □

Equation (8) actually provides a practical way of calculating the *pdf* of $\mathbf{X}$. Provided $pdf_i(x)$ for each $x_i \in \mathbf{X}$, we have the algorithm in pseudo-code below:
Algorithm 2: (*Finding a pdf for* $\mathbf{X}$)

```
Input: an n-item interval-valued dataset X;
       pdf_i(x) for every x_i ∈ X
Output: pdf(X)
# Initialization:
Concatenating x_ and x‾ as a list c
Sort c
For i from 1 to 2n − 1:
    segment_i = (c_i, c_{i+1}, 0)
End for
# Accumulating pdf on each segment:
For each x_i ∈ X find the j and k, such that
  c_j = x_i (lower) and c_k = x_i (upper)
    For l from j to k:
        segment_l.pdf += pdf_i
    End for
End for
# Calculating the pdf:
For i from 0 to 2n − 1:
    segment_i.pdf /= n
End for
Return segment_i for all i ∈ {1, 2, ..., 2n − 1}
```

Example 2. Find a *pdf* from the sample dataset $\mathbf{X}_0 = \{[1, 5], [1.5, 3.5], [2, 3], [2.5, 7], [4, 6]\}$. For simplicity, we assume a uniform distribution for each pdf_i's, i.e.,

$$pdf_i(x) = \begin{cases} \dfrac{1}{\overline{x}_i - \underline{x}_i} & \text{if } x \in x_i \\ 0, & \text{otherwise.} \end{cases}$$

Applying Algorithm 2, we have

$$f(X_0) = \frac{\sum_{i=1}^{5} pdf_i(x)}{5} = \begin{cases} 0.05 & \text{if } x \in [1, 1.5] \\ 0.15 & \text{if } x \in (1.5, 2] \\ 0.35 & \text{if } x \in (2, 2.5] \\ 0.39 & \text{if } x \in (2.5, 3] \\ 0.19 & \text{if } x \in (3, 3.5] \\ 0.09 & \text{if } x \in (3.5, 4] \\ 0.19 & \text{if } x \in (4, 5] \\ 0.14 & \text{if } x \in (5, 6] \\ 0.044 & \text{if } x \in (6, 7] \\ 0 & \text{otherwise.} \end{cases} \quad (9)$$

The *pdf* in the example is a stair function. This is because the uniform distribution assumption on each $x_i \in \mathbf{X}$. □

Here are few additional notes on finding a *pdf* for $\mathbf{X}$ with Algorithm 2 .

If assuming uniform distribution, how do we handle the case if $\exists i$ such that $\underline{x}_i = \overline{x}_i$? First of all, an interval element $\mathbf{x}_i$ is usually not degenerated as a constant. Even there is an i such that $\underline{x}_i = \overline{x}_i$, we can always assign an arbitrary non-negative *pdf* value at that point. This does not impact the calculation of probability in integrating the *pdf* function.

Algorithm 2 assumes $pdf_i(x) = 0, \forall x \notin x_i$. If it is not the case, the $2n$ numbers in $\underline{X}$ and $\overline{X}$ divide $\mathbb{R}$ in $2n + 1$ sub-intervals. They are $(-\infty, \min(\underline{X})), (\max(\overline{X}), \infty)$ together with the $2n - 1$ sub-intervals in $env(\mathbf{x})$. Therefore, the accumulation loop in Algorithm 2 should run through all of the $2n + 1$ sub-intervals, and then normalize them by dividing n.

Another implicit assumption of Theorem 1 is that, all $x_i \in \mathbf{X}$ are equally weighted. However, that is not necessary. If needed, one may place a positive weight w_i on each of pdf_i's as stated in the Corollary 2.

Corollary 2. *Let* $\mathbf{X} = (x_1, x_2, \ldots, x_n)$ *be an interval-valued dataset and* pdf_i *be the pdf of* $x_i \in X$*, then the function*

$$f(x) = \frac{\sum_{i=1}^{n} w_i \; pdf_i(x)}{\sum_{i=1}^{n} w_i} \quad \text{where } \forall i \; w_i > 0 \quad (10)$$

is a pdf of $\mathbf{X}$.

A proof of Corollary 2 is straightforward too. We have successfully applied the Corollary in computationally studying the stock market [12].

3.2 Probability Distribution of an Interval-Valued X Without Distribution Information for Any $x_i \in \mathbf{X}$

It is not necessary to assume the probability distribution for all $x_i \in \mathbf{X}$ to find a *pdf* of $\mathbf{X}$. An interval x is determined by its midpoint and radius. Let $u = mid(x)$ and $v = rad(x)$ be two point-valued random variables. Then, the *pdf* of x is a non-negative function $f(u,v) \geq 0$, such that $\int_{-\infty}^{\infty}\int_{-\infty}^{\infty} f(u,v)dudv = 1$. If we assume a normal distribution for $au + bv$, then $f(u,v)$ is a bivariate normal distribution [25]. The *pdf* of a bivariate normal distribution is:

$$p(u,v) = \frac{1}{2\pi\sigma_u\sigma_v\sqrt{1-\rho^2}} e^{\frac{-z}{2(1-\rho^2)}} \tag{11}$$

where $z = \frac{(u-\mu_u)^2}{\sigma_u^2} - \frac{2\rho(u-\mu_u)(v-\mu_v)}{\sigma_u\sigma_v} + \frac{(v-\mu_v)^2}{\sigma_v^2}$ and ρ is the normalized correlation between u and v, i.e., the ratio of their covariance and the product of σ_u and σ_v. Applying the *pdf*, we are able to estimate the probability over a region $u = [u_1, u_2]$, $v = [v_1, v_2]$ as

$$P(x) = \int_{v_1}^{v_2}\int_{u_1}^{u_2} p(u,v)dudv \tag{12}$$

To calculate the probability of an interval x, whose midpoint and radius are u_0 and v_0, we need a marginal *pdf* for either u or v. If we fix $u = u_0$, then the marginal *pdf* of v follows a single variable normal distribution. Thus,

$$p(v) = \frac{1}{\sigma_v\sqrt{2\pi}} e^{-\frac{1}{2}(\frac{v-\mu_v}{\sigma_v})^2}, \tag{13}$$

and the probability of x is

$$P(x) = \int_{-v_0}^{v_0} p(v)dv \tag{14}$$

An interval-valued dataset $\mathbf{X}$ provides us its $mid(X)$ and $rad(X)$. They are point-valued sample sets of u and v, respectively. All of $\mu_{mid(X)}, \mu_{rad(X)}, \sigma_{mid(X)}$, and $\sigma_{rad(X)}$ can be calculated as usual to estimate the μ_u, μ_v, σ_u, and σ_v in (11). For instance, from the sample $\mathbf{X}_0$, we have $\mu_{mid(X_0)} = 3.55$, $\mu_{rad(X_0)} = 1.35$, $\sigma_{mid(X_0)} = 1.1$, $\sigma_{rad(X_0)} = 0.66$, and $\rho = 0.404$, respectively. Furthermore, using $\mu_{rad(X_0)} = 1.35$ and $\sigma_{rad(X_0)} = 0.66$ in (13), we can estimate the probability of an arbitrary interval x with (14).

So far, we have established practical ways to calculate point-valued variance, standard deviation, and probability distribution for an interval-valued dataset X. With them, we are able to *directly apply commonly available inferential decision making schemes based on interval-valued dataset.*

4 Information Entropy of Interval-Valued Datasets

While it is out of the scope of this paper to discuss specific applications of inferential statistics on an interval-valued dataset, we are interested in measuring the amount of information in an interval-valued dataset. Information entropy is the average rate at which information is produced by a stochastic source of data [24]. Shannon introduced the concept of entropy in his seminal paper "A Mathematical Theory of Communication" [23]. The measure of information entropy associated with each possible data value is:

$$H(x) = -\sum_{i=1}^{n} p(x_i) \log p(x_i) \tag{15}$$

where $p(x_i)$ is the probability of $x_i \in X$.

An interval-valued dataset $\mathbf{X} = (x_1, x_2, \ldots, x_n)$ divides the real axis into $2n+1$ sub-intervals. Using $\mathcal{P}$ to denote the partition and $x^{(j)}$ to specify its j-th element, we have $\mathcal{P} = \left(x^{(1)}, x^{(2)}, \ldots, x^{(2n+1)}\right)$. As illustrated in Example 2, we can apply Algorithm 2 to find the pdf_j for each $x^{(j)} \in \mathcal{P}$. Then, the probability of $x^{(j)} = \int_{x^{(j)}} pdf_j(t)dt$ is available. Hence, we can apply (15) to calculate the entropy of an interval-valued dataset $\mathbf{X}$. For reader's convenience, we summarize the steps of finding the entropy of $\mathbf{X}$ as an algorithm below.

Algorithm 3: (*Finding the entropy for an interval-valued dataset* $\mathbf{X}$)

```
Input: an n-item interval dataset X
       pdf_i for all x_i ∈ X
Output: Entropy(X)

# Find the partition for the real axis:
Concatenating x̲ and x̄ as a list c
Sort c
The c forms a 2n+1 partition P of (−∞,∞)
# Find the probability for each x^(j) ∈ P:
For j from 1 to 2n+1
    Find a pdf_j on x^(j) with Algorithm 2
    Calculate p_j = ∫_{x^(j)} pdf_j(x)dx
End for
# Calculate the entropy:
Entropy(X) = 0
For j from 1 to 2n+1
    Entropy(X)  −= p_j log p_j
End for
Return Entropy(X)
```

The example below finds the entropy of the sample dataset $\mathbf{X}_0$ with the same assumption of uniform distribution in Example 2.

Example 3. Equation (9) in Example 2 provides the *pdf* of $\mathbf{X}_0$. Applying it, we obtain the probability of each interval $x^{(j)}$ as

$$p(x) = \int_{x^{(j)}} pdf(t)dt = \begin{cases} 0.025, & x^{(1)} = [1, 1.5] \\ 0.075, & x^{(2)} = [1.5, 2] \\ 0.175, & x^{(3)} = [2, 2.5] \\ 0.197, & x^{(4)} = [2.5, 3] \\ 0.098, & x^{(5)} = [3, 3.5] \\ 0.048, & x^{(6)} = [3.5, 4] \\ 0.194, & x^{(7)} = [4, 5] \\ 0.144, & x^{(8)} = [5, 6] \\ 0.044, & x^{(9)} = [6, 7] \\ 0, & otherwise \end{cases} \tag{16}$$

The entropy of $\mathbf{X}_0$ is $Entropy(X_0) = -\sum_i p_i \log p_i = 2.019$. □

Algorithm 3 provides us a much needed tool in studying point-valued information entropy of an interval-valued dataset. Applying it, we have investigated entropies of the real world financial dataset, which has used in the study of stock market forecasts [6,7], and [9], from the perspective of information theory. The results are reported in [12]. It not only reveals the deep reason of the significant quality improvements reported before, but also validates the concepts and algorithms presented here in this paper as a successful application.

5 Summary and Future Work

Recent advances have shown that using interval-valued data can significantly improve the quality and efficiency of information processing and uncertainty management. For interval-valued datasets, this work establishes much needed concepts of point-valued variational statistics, probability, and entropy for interval-valued datasets. Furthermore, this paper contains practical algorithms to find these point-valued measures. It provides additional theoretic foundations of applying point-valued methods in analyzing interval-valued datasets.

These point-valued measures enable us to directly apply currently available powerful point-valued statistic, probabilistic, theoretic results to interval-valued datasets. Applying these measures in various applications is definitely among a high priority of our future work. In fact, using this work as the theoretic foundation, we have successfully analyzed the entropies of the real world financial dataset related to the stock market forecasting mentioned in the introduction of this paper. The obtained results are reported in [12] and published in the same volume as this one. On a theoretic side, future work includes extending the concepts in this paper from single dimensional to multi-dimensional interval-valued datasets.

References

1. Bentkowska, U.: New types of aggregation functions for interval-valued fuzzy setting and preservation of pos-B and nec-B-transitivity in decision making problems. Inf. Sci. **424**(C), 385–399 (2018)

2. Billard, L., Diday, E.: Regression analysis for interval-valued data. In: Kiers, H.A.L., Rasson, J.P., Groenen, P.J.F., Schader, M. (eds.) Data Analysis, Classification, and Related Methods. STUDIES CLASS. Springer, Heidelberg (2000). https://doi.org/10.1007/978-3-642-59789-3_58
3. Dai, J., Wang, W., Mi, J.: Uncertainty measurement for interval-valued information systems. Inf. Sci. **251**, 63–78 (2013)
4. Gioia, F., Lauro, C.: Basic statistical methods for interval data. Statistica Applicata **17**(1), 75–104 (2005)
5. Grabisch, M., Marichal, J., Mesiar, R., Pap, E.: Aggregation Functions. Cambridge University Press, New York (2009)
6. He, L., Hu, C.: Midpoint method and accuracy of variability forecasting. J. Empir. Econ. **38**, 705–715 (2009). https://doi.org/10.1007/s00181-009-0286-6
7. He, L., Hu, C.: Impacts of interval computing on stock market forecasting. J. Comput. Econ. **33**(3), 263–276 (2009). https://doi.org/10.1007/s10614-008-9159-x
8. Hu, C., et al.: Knowledge Processing with Interval and Soft Computing. Springer, London (2008). https://doi.org/10.1007/978-1-84800-326-2
9. Hu, C., He, L.: An application of interval methods to stock market forecasting. J. Reliable Comput. **13**, 423–434 (2007). https://doi.org/10.1007/s11155-007-9039-4
10. Hu, C.: Using interval function approximation to estimate uncertainty. In: Interval/Probabilistic Uncertainty and Non-Classical Logics, pp. 341–352 (2008). https://doi.org/10.1007/978-3-540-77664-2_26
11. Hu, C.: A note on probabilistic confidence of the stock market ILS interval forecasts. J. Risk Finance **11**(4), 410–415 (2010)
12. Hu, C., and Hu, Z.: A computational study on the entropy of interval-valued datasets from the stock market. In: Lesot, M.-J., et al. (eds.) The Proceedings of the 18th International Conference on Information Processing and Management of Uncertainty in Knowledge-Based Systems (IPMU 2020), IPMU 2020, CCIS, vol. 1239, pp. 422–435. Springer (2020)
13. Huynh, V., Nakamori, Y., Hu, C., Kreinovich, V.: On decision making under interval uncertainty: a new justification of Hurwicz optimism-pessimism approach and its use in group decision making. In: 39th International Symposium on Multiple-Valued Logic, pp. 214–220 (2009)
14. IEEE Standard for Interval Arithmetic. IEEE Standards Association (2015). https://standards.ieee.org/standard/1788-2015.html
15. IEEE Standard for Interval Arithmetic (Simplified). IEEE Standards Association (2018). https://standards.ieee.org/standard/1788_1-2017.html
16. de Korvin, A., Hu, C., Chen, P.: Generating and applying rules for interval valued fuzzy observations. In: Yang, Z.R., Yin, H., Everson, R.M. (eds.) IDEAL 2004. LNCS, vol. 3177, pp. 279–284. Springer, Heidelberg (2004). https://doi.org/10.1007/978-3-540-28651-6_41
17. Lodwick, W.-A., Jamison, K.-D.: Interval-valued probability in the analysis of problems containing a mixture of possibilistic, probabilistic, and interval uncertainty. Fuzzy Sets Syst. **159**(21), 2845–2858 (2008)
18. Moore, R.E.: Methods and Applications of Interval Analysis. SIAM Studies in Applied Mathematics, Philadelphia (1979)
19. Marupally, P., Paruchuri, V., Hu, C.: Bandwidth variability prediction with rolling interval least squares (RILS). In: Proceedings of the 50th ACM SE Conference, Tuscaloosa, AL, USA, 29–31 March 2012, pp. 209–213. ACM (2012). https://doi.org/10.1145/2184512.2184562

20. Nordin, B., Hu, C., Chen, B., Sheng, V.S.: Interval-valued centroids in K-means algorithms. In: Proceedings of the 11th IEEE International Conference on Machine Learning and Applications (ICMLA), Boca Raton, FL, USA, pp. 478–481. IEEE (2012). https://doi.org/10.1109/ICMLA.2012.87
21. Pkala, B.: Uncertainty Data in Interval-Valued Fuzzy Set Theory: Properties, Algorithms and Applications, 1st edn. Springer, Cham (2018). https://doi.org/10.1007/978-3-319-93910-0
22. Rhodes, C., Lemon, J., Hu, C.: An interval-radial algorithm for hierarchical clustering analysis. In: 14th IEEE International Conference on Machine Learning and Applications (ICMLA), Miami, FL, USA, pp. 849–856. IEEE (2015)
23. Shannon, C.-E.: A mathematical theory of communication. Bell Syst. Tech. J. **27**, 379–423 (1948)
24. Wikipedia: Information entropy. https://en.wikipedia.org/wiki/Entropy_(information_theory)
25. Wolfram Mathworld. Binary normal distribution. http://mathworld.wolfram.com/BivariateNormalDistribution.html

A Computational Study on the Entropy of Interval-Valued Datasets from the Stock Market

Chenyi Hu[1(✉)] and Zhihui H. Hu[2]

[1] University of Central Arkansas, Conway, AR, USA
chu@uca.edu
[2] Edge Pursuit LLC, San Francisco, CA, USA

Abstract. Using interval-valued data and computing, researchers have reported significant quality improvements of the stock market annual variability forecasts recently. Through studying the entropy of interval-valued datasets, this work provides both information theoretic and empirical evidences on that the significant quality improvements are very likely come from interval-valued datasets. Therefore, using interval-valued samples rather than point-valued ones is preferable in making variability forecasts. This study also computationally investigates the impacts of data aggregation methods and probability distributions on the entropy of interval-valued datasets. Computational results suggest that both min-max and confidence intervals can work well in aggregating point-valued data into intervals. However, assuming uniform probability distribution should be a good practical choice in calculating the entropy of an interval-valued dataset in some applications at least.

Keywords: Interval-valued dataset · Stock market variability forecasting · Data aggregation · Probability distribution · Information entropy

1 Introduction

Recently, researchers have very successfully applied interval-valued data in information processing and uncertainty management. Related works on applications of interval-valued data include [13,21–25], and many more. With broad applications of interval computing, the IEEE Standard Association has released the IEEE Standards for Interval Arithmetic [19] and [20] recently.

This work is a continuation of the stock market interval-valued annual variability forecasts reported in [10,11,13,14,16], and [17]. In which, a real world six-dimensional point-valued monthly dataset is first aggregated into an interval-valued annual sample. Then, interval-valued annual predictions are made with interval least-squares (ILS) regression [15]. Comparing against the commonly used point-valued confidence interval predictions with ordinary least-squares (OLS), the interval approach increased the average accuracy ratio of annual stock market forecasts from 12.6% to 64.19%, and reduced the absolute mean error from 72.35% to 5.17% [14] with the same economical model [4] and the same raw dataset.

M.-J. Lesot et al. (Eds.): IPMU 2020, CCIS 1239, pp. 422–435, 2020.
https://doi.org/10.1007/978-3-030-50153-2_32

The quality improvements are significant. However, several questions arising from previous results still need to be answered. Among them are:

1. What is the theoretic reason for such a significant quality improvements?
2. What are the impacts of data aggregation methods on the results? and
3. What are the impacts of probability distributions on the entropy of an interval-valued dataset?

In this paper, we investigate these questions from the perspective of information theory [9]. To be able to calculate and compare entropies of interval-valued datasets, it is necessary to establish the concepts and algorithms on probability and entropy for interval-valued datasets. In our work [18], also published in this volume, we lay down both theoretic and algorithmic foundations for the investigation reported in this work. In which, point-valued statistic, probabilistic, and entropy measures for interval-valued datasets are established in details with practical algorithms. Interested readers should refer that article for a solid theoretical foundation.

In the rest of this paper, we briefly review related previous work, such as the stock market annual variability forecasting, the dataset, and information entropy in Sect. 2. We try to answer the question why interval-valued data leading better quality forecasts through comparing information entropy of interval-valued samples against point-valued ones in Sect. 3. We calculate and compare the impacts of two aggregation methods (min-max and confidence intervals) associated together with commonly used probability distributions (uniform, normal, and beta) in Sect. 4. We summarize the main results and possible future work in Sect. 5.

2 Related Previous Works

We first briefly review the dataset and the stock market annual variability forecasts; and then introduce related concepts and algorithms of calculating entropies of a point-valued dataset and of an interval-valued dataset.

2.1 The Stock Market Annual Variability Forecasting and the Dataset

The S & P 500 index is broadly used as a indicator for the overall stock market. The main challenge in studying the stock market is its volatility and uncertainty. Modeling the relationship between the stock market and relevant macroeconomic variables, Chen, Roll, and Ross [4] established a broadly accepted model in economics to forecast the overall level of the stock market. According to their model, the changes in the overall stock market value (SP_t) are linearly determined by the following five macroeconomic factors:

IP_t: the growth rate variations of adjusted Industrial Production Index,
DI_t: changes in expected inflation,

UI_t: and changes in unexpected inflation,
DF_t: default risk premiums, and
TM_t: unexpected changes in interest rates.

This relationship can be expressed as:

$$SP_t = a_t + I_t(IP_t) + U_t(UI_t) + D_t(DI_t) + F_t(DF_t) + T_t(TM_t) \quad (1)$$

By using historic data, one may estimate the coefficients of (1) to forecast changes of the overall stock market. The original dataset used in [14] and [17] consists of monthly data from January 1930 to December 2004 in 75 years for the six variables. Here are few sample lines of the data:

Yr-mth	UI	DI	SP	IP	DF	TM
30-Jan	-0.00897673	0	0.014382062	-0.003860512	0.0116	-0.0094
30-Feb	-0.00671673	-0.0023	0.060760088	-0.015592832	-0.0057	0.0115
30-Mar	-0.00834673	0.0016	0.037017628	-0.00788855	0.0055	0.0053
30-Apr	0.00295327	0.0005	0.061557893	-0.015966279	0.01	-0.0051
30-May	-0.00744673	-0.0014	-0.061557893	-0.028707502	-0.0082	0.0118
30-Jun	-0.00797673	0.0005	-0.106567965	-0.046763234	0.0059	0.0025
...	......	...	...	...	...	...
04-Jul	-0.00182673	0.0002	-0.024043354	0.00306212	0.0029	0.0147
04-Aug	0.00008127	0.0002	-0.015411102	-0.002424198	0	0.0385
04-Sep	0.00156327	0.0001	0.026033651	0.007217235	0.0005	0.0085
04-Oct	0.00470327	0	0.000368476	0.002001341	0.001	0.0143
04-Nov	-0.00002273	0	0.044493038	0.006654848	0.0034	-0.0245
04-Dec	-0.00461673	0.0004	0.025567309	0.001918659	0.0007	0.0235

To make an annual stock market forecast, a commonly used approach is to make a point-valued annual sample first, such as the end of each year, i.e., December data, or annual minimum for predicting the min, or annual maximum for estimating the max. Applying OLS to estimate the coefficients in (1), people are able to make a point-valued prediction. By adding and subtracting a factor (usually denoted as Z) of the standard deviation to the point-valued prediction, one form a confidence interval as an annual variability forecast. However, such confidence interval forecasting methods have never been widely used in the literature because of the poor forecasting quality [2] and [7] in forecasting the stock market. Normally, the forecasting intervals are so narrow that there is only a 50% chance, or even less, that a future point lies inside the interval [5] and [6]. In other cases, the forecasting intervals can be so wide that the forecasts are meaningless. This poor forecasting quality is deeply rooted in the methodology of point-based confidence interval forecasting.

Instead of commonly used point-valued approach, an interval-valued method has been proposed and applied for the annual stock market variability forecasts [14]. In which, the annual minimum and maximum form an interval-valued (min-max) sample of the year. By applying an interval least-squares algorithm [13] with the interval-valued sample, significant quality improvements of predictions are obtained. Figure 1 illustrates the interval-valued annual forecasts comparing against the actual variations of S & P 500 from 1940–2004. In which, a ten-year sliding window was used to make an out of sample forecast.

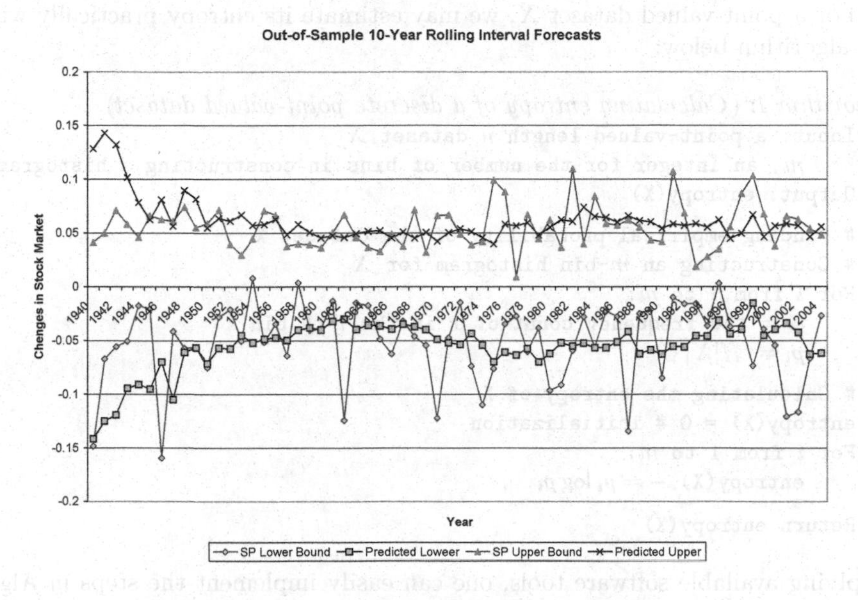

Fig. 1. Annual interval forecasts vs. actual market variations from 1940–2004

Further studies on forecasting the stock market [10] and [11], variability of mortgage rates [12], crude oi price prediction [29], and others, have consistently reported that the quality of variability forecasts with interval-valued samples and interval least-squares are significantly better than that of with point-valued samples and OLS.

As the main objective of this work, we want to investigate the major reason for such significant quality improvements through computing and comparing the entropies of point- and interval-valued samples.

2.2 Information Entropy of a Point- and an Interval-Valued Dataset

Our investigations are carried out through calculating and comparing information entropy, i.e., the average rate at which information produced by a stochastic source of data [28].

Shannon defines the entropy for a discrete dataset $X = \{x_1, x_2, \ldots, x_n\}$ in his seminal paper "A mathematical theory of communication" [26] as:

$$H(x) = -\sum_{i=1}^{n} p(x_i) \log p(x_i) \quad (2)$$

where $p(x_i)$ is the probability of event x_i. In information theory, Shannon's entropy has been referred as information entropy, and it has been used as a measure of information in data. Viewing the stock market as a stochastic source of data, we try to measure and compare the amount of information contained in datasets.

For a point-valued dataset X, we may estimate its entropy practically with the algorithm below:

Algorithm 1: (*Calculating entropy of a discrete point-valued dataset*)

```
Input: a point-valued length n dataset X
    m, an integer for the number of bins in constructing a histogram
Output: entropy(X)

# Finding empirical probability of the dataset X
# Constructing an m-bin histogram for X
For i from 1 to m:
    c_i = the frequency count of x in the i-th bin
    p_i = c_i/|X|

# Calculating the entropy of X
entropy(X) = 0 # initialization
For i from 1 to m:
    entropy(X) -= p_i log p_i .

Return entropy(X)
```

Applying available software tools, one can easily implement the steps in Algorithm 1 above. For example, calling the `histogram` method in Python `numpy` module returns the counts and bins in a histogram of a dataset. The rests are straightforward to implement.

However, it is not that straightforward to calculate information entropy of an interval-valued dataset. By the term interval, we mean a connected subset of $\mathbb{R}$. An interval-valued dataset is a collection of intervals. Using a boldfaced lowercase letter to denote an interval, and a boldfaced uppercase letter to specify an interval-valued dataset, we have $\mathbf{X} = (\boldsymbol{x}_1, \boldsymbol{x}_2, \ldots, \boldsymbol{x}_n)$ as an interval-valued dataset consisting of n intervals $\boldsymbol{x}_1, \boldsymbol{x}_2, \ldots \boldsymbol{x}_n$. Applying (2) to calculate the entropy of $\mathbf{X}$ demands a probability distribution of $\mathbf{X}$. Our paper [18] provides the theoretic and algorithmic foundations needed for calculating a point-valued probability of an interval-valued dataset. For readers' convenience, here are two related definitions and a theorem from that paper:

Definition 1. *A function $f(x)$ is called a probability density function, pdf of an interval-valued dataset $\boldsymbol{X}$ if and only if $f(x)$ satisfies all of the conditions:*

$$\begin{cases} f(x) \geq 0 \ \forall x \in (-\infty, \infty); \\ \sum_{i=1}^{n} \int_{\boldsymbol{x}_i \in \boldsymbol{X}} f(t)dt = 1. \end{cases} \tag{3}$$

Using pdf_i to denote the probability density function for $\boldsymbol{x}_i \in \mathbf{X}$, we have the theorem below to obtain a *pdf* for $\mathbf{X}$ practically.

Theorem 1. *Let $\mathbf{X} = (\boldsymbol{x}_1, \boldsymbol{x}_2, \ldots, \boldsymbol{x}_n)$ be an interval-valued dataset; and $pdf_i(x)$ be the pdf of $\boldsymbol{x}_i$ provided $i \in \{1, 2, \ldots, n\}$. Then,*

$$f(x) = \frac{\sum_{i=1}^{n} pdf_i(x)}{n} \tag{4}$$

is a pdf of $\mathbf{X}$.

With (4), we define the entropy for an interval-valued dataset $\mathbf{X}$ as

Definition 2. *Let $\mathcal{P}$ be an interval partition of the real axis and $pdf(x)$ be the probability density function of $\mathcal{P}$. Then, the probability of an interval $\boldsymbol{x}^{(j)} \in \mathcal{P}$ is $p_j = \int_{\boldsymbol{x}^{(j)}} pdf(t)dt$, and the entropy of $\mathcal{P}$ is*

$$entropy(X) = -\sum_{\mathcal{P}} p_j \log p_j \tag{5}$$

Example 1. Find a *pdf* and entropy for the interval-valued sample dataset $\mathbf{X}_0 = \{[1, 5], [1.5, 3.5], [2, 3], [2.5, 7], [4, 6]\}$.

For simplicity, we assume a uniform distribution for each $\boldsymbol{x}_i \in \mathbf{X}_0$, i.e.,

$$pdf_i(x) = \begin{cases} \dfrac{1}{\overline{x}_i - \underline{x}_i} & \text{if } x \in \boldsymbol{x}_i, \text{ and } \underline{x}_i \neq \overline{x}_i \\ \infty & \text{if } \underline{x}_i = \overline{x}_i \\ 0, & \text{otherwise.} \end{cases}$$

$$f(X_0) = \frac{\sum_{i=1}^{5} pdf_i(x)}{5} = \begin{cases} 0.05 & \text{if } x \in [1, 1.5] \\ 0.15 & \text{if } x \in (1.5, 2] \\ 0.35 & \text{if } x \in (2, 2.5] \\ 0.39 & \text{if } x \in (2.5, 3] \\ 0.19 & \text{if } x \in (3, 3.5] \\ 0.09 & \text{if } x \in (3.5, 4] \\ 0.19 & \text{if } x \in (4, 5] \\ 0.14 & \text{if } x \in (5, 6] \\ 0.044 & \text{if } x \in (6, 7] \\ 0 & \text{otherwise.} \end{cases} \tag{6}$$

The *pdf* of the example in (6) is a stair function. This is because of the uniform distribution assumption on each $\boldsymbol{x}_i \in \mathbf{X}_0$. The five intervals in $\mathbf{X}_0$ form a partition of $\mathbb{R}$ in eleven intervals including $(-\infty, 1)$ and $(7, \infty)$. Using (5), we have the entropy of the interval-valued sample dataset $entropy(X_0) = 2.019$ □

Example 1 illustrates the availability of a point-valued *pdf* for an interval-valued dataset. For more theoretic and algorithmic details, please refer [18]. We are ready now to investigate the question: why does the interval-valued approach significantly improve the quality of variability forecasts?

3 Why Does the Interval-Valued Approach Significantly Improve the Quality of Variability Forecasts?

Previous results have evidenced that the interval-valued approach can significantly improve the quality of forecasts in different areas (such as the stock market annual variability, the variability of the mortgage rate [12], and the variability of crude oil price [15]). However, using the same economical model and the same original dataset but point-valued samples, the quality of forecasts are much worse. To investigate the possible cause, we should examine the entropies of interval-valued and point-valued input datasets evidently.

Applying Algorithm 1 on point-valued annual samples of the six-dimensional financial dataset, we calculate their attribute-wise entropy. The four point-valued annual samples are December only, annual minimum, annual maximum, and annual midpoint[1]. With the Algorithm 3 in [18], we calculate the attribute-wise entropy of the annual min-max interval-valued sample. Table 1 summarizes the results. In which, the first row lists each of the six attributes in the dataset. The second to the last rows provide values of attribute-wise entropy of five different samples: December only, Annual minimum, Annual maximum, Annual midpoint, and Annual min-max interval, respectively.

Table 1. Entropy comparisons of different samples

	UI	DI	SP	IP	DF	TM
December only	2.32855	2.01183	2.12941	2.05978	2.39706	2.33573
Annual minimum	2.33076	2.16933	2.28035	2.09871	2.19422	2.62452
Annual maximum	1.88469	2.30266	1.53328	1.88045	2.34693	2.35843
Annual mean	2.04877	2.55961	2.31651	2.07323	2.09817	2.47341
Annual min-max intvl.	4.34192	3.06851	3.95838	4.30213	3.95359	4.31941

Figure 2 provides a visualized comparison of these entropy. From which, we can observe the followings:

The attribute-wise information entropies vary along with different samples. However, the attribute-wise entropies of the interval-valued sample are clearly much higher than that of any point-valued ones. Comparatively, the entropies of point-valued samples do not differ significantly. This indicates that the amount of information in these point-valued samples measured with entropies are somewhat similar. But, they are significantly less than that of the interval-valued ones. The greater the entropy is, the more information may possibly be extracted from. This is why the interval-valued forecasts can produce significantly better forecasts in [10,11,14], and others.

In both theory and practice, meaningless noises and irregularities may increase the entropy of a dataset too. However, it is not the case here in this study.

[1] The arithmetic average of annual min and annual max.

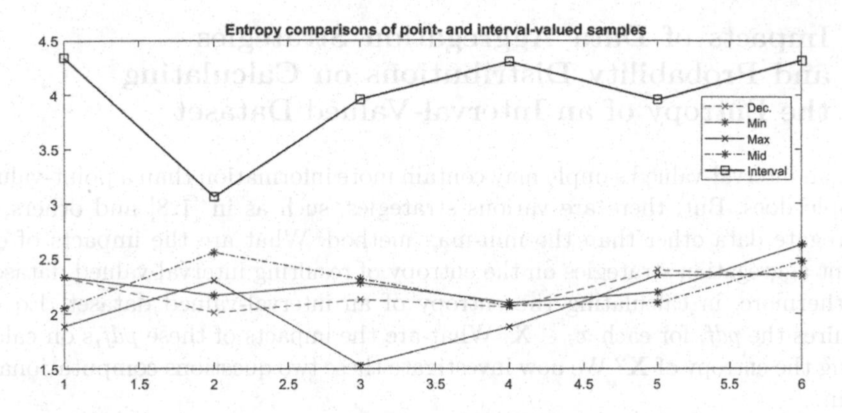

Fig. 2. Attribute-wise entropy comparison of point- and interval-valued samples

The interval rolling least-squares algorithm [16] has successfully extracted the additional information and made significant quality improvements. The advantages of using interval-valued samples instead of point-valued ones have also been observed in predicting variations of the mortgage rate [12], the crude oil price [15], and others. The interval-valued samples indeed contain more meaningful information. Therefore, in making variability forecasts like the stock market, it is preferable of using interval-valued samples rather than point-valued ones.

Here is an additional note. The attribute-wise entropies of the annual min-max interval-valued sample in Table 1 and the sum of entropies of the point-valued annual minimum and maximum are similar. If one uses the point-valued annual minimum and annual maximum separately, can he obtain quality forecasts similar to that of using the min-max interval-valued sample? Unfortunately, an empirical study show that is not the case. In [11], a comparison of the following two approaches is reported. One of the two is of applying the point-valued annual minimum and maximum samples to predict annual lower and upper bounds of the markct with the OLS separately. Then, confidence intervals are constructed as annual variability forecasts. The other applies the ILS with the min-max interval-valued sample. The quality of forecasts produced in the later approach is still much better than that of the former approach. In [10], using the sample of annual midpoints is studied for the same reason of performance comparison. The ILS with interval-valued annual sample still significantly outperform the point-valued approach in terms of higher average accuracy ratio, lower mean error, and a higher stability in terms of less standard deviation. This suggests that, to extract information from an interval-valued sample, one should use the ILS instead of OLS.

4 Impacts of Data Aggregation Strategies and Probability Distributions on Calculating the Entropy of an Interval-Valued Dataset

Yes, an interval-valued sample may contain more information than a point-valued sample does. But, there are various strategies, such as in [1,8] and others, to aggregate data other than the min-max method. What are the impacts of different aggregation strategies on the entropy of resulting interval-valued dataset? Furthermore, in calculating the entropy of an interval-valued dataset, Eq. (4) requires the pdf_i for each $\boldsymbol{x}_i \in \mathbf{X}$. What are the impacts of these pdf_is on calculating the entropy of $\mathbf{X}$? We now investigate these two questions computationally again.

In studying probability distribution of interval-valued annual stock market forecasts, point-valued data are aggregated with confidence intervals instead of annual min-max intervals [17]. In which, the points within a year are first fit with a normal distribution attribute-wise. Then, confidence intervals are formed at a selected level of probabilistic confidence with an intention of filtering out possible outliers. With different levels of confidence (by adjusting the Z-values), the interval-valued samples vary. So do the variability forecasts. However, we have observed that the variations are not very significant at all when Z is between 1.25 to 2, see [17]. Specifically, the average accuracy ratio associated with the Z-values are: 61.75% with $Z = 1.25$, 64.23% with $Z = 1.50$, 64.55% with $Z = 1.75$, and 62.94% with $Z = 2.00$. These accuracy ratios are very similar to 64.19% reported in [14] with the min-max aggregation.

In calculating the attribute-wise entropy of the annual min-max interval-valued sample with Algorithm 3 in [18] earlier, we have assumed a uniform distribution for each interval. In addition to uniform distribution, we consider both normal and beta distributions in this work because of their popularity in applications. In this study, we computationally investigate the impacts of a combination of an aggregation strategy associated with a probabilistic distribution on the entropy of resulting interval-valued data. We report our numerical results on each of the following four combinations:

(a) Min-max interval with uniform distribution;
(b) Fitting data with a normal distribution then forming confidence interval with $Z = 1.5$, using normal distribution in entropy calculation;
(c) Fitting data with a normal distribution then forming confidence interval with $Z = 1.5$, then assuming uniform distribution on each interval in entropy calculation; and
(d) Min-max interval fitting with a beta distribution.

Table 2 lists attribute-wise entropies for each of the four cases above. Figure 3 provides a visual comparison. Python modules `numpy` and `scipy` are used as the main software tools in carrying out the computational results.

We now analyze each of the outputs from (a)–(d) in Fig. 3.

Table 2. Entropy comparison of data aggregation methods and *pdf* selection

	UI	DI	SP	IP	DF	TM
(a) Min-max, unif.	4.34192	3.06851	3.95838	4.30213	3.95359	4.31941
(b) Conf. intvl, normal	2.69246	2.67623	2.61681	2.69736	2.73824	2.74129
(c) Conf. intvl, unif.	3.79327	3.76349	3.61804	3.80710	3.91177	3.91903
(d) Min-max, beta	1.96865	2.07197	2.04587	1.95605	2.08885	1.86871

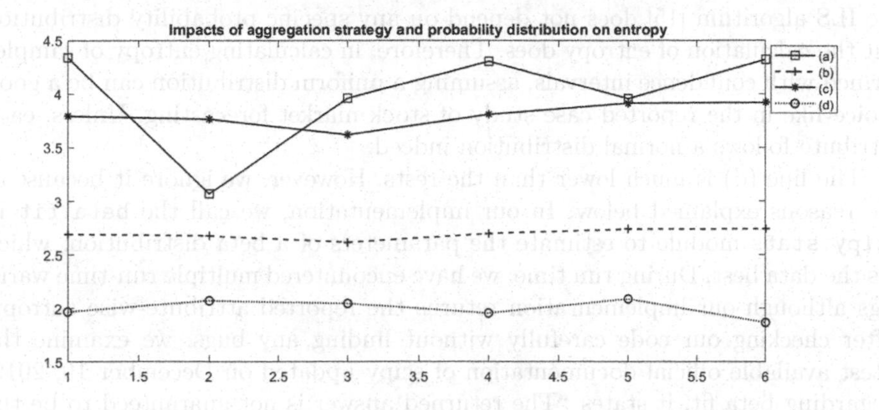

Fig. 3. Entropy comparison of data aggregation methods with *pdf* selection

The line (a) is exactly the same as the min-max interval line in Fig. 2. This is because of that we have already assumed uniform distribution in calculating the attribute-wise entropy for each of the min-max intervals.

The line (b) indicates that the entropies of the interval-valued sample formed with the method (b) are much less than that of the interval-valued one, i.e., the line (a). This is not by an accident. Equation (4) uses the arithmetic average of $\sum_i pdf_i$ as the *pdf* of an interval-valued dataset $\mathbf{X}$. As we know, the sum of normal random variables follows a normal distribution. Therefore, the resulting interval-valued dataset obtained with (b) follows a normal distribution, which is determined only by its mean and standard deviation with much less irregularity. Therefore, the calculated entropy is much less than that of (a). However, one should not abolish confidence interval aggregation at all. The only thing causing the relatively less entropy is the entropy calculation, in which, we assumed normal distribution for each pdf_i. This is further explained on the line (c) below.

The line (c) shows the results obtained with the same confidence intervals in (b) but then assuming a uniform distribution for each interval in calculating the entropy. The Corollary 2 in [18] makes this practically doable. Notice that the lines (c) and (a) are fairly close to each other comparing against (b) and (d). This means that using a confidence interval to aggregate points can still be a valid practical approach. Computational results in [17] repeated below further verify the claim as an evidence. By adjusting the Z-values of normal distribution, several interval-valued annual samples are formed at different levels of probabilistic confidence. Using them, that work reports some changes in overall quality of the stock market annual forecasts. The average accuracy ratio associated with the Z-values are: 61.75% with $Z = 1.25$, 64.23% with $Z = 1.50$, 64.55% with $Z = 1.75$, and 62.94% with $Z = 2.00$. They are very close to 64.19% reported in [14] with the min-max intervals. The relatively overall closeness of line (c) and (a) can be an explanation for the similarity of the average accuracy ratios. The closeness of (a) and (c) also implies that adjusting the Z-value in data aggregation may slightly improve the quality of forecasts but not significantly. Lastly, the ILS algorithm [15] does not depend on any specific probability distribution but the calculation of entropy does. Therefore, in calculating entropy of samples formed with confidence intervals, assuming a uniform distribution can be a good choice like in the reported case study of stock market forecasting. Unless, each attribute follows a normal distribution indeed.

The line (d) is much lower than the rests. However, we ignore it because of the reasons explained below. In our implementation, we call the `beta.fit` in `scipy.stats` module to estimate the parameters of a beta distribution, which fits the data best. During run time, we have encountered multiple run-time warnings although our implementation returns the reported attribute-wise entropy. After checking our code carefully without finding any bugs, we examine the latest available official documentation of scipy updated on December 19, 2019. Regarding beta fit, it states "The returned answer is not guaranteed to be the globally optimal MLE (Maximum Likelihood Estimate), it may only be locally optimal, or the optimization may fail altogether" [27]. We do not have any other explanations for the numerical results. Due to the run-time warnings and current software documentation, we accept that the specific computational results on (d) are not reliable as a fact.

5 Conclusions and Possible Future Work

Applying interval-valued data rather than point-valued ones, researchers have made very significant quality improvements of variability forecasts. This work strongly suggests that the significant quality improvements in previous studies very much likely come from the interval-valued inputs. Figure 2 clearly shows that the attribute-wise entropies of an interval-valued sample are much higher than that of those point-valued samples. The more information contained in the input data, the higher quality outputs could be expected. Furthermore, the interval least-squares algorithm [15] can be applied to successfully extract information from an interval-valued sample rather than using the traditional ordinary least-squares approaches as reported in [11] and others.

Computational results also conclude that both min-max and confidence intervals can be effectively used to aggregate point-valued data into intervals. Both of them may lead to similarly well quality variability forecasts with the evidence on the stock market reported in [3] and [17]. This is because of that they may result in interval-valued samples with similar entropies as illustrated in Fig. 3 lines (a) and (c). While the interval least-squares algorithm itself does not demand probability distribution information at all, calculating the entropy of an interval-valued dataset does. The lines (b) and (c) in Fig. 3 suggest that a uniform probability distribution on each interval can be a good choice in calculating the entropy of an interval-valued dataset.

In summary, this work provides information theoretic evidences, in addition to empirical results published previously, on the followings:

- Using interval-valued samples together with ILS is preferable than using point-valued ones with OLS in variability forecasts like predicting annual variability of the stock market and others.
- Applying min-max interval and/or confidence interval (at an appropriate level of confidence) to aggregate points into intervals may result in interval-valued samples containing similar amount of information.
- When estimating the entropy of an interval-valued dataset with (5), it can be a good choice of assuming a uniform distribution on each interval. Unless, it follows a normal distribution indeed.

The future work may consist of both sides of application and theory. With the information theoretic evidence, we have validated previously published results with interval-valued data and ILS. Therefore, applying interval methods in variability forecasts with uncertainty has a high priority. On the theoretic side, we should indicate that attribute-wise entropy is not exactly the same as the entropy of a multidimensional dataset. Investigating attribute-wise entropy in this study is not only because of its simplicity, but also because [18] only provides point-valued probability and entropy for single dimensional interval-valued datasets. Therefore, establishing point-valued probability and entropy for a multidimensional interval-valued dataset is among future works too.

Acknowledgment. The authors would very much like to express their sincere appreciations to the contributors of the freely available high quality Python software tools. Especially, the `spyder` IDLE, and the `numpy` and `scipy` modules have helped us greatly to implement our algorithms, and made our investigation much more effectively and efficiently.

References

1. Bouchon-Meunier, B.: Aggregation and Fusion of Imperfect Information. Springer, Heidelberg (2013)
2. Chatfield, C.: Prediction intervals for time-series forecasting. In: Armstrong, J.S. (ed.) Principles of Forecasting. International Series in Operations Research & Management Science, vol. 30, pp. 475–494. Springer, Boston (2001). https://doi.org/10.1007/978-0-306-47630-3_21
3. Chen, G., Hu, C.: A computational study on window-size selection in stock market RILS interval forecasting. In: 2009 World Congress on Computer Science and Information Engineering, Los Angeles, CA, pp. 297–301. IEEE (2009)
4. Chen, N.F., Roll, R., Ross, S.A.: Economic forces and the stock market. J. Bus. **59**(3), 383–403 (1986)
5. Gardner, E.: A simple method of computing prediction intervals for time series forecasts. Manage. Sci. **34**, 541–546 (1988)

6. Granger, C.: Can we improve the perceived quality of economic forecasts? J. Appl. Econom. **11**, 455–473 (1996)
7. Gooijer, J., Hyndman, R.: 25 years of time series forecasting. J. Forecast. **22**, 443–473 (2006)
8. Grabisch, M., Marichal, J., Mesiar, R., Pap, E.: Aggregation Functions. Cambridge University Press, New York (2009)
9. Gray, R.M.: Entropy and Information Theory, 1st edn., Corrected, Springer, New York (2013)
10. He, L.T., Hu, C.: Midpoint method and accuracy of variability forecasting. Empir. Econ. **38**, 705–715 (2009). https://doi.org/10.1007/s00181-009-0286-6
11. He, L.T., Hu, C.: Impacts of interval computing on stock market variability forecasting. Comput. Econ. **33**(3), 263–276 (2009). https://doi.org/10.1007/s10614-008-9159-x
12. He, L., Hu, C., Casey, M.: Prediction of variability in mortgage rates: interval computing solutions. J. Risk Finance **10**(2), 142–154 (2009)
13. Kreinovich, V., Korvin, A., Baker Kearfott, R., Hu, C. (eds.): Knowledge Processing with Interval and Soft Computing. AIKP. Springer, London (2008). https://doi.org/10.1007/978-1-84800-326-2
14. Hu, C., He, L.T.: An application of interval methods to stock market forecasting. Reliable Comput. **13**, 423–434 (2007). https://doi.org/10.1007/s11155-007-9039-4
15. Hu, C., He, L.T., Xu, S.: Interval function approximation and applications. In: Kreinovich, V., Korvin, A., Baker Kearfott, R., Hu, C. (eds.) Knowledge Processing with Interval and Soft Computing. AIKP, pp. 119–134. Springer, London (2008). https://doi.org/10.1007/978-1-84800-326-2_5
16. Hu, C.: Using interval function approximation to estimate uncertainty. In: Huynh, V.N., Nakamori, Y., Ono, H., Lawry, J., Kreinovich, V., Nguyen, H.T. (eds.) Interval / Probabilistic Uncertainty and Non-Classical Logics. AINSC, vol. 46, pp. 341–352. Springer, Heidelberg (2008). https://doi.org/10.1007/978-3-540-77664-2_26
17. Hu, C.: A note on probabilistic confidence of the stock market ILS interval forecasts. J. Risk Finance **11**(4), 410–415 (2010)
18. Hu, C., Hu, Z.: On statistics, probability, and entropy of interval-valued datasets. In: Lesot, M.-J., et al. (eds.) IPMU 2020. CCIS, vol. 1239, pp. 407–421. Springer, Cham (2020)
19. IEEE Standard for Interval Arithmetic. IEEE Standards Association (2015). https://standards.ieee.org/standard/1788-2015.html
20. IEEE Standard for Interval Arithmetic (Simplified). IEEE Standards Association (2018). https://standards.ieee.org/standard/1788_1-2017.html
21. de Korvin, A., Hu, C., Chen, P.: Generating and applying rules for interval valued fuzzy observations. In: Yang, Z.R., Yin, H., Everson, R.M. (eds.) IDEAL 2004. LNCS, vol. 3177, pp. 279–284. Springer, Heidelberg (2004). https://doi.org/10.1007/978-3-540-28651-6_41
22. Marupally, P., Paruchuri, V., Hu, C.: Bandwidth variability prediction with rolling interval least squares (RILS). In: Proceedings of the 50th ACM SE Conference, Tuscaloosa, AL, USA, 29–31 March 2012, pp. 209–213. ACM (2012). https://doi.org/10.1145/2184512.2184562
23. Nordin, B., Chen, B., Sheng, V.S., Hu, C.: Interval-valued centroids in K-Means algorithms. In: Proceedings of the 11th IEEE International Conference on Machine Learning and Applications (ICMLA), Boca Raton, FL, USA, pp. 478–481. IEEE (2012). https://doi.org/10.1109/ICMLA.2012.87

24. Pkekala, B.: Uncertainty Data in Interval-Valued Fuzzy Set Theory: Properties, Algorithms and Applications. SFSC, vol. 367, 1st edn. Springer, Cham (2019). https://doi.org/10.1007/978-3-319-93910-0
25. Rhodes, C., Lemon, J., Hu, C.: An interval-radial algorithm for hierarchical clustering analysis. In: Proceedings of the 14th IEEE International Conference on Machine Learning and Applications (ICMLA), Miami, FL, USA, pp. 849–856. IEEE (2015). https://doi.org/10.1109/ICMLA.2015.118
26. Shannon, C.-E.: A mathematical theory of communication. Bell Syst. Tech. J. **27**, 379–423 (1948)
27. Scipy.stats Documentation. https://docs.scipy.org/doc/scipy/reference/generated/scipy.stats.rv_continuous.fit.html. Updated 19 Dec 2019
28. Wikipedia, Information entropy. https://en.wikipedia.org/wiki/entropy_(information_theory). Edited 23 Dec 2019
29. Xu, S., Chen, X., Han, A.: Interval forecasting of crude oil price. In: Huynh, V.N., Nakamori, Y., Ono, H., Lawry, J., Kreinovich, V., Nguyen, H.T. (eds.) Interval / Probabilistic Uncertainty and Non-Classical Logics. AINSC, vol. 46, pp. 353–363. Springer, Heidelberg (2008). https://doi.org/10.1007/978-3-540-77664-2_27

Tolerance and Control Solutions of Two-Sided Interval Linear System and Their Applications

Worrawate Leela-apiradee[1(✉)], Phantipa Thipwiwatpotjana[2], and Artur Gorka[3,4]

[1] Department of Mathematics and Statistics, Faculty of Science and Technology, Thammasat University, Pathum Thani 12121, Thailand
worrawateleela@gmail.com

[2] Department of Mathematics and Computer Science, Faculty of Science, Chulalongkorn University, Bangkok 10330, Thailand
phantipa.t@chula.ac.th

[3] Department of Mathematics, Erskine College, Due West, SC 29639, USA

[4] Kamnoetvidya Science Academy, Wangchan, Rayong 21210, Thailand
art.gorka@erskine.edu

Abstract. This work investigates tolerance and control solutions to a two-sided interval linear system. Their semantics are different, even though, we would be able to interchange the role of the interval information algebraically. We present necessary and sufficient conditions of their solvabilities as the inequalities depending on center and radius of coefficient interval matrices on both sides of the system. In a situation when the vector of variables is nonnegative, the conditions can simply be modified as the inequalities depending on boundaries of the interval matrices. This result helps to find out the feasible solutions of a quadratic programming problem with two-sided interval linear equation constraints.

Keywords: Interval linear system · Tolerance solution · Control solution

1 Introduction

An interval linear system of equations is normally referred to as a system $\boldsymbol{A}x = \boldsymbol{b}$, where $\boldsymbol{A}$ is an interval matrix and $\boldsymbol{b}$ is an interval vector, while x is a vector of variables. As the matrix and the right hand side vector information of the system is not precise, it is impossible to provide a solution x to the system without any appropriate meaning.

Some literature presented different types of solutions of the system $\boldsymbol{A}x = \boldsymbol{b}$ depending on the purposes of the solutions. These solution types include weak, strong, tolerance and control solutions whose names reflect well on their mathematical definitions. For example, "x is a weak solution to $\boldsymbol{A}x = \boldsymbol{b}$" means

M.-J. Lesot et al. (Eds.): IPMU 2020, CCIS 1239, pp. 436–448, 2020.
https://doi.org/10.1007/978-3-030-50153-2_33

that

$$\exists A \in \boldsymbol{A}, \exists b \in \boldsymbol{b} \text{ such that } Ax = b,$$

while "x is a strong solution to $\boldsymbol{A}x = \boldsymbol{b}$" means that

$$\forall A \in \boldsymbol{A}, \forall b \in \boldsymbol{b} \text{ such that } Ax = b.$$

Beaumont presented in [1] an efficient method derived from the simplex algorithm to compute inner and outer inclusion of the weak (united) solution set. A full analysis of the solvability and the conditions for checking whether x is a particular solution type to the system $\boldsymbol{A}x = \boldsymbol{b}$ was also provided in the literature [2–4,8–12].

The background on a tolerance solution arose from the crane construction problem in [8] and the input-output planning problem with inexact data in [11]. The characteristic of a tolerance solution x is to make $\boldsymbol{A}x$ stay in the boundary of $\boldsymbol{b}$. Shary [12] first motivated an idea of a control solution, which is the opposite case of a tolerance solution. In addition, Tian et al. developed in [13] a tolerance-control solution for the case when each row index of the system $\boldsymbol{A}x = \boldsymbol{b}$ performs either tolerance or control. Recently, Leela-apiradee, [6] , have provided its solution set in terms of level set.

Instead of the system $\boldsymbol{A}x = \boldsymbol{b}$, the goal of this paper is to deal with a two-sided interval linear system. "Two-sided" means that the right hand side interval vector $\boldsymbol{b}$ is substituted by the term $\boldsymbol{B}y$. The paper then presents the tolerance and control solutions of a two-sided interval linear system together with their solvability conditions.

To lead to the main idea of the paper, let us first introduce some basic notation of an interval matrix and an interval vector that can be seen as a matrix and a vector of interval components as follows.

- An $m \times n$ interval matrix $\boldsymbol{A}$ is defined by

$$\boldsymbol{A} = \begin{pmatrix} [\underline{a}_{11}, \overline{a}_{11}] & [\underline{a}_{12}, \overline{a}_{12}] & \cdots & [\underline{a}_{1n}, \overline{a}_{1n}] \\ [\underline{a}_{21}, \overline{a}_{21}] & [\underline{a}_{22}, \overline{a}_{22}] & \cdots & [\underline{a}_{2n}, \overline{a}_{2n}] \\ \vdots & \vdots & \ddots & \vdots \\ [\underline{a}_{m1}, \overline{a}_{m1}] & [\underline{a}_{m2}, \overline{a}_{m2}] & \cdots & [\underline{a}_{mn}, \overline{a}_{mn}] \end{pmatrix},$$

  where $\underline{a}_{ij}$ and $\overline{a}_{ij}$ are real numbers such that $\underline{a}_{ij} \leq \overline{a}_{ij}$ for each $i \in \{1, 2, \ldots, m\}$ and $j \in \{1, 2, \ldots, n\}$.
- The interpretation of the interval matrix $\boldsymbol{A}$ can be written as the set of matrices, that is,

$$\boldsymbol{A} = [\underline{A}, \overline{A}] = \left\{ A \in \mathbb{R}^{m \times n} : \underline{A} \leq A \leq \overline{A} \right\},$$

  where

$$\underline{A} = \begin{pmatrix} \underline{a}_{11} & \underline{a}_{12} & \cdots & \underline{a}_{1n} \\ \underline{a}_{21} & \underline{a}_{22} & \cdots & \underline{a}_{2n} \\ \vdots & \vdots & \ddots & \vdots \\ \underline{a}_{m1} & \underline{a}_{m2} & \cdots & \underline{a}_{mn} \end{pmatrix} \quad \textit{and} \quad \overline{A} = \begin{pmatrix} \overline{a}_{11} & \overline{a}_{12} & \cdots & \overline{a}_{1n} \\ \overline{a}_{21} & \overline{a}_{22} & \cdots & \overline{a}_{2n} \\ \vdots & \vdots & \ddots & \vdots \\ \overline{a}_{m1} & \overline{a}_{m2} & \cdots & \overline{a}_{mn} \end{pmatrix}.$$

The ordering '$\leq$' between two matrices $\underline{A}$ and $\overline{A}$ is referred to componentwise inequality, i.e.,

$$\underline{A} \leq \overline{A} \text{ if and only if } \underline{a}_{ij} \leq \overline{a}_{ij} \ \forall i \in \{1,2,\ldots,m\}, \forall j \in \{1,2,\ldots,n\}.$$

Therefore, we note here that $A \in \boldsymbol{A}$ means $\underline{A} \leq A \leq \overline{A}$.

Moreover, we can use the following notation to represent the interval matrix $\boldsymbol{A}$.

$$\boldsymbol{A} = [\underline{A}, \overline{A}] = [A_c - \Delta_A, A_c + \Delta_A], \tag{1}$$

where $A_c = \frac{1}{2}(\overline{A} + \underline{A})$ and $\Delta_A = \frac{1}{2}(\overline{A} - \underline{A})$.

Fiedler et al. proved in [2] that the lower and upper bounds of the interval vector $\boldsymbol{A}x$, denoted by $\underline{\boldsymbol{A}x}$ and $\overline{\boldsymbol{A}x}$ for any $x = (x_1, x_2, \ldots, x_n)^T \in \mathbb{R}^n$, can be viewed by

$$\boldsymbol{A}x = [\underline{\boldsymbol{A}x}, \overline{\boldsymbol{A}x}] = [A_c x - \Delta_A|x|,\ A_c x + \Delta_A|x|], \tag{2}$$

where $|x|$ is defined as the absolute of vector x, i.e., $|x| = (|x_1|, |x_2|, \ldots, |x_n|)^T$.

Given $\boldsymbol{B}$ be another interval matrix and y be another vector of variables. In the situation when we have the term $\boldsymbol{B}y$ on the right hand side instead of the vector $\boldsymbol{b}$, the system "$\boldsymbol{A}x = \boldsymbol{b}$" would become "$\boldsymbol{A}x = \boldsymbol{B}y$", which is called a two-sided interval linear system. The dimensions of x and y do not need to be the same but the number of rows of interval matrices $\boldsymbol{A}$ and $\boldsymbol{B}$ does. The definitions of weak, strong, tolerance control and tolerance-control solutions of $\boldsymbol{A}x = \boldsymbol{B}y$ would mathematically be defined in the same fashion as the case of $\boldsymbol{A}x = \boldsymbol{b}$.

In this paper, we focus on the tolerance, control and tolerance-control solutions of the system $\boldsymbol{A}x = \boldsymbol{B}y$. In Sect. 2 we give the definitions and their characterizations by equivalent conditions. Usually, two sets of quantities are equal when the left and the right quantities are the same. However, there is often a situation with imprecise information that one set of quantities is being controlled by the other. This means that if the two sets of quantities are not precise with having interval information, then one set of quantities must be subset of the other. Moreover, one set of the interval information could be more important to the system than the other. The other set of the interval information must follow the semantics of the context. This leads to the interpretation of tolerance and control solutions of a two-sided interval linear system discussed in Sect. 3. A couple of application examples are demonstrated in Sect. 4, which can be modeled by a two-sided interval linear system and a quadratic programming problem with two-sided interval linear constraints. The conclusion is addressed in the last section.

2 Tolerance and Control Solutions of Two-Sided Interval System of Linear Equations

To see how the two-sided interval systems are motivated, consider two systems of standard linear equations below:

$$Ax = By, \tag{3}$$

$$Ax - By = 0, \tag{4}$$

where $A \in \mathbb{R}^{m \times n}$, $B \in \mathbb{R}^{m \times p}$, and m, n and p are positive integers. Any (x, y) in $\mathbb{R}^n \times \mathbb{R}^p$ satisfying System (3) (or (4)) is called a **solution** of the system. It is clear that Systems (3) and (4) are algebraically equivalent. The set of solutions of (3) is the same as the set of solutions of (4). However, if the entries of coefficient matrices A and B in (3) and (4) are interval data, Systems (3) and (4) turn into

$$\boldsymbol{A}x = \boldsymbol{B}y, \tag{5}$$

$$\boldsymbol{A}x - \boldsymbol{B}y = 0, \tag{6}$$

respectively, where coefficient terms $\boldsymbol{A}$ and $\boldsymbol{B}$ in (5) and (6) are interval matrices as defined in the introduction section. System (6) could not be well-defined in general since the left-hand side may be represented as interval vector with non-zero width as a result of Moore's standard interval arithmetic [7], while the right-hand side is a real zero vector with zero width. Therefore, we would not be able to move $\boldsymbol{B}y$ to the same side as $\boldsymbol{A}x$ of the equality as usual, when we deal with interval data. However, System (5) called a **two-sided interval system of linear equations** is well-defined since both sides of the equation are interval vectors. A solution $(x, y) \in \mathbb{R}^n \times \mathbb{R}^p$ to System (5) is not as simple as the case of standard matrices A and B, but it comes with its semantics.

In [2], Fiedler et al. defined tolerance and control solutions to an interval linear system $\boldsymbol{A}x = \boldsymbol{b}$. Based on the concepts of these solutions, Tian et al. later proposed a tolerance-control solution in [13].

Definition 1 (see [2] and [13]). *A vector $x \in \mathbb{R}^n$ is called*

1. *a* **tolerance solution** *of $\boldsymbol{A}x = \boldsymbol{b}$ if for each $A \in \boldsymbol{A}$ there exists $b \in \boldsymbol{b}$ such that $Ax = b$,*
2. *a* **control solution** *of $\boldsymbol{A}x = \boldsymbol{b}$ if for each $b \in \boldsymbol{b}$ there exists $A \in \boldsymbol{A}$ such that $Ax = b$,*
3. *a* **tolerance-control solution** *of $\boldsymbol{A}x = \boldsymbol{b}$ if each row index of the system is either tolerance or control.*

As we expand an interval linear system $\boldsymbol{A}x = \boldsymbol{b}$ to a two-sided interval linear system $\boldsymbol{A}x = \boldsymbol{B}y$, the types of solutions of $\boldsymbol{A}x = \boldsymbol{B}y$ presented in the following definition are developed in similar fashion as the solution concepts in $\boldsymbol{A}x = \boldsymbol{b}$.

Definition 2. *A vector $(x, y) \in \mathbb{R}^n \times \mathbb{R}^p$ is called*

1. *a* **tolerance solution** *of $\boldsymbol{A}x = \boldsymbol{B}y$ if for each $A \in \boldsymbol{A}$ there exists $B \in \boldsymbol{B}$ such that $Ax = By$,*

2. *a* **control solution** *of* $\boldsymbol{A}x = \boldsymbol{B}y$ *if for each* $B \in \boldsymbol{B}$ *there exists* $A \in \boldsymbol{A}$ *such that* $Ax = By$,
3. *a* **tolerance-control solution** *of* $\boldsymbol{A}x = \boldsymbol{B}y$ *if each row index of the system is either tolerance or control.*

Without semantics, the mathematical expression of tolerance and control solutions are the same. In Sect. 3, we will mention more about these two solutions and separately redefine them according to their semantics.

We now consecutively establish Theorems 1–3 based on Definition 2 to find necessary and sufficient conditions for checking tolerance and tolerance-control solvabilities of $\mathbf{A}x = \mathbf{B}y$. The conditions presented in Theorems 2 and 3 are in a form of inequalities depending on center and radius of the coefficient interval matrices $\boldsymbol{A}$ and $\boldsymbol{B}$ with absolute terms $|x|$ and $|y|$.

Theorem 1. *A vector* (x, y) *is a tolerance solution of* $\boldsymbol{A}x = \boldsymbol{B}y$ *if and only if it satisfies* $\boldsymbol{A}x \subseteq \boldsymbol{B}y$.

Proof. Assume that (x, y) is a tolerance solution of $\boldsymbol{A}x = \boldsymbol{B}y$. Let $A \in \boldsymbol{A}$. Then, there exists $B \in \boldsymbol{B}$ such that $Ax = By$. Thus,

$$Ax = By \in \boldsymbol{B}y = \left[\underline{\boldsymbol{B}y}, \overline{\boldsymbol{B}y}\right],$$

that is,

$$\underline{\boldsymbol{B}y} \leq Ax \leq \overline{\boldsymbol{B}y} \text{ for any } A \in \boldsymbol{A},$$

which concludes $\boldsymbol{A}x \subseteq \boldsymbol{B}y$. Conversely, we suppose that (x, y) satisfies $\boldsymbol{A}x \subseteq \boldsymbol{B}y$. Then, $Ax \in \boldsymbol{B}y$ for all $A \in \boldsymbol{A}$. Therefore, $Ax = By$ for all $A \in \boldsymbol{A}$ and for some $B \in \boldsymbol{B}$. Hence, (x, y) is a tolerance solution. □

Theorem 2. *A vector* (x, y) *is a tolerance solution of* $\boldsymbol{A}x = \boldsymbol{B}y$ *if and only if it satisfies*

$$|A_c x - B_c y| \leq \Delta_B |y| - \Delta_A |x|. \tag{7}$$

Proof. Let (x, y) be a tolerance solution of $\boldsymbol{A}x = \boldsymbol{B}y$. By Theorem 1 and (2),

$$[A_c x - \Delta_A |x|, A_c x + \Delta_A |x|\,] = \boldsymbol{A}x \subseteq \boldsymbol{B}y = [B_c y - \Delta_B |y|, B_c y + \Delta_B |y|\,].$$

Thus,

$$B_c y - \Delta_B |y| \leq A_c x - \Delta_A |x| \leq A_c x + \Delta_A |x| \leq B_c y + \Delta_B |y|,$$

which implies

$$-\left(\Delta_B |y| - \Delta_A |x|\right) \leq A_c x - B_c y \leq \Delta_B |y| - \Delta_A |x|. \tag{8}$$

Conversely, let (x, y) satisfy condition (7) Then, it gives (8), which means

$$\underline{\boldsymbol{B}y} = B_c y - \Delta_B |y| \leq A_c x - \Delta_A |x| = \underline{\boldsymbol{A}x}$$

and

$$\overline{\boldsymbol{A}x} = A_c x + \Delta_A |x| \leq B_c y + \Delta_B |y| = \overline{\boldsymbol{B}y}.$$

Therefore, $\boldsymbol{A}x \subseteq \boldsymbol{B}y$ and (x, y) becomes a tolerance solution. □

We can use inequality (7) to verify whether a given vector (x, y) is a tolerance solution to our two-sided interval linear system. The system of inequality (7) has the absolute terms $|x|$ and $|y|$, which means it is not the system of linear inequalities, in general. It depends on the signs of x and y components. However, when we consider the nonnegative domain of vector variables x and y, the inequality becomes a simple form as

$$\Delta_A x - \Delta_B y \leq A_c x - B_c y \leq \Delta_B y - \Delta_A x,$$

which is obtained by substituting $|x| = x$ and $|y| = y$. This turns into

$$(B_c - \Delta_B)y \leq (A_c - \Delta_A)x \text{ and } (A_c + \Delta_A)x \leq (\Delta_B + B_c)y. \tag{9}$$

According to (1), the inequalities (9) can be concluded as the corollary below.

Corollary 1. *Let x and y be nonnegative vector variables. A vector (x, y) is a tolerance solution of $\boldsymbol{A}x = \boldsymbol{B}y$ if and only if it satisfies*

$$\underline{B}y \leq \underline{A}x \text{ and } \overline{A}x \leq \overline{B}y.$$

The similar statements of Theorems 1–2 and Corollary 1 for a control solution can be done easily by interchanging the roles of "$\boldsymbol{A}$ and $\boldsymbol{B}$" and "x and y".

Theorem 3. *A vector (x, y) is a tolerance-control solution of $\boldsymbol{A}x = \boldsymbol{B}y$ if and only if it satisfies*

$$|A_c x - B_c y| \leq \Delta_A|x| + \Delta_B|y| - 2\delta, \tag{10}$$

where δ is a vector in $\mathbb{R}^m$ with the following components

$$\delta_i = \begin{cases} (\Delta_A|x|)_i, & \text{if } i \in P; \\ (\Delta_B|y|)_i, & \text{if } i \in M \setminus P, \end{cases}$$

and $P = \{i \in M : (\boldsymbol{A}x)_i \subseteq (\boldsymbol{B}y)_i\}$, $M = \{1, 2, \ldots, m\}$.

Proof. Assume that (x, y) is a tolerance-control solution of $\boldsymbol{A}x = \boldsymbol{B}y$. Let P be a subset of M such that row $i \in P$ of the system is tolerance. Then, the other $i \in M \setminus P$ of the system is control. Using Theorem 2, we have

$$|(A_c x)_i - (B_c y)_i| \leq (\Delta_B|y|)_i - (\Delta_A|x|)_i \text{ for each } i \in P \tag{11}$$

and

$$|(A_c x)_i - (B_c y)_i| \leq (\Delta_A|x|)_i - (\Delta_B|y|)_i \text{ for each } i \in M \setminus P \tag{12}$$

By putting Inequalities (11) and (12) together,

$$\begin{aligned} |A_c x - B_c y| &\leq \left(\Delta_A|x| - 2\sum_{i \in P}(\Delta_A|x|)_i e_i\right) + \left(\Delta_B|y| - 2\sum_{i \in M \setminus P}(\Delta_B|y|)_i e_i\right) \\ &= \Delta_A|x| + \Delta_B|y| - 2\left(\sum_{i \in P}(\Delta_A|x|)_i e_i + \sum_{i \in M \setminus P}\Delta_B|y|)_i e_i\right) \\ &= \Delta_A|x| + \Delta_B|y| - 2\delta, \end{aligned}$$

where e_i is a m-column vector containing 1 at the ith row and 0 elsewhere. Conversely, let (x, y) satisfy condition (10), which implies Inequalities (11) and (12). Therefore, (x, y) turns into a tolerance-control solution. □

3 Semantics of Tolerance and Control Solutions

Looking at the mathematical definitions in the previous section, it may seem that tolerance and control solutions are algebraically the same. So, why would we need to define them both? This is because one interval information could be more important than another. There are some missing details in Definition 2 about the priority of the interval information $\boldsymbol{A}$ and $\boldsymbol{B}$, which may not be represented clearly by just the mathematical quantification; "for all" and "for some".

To be able to achieve a control solution of the system $\boldsymbol{A}x = \boldsymbol{B}y$ and justify the semantics of the word "control", the boundary matrices $\underline{A}$ and $\overline{A}$ of $\boldsymbol{A}$ must be more important than the ones of $\boldsymbol{B}$. It could be the interval information that is given by an expert so that any quantity on the right must be controlled in the boundary of quantities on the left.

Similarly, to get the semantics of the word "tolerance" in a tolerance solution, the boundary matrices $\underline{A}$ and $\overline{A}$ of $\boldsymbol{A}$ must be more important. They provide the range of left hand side quantities. Moreover, the situation to come up with a tolerance solution is such that we need every element in the range to be under the control of the range of the right hand side quantities. In other words, the left hand side quantities tolerate themselves within the range of the right hand side quantities.

For those reasons, we cannot simply substitute $\boldsymbol{A}$ with $\boldsymbol{B}$ and x with y, and infer that tolerance and control solutions are the same. Therefore, the priority of the information $\boldsymbol{A}$ and $\boldsymbol{B}$ need to be stated in their definitions as rewritten below.

Definition 3. *Let an interval matrix $\boldsymbol{A}$ play more important role to the two-sided interval system of linear equations $\boldsymbol{A}x = \boldsymbol{B}y$. A vector $(x, y) \in \mathbb{R}^n \times \mathbb{R}^p$ is called*

1. *a* **tolerance solution** *of $\boldsymbol{A}x = \boldsymbol{B}y$ if for each $A \in \boldsymbol{A}$ there exists $B \in \boldsymbol{B}$ such that $Ax = By$. It is in the sense that the range of $\boldsymbol{A}x$ tolerates within the range of $\boldsymbol{B}y$,*
2. *a* **control solution** *of $\boldsymbol{A}x = \boldsymbol{B}y$ if for each $B \in \boldsymbol{B}$ there exists $A \in \boldsymbol{A}$ such that $Ax = By$. It is in the sense that the range of $\boldsymbol{A}x$ controls the range of $\boldsymbol{B}y$.*

4 Applications on Tolerance and Control Solutions of a Two-Sided Interval Linear System

In this section, we illustrate two small examples to show the difference between tolerance and control solutions. These examples could be parts of any relevant application systems.

- Problem Statement 1 is formulated as a tolerance solution to a two-sided interval linear system, which can be solved by system of inequalities.

– Problem Statement 2 is formulated as a quadratic programming with control solutions to two-sided interval linear constraints, whose numerical example is demonstrated in Example 1.

Problem Statement 1. *An animal food manufacturing company has its own quality control for its three formulas of chicken food: I, II, and III. These formulas must be checked for the level of carbohydrate, fiber, protein, and vitamins per kilogram per bag. The qualified bags must have the nutrients within the given boundaries as shown in the table below. The company also has a nutritionist who can advise customers about using these three formulas of chicken food for meat and egg chickens. To raise chickens for healthy meat and chickens for healthy eggs, those chickens should have nutrients within the range represented also in the table.*

Type of nutrients	Interval amount of nutrient (kg)				
	Formula I	Formula II	Formula III	Meat chicken	Egg chicken
1. Carbohydrate	$[\underline{a}_{11}, \overline{a}_{11}]$	$[\underline{a}_{12}, \overline{a}_{12}]$	$[\underline{a}_{13}, \overline{a}_{13}]$	$[\underline{b}_{11}, \overline{b}_{11}]$	$[\underline{b}_{12}, \overline{b}_{12}]$
2. Fiber	$[\underline{a}_{21}, \overline{a}_{21}]$	$[\underline{a}_{22}, \overline{a}_{22}]$	$[\underline{a}_{23}, \overline{a}_{23}]$	$[\underline{b}_{21}, \overline{b}_{21}]$	$[\underline{b}_{22}, \overline{b}_{22}]$
3. Protein	$[\underline{a}_{31}, \overline{a}_{31}]$	$[\underline{a}_{32}, \overline{a}_{32}]$	$[\underline{a}_{33}, \overline{a}_{33}]$	$[\underline{b}_{31}, \overline{b}_{31}]$	$[\underline{b}_{32}, \overline{b}_{32}]$
4. Vitamins	$[\underline{a}_{41}, \overline{a}_{41}]$	$[\underline{a}_{42}, \overline{a}_{42}]$	$[\underline{a}_{43}, \overline{a}_{43}]$	$[\underline{b}_{41}, \overline{b}_{41}]$	$[\underline{b}_{42}, \overline{b}_{42}]$

The nutritionist suggests customers to mix three formulas together before feeding. The total amount of nutrients in the mixed chicken food must be within the range of the total amount of all needed nutrients to guarantee that all chickens provide healthy products. This relationship can be represented as a two-sided interval linear system (13)

$$\begin{pmatrix} [\underline{a}_{11}, \overline{a}_{11}] & [\underline{a}_{12}, \overline{a}_{12}] & [\underline{a}_{13}, \overline{a}_{13}] \\ [\underline{a}_{21}, \overline{a}_{21}] & [\underline{a}_{22}, \overline{a}_{22}] & [\underline{a}_{23}, \overline{a}_{23}] \\ [\underline{a}_{31}, \overline{a}_{31}] & [\underline{a}_{32}, \overline{a}_{32}] & [\underline{a}_{33}, \overline{a}_{33}] \\ [\underline{a}_{41}, \overline{a}_{41}] & [\underline{a}_{42}, \overline{a}_{42}] & [\underline{a}_{43}, \overline{a}_{43}] \end{pmatrix} \begin{pmatrix} x_1 \\ x_2 \\ x_3 \end{pmatrix} = \begin{pmatrix} [\underline{b}_{11}, \overline{b}_{11}] & [\underline{b}_{12}, \overline{b}_{12}] \\ [\underline{b}_{21}, \overline{b}_{21}] & [\underline{b}_{22}, \overline{b}_{22}] \\ [\underline{b}_{31}, \overline{b}_{31}] & [\underline{b}_{32}, \overline{b}_{32}] \\ [\underline{b}_{41}, \overline{b}_{41}] & [\underline{b}_{42}, \overline{b}_{42}] \end{pmatrix} \begin{pmatrix} y_1 \\ y_2 \end{pmatrix}, \quad (13)$$

where x_j *is the amount of bags of animal food formula I, II and III that should be mixed, when* $j = 1, 2, 3$*, respectively,*
y_k *is the amount of meat and egg chickens that the customer should raise, when* $k = 1, 2$*, respectively.*

From the company's point of view, it is important to control the nutrients in each formula of chicken foods. It turns out that the set of solutions to System (13) is the set of tolerance solutions as the total of the nutrients created by the mixed food must be within the range of healthy nutrients. On the other hand, when considering the customer's side, the interval information about the nutrients needed for each chicken is the priority for the customer. In this case, the customer must control the nutrients in the mixed food by the range of the total amount of all needed nutrients for healthy chicken meat and eggs. Then, the set of solutions to System (13) can also be viewed as the tolerance solution set. Obviously, vector variables $x = (x_1, x_2, x_3)^T$ *and* $y = (y_1, y_2)^T$ *are nonnegative. From Corollary 1, the tolerance solution* (x, y) *is obtained by solving the system of inequalities as follows:*

$$\underline{b}_{i1}y_1 + \underline{b}_{i2}y_2 - \underline{a}_{i1}x_1 - \underline{a}_{i2}x_2 - \underline{a}_{i3}x_3 \leq 0$$

and

$$\overline{a}_{i1}x_1 + \overline{a}_{2}x_2 + \overline{a}_{i3}x_3 - \overline{b}_{i1}y_1 - \overline{b}_{i2}y_2 \leq 0,$$

for each $i \in \{1, 2, 3, 4\}$. □

Problem Statement 2. *A famous family bakery shop sells homemade fruit cakes and fruit tarts. The owner uses three grades of mixed dried berries: A, B, and C, by mixing them together to get the best dessert according to the family recipe. Suppose the owner determines the initial amount of fruit cakes and fruit tarts he/she wanted to make as* α_1 *and* α_2*, respectively, and sets up the initial amount of mixed dried berries grades A, B and C that he/she aims to buy as* β_1, β_2 *and* β_3*, respectively. One kilogram of each mixed dried fruit grade contains dried blueberries, dried cranberries, dried raspberries, and dried strawberries in the different uncertain quantities. The table below provides the interval quantities per kilogram of each mixed dried berries grade.*

Type of dried berries	Interval amount of berry (kg)		
	Grade A	Grade B	Grade C
1. Blueberries	$[\underline{b}_{11}, \overline{b}_{11}]$	$[\underline{b}_{12}, \overline{b}_{12}]$	$[\underline{b}_{13}, \overline{b}_{13}]$
2. Cranberries	$[\underline{b}_{21}, \overline{b}_{21}]$	$[\underline{b}_{22}, \overline{b}_{22}]$	$[\underline{b}_{23}, \overline{b}_{23}]$
3. Raspberries	$[\underline{b}_{31}, \overline{b}_{31}]$	$[\underline{b}_{32}, \overline{b}_{32}]$	$[\underline{b}_{33}, \overline{b}_{33}]$
4. Strawberries	$[\underline{b}_{41}, \overline{b}_{41}]$	$[\underline{b}_{42}, \overline{b}_{42}]$	$[\underline{b}_{43}, \overline{b}_{43}]$

To control the quality of the desserts, the recipe says that the dessert must contain each type of dried berries in a certain level as shown in the following table.

Type of dried berries	Interval amount of berry (kg)	
	Fruit cake	Fruit tart
1. Blueberries	$[\underline{a}_{11}, \overline{a}_{11}]$	$[\underline{a}_{12}, \overline{a}_{12}]$
2. Cranberries	$[\underline{a}_{21}, \overline{a}_{21}]$	$[\underline{a}_{22}, \overline{a}_{22}]$
3. Raspberries	$[\underline{a}_{31}, \overline{a}_{31}]$	$[\underline{a}_{32}, \overline{a}_{32}]$
4. Strawberries	$[\underline{a}_{41}, \overline{a}_{41}]$	$[\underline{a}_{42}, \overline{a}_{42}]$

The relationship of how many kilograms of the mixed dried berries grades A, B, and C that the shop should have and how many fruit cakes and fruit tarts that the shop should make to guarantee the overall quality of the dessert becomes a two-sided interval linear system as follows:

$$\begin{pmatrix} [\underline{a}_{11}, \overline{a}_{11}] & [\underline{a}_{12}, \overline{a}_{12}] \\ [\underline{a}_{21}, \overline{a}_{21}] & [\underline{a}_{22}, \overline{a}_{22}] \\ [\underline{a}_{31}, \overline{a}_{31}] & [\underline{a}_{32}, \overline{a}_{32}] \\ [\underline{a}_{41}, \overline{a}_{41}] & [\underline{a}_{42}, \overline{a}_{42}] \end{pmatrix} \begin{pmatrix} x_1 \\ x_2 \end{pmatrix} = \begin{pmatrix} [\underline{b}_{11}, \overline{b}_{11}] & [\underline{b}_{12}, \overline{b}_{12}] & [\underline{b}_{13}, \overline{b}_{13}] \\ [\underline{b}_{21}, \overline{b}_{21}] & [\underline{b}_{22}, \overline{b}_{22}] & [\underline{b}_{23}, \overline{b}_{23}] \\ [\underline{b}_{31}, \overline{b}_{31}] & [\underline{b}_{32}, \overline{b}_{32}] & [\underline{b}_{33}, \overline{b}_{33}] \\ [\underline{b}_{41}, \overline{b}_{41}] & [\underline{b}_{42}, \overline{b}_{42}] & [\underline{b}_{43}, \overline{b}_{43}] \end{pmatrix} \begin{pmatrix} y_1 \\ y_2 \\ y_3 \end{pmatrix}, \quad (14)$$

where x_j *is the amount of fruit cakes and fruit tarts that should be made, when* $j = 1, 2$*, respectively,*
y_k *is the amount of mixed dried berries grades A, B and C, when* $k = 1, 2, 3$*, respectively.*

The quality of desserts is very important to the bakery shop. The owner wants to re-evaluate the amount x_1, x_2 *and* y_1, y_2, y_3 *to make sure that the mixed dried berries the shop has would be covered by the range of the total amount of the mixed fruits in all dessert items. This way the shop would be able to guarantee the quality of the desserts. The set of solutions satisfies System (14) is considered to be the control solution set. In a situation when the owner wants the amount* x_1, x_2 *and* y_1, y_2, y_3 *as close as possible to the given values* α_1, α_2 *and* $\beta_1, \beta_2, \beta_3$*, respectively, it is the same as to minimize the function below:*

$$\|(x, y)^T - (\alpha, \beta)^T\|^2 = \sum_{j=1}^{2}(x_j - \alpha_j)^2 + \sum_{k=1}^{3}(y_k - \beta_k)^2$$
$$= \sum_{j=1}^{2}(x_j^2 - 2\alpha_j x_j + \alpha_j^2) + \sum_{k=1}^{3}(y_k^2 - 2\beta_k y_k + \beta_k^2). \quad (15)$$

It is sufficient to remove the constant terms α_j^2 *and* β_k^2 *from Eq. (15) for every* $j \in \{1, 2\}$ *and* $k \in \{1, 2, 3\}$*. Therefore, we can model this problem as a quadratic program* P_1 *with a two-sided interval linear constraint (14), whose constraints are obtained by using the statement of Corollary 1 for a control solution in the following way:*

$$P_1: \text{ minimize } f(x, y) = \sum_{j=1}^{2}(x_j^2 - 2\alpha_j x_j) + \sum_{k=1}^{3}(y_k^2 - 2\beta_k y_k)$$

$$\text{subject to} \quad \sum_{j=1}^{2}\underline{a}_{ij}x_j - \sum_{k=1}^{3}\underline{b}_{ik}y_k \le 0, \quad \forall i \in \{1, 2, 3, 4\}$$
$$-\sum_{j=1}^{2}\overline{a}_{ij}x_j + \sum_{k=1}^{3}\overline{b}_{ik}y_k \le 0, \quad \forall i \in \{1, 2, 3, 4\}$$
$$x_j, y_k \ge 0. \quad \forall j \in \{1, 2\}\ \forall k \in \{1, 2, 3\}$$

The Lagrangian function for Problem P_1 *can be written by*

$$L(z, \mu) = c^T z + \frac{1}{2}z^T Q z + \mu(A'z - b'), \quad (16)$$

where

$$c = \begin{pmatrix} -2\alpha_1 \\ -2\alpha_2 \\ -2\beta_1 \\ -2\beta_2 \\ -2\beta_3 \end{pmatrix}, \quad Q = \begin{pmatrix} 2\,0\,0\,0\,0 \\ 0\,2\,0\,0\,0 \\ 0\,0\,2\,0\,0 \\ 0\,0\,0\,2\,0 \\ 0\,0\,0\,0\,2 \end{pmatrix}, \quad z = \begin{pmatrix} x_1 \\ x_2 \\ y_1 \\ y_2 \\ y_3 \end{pmatrix}$$

and

$$A' = \begin{pmatrix} \underline{a}_{11} & \underline{a}_{12} & -\underline{b}_{11} & -\underline{b}_{12} & -\underline{b}_{13} \\ \underline{a}_{21} & \underline{a}_{22} & -\underline{b}_{21} & -\underline{b}_{22} & -\underline{b}_{23} \\ \underline{a}_{31} & \underline{a}_{32} & -\underline{b}_{31} & -\underline{b}_{32} & -\underline{b}_{33} \\ \underline{a}_{41} & \underline{a}_{42} & -\underline{b}_{41} & -\underline{b}_{42} & -\underline{b}_{43} \\ -\overline{a}_{11} & -\overline{a}_{12} & \overline{b}_{11} & \overline{b}_{12} & \overline{b}_{13} \\ -\overline{a}_{21} & -\overline{a}_{22} & \overline{b}_{21} & \overline{b}_{22} & \overline{b}_{23} \\ -\overline{a}_{31} & -\overline{a}_{32} & \overline{b}_{31} & \overline{b}_{32} & \overline{b}_{33} \\ -\overline{a}_{41} & -\overline{a}_{42} & \overline{b}_{41} & \overline{b}_{42} & \overline{b}_{43} \end{pmatrix}, \quad b' = \begin{pmatrix} 0 \\ 0 \\ 0 \\ 0 \\ 0 \\ 0 \\ 0 \\ 0 \end{pmatrix}$$

and $\mu = (\mu_1, \mu_2, \ldots, \mu_8) \geq 0$ *is the Lagrange multiplier with* 5*-dimensional row vector. Let* u *and* v *be surplus and slack variables to the inequalities*

$$c + Qz + (A')^T\mu^T \geq 0 \ \text{ and } \ A'z - b' \leq 0,$$

respectively. As shown in [5], we can represent (16) in the following linear constraints form:

$$Qz + (A')^T\mu^T - u = -c, \tag{17}$$

$$A'z + v = b', \tag{18}$$

$$z \geq 0, \ \ \mu \geq 0, \ \ u \geq 0, \ \ v \geq 0, \tag{19}$$

$$u^T z = 0, \ \ \mu v = 0, \tag{20}$$

where the equations shown in (20) prescribe complementary slackness. Since it can clearly be seen that the matrix Q *is positive definite, the conditions (17)–(20) are necessary and sufficient for a global optimum. To create the appropriate linear program, we add thirteen artificial variables* $a_1, a_2, \ldots, a_{13}$ *to each constraint of (17) and (18) together with minimizing their sum, that is,*

$$\begin{aligned} P_2: \ & \text{minimize } a_1 + a_2 + \cdots + a_{13} \\ & \text{subject to } \quad Qz + (A')^T\mu^T - u + a' = -c, \\ & \qquad\qquad A'z + v + a'' = b', \\ & \qquad\qquad z \geq 0, \ \ \mu \geq 0, \ \ u \geq 0, \ \ v \geq 0, \\ & \qquad\qquad u^T z = 0, \ \ \mu v = 0, \end{aligned}$$

where $a' = (a_1, a_2, \ldots, a_5)^T$ *and* $a'' = (a_6, a_7, \ldots, a_{13})^T$. *Hence, the optimal solution to the quadratic program* P_1 *is found out by solving the linear program* P_2. □

Type of dried berries	Interval amount of berry (kg)				
	Fruit cake	Fruit tart	Grade A	Grade B	Grade C
1. Blueberries	$[0.25, 0.32]$	$[0.20, 0.41]$	$[0.15, 0.30]$	$[0.60, 0.70]$	$[0.22, 0.25]$
2. Cranberries	$[0.05, 0.18]$	$[0.26, 0.35]$	$[0.20, 0.20]$	$[0.23, 0.35]$	$[0.08, 0.12]$
3. Raspberries	$[0.03, 0.15]$	$[0.24, 0.64]$	$[0.10, 0.16]$	$[0.04, 0.10]$	$[0.45, 0.55]$
4. Strawberries	$[0.04, 0.36]$	$[0.34, 0.45]$	$[0.32, 0.48]$	$[0.14, 0.28]$	$[0.18, 0.20]$

Example 1. According to Problem statement 2, we provide the numerical information in the following table.

- The initial amount of fruit cakes and fruit tarts is $\alpha_1 = 16$ and $\alpha_2 = 24$, respectively.
- The initial amount of mixed dried berries grade A, B and C is $\beta_1 = 15, \beta_2 = 12$ and $\beta_3 = 10$, respectively.

As explained in Problem statement 2 with the above information, our problem can be represented by Problem P_2, that is,

$$\text{minimize } a_1 + a_2 + a_3 + a_4 + a_5 + a_6 + a_7 + a_8 + a_9 + a_{10} + a_{11} + a_{12} + a_{13}$$

subject to

$$\begin{aligned}
2x_1 + 0.25\mu_1 + 0.05\mu_2 + 0.03\mu_3 + 0.04\mu_4 - 0.32\mu_5 - 0.18\mu_6 - 0.15\mu_7 - 0.36\mu_8 - u_1 + a_1 &= 32,\\
2x_2 + 0.20\mu_1 + 0.26\mu_2 + 0.24\mu_3 + 0.34\mu_4 - 0.41\mu_5 - 0.35\mu_6 - 0.64\mu_7 - 0.45\mu_8 - u_2 + a_2 &= 48,\\
2y_1 - 0.15\mu_1 - 0.20\mu_2 - 0.10\mu_3 - 0.32\mu_4 + 0.30\mu_5 + 0.20\mu_6 + 0.16\mu_7 + 0.48\mu_8 - u_3 + a_3 &= 30,\\
2y_2 - 0.60\mu_1 - 0.23\mu_2 - 0.04\mu_3 - 0.14\mu_4 + 0.70\mu_5 + 0.35\mu_6 + 0.10\mu_7 + 0.28\mu_8 - u_4 + a_4 &= 24,\\
2y_3 - 0.22\mu_1 - 0.08\mu_2 - 0.45\mu_3 - 0.18\mu_4 + 0.25\mu_5 + 0.12\mu_6 + 0.55\mu_7 + 0.20\mu_8 - u_5 + a_5 &= 20,\\
0.25x_1 + 0.20x_2 - 0.15y_1 - 0.60y_2 - 0.22y_3 + v_1 + a_6 &= 0,\\
0.05x_1 + 0.26x_2 - 0.20y_1 - 0.23y_2 - 0.08y_3 + v_2 + a_7 &= 0,\\
0.03x_1 + 0.24x_2 - 0.10y_1 - 0.04y_2 - 0.45y_3 + v_3 + a_8 &= 0,\\
0.04x_1 + 0.34x_2 - 0.32y_1 - 0.14y_2 - 0.18y_3 + v_4 + a_9 &= 0,\\
-0.32x_1 - 0.41x_2 + 0.30y_1 + 0.70y_2 + 0.25y_3 + v_5 + a_{10} &= 0,\\
-0.18x_1 - 0.35x_2 + 0.20y_1 + 0.35y_2 + 0.12y_3 + v_6 + a_{11} &= 0,\\
-0.15x_1 - 0.64x_2 + 0.16y_1 + 0.10y_2 + 0.55y_3 + v_7 + a_{12} &= 0,\\
-0.36x_1 - 0.45x_2 + 0.48y_1 + 0.28y_2 + 0.20y_3 + v_8 + a_{13} &= 0,
\end{aligned}$$

where all variables are nonnegative and complementary conditions are satisfied. By applying the simplex method, the optimal solution to this problem is eventually displayed as

$$\begin{pmatrix} x_1 \\ x_2 \end{pmatrix} = \begin{pmatrix} 18.49 \\ 22.16 \end{pmatrix} \text{ and } \begin{pmatrix} y_1 \\ y_2 \\ y_3 \end{pmatrix} = \begin{pmatrix} 17.49 \\ 11.01 \\ 8.19 \end{pmatrix}. \qquad \square$$

In this section we have presented two particular situations to show the difference between tolerance and control solutions together with the processes for solving them. The tolerance solution as Problem Statement 1 can directly be accomplished using system of inequalities. The quadratic programming problem constrained with control solutions as Problem Statement 2 was transformed to become an appropriate linear program. Then, the simplex method enables us to obtain the optimal solution of the problem as displayed in the above example.

5 Conclusion

This paper presents the concepts of tolerance and control solutions of a two-sided interval linear system together with their semantics to the system $\boldsymbol{A}x = \boldsymbol{B}y$. The conditions to verify that (x, y) is a tolerance or a control or a tolerance-control solution are also achieved in Theorems 1–3. In application problems, the vectors x and y are normally specified by nonnegative variables. Therefore, we simplify Theorem 2 to be Corollary 1 and applied it to implement a two-sided interval linear system and a quadratic programming with two-sided interval linear constraints. This work should be beneficial to any applications that have the restrictions in the format of two-sided interval linear system.

References

1. Beaumont, O.: Solving interval linear systems with linear programming techniques. Linear Algebra Appl. **281**(1–3), 293–309 (1998)
2. Fiedler, M., Nedoma, J., Ramik, J., Rohn, J., Zimmermann, K.: Linear Optimization Problems with Inexact Data. Springer, New York (2006). https://doi.org/10.1007/0-387-32698-7
3. Gerlach, W.: Zur lösung linearer ungleichungssysteme bei störimg der rechten seite und der koeffizientenmatrix. Optimization **12**(1), 41–43 (1981)
4. Hladík, M.: Weak and strong solvability of interval linear systems of equations and inequalities. Linear Algebra Appl. **438**(11), 4156–4165 (2013)
5. Jensen, P.A., Bard, J.F.: Operations Research Models and Methods, vol. 1. Wiley, Hoboken (2003)
6. Leela-apiradee, W.: New characterizations of tolerance-control and localized solutions to interval system of linear equations. J. Comput. Appl. Math. **355**, 11–22 (2019)
7. Moore, R.E.: Interval Analysis. Prince-Hall, Englewood Cliffs (1969)
8. Nuding, E., Wilhelm, J.: Über gleichungen und über lösungen. Zeitschrift für Angewandte Mathematik und Mechanik **52**, T188–T190 (1972)
9. Oettli, W., Prager, W.: Compatibility of approximate solution of linear equations with given error bounds for coeficients and right-hand sides. Numer. Math. **6**, 405–409 (1964). https://doi.org/10.1007/BF01386090
10. Prokopyev, O.A., Butenko, S., Trapp, A.: Checking solvability of systems of interval linear equations and inequalities via mixed integer programming. Eur. J. Oper. Res. **199**(1), 117–121 (2009)
11. Rohn, J.: Input-Output Planning with Inexact Data, vol. 9. Albert-Ludwigs-Universität, Freiburg (1978). Freiburger Intervall-Berichte
12. Shary, S.P.: On controlled solution set of interval algebraic systems. Interval Comput. **6**, 66–75 (1992)
13. Tian, L., Li, W., Wang, Q.: Tolerance-control solutions to interval linear equations. In: International Conference on Artificial Intelligence and Software Engineering, ICAISE 2013, pp. 18–22. Atlantis Press (2013)

Dealing with Inconsistent Measurements in Inverse Problems: An Approach Based on Sets and Intervals

Krushna Shinde[1(✉)], Pierre Feissel[1], and Sébastien Destercke[2]

[1] Roberval Laboratory, FRE-CNRS 2012, Université de Technologie de Compiègne, 60203 Compiègne, France
{krushna.shinde,pfeissel}@utc.fr

[2] Heudiasyc Laboratory, UMR CNRS 7253, Université de Technologie de Compiègne, 60203 Compiègne, France
sebastien.destercke@hds.utc.fr

https://roberval.utc.fr/, https://www.hds.utc.fr/

Abstract. We consider the (inverse) problem of finding back the parameter values of a physical model given a set of measurements. As the deterministic solution to this problem is sensitive to measurement error in the data, one way to resolve this issue is to take into account uncertainties in the data. In this paper, we explore how interval-based approaches can be used to obtain a solution to the inverse problem, in particular when measurements are inconsistent with one another. We show on a set of experiments, in which we compare the set-based approach with the Bayesian one, that this is particularly interesting when some measurements can be suspected of being outliers.

Keywords: Inverse problem · Interval uncertainty · Outlier detection

1 Introduction

Identifying the parameters of a physical model from a set of measurements is a common task in many fields such as image processing (tomographic reconstruction [1]), acoustic (source identification [2]), or mechanics (material properties identification [3]). Such a problem is known as the *inverse problem* and is the converse of the so-called forward problem. While the forward problem is usually well-posed, it is not the case of the inverse problem. Indeed, whenever there is noise in the measurements or error in the model, such a problem may well end-up having no solutions [4]. Common recourse to this issue that have been proposed in the literature is to consider either Least-square minimization techniques [5] or Bayesian approaches [6] modeling the noise in measurements.

M.-J. Lesot et al. (Eds.): IPMU 2020, CCIS 1239, pp. 449–462, 2020.
https://doi.org/10.1007/978-3-030-50153-2_34

Both these approaches, however, can be quite sensitive to outliers [7,8] or aberrant measurements. In addition to that, a lot of researchers argued that probabilistic methods such as Bayesian inverse methods are not well suited for representing and propagating uncertainty when information is missing or in case of partial ignorance [9–11]. In contrast, interval-valued methods [12–15] make a minimal amount of assumptions about the nature of the associated uncertainties, as they only require to define the region in which should be the measurement. In this paper, we propose an inverse strategy relying on interval analysis to deal with uncertain measurements and to detect inconsistent measurements (outliers). We apply the proposed strategy in experimentation concerning the identification of material elastic parameters in the presence of possibly inconsistent measurements (here, full-field displacements [16]).

This paper is organized as follows. Section 2 describes the identification strategy based on a set-valued approach and outlier detection method to select a set of consistent measurements as well as the numerical implementation of the identification strategy, including the discrete description of sets [11]. In Sect. 3, we present an application to static tensile tests of homogeneous plates to identify material parameters using the proposed outlier detection method and we compare it to the Bayesian inference method with sensitive data.

2 Identification Strategy and Outlier Detection Method

This section is composed of four parts. In Sect. 2.1, we introduce the inverse problem. The identification strategy with a set-valued approach based on intervals is described in Sect. 2.2. Section 2.3 introduces the outlier detection method to select a subset of consistent measurements. Section 2.4 describes the numerical implementation of the identification strategy with the discrete description of sets.

2.1 Inverse Problem Introduction

We consider an inverse problem where we want to identify some parameters of a model $y = f(\theta)$ from N measurements made on quantity y. The model f yields the relationship between the M model parameters $\theta \in \mathbb{R}^M$ and the measured quantity. We will denote by $\tilde{y} \in \mathbb{R}^N$ the measurements made on y. A typical example introduced in Sect. 3 is where θ corresponds to elastic Lamé parameters (λ and μ) and y is full-field displacement data obtained after applying a given load on the material specimen. In this paper, we consider the case where the discrepancy between $f(\theta)$ and $\tilde{y}$ is mainly due to measurement errors, i.e., we leave the issue of model error to future investigations.

2.2 Set-Valued Inverse Problem

In this Section, we propose a set-valued inverse problem strategy based on the interval-valued measurements.

Intervals to Model Uncertainty. Within the framework of Interval analysis, an interval $[x]$ in $\mathbb{R}$ is a closed set of connected real values noted by $[x] = [\underline{x}, \overline{x}] = \{x \in \mathbb{R} \mid \underline{x} \leq x \leq \overline{x}\}$ where $\underline{x} \in \mathbb{R}$ is the lower bound and $\overline{x}$ is the upper bound [17]. In our work, we choose to describe uncertainty on the measurements in interval form, as such a description requires almost no assumption regarding the nature and source of uncertainty [14]. To describe prior information about parameters, we use a multidimensional extension of intervals, i.e. hypercube or box of $\mathbb{R}^n$ defined as the Cartesian product of n intervals. For example, in the case of two parameters, x_1 and x_2, information on them is described by set $\mathbb{X}$ such that $\mathbb{X} = [x_1] \times [x_2] = [\underline{x_1}, \overline{x_1}] \times [\underline{x_2}, \overline{x_2}]$. Boxes are the easiest way to describe multidimensional sets.

Identification Strategy. In the proposed approach, intervals describe the uncertainty on the measurements and an hyper-cube describes the prior information about parameters. Hence, the solution of the inverse problem can be obtained thanks to a set inversion process [17]. The uncertainty in the measurements is described through the set $\mathbb{S}_y$.

$$\mathbb{S}_y = \prod_{k=1}^{N} [\underline{\tilde{y_k}}, \overline{\tilde{y_k}}] \subset \mathbb{R}^N \tag{1}$$

Each measurement is described with its lower bound $\underline{\tilde{y_k}}$ and an upper bound $\overline{\tilde{y_k}}$. Prior information about parameter is described through $\mathbb{S}_{0\theta} \subset \mathbb{R}^M$, i.e., with the box. Given a set $\mathbb{S}_y \subset \mathbb{R}^N$ describing the uncertainty on $\tilde{y}$ and prior parameter set $\mathbb{S}_{0\theta} \subset \mathbb{R}^M$, the set $\mathbb{S}_\theta \subset \mathbb{R}^M$ describing the solution of the inverse problem is defined as follows:

$$\mathbb{S}_\theta = \{\theta \in \mathbb{S}_{0\theta} \mid f(\theta) \in \mathbb{S}_y\} \tag{2}$$

In the current work, it is possible to obtain a solution set for each measurement as follows:

$$\mathbb{S}_\theta^k = \{\theta \in \mathbb{S}_{0\theta} \mid y_k(\theta) \in [\underline{\tilde{y_k}}, \overline{\tilde{y_k}}]\} \tag{3}$$

where $y_k(\theta)$ represents k^{th} response of the model $y = f(\theta)$ and then $\mathbb{S}_\theta$ can be obtained as the intersection of the $\mathbb{S}_\theta^k$:

$$\mathbb{S}_\theta = \bigcap_{k=1}^{N} \mathbb{S}_\theta^k \tag{4}$$

In case of inconsistent measurements, the set-valued inverse method gives an empty solution set $\mathbb{S}_\theta = \emptyset$ corresponding to $\bigcap_{k=1}^{N} \mathbb{S}_\theta^k = \emptyset$. There may be several reasons for the inconsistency of the measurements with respect to the model such as presence of measurement outliers or model error.

Example 1. We illustrate the set-valued inverse problem on a toy example. We consider a spring-mass system shown in Fig. 1 which can be described mathematically as

$$F/p = f(\theta) = y \tag{5}$$

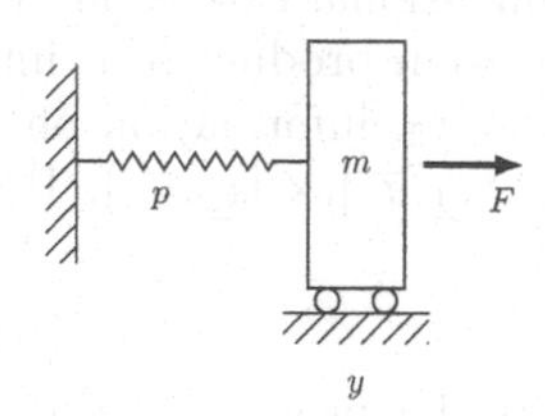

Fig. 1. Spring-mass system

where F represents the force applied on the spring in *Newton* (N), p is the spring stiffness constant (N/m) and the parameter to estimate, and y is the measured displacement of the spring in meter (m). We consider a case where a force F = 100 N is applied on the spring and a displacement, $\tilde{y}$ = 0.01 m is measured. Here, inverse problem consists of determining the parameter from the measurement $\tilde{y}$. To do this, we describe uncertainty on the prior knowledge about the parameter and the measurement in the interval form such that $\mathbb{S}_{0\theta}$ i.e., $p \in [P] = [8000, 12000]$ and $\mathbb{S}_y$ i.e., $\tilde{y} \in [\tilde{Y}] = [0.009, 0.0110]$. We solve the inverse problem numerically using Eq. (2), which gives a set of parameter which is included in set $[\mathbf{P}]$. We obtain the lower bound $\underline{\mathbf{P}} = 9152.5$ and upper bound $\overline{\mathbf{P}} = 11050.1$ of the interval $[\mathbf{P}]$ by taking $y = \overline{\tilde{Y}}$ and $y = \underline{\tilde{Y}}$ respectively. Hence, new interval of parameter is described by $[\mathbf{P}] = [\underline{\mathbf{P}}, \overline{\mathbf{P}}] = [9152.5, 11050.1] \subset [P]$. In the case of 1D, the length of the interval i.e., $\overline{\mathbf{P}} - \underline{\mathbf{P}}$ measures the area of $[\mathbf{P}]$, $\mathcal{A}([\mathbf{P}])$.

2.3 Outlier Detection Method

In case of inconsistency, a way to restore consistency is to remove incompatible measurements, i.e., possible outliers. To do this, our method relies on measures of consistency that we introduce now.

For any two solution sets $\mathbb{S}_\theta^k$ and $\mathbb{S}_\theta^{k'}$ corresponding to $\tilde{y_k}$ and $\tilde{y_{k'}}$ measurement respectively, $(k, k') \in \{1, ..., N\}^2$, we define the degree of inclusion (DOI) of one solution set $\mathbb{S}_\theta^k$ with respect to another $\mathbb{S}_\theta^{k'}$ as

$$DOI_{kk'} = \frac{\mathcal{A}(\mathbb{S}_\theta^k \cap \mathbb{S}_\theta^{k'})}{\mathcal{A}(\mathbb{S}_\theta^{k'})} \tag{6}$$

where $\mathcal{A}(\mathbb{S}_\theta^k)$ corresponds to the area of the set $\mathbb{S}_\theta^k$.

The DOI between two solution sets is non-symmetric, i.e., $DOI_{kk'} \neq DOI_{k'k}$. DOI reaches to its boundary values in the following situations as illustrated in Fig. 2.

$$DOI_{kk'} = \begin{cases} 1 & \text{iff } \mathbb{S}_\theta^{k'} \subseteq \mathbb{S}_\theta^k \\ 0 & \text{iff } \mathbb{S}_\theta^k \cap \mathbb{S}_\theta^{k'} = \emptyset \end{cases} \tag{7}$$

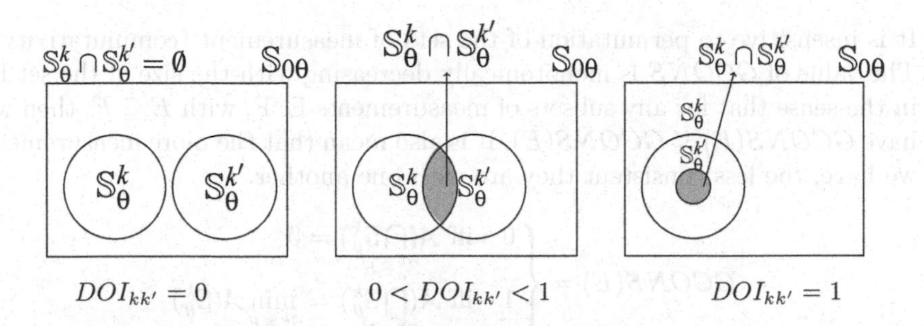

Fig. 2. DOI between two sets

Furthermore, the value of $DOI_{kk'}$ will always be between 0 and 1 when $\mathcal{A}(\mathbb{S}_\theta^k)$ is non-zero. The larger the value of DOI between one solution set and another, the higher the possibility of $\mathbb{S}_\theta^{k'}$ included in $\mathbb{S}_\theta^k$.

We now introduce a measurement-wise consistency degree from a set of measurements. By using the pairwise degree of inclusion (DOI) of the solution sets corresponding to the measurements, we define the global degree of consistency (GDOC) of any k^{th} measurement with respect to all other measurements as

$$GDOC(k) = \frac{\sum_{k'=1}^{N} \frac{\mathcal{A}(\mathbb{S}_\theta^k \cap \mathbb{S}_\theta^{k'})}{\mathcal{A}(\mathbb{S}_\theta^k)} + \sum_{k'=1}^{N} \frac{\mathcal{A}(\mathbb{S}_\theta^k \cap \mathbb{S}_\theta^{k'})}{\mathcal{A}(\mathbb{S}_\theta^{k'})}}{2N} \tag{8}$$

which reaches its boundary values in the following situations:

$$GDOC(k) = \begin{cases} 1 & \text{iff } \mathbb{S}_\theta^1 = \mathbb{S}_\theta^2, ..., = \mathbb{S}_\theta^N \\ 0 & \text{iff } \mathbb{S}_\theta^k \cap \mathbb{S}_\theta^{k'} = \emptyset, \ \forall\, k' \in \{1, ..., N\} \end{cases} \tag{9}$$

The value of $GDOC(k)$ will always be between 0 and 1. Note that the condition for $GDOC = 1$ is very strong, as it requires all sets to be identical. If $GDOC(k) = 0$ then the kth measurement is fully inconsistent with all other measurements. A high value of $GDOC$ for the kth measurement then indicates a high consistency with most of the other measurements.

Finally, we define a global consistency measure for a group of measurements. Let $S = \{\mathbb{S}_\theta^1, \ldots, \mathbb{S}_\theta^k, \ldots, \mathbb{S}_\theta^N\}$ with $\mathbb{S}_\theta^k \subseteq \mathbb{R}^M$ be the set of solutions to the inverse problems for the measurements $\{y_1, \ldots, y_N\}$. We define the general consistency

($GCONS$) for any subset $E \subset S$ of measurements as

$$GCONS(E) = \frac{\mathcal{A}(\bigcap_{\mathbb{S}_\theta^k \in E} \mathbb{S}_\theta^k)}{\min_{\mathbb{S}_\theta^k \in E} \mathcal{A}(\mathbb{S}_\theta^k)} \tag{10}$$

It has the following properties:

1. It is insensitive to permutation of the sets of measurement (commutativity).
2. The value of $GCONS$ is monotonically decreasing with the size of the set E, in the sense that for any subsets of measurements E, F, with $E \subseteq F$, then we have $GCONS(F) \leq GCONS(E)$. It is also mean that the more measurement we have, the less consistent they are with one another.
3. $$GCONS(E) = \begin{cases} 0 & \text{iff } \mathcal{A}(\bigcap_{\mathbb{S}_\theta^k \in E} \mathbb{S}_\theta^k) = \emptyset \\ 1 & \text{iff } \mathcal{A}(\bigcap_{\mathbb{S}_\theta^k \in E} \mathbb{S}_\theta^k) = \min_{\mathbb{S}_\theta^k \in E} \mathcal{A}(\mathbb{S}_\theta^k) \end{cases}$$

A good principle to choose a subset of consistent measurements would be to search for the biggest subset E (the maximal number of measurements) that has a reasonable consistency, that is for which $GCONS(E)$ is above some threshold. Yet, such a search could be exponential in N, which can be quite large, and therefore untractable. This is why we propose a greedy algorithm (Algorithm 1) that makes use of $GDOC$ measures to find a suitable subset E. The idea is quite simple: starting from the most consistent measurement according to $GDOC$ and ordering them according to their individual consistency, we iteratively add new measurements to E unless they bring the global consistency $GCONS$ under a pre-defined threshold, that is unless they introduce too much inconsistency.

Algorithm 1. GCONS outlier detection method

Require: $S = \{\mathbb{S}_\theta^1, ..., \mathbb{S}_\theta^N\}$, $GCONS_{threshold}$ ▷ Set $\mathbb{S}_\theta^1, \mathbb{S}_\theta^2, ...\mathbb{S}_\theta^N$ are arranged such that $GDOC(1) \geq GDOC(2).... \geq GDOC(N)$.

Ensure: Consistent set of solution sets corresponding to consistent measurements, S_{new} from S

1: $S_{new} = \{\mathbb{S}_\theta^1, \mathbb{S}_\theta^2\}$; ▷ Initial set
2: **for** $k \leftarrow 3$ to N **do**
3: $\quad E_k = \{S_{new}\} \cup \{\mathbb{S}_\theta^k\}$ ▷ $\mathbb{S}_\theta^k$ from S
4: $\quad$ **if** $GCONS(E_k) > GCONS_{threshold}$ **then**
5: $\quad\quad$ Accept $\mathbb{S}_\theta^k$
6: $\quad\quad S_{new} = \{S_{new}\} \cup \{\mathbb{S}_\theta^k\}$; ▷ $\mathbb{S}_\theta^k$ from S
7: $\quad$ **else**
8: $\quad\quad S_{new} = S_{new}$; ▷ Basically we are removing the kth measurement which gives solution set $\mathbb{S}_\theta^k$.

2.4 Implementation with Discrete Description of Sets

To solve the set-valued inverse problem, we need a discrete description of the sets. There are multiple ways to represent the sets in a discrete way, such as using boxes (SIVIA algorithm [15]) or a grid of points. Here, we use the same description as in [11], that is a grid of points, θ_i, $i \in \{1, ..., N_g\}$ as shown in Fig. 3a where N_g is the number of grid points. Such a description is convenient when comparing or intersecting the sets since the grid of points is the same for any set. Any set $\mathbb{S}_\theta \subset \mathbb{S}_{0\theta}$ is then characterized through its discrete characteristic function, defined at any point $\theta_i \in \mathbb{S}_{0\theta}$ of the grid as shown in Eq. (11) and Fig. 3b.

$$\chi_{\mathbb{S}_\theta}(\theta_i) = \begin{cases} 1 & \text{if } \theta_i \in \mathbb{S}_\theta \\ 0 & \text{otherwise} \end{cases} \tag{11}$$

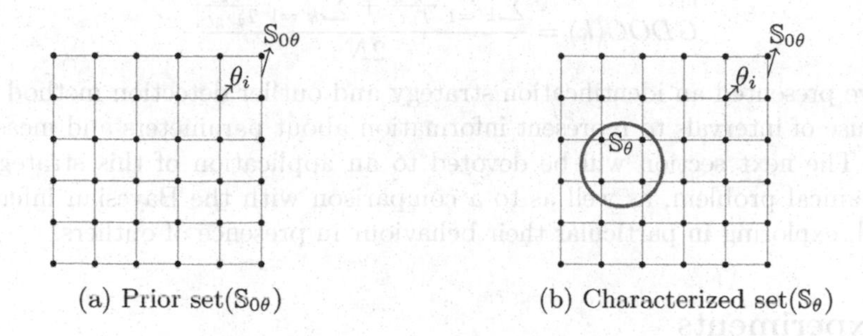

(a) Prior set($\mathbb{S}_{0\theta}$) (b) Characterized set($\mathbb{S}_\theta$)

Fig. 3. Discrete description of sets

In the current application, a uniform grid is chosen to describe the prior parameter set $\mathbb{S}_{0\theta}$, but it is not mandatory. In our method, each $\mathbb{S}_\theta^k$ is therefore described by its discrete characteristic function, defined at any point of the grid as

$$\chi_{\mathbb{S}_\theta^k}(\theta_i) = \begin{cases} 1 & \text{if } \underline{\tilde{y}_k} \leq f(\theta_i) \leq \overline{\tilde{y}_k} \\ 0 & \text{otherwise} \end{cases} \tag{12}$$

These discrete characteristic functions can be collected in a $N_g \times N$ matrix X as columns of boolean values as shown in Eq. (13). $N_g \times N$ matrix X is described as

$$X = \begin{bmatrix} 1\,1\,..\,1 \\ 0\,1\,..\,1 \\ ..\,..\,..\,.. \\ 1\,1\,..\,0 \end{bmatrix} \tag{13}$$

where $\chi_{\mathbb{S}_\theta^k}(\theta_i)$ is the element of column k and line i. Using matrix X, a $N \times N$ symmetric matrix $T = X^T X$ can be obtained, whose components are directly proportional to the inverse sets areas, and can therefore be used as an estimation of such areas:

$$T \propto \begin{bmatrix} \mathcal{A}(\mathbb{S}_\theta^1) & \mathcal{A}(\mathbb{S}_\theta^1 \cap \mathbb{S}_\theta^2) & .. & \mathcal{A}(\mathbb{S}_\theta^1 \cap \mathbb{S}_\theta^{k'}) \\ \mathcal{A}(\mathbb{S}_\theta^1 \cap \mathbb{S}_\theta^2) & \mathcal{A}(\mathbb{S}_\theta^2) & .. & \mathcal{A}(\mathbb{S}_\theta^2 \cap \mathbb{S}_\theta^{k'}) \\ .. & .. & .. & .. \\ \mathcal{A}(\mathbb{S}_\theta^1 \cap \mathbb{S}_\theta^{k'}) & \mathcal{A}(\mathbb{S}_\theta^2 \cap \mathbb{S}_\theta^{k'}) & .. & \mathcal{A}(\mathbb{S}_\theta^k) \end{bmatrix} \tag{14}$$

Indeed, the diagonal element T_{kk} of T represents the number of grid points for which the kth measurement is consistent and it is proportional to $\mathcal{A}(\mathbb{S}_\theta^k)$. The non-diagonal element $T_{kk'}$ of T represents the number of grid points for which both k^{th} and k'^{th} measurements are consistent and it is proportional to $\mathcal{A}(\mathbb{S}_\theta^k \cap \mathbb{S}_\theta^{k'})$. Hence, $GDOC$ can be computed from matrix T for any k^{th} measurement as follows

$$GDOC(k) = \frac{\sum_{k'=1}^{N} \frac{T_{k'k}}{T_{kk}} + \sum_{k'=1}^{N} \frac{T_{kk'}}{T_{k'k'}}}{2N} \tag{15}$$

We have presented an identification strategy and outlier detection method that makes use of intervals to represent information about parameters and measurements. The next section will be devoted to an application of this strategy to a mechanical problem, as well as to a comparison with the Bayesian inference method, exploring in particular their behaviour in presence of outliers.

3 Experiments

In this Section, we apply the set-valued inverse method to identify elastic properties (Lamé parameters: λ and μ) of a homogeneous 2D plate under plane strain as shown in Fig. 4a. The plate is clamped on the left side and loaded on the right side by a uniform traction $f = 1000$ N/m. To generate displacement measurement data $\tilde{y}$ (386 measurements), exact displacement data y^{Ref} is simulated by a Finite Element (FE) model (193 nodes, 336 elements) as shown in Fig. 4b considering the reference values $\lambda_0 = 1.15 \cdot 10^5$ MPa and $\mu_0 = 7.69 \cdot 10^4$ MPa. We also consider a possible Gaussian noise with 0 mean (no systematic bias) and with standard deviation σ. In the current work, σ was taken as 5% of the average of all the exact displacement values and in practical cases it can be assumed that σ can be deduced from the measurement technique.

For the set-valued inverse method, the uncertainty on the measurements has to be given in interval form. Therefore, each measurement is modelled as $[\tilde{y}_k - 2\sigma, \tilde{y}_k + 2\sigma]$. The width of 2σ ensures that sufficient measurements will be consistent with one another. Prior information about the parameters ($\mathbb{S}_{0\theta}$) is considered as a uniform $2D$ box $\lambda^p \times \mu^p$ with $\lambda^p = [0.72\ 10^5,\ 1.90\ 10^5]$ MPa and $\mu^p = [\ 7.2\ 10^4,\ 8.15\ 10^4]$ MPa .

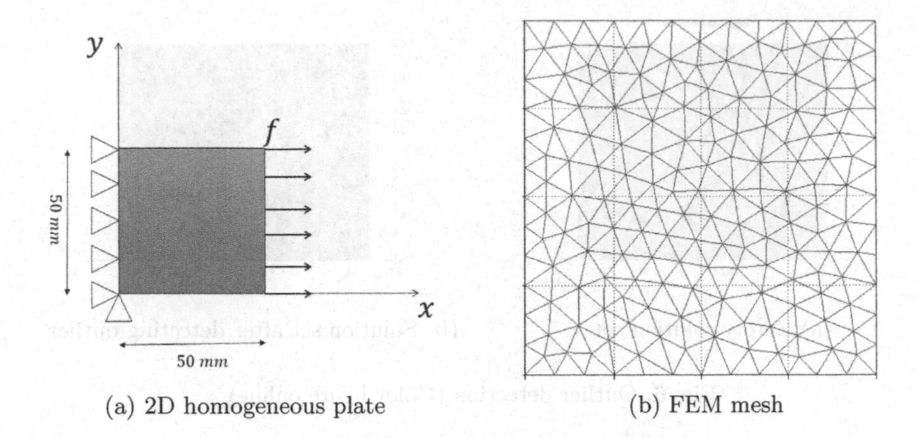

(a) 2D homogeneous plate (b) FEM mesh

Fig. 4. A homogeneous plate and its model

3.1 Application with the Set-Valued Inverse Method

We first apply the set-valued inverse method to identify the set of elastic parameters when there is no noise in the data. The measurement data was chosen such that $\tilde{y} = y^{Ref}$, and the information on the measurement $\tilde{y}$ was described in an interval form: $[\tilde{y} - 2\sigma, \quad \tilde{y} + 2\sigma]$.

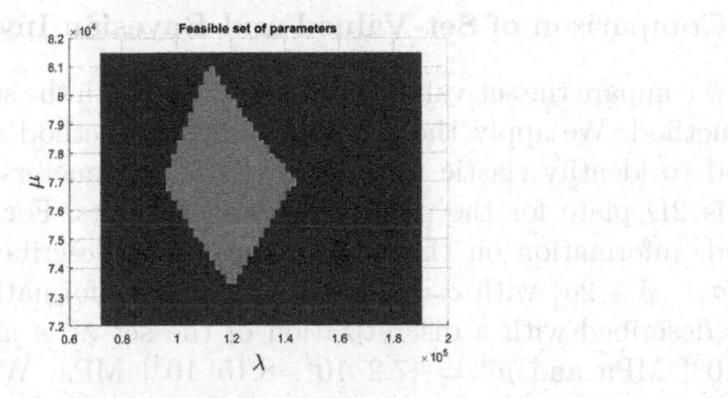

Fig. 5. Feasible set of identified parameters (Color figure online)

Figure 5 shows the feasible set (yellow color) of the identified parameter which is consistent with all 386 measurements using the set-valued inverse method. We can note that the exact value of the parameter included in the solution set.

We then apply the set-valued inverse method along with $GCONS$ outlier detection method (Algorithm 1) to identify the set of elastic parameters when there is random noise in the data. The measurement $\tilde{y}$ is created from y^{Ref} by adding to it a Gaussian white noise with standard deviation σ and the information on the measurement $\tilde{y}$ was described in an interval form: $[\tilde{y} - 2\sigma, \quad \tilde{y} + 2\sigma]$.

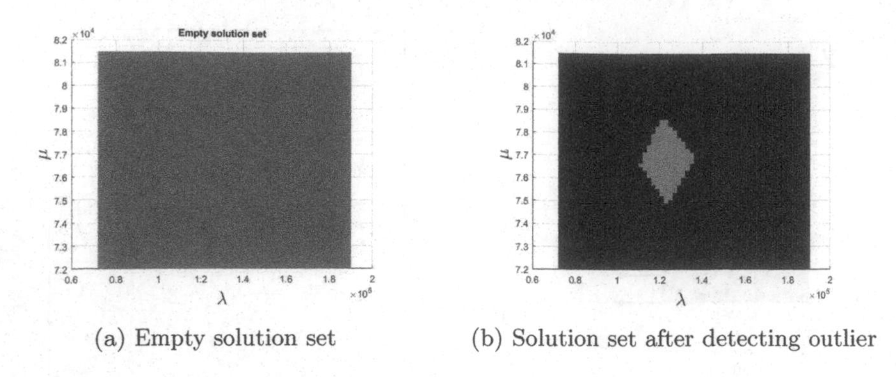

(a) Empty solution set (b) Solution set after detecting outlier

Fig. 6. Outlier detection (Color figure online)

Figure 6a shows that the identified set (green color) when taking all the measurements is empty due to inconsistency within the measurements. To obtain a non-empty solution set, we use our proposed solution and Algorithm 1 with the value of the $GCONS_{threshold}$ settled to 0.1. We use a low value of GCONS to ensure that a high enough number of measurements will be included. Figure 6b shows the feasible set (yellow color) of the identified parameter using GCONS method, with 55 measurements removed. We can note here again that the exact value of the parameter included in the solution set.

3.2 Comparison of Set-Valued and Bayesian Inverse Method

We now compare the set-valued inverse method with the standard Bayesian inference method. We apply the set-valued inverse method and Bayesian inference method to identify elastic properties (Lamé parameters: λ and μ) of a homogeneous 2D plate for the same 386 measurements. For the set-valued inverse method, information on the measurement $\tilde{y}$ is described in an interval form: $[\tilde{y} - 2\sigma, \quad \tilde{y} + 2\sigma]$ with $\sigma = 0.0020$ and prior information about the parameters is described with a discretization of the set $\lambda^p \times \mu^p$ with $\lambda^p = [0.72\ 10^5, 1.90\ 10^5]$ MPa and $\mu^p = [7.2\ 10^4, 8.15\ 10^4]$ MPa. With Bayesian inference method, error on the measurement $\tilde{y}$ is modelled by a Gaussian noise:$\sim \mathcal{N}(0, \sigma^2)$ with $\sigma = 0.0020$ and prior information about the parameter is modeled with a uniform distribution: $U_\lambda(0.72\ 10^5, 1.90\ 10^5)$ MPa, $U_\mu(7.2\ 10^4, 8.15\ 10^4)$ MPa. Rouhgly speaking, this means that the Bayesian model is not misspecified.

Figure 7 shows the feasible set (yellow color) of the identified parameter using the set-valued inverse method and the feasible set (red color) of the identified parameter using Bayesian inference method. In the case of Bayesian inference method, the feasible set (red color) corresponds to a credibility set having a probability of 90%. The results on this specific example indicate that both methods are consistent with each other, with the Bayesian approach delivering more precise inferences. This observation has been made on other simulations using a well-specified Bayesian model.

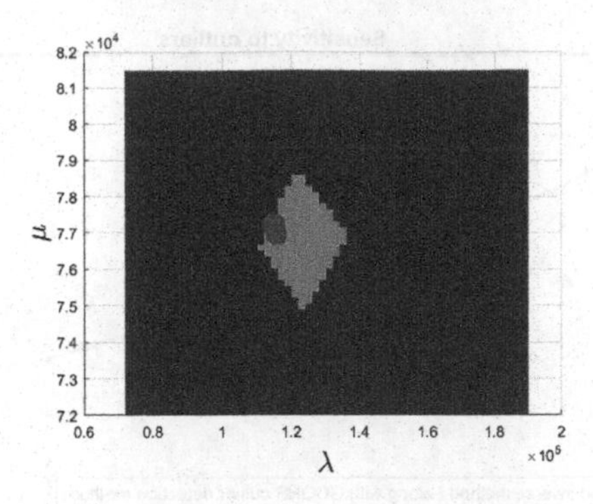

Fig. 7. Feasible set of parameters (Color figure online)

Now, we compare the set-valued inverse method and the Bayesian inference method in terms of their sensitivity to outliers i.e., how they perform when some data becomes aberrant, hence departing from the Bayesian assumptions. To do this, we use 8 sets of 100 experiments (each experiment with 386 measurements) in a way such that for each set the percentage of outlier measurements will increase. In practice, we use the following scheme

$$\tilde{y_0} = y^{Ref} + \epsilon \tag{16}$$

$$\tilde{y} = \tilde{y_0} + \alpha I \epsilon \tag{17}$$

where $\tilde{y_0}$ are noisy measurements, $\epsilon \sim \mathcal{N}(0, \sigma^2)$ is the initial noise, $\alpha = 5$ is a multiplicative constant applied to ϵ when a measurement is an outlier, and $I \sim \mathcal{B}(p_i)$ is a Bernoulli variable with parameter p_i depending on the experiment set, and indicating the average percentage of outlier measurements. In particular, we used the values 0%, 3%, 5%, 7%, 9%, 11%, 13%, 15% for p_i in our sets of experiment, starting from no outliers to an average of 15%.

For each experiment from the 8 sets (thus for 800 experiments), we have performed the identification using our set-valued inverse method and the Bayesian inference method to check their sensitivity towards outliers. For all experiments, we have chosen the value of $GCONS_{threshold} = 0.1$.

For each set of experiment, we have computed the average number of times that each method includes the true parameter values, denoted A_c in Fig. 8.

From this figure, it can be observed that when there is an increase in the percentage of over noisy data points per set, the A_C value starts to decrease in the case of Bayesian inference method but not with $GCONS$ method. So, while the Bayesian approach strongly suffers from a model misspecification, our method is robust to the presence of outliers, even in significant proportion. Hence, we can conclude that the two methods clearly follows different strategies and provide results that are qualitatively different in presence of outliers.

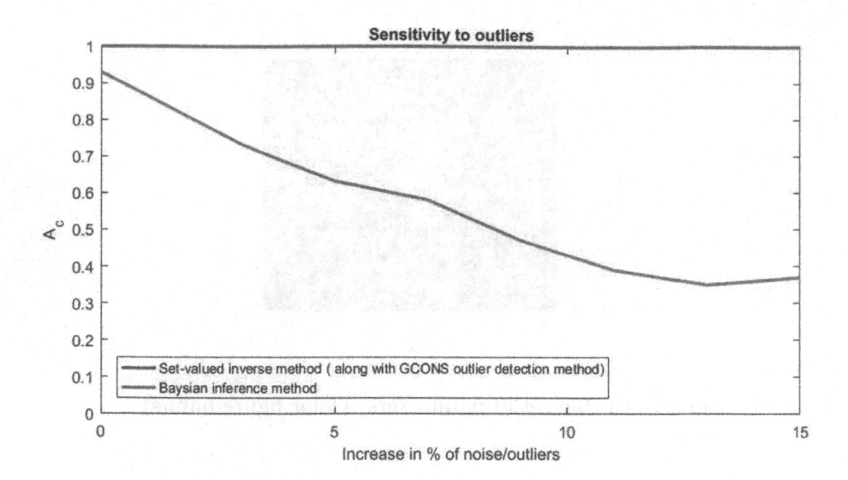

Fig. 8. Consistency with exact parameter values

4 Conclusions

In this paper we have presented a new parameter identification strategy relying on set theory and on interval measurements. In this approach, we have used intervals to describe uncertainty on measurements and parameters. In order to solve the inverse problem, we have proposed a discrete description of sets related to the information about the parameters. We have introduced indicators of consistency of measurements, using them to propose an outlier detection method, i.e., the *GCONS* method.

We applied this strategy to identify the elastic properties of homogenous isotropic material. The results showed that the identification strategy is not only helpful to obtain a feasible set of the parameters but is also able to detect the outliers in the noisy measurements. We also compared our identification strategy with the Bayesian inference method in terms of sensitivity to outliers and results showed that the Bayesian inference method can give a false prediction of the parameter when data is too noisy.

The application of the identification strategy considered in the current work concerns a relatively small number of measurements (at least for mechanical applications) and a 2D parameter identification. However, computational complexity in case of very high dimensions is an important issue that remains to be investigated. The next step in this work is to apply this strategy with high dimensional data as well as parameter identification.

Acknowledgements. The research reported in this paper has been supported by the project Labex MS2T, financed by the French Government through the program Investments for the future managed by the National Agency for Research (Reference ANR-11-IDEX-0004-02).

References

1. Chetih, N., Messali, Z.: Tomographic image reconstruction using filtered back projection (FBP) and algebraic reconstruction technique (ART). In: 2015 3rd International Conference on Control, Engineering Information Technology, CEIT, Tlemcen, pp. 1–6 (2015). https://doi.org/10.1109/CEIT.2015.7233031
2. Eller, M., Valdivia, N.P.: Acoustic source identification using multiple frequency information. Inverse Prob. **25**(11), 115005 (2009). https://doi.org/10.1088/0266-5611/25/11/115005
3. Tam, J.H., Ong, Z.C., Ismail, B.C., Khoo, S.Y.: Identification of material properties of composite materials using nondestructive vibrational evaluation approaches: a review. Mech. Adv. Mater. Struct. **24**(12), 971–986 (2017). https://doi.org/10.1080/15376494.2016.1196798
4. Tikhonov, A.N., Goncharsky, A., Stepanov, V.V., Yagola, A.G.: Numerical Methods for the Solution of Ill-Posed Problems. Mathematics and Its Applications, vol. 328. Springer, Netherlands (1995). https://doi.org/10.1007/978-94-015-8480-7
5. Teughels, A., Roeck, N.G.: Damage detection and parameter identification by finite element model updating. Revue Européenne de Génie Civil **9**(1–2), 109–158 (2005). https://doi.org/10.1080/17747120.2005.9692748
6. Gogu, C., Haftka, R., Riche, R.L., Molimard, J., Vautrin, A.: Introduction to the Bayesian approach applied to elastic constants identification. Mech. Adv. Mater. Struct. **8**(5), 893–903 (2010). https://doi.org/10.2514/1.40922
7. Blais, J.A.R.: Least Squares for Practitioners. Math. Probl. Eng. **2010**, 19 (2010)
8. Chen, Y., Breen, P., Andrew, N.L.: Impacts of outliers and mis-specification of priors on Bayesian fisheries-stock assessment. Can. J. Fish. Aquat. Sci. **57**, 2293–2305 (2000). https://doi.org/10.1139/f00-208
9. Ferson, S., Ginzburg, L.R.: Different methods are needed to propagate ignorance and variability. Reliab. Eng. Syst. Saf. **54**(2), 133–144 (1996). https://doi.org/10.1016/S0951-8320(96)00071-3
10. Elishakoff, I.: Essay on uncertanties in elastic and viscoelastic structures: from AM Freudenthal's criticisms to modern convex modeling. Comput. Struct. **56**(6), 871–895 (1995). https://doi.org/10.1016/0045-7949(94)00499-S
11. Sui, L., Feissel, P., Denæux, T.: Identification of elastic properties in the belief function framework. Int. J. Approximate Reasoning **101**, 69–87 (2018). https://doi.org/10.1016/j.ijar.2018.06.010
12. Helton, J.C., Johnson, J.D.: Quantification of margins and uncertainties: alternative representations of epistemic uncertainty. Reliab. Eng. Syst. Saf. **96**(9), 1034–1052 (2011). https://doi.org/10.1016/j.ress.2011.02.013
13. Moens, D., Hanss, M.: Non-probabilistic finite element analysis for parametric uncertainty treatment in applied mechanics: recent advances. Finite Elem. Anal. Des. **47**(1), 4–16 (2011). https://doi.org/10.1016/j.finel.2010.07.010
14. Zio, E., Pedroni, N.: Literature review of methods for representing uncertainty. 2013–03 of the Cahiers de la Sécurité Industrielle. Foundation for an Industrial Safety Culture, Toulouse (2013). http://www.foncsi.org/en/. ISSN 2100–3874
15. Jaulin, L., Walter, E.: Set inversion via interval analysis for nonlinear bounded-error estimation. Automatica **29**(4), 1053–1064 (1993). https://doi.org/10.1016/0005-1098(93)90106-4

16. Peters, W.H., Ranson, W.F.: Digital imaging techniques in experimental stress analysis. Opt. Eng. **21**(3), 427–431 (1982). https://doi.org/10.1117/12.7972925
17. Jaulin, L., Kieffer, M., Didrit, O., Walter, E.: Applied Interval Analysis. Software Engineering/Programming and Operating Systems. Springer, London (2001). https://doi.org/10.1007/978-1-4471-0249-6

Enhancing the Efficiency of the Interval-Valued Fuzzy Rule-Based Classifier with Tuning and Rule Selection

José Antonio Sanz[1,2](✉), Tiago da Cruz Asmus[1,3], Borja de la Osa[1], and Humberto Bustince[1,2]

[1] Universidad Publica de Navarra, Campus Arrosadia, Pamplona, Navarra, Spain
joseantonio.sanz@unavarra.es

[2] Institute of Smart Cities, Campus Arrosadia, Pamplona, Navarra, Spain

[3] Instituto de Matemática, Estatística e Física, Universidade Federal do Rio Grande, Rio Grande, Brazil

Abstract. Interval-Valued fuzzy rule-based classifier with TUning and Rule Selection, IVTURS, is a state-of-the-art fuzzy classifier. One of the key point of this method is the usage of interval-valued restricted equivalence functions because their parametrization allows one to tune them to each problem, which leads to obtaining accurate results. However, they require the application of the exponentiation several times to obtain a result, which is a time demanding operation implying an extra charge to the computational burden of the method.

In this contribution, we propose to reduce the number of exponentiation operations executed by the system, so that the efficiency of the method is enhanced with no alteration of the obtained results. Moreover, the new approach also allows for a reduction on the search space of the evolutionary method carried out in IVTURS. Consequently, we also propose four different approaches to take advantage of this reduction on the search space to study if it can imply an enhancement of the accuracy of the classifier. The experimental results prove: 1) the enhancement of the efficiency of IVTURS and 2) the accuracy of IVTURS is competitive versus that of the approaches using the reduced search space.

Keywords: Interval-valued fuzzy rule-based classification systems · Interval-valued fuzzy sets · Interval type-2 fuzzy sets · Evolutionary fuzzy systems

1 Introduction

Classification problems [10], which consist of assigning objects into predefined groups or classes based on the observed variables related to the objects, have been widely studied in machine learning. To tackle them, a mapping function from the input to the output space, called classifier, needs to be induced applying a learning algorithm. That is, a classifier is a model encoding a set of criteria

M.-J. Lesot et al. (Eds.): IPMU 2020, CCIS 1239, pp. 463–478, 2020.
https://doi.org/10.1007/978-3-030-50153-2_35

that allows a data instance to be assigned to a particular class depending on the value of certain variables.

Fuzzy Rule-Based Classification Systems (FRBCSs) [16] are applied to deal with classification problems, since they obtain accurate results while providing the user with a model composed of a set of rules formed of linguistic labels easily understood by humans. Interval-Valued FRBCSs (IVFRBCSs) [21], are an extension of FRBCSs where some (or all) linguistic labels are modelled by means of Interval-Valued Fuzzy Sets (IVFSs) [19].

IVTURS [22] is a state-of-the-art IVFRBCS built upon the basis of FARC-HD [1]. First, the two first steps of FARC-HD are applied to learn an initial fuzzy rule base, which is augmented with IVFSs to represent the inherent ignorance in the definition of the membership functions [20]. One of the key components of IVTURS is its Fuzzy Reasoning Method (FRM) [6], where all the steps consider intervals instead of numbers. When the matching degree between an example and the antecedent of a rule has to be computed, IVTURS makes usage of Interval-Valued Restricted Equivalence Functions (IV-REFs) [18]. These functions are introduced to measure the closeness between the interval membership degrees and the ideal ones, $[1, 1]$. Their interest resides in their parametric construction method, which allows them to be optimized for each specific problem. In fact, the last step of IVTURS applies an evolutionary algorithm to find the most appropriate values for the parameters used in their construction.

However, the accurate results obtained when using IV-REFs comes at the price of the computational cost. To use an IV-REF it is necessary to apply several exponentiation operations, which are very time demanding. Consequently, the aim of this contribution is to reduce the run-time of IVTURS by decreasing the number of exponentiation operations required to obtain the same results. To do so, we propose two modifications:

- A mathematical simplification of the construction method of IV-REFs, which allows one to reduce to half the number of exponentiation operations.
- Add a verification step to avoid making computations both with incompatible interval-valued fuzzy rules as well as with do not care labels.

Moreover, the mathematical simplification also offers the possibility of reducing the search space of the evolutionary process carried out in IVTURS. This reduction may imply a different behaviour of the classifier, which may derive to an enhancement of the results. In this contribution, we propose four different approaches to explore the reduced search space for the sake of studying whether they allow one to improve the system's performance or not.

We use the same experimental framework that was used in the paper where IVTURS was defined [22], which consist of twenty seven datasets selected from the KEEL data-set repository [2]. We will test whether our two modifications reduce the run-time of IVTURS and the reduction rate achieved as well as the performance of the four different approaches considered to explore the reduced search space. To support our conclusions, we conduct an appropriate statistical study as suggested in the literature [7,13].

The rest of the contribution is arranged as follows: in Sect. 2 we recall some preliminary concepts on IVFSs, IV-REFs and IVTURS. The proposals for speeding IVTURS up and those to explore the reduced search space are described in Sect. 3. Next, the experimental framework and the analysis of the results are presented in Sects. 4 and 5, respectively. Finally, the conclusions are drawn in Sect. 6.

2 Preliminaries

In this section, we review several preliminary concepts on IVFSs (Sect. 2.1), IV-REFs (Sect. 2.2) and IVFRBCSs (Sect. 2.3).

2.1 Interval-Valued Fuzzy Sets

This section is aimed at recalling the theoretical concepts related to IVFSs. We start showing the definition of IVFSs, whose history and relationship with other type of FSs as interval type-2 FSs can be found in [4].

Let $L([0,1])$ be the set of all closed subintervals in $[0,1]$:

$$L([0,1]) = \{\mathbf{x} = [\underline{x}, \overline{x}] | (\underline{x}, \overline{x}) \in [0,1]^2 \text{ and } \underline{x} \leq \overline{x}\}.$$

Definition 1. *[19] An interval-valued fuzzy set A on the universe $U \neq \emptyset$ is a mapping $A_{IV} : U \rightarrow L([0,1])$, so that*

$$A_{IV}(u_i) = [\underline{A}(u_i), \overline{A}(u_i)] \in L([0,1]), \textit{ for all } u_i \in U.$$

It is immediate that $[\underline{A}(u_i), \overline{A}(u_i)]$ is the interval membership degree of the element u_i to the IVFS A.

In order to model the conjunction among IVFSs we apply t-representable interval-valued t-norms [9] without zero divisors, that is, they verify that $\mathbf{T}(\mathbf{x}, \mathbf{y}) = 0_L$ if and only if $\mathbf{x} = 0_L$ or $\mathbf{y} = 0_L$. We denote them $\mathbf{T}_{T_a,T_b}$, since they are represented by T_a and T_b, which are the t-norms applied over the lower and the upper bounds, respectively. That is, $\mathbf{T}_{T_a,T_b}(\mathbf{x}, \mathbf{y}) = [\mathbf{T_a}(\underline{\mathbf{x}}, \underline{\mathbf{y}}), \mathbf{T_b}(\overline{\mathbf{x}}, \overline{\mathbf{y}})]$. Furthermore, we need to use interval arithmetical operations [8] to make some computations. Specifically, the interval arithmetic operations we need in the work are:

- Addition: $[\underline{x}, \overline{x}] + [\underline{y}, \overline{y}] = [\underline{x} + \underline{y}, \overline{x} + \overline{y}]$.
- Multiplication: $[\underline{x}, \overline{x}] * [\underline{y}, \overline{y}] = [\underline{x} * \underline{y}, \overline{x} * \overline{y}]$.
- Division: $\frac{[\underline{x}, \overline{x}]}{[\underline{y}, \overline{y}]} = [\min(\min(\frac{\underline{x}}{\underline{y}}, \frac{\overline{x}}{\overline{y}}), 1), \min(\max(\frac{\underline{x}}{\underline{y}}, \frac{\overline{x}}{\overline{y}}), 1)]$ with $\underline{y} \neq 0$.

where $[\underline{x}, \overline{x}]$, $[\underline{y}, \overline{y}]$ are two intervals in $\mathbb{R}^+$ so that $\mathbf{x}$ is larger than $\mathbf{y}$.

Finally, when a comparison between interval membership degrees is necessary, we use the total order relationship for intervals defined by Xu and Yager [23] (see Eq. (1)), which is also an admissible order [5].

$$[\underline{x}, \overline{x}] \leq [\underline{y}, \overline{y}] \text{ if and only if } \underline{x} + \overline{x} < \underline{y} + \overline{y} \text{ or } \underline{x} + \overline{x} = \underline{y} + \overline{y} \text{ and } \overline{x} - \underline{x} \geq \overline{y} - \underline{y} \quad (1)$$

Using Eq. (1) it is easy to observe that $0_L = [0,0]$ and $1_L = [1,1]$ are the smallest and largest elements in $L([0,1])$, respectively.

2.2 Interval-Valued Restricted Equivalence Functions

In IVTURS [22], one of the key components are the IV-REFs [11,18], whose aim is to quantify the equivalence degree between two intervals. They are the extension on IVFSs of REFs [3] and their definition is as follows:

Definition 2. *[11,18] An Interval-Valued Restricted Equivalence Function (IV-REF) associated with a interval-valued negation N is a function*

$$IV\text{-}REF : L([0,1])^2 \to L([0,1])$$

so that:

(IR1) $IV\text{-}REF(\boldsymbol{x}, \boldsymbol{y}) = IV\text{-}REF(\boldsymbol{y}, \boldsymbol{x})$ *for all* $\boldsymbol{x}, \boldsymbol{y} \in L([0,1])$*;*
(IR2) $IV\text{-}REF(\boldsymbol{x}, \boldsymbol{y}) = 1_L$ *if and only if* $\boldsymbol{x} = \boldsymbol{y}$*;*
(IR3) $IV\text{-}REF(\boldsymbol{x}, \boldsymbol{y}) = 0_L$ *if and only if* $\boldsymbol{x} = 1_L$ *and* $\boldsymbol{y} = 0_L$ *or* $\boldsymbol{x} = 0_L$ *and* $\boldsymbol{y} = 1_L$*;*
(IR4) $IV\text{-}REF(\boldsymbol{x}, \boldsymbol{y}) = IV\text{-}REF(N(\boldsymbol{x}), N(\boldsymbol{y}))$ *with N an involutive interval-valued negation;*
(IR5) For all $\boldsymbol{x}, \boldsymbol{y}, \boldsymbol{z} \in L([0,1])$*, if* $\boldsymbol{x} \leq_L \boldsymbol{y} \leq_L \boldsymbol{z}$*, then* $IV\text{-}REF(\boldsymbol{x}, \boldsymbol{y}) \geq_L IV\text{-}REF(\boldsymbol{x}, \boldsymbol{z})$ *and* $IV\text{-}REF(\boldsymbol{y}, \boldsymbol{z}) \geq_L IV\text{-}REF(\boldsymbol{x}, \boldsymbol{z})$*.*

In this work we use the standard negation, that is, $N(x) = 1 - x$.

An interesting feature of IV-REFs is the possibility of parametrize them by means of automorphisms as follows.

Definition 3. *An automorphism of the unit interval is any continuous and strictly increasing function* $\phi : [0,1] \to [0,1]$ *so that* $\phi(0) = 0$ *and* $\phi(1) = 1$*.*

An easy way of constructing automorphisms is by means of a parameter $\lambda \in (0, \infty)$: $\varphi(x) = x^{\lambda}$, and hence, $\varphi^{-1}(x) = x^{1/\lambda}$. Some automorphims constructed using different values of the parameter λ are shown in Fig. 1.

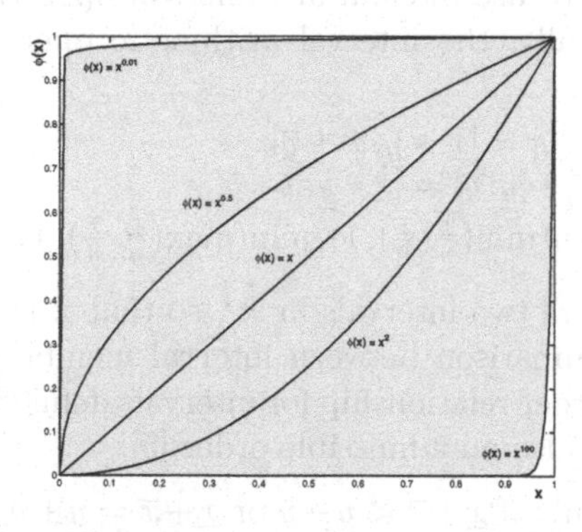

Fig. 1. Example of different automorphisms generated by different values of λ.

Then, the construction method of IV-REFs used in IVTURS can be seen in Eq. (2):

$$IV\text{-}REF(\mathbf{x},\mathbf{y}) = [T(\phi_1^{-1}(1-|\phi_2(\underline{x})-\phi_2(\underline{y})|),\phi_1^{-1}(1-|\phi_2(\overline{x})-\phi_2(\overline{y})|)), \\ S(\phi_1^{-1}(1-|\phi_2(\underline{x})-\phi_2(\underline{y})|),\phi_1^{-1}(1-|\phi_2(\overline{x})-\phi_2(\overline{y})|))] \quad (2)$$

where T is the minimum t-norm, S is the maximum t-conorm and φ_1, φ_2 are two automorphisms of the interval $[0,1]$ parametrized by λ_1 and λ_2, respectively. Therefore, the IV-REFs used in IVTURS are as follows:

$$IV\text{-}REF(\mathbf{x},\mathbf{y}) = [\min((1-|\underline{x}^{\lambda_2}-\underline{y}^{\lambda_2}|)^{1/\lambda_1},(1-|\overline{x}^{\lambda_2}-\overline{y}^{\lambda_2}|)^{1/\lambda_1}), \\ \max((1-|\underline{x}^{\lambda_2}-\underline{y}^{\lambda_2}|)^{1/\lambda_1},(1-|\overline{x}^{\lambda_2}-\overline{y}^{\lambda_2}|)^{1/\lambda_1})] \quad (3)$$

2.3 Interval-Valued Fuzzy Rule-Based Classification Systems

Solving a classification problem consists in learning a mapping function called *classifier* from a set of training examples, named *training set*, that allows new examples to be classified. The training set is composed of P examples, $x_p = (x_{p1},\ldots,x_{pn},y_p)$, where x_{pi} is the value of the i-th attribute ($i = 1,2,\ldots,n$) of the p-th training example. Each example belongs to a class $y_p \in \mathbb{C} = \{C_1, C_2, ..., C_m\}$, where m is the number of classes of the problem.

IVFRBCSs are a technique to deal with classification problems [20], where each of the n attributes is described by a set of linguistic terms modeled by their corresponding IVFSs. Consequently, they provide an interpretable model as the antecedent part of the fuzzy rules is composed of a subset of these linguistic terms as shown in Eq. (4).

$$\text{Rule } R_j: \text{ If } x_1 \text{ is } A_{j1} \text{ and } \ldots \text{ and } x_n \text{ is } A_{jn} \text{ then Class} = C_j \text{ with } RW_j \quad (4)$$

where R_j is the label of the jth rule, $x = (x_1,\ldots,x_n)$ is an n-dimensional pattern vector, A_{ji} is an antecedent IVFS representing a linguistic term, C_j is the class label, and RW_j is the rule weight [17].

IVTURS [22] is an state-of-the-art IVFRBCSs, whose learning process is composed of two steps:

1. To build an IV-FRBCS. This step involves the following tasks:
   - The generation of an initial FRBCS by applying FARC-HD [1].
   - Modelling the linguistic labels of the learned FRBCS by means of IVFSs.
   - The generation of an initial IV-REF for each variable of the problem.
2. To apply an optimization approach with a double purpose:
   - To learn the best values of the IV-REFs' parameters, that is, the values of the exponents of the automorphisms (λ_1 and λ_2).
   - To apply a rule selection process in order to decrease the system's complexity.

In order to be able to classify new examples, $x_p = (x_{p1},\ldots,x_{pn})$, IVTURS considers an Interval-Valued Fuzzy Reasoning Method [22] (IV-FRM), which uses the L interval-valued fuzzy rules composing the model as follows:

1. *Interval matching degree*: It quantifies the strength of activation of the if-part for all rules (L) in the system with the example x_p:

$$[\underline{A_j}(x_p), \overline{A_j}(x_p)] = \mathbf{T}_{T_a,T_b}(IV\text{-}REF([\underline{A_{j1}}(x_{p1}), \overline{A_{j1}}(x_{p1})], [1,1]), \ldots, IV\text{-}REF([\underline{A_{jn}}(x_{pn}), \overline{A_{jn}}(x_{pn})], [1,1])), \qquad j = 1, \ldots, L. \tag{5}$$

2. *Interval association degree*: for each rule, R_j, the interval matching degree is weighted by its rule weight $RW_j = [\underline{RW_j}, \overline{RW_j}]$::

$$[\underline{b_j}(x_p), \overline{b_j}(x_p)] = [\underline{\mu_{A_j}}(x_p), \overline{\mu_{A_j}}(x_p)] * [\underline{RW_j}, \overline{RW_j}] \quad j = 1, \ldots, L. \tag{6}$$

3. *Interval pattern classification soundness degree for all classes.* The positive interval association degrees are aggregated by class applying an aggregation function *f*.

$$[\underline{Y_k}, \overline{Y_k}] = f_{R_j \in RB;\, C_j = k}([\underline{b_j}(x_p), \overline{b_j}(x_p)] | [\underline{b_j}(x_p), \overline{b_j}(x_p)] > 0_L), \qquad k = 1, \ldots, m. \tag{7}$$

4. *Classification.* A decision function F is applied over the interval soundness degrees:

$$F([\underline{Y_1}, \overline{Y_1}], \ldots, [\underline{Y_m}, \overline{Y_m}]) = \arg \max_{k=1,\ldots,m}([\underline{Y_k}, \overline{Y_k}]) \tag{8}$$

3 Enhancing the Efficiency of IVTURS

IVTURS provides accurate results when tackling classification problems. However, we are concerned about its computational burden as it may be an obstacle to use it in real-world problems. The most computationally expensive operation in IVTURS is the exponentiation operation required when computing the IV-REFs, which are constantly used in the IV-FRM (Eq. (5)). Though there are twelve exponentiation operations in Eq. (3), only four of them need to be computed because: 1) $\underline{y} = \overline{y} = 1$, implying that the computation of $\underline{y}^{\lambda_2}$ and $\overline{y}^{\lambda_2}$ can be avoided as one raised to any number is one; 2) the lower and the upper bound of the resulting IV-REF are based on the minimum and maximum of the same operations, which reduces the number of operations to the half.

The aim of this contribution is to reduce the number of exponentiation operations needed to execute IVTURS, which will imply an enhancement of the system's efficiency. To do so, we propose two modifications to the original IVTURS: 1) to apply a mathematical simplification of the IV-REFs that reduces to half the number of exponentiation operations (Sect. 3.1) and 2) to avoid applying IV-REFs with both do not care labels and incompatible interval-valued fuzzy rules (Sect. 3.2).

Furthermore, the mathematical simplification of IV-REFs, besides reducing the number of exponentiation operations while obtaining the same results, would also allow us to also reduce the search space of the evolutionary algorithm, possibly implying in a different behavior in the system. We will study whether this reduction of the search space could result in a better performance of the system by using four different approaches to explore it (Sect. 3.3).

3.1 IV-REFs Simplification

IV-REFs are used to measure the degree of closeness (equivalence) between two intervals. In IVTURS, they are used to compute the equivalence between the interval membership degrees and the ideal membership degree, that is, IV-REF($[\underline{x}, \overline{x}], [1, 1]$). Precisely, because one of the input intervals is $[1, 1]$, we can apply the following mathematical simplification.

$$IV-REF([\underline{x}, \overline{x}], [1, 1]) = [(1-|\underline{x}^{\lambda_2} - 1^{\lambda_2}|)^{1/\lambda_1}, (1-|\overline{x}^{\lambda_2} - 1^{\lambda_2}|)^{1/\lambda_1}] = [\underline{x}^{\lambda_2/\lambda_1}, \overline{x}^{\lambda_2/\lambda_1}] \quad (9)$$

Therefore, we can obtain the same result by just raising the value of the interval membership degree to the division of both exponents (λ_2/λ_1), which imply reducing to half the number of operations.

3.2 Avoiding Incompatible Rules and Do Not Care Labels

When the inference process is applied to classify a new example, the interval matching degree has to be obtained for each rule of the system. The maximum number of antecedents of the interval-valued fuzzy rules used in IVTURS is limited to a certain hyper-parameter of the algorithm, k_t, whose default value is 3. This fact implies that in almost all the classification problems the usage of do not care labels is necessary, since the number of input attributes is greater than that of k_t. In order to program this feature of IVTURS, a do not care label is considered as an extra membership function that returns the neutral element for the t-representable interval-valued t-norm used ($[1, 1]$ in this case as the product is applied). In this manner, when performing the conjunction of the antecedents the usage of do not care labels do not change the obtained result. However, this fact implies that when having a do not care label, it returns $[1, 1]$ as interval membership degree and IV-REF ($[1, 1], [1, 1]$) needs to be computed ($[\underline{x}, \overline{x}] = [1, 1]$). Consequently, a large number of exponentiation operations can be saved if we avoid computing IV-REF in this situations as the result is always $[1, 1]$.

On the other hand, we also propose to avoid obtaining the interval matching degree and thus computing the associated IV-REFs when the example is not compatible with the antecedent of the interval-valued fuzzy rule. To do so, we need to perform an initial iteration where we check whether the example is compatible with the rule. Then, the interval matching degree is only computed when they are compatible. This may see to be an extra charge for the run-time but we take advantage of this first iteration to obtain the interval matching degrees, avoiding the do not care labels, and we send them to the function that computes the interval matching degree.

These two modifications could have a huge impact on the run-time of IVTURS because do not care labels are very common in the interval-valued fuzzy rules of the system and the proportion of compatible rules with an example is usually small.

3.3 Reducing the Search Space in the Evolutionary Process of IVTURS

In Sect. 3.1 we have presented a mathematical simplification of the IV-REFs that allows one to reduce the number of exponentiation operations obtaining exactly the same results than that obtained in the original formulation of IVTURS. However, according to Eq. (9) we can observe that both parameters of the simplified IV-REF (λ_1, λ_2) can be collapsed into a unique one (λ) as shown in Eq. (10).

$$IV - REF([\underline{x}, \overline{x}], [1, 1]) = [\underline{x}^{\lambda_2/\lambda_1}, \overline{x}^{\lambda_2/\lambda_1}] = [\underline{x}^{\lambda}, \overline{x}^{\lambda}] \tag{10}$$

In this manner, the search space of the evolutionary process carried out in IVTURS, where the values of λ_1 and λ_2 are tuned to each problem, can be also reduced to half because only the value of λ needs to be tuned. Consequently, the behaviour of the algorithm can change and we aim at studying whether this reduction is beneficial or not. Specifically, the structure of the chromosome is: $C_i = (g_{\lambda_1}, g_{\lambda_2}, \ldots, g_{\lambda_n})$, where $g_{\lambda_i}, i = 1, \ldots, n$, are the genes representing the value of λ_i and n is the number of input variables of the classification problem.

The parameter λ can vary theoretically between zero and infinity. However, in IVTURS, λ_1 and λ_2 are limited to the interval $[0.01, 100]$. On the other hand, in the evolutionary process, those genes used to encode them are codified in $[0.01, 1.99]$, $g_{\lambda_i} \in [0.01, 1.99]$, in such a way that the chances of learning values in $[0.01, 1]$ and in $(1, 100]$ are the same. Consequently, these genes have to be decoded so that they are in the range $[0.01, 100]$ when used in the corresponding IV-REF. The decoding process is driven by the following equation:

$$g_{\lambda_i} = \begin{cases} g_{\lambda_i}, & \text{if } 0 < g_{\lambda_i} \leq 1 \\ \frac{1}{2 - g_{\lambda_i}}, & \text{if } 1 < g_{\lambda_i} < 2 \end{cases} \tag{11}$$

In [12], Galar et al. use REFs (the numerical counterpart of IV-REFs) to deal with the problem of difficult classes applying the OVO decomposition strategy. In this method, on the one hand, those genes used for representing the parameter λ are coded in the range $(0, 1)$. On the other hand, the decoding process of the genes is driven by Eq. 12.

$$\lambda_i = \begin{cases} (2 \cdot g_{\lambda_i})^2 & \text{if } g_{\lambda_i} \leq 0.5 \\ \dfrac{1}{(1 - 2 \cdot (g_{\lambda_i} - 0.5))^2} & \text{otherwise.} \end{cases} \tag{12}$$

There are two main differences between these two methods: 1) the decoded value by Eq. (11) is in the range $[0.01, 100]$, whereas when using Eq. (12) the values are in $(0, \infty)$ and 2) the search space is explored in a different way as can be seen in the two first rows of Fig. 2, where the left and the right columns show how the final values when $\lambda_i \leq 1.0$ and $\lambda_i > 1.0$ are obtained, respectively.

Looking at these two methods, we propose another two new ones:

- Linear exploration of the search space: we encode all the genes in the range $[0.01, 1.99]$ and we decode them using a linear normalization in the ranges

[0.0001, 1.0] and (1.0, 10000] for the genes in [0.01, 1.0] and (1.0, 1.99], respectively.

- Mixture of Eq. (11) and Eq. (12). Genes are encoded in (0, 1) and they are decoded using Eq. (11) for genes in (0, 0.5] (linear decoding: $2 \cdot g_{\lambda_i}$)and Eq. (12) for genes in (0.5, 1.0].

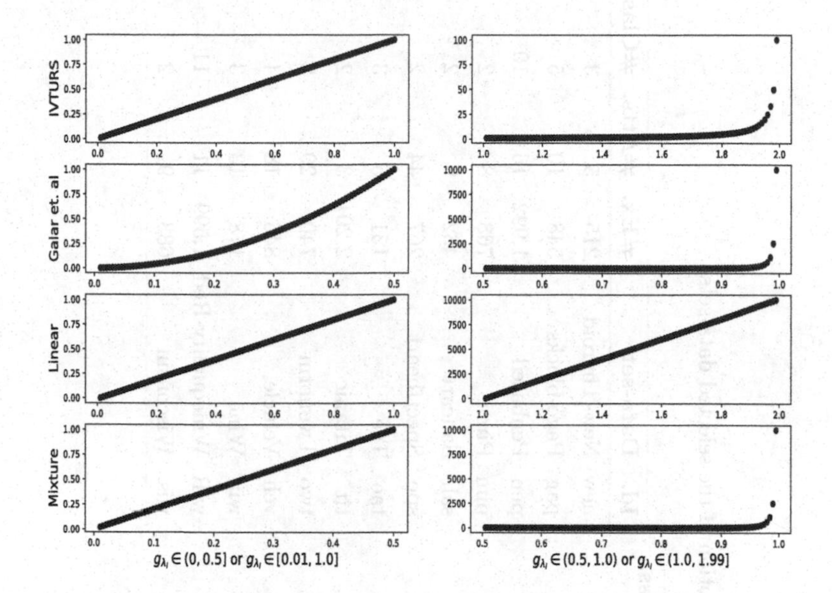

Fig. 2. Effect of the decoding method of the parameter g_λ on the way how the search space is explored.

4 Experimental Framework

We have considered the same datasets which were used in the paper where IVTURS was proposed. That is, we select twenty-seven real world data-sets from the KEEL data-set repository [2]. Table 1 summarizes their properties: number of examples (#Ex.), attributes (#Atts.) and classes (#Class.)[1]. We apply a *5-fold cross-validation model* using the standard accuracy rate to measure the performance of the classifiers.

In this contribution we use the configuration of IVTURS that was used in the paper were it was defined:

[1] We must recall that, as in the IVTURS' paper, the *magic, page-blocks, penbased, ring, satimage* and *shuttle* data-sets have been stratified sampled at 10% in order to reduce their size for training. In the case of missing values (*crx, dermatology* and *wisconsin*), those instances have been removed from the data-set.

Table 1. Description of the selected data-sets.

Id.	Data-set	#Ex.	#Atts.	#Class.
aus	Australian	690	14	2
bal	Balance	625	4	3
cle	Cleveland	297	13	5
con	Contraceptive	1,473	9	3
crx	Crx	653	15	2
der	Dermatology	358	34	6
eco	Ecoli	336	7	8
ger	German	1,000	20	2
hab	Haberman	306	3	2
hay	Hayes-Roth	160	4	3
hea	Heart	270	13	2
ion	Ionosphere	351	33	10
iri	Iris	150	4	3
mag	Magic	1,902	10	2

Id.	Data-set	#Ex.	#Atts.	#Class.
new	New-Thyroid	215	5	3
pag	Page-blocks	548	10	5
pen	Penbased	1,992	16	10
pim	Pima	768	8	2
sah	Saheart	462	9	2
spe	Spectfheart	267	44	2
tae	Tae	151	5	3
tit	Titanic	2,201	3	2
two	Twonorm	740	20	2
veh	Vehicle	846	18	4
win	Wine	178	13	3
wiR	Winequality-Red	1,599	11	11
wis	Wisconsin	683	9	2

- Fuzzy rule learning:
  - $Minsup$: 0.05.
  - $Maxconf$: 0.8.
  - $Depth_{max}$: 3.
  - k_t: 2.
- Evolutionary process
  - Population Size: 50 individuals.
  - Number of evaluations: 20,000.
  - Bits per gene for the Gray codification (for incest prevention): 30 bits.
- IVFSs construction:
  - Number of linguistic labels per variable: 5 labels.
  - Shape: Triangular membership functions.
  - Upper bound: 50% greater than the lower bound ($W = 0.25$).
- Configuration of the initial IV-REFs:
  - T-norm: minimum.
  - T-conorm: maximum.
  - First automorphism: $\phi_1(x) = x^1$ $(a = 1)$.
  - Second automorphism: $\phi_2(x) = x^1$ $(b = 1)$.
- Rule weight: fuzzy confidence (certainty factor) [17].
- Fuzzy reasoning method: additive combination [6].
- Conjunction operator: product interval-valued t-norm.
- Combination operator: product interval-valued t-norm.

5 Analysis of the Obtained Results

This section is aimed at showing the obtained results having a double aim:

1. To check whether the two modifications proposed for enhancing the run-time of IVTURS allow one to speed it up or not.
2. To study if the reduction of the search space made possible by the mathematical simplification of the IV-REFs allows one to improve the results of IVTURS.

In first place we show in Table 2 the run-time in seconds of the three versions of IVTURS[2], namely, the original IVTURS, IVTURS using the mathematical simplification of the IV-REFs ($IVTURS_{v1}$) and IVTURS avoiding the usage of incompatible interval-valued fuzzy rules and do not care labels ($IVTURS_{v2}$). For $IVTURS_{v1}$, the number in parentheses is the reduction rate achieved versus the original IVTURS, whereas in the case of $IVTURS_{v2}$ it is the reduction rate achieved with respect to $IVTURS_{v1}$.

[2] We do not show the accuracy of the methods because they obtain the same results.

Table 2. Run-time in seconds of IVTURS besides the two versions developed for speeding it up. The number in parentheses shows the reduction rate of the method in the column versus the method in its respective left column.

Dataset	IVTURS	$IVTURS_{v1}$	$IVTURS_{v2}$
aus	2032.22	1019.17 (x1.99)	288.22 (x3.54)
bal	909.98	463.19 (x1.96)	193.64 (x2.39)
cle	1460.90	720.98 (x2.03)	133.24 (x5.41)
con	5572.14	2896.43 (x1.92)	971.67 (x2.98)
crx	1781.01	885.32 (x2.01)	227.00 (x3.90)
der	2167.93	1087.80 (x1.99)	107.32 (x10.14)
eco	466.07	244.90 (x1.90)	87.32 (x2.80)
ger	9462.10	4894.81 (x1.93)	838.09 (x5.84)
hab	97.28	53.69 (x1.81)	31.78 (x1.69)
hay	99.16	52.54 (x1.89)	32.75 (x1.60)
hea	764.16	388.49 (x1.97)	102.06 (x3.81)
ion	1314.86	663.30 (x1.98)	81.15 (x8.17)
iri	35.10	19.61 (x1.79)	7.40 (x2.65)
mag	3840.45	1976.30 (x1.94)	581.34 (x3.40)
new	96.94	51.98 (x1.87)	22.77 (x2.28)
pag	571.19	293.44 (x1.95)	66.80 (x4.39)
pen	6417.29	3273.78 (x1.96)	511.53 (x6.40)
pim	1433.95	700.68 (x2.05)	306.43 (x2.29)
sah	963.71	485.20 (x1.99)	181.29 (x2.68)
spe	1319.11	647.94 (x2.04)	117.44 (x5.52)
tae	141.29	76.32 (x1.85)	40.33 (x1.89)
tit	490.51	261.57 (x1.88)	141.57 (x1.85)
two	2426.83	1203.81 (x2.02)	225.71 (x5.33)
veh	4264.57	2162.36 (x1.97)	416.60 (x5.19)
win	231.12	117.65 (x1.96)	23.41 (x5.02)
wiR	5491.86	2807.14 (x1.96)	842.27 (x3.33)
wis	726.39	360.51 (x2.01)	91.74 (x3.93)
Mean	2021.41	1029.96 (x1.95)	247.07 (x4.02)

Looking at the obtained results we can conclude that the two versions allow IVTURS to be more efficient. In fact, $IVTURS_{v1}$ allows one to reduce to half the run-time of IVTURS as expected, since the number of exponentiation operations is also reduced to half. On the other hand, $IVTURS_{v2}$ exhibits a huge reduction on the run-time with respect to that of the original IVTURS as it is 7.839 times faster (1.95*4.02). These modifications allow IVTURS to be applied in a wider range of classification problems as its efficiency has been notably enhanced. The code of the IVTURS method using the two modification for speeding it up can be found at: https://github.com/JoseanSanz/IVTURS.

The second part of the study is to analyze whether the reduction of the search space enabled by the mathematical simplification if the IV-REFs allows one to improve the accuracy of the system or not. As we have explained in Sect. 3.3, we propose four approaches to codify and explore the reduced search space: 1) the same approach than that used in the original IVTURS but using the reduced search space ($\text{IVTURS}_{Red.}$); 2) the approach defined by Galar et al. [12] but extended on IVFSs (IVTURS_{Galar}); 3) the mixture of the two previous approaches ($\text{IVTURS}_{Mix.}$) and 4) the linear exploration of the search space (IVTURS_{Linear}).

In Table 3 we show the results obtained in testing by these four approach besides those obtained by the original IVTURS. We stress in **bold-face** the best result for each dataset. Furthermore, we also show the averaged performance in the 27 datasets (Mean).

According to the results shown in Table 3, we can observe that both methods using the approach defined by Galar et al. (IVTURS_{Galar} and $\text{IVTURS}_{Mix.}$) allows one to improve the averaged accuracy of IVTURS. The reduction of the search space using the original approach defined in IVTURS, $\text{IVTURS}_{Red.}$, does not provide competitive results whereas the approach using a linear exploration of the search space also obtains worse results than those of the original IVTURS.

In order to give statistical support to our analysis we have carried out the Aligned Friedman's ranks test [14] to compare these five methods, whose obtained p-value is 1.21E-4 that implies the existence of statistical differences among them. For this reason, we have applied the Holm's post hoc test [15] to compare the control method (the one associated with the less rank) versus the remainder ones. In Table 4, we show both the ranks of the methods computed by the Aligned Friedman's test as well as the Adjusted P-Value (APV) obtained when applying the Holm's test.

Looking at the results of the statistical study we can conclude that IVTURS, $\text{IVTURS}_{Mix.}$ and $\text{IVTURS}_{Galar.}$ are statistically similar. However, there are statistical differences with respect to $\text{IVTURS}_{Red.}$ and a trend in favour to the three former methods when compared versus IVTURS_{Linear}. All in all, we can conclude that the approach defined in the original IVTURS provides competitive results even when compared against methods whose search space is reduced to half.

Table 3. Testing results provided by IVTURS and the four approaches using the reduced search space.

Dataset	IVTURS	$IVTURS_{Red.}$	$IVTURS_{Galar}$	$IVTURS_{Linear}$	$IVTURS_{Mix.}$
aus	**85.80**	85.07	84.20	84.64	85.36
bal	**85.76**	85.28	85.28	85.60	85.12
cle	**59.60**	58.24	58.58	56.22	58.93
con	53.36	53.02	53.16	**53.97**	53.57
crx	**87.14**	85.91	86.68	85.92	85.15
der	**94.42**	93.58	94.13	93.02	94.14
eco	78.58	78.28	80.96	**82.13**	80.07
ger	73.10	72.00	72.90	73.10	**73.30**
hab	72.85	73.17	72.19	71.55	**73.50**
hay	80.23	75.70	81.00	78.66	**81.77**
hea	**88.15**	85.93	87.41	86.67	**88.15**
ion	89.75	90.60	**92.60**	91.46	92.04
iri	96.00	**97.33**	96.00	96.00	96.00
mag	79.76	79.07	80.28	**80.91**	80.49
new	95.35	95.81	95.35	96.74	**97.21**
pag	95.07	94.16	94.70	**95.43**	94.89
pen	92.18	89.91	**92.64**	91.73	91.64
pim	75.90	74.48	74.87	**76.04**	74.61
sah	70.99	70.13	69.05	**71.20**	70.56
spe	80.52	79.39	**81.26**	80.15	80.16
tae	50.34	**58.30**	57.66	53.68	57.01
tit	**78.87**	**78.87**	**78.87**	**78.87**	**78.87**
two	92.30	90.95	92.43	91.22	**92.70**
veh	**67.38**	64.54	66.43	64.43	67.26
win	**97.19**	94.37	95.48	94.94	96.06
wiR	58.28	59.47	59.04	**59.66**	59.16
wis	96.49	**96.63**	**96.63**	96.34	96.34
Mean	80.57	80.01	80.73	80.38	**80.89**

Table 4. Results obtained by the Aligned Friedman's rank test and the Holm's test.

Method	Rank	APV
$IVTURS_{Mix.}$	52.48	
IVTURS	58.00	0.76
$IVTURS_{Galar}$	61.85	0.76
$IVTURS_{Linear}$	74.91	0.11
$IVTURS_{Red.}$	92.76	6.19E-4

6 Conclusion

In this contribution we have proposed two modifications over IVTURS aimed at enhancing its efficiency. On the one hand, we have used a mathematical simplification of the IV-REFs used in the inference process. On the other hand, we avoid making computations with both incompatible interval-valued fuzzy rules and do not care labels, since they do not affect the obtained results and they entail a charge to the computational burden of the method. Moreover, we have proposed a reduction of the search space of the evolutionary process carried out in IVTURS using four different approaches.

The experimental results have proven the improvement of the run-time of the method, since it is almost eight times faster that the original IVTURS when applying the two modifications. Regarding the reduction of the search space we have learned the following lessons: 1) the new methods based on the approach defined by Galar et al. allow one to improve the results without statistical differences versus IVTURS; 2) the simplification of the search space using the same setting defined in IVTURS does not provide competitive results, possibly due to the limited range where the genes are decoded when compared with respect the remainder approaches and 3) the linear exploration of the search space does not provide good results neither, which led us think that the most proper values are closer to one than to ∞.

Acknowledgments. This work was supported in part by the Spanish Ministry of Science and Technology (project TIN2016-77356-P (AEI/FEDER, UE)) and the Public University of Navarre under the project PJUPNA1926.

References

1. Alcala-Fdez, J., Alcala, R., Herrera, F.: A fuzzy association rule-based classification model for high-dimensional problems with genetic rule selection and lateral tuning. IEEE Trans. Fuzzy Syst. **19**(5), 857–872 (2011)
2. Alcalá-Fdez, J., et al.: KEEL data-mining software tool: data set repository, integration of algorithms and experimental analysis framework. J. Multiple-Valued Logic Soft Comput. **17**(2–3), 255–287 (2011)
3. Bustince, H., Barrenechea, E., Pagola, M.: Restricted equivalence functions. Fuzzy Sets Syst. **157**(17), 2333–2346 (2006)
4. Bustince, H., et al.: A historical account of types of fuzzy sets and their relationships. IEEE Trans. Fuzzy Syst. **24**(1), 179–194 (2016)
5. Bustince, H., Fernandez, J., Kolesárová, A., Mesiar, R.: Generation of linear orders for intervals by means of aggregation functions. Fuzzy Sets Syst. **220**(Suppl. C), 69–77 (2013)
6. Cordón, O., del Jesus, M.J., Herrera, F.: A proposal on reasoning methods in fuzzy rule-based classification systems. Int. J. Approximate Reason. **20**(1), 21–45 (1999)
7. Demšar, J.: Statistical comparisons of classifiers over multiple data sets. J. Mach. Learn. Res. **7**, 1–30 (2006)
8. Deschrijver, G.: Generalized arithmetic operators and their relationship to t-norms in interval-valued fuzzy set theory. Fuzzy Sets Syst. **160**(21), 3080–3102 (2009)

9. Deschrijver, G., Cornelis, C., Kerre, E.: On the representation of intuitionistic fuzzy t-norms and t-conorms. IEEE Trans. Fuzzy Syst. **12**(1), 45–61 (2004)
10. Duda, R.O., Hart, P.E., Stork, D.G.: Pattern Classification, 2nd edn. Wiley, Hoboken (2001)
11. Galar, M., Fernandez, J., Beliakov, G., Bustince, H.: Interval-valued fuzzy sets applied to stereo matching of color images. IEEE Trans. Image Process. **20**(7), 1949–1961 (2011)
12. Galar, M., Fernández, A., Barrenechea, E., Herrera, F.: Empowering difficult classes with a similarity-based aggregation in multi-class classification problems. Inf. Sci. **264**, 135–157 (2014)
13. García, S., Fernández, A., Luengo, J., Herrera, F.: Advanced nonparametric tests for multiple comparisons in the design of experiments in computational intelligence and data mining: experimental analysis of power. Inf. Sci. **180**(10), 2044–2064 (2010)
14. Hodges, J.L., Lehmann, E.L.: Ranks methods for combination of independent experiments in analysis of variance. Ann. Math. Stat. **33**, 482–497 (1962)
15. Holm, S.: A simple sequentially rejective multiple test procedure. Scand. J. Stat. **6**, 65–70 (1979)
16. Ishibuchi, H., Nakashima, T., Nii, M.: Classification and Modeling with Linguistic Information Granules: Advanced Approaches to Linguistic Data Mining. Springer, Berlin (2004)
17. Ishibuchi, H., Nakashima, T.: Effect of rule weights in fuzzy rule-based classification systems. IEEE Trans. Fuzzy Syst. **9**(4), 506–515 (2001)
18. Jurio, A., Pagola, M., Paternain, D., Lopez-Molina, C., Melo-Pinto, P.: Interval-valued restricted equivalence functions applied on clustering techniques. In: Carvalho, J., Kaymak, D., Sousa, J. (eds.) Proceedings of the Joint 2009 International Fuzzy Systems Association World Congress and 2009 European Society of Fuzzy Logic and Technology Conference, Lisbon, Portugal, 20–24 July 2009, pp. 831–836 (2009)
19. Sambuc, R.: Function Φ-Flous, Application a l'aide au Diagnostic en Pathologie Thyroidienne. Ph.D. thesis, University of Marseille (1975)
20. Sanz, J., Fernández, A., Bustince, H., Herrera, F.: A genetic tuning to improve the performance of fuzzy rule-based classification systems with interval-valued fuzzy sets: degree of ignorance and lateral position. Int. J. Approximate Reason. **52**(6), 751–766 (2011)
21. Sanz, J., Fernández, A., Bustince, H., Herrera, F.: Improving the performance of fuzzy rule-based classification systems with interval-valued fuzzy sets and genetic amplitude tuning. Inf. Sci. **180**(19), 3674–3685 (2010)
22. Sanz, J., Fernández, A., Bustince, H., Herrera, F.: IVTURS: a linguistic fuzzy rule-based classification system based on a new interval-valued fuzzy reasoning method with tuning and rule selection. IEEE Trans. Fuzzy Syst. **21**(3), 399–411 (2013)
23. Xu, Z.S., Yager, R.R.: Some geometric aggregation operators based on intuitionistic fuzzy sets. Int. J. General Syst. **35**(4), 417–433 (2006)

Robust Predictive-Reactive Scheduling: An Information-Based Decision Tree Model

Tom Portoleau[1,2](✉), Christian Artigues[1](✉), and Romain Guillaume[2](✉)

[1] LAAS-CNRS, Université de Toulouse, CNRS, Toulouse, France
{tom.portoleau,christian.artigues}@laas.fr
[2] ANITI, IRIT-CNRS, Université de Toulouse, Toulouse, France
romain.guillaume@irit.fr

Abstract. In this paper we introduce a proactive-reactive approach to deal with uncertain scheduling problems. The method constructs a robust decision tree for a decision maker that is reusable as long as the problem parameters remain in the uncertainty set. At each node of the tree we assume that the scheduler has access to some knowledge about the ongoing scenario, reducing the level of uncertainty and allowing the computation of less conservative solutions with robustness guarantees. However, obtaining information on the uncertain parameters can be costly and frequent rescheduling can be disturbing. We first formally define the robust decision tree and the information refining concepts in the context of uncertainty scenarios. Then we propose algorithms to build such a tree. Finally, focusing on a simple single machine scheduling problem, we provide experimental comparisons highlighting the potential of the decision tree approach compared with reactive algorithms for obtaining more robust solutions with fewer information updates and schedule changes.

1 Introduction

Dealing with uncertainty in scheduling problems is an issue that has been widely studied over the last decade. Many approaches emerged in order to cope with uncertainties. Proactive methods elaborate an initial baseline schedule while taking into account possible incoming breaks to make it as robust as possible. There exist several robustness criteria [7] that are widely used in these methods. However, one major flaw of proactive scheduling methods is their conservatism [9], the more robust solutions tend to be low-quality objective-wise. It is particularly the case when uncertainties are large and frequent. In these cases, reactive methods are more appropriate as they do not specially focus on the initial schedule but rather on how to modify it online, that is to say during the execution. They are usually based on priority rules [10] that allow decision maker to compute quickly new solutions according to the running scenario. However, on the contrary of proactive approaches, these is no guarantee regarding the objective value of solutions computed with a reactive method. In order to provide

M.-J. Lesot et al. (Eds.): IPMU 2020, CCIS 1239, pp. 479–492, 2020.
https://doi.org/10.1007/978-3-030-50153-2_36

solutions that are well balanced between robustness and quality, many hybrid methods have been suggested. The recoverable robustness framework [1] considers two stage decisions where robust decisions are taken at the first stage and where recovery algorithms are used to restore feasibility for a realised scenario on the second stage. This framework is linked to adjustable robust optimisation [12] that roughly transposes the concept of two-stage stochastic programming to robust optimisation: some recourse variables can be adjusted to the realised scenario.

In the scheduling literature, approaches that mix a proactive phase aiming at issuing a robust baseline schedule and a reactive phase that adapts the baseline scheduling in case of major disturbances have been widely studied under the proactive-reactive scheduling terminology [11]. Recently [3] remarked that in this series of approaches, the proactive and the reactive phases were rather treated separately while they should mutually influence each other. So in [3,4] the authors propose an integrated proactive reactive approach where they aim to find the best policy, which is in their case a robust initial schedule and a set of reactions giving transitions from a schedule to another schedule in response to a disruption, given a certain reaction cost. In a pioneering work, [6] proposed the so-called Just In Case approach, in which they compute a multiple contingent schedule, where transitions from a baseline schedule to alternative schedules were anticipated at some events having a high probability of break. This approach has been since then largely developed for AI planning problems, addressing among others incrementality and memory limits issues [5,8].

In this paper, we are interested in transposing contingency planning concepts to robust scheduling problems. We consider repeated scheduling problems where some parameters are known and constant at each scheduling iteration and some of them are known with imprecision, according to scenarios. In order to take advantage of known and constant parameters, we introduce a proactive reactive method in which we suppose that the decision maker may have access to some information about the imprecise parameters to reduce the uncertainty at predefined time points of the schedule execution, where the schedule can be changed. However, obtaining information on the uncertain parameters can be costly and frequent rescheduling can be disturbing. Hence at each decision time point the scheduler has the choice to use the information or not and to react or not depending on the impact of the reaction on the schedule robustness. Using intelligently these information, we build off-line a decision tree that will be used to schedule the problem at each repetition.

The problem we chose to validate our approach is the simple scheduling problem $1||L_{max}$, in which we suppose that the processing time are known and fixed, and the due dates are uncertain. We detail why we chose this problem in Sect. 4.

The outline of the paper is the following. Section 2 formally define the uncertainty and information models as well as the robust decision tree concept and its related problem. Then, we detail the algorithm we designed to solve these problems, namely the general algorithm for building the robust decision tree (Sect. 3) and the algorithm for solving the robust partitioning subproblem (Sect. 4). We then present some numerical results in Sect. 5 comparing our approach to a more standard proactive-reactive scheduling algorithm.

2 Notations and Definitions

2.1 Uncertainty Model

In practice, it is often easier for a decision-maker to establish bounds over uncertain parameters, like processing times or due dates, than to build an accurate probabilistic model which often requires a large amount of data. In this paper, we consider interval uncertainty. Given an uncertain parameter x we denote by $X = [x_{min}, x_{max}]$ the interval in which it takes its value. We make no assumption about which probabilistic law is followed by x in this interval. Given a set of uncertain parameters $(x_i)_{i \in I}$ we define the set of possible assignments of parameters by $\Omega = \prod_{i \in I} X_i$ (i.e. it is assumed that there is no correlation between them, all realisation x_i are independent). We call a scenario an assignment of each parameter $x_i, \forall i \in I$ such that $(x_i)_{i \in I} \in \Omega$.

From now on, when ω is a scenario, s a schedule and f the objective, we denote by $f(\omega, s)$ the objective value of s in scenario ω (note that for our application case, $f = L_{max}$).

In this paper we consider the $1||L_{max}$ problem, that is to say we want to schedule a set of I jobs with deterministic processing times $p_i, i \leq I$ and uncertain due dates $d_i \in D_i, i \leq I$ on a single machine so that the maximum lateness is minimum.

Example 1. *We consider a small instance of the problem* $1||L_{max}$ *with three tasks:*

- *task 1:* $p_1 = 10$, $d_1 \in D_1 = [10, 11]$
- *task 2:* $p_2 = 6$, $d_2 \in D_2 = [11, 17]$
- *task 3:* $p_3 = 4$, $d_3 \in D_3 = [13, 20]$

The set of scenarios for this instance is $\Omega = D_1 \times D_2 \times D_3$ *and* $\omega = (10, 12, 19)$ *is a scenario. We will keep this instance all along this paper to exemplify the notions and algorithms we introduce.*

2.2 Information Model

As explained in Sect. 1 we suppose that at some time during the execution of the schedule, some information become accessible. In our model, an information allows the scheduler to tighten the interval of uncertainty of a future realisation.

Definition 1. *For a given uncertain parameter* $x \in X$ *and a moment of decision* t *during the execution of the schedule, we call an information about* x *a value* k_x^t *and an operator in* $\{\leq, \geq\}$.

For instance, an information is $x \leq k_x^t$. So the decision maker, from time t on, is able to reduce the set of possible assignments by updating the interval X, $x \in X_{k_x^t}^{inf} = [x_{min}, k_x^t]$ or $x \in X_{k_x^t}^{sup} =]k_x^t, x_{max}]$. Note that the information depends on t, so the scheduler may have to ask for an information about the same data several times during the execution of the planning. Our model aims to make the best use of available information, and select the more relevant ones. For the problem considered in this paper we suppose that for a given moment of decision t and a task i, we have an information k_d^t if $\min(t + p_i, 2t) \in D_i$. If so, $k_d^t = \min(t + p_i, 2t)$. Otherwise we consider that we have no information about the task i. This hypothesis on the availability of information is arbitrary, but in fact it expresses two natural questions a scheduler may ask about uncertain due dates: "If task i started now, can it be completed without being late ? If so, is the due date d_i far from now ?" In any case, an answer to these questions allows the scheduler to bound the uncertainty of a due date d_i.

Example 2. *Let us look again at the instance from Example 1. We suppose that we are at a moment of decision* $t = 10$ *and that task 1 has been scheduled first. Given the hypothesis we made about information availability, we have:*

- *task 1 is completed, so there is no relevant information about it.*
- $\min(t + p_2, 2t) = 16 \in D_2$*, so* $k_{d_2}^t = 16$.
- $\min(t + p_3, 2t) = 14 \in D_3$*, so* $k_{d_3}^t = 14$.

Therefore, for any $\omega \in \Omega$*, the scheduler is able to determine, from time* $t = 10$*, if* $\omega_2 \leq 16$ *or if* $\omega_2 > 16$*, and if* $\omega_3 \leq 14$ *or if* $\omega_3 > 14$.

2.3 Robust Decision Tree

Definition 2. *A robust decision tree* T *is a tree where the nodes are labeled with a subset of* Ω*, and the arcs are labeled with a partial schedule. If* n *is a node of* T*,* Ω^n *denotes a subset of* Ω *associated to* n*. A robust decision tree satisfies the following properties:*

(i) *Let us denote by* $(n_j)_{j \leq J}$ *the children of node* n*. Then* $\bigcup_{j \leq J} \Omega^{n_j} = \Omega^n$.

(ii) *For any path* $(n_0, ..., n_m)$ *where* n_0 *is the root of* T *and* n_m *is a leaf, the schedule obtained by concatenating all the partial schedules on the arcs along the path is feasible.*

(iii) *Let* n *and* n' *be two nodes of* T*. The partial schedule* s' *on the arc* (n, n') *is robust:*

$$s' = \underset{s \in S}{\operatorname{argmin}} \max_{\omega \in \Omega^{n'}} f(\omega, s)$$

where S *is the set of admissible partial solutions.*

A robust decision tree can be seen as a compact representation of a set of solutions. Given that the decision maker has access to information that allow to split the set of scenarios, the tree makes it possible to retrieve a solution, with a robustness guarantee, for certain subsets of scenario. A generic illustration of

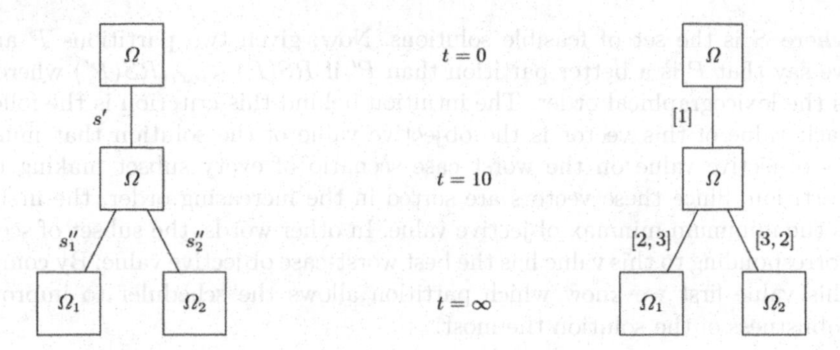

Fig. 1. Generic example of a Robust Decision Tree.

Fig. 2. Example of a robust decision for Example 3.

a robust decision tree is displayed in Fig. 1. In this case, if the decision maker is able to know if an ongoing scenario ω is in Ω_1 or in Ω_2, he can pick the more robust solution for each case.

In this paper our goal is, for a given set of scenario, to compute a robust decision tree that is actually usable by a decision maker during the execution of the schedule, allowing to adapt the schedule online and to make it as robust as possible, depending on some knowledge or information the decision maker has access to. For this purpose, we consider in our model that each level j in the tree coincide with a fixed moment of time t_j during the progress of the schedule. These moments are the points in time when the decision maker has access to some knowledge about uncertain parameters. An illustration of this is given in Example 3.

Example 3. *Still using data from Example 1, if we consider one moment of decision $t = 10$ we can use the information computed in Example 2. For this example, we set $\Omega_1 = D_1 \times [11, 16] \times D_3$ and $\Omega_2 = \Omega \setminus \Omega_1$. Then the robust decision tree shown in Fig. 2 is valid. The first task to be scheduled is task 1, regardless of the ongoing scenario, then thanks to the information accessible at time $t = 10$, the decision maker knows if the ongoing scenario ω is in Ω_1 or in Ω_2, and then can switch to the more robust solution (respectively scheduling task 2 before task 3 if $\omega \in \Omega_1$ and task 3 before task 2 otherwise).*

2.4 Partitioning the Scenarios

The core problem of our method is, for any node n, computing a robust partition of the scenario set, but how do one compare the robustness of two different partitions? We propose the following criterion. We define the Robustness Score (RS) of a partition P as the sorted vector of the worst case objective values of the optimal robust solution (considering absolute robustness criterion [7]) on each element of P:

$$RS(P) = (\min_{s \in S} \max_{\omega \in p} f(\omega, s))_{p \in P}$$

where S is the set of feasible solutions. Now, given two partitions P and P', we say that P is a better partition than P' if $RS(P) <_{lexi} RS(P')$ where $<_{lexi}$ is the lexicographical order. The intuition behind this criterion is the following: each value of this vector is the objective value of the solution that minimises its objective value on the worst case scenario of every subset making up the partition. Since these vectors are sorted in the increasing order, the first value is the minimum minmax objective value. In other words, the subset of scenarios corresponding to this value has the best worst-case objective value. By comparing this value first we know which partition allows the scheduler to improve the robustness of the solution the most.

Example 4. *Let us consider the partition P of $\Omega = \Omega_1 \cup \Omega_2$ from Example 3. Let us denote by $RV(\Omega)$ the minmax objective value on Ω, or more formally $RV(\Omega) = \min_{s \in S} \max_{\omega \in \Omega} f(\omega, s)$. We have:*
$RS(P) = (\min(RV(\Omega_1), RV(\Omega_2)), \max(RV(\Omega_1), RV(\Omega_2)))$
$RS(P) = (6, 7)$
Let now be P' another partition of $\Omega = \Omega'_1 \cup \Omega'_2$, with $\Omega'_1 = D_1 \times D_2 \times [13, 14]$ and $\Omega_2 = \Omega \setminus \Omega_1$. We compute $RS(P') = (4, 7)$. Since $RS(P') <_{lexi} RS(P)$, P' is a better partition considering our criterion.

As we have seen, an information k^t_x allows us to split in two the set of scenarios, since it enables us to distinguish scenarios where the data $x \in X^{inf}_{k^t_x}$ from those where $x \in X^{sup}_{k^t_x}$. More generally, if m information are available, we can split the set of scenarios in 2^m subsets. We denote by K^t the set of all information available at time t. Our goal is to use these subsets to create a size-limited partition of the set of scenarios and compute for each subset of scenarios within the partition a new robust solution. The idea is that diminishing the size of the set of scenarios necessarily improves the worst-case scenario and thus leads to better robust solutions. The maximum size of this new partition is a parameter L, decided by the decision maker. Moreover, the maximum number of information the partition is allowed to use is another parameter Q.

We express the core problem (our method involves solving it multiple times) of our approach, the Robust Partitioning Problem (RPP):

ROBUST PARTITIONING PROBLEM
INSTANCE: A set of scenarios Ω of dimension I, a set of information K^t and two integers Q and $L \leq 2^Q$.
SOLUTION: A partition P of Ω that verifies:

1- $|P| \leq L$
2- $\forall p \in P, \exists J_p \in \mathbb{N}, p = \bigcup_{j \leq J_p} \prod_{i \leq I} p_{i,j}$ with $p_{i,j} \in \{X^{inf}_{k^t_{x_i}}, X^{sup}_{k^t_{x_i}}, X_i\}$ if $k^t_{x_i} \in K^t$ and $p_{i,j} = X_i$ otherwise
3- $|\{k^t_{x_i} \in K^t | \exists p \in P, \exists j \leq J_p, \exists i \leq I, p_i \in \{X^{inf}_{k^t_{x_i}}, X^{sup}_{k^t_{x_i}}\}| \leq Q$

MEASURE: $RS(P)$ minimal for the lexicographical order.

Constraint 1 limits the size of the solution partition so it must be lower than L. Constraint 2 forces the partition to be composed of subsets formed by the information from K^t. In other words, a partition such that any of its subset cuts through the hyperplane defined by $\omega_i = k^t_{x_i}$ is not a feasible solution. And, finally, constraint 3 forces the maximum number of information to be lower than Q. Note that in the solution if we have $p_{i,j} = X_i$ for all $p_{j,i}$ and $k^t_{x_i} \in K^t$, for a parameter i, this means that despite the availability of a new information on parameter i at time t, this information has been ignored as the uncertainty set X_i is not partitioned according to $k^t_{x_i}$. This can be due to the limit on the number Q of information that can be used (Constraint 3) or to the absence of positive impact on this partition on the lexmin objective (i.e. the information is not locally relevant to improve the robustness).

Now that our model is set, the next sections introduce the algorithms we implemented to produce a robust decision tree and solve the RPP.

3 Robust Decision Tree Algorithm

In this section, we detail the general algorithm to build a robust decision tree.

Using the previous definitions we now propose a method to build a robust decision tree (see Definition 2). In this paper, we consider that the moments of decision (i.e. moments when the scheduler is able to access new information and change the schedule), denoted by $(t_j)_{j \in J}$ are fixed in advance. This may correspond in practice to special times, such as the end of a working day, or a shift change where the planning can be updated. Every decision moment corresponds to a level in the decision tree, such that t_1 corresponds to the first level, t_2 to the second one, etc. In that respect, the depth of the tree is controlled by the number of decision moments. At each fork at a level j in the tree, a new partial solution, consistent with the partial schedule that has been accomplished until t_j, is proposed according to the current set of scenarios. The root of the tree, that we consider being the level 0 corresponds to the time $t_0 = 0$, the beginning of the schedule. At this point no information is known, so only one robust solution is proposed. Thus, a single node is created at level 1. At this node, we retrieve all the information available at time t_1. Using up to Q information, we split the set of scenarios into -at most 2^Q- subsets forming a partition P. We then solve the Robust Partition Problem (we detail more about this solution procedure in Sect. 4), and obtain a robust partition P'. For each subsets in P' a new solution is proposed and a new branch is set up, leading to a new node at the next level. The set of scenarios considered in this node is the one from which it originated in P'. These steps are repeated until the last decision moment is reached.

4 Robust Partition Algorithm

In the general case, one can clearly see that this problem is highly combinatorial. Indeed we have to, for each combination of information, compute the best partition using these information. The complexity of the RPP depends on three factors. The first two are the number of tasks to schedule (let say n), since it gives an upper bound of the number of information available at each moment of decision, and Q the number of information we are allowed to use at each moment of decision. The number of possible combinations at a moment of decision is bounded by $\sum_{q=1}^{Q} \binom{n}{q}$. The third factor is the complexity of computing the min-max robust objective value $\min_{s \in S} \max_{\omega \in \Omega} f_\omega(s)$ for any Ω. Clearly, if the deterministic problem is already difficult, so is its robust variant. However, there are cases where a deterministic scheduling problem is polynomial, while its uncertain alternative is NP-Hard. We chose the problem $1||L_{max}$ to test our approach specifically because its robust min-max alternative is still polynomial [2].

Proposition 1. *If f admits a global worst case scenario, that is to say that there exists a scenario ω^{wc} such that:*

$$\forall \omega' \in \Omega, \forall s \in S, f(\omega^{wc}, s) \geq f(\omega', s)$$

then given a partition P of Ω and an integer L, it is possible to compute in polynomial time a partition P' so that $RS(P')$ is minimal and that satisfies:

(i) $|P'| = \min(L, |P|)$
(ii) for all $p \in P$ there exists $p' \in P'$ such that $p \subset p'$

Remark: For the $1||L_{max}$ problem, the global worst case scenario is $\omega = (d_{i_{min}})_{i \in I}$.

Proof. We prove the proposition by showing that any partition returned by Algorithm 1 verifies the properties from Proposition 1.
By construction, a partition P' returned by Algorithm 1 satisfies (i) and (ii). Now we must prove that $RS(P')$ is minimal. First we can observe that:

$$\min_{s \in S} \max_{\omega \in \bigcup_{j \in J} P_j} f(\omega, s) = \min_{s \in S} \max_{j \in J} \max_{\omega \in P_j} f(\omega, s)$$

for any family of disjoint sets $(P_j)_{j \in J}$. From that observation we can derive that for any partition P'' that satisfies (ii), the values contained in $RS(P'')$ are necessarily in $RS(P)$. Thus for any other partition P'' that verifies (ii) we have, for $i \leq L - 1$,

$$RS(P')_i \leq RS(P'')_i \tag{1}$$

because, by construction, $RS(P')_i$ is the i-th smallest possible value and RS vectors are sorted in the increasing order. In addition, since f admits a global worst case scenario ω^{wc}, there exists $p' \in P$ such that $\omega^{wc} \in p'$. Thus, for any union of subset of the form $\bigcup_{p \in P} p$ we have:

$$\min_{s \in S} \max_{\omega \in \bigcup_{p \in P} p} f(\omega, s) = \min_{s \in S} \max_{\omega \in p'} f(\omega, s) = \min_{s \in S} f(\omega^{wc}, s)$$

Thereby, for any partition P'' the last value in $RS(P'')$ is necessarily $\min_{s \in S} f(\omega^{wc}, s)$. Finally, from this result and (1) we conclude that $RS(P')$ is minimal. □

Algorithm 1

Require: $P = [P_1, P_2, ..., P_j]$ a partition of Ω and an integer L.
  for $i \leq j$ **do**
    $RS[i] \leftarrow \min_{s \in S} \max_{\omega \in p_i} f(\omega, s)$
  end for
  for $i \leq j$ **do**
    $RS_S[i] \leftarrow RS[\phi(i)]$
    $P_S[i] \leftarrow P[\phi(i)]$
    where ϕ is the permutation that sorts RS in the increasing order.
  end for
  $j' \leftarrow$ the index such that the global worst case scenario ω^{wc} is in $P[j']$
  $P_S[j], P_S[j'] \leftarrow P_s[j'], P_s[j]$
  if $|P_S| > L$ **then**
    $P_S \leftarrow [P_S[1], P_S[2], ..., P_S[L-1], \bigcup_{i=M}^{j} P_S[i]]$
  end if
  return P_S

An example of the execution of Algorithm 1 is given in Example 5.

Example 5. *Once again, let us consider instance from Example 1. Let us suppose that* $t = 10$ *is a moment of decision and we have to develop a node from the tree. In Example 2, we saw that* $K^{t=10} = \{k_{d_2}^{t=10}, k_{d_3}^{t=10}\}$ *with* $k_{d_2}^{t=10} = 16$ *and* $k_{d_3}^{t=10} = 14$. *We can split the set of scenarios into four subsets (as shown in Fig. 3).*
Let us apply Algorithm 1 with $P = [A, B, C, D]$ *and* $L = 3$. *Keeping the same notation, we have* $RSV = [7, 6, 4, 4]$, *then* $RSV_s = [4, 4, 6, 7]$ *and* $P_s = [C, D, B, A]$. *As* $|P_s| > L$, *we modify* P_s *to* $P_s = [C, D, A \cup B]$ *and its robustness score is* $[4, 4, 7]$.

Remark: Note that by definition of the shape of P any union of rectangles is feasible. In other words, even a non rectangle shape is acceptable. For instance, considering notations from Example 5, the partition $[A, B \cup C \cup D]$ is a feasible solution.

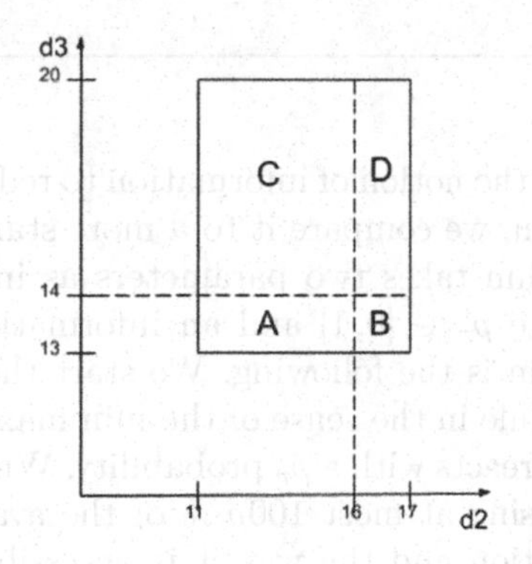

Fig. 3. Set of scenarios for Example 5

We now propose Algorithm 2 to solve the RPP problem. It is an exhaustive algorithm, that calls Algorithm 1 for each combination (smaller than Q) of information, so it has an exponential complexity.

We are now able to build a robust decision tree with the procedure introduced in Sect. 3.

5 Experimentations

The objective of the carried out experiments is to evaluate the robustness of our robust decision tree model, the quality of the selected information used for its construction and its stability in terms of number of reactions. For the numerical tests, we generated different types of instances for the $1||L_{max}$ problem with uncertainty on the due dates according to two parameters. The first parameter is the class of the instance. We distinguish two classes: the A class with small uncertainty intervals, and the B class with large uncertainty intervals. The second parameter is the number of tasks to schedule. For example, the instance A_{10} is an instance with 10 tasks, with small uncertainty intervals on the tasks' due dates.

Algorithm 2

Require: A set of scenarios Ω, a set of information K^t and two integers Q and L
  $P^* = [\Omega]$
  for $K \subseteq K^t$ **do**
    if $|K| \leq Q$ **then**
      $P = \{p|p = \bigcup_{j \in J_p} \prod_{i \in I} p_{i,j}$ with $p_{i,j} \in \{X^{inf}_{k^t_{x_i}}, X^{sup}_{k^t_{x_i}}, X_i\}$ if $k^t_{x_i} \in K^t$ and $p_{i,j} = X_i$ otherwise $\}$
      $P' = Algo1(P, L)$
      if $RS(P') \leq_{lexi} RS(P^*)$ **then**
        $P^* = P'$
      end if
    end if
  end for
  return P^*

As our approach uses the notion of information to reduce uncertainties to provide more robust solution, we compare it to a more standard proactive-reactive algorithms. This algorithm takes two parameters as input (in addition to the instance), a reaction rate $\rho_r \in [0, 1]$ and an information rate $\rho_i \in [0, 1]$. The principle of the algorithm is the following. We start the execution of the planning with a robust schedule in the sense of the min max criterion. At the end of each task, the algorithm reacts with a ρ_r probability. When it reacts, it computes a new robust solution using at most $100\rho_i\%$ of the available information. The definition of an information and the way it is accessible are strictly the same as the ones used to build a robust decision tree. To build a robust decision tree with our method, we need a couple of parameters: a list of moments of decision $(t_j)_{j \in J}$, the maximum number of information we are allowed to use at each moment of decision Q, and the maximum size of the partition we compute at each moment of decision L. In order to test different lists of moment of decision, we split the total schedule duration in T equal intervals. In our experiment we computed robust decision trees using the following values:

- Number of moment of decision $T \in [2, .., 10]$.
- Maximum number of information at each moment $Q \in [1, .., 10]$.
- Maximum size of partitions at each moment $L \in [2, .., 4]$.

As such, a tree computed with these parameter is denoted by $\text{Tree}_{Q,L,T}$. These values may seem small, but as we discuss it earlier, these parameters not only impact the computation time to solve the RPP, but also the size of the tree. In order to keep the total computation time of a tree reasonable, we had to keep these values not too high.

As we have seen before, the proactive-reactive algorithm takes two parameters, ρ_i and ρ_r. Since an execution of this algorithm is fast to simulate, we enumerate for ρ_i and ρ_r every values in [0, 1] with 0.1 steps.

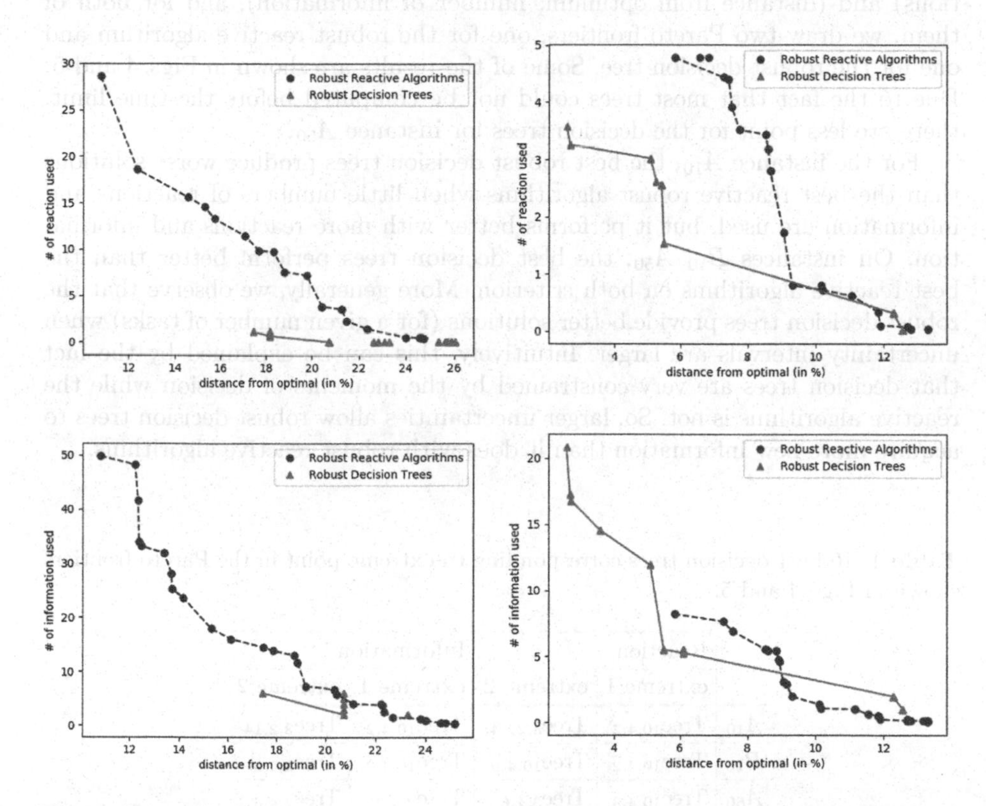

Fig. 4. Pareto frontier for both criterion on instance A_{50}

Fig. 5. Pareto frontier for both criterion on instance B_{50}

For each instance, we randomly pick 500 scenarios (with an uniform distribution). Then, for each set of parameters and each instance, we compute a robust decision tree, and go through the tree following the path corresponding to each random scenario. The same protocol is used with the robust reactive algorithms, we simulate them on each random scenarios.

5.1 Information/Reaction Efficiency

We assess the relevance of the information used to build the tree, and the stability of the planning proposed by the tree. To do so we collected, for each instance, the mean number (over the 500 random scenarios) of information used, the mean number of reactions (when a global solution changes between two moments of decisions) and the distance from optimum, which is, for a given scenario ω and a solution s:

$$100 \cdot \frac{L_{max}(\omega, s) - L_{max}(\omega, s^*)}{L_{max}(\omega, s^*)}$$

where s^* is the optimal solution minimising its L_{max} value on scenario ω. We consider two couples of criterion: (distance from optimum, number of reactions) and (distance from optimum, number of information), and for both of them, we draw two Pareto frontiers, one for the robust reactive algorithm and one for the robust decision tree. Some of the results are shown in Figs. 4 and 5. Due to the fact that most trees could not be computed before the time limit, there are less point for the decision trees for instance A_{50}.

For the instance A_{10}, the best robust decision trees produce worse solutions than the best reactive robust algorithms when little numbers of reactions and information are used, but it performs better with more reactions and information. On instances B_{10} A_{50}, the best decision trees perform better than the best reactive algorithms on both criterion. More generally, we observe that the robust decision trees provide better solutions (for a given number of tasks) when uncertainty intervals are larger. Intuitively, this can be explained by the fact that decision trees are very constrained by the moments of decision while the reactive algorithms is not. So, larger uncertainties allow robust decision trees to acquire more new information than it does with robust reactive algorithms.

Table 1. Robust decision trees corresponding to extreme point in the Pareto frontiers shown in Figs. 4 and 5.

	Reaction		Information	
	extreme 1	extreme 2	extreme 1	extreme 2
A_{10}	$\text{Tree}_{10,4,2}$	$\text{Tree}_{3,2,14}$	$\text{Tree}_{10,4,2}$	$\text{Tree}_{3,2,14}$
B_{10}	$\text{Tree}_{10,4,2}$	$\text{Tree}_{10,4,6}$	$\text{Tree}_{10,4,2}$	$\text{Tree}_{2,4,6}$
A_{50}	$\text{Tree}_{10,4,3}$	$\text{Tree}_{3,4,6}$	$\text{Tree}_{2,4,3}$	$\text{Tree}_{2,4,6}$

However, this way of presenting the results does not show which robust decision trees appear in the Pareto frontier. Table 1 shows which trees (and the set of parameters they were computed with) are the extreme points of the different Pareto frontiers. Quite intuitively, the decision trees propose the best solutions with few information and reactions, when the number of moments of decision is low and the maximum number of usable information is high. With the same idea, the decision trees built with a high number of moments of decision (i.e the deepest ones) but with few information available at each moment yield good solutions as well. Interestingly, this shows that the depth of the tree is more important to produce quality solutions than the maximum number of information. The table also shows that the maximum partition size was used for the best trees, which implies the results could be improved by increasing this value.

6 Conclusion

In this paper we introduce a proactive-reactive approach to deal with uncertain scheduling problems. The method constructs a robust decision tree for a decision maker that is reusable as long as the problem parameters belong to the uncertainty set. At each node of the tree we assume that the scheduler has access to some knowledge about the ongoing scenario, reducing the level of uncertainty and allowing the computation of less conservative solutions with robustness guarantees. However, obtaining information on the uncertain parameters can be costly and frequent rescheduling can be disturbing. We first formally define the robust decision tree and the information refining concepts in the context of uncertainty scenarios. We then introduce the Robust Partition Problem, the core problem of our approach. Then we propose algorithms to solve this problem and build such a tree. Finally, focusing on a simple single machine scheduling problem, we provide experimental comparisons highlighting the potential of the decision tree approach compared with reactive algorithms for obtaining more robust solutions with fewer information updates and schedule changes.

In the view of the encouraging results we believe that this method could be used in industrial cases. For our future works, we plan to apply and extend our method to hard problems, such as the Resource-Constrained Project Scheduling Problem.

Acknowledgments. This work has been partially funded by the ANR project PER4MANCE and the interdisciplinary institute in artificial intelligence ANITI.

References

1. van den Akker, M., Hoogeveen, H., Stoef, J.: Combining two-stage stochastic programming and recoverable robustness to minimize the number of late jobs in the case of uncertain processing times. J. Sched. **21**(6), 607–617 (2018). https://doi.org/10.1007/s10951-018-0559-z
2. Aloulou, M.A., Della Croce, F.: Complexity of single machine scheduling problems under scenario-based uncertainty. Oper. Res. Lett. **36**(3), 338–342 (2008)

3. Davari, M., Demeulemeester, E.: The proactive and reactive resource-constrained project scheduling problem. J. Sched. **22**(2), 211–237 (2017). https://doi.org/10.1007/s10951-017-0553-x
4. Davari, M., Demeulemeester, E.: Important classes of reactions for the proactive and reactive resource-constrained project scheduling problem. Ann. Oper. Res. **274**(1–2), 187–210 (2019)
5. Dearden, R., Meuleau, N., Ramakrishnan, S., Smith, D.E., Washington, R.: Incremental contingency planning. In: ICAPS-03 Workshop on Planning under Uncertainty (2003)
6. Drummond, M., Bresina, J., Swanson, K.: Just-in-case scheduling. In: AAAI, vol. 94, pp. 1098–1104 (1994)
7. Kouvelis, P., Yu, G.: Robust Discrete Optimization and Its Applications. Kluwer Academic Publishers, Dordrecht (1997)
8. Meuleau, N., Smith, D.E.: Optimal limited contingency planning. In: Proceedings of the Nineteenth Conference on Uncertainty in Artificial Intelligence, pp. 417–426. Morgan Kaufmann Publishers Inc. (2002)
9. Nikulin, Y.: Robustness in combinatorial optimization and scheduling theory: an extended annotated bibliography. Technical report. Manuskripte aus den Instituten für Betriebswirtschaftslehre der Universität Kiel (2006)
10. Rajendran, C., Holthaus, O.: A comparative study of dispatching rules in dynamic flowshops and jobshops. Eur. J. Oper. Res. **116**(1), 156–170 (1999). https://doi.org/10.1016/S0377-2217(98)00023-X
11. Van de Vonder, S., Demeulemeester, E., Herroelen, W.: A classification of predictive-reactive project scheduling procedures. J. Sched. **10**(3), 195–207 (2007)
12. Yanıkoğlu, İ., Gorissen, B.L., den Hertog, D.: A survey of adjustable robust optimization. Eur. J. Oper. Res. **277**(3), 799–813 (2019)

Orders Preserving Convexity Under Intersections for Interval-Valued Fuzzy Sets

Pedro Huidobro[1], Pedro Alonso[2], Vladimir Janiš[3], and Susana Montes[1(✉)]

[1] Department of Statistics and O.R., University of Oviedo, Oviedo, Spain
{huidobropedro,montes}@uniovi.es
[2] Department of Mathematics, University of Oviedo, Oviedo, Spain
palonso@uniovi.es
[3] Department of Mathematics, Matej Bel University, Banská Bystrica, Slovakia
vladimir.janis@umb.sk

Abstract. Convexity is a very important property in many areas and the studies of this property are frequent. In this paper, we have extended the notion of convexity for interval-valued fuzzy sets based on different order between intervals. The considered orders are related and their behavior analyzed. In particular, we study the preservation of the convexity under intersections, where again the chosen order is essential. After this study, we can conclude the appropriate behavior of the admissible orders for this purpose.

Keywords: Interval-valued fuzzy sets · Order between intervals · Intersection · Convexity

1 Introduction

Convexity is a basic mathematical concept that has been used as a tool in many different problems. It has important applications in many areas, like optimization [15], image processing [22], robotics [14] or geometry [13].

In real problems, the information we have to deal with is, in most of the cases, approximate. By this reason, the study of the convexity of a fuzzy set has been a very studied topic (see, for instance, Ammar and Metz [1], Diaz et al. [7], Ramik and Vlach [16], Sarkar [18], Syau and Lee [21] and Yang [25]).

Taking into account several real world problems, several extensions of the fuzzy sets have been introduced and studied in the last years. In particular, we

Authors would like to thank for the support of Spanish Ministry of Science and Technology project TIN-2017-87600-P (P. Alonso), Spanish Ministry of Science and Technology project PGC2018-098623-B-I00 (P. Huidobro and S. Montes), FICYT Project IDI/2018/000176 (P. Alonso, P. Huidobro and S. Montes) and Slovak grant agency VEGA project 1/0093/17 (V. Janiš).

M.-J. Lesot et al. (Eds.): IPMU 2020, CCIS 1239, pp. 493–505, 2020.
https://doi.org/10.1007/978-3-030-50153-2_37

are interested in interval-valued fuzzy sets. They were introduced independently by Zadeh [26], Grattan-Guiness [10], Jahn [12], Sambuc [17] in the seventies. From then, several concepts related to this extension have to be studied. Taking into account the previous comments, we are especially interested in the concept of a convex interval-valued fuzzy set. Since convexity is based on an order over the membership degrees and how the membership values are not numbers but intervals, we will obtain a different definition of convexity for each interval order on the set of intervals. The main aim of this paper is to introduce this general definition and study its dependence on the interval order considered. In particular, we are going to study in deep the preservation of the convexity under intersections, since it is a necessary property in many applications, as optimization (see [15]).

This paper is organized as follows. In Sect. 2, some basic concepts are introduced and the notation is fixed. Section 3 is devoted to the study of the different definitions we can consider for the intersection of two interval-valued fuzzy sets depending on the chosen order. In Sect. 4 we propose a definition of convexity for interval-valued fuzzy sets and we study the cases when the intersection of two convex sets remains convex. Finally, some conclusions and open problems are drawn in Sect. 5.

2 Basic Concepts

Let X denote the universe of discourse. An interval-valued fuzzy subset of X is a mapping $A : X \to L([0,1])$ such that $A(x) = [\underline{A}(x), \overline{A}(x)]$, where $L([0,1])$ denotes the family of closed intervals included in the unit interval $[0,1]$. Thus, an interval-valued fuzzy set A is totally characterized by two mapping, $\underline{A}$ and $\overline{A}$, from X into $[0,1]$ such that $\underline{A}(x) \leq \overline{A}(x), \forall x \in X$. These maps represent the lower and upper bound of the corresponding intervals. Let us notice that if $\underline{A}(x) = \overline{A}(x), \forall x \in X$, then A is a classical fuzzy sets. The collection of all the interval-valued fuzzy sets in X is denoted by $IVFS(X)$ and the subset formed by all the fuzzy sets in X is denoted by $FS(X)$.

For any pair of IVFS, it is usually considered that A is a subset of B if, and only if, $A(x)$ is lower than or equal to $B(x)$ for any $x \in X$. This definition is clear when we are dealing with fuzzy sets, since the usual order $\leq$ for the real number is used to define the inclusion. However, there is not a usual total order in $L([0,1])$ and so, several definitions of inclusion could be considered in accordance with the order considered in $L([0,1])$. The different usual orders for intervals are based on the specific points within the intervals which are considered as representatives.

Thus, if $a = [\underline{a}, \overline{a}]$ and $b = [\underline{b}, \overline{b}]$ are any two intervals in $L([0,1])$, we say that a is lower than or equal to b for the most usual orders between intervals if:

- Interval dominance [8]: $a \preceq_{ID} b$ if $\overline{a} \leq \underline{b}$.
- Lattice order [9]: $a \preceq_{Lo} b$ if $\underline{a} \leq \underline{b}$ and $\overline{a} \leq \overline{b}$, which is induced by the usual partial order in $\mathbb{R}^2$.
- Lexicographical order type 1 [5]: $a \preceq_{Lex1} b$ if $\underline{a} < \underline{b}$ or $\underline{a} = \underline{b}$ and $\overline{a} \leq \overline{b}$.

- Lexicographical order type 2 [5]: $a \preceq_{Lex2} b$ if $\overline{a} < \overline{b}$ or $\overline{a} = \overline{b}$ and $\underline{a} \leq \underline{b}$.
- The Xu and Yager order [24]: $a \preceq_{YX} b$ if $\underline{a} + \overline{a} < \underline{b} + \overline{b}$ or $\underline{a} + \overline{a} = \underline{b} + \overline{b}$ and $\overline{a} - \underline{a} < \overline{b} - \underline{b}$.
- Maximax order [19]: $a \preceq_{MM} b$ if $\overline{a} \leq \overline{b}$.
- Maximin order [20,23]: $a \preceq_{Mm} b$ if $\underline{a} \leq \underline{b}$.
- Hurwicz order [11]: $a \preceq_{H(\alpha)} b$ if $\alpha \cdot \underline{a} + (1 - \alpha) \cdot \overline{a} \leq \alpha \cdot \underline{b} + (1 - \alpha) \cdot \overline{b}$ with $\alpha \in [0, 1]$.
- Weak order [3]: $a \preceq_{wo} b$ if $\underline{a} \leq \overline{b}$.

Some of these orders are clearly related. Thus, it is well-known that if one interval a is lower than or equal to another interval b w.r.t. the order ID, a is also lower than or equal to b w.r.t. the lattice order. This also implies the same relation w.r.t. the lexicographical order type 1, which implies the same w.r.t. the maximax order and this implies that a is lower than or equal to b w.r.t. the weak order. All these implications and some other similar ones are summarized at the following figure.

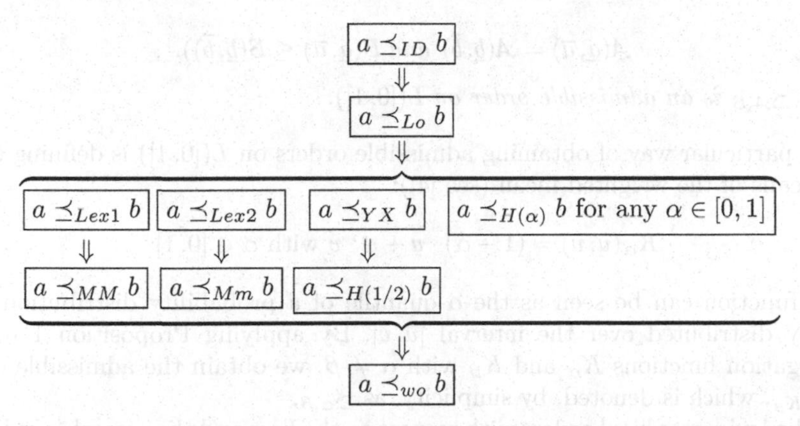

The implications represented here are fulfilled, but it is also known that the converse implications are not fulfilled in general.

In order to provide a total order that extends the usual orders between intervals, Bustince et al. introduced in [5] the concept of admissible order.

Definition 1. *[5] Let $(L([0,1]), \preceq)$ be a poset. The order $\preceq$ is called an admissible order if*

i) $\preceq$ is a linear order on $L([0,1])$,
ii) for all $[a,b], [c,d] \in L([0,1])$, $[a,b] \preceq [c,d]$ whenever $[a,b] \leq_{Lo} [c,d]$.

Once they introduced this definition, they also proposed a method to build these admissible orders in terms of two aggregation functions.

Definition 2. *[2] Let $\mathcal{A} : \bigcup_{i=1}^{n} [0,1]^i \to [0,1]$ such that*

- $\mathcal{A}(0, 0, \ldots, 0) = 0, \mathcal{A}(1, 1, \ldots, 1) = 1,$
- $\mathcal{A}(x) = x$ *for all* $x \in [0, 1]$,

– *$\mathcal{A}$ is monotone in each variable,*

then $\mathcal{A}$ is an aggregation function.

Notice that, there is a natural bijection between $L([0,1])$ and $K([0,1]) = \{(u,v) \in [0,1]^2 \mid u \leq v\}$ such that it identifies any interval $[\underline{a}, \overline{a}]$ to the point in $\mathbb{R}^2$ formed by its extremes, that is, $(\underline{a}, \overline{a})$ (see [5]). Thus, we can use aggregation functions to summarize the information given by an interval. Taking into account this idea, they obtained the following method to build admissible orders.

Proposition 1. *[5] Let $\mathcal{A}, \mathcal{B} : [0,1]^2 \to [0,1]$ be two continuous aggregation functions, such that for all $(u,v),(u',v') \in K([0,1])$, the equalities $\mathcal{A}(u,v) = \mathcal{A}(u',v')$ and $\mathcal{B}(u,v) = \mathcal{B}(u',v')$ can only hold if $(u,v) = (u',v')$. Define the relation $\preceq_{\mathcal{A},\mathcal{B}}$ on $L([0,1])$ by $a \preceq_{\mathcal{A},\mathcal{B}} b$ if and only if*

$$\mathcal{A}(\underline{a}, \overline{a}) < \mathcal{A}(\underline{b}, \overline{b})$$

or

$$\mathcal{A}(\underline{a}, \overline{a}) = \mathcal{A}(\underline{b}, \overline{b}) \text{ and } \mathcal{B}(\underline{a}, \overline{a}) \leq \mathcal{B}(\underline{b}, \overline{b})).$$

Then $\preceq_{\mathcal{A},\mathcal{B}}$ is an admissible order on $L([0,1])$.

A particular way of obtaining admissible orders on $L([0,1])$ is defining them by means of the weighted mean (see [4]):

$$K_\alpha(u,v) = (1-\alpha) \cdot u + \alpha \cdot v \text{ with } \alpha \in [0,1].$$

This function can be seen as the α-quantile of a probability distribution uniformly distributed over the interval $[u,v]$. By applying Proposition 1 to the aggregation functions K_α and K_β with $\alpha \neq \beta$, we obtain the admissible order $\preceq_{K_\alpha,K_\beta}$, which is denoted, by simplicity, as $\preceq_{\alpha,\beta}$.

The lexicographical orders with respect to the first and the second coordinate and the Xu and Yager order are particular cases of these admissible orders. Thus, $\preceq_{Lex1} \equiv \preceq_{0,1}$, $\preceq_{Lex2} \equiv \preceq_{1,0}$ and $\preceq_{YX} \equiv \preceq_{1/2,\beta}$ for any $\beta \in (1/2, 1]$ (see [5]).

3 Intersection of Interval-Valued Fuzzy Sets

From any order $\preceq_x$ in $L([0,1])$, we can deduce an order in $IVFS(X)$ given by the content relation obtained from this order. This relation in $IVFS(X)$ will be denoted by $\subseteq_x$. Thus, for instance,

$$A \subseteq_{YX} B \text{ iff } A(x) \preceq_{YX} B(x), \forall x \in X.$$

Let us notice that the inclusion between two interval-orders fuzzy sets does not imply the inclusion of the intervals which represent the membership degrees at any point in the referential, but that the interval associated to A is lower than or equal to w.r.t. the corresponding order to the interval associated to B.

Example 1. Let $X = \{x\}$ be the universe and let be $A, B, C \in IVFS(X)$ such that $A(x) = [0.4, 0.8]$, $B(x) = [0.2, 0.6]$ and $C(x) = [0.3, 0.9]$. We have that $B \subseteq_{Lo} A$ and $B \subseteq_{Lo} C$, since $[0.2, 0.6] \preceq_{Lo} [0.4, 0.8]$ and $[0.2, 0.6] \preceq_{Lo} [0.3, 0.9]$. However, $[0.2, 0.6] \not\subseteq [0.4, 0.8]$ and $[0.2, 0.6] \not\subseteq [0.3, 0.9]$. In fact, this is totally logical, since the membership degree of x to B is "lower" than the membership degree to A or C. In fact, we are using the same criteria as the usual one considered for fuzzy sets.

If the intersection of two sets is defined as the greatest set that is contained in both sets, then we have a different definition of intersection for each order we are considering in $IVFS(X)$.

Definition 3. *Let A, B be two sets in $IVFS(X)$ and let $\preceq_x$ an order in $L([0,1])$. We define the x-intersection of A and B, and we denote $A \cap_x B$ as the greatest interval-valued fuzzy set such that $A \cap_x B \subseteq_x A$ and $A \cap_x B \subseteq_x B$.*

So each order would have its own way to construct intersections between IVFS. To better understand this definition, we will see some examples after the general result of each order.

For any two interval orders $\preceq_x$ and $\preceq_y$ in $IVFS(X)$ such that $a \preceq_x b$ implies that $a \preceq_y b$, $\forall a, b \in L([0,1])$, we have that $A \cap_x B \subseteq_y A \cap_y B$ for any $A, B \in IVFS(X)$. Thus,

$$A \cap_{ID} B \subseteq_{Lo} A \cap_{Lo} B \subseteq_{Lex1} A \cap_{Lex1} B \subseteq_{MM} A \cap_{MM} B \subseteq_{wo} A \cap_{wo} B,$$

$$A \cap_{Lo} B \subseteq_{Lex2} A \cap_{Lex2} B \subseteq_{Mm} A \cap_{Mm} B \subseteq_{wo} A \cap_{wo} B,$$

$$A \cap_{Lo} B \subseteq_{YX} A \cap_{YX} B \subseteq_{H(1/2)} A \cap_{H(1/2)} B \subseteq_{wo} A \cap_{wo} B,$$

and

$$A \cap_{Lo} B \subseteq_{H(\alpha)} A \cap_{H(\alpha)} B \subseteq_{wo} A \cap_{wo} B.$$

Taking into account the relationship among the considered orders, we will study the definition that we obtain for the intersection for any of them, by considering some general behavior in those cases which is possible.

For the interval dominance we have that the intersection of two IVFS is a fuzzy set:

Proposition 2. *Let A, B be two sets in $IVFS(X)$. The ID-intersection of A and B is the interval-valued fuzzy set defined by*

$$A \cap_{ID} B(x) = \min\{\underline{A(x)}, \underline{B(x)}\}$$

for any $x \in X$.

Next, an example of the intersection of two IVFS is presented.

Example 2. In the same conditions of Example 1.

- The ID-intersection of A and B is the interval-valued fuzzy set $A \cap_{ID} B(x) = 0.2$.

- The ID-intersection of A and C is the interval-valued fuzzy set $A \cap_{ID} C(x) = 0.3$.

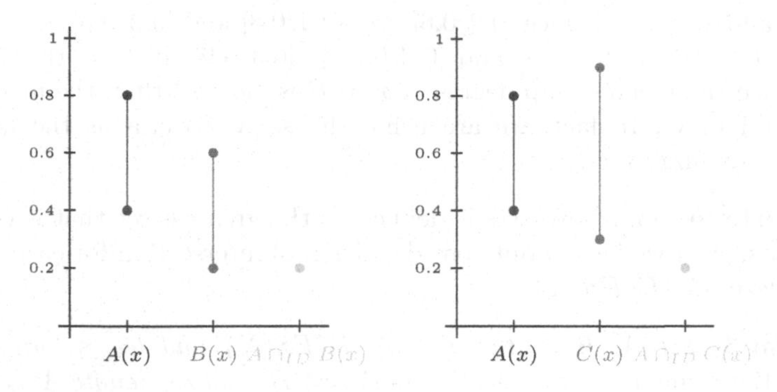

It is clear that $A \cap_{ID} B \subseteq_{ID} A$ since $0.2 \leq 0.4$ and $0.2 \leq 0.2$. Similarly, we can see that $A \cap_{ID} B \subseteq_{ID} B$. Both expressions are immediate, since the definition of intersection was given exactly as the set such that it is included in both of them and any other set included in them is a subset of the intersection. Of course, when we deal with the intersection and the content, the same order between intervals is considered.

However, for the less restrictive lattice ordering the intersection is not just a fuzzy set in general. In fact, with this order, we will obtain the usual way to define the intersection for two IVFS.

Proposition 3. *Let A, B be two sets in $IVFS(X)$. The Lo-intersection of A and B is the interval-valued fuzzy set defined by*

$$A \cap_{Lo} B(x) = [\min\{\underline{A(x)}, \underline{B(x)}\}, \min\{\overline{A(x)}, \overline{B(x)}\}]$$

for any $x \in X$.

This proposition coincides with the usual way to define the intersection between two interval-valued fuzzy sets, as it can be seen in [6], where intersection is studied in general for convolution lattices.

Following example shows what happens when we intersect two IVFS.

Example 3. In the same conditions of Example 1.

- The Lo-intersection of A and B is the interval-valued fuzzy set $A \cap_{Lo} B(x) = [0.2, 0.6]$.
- The Lo-intersection of A and C is the interval-valued fuzzy set $A \cap_{Lo} C(x) = [0.3, 0.8]$.

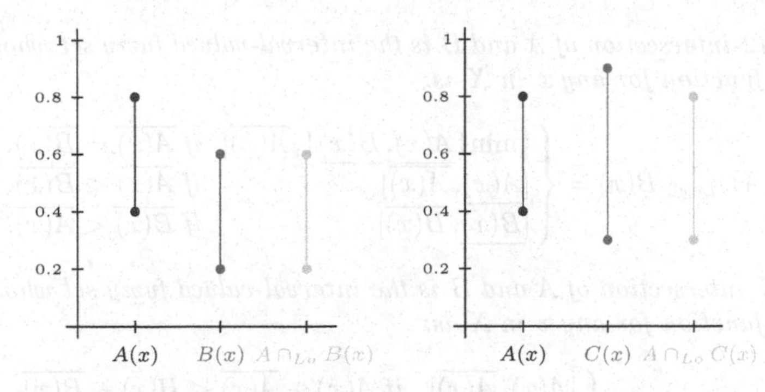

We can also use this example to remark that the intersection we are done is between two interval-valued fuzzy set, but it is not an intersection between intervals. Thus, since x has a membership degree in A given by the interval $[0.4, 0.8]$ and a membership degree in C given by the interval $[0.3, 0.9]$, we can say that the degree in which x is in their intersection is between 0.3 and 0.8. A value greater than 0.8 is impossible, since we have not this degree for A, but a value between 0.3 and 0.4 is possible, since x belongs to A at least in degree 0.4 and to B at least in degree 0.3.

The case of the lexicographical order and the Xu and Yager order, can be considered in a general way, since all of them are particular cases of admissible orders, as we commented previously.

Proposition 4. *Let $\mathcal{A}, \mathcal{B} : [0,1]^2 \to [0,1]$ be two continuous aggregation functions, such that for all $(u,v), (u',v') \in K([0,1])$, the equalities $\mathcal{A}(u,v) = \mathcal{A}(u',v')$ and $\mathcal{B}(u,v) = \mathcal{B}(u',v')$ can only hold if $(u,v) = (u',v')$. Let $\preceq_{\mathcal{A},\mathcal{B}}$ be the admissible order on $L([0,1])$ induced by them. For any $A, B \in IVFS(X)$, the $\mathcal{A},\mathcal{B}$-intersection of A and B is the interval-valued fuzzy set defined by:*

$$A \cap_{\mathcal{A},\mathcal{B}} B(x) = \begin{cases} [\underline{A(x)}, \overline{A(x)}] & \text{if } \mathcal{A}([\underline{A(x)}, \overline{A(x)}]) < \mathcal{A}([\underline{B(x)}, \overline{B(x)}]) \\ & \text{or } \Big\{\mathcal{A}([\underline{A(x)}, \overline{A(x)}]) = \mathcal{A}([\underline{B(x)}, \overline{B(x)}]) \\ & \text{and } \mathcal{B}([\underline{A(x)}, \overline{A(x)}]) \leq \mathcal{B}([\underline{B(x)}, \overline{B(x)}])\Big\}, \\ \left[\underline{B(x)}, \overline{B(x)}\right] & \text{if } \Big\{\mathcal{A}([\underline{A(x)}, \overline{A(x)}]) = \mathcal{A}([\underline{B(x)}, \overline{B(x)}]) \\ & \text{and } \mathcal{B}([\underline{B(x)}, \overline{B(x)}]) < \mathcal{B}([\underline{A(x)}, \overline{A(x)}])\Big\} \\ & \text{or } \mathcal{A}([\underline{B(x)}, \overline{B(x)}]) < \mathcal{A}([\underline{A(x)}, \overline{A(x)}]). \end{cases}$$

By applying this result to the specific admissible orders, we obtain that

Proposition 5. *Let A, B be two sets in $IVFS(X)$.*

– *The $Lex1$-intersection of A and B is the interval-valued fuzzy set whose membership function for any x in X is:*

$$A \cap_{Lex1} B(x) = \begin{cases} [\underline{A(x)}, \min\{\overline{A(x)}, \overline{B(x)}\}] & \text{if } \underline{A(x)} = \underline{B(x)}, \\ [\underline{A(x)}, \overline{A(x)}] & \text{if } \underline{A(x)} < \underline{B(x)}, \\ [\underline{B(x)}, \overline{B(x)}] & \text{if } \underline{B(x)} < \underline{A(x)}. \end{cases}$$

– *The Lex2-intersection of A and B is the interval-valued fuzzy set whose membership function for any x in X is:*

$$A \cap_{Lex2} B(x) = \begin{cases} [\min\{\underline{A}(x), \underline{B}(x)\}, \overline{A}(x)] & \text{if } \overline{A}(x) = \overline{B}(x), \\ [\underline{A}(x), \overline{A}(x)] & \text{if } \overline{A}(x) < \overline{B}(x), \\ [\underline{B}(x), \overline{B}(x)] & \text{if } \overline{B}(x) < \overline{A}(x). \end{cases}$$

– *The YX-intersection of A and B is the interval-valued fuzzy set whose membership function for any x in X is:*

$$A \cap_{YX} B(x) = \begin{cases} [\underline{A}(x), \overline{A}(x)] & \text{if } \underline{A}(x) + \overline{A}(x) < \underline{B}(x) + \overline{B}(x) \\ & \text{or } \{\underline{A}(x) + \overline{A}(x) = \underline{B}(x) + \overline{B}(x) \\ & \text{and } \overline{A}(x) - \underline{A}(x) \leq \overline{B}(x) - \underline{B}(x)\}, \\ [\underline{B}(x), \overline{B}(x)] & \text{if } \{\underline{A}(x) + \overline{A}(x) = \underline{B}(x) + \overline{B}(x), \\ & \text{and } \overline{B}(x) - \underline{B}(x) < \overline{A}(x) - \underline{A}(x)\} \\ & \text{or } \underline{B}(x) + \overline{B}(x) < \underline{A}(x) + \overline{A}(x). \end{cases}$$

In order to clarify this result, let us show some examples.

Example 4. In the same conditions of Example 1.

– The *Lex*1-intersection of A and B is the interval-valued fuzzy set $A \cap_{Lex1} B(x) = [0.2, 0.6]$.
– The *Lex*1-intersection of A and C is the interval-valued fuzzy set $A \cap_{Lex1} C(x) = [0.3, 0.9]$.

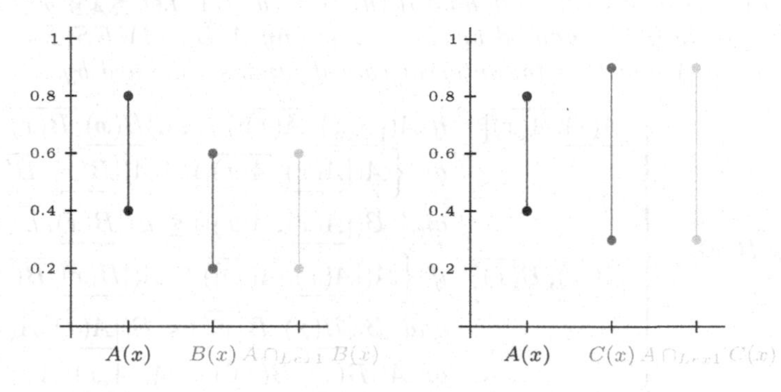

– The *Lex*2-intersection of A and B is the interval-valued fuzzy set $A \cap_{Lex2} B(x) = [0.2, 0.6]$.
– The *Lex*2-intersection of A and C is the interval-valued fuzzy set $A \cap_{Lex2} C(x) = [0.4, 0.8]$.

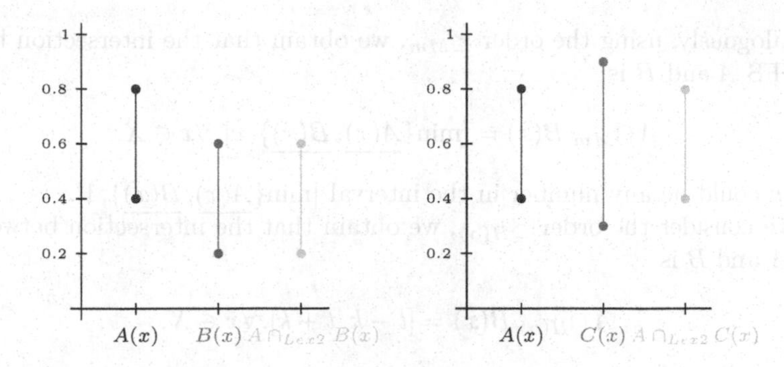

- The YX-intersection of A and B is the interval-valued fuzzy set $A \cap_{YX} B(x) = [0.2, 0.6]$.
- The YX-intersection of A and C is the interval-valued fuzzy set $A \cap_{YX} C(x) = [0.4, 0.8]$.
- On the other hand, if $D(x) = [0.2, 0.9]$, the YX-intersection of A and D is the interval-valued fuzzy set $A \cap_{YX} D(x) = [0.2, 0.9]$.

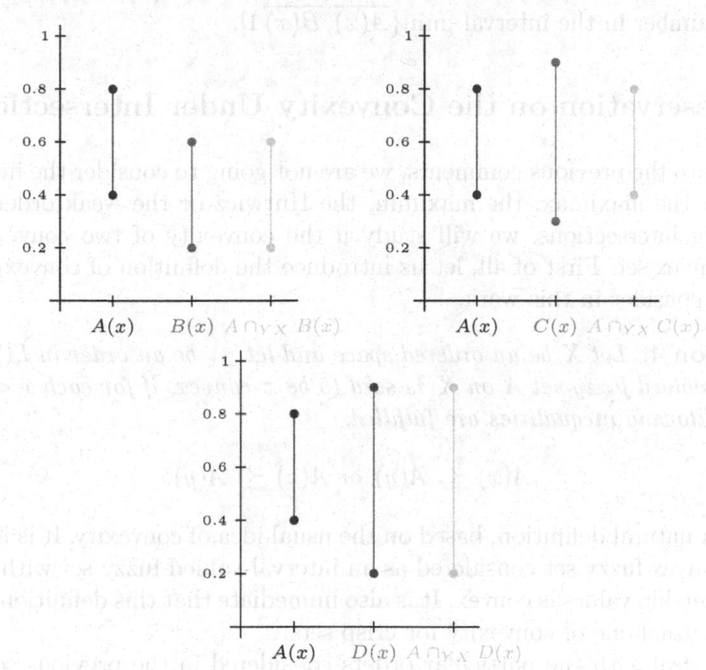

We are not going to consider the maximax, the maximin, the Hurwicz and the weak orders to define the intersection since it is not unique. Using the order $\preceq_{MM}$, we obtain that the intersection between two IVFS A and B is

$$A \cap_{MM} B(x) = [u, \min\{\overline{A(x)}, \overline{B(x)}\}], \forall x \in X,$$

where u could be any number in the interval $[0, \min\{\underline{A(x)}, \underline{B(x)}\}]$.

Analogously, using the order $\preceq_{Mm}$, we obtain that the intersection between two IVFS A and B is

$$A \cap_{Mm} B(x) = [\min\{\underline{A(x)}, \underline{B(x)}\}, v], \forall x \in X,$$

where v could be any number in the interval $[\min\{\underline{A(x)}, \underline{B(x)}\}, 1]$.

If we consider the order $\preceq_{H(\alpha)}$, we obtain that the intersection between two IVFS A and B is

$$A \cap_{H(\alpha)} B(x) = [t - k, t + k], \forall x \in X,$$

where $t = \min\{\alpha \cdot \underline{A(x)} + (1-\alpha) \cdot \overline{A(x)}, \alpha \cdot \underline{B(x)} + (1-\alpha) \cdot \overline{B(x)}\}$ and k could be any number in the interval $[0, \min\{t, 1-t\}]$.

For the weak order, the intersection is

$$A \cap_{MM} B(x) = [u, v], \forall x \in X,$$

where u could be any number in the interval $[0, \min\{\overline{A(x)}, \overline{B(x)}\}]$ and v could be any number in the interval $[\min\{\overline{A(x)}, \overline{B(x)}\ 1]$.

4 Preservation on the Convexity Under Intersections

Taking into the previous comments, we are not going to consider the intersection based on the maximax, the maximin, the Hurwicz or the weak order. For the remaining intersections, we will study if the convexity of two convex IVFS is still a convex set. First of all, let us introduce the definition of convexity we are going to consider in this work.

Definition 4. *Let X be an ordered space and let $\preceq_x$ be an order in $L([0,1])$. An interval-valued fuzzy set A on X is said to be x-convex, if for each $x < y < z$ in X the following inequalities are fulfilled:*

$$A(x) \preceq_x A(y) \text{ or } A(z) \preceq_x A(y).$$

It is a natural definition, based on the usual idea of convexity. It is immediate that a convex fuzzy set considered as an interval-valued fuzzy set with singleton as membership values is convex. It is also immediate that this definition coincides with the usual one of convexity for crisp sets.

If we deal with the particular orders considered in the previous section, we obtain that ID-convexity implies Lo-convexity and this implies $Lex1$-convexity, $Lex2$-convexity and YX-convexity.

About the important property of the preservation of the convexity under intersections, we have obtained the following results.

Proposition 6. *Let X be an ordered space, if $A, B \in IVFS(X)$ are ID-convex, then $A \cap_{ID} B$ is also ID-convex, whenever it is not empty.*

Unfortunately, the *Lo*-intersection of two IVFS which are *Lo*-convex is not always *Lo*-convex, as we can see at the following counterexample.

Example 5. Let $X = \{x, y, z\}$ with $x < y < z$. If we consider the IVFS A and B defined as follows

X	x	y	z
A	[0.1,0.7]	[0.2,0.8]	[0.3,0.5]
B	[0.1,0.7]	[0.4,0.6]	[0.3,0.5]
$A \cap_{Lo} B$	[0.1,0.7]	[0.2,0.6]	[0.3,0.5]

Then A is *Lo*-convex, since $[0.1, 0.7] \preceq_{Lo} [0.2, 0.8]$ and B is *Lo*-convex since $[0.3, 0.5] \preceq_{Lo} [0.4, 0.6]$. However, $A \cap_{Lo} B$ is not *Lo*-convex since $[0.2, 0.6]$ is not related with $[0.1, 0.7]$ or $[0.3, 0.5]$ by means of the order relation $\preceq_{Lo}$. This is the typical problem we can find any time we use partial orders. Thus, it is clear that convexity only makes sense for total orders as, for instance, admissible orders.

In fact, for admissible orders, we have been able to obtain a general results where we prove the good behaviour of them with respect to the convexity. Thus,

Proposition 7. *Let X be an ordered space and let $\preceq_{\mathcal{A},\mathcal{B}}$ an admissible order based on two aggregations functions $\mathcal{A}$ and $\mathcal{B}$. If $A, B \in IVFS(X)$ are $\mathcal{A}, \mathcal{B}$-convex, then $A \cap_{\mathcal{A},\mathcal{B}} B$ is also $\mathcal{A}, \mathcal{B}$-convex, whenever it is not empty.*

Thus, the case of the lexicographical orders and the Xu and Yager order are automatically solved.

Corollary 1. *Lex1-convexity, Lex2-convexity and YX-convexity are preserved under intersections.*

5 Concluding Remarks

In this paper we have proposed a definition of convexity for IVFS and we characterized the cases where it is preserved under intersections. The intersection of two IVFS is based on the chosen order between intervals and so several definitions of intersection are considered. It is not surprising that not all of the orders between intervals are appropriate for defining the intersection and that the lattice ordering defines the usual definition of intersection considered in the literature. However, this order has not a good behavior about preservation of convexity under intersections and admissible orders seem to be better for this purpose. An immediate pending work is the study of the cutworthy approach for this concept of convexity.

References

1. Ammar, E., Metz, J.: On fuzzy convexity and parametric fuzzy optimization. Fuzzy Sets Syst. **49**(2), 135–141 (1992). https://doi.org/10.1016/0165-0114(92)90319-Y
2. Beliakov, G., Bustince Sola, H., Calvo Sánchez, T.: A Practical Guide to Averaging Functions. SFSC, vol. 329. Springer, Cham (2016). https://doi.org/10.1007/978-3-319-24753-3
3. Bogart, K.P., Bonin, J., Mitas, J.: Interval orders based on weak orders. Discrete Appl. Math. **60**(1), 93–98 (1995). https://doi.org/10.1016/0166-218X(95)00110-D
4. Bustince, H., et al.: A class of aggregation functions encompassing two-dimensional OWA operators. Inf. Sci. **180**, 1977–1989 (2010). https://doi.org/10.1016/j.ins.2010.01.022
5. Bustince, H., Fernandez, J., Kolesarova, A., Mesiar, R.: Generation of linear orders for intervals by means of aggregation functions. Fuzzy Sets Syst. **220**(1), 69–77 (2013). https://doi.org/10.1016/j.fss.2012.07.015
6. De Miguel, L., Bustince, H., De Baets, B.: Convolution lattices. Fuzzy Sets Syst. **335**, 67–93 (2018). https://doi.org/10.1016/j.fss.2017.04.017
7. Diaz, S., Indurain, E., Janiš, V., Llinares, J.V., Montes, S.: Generalized convexities related to aggregation operators of fuzzy sets. Kybernetika **53**, 383–393 (2017). https://doi.org/10.14736/kyb-2017-3-0383
8. Fishburn, P.: Interval Orderings. Wiley, New York (1987)
9. Goguen, J.A.: L-fuzzy sets. J. Math. Anal. Appl. **18**(1), 145–174 (1967). https://doi.org/10.1016/0022-247X(67)90189-8
10. Grattan-Guinness, I.: Fuzzy membership mapped onto intervals and many-valued quantities-mathematical. Logic Q. **22**(1), 149–160 (1976). https://doi.org/10.1002/malq.19760220120
11. Hurwicz, L.: A class of criteria for decision-making under ignorance. Cowles Commission Discussion Paper: Statistics No. 356 (1951)
12. Jahn, K.U.: Intervall-wertige Mengen. Mathematische Nachrichten **68**, 115–132 (1975). https://doi.org/10.1002/mana.19750680109
13. Kim, Y., Xi, Z., Lien, J.M.: Disjoint convex shell and its applications in mesh unfolding. Comput.-Aided Des. **90**, 180–190 (2017). https://doi.org/10.1016/j.cad.2017.05.014
14. Miao, L., Wenyu, Y., Xiaoping, Z.: Projection on convex set and its application in testing force closure properties of robotic grasping. In: Liu, H., Ding, H., Xiong, Z., Zhu, X. (eds.) ICIRA 2010. LNCS (LNAI), vol. 6425, pp. 240–251. Springer, Heidelberg (2010). https://doi.org/10.1007/978-3-642-16587-0_22
15. Liberti, L.: Reformulation and convex relaxation techniques for global optimization. 4OR Q. J. Belgian French Italian Oper. Res. Soc. **2**(3), 255–258 (2004). https://doi.org/10.1007/s10288-004-0038-6
16. Ramik, S., Wlach, M.: Generalized Concavity in Fuzzy Optimization and Decision Analysis. Kluwer Academic Publishers, Boston (2002)
17. Sambuc, R.: Fonctions ϕ-floues. Application l'aide au diagnostic en pathologie thyroidienne, These de Doctorar en Merseille (1975)
18. Sarkar, D.: Concavoconvex fuzzy set. Fuzzy Sets Syst. **79**, 267–269 (1996). https://doi.org/10.1016/0165-0114(95)00089-5
19. Satia, J.K., Lave, R.E.: Markovian decision processes with uncertain transition probabilities. Oper. Res. **21**(3), 728–740 (1973). https://doi.org/10.1287/opre.21.3.728

20. Sniedovich, M.: Wald's maximin model: a treasure in disguise!. J. Risk Finance **9**(3), 287–291 (2008). https://doi.org/10.1108/15265940810875603
21. Syau, Y.R., Lee, E.S.:Fuzzy convexity and multiobjective convex optimization problems. Comput. Math. Appl. **52**(3), 351–362 (2006). https://doi.org/10.1016/j.camwa.2006.03.017
22. Tofighi, M., Yorulmaz, O., Kose, K., Kahraman, D.C., Cetin-Atalay, R., Cetin, A.E.: Phase and TV based convex sets for blind deconvolution of microscopic images. IEEE J. Sel. Topics Sig. Process. **10**(1), 81–91 (2015). https://doi.org/10.1109/JSTSP.2015.2502541
23. Wald, A.: Statistical decision functions which minimize the maximum risk. Ann. Math. **46**(2), 26 5–280 (1945). https://doi.org/10.1007/978-1-4612-0919-5_21
24. Xu, Z.S., Yager, R.R.: Some geometric aggregation operators based on intuitionistic fuzzy sets. Int. J. Gen. Syst. **35**(4), 417–433 (2006). https://doi.org/10.1080/03081070600574353
25. Yang, X.M.: A property on convex fuzzy sets. Fuzzy Sets Syst. **126**, 269–271 (2002). https://doi.org/10.1016/S0165-0114(01)00075-6
26. Zadeh, L.: The concept of a linguistic variable and its application to approximate reasoning-I. Inf. Sci. **8**, 199–249 (1975). https://doi.org/10.1016/0020-0255(75)90036-5

Discrete Models and Computational Intelligence

Improvements on the Convergence and Stability of Fuzzy Grey Cognitive Maps

István Á. Harmati[1(✉)] and László T. Kóczy[2,3]

[1] Department of Mathematics and Computational Sciences, Széchenyi István University, Egyetem tér 1, Győr 9026, Hungary
harmati@sze.hu

[2] Department of Information Technology, Széchenyi István University, Egyetem tér 1, Győr 9026, Hungary
koczy@sze.hu

[3] Department of Telecommunication and Media Informatics, Budapest University of Technology and Economics, Magyar tudósok körútja 2, Budapest 1117, Hungary

Abstract. Fuzzy grey cognitive maps (FGCMs) are extensions of fuzzy cognitive maps (FCMs), where the causal connections between the concepts are represented by so-called grey numbers. Just like in classical FCMs, the inference is determined by an iteration process, which may converge to an equilibrium point, but limit cycles or chaotic behaviour may also show up.

In this paper, based on network measures like in-degree, out-degree and connectivity, we provide new sufficient conditions for the existence and uniqueness of fixed points for FGCMs. Moreover, a tighter convergence condition is presented using the spectral radius of the modified weight matrix.

Keywords: Fuzzy cognitive map · Fuzzy grey cognitive map · Stability · Convergence · Equilibrium point

1 Introduction

Fuzzy cognitive maps are neural network-based decision support tools, where the neurons represent specific factors or characteristics of the modelled system [11]. Graphically, a fuzzy cognitive map is a weighted, directed graph. The constant weights assigned to the edges from the interval $[-1, 1]$ express the strength and direction of causal connections. The current states of the neurons (which are called concepts in FCM literature) are also characterized by numbers in the $[0, 1]$ interval (in some applications the interval $[-1, 1]$ is also applicable [12]). These are the activation values of the concepts [6].

M.-J. Lesot et al. (Eds.): IPMU 2020, CCIS 1239, pp. 509–523, 2020.
https://doi.org/10.1007/978-3-030-50153-2_38

Formally, the system can be described by the set of concepts $(C_1, C_2, \ldots, C_n)$; the current activation values of the concepts $(A_1, A_2, \ldots, A_n)$; the weight matrix W which assigns weight w_{ij} to each edge connecting the nodes C_i and C_j), expressing how strongly influenced is concept C_i by concept C_j. The sign of w_{ij} indicates whether the relationship between C_j and C_i is direct or inverse. So matrix W represents the weighted causal connections between the concepts. A transformation function $f : \mathbb{R} \to [0, 1]$ calculates the activation value of concepts at every time step of the iteration and the activation values in the allowed range (sometimes a function $f : \mathbb{R} \to [-1, 1]$ is applied).

The iteration rule which calculates the values of the concept at every step may or may not include self-feedback. In general form it can be written as

$$A_i(k) = f\left(\sum_{j=1, j\neq i}^{n} w_{ij}A_j(k-1) + d_iA_i(k-1)\right) \tag{1}$$

where $A_i(k)$ is the value of concept C_i at discrete time k, w_{ij} is the weight of the connection from concept C_j to concept C_i and $0 \leq d_i \leq 1$ expresses the possible self-feedback. If $d_i = 0$, then there is no self-feedback. If we include the d_is into the diagonal of weight matrix W, the iteration equation can be rewritten in more compact style:

$$A_i(k+1) = f\left(\sum_{j=1}^{n} w_{ij}A_j(k)\right) = f(w_iA(k)), \tag{2}$$

where $w_i = [w_{i1}, \ldots, w_{in}]$ is the ith row of W and $A(k) = [A_1(k), \ldots, A_n(k)]^T$ is the concept vector after k iterations. We apply dot product between them, so $w_iA^{(k)}$ is a real number.

Moreover, if we couple the coordinates of the concept vector together and denote by G the mapping $\mathbb{R}^n \to \mathbb{R}^n$ that generates the concept vector $A(k+1)$ from $A(k)$, then we have that:

$$A(k+1) = \begin{bmatrix} A_1(k+1) \\ \vdots \\ A_n(k+1) \end{bmatrix} = \begin{bmatrix} f(w_1A(k)) \\ \vdots \\ f(w_nA(k)) \end{bmatrix} = G(A(k)). \tag{3}$$

The iteration rule repeated until either the FCM converges to an equilibrium state (fixed point) or the maximal number of iterations is reached. Mathematically, the FCM may converge to a fixed point, may arrive to a limit cycle or shows chaotic pattern [2,5].

The weights of the connections are usually determined by human experts or by learning methods. In both of the cases there are some uncertainties about the exact values of the weights. This was the main motivation of Fuzzy Grey Cognitive Maps, where the weights and concept values are modelled by the so-called grey numbers [8–10,13].

A grey number (denoted by $\otimes g$) is a number whose accurate value is unknown, but we know the range within the value is included. A grey number with both a lower limit ($\underline{g}$) and an upper limit ($\overline{g}$) is called an interval grey number [4], so $\otimes g \in [\underline{g}, \overline{g}]$. In applications, a grey number is usually an interval. The basic arithmetic operations on grey numbers are the following [4]:

1. $\otimes g_1 + \otimes g_2 \in [\underline{g_1} + \underline{g_2}, \overline{g_1} + \overline{g_2}]$
2. $- \otimes g \in [-\overline{g}, -\underline{g}]$
3. $\otimes g_1 - \otimes g_2 \in [\underline{g_1} - \overline{g_2}, \overline{g_1} - \underline{g_2}]$
4. $\otimes g_1 \times \otimes g_2 \in [\min(S), \max(S)]$,
   where $S = \{\underline{g_1} \cdot \underline{g_2}, \underline{g_1} \cdot \overline{g_2}, \overline{g_1} \cdot \underline{g_2}, \overline{g_1} \cdot \overline{g_2}\}$
5. If $\lambda > 0$, $\lambda \in \mathbb{R}$, then $\lambda \cdot \otimes g \in [\lambda \underline{g}, \lambda \overline{g}]$

Beside the above defined operations, we have to provide a consistent definition for the generalization of any $f\colon \mathbb{R} \to \mathbb{R}$ function to grey numbers. The function of a grey number $\otimes g \in [\underline{g}, \overline{g}]$ is the grey number $f(\otimes g) \in [\underline{f(\otimes g)}, \overline{f(\otimes g)}]$, where

$$\underline{f(\otimes g)} = \inf\{f(\gamma)\colon \gamma \in [\underline{g}, \overline{g}]\} \tag{4}$$

$$\overline{f(\otimes g)} = \sup\{f(\gamma)\colon \gamma \in [\underline{g}, \overline{g}]\} \tag{5}$$

For a continuous and monotone increasing function f we have

$$\inf\{f(\gamma)\colon \gamma \in [\underline{g}, \overline{g}]\} = f(\underline{g}) \tag{6}$$

$$\sup\{f(\gamma)\colon \gamma \in [\underline{g}, \overline{g}]\} = f(\overline{g}) \tag{7}$$

Consequently, $\underline{f(\otimes g)} = f(\underline{g})$ and $\overline{f(\otimes g)} = f(\overline{g})$ and $f(\otimes g) \in [f(\underline{g}), f(\overline{g})]$.

The dynamics of an FGCM is similar to the original FCM's. It begins with an initial grey vector $A(0)$, which represents initial uncertainty. The elements of this vector are grey numbers, i.e. $A_i(0) \in [\underline{A_i(0)}, \overline{A_i(0)}]$ for every i. The activation values are computed by the iterative process, resulting grey numbers as concept values:

$$A_i(k) \in \left[\underline{f(w_i A(k-1))}, \overline{f(w_i A(k-1))}\right] \tag{8}$$

An FGCM with continuous threshold produces one of the following behaviours:

1. Fixed point: the FGCM converges to a grey fixed-point attractor. This fixed point is vector, whose coordinates are grey numbers (intervals). The convergence (stabilization) means that the endpoints of these intervals are stabilized after a certain number of iterations.
2. Limit cycle: the state values keep oscillating between several states. These states (elements of the limit cycle) are concept vectors with interval coordinates.
3. Chaotic behaviour: the FGCM produces different grey vector states for each iteration, without any pattern.

Usually, the behaviour of fuzzy cognitive maps is examined by trial-error methods. The main contribution of this paper is to present analytical conditions for the existence and uniqueness of attracting fixed points of FGCMs. It also ensures the global exponential stability of the system. Previously, Boutalis et al. [14] proved a condition for the convergence of a class of FCMs. Their result has been generalized in [2]. Knight et al. [15] studied the problem of fixed points of FCMs using only the topology, without the weights.

In this paper, we give several conditions for convergence and stability of fuzzy grey cognitive maps. In Sect. 2 different type of behaviours of FCMs and FGCMs are demonstrated by illustrative examples. In Sect. 3 we briefly summarize the mathematical background, in Sect. 4 some theorems are proved regarding to existence and uniqueness of fixed points of FGCMs. We illustrate the results with an example in Sect. 5, and shortly summarize them in Sect. 6.

2 Examples for Different Behaviour

Consider the following toy example to demonstrate the behaviour of FCMs and FGCMs (Fig. 1). Although this network is extremely simple, it is able to produce qualitatively different behaviours for different choice of weights. Let us apply the hyperbolic tangent function with parameter λ ($\tanh(\lambda x)$) as threshold function (for some properties of hyperbolic tangent FCMs see [3]).

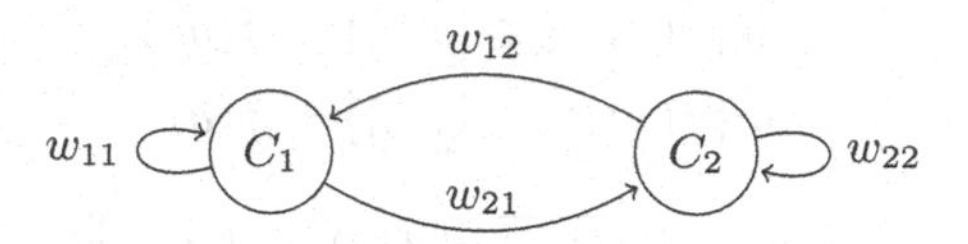

Fig. 1. The topology of the demonstrative example. The self-loops indicate the possible existence of self-feedback.

Different settings of weights and parameter λ yield completely different behaviour, although the topology remains the same.

For a certain set of parameters we may have a non-trivial fixed point (the trivial fixed point is the zero vector, since it is always a fixed point of hyperbolic tangent FCMs, but not always attractor [3]) (Fig. 2). Other setting yields oscillation, namely a quasiperiodic behaviour (Fig. 3).

Convergence of FGCMs means that the upper and lower endpoints of the intervals containing the activations values are stabilized. It can be observed in Fig. 4, while with different weights and parameter λ we can observe oscillating pattern (Fig. 5).

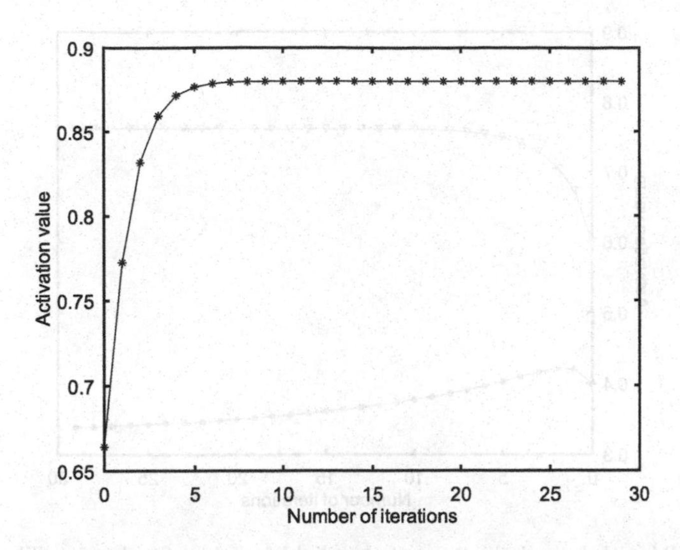

Fig. 2. FCM with hyperbolic tangent threshold function: fixed point. The activation value of concept C_1 vs. number of iterations. The parameters are $w_{11} = 1, w_{21} = 0.6, w_{21} = 0.4, w_{22} = 1, \lambda = 1$.

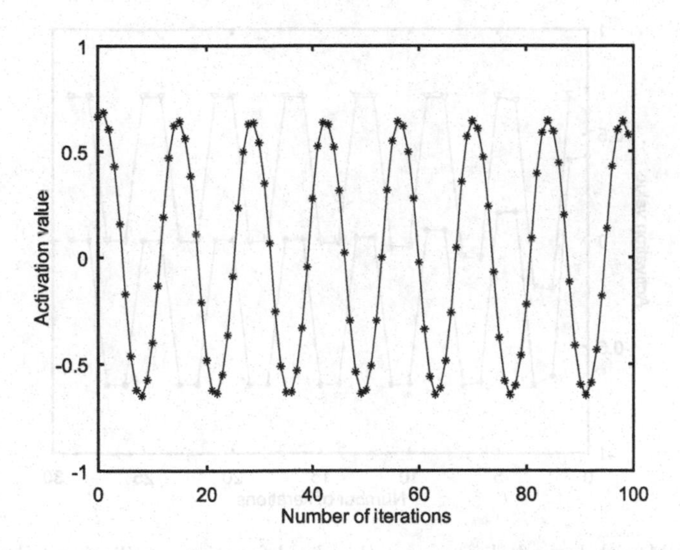

Fig. 3. FCM with hyperbolic tangent threshold function: quasiperiodic pattern. The activation value of concept C_1 vs. number of iterations. The parameters are $w_{11} = 1, w_{21} = 0.6, w_{21} = -0.4, w_{22} = 1, \lambda = 1$.

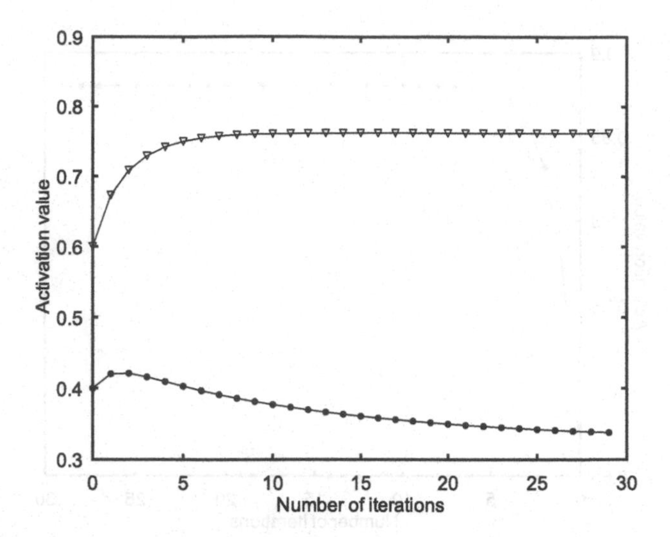

Fig. 4. FGCM with hyperbolic tangent threshold function: fixed point. The activation value (interval) of concept C_1 vs. number of iterations. The upper endpoint of the interval is denoted by ∇, the lower endpoint is denoted by $\bullet$. The parameters are $w_{11} = [0.9, 1]$, $w_{21} = [0.5, 0.7]$, $w_{21} = [0.3, 0.5]$, $w_{22} = [0.9, 1]$, $\lambda = 0.8$.

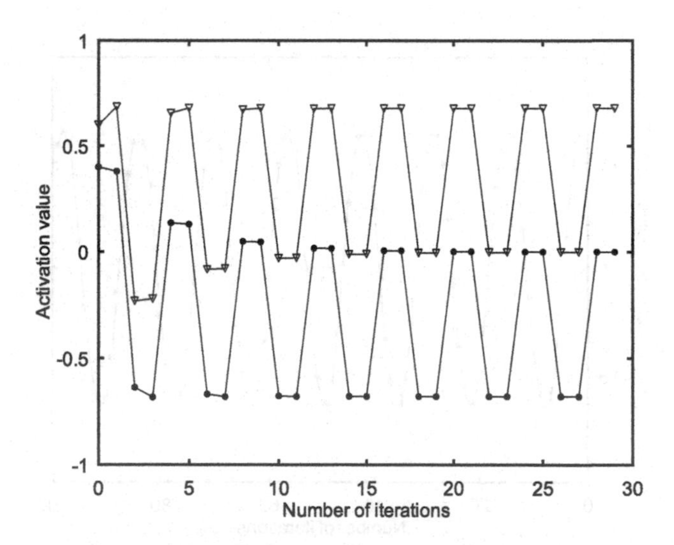

Fig. 5. FGCM with hyperbolic tangent threshold function: oscillating behaviour. The activation value (interval) of concept C_1 vs. number of iterations. The upper endpoint of the interval is denoted by ∇, the lower endpoint is denoted by $\bullet$. The parameters are $w_{11} = 0$, $w_{21} = [0.5, 0.7]$, $w_{21} = [-0.5, -0.3]$, $w_{22} = 0$, $\lambda = 2$.

3 Mathematical Background

The results presented in the next section are based on the contraction property of the mapping that generates the iteration. Here we recall the definition of contraction mapping [7]:

Definition 1. *Let (X, d) be a metric space. A mapping $G\colon X \to X$ is a* contraction mapping *or* contraction *if there exists a constant c (independent from x and y), with $0 \le c < 1$, such that*

$$d\left(G(x), G(y)\right) \le cd(x, y). \tag{9}$$

The notion of contraction is related to the distance metric d applied. It may happen that a function is a contraction w.r.t. one distance metric, but not a contraction w.r.t. another distance metric. The iterative process of an FCM may end at an equilibrium point, which is a so-called fixed point.

Let $G\colon X \to X$, then a point $x^* \in X$ such that $G(x^*) = x^*$ is a fixed point of G. The following theorem provides sufficient condition for the existence and uniqueness of a fixed point [7]. Moreover, if mapping that generates the iteration is a contraction, it ensures the stability of the iteration.

Theorem 1 *(Banach's fixed point theorem). If $G\colon X \to X$ is a contraction mapping on a nonempty complete metric space (X, d), then G has only one fixed point x^*. Moreover, x^* can be found as follows: start with an arbitrary $x_0 \in X$ and define the sequence $x_{n+1} = G(x_n)$, then $\lim_{n\to\infty} x_n = x^*$.*

Definition 2. *Let x^* be a fixed point of the iteration $x_{n+1} = G(x_n)$. x^* is* locally asymptotically stable *if there exist a neighborhood U of x^*, such that for each starting value $x_0 \in U$ we get that*

$$\lim_{n\to\infty} x_n = x^*. \tag{10}$$

If this neighborhood U is the entire domain of G, then x^* is a globally asymptotically stable *fixed point.*

Corollary 1. *If $G\colon X \to X$ is a contraction mapping on a nonempty complete metric space (X, d), then its unique fixed point x^* is globally asymptotically stable.*

In Sect. 4, the following property of the sigmoid function will be applied:

The derivative of the sigmoid function $f\colon \mathbb{R} \to \mathbb{R}$, $f(x) = 1/(1 + e^{-\lambda x})$, $(\lambda > 0)$ is bounded by $\lambda/4$. Moreover, for every $x, y \in \mathbb{R}$ the following inequality holds

$$|f(x) - f(y)| \le \lambda/4 \cdot |x - y|.$$

In [1] the following statements have been introduced about the convergence of fuzzy grey cognitive maps:

Theorem 2. *Let* $\otimes W$ *be the extended (including possible feedback) weight matrix of a fuzzy grey cognitive map (FGCM), where the weights* $\otimes w_{ij}$ *are nonnegative or nonpositive grey numbers and let* $\lambda > 0$ *be the parameter of the sigmoid function* $f(x) = 1/(1 + e^{-\lambda x})$ *applied for the iteration. Let* W^* *be a matrix defined by the absolute values of the weights, i.e.* $w^*_{ij} = \max\left\{|\underline{w_{ij}}|, |\overline{w_{ij}}|\right\}$. *If one of the inequalities*

$$\|W^*\|_1 < \frac{4}{\lambda} \tag{11}$$

$$\|W^*\|_\infty < \frac{4}{\lambda} \tag{12}$$

$$\|W^*\|_F < \frac{4}{\lambda} \tag{13}$$

hold, then the FGCM has one and only one grey fixed point, regardless of the initial concept values.

Here $\| * \|_1$, $\| * \|_\infty$ and $\| * \|_F$ denote the 1-norm, infinity norm and Frobenius norm of the matrix, respectively. Here fixed point $\otimes A^*$ is

$$\otimes A^* = [\otimes A^*_1, \ldots, \otimes A^*_n]^T \in \left[[\underline{A^*_1}, \overline{A^*_1}], \ldots, [\underline{A^*_n}, \overline{A^*_n}]\right]^T$$

The grey fixed point is unique in the sense that the endpoints of the intervals containing grey concept values are unique, i.e. the values $\underline{A^*_i}$ and $\overline{A^*_i}$ are unique for every i.

4 Convergence Conditions

In this section, we provide several theorems regarding the existence and uniqueness of attracting grey fixed point. The first three theorems are based on the structure of the FGCM, namely they are based in the so-called in-degree, out-degree and connectivity, which are widely used measures to describe the quality of the network. The last one is based on the spectral radius of the modified weight matrix W^* and it gives the better condition in the sense that it ensures the convergence for the largest set of parameter λ.

Definition 3. *The weighted in-degree of concept* C_j *equals the sum of the absolute values of the weights of in-coming edges:*

$$deg^{in}_j = \sum_{i=1}^{n} |w_{ij}| \tag{14}$$

which is the sum of the absolute values of the entries of the j*th column of* W.

Definition 4. *The weighted out-degree of concept C_i equals the sum of the absolute values of the weights of out-going edges:*

$$deg_i^{out} = \sum_{j=1}^{n} |w_{ij}| \tag{15}$$

which is the sum of the absolute values of the entries of the ith row of W.

We note that self-feedback means self-loop in the graph. So if self-feedbacks are applied in the concepts, then the weights of the feedback are counted in the in-degree and the out-degree, too. It is the reason that we did not exclude $i = j$ from the summations above.

Definition 5. *The connectivity of an FCM is the ratio of the number of connections between concepts to the maximum number of such possible connections.*

Connectivity measures the 'density' of the network. If self-feedback is allowed, then the maximum number of connections is n^2, if not allowed, then the maximum number of connections is $n(n-1)$.

The weighted in-degree, weighted out-degree and weighted connectivity can defined similarly for FGCMs, but instead of absolute values of real numbers (exact weights), we use the absolute values of grey numbers (intervals):

$$deg_j^{in} = \sum_{i=1}^{n} |\otimes w_{ij}| = \sum_{i=1}^{n} w_{ij}^*$$

$$deg_i^{out} = \sum_{j=1}^{n} |\otimes w_{ij}| = \sum_{j=1}^{n} w_{ij}^*$$

Definition 6. *The weighted connectivity of an FCM is the ratio of the sum of absolute values of weights of connections between concepts to the maximum number of such possible connections.*

If self-feedback is allowed, then the weighted connectivity is

$$Con_w = \frac{\sum_{i=1}^{n} \sum_{j=1}^{n} |w_{ij}|}{n^2}$$

If self-feedback is not allowed, then the weighted connectivity is

$$Con_w = \frac{\sum_{i=1}^{n} \sum_{j=1}^{n} |w_{ij}|}{n(n-1)}$$

For fuzzy grey cognitive maps, we apply the absolute values of the grey weights (w_{ij}^*– s), so the enumerator is the sum $\sum_{i=1}^{n} \sum_{j=1}^{n} w_{ij}^*$.

Theorem 3. *Let λ be the parameter of the sigmoid threshold function applied for every concept. If the maximal in-degree of the FGCM (including possible feedback) is less than $4/\lambda$, then the FGCM has one and only one fixed point.*

Proof. In [1] it has been shown that if $\|W^*\|_1 < 4/\lambda$, then the FGCM has one and only one grey fixed point. Moreover, since

$$\|W^*\|_1 = \max_{1\leq j\leq n} \sum_{i=1}^{n} w^*_{ij} = \max_{1\leq j\leq n} deg_j^{in} \tag{16}$$

this condition is equivalent to the requirement stated in the theorem.

Theorem 4. *Let λ be the parameter of the sigmoid threshold function applied for every concept. If the maximal out-degree of the FGCM (including possible feedback) is less than $4/\lambda$, then the FGCM has one and only one fixed point.*

Proof. The proof goes similarly to the previous one, but instead of 1-norm we use the infinity norm. In [1] it has been shown that if $\|W^*\|_\infty < 4/\lambda$, then the FGCM has one and only one grey fixed point. Moreover, since

$$\|W^*\|_\infty = \max_{1\leq i\leq n} \sum_{j=1}^{n} w^*_{ij} = \max_{1\leq i\leq n} deg_i^{out} \tag{17}$$

this condition is equivalent to the requirement stated in the theorem.

The theorems above are mathematically equivalent with the statements of Theorem 2, but they are easier to capture by the users of FCMs. While the users are not necessarily familiar with matrix norms, they can easily handle notions like in- and out-degree, which are graphically straightforward.

Theorem 5. *Let λ be the parameter of the sigmoid threshold function applied for every concept. If the weighted connectivity (Con_w) of the FGCM small enough, namely*

1. *if self-feedback is allowed:*

$$Con_w < \frac{4}{\lambda n^2},$$

2. *if self-feedback is not allowed:*

$$Con_w < \frac{4}{\lambda n(n-1)},$$

then the FGCM has one and only one fixed point.

Proof. We show that if $\sum_{i=1}^{n}\sum_{j=1}^{n} w^*_{ij} < 4/\lambda$, then mapping G is a contraction, so it has exactly one fixed point. Let us define the distance of grey concept vectors as

$$d(A, A') = \frac{1}{2}\left(\|\underline{A} - \underline{A}'\|_1 + \|\overline{A} - \overline{A}'\|_1\right) \tag{18}$$

We are going to show that with the distance measure above:

$$d(G(A), G(A')) \leq \frac{\lambda}{4} \sum_{i=1}^{n}\sum_{j=1}^{n} w^*_{ij} d(A, A')$$

By the definition of the distance of two grey-valued vectors, we have

$$d(G(A), G(A')) = \frac{1}{2}\left(\|\underline{G(A)} - \underline{G(A')}\|_1 + \|\overline{G(A)} - \overline{G(A')}\|_1\right)$$

It has been shown in [1] that the following upper estimation can be given for the difference of the ith coordinates (similar inequality holds for the difference of the upper endpoints):

$$\left|\underline{G(A)}_i - \underline{G(A')}_i\right| \leq \left|\frac{\lambda}{4} w_i^* |\underline{A} - \underline{A}'|\right|$$

where w_i^* is the ith row of matrix W^* and we apply dot product between w_i^* and $|\underline{A} - \underline{A}'| = (|A_1 - A_1'|, \ldots, |A_n - A_n'|)$. Moreover,

$$w_i^* |\underline{A} - \underline{A}'| \leq \|w_i^*\|_1 \cdot \|\underline{A} - \underline{A}'\|_1$$

Here $\|w_i^*\|_1 = \sum_{j=1}^n |w_{ij}^*| = \sum_{j=1}^n w_{ij}^*$. Use this inequality for the distance of $G(A)$ and $G(A')$:

$$d(G(A), G(A')) = \frac{1}{2}\left(\|\underline{G(A)} - \underline{G(A')}\|_1 + \|\overline{G(A)} - \overline{G(A')}\|_1\right) \tag{19}$$

$$= \frac{1}{2}\left(\sum_{i=1}^n \left|\underline{G(A)}_i - \underline{G(A')}_i\right| + \sum_{i=1}^n \left|\overline{G(A)}_i - \overline{G(A')}_i\right|\right) \tag{20}$$

$$\leq \frac{\lambda}{4}\frac{1}{2}\left(\sum_{i=1}^n \sum_{j=1}^n w_{ij}^* \|\underline{A} - \underline{A}'\|_1 + \sum_{i=1}^n \sum_{j=1}^n w_{ij}^* \|\overline{A} - \overline{A}'\|_1\right) \tag{21}$$

$$= \frac{\lambda}{4} \sum_{i=1}^n \sum_{j=1}^n w_{ij}^* \frac{1}{2}\left(\|\underline{A} - \underline{A}'\|_1 + \|\overline{A} - \overline{A}'\|_1\right) \tag{22}$$

$$= \frac{\lambda}{4} \sum_{i=1}^n \sum_{j=1}^n w_{ij}^* \cdot d(A, A') \tag{23}$$

If $\frac{\lambda}{4} \sum_{i=1}^n \sum_{j=1}^n w_{ij}^* < 1$, then the mapping is a contraction, so the iteration leads to a unique fixed point, regardless to the initial value. Rearanging this inequality and division both sides by n^2 (or $n(n-1)$) completes the proof.

Although Theorem 5 provides weaker condition, it has an important message expressed by connectivity: poorly connected FGCMs cannot produce complex behaviour (the term 'poorly' depends on λ and n).

Theorem 6. *Let $\otimes W$ be the extended (including possible feedback) weight matrix of a fuzzy grey cognitive map (FGCM), where the weights $\otimes w_{ij}$ are nonnegative or nonpositive grey numbers and let $\lambda > 0$ be the parameter of the sigmoid function $f(x) = 1/(1 + e^{-\lambda x})$ applied for the iteration. Let W^* be a* *matrix defined by the absolute values of the weights. If the spectral radius of* W^* *is less than* $4/\lambda$, *i.e if the inequality*

$$\rho(W^*) < \frac{4}{\lambda} \tag{24}$$

hold, then the FGCM has one and only one grey fixed point, regardless of the initial concept values.

Proof. Let us define the distance of two grey-valued vectors as the norm of their difference. At this stage we do not specify this norm:

$$d(A, A') = \|A - A'\| = \frac{1}{2}\left(\|\underline{A} - \underline{A}'\| + \|\overline{A} - \overline{A}'\|\right)$$

We are going to show that with the distance above and for a suitable matrix norm:

$$d(G(A), G(A')) \leq \frac{\lambda}{4}\|W^*\|d(A, A')$$

By the definition of the distance of two grey-valued vectors, we have

$$d(G(A), G(A')) = \frac{1}{2}\left(\|\underline{G(A)} - \underline{G(A')}\| + \|\overline{G(A)} - \overline{G(A')}\|\right)$$

It has been shown in [1] that the following upper estimation can be given for the difference of the ith coordinates (similar inequality holds for the difference of the upper endpoints):

$$\left|\underline{G(A)}_i - \underline{G(A')}_i\right| \leq \left|\frac{\lambda}{4}w_i^* |\underline{A} - \underline{A}'|\right|$$

where w_i^* is the ith row of matrix W^*. Since this inequality holds for every coordinates, we conclude to following inequality for the difference of the lower endpoint vectors:

$$\|\underline{G(A)} - \underline{G(A')}\| \leq \left\|\frac{\lambda}{4}W^* |\underline{A} - \underline{A}'|\right\|$$

Using this inequality (and the corresponding inequality for the upper endpoints) we provide upper estimation for the distance of $G(A)$ and $G(A')$:

$$\begin{aligned}
d(G(A), G(A')) &= \frac{1}{2}\left(\|\underline{G(A)} - \underline{G(A')}\| + \|\overline{G(A)} - \overline{G(A')}\|\right) && (25)\\
&\leq \frac{1}{2}\left(\left\|\frac{\lambda}{4}W^*|\underline{A} - \underline{A}'|\right\| + \left\|\frac{\lambda}{4}W^*|\overline{A} - \overline{A}'|\right\|\right) && (26)\\
&\leq \frac{\lambda}{4}\|W^*\|\frac{1}{2}\left(\|\underline{A} - \underline{A}'\| + \left\|\overline{A} - \overline{A}'\right\|\right) && (27)\\
&= \frac{\lambda}{4}\|W^*\| d(A, A') && (28)
\end{aligned}$$

In $\|W^*\|$, the matrix norm is induced by the vector norm. By the contraction mapping theorem, if the coefficient of $d(A, A')$ is less than one, then mapping

G is a contraction, consequently it has exactly one fixed point. Moreover, if the spectral radius of a matrix is less than one, then there exists a matrix norm, such that norm of the matrix is less then one, i.e. if $\rho\left(\frac{\lambda}{4}W^*\right) < 1$, then there exist a matrix norm, such that $\|\frac{\lambda}{4}W^*\| < 1$. Applying this matrix norm, mapping G is a contraction, which completes the proof.

Since $\rho(W^*) \leq \|W^*\|$ for any matrix norm, Theorem 6 gives the best condition expressed by W^*.

Remark 1. The results in Sect. 4: Theorem 3, Theorem 4, Theorem 5 and Theorem 6 are valid for fuzzy grey cognitive maps with hyperbolic tangent threshold function ($\tanh(\lambda x)$), too, but we have to replace $4/\lambda$ by $1/\lambda$, since the derivative of $\tanh(\lambda x)$ is bounded by λ (and not $\lambda/4$).

5 Example

Let us consider the following weight matrix with imprecise (grey) entries:

$$\otimes W = \begin{bmatrix} 0 & [0.1, 0.2] & [-0.6, -0.5] & 0 & 0 & 0 \\ 0 & 0 & [-0.7, -0.5] & 0 & 0 & [0.1, 0.3] \\ [0.6, 0.8] & 0 & 0 & [-0.6, -0.2] & 0 & 0 \\ [0.7, 0.9] & 0 & 0 & 0 & [0.6, 0.8] & 0 \\ 0 & 0 & [0.6, 0.7] & 0 & 0 & 0 \\ 0 & [0.1, 0.3] & 0 & [0.8, 1] & [-1, -0.8] & 0 \end{bmatrix} \quad (29)$$

Then the matrix W^* with the w^*_{ij} entries:

$$W^* = \begin{bmatrix} 0 & 0.2 & 0.6 & 0 & 0 & 0 \\ 0 & 0 & 0.7 & 0 & 0 & 0.3 \\ 0.8 & 0 & 0 & 0.6 & 0 & 0 \\ 0.9 & 0 & 0 & 0 & 0.8 & 0 \\ 0 & 0 & 0.7 & 0 & 0 & 0 \\ 0 & 0.3 & 0 & 1 & 1 & 0 \end{bmatrix} \quad (30)$$

The corresponding measures:

- Maximal weighted in-degree: 2.3
- Maximal weighted out-degree: 2
- Connectivity
  - without self-feedback: $Con_w = 0.2633$
  - with self-feedback: $Con_w = 0.2194$
- spectral radius: $\rho(W^*) = 1.1485$

According to Theorem 6, if $\lambda < 3.4827$, then this grey FCM has one and only one grey fixed point. It also means that in this case the FGCM produces globally asymptotically stable behaviour, since every initial grey vector leads to the same equilibrium state.

6 Summary

Fuzzy Grey Cognitive Maps are generalizations of classical FCMs, that can model the uncertainties of activation values and weights of causal connections.

In this paper, we provided some conditions for the convergence of FGCMs to a unique fixed point. The unicity of this attracting fixed point also ensures that the FGCM is globally exponentially stable, i.e. it converges to the same fixed point attractor regardless of the initial concept vector. Future work is focused on the effective detection of multiple fixed points scenarios and the prediction of oscillating patterns without simulations. The future goal is to provide exact analytical conditions for both of these behaviours.

Acknowledgment. The research presented in this paper was carried out as part of the EFOP-3.6.2-16-2017-00016 project in the framework of the New Széchenyi Plan. The completion of this project is funded by the European Union and co-financed by the European Social Fund.

This research was supported in part by National Research, Development and Innovation Office (NKFIH) K124055.

References

1. Harmati, I.Á., Kóczy, L.T.: On the convergence of sigmoidal fuzzy grey cognitive maps. Int. J. Appl. Math. Comput. Sci. **29**(3), 453–466 (2019)
2. Harmati, I.Á., Hatwágner, M.F., Kóczy, L.T.: On the existence and uniqueness of fixed points of fuzzy cognitive maps. In: Medina, J., et al. (eds.) IPMU 2018. CCIS, vol. 853, pp. 490–500. Springer, Cham (2018). https://doi.org/10.1007/978-3-319-91473-2_42
3. Harmati, I.A., Kóczy, L.T.: Notes on the dynamics of hyperbolic tangent fuzzy cognitive maps. In: 2019 IEEE International Conference on Fuzzy Systems (FUZZ-IEEE), New Orleans, LA, USA, pp. 1–6 (2019)
4. Liu, S., Lin, Y.: Grey Information. Springer, London (2006). https://doi.org/10.1007/1-84628-342-6
5. Nápoles, G., Papageorgiou, E., Bello, R., Vanhoof, K.: On the convergence of sigmoid fuzzy cognitive maps. Inf. Sci. **349–350**, 154–171 (2016)
6. Papageorgiou, E.I., Salmeron, J.L.: Methods and algorithms for fuzzy cognitive map-based decision support. In: Papageorgiou, E.I. (ed.) Fuzzy Cognitive Maps for Applied Sciences and Engineering (2013)
7. Rudin, W: Principles of Mathematical Analysis. McGraw-Hill Inc. (1964)
8. Salmeron, J.L.: Modelling grey uncertainty with fuzzy grey cognitive maps. Expert Syst. Appl. **37**(12), 7581–7588 (2010)
9. Salmeron, J.L., Papageorgiou, E.I.: A fuzzy grey cognitive maps-based decision support system for radiotherapy treatment planning. Knowl.-Based Syst. **30**, 151–160 (2012)
10. Salmeron, J.L., Palos-Sanchez, P.R.: Uncertainty propagation in fuzzy grey cognitive maps with Hebbian-like learning algorithms. IEEE Trans. Cybern. **49**(1), 211–220 (2017)
11. Stylios, C.D., Groumpos, P.P.: Modeling complex systems using fuzzy cognitive maps. IEEE Trans. Syst. Man Cybern.-Part A Syst. Hum. **34**(1), 155–162 (2004)

12. Tsadiras, A.K.: Comparing the inference capabilities of binary, trivalent and sigmoid fuzzy cognitive maps. Inf. Sci. **178**(20), 3880–3894 (2008)
13. Shojaei, P., Haeri, S.A.S.: Development of supply chain risk management approaches for construction projects: a grounded theory approach. Comput. Ind. Eng. **128**, 837–850 (2019)
14. Boutalis, Y., Kottas, T.L., Christodoulou, M.: Adaptive estimation of fuzzy cognitive maps with proven stability and parameter convergence. IEEE Trans. Fuzzy Syst. **17**(4), 874–889 (2009)
15. Knight, C.J., Lloyd, D.J., Penn, A.S.: Linear and sigmoidal fuzzy cognitive maps: an analysis of fixed points. Appl. Soft Comput. **15**, 193–202 (2014)

Group Definition Based on Flow in Community Detection

María Barroso[1(✉)], Inmaculada Gutiérrez[1], Daniel Gómez[1,2], Javier Castro[1,2], and Rosa Espínola[1,2]

[1] Faculty of Statistics, Complutense University, Avenida Puerta de Hierro, s/n, 28040 Madrid, Spain
{mbarro10,inmaguti}@ucm.es, {dagomez,jcastroc,rosaev}@estad.ucm.es
[2] Instituto de Evaluación Sanitaria, Complutense University, Madrid, Spain

Abstract. Community detection problems are one of the hottest disciplines in social network analysis. Nevertheless, most of the related algorithms are specific for non-directed networks, or are based on a density concept of group. In this paper, we deal with a new concept of community for directed networks that is based on the classical flow concept. A community is strong and cohesive if their members can communicate among them. With the aim of dealing with the identification of this new class of groups, in this work, we propose the use of fuzzy measures to represent the flow capacity of a group. We also provide a competitive community detection algorithm that focus on the identification of these new class of flow-based community.

Keywords: Directed networks · Flow · Fuzzy measures · Community detection problem · Louvain Algorithm

1 Introduction

Community detection problems are one of the most important topic in social network analysis [7,18]. The idea of finding communities is strongly related to the idea of finding clusters in data analysis. In general, a cluster can be considered as a set of items that are *closer each other* when they are compared to the rest of items of the problem. A good clusterization of a set of items is associated with the identification of a set of clusters that present internally high degree of intra-homogeneity, and high degree of inter-heterogeneity. In networks, the idea of intra-homogeneity of a cluster/community is usually associated with the density of the group. Then, a good community will be a dense set of nodes with many connections between the members of each group. The idea of high inter-heterogeneity is associated with the existence of lower relations between the clusters/groups. So, the more relationships among the groups, the greater the

This research has been partially supported by the Government of Spain, Grant Plan Nacional de I+D+i, MTM2015-70550-P, PGC2018096509-B-I00 and TIN2015-66471-P.

M.-J. Lesot et al. (Eds.): IPMU 2020, CCIS 1239, pp. 524–538, 2020.
https://doi.org/10.1007/978-3-030-50153-2_39

degree of inter-heterogeneity is. Taking this into account, an optimal community may be a set of nodes that induces a completed subgraph (so they are strongly connected), and are isolated from the rest of the nodes of the network (so they have no relation with any other node in the network).

However, this idea of community is not unanimously accepted when the graph is directed and valued. As noted in [15,18], the idea of community in directed networks can have different interpretations, so several definitions could be made. This is the reason why finding clusters in directed networks is a challenging task with several important applications, since many of the real networks are modelled in an undirected way.

Despite the importance of community detection problems in directed networks, this problem has been poorly studied in the literature. Nevertheless, we can find in [18] four different (but non-formalized) concepts about what could be understood as a community/group:

- The first notion is about a random walk. In this case, each group is formed by these nodes that are more likely to remain inside than outside. This kind of communities are usually obtained with random walk techniques. In the literature, we can find some algorithms that deal with these problems (see for example [7,19,24]).
- The second notion is about the density. In this context, the groups of nodes follow the traditional clustering definition, based on edge density characteristics. It is important to mention that, in this sense, the concept of modularity [20] has been redefined for directed networks. Taking this into account, new algorithms have been developed to deal with this problem. Many of them are adaptions of some well-known community detection algorithms (as the directed Louvain [2]) to the directed case.
- Co-citation groups. As mentioned in [18], edges density is not always the only criterion to identify a set of nodes that share many characteristics. In directed networks, the idea of co-citation group tries to identify groups of nodes (not necessarily connected) that follow or are followed by the same groups. In this sense, we could have a group of nodes that form a community because its followers' set is the same, even if they do not know each other. We also could have a group of nodes that form a community because the set of nodes from which they extract the information is the same. If we think of a citation network, for example, a community could be formed by those researchers who 'drink' their research from the same sources, or a community could be formed by those researchers who are cited by the same colleagues. Obviously, in this situation, two or more nodes may belong to the same community/cluster even if they are not directly connected by edges.
- The last one is related with the idea of flow. In this case, a group is as good as much information can be moved within it. Although this concept of community is clearly related with the idea of density (since the more relations between the members of a group there are, probably the higher the flow capacity will be), it is important to note that they differ in many respects. In flow problems, the structure and location of the edges can be decisive when we

have to distribute the information (by flow). Obviously, this is not reflected by the density of a group that only counts how many edges there are over the totals without specifying the way in which they are arranged.

In this work, we focus on this last idea of group, trying to identify groups in which the information moved by the flow is important [4]. It is easy to find many examples of this type of graphs in the field of social networks. For example, Twitter is a directed network in which each arrow may indicate the number of messages of i retweeted by j. Scientific reference pages such as Scopus, WOS or Google Academic, are also directed networks in which each arrow may indicate the number of times that author i has been cited by author j. In both cases, the flow measures the influence of i over j. A metro or a road network are also two examples of directed networks whose community structure depends on the flow.

The key is to find the way to incorporate this group definition into community detection problems. The use of fuzzy sets in social network analysis problems, and in particular, in community detection problems, is not new [10,11,13,23, 26,27]. Due to the way in which imprecision is modelled, fuzzy sets appear in a natural way when modelling real problems. In this sense, this paper proposes the use of fuzzy measures or capacity measures [7,25] to measure the relative strength of a group, according to the ability of their members to communicate among them. The more flow the members can send, the more cohesive the group will be.

Once the graph and the fuzzy measure are modelled, in this paper we provide a very efficient algorithm that combines the two class of information (the network and the fuzzy measure), allowing us to identify groups in which the idea of flow is considered. The proposed method may also be useful in the size reduction of large scale fuzzy cognitive maps [14], since their structure is a weighted directed digraph.

The rest of the paper is organized as follows. In Sect. 2 we introduce some basic definitions about community detection problems and fuzzy measures background. In Sect. 3 we introduce a new fuzzy measure related to the flow of a directed network. In Sect. 4 we propose an algorithm to deal with community detection problems with fuzzy measures in directed networks. Finally, some conclusions and future research are shown in Sect. 5.

2 Preliminaries

In this Section we introduce several concepts, definitions and algorithms necessary to have a proper understanding of this paper.

2.1 Community Detection Problems in Directed Networks Based on Density

Definition 1. Directed Network [18]. *A directed network is a set of individuals connected together, in which all the edges are directed from one individual to* *another. A directed network is usually represented by a graph $G = (V, E)$, where V is the set of individuals, called nodes or vertices, and $E = \{(i,j) | \ i, j \in V\}$ is the set of ordered pairs of $V \times V$, which are directed edges connecting pairs of nodes (i, j). Another way to represent directed graphs or networks is by means of its adjacency matrix, A, defined as follows:*

$$A_{ij} = \begin{cases} 1 & if \quad (i,j) \in E, \quad \forall i, j \in 1, \ldots, |V| \\ 0 & otherwise \end{cases}$$

where 1 represents the directed edge which connects i with j.

Then, let us recall the definition of community detection problems. Given a graph, this type of problem consists in finding a 'good' partition for the input set of individuals. The notion of 'good' may be different depending on the interests of each problem. Many measures have been proposed in the literature to quantify the goodness of a partition [16]. One of the most popular is the modularity, introduced by Girvan and Newman [22] for non-directed networks. This measure has been adapted to directed networks.

Definition 2. Directed Modularity $\mathbf{Q_d}$ [1]. *The modularity is a quality function to measure the goodness of a partition. Let G be a directed graph and P a partition of the nodes. The directed modularity is defined as:*

$$Q_d(G, P) = \frac{1}{m} \sum_{i,j} \left[A_{ij} - \frac{k_i^{in} k_j^{out}}{m} \right] \delta(c_i, c_j) \qquad (1)$$

where $\delta(c_i, c_j)$ is 1 if i belongs to the same group than j, and 0 otherwise, m is the amount of edges, and k_i^{in} and k_i^{out} are the in/out edges of node i.

There are many methods to deal with community detection problems [3,8,21]. Particularly, we focus on one of the most popular methods: Louvain Algorithm [2]. Because of its effectiveness and speed, it is one of the most used algorithms. It is based on modularity optimization, and works very well in large networks.

2.2 Fuzzy Measures, Directed Fuzzy Graphs, Extended Fuzzy Directed Graphs

Definition 3. Fuzzy Measure [25]. *Given a finite set V, a fuzzy measure is a function $\mu : 2^V \longrightarrow [0, 1]$ that is monotonous ($\forall A, B \subseteq V$ such that $A \subseteq B$) and normalized ($\mu(V) = 1$), and that satisfies the boundary condition ($\mu(\emptyset) = 0$).*

A characteristic of fuzzy measures is their k-additivity [12]. Particularly in this paper, we work with 2-additive fuzzy measures, so let us characterize them.

Definition 4. $2-$additive fuzzy measure [12]. *The fuzzy measure $\mu : 2^V \longrightarrow [0, 1]$ is said to be 2-additive if and only if, $\forall S \subseteq V$, it can be written as a linear combination $\mu(S) = \sum_{i=1}^{n} a_i x_i + \sum_{\{i,j\} \subset A} a_{ij} x_i x_j$, where $x_i = 1$ if $i \in S$ and $x_i = 0$ otherwise.*

Then, let us recall the notion of extended fuzzy graph, firstly introduced for non-directed networks in [13].

Definition 5. Extended Fuzzy Graph [13]. *Let $G = (V, E)$ be a graph, and let $\mu : 2^V \longrightarrow [0, 1]$ be a fuzzy measure defined over the set of nodes. The triplet $\widetilde{G} = (V, E, \mu)$ obtained from considering together the graph with the fuzzy measure, is called extended fuzzy graph.*

Note that this structure is much more complex than a fuzzy graph [23]. Fuzzy graphs could be somehow seen as weighted graphs, as the only available information is provided by their edges and their membership degree.

3 Fuzzy Measures from a Directed Networks: The Flow Capacity Measure

Classical community detection problems just consider the topological information provided by the adjacency matrix of networks. Other evidences, such as that given by the flow of the graph, have not been previously considered when dealing with this type of problems. Then, in this Section we propose a way to use the flow in community detection problems. To deal with it, we propose the use of a fuzzy measure which models the relative flow, by means of the weight of the edges. This weight represents the different degree of communication ability of each link, something obvious in real-life problems, in which different relations may have different importance. Then, we give a group idea related to the flow.

Definition 6. *Let $G = (V, E)$ be a directed graph, let (i, j) be an edge, and let f_{ij} be the flow between nodes i and j in G [9]. Then, $\forall S \subseteq V$, we define the function: $\mu^F(S) = \frac{\sum_{i,j \in S} f_{ij}}{\sum_{i,j \in V} f_{ij}}$.*

As a capacity measure, μ^F represents the communication capacity within a set of nodes of a directed network.

Proposition 1. *The function μ^F introduced in Definition 6 is a fuzzy measure.*

Proof. We will verify that μ^F meets the points mentioned in Definition 3.

1. $\mu^F(\emptyset) = 0$ Trivial.
2. $\mu^F(V) = 1$ due to normalization.
3. Let $A \subseteq B \subseteq V$. Then,
   $\mu^F(B) = \frac{\sum_{i,j \in B} f_{ij}}{\sum_{i,j \in V} f_{ij}} = \frac{\sum_{i,j \in A} f_{ij} + \sum_{i,j \in B \setminus A} f_{ij}}{\sum_{i,j \in V} f_{ij}} = \frac{\sum_{i,j \in A} f_{ij}}{\sum_{i,j \in V} f_{ij}} + \frac{\sum_{i,j \in B \setminus A} f_{ij}}{\sum_{i,j \in V} f_{ij}} =$
   $\mu^F(A) + \mu^F(B \setminus A) \geq \mu^F(A)$, since for all $i, j \in V$, $f_{ij} \geq 0$.

Proposition 2. *The fuzzy measure μ^F introduced in Definition 6 is a 2-additive fuzzy measure [12].*

Proof. We will verify that $\mu^F(S)$ can be defined as a linear combination:

$$\mu^F(S) = \sum_{i=1}^{n} a_i x_i + \sum_{\{i,j\} \in V} a_{ij} x_i x_j$$

where $x_i = 1$ if i $\in$ S and $x_i = 0$ otherwise.

Let us define: $a_i = \frac{f_{ii}}{\sum_{l,m \subseteq V} f_{lm}}$, and $a_{ij} = \frac{f_{ij}+f_{ji}}{\sum_{l,m \subseteq V} f_{lm}}$. Then, we can write:

$$\mu^F(S) = \sum_{i=1}^{n} \frac{f_{ii}}{\sum_{l,m \subseteq V} f_{lm}} x_i + \sum_{\{i,j\} \in V} \frac{f_{ij} + f_{ji}}{\sum_{l,m \subseteq V} f_{lm}} x_i x_j$$

Once we have the fuzzy measure μ^F which models the ability of the flow in a directed network, here we propose to build the graph associated with it, G_{μ^F}. To carry on with it, we work with the interaction index proposed by Grabisch [12]. Let us denote $\mu_S := \mu(S)$.

Definition 7. Interaction Index [12]*: Let V be a finite set, and let μ be a fuzzy measure defined over it. Let $\{i, j\} \in V$. The interaction index introduced by Grabisch, I_{ij} is defined as:*

$$I_{ij} = \sum_{k=0}^{n-2} \zeta_k \sum_{\substack{K \subset V \setminus \{i,j\} \\ |K|=k}} (\mu_{ijK} - \mu_{iK} - \mu_{jK} + \mu_K) \tag{2}$$

where $\zeta_k = \frac{(n-k-2)!k!}{(n-1)!} = \frac{1}{\binom{n-2}{k}(n-1)}$

Given two items i, j, the interaction index related to a fuzzy measure, represents a class of dependency/association in the global capacity. In this way, it is possible to construct a valued graph from a fuzzy measure which defines these dependencies. In [13], this was the way in which the fuzzy measure was taken into account for the community detection problem. We would like to emphasize that the capacity measure will be taken into account in the clustering problem thanks to the interaction index that will force some nodes to be in the same group while others separated.

Proposition 3. *Let $G = (V, E)$ be a directed graph whose related flow function is f, and let μ^F be the fuzzy measure introduced in Definition 6. Then:*

$$I_{ij} = \frac{f_{ij} + f_{ji}}{\sum_{l,m \in V} f_{lm}} \tag{3}$$

Proof. From equation (2), we can rewrite the components of μ^F as:

$$\mu^F_{ijK} = \sum_{\substack{l\in K \\ l\neq i,j}} \sum_{\substack{s\in K \\ s\neq i,j}} \frac{f_{ls}}{\sum_{r,m\subseteq V} f_{rm}} + \sum_{\substack{l\in K \\ l\neq i,j}} \frac{f_{li}+f_{il}}{\sum_{r,m\subseteq V} f_{rm}} + \sum_{\substack{l\in K \\ l\neq i,j}} \frac{f_{lj}+f_{jl}}{\sum_{r,m\subseteq V} f_{rm}} + \frac{f_{ij}+f_{ji}}{\sum_{r,m\subseteq V} f_{rm}}$$

$$\mu^F_{iK} = \sum_{\substack{l\in K \\ l\neq i,j}} \sum_{\substack{s\in K \\ s\neq i,j}} \frac{f_{ls}}{\sum_{r,m\subseteq V} f_{rm}} + \sum_{\substack{l\in K \\ l\neq i,j}} \frac{f_{li}+f_{il}}{\sum_{r,m\subseteq V} f_{rm}}$$

$$\mu^F_{jK} = \sum_{\substack{l\in K \\ l\neq i,j}} \sum_{\substack{s\in K \\ s\neq i,j}} \frac{f_{ls}}{\sum_{r,m\subseteq V} f_{rm}} + \sum_{\substack{l\in K \\ l\neq i,j}} \frac{f_{lj}+f_{jl}}{\sum_{r,m\subseteq V} f_{rm}}$$

$$\mu^F_{K} = \sum_{\substack{l\in K \\ l\neq i,j}} \sum_{\substack{s\in K \\ s\neq i,j}} \frac{f_{ls}}{\sum_{r,m\subseteq V} f_{rm}}$$

Hence, transcribing and reducing:

$$\begin{aligned} I_{ij} = {} & \sum_{k=0}^{n-2} \zeta_k \sum_{\substack{K\subset V\setminus\{i,j\} \\ |K|=k}} \Bigg[\sum_{\substack{l\in K \\ l\neq i,j}} \sum_{\substack{s\in K \\ s\neq i,j}} \frac{f_{ls}}{\sum_{r,m\subseteq V} f_{rm}} + \sum_{\substack{l\in K \\ l\neq i,j}} \frac{f_{li}+f_{il}}{\sum_{r,m\subseteq V} f_{rm}} \\ & + \sum_{\substack{l\in K \\ l\neq i,j}} \frac{f_{lj}+f_{jl}}{\sum_{r,m\subseteq V} f_{rm}} + \frac{f_{ij}+f_{ji}}{\sum_{r,m\subseteq V} f_{rm}} - \sum_{\substack{l\in K \\ l\neq i,j}} \sum_{\substack{s\in K \\ s\neq i,j}} \frac{f_{ls}}{\sum_{r,m\subseteq V} f_{rm}} \\ & - \sum_{\substack{l\in K \\ l\neq i,j}} \frac{f_{li}+f_{il}}{\sum_{r,m\subseteq V} f_{rm}} - \sum_{\substack{l\in K \\ l\neq i,j}} \sum_{\substack{s\in K \\ s\neq i,j}} \frac{f_{ls}}{\sum_{r,m\subseteq V} f_{rm}} - \sum_{\substack{l\in K \\ l\neq i,j}} \frac{f_{lj}+f_{jl}}{\sum_{r,m\subseteq V} f_{rm}} \\ & + \sum_{\substack{l\in K \\ l\neq i,j}} \sum_{\substack{s\in K \\ s\neq i,j}} \frac{f_{ls}}{\sum_{r,m\subseteq V} f_{rm}} \Bigg] = \sum_{k=0}^{n-2} \zeta_k \sum_{\substack{K\subset V\setminus\{i,j\} \\ |K|=k}} \frac{f_{ij}+f_{ji}}{\sum_{r,m\subseteq V} f_{rm}} \\ = {} & \sum_{k=0}^{n-2} \frac{1}{\binom{n-2}{k}(n-1)} \binom{n-2}{k} \frac{f_{ij}+f_{ji}}{\sum_{r,m\subseteq V} f_{rm}} = \sum_{k=0}^{n-2} \frac{1}{n-1} \frac{f_{ij}+f_{ji}}{\sum_{r,m\subseteq V} f_{rm}} \\ = {} & \frac{f_{ij}+f_{ji}}{\sum_{r,m\subseteq V} f_{rm}} = I_{ij} \end{aligned}$$

In the classical definition of the interaction index, the order of the elements i and j in the pair $\{i,j\}$ has no significance. Then, we propose an adaptation of it, in order to consider those cases in which this order is important.

Definition 8. Directed Interaction Index. *Let $G = (V,E)$ be a directed graph whose related flow function is f. Let μ be a fuzzy measure defined over the set of nodes, V. Given a pair of ordered nodes (i,j), where $i,j \in V$, we define the directed interaction index I^D_{ij} as:*

$$I^D_{ij} = \frac{f_{ij}}{\sum_{l,m\in V} f_{lm}} \tag{4}$$

From previous definition, we can trivially see that $\forall i,j,\ i\neq j,\ I_{ij} = I^D_{ij} + I^D_{ji}$. Then, the adjacency matrix of the G_{μ^F} is the matrix I^D.

Let us illustrate the calculation of μ^F and I^D with a toy example.

Fig. 1. Directed chain with 12 nodes.

Example 3.1. *We evaluate a simple example of a chain with 12 nodes, as it is drawn in Fig. 1. Let us assume that the weight of all the edges is 1. Then, we calculate* μ^F *and* I^D.

In this example we calculate the fuzzy measure μ^F *which, as it is shown previously, is 2-additive. Therefore, we only have to calculate it for those subsets of* V *with cardinality one and two. We also calculate the directed interactions,* I^D. *See Fig. 2.*

As it can be seen in matrix I^D, *the node 7 is the only that can communicate with the rest of nodes. At the same time, it is appreciable how this chain is divided by means of its flow values. The nodes 6, 5, 4, 3 and 2 only can communicate with the node with which the related flow reaches the lowest value. In the same way, 8, 9, 10, 11 can connect with the node with which the related flow reaches the highest value. On the other hand, 1 and 12 are isolated. Also, let us observe that in both matrices, the blanks mean 0, in Fig. 2.*

$$
A=\begin{pmatrix}
 & & & & & & & & & & & \\
1 & & & & & & & & & & & \\
 & 1 & & & & & & & & & & \\
 & & 1 & & & & & & & & & \\
 & & & 1 & & & & & & & & \\
 & & & & 1 & & & & & & & \\
 & & & & & 1 & 1 & & & & & \\
 & & & & & & & 1 & & & & \\
 & & & & & & & & 1 & & & \\
 & & & & & & & & & 1 & & \\
 & & & & & & & & & & 1 & \\
 & & & & & & & & & & & 1
\end{pmatrix}
\quad
\mu^F(\{i,j\})=\frac{1}{36}\begin{pmatrix}
1 & 1 & 1 & 1 & 1 & 1 & & & & & \\
 & 1 & 1 & 1 & 1 & 1 & & & & & \\
 & & 1 & 1 & 1 & 1 & & & & & \\
 & & & 1 & 1 & 1 & & & & & \\
 & & & & 1 & 1 & & & & & \\
 & & & & & 1 & & & & & \\
 & & & & & & 1 & 1 & 1 & 1 & 1 \\
 & & & & & & & 1 & 1 & 1 & 1 \\
 & & & & & & & & 1 & 1 & 1 \\
 & & & & & & & & & 1 & 1 \\
 & & & & & & & & & & 1
\end{pmatrix}
\quad
I^D_{ij}=\frac{1}{36}\begin{pmatrix}
1 & & & & & & & & & & & \\
1 & 1 & & & & & & & & & & \\
1 & 1 & 1 & & & & & & & & & \\
1 & 1 & 1 & 1 & & & & & & & & \\
1 & 1 & 1 & 1 & 1 & & & & & & & \\
1 & 1 & 1 & 1 & 1 & 1 & & 1 & 1 & 1 & 1 & 1 \\
 & & & & & & & & 1 & 1 & 1 & 1 \\
 & & & & & & & & & 1 & 1 & 1 \\
 & & & & & & & & & & 1 & 1 \\
 & & & & & & & & & & & 1
\end{pmatrix}
$$

Fig. 2. Adjacency matrix A, μ^F and interaction matrix I^D of directed chain.

4 Community Detection Problems with Capacity Measures in Directed Networks

In the previous Section, we have defined a fuzzy measure that represents the flow capacity of a group in a directed network. In this Section, we will take it into account to find communities in directed networks. The idea of using fuzzy measures in community detection problems was firstly introduced in [13] for non-directed graphs. There it is shown that the original concept of group/community change when a fuzzy measure is also considered, apart from the connections among nodes.

As we have pointed out in the introduction, our aim is to identify groups in which the idea of flow is considered. It is important to note that the modularity measure introduced in [20] for directed networks does not consider the flow. Then, as a natural consequence, any algorithm based on modularity optimization, will not be suitable for searching communities based on the flow. In order to show this fact with more emphasize, we propose another expression for the modularity formula.

Let $G = (V, E)$ be a directed graph, let $S \subseteq V$, and let $P = \{C_1, \ldots, C_L\}$ be a partition of the set of nodes. Here we introduce some notation:

- $K_S^{in} = \sum_{i \in S} k_i^{in}$, the number of links that goes to any node of S.
- $K_S^{out} = \sum_{i \in S} k_i^{out}$, the number of links that goes out from any node of S.
- $m_S = \sum_{i,j \in S} A_{ij}$, the number of links among the members of S.

Now, we can consider another expression of directed modularity [1] introduced by Newman for a given partition $P = \{C_1, \ldots, C_L\}$.

$$Q_d(G, P) = \frac{1}{m} \sum_{l=1}^{L} \sum_{i,j \in C_l} \left[A_{ij} - \frac{k_i^{in} k_j^{out}}{m} \right] = \frac{1}{m} \sum_{l=1}^{L} \left[m_{C_l} - \frac{K_{C_l}^{in} K_{C_l}^{out}}{m} \right].$$

From previous expression we can see that, fixed the edges in a group, the distribution and localization have not any impact in the modularity measure. The reason is that the important things of the measure are: the number of links inside the group, the number of links that goes from one element of the group to another (of the group or not), and the number of links that influences any member of the group (the origin of each link has no significance). In following example, we show it in detail.

Example 4.1. *Let us consider three directed graphs,* $G_i = (V_i, E_i)$ *for* $i = 1, \ldots, 3$, *where* $|V_1| = |V_2| = |V_3| = 6$ *and* $|E_1| = |E_2| = |E_3| = 5$.

Let us denote $V_1 = \{1, 2, \ldots, 6\}$, $V_2 = \{7, 8, \ldots, 12\}$, *and* $V_3 = \{13, 14, \ldots, 18\}$. *We assume that the graphs* G_1 *and* G_3 *are two directed stars (with hubs* 1 *and* 13*) and let us suppose that* G_2 *is a 6-directed chain.*

Let $E_1 = \{(1, 2), (1, 3), (1, 4), (1, 5), (1, 6)\}$; $E_2 = \{(7, 8), (8, 9), (9, 10), (10, 11), (11, 12)\}$ *and* $E_3 = \{(13, 14), (15, 13), (13, 16), (13, 17), (13, 18)\}$ *be the sets of edges of these graphs, respectively.*

Then, we present the following networks built from the aggregation of two of the previous graphs. $G_{12} = (V_1 \cup V_2, E_1 \cup E_2 \cup \{(6, 7)\})$; $G_{32} = (V_3 \cup V_2, E_3 \cup E_2 \cup \{(18, 7)\})$; $G_{13} = (V_1 \cup V_3, E_1 \cup E_3 \cup \{(6, 13)\})$ *(Fig. 3).*

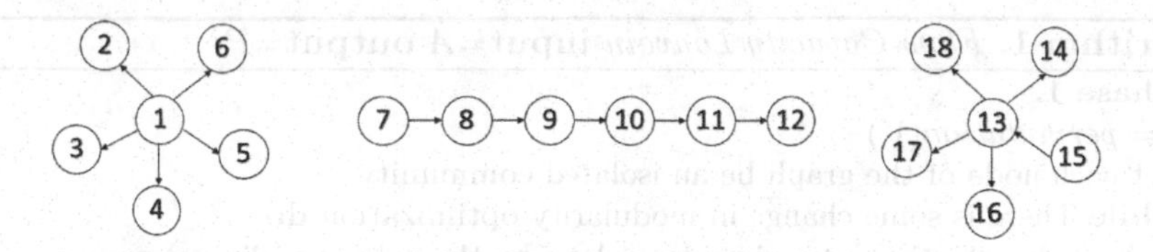

Fig. 3. Two directed stars and a 6-directed chain.

If we break these networks as $P_{12} = \{V_1, V_2\}$; $P_{32} = \{V_3, V_2\}$, $P_{13} = \{V_1, V_3\}$ *the modularity of each graph is:*

$$Q_d(G_{12}, P_{12}) = \frac{1}{m}\sum_{i,j\in V_1}\left[A_{ij} - \frac{k_i^{in}k_j^{out}}{m}\right] + \frac{1}{m}\sum_{i,j\in V_2}\left[A_{ij} - \frac{k_i^{in}k_j^{out}}{m}\right]$$

$$= \frac{1}{11}\left[5 - \frac{30}{11}\right] + \frac{1}{11}\left[5 - \frac{30}{11}\right] = \frac{25}{121} + \frac{25}{121}$$

$$Q_d(G_{32}, P_{32}) = \frac{1}{m}\sum_{i,j\in V_3}\left[A_{ij} - \frac{k_i^{in}k_j^{out}}{m}\right] + \frac{1}{m}\sum_{i,j\in V_2}\left[A_{ij} - \frac{k_i^{in}k_j^{out}}{m}\right]$$

$$= \frac{1}{11}\left[5 - \frac{30}{11}\right] + \frac{1}{11}\left[5 - \frac{30}{11}\right] = \frac{25}{121} + \frac{25}{121}$$

$$Q_d(G_{13}, P_{13}) = \frac{1}{m}\sum_{i,j\in V_1}\left[A_{ij} - \frac{k_i^{in}k_j^{out}}{m}\right] + \frac{1}{m}\sum_{i,j\in V_3}\left[A_{ij} - \frac{k_i^{in}k_j^{out}}{m}\right]$$

$$= \frac{1}{11}\left[5 - \frac{30}{11}\right] + \frac{1}{11}\left[5 - \frac{30}{11}\right] = \frac{25}{121} + \frac{25}{121}$$

Therefore, taking into account the inconveniences we found to get a 'good' partition with the current method, we introduce a capacity measure algorithm based on flow which will find clusters with maximum flow.

We propose a modification of Directed Louvain Algorithm [5,17] to work with extended fuzzy directed graphs denoted as Flow Capacity Louvain.

Let us define some concepts related to the Algorithm 1:

- $\Delta Q_i^d(j)$ is the increase in modularity when node j is incorporated into the community of i.
- A is the adjacency matrix, which has to guarantee the connections among the nodes.
- I^D is the directed interaction matrix.
- $\alpha \in [0,1]$ parameter of importance [13] which assigns a weight to each part of the extended fuzzy directed graph.
- $M = \alpha A + (1-\alpha)I^D$ is the matrix in which we search the partition, by maximizing its modularity.

Let us illustrate the performance of Flow Capacity Louvain Algorithm.

Algorithm 1. *Flow Capacity Louvain* **input**=A **output**=P

1: **Phase 1.**
2: $o = permutation(V)$
3: Let each node of the graph be an isolated community
4: **while** There is some change in modularity optimization **do**
5: According to the order given by o, let i be the corresponding element.
6: Then, find all out-edges j and in-edges j of i in A
7: Calculate $\Delta Q_i^d(j)$ in matrix $M = \alpha A + (1-\alpha)I^D$
8: Let j^* be the node for which ΔQ_i^d is maximum
9: **if** $\Delta Q_i^d(j^*) > 0$ **then**
10: Move node i to the community to which j^* belongs
11: **else**
12: i remains in its community
13: **end if**
14: **end while**
15: **Phase 1 Ends**
16: **Phase 2.**
17: A^* is the aggregated matrix obtained from A, whose nodes are the communities found in Phase 1
18: M^* is the aggregated matrix obtained from M, whose nodes are the communities found in Phase 1
19: While there is some change, apply Flow Capacity Louvain Algorithm, considering matrix A^* to find nodes and M^* to modularity optimization
20: **Phase 2 Ends**

Example 4.2. *Let us recall the graph introduced in Example 3.1. The partition P_1 obtained with the directed Louvain's algorithm divides this chain in three parts. The central cut $\{5,6,7,8\}$ has a bad behavior on flow (the modularity of I^D is not good). However, the partition P_2 obtained with Flow Capacity Louvain Algorithm (with $\alpha = 0.5$), defines 2 slices, both with good behavior on flow. All results can be seen in the Table 1.*

Table 1. Modularity of several partitions of the chain.

	Clustering classification according to chosen cut		
Directed Louvain	$P_1 = \{\{1,2,3,4\};\{5,6,7,8\};\{9,10,11,12\}\}$	$Q_d(A,P_1) = \mathbf{0.496}$	$Q_d(I^D,P_1) = \mathbf{0.222}$
Flow capacity	$P_2 = \{\{1,2,3,4,5,6\};\{7,8,9,10,11,12\}\}$	$Q_d(A,P_2) = \mathbf{0.413}$	$Q_d(I^D,P_2) = \mathbf{0.347}$

Example 4.3. *Let us consider three directed circles $(1,\ldots,6)$, $(1',\ldots,6')$ and $(1'',\ldots,6'')$. The graph of Fig. 4 is obtained by connecting each vertex of the first circle with its corresponding node of the second circle, and each vertex of the third circle with its corresponding node of the second circle. The interaction matrix I^D and the adjacency matrix A are represented in the Fig. 5. Let us consider that the weight of all the edges is 1. This structure can approach a wheel.*

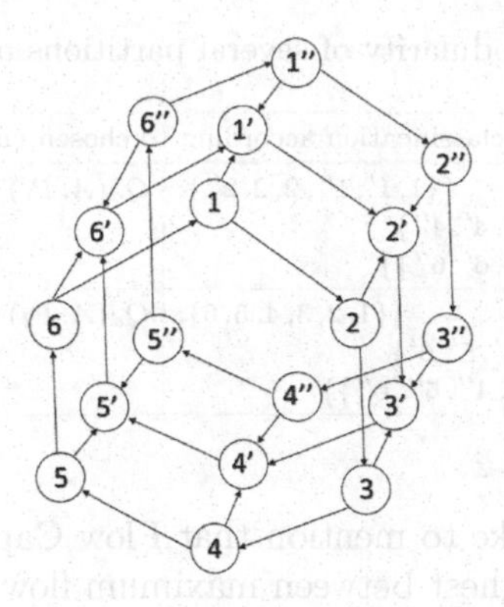

Fig. 4. Directed wheel with 18 nodes.

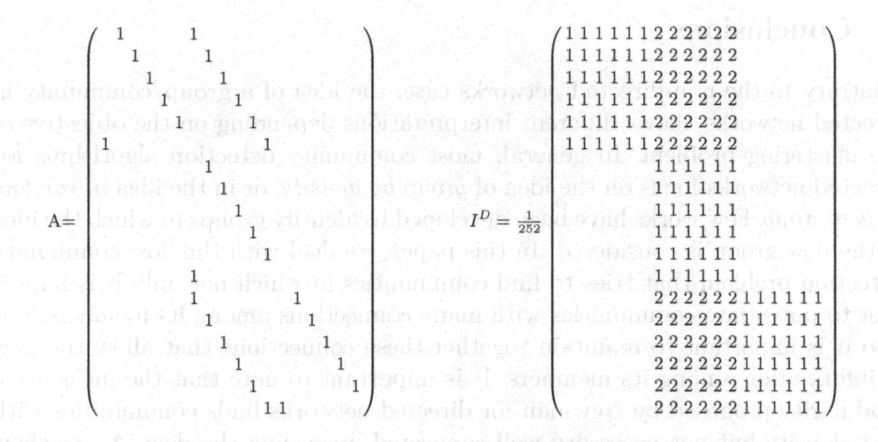

Fig. 5. Adjacency matrix A and Directed Interaction matrix I^D of directed wheel.

On this occasion, considering Louvain Algorithm [2], there are several partitions which maximize the modularity, cutting the wheel into three triangular zones. One of these partitions with maximum modularity in A is the cut P_1. In contrast, the modularity related to the flow (I^D) which is associated to P_1, is 0. Nevertheless, considering Flow Capacity Louvain Algorithm (with $\alpha = 0.5$), the obtained partition P_2 guarantees a high but not maximum modularity in A, and a maximum modularity in I^D. All the results can be seen in Table 2.

In this example, it is clearly seen how maximizing directed modularity does not achieve groups that maximize flow.

Let us remark that, in all these examples, we have assumed that the weight of all the edges is 1. Nevertheless, if we incorporate different weights, our algorithm will work correctly since Louvain Algorithm does.

Table 2. Modularity of several partitions of the wheel.

	Clustering classification according to chosen cut		
Directed Louvain	$P_1 = \{\{1, 1', 1'', 2, 2, 2''\}$; $\{3, 3', 3'', 4, 4', 4''\}$ $\{5, 5', 5'', 6, 6', 6''\}\}$	$Q_d(A, P_1) = 0.367$	$Q_d(I^D, P_1) = 0$
Flow Capacity	$P_2 = \{\{1, 2, 3, 4, 5, 6\}$; $\{1', 2', 3', 4', 5', 6'\}$ $\{1'', 2'', 3'', 4'', 5'', 6''\}\}$	$Q_d(A, P_2) = 0.320$	$Q_d(I^D, P_2) = 0.204$

Moreover, we would like to mention that Flow Capacity Louvain Algorithm complexity will be the highest between maximum flow among all pairs of nodes and Louvain Algorithm complexity [2,6].

5 Conclusions

Contrary to the non-directed networks case, the idea of a group/community in directed networks allows different interpretations depending on the objective of the clustering problem. In general, most community detection algorithms for directed networks focus on the idea of *group by density*, or in the idea of *random walker group*. Few works have been developed to identify groups in which the idea of the flow group is considered. In this paper, we deal with the flow community detection problem that tries to find communities in which not only it is important to potentiate communities with many connections among its members, but also it is important to maintain together these connections that allow the flow of information among its members. It is important to note that the measure of modularity proposed by Newman for directed networks finds communities with high density but not necessary well connected, regarding the flow. As we show with some examples in this work, modularity (and, as a consequence any optimization algorithm based on it), does not distinguish between different situations in which it is necessary to add some information to identify communities.

In order to take into account the flow capacity of a group, we incorporate to the community detection problem a 2-additive fuzzy measure that represents the relative flow capacity of each set of nodes. Then, following a similar methodology as that introduced in [13], we propose a modification of the Directed Louvain Algorithm [5,17] in order to incorporate the information provided by a fuzzy measure to the community detection algorithm in directed networks. Our proposal, Flow Capacity Louvain Algorithm, can consider, analyze and apply the information defined by a fuzzy measure when finding a partition in a directed network. Particularly, we propose to consider the fuzzy measure μ^F introduced in Sect. 3. Under the assumption of the new group definition based on the flow, we show that this algorithm provides very good results. As further work, we will develop an experimental study to test the efficiency of the algorithm here proposed, as well as an analysis of the processing time and memory usage of it. We will also work in some computational results, considering several benchmark models apart from the examples that we have included in this paper.

References

1. Arenas, A., Duch, J., Fernández, A., Gómez, S.: Size reduction of complex networks preserving modularity. New J. Phys. **9**(6), 176 (2007)
2. Blondel, V., Guillaume, J., Lambiotte, R., Lefevre, E.: Fast unfolding of communities in large networks. J. Stat. Mech.-Theory Exp. (2008). https://doi.org/10.1088/1742-5468/2008/10/P10008
3. Bollobás, B.: Modern Graph Theory, pp. 215–252. Springer, New York (1998). https://doi.org/10.1007/978-1-4612-0619-4
4. Borgatti, S.P.: Centrality and network flow. Soc. Netw. **27**(1), 55–71 (2005)
5. Dugué, N., Perez, A.: Directed Louvain: maximizing modularity in directed networks (2015)
6. Edmonds, J., Karp, R.M.: Theoretical improvements in algorithmic efficiency for network flow problems. J. ACM (JACM) **19**(2), 248–264 (1972)
7. Fortunato, S.: Community detection in graphs. Phys. Rep. **486**, 75–174 (2010)
8. Girvan, M., Newman, M.: Community structure in social and biological networks. Proc. Nat. Acad. Sci. **99**(12), 7821–7826 (2002)
9. Goldberg, A.V., Tarjan, R.E.: A new approach to the maximum-flow problem. J. ACM (JACM) **35**(4), 921–940 (1988)
10. Gómez, D., Rodríguez, J., Yáñez, J., Montero, J.: A new modularity measure for Fuzzy Community detection problems based on overlap and grouping functions. Int. J. Approx. Reason. **74**, 88–107 (2016)
11. Gómez, D., Zarrazola, E., Yáñez, J., Montero, J.: A divide-and-link algorithm for hierarchical clustering in networks. Inf. Sci. **316**, 308–328 (1997)
12. Grabisch, M.: K-order additive discrete fuzzy measures and their representation. Fuzzy Sets Syst. **92**(2), 167–189 (1997)
13. Gutiérrez, I., Gómez, D., Castro, J., Espínola, R.: A new community detection algorithm based on fuzzy measures. In: Kahraman, C., Cebi, S., Cevik Onar, S., Oztaysi, B., Tolga, A.C., Sari, I.U. (eds.) INFUS 2019. AISC, vol. 1029, pp. 133–140. Springer, Cham (2020). https://doi.org/10.1007/978-3-030-23756-1_18
14. Kosko, B., et al.: Fuzzy cognitive maps. Int. J. Man Mach. Stud. **24**(1), 65–75 (1986)
15. Lancichinetti, A., Fortunato, S.: Benchmarks for testing community detection algorithms on directed and weighted graphs with overlapping communities. Phys. Rev. E **80**(1), 016118 (2009)
16. Li, H., Xiang, J.: Explore of the fuzzy community structure integrating the directed line graph and likelihood optimization. J. Intell. Fuzzy Syst. **32**(6), 4503–4511 (2017)
17. Li, L., He, X., Yan, G.: Improved Louvain method for directed networks. In: Shi, Z., Mercier-Laurent, E., Li, J. (eds.) IIP 2018. IAICT, vol. 538, pp. 192–203. Springer, Cham (2018). https://doi.org/10.1007/978-3-030-00828-4_20
18. Malliaros, F., Vazirgiannis, M.: Clustering and community detection in directed networks: a survey. Phys. Rep. **533**(4), 95–142 (2013)
19. Masuda, N., Porter, M., Lambiotte, R.: Random walks and diffusion on networks (vol 716, pg 1, 2017). Phys. Rp-Rw Sect. Phys. Lett. **745**, 96 (2018)
20. Newman, M.: Finding community structure in networks using the eigenvectors of matrices. Phys. Rev. E **74**(3), 036104 (2006)
21. Newman, M.: Community detection and graph partitioning. EPL (Europhy. Lett.) **103**(2), 28003 (2013)

22. Newman, M., Girvan, M.: Finding and evaluating community structure in networks. Phys. Rev. **69**, 026113 (2004)
23. Rosenfeld, A.: Fuzzy graphs. Fuzzy Sets Appl. 77–95 (1975)
24. Rosvall, M., Bergstrom, C.: Multilevel compression of random walks on networks reveals hierarchical organization in large integrated systems. PloS One **6**(4), e18209 (2011)
25. Sugeno, M.: Fuzzy measures and fuzzy integrals–a survey. In: Readings in Fuzzy Sets for Intelligent Systems, pp. 251–257. Elsevier (1993)
26. Wu, T., Liu, X., Liu, F.: An interval type-2 fuzzy TOPSIS model for large scale group decision making problems with social network information. Inf. Sci. **432**, 392–410 (2018)
27. Zhang, D., Xie, F., Zhang, Y., Dong, F., Hirota, K.: Fuzzy analysis of community detection in complex networks. Phys. A **389**(22), 5319–5327 (2010)

Fuzzy Temporal Graphs and Sequence Modelling in Scheduling Problem

Margarita Knyazeva[1(✉)], Alexander Bozhenyuk[1(✉)], and Uzay Kaymak[2(✉)]

[1] Southern Federal University, Taganrog, Russian Federation
mknyazeva@sfedu.ru, avb002@yandex.ru
[2] School of Industrial Engineering, Eindhoven University of Technology, Eindhoven, The Netherlands
u.kaymak@ieee.org

Abstract. Processing sequential data and time-dependent data is a problem of constructing computational graph with a certain structure. A computational graph formalizes the structure of a set of computations including mapping temporal inputs and outputs. In this paper we apply graph theory and fuzzy interval representation of uncertain variables to indicate states of the temporal scheduling system. Descriptive model for temporal reasoning on graph, sequence modelling and ordering of fuzzy inputs for scheduling problem is introduced.

Keywords: Fuzzy sequence modelling · Computational graph · Fuzzy graph · Fuzzy temporal intervals · Temporal reasoning · State-transition system

1 Introduction

Temporal reasoning and temporal knowledge representation are the problems of introducing relations between elements or events in time with respect to known information and precedence relations between those events. Fuzzy temporal reasoning can be divided in two main approaches, depending on how time is represented: quantitative and qualitative. The quantitative approach is relevant when temporal data and temporal stamps are available and extracting new knowledge is necessary, while qualitative approach investigates relative fuzzy relations between elements or events - such as event A happens before event B, event C happens during event B and one needs to produce inferences on the known temporal facts.

Within both approaches there are two main problems to be solved: how to represent basic fuzzy units of time and handle it (how to "quantify time and measure it") and how to represent fuzzy relationships between basic units of time (how to "sequence events and locate them"). A number of different approaches exist to handle both of these problems: topological ordering techniques and algorithms, Allen's crisp interval algebra [1, 2], fuzzy intervals [3] and probabilistic intervals, computational graphs, state-variable representation [4] and sequence modelling techniques. These approaches are at the basis of proposed extensions to knowledge representation on, for example, the world wide web (see, e.g. [5]).

M.-J. Lesot et al. (Eds.): IPMU 2020, CCIS 1239, pp. 539–550, 2020.
https://doi.org/10.1007/978-3-030-50153-2_40

Measurement of time means the measurement of the uncertain temporal duration of events or “start-start” or “start-finish” fuzzy intervals, while specification means locating these events within the timeline and sequencing them. Locating an event in the time series means in the first instance locating it relative to other events.

2 Fuzzy Sequence Modelling and Temporal Interval Knowledge Representation

Usually we refer to sequence modelling when we use recurrent neural networks, or RNNs as a family of neural networks for processing sequential data. Sequence modeling in scheduling is the problem of predicting what event comes next: the current output is dependent on the previous input or state. In this paper we suggest temporal knowledge representation and sequence modelling technique for scheduling on a fuzzy graph.

The most traditional framework for handling the qualitative relations between time-dependent intervals is Allen’s Interval Algebra, or Interval Algebra (IA), formalized for the first time in [2]. The basic idea of IA is modelling temporal events as intervals, that have a starting and a finishing points in time. Based on this idea, 13 different temporal relations may be introduced between any given pair of events.

Table 1. Precedence relations between events according to Allen’s Interval Algebra.

Basic precedence relations		Graphical illustration	Formal notation
x before y y after x	< >	xxxx yyyy	$x^+ < y^-$
x meets y y is met by x	m m~	xxxx yyyy	$x^+ = y^-$
x overlaps y y is overlapped by x	o o~	xxxx yyyy	$x^- < y^- < x^+$ $x^+ < y^+$
x during y y includes x	d d~	xxx yyyyyyy	$x^- > y^-$ $x^+ < y^+$
x starts y y is started by x	s s~	xxx yyyyyyy	$x^- = y^-$ $x^+ < y^+$
x finish y y is finished by x	f f~	xxx yyyyyyy	$x^+ = y^+$ $x^- > y^-$
x equals y	≡	xxxx yyyy	$x^- = y^-$ $x^+ = y^+$

Table 1 shows the 13 relations between events performance, formalizing them with logical constraints (x^- denotes the left end of the time interval x and x^+ denotes the right end of x).

Temporal ordering of events may be expressed in terms of precedence relations and can be represented in the form of a graph, forming a type of graph that is called a temporal interval graph. However, in the real world, the actual start and finish times of many events may be hard to define, and so, in the fuzzy temporal interval graph, relations between events are defined in terms of degrees of truth, allowing multiple relations between elements at the same time. Figure 1 illustrates this kind of fuzzy graph relations between events.

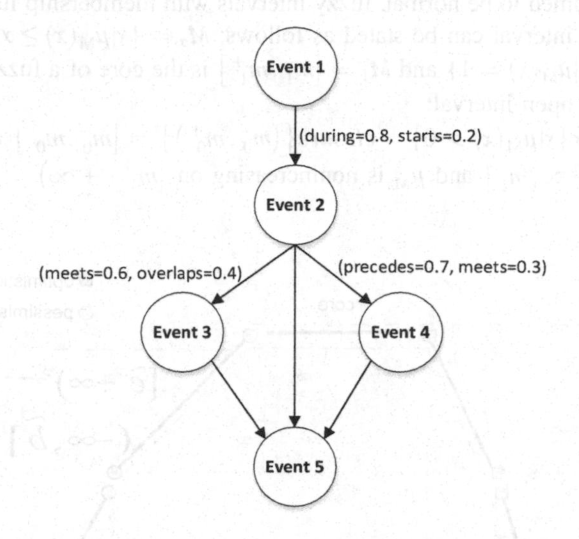

Fig. 1. Qualitative fuzzy relations between events in graph.

The truth value of a relation is the degree of "existence" of that relation, varying from 0 to 1. For example, In Fig. 1, Event-2 and Event-4 are related with propositions *precedes(Event-2, Event-4, 0.7)*, or relations like *meets(Event-2, Event-3, 0.6)*. Fuzzy edge between any two events is defined by a tuple that may be composed of multiple truth values for different fuzzy precedence relations.

The quantitative analysis for fuzzy temporal data can be done with the help of fuzzy interval representation of the areas near the edge of the interval that have less membership in the time-unit than the ones near the center, or simply fuzzy membership function. Gradual numbers as elements of fuzzy intervals were introduced as tools for computations on fuzzy sets, usually associated with combinatorial optimization and monotonic function evaluation.

Definition 1. Let X denote a set, then a *fuzzy set* $\tilde{A}$ on X is a set of ordered pairs $\tilde{A} = \{(x, \mu_{\tilde{A}}(x) : x \in X)\}$, where the membership function $\mu_{\tilde{A}}(x) : \mathrm{X} \to [0, 1]$ is a map from set X into the set of possible degrees of memberships with $\mu_{\tilde{A}}(x) = 1$ indicating full membership, $\mu_{\tilde{A}}(x) = 0$ indicating non-membership.

The α-cut of a fuzzy set is the set $\{x|\mu_{\tilde{A}}(x) \geq \alpha\}$ and is denoted as $\tilde{A}_{\alpha}$.

Definition 2. *A fuzzy interval M,* defined by its membership function $\mu_M(\cdot)$, is a fuzzy subset of the real line such that, if $\forall(x, y, z) \in \mathbb{R}^3,\ \ z \in [x, y]$, then $\mu_M(z) \geq \min(\mu_M(x), \mu_M(y))$ [6].

A fuzzy interval M is normal iff $\exists x \in \mathbb{R}$ such that $\mu_M(x) = 1$. A fuzzy interval is a normalized fuzzy set whose membership function is upper-semicontinuous and whose α-levels are convex.

Definition 3. The set $\{x|\mu_M(x) = 1\}$ is *the core of a fuzzy interval* (Fig. 2). As fuzzy intervals are assumed to be normal, fuzzy intervals with membership functions and the α-cut of a fuzzy interval can be stated as follows: $M_\alpha = \{x|\mu_M(x) \geq \alpha \geq 0\}$.

Let $M_1 = \{x|\mu_M(x) = 1\}$ and $M_1 = [m_1^-, m_1^+]$ is the core of a fuzzy interval, then its support is an open interval:

$M_0 = closure\{x|\mu_M(x) > 0\} = closure\{(m_0^-, m_0^+)\} = [m_0^-, m_0^+]$ and μ_M is nondecreasing on $(-\infty, m_1^-]$ and μ_M is nonincreasing on $[m_1^+, +\infty)$.

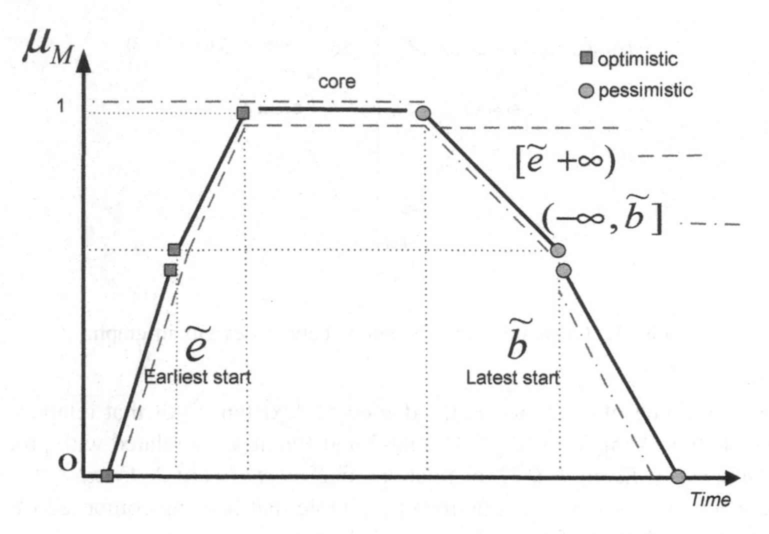

Fig. 2. Interval L-R representation for earliest and latest activity starting dates for a scheduling problem.

For practical purposes and ease of their use fuzzy intervals can be presented as left and right parts, or L-R representation of fuzzy interval [7]. Fuzzy interval can be defined by its membership function μ_M, core $[m^-, m^+]$, a support $[m^- - \alpha_M, m^+ + \beta_M]$ as well as L and R reference functions as follows:

$$\mu_M(x) = \begin{cases} 1 \; for \; x \in [m^-, m^+], \\ L\left(\frac{m^- - x}{\alpha_M}\right) for \; x < m^-, \\ R\left(\frac{x - m^+}{\beta_M}\right) for \; x > m^+. \end{cases} \quad (1)$$

In the literature of fuzzy optimization, the idea of fuzzy solution was introduced by Verdegay [9], who pays special attention to duality among fuzzy constraints and fuzzy objectives in fuzzy linear mathematical programming problems. Recently several investigations were done by Dubois, Fargier and others [8, 10, 11] about the notion of fuzzy elements and interval analysis for optimization methods. In this paper we suggest using fuzzy interval representation and gradual numbers for scheduling problems to deal with uncertainty in operational planning. Fuzzy interval here is the interval that reflects incomplete knowledge about some parameter that lies between two bounds, so that a value within the interval is possible and a value outside is impossible. Making uncertainty gradual means making the boundaries of the interval softer, and so fuzzy boundaries of fuzzy intervals should be gradual [6].

Definition 4. *A gradual number* $\tilde{r}$, is defined by a function, called the "assignment function" $A_{\tilde{r}} : (0, 1] \mapsto \mathbb{R}$. Then for each α, a real value r_α is given by $A_{\tilde{r}}(\alpha)$. A gradual real number can be understood as a real value parametrized by α.

Let μ_M^- and μ_M^+ be the parts of the membership function of a fuzzy interval *M*. They are functions from the real line to [0, 1], respectively, defined on $\left[m_0^-, m_1^-\right]$ and $\left[m_0^+, m_1^+\right]$. These functions are injective (μ_M^- is increasing and μ_M^+ is decreasing) and $(\mu_M^-)^{-1}(\alpha)$ and $(\mu_M^+)^{-1}(\alpha)$ are their inverse functions and the endpoints of the α-cut of M. The domain of a gradual number is defined as (0, 1] so that it represents all the possibility degrees for which the α-cuts are defined. Fortin, Dubois and Fargier treated fuzzy interval as a set of gradual numbers that lie between two gradual number endpoints in the same way that a real interval can be treated as a set of real numbers that lie between two real endpoints [6].

In this paper we suggest interpreting *fuzzy interval M* for earliest and latest starts of events as an interval with the α-cut mapping $\alpha \to M_\alpha$ as an assignment function from (0, 1] to the set of intervals.

Usually most algebraic properties of real numbers are preserved for gradual numbers, comparing with fuzzy intervals, but gradual real numbers are not totally ordered [6]. Using gradual numbers, we can introduce a fuzzy interval *M* for event performance by an ordered pair of gradual numbers $(\tilde{m}^-, \tilde{m}^+)$, where $\tilde{m}^-$ is called left profile or earliest starting time of event, while $\tilde{m}^+$ is the right profile or latest starting time (so-called lower and upper bounds). Such profiles of fuzzy sets are piecewise linear and can be easily implemented while estimating uncertain interval values, where a fuzzy bound is a gradual number. Algebraic operations and algebraic status of gradual numbers were discussed in several works [11–14] concerning graph-based scheduling problems. Other applications of gradual numbers can be implemented in path of maximal capacity problem, shortest path problem, allocation problems and other optimization problems, which can be formalized as linear programming problems.

3 Interval Temporal Modelling and Project Scheduling Problem

Usually a scheduling problem is characterized by precedence relations – an activity cannot start before its preceding activities are not finished; temporal constraints – an activity *i* cannot start before its earliest start e_i and b_i is the date after which activity cannot be started without delaying the end of the project (latest start time); capacity constraints – each activity *i* requires a certain level of resources for its execution at each moment t, those resources are limited and can be renewable and non-renewable. The critical path method (CPM) is an algorithm for scheduling a set of project activities, based on evaluation of temporal constraints. A critical path is determined by identifying the longest path of dependent precedence-related activities and measuring the time required to complete them from starting activity to finishing one.

Several main temporal variables need to be calculated based on temporal fuzzy intervals for CPM, gradual formalization and precedence relations for scheduling problem:

The earliest starting date e_i of an event i or activity is the date before which it can't be started without violation of a precedence constraint.

The latest starting date b_i is the date after which we cannot start the activity without delaying the end of the project.

The float is the difference between the latest starting date and the earliest starting date of the activity calculated as follows: $f_i = b_i - e_i$. An activity is considered to be critical iff its float is null.

According to Critical Path Method approach to resource-constrained scheduling problem we use a *forward graph propagation* to determine the earliest starting and finishing dates (finally project duration and the free floats) and a *backward graph propagation* to determine the latest starting and finishing dates.

P_{ij} denotes the set of all paths from activity i to activity j, while $T\ (p_{ij})$ is the temporal length of the path $p_{ij} \in P_{ij}$ in graph, while the length of longest path is the earliest starting date e_i form starting activity 1 (sink) to activity i (target) so that the following condition is true:

$$e_i = \max\{T(p_{ij})|p_{ij} \in P_{ij}\}. \tag{2}$$

The latest starting date is calculated as follows:

$$b_i = \max\{T(p_{ij})|p_{ij} \in P_{1,n}\} - \max\{T(p_{ij})|p_{ij} \in P_{i,n}\}. \tag{3}$$

CPM algorithm is based on the estimation of the longest possible continuous path in the graph, taken from the initial event to the terminal event so that:

$$e_i = \max\{e_j + d_j|j \in Pred_i\} \tag{4}$$

$$b_i = \min\{b_j - d_j|j \in Succ_i\}, \tag{5}$$

where d_j is the activity duration, $Pred_i$ and $Succ_i$ are the immediate predecessors and successors of the activity i. For each activity i in the graph, the three functions depending on the number of activities n are to be introduced: $e_i(\cdot)$, $b_i(\cdot)$ and $f_i(\cdot)$.

For example, earliest starting time calculated by forward graph propagation is increasing according to each argument:

$$\tilde{e}_i^- = \max\{\tilde{e}_i^- + \tilde{d}_i^- | j \in Pred_i\} \tag{6}$$

$$\tilde{e}_i^+ = \max\{\tilde{e}_i^+ + \tilde{d}_i^+ | j \in Pred_i\}. \tag{7}$$

Figure 3 illustrates fuzzy interval representation of temporal variable for earliest and latest starting times of activity *i*.

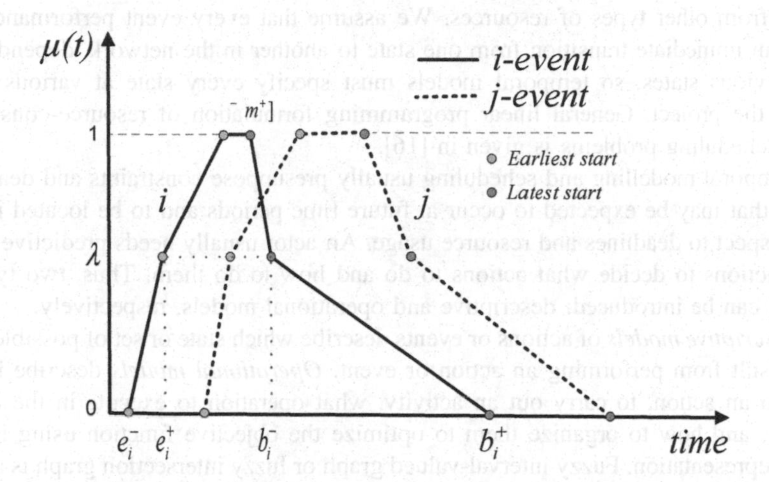

Fig. 3. Interval L-R profiles for a fuzzy intervals as a gradual number representation for a scheduling problem.

However the recursive backward graph propagation approach to determine the latest starting and finishing dates b_i is not appropriate due to the nature of the uncertainty that would be considered twice and the necessity to rank fuzzy numbers constantly due to the technique itself. Finally fuzzy subtraction increases the imprecision in the sense that the result of $\tilde{b}_i - \tilde{d}_i$ is more imprecise than $\tilde{b}_i$ or $\tilde{d}_i$ separately. Therefore variables $\tilde{b}_i$ and $\tilde{d}_i$ are non-interactive. Dubois and Prade introduced the concept of weakly non-interactive fuzzy numbers whose operations are based on the extension principle corresponding to each t-norm in place of the minimum operator [15]. Based on the idea of fuzzy interval calculations of floats and latest times, Zielinski introduced polynomial algorithms for determining the intervals of the latest starting times in scheduling networks and complexity results for floats [13].

Thus the fuzzy interval containing the float of activity *i* cannot be calculated by subtracting the fuzzy earliest starting time from the fuzzy latest starting time.

In this paper we consider arithmetic on intervals of gradual numbers [17], which uses the arithmetic on functions of the L-R profiles as a generalization of the real-valued interval analysis.

4 State-Transition and Time-Oriented Modelling

A project scheduling problem (PSP) can be defined by a number of events or activities representing different planning states of a project (or partial schedules) and a set of precedence relations between events. The problem can be presented as an event-network or a directed acyclic graph where vertices represent events while arcs illustrate precedence constraints. The idea of a project manager is generally to minimize the makespan of the project with respect to resources.

To perform an event, different kinds of resources may need to be assigned and consumed. Time is a resource required by every action or event to be performed, but it differs from other types of resources. We assume that every event performance produces an immediate transition from one state to another in the network, depending on the previous states, so temporal models must specify every state at various points during the project. General linear programming formulation of resource-constrained fuzzy scheduling problems is given in [16].

Temporal modelling and scheduling usually presuppose constraints and deadlines, events that may be expected to occur at future time periods and to be located in time with respect to deadlines and resource usage. An actor usually needs predictive model of its actions to decide what actions to do and how to do them. Thus, two types of models can be introduced: descriptive and operational models, respectively.

Descriptive models of actions or events describe which state or set of possible states may result from performing an action or event. *Operational models* describe how to perform an action, to carry out an activity, what operation to execute in the current context, and how to organize them to optimize the objective function using interval graph representation. Fuzzy interval-valued graph or fuzzy intersection graph is a fuzzy graph showing intersecting intervals on the real line so that each vertex is assigned an interval and two vertices are joined by an edge if and only if their corresponding intervals overlap.

Definition 6. By an *interval-valued directed fuzzy graph* of a graph $\tilde{G} = (V, E)$ we mean a pair $\tilde{G} = (A, R)$, where $A = \left[\mu_A^-, \mu_A^+\right]$ is a left-right interval-valued fuzzy set on the set of vertices V and $R = \left[\mu_R^-, \mu_R^+\right]$ is an interval-valued fuzzy relation on E.

Definition 7. A fuzzy interval M is defined by an ordered pair of gradual numbers $(\tilde{m}^-, \tilde{m}^+)$, where $\tilde{m}^-$ is called the fuzzy lower bound or left-profile and $\tilde{m}^+$ is called the fuzzy upper bound or right-profile, and $A_{\tilde{M}}$ is an assignment function.

Property 1. The domains of the assignment functions $A_{\tilde{m}^-}$ and $A_{\tilde{m}^+}$ must be (0, 1].

Property 2. Assignment function $A_{\tilde{m}^-}$ must be non-decreasing and assignment function $A_{\tilde{m}^+}$ must be non-increasing.

Property 3. Lower and upper bounds $\tilde{m}^-$ and $\tilde{m}^+$ must be such that $A_{\tilde{m}^-} \leq A_{\tilde{m}^+}$ for all α-cuts.

The precise organization of a hierarchy of data structures and state representations is a well-known area in computer science. Scheduling problems presuppose some decisions about when and how to perform a given set of actions with respect to time constraints, resource constraints and the objective function. They are typically NP-complete.

Definition 8. *A State-transition system or planning domain* is a 4-tuple $\Sigma = (S, A, \gamma, cost)$ where:

S – is a finite set of states for the system,
A – is a finite set of actions to perform,
γ – is a state-transition function, that gives the next state, or possible next states, after an action or event, $S \times A \rightarrow S$ with γ(s, a) being a predicted outcome.
Cost – is a partial function, $S \times A \rightarrow [0, \infty)$ having the same domain as γ. The cost function may represent monetary cost, time, or other resources to minimize

Each *state* of a system represents a time-oriented partial plan (schedule or interval-valued graph) that is associated with a certain timeline. Figure 4 illustrates two-dimensional state-time oriented model for a scheduling problem. State-oriented partial schedule keeps the notion of the global states and transition between the states (complete descriptions of the domain at some time point), while time-oriented profile represents the dynamics as a collection of partial intervals or primitives in time.

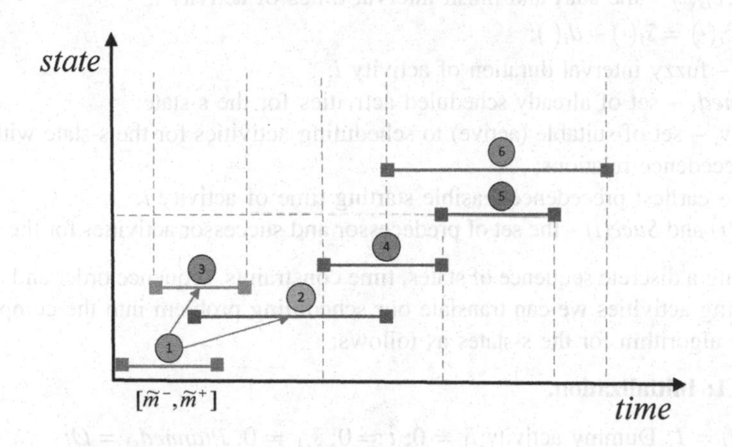

Fig. 4. Interval two-dimensional state-time oriented model.

Definition 9. *The planning computational graph* is a graph that consists of a number of states corresponding to timeline and computations or number of actions made according to the plan. Each state is associated with a number of inputs and outcome. Planning graph is an approximation of a complete enumeration tree of all possible states, actions and their results.

Figure 5 illustrates state-transition planning system, where each node represents the state at some time *t* and actions that are to be made to map the state at *t* to the state at *t + 1*. State-transition function γ gives the next state, or possible next states, after an action.

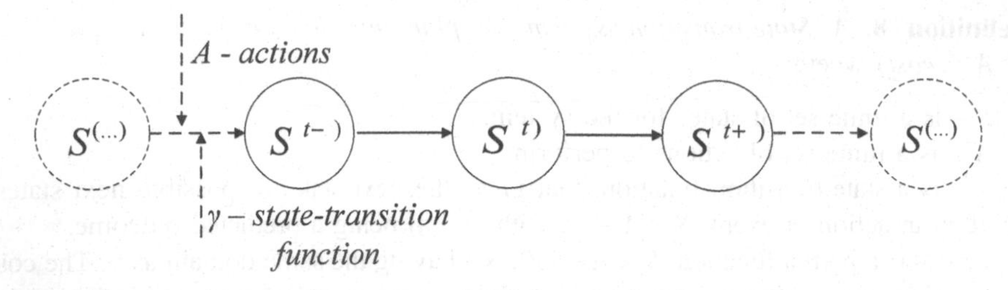

Fig. 5. The classical state-oriented dynamic system illustrated as an unfolded computational graph.

Let's consider the following notations:

S_t – the state of enumeration scheme for a certain partial schedule in moment *t*;
A_s – the set of activities for the certain *s*-state;
$\tilde{s}_{is}(\cdot), \tilde{c}_{is}(\cdot)$ – the start and finish interval times of activity *i*,
where $\tilde{c}_i(\cdot) = \tilde{s}_i(\cdot) + \tilde{d}_i(\cdot)$;
$\tilde{d}_i(\cdot)$ – fuzzy interval duration of activity *i*;
$Planned_s$ – set of already scheduled activities for the s-state;
$Ready_s$ – set of suitable (active) to scheduling activities for the s-state with respect to precedence relations;
$e_i(\cdot)$ – earliest precedence-feasible starting time of activity *i*;
Pred(i) and *Succ(i)* – the set of predecessor and successor activities for the *i*-activity

Having a discrete sequence of states, time constraints, sequence order and a number of planning activities we can translate our scheduling problem into the computational planning algorithm for the s-states as follows:

Step 1: Initialization.

Set $s_1 = 1$; Dummy activity $i_1 = 0$; $\tilde{t} := 0$; $\tilde{s}_{s1} = 0$; $Planned_{s1} = \emptyset$;

Step 2: Updating sets of activities.

Increase state number $s = s + 1$.
Update set of scheduled activities: $Planned_s = Planned_{s-1} \cup \{i_{s-1}\}$.
Compose the set of suitable activities:
$Ready_s = \{i \in A \backslash Planned_s | \text{pred}(i) \subseteq Planned_s\}$.
If the last dummy activity is active, then $n + 1 \in Ready_s$, then store the current solution and go to Step 5. Else go to step 3.

Step 3: Selecting the next activity from the set $Ready_s$ to be scheduled.

If there is no untested activity left in the set $Ready_s$ then go to step 5,
Else select the next activity $i_s \in Ready_s$.

Step 4: Computing the earliest precedence feasible starting time.

Compute the earliest precedence feasible starting time of the next activity:
$e_i(\cdot) = max\{\tilde{c}_{is}(\cdot) | i \in pred(i_s)\} + 1$. Go to Step 2.

Step 5: Backtracking.

Decrease the precedence state by 1: $s = s - 1$. If the precedence state s = 1, then STOP.

Else go to Step 4.

For each state of enumeration scheme evaluates all possible partial schedules and store the current solution until all the activities are scheduled. A detailed description of multi-mode resource-constrained project scheduling problems with non-preemptive activity splitting and the correspondent algorithms were given in [18].

5 Conclusion

In this work a new descriptive and operational model for temporal reasoning and temporal knowledge representation of relations between activities in computational graph have been introduced. In many situations involving computations with uncertain fuzzy variables, fuzzy arithmetic and a number of computations should be made to obtain feasible/optimal solution. In this paper we treat fuzzy intervals as crisp intervals of gradual numbers. A state-transition model for a planning system was introduced and the idea to apply unfolded computational graph to a planning system was proposed. On a future research, this approach could be used in topology: the unfolding of the system represents in a single branching structure all its possible computations, for example, in fuzzy graph transformation systems. The topological distribution of a system thus can be represented by a graph structure and the temporal dynamics of the system.

Acknowledgments. The reported study was funded by the Russian Foundation for Basic Research according to the research project #20-01-00197.

References

1. Allen, J.: Towards a general theory of action and time. Artif. Intell. **23**, 123–154 (1984). https://doi.org/10.1016/0004-3702(84)90008-0
2. Allen, J., Hayes, P.: A common-sense theory of time. In: Joshi, A. (ed.) Proceedings of the Ninth International Joint Conference on Artificial Intelligence (IJCAI 1985), San Francisco, pp. 528–531 (1985)
3. Dubois, D., Kerre, E., Mesiar, R., Prade, H.: Fuzzy interval analysis. In: Dubois, D., Prade, H. (eds.) Fundamentals of Fuzzy Sets: The Handbooks of Fuzzy Sets Series. FSHS, vol. 7, pp. 483–581. Springer, Boston (2000). https://doi.org/10.1007/978-1-4615-4429-6_11
4. Ghallab, M., Nau, D., Traverso, P.: Automated Planning and Acting. Cambridge University Press, Cambridge (2016)
5. Milea, V., Frasincar, F., Kaymak, U.: tOWL: a temporal web ontology language. IEEE Trans. Syst. Man Cybern. Part B (Cybern.) **42**(1), 268–281 (2012). https://doi.org/10.1109/TSMCB.2011.2162582
6. Fortin, J., Dubois, D., Fargier, H.: Gradual numbers and their application to fuzzy interval analysis. IEEE Trans. Fuzzy Syst. **16**(2), 388–402 (2008). https://doi.org/10.1109/TFUZZ.2006.890680

7. Dubois, D., Prade, H.: Operations on fuzzy numbers. Int. J. Syst. Sci. **9**(6), 613–626 (1978). https://doi.org/10.1080/00207727808941724
8. Dubois, D., Prade, H.: Gradual numbers and fuzzy solutions to fuzzy optimization problems. In: Pelta, D.A., Cruz Corona, C. (eds.) Soft Computing Based Optimization and Decision Models. SFSC, vol. 360, pp. 223–229. Springer, Cham (2018). https://doi.org/10.1007/978-3-319-64286-4_13
9. Verdegay, J.L.: A dual approach to solve the fuzzy linear programming problem. Fuzzy Sets Syst. **14**(2), 131–141 (1984). https://doi.org/10.1016/0165-0114(84)90096-4
10. Dubois, D., Fargier, H., Fortemps, P.: Fuzzy scheduling: modeling flexible constraints vs: coping with incomplete knowledge. Eur. J. Oper. Res. **147**, 231–252 (2003). https://doi.org/10.1016/S0377-2217(02)00558-1
11. Dubois, D., Fargier, H., Fortin, J.: A generalized vertex method for computing with fuzzy intervals. In: Proceedings of the IEEE International Conference on Fuzzy Systems FUZZ-IEEE, pp. 541–546 (2004). https://doi.org/10.1109/fuzzy.2004.1375793
12. Dubois, D., Fargier, H., Galvagnon, V.: On latest starting times and floats in activity networks with ill-known durations. Eur. J. Oper. Res. **147**, 266–280 (2003)
13. Zielinski, P.: On computing the latest starting times and floats of activities in a network with imprecise durations. Fuzzy Sets Syst. **150**, 53–76 (2005). https://doi.org/10.1016/j.fss.2004.08.007
14. Fortin, J., Dubois, D.: Solving fuzzy PERT using gradual real numbers. In: Proceedings of the 3rd European Starting AI Researcher Symposium (STAIRS 2006), pp. 196–207 (2006)
15. Dubois, D., Prade, H.: Additions of interactive fuzzy numbers. IEEE Trans. Autom. Control **26**(4), 99–135 (1981)
16. Knyazeva, M., Bozhenyuk, A., Kacprzyk, J.: Modeling decisions for project scheduling optimization problem based on type-2 fuzzy numbers. In: Batyrshin, I., Martínez-Villaseñor, M.L., Ponce Espinosa, H.E. (eds.) MICAI 2018. LNCS (LNAI), vol. 11288, pp. 357–368. Springer, Cham (2018). https://doi.org/10.1007/978-3-030-04491-6_27
17. Lodwick, W.A., Untiedt, E.A.: A comparison of interval analysis using constraint interval arithmetic and fuzzy interval analysis using gradual numbers. In: NAFIPS 2008: Annual Meeting of the North American Fuzzy Information Processing Society, New York City, NY, pp. 1–6 (2008)
18. Cheng, J., Fowler, J., Kempf, K., Mason, S.: Multi-mode resource-constrained project scheduling problems with non-preemptive activity splitting. Comput. Oper. Res. **53**, 275–287 (2015). https://doi.org/10.1016/j.cor.2014.04.018

Current Techniques to Model, Process and Describe Time Series

Predicting S&P500 Monthly Direction with Informed Machine Learning

David Romain Djoumbissie[1,2](✉) and Philippe Langlais[1]

[1] Department Computer Science and Operational Research, University of Montreal, Montreal, Canada
david.romain.djoumbissie@umontreal.ca, felipe@IRO.UMontreal.CA

[2] Canada Mortgage and Housing Corporation, Montreal, Canada

Abstract. We propose a systematic framework based on a dynamic functional causal graph in order to capture complexity and uncertainty on the financial markets, and then to predict the monthly direction of the S&P500 index. Our results highlight the relevance of (i) using the hierarchical causal graph model instead of modelling directly the S&P500 with its causal drivers (ii) taking into account different types of contexts (short and medium term) through latent variables (iii) using unstructured forward looking data from the Beige Book. The small size of our training data is compensated by the a priori knowledge on financial market. We obtain accuracy and F1-score of 70.9% and 67% compared to 64.1% and 50% for the industry benchmark on a period of over 25 years. By introducing a hierarchical interaction between drivers through a latent context variable, we improve performance of two recent works on same inputs.

Keywords: Financial knowledge representation · Functional causal graph · Prediction & informed machine learning

1 Introduction

Analyzing and predicting the dynamics of financial markets for investment decision-making over a monthly/quarterly horizon is an old challenge both in academy and in the asset management industry. The environment is complex, uncertain and modeling must take into account many factors, including incomplete, noisy and heterogeneous information with almost 80% in unstructured form [1,2].

The crucial parameter in this type of study is the prediction horizon. Indeed, it is strongly linked to the investment objectives/horizon; the paradigm used for modelling and the size of the training data. For the short-term (month/quarter) prediction, the losses recorded during the financial crisis (2008), in addition to all the previous ones have led many people to question the dominant paradigm. The latter is based essentially on a rational assumption and a direct relationship between S&P500 and a few causal drivers. In the literature, the solution for the short-term prediction might be classified into three groups.

M.-J. Lesot et al. (Eds.): IPMU 2020, CCIS 1239, pp. 553–566, 2020.
https://doi.org/10.1007/978-3-030-50153-2_41

The first group of studies are those from [3–7]. The foundations of their approach is based on pure rational argument and passive decision process without prediction. They assume that markets are efficient and it is difficult to predict the S&P500 index or to do better than a random walk. This solution serves as a benchmark in dynamic or active management segments of the industry.

The second group [8–14] proposes a solution based on a direct relationship (supervised algorithm) between S&P500 and a set of features from causal analysis or data mining. In [10,11], the authors found a direct relationship between S&P500 and four causal variables. The features obtained serve as input for a SVM model with innovation on the Kernel function in order to predict the monthly direction of S&P500 over 2006 to 2014. The main weaknesses from this group are: i) The weak predictions in an unstable environment; (ii) an adhoc approach to select the causal variables, the omission of the context and hierarchical interaction between drivers; (iii) The mismatch between the drivers and the prediction horizon.

The third group [13–17] is the most active at the moment and proposes: (i) to use all potential numerical/textual drivers; (ii) a deep architecture for learning representations directly on data. iii) and a prediction through deep supervised algorithms. In [15,16], the authors use NLP and deep learning on daily financial news to predict monthly direction of S&P500 without a priori knowledge. They learn features and predict directly from the data. The findings of this group are encouraging. However, a review we conducted on nearly 60 recent papers, the prediction horizon was less or equal to one day and more than 80% were tested on a very short period (less than 2 years). This prediction horizon (minutes, hours,..) has the advantage of providing a large training sample[1] but resolves a specific type of problem (high frequency transactions on financial markets), which are totally different from the problematic of monthly prediction.

The difference in terms of objectives, investment horizons, as well as the lack of validated studies over longer periods which will reflect the multiple changes in market regimes, make the notion of the state of art somewhat confused. Although there are a few names in the industry known for their ability to do better than the benchmark, recent studies and statistics [18,19] show it is difficult to conclude that one approach dominates the others.

In this paper, we propose a solution for a dynamic decision-making process based on the monthly prediction of the S&P500 index. The investor has an investment horizon of less than one year and uses a dynamic framework which is updated on a monthly basis. This frequency is also that of the publication and update of economic and financial information.

In order to reach our objectives, our contributions are threefold.

- Firstly, we combine our expertise with those of many studies in order to create the structure of a functional causal graph with four levels of the dynamics of the S&P500. Thus, we avoid learning this structure on small size and unstable data. Instead, we learn the distributional representation of latent variables

[1] data collected every second or minute.

(short/medium term context) from an unsupervised method (auto-encoder, similarity, rules). Level 1 includes observable causal variables, then a priori causal functional relationships allow the link with other 3 levels. The latent variables at the last level serve as features for a classifier.
- Secondly, we use unstructured forward looking data (1970–2019) in order to characterize the state of the business cycle. [15,16,20] confirmed the relevance of using unstructured daily data or events on companies published in 8-k form[2]. But the tests are conducted over short periods (24 months) and the aim was not to propose an effective decision-making process.
- Lastly, we perform a systematic validation and comparison with industry's benchmark over 25 years, as well as four sub-periods known as unstable and difficult to predict. We also make some comparison with other studies in the literature, which we formulated as special cases of our solution.

The remainder can be seen into four points: the description of the functional causal graph, the methodology for learning the representation of latent variables, the experiments with empirical results, finally the conclusion and future work.

2 Stock Market Dynamic via a Causal Functional Graph

Predicting S&P500 direction on the monthly horizon is formulated as a binary supervised classification task:

$$y_{t+1}^{S\&P500} = f(V_t) \tag{1}$$

where $y_{t+1}^{S\&P500}$ is the monthly price direction to be predicted (Up/Down), V_t the vector of features characterising the period t, derived from a functional causal graph (Fig. 1 and 2) of the dynamics of the S&P500 and f represents a classifier.

We describe the causal process of the dynamics of S&P500 through a causal functional graph with two essential goals: i) representing causal interactions (direct or hierarchic, linear or non-linear, static or dynamic,..) between short, medium and long term drivers, ii) learning dynamic embedding from temporal interactions between drivers. The a priori graph structure lies on two main source of knowledge. More than 50 years of literature on the financial markets (financial economic theory, behavioral finance, fundamental analysis, market microstructure, technical analysis), and our 15 years of experience in the conception/implementation of solutions for dynamic and tactical asset allocation.

Figure 1 describes different theories and the hierarchical top down interaction between long, medium, short term drivers and the stock market index. Figure 2 is a specific case based on three important medium/short term context (Business cycle, Market regime, Risk aversion). This choice is supported essentially by various works of two nobel prices in economic (Eugene Fama on empirical

[2] broad form used to notify investors in United States public companies of specified events that may be important to shareholders or the United States Securities and Exchange Commission.

analysis of asset prices[3], Daniel Kahneman on behavioral finance[4]) and one of the best portfolio manager of the century, Ray Dalio[5].

At time t, our biggest challenge is to characterize the current market environment (between t−k ...t) with a set of feature derived from Fig. 2 and use it to predict the S&P500 direction for time t+1. We use X_t to denote the realization of variable X at t and $X_{1:t}$ to denote the history of X between the period 1 to t. At time t, we are able to identify where was the market regime between 1..t−k, but we can only estimate the current market regime and we use the k most recent realisations to do. We will sometimes use $X_{1:t-k}$ and $X_{t-k:t}$.

The functional graph of Fig. 2, describe the dynamics of the S&P500. They have four levels and three main component: i) A set of 130 observable causal variables (ex: daily price index of 10 economics sectors); ii) A set of 6 latent context variables (ex: Risk aversion regime of Investors); and iii) 8 functions or algorithms that reflect a direct causal link between the variables (observable or latent). We suggest [21,22] for more details on functional causal graph in finance.

2.1 Graph Level 1: Observable Variables

The level 1 of the graph represents basic inputs organised around 4 groups of observable causal numerical variables and one group of textual information.

Observable Causal Numerical Variables: The variables in light blue (rectangular shape) designate observable numerical variables. All are available on the St. Louis Federal Reserve and Kenneth R. French websites.

S&P500$_{1:t-k}$: A numerical daily variable on the main U.S. equity market. At time t, the history from 1 to t-k (k represents the recent observations for which the regime is not known) serves as an input for ex-post identification algorithm of market regime (f_2, described in Sect. 3). The output of f_2 is an intermediate latent variable that characterizes the regime (bear/bull/range bound) in which the market was in the past (Market_Regime$_{1:t-k}$).

32_Risk_Factors$_{1:t}$: A set of 32 daily numerical variables denoting financial indexes. At time t, the history from 1 to t serves as an input for an unsupervised learning algorithm (f_3). The output is a set of intermediate latent variables characterizing the risk aversion of investors (Risk_Aversion$_t$).

80_Risk_Factors$_{1:t}$: A set of 80 numerical variables designating indices covering all asset classes and sectors of the economy. At time t, the history from 1 to t combined with Num_Repr_Beige_Book$_{1:t}$ (distributional representation of each Beige Book from 1..t) serves as an input for an unsupervised learning algorithm (f_4). The output of this algorithm is a set of intermediate latent variables that characterize the phase of the business cycle (Economic_Cycle$_t$).

---

[3] https://www.nobelprize.org/prizes/economic-sciences/2013/fama/facts/.
[4] https://www.nobelprize.org/prizes/economic\discretionary{-}{}{}sciences/2002/kahneman/biographical/.
[5] https://en.wikipedia.org/wiki/Ray_Dalio.

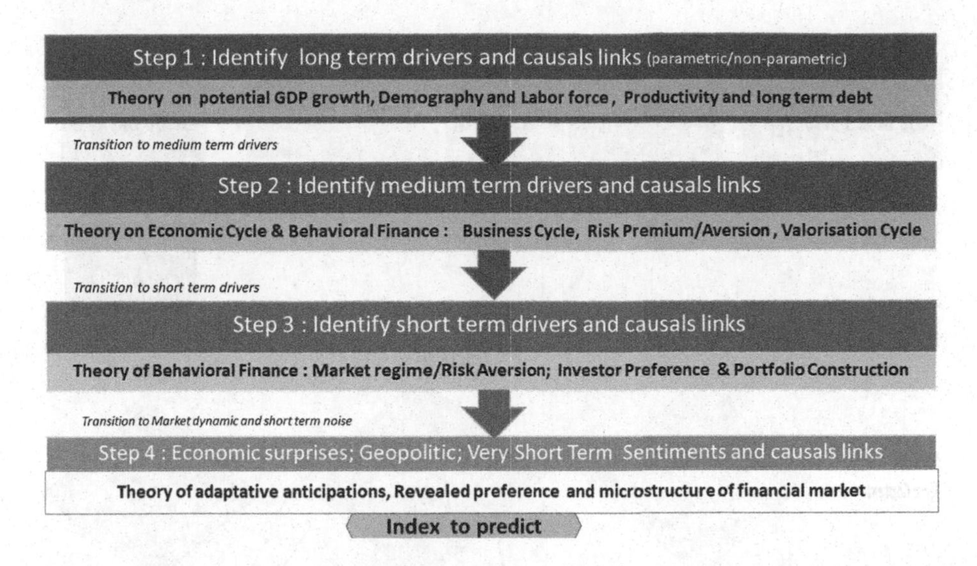

Fig. 1. Main component of the Causal hierachical top down dynamic of any stock index

S&P500_and_Rate$_{1:t}$: A set of the 3 numerical variables designating three of the most relevant indexes on US financial market. At time t, the history covering period 1 to t serves as an input with (Risk_Aversion$_{1:t}$, Economic_Cycle$_{1:t}$) for obtaining features via functions/algorithms or links (f_6, f_7) in the graph.

Observable Textual Causal Variables: It is shown in grey (rectangular shape with rounded side) on the graph (Textual_Data$_{1:t}$). It represent set of textual documents called the Beige Book, published 8 times a year by the U.S. Federal Reserve (≈2000 words for each edition of national summary) on highlights of economic activity, employment and prices. At time t, we use a function/algorithm (f_1, *Doc2vec*) to transform the most recent document into a set of p embedding denoting their distributional representation (Num_Repr_Beige_Book$_t$).

2.2 Graph Level 2: Latent Medium Term Context

Level 2 consists of three groups of light orange (lozenge shape) intermediate latent variables designating medium term context.

Market_Regime$_{1:t-k}$: it is a set of homogeneous cluster on historical price index S&P500$_{1:t-k}$. It summarizes ex-post the state or regime of the financial market for period 1..t-k via the function/algorithm or link f_2 in the graph.

Risk_Aversion$_{1:t}$: it summarizes other medium-term context. The risk aversion of investors on the markets for each period from 1 to t. It is obtained via the function/algorithm or link f_3 on the inputs 32_Risk_Factors$_{1:t}$.

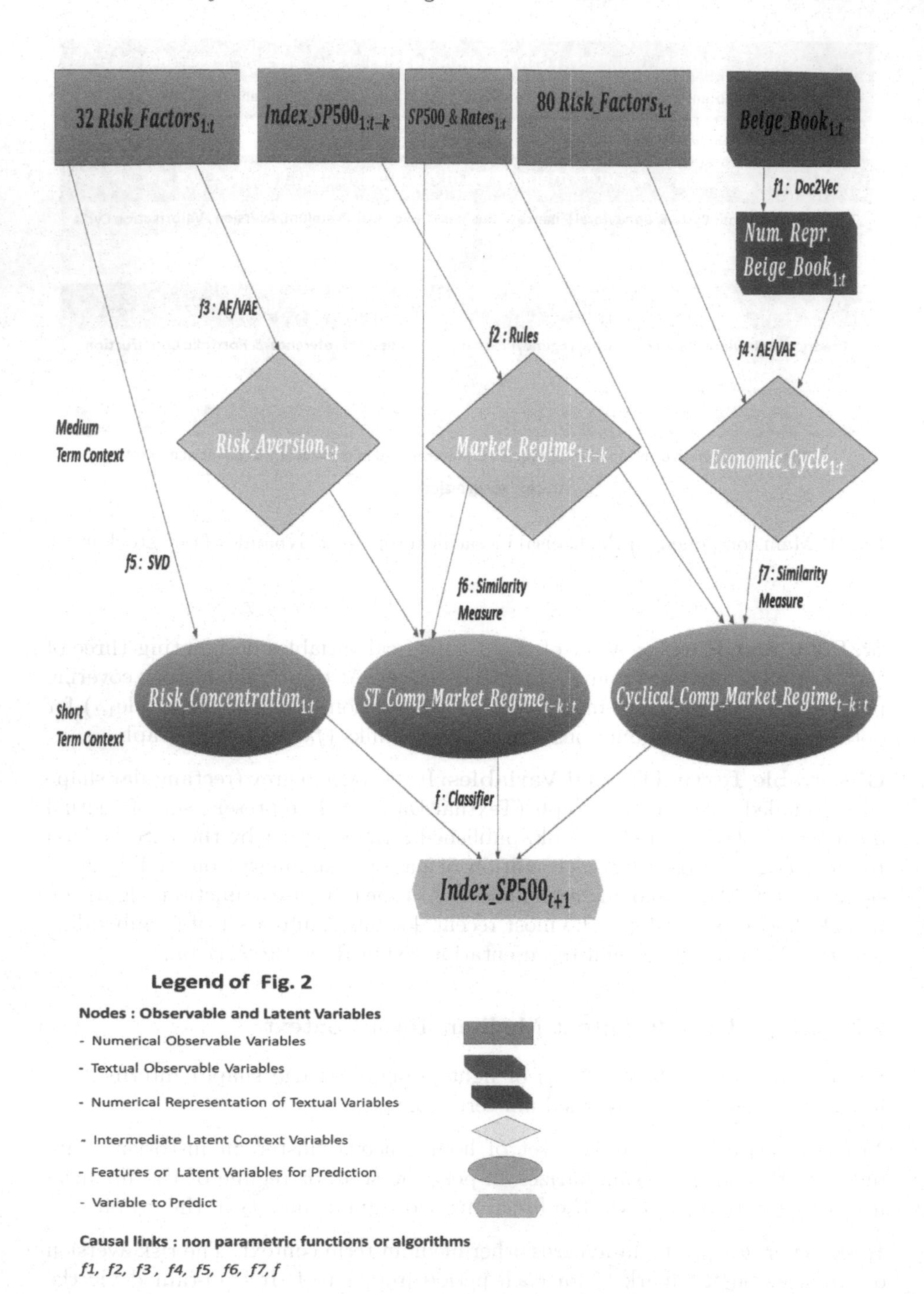

Fig. 2. Specific case of functional causal graph of the S&P500 dynamic

Economic_Cycle$_{1:t}$: it summarizes the last medium-term context, the phases of the economic cycle for each period 1 to t. It is obtained via the function/algorithm or link f_4 on the inputs (80_Risk_Factors$_{1:t}$ Num. Repr. Beige_Book$_{1:t}$).

2.3 Graph Level 3: Latent Short Term Drivers

Level 3 of the graph includes three other groups of latent variables, dark orange (Oval shape) used as features for a classifier on the S&P500.

ST_Comp_Market_Regime$_{t-k:t}$: A set of 8 latent variables designating the short-term component of the current market regime. For each of the recent periods between t-k to t, we obtain a statistical summary (median and asymmetry coefficient) of k measures of similarity with a function or link (f_6) in the graph. We measure the similarity between the recent k observations (t-k to t) and the historical observations (1 to t-k) of (Risk_Aversion$_{1:t}$, S&P500_and_Rate$_{1:t}$) of each of the homogeneous group obtained with Market_Regime$_{1:t-k}$.

MT_Comp_Market_Regime$_{t-k:t}$: A set of 8 latent variables designating the medium-term component of the current market regime. It is obtained in the same way, replacing the variable (Risk_Aversion$_{1:t}$) with (Economic_Cycle$_{1:t}$).

Risk_Concentration$_{1:t}$: A set of 4 latent variables denoting the concentration of sources of uncertainty in the markets. For each period from 1 to t, the percentage of the explained variance of the first 4 factors is obtained via a singular value decomposition (f_5) on the input 32_Risk_Factors$_{1:t}$.

3 Functions/Algorithms for the Latent Variables

We use a priori knowledge to going through the graph, learn separately the representation of each node and extract 20 business features (level 3) as inputs for a classifier. We validate this process on the reduced graph of Fig. 2 and the task of monthly prediction on S&P500. The generalization with a global graph in a unified embedding learning framework will be for the next step.

3.1 Algorithm f_2 for Intermediate Latent Variables Market_Regime

Market regime is identifiable ex-post. f_2 is a set of rules to separate the history of S&P500 into regimes (3 homogeneous groups). If $SP500_{t_0}...SP500_{t_n}$ is the sequence observed between $t_0,...t_n$, we define 3 market regimes :
Bull Market: Ex-post, the market was in a bullish mode between period $t_0,...t_h$ if starting to t_0, the S&P500 rises gradually to cross a certain threshold without returning below the initial price at t_0. Meaning the set of points
$\{t_0.., t_i, ..t_h;\ 0 \leq i \leq h and SP500_{t_i} \geq SP500_{t_0} \,\&\, SP500_{t_h} \geq (1+\lambda)SP500_{t_0}\}$
λ : Hyper-parameter based on empirical studies on risk premium.
Bear Market: Opposite of Bull Market (decrease trend)
Range Bound Market: Neither Bull, neither Bear

3.2 Algorithms f_3, f_4 for 2 Others Intermediate Latent Variables

f_3 *and* f_4 are two auto-encoders to learn the distributional representations of 2 others intermediate latent variables. At time t, a simple/variational auto-encoder (Fig. 3.) takes (32_Risk_Factors$_{1:t}$) as inputs and produces a representation of dimension q (Risk_Aversion$_t$), then takes (80_Risk_Factors$_{1:t}$; Num. Repr. Beige_Book$_{1:t}$) as inputs for other representation of dimension q (Economic_Cycle$_{1:t}$).

3.3 Algorithms f_5, f_6, f_7 and Features

We consider 3 others categories of latent variables to describe the current state of market and use distributional representations as features for a classifier.

8 Statistics Summarizing the Short-Term Component of the Current Market Regime. The current market regime is characterized by the similarity between the recent realisations of (Risk_Aversion$_{1:t}$; S&P500_and_Rate$_{1:t}$) and the historical observations organised in homogeneous groups.

Ex: Consider F_t, the similarity measure (Mahalanobis distance) in date t between recent (last month) observations (V_t^R) and Average/Variance of historical observations in the bullish regime (μ_t^H, S_t^H). We define F_t by :

$$F_t : R^n \times R^n \times R^{n \times n} \longrightarrow R^+$$
$$(V_t^R, \mu_t^H, \frac{1}{S_t^H}) = \sqrt{(V_t^R - \mu_t^H)^t \times \frac{1}{S_t^H} \times (V_t^R - \mu_t^H)}$$

Therefore, on a monthly horizon (20 days), we obtain a vector of 20 similarity measures that we aggregate by calculating a statistic like the median.

8 Statistics Summarizing the Cyclical Component of the Current Market Regime. They are also obtained by similarity measures as previously but replacing Risk_Aversion$_{1:t}$ by Economic_Cycle$_{1:t}$.

4 Factors Designating the Percentage of Explained Variance, obtained by singular values decomposition (f_5) and explaining more than 90% of the variance of key market risk factors (32_Risk_Factors$_{1:t}$).

These 20 features characterize the current market regime and constitute the main input for a classifier to predict the direction of the S&P500 index.

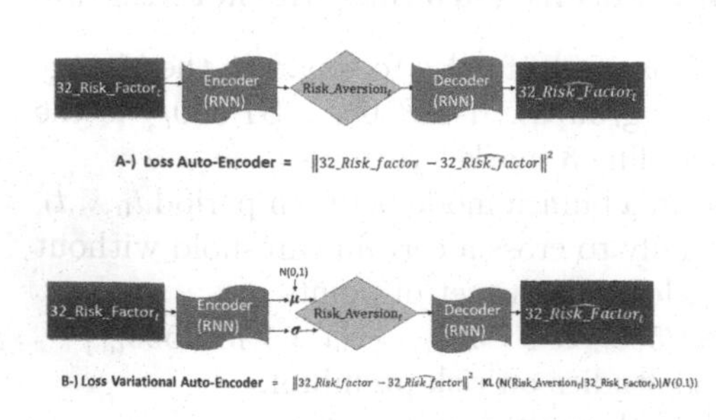

Fig. 3. Simple & Variational Auto-encodeur

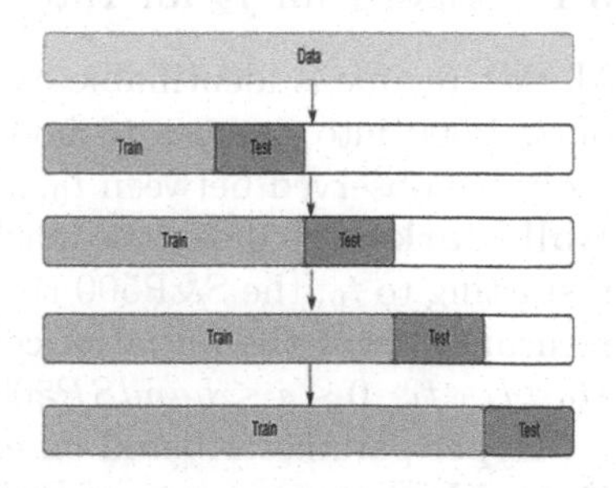

Fig. 4. Walk Forward Validation

4 Experiments and Empirical Results

4.1 Training, Validation and Testing Protocol

Time series have a certain dependence and the chronological order is a crucial element in the training and validation process. Walk Forward Validation (Fig. 4), based on an out-of-time dynamic validation that respects the chronological order, is widely used in finance and [27,28] provides additional details. The first 21 years (19 for training and 2 for validation) are used for initial training and to fix all hyper-parameters. Afterwards, at each date t, we obtain the features and predict the direction $\widehat{y_{t+1}}$ for the month t+1 on training sample $(y_{1...t}^{S\&P500}, V_{1...t})$. At t+1, we compare the prediction with realized $y_{t+1}^{S\&P500}$ and update two models. i) We update the parameters of the auto-encoder every ten years, which is enough to cover various market cycles and, ii) We update the parameters of the classifier every month with all training data until t+1 $(y_{1...t+1}^{S\&P500}, V_{1...t+1})$. We then iterate on the sample from 1992 to 2018 (Fig. 4).

We use three metrics detailed in [13] and adapted for Classification problems. The first is the accuracy (ACC) which is the total percentage of good predictions up and down. The second is the F1-score[6] and the last is the Matthews Correlation Coefficient[7] (MCC). The last two metrics allow a relevant analysis of the cost of errors. Indeed, the cost of bad decisions is high in the markets and the challenge is to have models with good *accuracy*, but especially an ability to limit false positives and false negatives. We make a comparison with the industry benchmark over the entire period, then in sub-periods known to be very unstable and difficult to predict. Our experiments are articulated into four points:

- We compare with the industry benchmark over the test period (1992 to 2018), then four unstable sub-periods (2000-02, 2007-08, 2011-12, 2015-16). This last comparison is typically not conducted in recent works.
- We Analyze the impact of features and latent variables by comparing 3 models of increasing complexity:
  i) **Model 1:** the link is direct between S&P500 and only the observable numerical variables of level 1 (no features, no latent variables). ii) **Model 2:** we consider the features, but they are obtained without two main medium-term context (Risk_Aversion and Cycle_Economic). iii) **Model 3:** we use all component of the graph (Fig. 2) and compare simple/variational auto-encoder to get the latent intermediate variables.
- The model with the textual data is compared to the model obtained only on numerical variables. The textual data from the Beige Book is replaced by traditional numerical Business cycle (Inflation, Industrial production).
- A comparison over the same test period and the same inputs deriving the work from [15] and [10,11] as specific cases of our solution. We identified five recent studies on monthly prediction of the direction of S&P50, then we selected the 2 best recent with available input. We transform the input to introduce a hierarchical interaction via a short term latent context.

[6] Harmonic average Precision and Recall.

[7] correlation coefficient between the observed and predicted binary classification.

4.2 API and Hyper-parameters Selection

We used the gensim implementation of Doc2vec to represent the Beige Book documents into vector of dimension k(hyper-parameter) and TensorFlow/Keras for training auto-encoders. For training the classifier, we used scikit-learn on python 3.6 (SVM, RandomForestClassifier and Ensemble.GradientBoostingClassifier).

We have two categories of Hyper-parameter: i) the first category (number of phases in the business cycle, number of market regimes, dimension of latent variables) are choosing based on our experience and some relative consensus on empirical studies on financial market [23–26]. ii) The second category for auto-encoder and classifier (dimension vector for Beige Book, learning rate, number of estimates, maximum depth of the trees, size of the sub-samples) are chosen to maximize output over the training (1970-89) and validation (1990–1991) period.

4.3 Performance and Comparison with the Industry Benchmark

The Table 1 shows our model consistently outperforms the benchmark over the test period (1992 to 2018) on all metrics. It highlights the limits of accuracy in predicting stock market. Indeed, the ACC of the long-term benchmark is around 64 and more when the markets are stable, but the cost of errors are better represented in the F1-score and MCC. We observe the absolute gain on all metrics with our model. The ACC, F1-Score and MCC are respectively 70.9%, 67%, 0.3 compare to 64.1%, 50% and 0 for the industry benchmark.

During the most unstable periods (2000-02, 2007-08), our model has an ACC of 72.2% and 70% versus 38.8% and 41.6% for the benchmark. The spread is more higher on f1-score (72% and 70% versus 22% and 25%) and MCC (0.52 and 0.54 vs 0 and 0). On the other relative unstable period, our model still outperforms the benchmark in (2011-12) but performs similarly in (2015–2016).

Globally, the benchmark has a good ACC on the long term, but masks the cost of error with poor f1-score and MCC and poor output during the unstable sub-periods.

4.4 Analysis of the Impact of the Short/Medium-Term Latent Context

We compared various auto-encoders, simple versus variational auto-encoder and feed-forward versus recurrent. The recurrent VAE gave us the best output. It was not possible to improve prediction with a convolutional auto-encoder. In order of importance, the 3 main points that Table 2 brings are :

- Overall, using the graph (Model 3) with all observable variables and latent context helps, and clearly outperforms model 1 in all test period and unstable sub-periods.
- Reccurent variational auto-encoder seems overall preferable to AE
- For the unstable periods, we don't have absolute conclusion and we need more investigation.

4.5 Impact of the Unstructured Data

The use of backward looking numerical data is considered as one limitation when analysing the financial markets. We explore and confirm the potential of the Beige Book to contain forward looking information for prediction.

The Table 3 shows the comparison with numerical data traditionally used to analyse the business cycle. On the test period (1992–2018), the ACC, f1-score and MCC of the final model are respectively 70.9%, 67%, 0.3 versus 68.5%, 64% and 0.25 for the model without unstructured data. This trend is the same on two highly unstable sub-periods, (2000-02, 2007-08) and one of the relative unstable sub-periods (2011-12). But it underperforms over the relative unstable sub-period of 2015-16 with statistic of (62.5%, 52%, 0.25) vs (54.1%, 46%, −0.04).

Table 1. Monthly prediction of S&P500 on different test periods

All test Period 1992-2018			
	ACC	**F1-S.**	**MCC**
Bench. Industry	64.1	50	0
This work	**70.9**	**67**	**0.3**
Sub Period 2000-2002			
Bench. Industry	38.8	22	0
This work	**72.2**	**72**	**0.52**
Sub Period 2007-2008			
Bench. Industry	41.6	25	0
This work	**70.8**	**70**	**0.54**
Sub Period 2011-2012			
Bench. Industry	58.3	43	0
This work	**70.8**	**70**	**0.39**
Sub Period 2015-2016			
Bench. Industry	**58.3**	43	0
This work	54.1	**46**	-0.04

Table 2. Impact of latent context

All test Period 1992-2018			
	ACC	**F1-S.**	**MCC**
Model 1	62	52	0
Model 2	65.7	60	0.16
Model 3 AE	68.8	65	0.26
Model 3 VAE	**70.9**	**67**	**0.3**
Sub Period 2000-2002			
Model 1	47.2	43	0.1
Model 2	55.6	56	0.14
Model 3 with AE	69.4	69	0.35
Model 3 with VAE	**72.2**	**72**	**0.52**
Sub Period 2007-2008			
Model 1	41.7	25	0
Model 2_F	54.1	54	0.07
Model 3 with AE	54.1	54	0.13
Model 3 with VAE	**70.8**	**70**	**0.54**
Sub Period 2011-2012			
Model 1	54.2	41	-0.18
Model 2	58.3	43	0
Model 3 with AE	**75**	**74**	**0.48**
Model 3 with VAE	70.8	70	0.39
Sub Period 2015-2016			
Model 1	54.2	41	-0.18
Model 2	**66.7**	**59**	**0.36**
Model 3 with AE	58.3	49	0.05
Model 3 with VAE	54.1	46	-0.04

Table 3. Impact of Textual Data

All test Period 1992-2018			
	ACC	**F1-S.**	**MCC**
Without Textual Data.	68.5	64	0.25
With Textual Data.	**70.9**	**67**	**0.3**
Sub Period 2000-2002			
Without Textual Data.	66.7	67	0.37
With Textual Data.	**72.2**	**72**	**0.52**
Sub Period 2007-2008			
Without Textual Data.	58.3	58	0.20
With Textual Data.	**70.8**	**70**	**0.54**
Sub Period 2011-2012			
Without Textual Data.	66.7	65	0.29
With Textual Data.	**70.8**	**70**	**0.39**
Sub Period 2015-2016			
Without Textual Data.	**62.5**	**52**	**0.25**
With Textual Data.	54.1	46	-0.04

Table 4. Comparison with related work

Ding Model : 2013	
	ACC
Benchmark Industry	75
Ding Model	55.9
This work	**80**
Pena Model : 2006-2014	
	ACC
Benchmark Industry	63.1
Pena Model	69.4
This work	**72**

4.6 Comparison with Two Works on the Same Inputs and Test Period

The metric available for comparison with two recent studies on monthly S&P500 prediction is the accuracy. We use the same inputs by formulating as specific cases of our result. [15] use neural networks on textual data and get a 55.9% accuracy over 12-month test period (Table 4). Although the test period is very short, the industry benchmark is 75% and the special case obtained from our solution is 80%. [10,11] use prior knowledge to select causal variables and innovate in the kernel function of an SVM algorithm. The accuracy is 69.4% for the period 2006–2014 (Table 4). We also obtain a specific case of our solution on the same causal variables with an accuracy of 72%. We show the importance of introducing a hierarchical interaction through a latent variable characterizing the short-term context.

5 Conclusion and Future Work

In this work, we tested several intuitions which should serve as a basis for generalizing of an integrated process with a small training sample of predicting financial markets on a monthly and quarterly basis. This is based on a framework of informed machine learning with an a priori functional causal graph of the S&P500 dynamics as the main input for predictive algorithms.

By combining our market experience, domain literature, we propose an a priori functional causal graph of the market dynamics. We learn separately the representation of each node, and then treat two similar work as special cases of our solution.

The proposed solution reconciles the theory, the selection of causal and context variables with great predictive powers, the domain knowledge features for monthly prediction on the small size of training data. The prediction are better than those of 5 similar works (including 2 studied here) and dominate the industry benchmark in all environments (stable and unstable).

The next step is to generalize using a global, dynamic functional causal graph with multiple unstructured data sources as the main input, then to automatically learn in a unified framework the embedding of all nodes and finally use it to predict the direction of various financial index and horizons (month, quarter,...).

References

1. Vuppala, K.: BlackRock, Text Analytics for Quant investing (2015)
2. Squirro: Use of unstructured data in financial services, White Paper (2014)
3. Fama, E.: Efficient capital market: a review of theory and empirical work. J. Finan. **25**, 383–417 (1970)
4. Sharpe, W.F.: Capital asset prices: a theory of market equilibrium under conditions of risk. J. Finan. **19**, 425–442 (1964)
5. Fama, E., French, K.: Efficient capital markets: II. J. Finan. **46**, 1575–1617 (1991)
6. Fama, E.F., French, K.R.: Common risk factors in the returns on stocks and bonds. J. Finan. Econ. **33**(1), 3–56 (1993)
7. Campbell, J.Y.: Empirical asset pricing: Eugene Fama, Lars Peter Hansen, and Robert Shiller. Working Paper, Department of Economics, Harvard University (2014)
8. Khaidem, L., Saha, S., Dey, S.R.: Predicting the direction of stock market prices using random forest (2016). arXiv:1605.00003
9. Harri, P.: Predicting the direction of US stock markets using industry returns. Empir.Finan. **52**, 1451–1480 (2017). https://doi.org/10.1007/s00181-016-1098-0
10. Peña, M., Arratia, A., Belanche, L.A.: Multivariate dynamic kernels for financial time series forecasting. In: Villa, A.E.P., Masulli, P., Pons Rivero, A.J. (eds.) ICANN 2016. LNCS, vol. 9887, pp. 336–344. Springer, Cham (2016). https://doi.org/10.1007/978-3-319-44781-0_40
11. Pena, M., Arratia, A., Belanche, L.: Forecasting financial time series with multivariate dynamic kernels. In: IWANN (2017)
12. Weng, B., Ahmed, M., Megahed, F.: Stock market one-day ahead movement prediction using disparate data sources. Expert Syst. Appl. **79**, 153–163 (2017)
13. Ican, O., Celik, T.: Stock market prediction performance of neural networks: a literature review. Int. J. Econ. Finan. **9**, 100–108 (2017)
14. Gu, S., Kelly, B.T., Xiu, D.: Empirical asset pricing via machine learning. Chicago Booth Research Paper No. 18–0 (2018)
15. Ding, X., Yue, Z., Liu, T., Duan, J.: Using structured events to predict stock price movement: an empirical investigation. Association for Computational Linguistics (2014)

16. Ding, X., Yue, Z., Liu, T., Duan, J.: Deep learning for event-driven stock prediction. In: Intelligence (IJCAI) (2015)
17. Chong, E., Han, C., Park, C.: Deep learning networks for stock market analysis and prediction: methodology, data representations, and case studies. Expert Syst. Appl. **83**, 187–2015 (2017)
18. Kenechukwu, A., Kruttli, M., McCabe, P., Osambela, E., Shin, C.: The shift from active to passive investing: potential risks to financial stability? Working Paper RPA 18–04 (2018)
19. Johnson, B.: Actively vs. Passively Managed Funds Performance. Morningstar Research Services LLC (2019)
20. Lee, H., Surdeanu, M., MacCartney, B., Jurafsky, D.: On the importance of text analysis for stock price prediction. In: Proceedings of the 9th International Conference on Language Resources and Evaluation, LREC (2014)
21. Denev, A.: Probabilistic Graphical Models : A New Way of Thinking in Financial Modelling. Risk Books, London (2015)
22. Denev, A., Papaioannou, A., Angelini, O.: A probabilistic graphical models approach to model interconnectedness. SSRN (2017)
23. Gonzalez, L., Powell, J.G., Shi, J., Wilson, A.: Two centuries of bull and bear market cycles. Int. Rev. Econ. Finan. **14**(4), 469–486 (2005)
24. Andrew, L.: Adaptive Markets, Financial Evolution at the Speed of Thought, Editions (2017)
25. Bry, G., Boschan, C.: Programmed selection of cyclical turning points. In: Cyclical Analysis of Time Series: Selected Procedures and Computer Programs, pp. 7–63 (1971)
26. Cotis, J.-P., Coppel, J.: Business cycle dynamics in OECD countries: evidence, causes and policy implications. In: RBA Annual Conference 2005: The Changing Nature of the Business Cycle Reserve Bank of Australia (2005)
27. Davide, F., Narayana, A.L., Turhan, B.: Preserving order of data when validation defect prediction model. ArXiv (2018)
28. Yao, M., et al.: Understanding hidden memories of recurrent neural networks. In: Conference on Visual Analytics Science and Technology (2017)

A Fuzzy Approach for Similarity Measurement in Time Series, Case Study for Stocks

Soheyla Mirshahi(✉) and Vilém Novák

Institute for Research and Applications of Fuzzy Modeling, NSC IT4Innovations,
University of Ostrava, 30. dubna 22, 701 03 Ostrava 1, Czech Republic
{soheyla.mirshahi,vilem.novak}@osu.cz

Abstract. In this paper, we tackle the issue of assessing similarity among time series under the assumption that a time series can be additively decomposed into a trend-cycle and an irregular fluctuation. It has been proved before that the former can be well estimated using the fuzzy transform. In the suggested method, first, we assign to each time series an adjoint one that consists of a sequence of trend-cycle of a time series estimated using fuzzy transform. Then we measure the distance between local trend-cycles. An experiment is conducted to demonstrate the advantages of the suggested method. This method is easy to calculate, well interpretable, and unlike standard euclidean distance, it is robust to outliers.

Keywords: Similarity measurements · Stock markets similarity · Time series analysis · Time series data mining

1 Introduction

Time series is a feasible way of representing data in many fields, including the finance sector. Financial crises in the 19th and early 20th caused a challenging situation for economies, and it led to a massive interest in economic and financial analysis. In this situation, any information that provides a better understanding to the behavior of markets is highly critical. Among many types of research concerning data mining in time series (see-[4, 7, 9, 10]); One of the key applications in this field [11] is stock data mining. Assessing time series similarity, i.e., the degree to which a given time series resembles another one is a core to many mining, retrieval, clustering, and classification tasks [18]. In the construction of financial portfolios (see [5]), diversification, which conveys investing in a variety of assets, is a key to reduce the risk of a chosen portfolio. Thus, identifying stocks that share similar behavior is vital. There is no straightforward approach, known as the best measure for assessing the similarities in time series. Surprisingly, many simple approaches like simple euclidean distance can outperform the most complicated approaches [18]. Wang et al., in 2013, perform an extensive comparison

M.-J. Lesot et al. (Eds.): IPMU 2020, CCIS 1239, pp. 567–577, 2020.
https://doi.org/10.1007/978-3-030-50153-2_42

between nine measurements across 38 data sets from various scientific domains (see [21]). One of their findings is that the euclidean distance remains an entirely accurate, robust, simple, and efficient way of measuring the similarity between two time series. However, stock markets have some properties which make the current similarity measures unfavorable. For instance, stocks react to a lot of exogenous factors such as news (see, e.g., [2]); thus, the presence of outliers in them is inevitable. Therefore, developing a measure that can react to the nature of stock markets seems essential.

A very effective technique in the analysis of time series is the fuzzy transform. Using it, we can extract trend-cycle (a low-frequency trend component) of the time series with high fidelity. The fuzzy transform provides not only the computed trend-cycle but also its analytic formula (cf. [16,17]). In this paper, using fuzzy transform, we first assign to each time series an adjoint one that consists of its local trend-cycle. Then we measure the distance between these approximate time series by a suggested formula.

There are several reasons to employ our fuzzy estimation of the trend-cycle for similarity measurement: Firstly, the trend-cycle in stocks tends to smoothen the price value and describes the behavior of the market concerning the changes in price values. Thus, it is more intuitive for experts than price values themself. It has been proven that we can successfully reach this goal using the fuzzy transform. Secondly, stock markets can be boisterous with outliers. Consequently, assessing similarities based on actual price values without any preprocessing can lead to unrealistic results. Using our method, we can easily "wipe out" the outliers without harming the basic characteristics of the time series. Finally, Our method is flexible and can answer the question of how we can find stocks that behave similarly in various time slots. For instance, experts can measure the similarity between stocks that behave similarly in a short to long term (e.g., one to several weeks).

The paper is structured as follows. After Introduction, we describe our method in Sects. 2 and 3. Section 4 is dedicated to an illustration of the purposed method and the evaluation of the results.

2 Preliminaries

2.1 Time Series Decomposition

Our techniques stem from the following characterization of a time series. It is understood as a stochastic process (see, e.g., [1,6]) $X : \mathbb{T} \times \Omega \to \mathbb{R}$ where Ω is a set of elementary random events and $\mathbb{T} = \{0, \ldots, p\} \subset \mathbb{N}$ is a finite set of numbers interpreted as time moments. Since financial time series typically posses no seasonality, we assume that they can be decomposed into components as follows:

$$X(t, \omega) = TC(t) + R(t, \omega), \qquad t \in \mathbb{T}, \tag{1}$$

where $TC(t) = Tr(t) + C(t)$ called *trend-cycle* and R is a random *noise*, i.e., a sequence of (possibly independent) random variables $R(t)$ such that for each $t \in \mathbb{T}$, the $R(t)$ has zero mean and finite variance.

2.2 Fuzzy Transform

Fuzzy transform (F-transform) is the fundamental theoretical tool for the suggested similarity measurement. Because of the lack of space, we will only briefly outline the main principles of the F-transform and refer the reader to the extensive literature, e.g., [15,16] and many others.

The F-transform is a procedure applied, in general, to a bounded real continuous function $f : [a, b] \to [c, d]$ where $a, b, c, d \in \mathbb{R}$. It is based on the concept of a *fuzzy partition* that is a set $\mathcal{A} = \{A_0, \ldots, A_n\}$, $n \geq 2$, of fuzzy sets fulfilling special axioms. The fuzzy sets are defined over nodes $a = c_0, \ldots, c_n = b$ in such a way that for each $k = 0, \ldots, n$, $A(c_k) = 1$ and $Supp(A_k) = [c_{k-1}, c_{k+1}]$[1]. The nodes are usually (but not necessarily) uniformly distributed, i.e., $c_{k+1} = c_k + h$ where $h > 0$ is a given value. To emphasize that the fuzzy partition is formed using the distance h, we will write $\mathcal{A}_h$.

The F-transform has two phases: direct and inverse. The *direct* F-transform assigns to each $A_k \in \mathcal{A}$ a component $F_k[f|\mathcal{A}]$. We distinguish *zero degree* F-transform whose components $F_k^0[f|\mathcal{A}]$ are numbers and first degree[2] F-transform whose components have the form $F_k^1[f|\mathcal{A}](x) = \beta_k^0[f] + \beta_k^1[f](x - c_k)$. The coefficient $\beta_k^1[f]$ provides estimation of an average value of the tangent (slope) of f over the area characterized by the fuzzy set $A_k \in \mathcal{A}$.

From the direct F-transform of f

$$\mathbf{F}[f|\mathcal{A}] = (F_0[f|\mathcal{A}], \ldots, F_n[f|\mathcal{A}])$$

we can form a function $\mathbf{I}[f|\mathcal{A}] : [a, b] \to [c, d]$ using the formula $\mathbf{I}[f|\mathcal{A}](x) = \sum_{k=0}^{n} (F_k[f|\mathcal{A}] \cdot A_k(x))$, $x \in [a, b]$. The function $\mathbf{I}[f|\mathcal{A}]$ is called the *inverse F-transform* of f and it approximates the original function f. It can be proved that this approximation is universal.

2.3 Application of the F-Transform to the Analysis of Time Series

The application of the F-transform to the time series analysis is based on the following result (cf. [14,16]). Let us now assume (without loss of generality) that the time series (1) contains periodic subcomponents with frequencies $\lambda_1 < \cdots < \lambda_r$. These frequencies correspond to periodicities

$$T_1 > \cdots > T_r, \tag{2}$$

respectively (via the equality $T = 2\pi/\lambda$).

[1] Of course, certain formal requirements must be fulfilled. They are omitted here and can be found in the cited literature.

[2] In general, higher degree F-transform.

Theorem 1. *Let $\{X(t) \mid t \in \mathbb{T}\}$ be a realization of the time series (1). Let us assume that all subcomponents with frequencies λ lower than λ_q are contained in the trend-cycle TC. If we construct a fuzzy partition $\mathcal{A}_h$ over the set of equidistant nodes with the distance $h = dT_q$ where $d \in \mathbb{N}$ and T_q is a periodicity corresponding to λ_q then the corresponding inverse F-transform $\mathbf{I}[X|\mathcal{A}]$ of $X(t)$ gives the following estimation of the trend-cycle:*

$$|\,\mathbf{I}[X|\mathcal{A}](t) - TC(t)| \leq 2\omega(h, TC) + D \tag{3}$$

for $t \in [c_1, c_{n-1}]$, where D is a certain small *number and $\omega(h, TC)$ is a modulus of continuity of TC w.r.t. h.*

The precise form of D and the detailed proof of this theorem can be found in [13,16]. It follows from this theorem that the F-transform makes it possible to filter out frequencies higher than a given threshold and also to reduce the noise R. Consequently, we have a tool for separation of the trend-cycle or trend. Theorem 1 tells us how the distance between nodes of the fuzzy partition should be set. This choice enables us to detect trend cycles for different time frames of interest. Of course, the estimation depends on the course of TC and it is the better the smaller is the modulus of continuity $\omega(h, TC)$ (which in case of the trend-cycle or trend is a natural assumption). The periodicities (2) can be found using the classical technique of periodogram—see [1,6].

Selection of T_q in Theorem 1 can be based on the following general OECD specification: *Trend (tendency)* is the component of a time series that represents variations of low frequency in a time series, the high and medium frequency fluctuations having been filtered out. *Trend-cycle* is the component of the time series that represents variations of low frequency, the high frequency fluctuations having been filtered out.

3 The Suggested Similarity Measurement

In this section, we will describe how our suggested method evaluates the pairwise similarity between time series.

Definition 1. *Let $X = \{X(t)|t = 1, \ldots, n\}$ and $Y = \{Y(t)|t = 1, \ldots, n\}$ be two time series of the length n and TC_X and TC_Y be estimations of trend cycles of X and Y respectively calculated based on Eq. (3). Then we define the similarity between these two time series as follows:*

$$S(X,Y) = 1 - \frac{1}{n}\sum_{t=1}^{n} \frac{|TC_X(t) - E(TC_X) - (TC_Y(t) - E(TC_Y))|}{|TC_X(t) - E(TC_X)| + |TC_Y(t) - E(TC_Y)|}, \tag{4}$$

where $E(TC_X)$ and $E(TC_Y)$ are mean values (averages) of TC_X and TC_Y, respectively and $|.|$ denotes absolute value. It is easy to show that $S(X,Y) \in [0,1]$ where it has certain features that is described on the following theorem and can be proved. In Definition 1, it is necessary to emphasize, that TC_X and TC_Y are estimation, not the real trend-cycles, since we do not know them (cf. formulas (1) and (3)).

Theorem 2. *$S(X,Y)$ is a fuzzy equality w.r.t. Lukasiewicz conjunction, i.e., it is: $reflexive$: $S(X,X) = 1$, $symmetric$: $S(X,Y) = S(Y,X)$ and $transitive$: $S(X,Y) \otimes S(Y,Z) \leq S(X,Z)$ where $\otimes$ is the Łukasiewicz conjunction defined by $a \otimes b = \{\max 0, a+b-1\}$.*

A stock can be seen as a time series $\{X(t)|t = 1,\ldots,n\}$ where $X(t)$ is closing price at time t within an interval $[0,T]$. For instance, let us consider closing price of a stock from Nasdaq INC[3], from 05.10.2008 to 30.09.2018 (522 weeks). In order to estimate its local trend-cycle, we first build a uniform fuzzy partition such that the length of each basic functions $A_2;\ldots;A_{n_1}$ is equal to a proper time slot. In our case, by setting the length of basic function to $four$, we obtain the approximation of the trend-cycles for one month. In other terms, the monthly behavior of this stock is our concern here. Figure 1 depicts the mentioned weekly stock and the fuzzy approximation of its local trend-cycle. The first and the last components of F-transform are subject to big error (because the corresponding basic functions (A_1 and A_n are incomplete). Regardless it is clear that F-transform has approximated the local-trend cycles of the stock successfully. As we mentioned before, stock markets react to many exogenous factors; thus, the presence of outliers is unavoidable. A red square in Fig. 1 shows one of these outliers for the mentioned stock. It is clear to see that F-transform has successfully wiped out the outlier while preserved the core behavior of the stock.

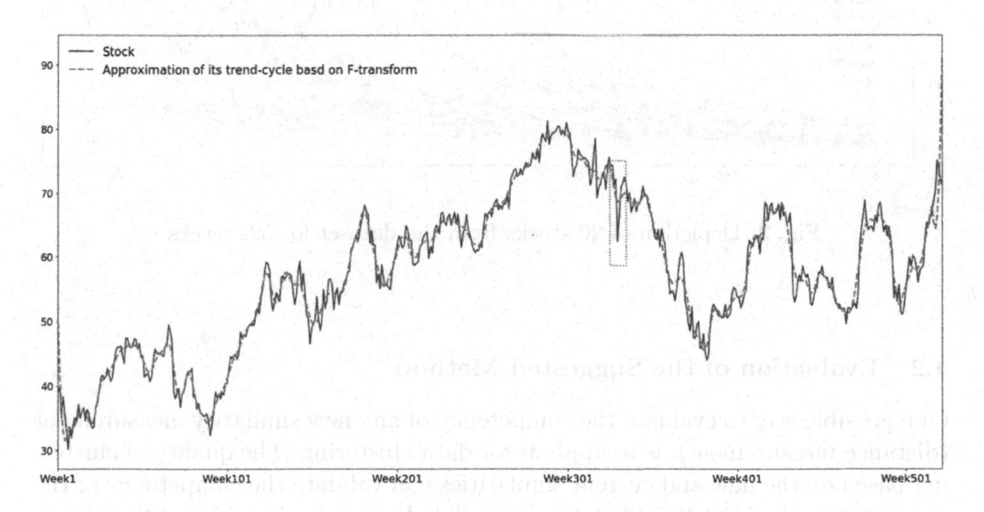

Fig. 1. A stock and its TC approximation based on F-transform.

The similarity from Definition 1 can be used for measuring the similarity between any number of stocks. We can measure using it also local behavior of them. In the next section, we will demonstrate how our suggested method works

[3] https://www.nasdaq.com/.

with a relatively large data set in conjunction with its comparison to standard the euclidean distance.

4 Illustration

4.1 Data Set

Our data set consists of a closing price of 92 stocks over 522 weeks obtained from Nasdaq INC. An example of twenty stocks from the mentioned data set is depicted in Fig. 2, where the x-axis and y-axis represent price values in dollars and number of weeks, respectively. From this figure, it is clear that any decision about the similarity between time series is impossible. Therefore it seems necessary to consider similarity between time series.

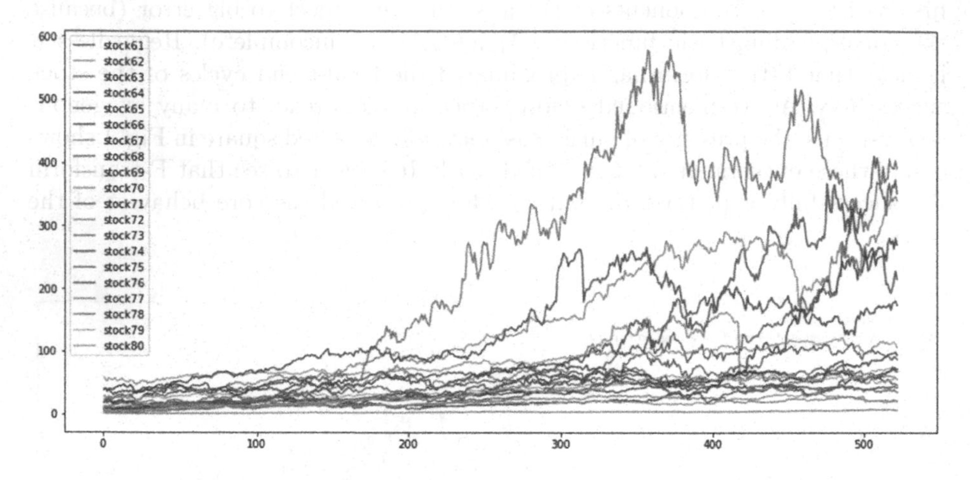

Fig. 2. Depiction of 20 stocks from the dataset for 522 weeks

4.2 Evaluation of the Suggested Method

One possible way to evaluate the competency of any new similarity measurement (distance measurement), is to apply it for data clustering. The quality of clustering based on the new and current similarities can validate the competency of the suggested method [12,19]. Therefore, we will below apply clustering of time series and compare the behavior of our similarity with the euclidean one. However, let us emphasize that time series clustering is not the primary goal of this research since our focus is on discovering the most similar pairs of stocks available in the database. As we mentioned before, the euclidean distance is an accurate, robust, simple, and efficient way to measure the similarity between two time series and, surprisingly, can outperform most of the more complex approaches (see [18,20]). Therefore we will compare our method with the euclidean distance by means of the quality of hierarchical clustering on a dataset. Hierarchical clustering is a method of cluster analysis which attempts at building a hierarchy of similar groups in data [8]. In this case, one problem to consider is the optimal number of clusters in a dataset. Overall, none of the methods for determining the optimal numbers of clusters is flawless, and none of the suggested similarities are fully satisfactory. Hierarchical clustering does not reveal an adequate number of clusters and estimation of the proper number of clusters is rather intuitive. Hence, there is a fair amount of subjectivity in determination of separate clusters. Figures 3 and 4, demonstrate the dendrogram of hierarchical clustering of the 92 stocks based on the suggested and euclidean similarity, respectively. The proper number of clusters for both similarities is equal to six. In these figures, the 92 stocks are represented in the x-axis, and their distances are depicted on the y-axis accordingly. Since the stocks are from various industries, they have different scales, and in the case of the clustering with the euclidean distance, we will eliminate the different scaling by normalizing the data. Nevertheless, this step is not demanded by the suggested method since the scale does not influence it.

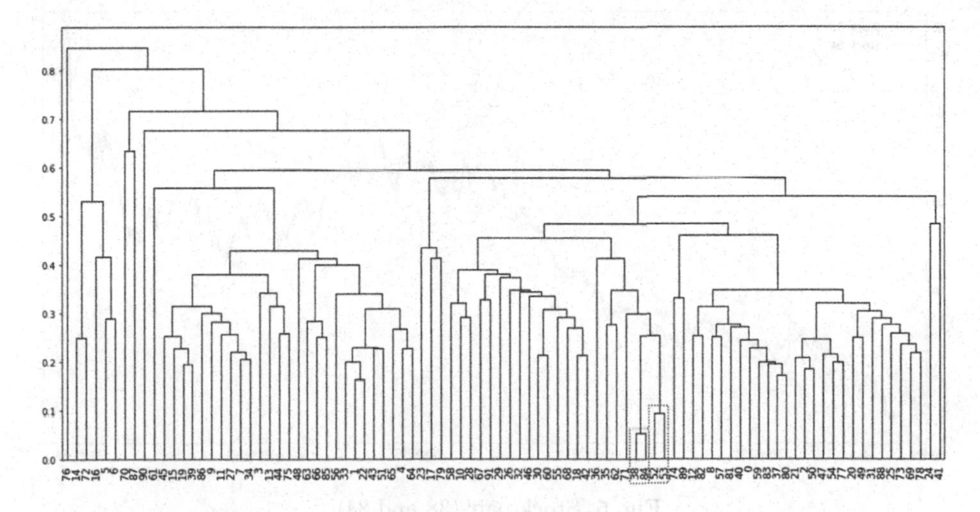

Fig. 3. Hierarchical clustering based on the suggested method (Color figure online)

Red dashed squares in 4.2 and 4.2 represent the most similar stock pairs, determined according to each method. Interestingly, both methods selected the same stock pairs; (38 and 84) and (52 and 53) as the most similar stocks. However, the suggested method, primarily determines stock pair (38 and 84) as the most similar stocks, following by stock pair (52 and 53) while the euclidean method suggests otherwise. Figure 5 and 6 shows the behaviour of theses stock pairs.

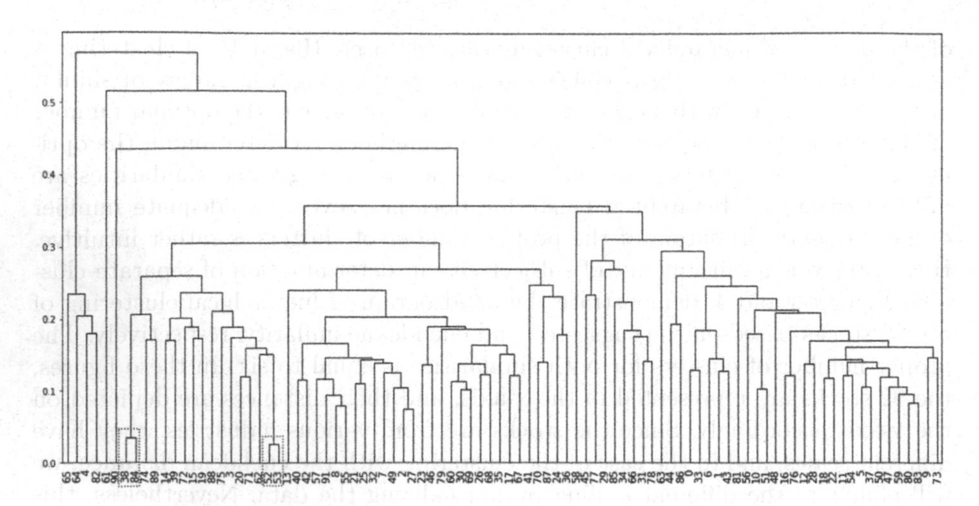

Fig. 4. Hierarchical clustering based on the Eucliden method

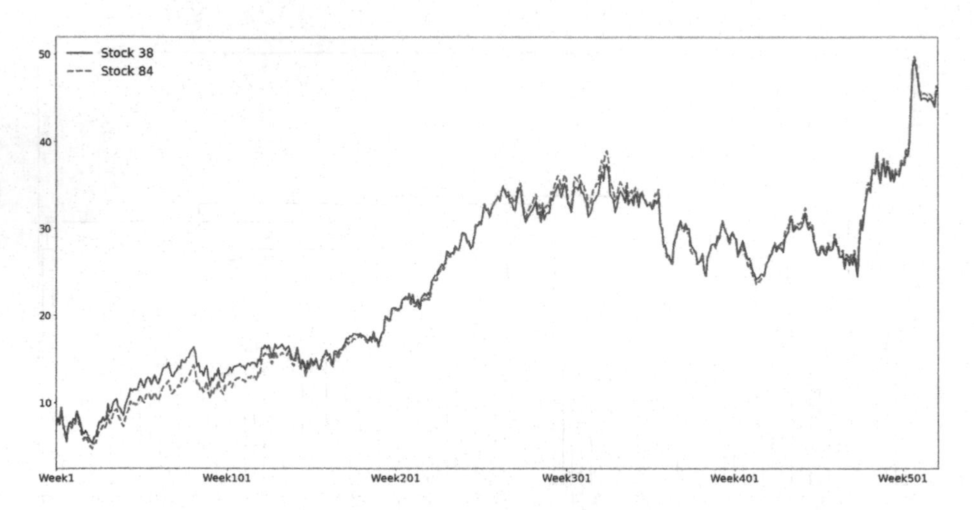

Fig. 5. Stock pair (38 and 84)

To measure the quality of clustering, we apply the Davies-Bouldin index, which is usually used in clustering. This measure evaluates intra-cluster similarity and inter-cluster differences [3]. Therefore, it can be a proper metric for clustering evaluation.

Table 1 demonstrates the Davies-Bouldin index for a different number of clusters based on the both similarities. Since the lower score indicates better quality of clustering, the, results reveal that not only is our method reasonably comparable to the euclidean method, but also it has provided more efficient clustering for these examples.

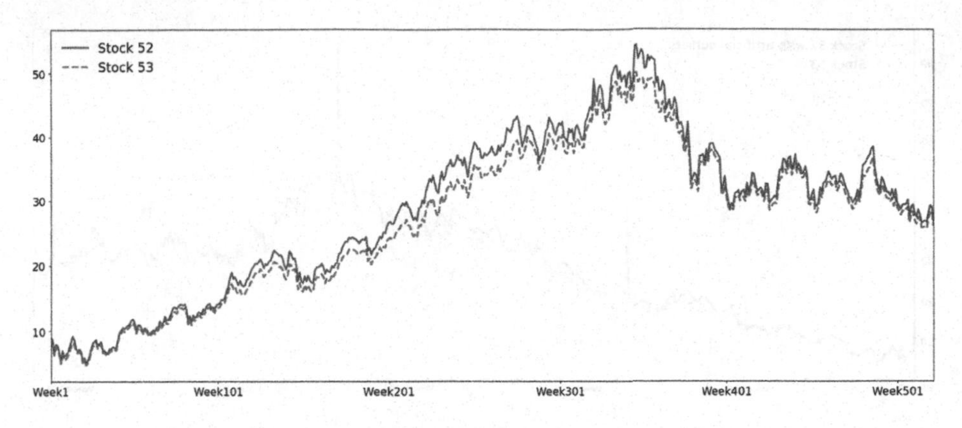

Fig. 6. Stock pair (52 and 53)

Table 1. The Davies-Bouldin index for clustering based on the proposed method and euclidean method

Method	6 clusters	8 clusters	10 clusters
The suggested method	0.61	0.64	0.72
The euclidean method	0.71	0.85	0.82

Furthermore, as we mentioned before, stock markets are prone to exogenous factors such as bad or good news (see e.g.,[2]). If a method pairs two stocks as similar, one can expect that after the occurrence of an outlier(s), the method would still evaluate these stocks alike. Hence, we will compare the performance of our method, and the euclidean distance metric for the stocks containing outliers. Recall from the previous section that based on both methods, stocks 52 and 53 are very similar to each other since their distance is minimal. Therefore, first, we will add some random artificial outliers to the stock 52, but we do not alter the stock 53 as shown in Fig. 7. Subsequently, we apply both methods to re-evaluate the similarity between these stocks.

Table 2 demonstrates the results. It is apparent, after including artificial outliers, that the euclidean distance has a dramatic jump (around 1800% increase). At the same time, the purposed method shows a minimal increase in distance (33%), which means that the suggested method is much less sensitive to the presence of outliers. Considering that the suggested method is based on the F-transform, it evaluates the similarity between the stocks concerning their local trend-cycles; therefore, it does not have the drawbacks of raw-data based approaches such as the euclidean distance. The latter methods are sensitive to noisy data [22]. One advantage of the euclidean method is its simplicity; however, the suggested method is also relatively simple since it has only one parameter to set (the length of the basic functions). Moreover, experts are able to adjust the suggested similarity measure, according to their time slot of interest.

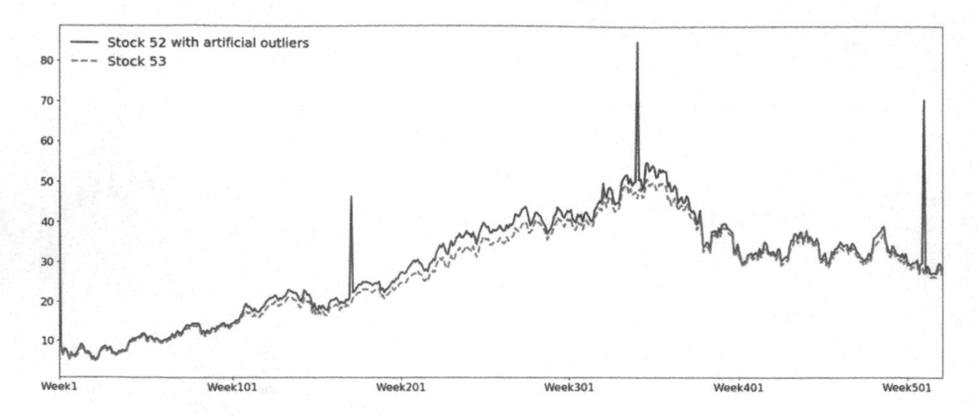

Fig. 7. Stock pair (52 and 53) containing artificial outliers

Table 2. The distance between stock 52 and 53, before and after outliers

Method	Distance before outliers	Distance after outliers
The suggested method	0.09	0.12
The euclidean method	0.17	3.33

5 Conclusion

In this paper, we developed a new method for pairwise similarity measurement. The method is based on the application of the fuzzy transform and a customized metric. The idea is based on the estimation of local trends using inverse fuzzy transform. The time series can then be paired together according to the similarity of the adjoint time series consisting of the local trends. We demonstrated the application of the suggested method in real life in addition to its comparison with the euclidean distance. Experimental results verify the capability of the suggested method for measuring the similarity between time series.

Further work will be focused on the application of this method in portfolio management and evaluation of its profitability in finance. Another addition to this work can be extending the method for time series of various lengths and compare the result with the so-called dynamic time warping (DTW) method.

Acknowledgment. The paper has been supported by the grant 18-13951S of GAČR, Czech Republic.

References

1. Anděl, J.: Statistical Analysis of Time Series. SNTL, Praha (1976). (in Czech)
2. Chan, W.S.: Stock price reaction to news and no-news: drift and reversal after headlines. J. Financ. Econ. **70**(2), 223–260 (2003)

3. Davies, D.L., Bouldin, D.W.: A cluster separation measure. IEEE Trans. Pattern Anal. Mach. Intell. **2**, 224–227 (1979)
4. Fu, T.C.: A review on time series data mining. Eng. Appl. Artif. Intell. **24**(1), 164–181 (2011)
5. Gilli, M., Maringer, D., Schumann, E.: Numerical Methods and Optimization in Finance. Academic Press, Cambridge (2019)
6. Hamilton, J.: Time Series Analysis. Princeton University Press, Princeton (1994)
7. Han, J., Pei, J., Kamber, M.: Data mining: concepts and techniques. Elsevier(2011)
8. Kaufman, L., Rousseeuw, P.J.: Finding Groups in Data: An Introduction to Cluster Analysis, vol. 344. Wiley, Hoboken (2009)
9. Keogh, E., Kasetty, S.: On the need for time series data mining benchmarks: a survey and empirical demonstration. Data Min. Knowl. Disc. **7**(4), 349–371 (2003)
10. Liao, T.W.: Clustering of time series data–a survey. Pattern Recogn. **38**(11), 1857–1874 (2005)
11. Mining, W.I.D.: Data Mining: Concepts and Techniques. Morgan Kaufinann, Burlington (2006)
12. Morse, M.D., Patel, J.M.: An efficient and accurate method for evaluating time series similarity. In: Proceedings of the 2007 ACM SIGMOD International Conference on Management of Data, pp. 569–580. ACM (2007)
13. Nguyen, L., Novák, V.: Filtering out high frequencies in time series using F-transform with respect to raised cosine generalized uniform fuzzy partition. In: Proceedings of International Conference on FUZZ-IEEE 2015. IEEE Computer Society, CPS, Istanbul (2015)
14. Nguyen, L., Novák, V.: Forecasting seasonal time series based on fuzzy techniques. Fuzzy Sets and Systems (to appear)
15. Novák, V., Perfilieva, I., Dvořák, A.: Insight into Fuzzy Modeling. Wiley, Hoboken (2016)
16. Novák, V., Perfilieva, I., Holčapek, M., Kreinovich, V.: Filtering out high frequencies in time series using F-transform. Inf. Sci. **274**, 192–209 (2014)
17. Novák, V., Štěpnička, M., Dvořák, A., Perfilieva, I., Pavliska, V., Vavříčková, L.: Analysis of seasonal time series using fuzzy approach. Int. J. Gen Syst **39**(3), 305–328 (2010)
18. Serra, J., Arcos, J.L.: An empirical evaluation of similarity measures for time series classification. Knowl.-Based Syst. **67**, 305–314 (2014)
19. Vlachos, M., Hadjieleftheriou, M., Gunopulos, D., Keogh, E.: Indexing multidimensional time-series. VLDB J. **15**(1), 1–20 (2006)
20. Wang, P.E. (ed.): Computing with Words. Wiley, New York (2001)
21. Wang, X., Mueen, A., Ding, H., Trajcevski, G., Scheuermann, P., Keogh, E.: Experimental comparison of representation methods and distance measures for time series data. Data Min. Knowl. Disc. **26**(2), 275–309 (2013)
22. Zervas, G., Ruger, S.M.: The curse of dimensionality and document clustering (1999)

Fuzzy k-NN Based Classifiers for Time Series with Soft Labels

Nicolas Wagner[1,2](✉), Violaine Antoine[1], Jonas Koko[1], and Romain Lardy[2]

[1] UCA, LIMOS, UMR 6158, CNRS, Clermont-Ferrand, France
nicolas.wagner@uca.fr

[2] UCA, INRAE, UMR Herbivores, 63122 Saint-Genès-Champanelle, France

Abstract. Time series are temporal ordered data available in many fields of science such as medicine, physics, astronomy, audio, etc. Various methods have been proposed to analyze time series. Amongst them, time series classification consists in predicting the class of a time series according to a set of already classified data. However, the performance of a time series classification algorithm depends on the quality of the known labels. In real applications, time series are often labeled by an expert or by an imprecise process, leading to noisy classes. Several algorithms have been developed to handle uncertain labels in case of non-temporal data sets. As an example, the fuzzy k-NN introduces for labeled objects a degree of membership to belong to classes. In this paper, we combine two popular time series classification algorithms, Bag of SFA Symbols (BOSS) and the Dynamic Time Warping (DTW) with the fuzzy k-NN. The new algorithms are called Fuzzy DTW and Fuzzy BOSS. Results show that our fuzzy time series classification algorithms outperform the non-soft algorithms especially when the level of noise is high.

Keywords: Time series classification · BOSS · Fuzzy k-NN · Soft labels

1 Introduction

Time series (TS) are data constrained with time order. Such data frequently appear in many fields such as economics, marketing, medicine, biology, physics... There exists a long-standing interest for time series analysis methods. Amongst the developed techniques, time series classification attract much attention since the need to accurately forecast and classify time series data spanned across a wide variety of application problems [2,9,20].

A majority of time series approaches consists in transforming time series and/or creating an alternative distance measure in order to finally employ a basic classifier. Thus, one of the most popular time series classifier is a *k-Nearest Neighbor* (k-NN) using a similarity measure called *Dynamic time warping* (DTW) [12] that allows nonlinear mapping. More recently, a *bag-of-words* model combined

M.-J. Lesot et al. (Eds.): IPMU 2020, CCIS 1239, pp. 578–589, 2020.
https://doi.org/10.1007/978-3-030-50153-2_43

with the *Symbolic Fourier Approximation* (SFA) algorithm [19] has been developed in order to deal with extraneous and erroneous data [18]. The algorithm, referred to as Bag of SFA Symbols (BOSS), converts time series into histograms. A distance is then proposed and applied to a k-NN classifier. The combinations of DTW and BOSS with a k-NN are simple and efficient approaches used as gold standards in the literature [1,8].

The k-NN algorithm is a lazy classifier employing labeled data to predict the class of a new data point. In time series, labels are specified for each timestamp and are obtained by an expert or by a combination of sensors. However, changing from one label to another can span multiple timestamps. For example, in animal health monitoring, an animal is more likely to become sick gradually than suddenly. As a consequence, using soft labels instead of hard labels to consider the animal state seems more intuitive.

The use of soft labels in classification for non-time series data sets has been studied and has shown robust prediction against label noise [7,21]. Several extensions of the k-NN algorithm have been proposed [6,10,14]. Amongst them, the fuzzy k-NN [11], which is the most popular algorithm [5], handles labels with probabilities membership for each class. The fuzzy k-NN has been applied in many domains: bioinformatics [22], image processing [13], fault detection [24], etc.

In this paper, we propose to replace the most popular time series classifiers, i.e. the k-NN algorithm, by a fuzzy k-NN. As a result, two new fuzzy classifiers are proposed: The Fuzzy DTW (F-DTW) and the Fuzzy BOSS (F-BOSS). The purpose is to tackle the problem of gradual labels in time series.

The rest of the work is organized as follows. Section 2 first recalls the DTW and BOSS algorithms. Then, the fuzzy k-NN classifier as well as the combinations between BOSS/DTW and fuzzy k-NN are detailed. Section 3 presents a comparison between hard and soft labels through several data sets.

2 Time Series Classifiers for Soft Labels

The most efficient way to deal with TS in classification is to use a specific metric such as DTW or to transform like BOSS the TS into non ordered data. A simple classifier can then be applied.

2.1 Dynamic Time Warping (DTW)

Dynamic Time Warping [3] is one of the most famous similarity measurement between two times series. It considers the fact that two similar times series may have different lengths due to various speed. The DTW measure allows then a non-linear mapping, which implies a time distortion. It has been shown that DTW is giving better comparisons than a Euclidean distance metric. In addition, the combination of the elastic measure with the 1-NN algorithm is a gold standard that produces competitive results [1], although DTW is not a distance function. Indeed, DTW does not respect the property of triangle inequality but in practice, this property is often respected [17]. Despite DTW has a quadratic complexity, the use of this measure with a simple classifier remains faster than other algorithms like neural networks. Moreover, using lower bound technique can decrease the complexity of the measure to a linear complexity [16].

2.2 The Bag of SFA Symbols (BOSS)

The bag of SFA Symbols algorithm (BOSS) [18] is a bag of words method using Fourier transform in order to reduce noise and to handle variable lengths. First, a sliding window of size w is applied on each time series of a data set. Then, windows from the same time series are converted into a word sequences according to the Symbolic Fourier Approximation (SFA) algorithm [19]. Words are composed of l symbols with an alphabet size of c. The time series is then represented by a histogram that corresponds to the number of word occurrences for each word. Finally, the 1-NN classifier can be used with distance computed between histograms. Given two histograms B_1 and B_2, the measure called d_{BOSS} is:

$$d_{BOSS}(B_1, B_2) = \sum_{a \in B_1; B_1(a)>0} [B_1(a) - B_2(a)]^2, \tag{1}$$

where a is a word and $B_i(a)$ the number of occurrences of a in the ith histogram. Note that the set of words are identical for B_1 and B_2, but the number of occurrences for some words can be equal to 0.

We propose to handle fuzzy labels in TS classification using the fuzzy k-NN algorithm.

2.3 Fuzzy k-NN

Let $\mathcal{D} = (\mathcal{X}, y)$ be a data set composed of $n = |\mathcal{X}|$ instances and $y_i \in \mathcal{C}$ be a label assigned to each instance $x_i \in \mathcal{X}$ with $\mathcal{C}$ the set of all possible labels.

For conventional hard classification algorithms, it is possible to compute a characteristic function $f_c : \mathcal{X} \rightarrow \{0, 1\}$ with $c \in \mathcal{C}$:

$$f_c(x_i) = \begin{cases} 1, \ c = y_i, \\ 0, \ c \neq y_i. \end{cases} \tag{2}$$

Rather than hard labels, soft labels allow to express a degree of confidence on the class membership of an object. Most of the time, this uncertainty is represented given by probabilistic distribution. In that case, soft labels correspond to fuzzy labels. Thereby, the concept of characteristic function is generalized to membership function $u_c : \mathcal{X} \rightarrow [0, 1]$ with $c \in \mathcal{C}$:

$$u_c(x_i) = \mathcal{P}(y_i = c), \tag{3}$$

such that

$$\sum_{c \in \mathcal{C}} u_c(x_i) = 1, \tag{4}$$

$$0 < \sum_{x \in \mathcal{X}} u_c(x) < n, \forall c \in \mathcal{C}. \quad (5)$$

There exists a wide range of k-NN variants using fuzzy labels in the literature [5]. The most famous and basic method, referred to as fuzzy k-NN [11], predicts the class membership of an object x_i using two steps. First, similarly to the hard k-NN algorithm, the k nearest neighbors $x_j \in \mathcal{K}$, $|\mathcal{K}| = k$ of x_i are retrieved. The second step differs from hard k-NN as it computes a membership degree for each class:

$$u_c(x_i) = \frac{\sum_{x_j \in \mathcal{K}} u_c(x_j) d(x_i, x_j)^{-2/(m-1)}}{\sum_{x_j \in \mathcal{K}} d(x_i, x_j)^{-2/(m-1)}}, \forall c \in \mathcal{C}, \quad (6)$$

with m a fixed coefficient controlling the fuzziness of the prediction, $d(x_i, x_j)$ the distance between instances x_i and x_j. Usually, $m = 2$ and the Euclidean distance is the most popular distance considered.

2.4 Fuzzy DTW and Fuzzy BOSS

In order to deal with time series and fuzzy labels, we propose two fuzzy classifiers called F-DTW and F-BOSS.

The F-DTW algorithm consists in using the fuzzy k-NN algorithm with DTW as distance function (see Fig. 1). It takes in entry a time series and computes the DTW distance with the labeled times series. Once the k closest time series found, the class membership is computed with Eq. (6).

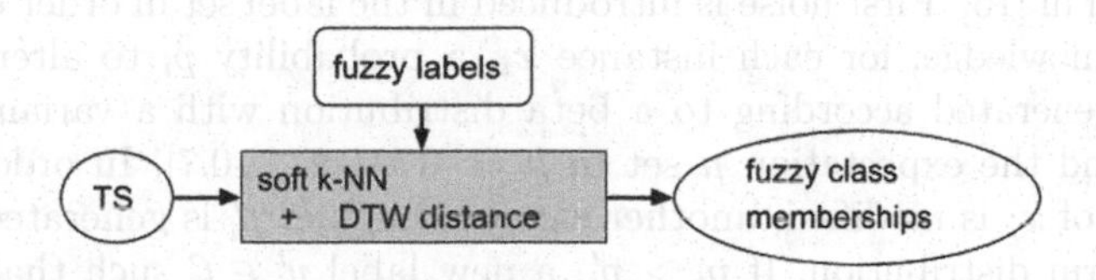

Fig. 1. F-DTW algorithm

The F-BOSS algorithm consists in first applying the BOSS algorithm in order to transform the time series into histograms. Then, the fuzzy k-NN is applied with BOSS distances. It generates fuzzy class memberships (see Fig. 2).

Once F-DTW and F-BOSS defined, experiments are carried out to show the interest of taking into account soft labels when there exists noise and/or uncertainties on the labels.

3 Experiments

Experiments consist in studying the parameters setting (i.e. the number of neighbors) and compare soft and hard methods when labels are noisy.

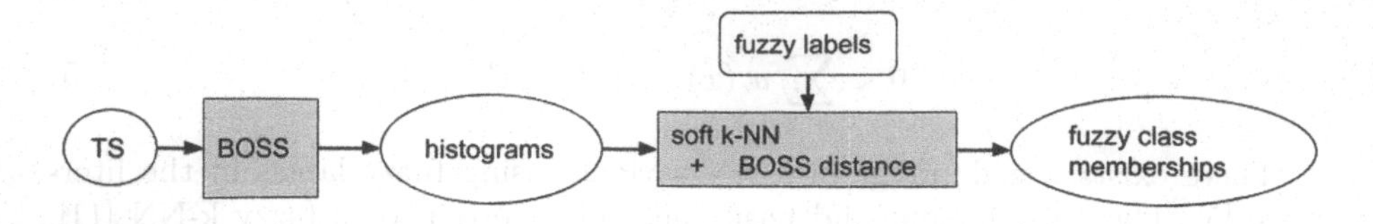

Fig. 2. F-BOSS algorithm

3.1 Experimental Protocol

We have selected five data sets from the University of California Riverside (UCR) archive [4]. Each data set have different characteristics detailed in Table 1.

Table 1. Characteristics of data sets.

Data set name	Size train	Size test	Size series	Nb classes	Type
WormsTwoClass	181	77	900	2	MOTION
Lightning2	60	61	637	2	SENSOR
ProximalPhalanxTW	400	205	80	6	IMAGE
Yoga	300	3000	426	2	IMAGE
MedicalImages	381	760	99	10	IMAGE

The hard labels are known for each data set. Thus, we generate fuzzy labels as described in [15]. First noise is introduced in the label set in order to represent uncertain knowledge: for each instance x_i, a probability p_i to alter label y_i is randomly generated according to a beta distribution with a variance σ set to $\sigma = 0.04$ and the expectation μ set to $\mu = [0.1, 0.2, ..., 0.7]$. In order to decide if the label of x_i is modified, another random number p_i' is generated according to an uniform distribution. If $p_i > p_i'$, a new label $y_i' \in \mathcal{C}$ such that $y_i' \neq y_i$ is randomly assigned to x_i. Second, fuzzy labels are deduced using p_i. Let $\Pi_c : \mathcal{X} \to [0,1]$ be a possibilistic function computed for each instance x_i and each class c:

$$\Pi_c(x_i) = \begin{cases} 1, & c = y_i', \\ p_i, & c \neq y_i'. \end{cases} \tag{7}$$

The possibilistic distribution allows to go from total certainty when $p_i = 0$ to total uncertainty when $p_i = 1$. Since our algorithms employ fuzzy labels, possibilities Π_i are converted into probabilities u_c by normalizing Eq. (7) with the sum of all possibilities:

$$u_c(x_i) = \frac{\Pi_c(x_i)}{\sum_{c \in \mathcal{C}} \Pi_c(x_i)}. \tag{8}$$

We propose to test and compare three strategies dealing with noisy labels. The two first ones are dedicated to classifiers taking in entry hard labels.

The first strategy, called strategy 1, considers that noise in labels is unknown. As a result soft labels are ignored and for each instance x_i, label y_i^* is chosen using the maximum probability membership rule, i.e. $\max(u_c(x_i))$.

The second strategy, called strategy 2, consists in discarding the most uncertain labels and transforming soft labels into hard labels. For each instance x_i the normalized entropy H_i is computed as follows:

$$H_i = \frac{1}{\log_2(|\mathcal{C}|)}(-\sum_{k \in \mathcal{C}} u_k(x_i) \log_2(u_k(x_i))). \tag{9}$$

Note that $H_i \in [0, 1]$ and $H_i = 0$ corresponds to a state of total certainty whereas $H_i = 1$ corresponds to a uniform distribution. If $H_i > \theta$ we consider the soft label of x_i as too uncertain and x_i is discarded from the fuzzy data set. In the experiments, we set the threshold θ to 0.95.

Finally, the third strategy, called strategy 3, keeps the whole fuzzy labels and apply a classifier able to handle such labels.

In order to compare strategies and since strategies 1 and 2 give hard labels whereas strategy 3 generates fuzzy labels, we convert fuzzy labels using the maximum membership rule, i.e. $\max(u_c(x_i))$, $\forall c \in \mathcal{C}$.

The best parameters of F-BOSS are found by a leave-one-out cross-validation on the training set. The values of the parameters are fixed as in [1]:

- window length $w = [10, ..., q]$, with $q = |x_i|$, the size of the series and $|w| = \min(200, \sqrt{q})$,
- alphabet size $\alpha = 4$,
- word length $l = [8, 10, 12, 14, 16]$.

Classifiers tested are soft k-NN, F-BOSS and F-DTW. For strategies 1 and 2, they correspond to k-NN, BOSS with k-NN and DTW with k-NN. For each classifier, different numbers of neighbors $k = [1, 2, ..., 10]$ and different values of μ, $\mu = [0, 0.1, 0.2, ..., 0.7]$ are analyzed. Note that $\mu = 0$ corresponds to the original data set without fuzzy processing. To compare the different classifiers and strategies, we choose to present the percentage of good classification, referred to as accuracy.

3.2 Influence of the Number of Neighbors in k-NN

Usually with DTW or BOSS with hard labels, the number of neighbors is set to 1. This experiment studies the influence of the parameter k when soft labels are used. Thus, we set $\mu = 0.3$ in order to represents a moderate level of noise that can exist in real applications and apply strategy 3 on all data sets. Figure 3 illustrates the result on the WormsTwoClass data set, i.e. the variation of the accuracy for the three classifiers according to k.

First, for all values of k the performance of the soft k-NN classifier is lower than the others. Such result has also been identified in other data sets. We

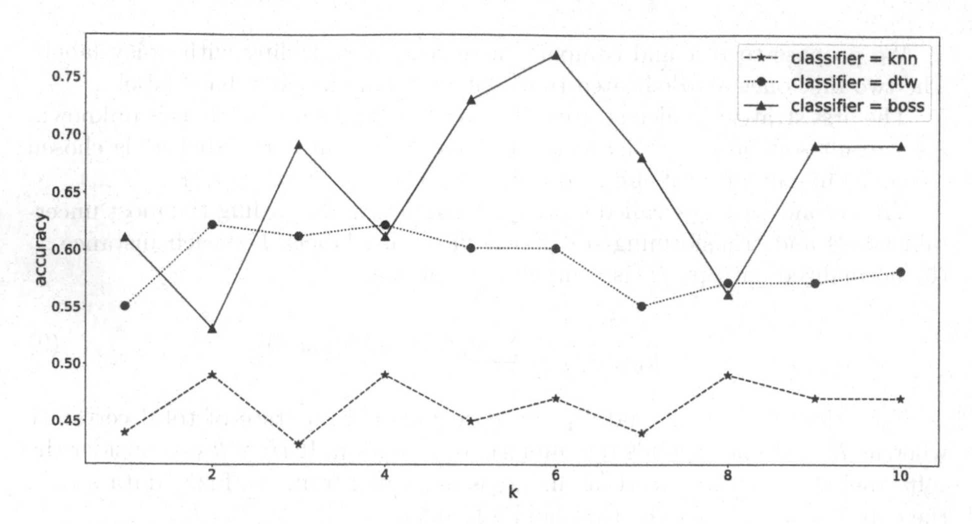

Fig. 3. Accuracy according to k for WormsTwoClass data set: $\mu = 0.3$ and strategy 3

Table 2. Accuracy for all data sets with $\mu = 0.3$ and $k = 5$.

	Strategy	Soft k-NN	F-DTW	F-BOSS
ProximalP.	1	0.32	0.33	0.36
	2	0.38	**0.41**	0.4
	3	0.38	0.39	**0.41**
Lightning2	1	0.43	0.67	0.56
	2	0.59	0.67	0.66
	3	0.56	**0.69**	0.56
WormsTwoC.	1	0.44	0.56	0.7
	2	0.48	0.58	0.68
	3	0.45	0.6	**0.73**
Yoga	1	0.64	0.68	0.67
	2	0.68	0.72	0.71
	3	0.68	**0.73**	0.7
MedicalI.	1	0.56	0.64	0.51
	2	0.58	**0.67**	0.54
	3	0.59	**0.67**	0.54

also observe on Fig. 3 that the F-BOSS algorithm is often better than F-DTW. However, the pattern of the F-BOSS curve is serrated that makes difficult the establishment of guidelines for the choice of k. In addition, the best k depends on the algorithm and the data set. Therefore, for the rest of the experiments section, we choose to set k to the median value $k = 5$.

3.3 Strategies and Algorithms Comparisons

Table 2 presents the results of all classifiers and all strategies on the five data sets for $k = 5$ and $\mu = 0.3$. F-BOSS and F-DTW outperform the k-NN classifier for all data sets. This result is expected since DTW and BOSS algorithms are specially developed for time series problems. The best algorithm between F-DTW and F-BOSS depends on the data set: F-DTW is the best one for Lightning2, Yoga and MedicalImages, and F-BOSS is the best one for WormsTwoClass. Note that for ProximalPhalanxTW, F-DTW is the best with strategy 2 and F-BOSS is the best for the strategy 3. Strategy 1 (i.e. hard labels) is most of the time worse than the two other strategies. This can be explained by the fact that the strategy 1 does not take the noise into account. For all best classifiers of all data sets, the strategy 3 is the best strategy even though for ProximalPhalanxTW and MedicalImages strategy 2 competes with strategy 3. The strategy 3 (i.e. soft) is therefore better than the strategy 2 (i.e. discard) one for five algorithms and equal for two algorithms. However, the best algorithm between F-BOSS and F-DTW depends on the data sets.

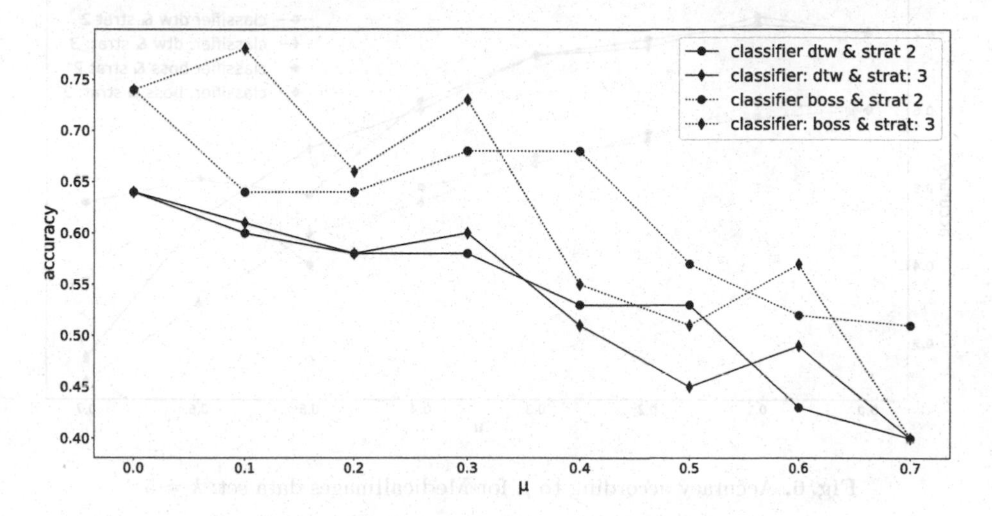

Fig. 4. Accuracy according to μ for WormsTwoClass data set: $k = 5$

3.4 Noise Impact on F-BOSS and F-DTW

To observe the impact of the μ parameter, Fig. 4, Fig. 5 and Fig. 6 illustrate respectively the accuracy variations for the WormsTwoClass, Lightning2 and MedicalImages data sets according to the value of μ. The k-NN classifier and the strategy 1 are not represented because their poor performance (see Sect. 3.3). The figures also include the value $\mu = 0$ that corresponds to the original data

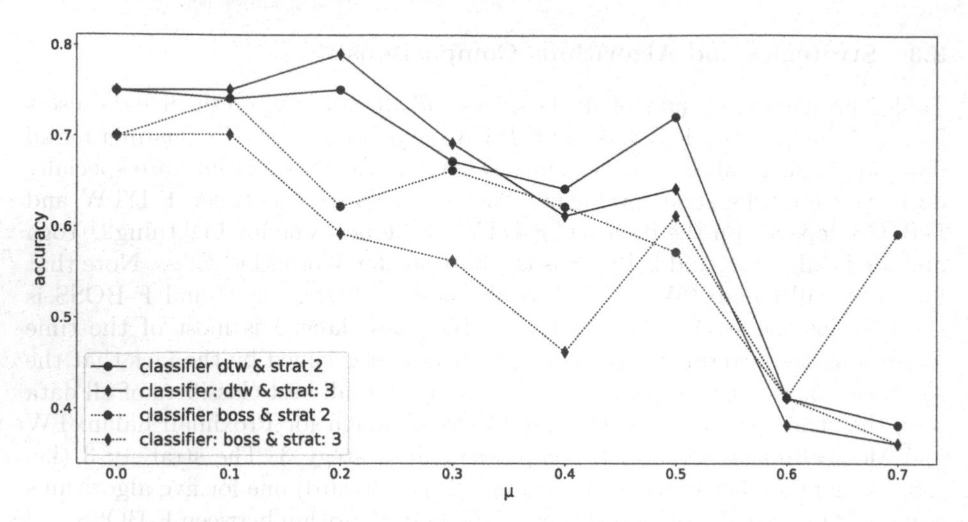

Fig. 5. Accuracy according to μ for Lightning2 data set: $k = 5$

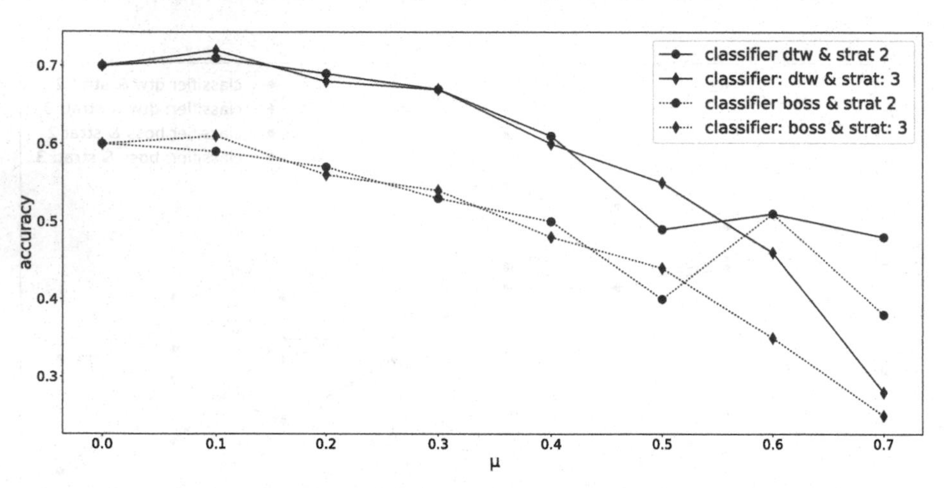

Fig. 6. Accuracy according to μ for MedicalImages data set: $k = 5$

without fuzzy processing. Results are not presented for the Yoga and ProximalPhalanxTW data sets because the accuracy differences between the strategies and the classifiers are not significant, especially when $\mu < 0.3$.

For WormsTwoClass, F-BOSS is better than F-DTW and inversely for Lightning2 and MedicalImages data sets. For the WormsTwoClass and Lightning2 data sets, with a low or moderate level of noise ($\mu < 0.3$), the third strategy is better than the second one. For the MedicalImages data set, the strategies 2 and 3 are quite equivalent, excepted for $\mu = 0.5$ where the third strategy is better. When $\mu > 0.5$, the strategy 2 is better. Higher levels of noise lead to better results with strategy 2. This can be explained as follows: strategy 2 is less disturbed by the important number of miss-classified instances since it removes them. On the opposite, with a moderate level of noise, the soft algorithms are more accurate because they keep informative labels.

Predicting soft labels instead of hard labels brings to the expert an extra information that can be analyzed. We propose to consider as uncertain all predicted fuzzy labels having a probability less than a threshold t. Figure 7 present the accuracy and the number of elements discarded varying with this threshold t for the WormsTwoClass data set. As it can be observed, the higher is t, the better is the accuracy and the more the number of predicted instances are discarded. Thus t is a tradeoff between good results and a sufficient number of predicted instances.

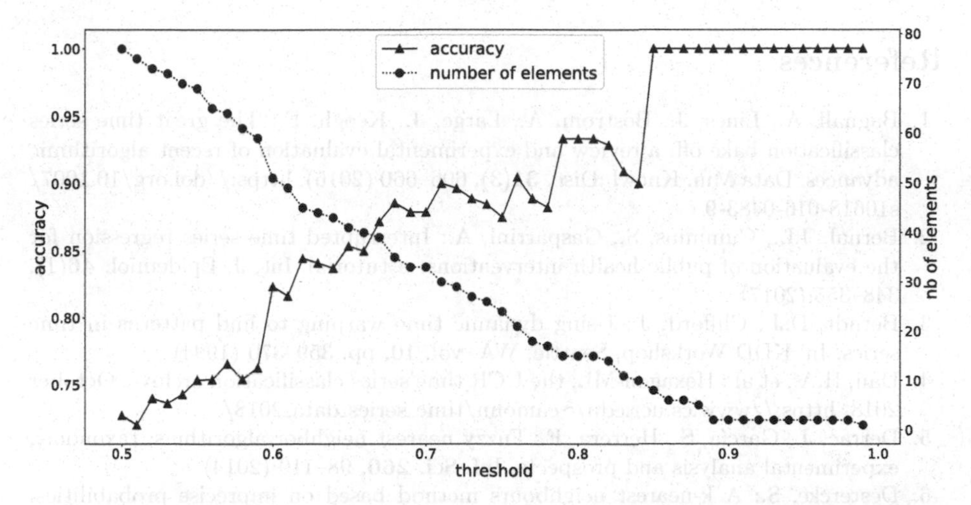

Fig. 7. Accuracy and number of elements according to the threshold t for WormsTwoClass data set: $\mu = 0.3$ and strategy 3

The results above show that methods designed for time series outperform the standard ones and the fuzzy strategies give a better performance for noisy labeled data.

4 Conclusion

This paper considers the classification problem of time series having fuzzy labels, i.e. labels with probabilities to belong to classes. We proposed two methods, F-BOSS and F-DTW, that are a combination of a fuzzy classifier (k-NN) and methods dedicated to times series (BOSS and DTW). The new algorithms are tested on five data sets coming from the UCR archives. With F-BOSS and F-DTW, integrating the information of uncertainty about the class memberships of the labeled instances outperforms strategies that does not take in account such information.

As perspectives we propose to modify the classification part of F-BOSS and F-DTW in order to attribute a weight on the neighbors depending on the distance to the object to predict. This strategy, inspired by some soft k-NN algorithms for non time series data sets, should improve the performances by giving less importance to far and uncertain labeled instances.

Another perspective consists in adapting the soft algorithms to possibilistic labels. Indeed, the possibilistic labels are more suitable for real applications as it allows an expert to assign a degree of uncertainty on an object to a class independently from the other classes. For instance, in a dairy cows application where the goal is to detect anomalies like diseases or estrus [23], the possibilistic labels are simple to retrieve and well appropriated because a cow can have two or more anomalies at the same time (e.g. a diseases and an estrus).

References

1. Bagnall, A., Lines, J., Bostrom, A., Large, J., Keogh, E.: The great time series classification bake off: a review and experimental evaluation of recent algorithmic advances. Data Min. Knowl. Disc. **31**(3), 606–660 (2016). https://doi.org/10.1007/s10618-016-0483-9
2. Bernal, J.L., Cummins, S., Gasparrini, A.: Interrupted time series regression for the evaluation of public health interventions: a tutorial. Int. J. Epidemiol. **46**(1), 348–355 (2017)
3. Berndt, D.J., Clifford, J.: Using dynamic time warping to find patterns in time series. In: KDD Workshop, Seattle, WA, vol. 10, pp. 359–370 (1994)
4. Dau, H.A., et al.: Hexagon-ML: the UCR time series classification archive, October 2018. https://www.cs.ucr.edu/~eamonn/time_series_data_2018/
5. Derrac, J., García, S., Herrera, F.: Fuzzy nearest neighbor algorithms: taxonomy, experimental analysis and prospects. Inf. Sci. **260**, 98–119 (2014)
6. Destercke, S.: A k-nearest neighbours method based on imprecise probabilities. Soft. Comput. **16**(5), 833–844 (2012). https://doi.org/10.1007/s00500-011-0773-5
7. El Gayar, N., Schwenker, F., Palm, G.: A study of the robustness of KNN classifiers trained using soft labels. In: Schwenker, F., Marinai, S. (eds.) ANNPR 2006. LNCS (LNAI), vol. 4087, pp. 67–80. Springer, Heidelberg (2006). https://doi.org/10.1007/11829898_7
8. Ismail Fawaz, H., Forestier, G., Weber, J., Idoumghar, L., Muller, P.-A.: Deep learning for time series classification: a review. Data Min. Knowl. Disc. **33**(4), 917–963 (2019). https://doi.org/10.1007/s10618-019-00619-1
9. Feyrer, J.: Trade and income–exploiting time series in geography. Am. Econ. J. Appl. Econ. **11**(4), 1–35 (2019)
10. Hüllermeier, E.: Possibilistic instance-based learning. Artif. Intell. **148**(1–2), 335–383 (2003)
11. Keller, J.M., Gray, M.R., Givens, J.A.: A fuzzy k-nearest neighbor algorithm. IEEE Trans. Syst. Man Cybern. **SMC-15**(4), 580–585 (1985)
12. Keogh, E., Ratanamahatana, C.A.: Exact indexing of dynamic time warping. Knowl. Inf. Syst. **7**(3), 358–386 (2004). https://doi.org/10.1007/s10115-004-0154-9
13. Machanje, D., Orero, J., Marsala, C.: A 2D-approach towards the detection of distress using fuzzy K-nearest neighbor. In: Medina, J., et al. (eds.) IPMU 2018. CCIS, vol. 853, pp. 762–773. Springer, Cham (2018). https://doi.org/10.1007/978-3-319-91473-2_64

14. Östermark, R.: A fuzzy vector valued knn-algorithm for automatic outlier detection. Appl. Soft Comput. **9**(4), 1263–1272 (2009)
15. Quost, B., Denœux, T., Li, S.: Parametric classification with soft labels using the evidential EM algorithm: linear discriminant analysis versus logistic regression. Adv. Data Anal. Classif. **11**(4), 659–690 (2017). https://doi.org/10.1007/s11634-017-0301-2
16. Ratanamahatana, C.A., Keogh, E.: Three myths about dynamic time warping data mining. In: Proceedings of the 2005 SIAM International Conference on Data Mining, pp. 506–510. SIAM (2005)
17. Ruiz, E.V., Nolla, F.C., Segovia, H.R.: Is the DTW "distance" really a metric? An algorithm reducing the number of DTW comparisons in isolated word recognition. Speech Commun. **4**(4), 333–344 (1985)
18. Schäfer, P.: The BOSS is concerned with time series classification in the presence of noise. Data Min. Knowl. Disc. **29**(6), 1505–1530 (2015). https://doi.org/10.1007/s10618-014-0377-7
19. Schäfer, P., Högqvist, M.: SFA: a symbolic fourier approximation and index for similarity search in high dimensional datasets. In: Proceedings of the 15th International Conference on Extending Database Technology, pp. 516–527. ACM (2012)
20. Susto, G.A., Cenedese, A., Terzi, M.: Time-series classification methods: review and applications to power systems data. In: Big Data Application in Power Systems, pp. 179–220. Elsevier (2018)
21. Thiel, C.: Classification on soft labels is robust against label noise. In: Lovrek, I., Howlett, R.J., Jain, L.C. (eds.) KES 2008. LNCS (LNAI), vol. 5177, pp. 65–73. Springer, Heidelberg (2008). https://doi.org/10.1007/978-3-540-85563-7_14
22. Tiwari, A.K., Srivastava, R.: An efficient approach for prediction of nuclear receptor and their subfamilies based on fuzzy k-nearest neighbor with maximum relevance minimum redundancy. Proc. Natl. Acad. Sci., India, Sect. A **88**(1), 129–136 (2018). https://doi.org/10.1007/s40010-016-0325-6
23. Wagner, N., et al.: Machine learning to detect behavioural anomalies in dairy cowsunder subacute ruminal acidosis. Comput. Electron. Agric. **170**, 105233 (2020). https://doi.org/10.1016/j.compag.2020.105233. http://www.sciencedirect.com/science/article/pii/S0168169919314905
24. Zhang, Y., Chen, J., Fang, Q., Ye, Z.: Fault analysis and prediction of transmission line based on fuzzy k-nearest neighbor algorithm. In: 2016 12th International Conference on Natural Computation, Fuzzy Systems and Knowledge Discovery (ICNC-FSKD), pp. 894–899. IEEE (2016)

Mathematical Fuzzy Logic and Graded Reasoning Models

Converting Possibilistic Networks by Using Uncertain Gates

Guillaume Petiot(✉)

CERES, Catholic Institute of Toulouse, 31 rue de la Fonderie, 31068 Toulouse, France
guillaume.petiot@ict-toulouse.fr
http://www.ict-toulouse.fr

Abstract. The purpose of this paper is to define a general frame to convert the Conditional Possibility Tables (CPT) of an existing possibilistic network into uncertain gates. In possibilistic networks, CPT parameters must be elicited by an expert but when the number of parents of a variable grows, the number of parameters to elicit grows exponentially. This problem generates difficulties for experts to elicit all parameters because it is time-consuming. One solution consists in using uncertain gates to compute automatically CPTs. This is useful in knowledge engineering. When possibilistic networks already exist, it can be interesting to transform them by using uncertain gates because we can highlight the combination behaviour of the variables. To illustrate our approach, we will present at first a simple example of the estimation for 3 test CPTs with behaviours MIN, MAX and weighted average. Then, we will perform a more significant experimentation which will consist in converting a set of Bayesian networks into possibilistic networks to perform the estimation of CPTs by uncertain gates.

Keywords: Possibilistic networks · Possibility theory · Uncertain logical gates · Estimation

1 Introduction

Knowledge engineering often needs to describe how information is combined. Experts' knowledge can be represented by a Directional Acyclic Graph (DAG). Unfortunately, this kind of knowledge is often imprecise and uncertain. The use of possibility theory [21] allows us to take into account these imperfections. Possibilistic networks [1,2], as Bayesian networks [17,20] in probability theory, can be used to propagate new information, called evidence, in a DAG. The main disadvantage of possibilistic networks is the need to elicit all parameters of Conditional Possibility Tables. When the number of parents of a variable grows, the number of parameters to elicit in the CPT grows exponentially. For example, if a variable of 2 states has 10 parents with 2 states, then we have $2^{11} = 2048$ parameters to elicit. Moreover, when experts define the parameters

M.-J. Lesot et al. (Eds.): IPMU 2020, CCIS 1239, pp. 593–606, 2020.
https://doi.org/10.1007/978-3-030-50153-2_44

of a CPT in possibilistic networks, they often define an implicit behaviour. The most common behaviours are those of Boolean logic, such as AND, OR, XOR, etc.

Noisy gates in probability theory [4] deal with this problem of parameters. They use a function to compute automatically the CPT. As a result, the number of parameters is reduced. Another advantage is the modelling of noise and incomplete knowledge. Indeed, it is often difficult to have all the knowledge during the task of modelling because of the complexity of the problem. The use of a variable which represents unknown knowledge leads to another kind of model. In possibility theory, authors Dubois et al. [5] proposed to use uncertain logical gates which present the same advantages as noisy gates. The behaviour of uncertain gates connectors depends on the choice of the function. The function can have the following behaviours: indulgent, compromise or severe.

Nowadays, there are several applications of possibilistic networks that exist where experts have elicited all CPT parameters. We propose in this paper a solution to convert an existing possibilistic network into a possibilistic network with uncertain gates. It is very interesting to see which connector corresponds to each CPT. This can provide a better understanding of information processing, which is not highlighted when an expert elicits CPT parameters. This is also useful in knowledge engineering.

This paper is structured as follows. In the first part, we will present possibility theory and uncertain gates. Then we will focus our interest on the estimation of uncertain gates and we will propose the examples of the estimation of uncertain gates for test CPTs and existing applications using Bayesian networks. So we will have to convert Bayesian networks into possibilistic networks and then we will perform the estimation of the CPTs. We will present the results of this experiment in the last part of this paper.

2 Uncertain Gates

Uncertain gates are an analogy of noisy gates in possibility theory [21]. Possibility theory proposes to model imprecise knowledge by using a possibility distribution. If Ω is the referential, then the possibility distribution π is defined from Ω to $[0,1]$ with $max_{v \in \Omega} \pi(v) = 1$. Dubois et al. [6] define the possibility measure Π and the necessity measure N from $P(\Omega)$ to $[0,1]$:

$$\forall A \in P(\Omega), \Pi(A) = \underset{x \in A}{sup}\ \pi(x) \text{ and } N(A) = 1 - \Pi(\neg A) = \underset{x \notin A}{inf}\ 1 - \pi(x). \quad (1)$$

The possibility theory is not additive but maxitive:

$$\forall A, B \in P(\Omega), \Pi(A \cup B) = \max(\Pi(A), \Pi(B)). \quad (2)$$

We can compute the possibility of a variable A by using the conditioning proposed by D. Dubois and H. Prade [6]:

$$\forall A, B \in P(\Omega), \Pi(A|B) = \begin{cases} \Pi(A, B) & \text{if } \Pi(A, B) < \Pi(B), \\ 1 & \text{if } \Pi(A, B) = \Pi(B). \end{cases} \quad (3)$$

Possibilistic networks [1,2] can be defined as follows for a DAG $G = (V, E)$ where V is the set of Variables and E the set of edges between the variables:

$$\Pi(V) = \bigotimes_{X \in V} \Pi(X/Pa(X)). \tag{4}$$

Pa stands for the parents of the variable X. There are two kinds of possibilistic networks: qualitative possibilistic networks (also called min-based possibilistic networks) where $\bigotimes$ is the function min, and quantitative possibilistic networks (also called product-based possibilistic networks) where $\bigotimes$ is the product. In this research, we consider the comparison of possibilistic values instead of the use of an intensity scale in [0,1] so we will use a min-based possibilistic network.

We can propose as in noisy gates to use the Independence of Causal Influence [4,9,25] to define a possibilistic model with the ICI. In fact, as in probability theory, there is a set of causal variables $X_1, ..., X_n$ which influence the result of an effect variable Y. We insert between each $X_i s$ and Y a variable Z_i to take into account the noise, i.e. uncertainty. We can also introduce another variable Z_l which represents the unknown knowledge [5]. The following figure presents the possibilistic model with ICI (Fig. 1):

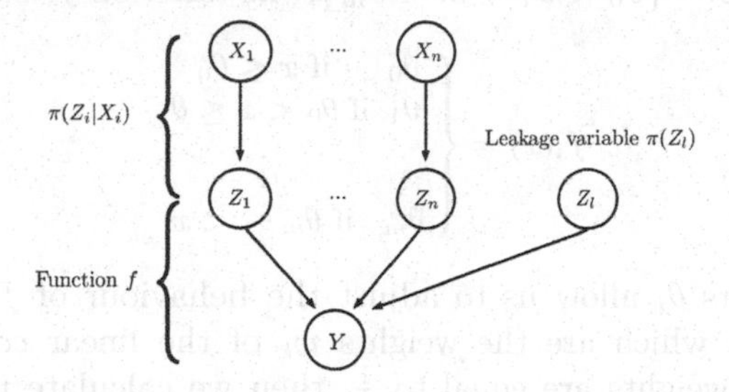

Fig. 1. Possibilistic model with ICI.

We propose to compute $\pi(Y|X_1, ..., X_n)$ by marginalizing the variables $Z_i s$:

$$\pi(y|x_1, ..., x_n) = \bigoplus_{z_1,...,z_n} \pi(y|z_1, ..., z_n) \otimes \bigotimes_{i=1}^{n} \pi(z_i|x_i) \tag{5}$$

The $\otimes$ is the minimum and the $\oplus$ is the maximum in possibility theory. If we have a deterministic model where the value of Y can be computed from the values of Z_is by using a function f then:

$$\text{where } \pi(y|z_1, ..., z_n) = \begin{cases} 1 \text{ if } y = f(z_1, ..., z_n) \\ 0 \text{ else} \end{cases} \tag{6}$$

As a result, we obtain the following equation:

$$\pi(y|x_1, ..., x_n) = \bigoplus_{z_1,...,z_n : y = f(z_1,...,z_n)} \bigotimes_{i=1}^{n} \pi(z_i|x_i) \tag{7}$$

If we add a leakage variable Z_l in the previous equation, we obtain:

$$\pi(y|x_1, ..., x_n) = \bigoplus_{z_1,...,z_n,z_l:y=f(z_1,...,z_n,z_l)} \bigotimes_{i=1}^{n} \pi(z_i|x_i) \otimes \pi(z_l) \quad (8)$$

The CPT is computed by using this formula. The function f can be AND, OR or their generalizations uncertain MIN and uncertain MAX as proposed by Dubois et al. [5]. These authors developed an optimization for the computation of uncertain MIN and MAX connectors. We can also use a weighted average function or the operator OWA [22], as described in our previous study [13,14]. For this experimentation, we will use the connectors uncertain MIN (UMIN), uncertain MAX (UMAX) and uncertain weighted average (UWAVG). The function f must have the same domain as the variable Y. We can see that the connectors uncertain MIN and uncertain MAX satisfy this property. The weighted average of the intensity level of the variables X_i can return a value outside the domain of Y. So in this case we have to combine a function noted g with a scaling function f_s which returns a value in the domain of Y. We can write $f = f_s \circ g$ where function g is $g(z_1, ..., z_n) = \omega_1 z_1 + ... + \omega_n z_n$. If the state of the variable Y defines an ordered scale $E = \{\vartheta_0 < \vartheta_1 < ... < \vartheta_m\}$, the function f_s can be for example:

$$f_s(x) = \begin{cases} \vartheta_0 & \text{if } x \leq \theta_0 \\ \vartheta_1 & \text{if } \theta_0 < x \leq \theta_1 \\ \vdots & \vdots \\ \vartheta_m & \text{if } \theta_{m-1} < x \end{cases} \quad (9)$$

The parameters θ_i allow us to adjust the behaviour of f_s. The function g has n parameters which are the weights w_i of the linear combination and n arguments. If all weights are equal to $\frac{1}{n}$, then we calculate the average of the intensities. If $\forall_{i \in [1,n]} \omega_i = 1$, then we make the sum of the intensities. To compute the Eq. 9, we have to define $\pi(Z_i|X_i)$, $\pi(Z_l)$, and the function f. If the states of a variable are ordered, we can assign an intensity level to each state as done by Dubois et al. [5]. In our experimentation with test CPTs we propose to use 3 states for the variables such as low, medium and high. The intensity levels are 0 for low, 1 for medium and 2 for high. The table of $\pi(Z_i|X_i)$ is the following (Table 1):

Table 1. Possibility table for 3 ordered states.

$\pi(Z_i\|X_i)$	$x_i = 2$	$x_i = 1$	$x_i = 0$
$z_i = 2$	1	$s_i^{2,1}$	$s_i^{2,0}$
$z_i = 1$	$\kappa_i^{1,2}$	1	$s_i^{1,0}$
$z_i = 0$	$\kappa_i^{0,2}$	$\kappa_i^{0,1}$	1

In the previous table, κ represents the possibility that an inhibitor exists if the cause is met and s_i the possibility that a substitute exists when the cause is not met. If a cause of weak intensity cannot produce a strong effect, then all

$s_i = 0$. So there are 6 parameters at the most per variable and 2 parameters for $\pi(Z_l)$. If all variables have the same intensity level and are graded variables with a normal state of intensity 0 (the first state is the normal state), there is no problem. In this case, the higher the intensity is, the more abnormal the situation becomes. But there are two other cases to discuss. The first case concerns the variables which are not graded variables. For example, if we have a variable where the domain is $(decrease, normal, accelerate)$, we can see that the normal speed is the neutral state i.e. with the intensity level 1. The second case concerns the incompatibilities of the intensity levels of the causal variables X_is and their parents. Indeed, if some causal variables do not share the same domain with the effect variable, we can imagine that the computation of the function f can give the results that are out of range. This remark leads to a constraint on Z_is, which is that Z_is must have the same domain as Y.

3 Estimation

The estimation of uncertain gates leads to two problems discussed in [24]. The first one is the simple case where we would like to perform an estimation of uncertain gates from an existing CPT. The second case concerns the estimation of the connector from the data without any CPT. Indeed, in possibilistic networks, the number of parameters to elicit grows exponentially with the number of parents of a variable, leading to problems of knowledge engineering. So if data is available, we propose to perform the estimation of the CPT instead of eliciting all parameters. Unfortunately, it is often difficult to have enough data to estimate all parameters.

In this paper, we will focus our interest on the first case. If a CPT already exists, it can be useful to know which uncertain gate corresponds to the CPT. The question can be which connector matches the target CPT among all existing connectors. To discover this, we have to compare the CPT generated by uncertain gates and the existing CPT. This leads to the problem of estimating the parameters of uncertain gates. We look for the closest CPT to a reference CPT. To do this, we must use a distance [8]. For our first experimentation we will use the Euclidean distance. The problems discussed in [24] for this distance are also present in our research. We propose to analyze them and to compare several distances in future works. With respect to this study, we will compare 4 estimation methods by using the distance of the CPTs as a cost function in order to improve the estimation: hill-climbing (HC), simulated annealing (SA), tabu search (TS), and genetic algorithm (GA). We propose to choose the connector with the smallest distance $\hat{\theta}$ to the target CPT.

We consider the example of hill-climbing and the estimation of the parameters of the uncertain MIN. We must at first initialize all parameters of the uncertain MIN to a random value in $[0, 1]$. Then we generate the CPT by using the connector and we calculate the initial distance between the generated CPT and the target CPT. Next, we evaluate all neighbours of the parameters by adding and subtracting a step (Function *GenNeighbour* in the following algorithm). We consider only the parameters with the smallest distance. These parameters become our temporary solution. We reiterate this process until the distance is lower than a constant, or a maximum number of iterations is reached. The algorithm for the estimation of the parameters of the uncertain MIN with hill-climbing is the following:

Algorithm 1: Hill climbing.

Input : Y a variable; $X_1, ..., X_n$ the n parents of Y; $\pi(Y|X_1, ..., X_n)$ the initial CPT; *max* the maximum number of iterations; ϵ the accepted distance; *step* a constant.
Output: The result is $\pi(Z_i|X_i)$.

1 **begin**
2 $\quad$ $iteration \leftarrow 0$; $error \leftarrow +\infty$; $current \leftarrow +\infty$; $Initialize(\pi(Z_i|X_i))$;
3 $\quad$ **while** $iteration <$ max *and* $error > \epsilon$ **do**
4 $\quad\quad$ $\pi'(Z_i|X_i) \leftarrow \pi(Z_i|X_i)$;
5 $\quad\quad$ $current \leftarrow error$;
6 $\quad\quad$ **forall** $\pi^*(Z_i|X_i) \in GenNeighbour(\pi(Z_i|X_i), step)$ **do**
7 $\quad\quad\quad$ $\pi^*(Y|X_1, ..., X_n) \leftarrow UMIN(Y, \pi^*(Z_i|X_i))$;
8 $\quad\quad\quad$ **if** $E(\pi^*(Y|X_1, ..., X_n), \pi(Y|X_1, ..., X_n)) < current$ **then**
9 $\quad\quad\quad\quad$ $current \leftarrow E(\pi^*(Y|X_1, ..., X_n), \pi(Y|X_1, ..., X_n))$;
10 $\quad\quad\quad\quad$ $\pi'(Z_i|X_i) \leftarrow \pi^*(Z_i|X_i)$;
11 $\quad\quad$ **if** $current < error$ **then**
12 $\quad\quad\quad$ $error \leftarrow current$;
13 $\quad\quad\quad$ $\pi(Z_i|X_i) \leftarrow \pi'(Z_i|X_i)$;
14 $\quad\quad$ $iteration \leftarrow iteration + 1$;

The simulated annealing algorithm was proposed [18] to describe a thermodynamic system. This is a probabilistic approach which leads to an approximation of a global optimum. We have proposed the following algorithm for the estimation of the parameters of the uncertain MIN:

Algorithm 2: Simulated annealing.

Input : Y a variable; $X_1, ..., X_n$ the n parents of Y; $\pi(Y|X_1, ..., X_n)$ the initial CPT; *max* the maximum number of iterations;ϵ the accepted distance;*step* a constant; $K_{\max}$ is the maximum of steps; T is the temperature.
Output: The result is $\pi(Z_i|X_i)$.

1 **begin**
2 $\quad$ $iteration \leftarrow 0$; $Initialize(\pi(Z_i|X_i))$; $SP \leftarrow +\infty$;
3 $\quad$ $\pi^p(Y|X_1, ..., X_n) \leftarrow UMIN(Y, \pi(Z_i|X_i))$;
4 $\quad$ **while** $iteration <$ max *and* $SP > \epsilon$ **do**
5 $\quad\quad$ $k \leftarrow 0$;
6 $\quad\quad$ **while** $k < k_{\max}$ **do**
7 $\quad\quad\quad$ $\pi^*(Z_i|X_i) \leftarrow GenNeighbour(\pi(Z_i|X_i), step)$
8 $\quad\quad\quad$ $\pi^*(Y|X_1, ..., X_n) \leftarrow UMIN(Y, \pi^*(Z_i|X_i))$;
9 $\quad\quad\quad$ $SP \leftarrow E(\pi^p(Y|X_1, ..., X_n), \pi(Y|X_1, ..., X_n))$;
10 $\quad\quad\quad$ $SN \leftarrow E(\pi^*(Y|X_1, ..., X_n), \pi(Y|X_1, ..., X_n))$;
11 $\quad\quad\quad$ $delta \leftarrow SN - SP$;
12 $\quad\quad\quad$ **if** $delta < 0$ **then**
13 $\quad\quad\quad\quad$ $\pi(Z_i|X_i) \leftarrow \pi^*(Z_i|X_i)$;
14 $\quad\quad\quad\quad$ $\pi^p(Y|X_1, ..., X_n) \leftarrow \pi^*(Y|X_1, ..., X_n)$;
15 $\quad\quad\quad$ $r \leftarrow RandomValue()$;
16 $\quad\quad\quad$ **if** $r < e^{-\frac{delta}{T}}$ **then**
17 $\quad\quad\quad\quad$ $\pi(Z_i|X_i) \leftarrow \pi^*(Z_i|X_i)$;
18 $\quad\quad\quad\quad$ $\pi^p(Y|X_1, ..., X_n) \leftarrow \pi^*(Y|X_1, ..., X_n)$;
19 $\quad\quad\quad$ $k \leftarrow k + 1$;
20 $\quad\quad$ $T \leftarrow \alpha \times T$;
21 $\quad\quad$ $iteration \leftarrow iteration + 1$;

We also propose to use a genetic algorithm to perform the estimation. The chromosomes gather all the parameters of the connector. The population of chromosomes will evolve by favouring the best individuals. A fitness function, which measures the distance between a CPT generated by one chromosome and the target CPT, is used to compare the individuals. The best chromosome is the one with the smallest fitness. The genetic algorithm performs several processing operations in a loop: the selection of the parents, the crossover, and the mutation. We can resume this with the following algorithm:

Algorithm 3: Genetic algorithm.

```
Input  : Y a variable; X1, ..., Xn the n parents of Y; π(Y|X1, ..., Xn) the initial CPT;
         max the maximum number of iterations; ε the accepted distance; step a
         constant; N is the number of chromosomes;
Output: The result is π(Zi|Xi).
1 begin
2     generation ← 0; bestFitness ← +∞; InitializePopulation();
3     while generation < max and bestFitness > ε do
4         ComputeFitness();
5         TournamentSelection();
6         Crossover();
7         Mutation();
8         generation ← generation + 1;
```

The tabu search was proposed by F. W. Glover [10–12]. This algorithm improves local search algorithms by using a list of previous moves to avoid processing a situation already explored. The Tabu search algorithm is the following:

Algorithm 4: Tabu search.

```
Input  : Y a variable; X1, ..., Xn the n parents of Y; π(Y|X1, ..., Xn) the initial CPT;
         max the maximum number of iterations; ε the accepted distance; step a
         constant; NC is the number of candidates; T is the temperature.
Output: The result is π(Zi|Xi).
1  begin
2      iteration ← 0; Initialize(π(Zi|Xi)); π^p(Y|X1, ..., Xn) ← UMIN(Y, π(Zi|Xi));
3      bestScore ← +∞;
4      while iteration < max and bestScore > ε do
5          k ← 0;
6          while k < NC do
7              π*(Zi|Xi) ← GenNeighbour(π(Zi|Xi), step)
8              π*(Y|X1, ..., Xn) ← UMIN(Y, π*(Zi|Xi));
9              score ← E(π*(Y|X1, ..., Xn), π(Y|X1, ..., Xn));
10             if π*(Zi|Xi) ∉ Candidate then
11                 Candidate ← Candidate ∪ π*(Zi|Xi);
12                 F(k) ← score;
13                 k ← k + 1;
14         S ← IndiceOfAscendingSort(F)
15         if (Candidate[S[0]] ∉ ListTabu) or ((Candidate[S[0]] ∈ ListTabu) and
             (F(S[0]) < bestScore)) then
16             ListTabu ← ListTabu ∪ Candidate[S[0]];
17             π(Zi|Xi) ← Candidate[S[0]];
18             currentScore ← F(S[0]);
19         else
20             b ← 0;
21             while ((b < NC) and (Candidate[S[0]] ∈ ListTabu)) do
22                 b ← b + 1;
23             if b < NC then
24                 ListTabu ← ListTabu ∪ Candidate[S[b]];
25                 π(Zi|Xi) ← Candidate[S[b]];
26                 currentScore ← F(S[b]);
27         if currentScore < bestScore then
28             bestScore ← currentScore;
29         iteration ← iteration + 1;
```

4 Experimentation

We now propose to test this approach on simulated CPTs for connectors MIN, MAX and weighted average. Next, we will compare the estimation of the CPTs. For this experimentation, we will use only two causal variables X_1 and X_2 and one effect variable Y. All the variables have states: low, medium and high with intensity levels 0, 1 and 2. We provide below the parameters used to compute the simulated CPTs (Table 2):

Table 2. Possibility table of $\pi(Z_i|X_i)$.

$\pi(Z_i\|X_i)$	$x_i = 2$	$x_i = 1$	$x_i = 0$
$z_i = 2$	1	0.3	0
$z_i = 1$	0.3	1	0
$z_i = 0$	0.2	0.3	1

We used the same parameters for X_1 and X_2. We defined $\pi(Z_l)$ with the values $\pi(Z_L = 0) = 1$ and $\pi(Z_L = 1) = \pi(Z_L = 2) = 0.1$. For the weighted average, we will simulate a mean behaviour by defining the weights equal to 0.5 for all variables but we will not take into account the leakage variable. We present the simulated CPTs obtained by using the above parameters in the following tables (Table 3):

Table 3. Simulated CPTs.

	X_1	Low			Medium			High		
	X_2	Low	Medium	High	Low	Medium	High	Low	Medium	High
Y	Low	1.0	1.0	1.0	1.0	0.3	0.3	1.0	0.3	0.2
	Medium	0.1	0.1	0.1	0.1	1.0	1.0	0.1	1.0	0.3
	High	0.1	0.1	0.1	0.1	0.3	0.3	0.1	0.3	1.0

(a) MIN CPT.

	X_1	Low			Medium			High		
	X_2	Low	Medium	High	Low	Medium	High	Low	Medium	High
Y	Low	1.0	0.3	0.2	0.3	0.3	0.2	0.2	0.2	0.2
	Medium	0.1	1.0	0.3	1.0	1.0	0.3	0.3	0.3	0.3
	High	0.1	0.3	1.0	0.3	0.3	1.0	1.0	1.0	1.0

(b) MAX CPT.

	X_1	Low			Medium			High		
	X_2	Low	Medium	High	Low	Medium	High	Low	Medium	High
Y	Low	1.0	1.0	0.3	1.0	0.3	0.3	0.3	0.3	0.2
	Medium	0.0	0.3	1.0	0.3	1.0	0.3	1.0	0.3	0.3
	High	0.0	0.0	0.0	0.0	0.3	1.0	0.0	1.0	1.0

(c) WAVG CPT.

We have performed the estimation with several steps 0.1, 0.01 and 0.001 and a limited number of iteration of 40000. The result of the estimation is the following (Table 4):

Table 4. Result of the estimation.

Connectors	Uncertain MIN				Uncertain MAX				Uncertain WAVG			
Estimation	HC	SA	TS	GA	HC	SA	TS	GA	HC	SA	TS	GA
MIN CPT	0.31	**0.0**	**0.0**	**0.0**	2.40	2.27	2.27	2.27	1.14	0.86	0.86	0.86
MAX CPT	2.22	2.05	2.11	2.17	0.62	**0.0**	**0.0**	0.22	1.09	0.94	0.86	0.86
WAVG CPT	1.84	1.72	1.72	1.73	2.08	1.96	1.96	1.98	0.10	**0.0**	0.22	**0.0**

(a) *step* = 0.1.

Connectors	Uncertain MIN				Uncertain MAX				Uncertain WAVG			
Estimation	HC	SA	TS	GA	HC	SA	TS	GA	HC	SA	TS	GA
MIN CPT	0.27	**0.0**	0.24	0.07	2.39	2.27	2.43	2.27	1.93	1.21	1.04	1.28
MAX CPT	2.18	2.05	2.08	2.08	0.26	**0.0**	**0.0**	**0.0**	1.63	2.21	1.29	0.86
WAVG CPT	1.89	1.72	1.72	1.92	2.22	2.01	1.96	1.96	0.56	1.24	**0.08**	0.22

(b) *step* = 0.01.

Connectors	Uncertain MIN				Uncertain MAX				Uncertain WAVG			
Estimation	HC	SA	TS	GA	HC	SA	TS	GA	HC	SA	TS	GA
MIN CPT	0.27	**0.0**	2.15	0.07	2.39	2.69	2.29	2.27	1.93	1.04	1.04	1.81
MAX CPT	2.11	2.08	2.25	2.10	0.20	0.58	**0.0**	0.22	1.86	1.80	1.63	0.95
WAVG CPT	1.89	1.73	2.57	1.73	2.22	1.97	1.96	1.97	0.58	0.22	**0.002**	0.08

(c) *step* = 0.001.

We can see in the above tables that all simulated CPTs are associated with the expected connector. We can see in bold in each line the smallest distance which corresponds to the best result for the estimation. By applying the decision rule, we select for each line the connector and the estimation algorithm which has produced the best result. For example, with step 0.001 we can see that the simulated MIN CPT was associated with the connector uncertain MIN and that the best result was provided by the simulated annealing algorithm with the final distance of 0. This means that the parameters are exactly estimated. The simulated MAX CPT was associated to the connector uncertain MAX and the best result was provided by tabu search. And finally, the simulated WAVG CPT was associated to uncertain WAVG connector as expected and the best result was provided by tabu search with a distance of 2×10^{-3}. If we compare the results, we can see that the best results are obtained with step 0.1. So we provide below the comparison of the estimation algorithms which lead to the best results for this step (Fig. 2):

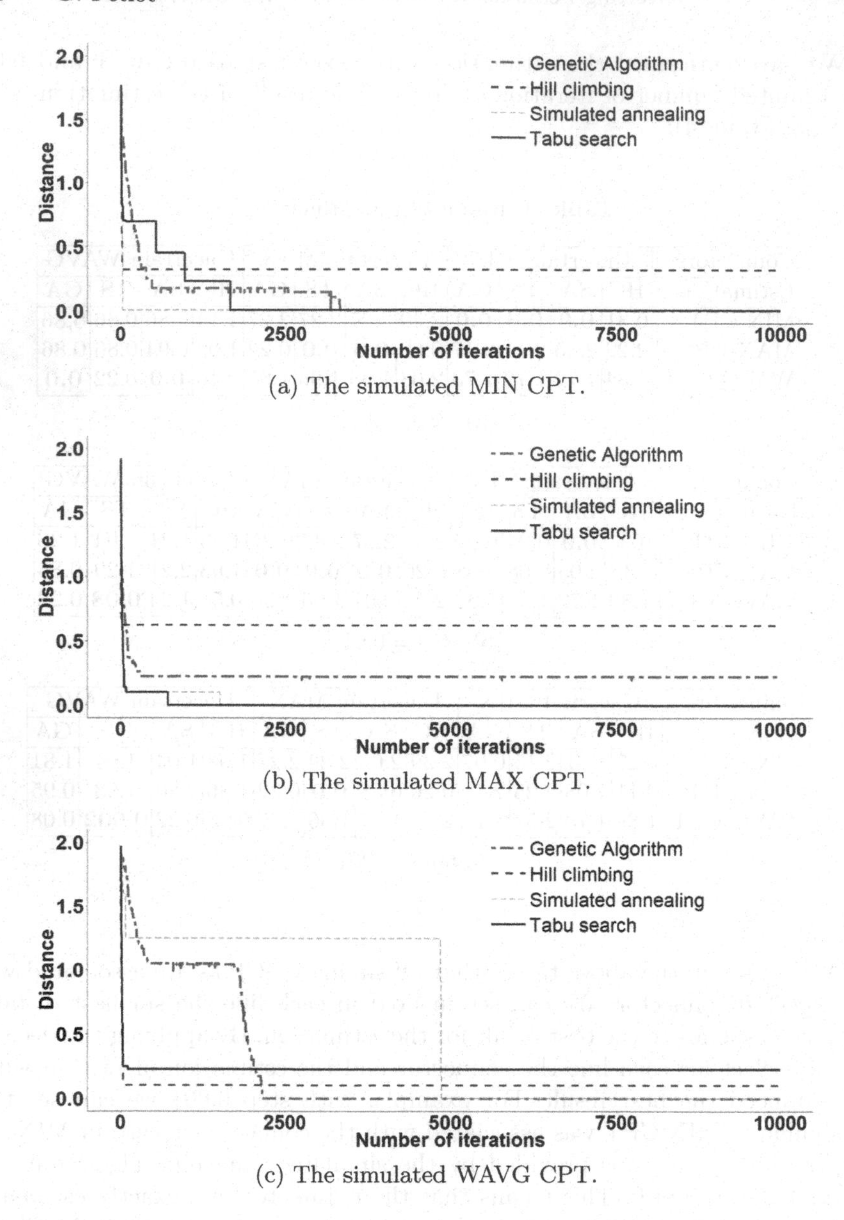

(a) The simulated MIN CPT.

(b) The simulated MAX CPT.

(c) The simulated WAVG CPT.

Fig. 2. Estimation of the simulated CPT with *step* = 0.1 for the first 10,000 iterations.

We can see in the above results that all algorithms did not converge to 0. The best result seems to be obtained by simulated annealing or tabu search but for step=0.001 and for the simulated MIN CPT the tabu algorithm didn't find the expected parameters. Nevertheless, we have tried several parameters for tabu search and we have obtained good results.

We have performed another experimentation by using existing Bayesian networks: Asia [16], Earthquake and Cancer [15], Survey [19], Sachs [23], Andes [3] because of the small number of data set available for possibilistic networks. We have transformed the conditional probability tables into conditional possibility tables to obtain possibilistic networks. The structure of the networks is the same in Bayesian networks and possibilistic networks. Then we have applied our approach to replace all CPTs by appropriate connectors. The problem of converting the probability table has already been discussed by Dubois et al. [7]. The temptation can be to perform the normalization of the probability but it is not sufficient to ensure that the possibility property $\Pi \geq P$ is satisfied. For example, if we have $p_1 > p_2 > ... > p_n$, we perform $p_i = \frac{p_i}{p_1}$ for all i. The correct solution proposed by Dubois et al. [7] consists in computing the possibilities as follows:

$$\pi_i = \begin{cases} 1 & \text{if } i = 1, \\ \sum_{j=i}^{n} p_j & \text{if } \pi_{i-1} > \pi_i, \\ \pi_{i-1} & \textit{otherwise}. \end{cases} \tag{10}$$

We present below the result of this experimentation. At first, we have performed the estimation for small networks. We can see in the first column the name of the data set, then the name of the node and the number of the parents of the node. In the fourth column, we present the size of the CPT, and in the following columns, the minimal distance for all connectors (Table 5).

Table 5. Results for small size networks (*step* = 0.01).

Name	Table	Parents	Size	MIN	MAX	WAVG
Asia	Dysp	2	8	1.238	**0.0707**	0.0709
	Either	2	8	1.581	**0.011**	0.012
Earthquake	Alarm	2	8	0.961	0.71	**0.204**
Cancer	Cancer	2	8	0.97	1.6856	**0.07**
Survey	E	2	12	1.56	**0.71**	0.71
	T	2	12	**0.70**	1.06	0.74

This first experiment allows us to distinguish two cases. The first one concerns the distance close to 0. In this case, the computation of the connector matches the target CPT. The connector can replace the CPT. The second case concerns the distance not close enough to 0, such as the tables alarm, E or T. We cannot replace the CPT by a connector, nevertheless the information provided is meaningful because we can deduce the behaviour of the CPT between severe and indulgent. We have also performed another experiment with Andes, which is a big network, and we obtain the following result for the first 30 nodes (Table 6):

Table 6. The 30 first nodes of Andes (*step* = 0.01).

Name	Table	Parents	Size	MIN	MAX	WAVG
Andes	RApp1	2	8	**0.01**	1.58	0.01
	SNode_8	2	8	1.52	**0.007**	0.008
	SNode_20	2	8	**0.005**	1.33	0.006
	SNode_21	2	8	**0.006**	1.33	0.003
	SNode_24	2	8	**0.009**	1.45	0.008
	SNode_25	2	8	**0.008**	1.45	0.001
	SNode_26	2	8	**0.006**	1.33	0.007
	SNode_47	2	8	**0.003**	1.33	0.007
	RApp3	3	16	**0.027**	2.598	0.013
	RApp4	2	8	**0.015**	1.581	0.01
	SNode_27	2	8	1.523	**0.006**	0.004
	GOAL_48	2	8	**0.006**	1.33	0.003
	GOAL_49	4	32	**0.008**	3.13	0.01
	GOAL_50	2	8	**0.006**	1.33	0.009
	SNode_51	3	16	**0.009**	2.36	0.009
	SNode_52	2	8	**0.004**	1.33	0.008
	GOAL_53	3	16	**0.01**	2.14	0.008
	SNode_28	3	16	**0.0104**	2.14	0.0105
	SNode_29	2	8	**0.006**	1.33	0.003
	SNode_31	2	8	**0.009**	1.45	0.008
	SNode_33	2	8	**0.009**	1.45	0.006
	SNode_34	2	8	**0.004**	1.33	0.007
	GOAL_56	3	16	**0.013**	2.14	0.009
	GOAL_57	2	8	**0.005**	1.33	0.003
	SNode_59	3	16	**0.006**	2.14	0.008
	SNode_60	2	8	**0.003**	1.33	0.006
	GOAL_61	3	16	**0.009**	2.14	0.007
	GOAL_62	2	8	**0.005**	1.33	0.006
	GOAL_63	3	16	**0.01**	2.14	0.008
	SNode_64	2	8	**0.007**	1.33	0.006

Having computed all nodes, we obtain the following results: for the uncertain MIN connector, 94% of the final distance is less than 0.1; for the uncertain MAX connector, only 6 results are below 0.1; for the uncertain WAVG connector most of the results have a distance less than 0.1 because it can replace the other connectors. Nevertheless, its computation time is high. The best results for the uncertain MIN connector were obtained by using the hill-climbing algorithm and the genetic algorithm. Nevertheless, the use of simulated annealing and tabu search can sometimes provide good results too. As for the uncertain MAX and uncertain WAVG connectors, all algorithms gave the same number of good results. The use of concurrent algorithms for the estimation has improved the final result.

5 Conclusion

In this paper, we have proposed a solution to convert the CPTs of an existing possibilistic network into uncertain gates. Uncertain gates provide a set of connectors with behaviours from severe to indulgent through the family of compromise (median, average, weighted average,...). Uncertain gates provide meaningful knowledge about how information is combined to compute the state of a variable. To convert the CPTs, we have used and compared optimization algorithms such as hill-climbing, simulated annealing, tabu search and the genetic algorithm. The goal was to find the closest CPT generated by using the connectors. The best result is provided by the connector which matches the best the initial CPT. To validate this approach, we have generated test CPTs with the behaviours MIN, MAX and weighted average and performed the estimation. The results correspond to our expectations. We have also proposed to test our solution on real datasets. To do this, we have converted Bayesian networks into possibilistic networks and performed the estimation of the connectors to select the most appropriate one for each table. For our future works we would like to improve the computation of the connectors by proposing a parallel algorithm. We would like to better analyze the problem of variables with different intensity scales. Also, it can be interesting to evaluate the effects on the decision of converting a CPT into uncertain gates.

References

1. Benferhat, S., Dubois, D., Garcia, L., Prade, H.: Possibilistic logic bases and possibilistic graphs. In: Proceedings of the Fifteenth Conference on Uncertainty in Artificial Intelligence, pp. 57–64, Morgan Kaufmann, Stockholm (1999)
2. Borgelt, C., Gebhardt, J., Kruse, R.: Possibilistic graphical models. In: Della Riccia, G., Kruse, R., Lenz, H.-J. (eds.) Computational Intelligence in Data Mining. ICMS, vol. 408, pp. 51–67. Springer, Vienna (2000). https://doi.org/10.1007/978-3-7091-2588-5_3
3. Conati, C., Gertner, A.S., VanLehn, K., Druzdzel, M.J.: On-line student modeling for coached problem solving using Bayesian networks. In: Jameson, A., Paris, C., Tasso, C. (eds.) User Modeling. ICMS, vol. 383, pp. 231–242. Springer, Vienna (1997). https://doi.org/10.1007/978-3-7091-2670-7_24
4. Dìez, F., Drudzel, M.: Canonical probabilistic models for knowledge engineering. Technical Report, CISIAD-06-01 (2007)
5. Dubois, D., Fusco, G., Prade, H., Tettamanzi, A.: Uncertain logical gates in possibilistic networks. an application to human geography. In: Beierle, C., Dekhtyar, A. (eds.) SUM 2015. LNCS (LNAI), vol. 9310, pp. 249–263. Springer, Cham (2015). https://doi.org/10.1007/978-3-319-23540-0_17

6. Dubois, D., Prade, H.: Possibility Theory: An Approach to Computerized Processing of Uncertainty. Plenum Press, New York (1988)
7. Dubois, D., Foulloy, L., Mauris, G., Prade, H.: Probability-possibility transformations, triangular fuzzy sets, and probabilistic inequalities. Reliable Comput. **10**, 273–297 (2004)
8. Lee, L.: On the effectiveness of the skew divergence for statistical language analysis. In: Proceedings of Artificial Intelligence and Statistics (AISTATS), pp. 65–72 (2001)
9. Heckerman, D., Breese, J.: A new look at causal independence. In: Proceedings of the Tenth Annual Conference on Uncertainty in Artificial Intelligence (UAI-94), pp. 286–292, Morgan Kaufmann, San Francisco (1994)
10. Glover, F.: Future paths for integer programming and links to artificial intelligence. Comput. Oper. Res. **13**(5), 533–549 (1986)
11. Glover, F.: Tabu Search-Part I. ORSA J. Comput. **1**(3), 190–206 (1989)
12. Glover, F.: Tabu Search-Part II. ORSA J. Comput. **2**(1), 4–32 (1990)
13. Petiot, G.: The calculation of educational indicators by uncertain gates. In: Proceedings of the 20th International Conference on Enterprise Information Systems, pp. 379–386, SCITEPRESS, Funchal, Madeira (2018)
14. Petiot, G.: Merging information using uncertain gates: an application to educational indicators. In: Medina, J., et al. (eds.) IPMU 2018. CCIS, vol. 853, pp. 183–194. Springer, Cham (2018). https://doi.org/10.1007/978-3-319-91473-2_16
15. Korb, K.B., Nicholson, A.E.: Bayesian Artificial Intelligence, 2nd edn. CRC Press, Boca Raton (2010). Section 2.2.2
16. Lauritzen, S.L., Spiegelhalter, D.J.: Local computations with probabilities on graphical structures and their application to expert systems. J. R. Stat. Soc.: Series B (Methodol.) **50**, 157–194 (1988)
17. Neapolitan, R.E.: Probabilistic Reasoning in Expert Systems: Theory and Algorithms. Wiley, New-York (1990)
18. Metropolis, N., Rosenbluth, A.W., Rosenbluth, M.N., Teller, A.H., Teller, E.: Equation of state calculations by fast computing machines. J. Chem. Phys. **21**(6), 1087–1092 (1953)
19. Scutari, M., Denis, J.-B.: Bayesian Networks: with Examples in R. Chapman & Hall (2014)
20. Pearl, J.: A new look at causal independence. In: Probabilistic Reasoning in Intelligent Systems: Networks of Plausible Inference. Morgan Kaufmann, San Mateo (1988)
21. Zadeh, L.A.: Fuzzy sets as a basis for a theory of possibility. Fuzzy Sets Syst. **1**, 3–28 (1978)
22. Yager, R.R.: On ordered weighted averaging aggregation operators in multicriteria decisionmaking. IEEE Trans. Syst. Man Cybern. **18**(1), 183–190 (1993)
23. Sachs, K.: Causal protein-signaling networks derived from multiparameter single-cell data. Science **309**, 1187–1187 (2005)
24. Zagorecki, A., Druzdzel, M.: Knowledge engineering for Bayesian networks: how common are noisy-MAX distributions in practice? IEEE Trans. Syst. Man. Cybern.: Syst. **43**, 186–195 (2013)
25. Zagorecki, A., Druzdzel, M.: Probabilistic independence of causal influences. In: Proceedings of the Third European Workshop on Probabilistic Graphical Models (PGM-2006), pp. 325–332 (2006)

Depth-Bounded Approximations of Probability

Paolo Baldi[(✉)], Marcello D'Agostino, and Hykel Hosni

Department of Philosophy, University of Milan, Milan, Italy
{paolo.baldi,marcello.dagostino,hykel.hosni}@unimi.it

Abstract. We introduce measures of uncertainty that are based on *Depth-Bounded Logics* [4] and resemble belief functions. We show that our measures can be seen as approximation of classical probability measures over classical logic, and that a variant of the PSAT [10] problem for them is solvable in polynomial time.

1 Introduction

In this work, we investigate the relation between belief functions (see the original [13], and, for a more recent survey, [5]) and probability, from a new logical perspective. Expanding on ideas first introduced in [1], we investigate measures of uncertainty which resemble Dempster-Shafer Belief Functions, but that, instead of being based on classical logic, are based on *Depth-Bounded Logics* (DB logics), a family of propositional logics approximating classical logic [4].

Our starting point is the observation that Belief Functions and Depth-Bounded Logics share a similar concern for the way *virtual* and *actual* information possessed by an agent is evaluated and manipulated.

Let us recall that belief functions can be uniquely determined from so-called mass functions (see e.g. [11]), i.e. probability distributions over the power sets of classical propositional evaluations. If such mass functions are non-zero only for singletons of evaluations, one obtains probability functions, as special cases. We will look at the mass functions behind the probability measures, as arising from the general mass functions (determining arbitrary belief functions) via a limiting process: agents originally assign masses to arbitrary sets of evaluations, reflecting their *actual information*, and they stepwise distribute such mass, only when requested to do so, by way of weighting additional *virtual information*, until they will have their say on the specific uncertainty associated with each single evaluation.

A related issue has been investigated in logic, where the family of DB logics [3,4] relies on the idea of separating two kinds of (classically valid) inferences: the inferences which only serve the purpose to make explicit the information

This research was funded by the Department of Philosophy "Piero Martinetti" of the University of Milan under the Project "Departments of Excellence 2018–2022" awarded by the Ministry of Education, University and Research (MIUR).

M.-J. Lesot et al. (Eds.): IPMU 2020, CCIS 1239, pp. 607–621, 2020.
https://doi.org/10.1007/978-3-030-50153-2_45

that agents already possess, i.e. those using only their actual information on the one hand, and those which make use of virtual information on the other. The latter type of inferences arises from the use of a single branching rule (see Fig. 1), reflecting the *principle of bivalence*, which allow agents to reason by cases, adding information not actually in their possession, and drawing further inferences thereon (see Sect. 2). The family of depth-bounded logics is then defined just by fixing maximal depths at which the application of such branching rule is allowed. Unbounded use of the rule results in (an alternative presentation of) classical logic, which can be thus seen as a limit of such family of weaker DB logics.

As an important consequence of the bounded use of the bivalence principle, it is shown in [4] that the consequence relation determined by each DB logic is decidable in polynomial time, hence we can realistically expect that (boundedly) rational agents would be able to recognize, in practice and not only in principle, whether a depth-bounded inference is actually correct. This contrasts with classical logic, which can be seen as the, computationally unfeasible, limit of the feasible DB logics.

The main contribution of this paper is twofold: first, we show that the measures of belief that we introduce, based on DB logics, provide approximations of classical probability measures over classical logic. Second, we prove that under certain reasonable conditions, the problem of finding whether there is any such measure satisfying a given set of linear constraints is solvable in polynomial time, in contrast with the analogous problem for classical logic and probability.

The rest of the paper is structured as follows. In Sect. 2 we recall some preliminaries about DB logics. In Sect. 3 we introduce our depth-bounded measure of uncertainty, based on DB logics, and in Sect. 4 we investigate computational issues. Section 5 contains conclusions and hints at future work.

2 Preliminaries

Let us fix a language $\mathcal{L}$, over a finite set $Var_{\mathcal{L}} = \{p_1, \ldots, p_n\}$ of propositional variables. We let $Fm_{\mathcal{L}}$ be the formulas built from the propositional variables by the usual classical connectives $\wedge, \vee, \neg$, and a constant $\curlywedge$ denoting contradiction. For each $p_i \in Var$ we denote by $\pm p_i$ any of the literals p_i and $\neg p_i$. For each set of formulas Γ we denote by $Sf(\Gamma)$ the subformulas of the formulas in Γ, and by $Var(\Gamma)$ the propositional variables occurring in Γ. Finally by $At_{\mathcal{L}} = \{\pm p_1 \wedge \pm p_2 \wedge \cdots \wedge \pm p_n \mid p_i \in Var_{\mathcal{L}}\}$ we denote the *atoms*, i.e. all the conjunctions of literals, formed from choosing (under the given order) exactly one literal for each of the (finitely many) variables of the language.

Let us now move to consider the family of DB Logics. We start from the 0-depth logic, that is the logic manipulating only actual information. Here we will limit ourselves to a proof-theoretic presentation, based on the Intelim (introduction and elimination) rules in Table 1. For a semantic characterization, see the nondeterministic truth tables, e.g. in [3,4].

The intelim rules determine a notion of 0-depth consequence relation $\vdash_0$, in the usual way.

Table 1. Introduction and elimination rules

$$\frac{\varphi \quad \psi}{\varphi \wedge \psi}\,(\wedge\mathcal{I}) \qquad \frac{\neg\varphi}{\neg(\varphi \wedge \psi)}\,(\neg\wedge\mathcal{I}1) \qquad \frac{\neg\psi}{\neg(\varphi \wedge \psi)}\,(\neg\wedge\mathcal{I}2)$$

$$\frac{\neg\varphi \quad \neg\psi}{\neg(\varphi \vee \psi)}\,(\neg\vee\mathcal{I}) \qquad \frac{\varphi}{\varphi \vee \psi}\,(\vee\mathcal{I}1) \qquad \frac{\psi}{\varphi \vee \psi}\,(\vee\mathcal{I}2)$$

$$\frac{\varphi \quad \neg\varphi}{\curlywedge}\,(\curlywedge\mathcal{I}) \qquad \frac{\varphi}{\neg\neg\varphi}\,(\neg\neg\mathcal{I})$$

$$\frac{\varphi \vee \psi \quad \neg\varphi}{\psi}\,(\vee\mathcal{E}1) \qquad \frac{\varphi \vee \psi \quad \neg\psi}{\varphi}\,(\vee\mathcal{E}2) \qquad \frac{\neg(\varphi \vee \psi)}{\neg\varphi}\,(\neg\vee\mathcal{E}1)$$

$$\frac{\neg(\varphi \vee \psi)}{\neg\psi}\,(\neg\vee\mathcal{E}2) \qquad \frac{\varphi \wedge \psi}{\varphi}\,(\wedge\mathcal{E}1) \qquad \frac{\varphi \wedge \psi}{\psi}\,(\wedge\mathcal{E}2)$$

$$\frac{\neg(\varphi \wedge \psi) \quad \varphi}{\neg\psi}\,(\neg\wedge\mathcal{E}1) \qquad \frac{\neg(\varphi \wedge \psi) \quad \psi}{\neg\varphi}\,(\neg\wedge\mathcal{E}2)$$

$$\frac{\neg\neg\varphi}{\varphi}\,(\neg\neg\mathcal{E}) \qquad \frac{\curlywedge}{\varphi}\,(\curlywedge\mathcal{E})$$

Definition 1. *For any set of formulas $\Gamma \cup \{\alpha\} \subseteq Fm_{\mathcal{L}}$, we let $\Gamma \vdash_0 \alpha$ iff there is a sequence of formulas $\alpha_1, \ldots, \alpha_m$ such that $\alpha_m = \alpha$ and each formula α_i is either in Γ or obtained by an application of the rules in Table 1 on the formulas α_j with $j < i$.*

The key feature of the consequence relation $\vdash_0$ is that only information actually possessed by an agent is allowed in a "0-depth deduction".

As already recalled in the introduction, the DB Logics for $k > 0$, will be defined via the amount k of virtual information which agents are allowed to use in their deductions. This leads to the recursive definition of the consequence relation $\vdash_k$, for $k > 0$, as follows.

Definition 2. *For each $k > 0$ and set of formulas $\Gamma \cup \{\alpha\} \subseteq Fm_{\mathcal{L}}$, we let $\Gamma \vdash_k \alpha$ iff there is a $\beta \in Sf(\Gamma \cup \{\alpha\})$ such that $\Gamma, \beta \vdash_{k-1} \alpha$ and $\Gamma, \neg\beta \vdash_{k-1} \alpha$.*

In other words, we suppose that β and $\neg\beta$ are pieces of "virtual information" which is not actually possessed by the agent, but which is used to derive α through case-based reasoning. While, according to Definition 1, the consequence $\vdash_0$ amounts to the existence of a suitable *sequence* of formulas, the derivability relation $\vdash_k$ amounts to the existence of a suitable *proof-tree*, where each node is labeled by a formula, which is either an assumption or obtained by formulas

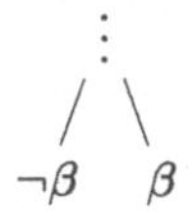

Fig. 1. The branching rule PB (principle of bivalence)

above it by means of an intelim rule, or of the branching rule (PB) in Fig. 1. The latter is then only allowed in a limited form: for $\vdash_k$ we are allowed at most k nested applications of (PB). Thus, $\Gamma \vdash_k \varphi$ can be equivalently taken to say that there is a proof-tree, as described above, so that φ is derivable from Γ in each branch, via the intelim rules, plus the additional virtual information introduced by the branching rules. One may run a proof-search procedure (see e.g. the algorithm in [4]), to verify whether such a proof-tree, deriving φ from Γ in each branch, exists. Even if this is not the case, i.e. if the proof-search procedure only produces proof-trees which derive φ in some (possibly none) and not all of the branches, we are still interested in the structure of such trees, in particular since they keep track of the virtual information that has been explored. This is the main inspiration behind our investigation of depth-bounded belief in the next section.

Before that, let us finish this section recalling two important properties of the DB logics, already mentioned in the introduction, and shown e.g. in [3,4]. First, DB logics provide a hierarchy of consequence relations approximating the classical one, that is, $\vdash_k \subseteq \vdash_{k+1}$ and $\lim_{k\to\infty} \vdash_k = \vdash$, where $\vdash$ stands for classical derivability. Finally, each $\vdash_k$ can be decided in polynomial time, and is thus feasible. This will be of particular use in Sect. 4 of the paper.

3 Depth-Bounded Proofs and Uncertain Reasoning

So far, we have recalled the definition of DB logics and given an idea of how proofs in such logics work, by distinguishing the use of actual and virtual information. Let us now assume that agents, whenever they add a piece of virtual information to their stock of assumptions, can also *weight* their belief on it, for extra-logical reasons. We will then take the belief that an agent commits to a formula φ to be the sum of all the weights assigned to the leaves of a depth-bounded proof-tree, that allow to derive φ. In particular, we request that, if all branches derive φ, which corresponds to φ being logically derivable, one would then obtain a degree of belief 1. We use these ideas as a bridge, from the realm of depth-bounded logic to that of depth-bounded *uncertain reasoning*. Let us recall that classical belief functions can be determined from mass functions, and that, when such mass functions are non-zero only for singletons, one obtain classical probabilities. Identifying formulas and sets of evaluations, one can reformulate this syntactically, by taking the mass functions behind belief functions to act over $Fm_{\mathcal{L}}$, and assume that those behind probabilities are non-zero only over $At_{\mathcal{L}}$ [11]. Our starting point towards depth-bounded uncertain reasoning, is to consider mass functions which are non-zero only over those formulas which keep track of the information (virtual and actual), used in each branch of a proof-search tree in DB logic. Before delving into our formal definition of depth-bounded belief functions, we will need to fix first various parameters.

- First, we will have a set $In \subseteq Fm_{\mathcal{L}} \cup \{*\}$, the *initial information*, which we assume to be finite. *In* stands for the formulas, for which an agent, for some extra-logical reasons, can assign a degree of belief, already at a shallow (0-depth) level. We can think of the values of such formulas as obtained from the available data, e.g. as information of statistical nature.
  In order to simplify the notation, we assume that *In* is nonempty, and we represent the case where no information at all is initially available by the symbol[1] $*$, which is not part of the language, and letting $In = \{*\}$. We adopt the convention that $* \vdash_k \varphi$ stands for $\vdash_k \varphi$.
  In our setting, we can consider a belief conditioned on a formula γ, by just assuming that, for each $\alpha \in In$, we have $\alpha \vdash_0 \gamma$. When this is the case, we say that *In* is γ-*based* and denote it by In_γ. Similarly we denote the case where $In = \{*\}$ by In_*.
- We have then a set Π, standing for the predictions that an agent wants to obtain, and that thus guide the weighting of her degree of belief. The idea is that an agent weights the uncertainty of virtual information and explores various possible scenarios, only in order to settle, eventually, the truth or falsity of all the formulas in Π.
- Finally, we have a set of virtual information V. This can be thought of as the set of questions that the agent may evaluate, in the process of assessing the formulas in Π. Typical example might be $V = Var(\Pi)$ or $V = Sf(\Pi)$.

Let us recapitulate our setting: starting from initial knowledge in *In*, agents ask themselves a number of questions about the formulas in V, thus specifying in more details the possible information states, which will be then used to settle the belief and make predictions about the formulas in Π.

We assume that the amount of questions the agents can ask themselves is bounded: the maximum number of questions an agent can ask corresponds, in a sense to be made precise later, to the depth of derivations in DB logic.

We are now ready to give our first formal definition of 0-depth mass functions, representing the initial evidence possessed by an agent. This is nothing else than a convex distribution over the set *In* of the initial information.

Definition 3. *A* 0*-depth mass function is a function* $m_0 \colon In \to [0,1]$ *such that* $\sum_{\alpha \in In} m_0(\alpha) = 1$ *and* $m_0(\alpha) = 0$ *if* $\alpha \vdash_0 \curlywedge$.

Note that, in case $In = \{\gamma\}$, we have $m_0(\gamma) = 1$ and $m_0(\alpha) = 0$ for any other formula α.

---

[1] We slightly depart from the notation in [4], where the state of no information is denoted by $\bot$, since the latter is often used as a constant for *falsum* in intuitionistic and various nonclassical logics.

Definition 4. *Given a 0-depth mass-function m_0, a 0-depth belief function is a function $B_0\colon Fm_{\mathcal{L}} \to [0,1]$ such that*

$$B_0(\varphi) = \sum_{\substack{\alpha \in In \\ \alpha \vdash_0 \varphi}} m_0(\alpha) \qquad B_0(\varphi) = 0 \text{ if for no } \alpha \in In, \alpha \vdash_0 \varphi.$$

If In is of the form In_γ for some $\gamma \in Fm_{\mathcal{L}}$, we will then have $B_0(\varphi) = 1$ if $\gamma \vdash_0 \varphi$ and $B_0(\varphi) = 0$ otherwise.

Remark 1. As in the case of classical belief and mass functions, $m_0(\varphi)$ represents a portion of belief committed *exclusively to* φ and to no other formula, while $B_0(\varphi)$ stands for the belief in φ, which is obtained by putting together all the basic pieces of belief leading to (i.e. 0-depth deriving) φ. Note that, while in principle 0-depth equivalent formulas can be assigned different values via a 0-depth mass, they will still be assigned the same 0-depth belief. Hence, in our framework, masses cannot be uniquely determined by belief functions. This is due to fact that we assign masses to formulas, rather than to equivalence classes of the corresponding Lindenbaum-Tarski algebra (see e.g. [11]).

The notions of 0-depth mass and 0-depth belief function encode the shallow information, which is provided to an agent. We will now introduce mass functions based on higher DB logics, corresponding to the setting where agents have both higher inferential and "imaginative" power, i.e. when they can weight the uncertainty of pieces of information going beyond what is originally given.

Let us fix a triplet $\mathcal{G} = \langle In, \Pi, V\rangle$ where $In \subseteq Fm_{\mathcal{L}} \cup \{*\}$, $\Pi \subseteq Fm_{\mathcal{L}}$, $V \subseteq Fm_{\mathcal{L}}$, with the intended meaning discussed above. We will represent the information evaluated by an agent, in the form of forests, with labels provided via the triplet $\mathcal{G}$. Let us recall that by a forest we just mean a disjoint union of trees, in graph-theoretic terms.

Definition 5. *Let F be a binary forest. A $\mathcal{G}$-label for F is a labeling of nodes of F into formulas in $Fm_{\mathcal{L}}$ such that:*

- *When restricted to the roots of the trees in F, the labeling is a bijection with the formulas in In.*
- *For each node labeled by α, the children nodes are labeled by $\alpha \wedge \beta$ and $\alpha \wedge \neg\beta$ for some $\beta \in V$.*

Before proceeding, we also need the following technical definition.

Definition 6. *Let F be any $\mathcal{G}$-labeled forest*

- *We say that a formula γ* k-decides *δ if $\gamma \vdash_k \delta$ or $\gamma \vdash_k \neg\delta$.*
- *We let $Lf(F)$ be the set of formulas that label the leaves of F.*
- *We say that a leaf labeled by α is* Π-closed *if $\alpha \vdash_0 \curlywedge$ or α k-decides δ, for each $\delta \in \Pi$. A leaf which is not Π-closed is said to be* open.

We will build now a set of $\mathcal{G}$- labeled forests of a given maximal depth. Each open node is expanded by two new children nodes, representing the addition of a certain piece of virtual information and its negation.

Definition 7. *Let $\mathcal{G} = \langle In, \Pi, V \rangle$. We define recursively the set of $\mathcal{G}$-labeled forests F_k of depth k, for any $k \in \mathbb{N}$, as follows :*

- *For $k = 0$ we let F_0 be a set of nodes with no edges, each labeled by a distinct formula in In. Clearly $Lf(F_0) = In$.*
- *The set F_k, for $k \geq 1$ is the set of all $\mathcal{G}$-labeled forests obtained as follows:*
  - *Pick a $\beta \in V$ and, for each $\mathcal{G}$-labeled forest $F' \in F_{k-1}$, expand each Π-open leaf labeled by α in F' with two nodes labeled by $\alpha \wedge \beta$ and $\alpha \wedge \neg\beta$.*
  - (MAX) *Among the resulting forests, add to F_k only those forests F such that the number of formulas in $Lf(F)$, which 0-depth derive $\pm\varphi$, for each $\varphi \in \Pi$, is maximal[2].*

In the following, for each $F \in F_k$ we call the forest $F' \in F_{k-1}$ from which it was obtained, via the construction above, the predecessor *of F.*

Definition 8. *Let F_k be the set of $\mathcal{G}$-labeled forests of depth k. For each forest $F \in F_k$, we let $m_k^F : Lf(F) \to [0,1]$ be any function such that:*

(i) $m_k^F(\gamma \wedge \alpha) + m_k^F(\gamma \wedge \neg\alpha) = m_{k-1}^{F'}(\gamma)$ *where $F' \in F_{k-1}$ is the predecessor of F, $\gamma \in Lf(F')$ and γ labels the parent node in F of $\gamma \wedge \alpha$ and $\gamma \wedge \neg\alpha$.*
(ii) $m_k^F(\gamma) = m_{k-1}^{F'}(\gamma)$ *if $F' \in F_{k-1}$ is the predecessor of F and $\gamma \in Lf(F') \cap Lf(F)$.*
(iii) $m_k^F(\gamma) = m_k^G(\delta)$ *for each $F, G \in F_k$, $\gamma \in Lf(F), \delta \in Lf(G)$ such that $\gamma \vdash_0 \delta$ and $\delta \vdash_0 \gamma$.*

Recalling Definition 3 and condition (i) in Definition 8, it is easy to see that, for each $F \in F_k$

$$\sum_{\alpha \in Lf(F)} m_k^F(\alpha) = 1$$

Each m_k^F is thus a mass functions, in the sense of Shafer's belief function [13], which is non-zero only over the leaves of the trees in F.

Definition 9. *Let $\mathcal{G} = \langle In, \Pi, V \rangle$ and F_k be a $\mathcal{G}$-labeled set of forests. For each $F \in F_k$, we define the F-*based *k-depth belief function B_k^F and the k-depth plausibility function Pl_k^F as follows:*

$$B_k^F(\varphi) = \sum_{\substack{\alpha \in Lf(F) \\ \alpha \vdash_0 \varphi}} m_k^F(\alpha) \qquad Pl_k^F(\varphi) = \sum_{\substack{\alpha \in Lf(F) \\ \alpha \nvdash_0 \neg\varphi}} m_k^F(\alpha)$$

[2] This condition might not seem intuitive, but actually plays an important conceptual role, given the motivations of our model. While we want to depart from unrealistic assumptions behind both classical inferences and probability, we still want our models to be *prescriptive*, rather than purely descriptive. In other words, we want to model how agents *should* weight their uncertainty, given their limited inferential ability. Therefore, even if it could be the case that agents use the *wrong* piece of virtual information (i.e. failing the condition (MAX)) we limit ourselves to the case where they only use the virtual information actually leading them to settle as many of their questions as possible, within their inferential abilities.

Finally, we define the $\mathcal{G}$-based k-depth belief *as a function* B_k *from the formulas in* $Fm_{\mathcal{L}}$ *to the interval subsets of* $[0,1]$, *associating to any formula* φ *the following interval:*

$$B_k(\varphi) = [\min_{F \in F_k} B_k^F(\varphi), \max_{F \in F_k} B_k^F(\varphi)]$$

Remark 2. It is immediate to see that $Pl_k^F(\varphi) = 1 - B_k^F(\neg\varphi)$, hence, for each forest F we can think that an exact measure of uncertainty of the formula φ lies within the interval $[B_k^F(\varphi), Pl_k^F(\varphi)]$. Any such interval is related with a single forest F. This should not be confused with the interval given by $[\min_{F \in F_k} B_k^F(\varphi), \max_{F \in F_k} B_k^F(\varphi)]$ which arises from considering various k-depth belief functions over different forests, that is, various proof-search strategies, involving different pieces of virtual information.

We now show some properties of our construction, which highlight its connection with belief functions on the one hand, and with DB logics on the other.

Proposition 1. *(a) Assume* $\mathcal{G} = (In_\gamma, \{\varphi\}, Sf(In \cup \{\varphi\}))$, *for some* $\gamma \in Fm_{\mathcal{L}} \cup \{*\}$, $\varphi \in Fm_{\mathcal{L}}$, *and let* F_k *be the set of* $\mathcal{G}$*-labeled* k*-depth forests. If* $\gamma \vdash_k \varphi$, *then for all the* $F \in F_k$, *we have* $B_k^F(\varphi) = 1$.

(b) Assume $\mathcal{G} = (In_\gamma, \{\varphi\}, Sf(In \cup \{\varphi\})$ *for some* $\gamma \in Fm_{\mathcal{L}} \cup \{*\}$,$\varphi \in Fm_{\mathcal{L}}$, *and let* F_k *be the set of* $\mathcal{G}$*-labeled* k*-depth forests. If* $\gamma \vdash_k \neg\varphi$, *then for all* $F \in F_k$, *we have* $B_k^F(\varphi) = 0$.

(c) Assume $\mathcal{G} = (In, \{\varphi, \psi\}, Sf(In \cup \{\varphi, \psi\}))$ *for some* $\varphi, \psi \in Fm_{\mathcal{L}}$, *and let* F_k *be the set of* $\mathcal{G}$*-labeled* k*-depth forests. If* $\varphi \vdash_k \psi$, *we have:*
- *There is an* $F \in F_k$ *such that* $B_k^F(\varphi) \leq B_k^F(\psi)$
- *There is an* $l \geq k$ *such that, for any forest* $F \in F_l$, *we get* $B_l^F(\varphi) \leq B_l^F(\psi)$.

(d) Assume $\mathcal{G} = (In, \Pi, V)$, *and let* F_k *be the set of* $\mathcal{G}$*-labeled* k*-depth forests. For each* $F \in F_k$,$\varphi_1, \ldots, \varphi_n \in Fm_{\mathcal{L}}$, *we have:*

$$B_k^F(\bigvee_{i=1}^{n} \varphi_i) \geq \sum_{\emptyset \neq S \subseteq 1,\ldots,n} (-1)^{|S|-1} B_k^F(\bigwedge_{i \in S} \varphi_i).$$

Proof. (a). Consider the forest G, obtained by attaching to any $\alpha \in In$, the tree containing the virtual information in a k-depth proof of φ from γ. Now, since $\alpha \vdash_0 \gamma$, and each $\beta \in Lf(G)$ contains all the virtual information in a k-depth proof of φ from γ, we will have that $\beta \vdash_0 \varphi$, for all $\beta \in Lf(G)$. By condition (MAX) since there is a forest, G, such that all its leaves derive φ, then all the forests $F \in F_k$ need to have the same property. Hence, we obtain that for each $F \in F_k$, $B_k^F(\varphi) = \sum_{\substack{\alpha \in Lf(F) \\ \alpha \vdash_0 \varphi}} m_k^F(\alpha) = 1$.

(b). By (a), for any forest F, we have $B_k^F(\neg\varphi) = 1$. This means that, for any $\alpha \in Lf(F)$ such that $m_k^F(\alpha) > 0$, $\alpha \vdash_0 \neg\varphi$. If $\alpha \vdash \varphi$ we would get $\alpha \vdash \curlywedge$, which by definition of m_k^F implies $m_k^F(\alpha) = 0$ in contradiction with our assumption. Hence, for any $\alpha \in Lf(F)$, $\alpha \not\vdash_0 \varphi$, and $B_k^F(\varphi) = 0$.

(c). The first claim holds, by just taking the forest F to be constituted of the virtual information used in a k-depth proof of ψ from φ. Then, any formula labeling a leaf 0-depth deriving φ, will derive ψ as well. For the second claim, take any forest $F \in F_k$, and consider all the leaves which are 0-depth deriving φ, but not deriving ψ. Expand such leaves with the virtual information contained in any k-depth proof of ψ from φ. This results in a forest of depth l for some $l \geq k$, where each leaf 0-depth deriving φ derives ψ as well. By the maximality condition (MAX) in Definition 7, all forests at depths l will have this property, since otherwise they cannot decide all the formulas in $\Pi = \{\varphi, \psi\}$.

(d). We straightforwardly adapt Theorem 4.1 in [11]. Pick a forest $F \in F_k$. First, for each $\alpha \in Lf(F)$ let $Ind(\alpha) = \{i \mid \alpha \vdash_k \varphi_i\}$. Note that one can show by induction on $|Ind(\alpha)|$, that if $Ind(\alpha) \neq \emptyset$, then $\sum_{\emptyset \neq S,\ S \subseteq Ind(\alpha)} (-1)^{|S|-1} = 1$. We thus get:

$$\begin{aligned}
B_k^F(\bigvee_{i=1}^{n} \varphi_i) &= \sum_{\substack{\alpha \in Lf(F) \\ \alpha \vdash_0 (\varphi_1 \vee \cdots \vee \varphi_n)}} m_k^F(\alpha) \\
&\geq \sum_{\substack{\alpha \in Lf(F) \\ \alpha \vdash_0 \varphi_1 \text{ or } \ldots \text{ or } \alpha \vdash_0 \varphi_n}} m_k^F(\alpha) \\
&= \sum_{Ind(\alpha) \neq \emptyset} m_k^F(\alpha) = \sum_{Ind(\alpha) \neq \emptyset} m_k^F(\alpha) \sum_{\emptyset \neq S \subseteq Ind(\alpha)} (-1)^{|S|-1} \\
&= \sum_{\emptyset \neq S \subseteq Ind(\alpha)} (-1)^{|S|-1} \sum_{\substack{\alpha \in Lf(F) \\ \alpha \vdash_0 \bigwedge_{i \in S} \varphi_i}} m_k^F(\alpha) \\
&= \sum_{\emptyset \neq S \subseteq \{1,\ldots,n\}} (-1)^{|S|-1} B_k(\bigwedge_{i \in S} \varphi_i).
\end{aligned}$$

Let us now discuss some examples.

Example 1. Let $\mathcal{G} = (\{*\}, \{\alpha \vee \beta\}, \{\alpha, \beta\})$.

At depth 0, we only have the tree with a single node labeled by $*$. We obtain $B_0(\alpha \vee \beta) = 0$ since $* \nvdash_0 \alpha \vee \beta$. At depth 1, our possible forests are actually just trees. Two trees satisfy the constraints in Definition 7, namely :

$$\overbrace{\alpha \quad \neg\alpha}^{*} \qquad\qquad \overbrace{\beta \quad \neg\beta}^{*}$$

Let us call the left tree above S and the right one T, and let $m_1^S(\alpha) = 0.5$ and $m_1^S(\neg\alpha) = 0.5$ while $m_1^T(\beta) = 0.4$ and $m_1^T(\neg\beta) = 0.6$. Applying Definition 9, we thus obtain $B_1^S(\alpha \vee \beta) = m_1^S(\alpha) = 0.5$ and $B_1^T(\alpha \vee \beta) = m_1^T(\beta) = 0.4$. Hence $B_1(\alpha \vee \beta) \in [B_1^T(\alpha \vee \beta), B_1^S(\alpha \vee \beta)] = [m_1^T(\beta), m_1^S(\alpha)] = [0.4, 0.5]$. Note that, on the other hand, $B_1(\alpha) \in [B_1^T(\alpha), B_1^S(\alpha)] = [m_1^T(\alpha), m_1^S(\alpha)] = [0, 0.5]$ and $B_1(\beta) \in [B_1^S(\beta), B_1^T(\beta)] = [m_0^S(\beta), m_0^T(\beta)] = [0, 0.4]$. Let us move now to depth

2. In S we only need to expand the node $\neg\alpha$, since α is $\{\alpha\vee\beta\}$-closed (it is sufficient to 0-depth derive $\alpha\vee\beta$). The same holds for $\neg\beta$ in the tree T. We get:

where for simplicity, we display only the piece of virtual information added by each node, rather than their actual labels (which can be read off by the conjunction of the formula displayed in the node with all its ancestors). Let us call the two trees above again S and T for simplicity. By condition (ii) in Definition 8, we have $m_2^S(\alpha) = m_1^S(\alpha) = 0.5$, and $m_2^T(\beta) = m_1^T(\beta) = 0.4$. For the remaining nodes of S, we let $m_2^S(\neg\alpha\wedge\beta) = 0.2$ and $m_2^S(\neg\alpha\wedge\neg\beta) = 0.3$, and for T, we let $m_2^T(\neg\beta\wedge\alpha) = 0.3$ and $m_2^T(\neg\beta\wedge\neg\alpha) = 0.3$. We obtain finally $B_2^T(\alpha\vee\beta) = m_2^T(\beta) + m_2^T(\neg\beta\wedge\alpha)$ and $B_2^S(\alpha\vee\beta) = (m_2^S(\alpha) + m_2^S(\neg\alpha\wedge\beta)]$, hence

$$B_2(\alpha\vee\beta) \in [B_2^T(\alpha\vee\beta), B_2^S(\alpha\vee\beta)] = [0.7, 0.7]$$

Remark 3. The example above can be generalized considering, for any $n \geq 2$, the triplet $\mathcal{G} = (\{*\}, \{\varphi_1\vee\cdots\vee\varphi_n\}, \{\varphi_1,\ldots,\varphi_n\})$. The corresponding $\mathcal{G}$- based belief function determines, for each $F \in F_n$ a corresponding permutation σ such that:

$$\begin{aligned} B_n^F(\varphi_1\vee\cdots\vee\varphi_n) = B_1^F(\varphi_{\sigma(1)}) + B_2^F(\neg\varphi_{\sigma(1)}\wedge\varphi_{\sigma(2)}) + \ldots \\ + B_n^F(\neg\varphi_{\sigma(1)}\wedge\cdots\wedge\neg\varphi_{\sigma(n-1)}\wedge\varphi_{\sigma(n)}) \end{aligned}$$

Example 2. Let us now consider the famous example of the Ellsberg urn [6]. We assume to have a language with propositional variables {Y,R,B} which stand for the proposition *the next extracted ball is Yellow— Red— Blue*, respectively. The initial knowledge is that 2/3 of the balls are either yellow or red and 1/3 are blue. The background theory is given by the conjunction γ of the formulas in the set

$$\{Y\rightarrow(\neg B\wedge\neg R), R\rightarrow(\neg B\wedge\neg Y), B\rightarrow(\neg R\wedge\neg Y)\}$$

which encode the information that any extracted ball has exactly one of the colors Y, B, R. We now consider the $\mathcal{G}$- based k-depth belief, with $\mathcal{G} = \langle In_\gamma, \Pi, V\rangle$, where

$$In_\gamma = \{(Y\vee R)\wedge\gamma, B\wedge\gamma\} \quad \Pi = \{Y, R, B\} \quad V = \{Y, R\}$$

We formalize the factual information about the proportion of the balls, together with the background theory, by letting: $m_0((Y\vee R)\wedge\gamma) = 2/3, m_0(B\wedge\gamma) = 1/3$.

This implies $B_0(Y\vee R) = 2/3, B_0(B) = 1/3, B_0(\gamma) = 1, B_0(Y) = B_0(R) = 0$. At depth 1, we are required to make use of virtual information. One can easily check that, via the node labeled $B\wedge\gamma$, we can already prove $B\wedge\gamma\vdash_0\neg Y$, that $B\wedge\gamma\vdash_0\neg R$ and that $B\wedge\gamma\vdash_0 B$. The node is thus Π-closed, and it should not be expanded. On the other hand, we will need to expand the node $(Y\vee R)\wedge\gamma$ with either the virtual information on Y or on R. We obtain thus a forest $F \in F_1$ of the form:

$$(Y \vee R) \wedge \gamma \qquad\qquad B \wedge \gamma$$

$$\widehat{Y \; \neg Y}$$

and a forest $G \in F_1$ of the form:

$$(Y \vee R) \wedge \gamma \qquad\qquad B \wedge \gamma$$

$$\widehat{R \; \neg R}$$

which have exactly the same structure. So at depth 1, the agent will assign $m_1^F(((Y \vee R) \wedge \gamma) \wedge Y)$ and $m_1^F(((Y \vee R) \wedge \gamma) \wedge \neg Y)$ such that their sum equals $m_0^F((Y \vee R) \wedge \gamma))$. Note that we easily obtain $((Y \vee R) \wedge \gamma) \wedge \neg Y \vdash_0 ((Y \vee R) \wedge \gamma) \wedge R$ and $((Y \vee R) \wedge \gamma) \wedge Y \vdash_0 ((Y \vee R) \wedge \gamma) \wedge \neg R$. The converse direction of the consequence holds as well, hence by condition (iii) in Definition 8, we have: $m_1^G(((Y \vee R) \wedge \gamma) \wedge R) = m_1^G(((Y \vee R) \wedge \gamma) \wedge \neg Y)$ and $m_1^G(((Y \vee R) \wedge \gamma) \wedge \neg R) = m_1^G(((Y \vee R) \wedge \gamma) \wedge Y)$.

At depth 1, considering that the information about the colors is completely symmetric, a natural assumption is now to adopt a uniform distribution, i.e. $m_1((Y \vee R) \wedge Y)) = m_1((Y \vee R) \wedge \neg Y)) = 1/3$. This means that $B_1(Y) = B_1(R) = B_1(B) = 1/3$.

To conclude this section, we now show that we can see usual classical probability functions as arising from sequences of depth-bounded belief functions. By classical probability in our setting, we just mean finitely additive measures, defined as functions $P\colon Fm_{\mathcal{L}} \to [0,1]$.

Theorem 1. *Let $P\colon Fm_{\mathcal{L}} \to [0,1]$ be a classical probability function. Then there is a sequence of $\mathcal{G}$-labeled depth bounded belief functions such that, for each formula φ, we have $P(\varphi) = \lim_{k \to \infty} B_k(\varphi)$.*

Proof. Let $\mathcal{G} = (\{*\}, Fm_{\mathcal{L}}, Var_{\mathcal{L}})$. Recalling that $Var_{\mathcal{L}} = \{p_1, \ldots, p_n\}$, we obtain that, for each forest $F \in F_n$, the set $Lf(F)$ coincides with $At_{\mathcal{L}}$, up to permutations of the literals in each atom. Let us consider the mass function m_n over the set $Lf(F)$ such that $m_n(\alpha) = P(\alpha)$ for each $\alpha \subset Lf(F)$. Now, we obtain that for any $F \in F_n$

$$P(\varphi) = \sum_{\substack{\alpha \in At_{\mathcal{L}} \\ \alpha \vdash \varphi}} P(\alpha) = \sum_{\substack{\alpha \in At_{\mathcal{L}} \\ \alpha \vdash_0 \varphi}} P(\alpha) = \sum_{\substack{\alpha \in Lf(F) \\ \alpha \vdash_0 \varphi}} m_n(\alpha) = B_n^F(\varphi)$$

All forests in F_n will have the same leaves, modulo a permutations of the literals appearing in the conjunction. Hence, by condition (iii) in Definition 8, for any $F, G \in F_k$, $\alpha \in Lf(F)$ and $\sigma(\alpha) \in G$, where σ is a permutation of the literals in α, we need to have $m_n^G(\sigma(\alpha)) = m_n^F(\alpha) = P(\alpha)$. Hence, by Definition 9, we have $B_n^F(\varphi) = B_n^G(\varphi) = P(\varphi)$ for each $\varphi \in Fm_{\mathcal{L}}$. This implies that the interval for $B_n(\varphi)$ in Definition 9 reduces to the single value $P(\varphi)$. On the other hand, all the leaves in the forests in F_n are $Fm_{\mathcal{L}}$-closed, since atoms decide all the formulas in $Fm_{\mathcal{L}}$, hence $F_k = F_n$ for any $k \geq n$, and $B_k(\varphi) = B_n(\varphi) = P(\varphi)$ for any $k \geq n$. From this the main claim immediately follows.

4 Complexity of Depth-Bounded Belief

In this section we investigate the conditions under which our approach provides a feasible model of reasoning under uncertainty. For concepts in complexity theory, we refer the reader e.g. to [14]. Following previous works based on classical probability, e.g. [7,8,10,11], we assume that an agent is provided n linear constraints over her belief on the formulas $\varphi_1, \ldots, \varphi_m$, of the form:

$$\sum_{j=1}^{m} a_{ij} B(\varphi_j) = w_i \quad i = 1, \ldots, n \quad a_{ij}, w_i \in \mathbb{Q}. \tag{1}$$

Our setup suggests then the following decision problem, which stands to our k-depth logic and k-depth belief functions as the GENPSAT problem (see e.g. [2]) stands to classical logic and classical probability functions:

GEN-B$_0$-SAT Problem

INPUT: The set of m formulas and n linear constraints in (1).

PROBLEM: Is there a 0-depth belief function B_0 over $In = \{\varphi_1, \ldots, \varphi_m\}$ satisfying the n constraints in (1)?

Recalling Definition 9, the problem boils down to finding a solution for the following system of linear inequalities in the unknowns $m_0(\varphi_1), \ldots, m_0(\varphi_m)$.

$$\sum_{j=1}^{m} a_{ij} \sum_{\substack{k=1,\ldots,m \\ \varphi_k \vdash_0 \varphi_j}} m_0(\varphi_k) = w_i \qquad \text{for each } i = 1, \ldots, n$$

$$m_0(\varphi_j) \geq 0 \qquad \text{for each } j = 1, \ldots, m$$

$$\sum_{j=1}^{m} m_0(\varphi_j) = 1$$

$$m_0(\varphi_j) = 0 \qquad \text{if } \varphi_j \vdash_0 \curlywedge$$

Let us denote by *size*(*In*) the number of symbols occurring in the formulas in *In*, and by *inc*(*In*) the number of inconsistent formulas among those in *In*. We recall from [4] that both, finding out whether $\varphi_j \vdash_0 \varphi_i$, and whether $\varphi_j \vdash_0 \curlywedge$ requires time polynomial in *size*(*In*). On the other hand, the system above has size $((n + m + 1 + inc(In)) \times m)$, and finding a solution is polynomial as well. Hence the problem above turns out to be in $\mathbf{PTIME}(size(In) + n)$.

Let us now consider the problem of finding out whether there is a k-depth belief function, for a given $k > 0$, satisfying the constraints in (1). Recalling the Definition 9, this problem amounts to solving a linear system as the one above, where the set *In* is replaced by the set $Lf(F)$, for all the various $F \in F_k$. Recall that the latter are determined by the parameters Π and V, discussed in the previous section. Let us still set $In = \{\varphi_1, \ldots, \varphi_m\}$, which is given as input to the problem, as the information initially provided to an agent. For the remaining two parameters, we take $V = Var(In)$ and $\Pi = f(In)$, for some computable function f. We fix a $k > 0$ and consider then the following:

GEN–B$_k$-SAT Problem

INPUT: A triplet $\mathcal{G} = (\{\varphi_1, \ldots, \varphi_m\}, f(\{\varphi_1 \ldots \varphi_m\}), Var(In))$ and the n constraints in (1).

PROBLEM: Is there a $\mathcal{G}$-based k-depth belief function B_k^F, with F in the $\mathcal{G}$-labeled set of forests F_k, which satisfies the n constraints in (1)?
Answering to this problem corresponds to finding an F in F_k, for which the following system, in the unknowns $m_k^F(\alpha)$, has a solution:

$$\sum_{j=1}^{m} a_{ij} \sum_{\substack{\alpha \in Lf(F) \\ \alpha \vdash_0 \varphi_j}} m_k^F(\alpha) = w_i \qquad \text{for each } i = 1, \ldots, n$$

$$m_k^F(\alpha) \geq 0 \qquad \text{for each } \alpha \in Lf(F)$$

$$\sum_{\alpha \in Lf(F)} m_k^F(\alpha) = 1$$

$$m_k^F(\alpha) = 0 \qquad \text{if } \alpha \vdash_0 \curlywedge$$

This problem also turns out to be polynomial, if the size of Π is polynomially bounded. We give a sketch of proof in the following.

Theorem 2. GEN-B$_k$-SAT *can be decided in* **PTIME**(size(In) + n + size(Π))*.*

Proof. By our construction, for any forest $F \in F_k$ the number of leaves in $Lf(F)$ is bounded above by $|In| \cdot 2^k$, which is linear in $|In|$ (once k is fixed, 2^k is constant). The number of possible forests, on the other hand, is bounded by the number of subsets of $|Var(In)|$ of cardinality k, which is polynomial in $|Var(In)|$ and, consequently, polynomial in $size(In)$. Indeed, we can safely disregard any permutation or repetitions of the same virtual information, due to condition (iii) of Definition 8 and condition (MAX) of Definition 7, respectively. We then need to do some "pruning" among all possible forests, by discarding the branches which are not Π-open and the forest that do not satisfy the maximality condition (MAX) in Definition 7. The latter is obtained then, by running, whenever necessary, for each formula in Π and its negation, the polynomial time algorithm, e.g. in [4]. Once we have determined the set F_k, we have then, for each $F \in F_k$ a set of formulas in $Lf(F)$. We have then to look for a solution to the system above. Each such system has size $(n+|Lf(F)|+1+inc(Lf(F)) \times |Lf(F)|$, hence it is still polynomially bounded. Finally, since solving each system requires polynomial time, we obtain the claim.

Finally, let us notice that, if we take $Size(\Pi) = f(Size(In))$, where f is a polynomially bounded function, then the GEN-B$_k$-SAT Problem is in **PTIME**$(Size(In) + n)$. As an example, this would hold if we take, as a reasonable choice $\Pi = Sf(In)$.

5 Conclusions and Future Work

In this work we have introduced feasible approximations of probability measures, based on Depth-Bounded logics. The resulting measures shed light on the connection between two approximation problems: the approximation of probability, as a limiting case of belief functions and that of classical logic as a limiting case of depth-bounded boolean logic. In future research, we plan to compare our approach with the Transferable Belief Model of [15], and similar works, which handle the relation between belief functions and probability. While the former are considered in [15] to be adequate to model the *credal*, i.e. purely mental, aspect of belief, the latter are taken as good models for its *pignistic* aspect, i.e. its role as a guide towards decisions. Decision-theoretic models are also a natural setting to evaluate and deepen our results. In particular, in the context of subjective expected utility, various weakenings of Savage axioms [12] have been considered in the literature (see e.g. [9] for an overview). We plan to investigate how these works relate to our approach, which weakens instead the logic.

References

1. Baldi, P., Hosni, H.: Depth-bounded belief functions. Int. J. Approximate Reasoning (2020). https://doi.org/10.1016/j.ijar.2020.05.001
2. Caleiro, C., Casal, F., Mordido, A.: Generalized probabilistic satisfiability. Inf. Comput. **332**, 39–56 (2017)
3. D'Agostino, M.: An informational view of classical logic. Theor. Comput. Sci. **606**, 79–97 (2015)
4. D'Agostino, M., Finger, M., Gabbay, D.M.: Semantics and proof-theory of depth-bounded boolean logics. Theor. Comput. Sci. **480**, 43–68 (2013)
5. Denoeux, T.: 40 years of Dempster-Shafer Theory. Int. J. Approximate Reasoning **79**, 1–6 (2016)
6. Ellsberg, D.: Risk, Ambiguity, and the Savage Axioms. Quart. J. Econ. **75**(4), 643–669 (1961)
7. Fagin, R., Halpern, J.Y., Megiddo, N.: A logic for reasoning about probabilities. Inf. Comput. **1–2**(87), 78–128 (1990)
8. Finger, M., De Bona, G.: Probabilistic satisfiability: algorithms with the presence and absence of a phase transition. Ann. Math. Artif. Intell. **75**(3–4), 351–389 (2015)
9. Gilboa, I., Marinacci, M.: Ambiguity and the Bayesian Paradigm, pp. 385–439. Springer, Cham (2016)
10. Hansen, P., Jaumard, B.: Probabilistic satisfiability. In: Kohlas, J., Moral, S. (eds.) Handbook of Defeasible Reasoning and Uncertainty Management Systems. Handbook of Defeasible Reasoning and Uncertainty Management Systems, vol. 5, pp. 321–367. Springer, Kluwer, Dordrecht (2000). https://doi.org/10.1007/978-94-017-1737-3_8
11. Paris, J.B.: The Uncertain Reasoner's Companion: A Mathematical Perspective. Cambridge University Press, Cambridge (1994)
12. Savage, L.J.: The Foundations of Statistics. 2nd edn., Dover (1972)

13. Shafer, G.: A Mathematical Theory of Evidence. Princeton University Press, Princeton (1976)
14. Sipser, M.: Introduction to the Theory of Computation. PWS Publishing Company (1997)
15. Smets, P., Kruse, R.: The Transferable Belief Model for Belief Representation. In: Motro, A., Smets, P. (eds.) Uncertainty Management in Information Systems, pp. 343–368. Springer, Boston (1997). https://doi.org/10.1007/978-1-4615-6245-0_12

Unification in Łukasiewicz Logic with a Finite Number of Variables

Marco Abbadini[1], Federica Di Stefano[2], and Luca Spada[2](✉)

[1] Dipartimento di Matematica "Federigo Enriques", Università degli Studi di Milano, via Cesare Saldini 50, 20133 Milan, Italy
marco.abbadini@unimi.it

[2] Dipartimento di Matematica, Università degli Studi di Salerno, Piazza Renato Caccioppoli, 2, 84084 Fisciano, SA, Italy
f.distefano3@studenti.unisa.it, lspada@unisa.it

Abstract. We prove that the unification type of Łukasiewicz logic with a finite number of variables is either infinitary or nullary. To achieve this result we use Ghilardi's categorical characterisation of unification types in terms of projective objects, the categorical duality between finitely presented MV-algebras and rational polyhedra, and a homotopy-theoretic argument.

Keywords: Łukasiewicz logic · MV-algebras · Unification · Universal cover

1 Introduction

The classical, syntactic unification problem is: given two terms s, t in a purely functional language, find a uniform replacement of the variables occurring in s and t by other terms that makes s and t identical. The substitution is then called *unifier*. When the latter syntactical identity is replaced by equality modulo a given equational theory E, one speaks of *E-unification*. The study of unification modulo an equational theory has acquired increasing significance in recent years (see e.g. [2,3]). The most basic piece of information one would like to have about E in connection with unification issues is its *unification type*. In order to define it precisely, let us recall some standard notions.

We consider a set $\mathscr{F}$ of function symbols along with a further set $\mathscr{V} = \{X_1, X_2, \ldots\}$ of *variables*. We then let $\mathsf{Term}_{\mathscr{V}}(\mathscr{F})$ be the *term algebra* built from $\mathscr{F}$ and $\mathscr{V}$ in the usual manner [5, Definition 10.1]. A *substitution* is a mapping $\sigma\colon \mathscr{V} \to \mathsf{Term}_{\mathscr{V}}(\mathscr{F})$ that acts identically but for a finite number of exceptions, i.e. is such that $\{X \in \mathscr{V} \mid \sigma(X) \neq X\}$ is a finite set. Any substitution extends in a unique way to the whole $\mathsf{Term}_{\mathscr{V}}(\mathscr{F})$ by requiring that it commutes with operations; hence it makes sense to speak of composition between substitutions.

Let E be a set of equations in the language $\mathscr{F}$. A *unification problem modulo E* is a finite set of pairs

$$\mathscr{E} = \{(s_j, t_j) \mid s_j, t_j \in \mathsf{Term}_{\mathscr{V}}(\mathscr{F}),\ j \in J\},$$

M.-J. Lesot et al. (Eds.): IPMU 2020, CCIS 1239, pp. 622–633, 2020.
https://doi.org/10.1007/978-3-030-50153-2_46

for some finite index set J. A *unifier* for $\mathscr{E}$ is a substitution σ such that

$$E \models \sigma(s_j) \approx \sigma(t_j)\,,$$

for each $j \in J$, i.e. such that the equality $\sigma(s_j) = \sigma(t_j)$ holds in every algebra of the variety $\mathbb{V}_E$ in the usual universal-algebraic sense [5, p. 78]. The problem $\mathscr{E}$ is *unifiable* if it admits at least one unifier. The set $U(\mathscr{E})$ of unifiers for $\mathscr{E}$ can be partially ordered as follows. If σ and τ are substitutions we say that σ is *more general* than τ (with respect to E), written $\tau \preceq_E \sigma$, if there exists a substitution ρ such that

$$E \models \tau(X) \approx (\rho \circ \sigma)(X)$$

holds for every $X \in \mathscr{V}$. This amounts to saying that τ is an instantiation of σ, up to E-equivalence. We endow $U(\mathscr{E})$ with the relation $\preceq_E$. The relation $\preceq_E$ is a pre-order. Let $\sim$ be the equivalence relation: $u \sim w$ if and only if both $u \preceq_E w$ and $w \preceq_E u$ hold. Then the quotient $\frac{U(\mathscr{E})}{\sim}$ carries a canonical partial order given by: $[\sigma] \leqslant_E [\tau]$ if, and only if, $\sigma \preceq_E \tau$.

The *unification type* of the unification problem $\mathscr{E}$ is:

1. *unitary*, if $\leqslant_E$ admits a maximum $[\mu] \in \frac{U(\mathscr{E})}{\sim}$ ($[\mu]$ is called a *most general unifier*);
2. *finitary*, if $\leqslant_E$ admits no maximum, but admits finitely many maximal elements $[\mu_1], \ldots, [\mu_u] \in \frac{U(\mathscr{E})}{\sim}$ such that every $[\sigma] \in \frac{U(\mathscr{E})}{\sim}$ lies below some $[\mu_i]$;
3. *infinitary*, if it is not finitary, and $\leqslant_E$ admits infinitely many maximal elements $\left\{[\mu_i] \in \frac{U(\mathscr{E})}{\sim} \mid i \in I\right\}$, for I an infinite index set, such that every $[\sigma] \in \frac{U(\mathscr{E})}{\sim}$ lies below some $[\mu_i]$;
4. *nullary*, if none of the preceding cases applies.

The unification types above are listed in order of desirability, with nullary being the worst possible case. The *unification type* of the equational theory E is now defined to be the worst unification type occurring among the unifiable problems $\mathscr{E}$ modulo E.

Unification has also found applications in the study of non classical logics for its connections with admissible rules [10,11,13,14]. For a propositional logic $\mathcal{L}$, a unification problem is simply a formula of $\mathcal{L}$ and a unifier is a substitution that makes that formula into a theorem. When a logic has an equivalent algebraic semantics, in the sense of [4], given by a class of algebras axiomatised by a set E of equations, the unification type of the logic and the unification type of E are the same.

This paper is devoted to an investigation of the unification type of fragments of *Łukasiewicz (infinite-valued propositional) logic* where only a finite number of variables are available. The standard references for Łukasiewicz logic are [7,18].

The unification type of Łukasiewicz logic is known to be the worst possible:

Theorem 1 ([17]). *The unification type of Łukasiewicz logic is nullary. Specifically, consider the unification problem*

$$p_1 \vee \neg p_1 \vee p_2 \vee \neg p_2, \qquad (\star)$$

where p_1 and p_2 are distinct propositional variables. Then the partially ordered set of unifiers for $\mathscr{E}$ contains a co-final chain of order-type ω.

By a *chain of order-type* ω we mean, as usual, a totally ordered set that is order-isomorphic to the natural numbers with their natural order. Recall also that a subset C of a partially ordered set $(P, \leqslant)$ is *co-final* if for every $p \in P$ there is $c \in C$ with $p \leqslant c$. In particular, Theorem 1 implies that *no unifier for the unifiable problem $\mathscr{E}$ is maximally general*—a condition that is strictly stronger than nullarity.

The proof of Theorem 1 requires an infinite amount of distinct propositional variables to be carried out—see Remark 4 for more details. Therefore the problem of establishing the unification type of fragments of Łukasiewicz logic with a finite number of variables was left open. The only case for which the unifications type was settled is the fragment of Łukasiewicz logic with only one variable. In this case the unification type is finitary (in a sense almost unitary):

Theorem 2 ([17, Theorem 4.1]). *Let $\varphi_i(X)$ be formulas of Łukasiewicz logic built from the single propositional variable X, for i ranging in some finite index set I. Then, if the unification problem $\mathscr{E} = \{\varphi_i(X) \mid i \in I\}$ is unifiable, it admits either one most general unifier, or two maximally general unifiers that are more general than any other unifier for $\mathscr{E}$. Further, each one of these cases is attained for some choice of $\mathscr{E}$.*

We shall see that when at least two variables are allowed the unification type becomes again non-finitary. The main result of this paper, Theorem 5, asserts that for every $n \geqslant 2$ the unification type of the fragment of Łukasiewicz logic with n distinct variables has unification type either infinitary of nullary. To establish such a result we first move to the equivalent algebraic semantics of Łukasiewicz logic, called MV-algebras; we then use Ghilardi's characterisation of the unification type in terms of projective and finitely presented objects [9]; as next step we use the duality between finitely presented MV-algebras and rational polyhedra [16]. In this latter category we build an infinite family of (dual) unifiers for the (dual of the) unification problem ($\star$) with the property that any infinite subfamily does not admit an upper bound in the order $\leqslant_E$. The proof of this last result rests upon the lifting property of the universal cover of the polyhedron associated with ($\star$).

We briefly discuss all these diverse tools used in the proof. More specifically, in Sect. 2 we spell out some preliminaries: in Subsect. 2.1 we summarise Ghilardi's approach to E-unification through projectivity; Subsect. 2.2 contains some basic information about MV-algebras and the background on polyhedral geometry required to state the duality theorem for finitely presented MV-algebras; in Subsect. 2.3 we give the needed background in algebraic topology. Finally, in Sect. 3 we prove the main theorem.

2 Background and Preliminaries

2.1 Ghilardi's Algebraic Unification Type

An object P in a category is called *projective* with respect to a class $\mathcal{E}$ of morphisms if for any $f: A \to B$ in $\mathcal{E}$ and any arrow $g: P \to B$, there exists an arrow $h: P \to A$ such that the following diagram commutes.

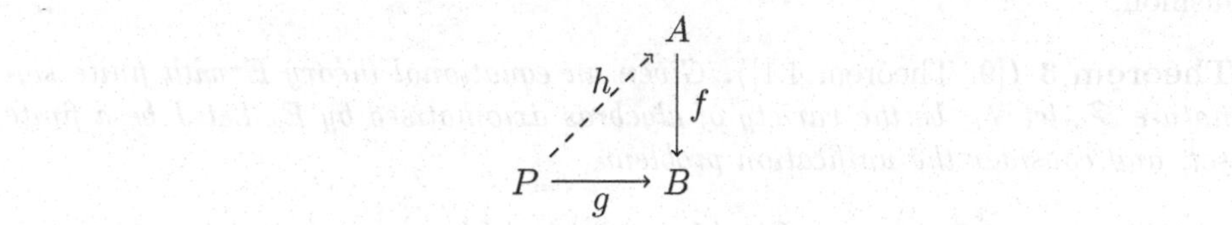

The class $\mathcal{E}$ may consist of all epimorphisms, regular epimorphisms, strong epimorphisms, etc. In this paper, the main objects are algebras in a variety, and the arrow $f: A \to B$ is always taken to be a surjection. It is well known that surjections in a variety are the same thing as regular epimorphisms, see e.g. [1, (vi) on p. 135].

An object A in a category is said to be a *retract* of an object B if there are arrows $s: A \to B$ and $r: B \to A$ such that $r \circ s$ is the identity on A. When this is the case, r is called a *retraction* (of s) and s a *section* (of r). If the category in question is a variety, it follows at once that r is surjective, and s is injective. One checks that on these definitions projective objects in any variety of algebras are stable under retractions, and they are precisely the retracts of free objects. In particular, free objects are projective.

Let us fix a variety $\mathbb{V}$ of algebras, and let us write $\mathcal{F}_I$ for the free object in $\mathbb{V}$ generated by a set I. Recall that an algebra A of $\mathbb{V}$ is *finitely presented* if it is a quotient of the form $A = \mathcal{F}_I/\theta$, with I finite and θ a finitely generated congruence. The elements of I are the *generators* of A, while any given set of pairs $(s, t) \in \theta$ that generates the congruence θ is traditionally called a *set of relators* for A.

Following [9], by an *algebraic unification problem* we mean a finitely presented algebra A of $\mathbb{V}$. An *algebraic unifier* for A is a homomorphism $u: A \to P$ with P a finitely presented projective algebra in $\mathbb{V}$; and A is *algebraically unifiable* if such an algebraic unifier exists.

Given another algebraic unifier $w: A \to Q$, we say that u *is more general than* w, written $w \preceq_{\mathbb{V}} u$, if there is a homomorphism $g: P \to Q$ making the following diagram commute.

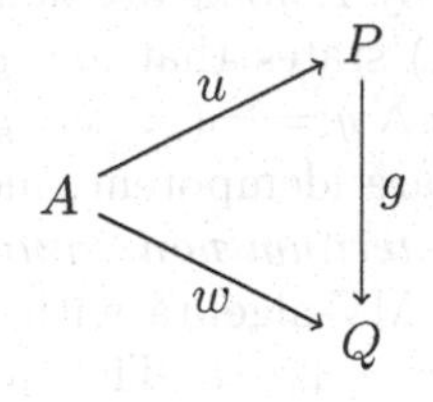

The relation $\preceq_{\mathbb{V}}$ is a pre-order on the set $U(A)$ of algebraic unifiers for A and can be canonically quotiented into a partial order $\left(\frac{U(A)}{\sim}, \leqslant_{\mathbb{V}}\right)$.

The *algebraic unification type* of an algebraically unifiable finitely presented algebra A in the variety $\mathbb{V}$ is now defined exactly as in the symbolic case (see the Introduction), using the partially ordered set $\left(\frac{U(A)}{\sim}, \leqslant_{\mathbb{V}}\right)$ in place of $\left(\frac{U(\mathscr{E})}{\sim}, \leqslant_E\right)$. One also defines the *algebraic unification type* of the variety $\mathbb{V}$ in the same fashion.

Theorem 3 ([9, Theorem 4.1]). *Given an equational theory E with finite signature $\mathscr{F}$, let $\mathbb{V}_E$ be the variety of algebras axiomatised by E. Let J be a finite set, and consider the unification problem*

$$\mathscr{E} = \{ (s_j , t_j) \mid i \in J \},$$

where $s_j, t_j \in \mathsf{Term}_{\mathscr{V}}(\mathscr{F})$ are terms. Let A be the algebra of $\mathbb{V}_E$ finitely presented by the relators $\mathscr{E}$.

Then $\mathscr{E}$ is unifiable if and only if A is algebraically unifiable. Further, the partially ordered sets $\left(\frac{U(A)}{\sim}, \leqslant_{\mathbb{V}_E}\right)$ of algebraic unifiers for A, and $\left(\frac{U(\mathscr{E})}{\sim}, \leqslant_E\right)$ of unifiers for $\mathscr{E}$, are isomorphic. In particular, the unification type of E and the algebraic unification type of $\mathbb{V}_E$ coincide.

Remark 1. Ghilardi's approach to unification goes far beyond the universal-algebraic contexts. Indeed, one readily sees that projectivity can be stated in any category, while thanks to the work of Gabriel and Ulmer [8] we know that the concept of 'finitely-presented' object makes sense in any locally small category. In particular this shows that the unification type is preserved under categorical equivalences.

2.2 MV-algebras and Rational Polyhedra

The equivalent algebraic semantics of Łukasiewicz logic, in the precise sense of Blok and Pigozzi [4], is given by *MV-algebras.* An MV-algebra is an algebraic structure $(M, \oplus, \neg, 0)$, where $0 \in M$ is a constant, $\neg$ is a unary operation satisfying $\neg\neg x = x$, $\oplus$ is a binary operation making $(M, \oplus, 0)$ a commutative monoid, the element 1 defined as $\neg 0$ satisfies $x \oplus 1 = 1$, and the law

$$\neg(\neg x \oplus y) \oplus y = \neg(\neg y \oplus x) \oplus x \tag{1}$$

holds. Any MV-algebra has an underlying structure of distributive lattice bounded below by 0 and above by 1. Joins are defined as $x \vee y = \neg(\neg x \oplus y) \oplus y$. Thus, the characteristic law (1) states that $x \vee y = y \vee x$. Meets are defined by the de Morgan condition $x \wedge y = \neg(\neg x \vee \neg y)$. Boolean algebras are precisely those MV-algebras that are idempotent, meaning that $x \oplus x = x$ holds, or equivalently, that satisfy the *tertium non datur* law $x \vee \neg x = 1$. The interval $[0, 1] \subseteq \mathbb{R}$ can be made into an MV-algebra with neutral element 0 by defining $x \oplus y = \min\{x + y, 1\}$ and $\neg x = 1 - x$. The underlying lattice order of this

MV-algebra coincides with the natural order that $[0,1]$ inherits from the real numbers. This MV-algebra is often referred as the *standard MV-algebra.*

Let us fix an integer $d \geqslant 0$ as the dimension of the real vector space $\mathbb{R}^d$. A *convex combination* of a finite set of vectors $v_1, \ldots, v_u \in \mathbb{R}^d$ is any vector of the form $\lambda_1 v_1 + \cdots + \lambda_u v_u$, for non-negative real numbers $\lambda_i \geqslant 0$ satisfying $\sum_{i=1}^{u} \lambda_i = 1$. If $S \subseteq \mathbb{R}^d$ is any subset, we let $\mathrm{conv}\, S$ denote the *convex hull* of S, i.e. the collection of all convex combinations of finite sets of vectors $v_1, \ldots, v_u \in S$. A *polytope* is any subset of $\mathbb{R}^d$ of the form $\mathrm{conv}\, S$, for some finite $S \subseteq \mathbb{R}^d$, and a (*compact*) *polyhedron* is a union of finitely many polytopes in $\mathbb{R}^d$. A polytope is *rational* if it may be written in the form $\mathrm{conv}\, S$ for some finite set $S \subseteq \mathbb{Q}^d \subseteq \mathbb{R}^d$ of vectors with rational coordinates. Similarly, a polyhedron is *rational* if it may be written as a union of finitely many rational polytopes.

Throughout, the adjective 'linear' is to be understood as 'affine linear'. A function $f \colon \mathbb{R}^d \to \mathbb{R}$ is *piecewise linear* if it is continuous (with respect to the Euclidean topology on $\mathbb{R}^d$ and $\mathbb{R}$), and there is a finite set of linear functions $l_1, \ldots, l_u$ such that for each $x \in \mathbb{R}^d$ one has $f(x) = l_i(x)$ for some choice of $i = 1, \ldots, u$. If, moreover, each l_i can be written as a linear polynomial with integer coefficients, then f is called *$\mathbb{Z}$-map.* For an integer $d' \geqslant 0$, a function $\lambda = (\lambda_1, \ldots, \lambda_{d'}) \colon \mathbb{R}^d \to \mathbb{R}^{d'}$ is a *piecewise linear map* (respectively, *a $\mathbb{Z}$-map*) if each one of its scalar components $\lambda_j \colon \mathbb{R}^d \to \mathbb{R}$ is a piecewise linear function (respectively, $\mathbb{Z}$-map). We now define piecewise linear maps ($\mathbb{Z}$-maps) $A \to B$ for arbitrary subsets $A \subseteq \mathbb{R}^d$, $B \subseteq \mathbb{R}^{d'}$ as the restriction and co-restriction of piecewise linear maps ($\mathbb{Z}$-maps) $\mathbb{R}^d \to \mathbb{R}^{d'}$.

When the spaces at issue are rational polyhedra, a useful equivalent to the preceding definition of $\mathbb{Z}$-map is available.

Lemma 1. *Let $P \subseteq \mathbb{R}^d$ be a rational polyhedron, and let $f \colon P \to \mathbb{R}$ be a continuous function. Then the following are equivalent.*

1. *f is a $\mathbb{Z}$-map.*
2. *There exist finitely many linear polynomials with integer coefficients $l_1, \ldots, l_u \colon \mathbb{R}^d \to \mathbb{R}$ such that, for each $p \in P$, $f(p) = l_{i_p}(p)$ for some $i_p \in \{1, \ldots, u\}$.*

Notice that in item 2 the result of gluing the linear polynomials $l_1, \ldots, l_u$ is only required to be continuous on P and not on the whole $\mathbb{R}^n$. For example, consider a polyhedron P which is a disjoint union of two polytopes A and B, and consider the map $f \colon P \to \mathbb{R}$ defined by $f(x) = 0$ for $x \in A$ and $f(x) = 1$ for $x \in B$. By lemma 1, f is automatically a $\mathbb{Z}$-map; we do not need to interpolate the two maps constantly equal to 0 and 1. To see that the hypothesis that P be a polyhedron is crucial, see [16, Remark 4.10].

It is not hard to show that the composition of $\mathbb{Z}$-maps between rational polyhedra is again a $\mathbb{Z}$-map. A $\mathbb{Z}$-map $\lambda \colon A \to B$ between rational polyhedra $A \subseteq \mathbb{R}^d$ and $B \subseteq \mathbb{R}^{d'}$ is a *$\mathbb{Z}$-homeomorphism* if there exists a $\mathbb{Z}$-map $\lambda' \colon B \to A$ such that $\lambda \circ \lambda' = \mathbf{1}_B$ and $\lambda' \circ \lambda = \mathbf{1}_A$. In other words, a $\mathbb{Z}$-map is a $\mathbb{Z}$-homeomorphism if it is a homeomorphism whose inverse is a $\mathbb{Z}$-map, too. With these definitions, rational polyhedra and $\mathbb{Z}$-maps form a category.

The following result is a particular case of a larger duality for semisimple MV-algebras (see [6,16] for more details).

Theorem 4 ([16, Corollary 4.12]). *There is a categorical equivalence between the category of finitely presented MV-algebras with MV-homomorphisms and the opposite of the category of rational polyhedra with $\mathbb{Z}$-maps.*

Remark 2. Theorem 4 provides a back-and-forth translation of algebraic concepts into geometric ones. Free n-generated MV-algebras correspond to $[0,1]^n$, an algebraic unification problem (=finitely presented MV-algebra) A becomes a rational polyhedron Q and an algebraic unifier for A (=finitely presented projective MV-algebra B with a homomorphism $f\colon A \to B$) becomes a polyhedron P, which is a retract by $\mathbb{Z}$-maps of $[0,1]^n$ for some $n \in \mathbb{N}$, together with a $\mathbb{Z}$-map $g\colon P \to Q$ (because the equivalence is contravariant). We shall call *co-unifier* such a pair (P, g).

2.3 The Universal Cover of the Circle

Let us recall some standard notions from algebraic topology; we refer to [12] for details.

A *path* in a space X is a continuous map $f\colon [0,1] \to X$; the *endpoints* of f are $f(0)$ and $f(1)$. A space X is *path-connected* if for any $x_0, x_1 \in X$ there is a path in X with endpoints x_0, x_1. On the other hand, X is *locally path-connected* if each point has arbitrarily small open neighbourhoods that are path-connected; that is, for each $y \in X$ and each neighbourhood U of y there is a path-connected open neighbourhood of y contained in U. It is not hard to prove that polyhedra are locally path-connected (in fact, locally contractible by [12, Proposition A.1]), and therefore that a polyhedron is connected if and only if it is path-connected.

A *loop* in X is a path p in X such that $p(0) = p(1)$. A space X is *simply-connected* if it is path-connected and, for every loop p in X, letting $x_0 := p(0)$, there is a continuous function $F\colon [0,1] \times [0,1] \to X$ such that, for every $x \in [0,1]$ we have $F(x,0) = p(x)$ and $F(x,1) = F(0,x) = F(1,x) = x_0$.

A *covering space* [12, Section 1.3] of a topological space X is a space $\tilde{X}$ together with a surjective continuous map $p\colon \tilde{X} \to X$, called a *covering map*, such that there is a open covering $\{O_i\}$ of X, with i ranging in some index set I, satisfying the following condition: for each $i \in I$ the inverse image $p^{-1}(O_i)$ is a disjoint union of open sets in $\tilde{X}$, each of which is mapped homeomorphically by p onto O_i.

If $p\colon \tilde{X} \to X$ is a covering map of the space X, and if Y is any space, a continuous map $f\colon Y \to X$ is said to *lift to* p (or, more informally, to $\tilde{X}$, when p is understood), if there is a continuous map $\tilde{f}\colon Y \to \tilde{X}$ such that $p \circ \tilde{f} = f$. Any such $\tilde{f}$ is then called a *lift* of f. In the next lemma we recall two important properties of covering maps with respect to lifts that we will use in Sect. 3.

Lemma 2 ([12, Proposition 1.33 and 1.34]). *Given topological spaces X and $\tilde{X}$, suppose that $p\colon \tilde{X} \to X$ is a covering map. Further, let Y be a topological space, and let $f\colon Y \to X$ be a continuous map. Then the following hold.*

1. (Unique lifting property.) *Assume Y is connected. If $\tilde{f}, \tilde{f}' \colon Y \to \tilde{X}$ are two lifts of f that agree at one point of Y, then $\tilde{f} = \tilde{f}'$.*
2. (Lifting property of simply-connected polyhedra.) *If, additionally, Y is a simply-connected locally path-connected space, then a lift of f does exist. In fact, for any point $y \in Y$, and for any point $\tilde{x} \in \tilde{X}$ lying in the fibre over $f(y)$, i.e. such that $p(\tilde{x}) = f(y)$, there is a lift $\tilde{f}$ of f such that $\tilde{f}(y) = \tilde{x}$.*

It is not hard to prove that polyhedra are locally path-connected (in fact, locally contractible by [12, Proposition A.1]), so item 2 applies to any simply-connected polyhedron Y.

Given a path-connected, locally path-connected space X, a covering map $p \colon \tilde{X} \to X$ is called a *universal covering map* if $\tilde{X}$ is simply-connected. In this case $\tilde{X}$ is called the *universal cover* of X. This name is due to the fact that, among the covering maps $t \colon Y \to X$ with Y connected, a universal covering map $p \colon \tilde{X} \to X$ is characterised by a certain universal property—namely, p factors through every such a t. A universal covering map (of a path-connected, locally path-connected space) is essentially unique. Under suitable conditions, a space X admits a universal cover (see [12, Theorem 1.38]). The following is an example of universal covering map.

Example 1. Let $\mathbb{S}^1 = \{(x, y) \in \mathbb{R}^2 \mid x^2 + y^2 = 1\}$ be the unit circle in the Euclidean plane $\mathbb{R}^2$, and let $\chi \colon \mathbb{R} \to \mathbb{S}^1$ be the continuous function given by

$$t \mapsto (\cos 2\pi t, \sin 2\pi t).$$

Upon embedding $\mathbb{R}$ into $\mathbb{R}^3$ as a helix H via $t \mapsto (\cos 2\pi t, \sin 2\pi t, t)$, χ acts on H as the orthogonal projection onto $\mathbb{S}^1$ along the z-axis. The surjective map

$$\chi \colon \mathbb{R} \to \mathbb{S}^1$$

is the *universal covering map of the circle*, and $\mathbb{R}$ is the *universal cover* of $\mathbb{S}^1$.

3 Main Result

Having set up all necessary background we turn to the main question of the paper: what is the unification type of the fragments of Łukasiewicz logic with at most n variables, with $n \geqslant 2$? So, we are interested in unification problems and unifiers that involve at most n variables. More precisely, using the algebraic notation introduced at the beginning of the paper, for any fixed $n \geqslant 2$, we let $\mathscr{V}_n := \{X_1 \ldots, X_n\}$ and $\mathscr{F}$ be the set of basic operations in the language of MV-algebras. We consider only unification problems whose terms range in $\mathsf{Term}_{\mathscr{V}}(\mathscr{F})$ and unifiers going from $\mathscr{V}$ into $\mathsf{Term}_{\mathscr{V}}(\mathscr{F})$. In terms of the framework presented in Sect. 2.1 this corresponds to restricting to finitely presented and projective algebras with up to n-generators, which we call n-generated.

Let us call a polyhedron *n-generated* if it is $\mathbb{Z}$-homeomorphic to a polyhedron inside $[0, 1]^n$. An easy inspection of [16] shows that the duality of Theorem 4 restricts as follows.

Corollary 1. *The category of n-generated MV-algebras and MV-homomorphisms among them is dually equivalent to the category of n-generated polyhedra and $\mathbb{Z}$-maps among them.*

We will concentrate on the co-unifiers of the rational polyhedron given by the border of the unit square:

$$\mathfrak{B} := \mathtt{conv}\{(0,0),(1,0)\} \cup \mathtt{conv}\{(1,0),(1,1)\} \cup \\ \mathtt{conv}\{(1,1),(0,1)\} \cup \mathtt{conv}\{(0,1),(0,0)\}. \tag{2}$$

Consider the map $\zeta\colon \mathbb{R} \to \mathfrak{B}$ which wraps $\mathbb{R}$ around $\mathfrak{B}$, counter-clockwise, at constant speed 1, sending 0 to $(0,0)$. More precisely,

$$\zeta\colon \mathbb{R} \longrightarrow \mathfrak{B}$$
$$x \longmapsto \begin{cases} (x - \lfloor x \rfloor, 0) & \text{if } \lfloor x \rfloor \equiv 0 \mod 4, \\ (1, x - \lfloor x \rfloor) & \text{if } \lfloor x \rfloor \equiv 1 \mod 4, \\ (1 - (x - \lfloor x \rfloor), 1) & \text{if } \lfloor x \rfloor \equiv 2 \mod 4, \\ (0, 1 - (x - \lfloor x \rfloor)) & \text{if } \lfloor x \rfloor \equiv 3 \mod 4, \end{cases} \tag{3}$$

where $\lfloor x \rfloor$ is the greatest integer below x.

Remark 3. Upon embedding $\mathbb{R}$ into $\mathbb{R}^3$ as a *squared* helix H, as depicted in Fig. 1, ζ acts on H as the orthogonal projection onto $\mathfrak{B}$ along the z-axis.

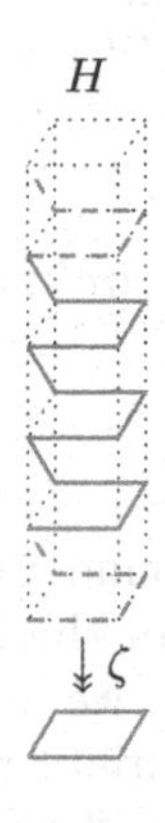

Fig. 1. The $\mathbb{Z}$-universal cover of $\mathfrak{B}$.

Lemma 3. *The map ζ is a universal cover of $\mathfrak{B}$.*

Proof. This is obvious as $\mathfrak{B}$ is homeomorphic to $\mathbb{S}^1$ and up to homeomorphism ζ maps $\mathbb{R}$ on $\mathfrak{B}$ as χ does on $\mathbb{S}^1$ in Example 1.

Notice that ζ is continuous but is not a $\mathbb{Z}$-map; however, the following holds.

Lemma 4. *The restriction $\zeta_{a,b}$ of ζ to any closed interval $[a, b]$ with $a, b \in \mathbb{Z}$ is a $\mathbb{Z}$-map.*

Proof. This can be seen using Lemma 1. Indeed, $\zeta_{a,b}$ is defined on every interval $[a+i, a+i+1]$, with $0 \leqslant i < |b-a|$, by one of the four cases of Eq. (3); each of those functions is linear with integer coefficients on $[a+i, a+i+1]$ because $x - \lfloor x \rfloor$ can be written on such an interval as $x - (a+i)$.

Remark 4. The co-final chain of Theorem 1 is obtained by taking increasingly larger parts of the piece-wise linear spiral depicted in Fig. 1, together with their projections onto $\mathfrak{B}$. An easy argument shows that $\mathbb{Z}$-homeomorphisms preserve the number of points with integer coordinates. As a consequence, it can be seen that there can be no finite bound on the dimensions of the unital cubes that embed the rational polyhedra in the above mentioned increasing sequence. Hence the proof strategy of Theorem 1 cannot be adopted for fragments of Łukasiewicz logic with a finite amounts of variables.

Lemma 5. *Let Y be a connected space and $f, g \colon Y \to \mathbb{R}$ be a pair of continuous functions. If $\zeta \circ f = \zeta \circ g$ then there exists $k \in \mathbb{Z}$ such that, for every $y \in Y$,*

$$g(y) = f(y) + 4k.$$

Proof. Let y_0 be any point of Y. By inspection of the definition (3) one easily sees that $\zeta(f(y_0)) = \zeta(g(y_0))$ entails the existence of some $k \in \mathbb{Z}$ such that $g(y_0) - f(y_0) = 4k$. Since by definition ζ has period 4, $\zeta \circ g = \zeta \circ f = \zeta \circ (f + 4k)$, so the maps $g \colon Y \to \mathbb{R}$ and $f + 4k \colon Y \to \mathbb{R}$ are both lifts to $\mathbb{R}$ of $\zeta \circ g$. By the choice of y_0, both g and $f + 4k$ attain the same value at y_0, because $g(y_0) = f(y_0) + 4k$. Thus, by item 1 in Lemma 2, $g = f + 4k$.

Definition 1. *Let Y be a connected space and $f \colon Y \to \mathfrak{B}$ be a continuous map which admits a lift to ζ. Every lift $\tilde{f} \colon Y \to \mathbb{R}$ has a connected image, whose length*[1]*, by Lemma 5, is independent of the choice of $\tilde{f}$. We denote the length of $\tilde{f}[Y]$ with $\mathtt{d}(Y, f)$ and we call it the* degree *of (Y, f).*

Lemma 6. *If (P, f) is a co-unifier then f admits a lift and $\mathtt{d}(P, f)$ is finite.*

Proof. It is straightforward to verify that, if A is a retract of B in the category of topological spaces and continuous functions, and B is simply-connected, then A is simply-connected. As a consequence, if (P, f) is a co-unifier, then P is simply-connected. So by item 2 in Lemma 2, f admits a lift. Moreover, $\mathtt{d}(P, f)$ is finite because P is compact.

Lemma 7. *Let (P, f) and (Q, g) be co-unifiers for $\mathfrak{B}$. If (P, f) is less general than (Q, g), then $\mathtt{d}(P, f) \leqslant \mathtt{d}(Q, g)$.*

Proof. If (P, f) is less general than (Q, g), then there exists $h \colon P \to Q$ such that $f = g \circ h$. Let $\tilde{g}$ be a lift of g. Then we have $f = \zeta \circ \tilde{g} \circ h$, so $\tilde{g} \circ h$ is a lift of f. The image of $\tilde{g}$ contains the image of $\tilde{g} \circ h$, so $\mathtt{d}(P, f) \leqslant \mathtt{d}(Q, g)$.

[1] By *length* of a connected subset of $\mathbb{R}$ we simply mean its Lebesgue measure.

Definition 2. *Consider the family $\{f_n \mid n \in \mathbb{N}\}$ of functions $f_n\colon [0,1] \to \mathfrak{B}$, where each f_n wraps the unit interval around $\mathfrak{B}$ n-times, counter-clockwise, at constant speed, starting at $(0,0)$. More precisely, for each $n \in \mathbb{N}$, set*

$$\begin{aligned} v_n\colon [0,1] &\longrightarrow \mathbb{R} \\ x &\longmapsto 4n \cdot x. \end{aligned}$$

and set

$$f_n := \zeta \circ v_n.$$

Lemma 8. *For every $n \in \mathbb{N}$, $([0,1], f_n)$ is a co-unifier of $\mathfrak{B}$.*

Proof. The polyhedron $[0,1]$ is obviously a retract of itself. Furthermore, notice that v_n is a $\mathbb{Z}$-map whose image is $[0,4n]$ and, by Lemma 4, the restriction of ζ to $[0,4n]$ is a $\mathbb{Z}$-map. Since composition of $\mathbb{Z}$-maps is a $\mathbb{Z}$-map, we have that each f_n is indeed a $\mathbb{Z}$-map.

Lemma 9. *For each $n \in \mathbb{N}$, the degree of $([0,1], f_n)$ is $4n$.*

Proof. For every $n \in \mathbb{N}$, v_n is a lift of f_n by definition. The image of v_n is $[0,4n]$, thus by Definition 1, $\mathtt{d}([0,1], f_n) = 4n$.

The following lemma gives us a picture of the relations among $([0,1], f_n)$, however the crucial property is contained in the subsequent lemma.

Lemma 10. *For every distinct, non-zero $m, n \in \mathbb{N}$ the co-unifiers $([0,1], f_m)$ and $([0,1], f_n)$ are incomparable.*

Proof. Suppose, without loss of generality, that $m < n$. By Lemma 9, for any $k \in \mathbb{N}$, $\mathtt{d}([0,1], f_k) = 4k$, so Lemma 7 implies that $([0,1], f_n)$ cannot be less general than $([0,1], f_m)$. To see that $([0,1], f_m)$ cannot be less general than $([0,1], f_n)$, suppose, by way of contradiction, that there exists $h\colon [0,1] \to [0,1]$ such that $f_n \circ h = f_m$. The functions $v_n \circ h$ and v_m are two lifts of the function $f_n \circ h = f_m$. By Lemma 5, there exists $k \in \mathbb{Z}$ such that $v_n \circ h = v_m + 4k$, i.e., for every $x \in [0,1]$, $4n \cdot h(x) = 4(mx + k)$, i.e., $h(x) = \frac{m}{n}x + \frac{k}{n}$. However, the hypothesis $m < n$ implies $\frac{m}{n} \in (0,1) \notin \mathbb{Z}$, so h is not a $\mathbb{Z}$-map: a contradiction.

Lemma 11. *Every infinite subset of the family of co-unifiers of Definition 2 does not admit an upper bound.*

Proof. Let $\mathcal{F}$ be such a subset. Since $\mathcal{F}$ is infinite, by Lemma 9, it must contain co-unifiers of arbitrarily large degree. Thus, by Lemma 7 an upper bound of such a family of co-unifiers cannot have finite degree and this contradicts Lemma 6.

Theorem 5. *For every $n \geqslant 2$ the fragments of Łukasiewicz logic with n distinct variables have unification type either infinitary or nullary.*

Proof. By way of contradiction, let us suppose that the fragment of Łukasiewicz logic with n distinct variables has finitary unification type. Then, by coupling Theorem 3 with Corollary 1, we get the the co-unification type of $\mathfrak{B}$ must be finitary. Let $(Q_1, g_1), \ldots, (Q_k, g_k)$ be the maximal co-unifiers of $\mathfrak{B}$. So each co-unifier $([0,1], f_m)$ must be less general than some (Q_i, g_i) for $i \leqslant k$. Since k is finite there must be at least an infinite family $\{([0,1], f_{m_j}) \mid j \in \mathbb{N}\}$ of co-unifiers that are all less general than some (Q_i, g_i). This contradicts Lemma 11.

References

1. Adámek, J., Rosický, J.: Locally Presentable and Accessible Categories. London Mathematical Society Lecture Note Series, vol. 189. Cambridge University Press, Cambridge (1994)
2. Baader, F., Ghilardi, S.: Unification in modal and description logics. Logic J. IGPL **19**(6), 705–730 (2011)
3. Baader, F., Snyder, W.: Unification theory. In: Robinson, J.A., Voronkov, A. eds. Handbook of Automated Reasoning, pp. 445–532. Elsevier/MIT Press (2001)
4. Blok, W.J., Pigozzi, D.: Algebraizable logics. Mem. Amer. Math. Soc. **77**(396), vi+78 (1989)
5. Burris, S., Sankappanavar, H.P.: A Course in Universal Algebra. Graduate Texts in Mathematics, vol. 78. Springer, New York (1981)
6. Cabrer, L.M., Spada, L.: MV-algebras, infinite dimensional polyhedra, and natural dualities. Arch. Math. Logic **56**, 21–42 (2016). https://doi.org/10.1007/s00153-016-0512-9
7. Cignoli, R.L.O., D'Ottaviano, I.M.L., Mundici, D.: Algebraic Foundations of Many-Valued Reasoning. Trends in Logic-Studia Logica Library, vol. 7. Kluwer Academic Publishers, Dordrecht (2000)
8. Gabriel, P., Ulmer, F.: Lokal präsentierbare Kategorien. LNM, vol. 221. Springer, Heidelberg (1971). https://doi.org/10.1007/BFb0059396
9. Ghilardi, S.: Unification through projectivity. J. Logic Comput. **7**(6), 733–752 (1997)
10. Ghilardi, S.: Unification in intuitionistic logic. J. Symb. Logic **64**(2), 859–880 (1999)
11. Ghilardi, S.: Best solving modal equations. Ann. Pure Appl. Logic **102**(3), 183–198 (2000)
12. Hatcher, A.: Algebraic Topology. Cambridge University Press, Cambridge (2002)
13. Iemhoff, R.: On the admissible rules of intuitionistic propositional logic. J. Symb. Logic **66**(1), 281–294 (2001)
14. Jeřábek, E.: Admissible rules of modal logics. J. Logic Comput. **15**, 411–431 (2005)
15. Mac Lane, S.: Categories for the Working Mathematician. GTM, vol. 5. Springer, New York (1978). https://doi.org/10.1007/978-1-4757-4721-8
16. Marra, V., Spada, L.: The dual adjunction between MV-algebras and Tychonoff spaces. Stud. Log. **100**(1–2), 253–278 (2012)
17. Marra, V., Spada, L.: Duality, projectivity, and unification in Łukasiewicz logic and MV-algebras. Ann. Pure Appl. Logic **164**(3), 192–210 (2013)
18. Mundici, D.: Advanced Łukasiewicz Calculus and MV-algebras. Trends in Logic-Studia Logica Library, vol. 35. Springer, New York (2011). https://doi.org/10.1007/978-94-007-0840-2

Two Dualities for Weakly Pseudo-complemented quasi-Kleene Algebras

Umberto Rivieccio[1], Ramon Jansana[2], and Thiago Nascimento[1](✉)

[1] Universidade Federal do Rio Grande do Norte, Natal, Brazil
urivieccio@dimap.ufrn.br, thiagnascsilva@gmail.com
[2] Universitat de Barcelona, Barcelona, Spain
jansana@ub.edu

Abstract. Quasi-Nelson algebras are a non-involutive generalisation of Nelson algebras that can be characterised in several ways, e.g. as (i) the variety of bounded commutative integral (not necessarily involutive) residuated lattices that satisfy the Nelson identity; (ii) the class of $(0,1)$-congruence orderable commutative integral residuated lattices; (iii) the algebraic counterpart of quasi-Nelson logic, i.e. the (algebraisable) extension of the substructural logic $\mathcal{FL}_{ew}$ by the Nelson axiom. In the present paper we focus on the subreducts of quasi-Nelson algebras obtained by eliding the implication while keeping the two term-definable negations. These form a variety that (following A. Sendlewski, who studied the corresponding fragment of Nelson algebras) we dub *weakly pseudo-complemented quasi-Kleene algebras*. We develop a Priestley-style duality for these algebras (in two different guises) which is essentially an application of the general approach proposed in the paper *A duality for two-sorted lattices* by A. Jung and U. Rivieccio.

Keywords: Quasi-Nelson algebras · Kleene algebras with weak pseudo-complement · Residuated lattices · Two-sorted duality · Semi-De Morgan

1 Introduction

Nelson's constructive logic with strong negation $\mathcal{N}$ (introduced in [10]; see also [14,20,25]) is a well-known non-classical logic that combines the constructive approach of (the $\{\wedge, \vee, \to\}$-fragment of) intuitionistic logic with a De Morgan

U. Rivieccio acknowledges partial funding by the Conselho Nacional de Desenvolvimento Científico e Tecnológico (CNPq, Brazil), under the grant 313643/2017-2 (Bolsas de Produtividade em Pesquisa - PQ). R. Jansana was financed by grant 2017 SGR 95 from the government of Catalonia and the research project MTM2016-74892-P from the government of Spain, which include FEDER funds from the European Union. T. Nascimento acknowledges partial support by the Coordenação de Aperfeiçoamento de Pessoal de Nível Superior (CAPES, Brazil) - Finance Code 001.

M.-J. Lesot et al. (Eds.): IPMU 2020, CCIS 1239, pp. 634–653, 2020.
https://doi.org/10.1007/978-3-030-50153-2_47

involutive negation. The algebraic models of $\mathcal{N}$ form the variety of *Nelson algebras* (alias *Nelson residuated lattices*), and have been studied since the late 1950's (first by Rasiowa; see [14] and references therein). One of the main algebraic insights on this variety is that every Nelson algebra can be represented as a special binary power (here called a *twist-algebra*) of a Heyting algebra. This correspondence was formulated as a categorical equivalence (first by A. Sendlewski) between Nelson algebras and a category of enriched Heyting algebras, and this made it possible to transfer a number of fundamental results from the theory of Heyting algebras to the Nelson realm.

More recent is the discovery (due to Spinks and Veroff [23,24]) that Nelson logic can be viewed as one of the so-called substructural logics. This entails that the class of Nelson algebras is term equivalent to a variety of bounded, commutative, integral residuated lattices [4]; hence the alternative name of *Nelson residuated lattices.* Given the recent flourish of studies on substructural logics and residuated structures, this alternative perspective also proved fruitful. Indeed, it made it possible, in the first place, to recover or recast a number of results on Nelson algebras by specialising more general ones about residuated structures. Furthermore, and maybe more interestingly, it allows us to formulate new questions on Nelson algebras/logic that can be best appreciated within the framework of residuated lattices. Among these is the problem that led to the introduction of *quasi-Nelson algebras*, which can be phrased as follows.

By the results of [23,24], Nelson algebras are term equivalent to a the class of (bounded, commutative, integral) residuated lattices that additionally satisfy the involutive law $(x \approx (x \Rightarrow 0) \Rightarrow 0)$ and the *Nelson identity*:

$$(x \Rightarrow (x \Rightarrow y)) \wedge ({\sim} y \Rightarrow ({\sim} y \Rightarrow {\sim} x)) \approx x \Rightarrow y.$$

Thus, all results that are specific to Nelson algebras (as opposed to general residuated lattices), including the connection with Heyting algebras given by twist-algebras, essentially depend on involutivity and on the Nelson identity.

The papers [17,18] are aimed at establishing to which extent the structure theory of Nelson algebras can be reconstructed (within the context of residuated lattices) in the presence of the Nelson identity but without relying on the involutive law. It turns out that some of the most characteristic results indeed do not depend on the involutive law. In particular, it is shown in [17,18] that (a suitable generalisation of) the twist-algebra construction can be performed in a not-necessarily involutive context: thus making it possible to recover the connection between Heyting algebras and 'non-involutive Nelson algebras', a variety we dubbed *quasi-Nelson algebras* (alias *quasi-Nelson residuated lattices*). This class can also be characterised by a purely congruence-theoretical property introduced in [22] under the name of (0, 1)-*congruence orderability*; the main result being that among (bounded, commutative, integral) residuated lattices, quasi-Nelson algebras are precisely the $(0, 1)$-congruence orderable ones. This generalises the result of [22] that, the $(0, 1)$-congruence orderable involutive residuated lattices are precisely the Nelson residuated lattices.

The very recent paper [15] extends the investigation of quasi-Nelson algebras initiated in [17,18] to the implication-free fragment; in turn, [15] relies on [16], in which the $\{\wedge, \vee, \sim\}$- fragment of quasi-Nelson logic was characterised. More precisely, [15] elides the implication operation $\rightarrow$ from the language while keeping the *two* negation operators (the primitive $\sim$ and a second one $\neg$ that is defined, in the full language of quasi-Nelson algebras, by $\neg x := x \rightarrow 0$). It turns out that a twist-algebra construction (see Definition 5 below) can be used to characterise the class of algebras corresponding to the two-negation fragment of quasi-Nelson algebras, dubbed in [15] *weakly pseudo-complemented quasi-Kleene algebras.*

In fact, [15] can be viewed as a non-involutive counterpart of Sendlewski's study on the two-negation subreducts of Nelson algebras [21]. Sendlewski shows that those subreducts form a variety (called *wp-Kleene algebras*) which corresponds via a twist-algebra construction to pseudo-complemented distributive lattices (i.e., the subreducts of Heyting algebras with negation but no implication). This entails that the functor between Nelson and Heyting algebras can be extended to one with similar properties relating the subreducts of both classes. Indeed, most results contained in [21] can be retrieved from [15] by restricting one's attention to involutive algebras.

In the present paper, we take advantage of the twist representation introduced in [15] to develop a Priestley-style duality for weakly pseudo-complemented quasi-Kleene algebras (thereby obtaining a duality for Sendlewski's *wp*-algebras as well). We present our duality in two guises based on the two twist representations introduced in [15]; both can be viewed as applications of the two-sorted approach to dualities proposed in [7].

2 WPQK-algebras and Their Representation

In this section we sum up the results from [15,16] that shall be needed for our present purposes. We begin by introducing quasi-Nelson algebras, the algebras in the full language (we refer the reader to [17,18] for further details and proofs; see also [4] for all unexplained algebraic and logical terminology). The most convenient way to do so is by taking the substructural route, starting from the notion of residuated lattice.

A *commutative integral bounded residuated lattice* (CIBRL) is an algebra $\mathbf{A} = \langle A; \wedge, \vee, *, \Rightarrow, 0, 1\rangle$ of type $\langle 2, 2, 2, 2, 0, 0\rangle$ such that:

(i) $\langle A; *, 1\rangle$ is a commutative monoid, (Mon)
(ii) $\langle A; \wedge, \vee, 0, 1\rangle$ is a bounded lattice (with order $\leq$), (Lat)
(iii) $a * b \leq c$ iff $a \leq b \Rightarrow c$ for all $a, b, c \in A$. (Res)

CIBRLs form a variety that is the algebraic counterpart of the logic $\mathcal{FL}_{ew}$, i.e. the extension of the Full Lambek Calculus $\mathcal{FL}$ obtained by adding the rules of *exchange* (*e*) and *weakening* (*w*), as well as a propositional constant (usually denoted $\perp$ or by 0) to be interpreted as the least element on the algebras. The *negation* connective/operation is defined by the term $\sim x := x \Rightarrow 0$.

Definition 1. *A* quasi-Nelson residuated lattice *or* quasi-Nelson algebra *is a CIBRL that satisfies the Nelson identity:*

$$(x \Rightarrow (x \Rightarrow y)) \wedge (\sim y \Rightarrow (\sim y \Rightarrow \sim x)) \approx x \Rightarrow y. \qquad \text{(Nelson)}$$

A Nelson residuated lattice *(or* Nelson algebra*) is a quasi-Nelson residuated lattice that additionally satisfies the involutive identity* $\sim\sim x \approx x$.

On every quasi-Nelson algebra, one can define a connective $\to$ (called *weak implication*, while the residuated $\Rightarrow$ is known as the *strong implication*) by the term $x \to y := x \Rightarrow (x \Rightarrow y)$. The weak implication is indeed a genuine implication; it is, in fact, the connective that gives (quasi-)Nelson logic a classical deduction-detachment theorem [9, Thm. 2]. As such, the weak implication can be used to introduce an alternative negation $\neg$ (distinct from $\sim$) given by the term $\neg x := x \to 0$. This observation led Sendlewski [21] to the study of *weakly pseudo-complemented Kleene algebras*, alias the $\{\wedge, \vee, \sim, \neg, 0, 1\}$-subreducts of Nelson algebras. The $\{\wedge, \vee, \sim, \neg, 0, 1\}$-subreducts of quasi-Nelson algebras are characterised in [15], and the corresponding variety dubbed *weakly pseudo-complemented quasi-Kleene algebras.* We proceed to introduce formally these classes of algebras, starting from the $\{\wedge, \vee, \sim\}$-fragment of quasi-Nelson algebras, which was studied in [16].

Definition 2 ([19])**.** *An algebra* $\mathbf{A} = \langle A; \wedge, \vee, \sim, 0, 1\rangle$ *of type* $\langle 2, 2, 1, 0, 0\rangle$ *is a* semi-De Morgan algebra *if the following properties and identities are satisfied:*

(SD1) $\langle A; \wedge, \vee, 0, 1\rangle$ *is a bounded distributive lattice,*
(SD2) $\sim 0 \approx 1$ *and* $\sim 1 \approx 0$,
(SD3) $\sim(x \vee y) \approx \sim x \wedge \sim y$,
(SD4) $\sim\sim(x \wedge y) \approx \sim\sim x \wedge \sim\sim y$,
(SD5) $\sim x \approx \sim\sim\sim x$.

A lower quasi-De Morgan algebra *is a semi-De Morgan algebra that satisfies:*

(QD) $x \ll \sim\sim x$,

(in the present paper, we take $\alpha \ll \beta$ *as an abbreviation for the formal identity* $\alpha \wedge \beta \approx \alpha$*). A* De Morgan algebra *can be defined as a semi-De Morgan algebra that further satisfies the involutive identity* $\sim\sim x \approx x$.

Besides De Morgan algebras, another well-known subvariety of semi-De Morgan algebras is the class of *pseudo-complemented distributive lattices* (also called *distributive p-algebras* or simply – as we will here – *p-lattices*). This class can be axiomatised, relative to semi-De Morgan algebras, by adding the lower quasi-De Morgan identity (QD) together with the following one [19, Cor. 2.8]: $\sim x \wedge \sim\sim x \approx 0$. The variety of p-lattices is precisely the class of $\{\wedge, \vee, \sim, 0, 1\}$-subreducts of Heyting algebras [1, Chapter VIII]. Alternatively, a p-lattice can be defined as a bounded distributive lattice $\langle A; \wedge, \vee, 0, 1\rangle$, with order $\leq$, bottom 0 and top 1, additionally satisfying the property that, for all $a, b \in A$,

(P) $\quad a \leq \sim b$ if and only if $a \wedge b = 0$. $\quad$ (pseudo-complement)

We shall refer to (P) as to the *property of the pseudo-complement.* It is useful to keep in mind that, on every distributive lattice A, the pseudo-complement $\sim b$ of each $b \in A$ (if it exists) is uniquely determined by the lattice structure in the following way: $\sim b = \bigvee\{a \in A : a \wedge b = 0\} = \max \bigvee\{a \in A : a \wedge b = 0\}$. Every p-lattice is a quasi-Kleene algebra (as defined below), but not necessarily a Kleene algebra.

Definition 3 **([16]).** *A* quasi-Kleene algebra *is a semi-De Morgan algebra* $\mathbf{A}$ *that additionally satisfies the following identities:*

(QK1) $x \wedge {\sim} x \ll y \vee {\sim} y$. *(the Kleene identity)*
(QK2) $x \ll {\sim}{\sim} x$ *(thus* $\mathbf{A}$ *is a lower quasi-De Morgan algebra)*
(QK3) ${\sim}{\sim} x \wedge {\sim}(x \wedge y) \ll {\sim} x \vee {\sim} y$.
(QK4) ${\sim}{\sim} x \wedge {\sim} x \ll x$.

A Kleene algebra *can be defined as a quasi-Kleene algebra that satisfies the involutive identity:* ${\sim}{\sim} x \approx x$.

Every Nelson algebra has a Kleene algebra reduct; indeed, Kalman's results [8] easily entail that Kleene algebras are precisely the $\{\wedge, \vee, {\sim}\}$-subreducts of Nelson algebras. Similarly, it is shown in [16, Cor. 6.6] that quasi-Kleene algebras are the $\{\wedge, \vee, {\sim}\}$-subreducts of quasi-Nelson algebras.

Given an algebra $\mathbf{A}$ having a quasi-Kleene algebra reduct and given $a, b \in A$, we write $a \preceq b$ as a shorthand for $a \leq {\sim} a \vee b$ and $a \equiv b$ as a shorthand for ($a \preceq b$ and $b \preceq a$). The binary relation associated with $\preceq$ is reflexive and transitive on every quasi-Kleene algebra. It is also clear that $a \leq b$ implies $a \preceq b$, for all $a, b \in A$. Thus, in particular, we have $0 \preceq a \preceq 1$ for all $a \in A$.

Definition 4. *A* weakly pseudo-complemented quasi-Kleene algebra *(WPQK-algebra) is an algebra* $\mathbf{A} = \langle A; \wedge, \vee, {\sim}, \neg, 0, 1\rangle$ *of type* $\langle 2, 2, 1, 1, 0, 0\rangle$ *such that:*

(i) $\langle A; \wedge, \vee, {\sim}, 0, 1\rangle$ *is a quasi-Kleene algebra,*
(ii) for all $a, b, c, d \in A$,
  1. $a \preceq \neg b$ *iff* $a \wedge b \preceq 0$ *(WP)*
  2. ${\sim}\neg a \equiv {\sim}{\sim} a$.

Item ii.1 in Definition 4 (the *property of the weak pseudo-complement*) can be equivalently replaced by the following conditions: for all $a, b \in A$,

(i) $\neg 1 = 0$,
(ii) $\neg(a \wedge {\sim} a) = 1$,
(iii) $a \wedge \neg(a \wedge b) \equiv a \wedge \neg b$.

Thus, the class of WPQK-algebras is a variety [15, Prop. 4.12]. The prime examples of WPQK-algebras are obviously the reducts of (quasi-)Nelson algebras [15, Prop. 4.4]. It is also easy to check that every p-lattice $\langle A; \wedge, \vee \neg, 0, 1\rangle$ forms a WPQK-algebra if we let ${\sim} x := \neg x$ (cf. [19, Cor. 2.8]). Sendlewski's wp-Kleene algebras are precisely the subvariety of WPQK-algebras satisfying the involutive identity ${\sim}{\sim} x \approx x$ [15, Prop. 4.15]. The reduct $\langle A; \wedge, \vee, \neg, 0, 1\rangle$ of a WPQK-algebra need not be a quasi-Kleene algebra, for the analogue of (QK2) for $\neg$ need not be satisfied. In fact, (SD4) and (SD5) may also fail, suggesting that $\langle A; \wedge, \vee, \neg, 0, 1\rangle$ may not even be a semi-De Morgan algebra. For further examples and properties of WPQK-algebras, see [15].

Proposition 1 **([15], Cor. 5.5).** *The class of* $\{\wedge, \vee, {\sim}, \neg, 0, 1\}$*-subreducts of quasi-Nelson algebras is precisely the variety of WPQK-algebras, and the class of* $\{\wedge, \vee, {\sim}, \neg, 0, 1\}$*-subreducts of Nelson algebras is the variety of wp-Kleene algebras.*

We note that Proposition 1 is informative also because, in general, the class of subreducts of a variety of algebras (in some proper subsignature) forms a quasi-variety but not necessarily a variety. The next and most fundamental result about WPQK-algebras is the representation as twist-algebras over pairs of p-lattices. Given a p-lattice $\mathbf{L} = \langle L; \wedge, \vee, \neg, 0, 1\rangle$ and a lattice filter $\nabla \subseteq L$, we say that ∇ is *dense* if $D(\mathbf{L}) \subseteq \nabla$, where $D(\mathbf{L}) := \{a \vee \neg a : a \in L\} = \{a \in L : \neg a = 0\}$. Notice that $D(\mathbf{L})$ is itself a lattice filter.

Definition 5. *A* WPQK-twist-structure *is a tuple* $\mathbb{L} = \langle \mathbf{L}_+, \mathbf{L}_-, n, p, \nabla\rangle$ *where* $\mathbf{L}_+ = \langle L_+; \wedge_+, \vee_+, \neg_+, 0_+, 1_+\rangle$ *is a p-lattice (with order* $\leq_+$*),* $\nabla \subseteq L_+$ *a dense filter,* $\mathbf{L}_- = \langle L_-; \wedge_-, \vee_-, 0_-, 1_-\rangle$ *a bounded distributive lattice (with order* $\leq_-$*), and* $n\colon L_+ \to L_-$ *and* $p\colon L_- \to L_+$ *are maps satisfying the following properties:*

(i) n *is a bounded lattice homomorphism,*
(ii) p *preserves finite meets and both lattice bounds,*
(iii) $n \circ p = Id_{L_-}$ *and* $Id_{L_+} \leq_+ p \circ n$.

The algebra $\mathbf{L}_+ \bowtie \mathbf{L}_- = \langle L_+ \times L_-; \wedge, \vee, \sim, \neg, 0, 1\rangle$ *is defined as follows. For all* $\langle a_+, a_-\rangle, \langle b_+, b_-\rangle \in L_+ \times L_-$,

$$1 := \langle 1_+, 0_-\rangle, \quad 0 := \langle 0_+, 1_-\rangle,$$
$$\sim\langle a_+, a_-\rangle := \langle p(a_-), n(a_+)\rangle, \quad \neg\langle a_+, a_-\rangle := \langle \neg_+ a_+, n(a_+)\rangle,$$
$$\langle a_+, a_-\rangle \wedge \langle b_+, b_-\rangle := \langle a_+ \wedge_+ b_+, a_- \vee_- b_-\rangle,$$
$$\langle a_+, a_-\rangle \vee \langle b_+, b_-\rangle := \langle a_+ \vee_+ b_+, a_- \wedge_- b_-\rangle.$$

The weakly pseudo-complemented quasi-Kleene twist-algebra *(WPQK twist-algebra)* $Tw(\mathbb{L})$ *is the* $\{\wedge, \vee, \sim, \neg, 0, 1\}$*-subreduct of* $\mathbf{L}_+ \bowtie \mathbf{L}_-$ *with universe:*

$$\{\langle a_+, a_-\rangle \in L_+ \times L_- : a_+ \vee_+ p(a_-) \in \nabla, a_+ \wedge_+ p(a_-) = 0_+\}.$$

While the whole algebra $\mathbf{L}_+ \bowtie \mathbf{L}_-$ does not need to be a WPQK-algebra, every WPQK twist-algebra is a WPQK-algebra [15, Prop. 4.3]. Moreover, every WPQK-algebra arises in this way [15, Thm. 6.2]. Before demonstrating this, let us comment on a few consequences of Definition 5 that will be useful in the next sections.

The maps n and p form an adjoint pair between the posets $\langle L_+, \leq_+\rangle$ and $\langle L_-, \leq_-\rangle$. As is well known, this entails that n preserves arbitrary existing joins and p arbitrary existing meets. Moreover, the lattice $\mathbf{L}_-$ is also pseudo-complemented, with the pseudo-complement given by $\neg_- a_- = n(\neg_+ p(a_-))$ for all $a_- \in A_-$. Both maps n and p preserve the pseudo-complement operation [15, Prop. 3.4].

Let $\mathbf{A} = \langle A; \wedge, \vee, \sim, \neg, 0, 1\rangle$, be a WPQK-algebra. Then the relation $\equiv$ introduced above is a congruence of the $\{\sim\}$-free reduct of $\mathbf{A}$ [15, Cor. 4.7], and the quotient algebra $\mathbf{A}_+ = \langle A/\equiv; \wedge, \vee, \neg, 0, 1\rangle$ is a p-lattice [15, Prop. 4.8]. This gives us the first factor for the twist representation. The second can be obtained as follows. Endow the set $A_- := \{[\sim a] : a \in A\} \subseteq A_+$ with operations given, for all $a, b \in A$, by:

$$[\sim a] \wedge_- [\sim b] := [\sim(a \vee b)] = [\sim a \wedge \sim b] = [\sim a] \wedge_+ [\sim b],$$
$$[\sim a] \vee_- [\sim b] := [\sim(a \wedge b)],$$
$$0_- := [\sim 1] = [0] = 0_+, \qquad 1_- := [\sim 0] = [1] = 1_+.$$

The pseudo-complement operation on A_- can be defined by $\neg_-[\sim a] := [\neg \sim a] = \neg_+[\sim a]$, obtaining a second p-lattice $\mathbf{A}_- = \langle A_-, \wedge_-, \vee_-, \neg_-, 0_-, 1_- \rangle$; see [15, Prop. 4.9]. The maps $p_{\mathbf{A}} \colon A_- \to A_+$ and $n_{\mathbf{A}} \colon A_+ \to A_-$ between $\mathbf{A}_+$ and $\mathbf{A}_-$ are given as follows: $p_{\mathbf{A}}$ is the identity map on A_-, and $n_{\mathbf{A}}([a]) := [\sim\sim a]$ for all $a \in A$. Note that $n_{\mathbf{A}}$ is well-defined because $a \equiv b$ entails $\sim\sim a \equiv \sim\sim b$ [16, Prop. 3.15.viii]. To obtain the dense filter $\nabla_{\mathbf{A}}$, we consider the set $F(\mathbf{A}) := \{a \in A : \sim a \leq a\}$ and we let $[F(\mathbf{A})] := \{[a] : a \equiv b \text{ for some } b \in F(\mathbf{A})\}$.

Theorem 1 (**[15], Thm. 6.2**). *Let* $\mathbf{A}$ *be a WPQK-algebra.*

(i) $\nabla_{\mathbf{A}} := [F(\mathbf{A})]$ *is a lattice filter of* $\mathbf{A}_+$ *and* $D(\mathbf{A}_+) \subseteq \nabla_{\mathbf{A}}$.
(ii) $\mathbf{A} \cong Tw(\langle \mathbf{A}_+, \mathbf{A}_-, n_{\mathbf{A}}, p_{\mathbf{A}}, \nabla_{\mathbf{A}} \rangle)$ *via the map* $\iota_{\mathbf{A}} \colon A \to A_+ \times A_-$ *given by* $\iota_{\mathbf{A}}(a) := \langle [a], [\sim a] \rangle$ *for all* $a \in A$.

By Theorem 1, every WPQK-algebra $\mathbf{A}$ is uniquely determined by a tuple $\langle \mathbf{A}_+, \mathbf{A}_-, n_{\mathbf{A}}, p_{\mathbf{A}}, \nabla_{\mathbf{A}} \rangle$. This correspondence at the object level can be extended to suitably defined morphisms, obtaining a (co-variant) categorical equivalence between the algebraic category WPQK of WPQK-algebras (with algebraic homomorphisms) and the category TW defined as follows.

Definition 6. *Let* TW *be the category having the WPQK-twist-structure given in Definition 5 as objects and as morphisms between objects* $\mathbb{L}$ *and* $\mathbb{L}'$ *the pairs* $\langle h_+, h_- \rangle$, *where* $h_+ \colon L_+ \to L'_+$ *is a p-lattice homomorphism such that* $h_+[\nabla] \subseteq \nabla'$, $h_- \colon L_- \to L'_-$ *is a bounded lattice homomorphism,* $h_+ \circ p = p' \circ h_-$ *and* $n' \circ h_+ = h_- \circ n$. *The composition of morphisms is given componentwise, that is, the composition of two composable morphisms* $\langle h_+, h_- \rangle$ *and* $\langle k_+, k_- \rangle$ *is* $\langle h_+ \circ k_+, h_- \circ k_- \rangle$. *The identity morphism for each* $\mathbb{L} \in$ TW *is the morphism* $\langle Id_{L_+}, Id_{L_-} \rangle$.

Checking that TW is indeed a category is straightforward. We define the functors $F \colon$ TW $\to$ WPQK and $G \colon$ WPQK $\to$ TW as follows. For every object $\mathbb{L} \in$ TW, we let $F(\mathbb{L}) := Tw(\mathbb{L})$. For a TW-morphism $\langle h_+, h_- \rangle$ from $\mathbb{L}$ to $\mathbb{L}'$, we let $F(\langle h_+, h_- \rangle)$ be given, for all $\langle a_+, a_- \rangle \in L_+ \times L_-$, by $F(\langle h_+, h_- \rangle)(\langle a_+, a_- \rangle) := \langle h_+(a_+), h_-(a_-) \rangle$. Conversely, for every WPQK-algebra $\mathbf{A}$, we let $G(\mathbf{A}) := \langle \mathbf{A}_+, \mathbf{A}_-, n_{\mathbf{A}}, p_{\mathbf{A}}, \nabla_{\mathbf{A}} \rangle$. For a WPQK-homomorphism $k \colon \mathbf{A} \to \mathbf{A}'$, we let $k_+ : \mathbf{A}_+ \to \mathbf{A}'_+$ and $k_- : \mathbf{A}_- \to \mathbf{A}'_-$ be the homomorphisms defined, respectively, by setting $k_+([a]) := [k(a)]$ and $k_-([\sim a]) := [k(\sim a)]$ for every $a \in A$. We then let $G(k) := \langle k_+, k_- \rangle$. It is easy to check that F and G are well-defined functors (concerning the role of the filter ∇, see e.g. [6]). Given $\mathbb{L} \in$ TW, we define the morphism $\iota_{\mathbb{L}} = \langle \iota_{L_+}, \iota_{L_-} \rangle$ from $\mathbb{L}$ to $G(F(\mathbb{L}))$ by setting $\iota_{L_+}(a_+) := [\langle a_+, n(\neg_+ a_+) \rangle]$ for every $a_+ \in L_+$ and $\iota_{L_-}(a_-) := [\sim\langle \neg_+ p(a_-), a_- \rangle]$ for every $a_- \in L_-$. The morphism $\iota_{\mathbb{L}}$ is an isomorphism between $\mathbb{L}$ and $GF(\mathbb{L})$. In this way we have a natural isomorphism from the identity functor on TW and the functor $G \circ F$.

Theorem 2. *The functors* F, G *establish an equivalence between* TW *and* WPQK.

3 Two-Sorted Duality

We are now going to introduce a Priestley-style duality for the category TW, which (in the light of Theorem 2) we view as a two-sorted alter ego of WPQK. We assume the reader is familiar with the basic duality results on distributive lattices [11], which we now briefly recall.

Priestley duality concerns the category D of bounded distributive lattices and bounded lattice homomorphisms. To every bounded distributive lattice $\mathbf{L}$, one associates the set $X(\mathbf{L})$ of its prime filters. On $X(\mathbf{L})$ one imposes the *Priestley topology* τ, generated by the sets $\phi(a) := \{P \in X(\mathbf{L}) : a \in P\}$ and $\phi'(a) := \{P \in X(\mathbf{L}) : a \notin P\}$, and the inclusion relation between prime filters as an order. The resulting ordered topological structures are called *Priestley spaces*[1]. A homomorphism h between bounded distributive lattices $\mathbf{L}$ and $\mathbf{L}'$ gives rise to a function $X(h) : X(\mathbf{L}') \to X(\mathbf{L})$, defined by $X(h)(P) = h^{-1}[P]$, that is continuous and order preserving. Taking functions with these properties (called *Priestley functions*) as morphisms between Priestley spaces, one obtains the category PrSp, and X is a contravariant functor from D to PrSp. For a functor in the opposite direction, one associates to every Priestley space $X = \langle X, \tau, \leq\rangle$ the set $L(X)$ of clopen up-sets. This is a bounded distributive lattice with respect to the set-theoretic operations $\cap, \cup, \emptyset$, and X. To a Priestley function $f : X \to X'$ one associates the function $L(f)$, given by $L(f)(U') = f^{-1}[U']$, which is a bounded lattice homomorphism from $L(X')$ to $L(X)$. Then L constitutes a contravariant functor from PrSp to D. The two functors are adjoint to each other with the units given by:

$$\Phi_{\mathbf{L}} : \mathbf{L} \to L(X(\mathbf{L})) \qquad \Phi_{\mathbf{L}}(a) := \{P \in X(\mathbf{L}) : a \in P\}$$
$$\Psi_X : X \to X(L(X)) \qquad \Psi_X(x) := \{U \in L(X) : x \in U\}.$$

These are the components of a natural transformation from the identity functor on D to $L \circ X$, and from the identity functor on PrSp to $X \circ L$, respectively. In particular, they are morphisms in their respective categories. Furthermore, they are *isomorphisms* and thus the central result of Priestley duality is obtained: the categories D and PrSp are dually equivalent.

A description of spaces dual to p-lattices can be found in [12,13]; see also [1]. We recall here the basic results that shall be needed. For a subset $Y \subseteq X$ of a Priestley space X, we let $\downarrow Y := \{x \in X : x \leq y \text{ for some } y \in Y\}$.

Proposition 2 ([12], Prop. 1). *A distributive lattice* $\mathbf{L}$ *is a p-lattice (i.e. can be endowed with a pseudo-complement operation) if and only if, for every clopen up-set* $U \in L(X(\mathbf{L}))$*, the set* $\downarrow U$ *is open in* $\langle X(\mathbf{L}), \subseteq, \tau_{\mathbf{L}}\rangle$.

Definition 7. *A* p-space *is a Priestley space* $\langle X, \leq, \tau\rangle$ *such that* $\downarrow U$ *is* τ*-open for all* $U \in L(X)$.

[1] Abstractly, a *Priestley space* is defined as a compact ordered topological space $\langle X, \tau, \leq\rangle$ such that, for all $x, y \in X$, if $x \nleq y$, then there is a clopen up-set $U \subseteq X$ with $x \in U$ and $y \notin U$. It follows that $\langle X, \tau\rangle$ is a Stone space.

Let $\langle X, \leq, \tau \rangle$ be a p-space and $U \in L(X)$. Then, defining

$$\neg U := X - \downarrow U = \{x \in X : \uparrow x \cap U = \emptyset\},$$

where $\uparrow x := \{y \in X : x \leq y\}$, we have that $\neg U \in L(X)$ is the pseudo-complement of U in $L(X)$, turning the distributive lattice of clopen up-sets into a p-lattice. Given a Priestley space $\langle X, \leq, \tau \rangle$, let $\max(X) := \{y \in X : y \text{ is } \leq\text{-maximal}\}$ and, for $x \in X$, $\max(x) := \{y \in X : x \leq y \text{ and } y \text{ is } \leq\text{-maximal}\}$.

Definition 8. *A* morphism *between p-spaces $\langle X, \leq, \tau \rangle$ and $\langle X', \leq', \tau' \rangle$ is a continuous order-preserving map $f : X \to X'$ such that $f[\max(x)] = \max(f(x))$ for all $x \in X$.*

Proposition 3 ([13], Prop. 3). *Let $\langle X, \leq, \tau \rangle$ and $\langle X', \leq', \tau' \rangle$ be p-spaces, and let $f : X \to X'$ be a continuous order-preserving map. Then f is a morphism of p-spaces if and only if $f^{-1} : L(X') \to L(X)$ preserves the pseudo-complement operation.*

The above results entail that the Priestley functors L and X establish a dual equivalence between the category of p-lattices (with algebraic homomorphisms) and the category of p-spaces with the morphisms given as per Definition 8. Upon this result we will build our two-sorted duality.

Let $\mathbb{L} = \langle \mathbf{L}_+, \mathbf{L}_-, n, p, \nabla \rangle \in \mathsf{TW}$. Then $\langle \mathbf{L}_+, \mathbf{L}_-, n, p \rangle$ is a *two-sorted lattice* in the sense of [7, Definition 4.1]. We thus follow [7] in defining its dual, the *two-sorted Priestley space* $X(\mathbb{L}) = \langle X(\mathbf{L}_+), X(\mathbf{L}_-), X(n), X(p), X(\nabla) \rangle$, as follows:

(i) $\langle X(\mathbf{L}_+), \tau_+, \leq_+ \rangle$ is the Priestley space of $\mathbf{L}_+$;
(ii) $\langle X(\mathbf{L}_-), \tau_-, \leq_- \rangle$ is the Priestley space of $\mathbf{L}_-$;
(iii) $X(p) \subseteq X(\mathbf{L}_+) \times X(\mathbf{L}_-)$ and $X(n) \subseteq X(\mathbf{L}_-) \times X(\mathbf{L}_+)$ are relations defined as follows:

$$X(n) := \{\langle P_-, P_+ \rangle \in X(\mathbf{L}_-) \times X(\mathbf{L}_+) : n^{-1}[P_-] \subseteq P_+\}$$

$$X(p) := \{\langle P_+, P_- \rangle \in X(\mathbf{L}_+) \times X(\mathbf{L}_-) : p^{-1}[P_+] \subseteq P_-\};$$

(iv) $X(\nabla) := \{P_+ \in X(\mathbf{L}_+) : \nabla \subseteq P_+\}$.

For item (iii), besides [7, Definition 4.1], we refer the reader to [3]. For (iv) and the subsequent treatment of $\nabla_{\mathbf{A}}$ (as well as its dual alter ego $\mathcal{C}_+$), see [6, Sec. 3.3]. Given sets X, Y, a relation $R \subseteq X \times Y$ and a subset $X' \subseteq X$, define: $R[X'] = \{y \in Y : \text{there is } x \in X' \text{ s.t. } \langle x, y \rangle \in R\}$. In particular, for $X' = \{x\}$, we write $R[x]$ instead of $R[\{x\}]$. For $Y' \subseteq Y$, let:

$$\Box_R Y' = \{x \in X : R[x] \subseteq Y'\}. \tag{BoxR}$$

The following proposition characterises the spaces that correspond to objects in TW.

Proposition 4. *Let $\mathbb{L} = \langle \mathbf{L}_+, \mathbf{L}_-, n, p, \nabla \rangle \in \mathsf{TW}$, and let the corresponding two-sorted Priestley space be $X(\mathbb{L}) = \langle X(\mathbf{L}_+), X(\mathbf{L}_-), X(n), X(p), X(\nabla) \rangle$. Then:*

(i) $X(n)[x_-] \subseteq X(\mathbf{L}_+)$ *and* $X(p)[x_+] \subseteq X(\mathbf{L}_-)$ *are non-empty closed up-sets;*
(ii) $\Box_{R_n} \circ \Phi_{\mathbf{L}_+} = \Phi_{\mathbf{L}_-} \circ n$ *and* $\Box_{R_p} \circ \Phi_{\mathbf{L}_-} = \Phi_{\mathbf{L}_+} \circ p$.
(iii) $X(n)$ *is* functional, *i.e., for all* $x_- \in X(\mathbf{L}_-)$ *there is* $x_+ \in X(\mathbf{L}_+)$ *such that* $\uparrow x_+ = X(n)[x_-]$.
(iv) $\leq_- = X(p) \circ X(n)$.
(v) $(X(n) \circ X(p)) \subseteq \leq_+$.
(vi) $X(\nabla_{\mathbf{A}})$ *is a* τ_+*-closed set such that* $X(\nabla) \subseteq \max(X(\mathbf{L}_+))$.

Proof. For items (i) and (ii), see [7, Prop. 4.2]. Item (iii) follows from [7, Prop. 5.1]. Items (iv) and (v) follow from [7, Prop. 5.3]. For (vi), see [6, Sec. 3.3].

We turn Proposition 4 into our official definition of two-sorted *p*-spaces.

Definition 9 (cf. [7]**, Def. 4.3).** *A* two-sorted *p*-space *is a structure* $X = \langle X_+, X_-, R_n, R_p, \mathcal{C}_+ \rangle$ *such that:*

(i) $X_+ = \langle X, \tau_+, \leq_+ \rangle$ *and* $X_- = \langle X, \tau_-, \leq_- \rangle$ *are p-spaces.*
(ii) $R_n \subseteq X_- \times X_+$, $R_p \subseteq X_+ \times X_-$ *satisfy:*
 1. $R_n[x_-]$ *and* $R_p[x_+]$ *are non-empty closed up-sets, for all* x_- *and* x_+*;*
 2. for all $U_- \in L(X_-)$, $U_+ \in L(X_+)$, *we have* $\Box_{R_p} U_- \in L(X_+)$ *and* $\Box_{R_n} U_+ \in L(X_-)$.
(iii) For all $x_- \in X_-$ *there is* $x_+ \in X_+$ *such that* $\uparrow x_+ = R_n[x_-]$.
(iv) $\leq_- = R_p \circ R_n$.
(v) $(R_n \circ R_p) \subseteq \leq_+$.
(vi) $\mathcal{C}_+$ *is a* τ_+*-closed set such that* $\mathcal{C}_+ \subseteq \max(X_+)$.

Given a two-sorted *p*-space $X = \langle X_+, X_-, R_n, R_p, \mathcal{C}_+ \rangle$, the dual $L(X) = \langle L(X_+), L(X_-), \Box_{R_n}, \Box_{R_p}, \nabla_{\mathcal{C}_+} \rangle$ is constructed in the expected way: $L(X_+)$ and $L(X_-)$ are as prescribed by Priestley duality (for *p*-lattices), $\Box_{R_n}, \Box_{R_p}$ are given as in (BoxR), and $\nabla_{\mathcal{C}_+} := \{U_+ \in \mathbf{L}(X_+) : \mathcal{C}_+ \subseteq U_+\}$.

Definition 10. *Let* $X = \langle X_+, X_-, R_n, R_p, \mathcal{C}_+ \rangle$ *and* $X' = \langle X'_+, X'_-, R'_n, R'_p, \mathcal{C}'_+ \rangle$ *be two-sorted p-spaces and let* $f_+ \colon X_+ \to X'_+$ *and* $f_- \colon X_- \to X'_-$ *be maps. The pair* $f = \langle f_+, f_- \rangle$ *is a* two-sorted *p*-space morphism *if the following conditions hold:*

(i) f_+ *and* f_- *are p-space morphisms.*
(ii) f *preserves* R_p *and* R_n, *that is, if* $\langle x_+, x_- \rangle \in R_p$, *then* $\langle f_+(x_+), f_-(x_-) \rangle \in R'_p$, *etc.*
(iii) f_+ *and* f_- *are bounded morphisms, that is,*
 1. if $\langle f_+(x_+), x'_- \rangle \in R'_p$, *then there is* $x_- \in X_-$ *such that* $f_-(x_-) \leq'_- x'_-$ *and* $\langle x_+, x_- \rangle \in R_p$.
 2. if $\langle f_-(x_-), x'_+ \rangle \in R'_n$, *then there is* $x_+ \in X_+$ *such that* $f_+(x_+) \leq'_+ x'_+$ *and* $\langle x_-, x_+ \rangle \in R_n$.
(iv) $f_+[\mathcal{C}_+] \subseteq \mathcal{C}'_+$.

We denote by 2pSP the category whose objects are two-sorted p-spaces and whose morphisms are given as in Definition 10. For every TW-morphism $h = \langle h_+, h_- \rangle$, the dual pair of maps $\langle X(h_+), X(h_-) \rangle$ form a 2pSP-morphism (see [7, Prop. 4.8], [6, Lemma 3.5]). Conversely, for every 2pSP-morphism $f = \langle f_+, f_- \rangle$, the dual pair of maps $\langle L(f_+), L(f_-) \rangle$ forms a TW-morphism according to Definition 6 (see [7, Prop. 4.8], [6, Lemma 3.6]). These observations, together with the above-mentioned results from Priestley duality (for D) easily entail the following.

Theorem 3. *The functors L and X establish a dual equivalence between* TW *and* 2pSP. *In consequence, the composite functors $F \circ L$ and $X \circ G$ establish a dual equivalence between* WPQK *and* 2pSP.

4 Nuclear Duality

In this section we propose an alternative duality based on the alternative representation for WPQK-algebras introduced in [15, Sec. 8], which in turn arises from the following observations.

Let $\mathbb{L} = \langle \mathbf{L}_+, \mathbf{L}_-, n, p, \nabla \rangle \in$ TW. Define the operation $\Box \colon L_+ \to L_+$ by $\Box a_+ := pn(a_+)$ for all $a_+ \in L_+$. Then $\Box$ is a *(dense) nucleus* on $\mathbf{L}_+$ in the sense of e.g. [2]. One can further show that $\mathbf{L}_-$ is isomorphic to the algebra $\mathbf{L}_+^{\Box} := \langle L_+^{\Box}; \wedge_+, \vee_+^{\Box}, 0_+, 1_+ \rangle$ with universe $L_+^{\Box} := \{\Box a_+ : a_+ \in A_+\}$ and operations given by the restrictions of those of $\mathbf{L}_+$ except for the join, which is defined as $a_+ \vee_+^{\Box} b_+ := \Box(a_+ \vee_+ b_+)$ for all $a_+, b_+ \in L_+^{\Box}$. This suggests that a tuple $\langle \mathbf{L}_+, \mathbf{L}_-, n, p, \nabla_{\mathbf{A}} \rangle$ can be represented by a pair $\langle \mathbf{L}, \nabla \rangle$, with $\mathbf{L}$ a p-lattice enriched with a nucleus (the maps n and p being replaced by, respectively, the map $\Box \colon L \to L^{\Box} = \{\Box a : a \in L\}$ and the identity map on $L^{\Box}$). In this way one obtains an alternative representation for WPQK-algebras [15, Sec. 8].

Definition 11. *A* nuclear p-lattice *(*np-lattice *for short) is an algebra* $\mathbf{L} = \langle L; \wedge, \vee, \neg, \Box, 0, 1 \rangle$ *of type* $\langle 2, 2, 1, 1, 0, 0 \rangle$ *such that:*

(i) $\langle L; \wedge, \vee, \neg, 0, 1 \rangle$ *is a p-lattice (with order* $\leq$*).*
(ii) The operator $\Box$ *is a* dense nucleus *on* L*, that is, for all* $a, b \in L$*,*
   1. $\Box 0 = 0$
   2. $\Box(a \wedge b) = \Box a \wedge \Box b$
   3. $a \leq \Box a = \Box\Box a$.

Given an np-lattice $\mathbf{L}$ and a dense filter $\nabla \subseteq L$, we can define a bounded distributive lattice $\mathbf{L}^{\Box}$ and maps $\Box \colon L \to L^{\Box}$ and $Id_{L^{\Box}} \colon L^{\Box} \to L$, so that $\langle \mathbf{L}, \mathbf{L}^{\Box}, \Box, Id_{L^{\Box}}, \nabla \rangle \in$ TW. This gives us the following representation for WPQK-algebras.

Theorem 4 ([15], Thm. 8.5). *Every WPQK-algebra* $\mathbf{A}$ *is isomorphic to the WPQK twist-algebra* $Tw\langle \mathbf{A}_+, \mathbf{A}_+^{\Box}, \Box, Id_{L_+^{\Box}}, \nabla_{\mathbf{A}} \rangle$*, where* $\mathbf{A}_+^{\Box}$ *arises from the np-lattice* $(\mathbf{A}_+, \Box)$ *obtained from* $\mathbf{A}_+$ *with the nucleus given by* $\Box[a] := [\sim\sim a]$ *for all* $a \in A$*, and* $\nabla_{\mathbf{A}} \subseteq \mathbf{A}_+$ *is the dense filter of the twist representation of* $\mathbf{A}$.

Relying on Theorem 4, we can proceed in parallel to Sect. 3.

Definition 12. *Let* NP *be the category whose objects are tuples* $\mathbb{L} = \langle \mathbf{L}, \nabla \rangle$, *with* $\mathbf{L}$ *an np-lattice and* $\nabla \subseteq L$ *a dense filter. A* morphism *between objects* $\mathbb{L} = \langle \mathbf{L}, \nabla \rangle$ *and* $\mathbb{L}' = \langle \mathbf{L}', \nabla' \rangle$ *is an np-lattice homomorphism* h *such that* $h[\nabla] \subseteq \nabla'$.

Given $\mathbb{L} = \langle \mathbf{L}, \nabla \rangle \in \mathsf{NP}$, the WPQK twist-algebra $Tw\langle \mathbf{L}, \mathbf{L}^{\Box}, \Box, Id_{L^{\Box}}, \nabla \rangle$ will be denoted by $Tw(\mathbb{L})$. The equivalence between NP and WPQK is given by the functors $H \colon \mathsf{NP} \to \mathsf{WPQK}$ and $K \colon \mathsf{WPQK} \to \mathsf{NP}$ defined as follows. For every $\mathbb{L} = \langle \mathbf{L}, \nabla \rangle \in \mathsf{NP}$, we let $H(\mathbb{L}) := Tw(\mathbb{L})$. For an NP-morphism $h \colon \mathbb{L} \to \mathbb{L}'$ between $\mathbb{L} = \langle \mathbf{L}, \nabla \rangle$ and $\mathbb{L}' = \langle \mathbf{L}', \nabla' \rangle$, we let $H(h) : H(\mathbb{L}) \to H(\mathbb{L}')$ be given, for all $\langle a, b \rangle \in L \times L^{\Box}$, by $H(h)(\langle a, b \rangle) := \langle h(a), h(b) \rangle$. Conversely, for every WPQK-algebra $\mathbf{A}$, let $K(\mathbf{A}) := \langle \langle \mathbf{A}_{+}, \Box \rangle, \nabla_{\mathbf{A}} \rangle$. For a WPQK-homomorphism $k \colon \mathbf{A} \to \mathbf{A}'$, let $K(k) : K(\mathbf{A}) \to K(\mathbf{A}')$ be defined by setting $K(k)([a]) := [k(a)]$ for every $a \in A$.

Theorem 5. *The functors* H *and* K *establish a co-variant equivalence between* NP *and* WPQK.

As with Theorem 2, we can rely on Theorem 5 to introduce a duality for the category NP viewed as another alter ego of WPQK. In doing so, since the nucleus is a modal-like operator, we shall rely on duality for distributive lattices with operators (see [5]).

As expected, the dual of $\mathbb{L} = \langle \mathbf{L}, \nabla \rangle$ is defined as a structure $X(\mathbb{L}) = \langle X(\mathbf{L}), X(\Box), X(\nabla) \rangle$, where:

(i) $\langle X(\mathbf{L}), \tau_{+}, \leq_{+} \rangle$ is the p-space of $\mathbf{L}$;
(ii) $X(\Box) \subseteq X(\mathbf{L}) \times X(\mathbf{L})$ is a relation given by:

$$X(\Box) := \{\langle P, Q \rangle \in X(\mathbf{L}) \times X(\mathbf{L}) : \Box^{-1}[P] \subseteq Q\};$$

(iii) $X(\nabla) := \{P \in X(\mathbf{L}) : \nabla \subseteq P\}$.

It is well known that the relation R corresponding to an operation $\Box$ that preserves (at least) finite meets and the top element satisfies the following:

(i) $\leq \circ R \circ \leq \subseteq R$, where $\leq$ is the Priestley order on $X(\mathbf{L})$;
(ii) $R[P] := \{Q \in X(\mathbf{L}) : \langle P, Q \rangle \in R\}$ is a closed set in the Priestley topology;
(iii) $R^{-1}[U] := \{P \in X(\mathbf{L}) : \langle P, Q \rangle \in R$ for some $Q \in U\}$ is clopen, for all $U \in L(X(\mathbf{L}))$.

We shall say that a relation R on a Priestley space satisfying the above properties is a $\Box$-relation, i.e., a relation corresponding to an operator $\Box$ on the distributive lattice dual to the space. We list below four further properties that the Priestley dual of every *np*-lattice satisfies.

Proposition 5. *Let* $\langle X(\mathbf{L}), X(\Box) \rangle$ *be the dual space of an np-lattice* $\mathbf{L}$ *and* $P, Q \in X(\mathbf{L})$.

(i) *There is* $Q' \in X(\mathbf{L})$ *such that* $\langle P, Q' \rangle \in X(\Box)$.
(ii) *If* $\langle P, Q \rangle \in X(\Box)$, *then there is* $Q' \in X(\mathbf{L})$ *s.t.* $\langle P, Q' \rangle \in X(\Box)$ *and* $\langle Q', Q \rangle \in X(\Box)$.
(iii) *If* $\langle P, Q \rangle \in X(\Box)$, *then* $P \subseteq Q$.
(iv) $X(\nabla)$ *is a closed set such that* $X(\nabla) \subseteq \max(X(\mathbf{L}))$.

Items (i)–(iii) of Proposition 5 are saying that the relation $X(\Box)$ is serial, dense and included in the Priestley order (regarding item (iv) see [6, Lemma 3.1]). We include these properties in our official definition of *np*-spaces.

Definition 13. *An np*-space *is a structure* $X = \langle X, R, \mathcal{C} \rangle$ *where* X *is a p-space,* $\mathcal{C}$ *is a closed set such that* $\mathcal{C} \subseteq \max(X)$, *and* R *is a* $\Box$*-relation which is serial, dense and included in the Priestley order.*

We note that it is possible to show that Properties (i) to (iii) in Definition 13 can be equivalently replaced by the following one (considered e.g. in [2]): for all $x, y \in X$, $\langle x, y \rangle \in R$ iff there is $z \in X$ s.t. $\langle z, z \rangle \in R$ and $x \leq z \leq y$.

Given an *np*-space $X = \langle X, R, \mathcal{C} \rangle$ and a clopen up-set $U \in L(X)$, we define $\Box_R U := \{x \in X : R[x] \subseteq U\}$. Defining the filter $\nabla_{\mathcal{C}} \subseteq L(X)$ as in the preceding section, it is easy to show that $\langle L(X), \Box_R, \nabla_{\mathcal{C}} \rangle \in \mathsf{NP}$. It follows from the duality for distributive lattices with a $\Box$-operator [5] that every *np*-lattice $\mathbf{L}$ is isomorphic to its double dual $\langle L(X(\mathbf{L})), \Box_{R_\Box} \rangle$. Conversely, for every *np*-space $X = \langle X, R, \mathcal{C} \rangle$, we have (by the duality for *p*-lattices) that the *p*-space X is homeomorphic to its double dual $X(L(X))$; furthermore, $\langle X, \leq, R \rangle$ is isomorphic, as a relational structure, to $\langle X(L(X)), \subseteq, R_{\Box_R} \rangle$. It is also easy to check that the Priestley isomorphisms respect ∇ and $\mathcal{C}$ [6, Lemmas 3.7 and 3.8].

Definition 14. *Let* $X = \langle X, R, \mathcal{C} \rangle$ *and* $X' = \langle X', R', \mathcal{C}' \rangle$ *be np-spaces, and let* $f \colon X \to X'$ *be a p-space morphism. We say that* f *is an np*-space morphism *if the following conditions hold:*

(i) *If* $\langle x, y \rangle \in R$, *then* $\langle f(x), f(y) \rangle \in R'$, *for all* $x, y \in X$.
(ii) *If* $\langle f(x), x' \rangle \in R'$, *then there is* $y \in X$ *such that* $f(y) \leq' x'$ *and* $\langle x, y \rangle \in R$, *for every* $x \in X$ *and* $x' \in X'$.
(iii) $f[\mathcal{C}] \subseteq \mathcal{C}'$.

Denote by npSP the category having as objects *np*-spaces and as morphisms the maps given in Definition 14. The following propositions are immediate consequences of Priestley duality for distributive lattices with a $\Box$-operator, together with [6, Lemmas 3.5 and 3.6].

Proposition 6. *Let* $\mathbb{L} = \langle \mathbf{L}, \nabla \rangle$, $\mathbb{L}' = \langle \mathbf{L}', \nabla' \rangle \in \mathsf{NP}$, *and let* $h \colon L \to L'$ *be an* NP*-morphism. Then* $h^{-1} \colon X(\mathbb{L}') \to X(\mathbb{L})$ *is an np-space morphism.*

Proposition 7. *Let* $X = \langle X, R, \mathcal{C} \rangle$ *and* $X' = \langle X', R', \mathcal{C}' \rangle$ *be np-spaces, and let* $f \colon X \to X'$ *be an np-morphism. Then* $f^{-1} \colon L(X') \to L(X)$ *is an* NP*-morphism.*

As in the previous cases, it is straightforward to check that the above-defined categories are dually equivalent via the Priestley functors.

Theorem 6. *The functors L, X establish a dual equivalence between* NP *and* npSP*. In consequence, the composite functors $F \circ L$ and $X \circ G$ establish a dual equivalence between* WPQK *and* npSP.

Appendix: Proofs of Theorems 2 and 5

Theorem 2. *The functors F, G establish a co-variant equivalence between* TW *and* WPQK.

Proof. First of all, we shall prove that F is a functor. Given $\mathbb{L} = \langle \mathbf{L}_+, \mathbf{L}_-, n, p, \nabla \rangle$ and $\mathbb{L}' = \langle \mathbf{L}'_+, \mathbf{L}'_-, n', p', \nabla' \rangle$ two objects in TW and $\langle h_+, h_- \rangle$ a TW-morphism from $\mathbb{L}$ to $\mathbb{L}'$, we will prove that $F(\langle h_+, h_- \rangle)$ is a WPQK-morphism from $F(\mathbb{L})$ to $F(\mathbb{L}')$. Let $\langle a_+, a_- \rangle, \langle b_+, b_- \rangle$ be elements of the universe of $F(\mathbb{L})$. For $\wedge$ we have:

$$\begin{aligned} F(\langle h_+, h_- \rangle)(\langle a_+, a_- \rangle \wedge \langle b_+, b_- \rangle) &= F(\langle h_+, h_- \rangle)(\langle a_+ \wedge b_+, a_- \vee b_- \rangle) \\ &= \langle h_+(a_+ \wedge b_+), h_-(a_- \vee b_-) \rangle \\ &= \langle h_+(a_+) \wedge h_+(b_+), h_-(a_-) \vee h_-(b_-) \rangle \\ &= \langle h_+(a_+), h_-(a_-) \rangle \wedge \langle h_+(b_+), h_-(b_-) \rangle \\ &= F(\langle h_+, h_- \rangle)(\langle a_+, a_- \rangle) \wedge F(\langle h_+, h_- \rangle)(\langle b_+, b_- \rangle). \end{aligned}$$

The proof for $\vee$ is analogous and will be omitted. For $\sim$ we have:

$$\begin{aligned} F(\langle h_+, h_- \rangle)(\sim\langle a_+, a_- \rangle) &= F(\langle h_+, h_- \rangle)\langle p(a_-), n(a_+) \rangle && (1) \\ &= \langle (h_+ \circ p)(a_-), (h_- \circ n)(a_+) \rangle && (2) \\ &= \langle (p' \circ h_-)(a_-), (n' \circ h_+)(a_+) \rangle && (3) \\ &= \sim\langle h_+(a_+), h_-(a_-) \rangle && (4) \\ &= \sim F(\langle h_+, h_- \rangle)(\langle a_+, a_- \rangle) && (5) \end{aligned}$$

From (3) to (4) we used the fact that $h_+ \circ p = p' \circ h_-$ and $h_- \circ n = n' \circ h_+$. For $\neg$ we have:

$$\begin{aligned} F(\langle h_+, h_- \rangle)(\neg\langle a_+, a_- \rangle) &= F(\langle h_+, h_- \rangle)(\langle \neg_+ a_+, n(a_+) \rangle) && (6) \\ &= \langle h_+(\neg_+ a_+), h_- \circ n(a+) \rangle && (7) \\ &= \langle \neg_+ h_+(a_+), n' \circ h_+(a_+) \rangle && (8) \\ &= \neg\langle h_+(a_+), h_-(a_-) \rangle && (9) \\ &= \neg F(\langle h_+, h_- \rangle)(\langle a_+, a_- \rangle) && (10) \end{aligned}$$

From (7) to (8) we used the identity $(h_- \circ n) = (n' \circ h_+)$. Now we move to prove that given an object $\mathbb{L} = \langle \mathbf{L}_+, \mathbf{L}_-, n, p, \nabla \rangle$ in TW and the identity morphism $Id_{\mathbb{L}} := \langle Id_{L_+}, Id_{L_-} \rangle$ for $\mathbb{L}$, $F(Id_{\mathbb{L}}) = Id_{F(\mathbb{L})}$, i.e. the identity homomorphism for $F(\mathbb{L})$. Notice that if $\langle a_+, a_- \rangle$ is an element of the universe of $F(\mathbb{L})$ and $F(\langle Id_{L_+}, Id_{L_-} \rangle)(\langle a_+, a_- \rangle) = \langle Id_{L_+}(a_+), Id_{L_-}(a_-) \rangle = \langle a_+, a_- \rangle$. Therefore,

$F(Id_{\mathbb{L}}) = Id_{F(\mathbb{L})}$. Finally, given two TW-morphisms $\langle h_+, h_- \rangle : \mathbb{L} \to \mathbb{L}'$ and $\langle f_+, f_- \rangle : \mathbb{L}' \to \mathbb{L}''$, we shall prove that $F(\langle h_+, h_- \rangle \circ \langle f_+, f_- \rangle) = F(\langle h_+, h_- \rangle) \circ F(\langle f_+, f_- \rangle)$. Let $\langle a_+, a_- \rangle$ be an element of the universe of $F(\mathbb{L})$. Then

$$\begin{aligned} F(\langle h_+, h_- \rangle \circ \langle f_+, f_- \rangle)(\langle a_+, a_- \rangle) &= F(\langle h_+ \circ f_+, h_- \circ f_- \rangle)(\langle a_+, a_- \rangle) \\ &= \langle (h_+ \circ f_+)(a_+), (h_- \circ f_-)(a_-) \rangle \\ &= F(\langle h_+, h_- \rangle)(\langle f_+(a), f_-(a) \rangle) \\ &= (F(\langle h_+, h_- \rangle) \circ F(\langle f_+, f_- \rangle))(\langle a_+, a_- \rangle). \end{aligned}$$

We now prove that G is a functor. Let $\mathbf{A}$ and $\mathbf{A}'$ be two WPQK-algebras and $k : \mathbf{A} \to \mathbf{A}'$ a homomorphism between them. In order to prove that $G(k)$ is a TW-morphism from $G(\mathbf{A})$ to $G(\mathbf{A}')$, we prove first that $k_+ : \mathbf{A}_+ \to \mathbf{A}'_+$ is a p-lattice homomorphism and $k_- : \mathbf{A}_- \to \mathbf{A}'_-$ is a bounded lattice homomorphism. Notice that for every $a \in A$, $k_+([a]) := [k(a)]$ and since k is a homomorphism of WPQK-algebras, the morphism $k_+ : A/{\equiv} \to A'/{\equiv}$ is a morphism of p-lattices. In order to prove that k_- is a bounded lattice homomorphism, let $a, b \in A$. We have:

$$\begin{aligned} k_-([{\sim} a] \wedge [{\sim} b]) = k_-([{\sim}(a \vee b)]) = [k({\sim}(a \vee b))] = [{\sim} k(a \vee b)] = \\ [{\sim}(k(a) \vee k(b))] = [{\sim} k(a)] \wedge [{\sim} k(b)] = \\ [k({\sim} a) \wedge k({\sim} b)] = [k({\sim} a)] \wedge [k({\sim} b)] = \\ k_-([{\sim} a]) \wedge k_-([{\sim} b]). \end{aligned}$$

For $\vee$ we have:

$$\begin{aligned} k_-([{\sim} a] \vee [{\sim} b]) = k_-([{\sim}(a \wedge b)]) = [k({\sim}(a \wedge b))] = [{\sim} k(a \wedge b)] = \\ [{\sim}(k(a) \wedge k(b))] = [{\sim} k(a)] \vee [{\sim} k(b)] = \\ [k({\sim} a)] \vee [k({\sim} b)] = \\ k_-([{\sim} a]) \vee k_-([{\sim} b]). \end{aligned}$$

We now prove that $k_+ \circ p_{\mathbf{A}} = p_{\mathbf{A}'} \circ k_-$ and $n_{\mathbf{A}'} \circ k_+ = k_- \circ n_{\mathbf{A}}$. Recall that $p_{\mathbf{A}} : A_- \to A_+$ is the identity map on A_- and that $n_{\mathbf{A}} : A_+ \to A_-$ is the function defined by $n_{\mathbf{A}}([a]) = [{\sim}{\sim} a]$ for all $a \in A$. Given $a \in A$, we have $(k_+ \circ p_{\mathbf{A}})([{\sim} a]) = k_+([{\sim} a])$, while $(p_{\mathbf{A}'} \circ k_-)([{\sim} a]) = p_{\mathbf{A}'}(k_-([{\sim} a])) = p_{\mathbf{A}'}([k({\sim} a)]) = k_+([{\sim} a])$. Therefore, $(k_+ \circ p_{\mathbf{A}})([{\sim} a]) = (p_{\mathbf{A}'} \circ k_-)([{\sim} a])$. We have that $(n_{\mathbf{A}'} \circ k_+)([a]) = n_{\mathbf{A}'}(k_+([a])) = [{\sim}{\sim} k(a)] = [k({\sim}{\sim} a)] = k_-([{\sim}{\sim} a])$, while $(k_- \circ n_{\mathbf{A}})([a]) = k_-([{\sim}{\sim} a])$. Hence, $(n_{\mathbf{A}'} \circ k_+)([a]) = (k_- \circ n_{\mathbf{A}})([a])$. We also need to prove that $k_+[\nabla_{\mathbf{A}}] \subseteq \nabla_{\mathbf{A}'}$. If $[a] \in \nabla_{\mathbf{A}}$, then ${\sim} a \leq a$. Since k is a WPQK-morphism, we have ${\sim} k(a) \leq k(a)$ and therefore $k(a) \in F(A)$, which implies $k_+([a]) \in \nabla_{\mathbf{A}'}$.

Given $\mathbf{A} \in$ WPQK, thanks to Theorem 1 we know that the morphism $\iota_{\mathbf{A}}$ defined in its statement is an isomorphism between $\mathbf{A}$ and $F(G(\mathbf{A}))$. Thus the morphisms $\iota_{\mathbf{A}}$ are the elements of a natural isomorphism from the identity functor on WPQK to the functor $F \circ G$.

We now prove that, for every $\mathbb{L} \in$ TW, one has that $\iota_{\mathbb{L}} = \langle \iota_{L_+}, \iota_{L_-} \rangle : \mathbb{L} \to G(F(\mathbb{L}))$ is a natural isomorphism. Let $\mathbb{L} = \langle \mathbf{L}_+, \mathbf{L}_-, n, p, \nabla \rangle \in$ TW. We have

$F(\mathbb{L}) = Tw(\langle \mathbf{L}_+, \mathbf{L}_-, n, p, \nabla \rangle)$. Let us denote by $\mathbf{A}$ the WPQK-algebra $F(\mathbb{L})$ so that then $G(F(\mathbb{L})) = \langle \mathbf{A}_+, \mathbf{A}_-, n_\mathbf{A}, p_\mathbf{A}, \nabla_\mathbf{A} \rangle$. Recall that, for every $a_+ \in L_+$, we have $\iota_{L_+}(a_+) = [\langle a_+, n(\neg_+ a_+) \rangle]$, and, for every $a_- \in L_-$, we have $\iota_{L_-}(a_-) = [\sim \langle \neg_+ p(a_-), a_- \rangle]$. First of all, we need to prove that the maps ι_{L_+} and ι_{L_-} are respectively maps from $\mathbf{L}_+$ to $\mathbf{A}_+$ and from $\mathbf{L}_-$ to $\mathbf{A}_-$. To this end, it is enough to prove that given $\langle a_+, n(\neg_+ a_+) \rangle$ and $\langle \neg_+ p(b_-), b_- \rangle$ in $L_+ \times L_-$, the pairs $\langle a_+, n(\neg_+ a_+) \rangle$ and $\langle \neg_+ p(b_-), b_- \rangle$ belong to $\mathbf{A}$, that is, that we have $a_+ \vee_+ p(n(\neg_+ a_+)) \in \nabla$, $\neg_+ p(b_-) \vee_+ p(b_-) \in \nabla$, $a_+ \wedge_+ p(n(\neg_+ a_+)) = 0_+$, and $\neg_+ p(b_-) \wedge_+ p(b_-) = 0_+$. We have that $a_+ \vee_+ \neg_+ a_+ \leq_+ a_+ \vee_+ p(n(\neg_+ a_+))$. Since ∇ is a filter and $a_+ \vee_+ \neg_+ a_+ \in \nabla$, then $a_+ \vee_+ p(n(\neg_+ a_+)) \in \nabla$. We have trivially that $\neg_+ p(b_-) \vee_+ p(b_-) \in \nabla$, given that ∇ is a dense filter. Now, since $Id_{\mathbf{A}_+} \leq_+ p \cdot n$ and p, n, preserve finite meets and the bounds, then $a_+ \wedge_+ p(n(\neg_+ a_+)) \leq_+ p(n(a_+)) \wedge_+ p(n(\neg_+ a_+)) = p(n(a_+ \wedge_+ \neg_+ a_+)) = p(n(0_+)) = 0_+$. Moreover, since $\neg_+$ is a pseudo-complement, then $\neg_+ p(b_-) \wedge_+ p(b_-) = 0_+$.

In order to prove that $\iota_\mathbb{L}$ is a morphism, we have to prove that ι_{L_+} is a p-lattice homomorphism from $\mathbf{L}_+$ to $\mathbf{A}_+$ and ι_{L_-} is a bounded lattice homomorphism from $\mathbf{L}_-$ to $\mathbf{A}_-$. Before we proceed, notice that given $a_+, b_+ \in L_+$ and $a_-, b_- \in L_-$, by [15, Lemma 3.5.iii] we have $[\langle a_+, a_- \rangle] = [\langle b_+, b_- \rangle]$ iff $a_+ = b_+$, and by [15, Lemma 3.5.ii] we have $[\sim\langle a_+, a_- \rangle] = [\sim\langle b_+, b_- \rangle]$ iff $a_- = b_-$. We will use these facts.

We first show that ι_{L_+} is a p-lattice homomorphism. Let $a_+, b_+ \in L_+$. Then:

($\wedge_+$) On the one hand, $\iota_{L_+}(a_+ \wedge_+ b_+) = [\langle a_+ \wedge_+ b_+, n(\neg_+(a_+ \wedge_+ b_+)) \rangle])$. On the other hand, $\iota_{L_+}(a_+) \wedge \iota_{L_+}(b_+) = [\langle a_+, n(\neg_+ a_+) \rangle] \wedge [\langle b_+, n(\neg_+ b_+) \rangle] = [\langle a_+ \wedge_+ b_+, n(\neg_+ a_+) \vee_- n(\neg_+ b_+) \rangle]$. It follows from [15, Lemma 3.5.iii] that $\iota_{L_+}(a_+ \wedge_+ b_+) = \iota_{L_+}(a_+) \wedge \iota_{L_+}(b_+)$.

($\vee_+$) On the one hand, $\iota_{L_+}(a_+ \vee_+ b_+) = [\langle a_+ \vee_+ b_+, n(\neg_+(a_+ \vee_+ b_+)) \rangle])$. On the othert hand, $\iota_{L_+}(a_+) \vee \iota_{L_+}(b_+) = [\langle a_+, n(\neg_+ a_+) \rangle] \vee [\langle b_+, n(\neg_+ b_+) \rangle] = [\langle a_+ \vee_+ b_+, n(\neg_+ a_+) \wedge_- n(\neg_+ b_+) \rangle]$. It follows from [15, Lemma 3.5.iii] that $\iota_{L_+}(a_+ \vee_+ b_+) = \iota_{L_+}(a_+) \vee \iota_{L_+}(b_+)$.

($\neg$) We have $\iota_{L_+}(\neg_+ a) = [\langle \neg_+ a, n(\neg_+ \neg_+ a) \rangle]$. Also, $\neg_+ \iota_{L_+}(a) = \neg[\langle a_+, n(\neg_+ a_+) \rangle] = [\langle \neg_+ a_+, n(n(\neg_+ a_+)) \rangle]$. It follows from [15, Lemma 3.5.iii] that $\iota_{L_+}(\neg_+ a) = \neg_+ \iota_{L_+}(a)$.

Now we prove that ι_{L_-} is a bounded lattice homomorphism. Let $a_-, b_- \in L_-$. Then:

($\wedge_-$) First notice that $\iota_{L_-}(a_- \wedge_- b_-) = [\sim\langle \neg_+ p(a_- \wedge_- b_-), a_- \wedge_- b_- \rangle]$. Secondly that $\iota_{L_-}(a_-) \wedge \iota_{L_-}(b_-) = [\sim\langle \neg_+ p(a_-), a_- \rangle] \wedge [\sim\langle \neg_+ p(b_-), b_- \rangle] = [\sim(\langle \neg_+ p(a_-), a_- \rangle \vee \langle \neg_+ p(b_-), b_- \rangle)] = [\sim(\neg_+ p(a_-) \vee_+ \neg_+ p(b_-), a_- \wedge_- b_-)]$. It follows from [15, Lemma 3.5.ii] that $\iota_{L_-}(a_- \wedge_- b_-) = \iota_{L_-}(a_-) \wedge \iota_{L_-}(b_-)$.

($\vee_-$) On the one hand, $\iota_{L_-}(a_- \vee_- b_-) = [\sim\langle \neg_+ p(a_- \vee_- b_-), a_- \vee_- b_- \rangle]$. On the other hand, $\iota_{L_-}(a_-) \vee \iota_{L_-}(b_-) = [\sim\langle - \neg_+ p(a_-), a_- \rangle] \vee [\sim\langle \neg_+ p(b_-), b_- \rangle] = [\sim(\langle \neg_+ p(a_-), a_- \rangle \wedge \langle \neg_+ p(b_-), b_- \rangle)] = [\sim(\neg_+ p(a_-) \wedge_+ \neg_+ p(b_-), a_- \vee_- b_-)]$. It follows from [15, Lemma 3.5.ii] that $\iota_{L_-}(a_- \vee b_-) = \iota_{L_-}(a_-) \vee \iota_{L_-}(b_-)$.

We proceed to prove that $\iota_{L_+} \circ p = p_{\mathbf{A}} \circ \iota_{L_-}$, $n_{\mathbf{A}} \circ \iota_{L_+} = \iota_{L_-} \circ n$ and that $\iota_{L_+}[\nabla] \subseteq \nabla_{\mathbf{A}}$. Notice that given $a_- \in L_-$, $\iota_{L_+}(p(a_-)) = [\langle p(a_-), n(\neg_+ p(a_-))\rangle]$ and $(p_{\mathbf{A}} \circ \iota_{L_-})(a_-) = [\sim\langle \neg_+ p(a_-), a_-\rangle] = [\langle p(a_-), n(\neg_+ p(a_-))\rangle]$. Thus we obtain that $\iota_{L_+} \circ p = p_{\mathbf{A}} \circ \iota_{L_-}$. To prove that $n_{\mathbf{A}} \circ \iota_{L_+} = \iota_{L_-} \circ n$, notice that given $a_+ \in L_+$, $n_{\mathbf{A}} \circ \iota_{L_+}(a_+) = [\sim\sim\langle a_+, n(\neg_+ a_+)\rangle] = [\sim\langle p(n(\neg_+ a_+)), n(a_+)\rangle]$ and $(\iota_{L_-} \circ n)(a_+) = [\sim\langle \neg_+ p(n(a_+)), n(a_+)\rangle]$. Thus, the desired equality follows. Finally, to prove that $\iota_{L_+}[\nabla] \subseteq \nabla_{\mathbf{A}}$, notice that $\nabla_{\mathbf{A}} = [F(A)]$ and thanks to [15, Prop. 6.1], $F(A) = \{\langle a_+, 0_-\rangle : \langle a_+, 0_-\rangle \in A\}$ and therefore $[F(A)] = \{[\langle a_+, 0_-\rangle] : \langle a_+, 0_-\rangle \in F(A)\}$. Let now $a_+ \in \nabla$. We prove that $\langle a_+, 0_-\rangle \in A$. To this end we need to show that $a_+ \vee_+ p(0_-) \in \nabla$ and $a_+ \wedge_+ p(0_-) = 0_-$. Notice that $a_+ \vee_+ p(0_-) = a_+ \vee_+ 0_+ = a_+$ and by hypothesis $a_+ \in \nabla$. Also, $a_+ \wedge_+ p(0_-) = a_+ \wedge_+ 0_+ = 0_+$. Since $\iota_{L_+}(a_+) = [\langle a_+, n(\neg_+ a_+)\rangle]$ using that $[\langle a_+, 0_-\rangle] = [\langle a_+, n(\neg_+ a_+)\rangle]$ ([15, Lemma 3.5.iii]), we obtain $\iota_{L_+}(a_+) \in \nabla_{\mathbf{A}}$.

It remains to prove that ι_{L_+} and ι_{L_-} are injective and surjective. We prove first injectivity.

ι_{L_+} If $\iota_{L_+}(a_+) = \iota_{L_+}(b_+)$, then $[\langle a_+, n(\neg_+ a_+)\rangle] = [\langle b_+, n(\neg_+ b_+)\rangle]$; and from [15, Lemma 3.5.iii] it follows that $a_+ = b_+$.

ι_{L_-} If $\iota_{L_-}(a_-) = \iota_{L_-}(b_-)$, then $[\sim\langle \neg_+ p(a_-), a_-\rangle] = [\sim\langle \neg_+ p(b_-), b_-\rangle]$, and from [15, Lemma 3.5.ii] it follows that $a_- = b_-$.

Now we prove that ι_{L_+} and ι_{L_-} are surjective:

ι_{L_+} Let $[\langle a_+, a_-\rangle] \in \mathbf{A}_+$ (so that $\langle a_+, a_-\rangle$ belongs to the universe of $F(\mathbb{L})$). By [15, Lemma 3.5.iii] $[\langle a_+, a_-\rangle] = [\langle a_+, n(\neg_+ a_+)\rangle]$. Therefore $\iota_{L_+}(a_+) = [\langle a_+, a_-\rangle]$.

ι_{L_-} Let $[\sim\langle a_+, a_-\rangle] \in \mathbf{A}_-$ (so that $\langle a_+, a_-\rangle$ belongs to the universe of $F(\mathbb{L})$). From [15, Lemma 3.5.ii], it follows that $[\sim\langle a_+, a_-\rangle] = [\sim\langle \neg_+ p(a_-), a_-\rangle]$. Therefore $\iota_{L_-}(a_-) = [\sim\langle a_+, a_-\rangle]$.

Theorem 5. *The functors H, K establish a co-variant equivalence between* NP *and* WPQK.

Proof. We prove first that H is a functor. Let $\mathbb{L} = \langle \mathbf{L}, \nabla\rangle$ and $\mathbb{L}' = \langle \mathbf{L}', \nabla'\rangle$ be two objects in NP and $h : \mathbb{L}' \to \mathbb{L}'$ a NP-morphism. The proof that $H(h) : H(\mathbb{L}) \to H(\mathbb{L}')$ is a WPQK-homomorphism is similar to the proof above for morphsims in TW and the functor F, because $H(\mathbb{L}) = Tw\langle \mathbf{L}, \mathbf{L}^{\Box}, \Box, Id_{L^{\Box}}, \nabla\rangle$ and $H(\mathbb{L}') = Tw\langle \mathbf{L}', \mathbf{L}'^{\Box'}, \Box, Id_{L'^{\Box'}}, \nabla'\rangle$.

Given $\mathbf{A}, \mathbf{A}'$ objects in WPQK and $k : \mathbf{A} \to \mathbf{A}'$ a WPQK-homomorphism, we shall prove that $K(k) : K(\mathbf{A}) \to K(\mathbf{A}')$ is a NP-morphism. Since k is a morphism of WPQK algebras, it easily follows that $K(k)$ is a homomorphism from $\langle \mathbf{A}_+, \Box\rangle$ to $\langle \mathbf{A}'_+, \Box'\rangle$. Let $a, b \in A$. Then:

$$\begin{aligned} K(k)([a] \wedge [b]) = K(k)([a \wedge b]) = ([k(a \wedge b)]) = \\ ([k(a) \wedge k(b)]) = [k(a)] \wedge [k(b)] = \\ K(k)([a]) \wedge K(k)([b]), \end{aligned}$$

$$K(k)([a] \vee [b]) = K(k)([a \vee b]) = ([k(a \vee b)]) = ([k(a) \vee k(b)]) = [k(a)] \vee [k(b)] = K(k)([a]) \vee K(k)([b]),$$

$$K(k)(\neg[a]) = [k(\neg a)] = [\neg k(a)] = \neg K(k)([a]),$$

$$K(k)(\Box[a]) = [k(\Box a)] = [k(\sim\sim a)] = [\sim\sim k(a)] = [\Box k(a)] = \Box K(k)([a]).$$

It remains to prove that $K(k)[\nabla_{\mathbf{A}}] \subseteq \nabla_{\mathbf{A}'}$. Given $[a] \in \nabla_{\mathbf{A}}$, we have $a \in F(A)$ and $\sim a \leq a$, since k is a morphism of WPQK-algebras, then $k(\sim a) \leq k(a)$ and it follows that $\sim k(a) \leq k(a)$. Therefore $[\sim k(a)] \leq [k(a)]$ and we conclude that $k[a] \in \nabla_{\mathbf{A}}$.

Given an object $\mathbf{A}$ in WPQK and the identity morphism $Id_{\mathbf{A}} : \mathbf{A} \to \mathbf{A}$, we shall prove that $K(Id_{\mathbf{A}}) = Id_{K(\mathbf{A})}$. Let $[a]$ be an element of the universe of $K(\mathbf{A})$. Notice that $K(Id_{\mathbf{A}})[a] = [Id_{\mathbf{A}}(a)] = [a]$. So we are done. Given WPQK-homomorphisms $f : \mathbf{A} \to \mathbf{A}'$ and $g : \mathbf{A}' \to \mathbf{A}''$, we shall prove that $K(g \circ f) = K(g) \circ K(f)$. Let $[a]$ be an element of the universe of $K(\mathbf{A})$.

$$\begin{aligned} K(g \circ f)([a]) &= [(g \circ f)(a)] \\ &= [g(f(a))] \\ &= K(g)([f(a)]) \\ &= K(g)(K(f)[a]) \\ &= K(f) \circ K(g)([a]). \end{aligned}$$

From [15, Theorem 7.11] it is easy to obtain the natural isomorphism between the identity functor on WPQK and the functor $H \circ K$.

We proceed to prove that, for every $\mathbb{L} \in$ NP, the map ι_{NP} defined by $\iota_{\mathsf{NP}}(a) = [\langle a, \Box\neg a\rangle]$ for all $a \in L$ is an isomorphism between $\mathbb{L}$ and $K(H(\mathbb{L}))$. This will provide the natural isomorphism between the identity functor on NP and the functor $K \circ H$.

Let $\mathbb{L} = \langle \mathbf{L}, \nabla\rangle \in$ NP. We have $H(\mathbb{L}) = Tw\langle \mathbf{L}, \mathbf{L}^{\Box}, \Box, Id_{L^{\Box}}, \nabla\rangle$. Let us denote by $\mathbf{A}$ the WPQK-algebra $H(\mathbb{L})$, with A its universe, so that $K(H(\mathbb{L})) = \langle\langle \mathbf{A}_{+}, \Box\rangle, \nabla_{\mathbf{A}}\rangle$. First of all, we need to prove that ι_{NP} is a map from $\mathbb{L}$ to $K(H(\mathbb{L}))$. To this end, it is enough to notice that for every $a \in L$, the element $\langle a, \Box\neg a\rangle$ belongs to the twist-algebra, that is, that $a \vee Id_{L^{\Box}}(\Box\neg a) \in \nabla$ and $a \wedge Id_{L^{\Box}}(\Box\neg a) = 0$. Since $Id_{L^{\Box}}$ is the identity map, we shall prove that $a \vee \Box\neg a \in \nabla$ and $a \wedge \Box\neg a = 0$. We have that, $a \vee \neg a \leq a \vee \Box\neg a$. Since ∇ is a filter and $a \vee \neg a \in \nabla$, we have $a \vee \Box\neg a \in \nabla$. We have that $a \wedge \Box\neg a \leq \Box a \wedge \Box\neg a = \Box(a \wedge \neg a) = \Box(0) = 0$. Therefore, $a \wedge \Box\neg a = 0$.

We now prove that ι_{NP} is a NP-morphism. Before we proceed, notice that given $a, b \in L$ and $c, d \in L^{\Box}$, by [15, Lemma 3.5.iii], we have $[\langle a, c\rangle] = [\langle b, d\rangle]$ iff $a = b$. Also note that given $\langle a, b\rangle \in A$, according to the definition of $\Box$ in Theorem 4, we have $\Box[\langle a, b\rangle] = [\sim\sim\langle a, b\rangle] = [\sim\langle b, \Box a\rangle] = [\langle \Box a, \Box b\rangle]$.

($\wedge$) $\iota_{\mathsf{NP}}(a \wedge b) = [\langle a \wedge b, \Box(\neg(a \wedge b))\rangle]$. While, $\iota_{\mathsf{NP}}(a) \wedge \iota_{\mathsf{NP}}(b) = [\langle a, \Box(\neg a)\rangle] \wedge [\langle b, \Box(\neg b)\rangle] = [\langle a \wedge b, \Box(\neg a) \vee \Box(\neg b)\rangle]$. It follows from [15, Lemma 3.5.iii] that $\iota_{\mathsf{NP}}(a \wedge b) = \iota_{\mathsf{NP}}(a) \wedge \iota_{\mathsf{NP}}(b)$.

($\vee$) $\iota_{\mathsf{NP}}(a \vee b) = [\langle a \vee b, \Box(\neg(a \vee b))\rangle]$. While, $\iota_{\mathsf{NP}}(a) \vee \iota_{\mathsf{NP}}(b) = [\langle a, \Box(\neg a)\rangle] \vee [\langle b, \Box(\neg b)\rangle] = [\langle a \vee b, \Box(\neg a) \wedge \Box(\neg b)\rangle]$. It follows from [15, Lemma 3.5.iii] that $\iota_{\mathsf{NP}}(a \vee b) = \iota_{\mathsf{NP}}(a) \vee \iota_{\mathsf{NP}}(b)$.

($\neg$) $\iota_{\mathsf{NP}}(\neg a) = [\langle \neg a, \Box(\neg(\neg a))\rangle]$. While $\neg\iota_{\mathsf{NP}}(a) = \neg[\langle a, \Box a\rangle] = [\langle \neg a, n(\Box(\neg a))\rangle]$. It follows from [15, Lemma 3.5.ii] that $\iota_{\mathsf{NP}}(\neg a) = \neg\iota_{\mathsf{NP}}(a)$.

($\Box$) $\iota_{\mathsf{NP}}(\Box a) = [\langle \Box a, \Box\neg\Box a\rangle]$. While $\Box\iota_{\mathsf{NP}}(a) = \Box[\langle a, \Box(\neg a)\rangle] = [\langle \Box a, \Box\Box\neg a\rangle]$ It follows from [15, Lemma 3.5.iii] that $\iota_{\mathsf{NP}}(\Box a) = \Box\iota_{\mathsf{NP}}(a)$.

We shall prove now that ι_{NP} is injective and surjective. Notice that if $\iota_{\mathsf{NP}}(a) = \iota_{\mathsf{NP}}(b)$, then $[\langle a, \Box(\neg a)\rangle] = [\langle b, \Box(\neg b)\rangle]$ and again from [15, Lemma 3.5.iii] it follows that $a = b$. Therefore ι_{NP} is injective. Let $\langle a, b\rangle \in A$ and notice that $\iota_{\mathsf{NP}}(a) = [\langle a, \Box(\neg a)\rangle] = [\langle a, b\rangle]$ by [15, Lemma 3.5.iii]; therefore ι_{NP} is surjective. It remains to prove that $\iota_{\mathsf{NP}}[\nabla] \subseteq \nabla_{\mathbf{A}}$. Notice that $\nabla_{\mathbf{A}} = [F(A)]$ and thanks to [15, Prop. 6.1], $F(A) = \{\langle a, 0\rangle : \langle a, 0\rangle \in A\}$ and therefore $[F(A)] = \{[\langle a, 0\rangle] : \langle a, 0\rangle \in F(A)\}$. Let $a \in \nabla$. In order to prove that $\iota_{\mathsf{NP}}(a) \in \nabla_{\mathbf{A}}$, we show first that $\langle a, 0\rangle \in A$. We have that $a \vee Id_{L^{\Box}}(0) = a \vee 0 = a$. Therefore, since by assumption $a \in \nabla$, $a \vee Id_{L^{\Box}}(0) \in \nabla$. Notice also that $a \wedge Id_{L^{\Box}}(0) = a \wedge 0 = 0$. We thus obtain $\langle a, 0\rangle \in A$. From [15, Lemma 3.5.iii] we have $[\langle a, \Box(\neg a)\rangle] = [\langle a, 0\rangle]$. Since $\iota_{\mathsf{NP}}(a) = [\langle a, \Box(\neg a)\rangle]$, it follows that $\iota_{\mathsf{NP}}(a) \in \nabla_{\mathbf{A}}$.

References

1. Balbes, R., Dwinger, P.: Distributive Lattices. University of Missouri Press, Columbia (1974)
2. Bezhanishvili, G., Ghilardi, S.: An algebraic approach to subframe logics. Intuitionistic case. Ann. Pure Appl. Log. **147**(1–2), 84–100 (2007)
3. Bezhanishvili, G., Jansana, R.: Priestley style duality for distributive meet-semilattices. Stud. Logica **98**(1–2), 83–122 (2011). https://doi.org/10.1007/s11225-011-9323-5
4. Galatos, N., Jipsen, P., Kowalski, T., Ono, H.: Residuated Lattices: An Algebraic Glimpse at Substructural Logics. Studies in Logic and the Foundations of Mathematics, vol. 151. Elsevier, Amsterdam (2007)
5. Goldblatt, R.: Varieties of complex algebras. Ann. Pure Appl. Log. **44**(3), 173–242 (1989)
6. Jansana, R., Rivieccio, U.: Dualities for modal N4-lattices. Log. J. IGPL **22**(4), 608–637 (2014)
7. Jung, A., Rivieccio, U.: A duality for two-sorted lattices (Submitted)
8. Kalman, J.A.: Lattices with involution. Trans. Am. Math. Soc. **87**, 485–491 (1958)
9. Liang, F., Nascimento, T.: Algebraic semantics for quasi-nelson logic. In: Iemhoff, R., Moortgat, M., de Queiroz, R. (eds.) WoLLIC 2019. LNCS, vol. 11541, pp. 450–466. Springer, Heidelberg (2019). https://doi.org/10.1007/978-3-662-59533-6_27
10. Nelson, D.: Constructible falsity. J. Symb. Log. **14**, 16–26 (1949)
11. Priestley, H.A.: Representation of distributive lattices by means of ordered Stone spaces. Bull. Lond. Math. Soc. **2**, 186–190 (1970)

12. Priestley, H.A.: Stone lattices: a topological approach. Fundam. Math. **84**(2), 127–143 (1974)
13. Priestley, H.A.: The construction of spaces dual to pseudocomplemented distributive lattices. Q. J. Math. **26**(1), 215–228 (1975)
14. Rasiowa, H.: An Algebraic Approach to Non-Classical Logics. Studies in Logic and the Foundations of Mathematics, vol. 78. North-Holland, Amsterdam (1974)
15. Rivieccio, U.: Fragments of quasi-Nelson: two negations (Submitted)
16. Rivieccio, U.: Representation of De Morgan and (semi-)Kleene lattices. Soft Comput. **24**(12), 8685–8716 (2020). https://doi.org/10.1007/s00500-020-04885-w
17. Rivieccio, U., Spinks, M.: Quasi-Nelson algebras. Electron. Notes Theor. Comput. Sci. **344**, 169–188 (2019)
18. Rivieccio, U., Spinks, M.: Quasi-Nelson; or, non-involutive Nelson algebras. Trends Log. (to appear)
19. Sankappanavar, H.P.: Semi-De Morgan algebras. J. Symb. Log. **52**, 712–724 (1987)
20. Sendlewski, A.: Some investigations of varieties of $\mathcal{N}$-lattices. Stud. Logica **43**, 257–280 (1984)
21. Sendlewski, A.: Topologicality of Kleene algebras with a weak pseudocomplementation over distributive p-algebras. Rep. Math. Log. **25**, 13–56 (1991)
22. Spinks, M., Rivieccio, U., Nascimento, T.: Compatibly involutive residuated lattices and the Nelson identity. Soft Comput. **23**(7), 2297–2320 (2018). https://doi.org/10.1007/s00500-018-3588-9
23. Spinks, M., Veroff, R.: Constructive logic with strong negation is a substructural logic. I. Stud. Logica **88**, 325–348 (2008). https://doi.org/10.1007/s11225-008-9113-x
24. Spinks, M., Veroff, R.: Constructive logic with strong negation is a substructural logic. II. Stud. Logica **89**, 401–425 (2008). https://doi.org/10.1007/s11225-008-9138-1
25. Vakarelov, D.: Notes on $\mathcal{N}$-lattices and constructive logic with strong negation. Stud. Logica **36**(1–2), 109–125 (1977)

On the Logic of Left-Continuous t-Norms and Right-Continuous t-Conorms

Lluís Godo[1(✉)], Martín Sócola-Ramos[2], and Francesc Esteva[1]

[1] Artificial Intelligence Research Institute (IIIA-CSIC), Campus de la UAB, Bellaterra, 08193 Barcelona, Spain
{godo,esteva}@iiia.csic.es

[2] Junivägen 13, 27172 Köpingebro, Sweden
martin1992.4@gmail.com

Abstract. Double residuated lattices are expansions of residuated lattices with an extra monoidal operator, playing the role of a strong disjunction operation, together with its dual residuum. They were introduced by Orłowska and Radzikowska. In this paper we consider the subclass of double residuated structures that are expansions of MTL-algebras, that is, prelinear, bounded, commutative and integral residuated lattices. MTL-algebras constitute the algebraic semantics for the MTL logic, the system of mathematical fuzzy logic that is complete w.r.t. the class of residuated lattices on the real unit interval $[0, 1]$ induced by left-continuous t-norms. Our aim is to axiomatise the logic whose intended semantics are commutative and integral double residuated structures on $[0, 1]$, that are induced by an arbitrary left-continuous t-norm, an arbitrary right-continuous t-conorm, and their corresponding residual operations.

Keywords: Mathematical fuzzy logic · Double residuated lattices · MTL · DMCTL · Semilinear logics · Standard completeness

1 Introduction

Residuated lattices [7] are a usual choice for general algebras of truth-values in systems of mathematical fuzzy logic [1–3, 9], as well as for valuation structures in lattice-valued fuzzy sets (L-fuzzy sets) [8] and in multiple-valued generalisations of information relations [11, 13, 14]. Actually, from this perspective, residuated lattices can be seen as generalizations of the algebraic structures on the real unit interval $[0, 1]$ induced by a (left-continuous) t-norm and its residuum. Indeed, in a residuated lattice, besides the lattice-meet $\wedge$ and lattice-join $\vee$ operations, that play the role of the min and max operations in $[0, 1]$, there is a monoidal operation $\odot$ that plays the role of a left-continuous t-norm, and its residuum $\rightarrow$ (or its left- and right- residua in case the monoidal operation is not commutative). In a logical setting, while the lattice-meet and lattice-join are used to model a *weak conjunction* and a *weak disjunction* respectively, the monoidal operation is

M.-J. Lesot et al. (Eds.): IPMU 2020, CCIS 1239, pp. 654–665, 2020.
https://doi.org/10.1007/978-3-030-50153-2_48

used to model a *strong conjunction* connective and its residuum to interpret an implication connective.

However, there is no primitive operation in a residuated lattice that properly accounts for a *strong disjunction* playing the role of a t-conorm in $[0,1]$. Of course, one can always define a disjunction operation by De Morgan duality from the monoidal operation $\odot$ and the definable residual negation ($\neg x = x \to 0$). However, depending on the properties of $\neg$, the resulting disjunction operation may not enjoy nice properties. But even if, for instance, $\neg$ is a well-behaved involutive negation, one is bound to a unique choice for a strong disjunction. To remedy this situation, Orłowska and Radzikowska introduce in [12] more general residuated structures by expanding residuated lattices with an extra monoidal operator playing the role of a t-conorm, together with its (dual) residuum as well. They call them *double residuated lattices*, DRLs for short.

In [12], the authors study basic properties of DRLs and their application to fuzzy information systems. In this paper we are interested in the particular subclass of double residuated structures that are counterpart of MTL-algebras [6], that is, prelinear, bounded, commutative and integral residuated lattices. MTL-algebras constitute the algebraic semantics for the MTL logic, the system of mathematical fuzzy logic that is complete w.r.t. the class of residuated lattices on the real unit interval $[0,1]$ induced by left-continuous t-norms [10]. Our aim is to axiomatise the logic whose intended semantics are double residuated structures on $[0,1]$, hence induced by an arbitrary pair of a left-continuous t-norm and a right-continuous t-conorm, together with their respective residual operations.

The paper is structured as follows. After this introduction, in Sect. 2 we recall basic definitions and facts about the MTL logic and on double residuated lattices. Then in Sect. 3, as a preliminary step, we axiomatise the logic dual to MTL, called dMTL, while in Sect. 4 we axiomatise the logic complete w.r.t. to the class of double residuated lattices on the real unit interval $[0,1]$. We finish in Sect. 5 with some conclusions and future work.[1]

2 Preliminaries

2.1 A Refresher on MTL

The language of MTL consists of three binary connectives $\{\wedge, \&, \to\}$ and a constant $\mathbf{0}$. Formulas are built from a countable set of propositional variables as usual. Further connectives are definable:

- $\neg\varphi := \varphi \to \mathbf{0}$
- $\varphi \vee \psi := ((\varphi \to \psi) \to \psi) \wedge ((\psi \to \varphi) \to \varphi)$
- $\varphi \equiv \psi := (\varphi \to \psi)\&(\psi \to \varphi)$
- $\mathbf{1} := \neg\mathbf{0}$

We present the MTL axiomatic system proposed in [6]:

[1] The results in this paper are based on the Master thesis [15].

(A1) $(\varphi \to \psi) \to ((\psi \to \chi) \to (\varphi \to \chi))$
(A2) $(\varphi \& \psi) \to \varphi$
(A3) $(\varphi \& \psi) \to (\psi \& \varphi)$
(A4) $(\varphi \wedge \psi) \to \varphi$
(A5) $(\varphi \wedge \psi) \to (\psi \wedge \varphi)$
(A6) $(\varphi \& (\varphi \to \psi)) \to (\varphi \wedge \psi)$
(A7) $(\varphi \to (\psi \to \chi)) \to ((\varphi \& \psi) \to \chi)$
(A8) $((\varphi \& \psi) \to \chi) \to (\varphi \to (\psi \to \chi))$
(A9) $((\varphi \to \psi) \to \chi) \to (((\psi \to \varphi) \to \chi) \to \chi)$
(A10) $\mathbf{0} \to \varphi$

The only inference rule for MTL is Modus Ponens:

$$(\mathrm{MP}) : \frac{\varphi \quad \varphi \to \psi}{\psi}$$

MTL is an algebraizable logic and its equivalent algebraic semantics is given by the variety of MTL-algebras. We recall their definition.

Definition 1. *A MTL-algebra is a bounded commutative integral residuated lattice*

$$\mathbf{A} = \langle A, \wedge, \vee, \odot, \to, 0, 1 \rangle,$$

where $\wedge, \vee$ *are the lattice meet and join operations, and* $\langle \odot, \to \rangle$ *is an adjoint pair satisfying the prelinearity condition.*

As a direct consequence of its algebraizability, we have the following general completeness result for MTL.

Theorem 1 (General completeness). $\Gamma \vdash_{MTL} \varphi$ *iff, for any MTL algebra* $\mathbf{A}$ *and for every* $\mathbf{A}$*-evaluation* e*,* $e(\psi) = 1$ *for all* $\psi \in \Gamma$ *implies* $e(\varphi) = 1$*.*

Thanks to the fact that MTL proves the prelinearity axiom $(\varphi \to \psi) \vee (\psi \to \varphi)$ (a direct consequence of (A9)), the above completeness result can be improved to get completeness with respect to linearly-ordered algebras, or chains.

Theorem 2 (Chain completeness). $\Gamma \vdash_{MTL} \varphi$ *iff, for any linearly-ordered MTL algebra* $\mathbf{A}$ *and for every* $\mathbf{A}$*-evaluation* e*,* $e(\psi) = 1$ *for all* $\psi \in \Gamma$ *implies* $e(\varphi) = 1$*.*

In other words, MTL is a *semilinear* logic [4]. As a matter of fact, semilinearity is inherited by many expansions of MTL with new axioms and (finitary) inference rules: the authors of [4,5] prove that an expansion S of MTL is semilinear iff for each newly added finitary inference rule

$$\textbf{(R):} \ \frac{\Gamma}{\varphi},$$

its corresponding $\vee$-form

$(\mathbf{R}^{\vee})$: $\dfrac{\Gamma \vee p}{\varphi \vee p}$

is derivable in S as well, where p is an arbitrary propositional variable not appearing in $\Gamma \cup \{\varphi\}$.

Moreover, going back to MTL, to check the validity of a deduction in MTL we can even restrict ourselves to check it on the class of MTL-chains defined on the real unit interval $[0,1]$, also known as *standard chains.* This was proved by Jenei and Montagna in [10].

Theorem 3 (Standard completeness). *$\Gamma \vdash_{MTL} \varphi$ iff, for any standard MTL-chain $\mathbf{A}$ and for every $\mathbf{A}$-evaluation e, $e(\psi) = 1$ for all $\psi \in \Gamma$ implies $e(\varphi) = 1$.*

In algebraic terms, this result indicates that the variety of MTL-algebras is generated by its standard chains. Note that if $\mathbf{A} = \langle [0,1], \wedge, \vee, \odot, \rightarrow, 1, 0\rangle$ is a MTL-chain, then necessarily $\odot$ is a left-continuous t-norm and $\rightarrow$ is its residuum. This is why MTL is known as the logic of left-continuous t-norms.

2.2 Double Residuated Lattices

As mentioned in the introduction, the class of double residuated lattices (DRL) was introduced by Orłowska and Radzikowska in [12], in order to come up with algebraic structures general enough to accommodate on a lattice not only a strong conjunction and its residuum as in the case of residuated lattices, but also a strong disjunction and its residuum.

Although the definition of double residuated lattice in [12] is very general, for the purpose of this paper we will restrict ourselves to the commutative, bounded and integral case.

Definition 2. *A (commutative, bounded, integral)* double *residuated lattice is a structure of the form*

$$\mathbf{A} = \langle A, \wedge, \vee, \odot, \rightarrow, \oplus, \leftarrow, 1, 0\rangle$$

where:

- $\langle A, \wedge, \vee, \odot, \rightarrow, 1, 0\rangle$ *is a (commutative, bounded, integral) residuated lattice,*
- $\langle A, \oplus, 0\rangle$ *is a commutative monoid,*
- $\leftarrow$ *is the dual residuum of $\oplus$, i.e. for all $x, y, z \in A$ the following dual residuation condition holds:*

$$y \leq x \oplus z \quad \textit{iff} \quad x \leftarrow y \leq z.$$

Analogously to (commutative, bounded, integral) residuated lattices, DRL-algebras form a variety. In a DRL-algebra, the lattice order relation can be recovered from both residua, indeed we have:

$$x \leq y \quad \text{iff} \quad x \rightarrow y = 1 \quad \text{iff} \quad y \leftarrow x = 0,$$

and, as a matter of fact, this common underlying order is the only link between both adjoint pairs of operations $(\odot, \rightarrow)$ and $(\oplus, \leftarrow)$.

DMCTL-algebras are also introduced in [12] as the subvariety of DRL-algebras satisfying a pre-linearity condition and its dual version.

Definition 3. *Let* $\mathbf{A} = \langle A, \wedge, \vee, \odot, \rightarrow, \oplus, \leftarrow, 1, 0\rangle$ *be a bounded, commutative and integral DRL. Then* $\mathbf{A}$ *is called:*

A double monoidal t-norm logic (DMTL) algebra iff for all $x, y, z \in A$ *it satisfies:*

$$(x \rightarrow y) \vee (y \rightarrow x) = 1 \qquad \text{(Prelinearity)}$$

A double monoidal t-conorm logic (DMCL) algebra iff for all $x, y, z \in A$ *it satisfies:*

$$(x \leftarrow y) \wedge (y \leftarrow x) = 0 \qquad \text{(Dual Prelinearity)}$$

If $\mathbf{A}$ *is both a DMTL- and a DMCL-algebra, then it is called a double monoidal t-conorm and t-norm logic (DMCTL) algebra.*

Natural (and inspiring) examples of DMCTL-algebras are structures on the unit real interval $[0, 1]$ of the form

$$\langle [0, 1], \min, \max, \odot, \rightarrow, \oplus, \leftarrow, 1, 0\rangle,$$

where $\odot$ is a left-continuous t-norm and $\oplus$ is a right-continuous t-conorm. Recall that a t-norm $\oplus$ has residuum iff it is left-continuous, and similarly, a t-conorm $\oplus$ has a dual residuum iff it is right-continuous. In such a case, the residuum of $\odot$ is given by

$$x \rightarrow y = \max\{z \in [0, 1] : x \odot z \leq y\},$$

and the dual residuum of $\oplus$ is given by

$$x \leftarrow y = \min\{z \in [0, 1] : x \oplus z \geq y\}.$$

And actually, any DMCTL-algebra on $[0, 1]$ is of that form.

3 dMTL: The Dual Logic of MTL

Before expanding the logic MTL with a strong disjunction and its residuum, it is convenient to start with a *dualised* version of MTL, that we call dMTL, with only a strong disjunction and its residuum. It is just a formal exercise, but it will help later when defining the whole logic.

The language of dMTL will be built from a countable set of propositional variables and the primitive set of connectives $\{\vee, \circledcirc, \leftarrow\}$ together with the constant $\mathbf{1}$, as usual:

- All propositional variables and $\mathbf{1}$ are formulas.
- If φ, ψ are formulas, then $\varphi \circledcirc \psi$, $\varphi \leftarrow \psi$ and $\varphi \vee \psi$ are formulas.

The following connectives are definable:

1. $\neg\varphi := \varphi \leftarrow \mathbf{1}$
2. $\varphi \wedge \psi := ((\varphi \leftarrow \psi) \leftarrow \psi) \vee ((\psi \leftarrow \varphi) \leftarrow \varphi)$
3. $\mathbf{0} := \neg\mathbf{1}$

The axiom system for dMTL, completely dual to that for MTL, is as follows:

(dA1) $(\varphi \leftarrow \psi) \leftarrow ((\psi \leftarrow \chi) \leftarrow (\varphi \leftarrow \chi))$
(dA2) $(\varphi \circledcirc \psi) \leftarrow \varphi$
(dA3) $(\varphi \circledcirc \psi) \leftarrow (\psi \circledcirc \varphi)$
(dA4) $(\varphi \vee \psi) \leftarrow \varphi$
(dA5) $(\varphi \vee \psi) \leftarrow (\psi \vee \varphi)$
(dA6) $(\varphi \circledcirc (\varphi \leftarrow \psi)) \leftarrow (\varphi \vee \psi)$
(dA7) $(\varphi \leftarrow (\psi \leftarrow \chi)) \leftarrow ((\varphi \circledcirc \psi) \leftarrow \chi)$
(dA8) $((\varphi \circledcirc \psi) \leftarrow \chi) \leftarrow (\varphi \leftarrow (\psi \leftarrow \chi))$
(dA9) $((\varphi \leftarrow \psi) \leftarrow \chi) \leftarrow (((\psi \leftarrow \varphi) \leftarrow \chi) \leftarrow \chi)$
(dA10) $\mathbf{1} \leftarrow \varphi$

The only inference rule for dMTL is the following dual form of Modus Ponens:

$$\text{(dMP)}: \frac{\psi \quad \psi \leftarrow \varphi}{\varphi}$$

Analogously to MTL, dMTL is also an algebraizable logic, whose equivalent algebraic semantics is given by the class of what we call dMTL-algebras. A *dMTL-algebra* is a structure $\mathbf{A} = \langle A, \wedge, \vee, \oplus, \leftarrow, 0, 1 \rangle$ such that:

- $\langle A, \wedge, \vee, 0, 1 \rangle$ is a lattice
- $\langle A, \oplus, 0 \rangle$ is a commutative monoid,
- $\leftarrow$ is the dual residuum of $\oplus$, i.e. for all $x, y, z \in L$ the following dual residuation condition holds:
$$y \leq x \oplus z \text{ iff } x \leftarrow y \leq z.$$
- The following dual prelinearity condition holds:
$$(x \leftarrow y) \wedge (y \leftarrow x) = 0$$

In fact, dMTL-algebras are the dual structures of MTL-algebras in the following sense. Let $\langle A, \wedge, \vee, 0, 1 \rangle$ be a lattice, and consider its dual version $\langle A, \wedge^d, \vee^d, 0^d, 1^d \rangle$, where $\wedge^d = \vee, \vee^d = \wedge, 0^d = 1$ and $1^d = 0$. Then, it is easy to check that $\mathbf{A} = \langle A, \wedge, \vee, \oplus, \leftarrow, 0, 1 \rangle$ is a dMTL-algebra iff $\mathbf{A}^d = \langle A, \wedge^d, \vee^d, \oplus, \leftarrow, 0^d, 1^d \rangle$ is a MTL-algebra. In other words, a dMTL-algebra is just a MTL-algebra over its dual lattice reduct.

As a consequence of this observation, it becomes clear that the logic dMTL is not a 1-preserving logic but a 0-preserving logic. Indeed, all the above axioms are evaluated to 0 by any evaluation on a dMTL-algebra and the inference rule (dMP) does not preserve the truth but the falsity: indeed, for any evaluation e, if $e(\psi) = 0$ and $e(\psi \leftarrow \varphi) = 0$, then $e(\varphi) = 0$ as well.

Theorem 4. *For any set of dMTL-formulas $\Gamma \cup \{\varphi\}$, $\Gamma \vdash_{dMTL} \varphi$ iff, for any dMTL algebra* **A** *and for every* **A***-evaluation e, $e(\psi) = 0$ for all $\psi \in \Gamma$ implies $e(\varphi) = 0$.*

Moreover, one can show that MTL and dMTL are equivalent deductive systems in the following strong sense. Let $^+$ be the map from dMTL-formulas to MTL-formulas such that

$$\mathbf{1}^+ = \mathbf{0},\ p^+ = p, \text{for each propositional variable } p, \text{and}$$
$$(\varphi \vee \psi)^+ = \varphi^+ \wedge \psi^+, (\varphi \leftarrow \psi)^+ = \varphi^+ \rightarrow \psi^+, (\varphi \circledcirc \psi)^+ = \varphi^+ \& \psi^+,$$

and let * be the converse map from MTL to dMT-formulas, i.e. the map such that

$$\mathbf{0}^* = \mathbf{1}, p^* = p, \text{for each propositional variable } p, \text{and}$$
$$(\varphi \wedge \psi)^* = \varphi^* \vee \psi^*, (\varphi \rightarrow \psi)^* = \varphi^* \leftarrow \psi^*, (\varphi \& \psi)^* = \varphi^* \circledcirc \psi^*.$$

Then one can easily prove that deductions can be translated from one logic to the other.

Lemma 1. *The following conditions hold:*

1. *For any set of MTL-formulas $\Gamma \cup \{\varphi\}$: if $\Gamma \vdash_{MTL} \varphi$, then $\Gamma^* \vdash_{dMTL} \varphi^*$.*
2. *For any set of dMTL-formulas $\Gamma \cup \{\varphi\}$: If $\Gamma \vdash_{dMTL} \varphi$, then $\Gamma^+ \vdash_{MTL} \varphi^+$.*

4 The Logic DMCTL: Putting Together MTL and dMTL

In this section we introduce the *double monoidal t-conorm and t-norm logic*, DMCTL for short, as the expansion of the logic MTL with a strong disjunction and its dual residuum.

The language of the logic DMCTL expands the language of MTL (built over primitive connectives $\wedge, \&, \rightarrow$ and $\mathbf{0}$) with two new connectives $\{\circledcirc, \leftarrow\}$, standing respectively for a strong disjunction and its dual residuum.

We will continue using the following definable connectives:

1. $\neg\varphi := \varphi \rightarrow \mathbf{0}$
2. $\varphi \vee \psi := ((\varphi \rightarrow \psi) \rightarrow \psi) \wedge ((\psi \rightarrow \varphi) \rightarrow \varphi)$
3. $\varphi \equiv \psi := (\varphi \rightarrow \psi) \& (\psi \rightarrow \varphi)$
4. $\ulcorner\varphi := \varphi \leftarrow \mathbf{1}$

From the syntactic point of view, we define the logic DMCTL by expanding MTL with, roughly speaking, the 'negation' of the axioms and rules of dMTL.

Definition 4. *DMCTL is the expansion of MTL with the following set of axioms*

$$\{\neg\Phi : \Phi \in \{(dA1), \ldots, (dA10)\}\}$$

and the rules

$$(\text{R1}) : \frac{\varphi \to \psi}{\neg(\psi \leftarrow \varphi)} \qquad (\text{R2}) : \frac{\neg(\psi \leftarrow \varphi)}{\varphi \to \psi}$$

Note that rules (R1) and (R2) enforce the requirement that the order relations induced by both implications coincide.

Remark 5. *Observe that the corresponding 'negation' of the dual of modus ponens* (dMP)*, i.e. the rule*

$$\frac{\neg\varphi \quad \neg(\varphi \leftarrow \psi)}{\neg\psi}$$

does not appear in the above definition of the DMCTL logic since this rule is derivable using (R2) *and modus tollens (from $\varphi \to \psi$ and $\neg\psi$, derive $\neg\varphi$), that is already derivable in MTL, and hence in DMCTL as well.*

It is not hard to show that DMCTL enjoys the following congruence properties for the new connectives.

Lemma 2. *DMCTL proves the following congruence properties for $\circledcirc$ and $\leftarrow$:*

(i) $\varphi \equiv \psi \vdash_{DMCTL} \varphi \circledcirc \chi \equiv \psi \circledcirc \chi$
(ii) $\varphi \equiv \psi \vdash_{DMCTL} (\psi \leftarrow \chi) \equiv (\varphi \leftarrow \chi)$
(iii) $\varphi \equiv \psi \vdash_{DMCTL} (\chi \leftarrow \varphi) \equiv (\chi \leftarrow \psi)$

These properties point out that DMCTL is in fact a *weakly implicative logic* and hence algebraizable as well [5], and it has the variety of DMCTL-algebras as its equivalent algebraic semantics. Therefore we already have for free the following general completeness result.

Theorem 6 (General completeness). *For any set of DMCTL-formulas $\Gamma \cup \{\varphi\}$, $\Gamma \vdash_{DMCTL} \varphi$ iff, for any DMCTL-algebra* **A** *and for every* **A***-evaluation e, $e(\psi) = 1$ for all $\psi \in \Gamma$ implies $e(\varphi) = 1$.*

By definition, DMCTL is a expansion of MTL, but in fact we can show more.

Lemma 3. *DMCTL is a conservative expansion of MTL.*

The relation between dMTL and DMCTL is obviously a bit different. In the following, we will use $\vdash^*_{MTL}$ and $\vdash^*_{dMTL}$ to denote derivations (in the language of DMCTL) using only axioms and rules of MTL and of dMTL respectively.[2]

Lemma 4. *For any set of dMTL formulas $\Gamma \cup \{\varphi\}$, we have:*
If $\Gamma \vdash^*_{dMTL} \varphi$ then $\neg\Gamma \vdash_{DMCTL} \neg\varphi$, where $\neg\Gamma = \{\neg\psi \mid \psi \in \Gamma\}$.

[2] This means that, e.g. in a derivation $\Gamma \vdash^*_{dMTL} \varphi$, any maximal subformula in $\Gamma \cup \{\varphi\}$ not belonging to the language of dMTL is treated as a new propositional variable.

4.1 DMCTL$^\ell$: The Semilinear Version of DMCTL

According to the observation made in Sect. 2.1, DMCTL, as an expansion of MTL, would be semilinear iff the following $\vee$-forms of the rules (R1) and (R2) following rules

$$(\mathrm{R1}^\vee): \frac{(\varphi \to \psi) \vee \gamma}{\neg(\psi \leftarrow \varphi) \vee \gamma} \qquad (\mathrm{R2}^\vee): \frac{\neg(\psi \leftarrow \varphi) \vee \gamma}{(\varphi \to \psi) \vee \gamma}$$

were derivable in DMCTL. However, we have not succeeded in proving this.

This leads us to consider the logic DMCTL$^\ell$ by replacing in DMCTL the rules (R1) and (R2) by the above stronger two rules (R1$^\vee$) and (R2$^\vee$) respectively. Then, it is clear that DMCTL$^\ell$ is indeed a semilinear logic whose algebraic semantics is given by the quasi-variety of DMCTL$^\ell$-algebras, where a DMCTL$^\ell$-algebra is just a DMCTL-algebra satisfying the two following additional condition (in fact it consists of two quasi-equations):

- $(x \to y) \vee z = 1$ iff $\neg(x \leftarrow y) \vee z = 1$

Since in a linearly ordered lattice, we have $x \vee y = 1$ iff either $x = 1$ or $y = 1$, an easy observation is that the set of linearly-ordered DMCTL$^\ell$-algebras and linearly-ordered DMCTL-algebras coincide.

Lemma 5. *$\mathbf{A}$ is a linearly ordered DMCTL$^\ell$-algebra iff $\mathbf{A}$ is a linearly ordered DMCTL-algebra.*

Therefore, we have the following chain-completeness result for DMCTL$^\ell$.

Theorem 7 (Chain completeness). *The logic DMCTL$^\ell$ is complete with respect to the class of linearly ordered DMCTL-algebras.*

In algebraic terms, this result says that the quasivariety of DMCTL$^\ell$-algebras is generated by the class of its linearly-ordered members.

4.2 Standard Completeness of DMCTL$^\ell$

Finally, in this section we show that, in fact, DMCTL$^\ell$ enjoys standard completeness, i.e. completeness with respect to the class of DMCTL-chains on the real unit interval $[0, 1]$, or in other words, completeness w.r.t. those algebras on $[0, 1]$ defined by a left-continuous t-norm and a right-continuous t-conorm and their residua. To do so, one has to show that if a formula φ does not get the value 1 for some evaluation on a countable linearly ordered DMCTL-algebra, then we can always find a linearly ordered DMCTL-algebra on the unit real interval [0, 1] and an evaluation e on this algebra such that $e(\varphi) < 1$.

Mimicking the constructions in the proof of the standard completeness for MTL in [10], we start with a countable DMCTL-chain

$$\mathbf{S} = \langle S, \odot, \to, \oplus, \leftarrow, \leq_S, 0_S, 1_S \rangle,$$

and we go through the steps we sketch below:

Step 1. We densify **S** in two different ways:

(a) Define the structure $\mathbf{X} = \langle X, \circ, \preceq, 0_X, 1_X \rangle$ as follows:

(i) $X = \{(s,1) : s \in S\} \cup \{(s,q) \mid s \in S \text{ and } \exists s', s = suc(s') > s', q \in \mathbb{Q} \cap (0,1)\}$,

(ii) $(s,q) \preceq (s',q')$ iff either $s <_S s'$, or $s = s'$ and $q \leq q'$,

(iii) $0_X = (0_S, 1)$ and $1_X = (1_S, 1)$,

(iv) $\circ$ is a binary operation defined as

$$(s,q) \circ (s',q') = \begin{cases} \min_X\{(s,q),(s',q')\}, & \text{if } s \odot s' = \min_S(s,s') \\ (s \odot s', 1), & \text{otherwise} \end{cases}$$

where $\min_X$ and $\min_S$ are the lattice meets with respect to $\preceq$ and $\leq_S$ respectively.

(b) Define the structure $\mathbf{Y} = \langle Y, \bullet, \preceq, 0_Y, 1_Y \rangle$ as follows:

(i) $Y = \{(s,0) : s \in S\} \cup \{(s,q) \mid s \in S \text{ and } \exists suc(s) > s, q \in \mathbb{Q} \cap (0,1)\}$,

(ii) $(s,q) \preceq (s',q')$ iff either $s <_S s'$, or $s = s'$ and $q \leq q'$,

(iii) $0_Y = (0_S, 0)$ and $1_Y = (1_S, 0)$,

(iv) $\bullet$ is a binary operation defined as

$$(s,q) \bullet (s',q') = \begin{cases} \max_X\{(s,q),(s',q')\}, & \text{if } s \oplus s' = \max_S(s,s') \\ (s \oplus s', 0), & \text{otherwise} \end{cases}$$

where $\max_X$ and $\max_S$ are the lattice joins with respect to $\preceq$ and $\leq_S$ respectively.

Note that while in **X** we have $\lim_{q \to 1}(s,q) = (s,1)$ and $\lim_{q \to 0}(s,q) = (s',1)$ whenever $s' = suc(s) < s$, in **Y** we have $\lim_{q \to 1}(s',q) = (s,1)$ and $\lim_{q \to 0}(s',q) = (s',0)$.

Step 2. Although they are different, $\langle X, \preceq, 0_X, 1_X \rangle$ and $\langle Y \preceq, 0_Y, 1_Y \rangle$ are order isomorphic through the mapping $h : X \to Y$ defined as

$$\begin{cases} h((s,1)) = (s,0) \\ h((s,r)) = (s',r) \text{ if } r < 1 \text{ and } s = suc(s') \end{cases}$$

Moreover, h preserves suprema and infima.

Step 3. One can show that $\langle X, \circ, \preceq, 0_X, 1_X \rangle$ is a commutative linearly ordered integral monoid with null element 0_X and that $\circ$ is in fact left-continuous with respect to the order topology on $\langle X, \preceq \rangle$.

Step 4. Analogously, one can show that $\langle Y, \bullet, \preceq, 0_Y, 1_Y \rangle$ is a commutative linearly ordered integral monoid with null element 1_Y and that $\bullet$ is right-continuous with respect to the order topology on $\langle Y, \preceq \rangle$.

As a consequence: the structure $\langle X, \bullet^*, \preceq, 0_X, 1_X \rangle$, where

$$(s,r) \bullet^* (s',r') = h^{-1}(h((s,r)) \bullet h((s',r'))),$$

is a commutative linearly ordered integral monoid with null element 1_X and show that $\bullet^*$ is right-continuous with respect to the order topology on $\langle X, \preceq \rangle$.

Step 5. The map $\Phi : S \to X$ defined as $\Phi(s) = (s, 1)$ is an embedding of the structure $\langle S, \odot, \oplus, \leq_S, 0_S, 1_S\rangle$ into the structure $\langle X, \circ, \bullet^*, \preceq, 0_X, 1_X\rangle$, and moreover, restricted to $\Phi(S)$:
 (1) The residuum of $\circ$ exists, call it $\Rightarrow$, and $\Phi(s) \Rightarrow \Phi(s') = \Phi(s \to s')$ for all $s, s' \in S$.
 (2) The residuum of $\bullet^*$ exists, call it $\Leftarrow$, and $\Phi'(s) \Leftarrow \Phi'(s') = \Phi'(s \leftarrow s')$ for all $s, s' \in S$.

Step 6. Since X is countable and dense, the structure $\langle X, \circ, \Rightarrow, \bullet^*, \Leftarrow, \preceq, 0_X, 1_X\rangle$ is order-isomorphic (through a mapping $\Psi : X \to \mathbb{Q} \cap [0,1]$) to a structure on the rationals in $[0, 1]$, $\langle \mathbb{Q} \cap [0, 1], \circ', \Rightarrow', \bullet', \Leftarrow', \leq, 0, 1\rangle$.

Step 7. The structure $\langle \mathbb{Q} \cap [0, 1], \circ', \bullet', \leq, 0, 1\rangle$ can be embedded into an analogous structure $\langle [0, 1], \hat{\circ}, \hat{\bullet}, \leq, 0, 1\rangle$ on the real unit interval, where:
 (1) $\hat{\circ}$ is the extension of $\circ'$ defined by means of left-continuity, i.e. $\alpha \hat{\circ} \beta = \sup\{x \circ' y \mid x, y \in \mathbb{Q} \cap [0, 1], x \leq \alpha, y \leq \beta\}$. Moreover $\hat{\circ}$ is left-continuous.
 (2) $\hat{\bullet}$ is the extension of $\bullet'$ defined by means of right-continuity, i.e. $\alpha \hat{\bullet} \beta = \inf\{x \bullet' y \mid x, y \in \mathbb{Q} \cap [0, 1], x \geq \alpha, y \geq \beta\}$. Moreover $\hat{\bullet}$ is right-continuous.

Step 8. Since $\hat{\circ}$ is left-continuous and $\hat{\bullet}$ is right-continuous on [0, 1], their residua and dual residua exist, call them $\Rightarrow_{\hat{\circ}}$ and $\Leftarrow_{\hat{\bullet}}$ respectively. Then $\langle [0, 1], \hat{\circ}, \Rightarrow_{\hat{\circ}}, \hat{\bullet}, \Leftarrow_{\hat{\bullet}}, \leq, 0, 1\rangle$ is a linearly ordered DMCTL-algebra where the initial DMCTL-chain $\mathbf{S}$ can be embedded. In particular, if $\Gamma : S \to [0,1]$ denotes the composition of the maps Φ and Ψ, it can be shown that $\Gamma(s \odot s') = \Gamma(s) \hat{\circ} \Gamma(s')$, $\Gamma(s \oplus s') = \Gamma(s) \hat{\bullet} \Gamma(s')$, $\Gamma(s \to s') = \Gamma(s) \Rightarrow_{\hat{\circ}} \Gamma(s')$, and $\Gamma(s \leftarrow s') = \Gamma(s) \Leftarrow_{\hat{\bullet}} \Gamma(s')$ for all $s, s' \in S$.

As a direct consequence of the above constructions, we can claim the standard completeness of the logic DMCTL$^{\ell}$.

Theorem 8 (Standard completeness). *The logic DMCTL$^{\ell}$ is standard complete. In other words, for any set of formulas $\Gamma \cup \{\varphi\}$, $\Gamma \vdash \varphi$ iff, for any standard DMCTL-algebra* $\mathbf{A}$ *and any* $\mathbf{A}$*-evaluation e, if $e(\psi) = 1$ for all $\psi \in \Gamma$, then $e(\varphi) = 1$.*

Proof. Suppose $\Gamma \not\vdash_{DMCTL^{\ell}} \varphi$, then by the chain-completeness result, there is a countable DMCTL-chain $\mathbf{S}$ and an $\mathbf{S}$-evaluation e that $e(\psi) = 1$ for all $\psi \in \Gamma$ while $e(\varphi) < 1$. Then, by the above construction, there is a DMCTL-chain on $[0, 1]$ where $\mathbf{A}$ can be embedded. Let h be such an embedding and let e' the evaluation of variables on $[0, 1]$ defined as $e'(p) = h(e(p)$ for every propositional variable p. Then one still has that $e'(\psi) = 1$ for all $\psi \in \Gamma$ while $e'(\varphi) < 1$.

5 Conclusions

In this paper we have been concerned with defining a logic complete with respect to the class of double residuated lattices on $[0, 1]$ induced by left-continuous t-norms and right-continuous t-conorms. Future work will focus on different extensions of the logic DMCTL$^{\ell}$, counterparts of well-known axiomatic extensions of MTL, like Involutive MTL (IMTL), Hájek's BL logic, or any of their extensions.

Acknowledgments. Esteva and Godo acknowledge partial support by the Spanish MINECO/FEDER project RASO (TIN2015- 71799-C2-1-P).

References

1. Cintula, P., Hájek, P., Noguera, C. (eds.): Handbook of Mathematical Fuzzy Logic - Volume 1, Studies in Logic, Mathematical Logic and Foundations, vol. 37. College Publications, London (2011)
2. Cintula, P., Hájek, P., Noguera, C. (eds.): Handbook of Mathematical Fuzzy Logic - Volume 2, Studies in Logic, Mathematical Logic and Foundations, vol. 38. College Publications, London (2011)
3. Cintula, P., Fermüller, C., Noguera, C. (eds.): Handbook of Mathematical Fuzzy Logic - Volume 3, Studies in Logic, Mathematical Logic and Foundations, vol. 58. College Publications, London (2016)
4. Cintula, P., Noguera, C.: Implicational (semilinear) logics I: a new hierarchy. Arch. Math. Log. **49**, 417–446 (2010). https://doi.org/10.1007/s00153-010-0178-7
5. Cintula, P., Noguera, C.: A general framework for mathematical fuzzy logic. In: Cintula, P., Hájek, P., Noguera, C. (eds.) Handbook of Mathematical Fuzzy Logic, Vol. 1, Volume 37 of Studies in Logic, Chap. 1, pp. 103–207. College Publications, London (2011)
6. Esteva, F., Godo, L.: Monoidal t-norm based logic: towards a logic for left-continuous t-norms. Fuzzy Sets Syst. **124**(3), 271–288 (2001)
7. Galatos, N., Jipsen, P., Kowalski, T., Ono, H.: Residuated Lattices: An Algebraic Glimpse at Substructural Logics. Elsevier, Amsterdam (2007)
8. Goguen, J.A.: The logic of inexact concepts. Synthese **19**, 325–373 (1968–1969)
9. Hájek, P.: Metamathematics of Fuzzy Logic. Kluwer, Dordrecht (1998)
10. Jenei, S., Montagna, F.: A proof of standard completeness for Esteva and Godo's logic MTL. Stud. Logica **70**(2), 183–192 (2002). https://doi.org/10.1023/A:1015122331293
11. Orłowska, E.: Many-valuedness and uncertainty. In: Multiple-Valued Logic, vol. 4, pp. 207–227 (1999)
12. Orłowska, E., Radzikowska, A.M.: Double residuated lattices and their applications. In: de Swart, H.C.M. (ed.) RelMiCS 2001. LNCS, vol. 2561, pp. 171–189. Springer, Heidelberg (2002). https://doi.org/10.1007/3-540-36280-0_12
13. Radzikowska, A.M., Kerre, E.: A fuzzy generalisation of information relations. In: Orłowska, E., Fitting, M. (eds.) Beyond Two: Theory and Applications of Multiple-Valued Logics. STUDFUZZ, vol. 114, pp. 287–312. Springer, Heidelberg (2003). https://doi.org/10.1007/978-3-7908-1769-0_13
14. Radzikowska, A.M., Kerre, E.E.: Fuzzy information relations and operators: an Algebraic approach based on residuated lattices. In: de Swart, H., Orłowska, E., Schmidt, G., Roubens, M. (eds.) Theory and Applications of Relational Structures as Knowledge Instruments II. LNCS (LNAI), vol. 4342, pp. 162–184. Springer, Heidelberg (2006). https://doi.org/10.1007/11964810_8
15. Sócola-Ramos, M.: Double residuated lattices. Master thesis, Universitat de Barcelona, September 2019

Automorphism Groups of Finite BL-Algebras

Stefano Aguzzoli[1] and Brunella Gerla[2](✉)

[1] DI, Università degli Studi di Milano, Via Celoria 18, Milan, Italy
aguzzoli@di.unimi.it

[2] DiSTA, Università degli Studi dell'Insubria, Via O. Rossi 9, Varese, Italy
brunella.gerla@uninsubria.it

Abstract. Using a category dual to finite $\mathbb{BL}$-algebras and their homomorphisms, in this paper we characterise the structure of the automorphism group of any given finite $\mathbb{BL}$-algebra. Further, we specialise our result to the case of the variety generated by the k-element $\mathbb{MV}$-algebra, for each $k > 1$.

Keywords: $\mathbb{BL}$-algebras, Automorphism group, Substitutions

1 Introduction

The variety of $\mathbb{BL}$-algebras constitutes the algebraic semantics of Hájek's Basic Fuzzy Logic BL, which in turn is the logic of all continuous t-norms and their residua. The characterisation of the automorphism group of an algebraic structure is a typical problem in algebra. When the algebraic structure is associated with a logical system, BL in our case, then automorphisms are related with substitutions. Indeed, a substitution acting on the first n propositional letters can be conceived as an endomorphism of the free n-generated $\mathbb{BL}$-algebra, and then automorphisms of the same algebra coincide with invertible substitutions. As any $\mathbb{BL}$-algebra is a quotient of some free $\mathbb{BL}$-algebra, an automorphism of a $\mathbb{BL}$-algebra is a substitution which is invertible over the equivalence classes determined by the quotient.

In this work we use a category dually equivalent to finite $\mathbb{BL}$-algebras with their homomorphisms, namely the category of *finite weighted forests*, to characterise the structure of the automorphism group of any given finite $\mathbb{BL}$-algebra.

We exploit the finite combinatorial description of finite weighted forests to decompose any such automorphism group by means of direct and semidirect products of symmetric groups. Our results constitute a generalisation of [6], where the same approach has been used to characterise the automorphism groups of finite Gödel algebras, and they are related with the forthcoming paper [1], where these combinatorial techniques are applied to several subvarieties of $\mathbb{MTL}$-algebras. The paper is structured as follows. In the first section we introduce the logic BL and $\mathbb{BL}$-algebras; further we recall the notion of automorphism group

M.-J. Lesot et al. (Eds.): IPMU 2020, CCIS 1239, pp. 666–679, 2020.
https://doi.org/10.1007/978-3-030-50153-2_49

of an algebra, and its relations with invertible substitutions in the associated logic. Then, as an example we apply our approach based on dual categorical equivalence to characterise the automorphism groups of finite Boolean algebras. We recall also the notion of semidirect product of groups. In the third section we introduce the category of *finite weighted forests* and the dual equivalence with finite $\mathbb{BL}$-algebras. In the fourth section we prove our main result about automorphism groups of finite $\mathbb{BL}$-algebras. Finally, we specialise our result to the case of the variety generated by the k-element $\mathbb{MV}$-algebra, for each $k > 1$.

Hájek's Basic Fuzzy Logic BL [15] is proven in [9] to be the logic of all continuous t-norms and their residua. We recall that a t-norm is an operator $*\colon [0,1]^2 \to [0,1]$ that is associative, commutative, monotonically non-decreasing in both arguments, having 0 as absorbent element, and 1 as unit. The t-norm $*$ is continuous if $*$ it is so in the standard euclidean topology of $[0,1]^2$. BL is axiomatised as follows.

(A1) $(\phi \to \chi) \to ((\chi \to \psi) \to (\phi \to \psi))$
(A2) $(\phi \odot \chi) \to \phi$
(A3) $(\phi \odot \chi) \to (\chi \odot \phi)$
(A4) $(\phi \odot (\phi \to \chi)) \to (\chi \odot (\chi \to \phi))$
(A5a) $(\phi \to (\chi \to \psi)) \to ((\phi \odot \chi) \to \psi)$
(A5b) $((\phi \odot \chi) \to \psi) \to (\phi \to (\chi \to \psi))$
(A6) $((\phi \to \chi) \to \psi) \to (((\chi \to \phi) \to \psi) \to \psi)$
(A7) $\bot \to \phi$

with modus ponens

$$\frac{\varphi \quad \varphi \to \psi}{\psi}$$

as the only inference rule.

The equivalent algebraic semantics of BL is given by the variety of $\mathbb{BL}$-algebras, as follows: A $\mathbb{BL}$-*algebra* is an algebra $(A, \odot, \to, \bot)$ of type $(2,2,0)$ such that upon defining $x \wedge y = x \odot (x \to y)$ (*divisibility*), $x \vee y = ((x \to y) \to y) \wedge ((y \to x) \to x)$, and $\top = a \to a$ for some arbitrarily fixed $a \in A$, the following holds:

(i) $(A, \odot, \top)$ is a commutative monoid;
(ii) $(A, \vee, \wedge, \top, \bot)$ is a bounded lattice;
(iii) *residuation* holds, that is, $x \odot z \leq y$ if and only if $z \leq x \to y$.

Gödel logic G is obtained by extending BL via the idempotency axiom $\varphi \to (\varphi \& \varphi)$. Łukasiewicz logic Ł is the extension of BL via the involutiveness of negation axiom $\neg\neg\varphi \to \varphi$ (where $\neg\varphi$ stands for $\varphi \to \bot$).

The equivalent algebraic semantics of G is given by the variety of Gödel algebras $\mathbb{G}$, that is the subvariety of $\mathbb{BL}$ formed by those algebras satisfying the identity $x = x \odot x$. $\mathbb{MV}$-algebras are those $\mathbb{BL}$-algebras satisfying $x = (x \to \bot) \to \bot$. The variety $\mathbb{MV}$ of $\mathbb{MV}$-algebras constitutes the equivalent algebraic semantics of Ł.

A variety $\mathbb{V}$ is locally finite if its n-generated free algebra $\mathbf{F}_n(\mathbb{V})$ is finite, for each $n \geq 0$.

We define $x^0 = \top$ and for all $n \geq 0$, $x^{n+1} = x^n \odot x$, and say that a $\mathbb{BL}$-algebra is n-contractive if it satisfies the identity $x^{n+1} = x^n$. While $\mathbb{G}$ is locally finite (and 1-contractive), $\mathbb{BL}$ and $\mathbb{MV}$ are not (see [2,13,19] for a description of finitely generated free algebras in these varieties), but their n-contractive subvarieties are. As a matter of fact, only the n-contractive subvarieties of $\mathbb{BL}$ (or $\mathbb{MV}$) are locally finite.

Let L be a schematic extension of BL having the variety $\mathbb{L}$ as equivalent algebraic semantics. Then its Lindenbaum algebra is the $\mathbb{L}$-algebra of the classes $\varphi/\equiv$ of logically equivalent formulas, that is $\varphi \equiv \psi$ iff L proves $\varphi \to \psi$ and $\psi \to \varphi$. The operations are defined through the connectives: for all binary connectives $*$, $(\varphi/\equiv) * (\psi/\equiv) := (\varphi * \psi)_{\equiv}$, the bottom element is $\bot/\equiv$.

The Lindenbaum algebra of L is isomorphic with the free $\mathbb{L}$-algebra over ω generators $\mathbf{F}_\omega(\mathbb{L})$. Analogously, for each $n \geq 0$, the Lindenbaum algebra of the formulas of L built using only the first n propositional letters $x_1, x_2, \ldots, x_n$ is isomorphic with the free $\mathbb{L}$-algebra over n many generators $\mathbf{F}_n(\mathbb{L})$.

A *substitution* σ over $\{x_1, \ldots, x_n\}$ is displayed as

$$x_1 \mapsto \varphi_1, \ldots, x_n \mapsto \varphi_n$$

for $\varphi_1, \ldots, \varphi_n$ formulas built over $\{x_1, \ldots, x_n\}$, with the obvious meaning that $\sigma(x_i) = \varphi_i$. The substitution σ extends naturally to each formula over $\{x_1, \ldots, x_n\}$, via the inductive definition $\sigma(*(\psi_1, \ldots, \psi_k)) = *(\sigma(\psi_1), \ldots, \sigma(\psi_k))$, for each k-ary connective $*$ and k-tuple of formulas $(\psi_1, \ldots, \psi_k)$. As it is clear that if $\varphi \equiv \psi$ then $\sigma(\varphi) \equiv \sigma(\psi)$, then the substitution σ can be identified with an endomorphism of the n-generated free algebra:

$$\sigma \colon \mathbf{F}_n(\mathbb{L}) \to \mathbf{F}_n(\mathbb{L}).$$

The set of all substitutions over $\{x_1, \ldots, x_n\}$, equipped with functional composition, forms the *monoid of endomorphisms* $\mathbf{End}(\mathbf{F}_n(\mathbb{L}))$ of $\mathbf{F}_n(\mathbb{L})$, having the identity $id\colon x_i \mapsto x_i$ as neutral element. The bijective endomorphisms in $\mathbf{End}(\mathbf{F}_n(\mathbb{L}))$ are clearly the same as isomorphisms of $\mathbf{F}_n(\mathbb{L})$ onto itself, and form the *group of automorphisms* $\mathbf{Aut}(\mathbf{F}_n(\mathbb{L}))$ of $\mathbf{F}_n(\mathbb{L})$. In terms of substitutions, $\mathbf{Aut}(\mathbf{F}_n(\mathbb{L}))$ is the group of invertible substitutions over $\{x_1, \ldots, x_n\}$, that is, those σ such that there exists a substitution σ^{-1} such that $\sigma \circ \sigma^{-1} = \sigma^{-1} \circ \sigma = id$.

In an earlier work [6] we have characterised the automorphism group of finite Gödel algebras—the algebraic semantics of propositional Gödel logic—by means of a dual categorical equivalence. In this paper we shall apply and generalise those techniques to finite $\mathbb{BL}$-algebras (and hence, to locally finite subvarieties of $\mathbb{BL}$, when the technique is applied to their free algebras). We take here the opportunity to correct a nasty mistake in the introduction of [6]: for a quirk of carelessness, there we erroneously declared that automorphisms preserve logical equivalence, which is clearly not the case (this mistake does not invalidate any technical result in the paper). In that paper we had in mind the more algebraic notion of equivalence given in the following Proposition, stressing the fact that automorphisms preserve all relevant algebraic information of logical equivalence between the formulas.

Proposition 1. *Let* $\mathbb{L}$ *be a locally finite variety constituting the algebraic semantics of a schematic extension L of BL. Then* σ *is an automorphism of* $\mathbf{F}_n(\mathbb{L})$ *if and only if for all pairs of formulas* φ, ψ*:*

$$\varphi \equiv \psi \qquad \textit{if and only if} \qquad \sigma(\varphi) \equiv \sigma(\psi).$$

Proof. Clearly the property holds for automorphisms. Pick then $\sigma \in \mathbf{End}(\mathbf{F}_n(\mathbb{L})) \setminus \mathbf{Aut}(\mathbf{F}_n(\mathbb{L}))$. Since $\mathbb{L}$ is locally finite, we assume σ is not injective. Then, there are $\varphi \not\equiv \psi$ such that $\sigma(\varphi) \equiv \sigma(\psi)$.

Clearly, every algebra in a variety comes with its automorphism group. Each algebra $\mathbf{A} \in \mathbb{L}$ is a quotient of some free algebra in $\mathbb{L}$. If $\mathbf{A} = \mathbf{F}_n(\mathbb{L})/\Theta$ for some congruence Θ of $\mathbf{F}_n(\mathbb{L})$, then we can interpret elements of $\mathbf{Aut}(\mathbf{A})$ as substitutions. As a matter of fact, $\sigma \in \mathbf{Aut}(\mathbf{A})$ if there is a substitution σ' defined over $\{x_1, \ldots, x_n\}$ such that $\sigma\colon \varphi/\Theta \mapsto (\sigma'(\varphi))/\Theta$ is bijective (and hence invertible). Whence σ is conceived as an invertible substitution over $\{x_1/\Theta, \ldots, x_n/\Theta\}$. In logical terms we have an analogous of Proposition 1. Let $\varphi \leftrightarrow \psi$ be a shortening of $(\varphi \to \psi) \wedge (\psi \to \varphi)$.

Proposition 2. *Let* $\mathbb{L}$ *be a locally finite variety constituting the algebraic semantics of a schematic extension L of BL. Let* $\mathbf{A} = \mathbf{F}_n(\mathbb{L})/\Theta$ *for some congruence* Θ*. Let* Γ_Θ *be the theory formed by all formulas* φ *such that* $\varphi\Theta\top$*. Then* σ *is an automorphism of* $\mathbf{A}$ *if and only if for all pair of formulas* φ, ψ*:*

$$\Gamma_\Theta \models \varphi \leftrightarrow \psi \qquad \textit{if and only if} \qquad \Gamma_\Theta \models \sigma(\varphi) \leftrightarrow \sigma(\psi).$$

Proof. The same proof of Proposition 1, *mutatis mutandis.*

In the paper [1], the same technique used in [6] is adapted to determine the structure of the group of automorphisms of finitely generated free algebras in several locally finite subvarieties of $\mathbb{MTL}$-algebras. We recall that $\mathbb{MTL}$ forms a supervariety of $\mathbb{BL}$ [12], which constitutes the equivalent algebraic semantics of the logic MTL, which in turns is the logic of all left-continuous t-norms and their residua. In [1] the algebra of automorphism invariant (equivalence classes of) formulas is introduced and studied (just for the case of Gödel logic).

2 Automorphism Groups

With each algebraic structure $\mathbf{A}$ we can associate its *monoid of endomorphisms* $\mathbf{End}(\mathbf{A}) = (\{f\colon \mathbf{A} \to \mathbf{A}\}, \circ, id)$, having as universe the set of all homomorphisms of $\mathbf{A}$ into itself, where $\circ$ is functional composition, and $id\colon a \mapsto a$ for each $a \in A$ is the identity. The invertible elements of $\mathbf{End}(\mathbf{A})$, that is, those f such that there exists $f^{-1} \in \mathbf{End}(\mathbf{A})$ with the property $f \circ f^{-1} = id = f^{-1} \circ f$, constitute the universe of the *group of automorphisms* $\mathbf{Aut}(\mathbf{A})$ of $\mathbf{A}$.

Let $\mathbf{Sym}(n)$ denote the symmetric group over n elements, that is, the group of all permutations of an n-element set.

Let $\mathbb{B}$ denote the variety of Boolean algebras. We prove the well-known fact stated in Proposition 4 by means of a dual categorical equivalence, since the same approach is used in [6] and [1], and we shall use it for the case of finite $\mathbb{BL}$-algebras.

We recall that two categories C and D are dually equivalent iff there exists a pair of contravariant functors $F: \mathsf{C} \to \mathsf{D}$ and $G: \mathsf{D} \to \mathsf{C}$ whose compositions FG and GF are naturally isomorphic with the identities in D and C.

Proposition 3. *The category $\mathbb{B}_{fin}$ of finite Boolean algebras and their homomorphisms is dually equivalent to the category Set_{fin} of finite sets and functions between them.*

Proof. This is just the restriction to finite objects of the well known Stone's duality between Boolean algebras and Stone spaces.

Let us call Sub: $\mathsf{Set}_{fin} \to \mathbb{B}_{fin}$ and Spec: $\mathbb{B}_{fin} \to \mathsf{Set}_{fin}$ the functors implementing the equivalence. It is folklore that Sub S is the Boolean algebra of the subsets of S, and Spec $\mathbf{A}$ is the set of maximal filters of $\mathbf{A}$. On arrows, Sub and Spec are defined by taking preimages.

Clearly, for each Boolean algebra $\mathbf{A}$, $\mathbf{Aut}(\mathbf{A}) \cong \mathbf{Aut}(\mathrm{Spec}\,\mathbf{A})$.

Proposition 4. $\mathbf{Aut}(\mathbf{F}_n(\mathbf{B})) \cong \mathbf{Sym}(2^n)$.

Proof. Just recall that Spec $\mathbf{F}_n(\mathbf{B})$ is the set of 2^n elements, and an automorphism of a finite set is just a permutation of its elements.

To deal with the structure of the automophism groups of finite $\mathbb{BL}$-algebras we shall introduce some constructions from group theory. We refer to [20] for background.

Definition 1. *Given two groups $\mathbf{H}$ and $\mathbf{K}$ and a group homomorphism $f: k \in \mathbf{K} \mapsto f_k \in \mathbf{Aut}(\mathbf{H})$, the* semidirect product *$\mathbf{H} \rtimes_f \mathbf{K}$ is the group obtained equipping $H \times K$ with the operation $(h,k) * (h',k') = (hf_k(h'), kk')$.*

Theorem 1. *Let $\mathbf{G}$ be a group with identity e and let $\mathbf{H}, \mathbf{K}$ be two subgroups of $\mathbf{G}$. If the following hold:*

- $\mathbf{K} \lhd \mathbf{G}$ *($\mathbf{K}$ is a normal subgroup of $\mathbf{G}$);*
- $G = H \times K$;
- $H \cap K = \{e\}$,

then $\mathbf{G}$ is isomorphic to the semidirect product of $\mathbf{H}$ and $\mathbf{K}$ with respect to the homomorphism $f: k \in \mathbf{K} \to f_k \in \mathbf{Aut}(\mathbf{H})$ where for each $h \in \mathbf{H}$, $f_k(h) = khk^{-1}$. Hence $|\mathbf{G}| = |\mathbf{H}| \cdot |\mathbf{K}|$.

In the following, we shall simply write $\mathbf{H} \rtimes \mathbf{K}$ instead of $\mathbf{H} \rtimes_f \mathbf{K}$, as in any usage we assume f is as in Theorem 1.

3 Finite $\mathbb{BL}$-Algebras and Finite Weighted Forests

In this Section we shall present the relevant facts for our purposes about the dual categorical equivalence between finite $\mathbb{BL}$-algebras (with homomorphisms) and the category of *finite weighted forests.* The duality is introduced in [4].

A *forest* is a poset such that the collection of lower bounds $\downarrow x = \{y \mid y \leq x\}$ of any given element x is totally ordered. A morphism of forests $f\colon F \to G$ is an order-preserving map that is *open*, that is whenever $z \leq f(x)$ for $z \in G$ and $x \in F$, then there is $y \leq x$ in F such that $f(y) = z$.

Finite forests and their morphisms form a category, denoted F_{fin}.

Let $\mathbb{N}^+$ denote the set of positive natural numbers. A *weighted forest* is a function $w\colon F \to \mathbb{N}^+$, where F is a forest, called the *underlying forest* of w. A weighted forest is *finite* if its underlying forest is. Consider two finite weighted forests $w\colon F \to \mathbb{N}^+$, $w'\colon F' \to \mathbb{N}^+$. By a *morphism* $g\colon w \to w'$ we mean an order-preserving map $g\colon F \to F'$ that is:

(M1) *open* (or is a *p-morphism*), that is, whenever $x' \leq g(x)$ for $x' \in F'$ and $x \in F$, then there is $y \leq x$ in F such that $g(y) = x'$, and
(M2) *respects weights*, meaning that for each $x \in F$, there exists $y \leq x$ in F such that $g(y) = g(x)$ and $w'(g(y))$ divides $w(y)$.

Contemplation of these definitions shows that finite weighted forests and their morphisms form a category. Let us write WF_{fin} for the latter category, and $\mathbb{BL}_{fin}$ for the category of finite $\mathbb{BL}$-algebras and their homomorphisms.

In the following we shall often manipulate the underlying forest F of a finite weighted forest $w\colon F \to \mathbb{N}^+$. Henceforth, in order to simplify exposition, we shall write (F, w) for such a finite weighted forest.

Theorem 2. *The category of finite $\mathbb{BL}$-algebras and their homomorphisms is dually equivalent to the category of weighted forests and their morphisms. That is, there are functors*

$$\mathrm{wSpec}\colon \mathbb{BL}_{fin} \to \mathsf{WF}_{fin} \quad \textit{and} \quad \mathrm{Sub}\colon \mathsf{WF}_{fin} \to \mathbb{BL}_{fin}$$

such that the composite functors wSpec ∘ Sub *and* Sub ∘ wSpec *are naturally isomorphic to the identity functors on* WF_{fin} *and* $\mathbb{BL}_{fin}$*, respectively.*

Proof. See [4] for the definition of the functors wSpec and Sub.

In particular, by [18, Thm. IV.4.1] the functor wSpec is essentially surjective, and this yields the following representation theorem for finite $\mathbb{BL}$-algebras.

Corollary 1. *Any finite $\mathbb{BL}$-algebra is isomorphic to* Sub(F, w) *for a weighted forest (F, w) that is unique to within an isomorphism of weighted forests.*

While the previous corollary has been already proved in [11, Sect. 5] and, as a special case of a more general construction, in [17, Sect. 6], the finite duality theorem is first introduced in [4].

4 The Automorphism Group of a Finite Weighted Forest

In this Section we shall study the automorphism group of a finite weighted forest (F, w) since by Theorem 2 and Corollary 1 this group is the same as the automorphism group of the dual finite $\mathbb{BL}$-algebra $\mathrm{Sub}\,(F, w)$.

An *order-preserving permutation* of F is a bijective map $\pi\colon F \to F$ such that if $x \leq y$ then $\pi(x) \leq \pi(y)$.

Theorem 3. *The automorphism group* $\mathbf{Aut}(F)$ *of a finite forest* F *is isomorphic with the group of order-preserving permutations of* F.

Proof. It is clear that $\pi\colon F \to F$ is an isomorphism iff π is a bijective morphism in F_{fin}. Since morphisms of finite forests do preserve order, then if $x \leq y$ it must be $\pi(x) \leq \pi(y)$.

Let (F, w) be a weighted forest. For each $x \in F$ we denote $\uparrow x = \{y \in F \mid x \leq y\}$ the *upset* of x and $\downarrow x = \{y \in F \mid y \leq x\}$ the *downset* of x.

If x is not minimal in F we denote by $p(x)$ its unique predecessor.

For any $x \in F$, let $h(x) = |\downarrow x|$ be the *height* of x. Let $h(F) = \max\{h(x) \mid x \in F\}$.

Let $f\colon F \to F$ be an order-preserving permutation of F. Then we say that f *preserves the weights* of the map $w\colon F \to \mathbb{N}^+$ when $w(f(x)) = w(x)$ for every $x \in F$.

Proposition 5. *An automorphism of a weighted forest* (F, w) *is an order-preserving bijection* f *from* F *to* F *preserving the weights of* w.

Proof. An automorphism $f\colon (F, W) \to (F, W)$ is a bijective morphism in WF_{fin}. Perusal of the definitions of morphism in WF_{fin} and in F_{fin}, shows that f is a map $f\colon F \to F$ which is a bijective morphism in F_{fin}, that is f is an order-preserving permutation of F, by Theorem 3. As $f\colon F \to F$ is invertible, condition (M2) holds iff $w(f(x)) = w(x)$ for any $x \in F$.

We denote by $\mathbf{Aut}(F, w)$ the group of all automorphisms of (F, w).

For every $i = 1, \ldots, h(F) - 1$, we consider the partition of F given by its *levels*:

$$A_i = \{x \in F \mid h(x) = i\}$$

and the relation $\mathcal{R}$ on F such that $x\mathcal{R}y$ if and only if $\uparrow x \cong \uparrow y$ as finite weighted forests, $w(x) = w(y)$ and either x and y are both minimal or $p(x) = p(y)$.

The relation $\mathcal{R}$ is an equivalence relation and we denote by $[x]$ its equivalence classes. Note that for every $x \in F$, if $x \in A_i$ then $[x] \subseteq A_i$ and the set $[A_i] = \{[x] \mid x \in A_i\}$ is a partition of A_i. Further, if $x \neq y$ and $x\mathcal{R}y$ then for every $x_1 \in \uparrow x$ there is no $y_1 \in \uparrow y$ such that $x_1\mathcal{R}y_1$.

If $\varphi \in \mathbf{Aut}(F, w)$ and $A \subseteq F$, by $\varphi \upharpoonright A$ we mean the restriction of φ to A.

Lemma 1. *The following hold:*

(i) For every $\varphi \in \mathbf{Aut}(F, w)$ *and* $x \in F$, $\varphi([x]) = [\varphi(x)]$;

(ii) if $\varphi, \psi \in \mathbf{Aut}(F, w)$ are such that $\varphi(p(x)) = \psi(p(x))$, then $\varphi([x]) = \psi([x])$.

In the following, we fix a strict order relation $\prec$ on F such that $\prec$ is total on each equivalence class and for any $x, y \in F$, x, y are incomparable if $[x] \neq [y]$. Hence $\prec$ is an order relation that compares just elements in the same class (hence at the same level, with the same weight and with an isomorphic upset).

Definition 2. *Let (F, w) be a weighted forest with $h(F) = h$. Then for $i = 1, \ldots, h$, an* i-permutation respecting $\prec$ *on (F, w) is an order-preserving bijection $\pi : F \to F$ also preserving the weight w and such that:*

(i) For every $j < i$ and $x \in A_j$, $\pi(x) = x$;
(ii) For every $k > i$ and $x, y \in A_k$, if $x \prec y$ then $\pi(x) \prec \pi(y)$;

Note that this definition is well given since if $x \prec y$ then $[x] = [y]$ and, being $\pi \in \mathbf{Aut}(F, w)$, by Lemma 1 we have $[\pi(x)] = [\pi(y)]$. If π is an i-permutation and $x \in A_i$ (with $i > 1$) then $\pi(p(x)) = p(x)$ hence $\pi(x) \in [x]$ and $\pi([x]) = [x]$.

By definition, i-permutations fix everything below the level i and permute (in such a way to respect order and weights) elements in the levels above i by keeping the same order $\prec$ in the classes. So, the order $\prec$ is needed to fix a canonical i-permutation π once we have defined π on A_i, as explained in the following lemma.

Lemma 2. *Let $\varphi \in \mathbf{Aut}(F, w)$ and $i \in \{1, \ldots, h(F)\}$. If $\varphi([x]) = [x]$ for every $x \in A_i$ then there is a unique i-permutation $(\varphi)_i$ respecting $\prec$ such that*

$$(\varphi)_i \upharpoonright A_i = \varphi \upharpoonright A_i.$$

Proof. For every $j < i$ and $x \in A_j$, set $(\varphi)_i(x) = x$ and for every $x \in A_i$ set $(\varphi)_i(x) = \varphi(x)$. Since $\varphi([x]) = [x]$, then $p(x) = p(\varphi(x))$ and for every $x \in A_i$,

$$\downarrow (\varphi)_i(x) = \varphi(x) \cup \{y \mid y \leq x\} = (\varphi)_i(\downarrow x).$$

Then $(\varphi)_i$ is an order-preserving permutation of $\downarrow A_i$. Further, for every $x \subset A_i$ and $y \in (\varphi)_i([x])$, $\uparrow x \cong \uparrow y$.

Let $x \in A_{i+1}$. Then φ bijectively maps $[x]$ onto $\varphi([x])$. Display $[x]$ as $\{x_1 \prec \cdots \prec x_s\}$ and $\varphi([x])$ as $\{y_1 \prec \cdots \prec y_s\}$. Then we set $(\varphi)_i(x_j) = y_j$ for $j = 1, \ldots, s$. Note that $(\varphi)_i$ is an order-preserving permutation of $\downarrow A_{i+1}$ and $\uparrow x \cong \uparrow y$ for every $y \in (\varphi)_i([x])$.

In general, let $k > i$ and suppose to have extended $(\varphi)_i$ to an order-preserving permutation of $\downarrow A_{k-1}$ such that $\uparrow x \cong \uparrow y$ for every $y \in (\varphi)_i([x])$. Let $x \in A_k$. Since $p(x) \in A_{k-1}$ then $\uparrow p(x) \cong \uparrow (\varphi)_i(p(x))$. Then there exists $y \in \uparrow (\varphi)_i(p(x)) \cap A_k$ such that $\uparrow [x] \cong \uparrow [y]$. We can hence display $[x]$ as $\{x_1 \prec \cdots \prec x_s\}$ and $[y]$ as $\{y_1 \prec \cdots \prec y_s\}$. Then we set $(\varphi)_i(x_j) = y_j$ for $j = 1, \ldots, s$. Clearly, $(\varphi)_i$ is an order-preserving permutation of A_k and $\uparrow x \cong \uparrow y$ for every $y \in (\varphi)_i([x])$.

We can hence define $(\varphi)_i$ on the whole F and it follows from the construction that $(\varphi)_i$ is the unique i-permutation respecting $\prec$ and coinciding with φ on A_i.

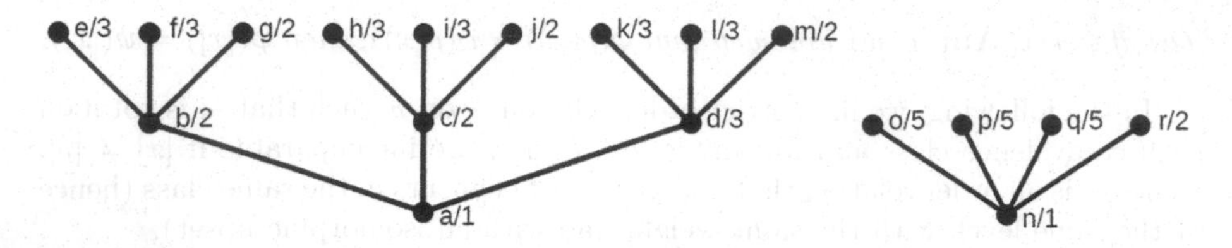

Fig. 1. Weighted forest (F, w) of Example 1

Example 1. Let (F, w) be as in Fig. 1, where each node is labeled with a letter and with its weight.

We have $h(F) = 3$ and levels are given by

$$A_1 = \{a, n\}$$
$$A_2 = \{b, c, d, o, p, q, r\}$$
$$A_3 = \{e, f, g, h, i, j, k, l, m\}.$$

Non-singleton classes are given by

$$[b] = [c] = \{b, c\} \subseteq A_2$$
$$[e] = [f] = \{e, f\} \subseteq A_3$$
$$[h] = [i] = \{h, i\} \subseteq A_3$$
$$[k] = [l] = \{k, l\} \subseteq A_3$$
$$[o] = [p] = [q] = \{o, p, q\} \subseteq A_2$$

while $[x] = \{x\}$ for every $x \in \{a, d, g, j, m, n, r\}$. Let us fix the order $b \prec c$, $e \prec f$, $h \prec i$, $k \prec l$ and $o \prec p \prec q$. The map φ_{21} defined by

x	a	b	c	d	e	f	g	h	i	j	k	l	m	n	o	p	q	r
$\varphi_{21}(x)$	a	c	b	d	h	i	j	e	f	g	k	l	m	n	o	p	q	r

maps b to c and viceversa and it is defined in accordance with $\prec$ on A_3, so it is a 2-permutation. Also φ_{22} defined by

x	a	b	c	d	e	f	g	h	i	j	k	l	m	n	o	p	q	r
$\varphi_{22}(x)$	a	b	c	d	e	f	g	h	i	j	k	l	m	n	p	q	o	r

is a 2-permutation since it fixes all elements but $\{o, p, q\}$. The composition $\varphi_2 = \varphi_{21}\varphi_{22} = \varphi_{22}\varphi_{21}$ is again a 2-permutation. Consider the automorphism φ defined by

x	a	b	c	d	e	f	g	h	i	j	k	l	m	n	o	p	q	r
$\varphi(x)$	a	c	b	d	i	h	j	f	e	g	k	l	m	n	p	q	o	r.

Note that φ is not a 2-permutation since $e \prec f$ but $\varphi(e) = i \not\prec \varphi(f) = h$. Nevertheless $\varphi \restriction A_2 = \varphi_2 \restriction A_2$. Note also that $\varphi \restriction A_3 \neq \varphi_2 \restriction A_3$ and we cannot apply Lemma 2 to level 3 since $\varphi([e]) \neq [e]$.

In the following, we shall omit the reference to the order $\prec$, and it will be tacitly assumed that all i-permutations are i-permutations respecting $\prec$.

The set P_i of all i-permutations is the domain of a subgroup $\mathbf{P}_i$ of $\mathbf{Aut}(F, w)$. We are going to describe it in terms of the symmetric groups $\mathbf{Sym}(n)$.

Definition 3. *Given $x \in A_i$, an $[x]$-permutation is an i-permutation π such that $\pi \restriction [x]$ is a permutation of $[x]$ and $\pi(y) = y$ for every $y \in A_i \setminus [x]$.*

Lemma 3. *Every i-permutation respecting $\prec$ is the composition of $[x]$-permutations for $[x] \in [A_i]$. Further,*

$$\mathbf{P}_i \cong \prod_{[x]\in[A_i]} \mathbf{Sym}(|[x]|).$$

Proof. Note that, since $[A_i] = \{[x] \mid x \in A_i\}$ is a partition of A_i, if π and π' are respectively an $[x]$-permutation and a $[y]$-permutation with $[x] \neq [y]$, then $\pi\pi' = \pi'\pi$. The claim then follows by the definition of i-permutation and by Lemma 2, noticing that for every i-permutation φ and $x \in A_i$, $\varphi([x]) = [x]$.

Example 2. Consider the weighted forest (F, w) from Example 1. We have

$$[A_1] = \{\{a\}, \{n\}\}$$
$$[A_2] = \{\{b, c\}, \{d\}, \{o, p, q\}, \{r\}\}$$
$$[A_3] = \{\{e, f\}, \{h, i\}, \{k, l\}, \{g\}, \{j\}, \{m\}\}.$$

If $x_1, \ldots, x_u \in A_i$, we use the notation $(x_1 x_2 \cdots x_u)$ to denote the unique i-permutation (in $\mathbf{Aut}(F, w)$) that maps $x_1 \mapsto x_2 \mapsto \cdots \mapsto x_u \mapsto x_1$ and fixes all the other elements of $\downarrow A_i$. Then we get

$$\mathbf{P}_3 = \{\mathrm{id}_F, (ef), (hi), (kl), (ef)(hi), (ef)(kl), (hi)(kl), (ef)(hi)(kl)\} \cong \mathbf{Sym}^3(2)$$

while

$$\mathbf{P}_2 = \{\mathrm{id}_F, (bc), (op), (oq), (pq), (opq), (oqp), (bc)(op), (bc)(oq), (bc)(pq), (bc)(opq), (bc)(oqp)\} \cong \mathbf{Sym}(2) \times \mathbf{Sym}(3)$$

and

$$\mathbf{P}_1 = \{\mathrm{id}_F\} \cong \mathbf{Sym}(1).$$

Let $\mathbf{N}_0 = \mathbf{Aut}(F, w)$ and for any $i = 1, \ldots, h(F)$,

$$\mathbf{N}_i = \{\varphi \in \mathbf{Aut}(F, w) \mid \varphi \restriction (\downarrow A_i) = id_{\downarrow A_i}\}.$$

Theorem 4. *Let $h = h(F)$. Then $\mathbf{N}_h = \{id_F\}$, $\mathbf{N}_{h-1} = \mathbf{P}_h$ and for any $i = 0, \ldots, h-2$*

$$\mathbf{N}_i \cong \mathbf{P}_{i+1} \rtimes \mathbf{N}_{i+1},$$

Proof. By definition, $\mathbf{N}_h = \{id_F\}$ and $\mathbf{N}_{h-1}$ is the set of order-preserving permutations that are distinct from the identity only on the level A_h. Since such a map π must satisfy $\pi([x]) = [x]$ for every $x \in A_h$, we have $\mathbf{N}_{h-1} = \mathbf{P}_h$.

It is easy to check that for every $i = 0, \ldots, h-2$, $\mathbf{P}_{i+1}$ and $\mathbf{N}_{i+1}$ are subgroups of $\mathbf{N}_i$. We first prove that $N_i = P_{i+1} \times N_{i+1}$ (as sets), that is for every element $\varphi \in \mathbf{N}_i$ there exist $\pi \in \mathbf{P}_{i+1}$ and $\psi \in \mathbf{N}_{i+1}$ such that $\varphi = \pi\psi$. Since for every $x \in A_{i+1}$, $\varphi([x]) = [x]$, then we can apply Lemma 2 and find $\pi = (\varphi)_{i+1} \in \mathbf{P}_{i+1}$ coinciding with φ over A_{i+1}. Let $\psi = (\varphi)_{i+1}^{-1}\varphi$. For every $x \in A_{i+1}$, $\varphi(x) \in A_{i+1}$ hence $\psi(x) = (\varphi)_i^{-1}(\varphi(x)) = \varphi^{-1}\varphi(x) = x$. For $x \in (\downarrow A_i)$, $\psi(x) = id_{A_i} id_{A_i}(x) = x$, hence $\psi \in N_{i+1}$. Clearly, $\varphi = \pi\psi$.

In order to prove that $\mathbf{N}_{i+1} \lhd \mathbf{N}_i$, let $\varphi \in \mathbf{N}_i$ and $\psi \in \mathbf{N}_{i+1}$. Then φ and ψ are both the identity function when restricted to $\downarrow A_i$ and further for every $x \in A_{i+1}$, $\varphi(\psi(\varphi^{-1}))(x) = \varphi\varphi^{-1}(x) = x$. Then $\varphi\psi\varphi^{-1}$ is the identity over $\downarrow A_{i+1}$ hence $\varphi\psi\varphi^{-1} \in N_{i+1}$.

Finally, we have to prove that $\mathbf{P}_{i+1} \cap \mathbf{N}_{i+1} = \{id_F\}$, but this is trivial since the only element in $\mathbf{P}_{i+1}$ coinciding with the identity over A_{i+1} is id_F.

The claim follows by Theorem 1.

Note that, by Theorem 1, the operation of the group $\mathbf{P}_{i+1} \rtimes \mathbf{N}_{i+1}$ is given by $(\pi, \psi) * (\pi', \psi') = (\pi\psi\pi'\psi^{-1}, \psi\psi')$.

Example 3. Consider the weighted forest (F, w) from Example 1 and the maps φ and φ_2. By definition, $\mathbf{N}_1 = \mathbf{Aut}(F, w) = \mathbf{N}_0$ since all the maps in $\mathbf{Aut}(F, w)$ are equal to the identity over the roots a and n. Further $\mathbf{N}_2 = \mathbf{P}_3$ and $\mathbf{N}_3 = \{\mathrm{id}_F\}$. By Theorem 4 we have $\mathbf{N}_1 \cong \mathbf{P}_2 \rtimes \mathbf{N}_2$. Consider for example the maps $\varphi \in \mathbf{Aut}(F, w)$ and $\varphi_2 \in \mathbf{P}_2$ from Example 1. Then $\varphi_2 = (bc)(opq) \in \mathbf{P}_2$ and $\varphi = \varphi_2 \circ (ef)(hi)(kl)$ where $(ef)(hi)(kl) \in \mathbf{P}_3 = \mathbf{N}_2$.

We are ready to state the structure of the automorphism group of a weighted forest (F, w).

Corollary 2. *Let (F, w) be a finite weighted forest. Let $h(F) = h$ and let $\mathbf{P}_i$ be the group of i-permutations of (F, w). Then*

$$\mathbf{Aut}(F, w) \cong \mathbf{P}_1 \rtimes (\mathbf{P}_2 \rtimes (\cdots \rtimes \mathbf{P}_h))$$

and

$$|\mathbf{Aut}(F, w)| = \prod_{i=1}^{h} \prod_{[x] \in [A_i]} |[x]|!.$$

Proof. The claim follows by Theorem 4 and Lemma 3.

Example 4. Let (F, w) be the weighted forest from Example 1. Then

$$\begin{aligned}\mathbf{Aut}(F, w) &\cong \mathbf{Sym}(1) \rtimes ((\mathbf{Sym}(2) \times \mathbf{Sym}(3)) \rtimes (\mathbf{Sym}^3(2))) \\ &\cong (\mathbf{Sym}(2) \times \mathbf{Sym}(3)) \rtimes (\mathbf{Sym}^3(2))\end{aligned}$$

and

$$|\mathbf{Aut}(F, w)| = \prod_{i=1}^{3} \prod_{[x] \in [A_i]} |[x]|! = 2 \cdot 3! \cdot 2 \cdot 2 \cdot 2 = 96.$$

We finally get,

Theorem 5. *Let* $\mathbf{A}$ *be a finite* $\mathbb{BL}$*-algebra. Let* $(F, w) = \mathrm{wSpec}\,\mathbf{A}$*, with* $h(F) = h$*. Let* $\mathbf{P}_i$ *be the group of i-permutations of* (F, w)*. Then*

$$\mathbf{Aut(A)} \cong \mathbf{P}_1 \rtimes (\mathbf{P}_2 \rtimes (\cdots \rtimes \mathbf{P}_h))$$

and

$$|\mathbf{Aut(A)}| = \prod_{i=1}^{h} \prod_{[x] \in [A_i]} |[x]|!.$$

Proof. Immediate, from Corollary 2, and the dual equivalence between WF_{fin} and $\mathbb{BL}_{fin}$.

5 Finite MV-Algebras and Finite Multisets of Natural Numbers

The following result is well-known [3,10,16].

Theorem 6. *The category of finite forests and order-preserving open maps is dually equivalent to the category of finite Gödel algebras and their homomorphisms.*

Notice that a finite forest F can be readily considered as a weighted forest $w \colon F \to \mathbb{N}^+$ such that w is a constant function. The most natural choice is $w(x) = 1$ for each $x \in F$, as the actual dual in $\mathbb{BL}_{fin}$ of a finite $\mathbb{BL}$-algebra that happens to be a Gödel algebra is a weighted forest where all nodes have weight 1. Whence, the automorphism group of a finite Gödel algebra is the same as the automorphism group of a weighted forest $w \colon F \to \mathbb{N}^+$, such that w is constant. The characterisation of these groups is precisely the subject of [6].

Another interesting case is when the weighted forest $w \colon F \to \mathbb{N}^+$ is such that all elements of F are uncomparable, that is F can be conceived as a set. Then (F, w) can be thought of as a multiset of positive natural numbers. This is precisely the case when (F, w) is the dual of a finite $\mathbb{MV}$-algebra.

The variety $\mathbb{MV}$ of MV-algebras constitutes the algebraic semantics of propositional Łukasiewicz logic [8]. $\mathbb{MV}$ is not locally finite, but the k-contractive $\mathbb{MV}$-algebras form a locally finite subvariety of $\mathbb{MV}$. Here we consider the subvariety $\mathbb{MV}_k$ generated by the MV-chain with k elements, which constitutes the algebraic semantics of k-valued Łukasiewicz logic. $\mathbb{MV}_k$ is axiomatised by imposing $(k-1)$-contractivity: $x^k = x^{k-1}$, and Grigolia's axioms [14] $k(x^h) = (h(x^{h-1}))^k$ for every integer $2 \leq h \leq k-2$ that does not divide $k-1$.

For any integer $d > 1$ let $\mathrm{Div}(d)$ be the set of coatoms in the lattice of divisors of d, and for any finite set of natural numbers X, let $\gcd(X)$ be the greatest common divisor of the numbers in X. Then let $\alpha(0, 1) = 1$, $\alpha(0, d) = 0$ for all $d > 1$, and for all $n \geq 1$,

$$\alpha(n, d) = (d+1)^n + \sum_{\emptyset \neq X \subseteq \mathrm{Div}(d)} (-1)^{|X|} (\gcd(X) + 1)^n.$$

Then $\alpha(n,d)$ counts the number of points in $[0,1]^n$ whose denominator is d. It is known that

$$\mathbf{F}_n(\mathbb{MV}_k) \cong \prod_{d|(k-1)} \text{Ł}_{d+1}^{\alpha(n,d)},$$

where Ł_m is the $\mathbb{MV}$-chain of cardinality m. Let $\mathsf{MN}_{\mathsf{k}fin}$ be the category whose objects are finite multisets of natural numbers dividing $k-1$ and whose arrows $f\colon M \to N$ are functions from M to N such that $f(x)$ divides x for any $x \in M$. Then $\mathsf{MN}_{\mathsf{k}fin}$ is dually equivalent to $\mathbb{MV}_k$. In particular, denoted $\mathrm{Spec}\colon \mathbb{MV}_k \to \mathsf{MN}_{\mathsf{k}fin}$ one of the pair of functors implementing the duality,

$$\mathrm{Spec}\,\mathbf{F}_n(\mathbb{MV}_k) \cong \bigcup_{d|(k-1)} \biguplus_{i=1}^{\alpha(n,d)} \{d\},$$

where $\biguplus_{i=1}^{m}\{t\}$ denotes the multiset formed by m copies of t.

It is clear that an automorphism $f\colon M \to M$ in $\mathsf{MN}_{\mathsf{k}fin}$ must be a bijection such that each copy of $x \in M$ is mapped to a copy of $x \in M$. Then

Theorem 7.

$$\mathbf{Aut}(\mathbf{F}_n(\mathbb{MV}_k)) \cong \prod_{d|(k-1)} \mathbf{Sym}(\alpha(n,d)),$$

and

$$|\mathbf{Aut}(\mathbf{F}_n(\mathbb{MV}_k))| = \prod_{d|(k-1)} (\alpha(n,d))!.$$

6 Conclusion

One can apply Theorem 5 to determine the automorphism group of all finitely generated free algebras in any given locally finite subvariety of $\mathbb{BL}$. Indeed, let $\mathbb{V}$ be such a subvariety. Then, by universal algebra, and by the dual equivalence between $\mathbb{BL}_{fin}$ and WF_{fin}, it holds that $\mathrm{wSpec}\,(\mathbf{F}_n(\mathbb{V})) \cong (\mathrm{wSpec}\,(\mathbf{F}_1(\mathbb{V})))^n$, where the power is computed using the recurrences for computing the products of finite forests (notice that in WF_{fin} and in F_{fin} the underlying set of the product is not the cartesian product of the underlying sets of the factors), see [4] and [5] for details. Once $\mathrm{wSpec}\,(\mathbf{F}_n(\mathbb{V}))$ is computed, Theorem 5 yields its automorphism group, which coincides with the automorphism group of $\mathbf{F}_n(\mathbb{V})$. Studying the structure of the automorphism group in subvarieties of $\mathbb{BL}$ that are not locally finite is much harder, as the combinatorial approach alone is not sufficient. The paper [7] provides a characterisation of the automorphism groups for a class of $\mathbb{MV}$-algebras containing members of infinite cardinality.

References

1. Aguzzoli, S.: Automorphism groups of Lindenbaum algebras of some propositional many-valued logics with locally finite algebraic semantics. In: Proceedings of FUZZ-IEEE - WCCI 2020 (2020, to appear)

2. Aguzzoli, S., Bova, S.: The free n-generated BL-algebra. Ann. Pure Appl. Logic **161**, 1144–1170 (2010)
3. Aguzzoli, S., Bova, S., Gerla, B.: Free algebras and functional representation for fuzzy logics. In: Handbook of Mathematical Fuzzy Logic. Studies in Logic. College Publications (2011)
4. Aguzzoli, S., Bova, S., Marra, V.: Applications of finite duality to locally finite varieties of BL-algebras. In: Artemov, S., Nerode, A. (eds.) LFCS 2009. LNCS, vol. 5407, pp. 1–15. Springer, Heidelberg (2008). https://doi.org/10.1007/978-3-540-92687-0_1
5. Aguzzoli, S., Codara, P.: Recursive formulas to compute coproducts of finite Gödel algebras and related structures. In: Proceedings of FUZZ-IEEE 2016, pp. 201–208. IEEE Computer Society Press (2016)
6. Aguzzoli, S., Gerla, B., Marra, V.: The automorphism group of finite Gödel algebras. In: Proceedings of the ISMVL 2000, pp. 21–26. IEEE Computer Society Press (2010)
7. Aguzzoli, S., Marra, V.: Finitely presented MV-algebras with finite automorpshism group. J. Logic Comput. **20**, 811–822 (2010)
8. Cignoli, R., D'Ottaviano, I., Mundici, D.: Algebraic Foundations of Many-Valued Reasoning. Trends in Logic, vol. 7. Springer, Dordrecht (1999). https://doi.org/10.1007/978-94-015-9480-6
9. Cignoli, R., Esteva, F., Godo, L., Torrens, A.: Basic fuzzy logic is the logic of continuous t-norms and their residua. Soft Comput. **4**(2), 106–112 (2000). https://doi.org/10.1007/s005000000044
10. D'Antona, O., Marra, V.: Computing coproducts of finitely presented Gödel algebras. Ann. Pure Appl. Logic **142**, 202–211 (2006)
11. Di Nola, A., Lettieri, A.: Finite BL-algebras. Discrete Math. **269**, 93–122 (2003)
12. Esteva, F., Godo, L.: Monoidal t-norm based logic: towards a logic for left-continuous t-norms. Fuzzy Sets Syst. **124**(3), 271–288 (2001)
13. Gerla, B.: A note on functions associated with Gödel formulas. Soft Comput. **4**(4), 206–209 (2000). https://doi.org/10.1007/PL00009890
14. Grigolia, R.: Algebraic analysis of Łukasiewicz-Tarski n-valued logical systems. In: Selected Papers on Łukasiewicz Sentential Calculi, pp. 81–91. Polish Academy of Science, Ossolineum (1977)
15. Hájek, P.: Metamathematics of Fuzzy Logic. Trends in Logic, vol. 4. Springer, Dordrecht (1998). https://doi.org/10.1007/978-94-011-5300-3
16. Horn, A.: Free L-algebras. J. Symb. Logic **34**, 475–480 (1969)
17. Jipsen, P., Montagna, F.: The Blok-Ferreirim theorem for normal GBL-algebras and its application. Algebra Univers. **60**, 381–404 (2009). https://doi.org/10.1007/s00012-009-2106-4
18. MacLane, S.: Categories for the Working Mathematician, 2nd edn. Springer (1998)
19. McNaughton, R.: A theorem about infinite-valued sentential logic. J. Symb. Logic **16**(1), 1–13 (1951)
20. Robinson, D.J.S: A Course in the Theory of Groups. Graduate Texts in Mathematics, vol. 80. Springer, New York (1995). https://doi.org/10.1007/978-1-4684-0128-8

Fuzzy Neighborhood Semantics for Multi-agent Probabilistic Reasoning in Games

Martina Daňková(✉) and Libor Běhounek

CE IT4Innovations–IRAFM, University of Ostrava, 30. dubna 22, 701 03 Ostrava, Czech Republic
{martina.dankova,libor.behounek}@osu.cz
https://ifm.osu.cz/

Abstract. In this contribution we apply fuzzy neighborhood semantics to multiple agents' reasoning about each other's subjective probabilities, especially in game-theoretic situations. The semantic model enables representing various game-theoretic notions such as payoff matrices or Nash equilibria, as well as higher-order probabilistic beliefs of the players about each other's choice of strategy. In the proposed framework, belief-dependent concepts such as the strategy with the best expected value are formally derivable in higher-order fuzzy logic for any finite matrix game with rational payoffs.

Keywords: Probabilistic reasoning · Fuzzy logic · Modal logic · Neighborhood semantics · Matrix game

1 Introduction

In this paper, we propose a semantics for multi-agent reasoning about uncertain beliefs. Using a suitable fuzzy logic for its representation makes it possible to formalize doxastic reasoning under uncertainty in a rather parsimonious way, which is of particular importance, e.g., in software modeling of rational agents.

As a prominent measure of uncertainty, we apply a fuzzy probability measure to fuzzy doxastic propositions. Fuzzy-logical modeling of probability started with [13]. Common approaches include two-layered expansions of suitable fuzzy logics by a fuzzy modality *probably,* states of MV-algebras, and probabilistic fuzzy description logics [10,15,16]. Here we generalize the fuzzy modal approach of [13], overcoming some of its limitations given by its two-layered syntax. Generally, though, we do not want to restrict the framework to perfectly rational agents. Therefore, we introduce a more general semantics that admits also probabilistically incoherent assignments of certainty degrees. This paves the way not only for accommodating the reasoning of probabilistically irrational agents, but also for modeling the agents' reasoning about the other agent's (ir)rationality and its potential exploitability.

M.-J. Lesot et al. (Eds.): IPMU 2020, CCIS 1239, pp. 680–693, 2020.
https://doi.org/10.1007/978-3-030-50153-2_50

As an illustration of the semantic framework, we apply it to probabilistic reasoning in game-theoretic situations. This application belongs to the broader research area of *logic in games* [4], which aims at a formal reconstruction of game-theoretical concepts by means of formal logic. As an interface between fuzzy logic and game theory, we employ the representation of (a broad class of) strategic games in fuzzy logic laid out in [2]. In the game-theoretic setting, the framework enables formalizing the player's beliefs about each other's choice of strategy, including higher-order beliefs (i.e., beliefs about the beliefs of others). In the game-theoretic setting, probabilistic beliefs are particularly important, due to the players' uncertainty about each others' choice of strategy, the possibility of using mixed (i.e., probabilistic) strategies, and is especially pronounced in games with incomplete information (such as most card games).

The paper is organized as follows. In Sect. 2, we gather prerequisites for developing fuzzy doxastic and probabilistic logic, including the logic ŁΠ and the notion of fuzzy probability measure. Fuzzy doxastic and probabilistic models for multi-agent reasoning are presented in Sect. 3. Next, in Sect. 4, we define fuzzy doxastic and probabilistic logic and discuss its relationships to various known logics. Section 5 provides an overview of game-theoretical notions formalized in fuzzy probabilistic or doxastic logic; subsequently, we apply these notions to represent probabilistic reasoning in a simple two-player game. Finally, the features of the introduced formalism and topics for future work are summarized in Sect. 6.

2 Preliminaries

For the formalization of probabilistic and doxastic reasoning in games, we will employ the expressively rich fuzzy logic ŁΠ. This choice is made for the sake of uniformity, even though many constructions described below can as well be carried out in some of its less expressive fragments such as $Ł_\triangle$ or $PŁ'_\triangle$. For details on the logic ŁΠ see [7,9]; here we just briefly recount the definition.

A salient feature of the logic ŁΠ is that it contains the connectives of many well-known fuzzy logics, including the three prominent t-norm based fuzzy logics (Gödel, Łukasiewicz, and product).

We use the symbols $\wedge, \vee, \neg, \otimes, \oplus, \sim, \Rightarrow, \odot, \oslash$, respectively, for the Gödel, Łukasiewicz, and product connectives $\&_G, \vee_G, \neg_G, \&_Ł, \vee_Ł, \neg_Ł, \Rightarrow_Ł, \&_\Pi, \Rightarrow_\Pi$ of ŁΠ. Of these, $\Rightarrow$, $\oslash$, $\odot$, and the truth constant 0 can be taken as the only primitives; the others are definable.

The *standard semantics* of ŁΠ, or the *standard* ŁΠ*-algebra* $[0,1]_{Ł\Pi}$, interprets the connectives by the following truth functions on $[0, 1]$:

$$
\begin{aligned}
x \wedge y &= \min(x, y) & x \otimes y &= \max(0, x + y - 1)\\
x \vee y &= \max(x, y) & x \Rightarrow y &= \min(1, 1 - x + y)\\
x \oplus y &= \min(1, x + y) & x \odot y &= x \cdot y\\
x \ominus y &= \max(0, x - y) & x \oslash y &= y/x \text{ if } x > y, \text{ otherwise } 1\\
{\sim}x &= 1 - x & \neg x &= 1 - \text{sign } x\\
x \Leftrightarrow y &= 1 - |x - y| & \triangle x &= 1 - \text{sign}(1 - x)
\end{aligned}
$$

An axiomatic system for ŁΠ consists of the axioms of Łukasiewicz and product logic respectively for Łukasiewicz and product connectives, the axioms $\triangle(\varphi \Rightarrow \psi) \Rightarrow (\varphi \oslash \psi)$, $\triangle(\varphi \oslash \psi) \Rightarrow (\varphi \Rightarrow \psi)$, $\varphi \odot (\psi \ominus \chi) \Leftrightarrow (\varphi \odot \psi) \ominus (\varphi \odot \chi)$, and the rules of modus ponens and $\triangle$-necessitation (from φ infer $\triangle\varphi$). The logic ŁΠ enjoys finite strong completeness of this axiomatic system w.r.t. its standard semantics on $[0, 1]$.

The general (linear) algebraic semantics of ŁΠ is given by the class of (linear) *ŁΠ-algebras* $\boldsymbol{L} = (L, \oplus, \sim, \oslash, \odot, 0, 1)$, where:

- $(L, \oplus, \sim, 0)$ is an MV-algebra,
- $(L, \vee, \wedge, \oslash, \odot, 0, 1)$ is a Π-algebra, and
- $x \odot (y \ominus z) = (x \odot y) \ominus (x \odot z)$ holds.

Like other fuzzy logics, ŁΠ also enjoys completeness w.r.t. the classes of linear and all ŁΠ-algebras. Except for the two-element ŁΠ-algebra $\{0, 1\}$, all non-trivial linear ŁΠ-algebras are isomorphic to the unit interval algebras of linearly ordered fields.

The first-order logic ŁΠ, denoted by ŁΠ∀, is defined as usual in fuzzy logic: in a first-order model $\boldsymbol{M} = (M, \boldsymbol{L}, I)$ over an ŁΠ-algebra $\boldsymbol{L}$, n-ary predicate symbols P are interpreted by $\boldsymbol{L}$-valued functions $I(P)\colon M^n \to \boldsymbol{L}$ and the quantifiers $\forall, \exists$ are evaluated as the infimum and supremum in $\boldsymbol{L}$. Safe ŁΠ∀-models (i.e., such that all required suprema and infima exist in $\boldsymbol{L}$) are axiomatized by the propositional axioms and rules of ŁΠ, generalization (from φ derive $(\forall x)\varphi$), and the axioms:

- $(\forall x)\varphi \Rightarrow \varphi(t)$, where t is free for x in φ, and
- $(\forall x)(\chi \Rightarrow \varphi) \Rightarrow (\chi \Rightarrow (\forall x)\varphi)$, where x is not free in χ.

First-order ŁΠ can be extended by the axioms for crisp identity, $x = x$ and $x = y \Rightarrow \triangle(\varphi(x) \Leftrightarrow \varphi(y))$, function symbols, and sorts of variables in a standard manner; see, e.g., [1].

The present paper will also make use of the logic ŁΠ of a higher order. Higher-order logic ŁΠ has been introduced in [1] and its Church-style notational variant in [20]. For the full description of higher-order ŁΠ we refer the reader to [1] or [3, Sect. A.3]; here we only highlight some of its features relevant to our purposes.

Of the full language of higher-order ŁΠ, in this paper we will only need its monadic fragment. Its syntax contains variables and constants for individuals $(x, y, \dots)$, first-order monadic predicates $(P, Q, \dots)$, second-order monadic predicates $(\mathcal{P}, \mathcal{Q}, \dots)$, etc. In a model over an ŁΠ-algebra $\boldsymbol{L}$, individual variables and constants are interpreted as elements of the model's domain X; first-order monadic predicates as fuzzy sets on X, i.e., elements of $\boldsymbol{L}^X$; second-order monadic predicates as fuzzy sets of fuzzy sets on X, i.e., elements of $\boldsymbol{L}^{\boldsymbol{L}^X}$; etc.

Besides the connectives of ŁΠ and the quantifiers of ŁΠ∀ (applicable to variables of any order), the language of monadic higher-order ŁΠ also contains *comprehension terms* $\{x^{(n)} \mid \varphi\}$, for all variables $x^{(n)}$ of any order n and all well-typed formulae φ. In $\boldsymbol{L}$-valued models, $\{x \mid \varphi(x)\}$ denotes the fuzzy set $A \in \boldsymbol{L}^X$

which assigns to each value of x the truth value of $\varphi(x)$. Analogously, $\{P \mid \varphi(P)\}$ denotes the second-order fuzzy set $\mathcal{A} \in \boldsymbol{L}^{\boldsymbol{L}^X}$ which assigns to each value of P the truth value of $\varphi(P)$, etc.

We will denote the higher-order logic ŁΠ by $\text{Ł}\Pi^\omega$. Its Henkin-style axiomatization in multi-sorted ŁΠ∀, consisting of the axioms of extensionality and comprehension for each type and complete w.r.t. Henkin models, can be found in [1,3].

In what follows we will need the following first-order fuzzy set operations (definable in $\text{Ł}\Pi^\omega$):

Definition 1. *Let X be a crisp set and $\boldsymbol{L}$ an ŁΠ-algebra. The fuzzy set operations $\Cap$, $\Cup$, $\smallsetminus$, and $\underline{0}$ are defined by setting for all $x \in X$ and $A, B \in \boldsymbol{L}^X$:*

$$(A \Cap B)(x) = A(x) \otimes B(x)$$
$$(A \Cup B)(x) = A(x) \oplus B(x)$$
$$(\smallsetminus A)(x) = {\sim}A(x)$$
$$\underline{0}(x) = 0$$

The following sections also refer to fuzzy probability measures, which have been extensively studied in the literature; for an overview see [10]. Fuzzy probability measures are usually defined as valued in the real unit interval $[0,1]$; here we use the definition generalized to any ŁΠ-algebra $\boldsymbol{L}$.

Definition 2. *Let $\boldsymbol{L}$ be an ŁΠ-algebra. A finitely additive $\boldsymbol{L}$-valued* fuzzy probability measure *on W is a function $\rho\colon \boldsymbol{L}^W \to \boldsymbol{L}$ such that the following conditions hold for all $A, B \in \boldsymbol{L}^W$:*

- $\rho(\underline{0}) = 0$
- $\rho(\smallsetminus A) = {\sim}\rho(A)$
- *If $\rho(A \Cap B) = 0$ then $\rho(A \Cup B) = \rho(A) \oplus \rho(B)$.*

Finally, let us briefly recall (multi-agent) *standard doxastic logic,* since fuzzy probabilistic and doxastic logics introduced below adapt its models to make them suitable for uncertain doxastic reasoning. For details on standard doxastic logic see, e.g., [18]. Standard multi-agent doxastic logic expands classical propositional logic by (freely nestable) unary modalities B_a for each agent a, where $\mathrm{B}_a\varphi$ is interpreted as "agent a believes that φ". Models for standard doxastic logic are given by Kripke-style (relational) semantics:

Definition 3. *A multi-agent* standard doxastic frame *is a tuple $\boldsymbol{F} = (W, A, \{R_a\}_{a \in A})$, where:*

- *$W \neq \emptyset$ is a set of* possible worlds.
- *$A \neq \emptyset$ is a set of* agents.
- *$R_a \subseteq W^2$ is the* accessibility relation *for each agent a. In standard doxastic logic it is assumed that all R_a are serial, transitive, and Euclidean.*

A multi-agent standard doxastic model *over the frame* $\boldsymbol{F}$ *is a pair* $\boldsymbol{M} = (\boldsymbol{F}, e)$, *where* $e\colon \mathit{Var} \times W \to \{0,1\}$ *is an* evaluation *of (countably many) propositional variables* $p_i \in \mathit{Var}$ *in each world* $w \in W$. *The truth value, or the* extension $\|\varphi\|_w$ *of a formula* φ *in a world* w *of the model* $\boldsymbol{M}$ *is defined by the recursive Tarski conditions:*

$$
\begin{aligned}
\|p\|_w = 1 &\quad \textit{iff} \quad e(p,w) = 1 \\
\|\neg\varphi\|_w = 1 &\quad \textit{iff} \quad \|\varphi\|_w = 0 \\
\|\varphi \Rightarrow \psi\|_w = 1 &\quad \textit{iff} \quad \|\varphi\|_w = 0 \textit{ or } \|\psi\|_w = 1 \\
\|\mathrm{B}_a\varphi\|_w = 1 &\quad \textit{iff} \quad R_a ww' \textit{ implies } \|\varphi\|_{w'} = 1 \textit{ for all } w' \in W
\end{aligned}
$$

The set $\|\varphi\| = \{w \in W\colon \|\varphi\|_w = 1\}$ *is called the* intension *of* φ *in* $\boldsymbol{M}$. *A formula is* valid *in* $\boldsymbol{M}$ *if* $\|\varphi\| = W$. *A formula is a doxastic* tautology *if it is valid in all doxastic models.*

Standard doxastic logic is axiomatized by adding the following axioms and rules to classical propositional logic, for all agents a:

(K)	$\mathrm{B}_a\varphi \wedge \mathrm{B}_a(\varphi \Rightarrow \psi) \Rightarrow \mathrm{B}_a\psi$	(logical rationality)
(D)	$\mathrm{B}_a\varphi \Rightarrow \neg\mathrm{B}_a\neg\varphi$	(consistency of beliefs)
(4)	$\mathrm{B}_a\varphi \Rightarrow \mathrm{B}_a\mathrm{B}_a\varphi$	(positive introspection)
(5)	$\neg\mathrm{B}_a\varphi \Rightarrow \mathrm{B}_a\neg\mathrm{B}_a\varphi$	(negative introspection)
(Nec)	from φ derive $\mathrm{B}_a\varphi$	(necessitation)

In standard doxastic frames, the intended role of accessibility relations R_a is such that the successor sets $wR_a = \{w' \mid R_a ww'\}$ comprise those worlds which the agent a in the world w does not rule out as being the actual world. The proposition "a believes that φ" is then considered true in w if φ is true in all worlds that a does not exclude in w, i.e., in all $w' \in wR_a$. The truth of $\mathrm{B}_a\varphi$ in w can thus be regarded as given by a (maxitive) two-valued measure $\beta_{a,w}$ on W:

$$
\beta_{a,w}(A) = \begin{cases} 1 & \text{if } wR_a \subseteq A \\ 0 & \text{otherwise,} \end{cases} \tag{1}
$$

for all $A \subseteq W$. Then the Tarski condition for B_a can be written as $\|B_a\varphi\|_w = \beta_{a,w}(\|\varphi\|)$. In the next section, the maxitive two-valued measure $\beta_{a,w}$ implicit in standard doxastic frames will be generalized to a finitely additive fuzzy probability measure suitable for doxastic reasoning under uncertainty.

3 Fuzzy Doxastic Models

For the modeling of doxastic or probabilistic reasoning of agents, we will employ a suitable fuzzy variant of possible-world (intensional) semantics (cf. [3,6,8,21]). The multi-agent fuzzy doxastic frames introduced in the following definition are a variant of similar structures that have already been employed for the semantics of probabilistic reasoning in the literature [5,11,12,22]. They also generalize (an equivalent reformulation of) the Kripke frames of standard doxastic logic [18].

Definition 4. *A* multi-agent fuzzy doxastic frame *is a tuple* $\boldsymbol{F} = (W, \boldsymbol{L}, A, \nu)$, *where:*

- $W \neq \emptyset$ *is a crisp set of* possible worlds *(or world states, situations).*
- $\boldsymbol{L}$ *is an* ŁΠ*-algebra of degrees.*
- $A \neq \emptyset$ *is a set of agents.*
- $\nu = \{\nu_{a,w}\}_{a \in A, w \in W}$, *where* $\nu_{a,w} \colon \boldsymbol{L}^W \to \boldsymbol{L}$ *for each* $a \in A$ *and* $w \in W$.

Fuzzy subsets of the set W of possible worlds are called *(fuzzy) propositions* or, synonymously, *(fuzzy) events.* In a multi-agent fuzzy doxastic frame $\boldsymbol{F}$, the functions $\nu_{a,w} \colon \boldsymbol{L}^W \to \boldsymbol{L}$ thus assign degrees to events. The value $\nu_{a,w}(E)$ is understood as the degree of the agent a's certainty in the world w about the event E. Each $\nu_{a,w}$ can also be regarded as a fuzzy set of events, $\nu_{a,w} \in \boldsymbol{L}^{\boldsymbol{L}^W}$.

The system $\nu = \{\nu_{a,w}\}_{a \in A, w \in W}$ can equivalently be viewed as assigning to each agent $a \in A$ a *fuzzy neighborhood function* $\nu_a \colon W \times \boldsymbol{L}^W \to \boldsymbol{L}$. These are known from the fuzzy neighborhood semantics of fuzzy modal logics [8,21], which is a fuzzy generalization of the well known Scott–Montague neighborhood semantics of modal logics [19,23]. The applicability of fuzzy neighborhood semantics to probabilistic and doxastic reasoning has been made explicit in [22].

A fuzzy neighborhood function ν_a assigns to each world w a fuzzy set of fuzzy "neighborhoods". The fuzzy neighborhoods of w will be understood as events that the agent a in the world w considers probable (to the degree assigned by $\nu_{a,w}$). We will interpret $\nu_{a,w}$ as measuring the subjective probability of events, as assessed by agent a in world w. In the general setting of multi-agent fuzzy doxastic frames we impose no restriction on $\nu_{a,w}$. In the following definition we specify additional conditions suitable for the probabilistic interpretation of $\nu_{a,w}$.

Definition 5. *Let* $\boldsymbol{F} = (W, \boldsymbol{L}, A, \nu)$ *be a multi-agent fuzzy doxastic frame. We say that* $\boldsymbol{F}$ *is a* (multi-agent) fuzzy probabilistic frame *if each* $\nu_{a,w}$ *is a finitely additive fuzzy probability measure.*

Thus, in fuzzy probabilistic frames, the subjective probability measures $\nu_{a,w}$ of all agents a and in all world states w are supposed to satisfy the axioms of fuzzy probability from Definition 2. This corresponds to the assumption of probabilistic rationality of all agents. In fuzzy doxastic frames, this condition is relaxed, which makes it possible to model agents with incomplete or incoherent assignments of probabilities.

In the probabilistic setting, the most common choice of $\boldsymbol{L}$ will be that of the standard ŁΠ-algebra, $\boldsymbol{L} = [0,1]_{\text{ŁΠ}}$; nevertheless, the definition also admits other ŁΠ-algebras of certainty degrees that may be suitable for probabilistic or doxastic reasoning, including the two-valued, rational-valued, or hyperreal-valued ones.

We will use fuzzy doxastic and probabilistic frames in the standard manner to define models for probabilistic and doxastic fuzzy modal logic. First we need to specify the modal language:

Definition 6. *Let Var be a countably infinite set of propositional variables and* A *a nonempty set of agents. By* $\mathscr{S}_A$ *we denote the propositional signature of the logic* ŁΠ *expanded by the unary modalities* P_a *for all* $a \in A$.

Thus, e.g., $(p \mathbin{\&} \mathrm{P}_a \triangle q) \Rightarrow \triangle \mathrm{P}_a \mathrm{P}_b (p \mathbin{\&} q)$ is a well-formed formula of $\mathscr{S}_{\{a,b\}}$.

Definition 7. *A (multi-agent) fuzzy doxastic model is a pair* $\boldsymbol{M} = (\boldsymbol{F}, e)$, *where* $\boldsymbol{F} = (W, \boldsymbol{L}, A, \nu)$ *is a multi-agent fuzzy probabilistic frame and* $e\colon \mathit{Var} \times W \to \boldsymbol{L}$ *is an* $\boldsymbol{L}$*-evaluation of propositional variables in each world.*

If $\boldsymbol{F}$ *is a fuzzy probabilistic frame, we speak of a* fuzzy probabilistic model.

As usual in intensional possible-world semantics, the semantic value of a formula φ in a model $\boldsymbol{M}$ is identified with its *intension* $\|\varphi\|\colon W \to \boldsymbol{L}$. The value of the intension $\|\varphi\|$ in a given world $w \in W$, i.e., the degree $\|\varphi\|(w) \in \boldsymbol{L}$, is called the *extension* of φ in w and denoted by $\|\varphi\|_w$. Note that in fuzzy doxastic frames, intensions of formulae are events and extensions are the degrees of the event's occurrence in particular worlds. Their values in fuzzy doxastic models are defined in a standard manner by recursive Tarski conditions (cf. [8,21]):

Definition 8. *The intensions* $\|\varphi\|\colon W \to \boldsymbol{L}$ *and extensions* $\|\varphi\|_w = \|\varphi\|(w)$ *of* $\mathscr{S}_A$*-formulae* φ *in the fuzzy doxastic model* $\boldsymbol{M}$ *are defined inductively as follows:*

$$\|p\|_w = e(p, w)$$
$$\|c(\varphi_1, \ldots, \varphi_n)\|_w = c^{\boldsymbol{L}}(\|\varphi_1\|_w, \ldots, \|\varphi_n\|_w)$$
$$\|\mathrm{P}_a \varphi\|_w = \nu_{a,w}(\|\varphi\|)$$

for all worlds $w \in W$, *agents* $a \in A$, *propositional variables* $p \in \mathit{Var}$, *all* $\mathscr{S}_A$*-formulae* $\varphi_1, \ldots, \varphi_n, \varphi$, *and each* n*-ary connective* c *of* ŁΠ, *where* $c^{\boldsymbol{L}}$ *is the truth function of* c *in the* ŁΠ*-algebra* $\boldsymbol{L}$.

It can be observed that standard doxastic frames (see Sect. 2) are special cases of fuzzy doxastic frames over the two-element ŁΠ-algebra $\{0,1\}$, when taking the maxitive two-valued measures $\beta_{a,w}$ of (1) for $\nu_{a,w}$.

Although the language $\mathscr{S}_A$ contains just a single graded probabilistic or doxastic modality P_a for each agent $a \in A$, various ranges and comparisons of probabilities (or certainty degrees) used in bivalent probabilistic logics are expressible by means of the connectives of ŁΠ. For instance, the qualitative probabilistic conditional $\varphi \succeq \psi$ of [12], "φ is at least as probable as ψ", is expressed for any agent a as $\triangle(\mathrm{P}_a \psi \Rightarrow \mathrm{P}_a \varphi)$. Similarly, the bivalent statement that "the probability assigned to φ by a is in the interval $\left[\frac{1}{3}, \frac{1}{2}\right]$" can be expressed by the $\mathscr{S}_A$-formula $\triangle(\mathrm{P}_a \varphi \oplus \mathrm{P}_a \varphi \oplus \mathrm{P}_a \varphi) \wedge \neg(\mathrm{P}_a \varphi \otimes \mathrm{P}_a \varphi)$. In general, by Proposition 1 below, any rational interval of a's probabilities for φ is expressible by an $\mathscr{S}_A$-formula. Note that this includes the threshold probabilistic modalities $\mathrm{P}_a^{\geq r} \varphi$, "$a$ believes that the probability of φ is at least r", for $r \in \mathbb{Q} \cap [0,1]$, used, e.g., in [11]. Since all infinite linear ŁΠ-algebras embed $\mathbb{Q} \cap [0,1]$, we can formulate the proposition more generally than just for standard models:

Proposition 1. *Let* $\boldsymbol{L}$ *be a linear* ŁΠ*-algebra,* $\boldsymbol{M} = (W, \boldsymbol{L}, A, \nu, e)$ *a fuzzy doxastic model,* $a \in A$, $w \in W$, *and* $r, s \in \mathbb{Q} \cap [0,1]$.

1. *There exist* ŁΠ*-formulae* $\chi_r, \chi_{\geq r}$ *in one propositional variable such that:*

$$\chi_r^{\boldsymbol{L}}(x) = \begin{cases} 1 & \text{if } x = r \\ 0 & \text{otherwise} \end{cases} \qquad \chi_{\geq r}^{\boldsymbol{L}}(x) = \begin{cases} 1 & \text{if } x \geq r \\ 0 & \text{otherwise,} \end{cases}$$

where $\chi_r^{\boldsymbol{L}}, \chi_{\geq r}^{\boldsymbol{L}}$ *are the truth functions of* $\chi_r, \chi_{\geq r}$ *in* $\boldsymbol{L}$.

2. $$\|\Delta(\mathrm{P}_a\psi \Rightarrow \mathrm{P}_a\varphi)\|_w = \begin{cases} 1 & \text{if } \nu_{a,w}(\|\varphi\|) \geq \nu_{a,w}(\|\psi\|) \\ 0 & \text{otherwise.} \end{cases}$$

3. $$\left\|\chi_{\geq r}(\mathrm{P}_a\varphi) \wedge \chi_{\geq 1-s}({\sim}\mathrm{P}_a\varphi)\right\|_w = \begin{cases} 1 & \text{if } \nu_{a,w}(\|\varphi\|) \in [r,s] \\ 0 & \text{otherwise.} \end{cases}$$

Proof. 1. If $r = 0$, take $\chi_r \equiv_{\mathrm{df}} \neg p$ and $\chi_{\geq r} \equiv_{\mathrm{df}} p \Rightarrow p$. If $r = 1$, take $\chi_r, \chi_{\geq r} \equiv_{\mathrm{df}} \Delta p$. If $r = \frac{m}{n} \in (0,1)$, $m, n \in \mathbb{N}$, then let $\varphi \equiv_{\mathrm{df}} p \wedge {\sim}p$; $\psi \equiv_{\mathrm{df}} \varphi \oslash (\varphi \oplus \varphi)$; $\vartheta \equiv_{\mathrm{df}} \bigodot_{i=1}^{\lceil \log_2(\max(m,n)) \rceil} \psi$; $\eta \equiv_{\mathrm{df}} (\bigoplus_{i=1}^{m} \vartheta) \oslash (\bigoplus_{i=1}^{n} \vartheta)$; $\chi_r \equiv_{\mathrm{df}} \Delta(\eta \Leftrightarrow p) \wedge \neg\Delta p$; and $\chi_{\geq r} \equiv_{\mathrm{df}} \Delta(\eta \Rightarrow p) \wedge \neg\neg p$. If $x \in \{0,1\}$, then it is easy to verify that $\eta^{\boldsymbol{L}}(x) = 1$, and thus $\chi_r^{\boldsymbol{L}}(0) = \chi_r^{\boldsymbol{L}}(1) = \chi_{\geq r}^{\boldsymbol{L}}(0) = 0$ and $\chi_{\geq r}^{\boldsymbol{L}}(1) = 1$ as desired. If $x \in \boldsymbol{L} \setminus \{0,1\}$, then $\varphi^{\boldsymbol{L}}(x) \leq \frac{1}{2}$, so $\psi^{\boldsymbol{L}}(x) = \frac{1}{2}$, $\vartheta^{\boldsymbol{L}}(x) = 2^{-\lceil \log_2(\max(m,n)) \rceil} \leq \max(\frac{1}{m}, \frac{1}{n})$, thus $\eta^{\boldsymbol{L}}(x) = \frac{m}{n}$, and then it is straightforward to verify that $\chi_r^{\boldsymbol{L}}(x), \chi_{\geq r}^{\boldsymbol{L}}(x)$ have the desired values.

Claims 2 and 3 follow directly from Claim 1 by Definition 8 and the semantics of propositional connectives in linear ŁΠ-algebras. □

4 Fuzzy Probabilistic and Doxastic Logic

The notions of truth, validity, tautologicity, and (local) entailment w.r.t. (classes of) fuzzy doxastic models are defined as usual in (fuzzy) intensional semantics (cf. [3,6,8]):

Definition 9. *Let* $\boldsymbol{M} = (W, \boldsymbol{L}, A, \nu, e)$ *be a fuzzy doxastic model. We say that an* $\mathscr{S}_A$*-formula* φ *is* true *in* $w \in W$ *if* $\|\varphi\|_w = 1^{\boldsymbol{L}}$. *We say that* φ *is* valid *in* $\boldsymbol{M}$ *if* $\|\varphi\|_w = 1^{\boldsymbol{L}}$ *for all* $w \in W$. *We say that* $\mathscr{S}_A$*-formulae* $\varphi_1, \ldots, \varphi_n$ *locally entail an* $\mathscr{S}_A$*-formula* φ *in* $\boldsymbol{M}$ *if* φ *is true in all worlds where all* $\varphi_1, \ldots, \varphi_n$ *are true.*

Let $\mathbf{K}$ *be a class of fuzzy doxastic models for* $\mathscr{S}_A$. *We say that an* $\mathscr{S}_A$*-formula* φ *is a* $\mathbf{K}$*-tautology, written* $\models_{\mathbf{K}} \varphi$, *if* φ *is valid in all models* $\boldsymbol{M} \in \mathbf{K}$. *We say that* $\mathscr{S}_A$*-formulae* $\varphi_1, \ldots, \varphi_n$ *locally entail an* $\mathscr{S}_A$*-formula* φ *in* $\mathbf{K}$, *written* $\varphi_1, \ldots, \varphi_n \models_{\mathbf{K}} \varphi$, *if* $\varphi_1, \ldots, \varphi_n$ *locally entail* φ *in every model* $\boldsymbol{M} \in \mathbf{K}$.

If $\mathbf{K}$ *is the class of* all *fuzzy doxastic models for* $\mathscr{S}_A$, *we denote* $\mathbf{K}$*-tautologies and entailment by* $\models_{\mathrm{FDL}_A}$ *and speak of* (multi-agent) fuzzy doxastic logic FDL_A. *Similarly if* $\mathbf{K}$ *is the class of all fuzzy* probabilistic *models for* $\mathscr{S}_A$, *we use* $\models_{\mathrm{FPL}_A}$ *and speak of* (multi-agent) fuzzy probabilistic logic FPL_A. *(For a generic set* A *of agents, we may drop the subscript and write just* FDL *or* FPL*).*

A sound and complete axiomatization, or at least an axiomatic approximation sufficiently strong for formalizing typical probabilistic or doxastic arguments, of

FDL and FPL in their own modal language $\mathscr{S}_A$ is a future work. Nevertheless, there is a standard translation (cf. [6, Prop. 4.18]) into higher-order ŁΠ, which provides a syntactic verification method for laws valid in FDL and FPL:

Definition 10. *The second-order predicate language $\mathcal{L}^2_A$ corresponding to the modal language $\mathscr{S}_A$ of Definition 6 consists of countably many monadic predicate symbols $P_1, P_2, \ldots$, one for each $p_1, p_2, \ldots \in Var$; individual variables $x, y, z, \ldots$; and a second-order monadic predicate symbol $\mathcal{N}_a$ for each $a \in A$.*

Let x be an individual variable of $\mathcal{L}^2_A$. The standard translation *of an $\mathscr{S}_A$-formula φ of* FDL *or* FPL *into an $\mathcal{L}^2_A$-formula* $\mathrm{tr}_x(\varphi)$ *of* $Ł\Pi^\omega$ *is defined recursively as follows:*

$$\begin{aligned}\mathrm{tr}_x(p_i) &= P_i(x)\\ \mathrm{tr}_x\big(c(\varphi_1,\ldots,\varphi_n)\big) &= c\big(\mathrm{tr}_x(\varphi_1),\ldots,\mathrm{tr}_x(\varphi_n)\big)\\ \mathrm{tr}_x(\mathrm{P}_a\varphi) &= \mathcal{N}_a(\{x \mid \mathrm{tr}_x(\varphi)\})\end{aligned}$$

for each $p_i \in Var$, each n-ary connective c of ŁΠ, *and each $a \in A$.*

It can be observed that every fuzzy doxastic model $\boldsymbol{M} = (\boldsymbol{F}, e)$ over a fuzzy doxastic frame $\boldsymbol{F} = (W, \boldsymbol{L}, A, \nu)$ can be regarded as an $\boldsymbol{L}$-valued $Ł\Pi^\omega$-model $\boldsymbol{M}' = (W, \boldsymbol{L}, I)$ with the domain W and the interpretation I of $\mathcal{L}^2_A$ such that $I(P_i) = e(p_i)$ for each $p_i \in Var$ and $I(\mathcal{N}_a)(w) = \nu_{a,w}$ for each $a \in A$ and $w \in W$. Vice versa, every $\boldsymbol{L}$-valued $Ł\Pi^\omega$-model $\boldsymbol{M}' = (W, \boldsymbol{L}, I)$ for $\mathcal{L}^2_A$ can be regarded as a fuzzy doxastic model $\boldsymbol{M} = ((W, \boldsymbol{L}, A, \nu), e)$, where $e(p_i) = I(P_i)$ and $\nu_{a,w} = I(\mathcal{N}_a)(w)$. The correspondence between the models is clearly one to one and the translation preserves the truth values of formulae:

Lemma 1. *Let $\boldsymbol{M}, \boldsymbol{M}'$ be as above. Then $\|\varphi\|^{\boldsymbol{M}}_w = \|\mathrm{tr}_x(\varphi)\|^{\boldsymbol{M}'}_{x\mapsto w}$, where $x \mapsto w$ denotes any $\boldsymbol{M}'$-evaluation η such that $\eta(x) = w$.*

Proof. Straightforward by definitions, analogously to [3, Th. 5.9].

Proposition 2. *Let $\varphi_1, \ldots, \varphi_n, \psi$ be $\mathscr{S}_A$-formulae and x an individual variable of $Ł\Pi^\omega$. Then:*

1. $\varphi_1, \ldots, \varphi_n \models_{\mathrm{FDL}} \psi$ *iff* $\mathrm{tr}_x(\varphi_1), \ldots, \mathrm{tr}_x(\varphi_n) \models_{Ł\Pi^\omega} \mathrm{tr}_x(\psi)$ *iff*
$$\models_{Ł\Pi^\omega} \Big(\bigwedge_{i=1}^{n} \triangle \mathrm{tr}_x(\varphi_i)\Big) \Rightarrow \mathrm{tr}_x(\psi).$$
2. $\varphi_1, \ldots, \varphi_n \models_{\mathrm{FPL}} \psi$ *iff* $\pi, \mathrm{tr}_x(\varphi_1), \ldots, \mathrm{tr}_x(\varphi_n) \models_{Ł\Pi^\omega} \mathrm{tr}_x(\psi)$ *iff*
$$\models_{Ł\Pi^\omega} \Big(\pi \wedge \bigwedge_{i=1}^{n} \triangle \mathrm{tr}_x(\varphi_i)\Big) \Rightarrow \mathrm{tr}_x(\psi),$$
where π is the $Ł\Pi^\omega$-formalization of the fuzzy probability axioms of Definition 2,
$$\begin{aligned}\pi \equiv_{\mathrm{df}} (\forall A)(\forall B) \bigwedge_{a\in A} \triangle\big[&\neg\mathcal{N}_a(\underline{0}) \wedge (\mathcal{N}_a(\setminus A) \Leftrightarrow {\sim}\mathcal{N}_a(A)) \wedge\\ &(\neg\mathcal{N}_a(A \Cap B) \Rightarrow (\mathcal{N}_a(A \Cup B) \Leftrightarrow (\mathcal{N}_a(A) \oplus \mathcal{N}_a(B))))\big].\end{aligned}$$

Proof. By inspection and easy modification of the proofs of [3, Th. 5.10, Cor. 5.12] and [3, Rem. 5.14]. □

5 Probabilistic and Doxastic Logic in Strategic Games

In this section we illustrate the apparatus of fuzzy probabilistic logic by applying it to formalization of uncertain doxastic reasoning in matrix games. In order to do so, we first need to have the game, determined by its payoff matrix, represented by formulae of fuzzy logic.

Logical representation of matrix games with only two strategies of each player and only two payoffs (*Boolean games*) was done in [14]. In [17], the representation was extended to finite strategic games with payoff values in finite MV-chains (*Łukasiewicz games*). A representation of a fairly broad class of matrix games in suitable fuzzy logics (including ŁΠ) was obtained in [2]. The latter representation covers all finite matrix games with rational payoffs, and also all n-player matrix games with strategies that can be mapped into rationals or reals from $[0, 1]$ and with each payoff function ŁΠ-representable. In this representation, the players' strategies and utilities (payoff values) are all encoded as elements of the standard ŁΠ-algebra $[0, 1]_{\text{Ł}\Pi}$ and the payoff function of a player a is expressed by an ŁΠ-formula v_a. It has been shown in [2] that for finite matrix games with rational payoffs, such game-theoretic concepts as the Nash equilibria in pure or mixed (i.e., probabilistic) strategies are expressible by ŁΠ-formulas.

Let us consider a finite matrix game $\mathbf{G}$ with a set $A = \{a_1, \ldots, a_n\}$ of players, where each player a_i is assigned a finite set of strategies S_{a_i} and a payoff function $u_{a_i}: \prod_{a \in A} S_a \to \mathbb{Q}$. By [2], the ŁΠ-representation of $\mathbf{G}$ encodes the strategies by elements of the standard ŁΠ-algebra $\boldsymbol{L} = [0, 1]_{\text{Ł}\Pi}$; without loss of generality we can assume that $|S_{a_i}| = m_i > 1$ and $S_{a_i} = \{\frac{j-1}{m_i-1} \mid 1 \leq j \leq m_i\}$ for each player $a_i \in A$. As shown in [2], the payoff functions u_a are affinely representable by ŁΠ-formulas: i.e., for each $a \in A$ there is an ŁΠ-formula v_a in n variables such that $v_a^{\boldsymbol{L}}(x_1, \ldots, x_n) = f(u_a(x_1, \ldots, x_n))$ whenever $x_i \in S_{a_i}$ for all $i \leq n$, where $v_a^{\boldsymbol{L}}$ is the truth function of v_a in the standard ŁΠ-algebra $\boldsymbol{L}$ and f is an affine function.

To model probabilistic beliefs of the players of $\mathbf{G}$, we will use a fuzzy probabilistic model $\boldsymbol{M} = (W, \boldsymbol{L}, A, \nu, e)$ over the standard ŁΠ-algebra $\boldsymbol{L}$. The events of interest are the players' chosen strategies; these will be represented by propositional variables $c^{a_1}, \ldots, c^{a_n} \in Var$. To ensure that $e(c^{a_i}, w) \in S_{a_i}$, we will characterize the strategies of $\mathbf{G}$ by a finite propositional theory $\Gamma_{\mathbf{G}} = \{\Gamma_{\mathbf{G}}^1, \ldots, \Gamma_{\mathbf{G}}^n\}$ in ŁΠ. The language of $\Gamma_{\mathbf{G}}$ consists of the variables c^{a_i} and additional variables $s_j^{a_i}$, representing the j-th strategy of player a_i (for all $i \leq n$ and $j \leq m_i$). The formulas $\Gamma_{\mathbf{G}}^i$ of $\Gamma_{\mathbf{G}}$ fix the values of $s_j^{a_i}$ as the elements of S_{a_i} and ensure that c^{a_i} are evaluated in S_{a_i}:

$$\Gamma_{\mathbf{G}}^i \equiv_{\text{df}} \bigwedge_{j=1}^{m_i} \chi_{\frac{j-1}{m_i-1}}(s_j^{a_i}) \wedge \bigvee_{j=1}^{m_i} (c^{a_i} \Leftrightarrow s_j^{a_i}),$$

where $\chi_{\frac{j-1}{m_i-1}}$ is the formula from Proposition 1. Since the values of s^a_j are fixed by $\Gamma_{\mathbf{G}}$ as elements of S_a (i.e., the $\boldsymbol{L}$-codes of a's strategies), by a slight abuse of language we will use s^a_j to refer directly to the elements of S_a and write, e.g., $s^a_j \in S_a$.

For every $a \in A$ and $j \leq m_i$, let c^a_j denote the formula $\triangle(c^a \Leftrightarrow s^a_j)$. The evaluation $e(c^a_j, w) \in \{0,1\} \subseteq \boldsymbol{L}$ indicates for each world w whether the player a chose the strategy s^a_j in w or not. The event $\|c^a_j\| \subseteq W$ is thus the (crisp) set of worlds where the player a chose to deploy the strategy $s^a_j \in S_a$. Player b's subjective probabilities (in w) of these events (i.e., player b's probabilistic beliefs about player a's choice of strategy) are the values $\|\mathrm{P}_a c^b_j\|_w \in \boldsymbol{L}$.

By the ŁΠ-representation of $\mathbf{G}$, the (affinely scaled) payoff of a player $a \in A$ is the value of the ŁΠ-formula $v_a(c^{a_1}, \ldots, c^{a_n})$. Given the choices $c^{a_i}_{j_i}$ (of the strategies $s^{a_i}_{j_i}$) by all players $a_i \in A$, the latter payoff formula is equivalent to $v_a(s^{a_1}_{j_1}, \ldots, s^{a_n}_{j_n})$. Thus, in each world $w \in \bigcap_{i=1}^n \|c^{a_i}_{j_i}\|$, the player a's payoff value is $\|v_a(s^{a_1}_{j_1}, \ldots, s^{a_n}_{j_n})\|_w \in \boldsymbol{L}$.

For simplicity, in the rest of the section we assume $A = \{a, b\}$.

Definition 11. *The* expected value *of a's i-th strategy $s^a_i \in S_a$ according to a's beliefs in w is the sum of a's payoffs weighted by a's probabilities for b's strategy choices, expressed by the $\mathscr{S}_A$-formula*

$$\eta_a(s^a_i) \equiv_{\mathrm{df}} \bigoplus_{j=1}^{|S_b|} (v_a(s^a_i, s^b_j) \odot \mathrm{P}_a c^b_j). \tag{2}$$

Observe that a player a's best-value strategy is indicated by the formula:

$$\sigma_a(s^a_m) \equiv_{\mathrm{df}} \triangle\Big(\Big(\bigvee_{i=1}^{|S_a|} \eta_a(s^a_i)\Big) \Rightarrow \eta_a(s^a_m)\Big) \tag{3}$$

Thus, for the optimal play according to the player's probabilistic beliefs about the other player's choice of strategy, the player a should choose the strategy s^a_m only in those worlds w where $\|\sigma_a(s^a_m)\|_w = 1$.

As an illustrative case study generalizable to any finite matrix game, the following example provides an analysis of strategy choices in the well known game of Stag Hunt.

Example 1 (Stag Hunt). The Stag Hunt game with 2 players ($\mathbf{SH}_2$) is specified as follows. To catch a stag, the two hunters $\{a, b\}$ need to cooperate (i.e., deploy the strategies $s^a_{\mathbf{C}}$ and $s^b_{\mathbf{C}}$); a hunter can also go for less valuable hares instead, i.e., defect ($s^a_{\mathbf{D}}$ or $s^b_{\mathbf{D}}$). The payoffs are given by the following payoff matrices:

u_a	$s^b_{\mathbf{C}}$	$s^b_{\mathbf{D}}$
$s^a_{\mathbf{C}}$	3	0
$s^a_{\mathbf{D}}$	2	1

u_b	$s^b_{\mathbf{C}}$	$s^b_{\mathbf{D}}$
$s^a_{\mathbf{C}}$	3	2
$s^a_{\mathbf{D}}$	0	1

The game has two pure Nash equilibria: either both players cooperate or both defect. Of the two, mutual defection is risk dominant (i.e., less risky), while the other is payoff dominant (i.e., yields better payoffs). Consequently, the more uncertainty about the other player's cooperation, the better to defect; however, if the player considers the other player's cooperation sufficiently probable to make it worth the risk, cooperation has a better expected value.

By [2], $\mathbf{SH}_2$ can be encoded in the logic $Ł\Pi$ as described above (e.g., with the payoff values $0, \frac{1}{3}, \frac{2}{3}, 1$). In fuzzy probabilistic logic FPL of Sect. 4 over $Ł\Pi$, the expected values of a's strategies $s \in \{s^a_{\mathbf{C}}, s^a_{\mathbf{D}}\}$ and a's best-value strategy are expressed by the formulas $\eta_a(s)$ and $\sigma_a(s)$ as in (2) and (3).

The following examples of $\mathscr{S}_A$-formulae are valid in FPL with $\Gamma_{\mathbf{SH}_2}$. In each world, they suggest the best-value strategy in $\mathbf{SH}_2$ under particular first- and higher-order beliefs of the player:

$$\triangle\big(\mathrm{P}_a c^b_{\mathbf{D}} \Rightarrow (\mathrm{P}_a c^b_{\mathbf{C}} \otimes \mathrm{P}_a c^b_{\mathbf{C}})\big) \Rightarrow \sigma_a(s^a_{\mathbf{C}}) \tag{4}$$

$$(\triangle \mathrm{P}_a(\triangle \mathrm{P}_b c^a_{\mathbf{D}} \Rightarrow c^b_{\mathbf{D}}) \otimes \triangle \mathrm{P}_a \triangle \mathrm{P}_b c^a_{\mathbf{D}}) \Rightarrow \sigma_a(s^a_{\mathbf{D}}) \tag{5}$$

The first formula says that a should cooperate (i.e., $s^a_{\mathbf{C}}$ is optimal) in worlds w where a believes that b will defect ($c^b_{\mathbf{D}}$) with probability at most $\frac{1}{3}$. Formula (5) says that if a believes that b plays rationally and that b believes that a is going to defect, then a's best-value strategy is to defect. It is straightforward to verify that the standard translations of both formulae by Definition 10 are indeed derivable in $Ł\Pi^\omega$ from the standard translation of $\Gamma_{\mathbf{SH}_2}$.

6 Conclusions

In this contribution, we proposed a logic for modeling probabilistic and doxastic multi-agent reasoning (Definition 9). The main feature of our approach is a parsimony of the presented formalism. We rendered propositions in the fuzzy logic $Ł\Pi$, which is expressively rich enough to provide the basic apparatus for formalizing game-theoretical notions therein.

In Sect. 3, we formulated a technical result related to a syntactic verification method for probabilistic and doxastic laws (Proposition 2). An open question remains the axiomatization of FDL and FPL in the modal language $\mathscr{S}_A$ itself, or at least an axiomatic approximation sufficiently strong for formalizing common probabilistic or doxastic reasoning.

Further, we showed in Sect. 5 that in the proposed logic, various important game-theoretic concepts (such as expected values and best strategy choices under uncertainty) can be expressed by formulas and derived by logical deduction. Similarly as the Stag Hunt game (Example 1), the framework can formalize uncertain reasoning in other simple matrix games such as the Prisoner's Dilemma, Chicken, Matching Pennies, Paper–Rock–Scissors, etc. The framework also naturally accommodates higher-order beliefs and, being based on fuzzy logic, also various graded concepts in games (such as the strength of a player's hand in Poker).

In Sect. 5 we assumed that the agents' beliefs are governed by the axioms of fuzzy probability (i.e., that $\nu_{a,w}$ are fuzzy probability measures). In future work, we want to model also agents with incoherent probability assignments, in order to formalize how to exploit their irrationality in games by Dutch-book strategies.

Acknowledgments. Supported by program NPU II project LQ1602 "IT4I XS" of MŠMT ČR. Based on joint work in progress with Tommaso Flaminio and Lluís Godo.

References

1. Běhounek, L., Cintula, P.: Fuzzy class theory. Fuzzy Sets Syst. **154**(1), 34–55 (2005)
2. Běhounek, L., Cintula, P., Fermüller, C., Kroupa, T.: Representing strategic games and their equilibria in many-valued logics. Logic J. IGPL **24**, 238–267 (2016)
3. Běhounek, L., Majer, O.: Fuzzy intensional semantics. J. Appl. Non-Class. Log. **28**, 348–388 (2018)
4. van Benthem, J.: Logic in Games. MIT Press, Cambridge (2014)
5. Bílková, M., Dostál, M.: Expressivity of many-valued modal logics, coalgebraically. In: Väänänen, J., Hirvonen, Å., de Queiroz, R. (eds.) WoLLIC 2016. LNCS, vol. 9803, pp. 109–124. Springer, Heidelberg (2016). https://doi.org/10.1007/978-3-662-52921-8_8
6. Chagrov, A., Zakharyaschev, M.: Modal Logic. Oxford University Press, Oxford (1997)
7. Cintula, P.: The ŁΠ and ŁΠ$\frac{1}{2}$ propositional and predicate logics. Fuzzy Sets Syst. **124**(3), 289–302 (2001)
8. Cintula, P., Noguera, C.: Neighborhood semantics for modal many-valued logics. Fuzzy Sets Syst. **345**, 99–112 (2018)
9. Esteva, F., Godo, L., Montagna, F.: The ŁΠ and ŁΠ$\frac{1}{2}$ logics: two complete fuzzy systems joining Łukasiewicz and product logics. Arch. Math. Log. **40**(1), 39–67 (2001)
10. Flaminio, T., Godo, L., Marchioni, E.: Reasoning about uncertainty of fuzzy events: an overview. In: Cintula, P., Fermüller, C., Godo, L., Hájek, P. (eds.) Understanding Vagueness: Logical, Philosophical, and Linguistic Perspectives, pp. 367–400. College Publications (2011)
11. Galeazzi, P., Lorini, E.: Epistemic logic meets epistemic game theory: a comparison between multi-agent Kripke models and type spaces. Synthese **193**, 2097–2127 (2016). https://doi.org/10.1007/s11229-015-0834-x
12. Gärdenfors, P.: Qualitative probability as an intensional logic. J. Philos. Log. **4**, 171–185 (1975)
13. Hájek, P., Godo, L., Esteva, F.: Fuzzy logic and probability. In: Proceedings of the 11th UAI Conference, Montreal, pp. 237–244 (1995)
14. Harrenstein, P., van der Hoek, W., Meyer, J.J.C., Witteveen, C.: Boolean games. In: van Benthem, J. (ed.) Proceedings of the 8th Conference on Theoretical Aspects of Rationality and Knowledge, TARK 2001, pp. 287–298. Morgan Kaufmann (2001)
15. Kroupa, T., Flaminio, T.: States of MV-algebras. In: Cintula, P., Fermüller, C., Noguera, C. (eds.) Handbook of Mathematical Fuzzy Logic, vol. 3, pp. 1183–1236. College Publications (2015)
16. Lukasiewicz, T.: Expressive probabilistic description logics. Artif. Intell. **172**, 852–883 (2008)

17. Marchioni, E., Wooldridge, M.J.: Łukasiewicz games: a logic-based approach to quantitative strategic interactions. ACM Trans. Comput. Logic **16**(4), 1–44 (2015). Article No. 33
18. Meyer, J.J.C.: Modal epistemic and doxastic logic. In: Gabbay, D., Guenthner, F. (eds.) Handbook of Philosophical Logic, vol. 10, pp. 1–38. Springer, Dordrecht (2003). https://doi.org/10.1007/978-94-017-4524-6_1
19. Montague, R.: Universal grammar. Theoria **36**, 373–398 (1970)
20. Novák, V.: Fuzzy type theory as higher order fuzzy logic. In: Proceedings of the 6th International Conference on Intelligent Technologies, InTech 2005, pp. 21–26 (2005)
21. Rodriguez, R.O., Godo, L.: Modal uncertainty logics with fuzzy neighbourhood semantics. In: Godo, L., Prade, H., Qi, G. (eds.) Working Papers of the IJCAI-2013 Workshop on Weighted Logics for Artificial Intelligence, WL4AI-2013, pp. 79–86 (2013)
22. Rodriguez, R.O., Godo, L.: On the fuzzy modal logics of belief KD45(A) and Prob($Ł_n$): axiomatization and neighbourhood semantics. In: Finger, M., Godo, L., Prade, H., Qi, G. (eds.) Working Papers of the IJCAI-2015 Workshop on Weighted Logics for Artificial Intelligence, WL4AI-2015, pp. 64–71 (2015)
23. Scott, D.: Advice on modal logic. In: Lambert, K. (ed.) Philosophical Problems in Logic. Synthese Library (Monographs on Epistemology, Logic, Methodology, Philosophy of Science, Sociology of Science and of Knowledge, and on the Mathematical Methods of Social and Behavioral Sciences), vol. 29, pp. 143–173. Springer, Dordrecht (1970). https://doi.org/10.1007/978-94-010-3272-8_7

Formal Concept Analysis, Rough Sets, General Operators and Related Topics

Towards a Logic-Based View of Some Approaches to Classification Tasks

Didier Dubois and Henri Prade(✉)

IRIT, CNRS & Univ. Paul Sabatier, 118 route de Narbonne, 31062 Toulouse Cedex 9, France
{dubois,prade}@irit.fr

Abstract. This paper is a plea for revisiting various existing approaches to the handling of data, for classification purposes, based on a set-theoretic view, such as version space learning, formal concept analysis, or analogical proportion-based inference, which rely on different paradigms and motivations and have been developed separately. The paper also exploits the notion of conditional object as a proper tool for modeling if-then rules. It also advocates possibility theory for handling uncertainty in such settings. It is a first, and preliminary, step towards a unified view of what these approaches contribute to machine learning.

Keywords: Data · Classification · Version space · Conditional object · If-then rule · Analogical proportion · Formal concept analysis · Possibility theory · Possibilistic logic · Bipolarity · Uncertainty

1 Introduction

It is an understatement to say that the current dominant paradigms in machine learning rely on neural nets and statistics; see, e.g., [1,8]. Yet, there have been quite a number of set theoretic- or logic-based views that have considered data sets from different perspectives: we can thus (at least) mention concept learning [24,25], formal concept analysis [19], rough sets [28], logical analysis of data [4], test theory [7], and GUHA method [22]. Still some other works, mentioned later, may be also relevant. These various paradigms can be related to logic, but have been developed independently. Strangely enough, little has been done to move towards a unified view of them.

This research note aims to be a first step in this direction. However, the result will remain modest, since we shall only outline connections between some settings, while other ones will be left aside for the moment. Moreover we shall mainly focus on Boolean data, even if some of what is said could be extended to nominal, or even numerical data. Still, we believe that it is of scientific interest to better understand the relationships between these different theoretical settings developed with various motivations and distinct paradigms, while all are starting from the same object: a set of data. In the long range, such a better understanding may contribute to some cooperation between these set theory-based views

M.-J. Lesot et al. (Eds.): IPMU 2020, CCIS 1239, pp. 697–711, 2020.
https://doi.org/10.1007/978-3-030-50153-2_51

and currently popular ones, such as neural nets or statistical approaches, perhaps providing tools for explanation capabilities; see, e.g., [6] for references and a tentative survey.

The paper is organized as follows. Section 2 states and discusses the problem of assigning an item to a class, given examples and counter-examples. Section 3 presents a simple propositional logic reading of the problem. Section 4 puts the discussion in a more appropriate setting using the notion of conditional object [12], which captures the idea of a rule, better than material implication. Moreover, a rule-based reading of analogical proportion-based classification [26] is also discussed in Sect. 5. Section 6 briefly recalls the version space characterization of the set of possible descriptions of a class on an example, emphasizing its bipolar nature. Section 7 advocates the interest of possibilistic logic [16] for handling uncertainty and coping with noisy data. Indeed, sensitivity to noise is a known drawback of pure set-theoretic approaches to data handling. Section 8 briefly surveys formal concept analysis and suggests its connection and potential relevance to classification. Section 9 mentions some other related matters and issues, pointing out lines for further research.

2 Classification Problem - A General View

Let us consider m pieces of data that describe items in terms of n attributes A_j. Namely an item i is represented by a vector $\boldsymbol{a}^i = (a_1^i, a_2^i, \cdots, a_n^i)$, with $i = 1, \ldots, m$, together with its class $cl(\boldsymbol{a}^i)$, where a_j^i denotes the value of the j-th attribute A_j for item $\boldsymbol{a}^i$, namely $A_j(\boldsymbol{a}^i) = a_j^i \in dom(A_j)$ ($dom(A_j)$ denotes the domain of attribute A_j). Each domain $dom(A_j)$ can be described using a set of propositional variables $\mathcal{V}_j$ specific to A_j, by means of logical formulas. If $|dom(A_j)| = 2$, one propositional variable $\mathcal{V}_j$ is enough and $dom(A_j) = \{v_j, \neg v_j\}$.

Let $\mathcal{C} = \{cl(\boldsymbol{a}^i)|i = 1, ..., m\}$ be a set of classes, where each object is supposed to belong to one and only one class. The classification problem amounts to predicting the class $cl(\boldsymbol{a}^*) \in \mathcal{C}$ of a new item $\boldsymbol{a}^*$ described in terms of the same attributes, on the basis of the m examples $(\boldsymbol{a}^i, cl(\boldsymbol{a}^i))$ consisting of classified objects.

There are other problems that are akin to classification, with different terminologies. Let us at least mention case-based decision, and diagnosis. In the first situation, we face a multiple criteria decision problem where one wants to predict the value of a new item on the basis of a collection of valued items (assuming that possible values belong to a finite scale), while in the second situation attribute values play the role of symptoms (present or not) and classes are replaced by diseases [13]. In both situations, the m examples constitute a repertory of reference cases already experienced. This is also true in case-based reasoning, where a *solution* is to be found for a new encountered *problem* on the basis of a collection of previously solved ones, for which the solution is known; however, case-based reasoning usually includes an adaptation step of the past solution selected, for a better adequacy with the new problem. Thus, ideas and methods developed in these different fields may be also of interest in a classification perspective.

Two further comments are in order here. First, for each class C, one may partition the whole set of m data in two parts: the set $\mathcal{E}$ of examples associated with this class, and the set $\mathcal{E}'$ of examples of other classes, which can be viewed as *counter-examples* for this class. The situation is pictured in Table 1 below. It highlights the fact that the whole set of items in class C is bracketed between $\mathcal{E}$ and $\overline{\mathcal{E}'}$ (where the overbar means complementation). If the table is contradiction-free, there is no item that is both in $\mathcal{E}$ and in $\mathcal{E}'$.

Second, the classification problem can be envisaged in two different manners:

1. as an *induction* problem, where one wants to build a plausible description of each class; it can be done in terms of if-then rules associating sets of attribute values with a class, these rules being used for prediction purposes;
2. as a *transduction* problem, where the prediction is made without the help of such descriptions, but by means of direct *comparisons* of the new item with the set of the m examples.

Table 1. Contradiction-free data table

	A_1	A_2	$\cdots$	A_n	cl	
$\mathbf{e}^1$	a_1^1	a_2^1	$\cdots$	a_n^1	C	
$\cdots$	$\cdots$	$\cdots$	$\cdots$	$\cdots$	C	$\mathcal{E}$
$\mathbf{e}^r$	a_1^r	a_2^r	$\cdots$	a_n^{1r}	C	
$\mathbf{e}'^1$	$a_1'^1$	$a_2'^1$	$\cdots$	$a_n'^1$	$\overline{C}$	
$\cdots$	$\cdots$	$\cdots$	$\cdots$	$\cdots$	$\overline{C}$	$\mathcal{E}'$
$\mathbf{e}'^s$	$a_1'^s$	$a_2'^s$	$\cdots$	$a_n'^s$	$\overline{C}$	
$\cdots$	$\cdots$	$\cdots$	$\cdots$	$\cdots$	?	
$\mathbf{e}^\star$	$a_1^\star$	$a_2^\star$	$\cdots$	$a_n^\star$	?	
$\cdots$	$\cdots$	$\cdots$	$\cdots$	$\cdots$	?	

3 A Simple Logical Reading

An elementary idea for characterizing a class C is to look for an attribute such that the subset of values taken for this attribute by the available examples of class C is *disjoint* from the subset of values taken by the examples of the other classes. If there exists at least one such attribute A_{j^*}, then one may inductively assume that belonging or not to class C, for any new item, can be predicted on the basis of its value for A_{j^*}. More generally, if a particular combination of attribute values can be encountered only for items of a class C, then a new item with this particular combination should also be put plausibly in class C. Let us now have a more systematic logical analysis of the data.

Let us consider a particular class $C \in \mathcal{C}$. Then the m items $\boldsymbol{a}^i$ can be partitioned into two subsets, the items $\boldsymbol{a}^i$ such that $cl(\boldsymbol{a}^i) = C$, and those such that $cl(\boldsymbol{a}^i) \neq C$ (we assume that $|\mathcal{C}| \geq 2$). Thus we have a set $\mathcal{E}$ of examples for C, namely $\boldsymbol{e}^i = (a_1^i, a_2^i, \cdots, a_n^i, 1) = (\boldsymbol{a}^i, 1)$, where '1' means that $cl(\boldsymbol{a}^i) = C$, and a set $\mathcal{E}'$ of counter-examples $\boldsymbol{e}'^j = (a_1'^j, a_2'^j, \cdots, a_n'^j, 0)$ where '0' means that $cl(\boldsymbol{a}'^j) \neq C$.

Let us assume that the domains $dom(A_j)$ for $j = 1, n$ are finite and denote by v_C the propositional variable associated to class C (v_C has truth-value 1 for elements of C and 0 otherwise). Using the attribute values as propositional logic symbols, an example $\boldsymbol{e}^i$ expresses the truth of the logical statement

$$a_1^i \wedge a_2^i \wedge \cdots \wedge a_n^i \rightarrow v_C$$

meaning that if it is an example, then it belongs to the class, while counter-examples $\boldsymbol{e}'^j$ are encoded by stating that the formula $a_1'^j \wedge a_2'^j \wedge \cdots \wedge a_n'^j \rightarrow \neg v_C$ is true, or equivalently

$$\models v_C \rightarrow \neg a_1'^j \vee \neg a_2'^j \vee \cdots \vee \neg a_n'^j.$$

Then any class (or concept) C that agrees with the m pieces of data is such that

$$\bigvee_{i: \boldsymbol{e}^i \in \mathcal{E}} (a_1^i \wedge a_2^i \wedge \cdots \wedge a_n^i) \models v_C \models \bigwedge_{j: \boldsymbol{e}'^j \in \mathcal{E}'} (\neg a_1'^j \vee \neg a_2'^j \vee \cdots \vee \neg a_n'^j). \qquad (1)$$

Letting $\mathcal{E}$ be the set of models of $\bigvee_i \boldsymbol{a}^i$ (the examples) and $\mathcal{E}'$ be the set of models of $\bigvee_j \boldsymbol{a}'_j$ (the counter-examples), (1) simply reads $\mathcal{E} \subseteq C \subseteq \overline{\mathcal{E}'}$ where the overbar denotes complementation. Note that the larger the number of counter-examples, the more specific the upper bound of C; the larger the number of examples, the more general the lower bound of C.

This logical expression states that if an item is identical to an example on all attributes then it is in the class, and that if an item is in the class then it should be different from all counter-examples on at least one attribute.

Let us assume Boolean attributes for simplicity, and let us suppose that $a_1^i = v_1$ is true for all the examples of class C and false for all the examples of other classes. Then it can be seen that (1) can be put under the form $v_1 \wedge L \models v_C \models v_1 \vee L'$ where L and L' are logical expressions that do not involve any propositional variable pertaining to attribute A_1. This provides a reasonable support for inducing that an item belongs to C as soon as v_1 is true for it. Such a remark can be generalized to a combination of attribute values and to nominal attributes.

Let us consider a toy example, small yet sufficient for an illustration of (1) and starting the discussion.

Example 1. It is an example with two Boolean attributes, two classes (C and $\overline{C}$), two examples and a counter-example. Namely, we have $\boldsymbol{e}^1 = (a_1^1, a_2^1, 1) =$

$(1,0,1) = (v_1, \neg v_2, v_C)$; $\mathbf{e^2} = (a_1^2, a_2^2, 1) = (0,1,1) = (\neg v_1, v_2, v_C)$; $\mathbf{e'^1} = (a_1'^1, a_2'^1, 0) = (0,0,0) = (\neg v_1, \neg v_2, \neg v_C)$.

We can easily see that $(v_1 \wedge \neg v_2) \vee (\neg v_1 \wedge v_2) \models v_C \models v_1 \vee v_2$, i.e., we have $v_1 \dot{\vee} v_2 \models v_C \models v_1 \vee v_2$, where $\dot{\vee}$ stands for exclusive *or*. Indeed depending on whether $(1,1)$ is an example or a counter-example, the class C will be described by $v_1 \vee v_2$, or by $v_1 \dot{\vee} v_2$ respectively.

Note that in the absence of any further information or principle, the two options for assigning $(1,1)$ to a class on the basis of $\mathbf{e^1}, \mathbf{e^2}$ and $\mathbf{e'^1}$, are equally possible here. □

Observe that if the bracketing of C in (1) is consistent, the conjunction of the lower bound expression and the upper bound expression yields the lower bound. But in case of an item which would appear both as an example and as a counter-example for C (noisy data), this conjunction would not be a contradiction, as we might expect in general, as shown by the example below.

Example 2. Assume we have $\mathbf{e^1} = (1,0,1)$; $\mathbf{e^2} = (1,1,1)$; $\mathbf{e'^1} = (1,1,0)$. The classes $\mathcal{E}$ and $\mathcal{E}'$ overlap since $\mathbf{e^2}$ and $\mathbf{e'^1}$ are the same item, classified differently. As a consequence we do not have that $\mathcal{E} \subseteq \overline{\mathcal{E}'}$. So equation (1) is not valid: we do not have that $v_1 = (v_1 \wedge \neg v_2) \vee (v_1 \wedge v_2) \models v_C \models \neg v_1 \vee \neg v_2$, i.e., $v_1 \models v_C \models \neg v_1 \vee \neg v_2$ is wrong even if $v_1 \wedge (\neg v_1 \vee \neg v_2) = v_1 \wedge \neg v_2 \neq \bot$. □

A more appropriate treatment of inconsistency will be proposed in the next section.

The two expressions bracketing C in (1) have a graded counterpart, proposed in [17], for assessing how satisfactory an item is, given a set of examples and a set of counter-examples supposed to describe what we are looking for. Then an item is all the better ranked as it is similar to at least one example on all important attributes, and that it is dissimilar to all counter-examples on at least one important attribute (where similarity, dissimilarity, and importance are matters of degrees). However, this ranking problem is somewhat different from the classification problem where each item should be assigned to a class. Here if an item is both close to an example and to a counter-example, it has a poor evaluation, just as it would be if it is close to a counter-example only.

Note that if one considers examples only, the graded counterpart amounts to searching for items that are similar to examples. In terms of classification, it means looking for the pieces of data that are sufficiently similar (on all attributes) to the item, the class of which one wants to predict, and to assign this item to the class shared by the majority of these such similar data. This is the k-nearest neighbor method. This is also very close to fuzzy case-based reasoning and instance-based learning [11,23].

4 Conditional Objects and Rules

A conditional object $b|a$, where a, b are propositions, is a three-valued entity, which is *true* if $a \wedge b$ is true; *false* if $a \wedge \neg b$ is true; *inapplicable* if a is false; see, e.g., [12]. It can be thought as the rule 'if a then b'. Indeed, the rule can be fired only if a is true; the examples of this rule are such that $a \wedge b$ is true, while its counter-examples are such that $a \wedge \neg b$ is true. This view of conditionals dates back to Bruno De Finetti 's works in the 1930's.

An (associative) quasi-conjunction & can be defined for conditional objects:

$$b|a \ \&\ d|c = (a \to b) \wedge (c \to d)|(a \vee c)$$

where $\to$ denotes the material implication. It fits with the intuition that a set of rules can be fired as soon as at least one rule can be fired, and when a rule is fired, the rule behaves like material implication. Moreover, entailment between conditional objects is defined by $b|a \vDash d|c$ iff $a \wedge b \vDash c \wedge d$ and $c \wedge \neg d \vDash a \wedge \neg b$, which expresses that the examples of rule 'if a then b' are examples of rule 'if c then d', and the counter-examples of rule 'if c then d' are counter-examples of rule 'if a then b'. It can be checked that $b|a = (a \wedge b)|a = (a \to b)|a$ since these three conditional objects have the same examples and the same counter-examples. It can be also shown that $a \wedge b|\top \vDash b|a \vDash a \to b|\top$ (where $\top$ denotes tautology), thus highlighting the fact that $b|a$ is bracketed by the conjunction $a \wedge b$ and the material implication $a \to b$.

Let us revisit expression (1) in this setting. For an example $\boldsymbol{e} = (\boldsymbol{a}, 1)$, and a counter-example $\boldsymbol{e'} = (\boldsymbol{a'}, 0)$ with respect to a class C, it leads to consider the conditional objects $v_C|\boldsymbol{a}$ and $\neg v_C|\boldsymbol{a'}$ respectively (if it is an example we are in the class, otherwise not).

For a collection of examples we have

$$\begin{aligned}(v_C|\boldsymbol{a^1}) \ \&\ \cdots \ \&\ (v_C|\boldsymbol{a^r}) &= ((\boldsymbol{a^1} \vee \cdots \vee \boldsymbol{a^r}) \to v_C)|(\boldsymbol{a^1} \vee \cdots \vee \boldsymbol{a^r}) \\ &= v_C|(\boldsymbol{a^1} \vee \cdots \vee \boldsymbol{a^r})\end{aligned}$$

Similarly, we have

$$\begin{aligned}(\neg v_C|\boldsymbol{a'^1}) \ \&\ \cdots \ \&\ (\neg v_C|\boldsymbol{a'^s}) &= ((\boldsymbol{a'^1} \vee \cdots \vee \boldsymbol{a'^s}) \to \neg v_C)|(\boldsymbol{a'^1} \vee \cdots \vee \boldsymbol{a'^s}) \\ &= \neg v_C|(\boldsymbol{a'^1} \vee \cdots \vee \boldsymbol{a'^s})\end{aligned}$$

Letting $\phi_E = \bigvee_{i=1}^{r} \boldsymbol{a^i}$ and $\phi_{E'} = \bigvee_{j=1}^{s} \boldsymbol{a'^j}$, we can join the two conditional expressions:

$$(v_C|\phi_E) \ \&\ (\neg v_C|\phi_{E'}) = (\phi_E \to v_C) \wedge (\phi_{E'} \to \neg v_C)|(\phi_E \vee \phi_{E'})$$

where

$$(\phi_E \wedge v_C) \vee (\phi_{E'} \wedge \neg v_C)|\top \vDash (v_C|\phi_E) \ \&\ (\neg v_C|\phi_{E'}) \vDash (\phi_E \to v_C) \wedge (\phi_{E'} \to \neg v_C)|\top$$

A set of conditional objects K is said to be consistent if and only if for no subset $S \subseteq K$ does the quasi-conjunction $Q(S)$ of the conditional objects in S entail a conditional contradiction of the form $\bot|\phi$. [12]. Contrary to material implication, the use of three-valued conditionals reveals the presence of contradictions in the data.

Example 3. (Example 2 continued) The data are $e^1 = (1,0,1)$; $e^2 = (1,1,1)$; $e'^1 = (1,1,0)$. In terms of conditional objects, considering the subset $\{e^2, e'^1\}$, we have

$$\begin{aligned} v_C|(v_1 \wedge v_2) \ \&\ \neg v_C|(v_1 \wedge v_2) &= (v_1 \wedge v_2) \rightarrow (v_C \wedge \neg v_C)|(v_1 \wedge v_2) \\ &= (v_C \wedge \neg v_C)|(v_1 \wedge v_2) = \bot|v_1 \wedge v_2, \end{aligned}$$

which is a conditional contradiction. □

5 Analogical Proportion-Based Transduction

Apart from the k-nearest neighbor method, there is another transduction approach to the classification problem which applies to Boolean, nominal and numerical attribute values [5]. For simplicity here, we only consider Boolean attributes. It relies on the notion of analogical proportion [26]. Analogical proportions are statements of the form "a is to b as c is to d", often denoted by $a : b :: c : d$, which express that "a differs from b as c differs from d and b differs from a as d differs from c". This statement can be encoded into a Boolean logical expression which is true only for the 6 following assignments $(0,0,0,0)$, $(1,1,1,1)$, $(1,0,1,0)$, $(0,1,0,1)$, $(1,1,0,0)$, and $(0,0,1,1)$ for (a,b,c,d). Boolean Analogical proportions straightforwardly extend to vectors of attributes values such as $\boldsymbol{a} = (a_1, ..., a_n)$, by stating $\boldsymbol{a} : \boldsymbol{b} :: \boldsymbol{c} : \boldsymbol{d}$ iff $\forall i \in [1,n]$, $a_i : b_i :: c_i : d_i$. The basic analogical inference pattern, is then

$$\frac{\forall i \in \{1, ..., p\}, \quad a_i : b_i :: c_i : d_i \text{ holds}}{\forall j \in \{p+1, ..., n\}, \quad a_j : b_j :: c_j : d_j \text{ holds}}$$

Thus analogical reasoning amounts to finding completely informed triples $(\boldsymbol{a}, \boldsymbol{b}, \boldsymbol{c})$ appropriate for inferring the missing value(s) in $\boldsymbol{d}$. When there exist several suitable triples, possibly leading to distinct conclusions, one may use a majority vote for concluding. This inference method is an extrapolation, which applies to classification (then the class $cl(\boldsymbol{x})$ is the unique solution, when it exists, such as $cl(\boldsymbol{a}) : cl(\boldsymbol{b}) :: cl(\boldsymbol{c}) : cl(\boldsymbol{x})$ holds).

Let us examine more carefully how it works. The inference in fact takes items pair by pair, and then puts two pairs in parallel. Let us first consider the case where three items belong to the same class ; the fourth item is the one, the class of which one wants to predict (denoted by 1 in the following). Considering a pair of items $\mathbf{a^i}$ and $\mathbf{a^j}$. There are attributes for which the two items are equal and attributes for which they differ. For simplicity, we assume that they differ only on the first attribute (the method easily extend to more attributes). So we have $\mathbf{a^i} = (a_1^i, a_2^i, \cdots, a_n^i, 1)$ and $\mathbf{a^j} = (a_1^j, a_2^j, \cdots, a_n^j, 1)$ with $a_1^j = \neg a_1^i$ and $a_t^j = a_t^i = v_t$ for $t = 2, \ldots, n$. This means that the change from a_1^i to a_1^j in context $(v_2, \cdots, v_n)$ does not change the class. Assume we have now another pair $\mathbf{a^k} = (v_1, a_2^k, \cdots, a_n^k, 1)$ and $\mathbf{a^\star} = (\neg v_1, a_2^\star, \cdots, a_n^\star, ?)$ involving the item, the class of which we have to predict, and exhibiting the same change on attribute

A_1 and being equal elsewhere, i.e., we have $a^k_t = a^\star_t = v^\sharp_t$ for $t = 2, \ldots, n$. Putting the two pairs in parallel, we obtain the following pattern

$(v_1, v_2, \cdots, v_n, 1)$
$(\neg v_1, v_2, \cdots, v_n, 1)$
$(v_1, v^\sharp_2, \cdots, v^\sharp_n, 1)$
$(\neg v_1, v^\sharp_2, \cdots, v^\sharp_n, ?)$.

It is not difficult to check that $\mathbf{a^i}$, $\mathbf{a^j}$, $\mathbf{a^k}$ and $\mathbf{a^\star}$ are in analogical proportion for each attribute. So $\mathbf{a^i} : \mathbf{a^j} :: \mathbf{a^k} : \mathbf{a^\star}$ holds. The solution of $1 : 1 :: 1 :?$ is obviously $? = 1$, so the prediction is $cl(\mathbf{a^\star}) = 1$. This conclusion is thus based on the idea that since the change from a^i_1 to a^j_1 in context $(v_2, \cdots, v_n)$ does not change the class, it is the same in the other context $(v^\sharp_2, \cdots, v^\sharp_n)$.

The case where $\mathbf{a^i}$ and $\mathbf{a^k}$ belong to class C while $\mathbf{a^j}$ is in $\neg C$ leads to another analogical pattern, where the change from a^i_1 to a^j_1 now changes the class in context $(v_2, \cdots, v_n)$. The pattern is

$(v_1, v_2, \cdots, v_n, 1)$
$(\neg v_1, v_2, \cdots, v_n, 0)$
$(v_1, v^\sharp_2, \cdots, v^\sharp_n, 1)$
$(\neg v_1, v^\sharp_2, \cdots, v^\sharp_n, ?)$

The conclusion is now $? = 0$, i.e., $\mathbf{a^\star}$ is not in C. This approach thus implements the idea that the switch from a^i_1 to a^j_1 that changes the class in context $(v_2, \cdots, v_n)$, also leads to the same change in context $(v^\sharp_2, \cdots, v^\sharp_n)$.

It has been theoretically established that analogical classifiers *always* yield exact prediction for Boolean affine functions describing the class (which includes x-or functions), and only for them [9]. Still, a majority vote among the predicting triples often yields the right prediction in other situations [5].

Let us see how it works on Example 1 and variants.

Example 4. In Example 1 we have: $\boldsymbol{e^1} = (1, 0, 1)$; $\boldsymbol{e^2} = (0, 1, 1)$; $\boldsymbol{e'^1} = (0, 0, 0)$. We can check that there is no analogical prediction in this case for $(1, 1, ?)$. Indeed, whatever the way we order the three vectors, either we get the 4-tuple $(1, 0, 0, 1)$ on one component, which is not a pattern in agreement with an analogical proportion, or the equation $0 : 1 :: 1 :?$ which has no solution. So analogy remains neutral in this case.

However, in the situation where would have $\boldsymbol{e^1} = (1, 0, 1)$; $\boldsymbol{e^2} = (1, 1, 1)$; $\boldsymbol{e'^1} = (0, 1, 0)$. Taking the triple $(\boldsymbol{e^2}, \boldsymbol{e^1}, \boldsymbol{e'^1})$, we can check that $(1, 1) : (1, 0) :: (0, 1) : (0, 0)$ holds on each of the two vector components. The solution of the equation $1 : 1 :: 0 :?$ is $? = 0$, which is the analogical prediction for $(0, 0, ?)$.

Similarly, in the case $\boldsymbol{e^1} = (1, 0, 1)$, $\boldsymbol{e^2} = (1, 1, 1)$ and $\boldsymbol{e^3} = (0, 1, 1)$, we would obtain $? = 1$ for $(0, 0, ?)$ as expected, using triple $(\boldsymbol{e^2}, \boldsymbol{e^1}, \boldsymbol{e^3})$. □

It is clear that the role of analogical reasoning here is to complete the data set with new examples or counter-examples obtained by transduction, assuming analogical inference patterns are valid in the case under study. It may be a first step prior to the induction of a classification model.

6 Concept Learning, Version Space and Logic

The version space setting, as proposed by Mitchell [24,25], offers an elegant elimination procedure, exploiting examples and counter-examples of a class, then called "concept", for restricting the hypotheses space and providing an approach to rule learning.

Let us recall the approach using a simple example, with 3 attributes:

- $A_1 = Sky$ (with possible values `Sunny`, `Cloudy`, and `Rainy`),
- $A_2 = Air\ Temp$ (with values `Warm` and `Cold`),
- $A_3 = Humidity$ (with values `Normal` and `High`).

The problem is to learn the concept of $C = Nice\ Day$ on the basis of examples and counter-examples. This means finding all hypotheses h, such that the implication $h \to v_C$ is compatible with the examples and the counter-examples.

Each hypothesis is described by a conjunction of constraints on the attributes, here *Sky*, *Air Temp*, and *Humidity*. Constraints may be ? (any value is acceptable), $\emptyset$ (no value is acceptable), a specific value, or a disjunction thereof. The target concept C, here *Nice Day*, is supposed to be represented by a disjunction of hypotheses (there may exist different h and h' such that $h \to v_C$ and $h' \to v_C$).

Descriptions of examples or counter-examples can be *ordered* according to their generality/specificity. Thus for instance, the following descriptions are ordered according to decreasing generality: $<?, ?, ?>$, $<\texttt{Sunny} \vee \texttt{Cloudy}, ?, ?>$, $<\texttt{Sunny}, ?, ?>$, $<\texttt{Sunny}, ?, \texttt{Normal}>$, $<\emptyset, \emptyset, \emptyset>$.

The version space is represented by its most general and least general members. The so-called general boundary G is the set of maximally general members of the hypothesis space that are consistent with the data. The specific boundary S is the set of maximally specific members of the hypothesis space that are consistent with the data. G and S are initialized as $G = <?, ?, ?>$ and $S = <\emptyset, \emptyset, \emptyset>$ (for 3 attributes as in the example).

The procedure amounts to finding a maximally specific hypothesis which covers the positive examples. Suppose we have two examples of *Nice Day*:

$$Ex1.\ <\texttt{Sunny, Warm, Normal}>\ ,\ Ex2.\ <\texttt{Sunny, Warm, High}>.$$

Then, taking into account *Ex1*, S is updated to $S_1 = <\texttt{Sunny,Warm,Normal}>$. Adding *Ex2*, S is improved into $S_2 = <\texttt{Sunny, Warm}, ?>$, which corresponds to the disjunction of *Ex1* and *Ex2*. The positive training examples force the S boundary of the version space to become increasingly general (S_2 is more general than S_1).

Although the version space approach was not cast in a logical setting, it is perfectly compatible with the logical encoding (1). Indeed here we have two examples of the form (v_1, v_2, v_3) and $(v_1, v_2, \neg v_3)$ (with $v_1 = \texttt{Sunny}; v_2 = \texttt{Warm}; v_3 = \texttt{Normal}, \neg v_3 = \texttt{High}$). A tuple of values such that $<v, v', v''>$ is to be understood as the conjunction $v \wedge v' \wedge v''$. So we obtain $(v_1 \wedge v_2 \wedge v_3) \vee (v_1 \wedge v_2 \wedge \neg v_3) \to v_C$. It corresponds to the left part of Eq. (1) for $n = 3$ and $|\mathcal{E}| = 2$, which yields

$(v_1 \wedge v_2) \wedge (v_3 \vee \neg v_3) \to v_C$, i.e., $(v_1 \wedge v_2) \to v_C$. So the more positive examples we have, the more general the lower bound of C in (1) (the set of models of a disjunction is larger than the set of models of each of its components). This lower bound, here $v_1 \wedge v_2$, is a maximally specific hypothesis h.

Negative examples play a complementary role. They force the G boundary to become increasingly specific. Consider we have the following counter-example for *Nice Day*: *cEx3*. <Rainy, Cold, High> .

The hypothesis in the G boundary must be specialized until it correctly classifies the new negative example. There are several alternative minimally more specific hypotheses. Indeed, the 3 attributes can be specialized for avoiding to cover *cEx3* by being ¬Rainy, or being ¬Cold, or being ¬High. This exactly corresponds to Equ(1), which here gives $v_C \to \neg\texttt{Rainy} \vee \neg\texttt{Cold} \vee \neg\texttt{High}$, i.e., $v_C \to \texttt{Sunny} \vee \texttt{Cloudy} \vee \texttt{Warm} \vee \texttt{Normal}$.

The elements of this disjunction correspond to maximally general potential hypotheses. But in fact we have only two new hypotheses in G: <Sunny, ?, ?> and <?, Warm, ?>, as explained now. Indeed, the hypothesis $h = (?, ?, \texttt{Normal})$ is not included in G, although it is a minimal specialization of G that correctly labels *cEx3* as a negative example. This is because example *Ex2* whose attribute value for A_3 is High, disagrees with the implication $\texttt{Normal} \to v_C$. So, hypothesis <?, ?, Normal> is excluded. Similarly, examples *Ex1* and *Ex2* (for which the attribute value for A_1 is Sunny) disagree with implication $\texttt{Cloudy} \to v_C$. This kind of elimination applies in Equation (1) as well. Indeed the expression $v \wedge L \models \neg v \vee L'$ can be simplified into $v \wedge L \models L'$.

We thus obtain upper and lower bounds from *Ex1*, *Ex2*, and *cEx3*

S_3: <Sunny, Warm, ? > G_3: {<Sunny, ?, ? > , <?, Warm, ? >}.

where $\{<v_1, v_1', v_1''>, <v_2, v_2', v_2''>\}$ logically reads $(v_1 \wedge v_1' \wedge v_1'') \vee (v_2 \wedge v_2' \wedge v_2'')$ (? stands for $\top$).

The S boundary of the version space thus summarizes the previously encountered positive examples. Any hypothesis more general than S will, by definition, cover any example that S covers and thus will cover any past positive example. In a dual fashion, the G boundary summarizes the information from previously encountered *negative* examples. Any hypothesis more specific than G is assured to be consistent with past negative examples. The set of all the hypotheses between S and G has a lattice structure. This in full agreement with Equation (1). The approach provides an iterative procedure that takes advantage of the examples and counter-examples progressively.

Thus, the general procedure for obtaining the bounds of the version space are as follows.

- If **e** is a positive example,
  1. remove from G any hypothesis inconsistent with **e**;
  2. substitute in S any minimal generalization h consistent with **e**.
- If **e** is a negative example,
  1. remove from S any hypothesis inconsistent with **e**;
  2. substitute in G any minimal specialization h consistent with **e**.

7 Towards a Possibilistic Variant of the Version Space

The main drawback of the version space approach is its sensitivity to noise. Indeed each example and each counter-example influence the result. In [18], the authors use rough set approximations to cope with this problem.

Here we make another suggestion using possibility theory. The idea is to associate each example and each counter-example with a certainty level, as in possibilistic logic (see, e.g., [16]) in order to express to what extent we consider it is certain that the corresponding piece of information is true (rather than false). This certainty level expresses our confidence in the piece of data as being exact. It can reflect the confidence we have in the source that provided it, or be the result of an analysis or filtering of the data that disqualifies outliers. In that respect we should remember that one semantics of possibility theory is in terms of (dis)similarity [29].

In other words, we have a multi-tiered set of examples and a multi-tiered set of counter-examples. So, considering all examples and all counter-examples whose certainty is above or equal to some given certainty level α yields a regular version space with classical bounds. Thus, for each α, it gives birth to a finite set of hypotheses to which α can be associated. We have thus a natural basis for rank-ordering hypotheses. The smaller α, the larger the numbers of examples and counter-examples taken into account, and the tighter the bounds.

This can be illustrated on the example of the previous section.

Example 5. Examples and counter-examples now come with certainty weights. Assume we have *Ex1*: (<Sunny, Warm, Normal>, 1); *cEx3*: (<Rainy, Cold, High> , α); *Ex2*: (<Sunny, Warm, High>, β), with $1 > \alpha > \beta$.

So, we obtain a layered version of the upper and lower bounds of the version space:

- at level 1, we have G_1 = <?, ?, ?> and S_1 = <Sunny, Warm, Normal>.
- at level α, we have G_α = {<Sunny, ?, ?>, <Cloudy, ?, ?>, <?, Warm, ?>} and S_α = <Sunny, Warm, Normal>.
- at level β, we have G_β = {<Sunny, ?, ?>, <?, Warm, ?>} and S_β = <Sunny, Warm, ?>.

□

The above syntactic view is simpler than the semantic one presented in [27] where the paper starts with a pair of possibility distributions over hypotheses, respectively induced by the examples and by the counter-examples.

8 Formal Concept Analysis

Formal concept analysis [19] is another setting where association rules between attributes can be extracted from a formal context $R \subseteq X \times Y$, which is nothing but a relation linking items in X with properties in Y. It provides a theoretical basis for data mining. Table 1 can be viewed as a context, restricting to rows $\mathcal{E} \cup \mathcal{E}'$ and considering the class of examples as just another attribute.

Let Rx and $R^{-1}y$ respectively denote the set of properties possessed by item x and the set of items having property y. Let $E \subseteq X$ and $A \subseteq Y$. The set of items having all properties in A is given by $A^{\downarrow} = \{x \mid A \subseteq Rx\}$ and the set of properties possessed by all items in E is given by $E^{\uparrow} = \{y \mid E \subseteq R^{-1}y\}$. A formal concept is then defined as a pair (E, A) such that $A^{\downarrow} = E$ and $E^{\uparrow} = A$ where E and A provides the extent and the intent of the formal concept respectively. Then, it can be shown that $E \times A \subseteq R$, and is maximal with respect to set inclusion, i.e., (E, A) defines a maximal rectangle in the formal context.

Let A and B be two subsets of Y. Then R satisfies the attribute implication $A \Rightarrow B$ if for every $x \in X$, such that $x \in A^{\downarrow}$, then $x \in B^{\downarrow}$. Formal concept analysis is not primarily oriented towards concept learning, but towards mining attribute implications (i.e., association rules). However, it might be interesting to consider formal contexts where Y also contains the names of classes, i.e., $\mathcal{C} \subseteq Y$. Then being able to find attribute implications of the form $A \Rightarrow C$ where $A \cap \mathcal{C} = \emptyset$ and $C \subseteq \mathcal{C}$, would be of a particular interest, especially if C is a singleton.

A construction dual to the theory of attribute implications has been proposed in [2], to extract disjunctive attribute implications $A \longrightarrow B$ which are satisfied by R if for every object x, if x possesses at least one property in A then it possesses at least one property in B. This approach interprets a zero in matrix R for object x and property a as the statement that x does not possess property a.

Disjunctive attribute implications can be extracted either by considering the complementary context $\overline{R}$, viewed as a standard context with negated attributes, and extracting attribute implications from it. Disjunctive attribute implications are then obtained by contraposition. Or we can derive them directly, replacing operator $A^{\downarrow}$ by a possibilistic operator $A^{\downarrow\Pi} = \{x \mid A \cup Rx \neq \emptyset\}$ introduced independently by several authors relating FCA and modal logic [21], rough sets [30] and possibility theory [10]. Then R satisfies the attribute implication $A \longrightarrow B$ if $A^{\downarrow\Pi} \subseteq B^{\downarrow\Pi}$.

It is interesting to notice that if Y also contains the names of classes, i.e., $\mathcal{C} \subseteq Y$, disjunctive attribute implications of the form $C \longrightarrow B$ where $B \cap \mathcal{C} = \emptyset$ and $C \subseteq \mathcal{C}$ corresponds to the logic rule $v_C \rightarrow \vee_{b \in B} b$, which is in agreement with the handling of exceptions in the logical reading of a classification task presented in Sect. 3. Indeed, the rule $v_C \rightarrow \vee_{b \in B} b$ can be read by contraposition: if an object violates all properties in B, then it is a counterexample. So there is a natural way of relating logical approaches to the classification problem and formal concept analysis, provided that a formal context is viewed as a set of examples and count examples (e.g., objects that satisfy a set of properties, vs. objects that do not).

Note finally that the rectangular nature of formal concepts expresses a form of convexity, which fits well with the ideas of Gärdenfors about conceptual spaces [20]. Moreover, using also operators other than $^{\downarrow}$ and $^{\uparrow}$ (see [14]) help characterizing independent sub-contexts and other noticeable structures. Formal concept analysis can be also related to the idea of clustering [15], where clusters are unions of overlapping concepts in independent sub-contexts. The idea of approximate concepts, i.e., rectangles with "holes", suggests a convexity-based completion principle, which might be useful in a classification perspective.

9 Concluding Remarks

This paper is clearly a preliminary step toward a unified, logical, study of set theory-based approaches in data management. It is preliminary in at least two respects: several of these approaches have been only cited in the introduction, while the others have been only briefly discussed. All these theoretical settings start with a Boolean table in the simplest case, and many of them extend to nominal, and possibly to numerical data. Still they have been motivated by different concerns such as describing a concept, predicting a class, or mining rules. Due to their set theory-based nature, they can be considered from a logical point of view, and a number of issues are common, such that handling incomplete information, missing values, inconsistent information, or non applicable attributes.

In a logical setting, the handling of uncertainty can be conveniently achieved using possibility theory and possibilistic logic [16]. We have suggested above how it can be applied to concept learning and how it may take into account uncertain pieces of data. Possibilistic logic can also handle default rules that can be obtained from Boolean data by looking for suitable probability distributions [3]; such rules provide useful summaries of data. The possible uses of possibilistic logic in data management is a general topic for further investigation.

Acknowledgements. The authors acknowledge a partial support of ANR-11-LABX-0040-CIMI (Centre International de Mathématiques et d'Informatique) within the program ANR-11-IDEX-0002-02, project ISIPA ("Intégrales de Sugeno, Interpolation, Proportions Analogiques").

References

1. Abu-Mostafa, Y.S., Magdon-Ismail, M., Lin, H.-T.: Learning from Data. A Short Course. AMLbook.com (2012)
2. Ait-Yakoub, Z., Djouadi, Y., Dubois, D., Prade, H.: Asymmetric composition of possibilistic operators in formal concept analysis: application to the extraction of attribute implications from incomplete contexts. Int. J. Intell. Syst. **32**(12), 1285–1311 (2017)
3. Benferhat, S., Dubois, D., Lagrue, S., Prade, H.: A big-stepped probability approach for discovering default rules. Int. J. Uncertainty Fuzziness Knowl.-Based Syst. **11**(1), 1–14 (2003)

4. Boros, E., Crama, Y., Hammer, P.L., Ibaraki, T., Kogan, A., Makino, K.: Logical analysis of data: classification with justification. Ann. OR **188**(1), 33–61 (2011)
5. Bounhas, M., Prade, H., Richard, G.: Analogy-based classifiers for nominal or numerical data. Int. J. Approx. Reason. **91**, 36–55 (2017)
6. Bouraoui, Z., et al.: From shallow to deep interactions between knowledge representation, reasoning and machine learning (Kay R. Amel group). CoRR abs/1912.06612 (2019)
7. Chikalov, I., et al.: Three Approaches to Data Analysis - Test Theory, Rough Sets and Logical Analysis of Data. Intelligent Systems Reference Library, vol. 41. Springer, Heidelberg (2013). https://doi.org/10.1007/978-3-642-28667-4
8. Cornuejols, A., Koriche, F., Nock, R.: Statistical computational learning. In: Marquis, P., Papini, O., Prade, H. (eds.) A Guided Tour of Artificial Intelligence Research, pp. 341–388. Springer, Cham (2020). https://doi.org/10.1007/978-3-030-06164-7_11
9. Couceiro, M., Hug, N., Prade, H., Richard, G.: Analogy-preserving functions: a way to extend Boolean samples. In: Proceedings of the IJCAI 2017, Stockholm, pp. 1575–1581 (2017)
10. Dubois, D., Dupin de Saint-Cyr, F., Prade, H.: A possibility-theoretic view of formal concept analysis. Fundamenta Informaticae **75**(1–4), 195–213 (2007)
11. Dubois, D., Hüllermeier, E., Prade, H.: Fuzzy methods for case-based recommendation and decision support. J. Intell. Inf. Syst. **27**(2), 95–115 (2006). https://doi.org/10.1007/s10844-006-0976-x
12. Dubois, D., Prade, H.: Conditional objects as nonmonotonic consequence relationships. IEEE Trans. Syst. Cybern. **24**(12), 1724–1740 (1994)
13. Dubois, D., Prade, H.: Fuzzy relation equations and causal reasoning. Fuzzy Sets Syst. **75**(2), 119–134 (1995)
14. Dubois, D., Prade, H.: Possibility theory and formal concept analysis: characterizing independent sub-contexts. Fuzzy Sets Syst. **196**, 4–16 (2012)
15. Dubois, D., Prade, H.: Bridging gaps between several forms of granular computing. Granul. Comput. **1**(2), 115–126 (2015). https://doi.org/10.1007/s41066-015-0008-8
16. Dubois, D., Prade, H.: Possibilistic logic: from certainty-qualified statements to two-tiered logics – a prospective survey. In: Calimeri, F., Leone, N., Manna, M. (eds.) JELIA 2019. LNCS (LNAI), vol. 11468, pp. 3–20. Springer, Cham (2019). https://doi.org/10.1007/978-3-030-19570-0_1
17. Dubois, D., Prade, H., Sédes, F.: Fuzzy logic techniques in multimedia database querying: a preliminary investigation of the potentials. IEEE Trans. Knowl. Data Eng. **13**(3), 383–392 (2001)
18. Dubois, V., Quafafou, M.: Concept learning with approximation: rough version spaces. In: Alpigini, J.J., Peters, J.F., Skowron, A., Zhong, N. (eds.) RSCTC 2002. LNCS (LNAI), vol. 2475, pp. 239–246. Springer, Heidelberg (2002). https://doi.org/10.1007/3-540-45813-1_31
19. Ganter, B., Wille, R.: Formal Concept Analysis: Mathematical Foundations. Springer, Heidelberg (1998). https://doi.org/10.1007/978-3-642-59830-2
20. Gärdenfors, P.: Conceptual Spaces. The Geometry of Thought. MIT Press, Cambridge (2000)
21. Gediga, G., Düntsch, I.: Modal-style operators in qualitative data analysis. In: Proceedings of the IEEE International Conference on Data Mining, pp. 155–162 (2002)
22. Hájek, P., Havránek, P.: Mechanising Hypothesis Formation - Mathematical Foundations for a General Theory. Springer, Heidelberg (1978). https://doi.org/10.1007/978-3-642-66943-9

23. Hüllermeier, E., Dubois, D., Prade, H.: Model adaptation in possibilistic instance-based reasoning. IEEE Trans. Fuzzy Syst. **10**(3), 333–339 (2002)
24. Mitchell, T.M.: Version spaces: a candidate elimination approach to rule learning. In: IJCAI, pp. 305–310 (1977)
25. Mitchell, T.M.: Version spaces: an approach to concept learning. Ph.D. thesis, Stanford University (1979)
26. Prade, H., Richard, G.: Analogical proportions and analogical reasoning - an introduction. In: Aha, D.W., Lieber, J. (eds.) ICCBR 2017. LNCS (LNAI), vol. 10339, pp. 16–32. Springer, Cham (2017). https://doi.org/10.1007/978-3-319-61030-6_2
27. Prade, H., Serrurier, M.: Bipolar version space learning. Int. J. Intell. Syst. **23**(10), 1135–1152 (2008)
28. Pawlak, Z.: Rough Sets. Theoretical Aspects of Reasoning about Data. Springer, Dordrecht (1991). https://doi.org/10.1007/978-94-011-3534-4
29. Sudkamp, T.: Similarity and the measurement of possibility. In: Actes Rencontres Francophones sur la Logique Floue et ses Applications (Montpellier, France), pp. 13–26. Cépadués Editions, Toulouse (2002)
30. Yao, Y.Y.: Concept lattices in rough set theory. In: Proceedings of Annual Meeting of the North American Fuzzy Information Processing Society, NAFIPS 2004, pp. 796–801 (2004)

Fuzzy Relational Mathematical Morphology: Erosion and Dilation

Alexander Šostak[1,2(✉)], Ingrīda Uljane[1,2(✉)], and Patrik Eklund[3]

[1] Institute of Mathematics, CS University of Latvia, Riga 1459, Latvia
aleksandrs.sostaks@lumii.lv
[2] Department of Mathematics, University of Latvia, Riga 1004, Latvia
ingrida.uljane@lu.lv
[3] Department of Computing Science, Umeå University, 90187 Umeå, Sweden
peklund@cs.umu.se

Abstract. In the recent years, the subject if fuzzy mathematical morphology entered the field of interest of many researchers. In our recent paper [23], we have developed the basis of the (unstructured) L-fuzzy relation mathematical morphology where L is a quantale. In this paper we extend it to the structured case. We introduce structured L-fuzzy relational erosion and dilation operators, study their basic properties, show that under some conditions these operators are dual and form an adjunction pair. Basing on the topological interpretation of these operators, we introduce the category of L-fuzzy relational morphological spaces and their continuous transformations.

Keywords: L-fuzzy relational erosion · L-fuzzy relational dilation · L-fuzzy relational morphological spaces · Duality · Adjointness · Continuous transformations

1 Introduction

Mathematical morphology has its origins in geological problems centered in the processes of erosion and dilation. The founders of mathematical morphology are engineer G. Matheron [18] and his student, engineer J. Serra [22]. The idea of the classical mathematical morphology can be explained as the process of modifying a subset A of a cube in an n-dimensional Euclidean space $\mathbb{R}^n$ by cutting out pieces of B from A (in case of erosion) or glueing them down to the set A (in case of dilation). The set B, intuitively, small if compared with A, is called the structuring set. In the first works on fuzzy morphology A and B were crisp sets, however soon the interest of some researchers was directed also to the case when A and B could be fuzzy. This allowed to describe gray scale processes of erosion and dilation. The first fundamental works on fuzzy mathematical morphology are the two papers by B. De Baets, E. Kerre and M. Gupta [7], [8].

The first and the second authors are thankful for the partial financial support from the project No. Lzp-2018/2-0338.

M.-J. Lesot et al. (Eds.): IPMU 2020, CCIS 1239, pp. 712–725, 2020.
https://doi.org/10.1007/978-3-030-50153-2_52

In the first period of the development of fuzzy mathematical morphology the domain where the operators of erosion and dilation were defined was restricted by Euclidean spaces $\mathbb{R}^n$, both in crisp and fuzzy approaches. This framework was usually adequate for studying practical problems in geology, it was appropriate also for applications of mathematical morphology in different applied sciences, in particular in pattern recognition, image processing, digital topology, etc. Additionally it was convenient since all basic constructions of fuzzy morphology, in particular, "cutting" and "glueing" pieces B from or to A were defined by means of the use of the linear structure in the Euclidean space. However, some researchers were attracted by the idea to extend basic concepts and constructions of fuzzy morphology from $\mathbb{R}^n$ to a more general context. This idea was interesting not only theoretically, but also in view of possible applications in some tasks beyond the classical ones. The principle how to find the "correct" extensions for the definitions were found in the interrelations between erosion and dilation operators that can be observed in almost all "classical" approaches. Namely, the principal features are that erosion ε and dilation δ are dual operators and the pair (ε, δ) is an adjunction.

This observation has lead to two mainstreams in the generalized approach to fuzzy mathematical morphology: the algebraic and the relational one. The algebraic approach, in its most general form, is based on two complete lattices L_1 and L_2 and two mappings: a mapping $\varepsilon : L_1 \to L_2$ that preserves arbitrary infima, and a mapping $\delta : L_2 \to L_1$ that preserves arbitrary joins. These mappings ε and δ should be related by Galois connection. This approach was initiated by Heijecman [12] and further developed by I. Bloch [3], [4] and some other authors. The second, less formal relational approach, considers a set X equipped with some relation R; this relation is used instead of linear transformations applied in the classical case, that is when X is the Euclidean space, see e.g. [19], see also [23].

In this paper, we start to develop an approach to L-fuzzy relational morphology in which, as different from e.g. [19] and [23], the erosion and dilation are structured by some L-fuzzy set B.

The paper consists of five sections. In the first one we recall and specify terminology related to quantales and L-fuzzy relations. In the second section images and preimages of L-fuzzy sets under L-fuzzy relations are considered; these images and preimages are closely related to the operators of erosion and dilation introduced and studied in Sect. 3. In the fourth section we consider interrelations between operators of erosion and dilation. Namely, we show that under some conditions they are dual and make an adjoint pair. In the next, fifth section, we introduce the category $\mathbb{M}(L)$ of L-fuzzy relational morphological spaces and its subcategory $\mathbb{M}^p(L)$ of unstructured fuzzy relational morphological spaces. We study some properties of this categories and compare these categories with certain categories of Fuzzy Topology. In the last, Conclusion section, we list some directions where our approach to the concept of fuzzy morphological spaces could be further developed.

2 Preliminaries

In this section, we recall some well-known concepts that will make the context of our work. The restricted volume of the paper does not allow to reproduce here all information used in the work. A reader is referred to the monographs [11,21] and other standard references sources for the remaining details.

2.1 Lattices, Quantales, Girard Monoids and MV-algebras

In our paper, $(L, \leq, \wedge, \vee)$ is a complete infinitely distributive lattice with bottom and top elements 0_L and 1_L respectively. Given a binary associative monotone operation $* : L \times L \to L$ on $(L, \leq, \wedge, \vee)$, the tuple $(L, \leq, \wedge, \vee, *)$ is called a quantale if $*$ commutes over arbitrary joins:

$$\alpha * \left(\bigvee_{i\in I} \beta_i\right) = \bigvee_{i\in I} (\alpha * \beta_i) \ \forall \alpha \in L, \ \forall \{\beta_i | i \in I\} \subseteq L.$$

The operation $*$ will be referred to as the product or the conjunction. A quantale is called commutative if the product $*$ is commutative. A quantale is called integral if the top element 1_L acts as the unit, that is $1_L * \alpha = \alpha$ for every $\alpha \in L$.

In a quantale a further binary operation $\mapsto: L \times L \to L$, the residuum, can be introduced as associated with operation $*$ of the quantale $(L, \leq, \wedge, \vee, *)$ via the Galois connection, that is $\alpha * \beta \leq \gamma \Longleftrightarrow \alpha \leq \beta \mapsto \gamma$ for all $\alpha, \beta, \gamma \in L$. Explicitly residium can be defined by $\alpha \mapsto \beta = \bigvee \{\lambda \in L | \lambda * \alpha \leq \beta\}$.

Further, a unary operator $^c : L \to L$ is called negation if it is an order reversing involution, that is $\alpha \leq \beta \Longrightarrow \beta^c \leq \alpha^c$ and $(\alpha^c)^c = \alpha$ for all $\alpha, \beta \in L$. For us it is important that negation ^{c} in a quantale is well-coordinated with the original quantale structure $(L, \leq, \wedge, \vee, *)$. Explicitly, this means that the negation should be defined according to the laws of fuzzy logic, that is $a^c = a \mapsto 0$. Therefore, to satisfy the properties of the negation, we have to request that

$$(\alpha \mapsto 0) \mapsto 0 = \alpha \ \forall \alpha \in L.$$

Quantales $(L, \leq, \wedge, \vee, *)$ satisfying this property are called Girard monoids [16] or Girard quantales. Girards quantales are a generalization of the concept of an MV-algebra, see e.g. [13], [14]: A quantale is called an MV-algebra if

$$(\alpha \mapsto \beta) \vee \beta = \alpha \vee \beta \text{ for all } \alpha, \beta \in L.$$

In an MV-algebra $(L, \leq, \wedge, \vee, *)$ operation $*$ distributes also over arbitrary meets see e.g. [13], [14]:

$$\alpha * \left(\bigwedge_{i\in I} \beta_i\right) = \bigwedge_{i\in I} (\alpha * \beta_i) \ \forall \alpha \in L, \ \forall \{\beta_i | i \in I\} \subseteq L.$$

In a Girard quantale $(L, \leq, \wedge, \vee, *)$ a further binary operation $\oplus$, so called coproduct or disjunction, can be defined by setting

$$\alpha \oplus \beta = (\alpha^c * \beta^c)^c \ \forall \, \alpha, \beta \in L.$$

Co-product is a commutative associative monotone operation and, in case $(L, \leq, \wedge, \vee, *)$ iks integral, 0_L acts as a zero, that is $\alpha \oplus 0_L = \alpha$ for every $\alpha \in L$. Important properties of operations in Girard quantales are given in the next Lemma:

Lemma 1. *Operation $\oplus$ in a Girard quantale $(L, \leq_L, \wedge_L, \vee_L, *)$ is distributive over arbitrary meets:*

$$\alpha \oplus \left(\bigwedge_{i\in I} \beta_i\right) = \bigwedge_{i\in I} (\alpha \oplus \beta_i) \ \ \forall \alpha \in L, \forall \{\beta_i | i \in I\} \subseteq L.$$

The proof follows from the following series of equalities justified by definitions: $\alpha \oplus (\bigwedge_{i\in I} \beta_i) = \left(\alpha^c * (\bigwedge_{i\in I} \beta_i)^c\right)^c = \left(\alpha^c * (\bigvee_{i\in I} \beta_i^c)\right)^c = \left(\bigvee_{i\in I} (\alpha^c * \beta_i^c)\right)^c = \bigwedge_{i\in I} \left(((\alpha^c)^c \oplus (\beta_i^c)^c)^c\right)^c = \bigwedge_{i\in I} (\alpha \oplus \beta_i)$. □

In a similar way one can prove the following Lemma.

Lemma 2. *If in a Girard quantale $(L, \leq, \wedge, \vee, *)$ operation $*$ distributes over arbitrary meets, then the corresponding operation $\oplus$ distributes over arbitrary joins:*

$$\alpha \oplus \left(\bigvee_{i\in I} \beta_i\right) = \bigvee_{i\in I} (\alpha \oplus \beta_i) \ \ \forall \alpha \in L, \ \forall \{\beta_i : i \in I\} \subseteq L.$$

We will need also the following Lemma, the proof of which can be found in [13, Lemme 1.4]; we reformulate it a way convenient for our use:

Lemma 3. *In a Girard quantale $(L, \leq, \wedge, \vee, *)$ the following equality holds:*

$$\alpha \mapsto \beta = (\alpha * (\beta \mapsto 0)) \mapsto 0 \qquad \forall \alpha, \beta \in L$$

A quantale $(L, \leq, \wedge, \vee, *)$ is called *divisible* if

$$a \leq b \Longleftrightarrow \text{exists } d \in L \text{ such that } a * d = b$$

see e.g. [15, p. 128]. It is known that every MV-algebra is divisible [15, p. 129]. We will need a stronger version of this property:

Definition 1. *A quantale is called strongly divisible if*

$$a * \lambda = b * c \Longleftrightarrow \exists \mu \in L \text{ such that } a * c * \mu = b \ \ \forall a, b, c, \lambda \in L.$$

Lemma 4. *In a strongly divisible quantale L the equality $a \mapsto b * c = (a \mapsto b) * c$ holds for all $a, b, c \in L$.*

Proof. $(a \mapsto b) * c = c * \bigvee\{\lambda \mid \lambda * a \leq b\} = \bigvee\{\lambda * c \mid \lambda * a \leq b\} \leq \bigvee\{\lambda * c \mid \lambda * a * c \leq b * c\} \leq \bigvee\{\mu \mid \mu \leq b * c\} = a \mapsto b * c$.

On the other hand, by strong divisibility $a \mapsto b * c = \bigvee\{\lambda \in L | \lambda * a \leq b\} \leq \bigvee\{\mu \in L | \mu * a * c \leq b\} = c * \bigvee\{\mu \in L | \mu * a \leq b\} = c * (a \mapsto b)$ □

In our work $(L, \leq, \wedge, \vee, *)$ *is always an integral commutative quantale,* sometimes satisfying additional, explicitly stated, conditions.

2.2 L-fuzzy Relations, L-fuzzy Relational Sets

Definition 2 (see e.g. [25]). *An L-fuzzy relation from a set X to a set Y is an L-fuzzy subset of the product $X \times Y$, that is a mapping $R : X \times Y \to L$. In case when $X = Y$ it is called an L-fuzzy relation on the set X. The triple (X, Y, R) where R is an L-fuzzy relation from X to Y is called an L-fuzzy relational triple and if $X = Y$ the pair (X, R) is called an L-fuzzy relational set.*

We will need some special properties of L-fuzzy relations specified below:

Definition 3 (see e.g. [24]). *An L-fuzzy relation $R : X \times Y \to L$ is called* left connected *if $\bigwedge_{y \in Y} \bigvee_{x \in X} R(x, y) = 1_L$. R is called strongly left connected if for every $y \in Y$ there exists $x \in X$ such that $R(x, y) = 1_L$. An L-fuzzy relation R is called* right connected *if $\bigwedge_{x \in X} \bigvee_{y \in Y} R(x, y) = 1_L$. R is called* strongly right connected *if for every $x \in X$ there exists $y \in Y$ such that $R(x, y) = 1_L$. An L- fuzzy relation R on a set X is called reflexive if $R(x, x) = 1_L$ for every $x \in X$. An L-fuzzy relation R on a set X is called symmetric, if $R(x, y) = R(y, x)$ for all $x, y \in X$. An L-fuzzy relation R on a set X is called transitive if $R(x, y) * R(y, z) \leq R(x, z)$ for all $x, y, z \in X$.*

3 Image and Preimage Operators on L-fuzzy Power-Sets Induced by L-fuzzy Relations

The subject of this section is, what we call, upper and lower image and preimage operators induced by an L-fuzzy relation $R : X \times Y \to L$. TAs we will see they are closely related to the operators of fuzzy relational erosion and dilation. These operators $R^{\to} : L^X \to L^Y$, and their basic properties can be found in different papers where L-fuzzy power-sets are involved. For reader's convenience we briefly discuss them here.

Definition 4. *The upper image of an L-fuzzy set $A \in L^X$ under L-fuzzy relation $R : X \times Y \to L$ is the L-fuzzy set $R^{\to}(A) \in L^Y$ defined by $R^{\to}(A)(y) = \bigvee_{x \in X} R(x, y) * A(x)$ for all $A \in L^X, y \in Y$.*

Definition 5. *The upper preimage of an L-fuzzy set $A \in L^Y$ under L-fuzzy relation $R : X \times Y \to L$ is the L-fuzzy set $R^{\leftarrow}(A) \in L^X$ defined by $R^{\leftarrow}(A)(x) = \bigvee_{y \in Y} R(x, y) * A(y)$ for all $A \in L^Y, x \in X$.*

Definition 6. *The lower image of an L-fuzzy set $A \in L^X$ under L-fuzzy relation $R : X \times Y \to L$ is the L-fuzzy set $R^{\Rightarrow}(A) \in L^Y$ defined by $R^{\Rightarrow}(A)(y) = \bigwedge_{x \in X} (R(x, y) \mapsto A(x))$ for all $A \in L^X, y \in Y$.*

Definition 7. *The lower preimage of an L-fuzzy set $A \in L^Y$ under L-fuzzy relation $R : X \times Y \to L$ is the L-fuzzy set $R^{\Leftarrow}(A) \in L^X$ defined by $R^{\Leftarrow}(A)(x) = \bigwedge_{y \in Y} (R(x, y) \mapsto A(y))$ for all $A \in L^Y, y \in Y$.*

Proposition 1. *If relation $R : X \times Y \to L$ is strongly left connected, then $R^{\Rightarrow}(A) \leq R^{\to}(A)$ for every $A \in L^X$.*

Proof. Given $y \in Y$, we take $x_y \in X$ such that $R(x_y, y) = 1_L$. Then $\bigwedge_{x\in X} R(x,y) \mapsto A(x) \leq R(x_y, y) \mapsto A(x_y) = 1_L \mapsto A(x_y) = A(x_y)$.
On the other hand, $\bigvee_{x\in X} R(x,y) * A(x) \geq R(x_y, y) * A(x_y) = A(x_y)$. □

Proposition 2. *If relation $R : X \times Y \to L$ is strongly right connected, then $R^{\Leftarrow}(A) \leq R^{\leftarrow}(A)$ for every $A \in L^Y$.*

Proof. Let $x \in X$ be fixed and let $y_x \in X$ satisfy $R(x, y_x) = 1_L$. Then $R^{\leftarrow}(A)(x) = \bigvee_y (R(x,y) * A(y)) \geq R(x, y_x) * A(y_x) = A(y_x)$. In its turn, $R^{\Leftarrow}(A)(x) = \bigwedge_y (R(x,y) \mapsto A(y)) \leq R(x, y_x) \to A(y_x) = A(y_x)$. □

4 Operators of Structured *L*-fuzzy Relational Erosion and Dilation

4.1 Structured Relational *L*-fuzzy Erosion

Modifying the definition of *L*-fuzzy relational erosion given in [19], see also [23] to the case when the fuzzy erosion of a fuzzy set $A \in L^X$ is structured by a fuzzy set $B \in L^Y$, we come to the following definition

Definition 8. *Given $A \in L^X$ and $B \in L^Y$, the erosion of A structured by B in a fuzzy relational triple (X, Y, R) is the L-fuzzy set $\varepsilon_R(A, B) \in L^Y$ defined by*

$$\varepsilon_R(A,B)(y) = \left(\bigwedge\nolimits_{x\in X} (R(x,y) \mapsto A(x))\right) * B^c(y).$$

Considering erosion for all $A \in L^X$ when $B \in L^Y$ is fixed, we get the operator of erosion $\varepsilon_R(\cdot, B) : L^X \to L^Y$.

Thus *L*-fuzzy erosion $\varepsilon_R(\cdot, 0_X) : L^X \to L^Y$ is actually the lower image operator $R^{\Rightarrow}$ induced by *L*-fuzzy relation $R : X \times Y \to L$.

In the next proposition we collect some properties of erosion operators.

Proposition 3. (1) $\varepsilon_R(1_X, B) = B^c$. *If R is left connected, then $\varepsilon_R(a_X, B) = a * B^c$ for every $a \in L$ where $a_X : X \to L$ is the constant function with value $a \in L$.*
(2) *Operator $\varepsilon_R(\cdot, B) : L^X \to L^Y$ is non-decreasing, that is if $A_1 \leq A_2 \in L^X$ then $\varepsilon_R(A_1, B) \leq \varepsilon_R(A_2, B)$.*
(3) *If $B_1 \leq B_2 \in L^Y$ then for every $A \in L^X$ $\varepsilon_R(A, B_1) \geq \varepsilon_R(A, B_2)$.*
(4) *If $*$ is distribute over arbitrary meets, then, given a family $\{A_i \mid i \in I\} \subseteq L^X$ and $B \in L^Y$, we have $\varepsilon_R\left(\bigwedge_{i\in I} A_i, B\right) = \bigwedge_{i\in I} \varepsilon_R(A_i, B)$.*

Proof. (1) From the definition it is clear that for every $y \in Y$ we have $\varepsilon_R(a_X, B)(y) = (\bigwedge_{x\in X} (R(x,y) \mapsto a) * B^c(y) = ((\bigvee_{x\in X} R(x,y)) \mapsto a) * B^c(y)$. In case $a = 1_L$, we have $\varepsilon_R(1_X, B)(y) = B^c(y)$ for every $y \in Y$. In its turn, if $R(x,y)$ is left connected, then $\bigvee_{x\in X} R(x,y) = 1_L$ for every $y \in Y$ and hence $\varepsilon_R(a_X, B)(y) = (1_L \mapsto a) * B^c(y) = a_Y * B^c(y)$.

The statements (2) and (3) are obvious.

(4) Given a family of L-fuzzy sets $\{A_i \mid i \in I\}$ and $y \in Y$, by meet-distributivity of $*$ we have $\varepsilon_R\left(\bigwedge_{i\in I} A_i, B\right)(y) = \bigwedge_{x\in X} \left(\bigwedge_{i\in I}(R(x,y) \mapsto A_i(x))\right) * B^c(y) = \bigwedge_{i\in I}\left(\bigwedge_{x\in X}(R(x,y) \mapsto A_i(x)) * B^c(y)\right) = \bigwedge_{i\in I} \varepsilon_R(A_i, B)(y)$. □

In the rest of this subsection, we consider the case when $X = Y$, that is when $R : X \times Y \to L$ is an L-fuzzy relation on a set X. In this case erosion has some important additional properties.

Proposition 4. *If L-fuzzy relation is reflexive, then for every $A \in L^X$ and every $B \in L^X$ it holds $\varepsilon_R(A, B) \leq A * B^c$. In particular, $\varepsilon_R(A, A) \leq A * A^c$.*

Proof Notice that for every $y \in Y$
$\varepsilon_R(A, B)(y) = \left(\bigwedge_{x\in X}(R(x,y) \mapsto A(x))\right) * B^c(y) \leq (R(y,y) \mapsto A(y)) * B^c(y) = (1_L \mapsto A(y)) * B^c(y) = A(y) * B^c(y) = (A * B^c)(y)$. □

Corollary 1. *If L-fuzzy relation R is reflexive, then $\varepsilon_R(A, B) \leq A$.*

To formulate the next proposition, we denote $\overline{B} = \sup\{B(y) : y \in Y\}$.

Proposition 5. *If L-fuzzy relation R is reflexive and symmetric and $*$ distributes over arbitrary meets, then for any L-fuzzy sets $A \in L^X$ and $B \in L^Y$ it holds*

$$\varepsilon_R(A, B) * (\overline{B})^c \leq \varepsilon_R(\varepsilon_R(A, B), B) \leq \varepsilon_R(A, B).$$

Proof. Inequality $\varepsilon_R(\varepsilon_R(A, B), B) \leq \varepsilon_R(A, B)$ follows from Corrolary 1. To show the other inequality let $X = Y = Z$ be sets, $R : X \times Y \to L$ be an L-fuzzy relation, define $R : Y \times Z \to L$ in the same way as the given L-fuzzy relation $R : X \times Y \to L$ and let $z \in Z$. Then by meet-distributivity of $*$ and symmetry of R and twice applying inequality $a \mapsto b * c \leq (a \mapsto b) * c$ we have

$$\varepsilon_R(\varepsilon_R(A, B), B)(z) = \left[\bigwedge_{y\in Y}(R(y,z) \mapsto \varepsilon_R(A, B)(y))\right] * B^c(z) =$$
$$\left[\bigwedge_{y\in Y}\left(R(y,z) \mapsto \left(\bigwedge_{x\in X}(R(x,y) \mapsto A(x))\right) * B^c(y)\right)\right] * B^c(z) \geq$$
$$\left[\bigwedge_{y\in Y}(R(y,z) \mapsto (R(z,y) \mapsto A(z)) * B^c(y))\right] * B^c(z) \geq$$
$$\left[\bigwedge_{y\in Y}(R(y,z) \mapsto (R(z,y) \mapsto A(z) * B^c(y))\right] * B^c(z) \geq$$
$$\left[\bigwedge_{y\in Y}((R(y,z) \mapsto A(z) * B^c(y))\right] * B^c(z) \geq$$
$$\bigwedge_{y\in Y}((R(y,z) \mapsto A(z)) * B^c(z) * B^c(y)) \geq$$
$$\bigwedge_{y\in Y}(R(y,z) \mapsto A(z)) * B^c(z) * (\overline{B})^c = \varepsilon_R(A, B)(z) * (\overline{B})^c.$$

□

In case $B = 0_Y$ we do not need to use meet-distributivity of $*$ and so we have:

Corollary 2. *If the L-fuzzy relation R is reflexive and symmetric and $B = 0_Y$, then for any $A \in L^X$ it holds $\varepsilon_R(\varepsilon_R(A, 0_Y), 0_Y) = \varepsilon_R(A, 0_Y)$. In particular this means that operator $\varepsilon_R(\cdot, 0_X) : L^X \to L^X$ is idempotent.*

4.2 Structured L-fuzzy Dilation

As before, let X, Y be sets, $R : X \times Y \to L$ an L-fuzzy relation and let $A \in L^X$ and $B \in L^Y$ be L-fuzzy sets. Generalizing definition of relational dilation of the L-fuzzy set given in [19], see also [23], for the situation when dilation of A is structured by B, we come to the following definition:

Definition 9. *Given $A \in L^X$, its L-fuzzy dilation structured by $B \in L^Y$ is an L-fuzzy set $\delta_R(A, B) \in L^Y$ defined by*

$$\delta_R(A, B)(y) = \Big(\bigvee\nolimits_{x\in X} R(y, x) * A(x)\Big) \oplus B(y).$$

Considering dilation for all $A \in L^X$ when the structuring L-fuzzy set B is fixed, we get the operator of dilation $\delta_R(\cdot, B) : L^X \to L^Y$.

In the next proposition we collect basic properties of dilation operator $\delta_R(\cdot, B)$.

Proposition 6. *Let $R : X \times Y \to L$ be an L-fuzzy relation. Then*

(1) *$\delta_R(0_X, B) = 0_Y$ and if R is right connected, then $\delta_R(a_X, B) = a_Y \oplus B$ for any $a \in L$, and in particular $\delta_R(1_X, B) = 1_Y$.*
(2) *$A_1 \le A_2 \in L^X \Longrightarrow \delta_R(A_1, B) \le \delta_R(A_2, B)$.*
(3) *If $B_1 \le B_2 \in L^Y$ then for every $A \in L^X$ $\delta_R(A, B_1) \le \delta_R(A, B_2)$.*
(4) *If operation $*$ is distributes over arbitrary joins, then given a family of L-fuzzy sets $\{A_i \mid i \in I\} \subseteq L^X$, it holds $\delta_R\left(\bigvee_{i\in I} A_i, B\right) = \bigvee_{i\in I} \delta_R(A_i, B)$.*

Proof. (1) For every $y \in Y$ we have
$\delta_R(a_X, B)(y) = (\bigvee_{x\in X} R(y, x) * a_X(x)) \oplus B(y) = (\bigvee_{x\in X} R(y, x) * a) \oplus B(y)$. Hence, $\delta_R(0_X, B)(y) = 0_Y$ for every $y \in Y$, that is $\delta_R(0_X, B) = 0_Y$. If R is right connected, then $\delta_R(a_X, B)(y) = a_Y \oplus B(y)$ for all $y \in Y$.

The proof of (2) and (3) is obvious.

(4) Let a family of L-fuzzy sets $\{A_i \mid i \in I\} \subseteq L^X$ and $y \in Y$ be given. Recalling that by Lemma 2 co-product distribures over arbitrary joins, we have
$\delta_R\left(\bigvee_{i\in I} A_i, B)\right)(y) = \left(\bigvee_{x\in X}\left(R(y, x) * \bigvee_{i\in I} A_i(x)\right)\right) \oplus B(y) =$
$\left(\bigvee_{x\in X}\left(\bigvee_{i\in I} (R(y, x) * A_i(x))\right)\right) \oplus B(y) =$
$\left(\bigvee_{i\in I}\left(\bigvee_{x\in X} (R(y, x) * A_i(x))\right)\right) \oplus B(y) =$
$\bigvee_{i\in I}\left(\bigvee_{x\in X} (R(y, x) * A_i(x)) \oplus B(y)\right) = \bigvee_{i\in I} \delta_R(A_i, B)(y)$. □

In the rest of this subsection, we consider the case when $X = Y$ that is when R is an L-fuzzy relation on the set X. In this case dilation has some additional properties.

Proposition 7. *If L-fuzzy relation R is reflexive, then for every $A \in L^X$ and every $B \in L^X$ it holds $\delta_R(A, B) \ge A \oplus B$. In particular, $\delta_R(A, A) \ge A \oplus A$.*

Proof. Given any point $y \in Y(= X)$, by reflexivity of R we have: $\delta_R(A, B)(y) = \left(\bigvee_{x\in X} R(y, x) * A(x)\right) \oplus B(y) \ge (R(y, y) * A(y)) \oplus B(y) = A(y) \oplus B(y)$. □

Corollary 3. *If L-fuzzy relation R is reflexive, then* $\delta_R(A,B) \geq A$ *for all* $A, B \in L^X$.

To formulate the next proposition we recall that $\overline{B} = \sup\{B(y) : y \in Y\}$

Proposition 8. *If the L-fuzzy relation is reflexive and symmetric and operation* $*$ *is distributes over arbitrary meets, then for all L-fuzzy sets* $A \in L^X$ *and* $B \in L^Y$ *the following inequality holds*

$$\delta_R(A,B) \oplus \overline{B} \geq \delta_R(\delta_R(A,B),B) \geq \delta_R(A,B).$$

Proof. The inequality $\delta_R(\delta_R(A,B),B) \geq \delta_R(A,B)$ follows from Corollary 3. We establish the second inequality as follows. Let $X = Y = Z$ be sets and take some $z \in Z$. Then

$\delta_R(\delta_R(A,B),B)(z) = \left(\bigvee_{y\in Y} R(z,y) * \delta_R(A,B)(y)\right) \oplus B(z) =$
$\left[\bigvee_{y\in Y} (R(z,y) * (\bigvee_{x\in X} R(y,x) * A(x)) \oplus B(y)))\right] \oplus B(z) \leq$
$\left[\bigvee_{y\in Y} (R(z,y) * (R(y,z) * A(z)) \oplus B(y))\right] \oplus B(z) \leq$
$\left[\bigvee_{y\in Y} (R(z,y) * (R(y,z) * A(z)) \oplus B(y))\right] \oplus B(z) =$
$\left[\bigvee_{y\in Y} (R(z,y) * (R(z,y) * A(z)) \oplus B(y))\right] \oplus B(z) \leq$
$\left[\bigvee_{y\in Y} (R(z,y) * A(z)) \oplus B(y))\right] \oplus B(z) =$ (by Lemma2)
$\bigvee_{y\in Y} ((R(z,y) * A(z)) \oplus B(z)) \oplus B(y)) \leq$
$\left[\bigvee_{y\in Y} (R(z,y) * A(z)) \oplus B(z)\right] \oplus \overline{B} = \delta_R(A,B)(y) \oplus \overline{B}.$ □

Since in case $B = 0_Y$ in the proof we do not need join-distributivity of the co-product $\oplus$, we get the following corollary from the previous theorem.

Corollary 4. *If the L-fuzzy relation R is reflexive and symmetric, then* $\delta_R(\delta_R(A, 0_Y), 0_Y) = \delta_R(A, 0_Y)$ *and hence operator* $\delta_R(\cdot, 0_Y)$ *is idempotent.*

5 Interrelations Between Fuzzy Relational Erosion and Dilation

One of the most important attributes of mathematical morphology is the interrelations between erosion and dilation which manifest in two ways: as the adjunction between erosion and dilation and as the duality between erosion and dilation. One or both of them exist in all approaches to fuzzy morphology known to us. It is the aim of this section to study the corresponding interrelation in our case. Unfortunately, to get the analogues of these interconnections in case of structured relational erosion and dilation, we have to assume additional conditions laid down on the quantale $(L, \leq, \wedge, \vee, *)$.

5.1 Duality Between Fuzzy Relational Erosion and Dilation

Theorem 1. *Let $(L, \leq, \vee, \wedge, *)$ be a Girard quantale. Then for every $B \in L^X$ operators $\varepsilon_R(\cdot, B)$ and $\delta_R(\cdot, B)$ make a dual pair:*

$$\varepsilon_R^c(A, B) = \delta_R(A^c, B) \ \forall A \in L^X.$$

Proof. We prove the theorem by a series of equivalent transitions which are justified by the definition of Girard quantale:
$\varepsilon_R^c(A, B)(y) =$
$((\bigwedge_{x \in X}(R(x, y) \mapsto A(x))) * B^c(y)) \mapsto 0 =$ (definition of a Girard quntale)
$(\bigwedge_{x \in X}((R(x, y) \mapsto A(x)) \mapsto 0)) \oplus (B^c(y) \mapsto 0) =$ (by Lemma 3)
$(\bigvee_{x \in X} R(x, y) * (A(x) \mapsto 0)) \oplus B(y) = \delta_R(A^c, B)(y).$ □

Corollary 5. *Let $(L, \leq, \vee, \wedge, *)$ be a Girard quantale. Then for every $B \in L^X$*

$$\delta_R^c(A, B) = \varepsilon_R(A^c, B) \ \forall A \in L^X.$$

5.2 Adjunction $(\varepsilon_R(\cdot, B), \delta_R(\cdot, B))$

When studying the problem of adjunction between operators $\varepsilon_R(\cdot, B)$ and $\delta_R(\cdot, B)$, we inevitably (?) have to assume that $*$ and $\oplus$ constitute an adjuction. In the next definition we specify what we mean by this.

Definition 10. *A pair $(*, \oplus)$ is called adjunctive if for any $\alpha, \beta, \gamma \in L$*

$$\alpha \oplus \beta \leq \gamma \iff \alpha \leq \gamma * \beta^c.$$

Unfortunately, at the moment we have only one example of an adjunctive pair $(*, \oplus)$- namely the one that corresponds to Łukasiewicz t-norm and its generalizations.

Theorem 2. *Let$(L, \leq, \wedge, \vee, *)$ be a strongly divisible Girard monoid, $*$ distribute over arbitrary meets and $(*, \oplus)$ be an adjunctive pair. Then $(\varepsilon_R(\cdot, B), \delta_R(\cdot, B))$ is an adjunctive pair.*

Proof. Since both functors $\varepsilon_R(\cdot, B)$ and $\delta_R(\cdot, B)$ are defined on the lattice L^X and take values in the same lattice L^X, the adjunction just means that these functors are related by Galois connection, that is for any $A, C \in L^X$:

$$\delta_R(A, B) \leq C \iff \varepsilon_R(C, B) \leq A.$$

We prove this by the following series of transitions: $\delta_R(A, B)(y) \leq C(y) \ \forall y \in X \iff$
$(\bigvee_{x \in X}(R(x, y) * A(y)) \oplus B(y) \leq C(y) \ \forall y \in Y \iff$ (by Lemma 2)
$(R(x, y) * A(y)) \oplus B(y) \leq C(y) \ \forall x, y \in X \iff$ (by adjunction $(*, \oplus)$)
$R(x, y) * A(y) \leq C(y) * B^c(y) \ \forall x, y \in X \iff$
$A(y) \leq R(x, y) \mapsto (C(y) * B^c(y)) \ \forall x, y \in X \iff$ (by Lemma 4)
$A(y) \leq (R(x, y) \mapsto C(y)) * B^c(y)) \ \forall x, y \in X \iff$
$A(y) \leq \bigwedge_{x \in X}(R(x, y) \mapsto C(y)) * B^c(y) \ \forall x, \in X \iff$
$A(y) \leq \varepsilon_R(C, B)(y) \ \forall y \in X.$ □

6 Fuzzy Morphological Spaces

Basing on the concepts of stratified relational erosion and dilation and the results obtained in the previous sections, in this section we introduce the concept of a fuzzy relational morphological space and consider its basic properties. Special attention is made to interpreting these properties from topological point of view. To make exposition more homogeneous, in this section we assume that $(L, \leq, \wedge, \vee, *)$ is a fixed quantale and operation $*: L \times L \to L$ distributes over arbitrary meets. Further, let X be a set and $R : X \times X \to L$ be a reflexive symmetric L-fuzzy relation on the set X. Now, properties of the erosian operator obtained in in Proposition 3 and Corollaries 1, 2 and properties of dilation operator obtained established in Proposition 6 and Corollaries 3, 4 allow to get the following list of properties:

(1) $\varepsilon_R(a_X, B) = a_X * B^c$ for every $a \in L$.
(2) If $A_1 \leq A_2 \in L^X$ then $\varepsilon_R(A_1, B) \leq \varepsilon_R(A_2, B)$.
(3) Given $\{A_i \mid i \in I\} \subseteq L^X$, we have $\varepsilon_R\left(\bigwedge_{i \in I} A_i, B\right) = \bigwedge_{i \in I} \varepsilon_R(A_i, B)$.
(4) $\varepsilon_R(A, B) \leq A$ for every $A \in L^X$.
(5) $\varepsilon_R(\varepsilon_R(A, B), B) \leq \varepsilon_R(A, B)$ for every $A \in L^X$
(5') $\varepsilon_R(\varepsilon_R(A, 0_Y), 0_Y) = \varepsilon_R(A, 0_Y)$ for every $A \in L^X$, and hence operator $\varepsilon_R(\cdot, 0_X)$ is idempotent.
(6) $\delta_R(a_X, B) = a_Y \oplus B$ for every $a \in L$.
(7) If $A_1 \leq A_2 \in L^X$ then $\delta_R(A_1, B) \leq \delta_R(A_2, B)$.
(8) Given $\{A_i \mid i \in I\} \subseteq L^X$, it holds $\delta_R\left(\bigvee_{i \in I} A, B\right) = \bigvee_{i \in I} \delta_R(A_i, B)$.
(9) $\delta_R(A, B) \geq A$ for every $A \in L^X$.
(10) $\delta_R(\delta_R(A, B), B) \geq \delta_R(A, B)$ for every $A \in L^X$.
(10') $\delta_R(\delta_R(A, 0_Y), 0_Y) = \delta_R(A, 0_Y)$ for every $A \in L^X$ and hence operator $\delta_R(\cdot, 0_Y)$ is idempotent.

Thus properties (1)–(5) remind basic properties of an L-fuzzy stratified pre-interior Alexandroff operator $int : L^X \to L^X$ [2, Appendix A]. Moreover, in case when (5) is replaced by (5′), they are just the axioms of an L-fuzzy stratified interior Alexandroff operator. In its turn, properties (6)–(10) of the dilation operator $\varepsilon_R(\cdot, B)$ remind basic properties of the L-fuzzy stratified Alexandroff pre-closure operator $cl : L^X \to L^X$ [2, Appendix A]. Moreover, in case when (10) is replaced by (10′), they are just the axioms of an L-fuzzy stratified closure Alexandroff operator.

Remark 1. Stratified interior means that $int(\alpha_X) = \alpha_X$ for all $\alpha \in L$ and *stratified closure* in our context mean that $cl(\alpha_X) = \alpha_X$ for all $\alpha \in L$ (and not only for $\alpha = 0$, see e.g. [17], [15]). The adjective *Alexandroff* means that the intersection axiom in the definition of interior and closure of an L-fuzzy topological space hold for arbitrary families (and not only finite) of (fuzzy) sets, see e.g. [1], [6]. Thus, the tuple $(X, R, \varepsilon_R, \delta_R)$ reminds the definition of a pre-di-topological space [5].

Definition 11. *A quadruple $(X, R, \varepsilon_R(\cdot, B), \delta_R(\cdot, B))$ is called an L-fuzzy relational morphological space.*

Further, in case $B = 0_Y$ operator $\varepsilon_R(\cdot, B)$ is idempotent by property (5′) and operator $\delta_R(\cdot, B)$ is idempotent by property (10′). Therefore by setting in an L-fuzzy relational morphological space $(X, R, \varepsilon_R(\cdot, B), \delta_R(\cdot, B))$ families of L-fuzzy sets $\mathcal{T}_R = \{U \in L^X | \varepsilon_R(U, B) = U\}$ and $\mathcal{S}_R = \{V \in L^X | \delta_R(V, B) = V\}$, we obtain a fuzzy di-topological [5] space $(X, R, \mathcal{T}_R, \mathcal{S}_R)$. However, wishing to view it as a special type of an L-fuzzy relational morphological space, we call such spaces *pure* L-fuzzy relational morphological spaces - "pure" in the sense that they were not influenced by structuring.

To view L-fuzzy morphological spaces as a category $\mathbb{M}(L)$ and the category $\mathbb{M}^p(L)$ of pure L-fuzzy morphological spaces as its subcategory, we must specify its morphisms. We do it patterned after the topological background of these categories.

Definition 12. *Let* $\mathcal{X}_1 = (X_1, R_1, \varepsilon_{R1}, \delta_{R1})$ *and* $\mathcal{X}_2 = (X_2, R_2, \varepsilon_{R2})$ *be two L-fuzzy relational morphological spaces. A mapping* $f : X_1 \to X_2$ *is called* a continuous transformation *from* $\mathcal{X}_1$ *to* $\mathcal{X}_2$ *if and only if the following conditions are satisfied:*

1. $R_1(x, y) \le R_2(f(x), f(y)) \ \forall \ x, y \in X_1$;
2. $f(\delta_{R_1}(A, B)) \le \delta_{R_2}(f(A), f(B)) \ \forall A, B \in L^{X_1}$.
3. $f^{-1}(\varepsilon_{R_2}(A, B)) \le \varepsilon_{R_1}(A, B) \ \forall A, B \in L^{X_2}$.

The proof of the following proposition is obvious

Proposition 9. *Given three L-fuzzy morphological spaces* $\mathcal{X}_1$, $\mathcal{X}_2$ *and* $\mathcal{X}_3$ *and continuous transformations* $f : \mathcal{X}_1 \longrightarrow \mathcal{X}_2$ *and* $g : \mathcal{X}_2 \longrightarrow \mathcal{X}_3$, *the composition* $g \circ f : \mathcal{X}_1 \to \mathcal{X}_3$ *is a continuous transformation. Given an L-fuzzy morphological space* $\mathcal{X}$, *the identity mapping* $id_X : X \to X$ *is continuous.*

Corollary 6. *L-fuzzy morphological spaces and their continuous transformations constitute a category* $\mathbb{M}(L)$.

Remark 2. In this section we did not assume any additional conditions on the quantale $(L, \le, \wedge, \vee, *)$ except of the conditions supposed throughout the paper and meet-semicontinuity of the operation $*$. However, in case when $(L, \le, \wedge, \vee, *)$ is a Girard quantale and/or satisfied conditions assumed in Theorem 3, some additional results, in particular, of categorical nature, can be obtained for the L-fuzzy relational morphological spaces. However, this will be the subject of the subsequent work.

7 Conclusion

In this paper, we have introduced the structured versions of L-fuzzy relational erosion and dilation operators defined on the L-power-set L^X of the relational set (X, R), generalizing (unstructured) L-fuzzy relational erosion and dilation counterparts introduced in [19] and further studied in [23]. After considering first separately and independently properties of L-fuzzy relational erosion and dilation we proceed with the study of interrelations between these operators. When developing the research in this direction we assume that L is a Girard quantale and in some cases impose additional conditions on the operation $*$ in the quantale L. The main result here is that under assumption of some conditions the operators ε_R and δ_R are dual and represent an adjunctive pair. In the last, fifth section we introduce category of L-fuzzy morphological spaces. Introducing these categories, we base on a certain analogy on behavior of erosion and dilation operators with topological operators of interior and closure.

As the main directions for the further research of structured L-fuzzy relational erosion and dilation operators and the corresponding categories of L-fuzzy morphological spaces we view the following.

- When studying the interrelations between structured L-fuzzy relational erosion and structured L-fuzzy relational dilation we had to impose some additional conditions on the quantale L, see e.g. Definitions 1 and 2. These conditions are sufficient but we do not know yet whether they are necessary. Probably these results can be obtained for some weaker conditions.
- In this paper, we address to the basic concepts of structured L-fuzzy relational mathematical morphology, namely erosion and dilation. Aiming to develop more or less full-bodied version of structured L-fuzzy relational mathematical morphology, as the second step we see the study of structured L-fuzzy relational opening and closing operators. At present we are working in this direction.
- As a challenging direction for the further research we consider the study of structured L-fuzzy relational morphological spaces, in particular to develop the categorical approach to L-fuzzy relational spaces.
- Quite interesting, especially from the point of possible application, will be to compare structured L-fuzzy relational morphological spaces with some kind of fuzzy rough approximation systems (cf [9,10,20], etc,). In particular, it could be useful in the study of big volumes of transformed data.
- One of the main directions of mathematical morphology is image processing. Probably, also our approach will have useful application in this area.

Acknowledgement. The authors express appreciation to the anonymous referees for reading the paper carefully and making useful criticisms.

References

1. Alexandroff, P.: Diskrete Räume. Sbornik **2**, 501–518 (1937)
2. Stadler, B.M.R., Stadler, P.F., Shpak, M., Wagner, G.P.: Recombination spaces, metrics and pretopologies. www.tbi.univie.ac.at/papers/Abstracts/01-02-011.pdf
3. Bloch, I.: Duality vs. adjunction for fuzzy mathematical morphology. Fuzzy Sets Syst. **160**, 1858–1867 (2009)
4. Bloch, I., Maitre, H.: Fuzzy mathematical morphology: a comparative study. Pattern Recogn. **28**, 1341–1387 (1995)
5. Brown, L.M., Ertürk, R., Dost, Ş.: Ditopological texture spaces and fuzzy topology, I. Basic concepts. Fuzzy Sets Syst. **147**, 171–199 (2004)

6. Chen, P., Zhang, D.: Alexandroff L-cotopological spaces. Fuzzy Sets Syst. **161**, 2505–2525 (2010)
7. De Baets, B., Kerre, E.E., Gupta, M.: The fundamentals of fuzzy mathematical morphology part I: basic concepts. Int. J. Gen. Syst. **23**, 155–171 (1995)
8. De Baets, B., Kerre, E.E., Gupta, M.: The fundamentals of fuzzy mathematical morphology part II: idempotence, convexity and decomposition. Int. J. Gen. Syst. **23**, 307–322 (1995)
9. Dubois, D., Prade, H.: Rough fuzzy sets and fuzzy rough sets. Int. J. Gen. Syst. **17**, 191–209 (1990)
10. Eļkins, A., Šostak, A., Uļjane, I.: On a category of extensional fuzzy rough approximation operators. Commun. Comput. Inform. Sci. **611**, 36–47 (2016)
11. Gierz, G., Hoffman, K.H., Keimel, K., Lawson, J.D., Mislove, M.W., Scott, D.S.: Continuous Lattices and Domains. Cambridge University Press, Cambridge (2003)
12. Heijcmans, H.J.A.M., Ronse, C.: The algebraic basis of mathematical morphology: dilations and erosions. Vis. Gr. Image Process **50**(3), 245–295 (1990)
13. Höhle, U.: M-valued sets and sheaves over integral commutative cl-monoids, Chapter 2. In: Rodabaugh, S.E., Klement, E.P., Höhle, U. (eds.) Applications of Category Theory to Fuzzy Subsets, pp. 33–73. Kluwer Academic Publishing (1992)
14. Höhle, U.: Many Valued Topology and its Application. Kluwer Academic Publisher, Boston (2001)
15. Höhle, U., Šostak, A.: Axiomatics for fixed-based fuzzy topologies, Chapter 3. In: Höhle, U., Rodabaugh, S.E. (eds.) Mathematics of Fuzzy Sets: Logic, Topology and Measure Theory - Handbook Series, vol. 3, pp. 123–272. Kluwer Academic Publisher (1999)
16. Jenei, S.: Structure of Girard monoid on [0,1], Chapter 10. In: Rodabaugh, S.E., Klement, E.P. (eds.) Topological and Algebraic Structures in Fuzzy Sets, pp. 277–308. Kluwer Academic Publishing, Boston (2003)
17. Liu, Y.M., Luo, M.K.: Fuzzy Topology. Advances in Fuzzy Systems - Applications and Topology. World Scientific. Singapore (1997)
18. Matheron, G.: Random Sets and Integral Geometry. Wiley, New York (1975)
19. Madrid, N., Ojeda-Aciego, M., Medina, J., Perfilieva, I.: L-fuzzy relational mathematical morphology based on adjoint triples. Inf. Sci. **474**, 75–89 (2019)
20. Pawlak, Z.: Rough sets. Int. J. Comput. Inf. Sci. **11**, 341–356 (1982)
21. Rosenthal, K.I.: Quantales and Their Applications. Pirman Research Notes in Mathematics 234. Longman Scientific & Technical (1990)
22. Serra, J.: Image Analysis and Mathematical Morphology. Academic Press, London (1982)
23. Šostak, A., Uļjane, I.: Some remarks on topological structure in the context of fuzzy relational mathematical morphology. Atlantis Series in Uncertainty Modelling, vol. 1, pp. 776–783 (2019). https://doi.org/10.2991/eusflat-19.2019.106
24. Valverde, L.: On the structure of F-indistinguishability operators. Fuzzy Sets Syst. **17**, 313–328 (1985)
25. Zadeh, L.A.: Similarity relations and fuzzy orderings. Inf. Sci. **3**, 177–200 (1971)

Isotone $\mathcal{L}$-Fuzzy Formal Concept Analysis and $\mathcal{L}$-Valued Fuzzy Measures and Integrals

Ondrej Krídlo(✉)

P.J. Šafárik University in Košice, Košice, Slovakia
ondrej.kridlo@upjs.sk

Abstract. The main idea of the paper is to generalize the concept of lattice valued fuzzy measures and integrals for data from complete residuated lattice where double negation law holds and then to show their relationship to isotone $\mathcal{L}$-fuzzy concept forming operators.

Keywords: Fuzzy measure · Fuzzy integral · Formal concept analysis

1 Introduction

Once I saw a very nice presentation about fuzzy sets and its applications and basic technics where very simple but nice toy example of fuzzy measure and fuzzy integral were presented. It was my almost first meeting with such an area where I found some small intuitive connection with fuzzy Formal concept analysis (FCA) [2,3,5,6,12,13] that is my main topic for years. After that experience my scientific curiosity led me to papers that seemed to me as a newcomer as most understandable [1,7,14–17] and also very useful was to read two papers that deals with both fuzzy integrals and fuzzy FCA [4,8].

Main topic of the paper is to first connect two notions, lattice valued fuzzy measures and integrals [17] and t-norm and t-conorm fuzzy measures and integrals [16] into one "complete residuated lattice"-valued fuzzy measures and integrals, where double negation law has to be preserved. Next step is to show their connection with isotone derivation (concept forming) operators of fuzzy formal context that are the most important part of FCA.

Second section is dedicated to basics of isotone FCA based on data from complete residuated lattice. Third section is about proposition of so called one-sided fuzzy concept-forming operators and the definition of new complete residuated lattice valued fuzzy measures and integrals. All needed and some new properties are proved.

This article was created in the framework of the National project IT Academy – Education for the 21st Century, which is supported by the European Social Fund and the European Regional Development Fund in the framework of the Operational Programme Human Resources. Supported also by the Slovak Research and Development Agency contract No. APVV-15-0091.

M.-J. Lesot et al. (Eds.): IPMU 2020, CCIS 1239, pp. 726–735, 2020.
https://doi.org/10.1007/978-3-030-50153-2_53

2 Isotone $\mathcal{L}$-fuzzy Formal Concept Analysis

Definition 1. *An algebra* $\mathcal{L} = \langle L, \wedge, \vee, 0, 1, \otimes, \rightarrow \rangle$ *is said to be a* complete residuated lattice *if*

1. $\langle L, \wedge, \vee, 0, 1 \rangle$ *is a complete lattice where* 0 *and* 1 *are the bottom and top elements (resp.).*
2. $\langle L, \otimes, 1 \rangle$ *is a commutative monoid.*
3. $\langle \otimes, \rightarrow \rangle$ *is an adjoint pair, i.e.* $k \otimes m \leq n$ *if and only if* $k \leq m \rightarrow n$*, for all* $k, m, n \in L$*, where* $\leq$ *is the ordering generated by* $\wedge$ *and* $\vee$*.*

It will be important to have $\mathcal{L}$ with double neggation law, ie. $\neg\neg k = k$ for any $k \in L$, where $\neg k = k \rightarrow 0$.

Definition 2. *A* Girard monoid *is a residuated lattice* $\mathcal{L}$ *satisfying the* law of double negation, *namely, the equality* $x = (x \rightarrow 0) \rightarrow 0$ *holds for all* $x \in L$*.*

This notion represents one of the several flavours in which one can find residuated lattices, it was used by Girard in his development programme for linear logics, and a study of its structure in the particular case of the unit interval can be found in [9]. Other well-known enriched versions of residuated lattices include, for instance, Heyting algebras (satisfying $x \otimes y = x \wedge y$), BL-algebras (satisfying divisibility, i.e. $x \wedge y = x \otimes (x \rightarrow y)$, and the prelinearity, i.e. $(x \rightarrow y) \vee (y \rightarrow x) = 1$), or MV-algebras (BL-algebra satisfying the law of double negation).

Definition 3. *The operator* $\oplus \colon L \times L \rightarrow L$ *is defined by*

$$a \oplus b = \neg a \rightarrow b = (a \rightarrow 0) \rightarrow b.$$

Assuming that we are working on a Girard monoid, it is not difficult to check that $\oplus$ is commutative and associative. Furthermore, the De Morgan laws between $\otimes$ and $\oplus$, and also between $\vee$ and $\wedge$, and contraposition law also hold. Hereafter, we will assume that $\mathcal{L}$ is a Girard monoid.

Let X be any final set. The ordered set (complete lattice) of all $\mathcal{L}$ sets over X will be denoted by $\mathcal{L}^X$.

Definition 4. $\mathcal{L}$*-Fuzzy* Formal Context *is a triple* $\langle B, A, \mathcal{L}, r \rangle$ *where* B *is the set of* objects, A *is the set of* attributes *and* $r : B \times A \rightarrow \mathcal{L}$ *is a binary relation between objects and attributes.*

Definition 5. *Let us define two pairs of so called* isotone derivation operators *between (*$\mathcal{L}$*-fuzzy) powerset complete lattices over the sets of objects and attributes as follows:*

1. – $\nearrow \colon \mathcal{L}^B \longleftarrow \mathcal{L}^A$ *and* $\swarrow \colon \mathcal{L}^A \longleftarrow \mathcal{L}^B$*.*
   – $\nearrow (f)(a) = \bigvee_{b \in B}(f(b) \otimes r(b,a))$ *for any* $f \in \mathcal{L}^B$*.*
   – $\swarrow (g)(b) = \bigwedge_{a \in A}(r(b,a) \rightarrow g(a))$ *for any* $g \in \mathcal{L}^A$*.*
2. – $\nwarrow \colon \mathcal{L}^B \longleftarrow \mathcal{L}^A$ *and* $\searrow \colon \mathcal{L}^A \longleftarrow \mathcal{L}^B$*.*
   – $\nwarrow (f)(a) = \bigwedge_{b \in B}(r(b,a) \rightarrow f(b))$ *for any* $f \in \mathcal{L}^B$*.*

- $\searrow(g)(b) = \bigvee_{a\in A}(g(a) \otimes r(b,a))$ *for any* $g \in \mathcal{L}^A$.

For above defined operators holds $(\swarrow, \nearrow)$ and $(\searrow, \nwarrow)$ form an isotone Galois connections between $\mathcal{L}^B$ and $\mathcal{L}^A$, i.e. for any $f \in \mathcal{L}^B$ and $g \in \mathcal{L}^A$ holds

$$\nearrow(f) \leq g \quad \Leftrightarrow \quad f \leq \swarrow(g) \quad \text{and} \quad g \leq \nwarrow(f) \quad \Leftrightarrow \quad \searrow(g) \leq f.$$

As a consequence of previous facts the compositions

- $\swarrow\nearrow$ and $\nwarrow\searrow$ forms closure operators on $\mathcal{L}^B$ or $\mathcal{L}^A$ respectively.
- $\nearrow\swarrow$ and $\searrow\nwarrow$ forms interior operator on $\mathcal{L}^A$ or $\mathcal{L}^B$ respectively.

That means that

- all compositions are monotone
- $\swarrow\nearrow$ and $\nwarrow\searrow$ are inflationary (i.e. $\nwarrow\searrow(g) \geq g$ for any $g \in \mathcal{L}^A$)
- $\nearrow\swarrow$ and $\searrow\nwarrow$ are deflationary (i.e. $\searrow\nwarrow(f) \leq f$ for any $f \in \mathcal{L}^B$).
- all compositions are idempotent (i.e. $\swarrow\nearrow\swarrow\nearrow(f) = \swarrow\nearrow(f)$ for any $f \in \mathcal{L}^B$)

Pairs $(f,g) \in \mathcal{L}^B \times \mathcal{L}^A$ such that $\nearrow(f) = g$ and $\swarrow(g) = f$ is called *$\mathcal{L}$-fuzzy formal concept* of $\mathcal{L}$-context $\langle B, A, \mathcal{L}, r\rangle$ with respect to operators $(\nearrow, \swarrow)$. f is then called *extent* and g is called *intent.* All extents will be denoted by $\mathrm{Ext}(B, A, \mathcal{L}, r, \nearrow, \swarrow)$ and all intents by $\mathrm{Int}(B, A, \mathcal{L}, r, \nearrow, \swarrow)$. Similarly for the other pair of operators $(\nwarrow, \searrow)$. Sets of all extents and intents ordered by $\mathcal{L}$-fuzzy set inclusion form complete lattices.

3 $\mathcal{L}$-fuzzy Isotone Derivation Operators and $\mathcal{L}$-fuzzy Measures and Integrals

Krajči in [10,11] defined a modification of antitone $\mathcal{L}$-fuzzy derivation operators $(\uparrow, \downarrow)$, mappings between $\mathcal{L}^B$ and $\mathcal{L}^A$ into the case of mappings between 2^B and $\mathcal{L}^A$, due to better understanding or interpretation of possible results.

So now the isotone derivation operators will be modified into one sided form:

Let $X \subseteq B$ be an arbitrary classical subset. χ_X is a characteristic function of X. Let us define a new mappings $\overline{\nearrow} : 2^B \to \mathcal{L}^A$ and $\overline{\nwarrow} : 2^B \to \mathcal{L}^A$

$$\begin{aligned}
\overline{\nearrow}(X)(a) &= \nearrow(\chi_X)(a) = \bigvee_{b\in B} \chi_X(b) \otimes r(b,a) \\
&= \bigvee_{b\in X} 1 \otimes r(b,a) \vee \bigvee_{b\in X^c} 0 \otimes r(b,a) \\
&= \bigvee_{b\in X} 1 \otimes r(b,a) \vee 0 = \bigvee_{b\in X} r(b,a)
\end{aligned}$$

$$\begin{aligned}\overline{\nwarrow}(X)(a) = \nwarrow(\chi_X)(a) &= \bigwedge_{b\in B}(r(b,a) \to \chi_X(b)) \\ &= \bigwedge_{b\in X}(r(b,a) \to 1) \wedge \bigwedge_{b\in X^c}(r(b,a) \to 0) \\ &= 1 \wedge \bigwedge_{b\in X^c}(r(b,a) \to 0) = \bigwedge_{b\in X^c} \neg r(b,a)\end{aligned}$$

Definition 6. *Let B be an arbitrary set and $\mathcal{L}$ be the lattice with bottom 0 and top element 1. The mapping $\mu : \mathcal{P}(B) \to \mathcal{L}$ is called $\mathcal{L}$-valued fuzzy measure iff μ is monotone, $\mu(\emptyset) = 0$ and $\mu(B) = 1$.*

Before the theorem about a new measure, let us define a notion of normal $\mathcal{L}$-set and normal $\mathcal{L}$-context.

Definition 7. *Any $\mathcal{L}$-set from $\mathcal{L}^X$ is called* normal *when its maximal membership value is equal to 1 (top of $\mathcal{L}$). Let $\langle B, A, \mathcal{L}, r\rangle$ be a $\mathcal{L}$-fuzzy formal context, such that its columns $r(-,a)$ are normal $\mathcal{L}$-fuzzy sets from $\mathcal{L}^B$. Such $\mathcal{L}$-context is also called normal.*

Theorem 1. *Let $\langle B, A, \mathcal{L}, r\rangle$ be a normal $\mathcal{L}$-fuzzy formal context. Let $a \in A$ be arbitrary attribute. The following upper and lower $\mathcal{L}$-valued mappings*

- $\underline{\mu}_a$ *where for any $X \subseteq B$ is $\underline{\mu}_a(X) = \overline{\nearrow}(X)(a)$*
- $\overline{\mu}_a$ *where for any $X \subseteq B$ is $\overline{\mu}_a(X) = \overline{\nwarrow}(X)(a)$.*

are $\mathcal{L}$-valued fuzzy measures.

Proof. It is well known that operators $\nearrow$ and $\nwarrow$ are monotone, hence also their one-sided fuzzy form will also be monotone. Moreover

$$\overline{\mu}_a(B) = \bigwedge_{b\in B^c} \neg r(b,a) = 1 \quad \text{and} \quad \underline{\mu}_a(\emptyset) = \bigvee_{b\in\emptyset} r(b,a) = 0$$

for any $a \in A$.

What can be little questionable are the following two facts. In general $\overline{\mu}_a(\emptyset)$ need not to be equal to 0 and $\underline{\mu}_a(B)$ need not to be equal to 1. This is why the precondition of normality is here, which says that there for any $a \in A$ there exists at least one $b \in B$ such that $r(b,a) = 1$. Hence

$$\overline{\mu}_a(\emptyset) = \bigwedge_{b\in\emptyset^c} \neg r(b,a) = \bigwedge_{b\in B} \neg r(b,a) = 0 \quad \text{and} \quad \underline{\mu}_a(B) = \bigvee_{b\in B} r(b,a) = 1$$

□

Before the definition of integral let us first define some auxiliary notation. Let $f \in \mathcal{L}^B$ be an arbitrary $\mathcal{L}$-set over B. Then

- $f_\alpha = \{b \in B | f(b) \geq \alpha\}$ is well known α-cut of $\mathcal{L}$-set for some $\alpha \in \mathcal{L}$
- $f^\alpha = \{b \in B | f(b) \not\leq \alpha\}$

The following definition is a generalisation of the definition of lower and upper-lattice valued fuzzy integrals from [17].

Theorem 2. *Let $\langle B, A, \mathcal{L}, r\rangle$ be the normal $\mathcal{L}$-fuzzy formal context and for each $a \in A$ there are corresponding $\underline{\mu}_a$ and $\overline{\mu}_a$ lower and upper $\mathcal{L}$-fuzzy measures. The following lower and upper $\mathcal{L}$-valued mappings defined for any $f \in \mathcal{L}^B$ and $X \subseteq B$ as follows*

$$\underline{\int}_X f d\underline{\mu}_a = \bigvee_{\alpha\in\mathcal{L}} \alpha \otimes \underline{\mu}_a(f_\alpha \cap X) \quad \text{and} \quad \overline{\int}_X f d\overline{\mu}_a = \bigwedge_{\alpha\in\mathcal{L}} \alpha \oplus \overline{\mu}_a(f^\alpha \cap X)$$

are lower and upper $\mathcal{L}$-valued fuzzy integrals.

Proof. In [14] where the definition of fuzzy integral says that it has to be idempotent homeomorphism. So we have to prove for above defined $\mathcal{L}$-valued mappings the following properties:

1. they are monotone
2. they map top $\mathcal{L}$-set from $\mathcal{L}^B$ to 1 and bottom to 0
3. for any $\gamma \in \mathcal{L}$ a constant $\mathcal{L}$-set $\overline{\gamma}$ defined as $\overline{\gamma}(b) = \gamma$ for any $b \in B$ holds:
   - $\underline{\int}_B \overline{\gamma} d\underline{\mu}_a = \gamma$
   - $\overline{\int}_B \overline{\gamma} d\overline{\mu}_a = \gamma$

1. The monotonicity follows directly from the definition of integrals.
2. From interior and closure operator properties of $\nearrow$ and $\swarrow$ from the Sect. 2 and the normality precondition the following holds.

$$\underline{\int}_B \chi_B d\underline{\mu}_a = \bigvee_{\alpha\in\mathcal{L}} \alpha \otimes \underline{\mu}_a(B) = \bigvee_{\alpha\in\mathcal{L}} \alpha \otimes 1 = 1 \otimes 1 = 1$$

$$\underline{\int}_B \chi_\emptyset d\underline{\mu}_a = \bigvee_{\alpha\in\mathcal{L}} \alpha \otimes \underline{\mu}_a(\emptyset) = \bigvee_{\alpha\in\mathcal{L}} \alpha \otimes 0 = 0$$

$$\overline{\int}_B \chi_B d\overline{\mu}_a(B) = \bigwedge_{\alpha\in LL} \alpha \oplus \overline{\mu}_a(B) = \bigwedge_{\alpha\in\mathcal{L}} \alpha \oplus 1 = 1$$

$$\overline{\int}_B \chi_B d\overline{\mu}_a(\emptyset) = \bigwedge_{\alpha\in LL} \alpha \oplus \overline{\mu}_a(\emptyset) = \bigwedge_{\alpha\in\mathcal{L}} \alpha \oplus 0 = 0 \oplus 0 = 0$$

3.

$$\begin{aligned}
\underline{\int_B} \overline{\gamma} d\underline{\mu}_a &= \bigvee_{\alpha\in\mathcal{L}} \alpha \otimes \underline{\mu}_a(B \cap \overline{\gamma}_\alpha) \\
&= \bigvee_{\alpha\in\mathcal{L};\alpha\leq\gamma} \alpha \otimes \underline{\mu}_a(B \cap \overline{\gamma}_\alpha) \vee \bigvee_{\alpha\in\mathcal{L};\alpha\nleq\gamma} \alpha \otimes \underline{\mu}_a(B \cap \overline{\gamma}_\alpha) \\
&= \bigvee_{\alpha\in\mathcal{L};\alpha\leq\gamma} \alpha \otimes \underline{\mu}_a(B) \vee \bigvee_{\alpha\in\mathcal{L};\alpha\nleq\gamma} \alpha \otimes \underline{\mu}_a(\emptyset) \\
&= \gamma \otimes \underline{\mu}_a(B) \vee \bigvee_{\alpha\in\mathcal{L};\alpha\nleq\gamma} \alpha \otimes 0 \\
&= (\gamma \otimes 1) \vee 0 = \gamma
\end{aligned}$$

$$\begin{aligned}
\overline{\int_B} \overline{\gamma} d\overline{\mu}_a &= \bigwedge_{\alpha\in\mathcal{L}} \alpha \oplus \overline{\mu}_a(B \cap \overline{\gamma}^\alpha) \\
&= \bigwedge_{\alpha\in\mathcal{L};\alpha\leq\gamma} \alpha \oplus \overline{\mu}_a(B \cap \overline{\gamma}^\alpha) \wedge \bigwedge_{\alpha\in\mathcal{L};\alpha\nleq\gamma} \alpha \oplus \overline{\mu}_a(B \cap \overline{\gamma}^\alpha) \\
&= \bigwedge_{\alpha\in\mathcal{L};\alpha\leq\gamma} \alpha \oplus \overline{\mu}_a(\emptyset) \wedge \bigwedge_{\alpha\in\mathcal{L};\alpha\nleq\gamma} \alpha \oplus \overline{\mu}_a(B) \\
&= \gamma \oplus \overline{\mu}_a(\emptyset) \wedge \bigwedge_{\alpha\in\mathcal{L};\alpha\nleq\gamma} \alpha \oplus 1 \\
&= \gamma \wedge 1 = \gamma
\end{aligned}$$

□

The following theorem will show a relationship between new integrals and concept forming operators.

Theorem 3. *Let $\langle B, A, \mathcal{L}, r\rangle$ be a $\mathcal{L}$-context and $\{\underline{\mu}_a | a \in A\}$ and $\{\overline{\mu}_a | a \in A\}$ be its collections of measures. Then for corresponding integrals holds:*

$$\underline{\int_B} f d\underline{\mu}_a = \nearrow (f)(a) \quad \textit{and} \quad \overline{\int_B} f d\overline{\mu}_a = \nwarrow (f)(a).$$

Proof. Let f be an arbitrary from $\mathcal{L}^B$.

$$\begin{aligned}
\underline{\int_B} f d\underline{\mu}_a &= \bigvee_{\alpha\in\mathcal{L}} \alpha \otimes \underline{\mu}_a(B \cap f_\alpha) \\
&= \bigvee_{\alpha\in\mathcal{L}} \alpha \otimes \bigvee_{b\in f_\alpha} r(b,a) \\
&= \bigvee_{\alpha\in\mathcal{L}} \alpha \otimes \bigvee_{b\in B; f(b)\geq\alpha} r(b,a)
\end{aligned}$$

$$= \bigvee_{\alpha \in \mathcal{L}} \bigvee_{b \in B; f(b) \geq \alpha} \alpha \otimes r(b,a)$$
$$= \bigvee_{b \in B} \bigvee_{\alpha \in \mathcal{L}; f(b) \geq \alpha} \alpha \otimes r(b,a)$$
$$= \bigvee_{b \in B} f(b) \otimes r(b,a) = \nearrow (f)(a)$$

$$\overline{\int}_B f d\overline{\mu}_a = \bigwedge_{\alpha \in \mathcal{L}} \alpha \oplus \overline{\mu}_a(B \cap f^\alpha)$$
$$= \bigwedge_{\alpha \in \mathcal{L}} \neg\alpha \to \bigwedge_{b \in (f^\alpha)^c} \neg r(b,a)$$
$$= \bigwedge_{\alpha \in \mathcal{L}} \bigwedge_{b \in B; f(b) \leq \alpha} \neg\alpha \to \neg r(b,a)$$
$$= \bigwedge_{\alpha \in \mathcal{L}} \bigwedge_{b \in B; f(b) \leq \alpha} r(b,a) \to \alpha$$
$$= \bigwedge_{b \in B} \bigwedge_{\alpha \in \mathcal{L}; f(b) \leq \alpha} r(b,a) \to \alpha$$
$$= \bigwedge_{b \in B} r(b,a) \to f(b) = \nwarrow (f)(a)$$

Theorem 4. *Let $\langle B, A, \mathcal{L}, r \rangle$ be a $\mathcal{L}$-context and $\{\underline{\mu}_a | a \in A\}$ and $\{\overline{\mu}_a | a \in A\}$ be its collections of measures. Then for corresponding integrals holds:*

$$\underline{\int}_X f d\underline{\mu}_a = \nearrow (f \cap \chi_X)(a) \quad \text{and} \quad \overline{\int}_X f d\overline{\mu}_a = \nwarrow (f \cap \chi_X)(a)$$

Proof. Let f be an arbitrary from $\mathcal{L}^B$ and $X \subseteq B$.

$$\underline{\int}_X f d\underline{\mu}_a = \bigvee_{\alpha \in \mathcal{L}} \alpha \otimes \underline{\mu}_a(f_\alpha \cap X)$$
$$= \bigvee_{\alpha \in \mathcal{L}} \alpha \otimes \bigvee_{b \in f_\alpha \cap X} r(b,a)$$
$$= \bigvee_{\alpha \in \mathcal{L}} \bigvee_{b \in f_\alpha \cap X} \alpha \otimes r(b,a)$$
$$= \bigvee_{\alpha \in \mathcal{L}} \bigvee_{b \in X, f(b) \geq \alpha} \alpha \otimes r(b,a)$$
$$= \bigvee_{b \in X} \bigvee_{\alpha \in \mathcal{L}, \alpha \leq f(b)} \alpha \otimes r(b,a)$$

$$
\begin{aligned}
&= \bigvee_{b\in X} f(b) \otimes r(b,a) \\
&= \bigvee_{b\in B} (f(b) \wedge \chi_X(b)) \otimes r(b,a) \\
&= \nearrow (f \cap \chi_X)(a)
\end{aligned}
$$

$$
\begin{aligned}
\overline{\int}_X f d\overline{\mu}_a &= \bigwedge_{\alpha\in\mathcal{L}} \alpha \oplus \overline{\mu}_a(f^\alpha \cap X) \\
&= \bigwedge_{\alpha\in\mathcal{L}} \alpha \oplus \bigwedge_{b\in(f^\alpha\cap X)^c} \neg r(b,a) \\
&= \bigwedge_{\alpha\in\mathcal{L}} \alpha \oplus \bigwedge_{b\in(f^\alpha)^c\cup X^c} \neg r(b,a) \\
&= \left(\bigwedge_{\alpha\in\mathcal{L}} \alpha \oplus \bigwedge_{b\in(f^\alpha)^c} \neg r(b,a)\right) \wedge \left(\bigwedge_{\alpha\in\mathcal{L}} \alpha \oplus \bigwedge_{b\in(X)^c} \neg r(b,a)\right) \\
&= \left(\bigwedge_{\alpha\in\mathcal{L}} \neg\alpha \to \bigwedge_{f(b)\leq\alpha} \neg r(b,a)\right) \wedge \left(\bigwedge_{\alpha\in\mathcal{L}} \neg\alpha \to \bigwedge_{b\in(X)^c} \neg r(b,a)\right) \\
&= \left(\bigwedge_{\alpha\in\mathcal{L}} \bigwedge_{f(b)\leq\alpha} r(b,a) \to \alpha\right) \wedge \left(\bigwedge_{\alpha\in\mathcal{L}} \bigwedge_{b\in(X)^c} r(b,a) \to \alpha\right) \\
&= \left(\bigwedge_{b\in B} \bigwedge_{\alpha\in\mathcal{L};\alpha\geq f(b)} r(b,a) \to \alpha\right) \wedge \left(\bigwedge_{b\in(X)^c} \bigwedge_{\alpha\in\mathcal{L}} r(b,a) \to \alpha\right) \\
&= \left(\bigwedge_{b\in B} r(b,a) \to f(b)\right) \wedge \left(\left(\bigwedge_{b\in(X)^c} r(b,a) \to 0\right) \wedge \left(\bigwedge_{b\in X} r(b,a) \to 1\right)\right) \\
&= \nwarrow (f)(a) \wedge \left(\bigwedge_{b\in B} (r(b,a) \to \chi_X(b))\right) \\
&= \nwarrow (f)(a) \wedge \nwarrow (\chi_X)(a) = \nwarrow (f \cap \chi_X)(a)
\end{aligned}
$$

□

Theorem 5. *Let $X \subseteq B$ and $f \in \mathcal{L}^B$ be arbitrary. Then*

$$
\overline{\int}_X f d\mu_a = \neg \underline{\int}_{X^c} \neg f d\mu_a
$$

Proof.

$$
\begin{aligned}
\overline{\int}_X f d\mu_a &= \searrow (f \cap \chi_X)(a) \\
&= \bigwedge_{b \in B} (r(b,a) \to (f \cap \chi_X)(b)) = \bigwedge_{b \in B} (\neg(f \cap \chi_X)(b) \to \neg r(b,a)) \\
&= \bigwedge_{b \in B} (\neg(f \cap \chi_X)(b) \to (r(b,a) \to 0)) \\
&= \bigwedge_{b \in B} ((\neg(f \cap \chi_X)(b) \otimes r(b,a)) \to 0) \\
&= (\bigvee_{b \in B} (\neg(f \cap \chi_X)(b) \otimes r(b,a)) \to 0) \\
&= \neg(\bigvee_{b \in B} (\neg(f \cap \chi_X)(b) \otimes r(b,a))) \\
&= \neg \nearrow (\neg(f \cap \chi_X))(a) = \neg \nearrow (\neg f \cap \chi_{X^c}))(a) \\
&= \neg \underline{\int}_{X^c} \neg f d\mu_a
\end{aligned}
$$

4 New Measures and Integrals Defined on Concept Lattices

And at the end we can also define a "new" measures and integrals for $\mathcal{L}$-context $\langle B, A, \mathcal{L}, r \rangle$ as follows:

- $\underline{\mu} : \mathcal{P}(B) \to \mathrm{Int}(B, A, \mathcal{L}, r, \nearrow, \swarrow)$ defined as $\underline{\mu}(X)(a) = \underline{\mu}_a(X)$
- $\overline{\overline{\mu}} : \mathcal{P}(B) \to \mathrm{Int}(B, A, \mathcal{L}, r, \nwarrow \searrow)$ defined as $\overline{\mu}(X)(a) = \overline{\mu}_a(X)$
- $\underline{\underline{\int}}_X f d\underline{\mu} : \mathcal{L}^B \to \mathrm{Int}(B, A, \mathcal{L}, r, \nearrow, \swarrow)$ defined as $\underline{\underline{\int}}_X f d\underline{\mu}(a) = \underline{\int}_X f d\underline{\mu}_a$
- $\overline{\int}_X f d\overline{\mu} : \mathcal{L}^B \to \mathrm{Int}(B, A, \mathcal{L}, r, \nearrow, \swarrow)$ defined as $\overline{\int}_X f d\overline{\mu}(a) = \underline{\int}_X f d\underline{\mu}_a$

In such a case the measures and integrals are mappings to closure and interior systems that are complete lattices

- $\mathrm{Int}(B, A, \mathcal{L}, r, \nearrow, \swarrow)$ with top $\nearrow (\chi_B)$ and bottom $\chi_\emptyset$
- $\mathrm{Int}(B, A, \mathcal{L}, r, \nwarrow, \searrow)$ with top χ_B and bottom $\nwarrow (\chi_\emptyset)$

Where all "inconveniences" with normality are hence solved.

Proposition 1. *Let $\underline{\mu}^B_a$ and $\overline{\mu}^B$ be $\mathcal{L}$-measures corresponded to $\mathcal{L}$-context $\langle B, A, \mathcal{L}, r \rangle$. Let $\underline{\mu}^A$ and $\overline{\mu}^A$ be $\mathcal{L}$-measures corresponded to $\mathcal{L}$-context $\langle A, B, \mathcal{L}, r^t \rangle$ where r^t is the transposition of r. Then*

$$
\underline{\int}_B f d\underline{\mu}^B \leq g \quad \Leftrightarrow \quad f \leq \overline{\int}_A g d\overline{\mu}^A
$$

and

$$
\underline{\int}_A g d\underline{\mu}^A \leq f \quad \Leftrightarrow \quad g \leq \overline{\int}_B f d\overline{\mu}^B
$$

Proof. From previous facts.

5 Conclusion

New lattice valued fuzzy measures and integrals are proposed by using data and operations from Girard monoid, i.e. complete residuated lattice with double negation law that are built from fuzzy formal context. Main result is to show the relationship between proposed measures and integrals and isotone concept-forming operators.

References

1. Ban, A., Fechete, I.: Componentwise decomposition of some lattice-valued fuzzy integrals. Inf. Sci. **177**(6), 1430–1440 (2007). ISSN 0020–0255
2. Bělohlávek, R.: Fuzzy Relational Systems. Springer, Heidelberg (2002). https://doi.org/10.1007/978-1-4615-0633-1
3. Bělohlávek, R.: Lattice generated by binary fuzzy relations (extended abstract). In: 4th International Conference on Fuzzy Sets Theory and Applications, p. 11 (1998)
4. Alcalde, C., Burusco, A.: On the use of Choquet integrals in the reduction of the size of L-fuzzy contexts. In: 2017 IEEE International Conference on Fuzzy Systems (FUZZ-IEEE), Naples, pp. 1–6 (2017)
5. Burusco, A., Fuentes-Gonzáles, R.: The study of the L-fuzzy concept lattice. Math. Soft Comput. **3**, 209–218 (1994)
6. Ganter, B., Wille, R.: Formal Concept Analysis - Mathematical Foundations. Springer, Heidelberg (1999). https://doi.org/10.1007/978-3-642-59830-2
7. Grabisch, M.: Fuzzy integral in multicriteria decision making. Fuzzy Sets Syst. **69**(3), 279–298 (1995)
8. Ilin, R.: Classification with concept lattice and Choquet integral. In: 2016 19th International Conference on Information Fusion (FUSION), Heidelberg, pp. 1554–1561 (2016)
9. Jenei, S.: Structure of girard monoids on [0,1]. In: Rodabaugh, S.E., Klement, E.P. (eds.) Topological and Algebraic Structures in Fuzzy Sets. Number 20 in Trends in Logic. Springer, Dordrecht (2003). https://doi.org/10.1007/978-94-017-0231-7_12
10. Krajči, S.: Cluster based efficient generation of fuzzy concepts. Neural Netw. World **13**(5), 521–530 (2003)
11. Krajči, S., Krajčiová, J.: Social network and one-sided fuzzy concept lattices. In: 2007 IEEE International Fuzzy Systems Conference, 23–26 July 2007, pp. 1–6 (2007). https://doi.org/10.1109/FUZZY.2007.4295369
12. Krídlo, O., Krajči, S., Ojeda-Aciego, M.: The category of L-Chu correspondences and the structure of L-bonds. Fundam. Inf. **115**(4), 297–325 (2012)
13. Krídlo, O., Krajči, S., Antoni, L.: Formal concept analysis of higher order. Int. J. Gener. Syst. **45**(2), 116–134 (2016)
14. Mesiar, R.: Fuzzy measures and integrals. Fuzzy Sets Syst. **156**(3), 365–370 (2005). ISSN 0165-0114
15. Mesiar, R., Mesiarová, A.: Fuzzy integrals and linearity. Int. J. Approx. Reason. **47**(3), 352–358 (2008). ISSN 0888-613X
16. Murofushi, T., Sugeno, M.: Fuzzy measures and fuzzy integrals
17. Xuecheng, L., Guangquan, Z.: Lattice-valued fuzzy measure and lattice-valued fuzzy integral. Fuzzy Sets Syst. **62**(3), 319–332 (1994)

Galois Connections Between Unbalanced Structures in a Fuzzy Framework

Inma P. Cabrera, Pablo Cordero, Emilio Muñoz-Velasco, and Manuel Ojeda-Aciego(✉)

Dept. Matemática Aplicada, Universidad de Málaga, Málaga, Spain
{ipcabrera,pcordero,ejmunoz,aciego}@uma.es

Abstract. The construction of Galois connections between unbalanced structures has received considerable attention in the recent years. In a nutshell, the problem is to find a right adjoint of a mapping defined between sets with unbalanced structure; in this paper we survey recent results obtained in this framework, focusing specially on the fuzzy structures that have been considered so far in this context: fuzzy preposets, fuzzy preordered structures, and fuzzy T-digraphs.

Keywords: Galois connection · Computational intelligence

1 Introduction

The notion of *Galois connection* (or its sibling, *adjunction*) has received considerable attention since its introduction [28], and it is common to find papers dealing with them either from a practical or a theoretical point of view, see [11] for a short survey. Galois connections (both in a crisp and in a fuzzy setting) have found applications in areas such as rough set theory [12,15,33]; (fuzzy) Mathematical Morphology in which the (fuzzy) operations of erosion and dilation are known to form a Galois connection [5,22,29,30]; another important source of applications of Galois connections is within the field of Formal Concept Analysis [7,13,16,31], where the concept-forming operators form either an antitone or isotone Galois connection (depending on the specific definition). Moreover, one can find applications in many other areas; for instance, Kycia [26] demonstrates how to construct a Galois connection between two systems with entropy; Brattka [6] considers a formal Galois connection in a certain lattice of representation spaces; Faul [17] uses adjunctions to study two apparently different approaches to broadcast domination of product graphs; Moraschini [27] introduces a logical and algebraic description of right adjoint functors between generalized quasi-varieties; Gibbons et al. [21] use adjunctions to elegantly explain relational algebra constructs.

Concerning the generalization to the fuzzy case of the notion of Galois connection, to the best of our knowledge, the first approach was due to Bělohlávek [3]. Later, a number of authors have considered different approaches to the so-called

M.-J. Lesot et al. (Eds.): IPMU 2020, CCIS 1239, pp. 736–747, 2020.
https://doi.org/10.1007/978-3-030-50153-2_54

fuzzy (isotone or antitone) Galois connections; see [4,14,18,20,23,24,32]. In [32], fuzzy Galois connections on fuzzy posets were introduced as a generalization of Bělohlávek's fuzzy Galois connection, and our approach is precisely based on this generalization.

In this paper, we survey recent results in our research line on the construction of Galois connections between sets with unbalanced structures initiated in [19], in which we attempt to characterize the existence of the right part of a Galois connection of a given mapping $f\colon \mathbb{A} \to B$ between sets with a different structure (it is precisely this condition of different structure that makes this problem to be outside the scope of Freyd's adjoint functor theorem). In [19], given a mapping from a crisp (pre-)ordered set $\mathbb{A} = (A, \leq_A)$ into an unstructured set B, we solved the problem of defining a suitable (pre-)ordering relation $\leq_B$ on B, for which there exists a mapping such that the pair of mappings forms an isotone Galois connection (or adjunction) between the (pre-)ordered sets $(A, \leq_A)$ and $(B, \leq_B)$.

Specifically, we consider the previous problem in different fuzzy frameworks: in Sect. 3 we focus on the case of a fuzzy preposet $\mathbb{A} = (A, \rho_A)$ and an unstructured B, see [8]; later, in Sect. 4, the work is extended by replacing crisp equality by a fuzzy equivalence relation, therefore the problem considers a mapping between a fuzzy preordered structure $\mathbb{A} = (A, \approx_A, \rho_A)$ and a fuzzy structure $(B, \approx_B)$, see [9]. Finally, in Sect. 5 we aim at obtaining a notion of Galois connection whose components are, in fact, relations between fuzzy T-digraphs [10].

2 Preliminary Definitions

The standard notion of Galois connection is defined between two partially ordered sets. However, not all the authors consider the same definition of Galois connection and it is remarkable that the definitions are not equivalent. In fact, there are four different notions of Galois connection, the most often used being the "right Galois connection" (also known as antitone Galois connection) and the "adjunction" (also known as isotone Galois connections).

Definition 1. *Let $\mathbb{A} = (A, \leq)$ and $\mathbb{B} = (B, \leq)$ be posets, $f\colon A \to B$ and $g\colon B \to A$ be two mappings. The pair (f,g) is called a*

- Right Galois Connection between $\mathbb{A}$ and $\mathbb{B}$, *denoted by $(f, g)\colon \mathbb{A} \leftharpoonup\!\rightharpoonup \mathbb{B}$ if, for all $a \in A$ and $b \in B$ it holds that $a \leq g(b)$ if only if $b \leq f(a)$.*
- Left Galois Connection between $\mathbb{A}$ and $\mathbb{B}$, *denoted by $(f, g)\colon \mathbb{A} \rightharpoondown\!\leftharpoondown \mathbb{B}$ if, for all $a \in A$ and $b \in B$ it holds that $g(b) \leq a$ if only if $f(a) \leq b$.*
- Adjunction between $\mathbb{A}$ and $\mathbb{B}$, *denoted by $(f, g)\colon \mathbb{A} \leftrightharpoons \mathbb{B}$ if, for all $a \in A$ and $b \in B$ it holds that $a \leq g(b)$ if only if $f(a) \leq b$.*
- Co-Adjunction between $\mathbb{A}$ and $\mathbb{B}$, *denoted by $(f, g)\colon \mathbb{A} \rightleftharpoons \mathbb{B}$ if, for all $a \in A$ and $b \in B$ it holds that $g(b) \leq a$ if only if $b \leq f(a)$.*

It is noteworthy that this definition is also compatible with the case of $\mathbb{A} = (A, \leq)$ and $\mathbb{B} = (B, \leq)$ being preordered sets.

Taking into account the dual order, $\mathbb{A}^{\partial} = (A, \geq)$, it is not difficult to check that the following conditions are equivalent:

1. $(f,g)\colon \mathbb{A} \leftharpoonup\!\!\!\rightharpoonup \mathbb{B}$.
2. $(f,g)\colon \mathbb{A}^\partial \leftharpoondown\!\!\!\rightharpoondown \mathbb{B}^\partial$.
3. $(f,g)\colon \mathbb{A} \leftrightharpoons \mathbb{B}^\partial$.
4. $(f,g)\colon \mathbb{A}^\partial \rightleftharpoons \mathbb{B}$.

It is worth mentioning that all the results can be stated both in terms of Galois connection or adjunctions, and either in terms of the existence and construction of right adjoints (or residual mappings, namely, the component g of the pair) or the existence and construction of left adjoints (or residuated mappings).

Galois Connections in the Fuzzy Case

As usual, we will consider a *complete residuated lattice* $\mathbb{L} = (L, \leq, \top, \bot, \otimes, \Rightarrow)$ as the underlying structure for considering the generalization to a fuzzy framework; supremum and infimum will be denoted by $\vee$ and $\wedge$, respectively.

An $\mathbb{L}$-*fuzzy set* is a mapping from the universe set, say X, to the lattice L, i.e. $X\colon U \to L$, where $X(u)$ means the degree in which u belongs to X. We will denote L^A to refer to the set of all mappings from A to L.

Given X and Y two $\mathbb{L}$-fuzzy sets, X is said to *be included in* Y, denoted as $X \subseteq Y$, if $X(u) \leq Y(u)$ for all $u \in U$. The *subsethood degree* $S(X,Y)$, by which X is a subset of Y, is defined by $S(X,Y) = \bigwedge_{u \in U} \big(X(u) \Rightarrow Y(u)\big)$.

The first notion of fuzzy Galois connection was given by Bělohlávek, and it can be rewritten as follows:

Definition 2 ([3]). *An ($\mathbb{L}$-)fuzzy Galois connection between A and B is a pair of mappings $f\colon L^A \to L^B$ and $g\colon L^B \to L^A$ such that, for all $X \in L^A$ and $Y \in L^B$ it holds that $S(X, g(Y)) = S(Y, f(X))$.*

An $\mathbb{L}$-*fuzzy binary relation on* U is an $\mathbb{L}$-fuzzy subset of $U \times U$, that is $\rho_U\colon U \times U \to L$, and it is said to be:

- *Reflexive* if $\rho_U(a,a) = \top$ for all $a \in U$.
- $\otimes$-*Transitive* if $\rho_U(a,b) \otimes \rho_U(b,c) \leq \rho_U(a,c)$ for all $a, b, c \in U$.
- *Symmetric* if $\rho_U(a,b) = \rho_U(b,a)$ for all $a, b \in U$.
- *Antisymmetric* if $\rho_U(a,b) = \rho_U(b,a) = \top$ implies $a = b$, for all $a, b \in U$.

We can now introduce the notions of *fuzzy poset* and *fuzzy preposet* as follows:

- An $\mathbb{L}$-*fuzzy poset* is a pair $\mathbb{U} = (U, \rho_U)$ in which ρ_U is a reflexive, antisymmetric and transitive $\mathbb{L}$-fuzzy relation on U.
- An $\mathbb{L}$-*fuzzy preposet* is a pair $\mathbb{U} = (U, \rho_U)$ in which ρ_U is a reflexive and transitive $\mathbb{L}$-fuzzy relation on U.

We will need the following order-related notions in the fuzzy framework:
Let $\mathbb{U} = \langle U, \rho_U \rangle$ be a fuzzy poset.

(i) The crisp set of *upper bounds* of a fuzzy set X on $\mathbb{U}$ is defined as

$$\mathrm{Up}(X) = \{a \in U \mid X(u) \leq \rho_A(u,a), \text{ for all } u \in U\}.$$

(ii) The *upset* and *downset* of an element $a \in U$ are defined as fuzzy sets $a^\uparrow, a^\downarrow \colon U \to L$, where $a^\downarrow(u) = \rho_U(u, a)$ and $a^\uparrow(u) = \rho_U(a, u)$ for all $u \in U$.
(iii) An element $a \in U$ is called a *maximum* of a fuzzy set X if $X(a) = \top$ and $X \subseteq a^\downarrow$. The definition of a minimum is similar.

In absence of antisymmetry it is possible that several maximum (resp. minimum) elements for X exist, which will be called *p-maximum* (resp. p-minimum). We will write p-max X (resp. p-min X) to denote the set of p-maxima (resp. p-minima) of X.

Remark 1. Although uniqueness is lost, given two p-maximum (resp. p-minimum) elements x and y, we have that $\rho_U(x, y) = \top$. This property will be relevant later in subsequent sections.

We can now recall the extension to the fuzzy case provided by Yao and Lu, also used in [8], which can be stated as follows:

Definition 3 ([32]). *Let $\mathbb{A} = \langle A, \rho_A \rangle$ and $\mathbb{B} = \langle B, \rho_B \rangle$ be fuzzy preposets. A pair of mappings $f \colon A \to B$ and $g \colon B \to A$ forms a* Galois connection *between $\mathbb{A}$ and $\mathbb{B}$, denoted $(f, g) : \mathbb{A} \rightleftharpoons \mathbb{B}$ if, for all $a \in A$ and $b \in B$, the equality $\rho_A(a, g(b)) = \rho_B(f(a), b)$ holds.*

Note that we have maintained the original term used by Yao and Lu, although it technically corresponds to an adjunction, not a Galois connection.

3 When the Domain Has the Structure of Fuzzy Preposet

In this section, we consider a mapping $f \colon \mathbb{A} \to B$ from a fuzzy preposet $\mathbb{A} = \langle A, \rho_A \rangle$ into an unstructured set B, and characterize those situations in which B can be endowed with a fuzzy preorder relation and an isotone mapping $g \colon B \to A$ can be built such that the pair (f, g) is an adjunction.

Let $\mathbb{A} = \langle A, \rho_A \rangle$ be a fuzzy preposet, and consider a mapping $f \colon A \to B$. The *fuzzy p-kernel* relation $\cong_A$ is the $\otimes$-transitive closure of the union of the fuzzy equivalence relations $\approx_A$ and $\equiv_f$, where

$$(a_1 \approx_A a_2) = \rho_A(a_1, a_2) \otimes \rho_A(a_2, a_1) \qquad \text{for all } a_1, a_2 \in A\,.$$

and

$$(a_1 \equiv_f a_2) = \begin{cases} \bot & \text{if } f(a_1) \neq f(a_2), \\ \top & \text{if } f(a_1) = f(a_2). \end{cases}$$

Note that $\cong_A$ is also a fuzzy equivalence relation and the fuzzy equivalence classes $[a]_{\cong_A} : A \to L$ are the fuzzy sets defined by

$$[a]_{\cong_A}(x) = (x \cong_A a). \tag{1}$$

In the definition of the inherited structure, and also in the right adjoint, we will make use of (some of) the following fuzzy powerings:

Given (A, ρ) and $X, Y \subseteq A$, we define the *Hoare, Smyth and full fuzzy powerings* as follows:

1. $\rho_H(X,Y) = \bigwedge_{x\in X} \bigvee_{y\in Y} \rho(x,y)$
2. $\rho_S(X,Y) = \bigwedge_{y\in Y} \bigvee_{x\in X} \rho(x,y)$
3. $\rho_\propto(X,Y) = \bigwedge_{x\in X} \bigwedge_{y\in Y} \rho(x,y)$

We can now state necessary and sufficient conditions for the existence of a right adjoint from a fuzzy preposet to an unstructured set.

Theorem 1. *Let $\mathbb{A} = \langle A, \rho_A\rangle$ be a fuzzy preposet, and consider a mapping $f\colon A \to B$, then there exist a fuzzy preorder relation ρ_B on B and a mapping $g\colon B \to A$ such that $(f,g) : \mathbb{A} \leftrightharpoons \mathbb{B}$ if and only if there exists a subset $S \subseteq A$ such that, for all $a, a_1, a_2 \in A$:*

1. $S \subseteq \bigcup_{a\in A} \text{p-max}[a]_{\cong_A}$.
2. $\text{p-min}(\text{Up}([a]_{\cong_A}) \cap S) \neq \varnothing$
3. $\rho_A(a_1,a_2) \leq \rho_H\big(\text{p-min}(\text{Up}([a_1]_{\cong_A}) \cap S), \text{p-min}(\text{Up}([a_2]_{\cong_A}) \cap S)\big)$.

The proof of the theorem is completely constructive, and the ordered structure on B is given as follows:

For any $a_0 \in A$, there exist a number of suitable definitions of $g\colon B \to A$, and all of them can be specified as follows:

- If $b \in f(A)$, then $g(b)$ is any element in $\text{p-min}(\text{Up}([x_b]_{\cong_A}) \cap S)$ for $x_b \in f^{-1}(b)$.
- If $b \notin f(A)$, then $g(b)$ is any element in $\text{p-min}(\text{Up}([a_0]_{\cong_A}) \cap S)$.

Finally, the fuzzy relation $\rho_B^{a_0} : B \times B \to L$ is defined as follows

$$\rho_B^{a_0}(b_1,b_2) = \rho_A(g(b_1), g(b_2)).$$

4 Changing Crisp Equality by a Fuzzy Equivalence Relation

A further step towards generalization to the fuzzy realm is possible when considering fuzzy equivalence relations in each of the involved sets instead of the mere equality relation. This leads to a notion of fuzzy Galois connection in which the mappings f and g can be seen, in some sense, as fuzzy mappings instead of being crisp ones.

In this section, we consider the case where there are two underlying fuzzy equivalence relations in both the domain and the codomain of the mapping f, more specifically, f is a *morphism* between the *fuzzy structures* $\langle A, \approx_A\rangle$ and $\langle B, \approx_B\rangle$ where, in addition, $\langle A, \approx_A\rangle$ is a fuzzy preordered structure.

The additional consideration of an underlying fuzzy equivalence relation suggests considering the following notions:

(i) A *fuzzy structure* $\mathcal{A} = \langle A, \approx_A \rangle$ is a set A endowed with a fuzzy equivalence relation $\approx_A$.
(ii) A *morphism* between two fuzzy structures $\mathcal{A}$ and $\mathcal{B}$ is a mapping $f : A \to B$ such that for all $a_1, a_2 \in A$ the following inequality holds: $(a_1 \approx_A a_2) \leq (f(a_1) \approx_B f(a_2))$. In this case, we write $f : \mathcal{A} \to \mathcal{B}$, and we say that f is *compatible* with $\approx_A$ and $\approx_B$.
(iii) A morphism between two fuzzy structures $\mathcal{A}$ and $\mathcal{B}$ is said to be $\approx$*-injective* if $(f(a_1) \approx_B f(a_2)) \leq (a_1 \approx_A a_2)$, for all $a_1, a_2 \in A$. $\approx$*-surjective* if for all $b \in B$ there exists $a \in A$ such that $(f(a) \approx_B b) = \top$.
(iv) Let $\mathcal{B} = \langle B, \approx_B \rangle$ be a fuzzy structure, and consider a crisp subset $X \subseteq B$. A mapping $h : B \to X$ is said to be a *contraction* if it is a morphism $h : \mathcal{B} \to \langle X, \approx_B \rangle$ and $h(x) = x$ for all $x \in X$.

Given a fuzzy structure $\mathcal{A} = \langle A, \approx_A \rangle$, we can now introduce the notion of *fuzzy preordered structure* as a pair $\mathbb{A} = \langle \mathcal{A}, \rho_A \rangle$ in which ρ_A is a fuzzy relation that is $\approx_A$-reflexive, $\otimes$-$\approx_A$-antisymmetric and $\otimes$-transitive, where

(i) $\approx_A$*-reflexive* means $(a_1 \approx_A a_2) \leq \rho_A(a_1, a_2)$ for all $a_1, a_2 \in A$.
(ii) $\otimes$-$\approx_A$*-antisymmetric* means $\rho_A(a_1, a_2) \otimes \rho_A(a_2, a_1) \leq (a_1 \approx_A a_2)$ for all $a_1, a_2 \in A$.

If the underlying fuzzy structure is not clear from the context, we will sometimes write a fuzzy preordered structure as a triplet $\mathbb{A} = \langle A, \approx_A, \rho_A \rangle$.

The formal notion of p-maximum (resp. p-minimum) in the context of fuzzy preordered structures is exactly the same as in the previous section; however, the use of the underlying fuzzy equivalence relation leads to different properties. Observe that, given two p-maxima x_1, x_2 of a fuzzy set X in a fuzzy preordered structure, we obtain $\rho_A(x_1, x_2) = \top = \rho_A(x_2, x_1)$ and by $\otimes$-$\approx_A$-antisymmetry, also $(x_1 \approx_A x_2) = \top$.

A reasonable approach to introduce the notion of Galois connection between fuzzy preordered structures $\mathbb{A}$ and $\mathbb{B}$ would be the following:

Definition 4 ([9]). *Let* $\mathbb{A}$ *and* $\mathbb{B}$ *be two fuzzy preordered structures. Given two morphisms* $f : \mathcal{A} \to \mathcal{B}$ *and* $g : \mathcal{B} \to \mathcal{A}$, *the pair* (f, g) *is said to be a* Galois connection *between* $\mathbb{A}$ *and* $\mathbb{B}$ *(briefly,* $(f, g) : \mathbb{A} \leftrightharpoons \mathbb{B}$*) if the following conditions hold for all* $a, a_1, a_2 \in A$ *and* $b, b_1, b_2 \in B$:

(G1) $(a_1 \approx_A a_2) \otimes \rho_A(a_2, g(b)) \leq \rho_B(f(a_1), b)$
(G2) $(b_1 \approx_B b_2) \otimes \rho_B(f(a), b_1) \leq \rho_A(a, g(b_2))$

The previous definition behaves as expected, namely, it is equivalent to the standard equality for Galois connections. More specifically, the pair (f, g) is a Galois connection between $\mathbb{A}$ and $\mathbb{B}$ if and only if both mappings are morphisms and $\rho_A(a, g(b)) = \rho_B(f(a), b)$ for all $a \in A$ and $b \in B$.

Once again, we need the corresponding version of the kernel relation and its equivalence classes. These definitions are given below:

Let $\mathcal{A}$ and $\mathcal{B}$ be two fuzzy structures and let $f\colon \mathcal{A} \to \mathcal{B}$ be a morphism. The *fuzzy kernel relation* $\equiv_f\colon A \times A \to L$ associated with f is defined as follows, for $a_1, a_2 \in A$,

$$(a_1 \equiv_f a_2) = \big(f(a_1) \approx_B f(a_2)\big).$$

The fuzzy kernel relation trivially is a fuzzy equivalence relation, and the equivalence class of an element $a \in A$ is the fuzzy set $[a]_f\colon A \to L$ defined by $[a]_f(u) = (f(a) \approx_B f(u))$ for all $u \in A$.

Given a fuzzy preordered structure $\mathbb{A} = \langle A, \approx_A, \rho_A\rangle$, and crisp subsets X, Y of A and . The fuzzy relations $\approx_A$ and ρ_A can be extended to the sets of p-maxima as follows:

$$\big(\text{p}-\max(X) \approx_A \text{p}-\max(Y)\big) \overset{\text{def}}{=} (x \approx_A y)$$

$$\rho_A\big(\text{p}-\max(X), \text{p}-\max(Y)\big) \overset{\text{def}}{=} \rho_A(x, y)$$

where x (resp. y) can be any element in $\text{p}-\max(X)$ (resp. $\text{p}-\max(Y)$). It is not difficult to prove that the definition does not depend on the choice of x and y.

The preceding notation allows us to state necessary conditions on f in order to have a right adjoint in a more compact form which essentially follows the scheme already obtained in [8] and [19].

Theorem 2 (Necessary conditions). *Consider two fuzzy preordered structures $\mathbb{A}$ and $\mathbb{B}$, together with two morphisms $f\colon \mathcal{A} \to \mathcal{B}$ and $g\colon \mathcal{B} \to \mathcal{A}$. If (f, g) is a Galois connection between $\mathbb{A}$ and $\mathbb{B}$, then*

1. $\text{p}-\max([a]_f)$ *is not empty for all* $a \in A$.
2. $\rho_A(a_1, a_2) \le \rho_A\big(\text{p}-\max([a_1]_f), \text{p}-\max([a_2]_f)\big)$, *for all* $a_1, a_2 \in A$.
3. $(a_1 \equiv_f a_2) \le \big(\text{p}-\max([a_1]_f) \approx_A \text{p}-\max([a_2]_f)\big)$, *for all* $a_1, a_2 \in A$.

We show now that the necessary conditions in Theorem 2 are sufficient in the case of a $\approx$-surjective mapping.

Theorem 3 (Sufficient conditions). *Consider a fuzzy preordered structure $\mathbb{A}$, a fuzzy structure $\mathcal{B} = \langle B, \approx_B\rangle$, and a $\approx$-surjective morphism $f\colon \mathcal{A} \to \mathcal{B}$. If the following conditions hold*

1. $\text{p}-\max([a]_f)$ *is not empty for all* $a \in A$;
2. $\rho_A(a_1, a_2) \le \rho_A\big(\text{p}-\max([a_1]_f), \text{p}-\max([a_2]_f)\big)$, *for all* $a_1, a_2 \in A$;
3. $(a_1 \equiv_f a_2) \le \big(\text{p}-\max([a_1]_f) \approx_A \text{p}-\max([a_2]_f)\big)$, *for all* $a_1, a_2 \in A$;

then there exists a $\approx_B$-reflexive, $\otimes$-$\approx_B$-antisymmetric and $\otimes$-transitive fuzzy relation ρ_B on B and a morphism $g\colon \mathcal{B} \to \mathcal{A}$ such that (f, g) is a Galois connection between the fuzzy preordered structures $\mathbb{A}$ and $\mathbb{B} = \langle \mathcal{B}, \rho_B\rangle$.

We also identify necessary and sufficient conditions in the case of a $\approx$-injective mapping.

Theorem 4. *Consider two fuzzy preordered structures $\mathbb{A} = \langle \mathcal{A}, \rho_A\rangle$ and $\mathbb{B} = \langle \mathcal{B}, \rho_B\rangle$. For a $\approx$-injective morphism $f\colon \mathcal{A} \to \mathcal{B}$, the following statements are equivalent:*

1. *There exists a morphism* $g\colon \mathcal{B} \to \mathcal{A}$ *such that* $(f,g)\colon \mathbb{A} \leftrightharpoons \mathbb{B}$.
2. *There exist a contraction* $h\colon \langle B, \approx_B\rangle \to \langle f(A), \approx_B\rangle$ *and a fuzzy relation* $\rho_{f(A)}$ *defined as* $\rho_{f(A)}(f(a_1), f(a_2)) = \rho_A(a_1, a_2)$ *such that the pair* (i,h) *is a Galois connection between* $\langle f(A), \approx_B, \rho_{f(A)}\rangle$ *and* $\langle B, \approx_B, \rho_B\rangle$, *where* $i\colon f(A) \to B$ *denotes the canonical embedding.*

The previous results lead to the systematic construction of the induced structure and the right adjoint in Algorithm 1.

Algorithm 1: Building Galois Connection

Data: A finite fuzzy preordered structure $\langle A, \approx_A, \rho_A\rangle$, a finite fuzzy structure $\langle B, \approx_B\rangle$ and a morphism $f\colon \langle A, \approx_A\rangle \to \langle B, \approx_B\rangle$.
Result: A morphism $g\colon \langle B, \approx_B\rangle \to \langle A, \approx_A\rangle$ and a $\approx_B$-reflexive, $\otimes$-$\approx_B$-antisymmetric and $\otimes$-transitive fuzzy relation ρ_B such that $(f,g) : \langle A, \approx_A, \rho_A\rangle \leftrightharpoons \langle B, \approx_B, \rho_B\rangle$ if they exist, or the message *"It is not possible to build a Galois connection"* otherwise.

1 Compute the relation $\equiv_f$ on A defined by $(a_1 \equiv_f a_2) := (f(a_1) \approx_B f(a_2))$
2 **foreach** $a \in A$ **do**
3 Compute p−max($[a]_f$) where $[a]_f$ is the equivalence class of a w.r.t. $\equiv_f$
4 **if** p−max($[a]_f$) $= \varnothing$ **then return** "It is not possible to build a Galois connection"
5 **else** Let $b = f(a)$ and consider an arbitrary element $\psi(b)$ from p−max($[a]_f$)
6 **foreach** $a_1, a_2 \in A$ **do**
7 **if** $\rho_A(a_1,a_2) \not\leq \rho_A(\psi f(a_1), \psi f(a_2))$ *or* $(a_1 \equiv_f a_2) \not\leq (\psi f(a_1) \approx_A \psi f(a_2))$ **then**
8 **return** "It is not possible to build a Galois connection"
9 Define $\rho_{f(A)}$ as $\rho_{f(A)}(b_1,b_2) := \rho_A(\psi(b_1), \psi(b_2))$ for each $b_1, b_2 \in f(A)$
10 **foreach** *contraction* $h\colon B \to f(A)$ **do**
11 Define μ_h in B as:
12 $\mu_h(b_1,b_2) := \rho_{f(A)}(b_1, h(b_2))$ if $b_1 \in f(A)$ and $\mu_h(b_1,b_2) := (b_1 \approx_B b_2)$ otherwise
13 Compute $\rho_B := \mu_h^2$ and $g := \psi \circ h$
14 **if** ρ_B *is* $\otimes$-$\approx_B$*-antisymmetric* **then return** g *and* ρ_B
15 **return** "It is not possible to build a Galois connection"

---

5 Relational Galois Connections Between Fuzzy T-Digraphs

We attempt here a first generalization of the notion of *relational* Galois connection to the fuzzy case. The focus is put on transitive fuzzy directed graphs, fuzzy T-digraphs for short, because of their interest for applications. One can find interesting theoretical applications of digraphs, for instance, Akram et al. [1] introduce the notion of fuzzy rough digraph and consider its application in decision making. In [2], Baykasoglu applies a fuzzy digraph model to quantify manufacturing flexibility. In [25], Koulouriotis and Ketipi develop a fuzzy digraph method for robot evaluation and selection, according to a given industrial application.

In this section, we focus specifically on providing an adequate notion of relational Galois connection between fuzzy T-digraphs which inherits most of the interesting equivalent characterizations of the notion of crisp Galois connection.

Our framework in this work is relational at the level of Galois connections (namely, the components of a Galois connection are crisp binary relations instead of functions) and fuzzy at the level of their domain and codomain.

We will use the following standard notions about relations: Given a binary relation $\mathcal{R} \subseteq A \times B$, the *afterset* $a^{\mathcal{R}}$ of an element $a \in A$ is defined as $\{b \in B \mid a\mathcal{R}b\}$.

Definition 5. *A pair* $\mathbb{A} = (A, \rho)$ *is said to be a* fuzzy T-digraph *if* ρ *is a* $\otimes$-*transitive fuzzy relation on* A.

The usual requirement that in a Galois condition both components should be antitone and their compositions inflationary leads to a preliminary approach to the definition of a relational Galois connection for fuzzy preposets.

Let us, firstly, fix the notions of antitone and inflationary relation in a fuzzy setting. Given (A, ρ) and (B, ρ):

1. A relation $\mathcal{R} \subseteq A \times B$ is *antitone* if $\rho(a_1, a_2) \le \rho(b_2, b_1)$ for all $b_1 \in a_1^{\mathcal{R}}$ and $b_2 \in a_2^{\mathcal{R}}$, or equivalently, $\rho(a_1, a_2) \le \rho_{\propto}(a_2^{\mathcal{R}}, a_1^{\mathcal{R}})$.
2. A relation $\mathcal{R} \subseteq A \times A$ is *inflationary* if $\rho(a_1, a_2) = \top$ for all $a_2 \in a_1^{\mathcal{R}}$ or, equivalently, $\rho_{\propto}(a, a^{\mathcal{R}}) = \top$.

We can obtain the following proposition which links the properties of antitone and inflationary to a pair of inequalities with a certain flavour to Galois condition.

Proposition 1. *Let* (A, ρ) *and* (B, ρ) *be fuzzy preposets and* $\mathcal{R} \subseteq A \times B$ *and* $\mathcal{S} \subseteq B \times A$ *be relations. Then* $\mathcal{R}$ *and* $\mathcal{S}$ *are antitone and* $\mathcal{R} \circ \mathcal{S}$ *and* $\mathcal{S} \circ \mathcal{R}$ *are inflationary if and only if the following inequalities hold:*

$$\rho_H(a, b^{\mathcal{S}}) \le \rho_S(b, a^{\mathcal{R}}) \qquad \text{and} \qquad \rho_H(b, a^{\mathcal{R}}) \le \rho_S(a, b^{\mathcal{S}}) \,. \tag{2}$$

This proposition suggests to consider inequalities (2) as a tentative definition of relational Galois connection between fuzzy T-digraphs. To begin with, we have the following result.

Proposition 2. *Let* (A, ρ) *and* (B, ρ) *be fuzzy T-digraphs and* $\mathcal{R} \subseteq A \times B$ *and* $\mathcal{S} \subseteq B \times A$ *be relations. If* $\mathcal{R}$ *and* $\mathcal{S}$ *are antitone and* $\mathcal{R} \circ \mathcal{S}$ *and* $\mathcal{S} \circ \mathcal{R}$ *are inflationary, then* $(\mathcal{R}, \mathcal{S})$ *satisfy condition* (2).

However, the following example shows that the converse does not hold.

Example 1. Consider the following fuzzy T-digraphs $\mathbb{A} = (\{a_1, a_2, a_3\}, \rho)$ and $\mathbb{B} = (\{b_1, b_2, b_3\}, \rho)$, and the relations $\mathcal{R} \subseteq A \times B$ and $\mathcal{S} \subseteq B \times A$ defined below:

ρ	a_1	a_2	a_3
a_1	1	1	1/2
a_2	0	0	0
a_3	0	1/2	1

ρ	b_1	b_2	b_3
b_1	1	0	0
b_2	1	0	0
b_3	1/2	0	1

x	$x^{\mathcal{R}}$
a_1	$\{b_1\}$
a_2	$\{b_2\}$
a_3	$\{b_3\}$

x	$x^{\mathcal{S}}$
b_1	$\{a_1\}$
b_2	$\{a_1, a_2\}$
b_3	$\{a_3\}$

It is routine to check that $(\mathcal{R}, \mathcal{S})$ satisfies condition (2). Nevertheless, $\mathcal{R} \circ \mathcal{S}$ is not inflationary, because $\{a_1\} \in a_2^{\mathcal{R} \circ \mathcal{S}}$, while $\rho(a_2, a_1) = 0$ and $\rho_S(a_2, a_2^{\mathcal{R} \circ \mathcal{S}}) = \rho_S(a_1, \{a_1, a_2\}) = 0$.

The question now is to discover some "missing" requirement which should be required in order to prove the converse of Proposition 2. Surprisingly, this requirement already appeared in previous sections as a property of the sets of p-minima or p-maxima; namely, all the elements in the aftersets should be related with degree $\top$. Formally, we have the following definition:

Definition 6. *Let (A, ρ) be a fuzzy T-digraph and $X \subseteq A$. We say that a nonempty set X is a clique if for all $x, y \in X$ it holds $\rho(x, y) = \top$ or, equivalently, $\rho_\propto(X, X) = \top$.*

Notice that given a fuzzy T-digraph (A, ρ), $X \subseteq A$ and $a \in A$, then if X is a clique, we have that $\rho_S(a, X) = \rho_\propto(a, X) = \rho_H(a, X)$. As a result, the inequalities in (2) collapse into the equality $\rho_\propto(a, b^{\mathcal{S}}) = \rho_\propto(b, a^{\mathcal{R}})$ and, furthermore, the following characterisation can be proved:

Theorem 5. *Let (A, ρ) and (B, ρ) be fuzzy T-digraphs. Given $\mathcal{R} \subseteq A \times B$ and $\mathcal{S} \subseteq B \times A$ then, $\mathcal{R}$ and $\mathcal{S}$ are antitone and $\mathcal{R} \circ \mathcal{S}$ and $\mathcal{S} \circ \mathcal{R}$ are inflationary between (A, ρ) and (B, ρ) if and only if the following conditions hold:*

(i) $\rho_\propto(a, b^{\mathcal{S}}) = \rho_\propto(b, a^{\mathcal{R}})$ for all $a \in A$ and $b \in B$,
(ii) $a^{\mathcal{R}}$ and $b^{\mathcal{S}}$ are cliques for all $a \in A$ and $b \in B$.

As a consequence, we can give an adequate definition of relational Galois connection between fuzzy T-digraphs which, on the one hand, generalizes the Galois condition and, on the other hand, guarantees the properties of the components of the connection:

Definition 7. *Let (A, ρ) and (B, ρ) be fuzzy T-digraphs and $\mathcal{R} \subseteq A \times B$ and $\mathcal{S} \subseteq B \times A$ be relations. We say that the pair $(\mathcal{R}, \mathcal{S})$ is a relational Galois connection if the following conditions hold:*

(i) $\rho_\propto(a, b^{\mathcal{S}}) = \rho_\propto(b, a^{\mathcal{R}})$ for all $a \in A$ and $b \in B$,
(ii) $a^{\mathcal{R}}$ and $b^{\mathcal{S}}$ are cliques for all $a \in A$ and $b \in B$.

6 Conclusions and Future Work

There are a number of possible options to extend the notion of a Galois connection to a fuzzy setting. We have surveyed some of the previous works in this area, and provided a somewhat unified presentation. In some cases, we have given a characterization theorem of the existence of a right adjoint for a given function. Moreover, we have provided the adequate notion of Galois connection between fuzzy T-digraphs, whilst the explicit construction of a right adjoint for a given relation is left for future work.

The relational generalization to fuzzy T-digraphs paves the way towards obtaining an operative notion of *fuzzy* relational Galois connection between *fuzzy* T-digraphs, and initiates the search for a characterization of the existence of a residual to a given fuzzy relation. On the other hand, it might enable a new approach to Formal Concept Analysis, provided that the definition of relational Galois connection is suitably adapted to formal contexts.

Acknowledgments. Partially supported by the Spanish Ministry of Science, Innovation, and Universities (MCIU), the State Agency of Research (AEI) and the European Social Fund (FEDER) through projects PGC2018-095869-B-I00 and TIN2017-89023-P, and Junta de Andalucía project UMA2018-FEDERJA-001.

References

1. Akram, M., Shumaiza, Arshad, M.: A new approach based on fuzzy rough digraphs for decision-making. J. Intell. Fuzzy Syst. **35**(2), 2105–2121 (2018)
2. Baykasoglu, A.: A practical fuzzy digraph model for modeling manufacturing flexibility. Cybern. Syst. **40**(6), 475–489 (2009)
3. Bělohlávek, R.: Fuzzy Galois connections. Math. Logic Q. **45**(4), 497–504 (1999)
4. Bělohlávek, R., Osička, P.: Triadic fuzzy Galois connections as ordinary connections. Fuzzy Sets Syst. **249**, 83–99 (2014)
5. Bloch, I.: Fuzzy sets for image processing and understanding. Fuzzy Sets Syst. **281**, 280–291 (2015)
6. Brattka, V.: A Galois connection between turing jumps and limits. Log. Methods Comput. Sci. **14**(3:13) (2018)
7. Butka, P., Pócs, J., Pócsová, J.: Isotone Galois connections and generalized one-sided concept lattices. In: Choroś, K., Kopel, M., Kukla, E., Siemiński, A. (eds.) MISSI 2018. AISC, vol. 833, pp. 151–160. Springer, Cham (2019). https://doi.org/10.1007/978-3-319-98678-4_17
8. Cabrera, I., Cordero, P., Garcia-Pardo, F., Ojeda-Aciego, M., De Baets, B.: On the construction of adjunctions between a fuzzy preposet and an unstructured set. Fuzzy Sets Syst. **320**, 81–92 (2017)
9. Cabrera, I., Cordero, P., Garcia-Pardo, F., Ojeda-Aciego, M., De Baets, B.: Galois connections between a fuzzy preordered structure and a general fuzzy structure. IEEE Trans. Fuzzy Syst. **26**(3), 1274–1287 (2018)
10. Cabrera, I., Cordero, P., Muñoz-Velasco, E., Ojeda-Aciego, M., De Baets, B.: Relational Galois connections between transitive fuzzy digraphs. Math. Methods Appl. Sci. **43**(9), 5673–5680 (2020)
11. Cabrera, I., Cordero, P., Ojeda-Aciego, M.: Galois connections in computational intelligence: a short survey. In: IEEE Symposium Series on Computational Intelligence (SSCI) (2017)
12. Cornelis, C., Medina, J., Verbiest, N.: Multi-adjoint fuzzy rough sets: definition, properties and attribute selection. Int. J. Approx. Reason. **55**(1), 412–426 (2014)

13. Denniston, J.T., Melton, A., Rodabaugh, S.E.: Formal contexts, formal concept analysis, and Galois connections. Electr. Proc. Theor. Comput. Sci. **129**, 105–120 (2013)
14. Djouadi, Y., Prade, H.: Interval-valued fuzzy Galois connections: algebraic requirements and concept lattice construction. Fundamenta Informaticae **99**(2), 169–186 (2010)
15. Dzik, W., Järvinen, J., Kondo, M.: Representing expansions of bounded distributive lattices with Galois connections in terms of rough sets. Int. J. Approx. Reason. **55**(1), 427–435 (2014)
16. Díaz, J., Medina, J., Ojeda-Aciego, M.: On basic conditions to generate multi-adjoint concept lattices via Galois connections. Int. J. Gen. Syst. **43**(2), 149–161 (2014)
17. Faul, P.F.: Adjunctions in the study of broadcast domination with a cost function. Aust. J. Comb. **72**, 70–81 (2018)
18. Frascella, A.: Fuzzy Galois connections under weak conditions. Fuzzy Sets Syst. **172**(1), 33–50 (2011)
19. García-Pardo, F., Cabrera, I., Cordero, P., Ojeda-Aciego, M., Rodríguez, F.: On the definition of suitable orderings to generate adjunctions over an unstructured codomain. Inf. Sci. **286**, 173–187 (2014)
20. Georgescu, G., Popescu, A.: Non-commutative fuzzy Galois connections. Soft Comput. **7**(7), 458–467 (2003)
21. Gibbons, J., Henglein, F., Hinze, R., Wu, N.: Relational algebra by way of adjunctions. Proc. ACM Program. Lang. **2**, 86:1–86:28 (2018)
22. González-Hidalgo, M., Massanet, S., Mir, A., Ruiz-Aguilera, D.: A fuzzy morphological hit-or-miss transform for grey-level images: a new approach. Fuzzy Sets Syst. **286**, 30–65 (2016)
23. Gutiérrez-García, J., Mardones-Pérez, I., de Prada-Vicente, M.A., Zhang, D.: Fuzzy Galois connections categorically. Math. Log. Q. **56**(2), 131–147 (2010)
24. Konecny, J.: Isotone fuzzy Galois connections with hedges. Inf. Sci. **181**, 1804–1817 (2011)
25. Koulouriotis, D.E., Ketipi, M.K.: A fuzzy digraph method for robot evaluation and selection. Expert Syst. Appl. **38**(9), 11901–11910 (2011)
26. Kycia, R.: Landauer's principle as a special case of Galois connection. Entropy **20**(12), 971 (2018)
27. Moraschini, T.: A logical and algebraic characterization of adjunctions between generalized quasi-varieties. J. Symb. Log. **83**(3), 899–919 (2018)
28. Ore, Ø.: Galois connexions. Trans. Am. Math. Soc. **55**, 493–513 (1944)
29. Shi, Y., Nachtegael, M., Ruan, D., Kerre, E.: Fuzzy adjunctions and fuzzy morphological operations based on implications. Int. J. Intell. Syst. **24**(12), 1280–1296 (2009)
30. Sussner, P.: Lattice fuzzy transforms from the perspective of mathematical morphology. Fuzzy Sets Syst. **288**, 115–128 (2016)
31. Yao, W., Han, S., Wang, R.: Lattice-theoretic contexts and their concept lattices via Galois ideals. Inf. Sci. **339**, 1–18 (2016)
32. Yao, W., Lu, L.-X.: Fuzzy Galois connections on fuzzy posets. Math. Log. Q. **55**(1), 105–112 (2009)
33. Yao, Y.: Rough-set concept analysis: interpreting RS-definable concepts based on ideas from formal concept analysis. Inf. Sci. **346–347**, 442–462 (2016)

Impact of Local Congruences in Attribute Reduction

Roberto G. Aragón, Jesús Medina, and Eloísa Ramírez-Poussa(✉)

Department of Mathematics, University of Cádiz, Cádiz, Spain
{roberto.aragon,jesus.medina,eloisa.ramirez}@uca.es

Abstract. Local congruences are equivalence relations whose equivalence classes are convex sublattices of the original lattice. In this paper, we present a study that relates local congruences to attribute reduction in FCA. Specifically, we will analyze the impact in the context of the use of local congruences, when they are used for complementing an attribute reduction.

Keywords: Formal Concept Analysis · Size reduction · Attribute reduction · Local congruence

1 Introduction

Formal Concept Analysis (FCA) is a mathematical framework to analyze datasets introduced by Ganter and Wille in eighties [12]. The main goals of FCA are the following: to obtain the knowledge from data, to represent the obtained knowledge by means of the mathematical structure called concept lattice and to discover dependencies in data. The applied potential of FCA has encouraged the development of different generalizations.

One of the most intensively studied research lines by the research community of FCA in the last years, consists on decreasing the number of attributes of a dataset, preserving the information provided by the dataset [1,2,7,8,10,11,13–18]. In [6], authors proved that every reduction of attributes of a formal context induces an equivalent relation whose equivalent classes are join-semilattices. In [3], local congruences were introduced and applied to this attribute reduction. Local congruences are equivalence relations on lattices whose equivalence classes are convex sublattices. The idea in [3] was to find the least local congruence containing the equivalent relation induced by an attribute reduction of a formal context, in order to group the concepts of the original concept lattice using closed structures.

Partially supported by the the 2014-2020 ERDF Operational Programme in collaboration with the State Research Agency (AEI) in project TIN2016-76653-P, and with the Department of Economy, Knowledge, Business and University of the Regional Government of Andalusia. in project FEDER-UCA18-108612, and by the European Cooperation in Science & Technology (COST) Action CA17124.

M.-J. Lesot et al. (Eds.): IPMU 2020, CCIS 1239, pp. 748–758, 2020.
https://doi.org/10.1007/978-3-030-50153-2_55

Sometimes, the induced equivalent relation by a reduction of the context is already a local congruence but sometime it is not. In the latter case, the fact of using a local congruence that contains the induced equivalence relation has an influence on the original reduction. In this paper, we present an initial study about the relationship between local congruences and the induced equivalent relation by an attribute reduction of a formal context. This study provides a first step to know the influence that this special kind of equivalence relations has on the reduction procedure. We will include several examples to illustrate the obtained result.

2 Preliminaries

We need to recall some basic notions used in this work. In order to present the preliminary notions as clearly as possible, we will divide this section into two parts. The first one will be devoted to recall those necessary notions of FCA and the second one to those related to local congruences.

2.1 Formal Concept Analysis

In FCA a context is a triple (A, B, R) where A is a set of attributes, B is a set of objects and $R\colon A \times B \to \{0, 1\}$ is a relationship, such that $R(a, x) = aRx = 1$, if the object $x \in B$ possesses the attribute $a \in A$, and $R(a, x) = 0$, otherwise. In addition, we call *concept-forming operators* to the mappings ${}^{\uparrow}\colon 2^B \to 2^A$ and ${}^{\downarrow}\colon 2^A \to 2^B$ defined for each $X \subseteq B$ and $Y \subseteq A$ as:

$$X^{\uparrow} = \{a \in A \mid \text{for all } x \in X, aRx\} \tag{1}$$

$$Y^{\downarrow} = \{x \in B \mid \text{for all } a \in Y, aRx\} \tag{2}$$

Taking into account the previous mappings, a *concept* is a pair (X, Y), with $X \subseteq B$ and $Y \subseteq A$ satisfying that $X^{\uparrow} = Y$ and $Y^{\downarrow} = X$. The subset X is called the *extent* of the concept and the subset Y is called the *intent*. The set of extents and intents are denoted by $\mathfrak{E}(A, B, R)$ and $\mathfrak{I}(A, B, R)$, respectively.

In addition, all the concepts together with the inclusion ordering on the left argument has the structure of a complete lattice, which is called *concept lattice* and it is denoted as $\mathcal{C}(A, B, R)$.

From now on, we will say that an *attribute-concept* is a concept generated by an attribute $a \in A$, that is $(a^{\downarrow}, a^{\downarrow\uparrow})$.

On the other hand, we need to recall the notion of meet-irreducible element of a lattice.

Definition 1. *Given a lattice $(L, \preceq)$, such that $\wedge$ is the meet operator, and an element $x \in L$ verifying*

1. *If L has a top element $\top$, then $x \neq \top$.*
2. *If $x = y \wedge z$, then $x = y$ or $x = z$, for all $y, z \in L$.*

we call x meet-irreducible ($\wedge$-irreducible) element *of L. Condition* (2) *is equivalent to*

2′. If $x < y$ and $x < z$, then $x < y \wedge z$, for all $y, z \in L$.

On the other hand, with respect to the attribute reduction in FCA, it is important to recall that when we reduce the set of attributes in a context, an equivalence relation on the set of concepts of the original concept lattice is induced. The following proposition was proved in [6] for the classical setting of FCA and it is recalled below.

Proposition 1 ([6]). *Given a context (A, B, R) and a subset $D \subseteq A$. The set $R_E = \{((X_1, Y_1), (X_2, Y_2)) \mid (X_1, Y_1), (X_2, Y_2) \in \mathcal{C}(A, B, R), X_1^{\uparrow_D\downarrow} = X_2^{\uparrow_D\downarrow}\}$ is an equivalence relation. Where $^{\uparrow_D}$ denotes the concept-forming operator $X^{\uparrow_D} = \{a \in D \mid \text{for all } x \in X, (a, x) \in R\}$ restricted to the subset of attributes $D \subseteq A$.*

In [6], the authors also proved that each equivalence class of the induced equivalence relation has a structure of join semilattice.

Proposition 2 ([6]). *Given a context (A, B, R), a subset $D \subseteq A$ and a class $[(X, Y)]_D$ of the quotient set $\mathcal{C}(A, B, R)/R_E$. The class $[(X, Y)]_D$ is a join semilattice with maximum element $(X^{\uparrow_D\downarrow}, X^{\uparrow_D\downarrow\uparrow})$.*

2.2 Local Congruences

The notion of local congruence arose with the goal of complementing attribute reduction in FCA. The purpose of local congruences is to obtain equivalence relations less-constraining than congruences [3] and with useful properties to be applied in size reduction processes of concept lattices. We recall the notion of local congruence in the next definition.

Definition 2. *Given a lattice $(L, \preceq)$, we say that an equivalence relation δ on L is a* local congruence *if each equivalence class of δ is a convex sublattice of L.*

The notion of local congruence can be characterized in terms of the equivalence relation, as the following result shows.

Proposition 3. *Given a lattice $(L, \preceq)$ and an equivalence relation δ on L, the relation δ is a local congruence on L if and only if, for each $a, b, c \in L$, the following properties hold:*

(i) If $(a, b) \in \delta$ and $a \preceq c \preceq b$, then $(a, c) \in \delta$.
(ii) $(a, b) \in \delta$ if and only if $(a \wedge b, a \vee b) \in \delta$.

Usually, we will look for a local congruence that contains a partition induced by an equivalence relation. When we say that a local congruence contain a partition provided by an equivalence relation, we are making use of the following definition of inclusion of equivalence relations.

Definition 3. *Let ρ_1 and ρ_2 be two equivalence relations on a lattice $(L, \preceq)$. We say that the equivalence relation ρ_1 is included in ρ_2, denoted as $\rho_1 \sqsubseteq \rho_2$, if for every equivalence class $[x]_{\rho_1} \in L/\rho_1$ there exists an equivalence class $[y]_{\rho_2} \in L/\rho_2$ such that $[x]_{\rho_1} \subseteq [y]_{\rho_2}$.*

3 Analyzing Local Congruences

In this section, we will present an initial study about the role of local congruences when they are used along or together with other mechanisms to attribute reduction. In particular, we will analyze the relationship between local congruences and the induced equivalence relation by an attribute reduction from the perspective of the attribute of the context as well as from the meet-irreducible elements of the concept lattices. We are interested in discovering under what conditions the induced equivalence relation is a local congruence. We are also interested in analyzing the influence of the use of local congruence in the reduction of attributes, when the induced equivalence relation is not a local congruence.

Firstly, in the first example we will illustrate the main idea of this study.

Example 1. Let us consider a formal context (A, B, R) composed of the attributes $A = \{a_1, a_2, a_3\}$ and the objects $B = \{b_1, b_2, b_3\}$, related by a relationship $R \subseteq A \times B$, which is shown in the left side of Table 1, together with the list of concepts which appears in the right side of the same table. The associated concept lattice is displayed in the left side of Fig. 1.

Table 1. Relation and list of concepts of the context of Example 1.

R	b_1	b_2	b_3
a_1	1	0	1
a_2	0	1	1
a_3	0	0	1

C_i	Extent			Intent		
	b_1	b_2	b_3	a_1	a_2	a_3
0	0	0	1	1	1	1
1	1	0	1	1	0	0
2	0	1	1	0	1	0
3	1	1	1	0	0	0

In order to analyze the influence of local congruences in the reduction of the set of attributes of the considered context, we include a list in which we show the attribute that generates each concept of the concept lattice:

$$C_0 = (a_3^{\downarrow}, a_3^{\downarrow\uparrow})$$
$$C_1 = (a_1^{\downarrow}, a_1^{\downarrow\uparrow})$$
$$C_2 = (a_2^{\downarrow}, a_2^{\downarrow\uparrow})$$

If we consider, for example, the subset $D_1 = \{a_2, a_3\}$ to carry out the reduction of the set of attributes, that is, we remove the attribute a_1, we obtain a partition of the concept lattice induced by the reduction that is highlighted by means of a dashed Venn diagram in the middle of Fig. 1. We obtain that the concepts C_1 and C_3 are grouped in the same class whereas the concepts C_0 and C_2 provide two different classes composed of a single concept each one. Therefore, according to Proposition 2, we can see that the obtained equivalence classes are

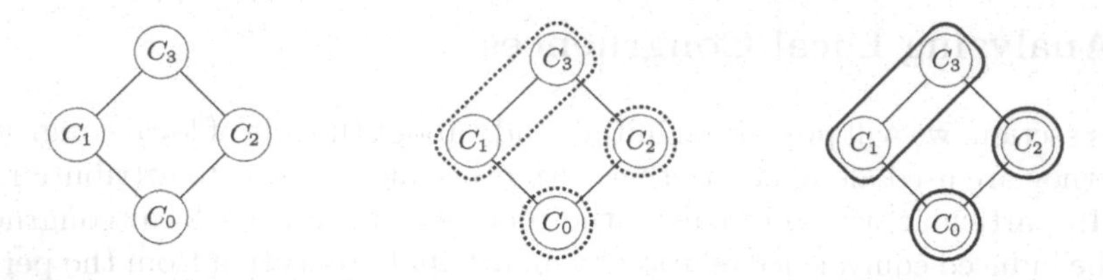

Fig. 1. Concept lattice of Example 1 (left), the partition induced by the subset D_1 (center) and the least local congruence containing the partition (right).

join semilattices. Indeed, all classes are convex sublattices of the original concept lattice.

As a consequence, the least local congruence containing such a reduction is the induced partition itself as it is shown in the right side of Fig. 1, where the local congruence is highlighted by means of a Venn diagram. In other words, the induced equivalence relation by the reduction is already a local congruence and, as a consequence, the consideration of local congruences does not alter the attribute reduction originally carried out on the set of attributes.

However, if the user decides to remove the attributes a_1 and a_2, that is, only the subset of attributes $D_2 = \{a_3\}$ is considered, the induced partition by the reduction is shown in the left side of Fig. 2.

Fig. 2. The partition induced by the elimination of the attributes a_1 and a_2 of Example 1 (left) and the least local congruence containing the induced partition (right).

The equivalence classes induced by the the reduction are the following:

$$[C_0]_{D_2} = \{C_0\}$$
$$[C_1]_{D_2} = [C_2]_{D_2} = [C_3]_{D_2} = \{C_1, C_2, C_3\}$$

In this case, the obtained equivalence classes are non-trivial join-semilattices since the concepts C_1, C_2 and C_3 do not form a convex sublattice of the original concept lattice. In this case, the infimum of the equivalence class $[C_1]_{D_2}$ is de concept C_0, which has not been included in $[C_1]_{D_2}$. This concept is not in $[C_1]_{D_2}$ because it is generated from the attribute a_3, which means that this attribute differences concept C_0 from the rest. Therefore, if this attribute is not removed in the reduction procedure, then it continues differentiating this concept from the rest and it cannot be in the same class of the rest.

If we compute the least local congruence containing the equivalence relation above, it groups all concepts in a single class, that is, the local congruence includes the infimum of the concepts C_1, C_2 and C_3, that is, the concept C_0, in the equivalence class $[C_1]_{D_2}$. This local congruence is depicted in the right side of Fig. 2. Clearly, this local congruence does not coincides with the equivalence relation induced by the attribute reduction which entails certain consequences with respect to the initial attribute reduction, since the inclusion of the concept C_0 in the equivalence class $[C_1]_{D_2}$, can be seen as a kind of elimination of the attribute a_3 (since the attribute a_3 generates the concept C_0). □

The previous example has shown different possibilities of applying local congruences for complementing an attribute reduction process. Hence, we have that some times the obtained equivalence relations is already a local congruence and other cases is not. In particular, we have seen a case that when the infimum of an induced equivalence class is generated by an attribute, which has not been removed during the reduction process, proper join semilattices arise and the induced equivalence relation is not a local congruence. In the following example, we will analyze another possible situations we can find when the set of attributes is reduced.

Example 2. We will consider a context composed of the set of attributes $A = \{a_1, a_2, a_3, a_4\}$ and the set of objects $B = \{b_1, b_2, b_3\}$, related by $R : A \times B \to \{0, 1\}$, defined on the left side of Table 2 together with the list of the corresponding concepts which appear in the right side of the same table. The associated concept lattice is given on the left side of Fig. 3.

Table 2. Relation and list of concepts of the context of Example 2.

R	b_1	b_2	b_3
a_1	1	1	0
a_2	1	0	1
a_3	0	1	1
a_4	0	0	1

C_i	Extent			Intent			
	b_1	b_2	b_3	a_1	a_2	a_3	a_4
0	0	0	0	1	1	1	1
1	1	0	0	1	1	0	0
2	0	1	0	1	0	1	0
3	0	0	1	0	1	1	1
4	1	1	0	1	0	0	0
5	1	0	1	0	1	0	0
6	0	1	1	0	0	1	0
7	1	1	1	0	0	0	0

From this context we obtain the following attribute-concepts:

$$C_3 = (a_4^{\downarrow}, a_4^{\downarrow\uparrow})$$
$$C_4 = (a_1^{\downarrow}, a_1^{\downarrow\uparrow})$$
$$C_5 = (a_2^{\downarrow}, a_2^{\downarrow\uparrow})$$
$$C_6 = (a_3^{\downarrow}, a_3^{\downarrow\uparrow})$$

For instance, if we are interested in considering the subset of attributes $D_1 = \{a_1, a_3\}$ and we carry out the corresponding reduction (removing the attributes a_2 and a_4), we obtain the partition induced by D_1, which is shown in the middle of Fig. 3. Once again, as in Example 1, the equivalence classes obtained from the reduction considering the subset D_1 are convex sublattices of the original concept lattice. Therefore, the least local congruence that contains such a reduction is the induced equivalence relation itself, as it can be seen in the right side of Fig. 3. Consequently, local congruences do not modify the considered attribute reduction.

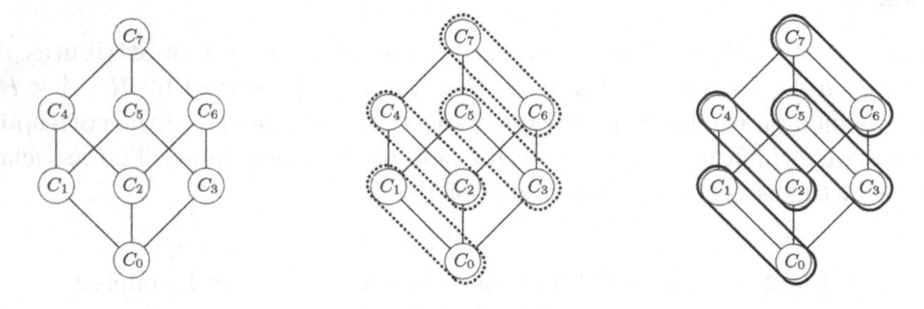

Fig. 3. Concept lattice of Example 2 (left), the partition induced by the subset D_1 (center) and the least local congruence containing the induced partition (right).

Now, if the attributes a_2 and a_3 are removed, i.e., only the subset of attributes $D_2 = \{a_1, a_4\}$ is considered, then the partition induced by the reduction is shown in the left side of Fig. 4 and the induced equivalence classes are listed below.

$$[C_0]_{D_2} = \{C_0\}$$
$$[C_1]_{D_2} = [C_2]_{D_2} = [C_4]_{D_2} = \{C_1, C_2, C_4\}$$
$$[C_3]_{D_2} = \{C_3\}$$
$$[C_5]_{D_2} = [C_6]_{D_2} = [C_7]_{D_2} = \{C_5, C_6, C_7\}$$

Notice that two of the obtained equivalence classes are not convex sublattices of the original concept lattice. The first one contains the concepts C_1, C_2, C_4 and the other one contains the concepts C_5, C_6, C_7. However, the reasons that make these classes are not convex sublattices are well differentiated.

On the one hand, with respect to the equivalence class of the concept $[C_5]_{D_2}$ we find a similar situation than the one shown in Example 1, that is, the infimum of the equivalence class $[C_5]_{D_2}$ is the concept C_3 which is generated from attribute a_4 that has not been removed in the reduction of the context.

On the other hand, the infimum of the equivalence class $[C_1]_{D_2}$ is the concept C_0 which is not generated by any attribute of the context. Nevertheless, $C_0 \notin [C_1]_{D_2}$ since in the decomposition of meet-irreducible concepts of the concept C_0, that is $C_0 = C_4 \wedge C_5 \wedge C_6$, we can find two meet-irreducible concepts C_5 and C_6 satisfying that $C_5, C_6 \notin [C_1]_{D_2}$.

In this case, the least local congruence whose equivalence classes contain the induced partition can be seen in the right side of Fig. 4. In this figure we have that the local congruence includes the infimum of the equivalence classes $[C_1]_{D_2}$ and $[C_5]_{D_2}$ in their respective classes. Thus, the least local congruence provides two different equivalence classes.

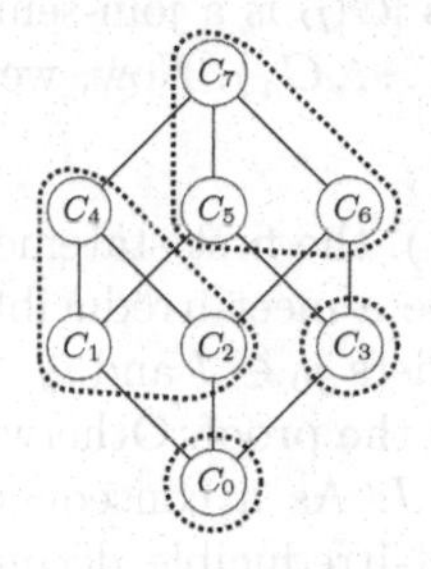

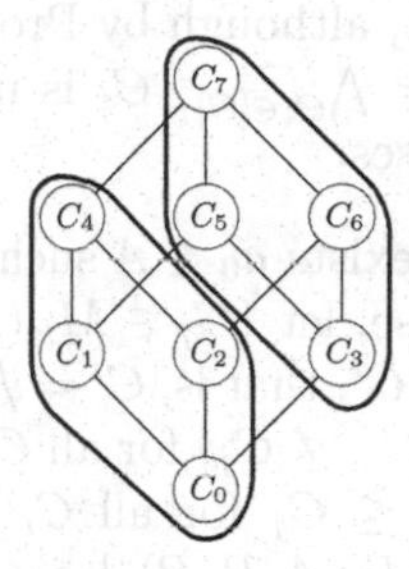

Fig. 4. The partition induced by the elimination of attributes a_2 and a_3 in Example 2 (left) and the least local congruence containing the induced partition (right).

Now, we will analyze how this local congruence influences in the reduction of the attributes. We can see that the inclusion of the concept C_3 in the equivalence $[C_5]_{D_2}$, is equivalent to the elimination of attribute a_4. We can also observe that the intension of the concept C_0 includes attribute a_4 which is ignored when C_0 is introduced in the equivalence class $[C_5]_{D_2}$. Hence, in spite of the reduction of the context was carried out originally from the elimination of attributes a_2 and a_3, somehow the consideration of the local congruence implies the elimination of attribute a_4. □

From the previous examples, we deduce that when the induced equivalence relation does not provide convex sublattices as equivalence classes, the use of local congruence relations alters the original attribute reduction, increasing the number of attributes to be removed. Moreover, it would be interesting to highlight these attributes, record its relationship with the removed attributes and the impact in attribute implications [4,5,9,19].

Next result relates the equivalence relations induced by an attribute reduction with the attributes-concepts and the meet-irreducible elements of the concept lattice. Due to the closely relation between the $\wedge$-irreducible concepts and the set of attributes of the context. This result summarizes the influence of local congruences in the attribute reduction of relational datasets.

Proposition 4. *Given a context (A, B, R), a subset of attributes $D \subseteq A$, an equivalence class $[C]_D$, with $C \in \mathcal{C}(A, B, R)$, of the induced equivalence relation which is not a convex sublattice and the concept $C' = \bigwedge_{C_i \in [C]_D} C_i$. Then, one of the following statements is satisfied:*

- *There exists at least one attribute $a \in A$ such that $C' = (a^{\downarrow}, a^{\downarrow\uparrow})$.*
- *There exists a concept $C^* \in M_F(A, B, R)$ in a meet-irreducible decomposition $\{C_j \in M_F(A, B, R) \mid j \in J\}$ of C', such as $C_{i_0} \not\leq C^*$ for a concept $C_{i_0} \in [C]_D$.*

Proof. Let us assume that we reduce the context (A, B, R), by considering a subset of attributes $D \subseteq A$, and that given $C \in \mathcal{C}(A, B, R)$, the induced equivalence class $[C]_D$ is not a convex sublattice of the original concept lattice.

Therefore, although by Proposition 2 the class $[C]_D$ is a join-semilattice, the concept $C' = \bigwedge_{C_i \in [C]_D} C_i$ is not in $[C]_D = \{C_1, \ldots, C_n\}$. Now, we will distinguish two cases:

(i) If there exists $a_0 \in A$ such that $C' = (a_0^{\downarrow}, a_0^{\downarrow\uparrow})$, the first statement holds.
(ii) Otherwise, let $\{C_j \in M_F(A, B, R) \mid j \in J\}$ be a meet-irreducible decomposition of C', that is, $C' = \bigwedge_{j \in J} C_j$. If there exists $j_0 \in J$ and $i_0 \in \{1, \ldots, n\}$, such as $C_{i_0} \not\leq C_{j_0}$ for all $C_i \in [C]_D$, we finish the proof. Otherwise, we have that $C_i \leq C_j$ for all $C_i \in [C]_D$ and $j \in J$. As a consequence, the set $\{C_j \in M_F(A, B, R) \mid j \in J\}$ is in the meet-irreducible decomposition of every concept in $[C]_D$, in particular in the maximum element of the class, denoted as C_M. Hence, we have that

$$C' < C_M \leq \bigwedge_{j \in J} C_j = C'$$

which leads us to a contradiction.

It is important to mention that the items exposed in the previous result are not exclusive, that is, we can find a concept C' satisfying simultaneously both conditions of the previous result. In this situation, this fact means that the intent of the concept C' has at least two different attributes, a_0 and a_1 such that these attributes do not belong to the intent of any concept $C_i \in [C]_D$ for all $i \in I$.

Notice also that the requirement "$C_{i_0} \not\leq C^*$ for a concept $C_{i_0} \in [C]_D$." in the second condition can be rewritten as C_{i_0} and C^* are incomparable, or $C^* < C_{i_0}$. This last inequality detects a possible non-distributivity lattice and discover the following consequences of Proposition 4.

Corollary 1. *Let (A, B, R) be a context where its concept lattice $\mathcal{C}(A, B, R)$ is distributive, $D \subseteq A$ a subset of attributes and $C \in M_F(A, B, R)$. If $C' = \bigwedge_{C_i \in [C]_D} C_i$ is not in $[C]_D$, then there exists an attribute $a \in A$ such that $C' = (a^{\downarrow}, a^{\downarrow\uparrow})$.*

These results show that the application of local congruences offers an advance and complemented procedure to reduce concept lattices, selecting and removing appropriate new attributes.

4 Conclusions and Future Work

In this paper, we have addressed an initial study about the relation between the equivalence classes provided by both an attribute reduction and the least local congruence containing such a reduction in FCA. In particular, we have analyzed more in detail the cases when the induced equivalence relation does not provide convex sublattices as equivalence classes and the behavior of the local congruence when we use it in these cases. As a consequence, we have observed that the use of local congruence relations modifies the subset of unconsidered attributes. Moreover, we have stated conditions on the attribute-concepts and the meet-irreducible elements of the concept lattice associated with a context in order to detect when an equivalence class is not a convex sublattice. All ideas presented in this paper have been illustrated by means of different examples.

As future work, we are interested in continuing the study of influence of local congruences in the attribute reduction of a dataset. For example, we will analyze the relationship of the use of local congruences with attribute implications and how the removed attributes can be recovered from the set of attribute implications associated with the context. Furthermore, we will explore the ideas presented in this paper in the fuzzy framework of the multi-adjoint concept lattices.

References

1. Alcalde, C., Burusco, A.: Study of the relevance of objects and attributes of L-fuzzy contexts using overlap indexes. In: Medina, J., et al. (eds.) IPMU 2018. CCIS, vol. 853, pp. 537–548. Springer, Cham (2018). https://doi.org/10.1007/978-3-319-91473-2_46
2. Antoni, L., Krajči, S., Krídlo, O.: Constraint heterogeneous concept lattices and concept lattices with heterogeneous hedges. Fuzzy Sets Syst. **303**, 21–37 (2016)
3. Aragón, R.G., Medina, J., Ramírez-Poussa, E.: Weaken the congruence notion to reduce concept lattices. Stud. Comput. Intell. 1–7 (2019, in press)
4. Belohlavek, R., Cordero, P., Enciso, M., Mora, A., Vychodil, V.: Automated prover for attribute dependencies in data with grades. Int. J. Approx. Reason. **70**, 51–67 (2016)
5. Belohlávek, R., Vychodil, V.: Attribute dependencies for data with grades II. Int. J. Gen Syst **46**(1), 66–92 (2017)
6. Benítez-Caballero, M.J., Medina, J., Ramírez-Poussa, E., Ślęzak, D.: A computational procedure for variable selection preserving different initial conditions. Int. J. Comput. Math. **97**(1–2), 387–404 (2019)
7. Chen, J., Mi, J., Xie, B., Lin, Y.: A fast attribute reduction method for large formal decision contexts. Int. J. Approx. Reason. **106**, 1–17 (2019)
8. Cornejo, M.E., Medina, J., Ramírez-Poussa, E.: Attribute reduction in multi-adjoint concept lattices. Inf. Sci. **294**, 41–56 (2015)

9. Cornejo, M.E., Medina, J., Ramírez-Poussa, E.: Computing the validity of attribute implications in multi-adjoint concept lattices. In: International Conference on Computational and Mathematical Methods in Science and Engineering, CMMSE 2016, Vol. II, pp. 414–423 (2016)
10. Cornejo, M.E., Medina, J., Ramírez-Poussa, E.: Attribute and size reduction mechanisms in multi-adjoint concept lattices. J. Comput. Appl. Math. **318**, 388–402 (2017). Computational and Mathematical Methods in Science and Engineering CMMSE-2015
11. Cornejo, M.E., Medina, J., Ramírez-Poussa, E.: Characterizing reducts in multi-adjoint concept lattices. Inf. Sci. **422**, 364–376 (2018)
12. Ganter, B., Wille, R.: Formal Concept Analysis: Mathematical Foundation. Springer, Heidelberg (1999). https://doi.org/10.1007/978-3-642-59830-2
13. Konecny, J., Krajča, P.: On attribute reduction in concept lattices: the polynomial time discernibility matrix-based method becomes the CR-method. Inf. Sci. **491**, 48–62 (2019)
14. Li, J., Kumar, C.A., Mei, C., Wang, X.: Comparison of reduction in formal decision contexts. Int. J. Approx. Reason. **80**, 100–122 (2017)
15. Liñeiro-Barea, V., Medina, J., Medina-Bulo, I.: Generating fuzzy attribute rules via fuzzy formal concept analysis. In: Kóczy, L.T., Medina, J. (eds.) Interactions Between Computational Intelligence and Mathematics. SCI, vol. 758, pp. 105–119. Springer, Cham (2018). https://doi.org/10.1007/978-3-319-74681-4_7
16. Medina, J.: Relating attribute reduction in formal, object-oriented and property-oriented concept lattices. Comput. Math. Appl. **64**(6), 1992–2002 (2012)
17. Ren, R., Wei, L.: The attribute reductions of three-way concept lattices. Know.-Based Syst. **99**(C), 92–102 (2016)
18. Shao, M.-W., Li, K.-W.: Attribute reduction in generalized one-sided formal contexts. Inf. Sci. **378**, 317–327 (2017)
19. Vychodil, V.: Computing sets of graded attribute implications with witnessed non-redundancy. Inf. Sci. **351**, 90–100 (2016)

Towards a Classification of Rough Set Bireducts

M. José Benítez-Caballero(✉), Jesús Medina, and Eloísa Ramírez-Poussa

Universidad de Cádiz, Cádiz, Spain
{mariajose.benitez,jesus.medina,eloisa.ramirez}@uca.es

Abstract. Size reduction mechanisms are very important in several mathematical fields. In rough set theory, bireducts arose to reduce simultaneously the set of attributes and the set of objects of the considered dataset, providing subsystems with the minimal sets of attributes that connect the maximum number of objects preserving the information of the original dataset. This paper presents the main properties of bireducts and how they can be used for removing inconsistencies.

Keywords: Rough set theory · Bireducts · Size reduction

1 Introduction

Rough Set Theory (RST) is one of the most useful mathematical tools to treat and manage datasets. In particular, RST was proposed by Pawlak in [7], to analyze datasets containing incomplete information. The main idea of this theory is to determine a set from two approximations. These approximations are called upper approximation and lower approximation.

In this theory, a relational database can be represented from two different point of view, as an information system or as a decision system. In the case of information system, the database is simulated by a set of objects and a set of attributes characterizing the objects. On the other hand, a decision system is a particular case of information system adding a new attribute that describes an action over the objects, which is called decision attribute.

Due to the size of databases has increased in late decades, size reduction mechanisms became into one of the main goals of different mathematical theories. In the particular case of RST, a reduct is a minimal subset of attributes preserving the same knowledge as the original set. This notion is deeply studied in many papers [4–6,11,12]. Therefore, reducts are focused on the reduction of the set of attributes. In order to reduce also the set of objects, the notion of

Partially supported by the the 2014-2020 ERDF Operational Programme in collaboration with the State Research Agency (AEI) in project TIN2016-76653-P, and with the Department of Economy, Knowledge, Business and University of the Regional Government of Andalusia. in project FEDER-UCA18-108612, and by the European Cooperation in Science & Technology (COST) Action CA17124.

M.-J. Lesot et al. (Eds.): IPMU 2020, CCIS 1239, pp. 759–770, 2020.
https://doi.org/10.1007/978-3-030-50153-2_56

bireduct arose [1–3,9,10]. In a general point of view, the main idea underlying bireducts is to choose the maximal consistent information subsystem. Moreover, as all the bireducts are computed, the user can choose the bireduct consistent subsystem that best suits their needs.

In this paper, we study some properties of bireducts. We will prove that the set of reducts can be obtained from the set of bireducts. We will analyze the relation between bireducts and the discernibility classes of the objects of the dataset. We will also inspect how bireducts can be used for detecting inconsistencies contained in the considered database. The presented study will be carried out for information systems, as well as for decision systems. All the presented results will be illustrated by means of examples.

The paper is organized as follows: the notions and results needed in this study are recalled in Sect. 2. Afterwards, Sect. 3 presents the contribution of this paper together with some examples. Finally, the conclusions and future works are presented in Sect. 4.

2 Preliminaries

This section recalls the main notions associated with information and decision systems and the characterizations of reducts and bireducts. More detailed information related to these notions can be found in [3].

2.1 Information Systems

We will recall the notions and results needed to carry out the attribute reduction in information systems. First of all, we present the definition of information system and the considered indiscernibility relation.

Definition 1. *An* information system $(U, \mathcal{A})$ *is a tuple, where* $U = \{x_1, x_2, \ldots, x_n\}$ *and* $\mathcal{A} = \{a_1, a_2, \ldots, a_m\}$ *are finite, non-empty sets of objects and attributes, respectively. Each* $a \in \mathcal{A}$ *corresponds to a mapping* $\bar{a} : U \to V_a$*, where* V_a *is the value set of the attribute a over U. For every subset D of* $\mathcal{A}$*, the D-*indiscernibility relation, *Ind(D), is defined by the following equivalence relation*

$$Ind(D) = \{(x_i, x_j) \in U \times U \mid \text{ for all } a \in D, \bar{a}(x_i) = \bar{a}(x_j)\}$$

where each equivalence class is written as $[x]_D = \{x_i \in U \mid (x, x_i) \in Ind(D)\}$*. These equivalence classes are called* indiscernibility class. *Ind(D) provides a partition on U denoted as* $U/Ind(D) = \{[x]_D \mid x \in U\}$*.*

In order to be able to reduce an information system, the notions of consistent set and reduct are fundamental.

Definition 2. *Let* $(U, \mathcal{A})$ *be an information system and a subset of attributes* $D \subseteq \mathcal{A}$*. The subset D is a* consistent set *of* $(U, \mathcal{A})$ *if* $Ind(D) = Ind(\mathcal{A})$*. Moreover, if for each* $a \in D$ *we have that* $Ind(D \setminus \{a\}) \neq Ind(\mathcal{A})$*, then D is a* reduct *of* $(U, \mathcal{A})$*.*

The following definition presents the idea of discernibility matrix and discernibility function. In particular, the discernibility matrix is a useful tool which is used to represent the attributes in which the objects differ.

Definition 3. *Given an information system* $(U, \mathcal{A})$, *its* discernibility matrix *is a matrix with order* $|U| \times |U|$, *denoted by* $M_{\mathcal{A}}$, *in which the element* $M_{\mathcal{A}}(x, y)$ *for each pair of objects* (x, y) *is defined by:*

$$M_{\mathcal{A}}(x, y) = \{a \in \mathcal{A} \mid \bar{a}(x_i) \neq \bar{a}(x_j)\}$$

and the discernibility function *of* $(U, \mathcal{A})$ *is defined by:*

$$\tau_{\mathcal{A}}^{uni} = \bigwedge \left\{ \bigvee (M_{\mathcal{A}}(x_i, x_j)) \mid x_i, x_j \in U \text{ and } M_{\mathcal{A}}(x_i, x_j) \neq \varnothing \right\}$$

The discernibility function of an information system is a powerful tool which is used in the following result in order to describe a method to obtain all reducts from an information system [3,8].

Theorem 1. *Let* $(U, \mathcal{A})$ *be a boolean information system. An arbitrary set* D, *where* $D \subseteq \mathcal{A}$, *is a reduct of the information system if and only if the cube* $\bigwedge_{a \in D} a$ *is a cube in the restricted disjunctive normal form.*

Next, we introduce an example, which will be developed throughout the paper.

Example 1. Let us consider the information system $(U, \mathcal{A})$, where the set of objects represents six patients $U = \{1, 2, 3, 4, 5, 6\}$, the set of attributes $\mathcal{A} = \{\text{fever(f)}, \text{cough(c)}, \text{tonsil inflam.(t)}, \text{muscle ache(a)}\}$ and the relation between them is shown in Table 1.

Table 1. Relation of Example 1.

R	fever(f)	cough(c)	tonsil inflam.(t)	muscle ache(a)
1	High	No	No	No
2	High	No	Yes	Yes
3	Low	Yes	Yes	No
4	Low	Yes	Yes	No
5	High	Yes	Yes	Yes
6	No	Yes	Yes	No

If we compare the objects, considering the indiscernibility relation presented in Definition 1, we can build the following discernibility matrix:

$$\begin{pmatrix} \varnothing & & & & & \\ \{t,a\} & \varnothing & & & & \\ \{f,c,t\} & \{f,c,a\} & \varnothing & & & \\ \{f,c,t\} & \{f,c,a\} & \varnothing & \varnothing & & \\ \{c,t,a\} & \{c\} & \{f,a\} & \{f,a\} & \varnothing & \\ \{f,c,t\} & \{f,c,a\} & \{f\} & \{f\} & \{f,a\} & \varnothing \end{pmatrix} \quad (1)$$

Now, we will use the discernibility matrix to build the unidimensional discernibility function:

$$\begin{aligned} \tau^{\text{uni}} &= \{t \vee a\} \wedge \{f \vee c \vee t\} \wedge \{c \vee t \vee a\} \wedge \{f \vee c \vee a\} \wedge \{c\} \wedge \{f \vee a\} \wedge \{f\} \\ &= \{f \wedge c \wedge a\} \vee \{f \wedge c \wedge t\} \end{aligned}$$

Therefore, by Theorem 1, we obtain two reducts:

$$\begin{aligned} D_1 &= \{\text{fever, cough, muscle ache}\} \\ D_2 &= \{\text{fever, cough, tonsil inflam.}\} \end{aligned}$$

□

The idea of bireducts arose as a path to prevent incompatibilities and eliminate noise in the original data by means of a reduction in the set of objects and the set of attributes, simultaneously.

Definition 4. *Given an information system $(U, \mathcal{A})$, we consider a pair (X, D), where $X \in U$ is a subset of objects and $D \in \mathcal{A}$ is a subset of attributes. We say that (X, D) is an information bireduct if and only if every pair of objects $i, j \in X$ are discernible by D and the following properties hold:*

- *There is no subset $C \subsetneq D$ such that C discerns every pair of objects of X.*
- *There is no subset of objects $X \subsetneq Y$ such that D discern every pair of objects of Y.*

Since we will work simultaneously with reducts and bireducts of an information system, we will use the notation (X, B) to denote bireducts in order to distinguish the subset of attributes from reducts and bireducts.

In order to generalize the mechanism to obtain reducts presented in Theorem 1, we need to improve the idea of discernibility function as follows:

$$\tau_{\mathcal{A}}^{\text{bi}} = \bigwedge \left\{ x_i \vee x_j \vee \bigvee (M(x_i, x_j)) \mid \text{ for all } x_i, x_j \in U, M(x_i, x_j) \neq \emptyset \right\}$$

Now, we can introduce the following theorem, in which a mechanism to obtain all the bireducts of an information system is presented.

Theorem 2 ([3])**.** *Given a boolean information system $(U, \mathcal{A})$. An arbitrary pair of sets (X, D), where $X \subseteq U$, $D \subseteq \mathcal{A}$, is a bireduct of the information system if and only if the cube $\bigwedge_{a \in D} a \wedge \bigwedge_{x_i \notin X} x_i$ is a cube in the restricted disjunctive normal form (RDNF) of $\tau_{U,\mathcal{A}}^{bi}$.*

Example 2. We are going to compute all the bireducts from the information system described in Example 1. In order to do that, we consider the discernibility matrix described in Expression (1), obtaining the following bidimensional discernibility function:

$$\begin{aligned}\tau^{\text{bi}} = {}& \{1 \vee 2 \vee \mathrm{t} \vee \mathrm{a}\} \wedge \{1 \vee 3 \vee \mathrm{f} \vee \mathrm{c} \vee \mathrm{t}\} \wedge \{1 \vee 4 \vee \mathrm{f} \vee \mathrm{c} \vee \mathrm{t}\} \wedge \{1 \vee 5 \vee \mathrm{c} \vee \mathrm{t} \vee \mathrm{a}\}\\ & \wedge \{1 \vee 6 \vee \mathrm{f} \vee \mathrm{c} \vee \mathrm{t}\} \wedge \{2 \vee 3 \vee \mathrm{f} \vee \mathrm{c} \vee \mathrm{a}\} \wedge \{2 \vee 4 \vee \mathrm{f} \vee \mathrm{c} \vee \mathrm{a}\} \wedge \{2 \vee 5 \vee \mathrm{c}\}\\ & \wedge \{2 \vee 6 \vee \mathrm{f} \vee \mathrm{c} \vee \mathrm{a}\} \wedge \{3 \vee 5 \vee \mathrm{f} \vee \mathrm{a}\} \wedge \{3 \vee 6 \vee \mathrm{f}\} \wedge \{4 \vee 5 \vee \mathrm{f} \vee \mathrm{a}\} \wedge \{4 \vee 6 \vee \mathrm{f}\}\\ & \wedge \{5 \vee 6 \vee \mathrm{f} \vee \mathrm{a}\}\end{aligned}$$

Now, we compute the reduced disjunctive normal form of τ^{bi} obtaining

$$\begin{aligned}= {}& \{1 \wedge 2 \wedge \mathrm{f}\} \vee \{1 \wedge 5 \wedge \mathrm{f}\} \vee \{1 \wedge \mathrm{f} \wedge \mathrm{c}\} \vee \{2 \wedge 5 \wedge \mathrm{f}\} \vee \{2 \wedge \mathrm{f} \wedge \mathrm{c}\} \vee \{2 \wedge \mathrm{f} \wedge \mathrm{t}\}\\ & \vee \{2 \wedge \mathrm{f} \wedge \mathrm{a}\} \vee \{5 \wedge \mathrm{f} \wedge \mathrm{t}\} \vee \{5 \wedge \mathrm{f} \wedge \mathrm{a}\} \vee \{6 \wedge \mathrm{c} \wedge \mathrm{a}\} \vee \{\mathrm{f} \wedge \mathrm{c} \wedge \mathrm{t}\} \vee \{\mathrm{f} \wedge \mathrm{c} \wedge \mathrm{a}\}\\ & \vee \{1 \wedge 2 \wedge 5 \wedge 6\} \vee \{1 \wedge 2 \wedge 6 \wedge \mathrm{a}\} \vee \{1 \wedge 5 \wedge 6 \wedge \mathrm{c}\} \vee \{1 \wedge 5 \wedge 6 \wedge \mathrm{a}\}\\ & \vee \{2 \wedge 3 \wedge 5 \wedge \mathrm{c}\} \vee \{2 \wedge 3 \wedge \mathrm{c} \wedge \mathrm{a}\} \vee \{2 \wedge 5 \wedge 6 \wedge \mathrm{c}\} \vee \{2 \wedge 5 \wedge 6 \wedge \mathrm{t}\}\\ & \vee \{2 \wedge 6 \wedge \mathrm{t} \wedge \mathrm{a}\} \vee \{5 \wedge 6 \wedge \mathrm{c} \wedge \mathrm{t}\} \vee \{4 \wedge \mathrm{c} \wedge \mathrm{t} \wedge \mathrm{a}\} \vee \{5 \wedge 6 \wedge \mathrm{t} \wedge \mathrm{a}\}\\ & \vee \{3 \wedge 4 \wedge \mathrm{c} \wedge \mathrm{a}\} \vee \{2 \wedge 3 \wedge \mathrm{f} \wedge \mathrm{a}\} \vee \{1 \wedge 2 \wedge 3 \wedge 4 \wedge 5\} \vee \{1 \wedge 2 \wedge 3 \wedge 4 \wedge 6\}\\ & \vee \{1 \wedge 2 \wedge 3 \wedge 4 \wedge \mathrm{a}\} \vee \{1 \wedge 3 \wedge 4 \wedge 5 \wedge 6\} \vee \{1 \wedge 3 \wedge 4 \wedge 5 \wedge \mathrm{c}\}\\ & \vee \{1 \wedge 3 \wedge 4 \wedge 5 \wedge \mathrm{a}\} \vee \{1 \wedge 3 \wedge 4 \wedge 6 \wedge \mathrm{c}\} \vee \{2 \wedge 3 \wedge 4 \wedge 5 \wedge 6\}\\ & \vee \{2 \wedge 3 \wedge 4 \wedge 5 \wedge \mathrm{c}\} \vee \{2 \wedge 3 \wedge 4 \wedge 5 \wedge \mathrm{t}\} \vee \{2 \wedge 3 \wedge 4 \wedge 6 \wedge \mathrm{a}\}\\ & \vee \{2 \wedge 3 \wedge 4 \wedge 6 \wedge \mathrm{c}\} \vee \{2 \wedge 3 \wedge 4 \wedge 6 \wedge \mathrm{t}\} \vee \{2 \wedge 3 \wedge 4 \wedge \mathrm{t} \wedge \mathrm{a}\}\\ & \vee \{3 \wedge 4 \wedge 5 \wedge 6 \wedge \mathrm{t}\} \vee \{3 \wedge 4 \wedge 5 \wedge \mathrm{c} \wedge \mathrm{t}\} \vee \{3 \wedge 4 \wedge 5 \wedge \mathrm{t} \wedge \mathrm{a}\}\\ & \vee \{3 \wedge 4 \wedge 6 \wedge \mathrm{c} \wedge \mathrm{t}\} \vee \{3 \wedge 4 \wedge 5 \wedge 6 \wedge \mathrm{a}\} \vee \{3 \wedge 4 \wedge 6 \wedge \mathrm{a} \wedge \mathrm{c}\}\end{aligned}$$

Therefore, there are 47 bireducts in the information system $(U, \mathcal{A})$. Some of them are described in Table 2 □

Table 2. Some bireducts of information system $(U, \mathcal{A})$ of Example 1

Bireduct	Subset of objects	Subset of attributes
(X_1, B_1)	$\{1, 2, 3, 4, 5, 6\}$	$\{\mathrm{f}, \mathrm{c}, \mathrm{t}\}$
(X_2, B_2)	$\{1, 2, 3, 4, 5, 6\}$	$\{\mathrm{f}, \mathrm{c}, \mathrm{a}\}$
(X_3, B_3)	$\{3, 4, 5, 6\}$	$\{\mathrm{f}\}$
(X_4, B_4)	$\{2, 3, 4, 5, 6\}$	$\{\mathrm{f}, \mathrm{c}\}$
(X_5, B_5)	$\{3, 4\}$	$\varnothing$
(X_6, B_6)	$\{1\}$	$\varnothing$
(X_7, B_7)	$\{2\}$	$\varnothing$
(X_8, B_8)	$\{5\}$	$\varnothing$
(X_9, B_9)	$\{6\}$	$\varnothing$

2.2 Decision Systems

In this section, we recall the main notions and results we will need in the framework of decision systems. First of all, we present the formal definition of a decision system.

Definition 5. *A* decision system $(U, \mathcal{A} \cup \{d\})$ *is a special kind of information system, in which* $d \notin \mathcal{A}$ *is called the decision attribute, and its equivalence class* $[x]_d$ *is called decision class.*

In this framework the role of reduct is a little bit different, since only objects with different decision attribute values are compared.

Definition 6. *The subset* $B \subseteq \mathcal{A}$ *is called a* decision reduct *for the decision system* $(U, \mathcal{A} \cup \{d\})$ *if it is an irreducible subset, such that* B *discerns all pairs* $x_i, x_j \in U$ *satisfying* $d(x_i) \neq d(x_j)$.

As we did in an information system, we use the notions of discernibility matrix an function in order to compute all reducts [8]. Therefore, the discernibility matrix of a decision system $(U, \mathcal{A} \cup \{d\})$ is a square and symmetric matrix defined as:

$$M(x_i, x_j) = \begin{cases} \varnothing & \text{if } d(x_i) = d(x_j) \\ \{a \in \mathcal{A} \mid a(x_i) \neq a(x_j)\} & \text{otherwise} \end{cases} \tag{2}$$

Therefore, there are two possibilities of obtaining the empty set: objects have the same decision value or are indiscernible by characteristic attributes. In addition, the *discernibility function* of $(U, \mathcal{A} \cup \{d\})$ is the map $\tau \colon \{0,1\}^m \to \{0,1\}$, defined by

$$\tau^{\text{uni}} = \bigwedge \Big\{ \bigvee M^*(x_i, x_j) \mid 1 \leq i < j \leq n \text{ and } M(x_i, x_j) \neq \varnothing \Big\} \tag{3}$$

It can be shown that the prime implicants of f constitute exactly all the decision reducts of $(U, \mathcal{A} \cup \{d\})$, as the generalization of Theorem 1 to a decision system.

Now, we present the definition of decision bireduct of a decision system:

Definition 7. *Given a decision system* $(U, A \cup \{d\})$*, the pair* (B, X)*, where* $B \subset A$ *and* $X \subset U$*, is called* decision bireduct *if and only if* B *discern every pair* $x_i, x_j \in X$ *where* $d(x_i) \neq d(x_j)$ *and the following properties are verified:*

1. *There is no subset* $C \subsetneq B$ *such that* C *discern every pair* $x_i, x_j \in X$ *where* $d(x_i) \neq d(x_j)$.
2. *There is no subset* $X \subsetneq Y$ *such that* B *discern every pair* $x_i, x_j \in Y$ *with* $d(x_i) \neq d(x_j)$.

In order to generalize the process to compute all bireducts, we consider the discernibility function:

$$\tau^{bir} = \bigwedge_{x_i, x_j \in U \mid d(x_i) \neq d(x_j)} \Big(x_i \vee x_j \vee \bigvee \{a \in A \mid a(x_i) \neq a(x_j)\} \Big) \tag{4}$$

The corresponding characterization theorem for bireducts from the discernibility function is as follows.

Theorem 3 ([3]). *Given a boolean information system $(U, \mathcal{A} \cup \{d\})$. An arbitrary pair of sets (X, D), where $X \subseteq U$, $D \subseteq \mathcal{A}$, is a decision bireduct of a decision system if and only if the cube $\bigwedge_{a \in D} a \wedge \bigwedge_{x_i \notin X} x_i$ is a cube in the restricted disjunctive normal form (RDNF) of τ^{bi}.*

Now that all the needed notions and results have been presented, we present the different kinds of bireducts that we can find in an information and decision systems.

3 Threshing Bireducts

This section highlights the main properties of bireducts. First of all, the following result shows that the reducts of an information system are particular cases of bireducts.

Proposition 1. *Let $(U, \mathcal{A})$ an information system and $(\mathcal{X}, \mathcal{B})$ the family of bireducts from the information system. If a bireduct (X, B) verifies that $X = \mathcal{A}$, then B is a reduct from the information system.*

Proof. This proof is straightly obtained from Definitions 2 and 4.

Moreover, we can assert that the decision reducts of a decision system are also decision bireducts due to the definitions of these notions.

The following example illustrates this result by means of the information system given in Example 1.

Example 3. Let us focus in the bireduct (X_1, B_1) and (X_2, B_2) obtained in Example 2, due to they have the whole set of objects. If we compare the set of attributes from the bireducts and the set of attributes described by the reducts of Example 1, we have that

$$D_1 = B_2 = \{\text{fever}, \text{cough}, \text{muscle ache}\}$$
$$D_2 = B_1 = \{\text{fever}, \text{cough}, \text{tonsil inflam.}\}$$

□

A special type of bireduct appears when the subset of attributes is the empty set, that is, there are no attribute to distinguish the elements in the subset of objects. Therefore, the objects are indiscernible. The following result formalizes this idea.

Proposition 2. *Let $(U, \mathcal{A})$ an information system and $(\mathcal{X}, \mathcal{B})$ the family of bireducts from the information system. If a bireduct (X, B) verifies that $B = \varnothing$, then X is the indiscernibility class of the objects $x \in X$.*

Proof. As (X, B) is a bireduct and $B = \varnothing$, we have that there is no attribute distinguishing all objects in the set X. Therefore, for every object $x \in X$, we have that $\bar{a}(x) = \bar{a}(x_j)$, for all attribute $a \in \mathcal{A}$ and any object $x_j \in X$. Consequently,

$$x \in \{x_j \in U \mid \text{ for all } a \in B, \bar{a}(x) = \bar{a}(x_j)\}$$

which is the definition of $[x]_{\mathcal{A}}$ presented in Definition 1. Therefore, $X \subseteq [x]_{\mathcal{A}}$. By the maximality of X, we obtain that X must be $[x]_{\mathcal{A}}$. □

In this example, we present the connection between the indiscernibility classes of the objects in an information system and the bireducts with no attributes.

Example 4. Let us continue the study of the information system in Example 1. If we consider the bireducts (X_i, B_i), with $i \in \{5, \dots, 9\}$, we have that $B_i = \varnothing$, for all $i \in \{5, \dots, 9\}$.

On the other hand, if we compute the indiscernibility classes of the objects of the considered information system, we obtain that:

$$[1]_{\mathcal{A}} = \{1\} \qquad [5]_{\mathcal{A}} = \{5\}$$
$$[2]_{\mathcal{A}} = \{2\} \qquad [6]_{\mathcal{A}} = \{6\}$$
$$[3]_{\mathcal{A}} = [4]_{\mathcal{A}} = \{3, 4\}$$

Comparing these subsets of objects with the bireducts (X_i, B_i), with $i \in \{5, \dots, 9\}$, we obtain a correspondence between the sets X_i, with $i \in \{5, \dots, 9\}$ and the indiscernibility classes $[x]_{\mathcal{A}}$, for all $x \in U$. □

In the particular case of a decision system, if the subset of attributes of a birreduct is empty, the subset of objects of that bireduct is the decision class provided by the decision attribute.

Proposition 3. *Given a decision system $(U, \mathcal{A} \cup \{d\})$ and its family of decision bireducts $(\mathcal{X}, \mathcal{B})$, if a bireduct (X, B) verifies that $B = \varnothing$, then $X = [x]_d$, for all $x \in X$.*

Proof. By definition of bireduct in a decision system, $B = \varnothing$ if and only if the value of the decision attribute is the same for all the objects in X. In this case, since two objects with the same decision attribute value are not compared further, the assumption $B = \varnothing$ automatically implies that all objects are in the same decision class, that is $X = [x]_d$, for every object $x \in X$.

Moreover, bireducts remove inconsistencies in the data, that is the cases when two objects have different decision value, but they are indiscernible by the attributes in $\mathcal{A}$.

Proposition 4. *Given a decision system $(U, \mathcal{A} \cup \{d\})$, if $x, y \in U$ satisfy that $d(x) \neq d(y)$ and $\bar{a}(x) = \bar{a}(y)$, for all $a \in \mathcal{A}$, then x and y do not belong to the same subset X, for any bireduct (X, B).*

Proof. By the definition of discernibility function of decision systems, given in Expression 4, if $x, y \in U$, such that $d(x) \neq d(y)$, the conjunctive normal form τ^{bir} will contain the clause

$$x \vee y \vee \bigvee \{a \in A \mid \bar{a}(x) \neq \bar{a}(y)\}$$

Since $\bar{a}(x) = \bar{a}(y)$, for all $a \in \mathcal{A}$, the set $\{a \in A \mid \bar{a}(x) \neq \bar{a}(y)\}$ is empty and so, the clause is $x \vee y$.

Therefore, every cube of the obtained reduced disjunctive normal form will contain x or y. As a consequence, by Theorem 3, we have that, for every bireduct (X, B), the set X cannot contain x and y simultaneously, which proves the result. □

As a consequence of this result, all bireducts are consistent and so, the obtained information from these subsystems is also consistent. The following example illustrates the previous notions and results in the particular case of a decision system.

Example 5. From the information system $(U, \mathcal{A})$ in Example 1, we will add a decision attribute. This decision attribute will represent whether a patient has flu or not. The relation is shown in Table 3.

Table 3. Relation of Example 5.

R	fever(f)	cough(c)	tonsil inflam.(t)	muscle ache(a)	flu?
1	High	No	No	No	No
2	High	No	Yes	Yes	Yes
3	Low	Yes	Yes	No	No
4	Low	Yes	Yes	No	Yes
5	High	Yes	Yes	Yes	Yes
6	No	Yes	Yes	No	No

As we can see in Table 3, the objects 3 and 4 have different values in the decision attribute but the values of these objects coincide for the rest of the attributes. Therefore, objects 3 and 4 represent an inconsistency in the data. Now, if we compare the objects considering the indiscernibility relation presented in Definition 1, we can build the following discernibility matrix:

$$\begin{pmatrix} \varnothing & & & & & \\ \{\mathrm{t},\mathrm{a}\} & \varnothing & & & & \\ \varnothing & \{\mathrm{f},\mathrm{c},\mathrm{a}\} & \varnothing & & & \\ \{\mathrm{f},\mathrm{c},\mathrm{t}\} & \varnothing & \varnothing & \varnothing & & \\ \{\mathrm{c},\mathrm{t},\mathrm{a}\} & \varnothing & \{\mathrm{f},\mathrm{a}\} & \varnothing & \varnothing & \\ \varnothing & \{\mathrm{f},\mathrm{c},\mathrm{a}\} & \varnothing & \{\mathrm{f}\} & \{\mathrm{f},\mathrm{a}\} & \varnothing \end{pmatrix} \tag{5}$$

There exist two cases to obtain the empty set as an element of the discernibility matrix: the objects have the same value in the decision attribute or, having different values in the decision attribute, the objects are indiscernible. Therefore, considering the discernibility matrix, we obtain the unidimensional discernibility function:

$$\begin{aligned} \tau^{\mathrm{uni}} &= \{\mathrm{t} \vee \mathrm{a}\} \wedge \{\mathrm{f} \vee \mathrm{c} \vee \mathrm{t}\} \wedge \{\mathrm{c} \vee \mathrm{t} \vee \mathrm{a}\} \wedge \{\mathrm{f} \vee \mathrm{c} \vee \mathrm{a}\} \wedge \{\mathrm{f} \vee \mathrm{a}\} \wedge \{\mathrm{f}\} \\ &= \{\mathrm{f} \wedge \mathrm{a}\} \vee \{\mathrm{f} \wedge \mathrm{t}\} \end{aligned}$$

Therefore, we obtain two decision reducts:

$$\begin{aligned} D_1 &= \{\text{fever}, \text{muscle ache}\} \\ D_2 &= \{\text{fever}, \text{tonsil inflam.}\} \end{aligned} \tag{6}$$

Now, we will compute all bireducts of the decision system $(U, \mathcal{A} \cup \{d\})$ from Theorem 2, that is, throughout the following bidimensional discernibility function.

$$\tau^{\text{bi}} = \{1 \vee 2 \vee \text{t} \vee \text{a}\} \wedge \{1 \vee 4 \vee \text{f} \vee \text{c} \vee \text{t}\} \wedge \{1 \vee 5 \vee \text{c} \vee \text{t} \vee \text{a}\} \wedge \{2 \vee 3 \vee \text{f} \vee \text{c} \vee \text{a}\} \\ \wedge \{2 \vee 6 \vee \text{f} \vee \text{c} \vee \text{a}\} \wedge \{3 \vee 4\} \wedge \{3 \vee 5 \vee \text{f} \vee \text{a}\} \wedge \{4 \vee 6 \vee \text{f}\} \wedge \{5 \vee 6 \vee \text{f} \vee \text{a}\}$$

From the formula above, the reduced disjunctive normal form is computed.

$$\tau^{\text{bi}} = \{4 \wedge \text{a}\} \vee \{1 \wedge 3 \wedge \text{f}\} \vee \{1 \wedge 4 \wedge \text{f}\} \vee \{3 \wedge \text{f} \wedge \text{t}\} \vee \{4 \wedge \text{f} \wedge \text{t}\} \vee \{3 \wedge \text{f} \wedge \text{a}\} \\ \vee \{1 \wedge 3 \wedge 6\} \vee \{2 \wedge 4 \wedge 5\} \vee \{3 \wedge 6 \wedge \text{c}\} \vee \{1 \wedge 4 \wedge 5 \wedge \text{c}\} \vee \{2 \wedge 4 \wedge \text{f} \wedge \text{c}\} \\ \vee \{2 \wedge 3 \wedge 5 \wedge \text{f}\} \vee \{2 \wedge 3 \wedge \text{f} \wedge \text{c}\} \vee \{2 \wedge 3 \wedge 6 \wedge \text{c}\} \vee \{4 \wedge 5 \wedge \text{c} \wedge \text{t}\} \\ \vee \{4 \wedge \text{f} \wedge \text{c} \wedge \text{t}\} \vee \{3 \wedge 6 \wedge \text{c} \wedge \text{a}\} \vee \{2 \wedge 3 \wedge 4 \wedge 6 \wedge \text{c}\}$$

Hence, we obtain 18 decision bireducts, some of them listed in Table 4.

Table 4. Several bireducts of the information system $(U, \mathcal{A} \cup \{d\})$ in Example 5

Bireduct	Subset of objects	Subset of attributes
(X_1, B_1)	$\{2,4,5,6\}$	{f}
(X_2, B_2)	$\{1,2,3,5,6\}$	{a}
(X_3, B_3)	$\{1,2,4,5,6\}$	{f, t}
(X_4, B_4)	$\{1,2,3,5,6\}$	{f, t}
(X_5, B_5)	$\{1,2,4,5,6\}$	{f, a}
(X_6, B_6)	$\{2,4,5\}$	$\varnothing$
(X_7, B_7)	$\{1,3,6\}$	$\varnothing$
(X_8, B_8)	$\{1,2,3,4,5\}$	{c, a}

If we observe the reducts in Table 2, we detect that bireducts (X_3, B_3), (X_4, B_4) and (X_5, B_5) have the same subsets of attributes as the reducts D_1 and D_2, listed in Expression 6. Notice that the subsets of objects in these three bireducts are not the whole set $\mathcal{A}$. This is due to objects 3 and 4 present an inconsistence in the data, as Proposition 4 asserts, they cannot belong to the same subset of objects of any bireduct. Therefore, when the considered dataset presents inconsistencies, a decision reduct is represented as a bireduct with the set of objects as large as possible without inconsistencies.

On the other hand, bireducts (X_6, B_6) and (X_7, B_7) do not consider any attribute. Comparing with the classes, we obtain that:

$$[2]_d = [4]_d = [5]_d = \{2,4,5\} = X_6$$
$$[1]_d = [3]_d = [6]_d = \{1,3,6\} = X_7$$

as Proposition 3 asserts. □

4 Conclusion and Future Work

In this paper, we have studied some properties of bireducts and highlighted specific obtained bireducts. Mainly, we have identified the bireducts that provide the indiscernibility classes of the objects of the considered dataset. Moreover, it has been proved that the reducts of information systems and decision systems can also be obtained from bireducts. Furthermore, in the particular case of decision systems, we have proven that inconsistencies can be detected with bireducts and that they consider the largest consistent subsets of objects.

As a future work, we will continue the study of the properties obtained from the reduction of a formal context by means of bireducts. Also, we will use this study in order to reduce the number of attribute implications in FCA. In addition, the notion of fuzzy bireduct will be investigated.

References

1. Benítez, M., Medina, J., Ślęzak, D.: Delta-information reducts and bireducts. In: Alonso, J.M., Bustince, H., Reformat, M. (eds.) 2015 Conference of the International Fuzzy Systems Association and the European Society for Fuzzy Logic and Technology (IFSA- EUSFLAT-2015), Gijón, Spain, pp. 1154 1160. Atlantis Press (2015)
2. Benítez, M., Medina, J., Ślęzak, D.: Reducing information systems considering similarity relations. In: Kacprzyk, J., Koczy, L., Medina, J. (eds.) 7th European Symposium on Computational Intelligence and Mathematices (ESCIM 2015), pp. 257–263 (2015)
3. Benítez-Caballero, M.J., Medina, J., Ramírez-Poussa, E., Ślęzak, D.: Bireducts with tolerance relations. Inf. Sci. **435**, 26–39 (2018). https://doi.org/10.1016/j.ins.2017.12.037
4. Cornelis, C., Jensen, R., Hurtado, G., Ślęzak, D.: Attribute selection with fuzzy decision reducts. Inf. Sci. **180**, 209–224 (2010)
5. Janusz, A., Ślęzak, D.: Rough set methods for attribute clustering and selection. Appl. Artif. Intell. **28**(3), 220–242 (2014). https://doi.org/10.1080/08839514.2014.883902
6. Medina, J.: Relating attribute reduction in formal, object-oriented and property-oriented concept lattices. Comput. Math. Appl. **64**(6), 1992–2002 (2012). https://doi.org/10.1016/j.camwa.2012.03.087
7. Pawlak, Z.: Rough sets. Int. J. Comput. Inf. Sci. **11**, 341–356 (1982). https://doi.org/10.1007/BF01001956
8. Skowron, A., Rauszer, C.: The discernibility matrices and functions in information systems. In: Słowiński, R. (ed.) Intelligent Decision Support: Handbook of Applications and Advances of the Rough Sets Theory, pp. 331–362. Kluwer Academic Publishers (1992)
9. Ślęzak, D., Janusz, A.: Ensembles of bireducts: towards robust classification and simple representation. In: Kim, T., et al. (eds.) FGIT 2011. LNCS, vol. 7105, pp. 64–77. Springer, Heidelberg (2011). https://doi.org/10.1007/978-3-642-27142-7_9
10. Stawicki, S., Ślęzak, D.: Recent advances in decision bireducts: complexity, heuristics and streams. In: Lingras, P., Wolski, M., Cornelis, C., Mitra, S., Wasilewski, P. (eds.) RSKT 2013. LNCS (LNAI), vol. 8171, pp. 200–212. Springer, Heidelberg (2013). https://doi.org/10.1007/978-3-642-41299-8_19

11. Yao, Y., Zhao, Y.: Attribute reduction in decision-theoretic rough set models. Inf. Sci. **178**(17), 3356–3373 (2008)
12. Zhao, Y., Yao, Y., Luo, F.: Data analysis based on discernibility and indiscernibility. Inf. Sci. **177**(22), 4959–4976 (2007). https://doi.org/10.1016/j.ins.2007.06.031. http://www.sciencedirect.com/science/article/pii/S0020025507003271

Computational Intelligence Methods in Information Modelling, Representation and Processing

Fast Convergence of Competitive Spiking Neural Networks with Sample-Based Weight Initialization

Paolo Gabriel Cachi[1(✉)], Sebastián Ventura[2], and Krzysztof Jozef Cios[1,3]

[1] Virginia Commonwealth University, Richmond, VA 23220, USA
{cachidelgadpg,kcios}@vcu.edu
[2] Universidad de Córdoba, Córdoba, Spain
sventura@uco.es
[3] Polish Academy of Sciences, Gliwice, Poland

Abstract. Recent work on spiking neural networks showed good progress towards unsupervised feature learning. In particular, networks called Competitive Spiking Neural Networks (CSNN) achieve reasonable accuracy in classification tasks. However, two major disadvantages limit their practical applications: high computational complexity and slow convergence. While the first problem has partially been addressed with the development of neuromorphic hardware, no work has addressed the latter problem. In this paper we show that the number of samples the CSNN needs to converge can be reduced significantly by a proposed new weight initialization. The proposed method uses input samples as initial values for the connection weights. Surprisingly, this simple initialization reduces the number of training samples needed for convergence by an order of magnitude without loss of accuracy. We use the MNIST dataset to show that the method is robust even when not all classes are seen during initialization.

Keywords: Spiking Neural Networks · Competitive learning · Unsupervised feature learning

1 Introduction

The competitive learning paradigm has been successful in dealing with unsupervised data [7,24,33]. In competitive learning, units/neurons compete with each other for the right to respond to the given input. The winner units are then updated and become more specialized. At the end of training, all units are tuned to respond to specific input patterns and their activation is used to classify new unseen samples [29,33].

Competitive learning inspired design of several clustering and unsupervised feature learning algorithms, such as Vector Quantization [22], Self Organizing Maps (SOM) [17], and Deep Self Organizing Maps (DSOM) [39]. While these algorithms are good for extracting spatial information from unlabeled data, their use for classification tasks is limited by their performance. For example, classification accuracy achieved by DSOM on the MNIST dataset was 87.12% [39], compared with 99.79% achieved by current state of the art fully supervised algorithms [5,34,38].

M.-J. Lesot et al. (Eds.): IPMU 2020, CCIS 1239, pp. 773–786, 2020.
https://doi.org/10.1007/978-3-030-50153-2_57

A considerable increase in classification performance has been achieved by competitive learning networks using spiking neurons. Spiking neurons are dynamic units that respond not only to the current state of their inputs, as traditional neural networks do [11, 19, 33], but also take into account their previous states [2, 23, 32]. These networks, named Competitive Spiking Neural Networks (CSNN), achieved 95% accuracy on the MNIST dataset, almost 8% increase over DSOM [6, 39].

The CSNNs, however, are limited by two factors: high computational complexity and slow convergence. The first problem is due to the fact that the spiking neurons are implemented as independent units, which requires using highly parallel processors. The neuromorphic processors provide parallel architecture needed for these networks and that, at the same time, considerably reduce energy consumption when compared with traditional deep neural network implementations [8, 27]. On the other hand, there is little, if any, work on reducing the network convergence time. The state of the art CSNN [6] needs around 20,000 samples to converge which in computational time represents more than 2 h of training using a single thread implementation on an Intel Core i9-9900K processor. Thus, developing a method that would require less number of samples for training is urgently needed to expand their real world applications [14, 25, 26].

Here, we show that using the input samples as initial weights in the CSNN reduces the number of training samples needed for convergence by an order of magnitude, and with no loss in accuracy. We use different combinations of the initial weights to check the method's robustness to cases where the samples used for initialization do not represent all classes in the data. The method is evaluated on the MNIST dataset.

The rest of the paper is organized as follows. Section 2 presents a review of work on the CSNNs. A general overview of the network topology and its main characteristics are presented in Sect. 3. Section 4 introduces the proposed initialization method. Section 5 describes the dataset used, experimental settings, and evaluation metrics. Section 6 discusses the results. Section 7 ends with conclusions.

2 Related Work

The use of CSNN for unsupervised feature learning was originally proposed in [31]. The authors used an array of memrirstors to implement a CSNN for unsupervised feature learning. Their network used 300 spiking-like units to achieve 93% accuracy on the MNIST dataset and showed robustness to parameter variations. An extension of this work achieved accuracy of 95% but at the cost of using 6,400 complex spiking neurons [6]. In terms of the convergence time, both implementations converged only after 20,000 sample presentations, which in combination with the high computational cost undermines their practical applications.

Other authors used self-organizing and convolutional spiking network implementations. In [12] the authors reported accuracy of 94.07% using 1,600 neurons, however, we believe that this increase of performance was due more to the use of a specific classification method. In fact, when using the same classification method as in [6], this CSNN achieves only 92.96% accuracy. The convolutional spiking network had only 84.3% accuracy [35].

3 Competitive Spiking Neural Networks

The CSNN uses a spiking neuron layer with Spike Time Dependence Plasticity (STDP), lateral inhibition, and homeostasis to learn input data patterns in an unsupervised way. At any given time, the output neuron that is most active (spikes the most) represents the current data input.

3.1 Network's Topology

The detailed network topology is shown in Fig. 1. The Sensory layer first transforms an N-dimensional input vector, via N Poisson units (using Poisson distribution), into a series of spikes, which are fed into a layer of M spiking neurons with lateral inhibition.

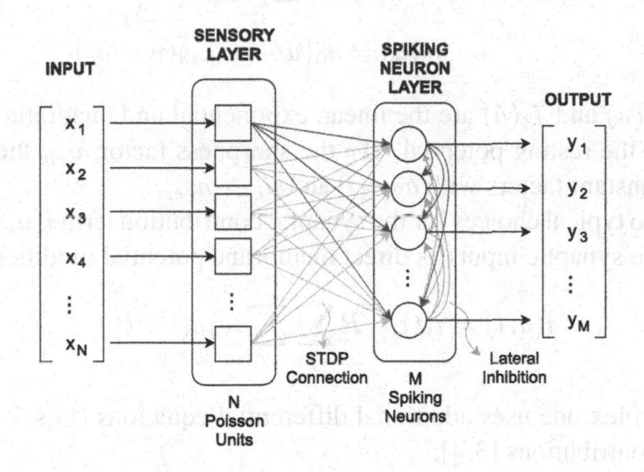

Fig. 1. Competitive spiking neural network topology

The N sensory units are fully connected to the M spiking neurons. The learning process uses the STDP implementation of the Konorski/Hebb rule [13,18]. All spiking neurons are connected with all others by fixed inhibitory weights; this is known as lateral inhibition used to ensure that only one neuron fires for a given input. The specific mechanisms used for both connections are described in detail in Sects. 3.3 and 3.4.

3.2 Spiking Neuron Model

In this paper we use a spiking neuron model known as Integrate and Fire [10,15,16]. This model uses a differential equation and a threshold mechanism to define the neuron behavior:

$$\tau \frac{du}{dt} = f(u) + i(u,t) \tag{1}$$

$$t^f: \quad u(t^f) = \theta_{reset} \quad and \quad \left.\frac{du(t)}{dt}\right|_{t=t^f} > 0 \tag{2}$$

Equation 1 describes the evolution of the membrane potential u in terms of a linear/ non-linear function $f(u)$ and a synaptic input term $i(u,t)$. Equation 2 defines the fire time t^f as the moment the membrane potential u crosses, from below, its threshold value θ_{reset}. When this happens, a spike is generated and the neuron enters a refractory period, during which it cannot respond to any input, after that the neuron's membrane potential is reset.

The choice of $f(u)$ gives rise to different variations of the Integrate and Fire neuron model [10]. A linear choice (Eq. 3) defines the Leaky Integrate and Fire Model. Non linear choice gives rise to the Exponential Integrate and Fire Model (Eq. 4) and the Quadratic Integrate and Fire Model (Eq. 5):

$$f_1(u) = -(u - u_{rest}) \tag{3}$$

$$f_2(u) = -(u - u_{rest}) + \Delta_T \exp\frac{(u - \vartheta_{rh})}{\Delta_T} \tag{4}$$

$$f_3(u) = a_0(u - u_{rest})(u - u_c) \tag{5}$$

where $f_1(u)$, $f_2(u)$ and $f_3(u)$ are the linear, exponential and quadratic function terms, u_{rest} represents the resting potential, Δ_T the sharpness factor, ϑ_{rh} the threshold, and a_0 and u_c are constant factors with $a_0 > 0$ and $u_c > u_{rest}$.

There are two typical choices for the synaptic contribution term $i(u,t)$. The simplest one considers the synaptic inputs as direct membrane potential modifiers, Eq. 6:

$$i(u,t) = i(t) = R\sum_j \sum_{t^f} w_j \rho(t - t_j^f) \tag{6}$$

A more complex one uses additional differential equations (Eqs. 7 and 8) for conductance level contributions [3,4]:

$$i(u,t) = g(t)(u_{input} - u) \tag{7}$$

$$\frac{dg}{dt} = -\frac{g}{\tau_g} + \sum_j \sum_{t^f} w_j \delta(t - t_j^f) \tag{8}$$

where $g(t)$ represents the conductance contribution, τ_g the conductance time constant, w_j the synaptic weight of connection j, and t_j^f is the firing time of the input neuron j.

3.3 Learning Rule

The CSNN uses STDP to modify the connection weights between the Poisson units and the spiking neurons [31]. In STDP, the adjustment of the strength of each weight depends on the relative activity between the pre- and post-synaptic neurons, Eq. 9:

$$STDP(\Delta t) = \begin{cases} \alpha_+ \exp(-\Delta t/\tau_+) & \text{if } \Delta t > 0 \\ -\alpha_- \exp(\Delta t/\tau_-) & \text{if } \Delta t \leq 0 \end{cases} \tag{9}$$

where δt is the time between pre- and post-synaptic neuron firings, α_+ and α_- are learning rates, and τ_+ and τ_- are time constants [1,37]. Over time, and because STPD takes effect only after a spiking event, the synaptic weights become selective to specific input patterns. In contrast to the gradient descent learning, STDP is a local rule that uses information only from the pre- and post-synaptic neuron firings, while gradient descent updates all weights based on the minimization of a global loss function.

STDP can be implemented using exponential decay variables to keep track of the weight update values a_{pre} and a_{post}, Eqs. 10 and 11 [28]. When a pre-synaptic spike is registered, the synaptic weight is decreased by a_{post} and a_{pre} is updated to a constant value A_{pre}. In contrast, when a post-synaptic spike is registered, the weight value is increased by a_{pre} and at the same time a_{post} is updated to A_{post}.

$$\tau_{pre}\frac{da_{pre}}{dt} = -a_{pre} \tag{10}$$

$$\tau_{post}\frac{da_{post}}{dt} = -a_{post} \tag{11}$$

where a_{pre} and a_{post} are the pre- and post-synaptic trace values used to update connection weights in the event of a pre- or post-synaptic neuron spikes. τ_{pre} and τ_{post} are the constant time factors for each exponential decay variable.

3.4 Lateral Inhibition

All spiking neurons at the spiking layer are connected to each other via direct inhibitory synapses, with the purpose of feedback regulation [6,31]. When a neuron produces a spike, all neurons connected to it receive a negative potentiation (their membrane potentials are decreased) which reduces the neuron's probability of reaching its firing threshold to generate a spike, see Fig. 1.

3.5 Homeostasis

It is important that the membrane firing threshold $\theta_{threshold}$ is adaptive to make sure all neurons have a chance to fire during training. The threshold value is defined by an exponential decay function with a constant increase every time the neuron fires. Thus, a neuron that fired recently is less able to fire again because of its higher threshold value. Equation 12 describes the membrane reset value θ_{reset} as a function of a dynamic variable θ, Eq. 13:

$$\theta_{reset} = \theta_{offset} + \theta \tag{12}$$

$$\frac{d\theta}{dt} = \frac{-\theta}{\tau_\theta} + \sum_{t^f} \alpha\delta(t - t_j^f) \tag{13}$$

where θ_{offset} is the offset value when $\theta = 0$, τ_θ is the time constant and α is the increase constant value [30,40].

3.6 Weight Normalization

The purpose of weight normalization is to limit the total input a neuron receives. To do so, each input connection is normalized according to Eq. 14:

$$w_{ij}^{norm} = w_{ij}\frac{\lambda}{\sum_i w_{ij}} \tag{14}$$

where w_{ij} is the weight value for connection i of neuron j, and λ is the total sum of weights [9,21].

While a straightforward effect of normalization is to balance all input connections, a not so obvious effect can be stated as helping to spread "information" to all active inputs. This means that if one synapse is increased/decreased by STDP, the normalization will average the change in all the incoming inputs. In that way, not only one input is modified, but all of them.

4 Sample-Based Weight Initialization

What are the effects of using STDP, normalization, and lateral inhibition on the network operation. If a neuron, using STDP learning, is excited with a single input image for a long period of time, its synapse weights will increase/decrease proportionally to each pixel input activation rate. The weights corresponding to high pixel values will increase the most. Performing weight normalization bounds the weight changes so the system will not become unstable. These two operations result in the weights trying to copy its input. If more images are used, then the changes are averaged and the final weights are "finding" single prototypes among all the input images. Finally, using lateral inhibition makes sure that the neuron only updates in response to the inputs that are close to its current prototype. For example, a neuron that is following the prototype for number "2" will be only updated with inputs of this class (other inputs of "2") thus increasing its selectiveness.

We use the above analysis to reduce the network's training convergence time as follows. If each neuron strives to find prototypes among the input images, we can reduce the training time by initializing its weights with the input pixel values, which are closer to some of the final prototypes than a random initialization. We thus use the first M (out of P) training samples to serve as initial connection weights between the sensory layer and the M neurons at the spiking neuron layer. Since we use weight normalization, there is no need to re-scale the pixel values. We also tested the effects of using different degrees of blurring filters to soften the contrast in the input images; for that purpose, the OpenCV's blurring filter was used.

The pseudo-code for Sample-Based Weight Initialization is shown in Algorithm 1. The competitive spiking network is instantiated in line 2. Line 3 initializes the connection weights with the resulted images after passing the first M training samples through a 5×5 blurring filter. Line 4 creates a Spike Monitor instance used to keep track of each neuron's firing events. The FOR loop in lines 5 through 9 presents all the $P - M$ remaining training samples. First, the connection weights are normalized. Then, the firing rates of the Poisson neurons are set based on the input image. Each sample is presented for 350 ms.

After training, a new run over all training samples, with STDP turned off, is done again to associate each spiking neuron with a unique class label. Algorithm 2 describes the pseudo-code for the labeling process.

Line 2 loads the resulted network from the training process and line 3 turns STDP off so the network connections are not any longer modified. Line 4 creates a spiking counter to save each neuron's firing pattern. As in the training process, a FOR loop is

Algorithm 1. Training - Sample-Based Weight Initialization

```
1: trainingSet = load(MNIST-training)
2: spikingNetwork = CompetitiveSpikingNetwork()
3: spikingNetwork.STDPconnection[:, :] = CV2.blur(trainingSet[: M], (5,5))
4: spikeMonitor = SpikeCounter(spikingNetwork['Spiking'])
5: for iterator = m, m + 1, ..., P do
6:     normalizeSTDPConnection(78.0)
7:     spikingNetwork['Poisson'].rate = trainingSet.data[iterator]
8:     run(spikingNetwork, 350ms)
9: end for
10: saveSpikingNetwork(spikingNetwork)
```

Algorithm 2. Labeling

```
1: trainingSet = load(MNIST-training)
2: spikingNetwork = loadSpikingNetwork()
3: spikingNetwork.disableSTDP()
4: spikeMonitor = SpikeCounter(spikingNetwork['Spiking'])
5: for iterator = 1, 2, ..., P do
6:     spikingNetwork['Poisson'].rate = trainingSet.data[iterator]
7:     run(spikingNetwork, 350ms)
8: end for
9: labels = getSpikingNeuronLabels(spikeMonitor, trainigSet.labels)
10: saveLabels(labels)
```

used to present all training samples (lines 5 to 8) but the difference is that normalization is no longer needed since all connections are fixed. The spiking counter and the training labels are used to decide each neuron's label in line 9.

The already assigned labels are used to classify new unseen samples via a voting process, such as maximum, confidence, or distant-based [6,12,35]. Additionally, the firing pattern can be used directly for predictions through some decision function, which can be predefined [12,36], or learned by using the firing pattern matrix as input to any add-on machine learning classifier [31], such as a conventional neural network.

5 Experiments

5.1 Dataset

All experiments are performed on the MNIST dataset, which consists of 70,000 samples of hand written 28×28 pixel images divided into 60,000 samples for training and 10,000 samples for testing [20]. The raw images are first flattened (turned into column vectors) and scaled to the range from 0 to 63.75, and are used as input to the sensory layer to determine the firing rates of the Poisson units.

5.2 Experimental Settings

To analyze the performance of the sample-based initialization, three different experiments are performed. First, we compare the training convergence and testing accuracy of random initialization with our initialization method. Second, we evaluate our method's robustness using samples from only one class (from 10 total) as the initial weights. Third, we compare the prediction results of the CSNN with a fully supervised traditional neural network, using the same topology and number of neurons.

Two CSNNs with 400 spiking neurons are used: the state of the art CSNN [6], and another one simplified by us. The state of the art CSNN uses 784 sensory layer Poisson units, 400 Leaky Integrate and Fire neurons with conductance-based stimulation input to the spiking neuron layer, trace-based STDP, indirect inhibition, weight normalization, and resting period of 150 ms between each sample presentation. The simplified CSNN uses the same spiking neuron model, learning rule and weight normalization but differs in the use of direct inhibition, with no resting period, and a different value of the membrane constant time ($3 \cdot 10^6$ ms instead of $1 \cdot 10^7$ ms). Importantly, the simplified CSNN trains in half the time than the state of the art CSNN.

All simulations were carried out using the Python's Brian Simulator package on an Intel Core i9-9900K with 64GB RAM computer (the code is publicly available on GitHub).[1]

5.3 Evaluation Metrics

The training convergence and testing accuracy are used to evaluate all experiments. The training convergence is based on the number of samples needed to reach a stable state, which is defined as the number of samples needed to reach 80% accuracy. The training accuracy is calculated after every 1,000 sample presentations in a two step process. First, the neuron labels are assigned based on the maximum firing rate of the previous 1,000 samples. Second, the assigned labels are used to predict the classes for the next 1,000 samples using maximum voting.

The testing accuracy for all 10,000 testing samples is calculated using three different methods: the maximum- and confidence-based voting, and using an add-on two layer neural network classifier. The latter uses 200 neurons with Relu activation in its first layer, 10 neurons with soft max activation in the output layer, dropout of 0.2 between the layers, and cross entropy loss function. All results are reported as average of 10 runs.

6 Results and Discussion

6.1 Convergence Time

The accuracies for 60K training samples using random initialization and the sample-based initialization are shown in Fig. 2. Figure 2a shows the accuracy on the first 20K sample presentations, and Fig. 2b shows the result on the next 40K samples. Five lines are plotted: one for the current state of the art CSNN with random initialization (Base case) [6], one for the simplified by us CSNN with random initialization (Random), and three for sample-based initialization with different degrees of image blurring. The plot starts after 1K iterations since we estimate the training accuracy using a 1K window.

[1] https://github.com/PaoloGCD/fastCSNN.

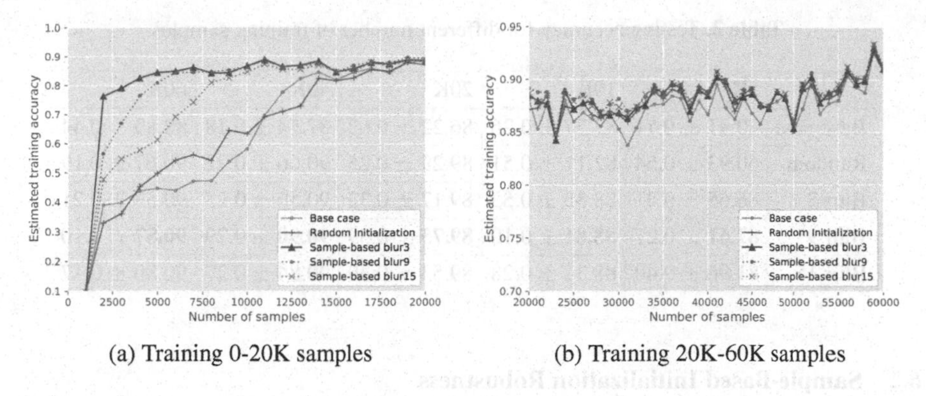

(a) Training 0-20K samples (b) Training 20K-60K samples

Fig. 2. Training accuracy vs. number of samples, using samples from all classes for initialization

The convergence time for our initialization method is faster than for random initialization (both base case and random). Specifically, using sample-based initialization with blurring of 3 achieves 80% accuracy using less than 3K samples. Bigger blurring factors reduce the convergence time (blurring of 9 and 15), but are still faster than the base case and the random initialization that need around 12.5K samples each to reach 80% accuracy. Blurring of 9 reaches 80% accuracy after around 5K samples and blurring of 15 after 8K samples.

Table 1 shows results using maximum and confidence voting, and using an add-on neural network classifier for classification prediction of the test set (trained on 60K samples). We see that sample-based initialization with blurring of 9 achieves the best accuracy in all three methods. Namely, it achieves 90.87%, 91.27% and 92.54% accuracies, which are higher than for the base case (88.89%, 90.37% and 91.73%), and higher than for random initialization (90.74%, 91.17% and 92.43%).

Table 1. Testing accuracy using different decision methods.

	Max voting	Confidence voting	Neural network
Base case	88.89 ± 0.44	90.37 ± 0.31	91.73 ± 0.16
Random	90.67 ± 0.19	91.14 ± 0.12	92.43 ± 0.06
Blur 3	90.53 ± 0.21	91.08 ± 0.16	92.37 ± 0.12
Blur 9	**90.87** ± 0.10	**91.27** ± 0.11	**92.54** ± 0.11
Blur 15	90.80 ± 0.17	91.18 ± 0.12	92.54 ± 0.12

Table 2 shows accuracy results on the testing set after training with 5K, 10K, 20K, 40K and 60K sample presentations, using maximum voting.

Before convergence (5K and 10K) sample-based initialization produces better results than random initialization. While at convergence (20K, 40K, and 60K) the results are about the same. The results with blurring of 9 are consistently the best in all cases.

Table 2. Testing accuracy for different number of training samples.

	5K	10K	20K	40K	60K
Base case	70.91 ± 0.71	83.37 ± 0.26	86.22 ± 0.62	87.54 ± 0.18	88.89 ± 0.44
Random	60.93 ± 0.54	82.11 ± 0.51	89.29 ± 0.25	90.66 ± 0.22	90.67 ± 0.19
Blur 3	86.65 ± 0.41	88.35 ± 0.53	89.17 ± 0.32	90.56 ± 0.23	90.53 ± 0.21
Blur 9	**87.61** ± 0.27	**88.81** ± 0.10	**89.73** ± 0.39	**90.90** ± 0.29	**90.87** ± 0.10
Blur 15	81.96 ± 0.40	88.31 ± 0.28	89.53 ± 0.39	90.84 ± 0.27	90.80 ± 0.17

6.2 Sample-Based Initialization Robustness

When training, often not all classes are seen in the first M samples, which are used to set the sample-based initial weights. Thus, we initialize the connection weights using samples of just one class (out of 10). Although, all classes were tested, we discuss here results only for training with classes 1, 5, and 7. Figure 3 shows training accuracies using initialization with these classes. The base and random cases are shown for reference using results from Fig. 2.

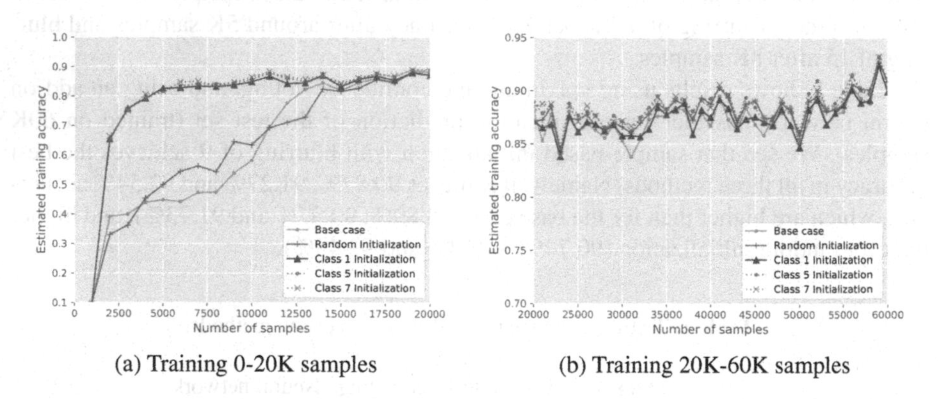

(a) Training 0-20K samples (b) Training 20K-60K samples

Fig. 3. Training accuracy vs. number of samples, using samples from only one class for initialization

We see that the convergence times for all sample-based initialization cases are still faster than for random initializations even when only one class is used to initialize the connection values. All these cases reach 80% accuracy after 4.5K sample presentations, while random initialization reaches 80% accuracy after 12K samples.

Table 3 shows testing accuracy results for maximum and confidence voting and for the add-on neural network classifier.

Overall, sample-based initialization of class 5 achieved the best result for all three methods (91.06%, 91.41% and 92.66%), while class 1 initialization was the worst (89.85%, 90.52% and 91.86%), but is still higher than the base case. The variance in all cases is less than 0.5% which indicates consistency across all cases.

Table 3. Testing accuracy using samples from only one class for initialization.

	Max voting	Confidence voting	Neural network
Base case	88.89 ± 0.441	90.37 ± 0.308	91.73 ± 0.157
Random	90.67 ± 0.190	91.14 ± 0.115	92.43 ± 0.085
Class 1	89.85 ± 0.470	90.52 ± 0.335	91.86 ± 0.147
Class 5	**91.06** ± 0.095	**91.41** ± 0.152	**92.66** ± 0.122
Class 7	90.74 ± 0.258	91.23 ± 0.190	92.52 ± 0.087

6.3 CSNN and Fully Supervised Neural Network Comparison

We compared the CSNN's performance with a fully supervised classical neural network. Table 4 shows testing accuracy for two best CSNNs, namely, sample-based initialization using blurring of 5 (fastest convergence) and blurring of 9 (best accuracy). The used NN is a 3 layer feed-forward neural network with 400, 200, and 10 neurons.

Table 4. Testing accuracy, CSNN and Fully Supervised NN comparison.

	1 epoch	3 epochs	5 epochs	10 epochs
CSNN-blurr5-NN	92.26	92.52	92.57	92.41
CSNN-blurr9-NN	**92.55**	92.89	92.97	92.85
Fully supervised NN	91.81	**94.73**	**95.49**	**96.24**

Importantly, we observe that both CSNNs achieve better accuracy after just 1 epoch of training, which can be advantageous in many real world applications. The testing accuracy for the CSNN improves slightly after 3 and 5 epochs but starts decaying at 10 epochs.

7 Conclusions

In this paper we introduced a new initialization method that uses the training samples as initial values for the as connection weights, between the Poisson units and the spiking neurons in Competitive Spiking Neural Networks. This method reduces the amount of training samples needed to achieve convergence and increases accuracy. Specifically, it significantly reduced the convergence time to around 3K samples as compared with random initialization that needed around 12.5K samples on the MNIST dataset. It also achieved a slight increase of accuracy using maximum voting, confidence voting, as well as using an add-on neural network classifier. We also showed that the convergence time and accuracy gains are about the same regardless of the class distribution in the samples used to initialize the connection weights. Importantly, we compared the CSNN with a fully supervised feed forward neural network and have shown that it performed better for small number of sample presentations, which is a strongly desired characteristic for real world applications.

References

1. Bi, G., Poo, M.: Synaptic modification by correlated activity: Hebb's postulate revisited. Ann. Rev. Neurosci. **24**(1), 139–166 (2001). https://doi.org/10.1146/annurev.neuro.24.1.139. pMID: 11283308
2. Brette, R., Gerstner, W.: Adaptive exponential integrate-and-fire model as an effective description of neuronal activity. J. Neurophysiol. **94**(5), 3637–3642 (2005). https://doi.org/10.1152/jn.00686.2005. pMID: 16014787
3. Cavallari, S., Panzeri, S., Mazzoni, A.: Comparison of the dynamics of neural interactions between current-based and conductance-based integrate-and-fire recurrent networks. Front. Neural Circuits **8**, 12 (2014). https://doi.org/10.3389/fncir.2014.00012. https://www.frontiersin.org/article/10.3389/fncir.2014.00012
4. Cessac, B., Viéville, T.: On dynamics of integrate-and-fire neural networks with conductance based synapses. Front. Comput. Neurosci. **2**, 2 (2008). https://doi.org/10.3389/neuro.10.002.2008. https://www.frontiersin.org/article/10.3389/neuro.10.002.2008
5. Ciregan, D., Meier, U., Schmidhuber, J.: Multi-column deep neural networks for image classification. In: 2012 IEEE Conference on Computer Vision and Pattern Recognition, pp. 3642–3649, June 2012. https://doi.org/10.1109/CVPR.2012.6248110
6. Diehl, P., Cook, M.: Unsupervised learning of digit recognition using spike-timing-dependent plasticity. Front. Comput. Neurosci. **9**, 99 (2015). https://doi.org/10.3389/fncom.2015.00099. https://www.frontiersin.org/article/10.3389/fncom.2015.00099
7. Fukushima, K.: Cognitron: a self-organizing multilayered neural network. Biol. Cybern. **20**(3), 121–136 (1975). https://doi.org/10.1007/BF00342633
8. Furber, S.B., Galluppi, F., Temple, S., Plana, L.A.: The spinnaker project. Proc. IEEE **102**(5), 652–665 (2014). https://doi.org/10.1109/JPROC.2014.2304638
9. Gerstner, W., Kistler, W.M.: Mathematical formulations of Hebbian learning. Biol. Cybern. **87**(5), 404–415 (2002). https://doi.org/10.1007/s00422-002-0353-y
10. Gerstner, W., Kistler, W.M., Naud, R., Paninski, L.: Neuronal Dynamics: From Single Neurons to Networks and Models of Cognition. Cambridge University Press, New York (2014)
11. Goodfellow, I., et al.: Generative adversarial nets. In: Ghahramani, Z., Welling, M., Cortes, C., Lawrence, N.D., Weinberger, K.Q. (eds.) Advances in Neural Information Processing Systems, vol. 27, pp. 2672–2680. Curran Associates, Inc. (2014). http://papers.nips.cc/paper/5423-generative-adversarial-nets.pdf
12. Hazan, H., Saunders, D., Sanghavi, D.T., Siegelmann, H., Kozma, R.: Unsupervised learning with self-organizing spiking neural networks. In: 2018 International Joint Conference on Neural Networks (IJCNN), July 2018. https://doi.org/10.1109/ijcnn.2018.8489673
13. Hebb, D.O.: The Organization of Behavior: A Neuropsychological Theory. Wiley/Chapman & Hall, Hoboken (1949)
14. Jia, Y., Huang, C., Darrell, T.: Beyond spatial pyramids: receptive field learning for pooled image features. In: 2012 IEEE Conference on Computer Vision and Pattern Recognition, pp. 3370–3377, June 2012. https://doi.org/10.1109/CVPR.2012.6248076
15. Kandel, E.R., et al.: Principles of Neural Science, vol. 5. McGraw-hill, New York (2013)
16. Koch, C., Segev, I.: Methods in Neuronal Modeling: from Ions to Networks. MIT Press, Cambridge (1998)
17. Kohonen, T.: Self-organized formation of topologically correct feature maps. Biol. Cybern. **43**(1), 59–69 (1982). https://doi.org/10.1007/BF00337288
18. Konorski, J.: Conditioned Reflexes and Neuron Organization. Cambridge University Press, Cambridge (1948)
19. LeCun, Y., Bengio, Y., Hinton, G.: Deep learning. Nature **521**(7553), 436–444 (2015). https://doi.org/10.1038/nature14539

20. LeCun, Y., Cortes, C.: MNIST handwritten digit database (2010). http://yann.lecun.com/exdb/mnist/
21. Liang, Z., Schwartz, D., Ditzler, G., Koyluoglu, O.O.: The impact of encoding-decoding schemes and weight normalization in spiking neural networks. Neural Netw. **108**, 365–378 (2018). https://doi.org/10.1016/j.neunet.2018.08.024. http://www.sciencedirect.com/science/article/pii/S0893608018302508
22. Linde, Y., Buzo, A., Gray, R.: An algorithm for vector quantizer design. IEEE Trans. Commun. **28**(1), 84–95 (1980)
23. Maass, W.: Networks of spiking neurons: the third generation of neural network models. Neural Netw. **10**(9), 1659–1671 (1997). https://doi.org/10.1016/S0893-6080(97)00011-7. http://www.sciencedirect.com/science/article/pii/S0893608097000117
24. von der Malsburg, C.: Self-organization of orientation sensitive cells in the striate cortex. Kybernetik **14**(2), 85–100 (1973). https://doi.org/10.1007/BF00288907
25. McDonnell, M.D., Vladusich, T.: Enhanced image classification with a fast-learning shallow convolutional neural network. In: 2015 International Joint Conference on Neural Networks (IJCNN), pp. 1–7, July 2015. https://doi.org/10.1109/IJCNN.2015.7280796
26. Mishkin, D., Matas, J.: All you need is a good init. arXiv preprint arXiv:1511.06422 (2015)
27. Monroe, D.: Neuromorphic computing gets ready for the (really) big time. Commun. ACM **57**(6), 13–15 (2014). https://doi.org/10.1145/2601069
28. Morrison, A., Diesmann, M., Gerstner, W.: Phenomenological models of synaptic plasticity based on spike timing. Biol. Cybern. **98**(6), 459–478 (2008). https://doi.org/10.1007/s00422-008-0233-1. https://pubmed.ncbi.nlm.nih.gov/18491160. 18491160[pmid]
29. Nowlan, S.J.: Maximum likelihood competitive learning. In: Touretzky, D.S. (ed.) Advances in Neural Information Processing Systems, vol. 2, pp. 574–582. Morgan-Kaufmann (1990). http://papers.nips.cc/paper/225-maximum-likelihood-competitive-learning.pdf
30. Pfister, J.P., Gerstner, W.: Triplets of spikes in a model of spike timing-dependent plasticity. J. Neurosci. **26**(38), 9673–9682 (2006). https://doi.org/10.1523/JNEUROSCI.1425-06.2006
31. Querlioz, D., Bichler, O., Dollfus, P., Gamrat, C.: Immunity to device variations in a spiking neural network with memristive nanodevices. IEEE Trans. Nanotechnol. **12**(3), 288–295 (2013). https://doi.org/10.1109/TNANO.2013.2250995
32. Querlioz, D., Bichler, O., Gamrat, C.: Simulation of a memristor-based spiking neural network immune to device variations. In: The 2011 International Joint Conference on Neural Networks, pp. 1775–1781, July 2011. https://doi.org/10.1109/IJCNN.2011.6033439
33. Rumelhart, D.E., Zipser, D.: Feature discovery by competitive learning. Cognit. Sci. **9**(1), 75–112 (1985). https://doi.org/10.1016/S0364-0213(85)80010-0. http://www.sciencedirect.com/science/article/pii/S0364021385800100
34. Sato, I., Nishimura, H., Yokoi, K.: APAC: augmented pattern classification with neural networks. arXiv preprint arXiv:1505.03229 (2015)
35. Saunders, D.J., Siegelmann, H.T., Kozma, R., et al.: STDP learning of image patches with convolutional spiking neural networks. In: 2018 International Joint Conference on Neural Networks (IJCNN), pp. 1–7. IEEE (2018). https://doi.org/10.1109/IJCNN.2018.8489684
36. Shin, J., et al.: Recognition of partially occluded and rotated images with a network of spiking neurons. IEEE Trans. Neural Netw. **21**(11), 1697–1709 (2010). https://doi.org/10.1109/TNN.2010.2050600
37. Sjöström, P.J., Rancz, E.A., Roth, A., Häusser, M.: Dendritic excitability and synaptic plasticity. Physiol. Rev. **88**(2), 769–840 (2008). https://doi.org/10.1152/physrev.00016.2007. pMID: 18391179
38. Wan, L., Zeiler, M., Zhang, S., Cun, Y.L., Fergus, R.: Regularization of neural networks using dropconnect. In: Dasgupta, S., McAllester, D. (eds.) Proceedings of the 30th International Conference on Machine Learning. Proceedings of Machine Learning Research, vol. 28, pp. 1058–1066. PMLR, Atlanta, June 2013. http://proceedings.mlr.press/v28/wan13.html

39. Wicramasinghe, C.S., Amarasinghe, K., Manic, M.: Deep self-organizing maps for unsupervised image classification. IEEE Trans. Ind. Inf. **15**(11), 5837–5845 (2019). https://doi.org/10.1109/TII.2019.2906083
40. Zierenberg, J., Wilting, J., Priesemann, V.: Homeostatic plasticity and external input shape neural network dynamics. Phys. Rev. X **8**, 031018 (2018). https://doi.org/10.1103/PhysRevX.8.031018

Intelligent Detection of Information Outliers Using Linguistic Summaries with Non-monotonic Quantifiers

Agnieszka Duraj(✉), Piotr S. Szczepaniak, and Lukasz Chomatek

Institute of Information Technology, Lodz University of Technology, ul. Wolczanska 215, 90-924 Lodz, Poland
{agnieszka.duraj,piotr.szczepaniak,lukasz.chomatek}@p.lodz.pl

Abstract. In the processing of imprecise information, principally in big data analysis, it is very advantageous to transform numerical values into the standard form of linguistic statements. This paper deals with a novel method of outlier detection using linguistic summaries. Particular attention is devoted to examining the usefulness of non-monotonic quantifiers, which represent a fuzzy determination of the amount of analyzed data. The answer is positive. The use of non-monotonic quantifiers in the detection of outliers can provide a more significant value of the degree of truth of a linguistic summary. At the end, this paper provides a computational example of practical importance.

Keywords: Intelligent data analysis · Linguistic summaries · Monotonic and non-monotonic quantifiers · Intelligent outlier detection

1 Introduction

Outliers represent objects whose attributes (or certain attributes) exhibit abnormal behavior in a particular or examined context. Outliers may include unexpected values for all the parameters that describe the object. They may additionally express unexpected values for a particular feature, attribute, or parameter. The customarily used definitions and recent concepts are the following:

- The formal proposition of Hawkins [20] is as follows: "An outlier is an observation which deviates so much from the other observations as to arouse suspicions that it was generated by a different mechanism".
- Barnett and Lewis [3]: "An observation (or subset of observations) which appears to be inconsistent with the remainder of that set of data".
- A collection of objects – the subjects of a linguistic summary, is be called outliers if Q objects having the feature S is a true statement in the sense of fuzzy logic, where Q is a selected relative quantifier (e.g.*very few*), and S is a finite, non-empty set of attributes (features) of the set of examined objects, cf. [12,13].

M.-J. Lesot et al. (Eds.): IPMU 2020, CCIS 1239, pp. 787–799, 2020.
https://doi.org/10.1007/978-3-030-50153-2_58

- In the field of Knowledge Discovery in Databases (KDD), or more specific in KDD in Data Mining, outliers are detected as a degree of a deviation from a specified pattern.

The decision to identify outliers is considered when developing decision-making systems, performing intelligent data analysis, and in other situations wherever any impurity or noise affects the proper functioning of systems and may or may not lead to application errors. Therefore, they must be detected and checked to determine if they will be a significant factor or not. Predominantly, there are two distinct approaches to detecting outliers. One way is the case when an object detected as an outlier can be eliminated and deleted at the data preparation stage [18,39]. The second approach assumes the "unique" objects are identified as distinct, retaining an unclear meaning for the processed data [23], and therefore they are not removed. When using artificial intelligence or soft-computing, the methods of detecting outliers are considered to be a part of Intelligent Data Analysis (IDA).

In this paper, the authors show in detail how the use of linguistic summaries given in natural language becomes a method for detecting outliers. The basis is Yager's [40–42] idea of linguistic summaries and some of the numerous extensions and modifications introduced by Kacprzyk and Zadrozny [25,26,29–32]. The innovative aspect of this work lies in the use and examination of non-monotonic quantifiers what reflects situations appearing in practice.

The paper is organized as follows. In Sect. 2, the scopes of related works are briefly presented. Basic definitions of a linguistic variable and non-monotonic quantifiers related to classic fuzzy sets are given in Sect. 3. In the next section, the concept of a linguistic summary and the way of its generation is explained. The practical rules for determining the degree of truth for monotonic and non-monotonic quantifiers are given. In Sects. 5 and 6, the formal definition of an outlier based on the concept of linguistic summary is formulated and the practice of outliers' detection is presented. The ending of the work is constituted by conclusions.

2 Related Works

The scope of applicability of methods of outlier detection in applications is very wide and varied. Numerous works are strictly focus on aimed towards specific applications, e.g. detection of production defects [19], hacker attacks on the computer network [21], fraudulent credit card transactions on credit cards or their abuse [34], public monitoring systems [33], and climate change [2]. There are works that deal with the detection of outliers in networks [16], chat activity, text messages, and the identification of illegal activities [22] in this regard.

There are works on detecting outliers in medical research and applications, e.g. personalized medicine, breast cancer, arrhythmia, monitoring of performance and endurance of athletes, or where outliers are pathogens or anomalies, e.g [1,8,35].

Outliers are distinct, they operate in separate dimensions. Outlier detection methods must, therefore, adapt to both the type of data they work on and the context in which they are operated. Numerous studies indicate an excessive interest in the issue of outlier detection, and the number of approaches increases. This is because we need utilizing a variety of methods adapted to the specific type of data we will analyze. Considering the aforementioned examples into consideration, it should be stated that tasks related to outlier to detection focus on the use of methods dedicated specific sets of data. For example, numerical and textual data - outliers are detected by using linguistic summaries based on classic and interval-valued fuzzy sets [12,13]. Another new approach is application of multiobjective genetic algorithms [7,11].

At present, the complexity of decision problems is constantly increasing. Therefore, authors of many works [6,24–30,32,38] describe not only the implementation and use of linguistic summaries but also emphasize the significance of linguistic summaries in decision-making processes. Moreover, according to Kacprzyk and Zadrozny [26,31] systems based on natural language will continue to develop.

3 Non-monotonic Quantifiers

The idea of a linguistic variable introduced by Zadeh [43,44]. The ideas used in natural language, such as *less than, almost half, about, hardly, few*, etc. can be interpreted as mathematically fuzzy linguistic concepts determining the number of items that fulfill a given criterion. It is worth noting that the relative quantifiers are defined on the interval of real numbers [0; 1]. They describe the relationship of the objects that meet the summary feature for all items in the analyzed dataset. Absolute quantifiers are defined on the set of non-negative real numbers. They describe the exact number of objects that meet the summary feature. A linguistic quantifier represents a determination of the cardinality. This means that it is a fuzzy set or a single value of the linguistic variable describing the number of objects that meet specific characteristics.

In practical solutions, monotone quantifiers are defined as classic fuzzy sets. For example, the linguistic variable $Q =$ *"few"* can be defined as a membership function in the form of a fuzzy set in the classical form as a trapezoidal or triangular function, etc. However, monotonic quantifiers do not include all possible situations.

The monotonic logic follows an intuitive indication that new knowledge does not reduce the existing set of rules and conclusions. However, it is unable to cope in cases or tasks where some rules must be removed as a consequence of further reasoning and concluding. Problems of non-monotonic logic were introduced in the 1980s. Non-monotonic logic is used to represent formalism, to describe phenomena that cannot be calculated and clearly defined. It has been pointed out that non-monotonicity remain a property of consequence. The logic system is considered to be non-monotonic if its consequence relation possesses a non-monotonic property.

In other words, non-monotonic logic is designed to represent possible conclusions, while initial conclusions may be withdrawn based on other further evidence. Non-monotonicity is closely related to the default conclusions. Non-monotonic formalism is often used in systems based on natural language and many papers present its usefulness, e.g. [4,5,17,36]. In works [12,13], the detection of outliers for monotonic quantifiers was considered. It was observed that the determination of the number used to detect outliers may not always be based on monotonic logic. The *"few"* and *"very few"* quantifiers are of particular importance in this context. However, not all quantifiers meet the condition of monotonicity [37]. The quantifiers should be normal and convex. Normal, because the height of the fuzzy set representing the quantifiers is equal to 1. Convex, because for any $\lambda \in [0,1]$, $\mu_Q(\lambda x_1 + (\lambda - 1)x_2) \geq \min(\mu_Q(x_1) + \mu_Q(x_2))$ We will use the $L-R$ fuzzy number to model the quantifiers with the membership function, where $L, R : [0,1] \longrightarrow [0,1]$ nondecreasing shape functions and $L(0) = R(0) = 0$, $L(1) = R(1) = 1$. The term *"few"*, particularly, is a non-monotonic quantifier, so such linguistic variables can be defined as membership functions in the form of (1).

$$\mu_Q(r) = \begin{cases} L(\frac{r-a}{b-a}) & r \in [a,b] \\ 1 & r \in [b,c] \\ 0 & otherwise \\ R(\frac{d-r}{d-c}) & r \in [c,d] \end{cases} \tag{1}$$

The function (1) can be written as a combination of functions L and R defined by Eqs. (2) and (3).

$$\mu_{Q_1}(r) = \begin{cases} 0 & r < a \\ L(\frac{r-a}{b-a}) & r \in [a,b] \\ 1 & r > b \end{cases} \tag{2}$$

$$\mu_{Q_2}(r) = \begin{cases} 0 & r < c \\ R(\frac{r-c}{d-c}) & r \in [c,d] \\ 1 & r > d \end{cases} \tag{3}$$

In the following section, the non-monotonic quantifiers defined above will be used in linguistic summaries. Both, the monotonic and non-monotonic quantifiers are applied to the detection of exceptions, and the results are compared.

4 Determining the Degree of Truth T_1 in a Linguistic Summary

The definition of linguistic summary introduce by R. Yager [41,42] is as follows.

Definition 1 *The ordered form of four elements [41,42], <Q; P; S; T₁> is called a linguistic summary. Here*
Q - a linguistic quantifier, or quantity in agreement, which is a fuzzy determination of the amount. Quantifier Q determines how many records in an analyzed

database fulfill the following required condition - has the characteristic S.
P - *the subject of the summary; it means the actual objects stored in the records of database;*
S - *the summarizer, the feature by which the database is scanned;*
R - *the subject's description of the summary;*
T_1- *the degree of truth; it determines the extent to which the result of the summary, expressed in a natural language, is true.*

According to the definition of linguistic summaries, we get the response in the natural language of the form:
Q objects being P are (have a feature) S [the degree of truth of this statement is $[T_1]$;
or the extended version:
Q P being R are/have S T_1
where R is the subject's description of the summary;
or in short:
Q P are/have the property S $[T_1]$.
Generating natural language responses as Yager's summaries consist of creating all possible expressions for the predefined quantifiers and summarizers of the analyzed set of objects. The value of the degree of truth for each summary is determined according to $T_1 = \mu_Q(r)$, where $r = \frac{1}{n}\sum_{i=1}^{n}\mu(a_i)$. The value r is determined for each attribute $a_i \in A$. We determine the membership function $\mu_Q(a_i)$, thus defining how well attribute a_i matches the feature given in summarizer S.

Yager's basic linguistic summary takes into consideration only a simple feature that operates on the values of one attribute. The subject is then always a set of analyzed objects in the information system, and the summarizer S denotes that the objects belong to one of the classes of the linguistic variable. Nowadays we can observe numerous extensions of Yager's method. For example, the extension of George and Srikanth [15] which proposed a family of fuzzy sets for features S, R as (4). For multiple attributes (Kacprzyk and Zadrozny's modification [27]) r is defined as (5).

$$r = \frac{1}{n}\sum_{i=1}^{n}(\mu_R(a_i)\cdot\mu_S(b_i)) \tag{4}$$

$$r = \frac{\sum_{i=1}^{n}(\mu_R(a_i)\cdot\mu_S(b_i))}{\sum_{i=1}^{n}\mu_R(a_i)} \tag{5}$$

Example 1. Let's assume we're analyzing a set of data with the attributes: age, blood sugar. If we ask:
How many middle-aged patients have a blood sugar level above average?
The resulting summary could be:
Few middle-aged patients have a blood sugar level above average [0.60].
Many middle-aged patients have a blood sugar level above average [0.25].
Almost all middle-aged patients have a blood sugar level above average [0.15].
The numbers [0.60], [0.25], and [0.15] represent the obtained degrees of truth.

The degree of truth of the linguistic summary with the use of non-monotonic quantifiers is calculated as follows (6) or (7). (cf. Sect. 4 cf. Sect. 3):

$$T_1(Q\ P \text{ is has } S) = T_1(Q_1\ P \text{ is (has) } S) - T_1(Q_2\ P \text{ is (has) } S) \tag{6}$$

$$T_1 = \mu_Q(r) = \mu_{Q_1}(r) - \mu_{Q_2}(r) \tag{7}$$

5 Detection of Outliers

Let us define the concept of an outlier using a linguistic summary.

Definition 2 *Let $X = \{x_1, x_2, ..., x_N\}$ for $N \in \mathbb{N}$ be a finite, non-empty set of objects. Let S be a finite, non-empty set of attributes (features) of the set of objects X. $S = \{s_1, s_2, ..., s_n\}$.*
Let Q be relative quantifiers.
A collection of objects, which are the subjects of a linguistic summary, will be called outliers if Q objects having the feature S is a true statement in the sense of fuzzy logic.
If the linguistic summary of Q objects in P are/have S, $[T_1]$ and $T_1 > 0$ (therefore, it is true in the sense of fuzzy logic), than outliers were found.

The procedure for detecting outliers using linguistic summaries according to Definition 2 begins with defining a set of linguistic values $X = \{Q_1, Q_2, ..., Q_n\}$. The next step is to calculate the value of r according to the procedure of generating linguistic summary described in Sect. 4. We determine T_1 for classic fuzzy sets. If we used non-monotonic quantifiers in the form of classic fuzzy sets, the degree of the truth T_1 can be determined according to (7).

One obtains
Q_1 P is (has) S $[T_1]$
Q_2 P is (has) S $[T_1]$
...
Q_N P is (has) S $[T_1]$

It is known that if $T_1 > 0$, one obtains a true sentence in the Zadeh's sense. Outliers are found if $T_1 > 0$ for q_i, where q_i is defined as: *very few, few, almost none*, and the like. For example, if for the linguistic variable $Q_1 = \{few\} T_1 > 0$ then one can expect that outliers are present.

If the set of linguistic variables is composed of several values like *"very few"*, *"few"*, *"almost none"*, then all summaries generated for those variables whose values are $T_1 > 0$ should be taken into consideration. In practical applications [10,12,14], the authors take into account the maximum to variables characterizing outliers. There exist four sets of possible responses, which are given in Table 1. Consequently, the use of linguistic summaries enables to generate information if outliers exist in the data bases under consideration. Note, that for the companies' management, the information provided in the linguistic form is of preferable form. The non-trivial example is examined in the following section.

Table 1. Types of responses of a linguistic summary indicating the existence of outliers.

Q_1 P is (has) S $[T_1]$	$T_1 = 0$	Outliers have not been found
Q_2 P is (has) S $[T_1]$	$T_1 > 0$	Outliers have been found
Q_1 P is (has) S $[T_1]$	$T_1 > 0$	Outliers have been found
Q_2 P is (has) S $[T_1]$	$T_1 = 0$	Outliers have not been found
Q_1 P is (has) S $[T_1]$	$T_1 > 0$	Outliers have been found
Q_2 P is (has) S $[T_1]$	$T_1 > 0$	Outliers have been found
Q_1 P is (has) S $[T_1]$	$T_1 = 0$	Outliers have not been found
Q_2 P is (has) S $[T_1]$	$T_1 = 0$	Outliers have not been found

6 The Practice of Outliers' Detection

Let us consider a set describing the activities of enterprises. The dataset was composed of publicly available data from Statistics Poland [9]. The examined set consists of many attributes which allow to reason about the accounting liquidity of enterprises. Attributes include, among others: company size, short-term liabilities, long-term liabilities, company assets, number of employees, financial liquidity ratio and bankruptcy risk.

Example of the data is presented in the Table 2. Current Ratio measures whether resources owned by a company are enough to meet its short-term obligations. All the calculations were performed in Java and R environment.

Table 2. Sample records of the dataset analyzed in the paper

ID	Current ratio	...	Bankruptcy risk
0001	0.148	...	0.2
0002	3.65	...	1.45
...	...	...	...
0298	2.44	...	0.54
0299	4.39	...	0.12
...	...	...	...
1587	1.74	...	2.13
1588	0.43	...	0.73
...	...	...	...

Let us consider the two following questions.
Query 1: How many enterprises with a high current ratio are in the high risk of bankruptcy group?
Query 2: How many enterprises with low profitability are in the high-risk group?

Let for the linguistic variables describing the risk of bankruptcy, the considered values are: *low, medium and high.* For the current ratio of a company, the assumed values are: *very low, low, medium, and high.*

For the each values (*low,medium, high*) the risk of bankruptcy is determined using trapezoidal membership functions
$Trap[x, a, b, c, d] = 0 \vee (1 \wedge \frac{x-a}{b-a} \wedge \frac{d-x}{d-c}), a < b \leq c < d, x \in X$:
$Trap_{low}[0, 0, 0, 2, 0.4]$, $Trap_{medium}[0.3, 0.5, 0.7, 0.9]$ and $Trap_{high}[0.6, 0.8, 1, 1]$. Similarly, the membership functions of the current liquidity indicator can be defined.

6.1 Monotonic Quantifiers

According to the procedure for detecting outliers using linguistic summaries, the set of linguistic values must be defined, here Q={ *"very few"*, *"few"*, *"many"*, *"almost all"*} and the trapezoidal form is chosen:
$Trap_{veryfew}[0, 0.1, 0.2, 0.3]$, $Trap_{few}[0.15, 0.3, 0.45, 0.6]$,
$Trap_{many}[0.5, 0.65, 0.8, 0.95]$, $Trap_{almostall}[0.75, 0.9, 1, 1]$.

On the basis of Eq. (5), the values of the coefficient of r for the two queries of interest are calculated as (8), where cls is the current liquidity indicator and $risk$ indicates the risk of bankruptcy.

$$r_{Query1} = \frac{\sum_{i=1}^{n}(\mu_{risk}(a_i) \cdot \mu_{cli}(b_i))}{\sum_{i=1}^{n} \mu_{risk}(a_i)} = 0.28 \tag{8}$$

$$r_{Query2} = \frac{\sum_{i=1}^{n}(\mu_{risk}(a_i) \cdot \mu_{prof}(b_i))}{\sum_{i=1}^{n} \mu_{risk}(a_i)} = 0.34 \tag{9}$$

The obtained linguistic summaries are of the form: Query No. 1:
Very few enterprises with a high current ratio are in the high risk of bankruptcy group; $T_1[0.2]$.
Few enterprises with a high current ratio are in the high risk of bankruptcy group; $T_1[0, 86]$.
Many enterprises with a high current ratio are in the high risk of bankruptcy group; $T_1[0]$.
Almost all enterprises with a high current ratio are in the high risk of bankruptcy group; $T_1[0]$.
According to the Definition 2 outliers were detected – see the values of the degree of truth T_1 for *few* and *very few*.

Query No. 2:
Very few enterprises with low profitability are in the high risk group; $T_1[0]$.
Few enterprises with low profitability are in the high risk group; $T_1[1]$.
Many enterprises with low profitability are in the high risk group; $T_1[0]$.
Almost all enterprises with low profitability are in the high risk group; $T_1[0]$.
Outliers were not detected because $T_1 = 0$ for all quantifiers.

6.2 Non-monotonic Quantifiers

Let the linguistic variables Q_1= *"very few"* and Q_2= *"few"* now be non-monotonic classic fuzzy sets. According to the Eq. (1), in this case the membership function Q_1 is transformed into two functions (2) and (3), and one obtains (10) and (13). Similarly for Q_2 we have (12) and (11).

$$\mu_{Q_{vf1}}(r) = \begin{cases} 0 & r < 0 \\ L(\frac{r}{0.1}) & r \in [0, 0.1] \\ 1 & r > 0.1 \end{cases} \tag{10}$$

$$\mu_{Q_{vf2}}(r) = \begin{cases} 0 & r < 0.2 \\ R(\frac{r-0.2}{0.1}) & r \in [0.2, 0.3] \\ 1 & r > 0.3 \end{cases} \tag{11}$$

$$\mu_{Q_{f1}}(r) = \begin{cases} 0 & r < 0.15 \\ L(\frac{r-0.15}{0.15}) & r \in [0.15, 0.3] \\ 1 & r > 0.3 \end{cases} \tag{12}$$

$$\mu_{Q_{f2}}(r) = \begin{cases} 0 & r < 0.45 \\ R(\frac{r-0.45}{0.15}) & r \in [0.45, 0.6] \\ 1 & r > 0.6 \end{cases} \tag{13}$$

The next step in the procedure of detecting outliers is to calculate the value of the coefficient of r. We use the Eq. (5) for Query No. 1 (8), Eq. (9) for Query No. 2. In the case of non-monotonic quantifiers T_1, we designate with (7).

We received the following generated sentences:

Query No. 1:

Very few enterprises with a high current ratio are in the high risk of bankruptcy group. T_1 [0.7]
because:

$$T_1(\mu_{Q_{veryfew}}) = T_1(\mu_{Q_{vf1}}) - T_1(\mu_{Q_{vf2}}) = 1 - 0.3 = 0.7$$

Few enterprises with a high current ratio are in the high risk of bankruptcy group. T_1[0.86]
because:

$$T_1(\mu_{Q_{few}}) = T_1(\mu_{Q_{f1}}) - T_1(\mu_{Q_{f2}}) = 0.86 - 0 = 0.86$$

Many enterprises with a high current ratio are in the high risk of bankruptcy group. T_1[0]

Almost all enterprises with a high current ratio are in the high risk of bankruptcy group. T_1[0]

Query No. 2:
Very few enterprises with low profitability are in the high risk group T_1 [0.1].
because:

$$T_1(\mu_{Q_{veryfew}}) = T_1(\mu_{Q_{vf1}}) - T_1(\mu_{Q_{vf2}}) = 1 - 0.9 = 0.1$$

Few enterprises with low profitability are in the high risk group T_1[1].
because:

$$T_1(\mu_{Q_{few}}) = T_1(\mu_{Q_{f1}}) - T_1(\mu_{Q_{f2}}) = 1 - 0 = 1$$

Many enterprises with low profitability are in the high risk group T_1[0].
Almost all enterprises with low profitability are in the high risk group T_1[0].

Table 3. The results of the degree of truth for the monotonic and non-monotonic quantifiers *very few* and *few*

Query	Quantificator	Monotonic	Non-monotonic
No. 1	Very few	0.2	0.7
No. 1	Few	0.86	0.86
No. 2	Very few	0.0	0.1
No. 2	Few	1	1

In Table 3, illustration of the degree of truth obtained for both monotonic and non-monotonic quantifiers is given. Application of non-monotonic quantifiers also indicates the existence of outliers but the value of the degree of truth is bigger. This fact can be interpreted that non-monotonic quantifiers give higher reliability of the result.

7 Conclusions

The aim of this study was to present a non-standard approach to the detection of outliers using linguistic summaries. It is a practical solution to the mentioned problem when a dataset is of numeric, or both numeric and linguistic character. However, the text attributes should be partially standardized. The presented idea is based on the summaries introduced by Yager. Other well-known standard approaches cannot directly be used for the analysis of textual or mixed data, and this is a significant advantage of the method which can operate in the case of big data evaluation as well. The results obtained in the form of sentences in a natural language are understandable and user friendly. This paper has introduced an algorithm for detecting outliers using non-monotonic quantifiers in linguistic summaries based on classic fuzzy sets. The non-monotonic quantifiers has not been considered in any of the previous studies on outlier detection with the use of linguistic summaries. In Sect. 6, the performance of the algorithm was illustrated. The conducted research and experiments confirm, that it is possible to detect outliers using linguistic summaries. To be specific, the work verified the correct functioning of the proposed method for non-monotonic quantifiers. This method enhances database analysis and decision-making processes, and it is useful for managers and data science experts.

References

1. Aggarwal, C.C.: Toward exploratory test-instance-centered diagnosis in high-dimensional classification. IEEE Trans. Knowl. Data Eng. **19**(8), 1001–1015 (2007)
2. Angiulli, F., Basta, S., Pizzuti, C.: Distance-based detection and prediction of outliers. IEEE Trans. Knowl. Data Eng. **18**(2), 145–160 (2006)
3. Barnett, V., Lewis, T.: Outliers in Statistical Data, 584 p. Wiley, Chichester (1964)
4. Benferhat, S., Dubois, D., Prade, H.: Nonmonotonic reasoning, conditional objects and possibility theory. Artif. Intell. **92**(1–2), 259–276 (1997)
5. van Benthem, J., Ter Meulen, A.: Handbook of Logic and Language. Elsevier, Amsterdam (1996)
6. Boran, F.E., Akay, D., Yager, R.R.: A probabilistic framework for interval type-2 fuzzy linguistic summarization. IEEE Trans. Fuzzy Syst. **22**(6), 1640–1653 (2014)
7. Chomatek, L., Duraj, A.: Multiobjective genetic algorithm for outliers detection. In: 2017 IEEE International Conference on INnovations in Intelligent SysTems and Applications (INISTA), pp. 379–384. IEEE (2017)
8. Cramer, J.A., Shah, S.S., Battaglia, T.M., Banerji, S.N., Obando, L.A., Booksh, K.S.: Outlier detection in chemical data by fractal analysis. J. Chemom. **18**(7–8), 317–326 (2004)
9. Databases: Statistic Poland. https://stat.gov.pl/en/databases/
10. Duraj, A.: Outlier detection in medical data using linguistic summaries. In: 2017 IEEE International Conference on INnovations in Intelligent SysTems and Applications (INISTA), pp. 385–390. IEEE (2017)
11. Duraj, A., Chomatek, L.: Supporting breast cancer diagnosis with multi-objective genetic algorithm for outlier detection. In: Kościelny, J.M., Syfert, M., Sztyber, A. (eds.) DPS 2017. AISC, vol. 635, pp. 304–315. Springer, Cham (2018). https://doi.org/10.1007/978-3-319-64474-5_25
12. Duraj, A., Niewiadomski, A., Szczepaniak, P.S.: Outlier detection using linguistically quantified statements. Int. J. Intell. Syst. **33**(9), 1858–1868 (2018)
13. Duraj, A., Niewiadomski, A., Szczepaniak, P.S.: Detection of outlier information by the use of linguistic summaries based on classic and interval-valued fuzzy sets. Int. J. Intell. Syst. **34**(3), 415–438 (2019)
14. Duraj, A., Szczepaniak, P.S.: Information outliers and their detection. In: Burgin, M., Hofkirchner, W. (eds.) Information Studies and the Quest for Transdisciplinarity, vol. 9, Chapter 15, pp. 413–437. World Scientific Publishing Company (2017)
15. George, R., Srikanth, R.: Data summarization using genetic algorithms and fuzzy logic. In: Genetic Algorithms and Soft Computing, pp. 599–611 (1996)
16. Giatrakos, N., Kotidis, Y., Deligiannakis, A., Vassalos, V., Theodoridis, Y.: In-network approximate computation of outliers with quality guarantees. Inf. Syst. **38**(8), 1285–1308 (2013)
17. Giordano, L., Gliozzi, V., Olivetti, N., Pozzato, G.L.: A non-monotonic description logic for reasoning about typicality. Artif. Intell. **195**, 165–202 (2013)

18. Guevara, J., Canu, S., Hirata, R.: Support measure data description for group anomaly detection. In: ODDx3 Workshop on Outlier Definition, Detection, and Description at the 21st ACM SIGKDD International Conference On Knowledge Discovery And Data Mining (KDD2015) (2015)
19. Guo, Q., Wu, K., Li, W.: Fault forecast and diagnosis of steam turbine based on fuzzy rough set theory. In: Second International Conference on Innovative Computing, Information and Control, ICICIC 2007, pp. 501–501. IEEE (2007)
20. Hawkins, D.M.: Identification of Outliers, vol. 11. Springer, Heidelberg (1980). https://doi.org/10.1007/978-94-015-3994-4
21. Hawkins, S., He, H., Williams, G., Baxter, R.: Outlier detection using replicator neural networks. In: Kambayashi, Y., Winiwarter, W., Arikawa, M. (eds.) DaWaK 2002. LNCS, vol. 2454, pp. 170–180. Springer, Heidelberg (2002). https://doi.org/10.1007/3-540-46145-0_17
22. He, Z., Xu, X., Deng, S.: Discovering cluster-based local outliers. Pattern Recogn. Lett. **24**(9), 1641–1650 (2003)
23. Jayakumar, G., Thomas, B.J.: A new procedure of clustering based on multivariate outlier detection. J. Data Sci. **11**(1), 69–84 (2013)
24. Kacprzyk, J., Wilbik, A., Zadrożny, S.: Linguistic summarization of time series using a fuzzy quantifier driven aggregation. Fuzzy Sets Syst. **159**(12), 1485–1499 (2008)
25. Kacprzyk, J., Wilbik, A., Zadrozny, S.: Linguistic summaries of time series via a quantifier based aggregation using the Sugeno integral. In: 2006 IEEE International Conference on Fuzzy Systems, pp. 713–719. IEEE (2006)
26. Kacprzyk, J., Wilbik, A., Zadrożny, S.: An approach to the linguistic summarization of time series using a fuzzy quantifier driven aggregation. Int. J. Intell. Syst. **25**(5), 411–439 (2010)
27. Kacprzyk, J., Yager, R.R.: Linguistic summaries of data using fuzzy logic. Int. J. Gen. Syst. **30**(2), 133–154 (2001)
28. Kacprzyk, J., Yager, R.R., Zadrożny, S.: A fuzzy logic based approach to linguistic summaries of databases. Int. J. Appl. Math. Comput. Sci. **10**(4), 813–834 (2000)
29. Kacprzyk, J., Yager, R.R., Zadrozny, S.: Fuzzy linguistic summaries of databases for an efficient business data analysis and decision support. In: Abramowicz, W., Zurada, J. (eds.) Knowledge Discovery for Business Information Systems. SECS, vol. 600, pp. 129–152. Springer, Heidelberg (2002). https://doi.org/10.1007/0-306-46991-X_6
30. Kacprzyk, J., Zadrożny, S.: Linguistic database summaries and their protoforms: towards natural language based knowledge discovery tools. Inf. Sci. **173**(4), 281–304 (2005)
31. Kacprzyk, J., Zadrozny, S.: Protoforms of linguistic database summaries as a human consistent tool for using natural language in data mining. Int. J. Softw. Sci. Comput. Intell. (IJSSCI) **1**(1), 100–111 (2009)
32. Kacprzyk, J., Zadrożny, S.: Computing with words is an implementable paradigm: fuzzy queries, linguistic data summaries, and natural-language generation. IEEE Trans. Fuzzy Syst. **18**(3), 461–472 (2010)
33. Knorr, E.M., Ng, R.T., Tucakov, V.: Distance-based outliers: algorithms and applications. VLDB J. Int. J. Very Large Data Bases **8**(3–4), 237–253 (2000). https://doi.org/10.1007/s007780050006
34. Last, M., Kandel, A.: Automated detection of outliers in real-world data. In: Proceedings of the Second International Conference on Intelligent Technologies, pp. 292–301 (2001)

35. Ng, R.: Outlier detection in personalized medicine. In: Proceedings of the ACM SIGKDD Workshop on Outlier Detection and Description, p. 7. ACM (2013)
36. Schulz, K., Van Rooij, R.: Pragmatic meaning and non-monotonic reasoning: the case of exhaustive interpretation. Linguist. Philos. **29**(2), 205–250 (2006). https://doi.org/10.1007/s10988-005-3760-4
37. Wilbik, A., Kaymak, U., Keller, J.M., Popescu, M.: Evaluation of the truth value of linguistic summaries – case with non-monotonic quantifiers. In: Angelov, P., et al. (eds.) Intelligent Systems 2014. AISC, vol. 322, pp. 69–79. Springer, Cham (2015). https://doi.org/10.1007/978-3-319-11313-5_7
38. Wilbik, A., Keller, J.M.: A fuzzy measure similarity between sets of linguistic summaries. IEEE Trans. Fuzzy Syst. **21**(1), 183–189 (2013)
39. Xiong, L., Póczos, B., Schneider, J., Connolly, A., Vander Plas, J.: Hierarchical probabilistic models for group anomaly detection. In: International Conference on Artificial Intelligence and Statistics 2011, pp. 789–797 (2011)
40. Yager, R.: Linguistic summaries as a tool for databases discovery. In: Workshop on Fuzzy Databases System and Information Retrieval (1995)
41. Yager, R.R.: A new approach to the summarization of data. Inf. Sci. **28**(1), 69–86 (1982)
42. Yager, R.R.: Linguistic summaries as a tool for database discovery. In: FQAS, pp. 17–22 (1994)
43. Zadeh, L.A.: Fuzzy sets. Inf. Control **8**(3), 338–353 (1965)
44. Zadeh, L.A.: The concept of a linguistic variable and its application to approximate reasoning-iii. Inf. Sci. **9**(1), 43–80 (1975)

Network of Fuzzy Comparators for Ovulation Window Prediction

Łukasz Sosnowski[1(✉)], Iwona Szymusik[2], and Tomasz Penza[3]

[1] Systems Research Institute, Polish Academy of Sciences, Newelska 6, 01-447 Warsaw, Poland
sosnowsl@ibspan.waw.pl

[2] Department of Obstetrics and Gynecology, Medical University of Warsaw, Żwirki i Wigury 61, 02-091 Warsaw, Poland
iwona.szymusik@gmail.com

[3] OvuFriend Sp. z o.o., Złota 61/100, 00-819 Warsaw, Poland
tomasz.penza@ovufriend.com

Abstract. This paper presents the problem and the solution of ovulation date prediction based on simple data acquired by a woman in home environment. It describes a method of processing collected data as a multivariate time series. The novelty of this algorithm lies in its ability to predict the ovulation date and not only to retrospectively detect it. This is achieved by applying the fuzzy network of comparators (NoC) to compare the menstrual cycle being analyzed with the reference set of historical cycles.

Keywords: Ovulation prediction · Fertility · Network of comparators · Compound objects · Time series data · comparators · Similarity · Fuzzy sets · Classifiers

1 Introduction

Infertility is one of the most challenging contemporary problems as the population is aging. The procreation window of a woman is constantly being shifted towards the age of about 35 years. It differs slightly across the world, but the tendency is uniform. Statistics say that every fifth couple that is trying to conceive (TTC) has a problem to achieve pregnancy in the first 12 months of efforts [1]. This effect can be reinforced by polluted environment, frequent travels, stress or an unhealthy lifestyle, all of which may affect the length of the fertility window.

In this situation, the ability to precisely determine the date of ovulation becomes crucial. Knowing the date of ovulation allows the couple to plan their intercourses (in the immediate vicinity of ovulation) in order to maximize

Co-financed by the EU Smart Growth Operational Programme 2014–2020 under the project "Development of New World Scale Solutions in the Field of Machine Learning Supporting Family Planning and Overcoming the Infertility Problem", POIR.01.01.01-00-0831/17-00.

M.-J. Lesot et al. (Eds.): IPMU 2020, CCIS 1239, pp. 800–813, 2020.
https://doi.org/10.1007/978-3-030-50153-2_59

chances of successful conception. In case of an anovulatory cycle, the knowledge that it is anovulatory reduces the stress associated with waiting for possible pregnancy confirmation and allows women to concentrate on better preparation for their next cycle. Accurate designation of the ovulation day may also reduce some equally important problems on the male side.

From the medical point of view, assessing the Graafian follicle using ultrasound (its presence, size and proof of rupture) and measuring the serum progesterone concentration (it should be over 5 ng/ml in the luteal phase) are the only reliable methods that are considered to prove that ovulation has occurred. In natural family planning there are other methods that are less reliable, but they nonetheless allow couples to obtain some information on ovulation at home. One of them is measuring the basal body temperature (BBT). An increase in BBT of at least 0.2 °C above the baseline taken at the same time every morning indicates that ovulation has occurred [8,9]. Additionally, the measurements have to be taken in the same body area and before performing any activities (ideally immediately after waking up). Moreover, it is required to measure BBT using the same device everyday. These limitations can impose great difficulties and sometimes cannot be accepted by women trying to conceive. It also requires a lot of determination. Another method is analyzing the variability of cervical mucus (the amount and consistency of cervical secretions) [3,9]. This method is also associated with various inconveniences related to measurements. The observation of the uterine cervix position, texture and opening constitute the third possibility in this field [9]. However this method requires a lot of experience and knowledge of one's own anatomical structure. It is very individual and not every woman will be able to reliably make the appropriate measurements. All these three methods belong to the retrospective group of indicators, which means that they cannot predict the date of the ovulation before it occurs, they are only able to confirm (to some extent) that ovulation has occurred in the recent past. Although their effectiveness is quite high (especially when using two methods at the same time), the aforementioned inconveniences amount to serious drawbacks of these methods.

Another group of ovulation indicators is based on LH hormone concentration [8,15]. There are many different urine or saliva tests and fertility monitoring devices that measure LH. These methods are able to predict ovulation 24 to 72 h in the future, unfortunately without specifying exactly which of the upcoming days is the most likely to be the ovulation day. Such tests operate by checking whether the given average concentration threshold calculated for the population is exceeded. Therefore it may happen, when someone's usual concentration varies from the norm, that the test will give incorrect results. Disadvantages of these devices include difficulties with interpretation of the results and consequent discrepancies, as well as difficulties with determining the appropriate period in which they should be performed. In addition, the need to buy test strips can often discourage women from using them. Nevertheless, they have the desirable feature of prediction before the actual occurrence of ovulation.

In order to be able to analyze women's menstrual cycles and to draw conclusions that facilitate detection of ovulation for more general purposes and in other future situations, it is necessary to collect large amounts of data from women with diverse fertility characteristics: e.g. age, occupation, lifestyle, weight, etc. Such data is collected by specialized portals that allow women to record their fertility parameters and provide other services to help couples plan their family. One of them is ovufriend.pl[1], the leading portal in Poland dealing with this topic. The previous publication presents the architecture of information processing and of its effective storage developed for this portal. The central point of the system is a *data warehouse* that provides data storage services in the sense of *Big Data*.

The purpose of creating a new ovulation detection algorithm, is to address the need for a universal, easily accessible solution that abolishes or reduces the existing barriers for users. Additionally, it is to use the existing knowledge and science, skillfully adjusting it to fit the real-life situations (e.g. combining fragmentary information obtained with various ovulation detection methods) and to support new methods of automatic data acquisition, e.g. wearable devices like fit bracelets [4,7,8]. The solution is also meant to be effective by economically processing large amounts of data on different scales and from many diverse users that represent different features present in the population. This solution utilizes data of many women to learn possible solutions for currently processed cycles (by computing complex multi-aspect similarities to other cycles). It creates the possibility of forecasting (determining in the future) the date of ovulation, as opposed to only confirming its occurrence in the recent past. This last point is the most important, because it is a kind of breakthrough in the current approach to fertility designation in the home environment. To achieve this goal, methods of artificial intelligence known as *similarity-based reasoning* were applied.

This paper is an overview of the solution that was developed for a commercial application. For this reason the details of the implementation are omitted and the focus is on the ideas behind the solution. The paper is organized as follows. Section two describes the data collected and its interpretation as compound objects. Sections three and four respectively describe solutions and methods applied during the process of research and development of the novel method of forecasting the date of ovulation. Section five provides evaluation results for different phases of the menstrual cycles. The last section provides elements of discussion and concludes our article.

2 Menstrual Cycle as a Compound Object

The central object of interest is the menstrual cycle described by the ensembles of time series inter-correlated with each other, constructed from observations taken by women. The individual components are indexed with the same time quanta representing particular days of the cycle. Depending on the cycle and the woman's behavior there are many possible combinations of data types that

[1] https://www.ovufriend.pl.

constitute this multivariate time series [2]. The set of features contains: BBT, cervical mucus, cervix parameters, LH urinary tests, pregnancy tests, statistics and occurrence of user-specific symptoms that may signify approaching ovulation.

BBT data consists of temperature values. The measurements are compared to the mean temperature of the previous 6 days. At the same time other factors are computed (eg. mean, relative difference, etc.) and stored together with the BBT time series for later processing.

Cervical mucus is defined by one of five possible values taken from the enumerative scale: dry, sticky, creamy, watery, stretchy. Each value describes different state of the mucus. Making use of this parameter requires detection of patterns in its variability. Therefore it is not enough to get a single measurement. The data should be collected day by day in a certain range.

Cervix has three parameters that can be tracked: opening, position and texture. Each of them has three values respectively: {open, medium, closed}, {high, medium, low}, {soft, medium, hard}. The observations are collected independently, but the interpretation of the whole state depends on all these values combined (at least two of them). These data create an additional nested three dimensional time series that describes one feature.

Ovulation test has a binary value: positive or negative. However there are some difficulties with interpreting its result which sometimes leads to wrong classification as one of these two states on part of the user. In this type of data a series of measurements is also required, in particular one containing a transition from negative to positive values. A single positive measurement is often not enough to accurately determine ovulation day.

Pregnancy test also has binary positive and negative values. If a woman got pregnant during the cycle, the pregnancy test will come out positive, but only if it was taken an appropriate amount of time after the ovulation. Thus both positive and negative values of pregnancy test in such a cycle may convey some information on the date of the ovulation.

Statistics are useful, because the length of the luteal phase is expected to be constant across a given woman's menstrual cycles. Simple statistical data particular to the user are used: average cycle and luteal phase lengths.

Symptoms are the most complex feature in terms of stored information. It consists of more than 80 elements which describe symptoms (e.g. various pains, mental states, infections, libido, etc.) on a single day of the cycle. Most of them are binary, but together they create a complex structure. Elementary symptoms are granulated and combined into groups of similar elements.

A representation of a single menstrual cycle can be any combination of these data. Moreover, each of the time series independently may require handling of missing values and of imprecision of processed values [5]. An important element is a correlation of the particular sub-time-series. Thus there arises the need to create a representation whose values are determined by the mutual influence of the individual parts of the multivariate time series.

Such combined time series for each cycle constitutes a *compound object* described by various features and consisting of many sub-objects in the sense of the definition in [12]. Having a compound object described in this way, allows for the use of the methodology of a network of compound object comparators (*NoC*) as a tool to perform reasoning based on similarity [12].

3 Fuzzy Networks of Comparators (NoC)

Networks of Comparators (NoC) are described in detail in [11] and [12]. The reader interested in theoretical or practical aspects of NoC like precise mathematical formulas or training methods should consult these sources. NoC is a general approach to reasoning about compound objects based on their multi-aspect similarity to other objects. In a sense NoC models analogies between objects and their mutual relations. It is most useful in situations that involve information granularity (see [13]). NoC's operation can be interpreted as the calculation of the following function:

$$\mu_{\text{net}}^{\text{ref}_{\text{out}}} : X \to [0,1]^{|\text{ref}_{\text{out}}|}, \tag{1}$$

where the argument is the input object $x \in X$ and ref_{out} is the reference set for the network's output layer. The target set (co-domain) of $\mu_{\text{net}}^{\text{ref}_{\text{out}}}$ is the space of proximity vectors. These vectors encapsulate information about similarities between the given input object x and objects from the reference set $\text{ref}_{\text{out}} = \{y_1, \ldots, y_n\}$. The value of the network's function is given by formula

$$\mu_{\text{net}}^{\text{ref}_{\text{out}}}(x) = \langle \text{SIM}(x, y_1), \ldots, \text{SIM}(x, y_n) \rangle, \tag{2}$$

where $\text{SIM}(x, y_i)$ is the value of *global similarity* established by the network for the input object x and a reference object y_i. Global similarity depends on partial (local) similarities calculated by the elements of the network: layers, comparators, local aggregators, translators, projection modules and global aggregators.

In the NoC there are three types of layers: input, intermediate and output. The similarity functions for each type of layer are denoted respectively by $\mu_{\text{layer-in}}^{\text{ref}_1}$, $\mu_{\text{layer-int}}^{\text{ref}_i}$ and $\mu_{\text{layer-out}}^{\text{ref}_{\text{out}}}$, where ref_i is the reference set corresponding to layer i. Further algebraic details are provided in [12]. A given network may have several intermediate layers. Each layer consists of comparators that are grouped together by the common purpose of processing a particular piece of information (attributes) about the object in question. Each layer contains a set of comparators working in parallel and a specific translating/aggregating mechanisms that facilitate the flow of information (proximity vectors) between layers. As sets of comparators in a particular layer corresponds to a specific combination of attributes and each returns own similarity vector, local aggregator responsible for converting a similarity matrix into one vector for whole layer is needed. If reference sets are different in consecutive layers, then t the output of the previous layer has to be translated first. Additionally a projection module can be used in a layer if there is a need to select a subset of coordinates of a proximity vector to be preserved (e.g. above a threshold, top N, etc.).

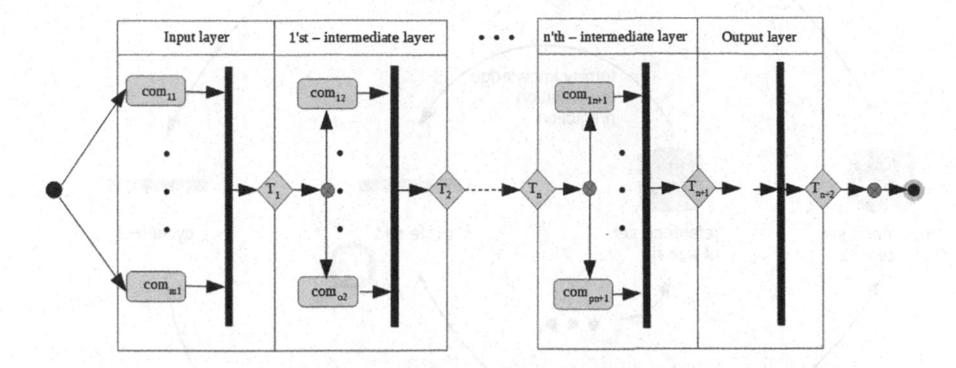

Fig. 1. General scheme of a comparator network in UML-like representation. Notation: com_{ij} – comparators, T_i – translators. Symbols: oval – comparator, thick vertical line – aggregator, rhombus – translator, encircled cross – projection module.

The global aggregator is a compulsory element of the output layer. Unlike local aggregators, which process results within a single layer, the global aggregator may process values resulting from all layers at the same time. This element implements methods of consensus reaching [6]. The comparator network function can be expressed as a composition of functions from subsequent layers:

$$\mu_{\mathrm{net}}^{\mathrm{ref_{out}}}(x) = \mu_{\mathrm{layer-out}}^{\mathrm{ref_{out}}}(\mu_{\mathrm{layer-int}}^{\mathrm{ref}_{k-1}} \cdots (\mu_{\mathrm{layer-in}}^{\mathrm{ref}_1}(x)) \ldots). \tag{3}$$

The graphical interpretation is presented in Fig. 1.

The final result of the NoC is fuzzy set, that's why there is required a defuzzificator to be used, to get crisp result at the end. This element for NoC was described in details in [14].

4 Proposed Solution

The designed solution is based on multistage classifiers that accumulate information throughout multiple menstrual cycles. A given cycle is thus important not only for its duration, but it also provides ovulation prediction in the user's future cycles. Multistage classifier is a general concept that describes processing information in parts and in a time sequence.

The algorithm consists of a set of sub-algorithms (called detectors) that independently analyze the available information about the menstrual cycle. They are divided into two classes - prognostic and retrospective. Each of them tries to detect clues to the occurrence of ovulation contained in the type of information that it processes. Every detector returns a fuzzy set whose universe is the set of

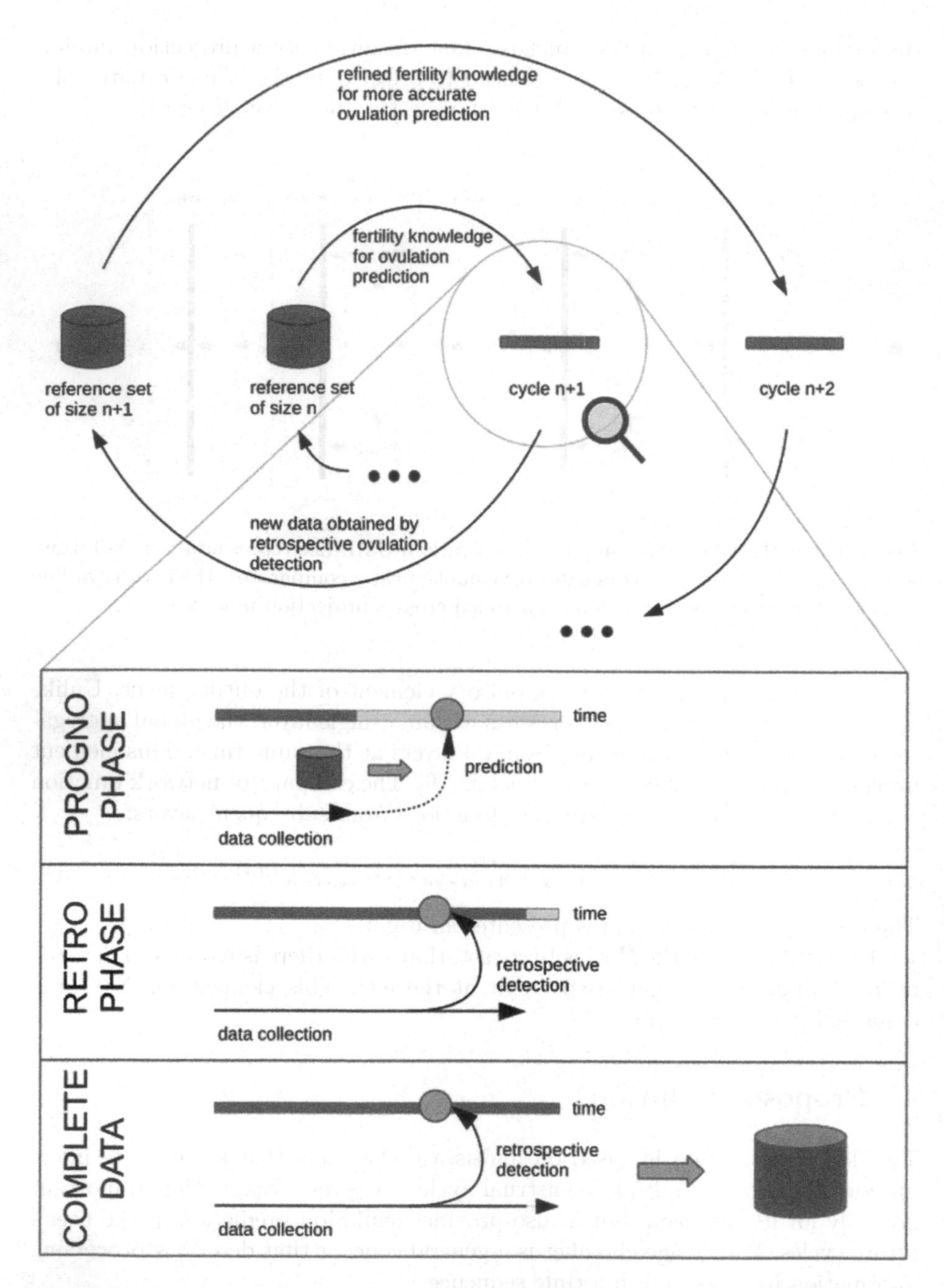

Fig. 2. General scheme of ovulation window prediction algorithm. Upper part presents general look at the data flow process in the algorithm which is arranged in the form of a spiral, where data is processed repeatedly in subsequent iterations on subsequent levels. The bottom part shows details of ovulation detection divided into *progno* and *retro* phases.

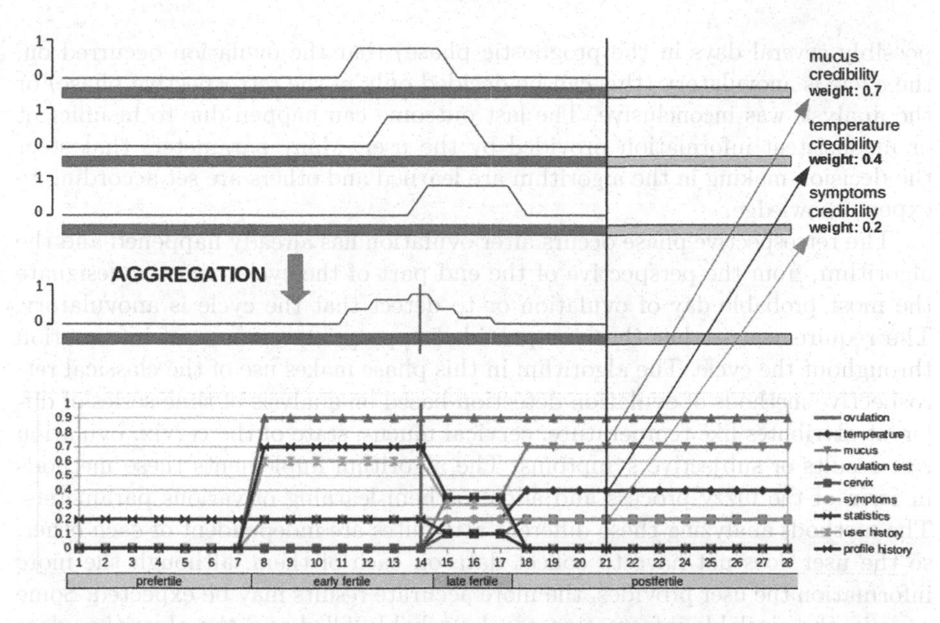

Fig. 3. As time progresses, the cycle advances through four successive sub-phases. Each of them has a different set of weights that determine the impact of the results of different detectors on the aggregated result. In this example results of three detectors are aggregated in the postfertile sub-phase. The red line marks the day to which the algorithm assigned the highest credibility of ovulation. (Color figure online)

all days of the cycle. This fuzzy set describes the ovulation process where the value of the membership function on a given day, represents the credibility that in given day the ovulation exists. The outputs of all detectors are aggregated to obtain a single fuzzy set. The aggregation process is dependent on the current phase of the ongoing cycle. The two main phases of ovulation detection during the cycle are the prognostic phase and the retrospective phase. They are subdivided into four sub-phases: prefertile, early fertile, late fertile and postfertile. Each is characterized by different weights given during aggregation to the particular detectors of prognostic and retrospective classes (including a zero weight which disables a detector), so that the ovulation detection is catered to the specific sub-phase. In the successive sub-phases the following approaches are realized respectively: purely statistical prediction, prediction based on fuzzy NoC, mix of predictive and retrospective methods and finally only retrospective detection. The weight of a detector during aggregation can also be decreased if data of the particular type is found to be of poor quality (inconsistent, unbelievable) in the given cycle. The process of aggregation is presented in Fig. 3. The aggregated fuzzy set is subjected to some additional processing to filter out improbable days. The analysis and defuzzification of this fuzzy set results in the final crisp decision of the algorithm. There can be used various defuzzifiers but here in this paper the *the last max* was used. The decision is one of the possible outcomes: the cycle is ovulatory and the algorithm designated the most probable day (or possibly several days in the prognostic phase) that the ovulation occurred on, the cycle is anovulatory (this can be decided only in the retrospective phase) or the analysis was inconclusive. The last outcome can happen due to insufficient or inconsistent information provided by the user. Many parameters that steer the decision-making in the algorithm are learned and others are set according to expert knowledge.

The retrospective phase occurs after ovulation has already happened and the algorithm, from the perspective of the end part of the cycle, aims to designate the most probable day of ovulation or to detect that the cycle is anovulatory. The requirement is that the user provided appropriate amount of information throughout the cycle. The algorithm in this phase makes use of the classical retrospective methods of ovulation detection based on analysis of time-series of different attributes like temperature, cervical mucus, state of the cervix, ovulation test results or subjective symptoms. The algorithm implements these methods in form of the fuzzy process and adds to them learning of various parameters. The methods analyzing these different attributes are independent of each other, so the user does not have to collect data on each of them, although the more information the user provides, the more accurate results may be expected. Some gaps in the available information can be reliably filled and the algorithm does that during the preprocessing step. The strength of the retrospective approach is its high efficacy under condition of access to reliable and fairly complete data on the menstrual cycle. It allows to designate ovulation in past menstrual cycles collected in the data base to obtain a vast reference set that makes it possible for the prognostic detectors to perform ovulation prediction in ongoing cycles.

After completing first menstrual cycle the system has some information about the user's fertility. At this stage the real prognostic phase of ovulation detection is enabled. In the prognostic phase the algorithm aims to indicate a future day or a few days when the ovulation will most probably occur. The ovulation prediction is achieved using the fuzzy NoC approach. The network structure is related to the set of attributes that describe the menstrual cycle. It has three layers and it compares time series of five types of data between the current cycle and the historical reference cycles using dedicated similarity measures. Comparisons are made for the time series of BBT, mucus type, cervix state, ovulation test results and symptoms. The algorithm uses various time series distance measures, such as DTW, $LCSS$ and other ovulation case specific ones [10]. The two main prognostic detectors use the same structure of the network, but they work on different reference sets. One of them uses as the reference elements all the previous ovulatory cycles of the user. The other utilizes the clustering of menstrual cycles in the database and creating profiles of similar users which are used to construct the reference set. When the reference set is determined, the detector compares the available part of the ongoing cycle to the reference cycles. If enough similarity is found to a reference cycle, then the reference ovulation day influences the fuzzy set which is an output of the detector. In this phase the algorithm does not mark cycles as anovulatory, although the fact that no similar historical cycle that was ovulatory was found can be interpreted as a possible anovulatory indication. This approach was used with success during the evaluation of the algorithm performance in the prognostic phase, however the system is designed to promote fertility and discouraging the user from attempting conception by a possibly inaccurate prediction that the cycle is anovulatory would go against its purpose.

There are three more prognostic detectors in addition to those using NoC. The simplest one uses statistics of user cycles (or of population if those are unavailable), i.e. the average cycle length and average luteal phase length. Its limited usefulness is thanks to (relative) constancy of the luteal phase length. Next one is the detector analyzing ovulation test results. It can predict ovulation in the near future based on the pattern of these results. The detector analyzing symptoms reported by the user is the most complex of the three. It relies on the detection of symptom rules periodically performed on the entire collection of menstrual cycles gathered in the data base. A rule is a combination of symptoms observed on a single day. It has a strength value assigned that measures how strong was the correlation between the first occurrence of this particular combination of symptoms in the cycle and the occurrence of ovulation in the near future. Each rule has an associated fuzzy set pattern that, when the rule is triggered on a given day of the cycle, is used to construct a fuzzy set in the neighbourhood of this day – according to the pattern – that conveys to what degree different future days are likely to be the ovulation day. This allows different rules to predict ovulation different amounts of days into the future.

We may now look holistically at the flow of information through the system. When a new cycle begins, the algorithm is in the prognostic phase and it uses the information accumulated throughout the previous cycles to predict ovulation date in the current cycle. At the very beginning of the cycle, the algorithm is in the prefertile sub-phase when the menstruation probably still lasts. In this time when there is virtually no information on the current cycle, the algorithm offers only simple predictions based on statistics of the user (if available) or the population - it gives the range of seven, five or three days (depending on accuracy) where the ovulation will statistically most probably appear. Next, when the cycle enters into the assumed fertile period, the algorithm proceeds to the early fertile sub-phase. Now that there is some information available on the current cycle, in addition to statistics it starts using the ovulation test detector, the symptom rules and the fuzzy NoC to make much more accurate predictions. From the practical point of view this is one of the most valuable functionalities that the system offers, as it gives comfort of better planning for couples hoping for pregnancy. After the gradual accumulation of individual signs of ovulation from retrospective detectors, the algorithm enters the late fertile sub-phase and starts combining prognostic and retrospective methods to indicate the ovulation day. In this period ovulation is in the near past and a quick confirmation (3–4 days) that the forecast from the previous sub-phase was good is possible. Subsequently the algorithm transitions to the postfertile sub-phase which corresponds completely to the retrospective phase. Having data on most of the cycle it can reliably designate ovulation or determine that the cycle is anovulatory. When the cycle concludes, it is added to the reference set of closed cycles and it will help refine ovulation prediction for the cycles that will follow it.

In general, this solution assumes that each new cycle feeds the knowledge base extending the system's capabilities. This way it can adapt to new cases and have the ability to predict ovulation for new types of cycles. Also additional cycles coming into the system help dilute the outliers and smooth the distribution of different types of situations so that general trends become more prominent. In this regard it has some analogy to the Kalman filter. The succession of the two phases - retrospective which allows gathering of new information and prognostic which uses that information to make more accurate predictions - in a way creates a spiral through time. This process is presented in Fig. 2.

5 Evaluation and Results

The data set used in the experiment contains 1097 menstrual cycles. Every cycle in the set was tagged by medical experts with a label signifying occurrence of ovulation or that the cycle is anovulatory. In the former case the experts also selected the day on which ovulation most probably occurred (or several days if the situation was too vague to reliably select a single day). Additionally, the experts recorded their certainty when making a decision (as a number from the unit interval $[0,1]$).

In the data set there are represented cycles with different levels of completeness of data. Those that have a high level of coverage with data points and those that have noticeably less information available. The cycles can also be classified in terms of types of measurements (time series) provided by the user, e.g. cycles with no data on temperature, cycles with no data on ovulation test, etc. This assures that cycles in the data set are diversified and the data set is balanced in terms of different features.

To designate ovulation or to mark the cycle as anovulatory, learning of a set of coefficients is necessary. The *ReSample* method was used which randomly selects 33% of cycles from the data set to the training set and the remaining cycles constitute the testing set. This procedure was performed k times and the counts of particular categories (TP, TN, FP, FN) were summed to compute the average efficiency of the algorithm. It was measured for closed cycles, so the ones that are in the post-fertile phase when the learned set of coefficients is used (only in this sub-phase the cycles can be marked as anovulatory). The results of this experiment (ran $k = 10$ times) are shown in Table 1.

Table 1. The averaged results achieved for 1097 cycles divided into 2 subsets (learning (33%) and testing (67%)) with $k = 10$ repeats.

TP	TN	FP	FN	Pr	Re	F1	Acc
6097	402	138	707	0.90	0.98	0.94	0.89

We also simulated the behaviour of the algorithm in the earlier sub-phases: prefertile, early fertile and late fertile. Since these sub-phases do not use the learned coefficients (because it is not needed to decide between ovulatory and anovulatory), these experiments were ran only once on the appropriately modified cycles.

To analyze efficiency of the algorithm in the pre-fertile sub-phase, each cycle was cut to its first 2 days, so to its very beginning. So prepared, the cycles were processed by the algorithm. Its task was to find the interval in which the ovulation will most probably occur if the cycle will turn out to have ovulation.

Table 2. Results for prefertile sub-phase with two types of sets: F - full set (1097), O - only ovulation cycles (1033). Cycles where cut to length of 2 first days. W - window in which the result was considered.

Set	W	TP	TN	FP	FN	Pr	Re	F1	Acc
F	1	488	0	66	543	0.88	0.47	0.62	0.44
F	2	687	0	66	344	0.91	0.67	0.77	0.63
F	3	823	0	66	208	0.93	0.80	0.86	0.75
O	1	488	0	2	543	0.99	0.47	0.64	0.47
O	2	687	0	2	344	0.99	0.67	0.80	0.67
O	3	823	0	2	208	0.99	0.80	0.89	0.80

Table 2 contains results for all cycles in the reference set of cycles tagged by medical experts (both these with ovulation and these that are anovulatory). Of course, since in this sub-phase the algorithm does not mark cycles as anovulatory, all anovulatory cycles were classified as mistakes. Next, the experiment was performed on the reference set with anovulatory cycles removed. In this case, the efficiency results take into account only the actual efficiency for cycles that had ovulation. The results are also contained in Table 2. They are listed depending on the interval width: 7 (window 3), 5 (window 2), or 3 (window 1) days. The narrower the interval, the lower the efficiency.

To simulate the early fertile sub-phase, with the assumption that the ovulation is to be predicted before it happens, the cycles were cut to one day before ovulation for cycles with ovulation and to half-length of the cycle for anovulatory cycles. Two experiments were performed whose results are presented in Table 3. In the first one if the algorithm did not return any prediction, we interpreted it as the decision that the cycle is anovulatory. This interpretation is justified by the fact that these detectors operate on cycle similarity, so not finding any cycle with ovulation that is similar enough to the cycle we are analyzing, indicates that it may be anovulatory. The second experiment was performed only on ovulatory cycles.

The third sub-phase is such that the ovulation has already occurred, but only in the near past. The simulation of this sub-phase was done by cutting the cycle to the fourth day after the ovulation or in the case of an anovulatory cycle by cutting it to five days before the end. In this sub-phase the algorithm uses both the prognostic and the retrospective detectors. Again, two experiments were performed with the same rules as in the previous sub-phase. Their results are presented in Table 3.

Table 3. Results for early fertile and late fertile sub-phases using two types of sets: F - full set, O - only ovulation cycles. Results for early fertile phase are placed in columns with ∗. Early fertile cycles were cut to the length of one 1 day before and late fertile cut to the 4 days after the ovulation day. In case of no ovulation half of original length or 5 days before end of cycle was set respectively.

Set	W	TP*	TN*	FP*	FN*	Pr*	Re*	F1*	Acc*	TP	TN	FP	FN	Pr	Re	F1	Acc
F	1	730	68	1	297	0.99	0.71	0.83	0.73	850	65	0	182	1.00	0.82	0.90	0.83
F	2	836	68	1	191	0.99	0.81	0.90	0.82	928	65	0	104	1.00	0.90	0.95	0.91
F	3	910	68	1	117	0.99	0.89	0.94	0.89	973	65	0	59	1.00	0.94	0.97	0.95
O	1	730	5	0	297	1.00	0.71	0.83	0.71	850	1	0	182	1.00	0.82	0.90	0.82
O	2	836	5	0	191	1.00	0.81	0.90	0.81	928	1	0	104	1.00	0.90	0.95	0.90
O	3	910	5	0	117	1.00	0.89	0.94	0.89	973	1	0	59	1.00	0.94	0.97	0.94

6 Summary

In this paper we presented a novel approach to designating the ovulation day in women's menstrual cycles using standard declarative data, which can be observed at home without specialized equipment. Particularly noteworthy is the fact that the algorithm allows for prediction of the day of ovulation and not only for its confirmation as is the case with most classic algorithms in this area.

The experiment confirmed the very good efficiency of the algorithm in the different phases of the cycle. The average F1 score value for the retrospective phase is 0.94, and for the earlier phases it is not much lower. The comparison of efficiency results achieved in different phases of the cycle shows that throughout all of the cycle, the algorithm is able to provide valuable information to a woman that is trying to conceive. Even at the beginning of the cycle, the algorithm is able to indicate an interval in which the ovulation is most likely to occur with relatively high reliability. When the cycle advances and the future ovulation gets closer in time, it is able to indicate the correct ovulation day with very high credibility. These results confirm the effectiveness of the solution and in particular of the networks of comparators used here to process such complex multidimensional time series.

The immediate next step in our work will be focused on application of the solution in the Ovufriend platform products, where real-time users will use the implemented algorithms. The presented algorithm will increase the chance of pregnancy even for couples with a narrow fertility window, because it will more accurately determine the ovulation day and, above all, it will predict it and not only confirm.

References

1. Bablok, L., Dziadecki, W., Szymusik, I., et al.: Patterns of infertility in Poland - multicenter study. Neuro Endocrinol Lett. **32**(6), 799–804 (2011)
2. Fedorowicz, J., et al.: Multivariate ovulation window detection at OvuFriend. In: Mihálydeák, T., et al. (eds.) IJCRS 2019. LNCS (LNAI), vol. 11499, pp. 395–408. Springer, Cham (2019). https://doi.org/10.1007/978-3-030-22815-6_31

3. Fehring, R., Schneider, M., Raviele, K., Barron, M.: Efficacy of cervical mucus observations plus electronic hormonal fertility monitoring as a method of natural family planning. Obstet. Gynecol. Neonatal Nurs. **36**(2), 152–160 (2007)
4. Goodale, B.M., Shilaih, M., Falco, L., Dammeier, F., Hamvas, G., Leeners, B.: Wearable sensors reveal menses-driven changes in physiology and enable prediction of the fertile window: observational study. J. Med. Internet Res. **21**(4) (2019)
5. Kacprzyk, J., Zadrożny, S.: Fuzzy logic-based linguistic summaries of time series: a powerful tool for discovering knowledge on time varying processes and systems under imprecision. Wiley Interdiscip. Rev. Data Min. Knowl. Discov. **6**(1), 37–46 (2016)
6. Kacprzyk, J., Zadrożny, S.: Towards a fairness-oriented approach to consensus reaching support under fuzzy preferences and a fuzzy majority via linguistic summaries. In: Nguyen, N.T., Kowalczyk, R., Mercik, J. (eds.) Transactions on Computational Collective Intelligence XXIII. LNCS, vol. 9760, pp. 189–211. Springer, Heidelberg (2016). https://doi.org/10.1007/978-3-662-52886-0_13
7. Shilaih, M., Clerck, V., Falco, L., Kübler, F., Leeners, B.: Pulse rate measurement during sleep using wearable sensors, and its correlation with the menstrual cycle phases, a prospective observational study. Sci. Rep. **7**(1) (2017)
8. Shilaih, M., Goodale, B., Falco, L., Kübler, F., De Clerck, V., Leeners, B.: Modern fertility awareness methods: wrist wearables capture the changes in temperature associated with the menstrual cycle. Biosci. Rep. **38**(6) (2018)
9. Smoley, B., Robinson, C.: Natural family planning. Am. Fam. Physician **86**(10), 924–928 (2012)
10. Soleimany, G., Abessi, M.: A new similarity measure for time series data mining based on longest common subsequence (2019)
11. Sosnowski, Ł.: Framework of compound object comparators. Intell. Decis. Technol. **9**(4), 343–363 (2015)
12. Sosnowski, Ł.: Compound objects comparators in application to similarity detection and object recognition. In: Peters, J.F., Skowron, A. (eds.) Transactions on Rough Sets XXI. LNCS, vol. 10810, pp. 169–300. Springer, Heidelberg (2019). https://doi.org/10.1007/978-3-662-58768-3_6
13. Sosnowski, Ł., Szczuka, M., Ślęzak, D.: Granular modeling with fuzzy comparators. In: 2015 IEEE International Conference on Big Data, Big Data 2015, Santa Clara, CA, USA, 29 October–1 November 2015, pp. 1550–1555. IEEE (2015)
14. Sosnowski, Ł., Szczuka, M.: Defuzzyfication in interpretation of comparator networks. In: Medina, J., et al. (eds.) IPMU 2018. CCIS, vol. 854, pp. 467–479. Springer, Cham (2018). https://doi.org/10.1007/978-3-319-91476-3_39
15. Su, H., Yi, Y., Wei, T., Chang, T., Cheng, C.: Detection of ovulation, a review of currently available methods. Bioeng. Transl. Med. **2**(3), 238–246 (2017)

Contextualizing Naive Bayes Predictions

Marcelo Loor[1,2](✉) and Guy De Tré[1]

[1] Department of Telecommunications and Information Processing, Ghent University, Sint-Pietersnieuwstraat 41, 9000 Ghent, Belgium
{Marcelo.Loor,Guy.DeTre}@UGent.be

[2] Department of Electrical and Computer Engineering, ESPOL University, Campus Gustavo Galindo V., Km. 30.5 Via Perimetral, Guayaquil, Ecuador

Abstract. A classification process can be seen as a set of actions by which several objects are evaluated in order to predict the class(es) those objects belong to. In situations where transparency is a necessary condition, predictions resulting from a classification process are needed to be interpretable. In this paper, we propose a novel variant of a naive Bayes (NB) classification process that yields such interpretable predictions. In the proposed variant, augmented appraisal degrees (AADs) are used for the contextualization of the evaluations carried out to make the predictions. Since an AAD has been conceived as a mathematical representation of the connotative meaning in an experience-based evaluation, the incorporation of AADs into a NB classification process helps to put the resulting predictions in context. An illustrative example, in which the proposed version of NB classification is used for the categorization of newswire articles, shows how such contextualized predictions can favor their interpretability.

Keywords: Explainable artificial intelligence · Augmented appraisal degrees · Naive Bayes classification · Context handling

1 Introduction

Computer applications like scoring tools that make judgments about individuals, or graphical applications that incorporate scene recognition to get stunning photos, can be driven by *artificial intelligence* (AI). Although such systems can be very convenient, they might be restricted or avoided in situations where transparency and accountability are highly important. For example, systems that predict the degree to which individuals are suitable (or unsuitable) for a job without explaining their predictions can be banned from using in the European Union according to the *General Data Protection Regulation* (GDPR) [6]. An ongoing challenge in this regard is to find appropriate mechanisms to explain such predictions.

In a previous work [14], we proposed a method to address that challenge in predictions made by a *support vector machine* (SVM) classification process [20,21]. In that method, an evaluation performed to predict whether an object

M.-J. Lesot et al. (Eds.): IPMU 2020, CCIS 1239, pp. 814–827, 2020.
https://doi.org/10.1007/978-3-030-50153-2_60

belongs to a given class or not is augmented in such a way that the object's features supporting the evaluation are also recorded. It has been shown how such an augmentation, which is represented by means of an *augmented appraisal degree* (AAD) [12], can favor the interpretability of SVM predictions.

As a sequel to [14], in this paper we propose a novel version of a naive Bayes (NB) classification process [9], in which AADs are incorporated to contextualize the evaluations performed to predict the class(es) an object belongs to. Our motivation here is that, while the context of evaluations performed by a person can sometimes be inferred from factors like situational or environmental aspects, the context of evaluations carried out by a machine might be difficult to infer. Thus, an explicit representation of the context of evaluations through AADs can help a *NB classifier* (NBC) to offer predictions that are better interpretable.

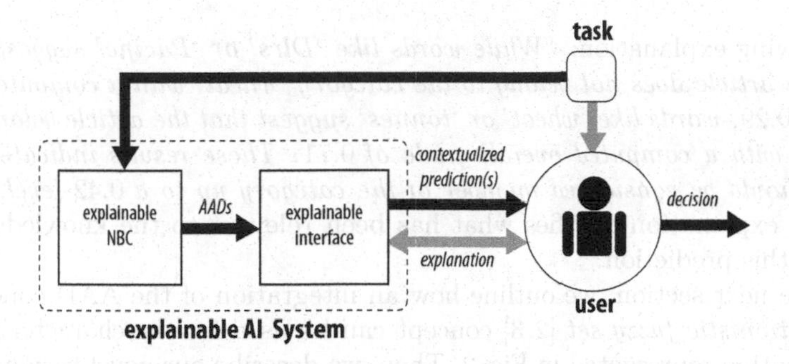

Fig. 1. A general view of the proposed version of NBC in the context of the explanation framework included in the 2016 DARPA report [5].

Contextualized predictions can be useful in situations where informed decisions are needed. In this regard, the proposed NBC, named *explainable NBC* (XNBC), can be included within an *explainable artificial intelligence* (XAI) system [5], by which a user can receive those contextualized predictions to make a decision as shown in Fig. 1. In addition, contextualized predictions can provide direct insights about what is deemed to be relevant to the (knowledge) model used by a classification process. This means that such contextualized predictions can also be used by, say, an AI practitioner to assess the quality of models that result from different learning scenarios.

To illustrate how the novel XNBC works, we develop a *text categorization* process (cf. [10]) by which newswire articles included in the *Reuters-21578* collection [8] are evaluated to predict the class(es) those articles belong to. Figure 2 shows a resulting visual representation where it is indicated why and why not XNBC predicts that a newswire article belongs to a given class up to a specific level: while the size of a word denotes its influence on the classification, its typographical style denotes whether the word is *in favor of* or against the membership in that class. The evaluation behind such a visual representation can also be used by, say, the explainable interface of an XAI system to provide

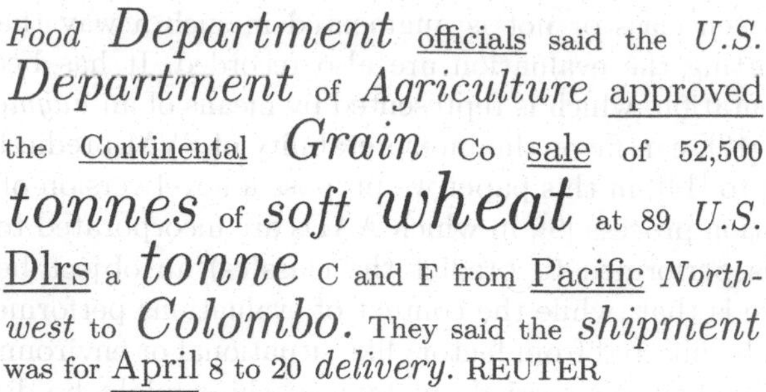

Food Department officials said the U.S. Department of Agriculture approved the Continental Grain Co sale of 52,500 tonnes of soft wheat at 89 U.S. Dlrs a tonne C and F from Pacific North-west to Colombo. They said the shipment was for April 8 to 20 delivery. REUTER

Fig. 2. A visual representation of the reasons that justify why and why not a newswire article belongs to the category 'wheat' up to a specific level.

the following explanation: *"While words like 'Dlrs' or 'Pacific' suggest that the newswire article does not belong to the category 'wheat' with a computed overall grade of* 0.29, *words like 'wheat' or 'tonnes' suggest that the article belongs to the category with a computed overall grade of* 0.71. *These results indicate that the article should be considered member of the category up to a* 0.42*-level."* Notice how this explanation clarifies what has been relevant to the knowledge model used for this prediction.

In the next section, we outline how an integration of the AAD concept into the *intuitionistic fuzzy set* [2,3] concept can be used for the characterization of the evaluation represented in Fig. 2. Then, we describe our novel variant of NBC in Sect. 3 and illustrate how it works in Sect. 4. After that, we present some related work in Sect. 5. The paper is concluded in Sect. 6.

2 Preliminaries

As previously stated, a classification can be seen as a process in which one or more objects are evaluated in order to predict whether those objects can be situated in one or more classes. In situations where an object, say x, has features suggesting that it partially belongs to a given class, say A, a classification algorithm can use the framework of *fuzzy set theory* [23] to model in mathematical terms the evaluation of the level to which x is a member of A. In this framework, such an evaluation can be characterized by a *membership grade*, which is a number $\mu_A(x)$ in the unit interval $[0, 1]$, where 0 and 1 represent in that order the lowest and the highest membership grades. For example, if the newswire article shown in Fig. 2 is denoted by x, and (what has been learned about) the category 'wheat' is represented by A, then $\mu_A(x)$ indicates the level to which x belongs to A. In this regard, if another category, say 'corn', is denoted by B, the expression $\mu_B(x) < \mu_A(x)$ indicates that the level to which (the newswire article) x belongs to B is less than the level to which x belongs to A.

An object can also have features suggesting that it does not belong to a class. Notice in Fig. 2 that, while words such as 'wheat' or 'grain' are in favor of the membership in the category 'wheat', words like 'Dlrs' or 'Pacific' are against that membership. In this case, a classification algorithm can make use of the *intuitionistic fuzzy set* (IFS) [2,3] framework to model the evaluation of an object x by means of an *IFS element.* An IFS element, say $\langle x, \mu_A(x), \nu_A(x)\rangle$, is constituted by the evaluated object x, a *membership grade* $\mu_A(x)$ and a *nonmembership grade* $\nu_A(x)$, where $\mu_A(x)$ and $\nu_A(x)$ are two numbers in the unit interval $[0,1]$ that satisfy the *consistency condition* $0 \leq \mu_A(x) + \nu_A(x) \leq 1$. The *buoyancy* [15] of $\langle x, \mu_A(x), \nu_A(x)\rangle$, i.e., $\rho_A(x) = \mu_A(x) - \nu_A(x)$, can be used for comparing this element to another. For example, if $\langle x, \mu_A(x), \nu_A(x)\rangle$ and $\langle x, \mu_B(x), \nu_B(x)\rangle$ denote the evaluations of the membership and nonmembership of x in categories A and B respectively, the expression $\rho_A(x) > \rho_B(x)$ will suggest that x belongs to a larger extent to A than to B.

As can be noticed, neither a membership grade, nor an IFS element can be used to record the object's characteristics that lead to the level to which the object belongs or not to a given class. To record those characteristics, the notion of *augmented appraisal degrees* (AADs) has been proposed in [12]. An AAD, say $\hat{\mu}_{A@K}(x)$, is a pair $\langle \mu_{A@K}(x), F_{\mu_{A@K}}(x)\rangle$ that represents the level $\mu_{A@K}(x)$ to which x belongs to A, as well as the particular collection of x's features $F_{\mu_{A@K}}(x)$ that have been taken into account to determine (the value of) $\mu_{A@K}(x)$ based on the knowledge K. Here, $A@K$ denotes what has been learned about A after following a learning process that yields K as a result. For example, consider that A and x denote the category 'wheat' and the newswire article shown in Fig. 2 respectively. With this consideration, one can use an AAD, say $\hat{\mu}_{A@K}(x) = \langle \mu_{A@K}(x), F_{\mu_{A@K}}(x)\rangle$, to represent the evaluation of the proposition 'x is member of A' according to what has been learned about the category 'wheat' after following a learning process that produces K as a result. In this case, while $\mu_{A@K}(x)$ represents the level to which x belongs to the category 'wheat', $F_{\mu_{A@K}}(x)$ represents the collection of x's words such as 'agriculture', 'grain', or 'wheat' that have been considered for quantifying the value of $\mu_{A@K}(x)$ according to (the knowledge) K.

As has been mentioned above, the newswire article x can also contain words suggesting that it does not belong to the category 'wheat'. To characterize the context of this kind of evaluations, the idea of an *augmented IFS element*, say $\langle x, \hat{\mu}_{A@K}(x), \hat{\nu}_{A@K}(x)\rangle$, has been introduced in [12]. As noticed, an augmented IFS element consists of a membership AAD, $\hat{\mu}_{A@K}(x)$, and a nonmembership AAD, $\hat{\nu}_{A@K}(x)$. Hence, the evaluation of the previous example can be better characterized by $\langle x, \hat{\mu}_{A@K}(x), \hat{\nu}_{A@K}(x)\rangle$, where the meaning of $\hat{\nu}_{A@K}(x)$ is analogous to the meaning of $\hat{\mu}_{A@K}(x)$, i.e., $\hat{\nu}_{A@K}(x)$ is a pair $\langle \nu_{A@K}(x), F_{\nu_{A@K}}(x)\rangle$ such that $\nu_{A@K}(x)$ represents the level to which x does not belong to the category 'wheat' and $F_{\nu_{A@K}}(x)$ is the collection of features that have been considered for quantifying the value of $\nu_{A@K}(x)$ according to K.

In the next section, we describe how to use AADs to contextualize predictions made by our novel variant of a naive Bayes classification process.

3 Explainable Naive Bayes Classification

Let F be the set of features under consideration. In *naive Bayes classification* [9,24], the probability $P(A|x)$ of an object, say x, being in a class (or category), say A, is given by

$$P(A|x) \propto P(A) \prod_{f \in x} P(f|A), \tag{1}$$

where $P(A)$ is the prior probability of x being member of A, and $P(f|A)$ is the conditional probability of a feature $f \in F$ occurring in an object x that belongs to A. This expression takes into account the "naive" assumption made in naive Bayes classification, which states that all features in x are *mutually independent*.

The actual value of $P(A|x)$ might be unknown. However, one can compute an approximation, say $\tilde{P}(A|x) = \tilde{P}(A) \prod_{f \in x} \tilde{P}(f|A)$, through a (knowledge) model obtained from a training set, say X_0. In this regard, $\tilde{P}(A)$ can be computed by means of

$$\tilde{P}(A) = \frac{|X_A|}{|X_A| + |X_{\bar{A}}|}, \tag{2}$$

where $|X_A|$ and $|X_{\bar{A}}|$ represent, in that order, the number of objects in X_0 that belong to A and the number of objects in X_0 that do not belong to A. Likewise, $\tilde{P}(f|A)$ can be computed by means of

$$\tilde{P}(f|A) = \frac{|F_A[f]|}{|F_A[f]| + |F_{\bar{A}}[f]|}, \tag{3}$$

where f denotes any of the x's features, $|F_A[f]|$ represents the number of occurrences of f in training objects that belong to A, and $|F_{\bar{A}}[f]|$ represents the number of occurrences of f in training objects that do not belong to A. In this regard, $\tilde{P}(f|A)$ can be seen as a quantification of the level to which f favors the membership of x in A.

Instead of multiplying many conditional probabilities in Eq. 1, performing the computation by summing logarithms of probabilities is preferred. Hence, the logarithm of $\tilde{P}(A|x)$ can be computed by

$$\log \tilde{P}(A|x) = \log \tilde{P}(A) + \sum_{f \in x} \log \tilde{P}(f|A). \tag{4}$$

Additionally, to avoid zeros, one can use *Laplace smoothing* [16], which adds one to each count. Thus, Eq. 4 can be rewritten as

$$\log \tilde{P}(A|x) \propto \log \frac{|X_A| + 1}{(|X_A| + |X_{\bar{A}}|) + 1} + \sum_{f \in x} \log \frac{|F_A[f]| + 1}{(|F_A[f]| + |F_{\bar{A}}[f]|) + |F_{X_0}|}, \tag{5}$$

where $|F_{X_0}|$ denotes the number of features detected in the training objects.

Given a collection of well-known classes, say $\mathcal{C}$, one can use Eq. 5 to predict the best class C for an object x by means of

$$C = \underset{A \in \mathcal{C}}{\operatorname{argmax}}(\log \tilde{P}(A|x)). \tag{6}$$

As can be noticed, Eq. 6 computes the predicted category without giving any explanation of what has been taken into account to make that prediction. For this reason, we consider that an explicit representation of the context of the evaluations made by Eq. 5 is strongly recommended. Hence, we propose our novel version of naive Bayes classification (NBC), named *explainable NBC* (XNBC), which main components: a learning process, an evaluation process and a prediction step, are described next.

3.1 Learning Process

The purpose of the learning process in XNBC is to obtain a model of what is known about a given category. Hence, a *feature-influence model* [13], which allows for the representation of the influence of features on the classification, is built with Algorithm 1. This algorithm uses a training set, X_0, and an identifier of the category, A, as input, and returns a model $K_A = \langle \hat{\mathbf{u}}_A, t_A \rangle$ as output. The model K_A is characterized by both a *directional vector* $\hat{\mathbf{u}}_A = \omega_1 \hat{\mathbf{f}}_1 + \cdots + \omega_m \hat{\mathbf{f}}_m$ and a *threshold point* t_A in a m-dimensional feature space $\mathcal{M}$, where ω_i denotes the influence of a feature f_i, which is represented by a unit vector $\hat{\mathbf{f}}_i$ in $\mathcal{M}$. As shown in Fig. 3, the model K_A can be seen as a *line* defined by $\hat{\mathbf{u}}_A$ and t_A: while the direction of $\hat{\mathbf{u}}_A$ points towards a place where the membership in A is favored, the location of t_A identifies a point where the membership in A is neither favored nor disfavored.

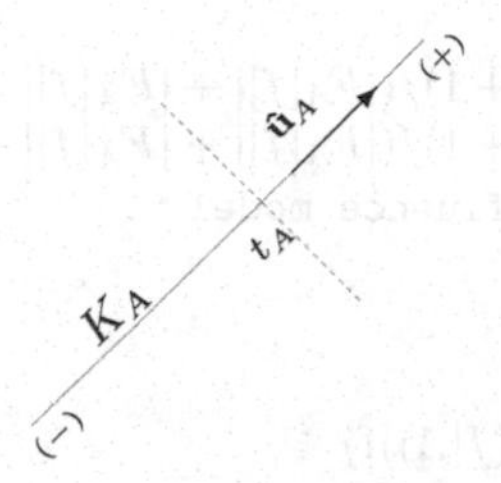

Fig. 3. Characterization of the knowledge model K_A.

To build the model, Algorithm 1 explores the objects included in the training set X_0 in order to determine the prior probability of a given object being a member of A, as well as the conditional probabilities of the features occurring in objects that belong to A. It is worth recalling that in NBC the best class for an object is considered to be the most likely. For this reason, Algorithm 1 first updates the following counters (see Lines 3–13): (i) $|X_A|$, which counts how many objects belong to the category A; (ii) $|X_{\bar{A}}|$, which counts how many objects do not belong to A; (iii) $|F_A[f]|$, which counts the occurrence of the feature f in objects that belong to A; (iv) $|F_{\bar{A}}[f]|$, which counts the occurrence of f in objects that do not belong to A. Then, the algorithm uses these counters to compute the following probabilities: (i) the prior probability $P(A)$ of an object

Algorithm 1: XNBC - Learning Process.

Data: A, X_0 /* category, training set */

Result: K_A /* knowledge model $K_A = \langle \hat{\mathbf{u}}_A, t_A \rangle$ */

1 $|X_A| \leftarrow 0$ /* number of objects that are member of A */
2 $|X_{\bar{A}}| \leftarrow 0$ /* number of objects that are nonmember of A */
3 **foreach** $x \in X_0$ **do**
4   **if** $x \in A$ **then** /* if x is member */
    /* ...increase the number of members */
5     $|X_A| \leftarrow |X_A| + 1$
6   **foreach** $f \in x$ **do** /* for each f in x's features */
      /* ..increase the occurrence of f in members */
7       $|F_A[f]| \leftarrow |F_A[f]| + \text{count}(f, x)$
8       $F_{X_0} \leftarrow F_{X_0} \cup \{f\}$
9   **else** /* if x is nonmember */
    /* ..increase the number of nonmembers */
10     $|X_{\bar{A}}| \leftarrow |X_{\bar{A}}| + 1$
11   **foreach** $f \in x$ **do** /* for each f in x's features */
      /* ..increase the occurrence of f in nonmembers */
12       $|F_{\bar{A}}[f]| \leftarrow |F_{\bar{A}}[f]| + \text{count}(f, x)$
13       $F_{X_0} \leftarrow F_{X_0} \cup \{f\}$
/* compute the prior probabilities */
14 $|X| \leftarrow |X_A| + |X_{\bar{A}}|$
15 $P(A) \leftarrow \log((|X_A| + 1)/(|X| + 1)$
16 $P(\bar{A}) \leftarrow \log((|X_{\bar{A}}| + 1)/(|X| + 1))$
/* compute the conditional probabilities */
17 **foreach** $f \in F_{X_0}$ **do**
18   $P(f|A) \leftarrow \log((|F_A[f]| + 1)/(|F_A[f]| + |F_{\bar{A}}[f]| + |F_{X_0}|))$
19   $P(f|\bar{A}) \leftarrow \log((|F_{\bar{A}}[f]| + 1)/(|F_A[f]| + |F_{\bar{A}}[f]| + |F_{X_0}|))$
/* build the feature-influence model */
20 $b \leftarrow P(A) - P(\bar{A})$
21 $\mathbf{w} \leftarrow \mathbf{0}$
22 **foreach** $f \in F_{X_0}$ **do**
23   $\mathbf{w} \leftarrow \mathbf{w} + (P(f|A) - P(f|\bar{A}))\hat{\mathbf{f}}_f$
24 $\hat{\mathbf{u}}_A \leftarrow \mathbf{w}/||\mathbf{w}||$
25 $t_A \leftarrow -b/||\mathbf{w}||$
26 $K_A \leftarrow \langle \hat{\mathbf{u}}_A, t_A \rangle$
27 **return** K_A

x being in A (see Line 15); (ii) the prior probability $P(\bar{A})$ of an object x not being in A (see Line 16); (iii) the conditional probability $P(f|A)$ of a feature f occurring in an object that belongs to A (see Line 18); and (iv) the conditional probability $P(f|\bar{A})$ of a feature f occurring in an object that does not belong to A (see Line 19). These probabilities are used for computing the components of K_A, i.e., $\hat{\mathbf{u}}_A$ and t_A (see Lines 20–25). As noticed, the conditional probability of each feature is used as an indicator of its relative influence on the classification.

Algorithm 2: XNBC - Evaluation Process.

Data: x, K_A `/* object, knowledge model` $K_A = \langle \hat{\mathbf{u}}_A, t_A \rangle$ `*/`
Result: $\langle x, \hat{\mu}_A(x), \hat{\nu}_A(x) \rangle$ `/* augmented IFS element */`

1 $\tilde{F}_{\mu_A}(x) \leftarrow \{\}$ `/* pro-membership` x`'s features */`
2 $\tilde{F}_{\nu_A}(x) \leftarrow \{\}$ `/* pro-nonmembership` x`'s features */`
3 $\tilde{\mu}_A(x) \leftarrow 0$ `/* pro-membership` x`'s score */`
4 $\tilde{\nu}_A(x) \leftarrow 0$ `/* pro-nonmembership` x`'s score */`
5 **if** $t_A < 0$ **then** `/* a negative threshold favors the score */`
6 | $\tilde{\mu}_A(x) \leftarrow \tilde{\mu}_A(x) + \text{abs}(t_A)$ `/* increase the positive score of` x `*/`
7 **else** `/* a positive threshold disfavors the score */`
8 | $\tilde{\nu}_A(x) \leftarrow \tilde{\nu}_A(x) + t_A$ `/* increase the negative score of` x `*/`
`/* recall that` $\hat{\mathbf{u}}_A = \sum_{f \in F_{X_0}} \omega_f \hat{\mathbf{f}}_f$ `*/`
9 **foreach** $f \in x$ **do** `/* for each` f `in` x`'s features */`
10 | $s_f \leftarrow \text{count}(f, x) * \omega_f$ `/* compute` f`'s influence */`
11 | **if** $s_f > 0$ **then** `/* if` f `is in favor of` $x \in A$ `*/`
12 | | $\tilde{\mu}_A(x) \leftarrow \tilde{\mu}_A(x) + s_f$ `/* increase` x`'s positive score */`
13 | | $\tilde{F}_{\mu_A}(x) \leftarrow \tilde{F}_{\mu_A}(x) \cup \{\langle f, s_f \rangle\}$ `/* and record` f`'s influence */`
14 | **else** `/*` f `is against` $x \in A$ `*/`
15 | | $\tilde{\nu}_A(x) \leftarrow \tilde{\nu}_A(x) + \text{abs}(s_f)$ `/* increase` x`'s negative score */`
16 | | $\tilde{F}_{\nu_A}(x) \leftarrow \tilde{F}_{\nu_A}(x) \cup \{\langle f, \text{abs}(s_f) \rangle\}$ `/* and record` f`'s influence */`
`/* handle the consistency condition` $0 \leq \mu_A(x) + \nu_A(x) \leq 1$ `*/`
17 $\text{maxLevel} \leftarrow \max(1, \tilde{\mu}_A(x) + \tilde{\nu}_A(x))$
18 **foreach** $\langle f, s_f \rangle \in \tilde{F}_{\mu_A}(x)$ **do**
19 | $F_{\mu_A}(x) \leftarrow F_{\mu_A}(x) \cup \{\langle f, (s_f/\text{maxLevel}) \rangle\}$
20 **foreach** $\langle f, s_f \rangle \in \tilde{F}_{\nu_A}(x)$ **do**
21 | $F_{\nu_A}(x) \leftarrow F_{\nu_A}(x) \cup \{\langle f, (s_f/\text{maxLevel}) \rangle\}$
22 $\mu_A(x) \leftarrow \tilde{\mu}_A(x)/\text{maxLevel}$
23 $\nu_A(x) \leftarrow \tilde{\nu}_A(x)/\text{maxLevel}$
`/* finally, build the augmented IFS element */`
24 $\hat{\mu}_A(x) \leftarrow \langle \mu_A(x), F_{\mu_A}(x) \rangle$
25 $\hat{\nu}_A(x) \leftarrow \langle \nu_A(x), F_{\nu_A}(x) \rangle$
26 **return** $\langle x, \hat{\mu}_A(x), \hat{\nu}_A(x) \rangle$

3.2 Evaluation Process

The purpose of the evaluation process is to obtain a contextualized evaluation of the membership of a given object in a given category. The steps of this process are described in Algorithm 2. This algorithm uses an object x and the knowledge model $K_A = \langle \hat{\mathbf{u}}_A, t_A \rangle$ for a category A as input, and builds an augmented IFS element[1] $\langle x, \hat{\mu}_A(x), \hat{\nu}_A(x) \rangle$ representing a contextualized evaluation that is returned as output.

[1] To be consistent with the notation introduced in Sect. 2, we should write $\langle x, \hat{\mu}_{A@X_0}(x), \hat{\nu}_{A@X_0}(x) \rangle$. However, for the sake of readability, we use this simplified notation hereafter.

To build a contextualized evaluation, Algorithm 2 computes both a positive score $\tilde{\mu}_A(x)$ and a negative score $\tilde{\nu}_A(x)$ of x being in category A based on the threshold point t_A and the influence of the features in the directional vector $\hat{\mathbf{u}}_A$. The positive score is increased in two cases: if t_A is negative (see Line 6); and if the influence of a feature is positive (see Line 12). Likewise, the negative score is increased in two cases: if t_A is positive (see Line 8); and if the influence of a feature is negative (see Line 15). While the conditions that arise when a positive score is increased are recorded in $\tilde{F}_{\mu_A}(x)$ (see Line 13), the conditions that arise when a negative score is increased are recorded in $\tilde{F}_{\nu_A}(x)$ (see Line 16).

The consistency condition of an IFS element, i.e., $0 \leq \mu_A(x) + \nu_A(x) \leq 1$, is guaranteed by Algorithm 2 in Lines 17–23. After this, the algorithm records the components of the augmented IFS element in Lines 24–25.

3.3 Predicting the Best Class(es)

To predict the best class $C \in \mathcal{C}$ for an object x, an XAI system (see Sect. 1) can first use Algorithm 1 for building a knowledge model for each class in $\mathcal{C}$. Then, that system can use Algorithm 2 to obtain the contextualized evaluation of the membership of x in each class using these models. After that, the system can use the buoyancy of those contextualized evaluations (see Sect. 2) to sort them in descending order. Then, the system can, say, list the top-k of the contextualized evaluations so that a user can be offered the k best classes with the best context. For each class an augmented IFS element, expressing the context of the evaluation of x belonging to the class or not, is provided. Together these explain to users why x has been classified in this way. Hence, with XNBC users and applications have extra information for giving preference to those classes with the best credible justification.

4 Illustrative Example

In this section, we present an example where our novel version of naive Bayes classification is used for predicting the classes of newswire articles. In this example, the *Reuters-21578* collection [8], which consists of 21578 newswire articles provided by Reuters, Ltd, has been used. Specifically, we made use of the articles established in the *"modified Apte split"* (ModApte) of this collection.

To use Algorithm 1, each article had to be modeled as a feature-influence vector whose components are the words in the article. Hence, each article was first split into words using separators such as commas or blank-spaces. Then, *stop words*, i.e., words like prepositions or conjunctions that have a negligible impact on the classification [11], were removed from the previous list of words. Additionally, words having a common stem were tokenized using the *Porter Stemming Algorithm* [18]. After that, Algorithm 1 was used with the feature-influence vectors corresponding to the 9603 articles included in the training set of the ModApte split for building a knowledge model for each of the following categories: earn, acq, money-fx, grain, crude, trade, interest, ship, wheat, corn.

For the sake of illustration in this paper we consider one article from the test set of the ModApte split, namely the newswire article identified by 14841. Algorithm 2 was used with the resulting knowledge models to evaluate the membership of this article to each of the aforementioned categories. This means that, augmented IFS elements like $\langle x, \hat{\mu}_{earn}(x), \hat{\nu}_{earn}(x)\rangle$ or $\langle x, \hat{\mu}_{grain}(x), \hat{\nu}_{grain}(x)\rangle$ were obtained as output – here x represents the article identified by 14841.

The resulting augmented IFS elements were used for building visual representations like the ones depicted in Fig. 4. For instance, $\mu_{grain}(x)$ and $\nu_{grain}(x)$, which are parts of $\hat{\mu}_{grain}(x)$ and $\hat{\nu}_{grain}(x)$ respectively, were used for computing the buoyancy $\rho_{grain}(x) = 0.60$ of the article in category 'grain' (see Fig. 4(a)). Analogously, the positive influence of the word 'wheat' on the membership of this article in category 'grain', namely $\langle$'wheat', $0.15\rangle \in F_{\mu_{grain}}(x)$, was used for setting both the size and the typographical style of this word. Herein, while the size of the word denotes the influence of this word on the classification, the typographical style denotes whether this influence is *positive* or negative.

Food *Department* *officials* said the *U.S.* *Department* of *Agriculture* approved the *Continental* *Grain* Co sale of 52,500 *tonnes* of *soft* *wheat* at 89 *U.S.* Dlrs a *tonne* C and F from Pacific *Northwest* to *Colombo.* They said the *shipment* was for April 8 to 20 *delivery.* REUTER

(a) $\rho_{grain}(x) = 0.60$

Food *Department* officials said the *U.S.* *Department* of *Agriculture* approved the Continental *Grain* Co sale of 52,500 *tonnes* of *soft* *wheat* at 89 *U.S.* Dlrs a *tonne* C and F from Pacific *Northwest* to *Colombo.* They said the *shipment* was for April 8 to 20 *delivery.* REUTER

(b) $\rho_{wheat}(x) = 0.42$

Food *Department* officials said the *U.S.* *Department* of *Agriculture* approved the Continental *Grain* Co sale of 52,500 *tonnes* of *soft* *wheat* at 89 *U.S.* Dlrs a *tonne* C and F from Pacific *Northwest* to *Colombo.* They said the *shipment* was for April 8 to 20 *delivery.* REUTER

(c) $\rho_{corn}(x) = 0.23$

Food Department *officials* said the U.S. Department of Agriculture approved the *Continental* *Grain* Co sale of 52,500 *tonnes* of *soft* *wheat* at 89 U.S. Dlrs a *tonne* C and F from Pacific *Northwest* to *Colombo.* They said the *shipment* was for April 8 to 20 delivery. REUTER

(d) $\rho_{ship}(x) = -0.47$

Fig. 4. The four best evaluated categories for a newswire article x.

Those augmented IFS elements were also used for building explanations like the following: *"While words like 'Dlrs', 'April' or 'Pacific' suggest that article* 14841 *does not belong to category 'grain' with a computed overall grade of* 0.20, *words like 'grain', 'wheat' or 'tonnes' suggest that the article belongs to the category with a computed overall grade of* 0.80. *These results indicate that article* 14841 *should be considered member of category 'grain' up to a* 0.60*-level."* Notice that this explanation indicates not only the level to which this article belongs to the category 'grain' but also provides practical information about what words (features) have been focused on during the evaluation. We foresee that this kind of explanation can help, say, an AI practitioner to improve the knowledge model used for the evaluation. For instance, if an AI practitioner considers that 'Dlrs' and 'April' are irrelevant to the evaluation, he/she might exclude these words from the list that is used during the learning process. Notice also that only the six most influential words (three with positive influence and three with negative influence) have been included in the explanation in order to keep it simple and interpretable. A future work will reveal how this simplification could be used for improving knowledge models that result from training sets having imperfect or scarce data.

Regarding the prediction of the best category (or categories) for article 14841, the contextualized evaluations were first sorted in descending order according to the computed buoyancy. After that, the four best evaluated categories (see Fig. 4) were presented as the most optimistic predictions. As noticed, these predictions reuse the context of the evaluations and, thus, they can be easily interpreted. Hence, a user can choose the category which prediction has the most adequate justification according to his/her perspective. In this regard, experimental studies about the interpretability and usability of such predictions are considered and highly suggested.

5 Related Work

Methods aiming to produce a set of rules that explain predictions can be found in the literature. For instance, a Bayesian method for learning rules that provide explanations of the predictions according to prior parameters fixed by a user is proposed in [22]. Another example is the method proposed in [7] for building Bayesian rules that discretize a high-dimensional feature space into a series of interpretable decision statements. In the framework of fuzzy set theory, an example is the variant of the neuro-fuzzy classification method presented in [17]. This variant tries to produce a small set of interpretable fuzzy rules for the diagnosis of patients.

A comprehensive survey of methods proposed for explaining computer predictions can be found in [4]. This survey has identified two main approaches of the works found in the literature: one trying to describe how 'black box' machine learning approaches work, and the other trying to explain the result of such approaches without knowing the details on how these work. In the first approach, the goal is to make "transparent classifiers" by training interpretable models that can be used for yielding satisfactory explanations. In the second approach, the purpose is to understand the reasons for the classification or how a model behaves by, say, changing one or more inputs. In this regard, while our novel XNBC can be considered to belong to the works following the first approach, the explanation technique proposed in [19] is an example of the second approach. It is worth mentioning that techniques based on the second approach try to explain only the reasons for a specific prediction. In contrast, techniques like XNBC try to explain what has been relevant to the knowledge model and is applicable for all the possible predictions.

Contributions proposed by the fuzzy logic community for explaining computer predictions are analyzed in [1]. This analysis suggests that efforts made by the non-fuzzy community and by the fuzzy logic community can be linked to solve problems related to the interpretability of computer predictions.

6 Conclusions

In this paper, we have proposed a novel variant of a naive Bayes classification (NBC) process that produces contextualized predictions. The novel NBC process, named *explainable NBC* or XNBC, consists of a learning process, an evaluation process and a prediction step: while the purpose of the first is to obtain a model of what is known about a particular class, the purpose of the second is to obtain contextualized evaluations of the level to which other objects belong to that class, these evaluations can then be used in the third for offering users the k best classes with the best context.

The learning process looks into the objects included into a training collection to build a knowledge model in which the influence of the features on the contextualized evaluations is represented. In this process, the influence of a feature is determined by the conditional probability of the feature occurring in objects that belong to the analyzed class.

The evaluation process uses such a knowledge model as input to quantify the influence of the features on the classification of other objects. *Augmented appraisal degrees* (AADs), which are mathematical representations of the context of experienced-based evaluations, are used for handling the evaluations performed during this process. Hence, the evaluation process produces contextualized evaluations that put the forthcoming predictions in context.

In the prediction step, the k best classes corresponding to the top-k of the resulting contextualized evaluations are presented in such a way that users have additional information for giving preference to the class(es) with the best credible justification.

By means of an example in which the categories of newswire articles are predicted, we have illustrated how the proposed XNBC process can produce contextualized predictions. We have also explained how those contextualized predictions can help a user to decide which prediction is the most appropriate according to his/her perspective and, thus, make an informed (classification) decision. In spite of that, further study is needed to demonstrate the interpretability and usability of such contextualized predictions.

References

1. Alonso, J.M., Castiello, C., Mencar, C.: A bibliometric analysis of the explainable artificial intelligence research field. In: Medina, J., et al. (eds.) IPMU 2018. CCIS, vol. 853, pp. 3–15. Springer, Cham (2018). https://doi.org/10.1007/978-3-319-91473-2_1

2. Atanassov, K.T.: Intuitionistic fuzzy sets. Fuzzy Sets Syst. **20**(1), 87–96 (1986). https://doi.org/10.1016/S0165-0114(86)80034-3
3. Atanassov, K.T.: On Intuitionistic Fuzzy Sets Theory. Studies in Fuzziness and Soft Computing, vol. 283. Springer, Heidelberg (2012). https://doi.org/10.1007/978-3-642-29127-2
4. Guidotti, R., Monreale, A., Ruggieri, S., Turini, F., Giannotti, F., Pedreschi, D.: A survey of methods for explaining black box models. ACM Comput. Surv. **51**(5) (2018). https://doi.org/10.1145/3236009
5. Gunning, D.: Explainable Artificial Intelligence (XAI) (2017). www.darpa.mil/attachments/XAIIndustryDay_Final.pptx
6. Kaminski, M.E.: The right to explanation, explained. Berkeley Tech. LJ 34, 189 (2019). https://doi.org/10.15779/Z38TD9N83H
7. Letham, B., Rudin, C., McCormick, T.H., Madigan, D.: Interpretable classifiers using rules and Bayesian analysis: building a better stroke prediction model. Ann. Appl. Stat. **9**(3), 1350–1371 (2015). https://doi.org/10.1214/15-AOAS848
8. Lewis, D.D.: Reuters-21578 Text Categorization Collection. http://kdd.ics.uci.edu/databases/reuters21578/reuters21578.html
9. Lewis, D.D.: Naive (Bayes) at forty: the independence assumption in information retrieval. In: Nédellec, C., Rouveirol, C. (eds.) ECML 1998. LNCS, vol. 1398, pp. 4–15. Springer, Heidelberg (1998). https://doi.org/10.1007/BFb0026666
10. Lewis, D.D., Jones, K.S.: Natural language processing for information retrieval. Commun. ACM **39**(1), 92–101 (1996). https://doi.org/10.1145/234173.234210
11. Liu, B.: Web Data Mining: Exploring Hyperlinks, Contents, and Usage Data. Springer, Heidelberg (2007). https://doi.org/10.1007/978-3-642-19460-3
12. Loor, M., De Tré, G.: On the need for augmented appraisal degrees to handle experience-based evaluations. Appl. Soft Comput. **54**, 284–295 (2017). https://doi.org/10.1016/j.asoc.2017.01.009
13. Loor, M., De Tré, G.: Identifying and properly handling context in crowdsourcing. Appl. Soft Comput. **73**, 203–214 (2018). https://doi.org/10.1016/j.asoc.2018.04.062
14. Loor, M., De Tré, G.: Explaining computer predictions with augmented appraisal degrees. In: 2019 Conference of the International Fuzzy Systems Association and the European Society for Fuzzy Logic and Technology, EUSFLAT 2019. Atlantis Press, August 2019. https://doi.org/10.2991/eusflat-19.2019.24
15. Loor, M., Tapia-Rosero, A., De Tré, G.: Usability of concordance indices in FAST-GDM problems. In: Proceedings of the 10th International Joint Conference on Computational Intelligence, IJCCI 2018, pp. 67–78 (2018). https://doi.org/10.5220/0006956500670078
16. Manning, C.D., Raghavan, P., Schütze, H.: Introduction to Information Retrieval. Cambridge University Press, Cambridge (2008)
17. Nauck, D., Kruse, R.: Obtaining interpretable fuzzy classification rules from medical data. Artif. Intell. Med. **16**(2), 149–169 (1999). https://doi.org/10.1016/S0933-3657(98)00070-0. Fuzzy Diagnosis
18. Porter, M.F., et al.: An algorithm for suffix stripping. Program **14**(3), 130–137 (1980). https://doi.org/10.1108/00330330610681286
19. Ribeiro, M.T., Singh, S., Guestrin, C.: "Why Should I Trust You?": Explaining the predictions of any classifier. In: Proceedings of the 22nd ACM SIGKDD International Conference on Knowledge Discovery and Data Mining, KDD 2016, pp. 1135–1144. ACM, New York (2016). https://doi.org/10.1145/2939672.2939778
20. Vapnik, V.N.: The Nature of Statistical Learning Theory. Springer, New York (1995). https://doi.org/10.1007/978-1-4757-2440-0

21. Vapnik, V.N., Vapnik, V.: Statistical Learning Theory, vol. 1. Wiley, New York (1998)
22. Wang, T., Rudin, C., Velez-Doshi, F., Liu, Y., Klampfl, E., MacNeille, P.: Bayesian rule sets for interpretable classification. In: 2016 IEEE 16th International Conference on Data Mining (ICDM), pp. 1269–1274, December 2016. https://doi.org/10.1109/ICDM.2016.0171
23. Zadeh, L.: Fuzzy sets. Information and control **8**(3), 338–353 (1965). https://doi.org/10.1016/S0019-9958(65)90241-X
24. Zhang, H.: The optimality of Naive Bayes. In: Proceedings of the Seventeenth International Florida Artificial Intelligence Research Society Conference, pp. 562–567. The AAAI Press, Menlo Park (2004). https://aaai.org/Papers/FLAIRS/2004/Flairs04-097.pdf

Author Index

Printed in the United States
By Bookmasters